T0189099

Springer Proceedings in Advanced Robotics 13

The Springer Proceedings in Advanced Robotics (SPAR) publishes new developments and advances in the fields of robotics research, rapidly and informally but with a high quality.

The intent is to cover all the technical contents, applications, and multidisciplinary aspects of robotics, embedded in the fields of Mechanical Engineering, Computer Science, Electrical Engineering, Mechatronics, Control, and Life Sciences, as well as the methodologies behind them.

The publications within the "Springer Proceedings in Advanced Robotics" are primarily proceedings and post-proceedings of important conferences, symposia and congresses. They cover significant recent developments in the field, both of a foundational and applicable character. Also considered for publication are edited monographs, contributed volumes and lecture notes of exceptionally high quality and interest.

An important characteristic feature of the series is the short publication time and world-wide distribution. This permits a rapid and broad dissemination of research results.

More information about this series at http://www.springer.com/series/15556

Ken Goldberg · Pieter Abbeel ·
Kostas Bekris · Lauren Miller
Editors

Algorithmic Foundations of Robotics XII

Proceedings of the Twelfth Workshop
on the Algorithmic Foundations of Robotics

 Springer

Editors
Ken Goldberg
Industrial Engineering
and Operations Research
University of California
Berkeley, CA, USA

Kostas Bekris
Computer Science
Rutgers University
Piscataway, NJ, USA

Pieter Abbeel
Electrical Engineering
and Computer Sciences
University of California
Berkeley, CA, USA

Lauren Miller
Electrical Engineering
and Computer Sciences
University of California
Berkeley, CA, USA

ISSN 2511-1256   ISSN 2511-1264   (electronic)
Springer Proceedings in Advanced Robotics
ISBN 978-3-030-43091-7   ISBN 978-3-030-43089-4   (eBook)
https://doi.org/10.1007/978-3-030-43089-4

This Springer imprint is published by the registered company Springer Nature Switzerland AG
The registered company address is: Gewerbestrasse 11, 6330 Cham, Switzerland

Foreword

At the dawn of the century's third decade, robotics is reaching an elevated level of maturity and continues to benefit from the advances and innovations in its enabling technologies. These all are contributing to an unprecedented effort to bringing robots to the human environment in hospitals and homes, factories, and schools, in the field for robots fighting fires, making goods and products, picking fruits and watering the farmland, saving time and lives. Robots today hold the promise for making a considerable impact in a wide range of real-world applications from industrial manufacturing to health care, transportation, and exploration of the deep space and sea. Tomorrow, robots will become pervasive and touch upon many aspects of modern life.

The *Springer Tracts in Advanced Robotics (STAR)* was launched in 2002 with the goal of bringing to the research community the latest advances in the robotics field based on their significance and quality. During the latest fifteen years, the STAR series has featured publication of both monographs and edited collections. Among the latter, the proceedings of thematic symposia devoted to excellence in robotics research, such as ISRR, ISER, FSR, and WAFR, have been regularly included in STAR.

The expansion of our field as well as the emergence of new research areas has motivated us to enlarge the pool of proceedings in the STAR series in the past few years. This has ultimately led to launching a sister series in parallel to STAR. The *Springer Proceedings in Advanced Robotics (SPAR)* is dedicated to the timely dissemination of the latest research results presented in selected symposia and workshops.

This volume of the SPAR series brings the proceedings of the twelfth edition of the Workshop Algorithmic Foundations of Robotics (WAFR). WAFR went back to its roots and was held from December 18 to 20, 2016, in San Francisco, California, the same city in which the very first WAFR was held in 1994.

The volume edited by Ken Goldberg, Pieter Abbeel, Kostas Bekris, and Lauren Miller is a collection of 58 contributions spanning a wide range of applications in manufacturing, medicine, distributed robotics, human–robot interaction, intelligent prosthetics, computer animation, computational biology, and many other areas.

Validation of algorithms, design concepts, or techniques is the common thread running through this focused collection.

Rich by topics and authoritative contributors, WAFR culminates with this unique reference on the current developments and new directions in the field of algorithmic foundations. A very fine addition to the series!

More information about the conference, including videos of the presentations can be found under the conference's website: http://wafr2016.berkeley.edu

January 2020   Bruno Siciliano
                                                       Oussama Khatib
                                                        SPAR Editors

Contents

Multiple Start Branch and Prune Filtering Algorithm for Nonconvex Optimization

Rangaprasad Arun Srivatsan[1] and Howie Choset[1]

Robotics Institute at Carnegie Mellon University, Pittsburgh, PA 15213, USA,
(rarunsrivatsan@,choset@cs.) cmu.edu

Abstract. Automatic control systems, electronic circuit design, image registration, SLAM and several other engineering problems all require nonconvex optimization. Many approaches have been developed to carry out such nonconvex optimization, but they suffer drawbacks including large computation time, require tuning of multiple unintuitive parameters and are unable to find multiple local/global minima. In this work we introduce multiple start branch and prune filtering algorithm (MSBP), a Kalman filtering-based method for solving nonconvex optimization problems. MSBP starts off with a number of initial state estimates, which are branched and pruned based on the state uncertainty and innovation respectively. We show that compared to popular methods used for solving nonconvex optimization problems, MSBP has fewer parameters to tune, making it easier to use. Through a case study of point set registration, we demonstrate the efficiency of MSBP in estimating multiple global minima, and show that MSBP is robust to initial estimation error in the presence of noise and incomplete data. The results are compared to other popular methods for nonconvex optimization using standard datasets. Overall MSBP offers a better success rate at finding the optimal solution with less computation time.

1 Introduction

In various engineering applications such as automatic control systems, signal processing, mechanical systems design, image registration, etc., we encounter problems that require optimization of some objective function. While many efficient algorithms have been developed for convex optimization, dealing with nonconvex optimization remains an open question [17]. In this work, we introduce a new method for nonconvex optimization, called *multiple start branch and prune filtering algorithm* (MSBP). Compared to popular methods, branch and bound [12], simulated annealing [2], genetic algorithms [5], etc., MSBP only has a few parameters to tune and can provide fast online estimates of the optimal solutions.

We believe that Kalman filter-based methods for nonconvex optimization [28] suffer less from issues surrounding computational efficiency and parameter tuning. Multi-hypothesis filtering [20] and the heuristic Kalman algorithm (HKA) [28,

© Springer Nature Switzerland AG 2020
K. Goldberg et al. (Eds.): *Algorithmic Foundations of Robotics XII*, SPAR 13, pp. 1–16, 2020.
https://doi.org/10.1007/978-3-030-43089-4_1

29] are two popular choices for filtering based methods for nonconvex optimization. Both these methods, as well as MSBP, fall under the category of population based stochastic optimization techniques. MSBP was developed for nonconvex optimization problems where the objective function is available in an analytical form and yet is expensive to evaluate (for example the case of point registration).

Unlike the HKA which starts with one initial state estimate, MSBP starts with multiple such estimates. These are further branched, updated and then pruned to explore the search space efficiently while avoiding premature convergence to a local minimum. A major advantage of MSBP over other methods is the high success rate of estimating all the minima in problems with multiple local/global minima. The MSBP requires tuning of only three intuitive parameters, which makes it easy for a non-expert to use the method.

In this work we evaluate and compare the efficiency of MSBP to other methods on the Griewank function, which is a standard test for nonconvex optimization methods. We also test MSBP on point set registration. This application is specifically chosen to test our algorithm because of its analytical and yet expensive function evaluation which offers practical challenges to most of the existing algorithms for nonconvex optimization. MSBP is tested in the presence of high initial error, multiple global minima, noisy data and incomplete data. In all these cases, MSBP accurately estimates the global minima with a high success rate over multiple runs of the algorithm.

2   Background

In a general setting, an optimization problem consists of finding input variables within a valid domain that minimize a function of those variables. An optimization problem can be represented as

$$
\begin{aligned}
\text{minimize} \quad & h(\boldsymbol{x}), & \boldsymbol{x} \in \boldsymbol{R}^{n_x} \\
\text{subject to} \quad & g_i(\boldsymbol{x}) \leq 0, & i = 1, \ldots, n_c \\
& e_j(\boldsymbol{x}) = 0, & j = 1, \ldots, n_e.
\end{aligned} \tag{1}
$$

In Eq. 1, $\boldsymbol{x}$ is the n_x dimensional input variable, also known as the optimization variable, h is the objective function to be minimized, $g_i(\boldsymbol{x})$ and $e_j(\boldsymbol{x})$ are the inequality and equality constraints respectively and n_c and n_e are the number of inequality and equality constraints respectively.

2.1   Nonconvex optimization problems

We often encounter optimization problems that have a number of locally optimal solutions which are optimal only within a small neighborhood but do not correspond to the globally optimal solution that minimizes the function in the function domain. Such problems are termed "nonconvex" optimization problems, in contrast to "convex" optimization problems where any local minimum is also a global minimum. Nonconvex optimization problems are in general non-trivial to solve because it is difficult to guarantee that the solution returned by the optimizer is global rather than local.

For these problems, a standard approach is to use convex optimizers that employ different randomization techniques to choose multiple initial starts [22]. The drawback of this approach is that for problems with a large number of local minima solutions, a lot of computational effort may be needed to find the global optimum [17]. Branch and bound methods are also commonly used, but the curse of dimensionality leaves them ineffective in cases with many optimization variables [12].

2.2   Heuristic methods for nonconvex optimization problems

Several heuristic methods have been developed to estimate global minima in nonconvex optimization problems such as simulated annealing (SA) [2], particle swarm optimization (PSO) [18], genetic algorithms (GA) [5] and more recently recursive decomposition (RD) [3]. SA is widely considered as versatile and easy to implement, but there are two major drawbacks: 1) there are multiple unintuitive parameters that require tuning, and the results are known to be sensitive to the choice of these parameters [7]; 2) the computation time is generally high for most practical applications [19]. PSO and GA are both categorized as population-based random-search methods. PSO is more sensitive than GA to the choice of parameters, and is known to prematurely converge unless the parameters are tuned well. Also, GA is known to be computationally intractable for many high dimensional problems [25]. In contrast, RD decomposes the objective function into approximately independent sub-functions, and then optimizes the simpler sub-functions using gradient based techniques. The drawback of such a method is that not all functions can be decomposed into sub-functions, in which case RD would perform similarly to a gradient descent with multiple starts.

2.3   Filtering-based methods for nonconvex optimization problems

Due to their ease of use and small number of tuning parameters, Kalman filter-based methods have also been used in optimization [4, 24, 29]. Typically such methods adapt a Kalman filter to have a static process model with the state vector comprised of the optimization variables x and an initial state uncertainty Σ spanning the domain of the search space. The measurement model is taken to be an evaluation of the objective function. The measurement is chosen to be the value of the minimum that we want the objective function to attain. By definition, with each iteration of the Kalman filter, the state vector is updated such that the difference between the measurement and the measurement model is decreased [10], thus ensuring that the objective function is minimized. The corresponding covariance also decreases as the number of iterations increases. When the mean of the state stops changing over iterations, or when the uncertainty decreases below a set threshold, we consider the state to be the optimal estimate.

A Kalman filter can provide the optimal estimate of $\boldsymbol{x}_{k+1|k+1}$ [1] such that

$$\boldsymbol{x}_{k+1|k+1} = \underset{\boldsymbol{x}}{\operatorname{argmax}}\; prob(\boldsymbol{x}|\boldsymbol{z}_{k+1}, \boldsymbol{x}_{k+1|k}), \tag{2}$$

where $\boldsymbol{z}_{k+1}$ is the current measurement and $\boldsymbol{x}_{k+1|k}$ is the mean of the predicted state estimate. The solution to Eq. 2 is

$$\boldsymbol{x}_{k+1|k+1} = \underset{\boldsymbol{x}}{\operatorname{argmin}}\; (\boldsymbol{x} - \boldsymbol{x}_{k+1|k})^T \boldsymbol{\Sigma}_{k+1|k}^{-1}(\boldsymbol{x} - \boldsymbol{x}_{k+1|k})+$$
$$(\boldsymbol{z}_{k+1} - h(\boldsymbol{x}))^T \boldsymbol{R}^{-1}(\boldsymbol{z}_{k+1} - h(\boldsymbol{x})), \tag{3}$$

where $\boldsymbol{\Sigma}_{k+1|k}$ is the uncertainty of the predicted state, and h is the unconstrained objective function as defined in Eq. 1.

Let h_{min} be the smallest value that h can attain, which is attained at $\boldsymbol{x} = \boldsymbol{x}_{min}$. Since there is uncertainty associated with $\boldsymbol{x}$, we have $h_{min} = h(\boldsymbol{x}_{min}) = h(\boldsymbol{x}) + \boldsymbol{v}$, where $\boldsymbol{v} \sim \mathcal{N}(0, \boldsymbol{R}(\boldsymbol{x}))$ is state dependent measurement noise drawn from a zero mean distribution with covariance $\boldsymbol{R}$. For an optimization problem as shown in Eq. 1, we choose the following measurement model

$$\boldsymbol{z}_k = h(\boldsymbol{x}) + \boldsymbol{v}_k(\boldsymbol{x}).$$

We set the measurement $\boldsymbol{z}_k = h_{min}$, as the state $\boldsymbol{x}$ will then be updated such that the value of h is close to h_{min}. If h_{min} is not known *a priori*, $\boldsymbol{z}_{k+1}$ can be set to an arbitrarily small value. The uncertainty $\boldsymbol{R}(\boldsymbol{x})$ can be computed analytically as shown in [9, pp. 90–91]. The resulting filter would provide an optimal estimate of $\boldsymbol{x}$ as long as h is linear [10]. The following are the Kalman filter equations modified for an optimization problem:

$$\boldsymbol{x}_{k+1} = \boldsymbol{x}_k + \boldsymbol{K}_{k+1}(h_{min} - h(\boldsymbol{x}_k)), \tag{4}$$
$$\boldsymbol{\Sigma}_{k+1} = \boldsymbol{\Sigma}_k - \boldsymbol{K}_{k+1}\boldsymbol{H}_{k+1}\boldsymbol{\Sigma}_k, \tag{5}$$

where the Kalman gain $\boldsymbol{K}_{k+1} = \boldsymbol{\Sigma}_k \boldsymbol{H}_{k+1}^T(\boldsymbol{H}_{k+1}\boldsymbol{\Sigma}_k\boldsymbol{H}_{k+1}^T + \boldsymbol{R})^{-1}$ and $h(\boldsymbol{x}_k) = \boldsymbol{H}_k\boldsymbol{x}_k$. $\boldsymbol{H}$ is the Jacobian of the objective function $h(\boldsymbol{x})$. If h is nonlinear, variants such as the extended Kalman filter (EKF), unscented Kalman filter (UKF), etc. can be used. In the presence of constraint functions that must be satisfied as shown in Eq. 1, equality or inequality constrained Kalman filtering techniques can be applied [4, 26].

In general, the Kalman filter only estimates the local minimum. A popular approach for nonconvex optimization problems is multi-start or multi-hypothesis filtering as shown in Fig. 1(a) [20]. Multiple filters each having a different randomly chosen initial start, are run in parallel, and after each iteration the estimate with the maximum likelihood is chosen as the current best estimate. Such an implementation has a good chance of finding global minima but at the expense of increased computation time.

Particle filters have also been adapted as a smart alternative to multi-hypothesis filtering [15]. The resampling step in a particle filter ensures that states with low

---

[1] $\boldsymbol{v}_{a|b}$ is the estimate of $\boldsymbol{v}$ at the ath iteration given measurements upto b iterations.

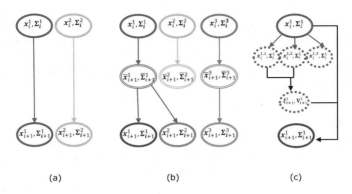

Fig. 1. (a) Steps involved in one iteration of a multi-hypothesis filter with 2 initial start states. After each iteration the state with maximum likelihood estimate is chosen as the best current estimate. (b) Steps involved in a particle filter with 3 particles. After updating the particles based on the measurement, resampling is performed to remove particles with low weights. (c) Steps involved in one iteration of the heuristic Kalman algorithm. In this example, the parent's state is divided into 3 child states. The weighted sum of 2 child states with the lowest objective value is used to obtain the pseudo measurement $\boldsymbol{\xi}_{i+1}$.

weights are pruned while the others are retained (see Fig. 1(b)). Particle filters and multi-hypothesis filters both suffer from the curse of dimentionality. When estimating high dimensional parameters (> 4), a large number of particles are needed to span the search space to find the global optimum, which can be computationally expensive especially if the function evaluation is not cheap.

The heuristic Kalman algorithm (HKA), introduced by Toscana et al. [28], is a combination of Kalman filtering and population-based random-search methods (see Fig. 1(c)). Starting with a parent state, HKA spans child states and evaluates the function at the child states. A pseudo measurement and its uncertainty ($\boldsymbol{\xi}_{i+1}, \boldsymbol{V}_{i+1}$) are then obtained from the n best states with the smallest function value, and the state ($\boldsymbol{x}_i, \boldsymbol{\Sigma}_i$) is updated using the pseudo measurement. Even though the parent state is divided into a number of child states, in each iteration of the algorithm only a single state, is updated. Such an approach has been shown to be suitable in situations where the function can only be evaluated using experimental simulations and not analytically. For such problems, HKA is a good optimization tool with very few parameters to tune and a good success rate of finding global minimum [29]. However, when an analytical form of the objective function is available, other methods perform much better than HKA.

In this work, we introduce the multiple start branch and prune filtering algorithm (MSBP), which is similar to HKA in that at every iteration the parent state is divided into several child states which are then updated using the current measurement. However, instead of using a weighted mean of the n selected child states to estimate a single next state, we update the m child states based on the measurements and select the n best updated states as parent states for the next iteration and repeat the process until convergence. MSBP also incorporates a procedure to prune the parent states if they are within a threshold of each other. This prevents duplication of computation on similar states and encourages exploration. Such an approach allows us to find multiple global/local minima solutions.

3   Modeling

The basic framework of the MSBP is shown in Fig. 2. The various steps involved in the MSBP implementation are as follows:

1. The algorithm is initialized with n initial parent states $(\boldsymbol{x}_k^i, \boldsymbol{\Sigma}_k^i)$, $i = 1, 2, \ldots, n$, where k denotes the iteration index (see Section 3.1 for information on how to choose the initial states).
2. Each parent state is divided into m child states $(\boldsymbol{x}_k^{i,j}, \boldsymbol{\Sigma}_k^i)$, $(j = 1, 2, \ldots, m)$, by sampling from the distribution $(\boldsymbol{x}_k^i, \boldsymbol{\Sigma}_k^i)$. The parent state is always retained as one of the m child states. The child states that are generated from the parent state can be viewed as perturbations being added to the states to overcome local minima and to encourage exploration.
3. The child states are then updated using Eq. 4 to obtain $(\boldsymbol{x}_{k+1}^{i,j}, \boldsymbol{\Sigma}_{k+1}^{i,j})$.
4. From the $n \times m$ child states, the n states with the lowest innovation, i.e., $\boldsymbol{z}_{k+1} - h(\boldsymbol{x}_{k|k+1})$ from Eq. 3 are chosen as parents for the next iteration.
5. Among the n parent states chosen, if the means of any states come within an ϵ threshold of each other, the state with the lower innovation is retained and the others are pruned (n decreases every time pruning happens).
6. Steps 2-5 are repeated until convergence or up to a fixed number of iterations.

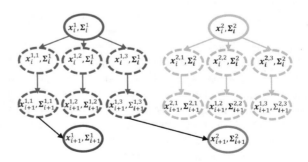

Fig. 2. Steps involved in one iteration of the MSBP. Parent states are shown in bold ellipses and child states are shown in dashed ellipses. In this example, $n = 2$, $m = 3$.

From Fig. 1(a) it can be noted that multi-hypothesis filtering is a special case of MSBP with $m = 1$ and $\epsilon = 0$. The multi-hypothesis filter requires a large number of initial states to converge onto a global minimum, as a lack of perturbation can result in premature convergence to a local minimum. Also lack of a pruning step in the multi-hypothesis filter often results in duplication of estimates by multiple filters. Particle filters prune states with lower probability during the resampling step and offer an advantage over multi-hypothesis filter. However, particle filters lack the perturbation step and the state update step present in MSBP, which helps in over coming local minima and quick convergence to the optimal solutions. In comparison to other methods, such as GA, SA, PSO, etc., at each iteration in addition to evaluation of the objective function at multiple states, the states themselves are updated by the update model of the MSBP. While this could be viewed as additional computation, the update step allows us to minimize the function faster and quickly identify the minima compared to the other methods. MSBP provides a maximum of n estimates after each iteration as opposed to a single estimate provided by HKA (see Fig. 1(c)). This is a drawback for HKA in problems that have multiple global minima, as HKA would tend to return an estimate that is at a location intermediate to both the minima. Running the HKA multiple times with different initial states can improve the success rate of finding the global minimum, but at the cost of increased computation time.

Thus, the MSBP offers the advantage of reduced computational load and memory storage in addition to a higher success rate of estimating the global minima, for problems with analytical objective functions. The only shortcoming is that when dealing with very high dimensional systems (typically > 20), the update step of the Kalman filter can become expensive as it would involve inverting a high-dimensional matrix.

3.1 Choice of initial state and parameters

In addition to the choice of initial states, there are three parameters that require tuning in the MSBP: n, m and ϵ. This section describes the intuition behind selecting these parameters and the initial states.

Initial state: In most practical problems the domain of the search space for optimization is known. Without loss of generality, the uncertainty of all the initial states is chosen to be equal to each other. The uncertainty is chosen to be a diagonal matrix with each diagonal element set to be equal to σ^2, such that 6σ equals the span of the domain in that dimension. Such a choice for Σ_0^i is generally conservative, and restricts the uncertainty in each of the parent state to the search domain. The mean of the states x_0^i are randomly chosen from the valid search domain.

Number of parent states n: n can be chosen based on prior knowledge of the number of global minima present in the problem. If that number is not known *a priori*, then a conservative estimate can be made. In practice we observe that choosing a value of n greater than the number of global and local minima present in the search domain improves the success rate of the algorithm. However, increasing the value of n also increases the computation time.

Number of child states m: m is the number of child states per parent state. If the estimator is stuck at a local minima, the perturbations help get it out of the local minima. Hence, the greater the value of m, the greater the chances of MSBP capturing the global minima. However a higher m would also mean increased computation time. As result m should be chosen depending on the allowable computation time for the application.

Choice of ϵ: ϵ is the parameter that decides the threshold between the parent states. ϵ helps prevent unnecessary computation and encourages exploration. A large value of ϵ can prune several parent states at once and can result in missing some solutions. $\epsilon = 0$ would not prune any parent state, resulting in unwanted computation in cases where multiple parent states are identical. Depending on the application, ϵ can be chosen to be a fixed number for all iterations or its value can be varied over the iterations.

Section 4.2 describes in more detail how these parameters are tuned for a case study on point set registration.

4   Results

In this section, we first demonstrate the performance of MSBP by testing it on the Griewank function. Following that, we do a case study of point set registration problem.

4.1   Numerical experiment with Griewank function

A number of standard functions are used to test the performance of nonconvex optimization methods [23]. In this work, we choose to test the MSBP on the Griewank function. Fig. 3(a) shows the plot of Griewank function for $x \in [-60, 60]$. In the chosen domain, the function is known to have a global minima at $x = 0$ and twenty local minima at ± 6.28, ± 12.56, ± 18.84, ± 25.12, ± 31.45, ± 37.55, ± 43.93, ± 50.3, ± 56.67.

As mentioned in Section 3.1 in order to ensure that 99% of samples fall within the search domain we choose the uncertainty of the initial states, $\Sigma_0 = \sigma_0^2$, such that $6\sigma_0 = 120$. The mean of the initial parent states are sampled from the normal distribution $\mathcal{N}(\mu, \Sigma_0)$, where $\mu = 5$. We choose $\mu = 5$, as it is closer to the local minima at $x = 6.28$ than the global minima at $x = 0$, and would be a more challenging test for the optimization algorithm. For our implementation of MSBP, we use an EKF since the function is non-linear. In addition we choose $n = 21, m = 10, \epsilon = 2$. We run all the algorithms until convergence. The algorithm is set to have converged when the change in the estimate of the minima is $< 10^{-6}$. We observe that the maximum number of iterations required by any algorithm is generally under 20. For the sake of a fair comparison, the values of the parameters for all the algorithms were tuned as per the recommendation in [28] and the best results have been reported.

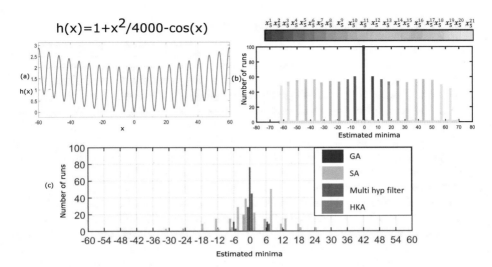

Fig. 3. (a) A plot of the Griewank function. (b) A histogram showing the values estimated by 21 parent states of MSBP over 100 runs. The Y axis of the plot shows the number of runs that estimate a particular state and the X axis shows the estimated value. A histogram of the estimated value over 100 runs is shown for the following algorithms (c) Histogram showing values estimated by Genetic algorithm, Simulated annealing, Multi-hypothesis filter, and HKA.

We repeat the experiment 100 times to observe the performance of the method over multiple runs. Fig. 3(b) shows the histogram of the values estimated by MSBP over 100 runs, all of which converged within five iterations. The global minimum is estimated correctly at $x = 0$ each time, while the local minima solutions are accurately predicted by the remaining twenty parent states. The order in which the other parent states estimate the local minima varies in each run of the algorithm, but they are tracked in all of them.

In comparison, HKA was implemented with initial state $(x_0, \Sigma_0) = (5, 400)$, 20 divisions, 2 best candidates, and a slow down coefficient of 0.7 . Fig. 3(c) shows the histogram of values estimated by HKA over 100 runs. We observe that the algorithm correctly estimates the local minima only 10% of the time. In 8% of the runs, HKA estimates the local minima at $x = 6.28$ and $x = -6.28$, which are closest to the mean of the initial state, $x_0 = 5$. SA also estimates the global minimum only 15% of times and rest of the times it gets stuck at nearby local minima (see Fig. 3(c)).

A multi-hypothesis filter was also implemented, where we choose the same initial states as those for MSBP with $n = 21, m = 1, \epsilon = 0$. Fig. 3(c) shows the histogram of estimated values over 100 repeated runs. More than 50% of the time, the algorithm estimates the global minimum correctly. The rest of the times it estimates values close to the global minimum or one of the two local minima closest to the global minimum similar to the GA (Fig. 3(c)).

4.2   Point set registration: A case study

Point set registration is the process of finding a spatial transformation that aligns the elements of two point sets. Point set registration is frequently encountered in robotic applications, such as computer vision [13], localization and mapping [8], surgical guidance [15], etc. When the correspondence between the points in the two point sets is known, rigid registration can be solved analytically as shown in [6]. However, when point correspondences are unknown, finding the optimal transformation becomes a nonconvex optimization problem with several local minima solutions. Besl et al. came up with the popular iterative closest point (ICP) method that recursively finds correspondences and minimizes the alignment difference between point sets [1]. Over the years several variants of the ICP have been developed [21], and also filtering based solutions have been developed that are better at handling noise in the data and provide online estimates [16].

Most of the point registration methods mentioned above use tools that are not designed for nonconvex optimization and so often converge to local minima. Branch and bound based technique has been developed to avoid this problem [31]. However, this methods has high computation time and is not suitable in real time applications. In this work, we use the MSBP for registration of point sets and demonstrate that it is able to find accurate estimates with low computation times. We perform multiple case studies with different conditions using different standard 3D shape datasets to show the versatility of our algorithm.

We use a dual-quaternion based Kalman filter (DQF) for estimating the registration parameters [27]. The MSBP can also be used with other filtering implementation for registration estimation such as [11, 16]. The DQF uses dual-quaternion representation for pose and reposes the originally nonlinear estimation problem as a linear estimation problem and hence Eq. 4 is readily used for estimating the optimal registration parameters (see [27] for more information on the expression for objective function used). In each iteration of the MSBP, closest point correspondence is found between a pair of points (as opposed to finding correspondence for all the points in the case of methods such as ICP). The correspondence found using the parent states is retained for the child states as well. Since the number of different correspondences that can be formed between the two point sets is combinatorial, we expect many local minima solutions. Hence, we choose a large value for n in all the applications below. When the state uncertainty reduces below a desired threshold, we end the estimation process and stop collecting measurements. Thus, compared to batch processing methods such as the ICP (in which we wait for all the measurements to be collected before estimating the optimal registration), the DQF has the advantage of faster computation using fewer point measurements [27].

Large initial transformation error Fig. 4(a) shows the CAD model of a Stanford bunny [30]. The CAD model is geometrically discretized using a triangular mesh with 43318 triangle vertices. We collect 1000 random samples of points from the CAD model and apply a known transformation to those points. We then estimate the applied transformation with the MSBP. The values of

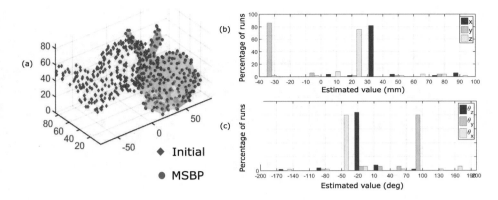

Fig. 4. (a)CAD model of a Stanford bunny. The initial position of 1000 points in shown in blue-diamond markers, the position estimated by MSBP is shown in red-circular markers. (b) Histogram of the estimated translation parameters, (c) histogram of the estimated rotation parameters over 100 runs of the algorithm. In (b) and (c), the Y axis shows the percentage of runs that return a particular value and the X axis shows the estimated value returned by the parent state with the smallest innovation. MSBP has a high success rate of estimating the optimal parameters.

various parameters used are $n = 40, m = 10, \epsilon = 1$. The experiment is repeated 100 times to note the statistical performance of our method. Fig. 4(a) shows the MSBP estimated points lie on top of the CAD model indicating accurate registration. On an average our algorithm converges after using 120 measurements.

Table 1 shows the actual registration parameters and the estimated values. The algorithm is compared with HKA, multi-hypothesis filtering, ICP, SA and GA. The SA and GA implementation we use for the sake of comparison are as described in [14, 25], which are a modified form of the original implementations of SA and GA with internal ICP computations. The authors of [14, 25] show that even though their approaches are expensive per iteration, they result is requirement of fewer iterations over all for convergence, and hence are faster and more accurate at estimating the registration parameters (These observations have been independently verified by us and hence we do not report results for vanilla implementations of SA and GA in this work).

While MSBP and multi-hypothesis filter estimate the registration parameters in a dual-quaternion space, we convert the estimated values into Cartesian coordinates and Euler angles for easy comparison with other methods. The penultimate column and the last column of Table 1 show the RMS error and time taken for various algorithms [2].

For multi-hypothesis filtering, we use the same set of initial states as MSBP. For HKA, ICP and SA we use a 4×4 identity matrix as the initial transformation.

---

[2] The computational time taken is calculated for script written in MATLAB R2015a software from MathWorks, running on a ThinkPad T450s (20BX0011GE) laptop from Lenovo with 8 GB RAM and intel i7 processor.

Table 1. Registration for large initial transformation error

Case1	x (mm)	y (mm)	z (mm)	θ_x (deg)	θ_y (deg)	θ_z (deg)	RMS (mm)	Time (sec)
Actual	30	-40	15	-55	80	-20	–	–
MSBP	29.89	-39.84	14.67	-58.57	80.59	-23.31	0.48	28
ICP	42.04	-35.22	8.52	17.83	19.21	33.26	35.06	5.82
Multi-hyp	59.79	-20.66	15.26	53.08	-45.58	30.29	18.25	404.62
HKA	-3.97	-17.69	17.45	31.39	31.69	-22.05	53.44	201.97
SA	29.16	-38.34	13.97	-51.75	81.67	-14.49	2.36	353.67
GA	30.08	-39.93	15.05	-54.59	79.92	-19.51	0.08	1051.00

The bounds on the search space are $[-100, 100]$ for translation and $[-\pi, \pi]$ for rotation around each axis. For HKA we use 40 divisions, 4 best candidates and a slow down coefficient of 0.4. For GA we use an initial population of 100, crossover probability of 0.7 and mutation probability of 0.2. These values for the parameters are tuned as per [29] and the best results are reported.

Fig. 4(a) shows that the displacement between the initial position of the points and their true position on the CAD model, is quite high and ICP does not perform well for such high initial errors. We notice that HKA also does not estimate the transformation accurately, presumably because it gets stuck at a local minimum. MSBP, SA and GA accurately estimate the transformation, however, SA and GA take much more time than MSBP to estimate. Since each function evaluation consists of an iteration of ICP internally, SA and GA both have higher estimation time than MSBP.

The MSBP algorithm is run 100 times and a histogram of the estimated translation and rotation parameters are shown in Fig. 4(b) and Fig. 4(c), respectively, which show that there is a $> 85\%$ chance of MSBP converging to the correct value. In comparison with MSBP, GA has a success rate of 10% and SA has a success rate of 20%. Thus, the MSBP produces accurate and repeatable results with high success rate, despite large errors in initial registration.

Multiple global minima In this example, we consider a snowflake as shown in Fig. 5(a), which has rotational symmetry about an axis passing through its center and perpendicular to its plane. The object is symmetric to its original shape when rotated about this axis by $\pm 60°, \pm 120°$ and $180°$. We sample 100 points from the CAD model of the snowflake and transform those points by a known transformation: $(x, y, \theta_z) = (15\text{mm}, 30\text{mm}, 45°)$. We then use MSBP to estimate the applied transformation. Since the snowflake is 2-dimensional, we restrict ourselves to in-plane registration. We use the following parameter values for MSBP: $n = 100, m = 10, \epsilon = 5$. After 100 iterations, the number of surviving parent states is 16. Fig. 5(b)-(g) show the position of the points after applying a transformation given by the first six parent states as estimated by the MSBP. The first six parent states of the MSBP accurately capture the six global minima

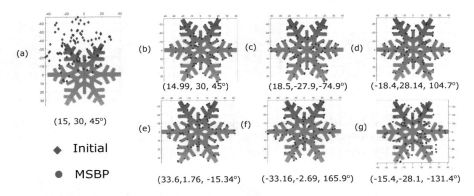

Fig. 5. (a) CAD model of a snowflake.The initial position of 100 points and the position estimated by MSBP are shown in blue-diamond and red-circular markers respectively. The actual transformation between the points and the CAD model is $(15, 30, 45°)$. (b)-(g) The first six parent states of MSBP. The estimated registration parameters are given below the figure. Note how the rotation angles are $45° \pm n \times 60°$, $(n = 0, 1, 2)$ due to the 6 way symmetry in the shape of the snowflake. Snowflake CAD model courtesy of Thingiverse CAD model repository

(Note that we limit our search domain to $[-180°,\ 180°]$ and hence there are 6 global minima in the search domain upto the rotational periodicity).

Noise in the input data In order to test the robustness of the registration using MSBP in the presence of noise in one of the point sets, we consider the example of Fertility as shown in Fig. 6. 200 Points are sampled from the CAD model and a Gaussian noise $\mathcal{N}(0, \sigma_n^2)$ is applied to each of the points. The standard deviation σ_n is kept constant for all the points, but is gradually increased from 1 to 20 in increments of 1 over several runs (For reference, the CAD model can be fit in a box of size $300 \times 200 \times 100$ units). Left hand side of Fig. 6(a)-(c) shows that CAD model and the initial position of the points in blue-diamond markers for 3 different values of σ_n. The right hand side of Fig. 6(a)-(c) shows the CAD model and the location of the points after applying the transformation as estimated by MSBP.

Note how the MSBP is able to successfully register the points for all the three cases shown in the figure. Also note how after registration, the points appear to be lying on the CAD model for lower σ_n and appear to be spread out of the CAD model for the case with higher σ_n.

Robustness to incomplete data A number of practical applications that require registration involve partial or incomplete datasets [8]. In order to test the performance of our approach for such applications, we consider an example of Stanford Armadillo man [30] (see Fig. 7). 500 points are sampled from the CAD model. In each run of the algorithm, one point is picked from the point

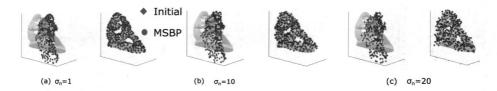

Fig. 6. CAD model of Fertility and 100 points sampled from it and a noise $\mathcal{N}(0, \sigma_n^2)$ is added to the points. (a) the plot for $\sigma_n = 1$ (b) the plot for $\sigma_n = 10$ (c) the plot for $\sigma_n = 20$. CAD model of Fertility courtesy of AIM@SHAPE model repository

Fig. 7. CAD model of a Stanford Armadillo man [30] and set of initial points sampled from parts of the model. The points are not sampled uniformly from all over the CAD, but have regions of missing information. (a) and (b) show two instances of incomplete data registered accurately to the CAD model using MSBP.

set at random and the selected point along with 250 of its nearest neighbors are removed from the point set. The rest of the points are then used for registration with the original CAD model.

We observe that in spite of the lack of complete point set information, MSBP is able to correctly register the points to the CAD model. Fig. 7 shows two arbitrary runs of the algorithm with different sets of points missing in each. In both the cases, the MSBP correctly registers the points to the CAD model as shown in Fig. 7(a)-(b).

5 Conclusion

In this work, we developed the multiple start branch and prune filtering algorithm (MSBP), a Kalman filter based method for nonconvex optimization. We show that using multiple initial states along with branching, updating and pruning, allows us to efficiently search for the optimal solution(s) in the domain of the search space without prematurely converging to a locally optimal solution. MSBP requires tuning of three parameters, the intuition behind which has been described and empirically verified with several examples. We show that the standard multi-hypothesis filter is a computationally less efficient, special case of the MSBP. With an example of point registration, MSBP is also compared with popular methods for nonconvex optimization and is found to estimate the optimal solutions accurately with a higher success rate especially when: 1) the objective function is available in an analytical form, 2) each function evaluation is expensive, 3) there are multiple global/local minima, and 4) the parameter space is relatively low dimensional (< 20).

Future work will involve an intermediate step to cluster the updated child states instead of using an ϵ threshold. By using an information filter instead of a Kalman filter, the expensive matrix inversion operation step in the state update can be avoided. This would allow us to extend the MSBP for problems involving high dimensional parameter spaces. Validating the effectiveness of MSBP on a variety of nonconvex problems with different functional complexities, different number of parameters, and studying parameter sensitivity, will be a part of our future publication.

Acknowledgments

This work has been funded through the National Robotics Initiative by NSF grant IIS-1426655.

References

1. Besl, P.: A Method for Registration of 3-D Shapes. IEEE Transactions on Pattern Analysis and Machine Intelligence 14(2), 239–256 (1992)
2. Brooks, S.P., Morgan, B.J.: Optimization using simulated annealing. The Statistician pp. 241–257 (1995)
3. Friesen, A.L., Domingos, P.: Recursive decomposition for nonconvex optimization. In: Proceedings of IJCAI (2015)
4. Gupta, N., Hauser, R.: Kalman filtering with equality and inequality state constraints. arXiv preprint arXiv:0709.2791 (2007)
5. Holland, J.H.: Outline for a logical theory of adaptive systems. Journal of the ACM 9(3), 297–314 (1962)
6. Horn, B.: Closed-form solution of absolute orientation using unit quaternions. Journal of the Optical Society of America A 4, 629–642 (1987)
7. Ingber, L.: Simulated annealing: Practice versus theory. Mathematical and computer modelling 18(11), 29–57 (1993)
8. Izadi, S., Kim, D., Hilliges, O., Molyneaux, D., Newcombe, R., Kohli, P., Shotton, J., Hodges, S., Freeman, D., Davison, A., et al.: KinectFusion: real-time 3D reconstruction and interaction using a moving depth camera. In: Proceedings of the 24th annual ACM symposium on User interface software and technology. pp. 559–568 (2011)
9. Jazwinski, A.H.: Stochastic processes and filtering theory. Courier Corp. (2007)
10. Kalman, R.E.: A new approach to linear filtering and prediction problems. Journal of Basic Engineering 82(1), 35–45 (1960)
11. Kang, Z., Chen, J., Wang, B.: Global Registration of Subway Tunnel Point Clouds Using an Augmented Extended Kalman Filter and Central-Axis Constraint. PloS one 10(5) (2015)
12. Lawler, E.L., Wood, D.E.: Branch-and-bound methods: A survey. Operations research 14(4), 699–719 (1966)
13. Lucas, B.D., Kanade, T., et al.: An iterative image registration technique with an application to stereo vision. In: International Joint Conference on Artificial Intelligence. vol. 81, pp. 674–679 (1981)

14. Luck, J., Little, C., Hoff, W.: Registration of range data using a hybrid simulated annealing and iterative closest point algorithm. In: Proceedings of IEEE International Conference on Robotics and Automation. pp. 3739–3744. IEEE (2000)
15. Ma, B., Ellis, R.E.: Surface-based registration with a particle filter. In: International Conference on Medical Image Computing and Computer-Assisted Intervention. pp. 566–573. Springer (2004)
16. Moghari, M.H., Abolmaesumi, P.: Point-based rigid-body registration using an unscented Kalman filter. IEEE Transactions on Medical Imaging 26(12), 1708–1728 (2007)
17. Neumaier, A., Shcherbina, O., Huyer, W., Vinkó, T.: A comparison of complete global optimization solvers. Mathematical programming 103(2), 335–356 (2005)
18. Poli, R., Kennedy, J., Blackwell, T.: Particle swarm optimization. Swarm intelligence 1(1), 33–57 (2007)
19. Ram, D.J., Sreenivas, T., Subramaniam, K.G.: Parallel simulated annealing algorithms. Journal of parallel and distributed computing 37(2), 207–212 (1996)
20. Reid, D.B.: An algorithm for tracking multiple targets. IEEE Transactions on Automatic Control 24(6), 843–854 (1979)
21. Rusinkiewicz, S., Levoy, M.: Efficient variants of the ICP algorithm. In: Proceedings. Third International Conference on 3-D Digital Imaging and Modeling. pp. 145–152 (2001)
22. Schoen, F.: Stochastic techniques for global optimization: A survey of recent advances. Journal of Global Optimization 1(3), 207–228 (1991)
23. Schoen, F.: A wide class of test functions for global optimization. Journal of Global Optimization 3(2), 133–137 (1993)
24. Schön, T., Gustafsson, F., Hansson, A.: A note on state estimation as a convex optimization problem. In: Proceedings of the IEEE International Conference on Acoustics, Speech, and Signal Processing. vol. 6, pp. VI–61. IEEE (2003)
25. Seixas, F.L., Ochi, L.S., Conci, A., Saade, D.M.: Image registration using genetic algorithms. In: Proceedings of the 10th annual conference on Genetic and evolutionary computation. pp. 1145–1146. ACM (2008)
26. Simon, D., Chia, T.L.: Kalman filtering with state equality constraints. IEEE transactions on Aerospace and Electronic Systems 38(1), 128–136 (2002)
27. Srivatsan, R.A., Rosen, G.T., Naina, F.M., Choset, H.: Estimating SE(3) elements using a dual-quaternion based linear Kalman filter. in the proceedings of Robotics Science and Systems (2016)
28. Toscano, R., Lyonnet, P.: Heuristic Kalman algorithm for solving optimization problems. IEEE Transactions on Systems, Man, and Cybernetics 39(5), 1231–1244 (2009)
29. Toscano, R., Lyonnet, P.: A Kalman Optimization Approach for Solving Some Industrial Electronics Problems. IEEE Transactions on Industrial Electronics 11(59), 4456–4464 (2012)
30. Turk, G., Levoy, M.: The Stanford 3D Scanning Repository. Stanford University Computer Graphics Laboratory http://graphics.stanford.edu/data/3Dscanrep
31. Yang, J., Li, H., Jia, Y.: Go-ICP: Solving 3D Registration Efficiently and Globally Optimally. In: 2013 IEEE International Conference on Computer Vision (ICCV). pp. 1457–1464 (2013)

Designing Sparse Reliable Pose-Graph SLAM: A Graph-Theoretic Approach

Kasra Khosoussi[1], Gaurav S. Sukhatme[2]
Shoudong Huang[1], and Gamini Dissanayake[1]

[1] Centre for Autonomous Systems
University of Technology Sydney
Sydney, NSW 2007, Australia
`kasra.khosoussi@uts.edu.au`

[2] Department of Computer Science
University of Southern California
Los Angeles, CA 90089, USA

Abstract. In this paper, we aim to design sparse D-optimal (determinant-optimal) pose-graph SLAM problems through the synthesis of sparse graphs with the maximum weighted number of spanning trees. Characterizing graphs with the maximum number of spanning trees is an open problem in general. To tackle this problem, several new theoretical results are established in this paper, including the monotone log-submodularity of the weighted number of spanning trees. By exploiting these structures, we design a complementary pair of near-optimal efficient approximation algorithms with provable guarantees. Our theoretical results are validated using random graphs and a publicly available pose-graph SLAM dataset.

Keywords: Number of Spanning Trees, D-Optimal Pose-Graph SLAM, Approximation Algorithms

1 Introduction

Graphs arise in modelling numerous phenomena across science and engineering. In particular, estimation-on-graph (EoG) is a class of (maximum likelihood) estimation problems with a natural graphical representation that arise especially in robotics and sensor networks. In such problems, each vertex corresponds to an unknown state, and each edge corresponds to a relative noisy measurement between the corresponding states. Simultaneous localization and mapping (SLAM) and sensor network localization (SNL) are two well-studied EoGs.

Designing sparse, yet "well-connected" graphs is a subtle task that frequently arises in various domains. First, note that graph sparsity—in EoGs and many other contexts—lead to computational efficiency. Hence, maintaining sparsity is often crucial. It is useful to see graph connectivity as a spectrum, as we often need to compare the connectivity of connected graphs. In engineering, well-connected

© Springer Nature Switzerland AG 2020
K. Goldberg et al. (Eds.): *Algorithmic Foundations of Robotics XII*, SPAR 13, pp. 17–32, 2020.
https://doi.org/10.1007/978-3-030-43089-4_2

graphs often exhibit desirable qualities such as *reliability*, *robustness*, and *resilience* to noise, outliers, and link failures. More specifically, a well-connected EoG is more resilient to a fixed noise level and results in a more reliable estimate (i.e., smaller estimation-error covariance in the Loewner ordering sense). Consequently, maintaining a sufficient connectivity is also critical. Needless to say, sparsity is, by its very essence, at odds with well-connectivity. This is the case in SLAM, where there is a trade-off between the cost of inference and the reliability of the resulting estimate. This problem is not new. Measurement selection and pose-graph pruning have been extensively studied in the SLAM literature (see, e.g., [21, 12]). However, in this paper we take a novel graph-theoretic approach by reducing the problem of designing sparse reliable SLAM problems to the purely combinatorial problem of synthesizing sparse, yet well-connected graphs. In what follows, we briefly justify this approximate reduction.

First, note that by estimation reliability we refer to the standard D-optimality criterion, defined as the determinant of the (asymptotic) maximum likelihood estimator covariance matrix. D-optimality is a standard and popular design criterion; see, e.g., [13, 16] and the references therein. Next, we have to specify how we measure graph connectivity. Among the existing combinatorial and spectral graph connectivity criteria, the number of spanning trees (sometimes referred to as *graph complexity* or *tree-connectivity*) stands out: despite its combinatorial origin, it can also be characterized solely by the spectrum of the graph Laplacian [9]. In [16, 17, 18], we shed light on the connection between the D-criterion in SLAM—and some other EoGs—and the tree-connectivity of the underlying graph. Our theoretical and empirical results demonstrate that, under some standard conditions, D-optimality in SLAM is *significantly* influenced by the tree-connectivity of the graph underneath. Therefore, one can accurately estimate the D-criterion without using any information about the robot's trajectory or realized measurements (see Section 3). Intuitively speaking, our approach can be seen as a dimensionality reduction scheme for designing D-optimal SLAM problems from the joint space of trajectories and graph topologies to only the space of graph topologies [18].

Although this work is specifically motivated by the SLAM problem, designing sparse graphs with the maximum tree-connectivity has several other important applications. For example, it has been shown that tree-connectivity is associated with the D-optimal incomplete block designs [7, 5, 1]. Moreover, tree-connectivity is a major factor in maximizing the connectivity of certain random graphs that model unreliable networks under random link failure (*all-terminal network reliability*) [15, 30]. In particular, a classic result in network reliability theory states that if the *uniformly-most reliable* network exits, it must have the maximum tree-connectivity among all graphs with the same size [2, 23, 3].

Known Results. Graphs with the maximum weighted number of spanning trees among a family of graphs with the same vertex set are called *t-optimal*. The problem of characterizing unweighted *t*-optimal graphs among the set of graphs with n vertices and m edges remains open and has been solved *only* for specific pairs of n and m; see, e.g., [27, 5, 14, 25]. The span of these special cases is too narrow for the types of graphs that typically arise in SLAM and sensor networks. Furthermore, in many cases the (n,m) constraint alone is insufficient for describing the true set of feasible graphs and cannot capture implicit practical constraints that exist in SLAM. Finally, it is not clear how these results can be extended to the case of (edge) weighted graphs, which are essential for representing SLAM problems, where the weight of each edge represents the precision of the corresponding pairwise measurement [18].

Contributions. This paper addresses the problem of designing sparse t-optimal graphs with the ultimate goal of designing D-optimal pose-graph SLAM problems. First and foremost, we formulate a combinatorial optimization problem that captures the measurement selection and measurement pruning scenarios in SLAM. Next, we prove that the weighted number of spanning trees, under certain conditions, is a monotone log-submodular function of the edge set. To the best of our knowledge, this is a new result in graph theory. Using this result, we prove that the greedy algorithm is near-optimal. In our second approximation algorithm, we formulate this problem as an integer program that admits a straightforward convex relaxation. Our analysis sheds light on the performance of a simple deterministic rounding procedure that have also been used in more general contexts. The proposed approximation algorithms provide near-optimality certificates. The proposed graph synthesis framework can be readily applied to any application where maximizing tree-connectivity is desired.

Notation. Throughout this paper, bold lower-case and upper-case letters are reserved for vectors and matrices, respectively. The standard basis for $\mathbb{R}^n$ is denoted by $\{\mathbf{e}_i^n\}_{i=1}^n$. Sets are shown by upper-case letters. $|\cdot|$ denotes the set cardinality. For any finite set $\mathcal{W}$, $\binom{\mathcal{W}}{k}$ is the set of all k-subsets of $\mathcal{W}$. We use $[n]$ to denote the set $\{1,2,\ldots,n\}$. The eigenvalues of symmetric matrix $\mathbf{M}$ are denoted by $\lambda_1(\mathbf{M}) \leq \cdots \leq \lambda_n(\mathbf{M})$. $\mathbf{1}$, $\mathbf{I}$ and $\mathbf{0}$ denote the vector of all ones, the identity and the zero matrices with appropriate sizes, respectively. $\mathbf{S}_1 \succ \mathbf{S}_2$ (resp. $\mathbf{S}_1 \succeq \mathbf{S}_2$) means $\mathbf{S}_1 - \mathbf{S}_2$ is positive definite (resp. positive semidefinite). Finally, $\mathrm{diag}(\mathbf{W}_1,\ldots,\mathbf{W}_k)$ is the block-diagonal matrix whose main diagonal blocks are $\mathbf{W}_1,\ldots,\mathbf{W}_k$.

2   Preliminaries

Graph Matrices. Throughout this paper, we usually refer to undirected graphs $\mathcal{G} = (\mathcal{V},\mathcal{E})$ with n vertices (labeled with $[n]$) and m edges. With a little abuse of notation, we call $\widetilde{\mathbf{A}} \in \{-1,0,1\}^{n \times m}$ the incidence matrix of $\mathcal{G}$ after choosing an arbitrary orientation for its edges. The Laplacian matrix of $\mathcal{G}$ is defined as $\widetilde{\mathbf{L}} \triangleq \widetilde{\mathbf{A}}\widetilde{\mathbf{A}}^\top$. $\widetilde{\mathbf{L}}$ can be written as $\sum_{i=1}^m \widetilde{\mathbf{L}}_{e_i}$ in which $\widetilde{\mathbf{L}}_{e_i}$ is the *elementary Laplacian* associated with edge $e_i = \{u_i,v_i\}$, where the (u_i,u_i) and (v_i,v_i) entries are 1, and the (u_i,v_i) and (v_i,u_i) entries are -1. Anchoring $v_0 \in \mathcal{V}$ is equivalent to removing the row associated with v_0 from $\widetilde{\mathbf{A}}$. Anchoring v_0 results in the *reduced* incidence matrix $\mathbf{A}$ and the *reduced* Laplacian matrix $\mathbf{L} \triangleq \mathbf{A}\mathbf{A}^\top$. $\mathbf{L}$ is also known as the *Dirichlet*. We may assign positive weights to the edges of $\mathcal{G}$ via $w : \mathcal{E} \to \mathbb{R}_{>0}$. Let $\mathbf{W} \in \mathbb{R}^{m \times m}$ be the diagonal matrix whose (i,i) entry is equal to the weight of the ith edge. The *weighted* Laplacian (resp. reduced *weighted* Laplacian) is then defined as $\widetilde{\mathbf{L}}_w \triangleq \widetilde{\mathbf{A}} \mathbf{W} \widetilde{\mathbf{A}}^\top$ (resp. $\mathbf{L}_w \triangleq \mathbf{A} \mathbf{W} \mathbf{A}^\top$). Note that the (reduced) unweighted Laplacian is a special case of the (reduced) weighted Laplacian with $\mathbf{W} = \mathbf{I}_m$ (i.e., when all edges have unit weight).

Spanning Trees. A spanning tree of $\mathcal{G}$ is a spanning subgraph of $\mathcal{G}$ that is also a tree. Let $\mathbb{T}_{\mathcal{G}}$ denote the set of all spanning trees of $\mathcal{G}$. $t(\mathcal{G}) \triangleq |\mathbb{T}_{\mathcal{G}}|$ denotes the number of spanning trees in $\mathcal{G}$. As a generalization, for graphs whose edges are weighted by $w : \mathcal{E} \to \mathbb{R}_{>0}$, we define the *weighted number of spanning trees*,

$$t_w(\mathcal{G}) \triangleq \sum_{\mathcal{T} \in \mathbb{T}_{\mathcal{G}}} \mathbb{V}_w(\mathcal{T}). \tag{1}$$

We call $\mathbb{V}_w : \mathbb{T}_{\mathcal{G}} \to \mathbb{R}_{>0}$ the *tree value function* and define it as the product of the edge weights along a spanning tree. Notice that for unit edge weights, $t_w(\mathcal{G})$ coincides with $t(\mathcal{G})$. Thus, unless explicitly stated otherwise, we generally assume the graph is weighted. To prevent overflow and underflow, it is more convenient to work with $\log t_w(\mathcal{G})$. We formally define *tree-connectivity* as,

$$\tau_w(\mathcal{G}) \triangleq \begin{cases} \log t_w(\mathcal{G}) & \text{if } \mathcal{G} \text{ is connected,} \\ 0 & \text{otherwise.} \end{cases} \tag{2}$$

For the purpose of this work, without loss of generality we can assume $w(e) \geq 1$ for all $e \in \mathcal{E}$, and thus $\tau_w(\mathcal{G}) \geq 0$.[3] The equality occurs only when either $\mathcal{G}$ is not connected, or when $\mathcal{G}$ is a tree whose all edges have unit weight. Kirchhoff's seminal matrix-tree theorem is a classic result in spectral graph theory. This theorem relates the spectrum of the Laplacian matrix of graph to its number of spanning trees. The original matrix-tree theorem states that,

$$t(\mathcal{G}) = \det \mathbf{L} \tag{3}$$

$$= \frac{1}{n} \prod_{i=2}^{n} \lambda_i(\widetilde{\mathbf{L}}). \tag{4}$$

Here $\mathbf{L}$ is the reduced Laplacian after anchoring an arbitrary vertex. Kirchhoff's matrix-tree theorem has been generalized to the case of edge-weighted graphs. According to the generalized theorem, $t_w(\mathcal{G}) = \det \mathbf{L}_w = \frac{1}{n} \prod_{i=2}^{n} \lambda_i(\widetilde{\mathbf{L}}_w)$.

Submodularity. Suppose $\mathcal{W}$ is a finite set. Consider a set function $\xi : 2^{\mathcal{W}} \to \mathbb{R}$. ξ is called:

1. *normalized* iff $\xi(\varnothing) = 0$.
2. *monotone* iff $\xi(\mathcal{B}) \geq \xi(\mathcal{A})$ for every $\mathcal{A}$ and $\mathcal{B}$ s.t. $\mathcal{A} \subseteq \mathcal{B} \subseteq \mathcal{W}$.
3. *submodular* iff for every $\mathcal{A}$ and $\mathcal{B}$ s.t. $\mathcal{A} \subseteq \mathcal{B} \subseteq \mathcal{W}$ and $\forall s \in \mathcal{W} \setminus \mathcal{B}$ we have,

$$\xi(\mathcal{A} \cup \{s\}) - \xi(\mathcal{A}) \geq \xi(\mathcal{B} \cup \{s\}) - \xi(\mathcal{B}). \tag{5}$$

---

[3] Replacing any $w : \mathcal{E} \to \mathbb{R}_{\geq 0}$ with $w' : \mathcal{E} \to \mathbb{R}_{\geq 1} : e \mapsto \alpha_w w(e)$ for a sufficiently large constant α_w does not affect the set of t-optimal graphs.

3   D-Optimality via Graph Synthesis

In this section, we discuss the connection between D-optimality and t-optimality in SLAM by briefly reviewing the results in [16, 17, 18]. Consider the 2-D pose-graph SLAM problem where each measurement consists of the rotation (angle) and translation between a pair of robot poses over time, corrupted by an independently-drawn additive zero-mean Gaussian noise. According to our model, the covariance matrix of the noise vector corrupting the ith measurement can be written as $\mathrm{diag}(\sigma_{p_i}^2 \mathbf{I}_2, \sigma_{\theta_i}^2)$, where $\sigma_{p_i}^2$ and $\sigma_{\theta_i}^2$ denote the translational and rotational noise variances, respectively. As mentioned earlier, SLAM, as an EoG problem, admits a natural graphical representation $\mathcal{G} = (\mathcal{V}, \mathcal{E})$ in which poses correspond to graph vertices and edges correspond to the relative measurements between the corresponding poses. Furthermore, measurement precisions are incorporated into our model by assigning positive weights to the edges of $\mathcal{G}$. Note that for each edge we have two separate weight functions w_p and w_θ, defined as $w_p : e_i \mapsto \sigma_{p_i}^{-2}$ and $w_\theta : e_i \mapsto \sigma_{\theta_i}^{-2}$.

Let $\mathbb{V}\mathrm{ar}[\hat{\mathbf{x}}_{\mathsf{mle}}]$ be the asymptotic covariance matrix of the maximum likelihood estimator (Cramér-Rao lower bound) for estimating the trajectory $\mathbf{x}$. In [16, 17, 18], we investigated the impact of graph topology on the D-optimality criterion $(\det \mathbb{V}\mathrm{ar}[\hat{\mathbf{x}}_{\mathsf{mle}}])$ in SLAM. The results presented in [18] are threefold. First, in [18, Proposition 2] it is proved that

$$-2\,\tau_{w_p}(\mathcal{G}) - \log \det(\mathbf{L}_{w_\theta} + \gamma \mathbf{I}) \le \log \det \mathbb{V}\mathrm{ar}[\hat{\mathbf{x}}_{\mathsf{mle}}] \le -2\,\tau_{w_p}(\mathcal{G}) - \tau_{w_\theta}(\mathcal{G}) \quad (6)$$

in which γ is a parameter whose value depends on the maximum distance between the neighbouring robot poses normalized by $\sigma_{p_i}^2$'s; e.g., this parameter shrinks by reducing the distance between the neighbouring poses, or by reducing the precision of the translational measurements (see [18, Remark 2]). Next, based on this result, it is easy to see that [18, Theorem 5],

$$\lim_{\gamma \to 0^+} \log \det \mathbb{V}\mathrm{ar}[\hat{\mathbf{x}}_{\mathsf{mle}}] = -2\,\tau_{w_p}(\mathcal{G}) - \tau_{w_\theta}(\mathcal{G}). \quad (7)$$

Note that the expression above depends only on the graphical representation of the problem. Finally, the empirical observations and Monte Carlo simulations based on a number of synthetic and real datasets indicate that the RHS of (7) provides a reasonable estimate for $\log \det \mathbb{V}\mathrm{ar}[\hat{\mathbf{x}}_{\mathsf{mle}}]$ even in the non-asymptotic regime where γ is not negligible. In what follows, we demonstrate how these results can be used in a graph-theoretic approach to the D-optimal measurement selection and pruning problems.

Measurement Selection. Maintaining sparsity is essential for computational efficiency in SLAM, especially in long-term autonomy. Sparsity can be preserved by implementing a measurement selection policy to asses the significance of new or existing measurements. Such a vetting process can be realized by (i) assessing the significance of any new measurement before adding it to the graph, and/or (ii) pruning a subset of the acquired measurements if their contribution is deemed to be insufficient. These ideas have been investigated in the literature; for the former approach see, e.g., [13, 26], and see, e.g., [21, 12] for the latter.

Now consider the D-optimal measurement selection problem whose goal is to select the optimal k-subset of measurements such that the resulting $\log \det \mathbb{V}\mathrm{ar}[\hat{\mathbf{x}}_{\mathsf{mle}}]$ is minimized. This problem is closely related to the D-optimal sensor selection problem for which two successful approximation algorithms have been proposed in [13] and [26] under the assumption of linear sensor models. The measurement models in SLAM are nonlinear. Nevertheless, we can still use [13, 26] after linearizing the measurement model. Note that the Fisher information matrix and $\log \det \mathbb{V}\mathrm{ar}[\hat{\mathbf{x}}_{\mathsf{mle}}]$ in SLAM depend on the true $\mathbf{x}$. Since the true value of $\mathbf{x}$ is not available, in practice these terms are approximated by evaluating the Jacobian matrix at the estimate obtained by maximizing the log-likelihood function using an iterative solver.

An alternative approach would be to replace $\log \det \mathbb{V}\mathrm{ar}[\hat{\mathbf{x}}_{\mathsf{mle}}]$ with a graph-theoretic objective function based on (7). Note that this is equivalent to reducing the original problem into a graph synthesis problem. The graphical approach has the following advantages:

1. *Robustness*: Maximum likelihood estimation in SLAM boils down to solving a non-convex optimization problem via iterative solvers. These solvers are subject to local minima. Hence, the approximated $\log \det \mathbb{V}\mathrm{ar}[\hat{\mathbf{x}}_{\mathsf{mle}}]$ can be highly inaccurate and lead to misleading designs if the Jacobian is evaluated at a local minimum (see [18, Section VI] for an example). The graph-theoretic objective function based on (7), however, is independent of the trajectory $\mathbf{x}$ and, therefore, is robust to such convergence errors.

2. *Flexibility*: To directly compute $\log \det \mathbb{V}\mathrm{ar}[\hat{\mathbf{x}}_{\mathsf{mle}}]$, we first need a nominal or estimated trajectory $\mathbf{x}$. Furthermore, for the latter we also need to know the realization of relative measurements. Therefore, any design or decisions made in this way will be confined to a particular trajectory. On the contrary, the graphical approach requires only the knowledge of the topology of the graph, and thus is more flexible. Note that the t-optimal topology corresponds to a range of trajectories. Therefore, the graphical approach enables us to assess the D-optimality of a particular design with minimum information and without relying on any particular—planned, nominal or estimated—trajectory.

We will investigate the problem of designing t-optimal graphs in Section 4.

4   Synthesis of Near-t-Optimal Graphs

Problem Formulation. In this section, we formulate and tackle the combinatorial optimization problem of designing sparse graphs with the maximum weighted tree-connectivity. Since the decision variables are the edges of the graph, it is more convenient to treat the weighted tree-connectivity as a function of the edge set of the graph for a given set of vertices ($\mathcal{V} = [n]$) and a positive weight function $w : \binom{[n]}{2} \to \mathbb{R}_{\geq 1}$. $\mathsf{tree}_{n,w} : 2^{\binom{[n]}{2}} \to \mathbb{R}_{\geq 0} : \mathcal{E} \mapsto \tau_w([n], \mathcal{E})$ takes as input a set of edges $\mathcal{E}$ and returns the weighted tree-connectivity of graph $([n], \mathcal{E})$. To simplify our notation, hereafter we drop n and/or w from $\mathsf{tree}_{n,w}$ (and similar terms) whenever n and/or w are clear from the context.

Problem 1 (k-ESP). Suppose the following are given:

- a *base graph* $\mathcal{G}_{\text{init}} = ([n], \mathcal{E}_{\text{init}})$
- a weight function $w : \binom{[n]}{2} \to \mathbb{R}_{\geq 1}$
- a set of c *candidate* edges (either $\mathcal{C}^+$ or $\mathcal{C}^-$)
- an integer $k \leq c$

Consider the following edge selection problems (ESP):

⬦ k-ESP$^+$

$$\underset{\mathcal{E} \subseteq \mathcal{C}^+ \subseteq \binom{[n]}{2} \setminus \mathcal{E}_{\text{init}}}{\text{maximize}} \quad \text{tree}(\mathcal{E}_{\text{init}} \cup \mathcal{E}) \quad \text{subject to} \quad |\mathcal{E}| = k. \tag{8}$$

⬦ k-ESP$^-$

$$\underset{\mathcal{E} \subseteq \mathcal{C}^- \subseteq \mathcal{E}_{\text{init}}}{\text{maximize}} \quad \text{tree}(\mathcal{E}_{\text{init}} \setminus \mathcal{E}) \quad \text{subject to} \quad |\mathcal{E}| = k. \tag{9}$$

Remark 1. It is easy to see that any instance of (8) can be expressed as an instance of (9) and vice versa. Therefore, without loss of generality, in this work we only consider k-ESP$^+$.

1-ESP$^+$. Consider the simple case of $k = 1$. $\Delta_{uv} \triangleq \mathbf{a}_{uv} \mathbf{L}^{-1} \mathbf{a}_{uv}$ is known as the *effective resistance* between vertices u and v. Here $\mathbf{a}_{uv} \in \{-1,0,1\}^{n-1}$ is the vector $\mathbf{e}_u^n - \mathbf{e}_v^n$ after crossing out the entry that corresponds to the anchored vertex. Effective resistance has emerged from several other contexts as a key factor; see, e.g., [8]. In [19, Lemma 3.1] it is shown that the optimal choice in 1-ESP$^+$ is the candidate edge with the maximum $w(e)\Delta_e$. The effective resistance can be efficiently computed by performing a Cholesky decomposition on the reduced weighted Laplacian matrix of the base graph $\mathbf{L}_{\text{init}}$ and solving a triangular linear system (see [19]). In the worst case and for a dense base graph 1-ESP$^+$ can be solved in $O(n^3 + cn^2)$ time.

4.1 Approximation Algorithms for k-ESP$^+$

Solving the general case of k-ESP$^+$ by exhaustive search requires examining $\binom{c}{k}$ graphs. This is not practical even when c is bounded (e.g., for $c = 30$ and $k = 10$ we need to perform more than 3×10^7 Cholesky factorizations). Here we propose a complementary pair of approximation algorithms.

I: Greedy. The greedy algorithm finds an approximate solution to k-ESP$^+$ by decomposing it into a sequence of k 1-ESP$^+$ problems, each of which can be solved using the procedure outlined above. After solving each subproblem, the optimal edge is moved from the candidate set to the base graph. The next 1-ESP$^+$ subproblem is defined using the updated candidate set and the updated base graph. If the graph is dense, a naive implementation of the greedy algorithm requires less than $O(kcn^3)$ operations. An efficient implementation of this approach that requires $O(n^3 + kcn^2)$ time is described in [19, Algorithm 1].

Analysis. Let $\mathcal{G}_{\text{init}} = ([n], \mathcal{E}_{\text{init}})$ be a *connected* base graph and $w : \binom{[n]}{2} \to \mathbb{R}_{\geq 1}$. Consider the following function.

$$\mathcal{X}_w : \mathcal{E} \mapsto \text{tree}(\mathcal{E} \cup \mathcal{E}_{\text{init}}) - \text{tree}(\mathcal{E}_{\text{init}}). \tag{10}$$

In k-ESP⁺, we restrict the domain of $\mathcal{X}_w$ to $2^{\mathcal{C}^+}$. Note that $\text{tree}(\mathcal{E}_{\text{init}})$ is a constant and, therefore, we can express the objective function in k-ESP⁺ using $\mathcal{X}_w$,

$$\underset{\mathcal{E} \subseteq \mathcal{C}^+}{\text{maximize}} \quad \mathcal{X}_w(\mathcal{E}) \quad \text{subject to} \quad |\mathcal{E}| = k. \tag{11}$$

Theorem 1. $\mathcal{X}_w$ *is normalized, monotone and submodular.*

Proof. Omitted due to space limitation—see the technical report [19].

Maximizing an arbitrary monotone submodular function subject to a cardinality constraint can be NP-hard in general (see, e.g., the Maximum Coverage problem [11]). Nemhauser et al. [24] in their seminal work have shown that the greedy algorithm is a constant-factor approximation algorithm with a factor of $\eta \triangleq (1 - 1/e) \approx 0.63$ for any (normalized) monotone submodular function subject to a cardinality constraint. Let OPT be the optimum value of (8), $\mathcal{E}_{\text{greedy}}$ be the edges selected by the greedy algorithm, $\tau_{\text{greedy}} \triangleq \text{tree}(\mathcal{E}_{\text{greedy}} \cup \mathcal{E}_{\text{init}})$ and $\tau_{\text{init}} \triangleq \text{tree}(\mathcal{E}_{\text{init}})$.

Corollary 1. $\tau_{\text{greedy}} \geq \eta \, \text{OPT} + (1 - \eta) \, \tau_{\text{init}}.$

II: Convex Relaxation. In this section, we design an approximation algorithm for k-ESP⁺ through convex relaxation. We begin by assigning an auxiliary variable $0 \leq \pi_i \leq 1$ to each candidate edge $e_i \in \mathcal{C}^+$. The idea is to reformulate the problem such that finding the optimal set of candidate edges is equivalent to finding the optimal value for π_i's. Let $\boldsymbol{\pi} \triangleq [\pi_1 \ \pi_2 \ \cdots \ \pi_c]^\top$ be the stacked vector of auxiliary variables. Define,

$$\mathbf{L}_w(\boldsymbol{\pi}) \triangleq \mathbf{L}_{\text{init}} + \sum_{e_i \in \mathcal{C}^+} \pi_i w(e_i) \mathbf{L}_{e_i} = \mathbf{A}\mathbf{W}^\pi \mathbf{A}^\top, \tag{12}$$

where $\mathbf{L}_{e_i}$ is the reduced elementary Laplacian, $\mathbf{A}$ is the reduced incidence matrix of $\mathcal{G}_\bullet \triangleq ([n], \mathcal{E}_{\text{init}} \cup \mathcal{C}^+)$, and $\mathbf{W}^\pi$ is the diagonal matrix of edge weights assigned by the following weight function,

$$w^\pi(e_i) = \begin{cases} \pi_i w(e_i) & e_i \in \mathcal{C}^+, \\ w(e_i) & e_i \notin \mathcal{C}^+. \end{cases} \tag{13}$$

Lemma 1. *If $\mathcal{G}_{\text{init}}$ is connected, $\mathbf{L}_w(\boldsymbol{\pi})$ is positive definite for any $\boldsymbol{\pi} \in [0,1]^c$.*

As before, for convenience we assume $\mathcal{G}_{\text{init}}$ is connected. Consider the following optimization problems over $\boldsymbol{\pi}$.

$$
\begin{aligned}
\underset{\boldsymbol{\pi}}{\text{maximize}} \quad & \log \det \mathbf{L}_w(\boldsymbol{\pi}) \\
\text{subject to} \quad & \|\boldsymbol{\pi}\|_0 = k, \\
& 0 \le \pi_i \le 1, \ \forall i \in [c].
\end{aligned} \tag{P_1}
$$

$$
\begin{aligned}
\underset{\boldsymbol{\pi}}{\text{maximize}} \quad & \log \det \mathbf{L}_w(\boldsymbol{\pi}) \\
\text{subject to} \quad & \|\boldsymbol{\pi}\|_1 = k, \\
& \pi_i \in \{0,1\}, \ \forall i \in [c].
\end{aligned} \tag{P_1'}
$$

P_1 is equivalent to our original definition of k-ESP$^+$. First, note that from the generalized matrix-tree theorem we know that the objective function is equal to the weighted tree-connectivity of graph $\mathcal{G}_\bullet = ([n], \mathcal{E}_{\text{init}} \cup \mathcal{C}^+)$ whose edges are weighted by w^π. The auxiliary variables act as selectors: the ith candidate edge is selected iff $\pi_i = 1$. The combinatorial difficulty of k-ESP$^+$ here is embodied in the non-convex ℓ_0-norm constraint. It is easy to see that in P_1, at the optimal solution, auxiliary variables take binary values. This is why the integer program P_1' is equivalent to P_1. A natural choice for relaxing P_1' is to replace $\pi_i \in \{0,1\}$ with $0 \le \pi_i \le 1$; i.e.,

$$
\begin{aligned}
\underset{\boldsymbol{\pi}}{\text{maximize}} \quad & \log \det \mathbf{L}_w(\boldsymbol{\pi}) \\
\text{subject to} \quad & \|\boldsymbol{\pi}\|_1 = k, \\
& 0 \le \pi_i \le 1, \ \forall i \in [c].
\end{aligned} \tag{P_2}
$$

The feasible set of P_2 contains that of P_1'. Hence, the optimum value of P_2 is an upper bound for the optimum of P_1 (or, equivalently, P_1'). Note that the ℓ_1-norm constraint here is identical to $\sum_{i=1}^c \pi_i = k$. P_2 is a convex optimization problem since the objective function (tree-connectivity) is concave and the constraints are linear and affine in $\boldsymbol{\pi}$. In fact, P_2 is an instance of the MAXDET problem [29] subject to additional affine constraints on $\boldsymbol{\pi}$. It is worth noting that P_2 can be reached also by relaxing the non-convex ℓ_0-norm constraint in P_1 into the convex ℓ_1-norm constraint $\|\boldsymbol{\pi}\|_1 = k$. Furthermore, P_2 is also closely related to a ℓ_1-regularised variant of MAXDET,

$$
\begin{aligned}
\underset{\boldsymbol{\pi}}{\text{maximize}} \quad & \log \det \mathbf{L}_w(\boldsymbol{\pi}) - \lambda \|\boldsymbol{\pi}\|_1 \\
\text{subject to} \quad & 0 \le \pi_i \le 1, \ \forall i \in [c].
\end{aligned} \tag{P_3}
$$

This problem is a penalized form of P_2; these two problems are equivalent for some positive value of λ. Problem P_3 is also a convex optimization problem for any non-negative λ. The ℓ_1-norm in P_3 penalizes the loss of sparsity, while the log-determinant rewards stronger tree-connectivity. λ is a parameter that controls the sparsity of the resulting graph; i.e., a larger λ yields a sparser vector of selectors $\boldsymbol{\pi}$. P_3 is closely related to graphical lasso [6]. P_2 (and P_3) can be solved globally in polynomial time using interior-point methods [4, 13]. After finding a globally optimal solution $\boldsymbol{\pi}^\star$ for the relaxed problem P_2, we ultimately need to map it into a feasible $\boldsymbol{\pi}$ for P_1'; i.e., choosing k edges from the candidate set $\mathcal{C}^+$.

Lemma 2. $\boldsymbol{\pi}^\star$ *is an optimal solution for* k-ESP$^+$ *iff* $\boldsymbol{\pi}^\star \in \{0,1\}^c$.

Rounding. In general, $\pi^\star$ may contain fractional values that need to be mapped into feasible integral values for P$_1'$ by a *rounding procedure* that sets k auxiliary variables to one and others to zero. The most intuitive deterministic rounding policy is to pick the k edges with the largest $\pi_i^\star$'s.

The idea behind the convex relaxation technique described so far can be seen as a graph-theoretic special case of the algorithm proposed in [13]. However, it is not clear yet how the solution of the relaxed convex problem P$_2$ is related to that of the original non-convex k-ESP$^+$ in the integer program P$_1'$. To answer this question, consider the following randomized strategy. We may attempt to find a suboptimal solution for k-ESP$^+$ by randomly sampling candidates. In this case, for the ith candidate edge, we flip a coin whose probability of heads is π_i (independent of other candidates). We then select that candidate edge if the coin lands on head.

Theorem 2. *Let the random variables k^* and t_w^* denote, respectively, the number of chosen candidate edges and the corresponding weighted number of spanning trees achieved by the above randomized algorithm. Then,*

$$\mathbb{E}\left[k^*\right] = \sum_{i=1}^{c} \pi_i, \tag{14}$$

$$\mathbb{E}\left[t_w^*\right] = \det \mathbf{L}_w(\pi). \tag{15}$$

Proof. See [19] for the proof.[4]

According to Theorem 2, the randomized algorithm described above on average selects $\sum_{i=1}^{c} \pi_i$ candidate edges and achieves $\det \mathbf{L}_w(\pi)$ weighted number of spanning trees in expectation. Note that these two terms appear in the constraints and objective of the relaxed problem P$_2$, respectively. Therefore, the relaxed problem can be interpreted as the problem of finding the optimal sampling probabilities π for the randomized algorithm described above. This offers a new narrative:

Corollary 2. *The objective in P$_2$ is to find the optimal probabilities $\pi^\star$ for sampling edges from $\mathcal{C}^+$ such that the weighted number of spanning trees is maximized in expectation, while the expected number of newly selected edges is equal to k.*

In other words, P$_2$ can be seen as a convex relaxation of P$_1$ at the expense of maximizing the objective and satisfying the constraint, both *in expectation*. This new interpretation can be used as a basis for designing randomized rounding procedures based on the randomized technique described above. If one uses $\pi^\star$ (the fractional solution of the relaxed problem P$_2$) in the aforementioned randomized rounding scheme, Theorem 2 ensures that, on average, such a method attains $\det \mathbf{L}(\pi^\star)$ by picking k new edges in expectation. Finally, we note that this new

---

[4] A generalized version of this theorem that covers the more general case of [13] is proved in [19].

interpretation sheds light on why the deterministic rounding policy described earlier performs well in practice. Note that randomly sampling candidate edges with the probabilities in $\boldsymbol{\pi}^\star$ does not necessarily result in a feasible solution for P'_1. That being said, consider every feasible outcome in which exactly k candidate edges are selected by the randomized algorithm with probabilities in $\boldsymbol{\pi}^\star$. It is easy to show that the deterministic procedure described earlier (picking k candidates with the largest $\pi_i^\star$'s) is in fact selecting the most probable feasible outcome (given that exactly k candidates have been selected).

Near-Optimality Certificates. It is impractical to compute OPT via exhaustive search in large problems. Nevertheless, the approximation algorithms described above yield lower and upper bounds for OPT that can be quite tight in practice. Let $\tau_{\text{cvx}}^\star$ be the optimum value of P_2. Moreover, let τ_{cvx} be the suboptimal value obtained after rounding the solution of P_2 (e.g., picking the k largest $\pi_i^\star$'s). The following corollary readily follows from the analysis of the greedy and convex approximation algorithms.

Corollary 3.

$$\max\left\{\tau_{\text{greedy}},\tau_{\text{cvx}}\right\} \leq \text{OPT} \leq \min\left\{\mathcal{U}_{\text{greedy}},\tau_{\text{cvx}}^\star\right\} \tag{16}$$

where $\mathcal{U}_{\text{greedy}} \triangleq \zeta\tau_{\text{greedy}} + (1-\zeta)\tau_{\text{init}}$ *in which* $\zeta \triangleq \eta^{-1} \approx 1.58$.

The bounds in Corollary 3 can be computed by running the greedy and convex relaxation algorithms. Whenever OPT is beyond reach, the upper bound can be used to asses the quality of any feasible design. Let $\mathcal{S}$ be an arbitrary k-subset of $\mathcal{C}^+$ and $\tau_\mathcal{S} \triangleq \text{tree}(\mathcal{S} \cup \mathcal{E}_{\text{init}})$. $\mathcal{S}$ can be, e.g., the solution of greedy algorithm, the solution of P_2 after rounding, an existing design (e.g., an existing pose-graph problem) or a suboptimal solution proposed by a third party. Let $\mathcal{L}$ and $\mathcal{U}$ denote the lower and upper bounds in (16), respectively. From Corollary 3 we have,

$$\max\left\{0,\mathcal{L}-\tau_\mathcal{S}\right\} \leq \underbrace{\text{OPT}-\tau_\mathcal{S}}_{\text{optimality gap}} \leq \mathcal{U}-\tau_\mathcal{S}. \tag{17}$$

Therefore, $\mathcal{U}-\tau_\mathcal{S}$ (or similarly, $\mathcal{U}/\tau_\mathcal{S} \geq \text{OPT}/\tau_\mathcal{S}$) can be used as a near-optimality certificate for an arbitrary design $\mathcal{S}$.

Two Weight Functions. In the synthesis problem studied so far, it was implicitly assumed that each edge is weighted by a single weight function. This is not necessarily the case in SLAM, where each measurement has two components, each of which has its own precision, i.e., w_p and w_θ in (7). Hence, we need to revisit the synthesis problem in a more general setting, where multiple weight functions assign weights, simultaneously, to a single edge. It turns out that the proposed approximation algorithms and their analyses can be easily generalized to handle this case.

1. *Greedy Algorithm*: For the greedy algorithm, we just need to replace $\mathcal{X}_w$ with $\mathcal{Y}_w : \mathcal{E} \mapsto 2\,\mathcal{X}_{w_p}(\mathcal{E}) + \mathcal{X}_{w_\theta}(\mathcal{E})$; see (7). Note that $\mathcal{Y}_w$ is a linear combination of

normalized monotone submodular functions with positive weights, and therefore is also normalized, monotone and submodular.

2. *Convex Relaxation*: The convex relaxation technique can be generalized to the case of multi-weighted edges by replacing the concave objective function $\log \det \mathbf{L}_w(\boldsymbol{\pi})$ with $2 \log \det \mathbf{L}_{w_p}(\boldsymbol{\pi}) + \log \det \mathbf{L}_{w_\theta}(\boldsymbol{\pi})$, which is also concave.

Remark 2. Recall that our goal was to design sparse, yet reliable SLAM problems. So far we considered the problem of designing D-optimal SLAM problems with a given number of edges. The dual approach would be to find the sparsest SLAM problem such that the determinant of the estimation-error covariance is less than a desired threshold. Take for example the following scenario: find the sparsest SLAM problem by selecting loop-closure measurements from a given set of candidates such that the resulting D-criterion is 50% smaller than that of dead reckoning. The dual problem can be written as,

$$\underset{\mathcal{E} \subseteq \mathcal{C}^+}{\text{minimize}} \quad |\mathcal{E}| \quad \text{subject to} \quad \mathfrak{X}_w(\mathcal{E}) \geq \tau_{\min}. \qquad (18)$$

in which $\tau_{\min}$ is given. In [19] we have shown that our proposed approximation algorithms and their analyses can be easily modified to address the dual problem. Due to space limitation, we have to refrain from discussing the dual problem in this paper.

4.2   Experimental Results

The proposed algorithms were implemented in MATLAB. P_2 is modelled using YALMIP [22] and solved using SDPT3 [28].

Random Graphs. Figure 1 illustrates the performance of our approximate algorithms in randomly generated graphs. The set of candidates in these experiments is $\mathcal{C}^+ = \binom{[n]}{2} \setminus \mathcal{E}_{\text{init}}$. Figures 1a and 1b show the resulting tree-connectivity as a function of the number of randomly generated edges for a fixed $k = 5$ and, respectively, $n = 20$ and $n = 50$. Our results indicate that both algorithms exhibit remarkable performances for $k = 5$. Note that computing OPT by exhaustive search is only feasible in small instances such as Figure 1a. Nevertheless, computing the exact OPT is not crucial for evaluating our approximate algorithms, as Corollary 3 guarantees that $\tau^\star_{\text{greedy}} \leq \text{OPT} \leq \tau^\star_{\text{cvx}}$; i.e., the space between each black $\cdot$ and the corresponding green $\times$. Figure 1c shows the results obtained for varying k. The optimality gap for τ_{cvx} gradually grows as the planning horizon k increases. Our greedy algorithm, however, still yields a near-optimal approximation.

Real Pose-Graph Dataset. We also evaluated the proposed algorithms on the Intel Research Lab dataset as a popular pose-graph SLAM benchmark.[5] In this scenario, $\mathcal{E}_{\text{init}}$ is chosen to be the set of odometry edges, and $\mathcal{C}^+$ is the set

---

[5] https://svn.openslam.org/data/svn/g2o/trunk/data/2d/intel/intel.g2o

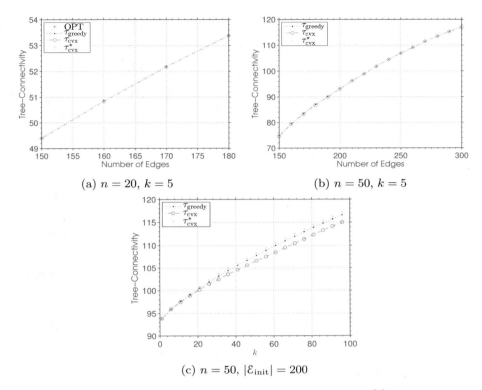

(a) $n = 20$, $k = 5$
(b) $n = 50$, $k = 5$

(c) $n = 50$, $|\mathcal{E}_{\text{init}}| = 200$

Fig. 1: k-ESP$^+$ on randomly generated graphs with $\mathcal{C}^+ = \binom{[n]}{2} \setminus \mathcal{E}_{\text{init}}$.

of loop closures. The parameters in this graph are $n = 943$, $|\mathcal{E}_{\text{init}}| = 942$ and $|\mathcal{C}^+| = 895$. Note that computing the true OPT via exhaustive search is clearly impractical; e.g., for $k = 100$, there are more than 10^{134} possible graphs. For the edge weights, we are using the original information (precisions) reported in the dataset. Since the translational and rotational measurements have different precisions, two weight functions—w_p and w_θ—assign weights to each edge of the graph, and the objective is to maximize $2\tau_{w_p}(\mathcal{G}) + \tau_{w_\theta}(\mathcal{G})$. Figure 2 shows the resulting objective value for the greedy and convex relaxation approximation algorithms, as well as the upper bounds ($\mathcal{U}$) in Corollary 3.[6] According to Figure 2, both algorithms have successfully found near-t-optimal (near-D-optimal) designs. The greedy algorithm has outperformed the convex relaxation with the simple deterministic (sorting) rounding procedure. For small values of k, the upper bound $\mathcal{U}$ on OPT is given by $\mathcal{U}_{\text{greedy}}$ (blue curve). However, for $k \geq 60$, the convex relaxation provides a significantly tighter upper bound on OPT (green curve). In this dataset, YALMIP+SDPT3 on an Intel Core i5-2400 operating at 3.1 GHz can solve the convex program in about 20-50 seconds, while a naive

---

[6] See also https://youtu.be/5JZF2QiRbDE for a visualization.

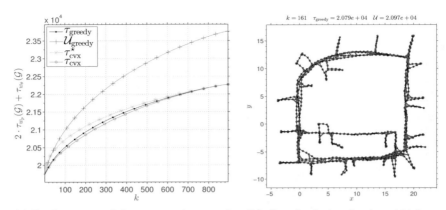

(a) Performance of the proposed approximation algorithms.

(b) Greedy design for $k = 161$ loop closures (out of 895). Loop-closure edges are shown in blue.

Fig. 2: k-ESP$^+$ for pose-graph SLAM on the Intel Research Lab dataset.

implementation of the greedy algorithm (without using rank-one updates) can solve the case with $k = 400$ in about 25 seconds.

5   Conclusion

We presented a graph-theoretic approach to the problem of designing sparse reliable (i.e., near-D-optimal) pose-graph SLAM. This paper demonstrated that this problem boils down to a combinatorial optimization problem whose goal is to find a sparse graph with the maximum weighted number of spanning trees. The problem of characterizing t-optimal graphs is an open problem with—to the best of our knowledge—no known efficient algorithm. We designed two efficient approximation algorithms with provable guarantees and near-optimality certificates. First and foremost, we introduced a new submodular graph invariant, i.e., weighted tree-connectivity. This was used to guarantee that the greedy algorithm is a constant-factor approximation algorithm for this problem with a factor of $(1 - 1/e)$ (up to a constant normalizer). In another approach, we formulated the original combinatorial optimization problem as an integer program that admits a natural convex relaxation. We discussed deterministic and randomized rounding schemes. Our analysis sheds light on the connection between the original and the relaxed problems. Finally, we evaluated the performance of the proposed approximation algorithms using random graphs and a real pose-graph SLAM dataset. Although this paper specifically targeted SLAM, we note that the proposed algorithms can be readily used to synthesize near-t-optimal graphs in any domain where maximizing tree-connectivity is useful. See, e.g., [10, 20, 3, 19] for applications in Chemistry, RNA modelling, network reliability under random link failure and estimation over sensor networks, respectively.

Bibliography

[1] Bailey, R.A., Cameron, P.J.: Combinatorics of optimal designs. Surveys in Combinatorics **365** (2009) 19–73

[2] Bauer, D., Boesch, F.T., Suffel, C., Van Slyke, R.: On the validity of a reduction of reliable network design to a graph extremal problem. Circuits and Systems, IEEE Transactions on **34**(12) (1987) 1579–1581

[3] Boesch, F.T., Satyanarayana, A., Suffel, C.L.: A survey of some network reliability analysis and synthesis results. Networks **54**(2) (2009) 99–107

[4] Boyd, S., Vandenberghe, L.: Convex optimization. Cambridge university press (2004)

[5] Cheng, C.S.: Maximizing the total number of spanning trees in a graph: two related problems in graph theory and optimum design theory. Journal of Combinatorial Theory, Series B **31**(2) (1981) 240–248

[6] Friedman, J., Hastie, T., Tibshirani, R.: Sparse inverse covariance estimation with the graphical lasso. Biostatistics **9**(3) (2008) 432–441

[7] Gaffke, N.: D-optimal block designs with at most six varieties. Journal of Statistical Planning and Inference **6**(2) (1982) 183–200

[8] Ghosh, A., Boyd, S., Saberi, A.: Minimizing effective resistance of a graph. SIAM review **50**(1) (2008) 37–66

[9] Godsil, C., Royle, G.: Algebraic graph theory. Graduate Texts in Mathematics Series. Springer London, Limited (2001)

[10] Gutman, I., Mallion, R., Essam, J.: Counting the spanning trees of a labelled molecular-graph. Molecular Physics **50**(4) (1983) 859–877

[11] Hochbaum, D.S.: Approximation algorithms for NP-hard problems. PWS Publishing Co. (1996)

[12] Huang, G., Kaess, M., Leonard, J.J.: Consistent sparsification for graph optimization. In: Mobile Robots (ECMR), 2013 European Conference on, IEEE (2013) 150–157

[13] Joshi, S., Boyd, S.: Sensor selection via convex optimization. Signal Processing, IEEE Transactions on **57**(2) (2009) 451–462

[14] Kelmans, A.K.: On graphs with the maximum number of spanning trees. Random Structures & Algorithms **9**(1-2) (1996) 177–192

[15] Kelmans, A.K., Kimelfeld, B.: Multiplicative submodularity of a matrix's principal minor as a function of the set of its rows and some combinatorial applications. Discrete Mathematics **44**(1) (1983) 113–116

[16] Khosoussi, K., Huang, S., Dissanayake, G.: Novel insights into the impact of graph structure on SLAM. In: Proceedings of IEEE/RSJ International Conference on Intelligent Robots and Systems (IROS), 2014. (2014) 2707–2714

[17] Khosoussi, K., Huang, S., Dissanayake, G.: Good, bad and ugly graphs for SLAM. RSS Workshop on the problem of mobile sensors (2015)

[18] Khosoussi, K., Huang, S., Dissanayake, G.: Tree-connectivity: A metric to evaluate the graphical structure of SLAM problems. Proceedings of the IEEE International Conference on Robotics and Automation (ICRA) (2016)

[19] Khosoussi, K., Sukhatme, G.S., Huang, S., Dissanayake, G.: Maximizing the weighted number of spanning trees: Near-t-optimal graphs. arXiv:1604.01116 (2016)

[20] Kim, N., Petingi, L., Schlick, T.: Network theory tools for RNA modeling. WSEAS transactions on mathematics **9**(12) (2013) 941

[21] Kretzschmar, H., Stachniss, C., Grisetti, G.: Efficient information-theoretic graph pruning for graph-based slam with laser range finders. In: Intelligent Robots and Systems (IROS), 2011 IEEE/RSJ International Conference on. (2011) 865 –871

[22] Löfberg, J.: Yalmip : A toolbox for modeling and optimization in MATLAB. In: Proceedings of the CACSD Conference, Taipei, Taiwan (2004)

[23] Myrvold, W.: Reliable network synthesis: Some recent developments. In: Proceedings of International Conference on Graph Theory, Combinatorics, Algorithms, and Applications. (1996)

[24] Nemhauser, G.L., Wolsey, L.A., Fisher, M.L.: An analysis of approximations for maximizing submodular set functions - I. Mathematical Programming **14**(1) (1978) 265–294

[25] Petingi, L., Rodriguez, J.: A new technique for the characterization of graphs with a maximum number of spanning trees. Discrete mathematics **244**(1) (2002) 351–373

[26] Shamaiah, M., Banerjee, S., Vikalo, H.: Greedy sensor selection: Leveraging submodularity. In: 49th IEEE conference on decision and control (CDC), IEEE (2010) 2572–2577

[27] Shier, D.: Maximizing the number of spanning trees in a graph with n nodes and m edges. Journal Research National Bureau of Standards, Section B **78** (1974) 193–196

[28] Tütüncü, R.H., Toh, K.C., Todd, M.J.: Solving semidefinite-quadratic-linear programs using sdpt3. Mathematical programming **95**(2) (2003) 189–217

[29] Vandenberghe, L., Boyd, S., Wu, S.P.: Determinant maximization with linear matrix inequality constraints. SIAM journal on matrix analysis and applications **19**(2) (1998) 499–533

[30] Weichenberg, G., Chan, V.W., Médard, M.: High-reliability topological architectures for networks under stress. Selected Areas in Communications, IEEE Journal on **22**(9) (2004) 1830–1845

Batch Misalignment Calibration of Multiple Three-Axis Sensors

Gabriel Hugh Elkaim[1]

Author is with the Autonomous Systems Lab at the University of California in Santa Cruz, Santa Cruz, CA 95060, USA. elkaim at soe.ucsc.edu

Abstract. When using multiple three-axis sensors, misalignments between the sensors can corrupt the measurements even when each individual sensor is itself well calibrated. An algorithm was developed to determine the misalignment between any two three-axis sensors whose unit vectors are known in the inertial frame and can be measured by the sensor in body coordinates. This is a batch algorithm, assuming each sensor has been individually calibrated, and requires no extraneous equipment. The input to the misalignment algorithm is a locus of measurements, the paired body-fixed measurements of both sensors, and the known unit vectors for each in inertial coordinates. The output of the algorithm is the misalignment direction cosine matrix matrix such that the second (slave) sensor can be rotated into a consistent coordinate frame as the first (master) sensor. The algorithm works both with heterogenous and homogenous sensors (e.g., accelerometer and magnetometer or multiple magnetometers). The algorithm was validated using Monte Carlo simulations for both large and small misalignments, with and without realistic sensor noise, and shows excellent convergence properties. The algorithm is demonstrated experimentally on a small UAV sensor package and on a small satellite equipped with three high quality magnetometers. In both cases, the algorithm identifies large misalignments created during installation, as well as residual small misalignments when sensing axes are aligned.

1 Introduction

The problem of estimating attitude from vector measurements is well studied for aircraft and satellites. Often known as Wahba's problem due to her 1965 problem in the SIAM Journal [1], there have been myriad solutions proposed and implemented [2]. While these methods all work very well in practice, they assume that the sensors are in fact perfectly aligned and calibrated. That is, the assumption is that the two independent sensors are rigidly coupled and have collinear sensing axes. Furthermore, all null shifts, scale factors, and cross coupling on the axes are also assumed to be fully calibrated.

The literature is rich with calibration solutions for determining the null shifts, scale factor errors, and cross coupling [3][4][5]. However, there are far fewer works specifically about calibrating the misalignment of one three axis sensor

© Springer Nature Switzerland AG 2020
K. Goldberg et al. (Eds.): *Algorithmic Foundations of Robotics XII*, SPAR 13, pp. 33–47, 2020.
https://doi.org/10.1007/978-3-030-43089-4_3

to another so that the sensor frames are coincident. [6] solves a misalignment problem for multiple IMUs using a manufactured platonic solid; [7] (and its related works) proposes three different related methods for "harmonizing" the sensor axes of homogenous sensors; [8] addresses a similar problem with Rotation Averaging for cameras in the Computer Vision domain. It can be assumed that the sensor suite is *never* perfectly aligned with the body axes. This is also true of robotics where accelerometers mounted at the joints are misaligned due to inherent manufacturing variations.

All of the existing prior art in misalignment calibration of vector measurements rely on batch processing in order to compute the misalignment matrix. Indeed, the conventional calibration of the sensors themselves individually can increase the misalignment depending on the results of the scale factor and bias errors [9][10][11]. As seen later in the Section 2, the equations are non-linear, and thus a recursive or online method would require an Extended Kalman Filter or other non-linear estimation scheme whose convergence to local minima is well documented. Indeed both [12] and [9] show results of misalignments (inter-axis calibration in their terminology) and calibration errors together. [12] uses a parameter search using the particle gradient descent algorithm (PGDA); the results of their misalignment calibration is similar to the results presented in this paper, but the method differs in choosing a stopping criteria in the minimum standard deviation for the solid angle between the magnetometer measurements and accelerometer measurements.

[9] use an direct external measurement of body attitude using a precise reference to generate known attitudes, and demonstrate superior results in base calibration. Since both accelerometer and magnetometer triads are being calibrated to a known external reference using the "Dot Product Invariance" method, the misalignment is directly embedded in the results of their calibration. This method is fundamentally different from the one proposed here due to the requirement for an external reference of attitude.

Lastly, [13] shows again a simultaneous calibration of the magnetometer and its misalignment of homogenous sensors, and demonstrates an iterated least squares algorithm that is shown in simulation work to decrease the cost function of the optimization over both small and large angles of misalignment.

There are many applications where three axis sensors are used for attitude measurements (e.g., aircraft, satellites, UAVs, and underwater vehicles). For example, small UAVs will often use a combination of accelerometers and magnetometers to provide aiding which is then used to estimate biases on the gyros. Previous work on calibrating the three axis sensors for inherent biases and scale factor errors based on a two-step solution [3][4] or an iterated least squares solution [13] have proven effective. Note that while these solutions calibrate the individual sensor triads, *they provide no information about relative sensing axes*.

When considering the *ensemble averaging* of sensor measurements (using multiple of the same sensors to improve the signal to noise), there is an inherent assumption of a common coordinate frame for each of the sensors. If the sensor axes are misaligned, then the averaging of each axis will degrade the signal to noise ratio (SNR). When using multiple sensors to measure the *gradient* of the field as an additional measurement, again each sensor must be in a consistent coherent axis system or the results will be off [14].

The contribution of this new algorithms is to extract the misalignment between two sensors without any need for an external reference. This greatly simplifies the calibration procedures, and can easily be implemented in the field.

Note that this same algorithm can be applied to align heterogenous sensors (e.g., magnetometer and accelerometer), to align multiples of the same sensor (for ensemble averaging or gradient measurement), or to align the sensor suite to the vehicle body frame. The paper proceeds as follows: Sec. 2 will develop the theory of the new algorithm and its application, Sec. 3 will explore numerical simulations of the algorithm showing convergence metrics and noise sensitivity, Sec. 4 will describe two experiments that were used to test the algorithm on real world data and the results of our new algorithm, and Sec. 5 presents conclusions and future directions.

2 Theory

Of the many solutions to Wahba's Problem, we will use the formulation by Markley in [15] which uses the singular value decomposition (*SVD*) to solve for the direction cosine matrix (*DCM*). For completeness, that solution is presented here:

$$\min_{\mathcal{R}} J(\mathcal{R}) = \frac{1}{2} \sum_{i=1}^{n} a_i \left\| \boldsymbol{w}_i - \mathcal{R} \boldsymbol{v}_i \right\|^2 \tag{1}$$

where $\boldsymbol{w}_i$ are a set of vectors in the inertial frame, $\boldsymbol{v}_i$ are the corresponding set of vectors in the body frame, and $\mathcal{R}$ is the *DCM* that transforms a vector from the body to the inertial frame. The a_i's are an optional set of weights, which for normalization purposes, $\sum a_i = 1$.

From the original solution to Wahba's Problem in [16], we have the formulation that $J(\mathcal{R})$ can be expressed as:

$$J(\mathcal{R}) = \frac{1}{2} \sum_{i=1}^{n} a_i \left\| \boldsymbol{w}_i - R \boldsymbol{v}_i \right\|^2 \tag{2}$$

$$= \frac{1}{2} tr(W - \mathcal{R}V)^T (W - \mathcal{R}V) \tag{3}$$

$$= 1 - tr(W \mathcal{R} V^T) \tag{4}$$

$$J(\mathcal{R}) = 1 - \sum_{i=1}^{n} a_i w_i^T \mathcal{R} v_i \tag{5}$$

Markley's solution [15] is based on the *SVD* decomposition, and proceeds as follows:

$$\mathcal{B} = \frac{1}{2} \sum_{i=1}^{n} a_i \boldsymbol{w}_i \boldsymbol{v}_i^T \tag{6}$$

$$\text{svd}(\mathcal{B}) = U \Sigma V^T \tag{7}$$

$$\mathcal{R} = UMV^T \tag{8}$$

where $M = diag(\begin{bmatrix} 1 & 1 & \det(U)\det(V) \end{bmatrix})$ and is used to enforce that R is a rotation matrix for a right handed coordinate frame. That is:

$$\mathcal{R}_{opt} = U \begin{bmatrix} 1 & 0 & 0 \\ 0 & 1 & 0 \\ 0 & 0 & \det(U)\det(V) \end{bmatrix} V^T \tag{9}$$

From this solution, the optimal rotation matrix, $\mathcal{R}$, can be extracted for any set of non-collinear vectors. Markley's solution will be used extensively throughout the development of the algorithm.

Before continuing with the solution, it is useful to visualize the problem. Consider two unit vectors that are known in inertial coordinates, defined as $\hat{\mathbf{m}}$ and $\hat{\mathbf{s}}$ for *master* and *slave* respectively. The notation we will use shows the coordinate frame to the right of the vector such that $^I\hat{\mathbf{m}}$ is the inertial master unit vector, expressed in the inertial frame, and $^B\hat{\mathbf{m}}$ is the same unit vector expressed in the body frame.

The key observation is that while the individual unit vectors may be pointed anywhere (e.g., the body frame is not aligned to the inertial frame), there is a constant solid angle between the two (master and slave) unit vectors *in the body frame*. When rotating the body through a set of orientations, the error in that solid angle will project onto each axis of the basis set and make the misalignment matrix observable. That is, a rich set of paired measurements in the body frame is sufficient to reconstruct the misalignment matrix between master and slave sensors.

Due to the nature of the Markley solution, the full attitude can be reconstructed given that $\mathcal{B}$ is full rank, which implies that both master and slave inertial vectors are non-collinear, and that one has at least one simultaneous measurement of both vectors in the body frame. In the case with noise, however, more measurements give a better solution as noise on the measurements directly distorts the attitude solution.

2.1   Misalignment Calculation

Consider a rigid body with two three axis sensors (one master, $\mathbf{m}$, and one slave, $\mathbf{s}$) which measure the three components of their respective signals. That is, for any arbitrary rotation of the rigid body, $\mathcal{R}$, the sensors will measure:

$$^B\hat{\mathbf{m}}_i = \mathcal{R}_i^T [^I\hat{\mathbf{m}}] + \nu_m \tag{10}$$

$$^B\hat{\mathbf{s}}_i = \mathcal{R}_i^T [^I\hat{\mathbf{s}}] + \nu_s \tag{11}$$

where $\mathcal{R}_i$ is the rotation matrix for that specific measurement, and is not known, and ν is sensor noise. Note that in this case, Wahba's problem can be used to find $\mathcal{R}_i$, but that is only because the sensors both share a common coordinate frame. If, however, the master and slave do *not* have the same coordinate frames, then there will be a misalignment matrix, $\mathcal{R}_{mis}$ that effectively rotates measurements in the *body-slave* frame to the *body-master*. That is:

$$^{BM}\hat{\mathbf{s}}_i = \mathcal{R}_{mis}{}^{BS}\hat{\mathbf{s}}_i \tag{12}$$

Where BM and BS refer to body-master and body-slave coordinates. The misalignment matrix $\mathcal{R}_{mis}$ is *constant in the body frame*, then Eq. 10 becomes:

$$^{B}\hat{\mathbf{m}}_i = \mathcal{R}_i{}^T[{}^I\hat{\mathbf{m}}] + \nu_m \tag{13}$$
$$^{B}\hat{\mathbf{s}}_i = \mathcal{R}_{mis}\mathcal{R}_i{}^T[{}^I\hat{\mathbf{s}}] + \nu_s \tag{14}$$

This constant, unknown misalignment between the coordinate frame of the master and the slave must be estimated before using the sensors (either for ensemble averaging or for attitude estimation). Data from each sensor is collected at different attitudes of the body to generate the paired measurements which can be collected into matrix form:

$$^{B}\mathbf{M} = \begin{bmatrix} ^{B}\hat{\mathbf{m}}_1 & ^{B}\hat{\mathbf{m}}_2 & \cdots & ^{B}\hat{\mathbf{m}}_n \end{bmatrix} \tag{15}$$
$$^{B}\mathbf{S} = \begin{bmatrix} ^{B}\hat{\mathbf{s}}_1 & ^{B}\hat{\mathbf{s}}_2 & \cdots & ^{B}\hat{\mathbf{s}}_n \end{bmatrix} \tag{16}$$

The algorithm for estimating $\mathcal{R}_{mis}$ uses an iterative approach through all of the paired measurements. $\widehat{\mathcal{R}}_{mis}$ is initialized to the identity matrix (a valid rotation matrix). We rotate the body fixed measurements of the slave by the estimate of the misalignment matrix:

$$\widehat{{}^{B}\mathbf{S}} = \widehat{\mathcal{R}}_{mis}{}^{B}\mathbf{S} = \widehat{\mathcal{R}}_{mis}\begin{bmatrix} ^{B}\hat{\mathbf{s}}_1 & ^{B}\hat{\mathbf{s}}_2 & \cdots & ^{B}\hat{\mathbf{s}}_n \end{bmatrix} \tag{17}$$

Eq. 17 is used to rotate all slave measurements by the estimate of $\mathcal{R}_{mis}$. Each measurement pair is then plugged into a Wahba's problem to solve for the individual rotation matrix $\mathcal{R}_i$ for each measurement (see Eq. 9). Thus, for each measurement i we compute $\mathcal{R}_i$ from $\widehat{{}^{B}\mathbf{S}}_i$ and $^{B}\hat{\mathbf{m}}_i$. With this $\mathcal{R}_i$ estimate, we calculate a new $^{B}\hat{\mathbf{s}}$ by rotating the slave unit vector $^I\hat{\mathbf{s}}$ into the body frame. That is:

$$^{B}\mathbf{s}_i = \mathcal{R}_i{}^T[{}^I\hat{\mathbf{s}}] \tag{18}$$

which are aggregated into a larger $3 \times n$ matrix, $^{B}\mathbf{S}$. We then again use the solution to Wahba's problem, this time with $^{B}\mathbf{S}$ and $\widehat{{}^{B}\mathbf{S}}$ with the resulting solution being a new estimate of the misalignment matrix, $\widehat{\mathcal{R}}_{mis}$. This is repeated until the misalignment matrix converges.

When $\widehat{\mathcal{R}}_{mis}$ is equal to $\mathcal{R}_{mis}$, then by substituting Eq. 17, Eq. 18, and Eq. 14 we show that $^{B}\mathbf{S}$ will equal $\widehat{{}^{B}\mathbf{S}}$, because $\widehat{\mathcal{R}}_{mis}^T\mathcal{R}_{mis} = I_{3\times3}$. We check the convergence of $\widehat{\mathcal{R}}_{mis}$ by checking the Frobenius norm of the difference between subsequent steps and stop when it is below some tolerance (generally set at 10^{-15}).

In summation, the algorithm uses all of the measurements (along with the estimate of the misalignment matrix) to develop the rotation matrix for each measurement, and then uses that set of rotation matrices and *only* the slave measurements to generate a new estimate of the misalignment matrix. This is then iterated to converge on a minimum.

Indeed, the algorithm can be shown to always converge to a local minima by expanding Eq. 1 with the definitions of the body measurements, Eq. 14, ignoring the noise terms and rearranging:

$$J(\mathcal{R}, \mathcal{R}_{mis}) = \frac{1}{2} \sum_{i=1}^{n} \left\| {}^I\mathbf{s}[\mathbf{I} - \mathcal{R}_i \mathcal{R}_{mis} \mathcal{R}_i{}^T] \right\|^2 \tag{19}$$

The cost function is everywhere differentiable, thus the algorithm breaks down into a block-coordinate descent approach which alternately optimizes for the $\mathcal{R}_i$'s with $\mathcal{R}_{mis}$ held fixed (which decouples into n Whaba's problems), and then for $\mathcal{R}_{mis}$ with the $\mathcal{R}_i$'s fixed (which is a simple Procrustes problem on the vectors $\mathbf{s}$). Since each sub-problem can be optimally solved, the block coordinate descent always converges to a local minima.

Algorithm 1 Compute $\mathcal{R}_{mis}$

1: $\widehat{\mathcal{R}}_{mis} \leftarrow I_{3 \times 3}$
2: **if** $\widehat{\mathcal{R}}_{mis}$ not converged **then**
3: $\quad {}^{\widehat{B}}\mathbf{S} \leftarrow \widehat{\mathcal{R}}_{mis} \begin{bmatrix} {}^B\hat{\mathbf{s}}_1 & {}^B\hat{\mathbf{s}}_2 & \cdots & {}^B\hat{\mathbf{s}}_n \end{bmatrix}$
4: $\quad$ **for all** measurements i in the body frame **do**
5: $\qquad$ Solve Wahba's Problem using $\boldsymbol{v} = \begin{bmatrix} {}^B\hat{\mathbf{m}}_i & {}^{\widehat{B}}\mathbf{S}_i \end{bmatrix}$ and $\boldsymbol{w} = \begin{bmatrix} {}^i\hat{\mathbf{m}} & {}^i\hat{\mathbf{s}} \end{bmatrix}$ for $\mathcal{R}_i$
6: $\qquad {}^B\mathbf{s}_i \leftarrow \mathcal{R}_i{}^T[{}^i\hat{\mathbf{s}}]$
7: $\quad$ Collect all ${}^B\mathbf{s}_i$ into ${}^B\mathbf{S}$
8: $\quad$ Solve Wahba's Problem using $\boldsymbol{V} = {}^{\widehat{B}}\mathbf{S}$ and $\boldsymbol{W} = {}^B\mathbf{S}$ for $\widehat{\mathcal{R}}_{mis}$
9: $\quad$ **if** $\left\| \widehat{\mathcal{R}}_{mis} - \widehat{\mathcal{R}}_{mis}^{prev} \right\|_{Frobenius} \leqslant tol$ **then**
10: $\qquad$ done
11: $\quad$ **else**
12: $\qquad$ goto step [3]
13: $\mathcal{R}_{mis} \leftarrow \widehat{\mathcal{R}}_{mis}$

Note that the algorithm still works when $\hat{\mathbf{m}}$ and $\hat{\mathbf{s}}$ are the same (e.g., homogenous sensors). This is due to the *SVD* nature of Markley's solution to Wahba's Problem, which returns the minimal answer (Eq. 9) even when Rank $\mathcal{B} = 2$; as long as there are sufficiently diverse measurements observe the misalignment, the estimate for $\mathcal{R}_{mis}$ will converge to a local minima.

In order for the $\mathcal{R}_{mis}$ to be observable, then ${}^B\mathbf{S}$ must be full rank (3). Depending on the exact misalignment, the minimum of three sensor measurements (at different attitudes) must include sufficient span that the projection of the misalignment matrix onto the measurement axes to achieve full rank. Thus including the measurement for attitude estimation, a minimum of three distinct measurements must be used to solve the misalignment problem (in the noise free case). Future work will determine the observability of the misalignment matrix with respect to the measurements.

3   Simulation

In order to determine the convergence and stability of the algorithm, as well quantify the error in estimates of $\mathcal{R}_{mis}$, a set of Monte Carlo numerical simulations were performed.

Two unit vectors, $\hat{\mathbf{m}}$ and $\hat{\mathbf{s}}$, and $\mathcal{R}_{mis}$ were chosen at random, using $\mathcal{R}_{mis} = e^{[\omega \times]}$ where ω is a randomly chosen rotation $[3 \times 1]$ and $[\cdot \times]$ is defined as the skew symmetric matrix. This is a valid rotation matrix.

An additional number of random rigid body rotations, $\mathcal{R}_i$ (varied from 3 to 50) were used to generate $^B\mathbf{M}$ and $^B\mathbf{S}$. The simulations were run with both with and without Gaussian white noise added to the body measurements. The noise variance on each measurement was set to 0.01 and added to each component of the sensor measurement. Anecdotal observations show the algorithm is quite tolerant of larger noise variance in simulation.

$$\nu_m \sim \mathcal{N}(0, 0.01) \tag{20}$$

$$\nu_s \sim \mathcal{N}(0, 0.01) \tag{21}$$

For observability of the $\widehat{\mathcal{R}}_{mis}$ a minimum of three non-coplanar body fixed measurements are required. However, with only three measurements, the likelihood of converging to the correct $\widehat{\mathcal{R}}_{mis}$ is poor. More measurements increase the probability of converging to the correct answer.

Fig. 1 shows a typical run for a scenario with $n = 4$ unique $\mathcal{R}_i$'s. The two true inertial unit vectors (orange), master (black) and true slave (blue) unit vector triads are shown in the body coordinates. The misaligned slave (green dashed) are shown as well (green dots show the misalignment iterations for clarity). The solution for the estimated slave (blue dots) are shown at each iteration. In this case, $\widehat{\mathcal{R}}_{mis}$ does indeed converge to the true value. As another visualization, the axis of $\widehat{\mathcal{R}}_{mis}$ (cyan, in an axis and angle representation) demonstrates the estimate converging to the correct one (red). In this case, the difference between the true and estimated misalignment matrix (Frobenius norm) was 4.29×10^{-15}.

The true misalignment matrix had an axis of $\begin{bmatrix} -0.3554 & -0.9850 & -0.2695 \end{bmatrix}$ and an angle of 51.11°. Which is to note that this is *not* a particularly small misalignment. This was a noise free case, and converged to the true value in 208 iterations.

Fig. 2 shows another simulation with $n = 12$ measurements; again the unit vectors (in orange), along with the body measured vectors (in black with triangles for the master and blue for the slave). The misaligned measurements are

Misalignment Simulation, n=4

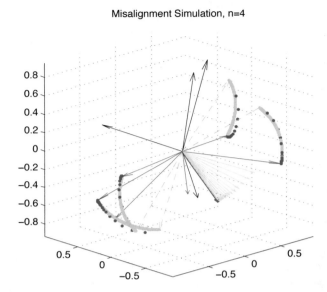

Fig. 1: Simulation of the misalignment problem with $n = 4$ measurements, showing the master (black) and true slave (blue) unit vector triads. The misaligned slave unit vector triads (green dashed), and the iterations that walk in the solutions to $\mathcal{R}_{mis}$, showing convergence to the true axes. The inertial sensor vectors are shown in orange, and the axis of $\mathcal{R}_{mis}$ is shown in red, with the steps shown in cyan.

shown in green dashed line with triangles with the dots showing the evolution of the $\widehat{\mathcal{R}}_{mis}$. Again, here the cyan vectors show the axis of the estimated misalignment and the blue dots show the convergence of the estimate. In this simulation, noise has been added to *all* body measurements. The estimate converges to within 10^{-2} of the exact misalignment, even in the presence of Gaussian white noise (see Eq. 20). This corresponds to an angular error of less than $\frac{1}{2}^\circ$.

The tolerance was set to 1×10^{-15}. Note that due to the use of the Markley *SVD* solution, the covariance of the misalignment matrix is also available, and shows in simulation that the converged misalignment is well within the noise covariance of true even when using noisy data. See [15] for details of the covariance matrix.

Using Monte Carlo simulations with 10000 runs for each of $n = [3, 4, 5, 10, 20, 50]$ demonstrates that the more data available in terms of measurements, the more likely convergence will occur. There are two available metrics to quantify convergence: (i) $\left\| \mathcal{R}_{mis} - \widehat{\mathcal{R}}_{mis} \right\|_F$ and (ii) the dot product of the true axis of rotation and the estimated one: $1 - \mathbf{v}^T \hat{\mathbf{v}}$. Since both of these measures are very small when the algorithm converges, the log of both is taken to quantify convergence;

Misalignment Simulation, n=12, noise added

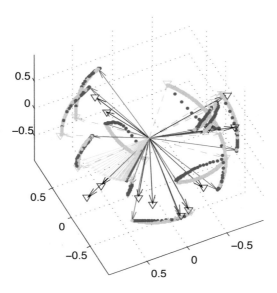

Fig. 2: Simulation of the misalignment problem with noise and $n = 12$, showing that the misaligned slave vectors (green dashed) converge to their true values (blue) even in the presence of noise on all measurements.

anything less than -6 is for all intents and purposes converged (corresponding to angular errors of less than $\frac{1}{100}^{\circ}$). Fig. 3 shows the cumulative probability distribution from the Monte Carlo runs for various measurement numbers (in the noise free case). The probability of converging on the correct misalignment is greater than 98% when $n = 50$. The small number of iterations that failed to converge were due to the measurement rotations $\mathcal{R}_i$ being insufficiently diverse ($^B\mathbf{S}$ being rank deficient) to allow reconstruction of $\mathcal{R}_i$ from Whaba's problem.

It is noted from the simulation results that the estimate of the misalignment matrix can become trapped in a local minima. If the algorithm is restarted with a different initial condition ($\mathcal{R}_0 \neq I_{3\times3}$) then it will often converge to the correct solution. This suggested how to validate the global nature of the estimate in practice. Segregate the data into two or more segments of at least n points each. In theory, n could be 3 or 4, however given the results of the Monte Carlo simulations, a higher n of 20-50 proved better in practice. Run the algorithm on each from at least two separate $\mathcal{R}_0$. If all converge to the same estimate, then the estimate has converged to the true value with high confidence.

Fig. 4 shows the CDF for several variants of the $n = 20$ Monte Carlo simulation (note that this is a relatively small number of individual measurements). The nominal noise free case (blue) has a 96.5% chance of success. If a simple

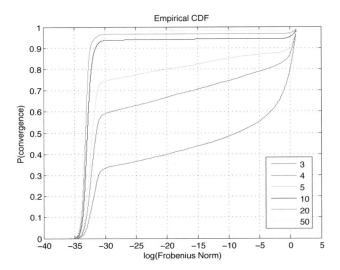

Fig. 3: Cumulative probability distribution of convergence for various measurements (noise free), showing the error between true $\mathcal{R}_{mis}$ and the estimate.

retry of resetting the initial guess on $\mathcal{R}_0$ is allowed, then the probability goes to 99.2% (red). In the case of noise added to the measurements, it simply moves the lowest Frobenius norm, but does not actually affect convergence (black). This is because the noise is uniformly distributed and thus averages out through the Wahba's problem solution. When the misalignment matrix is constrained to be small ($< 6°$ on any axis), then the convergence in 100% for both the noise free and noisy cases (green and cyan respectively). Thus, for small misalignments, this algorithm always finds the true solution in simulation.

4   Experimental Results

In order to demonstrate applications of the algorithm, two different experiments were performed. The first was from a small satellite experiment where three high quality magnetometers are used to take an ensemble average to generate better data for a *space weather* application. The second was from a small UAV autopilot (see Fig. 6) from its accelerometer and magnetometer sensors, where the data comes from a tumble test used to calibrate the sensors for scale factor, biases, and non-orthogonality [17].

4.1   SmallSat Experiment

For the SmallSat experiment, data was provided from three high quality flux-gate magnetometers mounted on a rigid platform. Rather than rotating the rigid platform, it was inserted into a Helmholz coil which could manipulate the magnetic

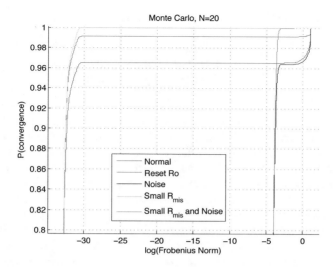

Fig. 4: Cumulative probability distribution for $n = 20$

field simulating body rotations in a very controlled manner. Three experiments were run, with data collected at each instance. Before any calibration was performed, the magnitude of the total magnetic field was calculated for each of the three sensors showing a mean of ~$900nT$ and a standard deviation of ~5-10nT. Note that this is quite small compared to Earth's magnetic field at $50000nT$.

The first calibration pass was to correct the magnetometers for bias, scale factor, and non-orthogonality errors using our previous *two step* techniques outlined in [3] and [4]. This forced the mean to exactly $900nT$ and resulted in a standard deviation of ~$1.35nT$.

When plotting the post calibration data individually (see Fig. 5a), it can be easily seen that this is a general lack of agreement between some of the axes of the three magenetometers (this was not known until the algorithm was run to estimate $\mathcal{R}_{mis}$). That is, there is a large rotation between one of the magnetometers and the other two due to the installation of the magnetometers on the platform. In the post-calibration image, it can be seen that the measurements of the three different magnetometers each line up on top of one another (as expected).

Using Magnetometer "A" as the master, the misalignment matrix between A and B is computed as:

$$\overset{A \to B}{\mathcal{R}_{mis}} = \begin{bmatrix} -0.00287 & 0.99998 & 0.00496 \\ -0.99993 & -0.00292 & 0.0119 \\ 0.0119 & -0.00492 & 0.99992 \end{bmatrix}$$

and the misalignment between A and C is computed as:

$$
\mathcal{R}_{mis}^{A \to C} = \begin{bmatrix} -0.99992 & -0.01257 & -0.00214 \\ -0.01256 & 0.9999 & -0.0071 \\ 0.00223 & -0.0071 & -0.99997 \end{bmatrix}
$$

Note that while $\mathcal{R}_{mis}^{A \to C}$ is somewhat close to a $180°$ rotation (x and z axes reversed), the one from $A \to B$ is not, indicating that it is mounted differently (the third magnetometer is mounted underneath the test platform). Finally, magnetometers B and C are rotated into the coordinate frame of A, and an ensemble average is taken. For the ensemble measurement, the norm of the magnetic field has a mean of $900nT$ and a standard deviation of $1.31nT$. While this is only a small improvement, it nevertheless caught the large misalignment of magnetometer B which was mounted on its side. Fig. 5b shows the aligned data, and shows a much better match on the data then the pre-alignment data.

The conclusion from the small improvement from the ensemble average (theoretically the signal to noise should be an improved by $\sqrt{3}$) is that the noise is not independent, but rather from the Helmholz coil itself. This means that the magnetometers are actually capable of better performance than is indicated by this experiment. This is a result that would not have been possible to determine without the misalignment correction.

4.2   UAV Tumble Experiment

The second experiment is based on data from the SLUGS autopilot developed at UCSC for UAV research into guidance, navigation, and control (GNC). The SLUGS has a MEMS-based Analog Devices three axis accelerometer, and a Honeywell three axis magnetometer on its circuit board (Fig. 6). While every care has been made to align the axes during manufacture, assembly, and mounting to the aircraft, this alignment cannot be trusted to be exact. Furthermore, the manufacturers of the sensors disagree on coordinate frames for sensing axes resulting in ambiguity of the polarity of the measurements. The magnetometer is run through the misalignment calibration process twice: first to correct the coordinate frame ambiguity, and then again with the appropriate axes flipped to do the fine alignment.

In order to calibrate these sensors, the aircraft was tumbled in a field by hand to collect data for processing in our *two step* calibration routines. This same data was then used to run the misalignment algorithm to determine the actual misalignment. Note that the final "realigned" sensor will still be misaligned from the body axes of the aircraft, but both magnetometer and accelerometer will be on a single coherent "sensor" frame.

The data from the tumble test was $25,633$ pairs of accelerometer and magnetometer data that were run through the algorithm. The algorithm converged after 241 iterations. The final misalignment DCM was:

$$\mathcal{R}_{mis} = \begin{bmatrix} 0.9952 & -0822 & 0.0537 \\ 0.0816 & 0.9966 & 0.0129 \\ -0.0546 & -0.0085 & 0.9985 \end{bmatrix}$$

From the misalignment matrix, it is observed to have pitch and roll misalignment on the order of $3-4°$. More interesting still is the covariance of the misalignment estimate:

$$\mathcal{P}_{body} = \begin{bmatrix} 1.603 & 0.7256 & -4.576 \\ 0.7256 & 3.013 & -9.802 \\ -4.576 & -9.802 & 60.42 \end{bmatrix}$$

Which shows that the algorithm had the hardest part in determining the $[3,3]$ term of the matrix. That is, the relative confidence of the algorithm in errors within the misalignment matrix shows the greatest difficulty in determining the $[3,3]$ term. For a complete discussion of the attitude matrix covariance, see [15]. Rerunning the algorithm using decimated data and different initial $\mathcal{R}_0$ always converges to the same value. This gives confidence that this is the true misalignment matrix for the magnetometer relative to the accelerometer.

5   Conclusions

A batch algorithm for estimating the misalignment between multiple three axis sensors from sensor measurements in the body frame has been developed. This is done via repeatedly applying Wahba's Problem solution to the data and iterating until converged. The algorithm works with either heterogenous or homogenous sensors, and is shown to converge well with sufficient number of points. The misalignment matrix is from one *master* to a *slave* sensor. Monte Carlo simulations show convergence probabilities in both the noise free and noisy data cases. The algorithm was run on two real experiments: a SmallSat (homogenous) and a UAV (heterogenous). In both cases, the algorithm was able to identify both large and small misalignments and quantify them in a mathematically rigorous manner.

References

1. Wahba, G., et al.: Problem 65-1: A least squares estimate of satellite attitude. Siam Review **7**(3) (1965) 409
2. Crassidis, J., Markley, F., Cheng, Y.: Survey of nonlinear attitude estimation methods. Journal of Guidance Control and Dynamics **30**(1) (2007) 12
3. Elkaim, G.H., Foster, C.: Extension of a non-linear, two-step calibration methodology to include non-orthogonal sensor axes. IEEE Transactions on Aerospace Electronic Systems **44** (07/2008 2008)
4. Vasconcelos, J., Elkaim, G.H., Silvestre, C., Oliveira, P., Cardeira, B.: A geometric approach to strapdown magnetometer calibration in sensor frame. IEEE Transactions on Aerospace Electronic Systems (2010)

5. Dorveaux, E.: Magneto-inertial navigation: Principles and application to an indoor pedometer. PhD thesis, École Nationale Supérieure des Mines de Paris (2011)
6. Nilsson, J.O., Skog, I., Handel, P.: Aligning the forceseliminating the misalignments in imu arrays. IEEE Transactions on Instrumentation and Measurement **63**(10) (2014) 2498–2500
7. Dorveaux, E., Vissiere, D., Petit, N.: On-the-field calibration of an array of sensors. In: American Control Conference (ACC), 2010, IEEE (2010) 6795–6802
8. Hartley, R., Trumpf, J., Dai, Y., Li, H.: Rotation averaging. International Journal of Computer Vision **103**(3) (2013)
9. Li, X., Li, Z.: A new calibration method for tri-axial field sensors in strap-down navigation systems. Measurement Science and Technology **23**(10) (2012) 105105
10. Včelák, J., Ripka, P., Kubik, J., Platil, A., Kašpar, P.: Amr navigation systems and methods of their calibration. Sensors and Actuators A: Physical **123** (2005) 122–128
11. Bonnet, S., Bassompierre, C., Godin, C., Lesecq, S., Barraud, A.: Calibration methods for inertial and magnetic sensors. Sensors and Actuators A: Physical **156**(2) (2009) 302–311
12. Mizutani, A., Rosser, K., Chahl, J.: Semiautomatic calibration and alignment of a low cost, 9 sensor inertial magnetic measurement sensor. Volume 7975. (2011) 797517–797517–8
13. Dorveaux, E., Vissiere, D., Martin, A.P., Petit, N.: Iterative calibration method for inertial and magnetic sensors. In: Decision and Control, 2009 held jointly with the 2009 28th Chinese Control Conference. CDC/CCC 2009. Proceedings of the 48th IEEE Conference on, IEEE (2009) 8296–8303
14. Dorveaux, E., Boudot, T., Hillion, M., Petit, N.: Combining inertial measurements and distributed magnetometry for motion estimation. In: American Control Conference (ACC), 2011, IEEE (2011) 4249–4256
15. Markley, F.: Attitude determination using vector observations and the singular value decomposition. Journal of the Astronautical Sciences **36**(3) (July-September 1988) 245–258
16. Farrell, J., Stuelpnagel, J., Wessner, R., Velman, J., Brook, J.: A least squares estimate of satellite attitude (grace wahba). SIAM Review **8**(3) (1966) 384–386
17. Lizarraga, M., Elkaim, G.H., Curry, R.: Slugs uav: A flexible and versatile hardware/software platform for guidance navigation and control research. In: American Control Conference (ACC), 2013, IEEE (2013) 674–679

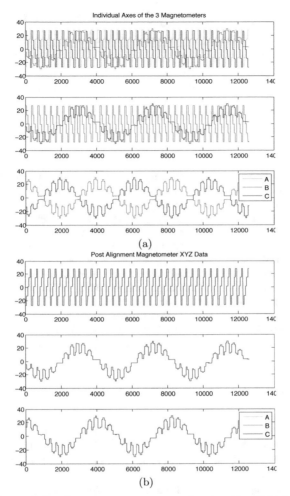

Fig. 5: XYZ data of the three magnetometers showing the Helmholtz coil magnetic field as a function of time: (a) Pre-Alignment and (b) Post-Alignment

Fig. 6: The SLUGS autopilot

High-Accuracy Preintegration for Visual-Inertial Navigation

Kevin Eckenhoff, Patrick Geneva, and Guoquan Huang*

Department of Mechanical Engineering, University of Delaware, Newark, DE
{keck,pgeneva,ghuang}@udel.edu

Abstract. Visual-inertial navigation that is able to provide accurate 3D localization in GPS-denied environments has seen popularity in recent years due to the proliferation of cost-effective cameras and inertial measurement units (IMUs). While an extended Kalman filter (EKF) is often used for sensor fusion, factor graph-based optimization has recently revealed its superior performance, which, however, is still compromised by the lack of rigorous IMU preintegration (i.e., integrating IMU measurements in a local frame of reference). To address this issue, in this paper, we analytically derive preintegration based on closed-form solutions of the continuous IMU measurement equations. These expressions allow us to analytically compute the mean, covariance, and bias Jacobians for a set of IMU preintegration factors. These accurate factors are subsequently fused with the visual information via visual-inertial factor graph optimization to provide high-precision trajectory estimates. The proposed method is validated on both Monte Carlo simulations and real-world experiments.

1 Introduction

Visual-inertial navigation systems (VINS) that fuse visual and inertial information to provide accurate localization, have become nearly ubiquitous in part because of their low cost and light weight (e.g., see [1, 2, 3]). IMUs provide local angular velocity and linear acceleration measurements, while cameras are a cheap yet informative means for sensing the surrounding environment and thus an ideal aiding source for inertial navigation. In particular, these benefits have made VINS popular in resource-constrained systems such as micro aerial vehicles (MAVs) [4]. Traditionally, navigation solutions have been achieved via extended Kalman Filters (EKFs), where incoming proprioceptive (IMU) and exteroceptive (camera) measurements are processed to propagate and update state estimates and covariances, respectively. These filtering methods do not update past state estimates that have been marginalized out, thus causing them to be susceptible to drift due to the compounding of errors.

Graph-based optimization methods, by contrast, process all measurements taken over a trajectory simultaneously to estimate a smooth history of sensor states. These methods achieve higher accuracy due to the ability to relinearize nonlinear measurement functions and correct previous state estimates while taking advantage of the sparse

---

* This work was partially supported by the University of Delaware College of Engineering, UD Cybersecurity Initiative, the Delaware NASA/EPSCoR Seed Grant, the NSF (IIS-1566129), and the DTRA (HDTRA1-16-1-0039).

K. Goldberg et al. (Eds.): *Algorithmic Foundations of Robotics XII*, SPAR 13, pp. 48–63, 2020.
https://doi.org/10.1007/978-3-030-43089-4_4

structure of the problem [5]. Recently, graph-based formulations have been introduced that allow the incorporation of IMU measurements into "preintegrated" factors by performing integration of the system dynamics in a *local* frame of reference [6, 7, 8]. However, these methods often simplify the required preintegration by resorting to discrete solutions under the approximation of piece-wise constant global accelerations. To improve this IMU preintegration, in this paper, we instead model the local IMU measurements as piece-wise constant and rigorously derive closed-form solutions of the integration equations. These solutions precisely model the underlying continuous dynamics of the preintegrated measurements. Based on these expressions, we offer analytical computations of the mean, covariance, and bias Jacobians, which have historically been solved using the discretized, rather than continuous, integrations.

After reviewing past literature in Section 2, we briefly review graph-based batch optimization. Following this, Section 4 presents our rigorous derivations of the analytical IMU preintegration and their respective Jacobians needed for graph-based VINS. In Section 5, we then explain the sliding window based visual tracking. We benchmark our analytical preintegration against the state-of-the-art discrete preintegration in Monte-Carlo simulations in Section 6, where we also offer real-world evaluations of the proposed VINS on a publicly available dataset. Finally, Section 7 concludes the paper and provides possible directions for future research.

2   Related Work

Filtering formulations for the VINS have been dominated by sliding window filters. Mourikis et al. [9] introduced the Multi-State Constraint Kalman Filter (MSCKF). This method was based on the idea of stochastic cloning [10], where the current state is augmented with copies of the past n sensor poses. This allowed for short term smoothing and past error correction, as well as processing exteroceptive sensor measurements without needing to store features in the state vector. This was achieved by projecting the measurement residuals onto the nullspace of the measurement Jacobians corresponding to features. Later filters [1, 3] improved the consistency of the MSCKF by enforcing the correct observability properties on the nonlinear system. In a similar vein, the Optimal-State Constraint (OSC)-EKF proposed in our prior work [11] also stores a sliding window of historical IMU poses. Rather than the projection method used in MSCKF, the OSC-EKF generates a multi-state constraint using a local batch optimization across the camera measurements in the window, and removes the dependency on the environment by marginalizing out features. This idea is also used in our visual tracking front-end in this work (see Section 5). These filtering formulations historically utilize the highly-accurate continuous view of the dynamics when performing propagation. This view of the nonlinear system motivates our derivation of closed-form expressions for preintegration. The error associated in discretization becomes especially detrimental when using low-cost sensors which cannot sample fast enough to mitigate linearization errors. These EKFs, while being computationally inexpensive, suffer from the inability to correct past states, leading to large estimation drift.

Graph-based formulations improve accuracy over their filtering counterparts by processing all measurements at once to estimate an entire trajectory [5, 12]. While these methods are well suited for the case of relative pose measurements, they have histor-ically had difficulty processing IMU measurements. The theory of preintegration was first introduced by Lupton et al [6]. By integrating multiple IMU measurements in a *local* frame of reference, initial conditions (velocity and gravity) can be recovered. The dependency of these integrations on bias was removed via a first-order Taylor series expansion, linearized about the current bias estimate. These techniques allowed for IMU measurements to be processed with a graph in a tractable fashion. Note that, in that work [6], rotation was parametrized using Euler angles, which are known to suffer from singularities. This was mitigated by Forster et al [7] who used the Lie algebra representation of rotation to analytically derive the involved Jacobians by exploiting properties of the manifold. However, both of these techniques discretize the preintegra-tion equations. By contrast our solutions are based on the continuous dynamics which properly model the underlying system. Although preintegration with quaternions has been used by Ling et al [8], their preintegration was set up in a continuous manner, but discretization was still used in the solution. In addition, bias updates were not used to correct preintegrated measurements. Other works that utilized preintegrated measure-ments without bias update can also be found in [13, 14]. Without using bias Jacobians, these methods are unable to correct their preintegrated measurements when the bias linearization point changes, thereby introducing avoidable errors into the system. All these issues will be rigorously addressed in our proposed preintegration.

3   Graph-based Batch Optimization

Batch optimization techniques provide more accurate estimation over their filtering counterparts by processing all measurements taken during a trajectory simultaneously to build a consistent and smooth estimate over the entire path. Measurements are com-monly modeled as being corrupted by Gaussian noise in the form:[1]

$$\mathbf{z}_i = \mathbf{h}_i\left(\mathbf{x}\right) + \mathbf{n}_i \, , \quad \mathbf{n}_i \sim \mathcal{N}\left(\mathbf{0}, \Lambda_i^{-1}\right) \, , \tag{1}$$

where $\mathbf{z}_i$ is the measurement vector, $\mathbf{x}$ is the true value of the state that is being esti-mated, $\mathbf{h}_i$ is a function that maps our state into the measurement, and $\mathbf{n}_i$ is zero-mean Gaussian noise with information matrix Λ_i. Maximum Likelihood Estimation (MLE) for our state can be formulated as the following Nonlinear Least-Squares (NLS) opti-mization problem [5, 12]:

$$\hat{\mathbf{x}} = \arg\min_{\mathbf{x}} \sum_i ||\mathbf{z_i} - \mathbf{h}_i\left(\mathbf{x}\right)||^2_{\Lambda_i} = \arg\min_{\mathbf{x}} \sum_i ||\mathbf{r}_i\left(\mathbf{x}\right)||^2_{\Lambda_i} \, , \tag{2}$$

where $||\mathbf{w}||^2_{\Lambda} = \mathbf{w}^\top \Lambda \mathbf{w}$. We define $\mathbf{r}_i$ as the residual associated with measurement i. This problem is often solved iteratively via Gauss-Newton approximations of the cost

---

[1] Throughout this paper $\hat{x}$ is used to denote the estimate of a random variable x, while $\tilde{x} = x - \hat{x}$ is the error in this estimate. $\mathbf{I}_{n \times n}$ and $\mathbf{0}_{n \times n}$ are the $n \times n$ identity and zero matrices, respectively. $\mathbf{e}_1, \mathbf{e}_2$ and $\mathbf{e}_3 \in \mathbb{R}^3$ are the unit vectors along $x-$, $y-$ and $z-$axes. The left superscript denotes the frame of reference which the vector is expressed with respect to.

about the current estimate.

$$\delta\mathbf{x} = \arg\min_{\delta\mathbf{x}} \sum_i ||\mathbf{r}_i(\hat{\mathbf{x}} \boxplus \delta\mathbf{x})||_{\boldsymbol{\Lambda}_i}^2 \simeq \arg\min_{\delta\mathbf{x}} \sum_i ||\mathbf{r}_i(\hat{\mathbf{x}}) + \mathbf{J}_i\delta\mathbf{x}||_{\boldsymbol{\Lambda}_i}^2 , \qquad (3)$$

where $\mathbf{J}_i := \left.\frac{\partial \mathbf{r}_i(\hat{\mathbf{x}} \boxplus \delta\mathbf{x})}{\partial \delta\mathbf{x}}\right|_{\delta\mathbf{x}=0}$ is the Jacobian of the residual with respect to the error state, evaluated at the error state being zero. We also define a corrective operator, $\boxplus$, that applies a correction term to the current estimate to yield a new estimate. This is a generalization of the additive noise associated with vector spaces, and allows the use of states restricted to a manifold, such as that of the unit quaternions. By stacking our measurements, the total corrective term can be obtained via the normal equation:

$$\delta\mathbf{x} = -\left(\mathbf{J}^\top \boldsymbol{\Lambda} \mathbf{J}\right)^{-1} \mathbf{J}^\top \boldsymbol{\Lambda} \mathbf{r} . \qquad (4)$$

Our estimate at the next iteration is updated as $\hat{\mathbf{x}}^{k+1} = \hat{\mathbf{x}}^k \boxplus \delta\mathbf{x}^k$. This process is repeated until convergence. The resulting distribution is then approximated with the following Gaussian:

$$\mathbf{x} = \hat{\mathbf{x}} \boxplus \delta\mathbf{x} , \quad \delta\mathbf{x} \sim \mathcal{N}\left(\mathbf{0}, (\mathbf{J}^\top \boldsymbol{\Lambda} \mathbf{J})^{-1}\right) . \qquad (5)$$

That is, our true state is modeled as being achievable from our current state estimate with a random error correction which is pulled from a Gaussian distribution with zero mean and information matrix of $\mathbf{J}^\top \boldsymbol{\Lambda} \mathbf{J}$.

4   Analytical IMU Preintegration

An IMU typically measures the local angular velocity $\boldsymbol{\omega}$ and linear acceleration $\boldsymbol{a}$ of its body, which are assumed to be corrupted by the Gaussian white noise ($\mathbf{n}_w$ and $\mathbf{n}_a$) and the random-walk biases ($\mathbf{b}_w$ and $\mathbf{b}_a$) [15]:

$$\boldsymbol{\omega}_m = \boldsymbol{\omega} + \mathbf{b}_w + \mathbf{n}_w , \quad \mathbf{a}_m = \mathbf{a} + \mathbf{g} + \mathbf{b}_a + \mathbf{n}_a , \quad \dot{\mathbf{b}}_w = \mathbf{n}_{wg} , \quad \dot{\mathbf{b}}_a = \mathbf{n}_{wa} , \qquad (6)$$

where $\mathbf{g}$ is the gravity vector in the local frame whose global counterpart is constant (e.g., $^G\mathbf{g} = [0 \ 0 \ 9.8]^T$). The navigation state at time-step k is given by:

$$\mathbf{x}_k = \begin{bmatrix} {}^{L_k}_G\bar{q}^T & \mathbf{b}_{w_k}^T & {}^G\mathbf{v}_k^T & \mathbf{b}_{a_k}^T & {}^G\mathbf{p}_k^T \end{bmatrix}^T , \qquad (7)$$

where ${}^{L_k}_G\bar{q} := \begin{bmatrix} \mathbf{q}^T & q_4 \end{bmatrix}^T$ is the JPL convention, [15], that describes the rotation from frame $\{G\}$ to frame $\{L_k\}$, and ${}^G\mathbf{v}_k$ and ${}^G\mathbf{p}_k$ are the velocity and position of the k-th local frame in the global frame, respectively. The corresponding error state and $\boxplus$ operation used in batch optimization can be written as (note that hereafter the transpose has been omitted for brevity):

$$\tilde{\mathbf{x}}_k = \begin{bmatrix} {}^{L_k}\delta\boldsymbol{\theta}_G & \tilde{\mathbf{b}}_{w_k} & {}^G\tilde{\mathbf{v}}_k & \tilde{\mathbf{b}}_{a_k} & {}^G\tilde{\mathbf{p}}_k \end{bmatrix} , \qquad (8)$$

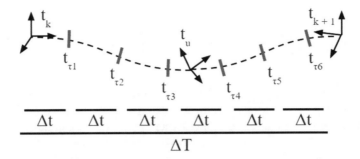

Fig. 1: IMU measurements are collected at discrete times t_τ with period Δt. This sampling occurs between image times t_k and t_{k+1}. We refer to the step k with a time of t_k, with t_u describing a generic time in the continuous domain.

$$\mathbf{x}_k = \hat{\mathbf{x}}_k \boxplus \tilde{\mathbf{x}}_k = \begin{bmatrix} {}^{L_k}_{\hat{L}_k}\delta\bar{q} \otimes {}^{\hat{L}_k}_{G}\hat{\bar{q}} \\ \hat{\mathbf{b}}_{w_k} + \tilde{\mathbf{b}}_{w_k} \\ {}^{G}\hat{\mathbf{v}}_k + {}^{G}\tilde{\mathbf{v}}_k \\ \hat{\mathbf{b}}_{a_k} + \tilde{\mathbf{b}}_{a_k} \\ {}^{G}\hat{\mathbf{p}}_k + {}^{G}\tilde{\mathbf{p}}_k \end{bmatrix} , \tag{9}$$

where the operator $\otimes$ denotes quaternion multiplication, and ${}^{L_k}_{\hat{L}_k}\delta\bar{q}$ is the error quaternion whose vector portion is half the error angle, ${}^{L_k}\delta\boldsymbol{\theta}_G = 2\mathbf{vec}\left({}^{L_k}_{\hat{L}_k}\delta\bar{q}\right) := 2{}^{L_k}_{\hat{L}_k}\delta\mathbf{q}$.

Between two imaging steps, k to $k+1$, IMU measurements are collected and integrated at times $t_k < t_\tau < t_{k+1}$ without accessing the state estimates (in particular, the orientation). This is done by performing the following factorization of the *current* rotation matrix and integration of the measurements [16]:

$$
{}^{G}\mathbf{p}_{k+1} = {}^{G}\mathbf{p}_k + {}^{G}\mathbf{v}_k\Delta T - \frac{1}{2}{}^{G}\mathbf{g}\Delta T^2 + {}^{G}_k\mathbf{R}\underbrace{\int_{t_k}^{t_{k+1}}\int_{t_k}^{s}{}^{k}_u\mathbf{R}\left({}^{u}\mathbf{a}_m - \mathbf{b}_a - \mathbf{n}_a\right)duds}_{{}^{k}\boldsymbol{\alpha}_{k+1}}
$$

$$
=: {}^{G}\mathbf{p}_k + {}^{G}\mathbf{v}_k\Delta T - \frac{1}{2}{}^{G}\mathbf{g}\Delta T^2 + {}^{G}_k\mathbf{R}{}^{k}\boldsymbol{\alpha}_{k+1} , \tag{10}
$$

$$
{}^{G}\mathbf{v}_{k+1} = {}^{G}\mathbf{v}_k - {}^{G}\mathbf{g}\Delta T + {}^{G}_k\mathbf{R}\underbrace{\int_{t_k}^{t_{k+1}}{}^{k}_u\mathbf{R}\left({}^{u}\mathbf{a}_m - \mathbf{b}_a - \mathbf{n}_a\right)du}_{{}^{k}\boldsymbol{\beta}_{k+1}}
$$

$$
=: {}^{G}\mathbf{v}_k - {}^{G}\mathbf{g}\Delta T + {}^{G}_k\mathbf{R}{}^{k}\boldsymbol{\beta}_{k+1} , \tag{11}
$$

$$
{}^{k+1}_{G}\mathbf{R} = {}^{k+1}_{k}\mathbf{R}{}^{k}_{G}\mathbf{R} , \tag{12}
$$

where $\Delta T = t_{k+1} - t_k$, and s and u are dummy variables used for integration (see Figure 1). Following the notation of Ling et al. [8], it becomes clear that the above integrals have been collected into the *preintegrated* measurements, ${}^{k}\boldsymbol{\alpha}_{k+1}$ and ${}^{k}\boldsymbol{\beta}_{k+1}$, which are expressed in the k-th local frame. Rearrangement of these equations yields:

$$
{}_{G}^{k}\mathbf{R}\left({}^{G}\mathbf{p}_{k+1} - {}^{G}\mathbf{p}_k - {}^{G}\mathbf{v}_k \varDelta T + \frac{1}{2}{}^{G}\mathbf{g}\varDelta T^2\right) = {}^{k}\boldsymbol{\alpha}_{k+1}\left(\mathcal{I}, \mathbf{n}_a, \mathbf{n}_w, \mathbf{b}_a, \mathbf{b}_w\right), \tag{13}
$$

$$
{}_{G}^{k}\mathbf{R}\left({}^{G}\mathbf{v}_{k+1} - {}^{G}\mathbf{v}_k + {}^{G}\mathbf{g}\varDelta T\right) = {}^{k}\boldsymbol{\beta}_{k+1}\left(\mathcal{I}, \mathbf{n}_a, \mathbf{n}_w, \mathbf{b}_a, \mathbf{b}_w\right), \tag{14}
$$

$$
{}_{G}^{k+1}\mathbf{R}\,{}_{G}^{k}\mathbf{R}^{\top} = {}_{k}^{k+1}\mathbf{R}\left(\mathcal{I}, \mathbf{n}_w, \mathbf{b}_w\right), \tag{15}
$$

where $\mathcal{I}$ is the set of all discrete IMU measurements collected between times t_k and t_{k+1} i.e., $\{{}^{\tau}\mathbf{a}_m, {}^{\tau}\boldsymbol{\omega}_m\}$. For the remainder of this paper the biases will refer to the those of state k, and are approximated as constant over the preintegration interval. It is important to note that the above equations (13)–(15) are dependent on the true biases which will causes exact preintegration to be intractable. In particular, they naively require the re-computation of the preintegration terms every time the bias linearization point changes. To address this issue, we employ the following first-order Taylor series expansion with respect to the biases:

$$
{}_{G}^{k}\mathbf{R}\left({}^{G}\mathbf{p}_{k+1} - {}^{G}\mathbf{p}_k - {}^{G}\mathbf{v}_k \varDelta T + \frac{1}{2}{}^{G}\mathbf{g}\varDelta T^2\right) \simeq \tag{16}
$$

$$
{}^{k}\boldsymbol{\alpha}_{k+1}\left(\mathcal{I}, \mathbf{n}_a, \mathbf{n}_w, \bar{\mathbf{b}}_a, \bar{\mathbf{b}}_w\right) + \left.\frac{\partial\boldsymbol{\alpha}}{\partial\mathbf{b}_a}\right|_{\bar{\mathbf{b}}_a}\varDelta\mathbf{b}_a + \left.\frac{\partial\boldsymbol{\alpha}}{\partial\mathbf{b}_w}\right|_{\bar{\mathbf{b}}_w}\varDelta\mathbf{b}_w,
$$

$$
{}_{G}^{k}\mathbf{R}\left({}^{G}\mathbf{v}_{k+1} - {}^{G}\mathbf{v}_k + {}^{G}\mathbf{g}\varDelta T\right) \simeq \tag{17}
$$

$$
{}^{k}\boldsymbol{\beta}_{k+1}\left(\mathcal{I}, \mathbf{n}_a, \mathbf{n}_w, \bar{\mathbf{b}}_a, \bar{\mathbf{b}}_w\right) + \left.\frac{\partial\boldsymbol{\beta}}{\partial\mathbf{b}_a}\right|_{\bar{\mathbf{b}}_a}\varDelta\mathbf{b}_a + \left.\frac{\partial\boldsymbol{\beta}}{\partial\mathbf{b}_w}\right|_{\bar{\mathbf{b}}_w}\varDelta\mathbf{b}_w,
$$

$$
{}_{G}^{k+1}\mathbf{R}\,{}_{G}^{k}\mathbf{R}^{T} \simeq \mathbf{R}\left(\varDelta\mathbf{b}_w\right)\,{}_{k}^{k+1}\mathbf{R}\left(\mathcal{I}, \mathbf{n}_w, \bar{\mathbf{b}}_w\right), \tag{18}
$$

where the preintegration functions have been linearized about the current bias estimates, $\bar{\mathbf{b}}_w$ and $\bar{\mathbf{b}}_a$, and $\varDelta\mathbf{b}_w := \mathbf{b}_w - \bar{\mathbf{b}}_w$ and $\varDelta\mathbf{b}_a := \mathbf{b}_a - \bar{\mathbf{b}}_a$ are the difference between the true biases and their linearization points. Note that in the case of the relative rotations, a change in bias is modeled as inducing a further rotation on our preintegrated relative rotation. This linearization process allows for the computation of the preintegration *once*, while still allowing approximate updates when the bias linearization point changes. The corresponding residuals of these preintegrated measurements for use in graph optimization are given by:

$$
{}^{k}\delta\boldsymbol{\alpha}_{k+1} = {}_{G}^{k}\mathbf{R}\left({}^{G}\mathbf{p}_{k+1} - {}^{G}\mathbf{p}_k - {}^{G}\mathbf{v}_k\varDelta T + \frac{1}{2}{}^{G}\mathbf{g}\varDelta T^2\right) - \frac{\partial\boldsymbol{\alpha}}{\partial\mathbf{b}_a}\varDelta\mathbf{b}_a - \frac{\partial\boldsymbol{\alpha}}{\partial\mathbf{b}_w}\varDelta\mathbf{b}_w - {}^{k}\breve{\boldsymbol{\alpha}}_{k+1},
$$

$$
{}^{k}\delta\boldsymbol{\beta}_{k+1} = {}_{G}^{k}\mathbf{R}\left({}^{G}\mathbf{v}_{k+1} - {}^{G}\mathbf{v}_k + {}^{G}\mathbf{g}\varDelta T\right) - \frac{\partial\boldsymbol{\beta}}{\partial\mathbf{b}_a}\varDelta\mathbf{b}_a - \frac{\partial\boldsymbol{\beta}}{\partial\mathbf{b}_w}\varDelta\mathbf{b}_w - {}^{k}\breve{\boldsymbol{\beta}}_{k+1},
$$

$$
{}^{k+1}\delta\boldsymbol{\theta}_k = 2\mathrm{vec}\left({}_{G}^{k+1}\bar{q} \otimes {}_{G}^{k}\bar{q}^{-1} \otimes {}_{k}^{k+1}\breve{\bar{q}}^{-1} \otimes \bar{q}\left(\varDelta\mathbf{b}_w\right)^{-1}\right), \tag{19}
$$

where ${}^{k}\breve{\boldsymbol{\alpha}}_{k+1}$, ${}^{k}\breve{\boldsymbol{\beta}}_{k+1}$ and ${}_{k}^{k+1}\breve{\bar{q}}$ are the preintegrated measurements with the quaternion being associated with the preintegrated rotation.

4.1 Compute Preintegration Mean and Covariance via Linear Systems

Before we use the IMU preintegrated measurement residuals (19) in batch optimization, we need to find their mean and covariance. To this end, we first note that the rotation (quaternion) time evolution is given by:

$$
{}_k^u \dot{\bar{q}} = \frac{1}{2} \boldsymbol{\Omega}({}^u\boldsymbol{\omega}_m - \bar{\mathbf{b}}_w - \mathbf{n}_w){}_k^u\bar{q}\,, \quad \boldsymbol{\Omega}(\boldsymbol{\omega}) = \begin{bmatrix} -\lfloor \boldsymbol{\omega}\times \rfloor & \boldsymbol{\omega} \\ -\boldsymbol{\omega}^\top & 0 \end{bmatrix}. \tag{20}
$$

This can be solved using the zeroth order quaternion integrator (see [15]). Based on (10) and (11), we have the following continuous measurement dynamics:

$$
{}^k\dot{\boldsymbol{\alpha}}_u = {}^k\boldsymbol{\beta}_u\,, \tag{21}
$$

$$
{}^k\dot{\boldsymbol{\beta}}_u = {}_u^k\mathbf{R}\left({}^u\mathbf{a}_m - \bar{\mathbf{b}}_a - \mathbf{n}_a\right). \tag{22}
$$

From these, we formulate the following *linear* system that describes the time evolution of the preintegrated measurements by taking the (approximate) expectation of the dynamic equations (21) and (22):

$$
\begin{bmatrix} {}^k\dot{\hat{\boldsymbol{\alpha}}}_u \\ {}^k\dot{\hat{\boldsymbol{\beta}}}_u \end{bmatrix} = \begin{bmatrix} \mathbf{0} & \mathbf{I}_{3\times3} \\ \mathbf{0} & \mathbf{0} \end{bmatrix} \begin{bmatrix} {}^k\hat{\boldsymbol{\alpha}}_u \\ {}^k\hat{\boldsymbol{\beta}}_u \end{bmatrix} + \begin{bmatrix} \mathbf{0} \\ {}_u^k\hat{\mathbf{R}} \end{bmatrix} \left({}^u\mathbf{a}_m - \bar{\mathbf{b}}_a\right). \tag{23}
$$

To *analytically* compute the mean of the preintegration measurement, we perform direct integration between two sample steps τ and $\tau + 1$, which correspond to IMU measurement times t_τ to $t_{\tau+1}$. With a little abuse of notation, we define $\hat{\mathbf{a}} = {}^\tau\mathbf{a}_m - \bar{\mathbf{b}}_a$, $\hat{\boldsymbol{\omega}} = {}^\tau\mathbf{w}_m - \bar{\mathbf{b}}_\omega$, and $\Delta t = t_{\tau+1} - t_\tau$.

$$
\begin{bmatrix} {}^k\hat{\boldsymbol{\alpha}}_{\tau+1} \\ {}^k\hat{\boldsymbol{\beta}}_{\tau+1} \end{bmatrix} = \begin{bmatrix} {}^k\hat{\boldsymbol{\alpha}}_\tau + {}^k\hat{\boldsymbol{\beta}}_\tau \Delta t \\ {}^k\hat{\boldsymbol{\beta}}_\tau \end{bmatrix} + \begin{bmatrix} {}_{\tau+1}^k\hat{\mathbf{R}}\left(\frac{(\Delta t^2)}{2}\mathbf{I}_{3\times3} + \frac{|\hat{\boldsymbol{\omega}}|\Delta t\cos(|\hat{\boldsymbol{\omega}}|\Delta t)-\sin(|\hat{\boldsymbol{\omega}}|\Delta t)}{|\hat{\boldsymbol{\omega}}|^3}\lfloor\hat{\boldsymbol{\omega}}\times\rfloor \\ \qquad + \frac{(|\hat{\boldsymbol{\omega}}|\Delta t)^2 - 2\cos(|\hat{\boldsymbol{\omega}}|\Delta t) - 2(|\hat{\boldsymbol{\omega}}|\Delta t)\sin(|\hat{\boldsymbol{\omega}}|\Delta t)+2}{2|\hat{\boldsymbol{\omega}}|^4}\lfloor\hat{\boldsymbol{\omega}}\times\rfloor^2\right)(\hat{\mathbf{a}}) \\ {}_{\tau+1}^k\hat{\mathbf{R}}\left(\Delta t\mathbf{I}_{3\times3} - \frac{1-\cos(|\hat{\boldsymbol{\omega}}|(\Delta t))}{|\hat{\boldsymbol{\omega}}|^2}\lfloor\hat{\boldsymbol{\omega}}\times\rfloor \\ \qquad + \frac{(|\hat{\boldsymbol{\omega}}|\Delta t)-\sin(|\boldsymbol{\omega}|\Delta t)}{|\hat{\boldsymbol{\omega}}|^3}\lfloor\hat{\boldsymbol{\omega}}\times\rfloor^2\right)(\hat{\mathbf{a}}) \end{bmatrix}. \tag{24}
$$

The final preintegrated measurements (${}^k\breve{\boldsymbol{\alpha}}_{k+1}$, ${}^k\breve{\boldsymbol{\beta}}_{k+1}$ and ${}_k^{k+1}\breve{q}$) can be computed incrementally by applying the above expression and the zeroth order quaternion integrator as new IMU measurements arrive. At the end of an integration window, we set the preintegrated measurements as the expected values for the total preintegrated quantities across the interval (e.g., ${}^k\breve{\boldsymbol{\alpha}}_{k+1} = {}^k\hat{\boldsymbol{\alpha}}_{k+1}$). The complete derivations can be found in our technical report [17].

To find the covariance of the preintegrated measurements, we first write out the dynamics of the corresponding error-states as follows:

$$
{}^k\delta\dot{\boldsymbol{\alpha}}_u = {}^k\delta\boldsymbol{\beta}_u\,, \tag{25}
$$

$$
{}^k\delta\dot{\boldsymbol{\beta}}_u = {}_u^k\hat{\mathbf{R}}(\mathbf{I}_{3\times3} + \lfloor{}^u\delta\boldsymbol{\theta}_k\times\rfloor)\left({}^u\mathbf{a}_m - \bar{\mathbf{b}}_a - \mathbf{n}_a\right) - {}_u^k\hat{\mathbf{R}}\left({}^u\mathbf{a}_m - \bar{\mathbf{b}}_a\right)
$$

$$
= {}_u^k\hat{\mathbf{R}}\left(-\mathbf{n}_a\right) + {}_u^k\hat{\mathbf{R}}\lfloor{}^u\delta\boldsymbol{\theta}_k\times\rfloor\left({}^u\mathbf{a}_m - \bar{\mathbf{b}}_a\right), \tag{26}
$$

$$
{}^u\delta\dot{\boldsymbol{\theta}}_k = -\lfloor\left({}^u\boldsymbol{\omega}_m - \bar{\mathbf{b}}_w\right)\times\rfloor\,{}^u\delta\boldsymbol{\theta}_k - \mathbf{n}_w\,. \tag{27}
$$

This immediately yields the following linearized system of the error states:

$$\begin{bmatrix} {}^k\delta\dot{\boldsymbol{\alpha}}_u \\ {}^k\delta\dot{\boldsymbol{\beta}}_u \\ {}^u\delta\dot{\boldsymbol{\theta}}_k \end{bmatrix} = \begin{bmatrix} \mathbf{0} & \mathbf{I}_{3\times3} & \mathbf{0} \\ \mathbf{0} & \mathbf{0} & -{}^k_u\hat{\mathbf{R}}\lfloor({}^u\mathbf{a}_m - \bar{\mathbf{b}}_a)\times\rfloor \\ \mathbf{0} & \mathbf{0} & -\lfloor({}^u\boldsymbol{\omega}_m - \bar{\mathbf{b}}_w)\times\rfloor \end{bmatrix} \begin{bmatrix} {}^k\delta\boldsymbol{\alpha}_u \\ {}^k\delta\boldsymbol{\beta}_u \\ {}^u\delta\boldsymbol{\theta}_k \end{bmatrix} + \begin{bmatrix} \mathbf{0} & \mathbf{0} \\ -{}^k_u\hat{\mathbf{R}} & \mathbf{0} \\ \mathbf{0} & -\mathbf{I}_{3\times3} \end{bmatrix} \begin{bmatrix} \mathbf{n}_a \\ \mathbf{n}_w \end{bmatrix},$$

$$\Rightarrow \dot{\mathbf{r}} = \mathbf{F}\mathbf{r} + \mathbf{G}\mathbf{n}. \tag{28}$$

Therefore, the noise covariance, $\mathbf{P}$, can be found as follows:

$$\mathbf{P}_k = \mathbf{0}_{9\times9} \tag{29}$$

$$\mathbf{P}_{\tau+1} = \boldsymbol{\Phi}(t_{\tau+1}, t_\tau)\mathbf{P}_\tau\boldsymbol{\Phi}(t_{\tau+1}, t_\tau)^\top + \mathbf{Q}_d \tag{30}$$

$$\mathbf{Q}_d = \int_{t_\tau}^{t_{\tau+1}} \boldsymbol{\Phi}(t_{\tau+1}, u)\mathbf{G}(u)\mathbf{Q}_c\mathbf{G}^\top(u)\boldsymbol{\Phi}^\top(t_{\tau+1}, u)du, \quad \mathbf{Q}_c = \begin{bmatrix} \sigma_a^2\mathbf{I}_{3\times3} & \mathbf{0}_{3\times3} \\ \mathbf{0}_{3\times3} & \sigma_w^2\mathbf{I}_{3\times3} \end{bmatrix}. \tag{31}$$

We want to stress that rather than using the discrete covariance approximation as in [8], we analytically compute the state-transition matrix, $\boldsymbol{\Phi}$, and discrete-time noise covariance, $\mathbf{Q}_d$, as derived in our tech report [17]. Although these expressions are more complicated than those found in discrete methods, we find that they do not prevent real-time processing of IMU data due their closed forms.

4.2 Compute Preintegration Bias Jacobians

As shown earlier, changes in bias are modeled as adding corrections to our preintegration measurements through the use of bias Jacobians [see (16) and (17)]. In particular, as seen from (24), each update term is linear in the estimated acceleration, $\hat{\mathbf{a}} = {}^\tau\mathbf{a}_m - \bar{\mathbf{b}}_a$, thus we find the bias Jacobians of ${}^k\boldsymbol{\alpha}_{k+1}$ and ${}^k\boldsymbol{\beta}_{k+1}$ with respect to $\mathbf{b}_a$ as follows:

$$\begin{bmatrix} \frac{\partial\boldsymbol{\alpha}}{\partial\mathbf{b}_a} \\ \frac{\partial\boldsymbol{\beta}}{\partial\mathbf{b}_a} \end{bmatrix} =: \begin{bmatrix} \mathbf{H}_\alpha(\tau+1) \\ \mathbf{H}_\beta(\tau+1) \end{bmatrix} = \begin{bmatrix} \mathbf{H}_\alpha(\tau) + \mathbf{H}_\beta(\tau)\Delta t \\ \mathbf{H}_\beta(\tau) \end{bmatrix} - \tag{32}$$

$$\begin{bmatrix} {}^k_{\tau+1}\mathbf{R}\left(\frac{\Delta t^2}{2}\mathbf{I}_{3\times3} + \frac{|\hat{\boldsymbol{\omega}}|\Delta t\cos(|\hat{\boldsymbol{\omega}}|\Delta t) - \sin(|\hat{\boldsymbol{\omega}}|\Delta t)}{|\hat{\boldsymbol{\omega}}|^3}\lfloor\hat{\boldsymbol{\omega}}\times\rfloor + \frac{(|\hat{\boldsymbol{\omega}}|\Delta t)^2 - 2\cos(|\hat{\boldsymbol{\omega}}|\Delta t) - 2(|\hat{\boldsymbol{\omega}}|\Delta t)\sin(|\hat{\boldsymbol{\omega}}|\Delta t) + 2}{2|\hat{\boldsymbol{\omega}}|^4}\lfloor\hat{\boldsymbol{\omega}}\times\rfloor^2\right) \\ {}^k_{\tau+1}\mathbf{R}\left(\Delta t\mathbf{I}_{3\times3} - \frac{1 - \cos(|\hat{\boldsymbol{\omega}}|(\Delta t))}{|\hat{\boldsymbol{\omega}}|^2}\lfloor\hat{\boldsymbol{\omega}}\times\rfloor + \frac{(|\hat{\boldsymbol{\omega}}|\Delta t) - \sin(|\hat{\boldsymbol{\omega}}|\Delta t)}{|\hat{\boldsymbol{\omega}}|^3}\lfloor\hat{\boldsymbol{\omega}}\times\rfloor^2\right) \end{bmatrix}.$$

Similarly, the Jacobians with respect to the gyro bias $\frac{\partial\boldsymbol{\alpha}}{\partial\mathbf{b}_w} =: \mathbf{J}_\alpha$, $\frac{\partial\boldsymbol{\beta}}{\partial\mathbf{b}_w} =: \mathbf{J}_\beta$ can be found. The detailed derivations are omitted here for brevity but are included in our companion technical report [17].

Now consider the relative-rotation measurement, ${}^{k+1}_k\mathbf{R}$. The updated rotation can be approximated as [7]:[2]

$$ {}^{k+1}_k\mathbf{R}_\oplus = \exp\left(\lfloor\mathbf{J}_q(k+1)(\mathbf{b}_w - \bar{\mathbf{b}}_w)\times\rfloor\right){}^{k+1}_k\mathbf{R}_\ominus, \tag{33}$$

where $\exp(\cdot)$ is the matrix exponential. It should be pointed out that in the above expression, that the rotational bias Jacobian, $\mathbf{J}_q$, can be computed *incrementally* using the

---

[2] We use the symbol $\oplus$ to denote an estimate after update and $\ominus$ before update.

right Jacobian of SO(3), $\mathbf{J}_{r_{\tau+1}} = \mathbf{J}_r(\boldsymbol{\omega}_\tau(t_{\tau+1} - t_\tau))$ (see [7, 17, 18]):

$$\mathbf{J}_q(\tau + 1) = {}_{\tau}^{\tau+1}\hat{\mathbf{R}} \sum_{u=k}^{\tau} {}_{u}^{\tau}\hat{\mathbf{R}}\mathbf{J}_{r_u}\Delta t + \mathbf{J}_{r_{\tau+1}}\Delta t = {}_{\tau}^{\tau+1}\hat{\mathbf{R}}\mathbf{J}_q(\tau) + \mathbf{J}_{r_{\tau+1}}\Delta t \,. \qquad (34)$$

The angle measurement residual can then be written as:

$$^{k+1}\delta\boldsymbol{\theta}_k = 2\mathbf{vec}\left({}_G^{k+1}\bar{q} \otimes {}_G^k\bar{q}^{-1} \otimes {}_k^{k+1}\breve{q}^{-1} \otimes \mathbf{quat}(\exp(-\lfloor\mathbf{J}_q(k+1)(\mathbf{b}_w - \bar{\mathbf{b}}_w)\times\rfloor))\right) , \quad (35)$$

$$= 2\mathbf{vec}\left({}_G^{k+1}\bar{q} \otimes {}_G^k\bar{q}^{-1} \otimes {}_k^{k+1}\breve{q}^{-1} \otimes \begin{bmatrix} \frac{\boldsymbol{\theta}}{|\boldsymbol{\theta}|}\sin(\frac{|\boldsymbol{\theta}|}{2}) \\ \cos(\frac{|\boldsymbol{\theta}|}{2}) \end{bmatrix}\right), \boldsymbol{\theta} = \mathbf{J}_q(\mathbf{b}_w - \bar{\mathbf{b}}_w) \qquad (36)$$

$$\simeq 2\mathbf{vec}\left({}_G^{k+1}\bar{q} \otimes {}_G^k\bar{q}^{-1} \otimes {}_k^{k+1}\breve{q}^{-1} \otimes \begin{bmatrix} \frac{1}{2}(\mathbf{J}_q(\mathbf{b}_w - \bar{\mathbf{b}}_w)) \\ 1 \end{bmatrix}\right) , \qquad (37)$$

where $\mathbf{quat}(\cdot)$ denotes the transformation of a rotation matrix to the corresponding quaternion. In the above expression, we have also used the common assumption that $(\mathbf{b}_w - \bar{\mathbf{b}}_w)$ is small. Note that we only use this approximation for the computation of Jacobians, while the more accurate (35) is used for the evaluation of actual residuals.

4.3   Compute Preintegration Measurement Jacobians

Our total preintegrated measurement residuals can now be written as:

$$\mathbf{r} = \begin{bmatrix} {}_G^k\mathbf{R}\left({}^G\mathbf{p}_{k+1} - {}^G\mathbf{p}_k - {}^G\mathbf{v}_k\Delta T + \frac{1}{2}{}^G\mathbf{g}\Delta T^2\right) - \mathbf{J}_\alpha(\mathbf{b}_w - \bar{\mathbf{b}}_w) - \mathbf{H}_\alpha(\mathbf{b}_a - \bar{\mathbf{b}}_a) - {}^k\breve{\boldsymbol{\alpha}}_{k+1} \\ {}_G^k\mathbf{R}\left({}^G\mathbf{v}_{k+1} - {}^G\mathbf{v}_k + {}^G\mathbf{g}\Delta T\right) - \mathbf{J}_\beta(\mathbf{b}_w - \bar{\mathbf{b}}_w) - \mathbf{H}_\beta(\mathbf{b}_a - \bar{\mathbf{b}}_a) - {}^k\breve{\boldsymbol{\beta}}_{k+1} \\ 2\mathbf{vec}\left({}_G^{k+1}\bar{q} \otimes {}_G^k\bar{q}^{-1} \otimes {}_k^{k+1}\breve{q}^{-1} \otimes \mathbf{quat}(\exp(-\lfloor\mathbf{J}_q(\mathbf{b}_w - \bar{\mathbf{b}}_w)\times\rfloor))\right) \end{bmatrix}$$

$$(38)$$

In order to use these residuals in graph-based optimization (3), the corresponding Jacobians with respect to the optimization variables are necessary. To this end, we first rewrite the relative-rotation measurement residual as:

$$^{k+1}\delta\boldsymbol{\theta}_k = 2\mathbf{vec}\left({}_G^{k+1}\bar{q} \otimes {}_G^k\bar{q}^{-1} \otimes {}_k^{k+1}\breve{q}^{-1} \otimes \bar{q}_b\right) , \qquad (39)$$

where, for ease of notation, $\bar{q}_b$ is the quaternion induced by a change in gyro bias. The measurement Jacobian with respect to one element of the state vector can be found by perturbing the residual function by the corresponding element. For example, the relative-rotation measurement residual is perturbed by a change in gyro bias around the current estimate (i.e., $\mathbf{b}_w - \bar{\mathbf{b}}_w = \hat{\mathbf{b}}_w + \tilde{\mathbf{b}}_w - \bar{\mathbf{b}}_w$):

$$^{k+1}\delta\boldsymbol{\theta}_k = 2\mathbf{vec}\left(\underbrace{{}_G^{k+1}\hat{\bar{q}} \otimes {}_G^{k+1}\hat{\bar{q}}^{-1} \otimes {}_k^{k+1}\breve{q}^{-1}}_{\hat{\bar{q}}_r} \otimes \begin{bmatrix} \frac{\mathbf{J}_q(\hat{\mathbf{b}}_w + \tilde{\mathbf{b}}_w - \bar{\mathbf{b}}_w)}{2} \\ 1 \end{bmatrix}\right)$$

$$= 2\mathbf{vec}\left(\begin{bmatrix} \hat{q}_{r,4}\mathbf{I}_{3\times3} - \lfloor\hat{\mathbf{q}}_r\times\rfloor & \hat{\mathbf{q}}_r \\ -\hat{\mathbf{q}}_r^\top & \hat{q}_{r,4} \end{bmatrix}\begin{bmatrix} \frac{\mathbf{J}_q(\hat{\mathbf{b}}_w + \tilde{\mathbf{b}}_w - \bar{\mathbf{b}}_w)}{2} \\ 1 \end{bmatrix}\right)$$

$$= (\hat{q}_{r,4}\mathbf{I}_{3\times3} - \lfloor\hat{\mathbf{q}}_r\times\rfloor)\mathbf{J}_q(\mathbf{b}_w + \tilde{\mathbf{b}}_w - \bar{\mathbf{b}}_w) + \text{other terms} ,$$

$$\Rightarrow \quad \frac{\partial^{k+1}\delta\boldsymbol{\theta}_k}{\partial\tilde{\mathbf{b}}_w} = (\hat{q}_{r,4}\mathbf{I}_{3\times3} - \lfloor\hat{\mathbf{q}}_r\times\rfloor)\mathbf{J}_q . \tag{40}$$

Similarly, the Jacobian with respect to $^{k+1}\delta\boldsymbol{\theta}_G$ can be found as follows:

$$^{k+1}\delta\boldsymbol{\theta}_k = 2\mathbf{vec}\left(\begin{bmatrix}\frac{1}{2}{}^{k+1}\delta\boldsymbol{\theta}_G \\ 1\end{bmatrix} \otimes \underbrace{{}^{k+1}_G\hat{\bar{q}} \otimes {}^k_G\hat{\bar{q}}^{-1} \otimes {}^{k+1}_k\bar{q}^{-1} \otimes \hat{\bar{q}}_b}_{\hat{\mathbf{q}}_{rb}}\right)$$

$$= 2\mathbf{vec}\left(\begin{bmatrix}\hat{q}_{rb,4}\mathbf{I}_{3\times3} + \lfloor\hat{\mathbf{q}}_{rb}\times\rfloor & \hat{\mathbf{q}}_{rb} \\ -\hat{\mathbf{q}}_{rb}^\top & \hat{q}_{rb,4}\end{bmatrix}\begin{bmatrix}\frac{1}{2}{}^{k+1}\delta\boldsymbol{\theta}_G \\ 1\end{bmatrix}\right)$$

$$= (\hat{q}_{rb,4}\mathbf{I}_{3\times3} + \lfloor\hat{\mathbf{q}}_{rb}\times\rfloor)^{k+1}\delta\boldsymbol{\theta}_G + \text{ other terms} ,$$

$$\Rightarrow \quad \frac{\partial^{k+1}\delta\boldsymbol{\theta}_k}{\partial^{k+1}\delta\boldsymbol{\theta}_G} = \hat{q}_{rb,4}\mathbf{I}_{3\times3} + \lfloor\hat{\mathbf{q}}_{rb}\times\rfloor . \tag{41}$$

The Jacobian with respect to $^k\delta\boldsymbol{\theta}_G$ is given by:

$$^{k+1}\delta\boldsymbol{\theta}_k = 2\mathbf{vec}\left(\underbrace{{}^{k+1}_G\hat{\bar{q}} \otimes {}^k_G\hat{\bar{q}}^{-1}}_{\hat{q}_n} \otimes \begin{bmatrix}-\frac{{}^k\delta\boldsymbol{\theta}_G}{2} \\ 1\end{bmatrix} \otimes \underbrace{{}^{k+1}_k\bar{q}^{-1} \otimes \hat{\bar{q}}_b}_{\hat{q}_{mb}^{-1}}\right)$$

$$= 2\mathbf{vec}\left(\begin{bmatrix}\hat{q}_{n,4}\mathbf{I}_{3\times3} - \lfloor\hat{\mathbf{q}}_n\times\rfloor & \hat{\mathbf{q}}_n \\ -\hat{\mathbf{q}}_n^\top & \hat{q}_{n,4}\end{bmatrix}\begin{bmatrix}\bar{q}_{mb,4}\mathbf{I}_{3\times3} - \lfloor\bar{\mathbf{q}}_{mb}\times\rfloor & -\mathbf{q}_{mb} \\ \mathbf{q}_{mb}^\top & \bar{q}_{mb,4}\end{bmatrix}\begin{bmatrix}-\frac{{}^k\delta\boldsymbol{\theta}_G}{2} \\ 1\end{bmatrix}\right)$$

$$= -((\hat{q}_{n,4}\mathbf{I}_{3\times3} - \lfloor\hat{\mathbf{q}}_n\times\rfloor)(q_{mb,4}\mathbf{I}_{3\times3} - \lfloor\mathbf{q}_{mb}\times\rfloor) + \hat{\mathbf{q}}_n\mathbf{q}_{mb}^\top)^k\delta\boldsymbol{\theta}_G + \text{ other terms} ,$$

$$\Rightarrow \quad \frac{\partial^{k+1}\delta\boldsymbol{\theta}_k}{\partial^k\delta\boldsymbol{\theta}_G} = -((\hat{q}_{n,4}\mathbf{I}_{3\times3} - \lfloor\hat{\mathbf{q}}_n\times\rfloor)(\bar{q}_{mb,4}\mathbf{I}_{3\times3} - \lfloor\mathbf{q}_{mb}\times\rfloor) + \hat{\mathbf{q}}_n\bar{\mathbf{q}}_{mb}^\top) . \tag{42}$$

Note than in the preceding Jacobians, we have defined several intermediate quaternions $(\hat{\bar{q}}_r, \hat{\bar{q}}_{rb}, \hat{\bar{q}}_n, \text{ and } \hat{\bar{q}}_{mb})$ for ease of notation which can easily be interpreted from context. Following the same methodology, we can find the Jacobians of the $^k\boldsymbol{\alpha}_{k+1}$ measurement with respect to the position, velocity and bias.

$$^k\boldsymbol{\alpha}_{k+1} = {}^k_G\mathbf{R}\left({}^G\mathbf{p}_{k+1} - {}^G\mathbf{p}_k - {}^G\mathbf{v}_k\Delta T + \frac{1}{2}{}^G\mathbf{g}\Delta T^2\right) - \mathbf{J}_\alpha(\mathbf{b}_w - \bar{\mathbf{b}}_w) - \mathbf{H}_\alpha(\mathbf{b}_a - \bar{\mathbf{b}}_a)$$

$$\simeq (\mathbf{I}_{3\times3} - \lfloor{}^k\delta\boldsymbol{\theta}_G\times\rfloor)^k_G\hat{\mathbf{R}}\left({}^G\hat{\mathbf{p}}_{k+1} + {}^G\tilde{\mathbf{p}}_{k+1} - {}^G\hat{\mathbf{p}}_k - {}^G\tilde{\mathbf{p}}_k - {}^G\hat{\mathbf{v}}_k\Delta T - {}^G\tilde{\mathbf{v}}_k\Delta T\right.$$
$$\left. + \frac{1}{2}{}^G\mathbf{g}\Delta T^2\right) - \mathbf{J}_\alpha\left(\hat{\mathbf{b}}_w + \tilde{\mathbf{b}}_w - \bar{\mathbf{b}}_w\right) - \mathbf{H}_\alpha\left(\hat{\mathbf{b}}_a + \tilde{\mathbf{b}}_a - \bar{\mathbf{b}}_a\right) . \tag{43}$$

Then the following Jacobians immediately become available:

$$\frac{\partial^k\boldsymbol{\alpha}_{k+1}}{\partial^k\delta\boldsymbol{\theta}_G} = \left\lfloor{}^k_G\hat{\mathbf{R}}\left({}^G\hat{\mathbf{p}}_{k+1} - {}^G\hat{\mathbf{p}}_k - {}^G\hat{\mathbf{v}}_k\Delta T + \frac{1}{2}{}^G\mathbf{g}\Delta T^2\right)\times\right\rfloor ,$$

$$\frac{\partial^k\boldsymbol{\alpha}_{k+1}}{\partial^G\tilde{\mathbf{p}}_k} = -{}^k_G\hat{\mathbf{R}} , \quad \frac{\partial^k\boldsymbol{\alpha}_{k+1}}{\partial^G\tilde{\mathbf{p}}_{k+1}} = {}^k_G\hat{\mathbf{R}} , \quad \frac{\partial^k\boldsymbol{\alpha}_{k+1}}{\partial^G\tilde{\mathbf{v}}_k} = -{}^k_G\hat{\mathbf{R}}\Delta T ,$$

$$\frac{\partial^k \boldsymbol{\alpha}_{k+1}}{\partial \tilde{\mathbf{b}}_w} = -\mathbf{J}_\alpha \, , \, \frac{\partial^k \boldsymbol{\alpha}_{k+1}}{\partial \tilde{\mathbf{b}}_a} = -\mathbf{H}_\alpha \, . \tag{44}$$

Similarly, we can write our $^k\boldsymbol{\beta}_{k+1}$ measurement with respect to the position, velocity and bias as:

$$
\begin{aligned}
^k\boldsymbol{\beta}_{k+1} &= {}^k_G\mathbf{R}\left({}^G\mathbf{v}_{k+1} - {}^G\mathbf{v}_k + {}^G\mathbf{g}\Delta T\right) - \mathbf{J}_\beta(\mathbf{b}_w - \bar{\mathbf{b}}_w) - \mathbf{H}_\beta(\mathbf{b}_a - \bar{\mathbf{b}}_a) \\
&\simeq (\mathbf{I}_{3\times 3} - \lfloor {}^k\delta\boldsymbol{\theta}_G \times \rfloor){}^k_G\hat{\mathbf{R}}\left({}^G\hat{\mathbf{v}}_{k+1} + {}^G\tilde{\mathbf{v}}_{k+1} - {}^G\hat{\mathbf{v}}_k - {}^G\tilde{\mathbf{v}}_k + {}^G\mathbf{g}\Delta T\right) \\
&\quad - \mathbf{J}_\beta(\hat{\mathbf{b}}_w + \tilde{\mathbf{b}}_w - \bar{\mathbf{b}}_w) - \mathbf{H}_\beta(\hat{\mathbf{b}}_a + \tilde{\mathbf{b}}_a - \bar{\mathbf{b}}_a) \, ,
\end{aligned}
\tag{45}
$$

which leads to the following Jacobians:

$$
\begin{aligned}
\frac{\partial^k\boldsymbol{\beta}_{k+1}}{\partial^k\delta\boldsymbol{\theta}_G} &= \left\lfloor {}^k_G\hat{\mathbf{R}}({}^G\hat{\mathbf{v}}_{k+1} - {}^G\hat{\mathbf{v}}_k + {}^G\mathbf{g}\Delta T)\times \right\rfloor \, , \\
\frac{\partial^k\boldsymbol{\beta}_{k+1}}{\partial^G\tilde{\mathbf{v}}_k} &= -{}^k_G\hat{\mathbf{R}} \, , \, \frac{\partial^k\boldsymbol{\beta}_{k+1}}{\partial^G\tilde{\mathbf{v}}_{k+1}} = {}^k_G\hat{\mathbf{R}} \, , \, \frac{\partial^k\boldsymbol{\beta}_{k+1}}{\partial\tilde{\mathbf{b}}_w} = -\mathbf{J}_\beta \, , \, \frac{\partial^k\boldsymbol{\beta}_{k+1}}{\partial\tilde{\mathbf{b}}_a} = -\mathbf{H}_\beta \, .
\end{aligned}
\tag{46}
$$

5 Sliding-Window Visual Tracking

Reliance on pure inertial measurements causes large drift over time due to the high noise factors, thus we rely on additional visual measurements from a camera. As an IMU-camera sensor suite moves throughout the environment, images are taken from the mounted stereo camera and features are extracted and tracked over a window of historical camera poses. Naively, we could add these features into our state vector, thereby greatly increasing the computational burden. Instead, we seek to extract all the information contained in these measurements about the sensor suite's states. To this end, we add these features into a *local* graph containing a sliding window of states and the corresponding tracked features, and perform a local *marginalization* across these features.

In particular, the measurement model associated with feature factors involves a transformation into a camera frame, followed by a projection onto the corresponding image plane. This function for a feature j detected on an image i is given by:

$$\mathbf{z}_{ij} = \begin{bmatrix} \frac{{}^{C_i}\mathbf{p}_j(1)}{{}^{C_i}\mathbf{p}_j(3)} \\ \frac{{}^{C_i}\mathbf{p}_j(2)}{{}^{C_i}\mathbf{p}_j(3)} \end{bmatrix} + \mathbf{n}_{ij} \, , \tag{47}$$

$$^{C_i}\mathbf{p}_j = {}^C_I\mathbf{R}\,{}^{I_i}_G\mathbf{R}\left({}^G\mathbf{p}_j - {}^G\mathbf{p}_{I_i}\right) + {}^C\mathbf{p}_I \, , \tag{48}$$

where $\mathbf{n}_{ij} \sim \mathcal{N}\left(\mathbf{0}, \boldsymbol{\Lambda}_{ij}^{-1}\right)$ is the zero-mean white Gaussian noise, and $\{{}^C_I\mathbf{R}, {}^C\mathbf{p}_I\}$ is the extrinsic calibration between the IMU and the camera, which is assumed to be known and constant over time.

Given the set of feature measurements, we seek the MLE for all the IMU poses in the window as well as the corresponding features (see Section 3). This optimization yields a normal distribution in the form (see [11, 19, 20]):

$$\begin{bmatrix} {}^G\mathbf{x} \\ {}^G\mathbf{p}_f \end{bmatrix} = \begin{bmatrix} {}^G\check{\mathbf{x}} \\ {}^G\check{\mathbf{p}}_f \end{bmatrix} \boxplus \begin{bmatrix} {}^G\delta\mathbf{x} \\ {}^G\delta\mathbf{p}_f \end{bmatrix} \, , \text{ with } \begin{bmatrix} {}^G\delta\mathbf{x} \\ {}^G\delta\mathbf{p}_f \end{bmatrix} \sim \mathcal{N}\left(\begin{bmatrix} \mathbf{0} \\ \mathbf{0} \end{bmatrix}, \boldsymbol{\Lambda}^{-1}\right) \, , \tag{49}$$

where $^G\mathbf{x}$ and $^G\mathbf{p}_f$ are the stacked sensor poses and feature positions respectively, while $(\check{*})$ refers to the estimates achieved by the MLE optimization. Defining $\mathbf{J}_{ij}$ as the Jacobian of the ij-th measurement, the information matrix is computed as:

$$\Lambda = \sum_{(i,j)} \mathbf{J}_{ij}^\top \Lambda_{ij} \mathbf{J}_{ij} = \begin{bmatrix} \Lambda_{ss} & \Lambda_{sf} \\ \Lambda_{fs} & \Lambda_{ff} \end{bmatrix} . \tag{50}$$

Note that the information matrix is partitioned according to the dimensions of the states and features (s and f respectively). Marginalizing the features yields the following normal distribution of our error state:

$$^G\delta\mathbf{x} \sim \mathcal{N}\left(\mathbf{0}, (\Lambda_{ss} - \Lambda_{sf}\Lambda_{ff}^{-1}\Lambda_{fs})^{-1}\right) . \tag{51}$$

It is important to note that this distribution encapsulates all the information in the measurements about the window nodes (up to linearization errors) [11]. Due to the lack of measurements anchoring the graph to the global frame, this optimization problem will typically be under-constrained. We therefore shift the frame of reference of the optimization problem into that of the oldest node in the window, and thus have the distribution on the *relative* states denoted by the left superscript "L" [see (49)]:

$$\begin{bmatrix} ^L\mathbf{x} \\ ^L\mathbf{p}_f \end{bmatrix} = \begin{bmatrix} ^L\check{\mathbf{x}} \\ ^L\check{\mathbf{p}}_f \end{bmatrix} \boxplus \begin{bmatrix} ^L\delta\mathbf{x} \\ ^L\delta\mathbf{p}_f \end{bmatrix}, \text{ with } \begin{bmatrix} ^L\delta\mathbf{x} \\ ^L\delta\mathbf{p}_f \end{bmatrix} \sim \mathcal{N}\left(\begin{bmatrix} \mathbf{0} \\ \mathbf{0} \end{bmatrix}, {}^L\Lambda^{-1}\right) . \tag{52}$$

Insertion back in the graph gives the following residuals for an example window of states $\{\mathbf{x}_k\}_{k=0}^n$, which will be used along with the IMU preintegration measurements in the graph optimization:

$$\mathbf{r}_f = \begin{bmatrix} 2\mathbf{vec}\left(_G^1\bar{q} \otimes {}_G^0\bar{q}^{-1} \otimes {}_0^1\check{q}^{-1}\right) \\ {}_G^0\mathbf{R}\left(^G\mathbf{p}_1 - {}^G\mathbf{p}_0\right) - {}^0\check{\mathbf{p}}_1 \\ \vdots \\ 2\mathbf{vec}\left(_G^n\bar{q} \otimes {}_G^0\bar{q}^{-1} \otimes {}_0^n\check{q}^{-1}\right) \\ {}_G^0\mathbf{R}\left(^G\mathbf{p}_n - {}^G\mathbf{p}_0\right) - {}^0\check{\mathbf{p}}_n \end{bmatrix} . \tag{53}$$

We therefore have extracted the information contained in the feature measurements about the states in the window while not storing features in the global graph.

6 Experimental Results

6.1 Monte-Carlo Simulations

The proposed method was implemented in C++ and tested on a MATLAB generated simulation in order to compare the proposed analytical preintegration against the discrete one [7]. A dynamic trajectory of approximately 107 meters traversed in 100 seconds, as well as a set of random 3D features were generated. Ten sets of noisy IMU and synthetic stereo image outputs were collected for Monte Carlo evaluation. Realistic noise levels and camera calibration parameters from the dataset below were used (see Section 6.2), while feature projections were corrupted by one pixel noise. IMU measurements were generated at a rate of 100 Hz, while synthetic images were created at 10 Hz. New state nodes were created every time a synthetic image pair was collected with features being "tracked" across a sliding window of six images. The local graph problem was solved to compute the vision factors (see Section 5) when the window reached its full size. In addition to the vision and IMU preintegration measurements, bias drift factors [21] were added into the global graph to constrain the difference in biases between nodes. Both the local and full graph optimizations were performed using the GTSAM library [22] and the discrete preintegration [7] was implemented using the open source code available within GTSAM. Figure 2 shows the generated path, while the root mean square errors (RMSE) are shown in Figure 3. The proposed method achieved an ending position RMSE of 0.64m (0.6% of the distance traveled) with an average RMSE of 0.36 m and 0.35 deg across the entire path. By contrast, the same system using discrete preintegration achieved an ending RMSE of 0.74 m (0.7% of the total path) with an average of 0.41 m and 0.38 deg across the entire path. This clearly demonstrates the improvements offered by our analytical preintegration.

6.2   Real-World Experiments

The proposed algorithm was validated on one of the "EuRoC MAV Datasets" that are publicly available [23]. The datasets use two Aptina MT9V034 global shutter cameras at 20 FPS with an image resolution of 752×480 pixels and the MEMS ADIS16448 IMU at a rate of 200 Hz. A MAV where the sensor suite is mounted, flies through the environment in a dynamic motion. The left to right images and IMU timestamps come synchronized, allowing for the inherent sensor time delay to be ignored. All extrinsic calibrations are provided including the camera to camera and IMU to camera transformations.

All images are histogram equalized to allow for better feature extraction and rectified using the OpenCV library [24]. New IMU measurements were preintegrated by "stacking" the readings over time [see (24)]. The current state was stored in a node and linked with this preintegration edge when a new stereo image pair is received. After ten images, a new feature factor was created through matching SIFT features across the first stereo pair and matching to older images using the KLT tracking method. To reject outliers, epipolar constraints are then enforced in each of the stereo pairs. In this experiment, g2o [12] was used as the graph solver, and the sliding window of ten images was chosen. The rest of the implementation was performed as explained in Section 6.1.

The Machine Hall 01 recording is 140 seconds long and provides dynamic aerial motion in an indoor area. The estimated trajectory verses the ground truth is shown in Figure 4, and the estimation errors are depicted in Figure 5. In this test, the proposed approach attains the position error of 0.5m, approximately 0.7% of the total distance traveled. Note that we present here only position errors, as highly accurate ground-truth orientation measurements were not available for the given dataset. Along with the Monte-Carlo simulation results, these real-world experiments clearly validate the proposed analytical preintegration for graph-based VINS.

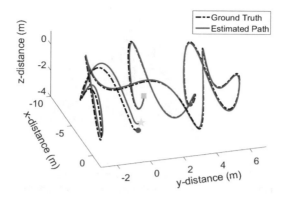

Fig. 2: Simulation results: Estimated trajectory versus the ground truth for an example Monte Carlo run. The initial start is show with a green square, with ground truth ending with a blue circle, and the estimation a teal star. Note that the discrete trajectory is not shown.

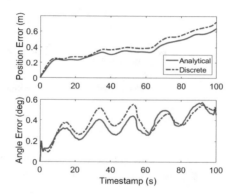

Fig. 3: Simulation results: Position and orientation RMSE for 10 Monte-Carlo simulations.

7   Conclusions and Future Work

We have introduced a high-accuracy preintegration theory based on *closed-form* solutions of the IMU integration equations, which allow us to accurately compute the mean, covariance, and bias Jacobians for each of the preintegration factors. This theory was integrated into graph-based VINS system and validated on both synthetic and real data. As currently our system implementation does not include large loop closures and thus does not fully gain the benefits of the batch optimization, in the future, we will include these constraints, which would allow us to further improve accuracy by taking full advantage of the graph formulation. In addition, we will seek to reduce the computational complexity of our system by intelligently sparsifying the graph so as to enable long-term and large-scale robot navigation. Lastly, we plan to incorporate our inertial processing with a variety of different vision techniques. For example, results for analytical preintegration with *direct* visual odometry can be found in [25].

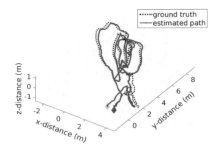

Fig. 4: Experimental results: Estimated trajectory versus the ground truth. The initial start is show with a red diamond and the ending location is shown with a green diamond.

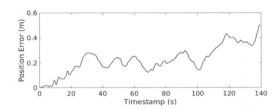

Fig. 5: Experimental results: Position estimation errors over the Machine Hall 01 dataset.

References

[1] Hesch, J., Kottas, D., Bowman, S., Roumeliotis, S.: Camera-IMU-based localization: Observability analysis and consistency improvement. International Journal of Robotics Research **33** (2014) 182–201

[2] Li, M., Mourikis, A.: High-precision, consistent EKF-based visual-inertial odometry. International Journal of Robotics Research **32**(6) (2013) 690–711

[3] Huang, G., Kaess, M., Leonard, J.: Towards consistent visual-inertial navigation. In: Proc. of the IEEE International Conference on Robotics and Automation, Hong Kong, China (May 31–June 7, 2014) 4926–4933

[4] Kumar, V., Michael, N.: Opportunities and challenges with autonomous micro aerial vehicles. International Journal of Robotics Research **31**(11) (September 2012) 1279–1291

[5] Dellaert, F., Kaess, M.: Square root SAM: Simultaneous localization and mapping via square root information smoothing. International Journal of Robotics Research **25**(12) (Dec. 2006) 1181–1203

[6] Lupton, T., Sukkarieh, S.: Visual-inertial-aided navigation for high-dynamic motion in built environments without initial conditions. IEEE Transactions on Robotics **28**(1) (February 2012) 61 –76

[7] Forster, C., Carlone, L., Dellaert, F., Scaramuzza, D.: Imu preintegration on manifold for efficient visual-inertial maximum-a-posteriori estimation. In: Robotics: Science and Systems XI. Number EPFL-CONF-214687 (2015)

[8] Ling, Y., Liu, T., Shen, S.: Aggressive quadrotor flight using dense visual-inertial fusion. In: 2016 IEEE International Conference on Robotics and Automation (ICRA). (May 2016) 1499–1506

[9] Mourikis, A.I., Roumeliotis, S.I.: A multi-state constraint Kalman filter for vision-aided inertial navigation. In: Proceedings of the IEEE International Conference on Robotics and Automation, Rome, Italy (April 10–14, 2007) 3565–3572

[10] Roumeliotis, S.I., Burdick, J.W.: Stochastic cloning: A generalized framework for processing relative state measurements. In: Proceedings of the IEEE International Conference on Robotics and Automation, Washington, DC (May 11-15 2002) 1788–1795

[11] Huang, G., Eckenhoff, K., Leonard, J.: Optimal-state-constraint EKF for visual-inertial navigation. In: Proc. of the International Symposium on Robotics Research, Sestri Levante, Italy (September 12–15, 2015)

[12] Kümmerle, R., Grisetti, G., Strasdat, H., Konolige, K., Burgard, W.: g2o: A general framework for graph optimization. In: Proc. of the IEEE International Conference on Robotics and Automation, Shanghai, China (May 9–13, 2011) 3607–3613

[13] Indelman, V., Williams, S., Kaess, M., Dellaert, F.: Factor graph based incremental smoothing in inertial navigation systems. In: Proc. of the International Conference on Information Fusion, Singapore (July 2012)

[14] Indelman, V., Melim, A., Dellaert, F.: Incremental light bundle adjustment for robotics navigation. In: 2013 IEEE/RSJ International Conference on Intelligent Robots and Systems, IEEE (2013) 1952–1959

[15] Trawny, N., Roumeliotis, S.I.: Indirect Kalman filter for 3D attitude estimation. Technical report, University of Minnesota, Dept. of Comp. Sci. & Eng. (March 2005)

[16] Lupton, T., Sukkarieh, S.: Visual-inertial-aided navigation for high-dynamic motion in built environments without initial conditions. IEEE Transactions on Robotics **28**(1) (Feb 2012) 61–76

[17] Eckenhoff, K., Geneva, P., Huang, G.: High-accuracy preintegration for visual inertial navigation. Technical Report RPNG-2016-001, University of Delaware (2016) Available: http://udel.edu/~ghuang/papers/tr_hapi.pdf.

[18] Chirikjian, G.S.: Stochastic Models, Information Theory, and Lie Groups, Volume 2: Analytic Methods and Modern Applications. Volume 2. Springer Science & Business Media (2011)

[19] Huang, G.P., Mourikis, A.I., Roumeliotis, S.I.: An observability constrained sliding window filter for SLAM. In: Proc. of the IEEE/RSJ International Conference on Intelligent Robots and Systems, San Francisco, CA (September 25–30, 2011) 65–72

[20] Huang, G., Kaess, M., Leonard, J.: Consistent sparsification for graph optimization. In: Proc. of the European Conference on Mobile Robots, Barcelona, Spain (September 25–27, 2013) 150–157

[21] Forster, C., Pizzoli, M., Scaramuzza, D.: SVO: Fast semi-direct monocular visual odometry. In: Proc. of the IEEE International Conference on Robotics and Automation, Hong Kong, China (May 2014)

[22] Kummerle, R., Grisetti, G., Strasdat, H., Konolige, K., Burgard, W.: g2o: A general framework for graph optimization. In: Proc. of the IEEE International Conference on Robotics and Automation, Shanghai, China (May 9–13, 2011) 3607–3613

[23] Burri, M., Nikolic, J., Gohl, P., Schneider, T., Rehder, J., Omari, S., Achtelik, M.W., Siegwart, R.: The euroc micro aerial vehicle datasets. The International Journal of Robotics Research (2016)

[24] OpenCV Developers Team: Open source computer vision (OpenCV) library. Available: http://opencv.org

[25] Eckenhoff, K., Geneva, P., Huang, G.: Direct visual-inertial navigation with analytical preintegration. In: 2017 IEEE International Conference on Robotics and Automation. (May 2017)

A Certifiably Correct Algorithm for Synchronization over the Special Euclidean Group

David M. Rosen$^{\star 1}$, Luca Carlone^2, Afonso S. Bandeira^3, and
John J. Leonard^1

^{1} Computer Science and Artificial Intelligence Laboratory, Massachusetts Institute of
Technology, Cambridge, MA 02139, USA
^{2} Laboratory for Information and Decision Systems, Massachusetts Institute of
Technology, Cambridge, MA 02139, USA
^{3} Department of Mathematics and Center for Data Science, Courant Institute of
Mathematical Sciences, New York University, New York, NY 10012, USA

Abstract. Many geometric estimation problems naturally take the form
of *synchronization over the special Euclidean group*: estimate the values of a set of unknown poses $x_1, \ldots, x_n \in \mathrm{SE}(d)$ given noisy measurements of a subset of their pairwise relative transforms $x_i^{-1} x_j$. Examples
of this class include the foundational problems of pose-graph simultaneous localization and mapping (SLAM) (in robotics) and camera motion
estimation (in computer vision), among others. This problem is typically formulated as a nonconvex maximum-likelihood estimation that is
computationally hard to solve in general. Nevertheless, in this paper we
present an algorithm that is able to efficiently recover *certifiably globally
optimal* solutions of the special Euclidean synchronization problem in a
non-adversarial noise regime. The crux of our approach is the development of a semidefinite relaxation of the maximum-likelihood estimation
whose minimizer provides the *exact* MLE so long as the magnitude of
the noise corrupting the available measurements falls below a certain
critical threshold; furthermore, whenever exactness obtains, it is possible to *verify* this fact a posteriori, thereby *certifying* the optimality of
the recovered estimate. We develop a specialized optimization scheme for
solving large-scale instances of this semidefinite relaxation by exploiting
its low-rank, geometric, and graph-theoretic structure to reduce it to
an equivalent optimization problem on a low-dimensional Riemannian
manifold, and then design a Riemannian truncated-Newton trust-region
method to solve this reduction efficiently. We combine this fast optimization approach with a simple rounding procedure to produce our
algorithm, *SE-Sync*. Experimental evaluation on a variety of simulated
and real-world pose-graph SLAM datasets shows that SE-Sync is capable
of recovering globally optimal solutions when the available measurements
are corrupted by noise up to an order of magnitude greater than that typically encountered in robotics and computer vision applications, and does
so more than an order of magnitude faster than the Gauss-Newton-based
approach that forms the basis of current state-of-the-art techniques.

---

* Corresponding author. Email: dmrosen@mit.edu

© Springer Nature Switzerland AG 2020
K. Goldberg et al. (Eds.): *Algorithmic Foundations of Robotics XII*, SPAR 13, pp. 64–79, 2020.
https://doi.org/10.1007/978-3-030-43089-4_5

1   Introduction

Over the coming decades, the increasingly widespread adoption of robotic technology in areas such as transportation, medicine, and disaster response has tremendous potential to increase productivity, alleviate suffering, and preserve life. At the same time, however, these high-impact applications often place autonomous systems in safety- and life-critical roles, where misbehavior or undetected failures can carry grave consequences. While empirical evaluation has traditionally been a driving force in the design and implementation of autonomous systems, safety-critical applications such as these necessitate algorithms that come with clearly-delineated performance guarantees. This paper presents one such algorithm, *SE-Sync*, an efficient and *certifiably correct* method for solving the fundamental problem of *pose estimation*.[1]

Formally, we consider the *synchronization problem* of estimating a collection of unknown poses based upon a set of relative measurements between them. This estimation problem lies at the core of many fundamental perceptual problems in robotics; for example, simultaneous localization and mapping (SLAM) [30] and multi-robot localization [3]. Closely-related formulations also arise in structure from motion [25, 32] and camera network calibration [42] (in computer vision), sensor network localization [35], and cryo-electron microscopy [39]. These synchronization problems are typically formulated as instances of maximum-likelihood estimation under an assumed probability distribution for the measurement noise. This formulation is attractive from a theoretical standpoint due to the powerful analytical framework and strong performance guarantees that maximum-likelihood estimation affords [20]. However, this formal rigor comes at the expense of computational tractability, as the maximum-likelihood formulation leads to a nonconvex estimation problem that is difficult to solve in general.

Related Work. In the context of SLAM, the pose synchronization problem is commonly solved using iterative numerical optimization methods, e.g. Gauss-Newton [28–30], gradient descent [23, 34], or trust region methods [36]. This approach is attractive because the rapid convergence speed of second-order numerical optimization methods [33], together with their ability to exploit the measurement sparsity that typically occurs in naturalistic problem instances [18], enables these techniques to scale efficiently to large problem sizes while maintaining real-time operation. However, this computational expedience comes at the expense of *reliability*, as their restriction to *local* search renders these methods vulnerable to convergence to suboptimal critical points, even for relatively small noise levels [14, 37]. This observation, together with the fact that suboptimal critical points usually correspond to egregiously wrong trajectory estimates, has motivated two general lines of research. The first addresses the *initialization problem*, i.e., how to compute a suitable initial guess for iterative refinement; examples of this effort are [12, 15, 37]. The second aims at a deeper understanding

---

[1] A *pose* is a position and orientation in d-dimensional Euclidean space; this is an element of the *special Euclidean group* $\mathrm{SE}(d) \cong \mathbb{R}^d \rtimes \mathrm{SO}(d)$.

of the global structure of the pose synchronization problem (e.g. number of local minima, convergence basin), see for example [26, 27, 45].

Contribution. In our previous work [13, 14, 16], we demonstrated that Lagrangian duality provides an effective means of *certifying* the optimality of an in-hand solution for the pose synchronization problem, and could in principle be used to *directly compute* such certifiably optimal solutions by solving a Lagrangian relaxation. However, this relaxation turns out to be a *semidefinite program* (SDP), and while there do exist mature general-purpose SDP solvers, their high per-iteration computational cost limits their practical utility to instances of this relaxation involving only a few hundred poses,[2] whereas real-world pose synchronization problems are typically one to two orders of magnitude larger.

The main contribution of this paper is the development of a specialized structure-exploiting optimization procedure that is capable of efficiently solving large-scale instances of the semidefinite relaxation in practice. This procedure enables the recovery of *certifiably globally optimal* solutions of the pose synchronization problem by means of solving the semidefinite relaxation within a non-adversarial noise regime in which minimizers of the latter correspond to *exact* solutions of the former. Our overall pose synchronization method, *SE-Sync*, is thus a *certifiably correct* algorithm [4], meaning that it is able to efficiently compute *globally* optimal solutions of generally intractable problems within a restricted range of operation, and *certify* the optimality of the solutions so obtained. In the case of our algorithm, experimental evaluation on a variety of simulated and real-world pose-graph SLAM datasets shows that SE-Sync is capable of recovering certifiably globally optimal solutions when the available measurements are corrupted by noise up to an order of magnitude greater than that typically encountered in robotics and computer vision applications, and does so more than an order of magnitude faster than the Gauss-Newton-based approach that forms the basis of current state-of-the-art techniques.

2 Problem formulation

The SE(d) synchronization problem consists of estimating the values of a set of n unknown poses $x_1, \ldots, x_n \in \mathrm{SE}(d)$ given noisy measurements of m of their pairwise relative transforms $x_{ij} \triangleq x_i^{-1} x_j$ ($i \neq j$). We model the set of available measurements using an undirected graph $G = (\mathcal{V}, \mathcal{E})$ in which the nodes $i \in \mathcal{V}$ are in one-to-one correspondence with the unknown poses x_i and the edges $\{i, j\} \in \mathcal{E}$ are in one-to-one correspondence with the set of available measurements, and we assume without loss of generality that G is connected.[3] We let $\overrightarrow{G} = (\mathcal{V}, \overrightarrow{\mathcal{E}})$ be a directed graph obtained from G by fixing an orientation for each of its edges,

---

[2] This encompasses the most commonly-used interior-point-based SDP software libraries, including SDPA, SeDuMi, SDPT3, CSDP, and DSDP.

[3] If G is not connected, then the problem of estimating the unknown poses $x_1, \ldots, x_n$ decomposes into a set of independent estimation problems that are in one-to-one correspondence with the connected components of G; thus, the general case is always reducible to the case of connected graphs.

and assume that a noisy measurement $\tilde{x}_{ij}$ of each relative pose $x_{ij} = (t_{ij}, R_{ij})$ is obtained by sampling from the following probabilistic generative model:[4]

$$\tilde{t}_{ij} = \underline{t}_{ij} + t_{ij}^\epsilon, \qquad t_{ij}^\epsilon \sim \mathcal{N}\left(0, \tau_{ij}^{-1} I_d\right),$$
$$\tilde{R}_{ij} = \underline{R}_{ij} R_{ij}^\epsilon, \qquad R_{ij}^\epsilon \sim \text{Langevin}\left(I_d, \kappa_{ij}\right), \qquad \forall (i,j) \in \overrightarrow{\mathcal{E}}. \tag{1}$$

Here $\underline{x}_{ij} = (\underline{t}_{ij}, \underline{R}_{ij})$ is the true (latent) value of x_{ij}, $\mathcal{N}(\mu, \Sigma)$ denotes the standard multivariate Gaussian distribution with mean $\mu \in \mathbb{R}^d$ and covariance $\Sigma \succeq 0$, and Langevin(M, κ) denotes the *isotropic Langevin distribution* on $\text{SO}(d)$ with mode $M \in \text{SO}(d)$ and concentration parameter $\kappa \geq 0$ (this is the distribution on $\text{SO}(d)$ whose probability density function is given by:

$$p(R; M, \kappa) = \frac{1}{c_d(\kappa)} \exp\left(\kappa \operatorname{tr}(M^\mathsf{T} R)\right) \tag{2}$$

with respect to the Haar measure [17]). Finally, we define $\tilde{x}_{ji} \triangleq \tilde{x}_{ij}^{-1}$, $\kappa_{ji} \triangleq \kappa_{ij}$, $\tau_{ji} \triangleq \tau_{ij}$, and $\tilde{R}_{ji} \triangleq \tilde{R}_{ij}^\mathsf{T}$ for all $(i,j) \in \overrightarrow{\mathcal{E}}$.

Given a set of noisy measurements $\tilde{x}_{ij}$ sampled from the generative model (1), a straightforward computation shows that a maximum-likelihood estimate $\hat{x}_{\text{MLE}} \in \text{SE}(d)^n$ for the poses $x_1, \ldots, x_n$ is obtained as a minimizer of:

Problem 1 (Maximum-likelihood estimation for $\text{SE}(d)$ *synchronization).*

$$p_{\text{MLE}}^* = \min_{\substack{t_i \in \mathbb{R}^d \\ R_i \in \text{SO}(d)}} \sum_{(i,j) \in \overrightarrow{\mathcal{E}}} \kappa_{ij} \|R_j - R_i \tilde{R}_{ij}\|_F^2 + \tau_{ij} \left\|t_j - t_i - R_i \tilde{t}_{ij}\right\|_2^2 \tag{3}$$

Unfortunately, Problem 1 is a high-dimensional, nonconvex nonlinear program, and is therefore computationally hard to solve in general. Consequently, in this paper our strategy will be to *approximate* this problem using a (convex) *semidefinite relaxation* [44], and then exploit this relaxation to search for solutions of the original (hard) problem.

3   Forming the semidefinite relaxation

In this section we develop the semidefinite relaxation that we will solve in place of the maximum-likelihood estimation Problem 1.[5] To that end, our first step will be to rewrite Problem 1 in a simpler and more compact form that emphasizes the structural correspondences between the optimization problem (3) and several simple graph-theoretic objects that can be constructed from the set of available measurements $\tilde{x}_{ij}$ and the graphs G and $\overrightarrow{G}$.

---

[4] We use a directed graph to model the measurements $\tilde{x}_{ij}$ sampled from (1) because the distribution of the noise corrupting the latent values $\underline{x}_{ij}$ is not invariant under $\text{SE}(d)$'s group inverse operation. Consequently, we must keep track of which state x_i was the "base frame" for each measurement.

[5] Due to space limitations, we omit all derivations and proofs; please see the extended version of this paper [38] for detailed derivations and additional results.

We define the *translational weight graph* $W^\tau = (\mathcal{V}, \mathcal{E}, \{\tau_{ij}\})$ to be the weighted undirected graph with node set $\mathcal{V}$, edge set $\mathcal{E}$, and edge weights τ_{ij} for $\{i,j\} \in \mathcal{E}$. The Laplacian $L(W^\tau)$ of W^τ is then:

$$L(W^\tau)_{ij} = \begin{cases} \sum_{\{i,k\} \in \mathcal{E}} \tau_{ik}, & i = j, \\ -\tau_{ij}, & \{i,j\} \in \mathcal{E}, \\ 0, & \{i,j\} \notin \mathcal{E}. \end{cases} \tag{4}$$

Similarly, let $L(\tilde{G}^\rho)$ denote the *connection Laplacian* for the rotational synchronization problem determined by the measurements $\tilde{R}_{ij}$ and measurement weights κ_{ij} for $(i,j) \in \overrightarrow{\mathcal{E}}$; this is the symmetric $(d \times d)$-block-structured matrix determined by (cf. [40, 46]):

$$L(\tilde{G}^\rho) \in \mathrm{Sym}(dn)$$

$$L(\tilde{G}^\rho)_{ij} \triangleq \begin{cases} \left(\sum_{\{i,k\} \in \mathcal{E}} \kappa_{ik} \right) I_d, & i = j, \\ -\kappa_{ij} \tilde{R}_{ij}, & \{i,j\} \in \mathcal{E}, \\ 0_{d \times d}, & \{i,j\} \notin \mathcal{E}. \end{cases} \tag{5}$$

We also define a few matrices constructed from the set of translational observations $\tilde{t}_{ij}$. We let $\tilde{V} \in \mathbb{R}^{n \times dn}$ be the $(1 \times d)$-block-structured matrix with (i,j)-blocks determined by:

$$\tilde{V}_{ij} \triangleq \begin{cases} \sum_{\{k \in \mathcal{V} | (j,k) \in \overrightarrow{\mathcal{E}}\}} \tau_{jk} \tilde{t}_{jk}^{\mathsf{T}}, & i = j, \\ -\tau_{ji} \tilde{t}_{ji}^{\mathsf{T}}, & (j,i) \in \overrightarrow{\mathcal{E}}, \\ 0_{1 \times d}, & \text{otherwise}, \end{cases} \tag{6}$$

$\tilde{T} \in \mathbb{R}^{m \times dn}$ denote the $(1 \times d)$-block-structured matrix with rows and columns indexed by $e \in \overrightarrow{\mathcal{E}}$ and $k \in \mathcal{V}$, respectively, and whose (e,k)-block is given by:

$$\tilde{T}_{ek} \triangleq \begin{cases} -\tilde{t}_{kj}^{\mathsf{T}}, & e = (k,j) \in \overrightarrow{\mathcal{E}}, \\ 0_{1 \times d}, & \text{otherwise}, \end{cases} \tag{7}$$

and $\Omega \triangleq \mathrm{Diag}(\tau_{e_1}, \ldots, \tau_{e_m}) \in \mathrm{Sym}(m)$ denote the diagonal matrix indexed by the directed edges $e \in \overrightarrow{\mathcal{E}}$ whose eth element gives the precision of the translational measurement $\tilde{t}_e$ corresponding to that edge. Finally, we also aggregate the rotational and translational state estimates into the block matrices $R \triangleq \begin{pmatrix} R_1 & \cdots & R_n \end{pmatrix} \in \mathrm{SO}(d)^n \subset \mathbb{R}^{d \times dn}$ and $t \triangleq (t_1, \ldots, t_n) \in \mathbb{R}^{dn}$.

With these definitions in hand, let us return to Problem 1. We observe that for a *fixed* value of the rotational states $R_1, \ldots, R_n$, this problem reduces to the *unconstrained* minimization of a quadratic form in the translational variables $t_1, \ldots, t_n \in \mathbb{R}^d$, for which we can find a closed-form solution using a generalized Schur complement [21, Proposition 4.2]. This observation enables us to *analytically eliminate* the translational states from the optimization problem (3), thereby obtaining the simplified but equivalent maximum-likelihood estimation:

Problem 2 (Simplified maximum-likelihood estimation).

$$p_{\mathrm{MLE}}^* = \min_{R \in \mathrm{SO}(d)^n} \mathrm{tr}(\tilde{Q}R^\mathsf{T}R) \tag{8a}$$

$$\tilde{Q} = L(\tilde{G}^\rho) + \tilde{T}^\mathsf{T} \Omega^{\frac{1}{2}} \Pi \Omega^{\frac{1}{2}} \tilde{T}. \tag{8b}$$

Furthermore, given any minimizer R^* of Problem 2, we can recover a corresponding optimal value t^* for t according to:

$$t^* = -\mathrm{vec}\left(R^* \tilde{V}^\mathsf{T} L(W^\tau)^\dagger\right). \tag{9}$$

In (8b) $\Pi \in \mathbb{R}^{m \times m}$ is the matrix of the orthogonal projection operator onto $\ker(A(\overrightarrow{G})\Omega^{\frac{1}{2}}) \subseteq \mathbb{R}^{m \times m}$, where $A(\overrightarrow{G}) \in \mathbb{R}^{n \times m}$ is the *incidence matrix* of the directed graph $\overrightarrow{G}$. Although Π is generically dense, by exploiting the fact that it is derived from a sparse graph, we have been able to show that it admits the sparse decomposition:

$$\Pi = I_m - \Omega^{\frac{1}{2}} A(\overrightarrow{G})^\mathsf{T} L^{-\mathsf{T}} L^{-1} A(\overrightarrow{G}) \Omega^{\frac{1}{2}} \tag{10}$$

where $\underline{A}(\overrightarrow{G})$ is the *reduced incidence matrix* of $\overrightarrow{G}$ and $\underline{A}(\overrightarrow{G})\Omega^{\frac{1}{2}} = LQ_1$ is a thin LQ decomposition of $\underline{A}(\overrightarrow{G})\Omega^{\frac{1}{2}}$. Note that expression (10) requires only the sparse lower-triangular factor L, which can be efficiently obtained in practice (e.g. by applying Givens rotations [22, Sec. 5.2.1]). Decomposition (10) will play a critical role in the implementation of our efficient optimization procedure.

Now we derive the semidefinite relaxation of Problem 2 that we will solve in practice, exploiting the simplified form (8). We begin by relaxing the condition $R \in \mathrm{SO}(d)^n$ to $R \in \mathrm{O}(d)^n$. The advantage of this latter version is that $\mathrm{O}(d)$ is defined by a set of (quadratic) orthonormality constraints, so the orthogonally-relaxed version of Problem 2 is a *quadratically constrained quadratic program*; consequently, its Lagrangian dual relaxation is a *semidefinite program* [31]:

Problem 3 (Semidefinite relaxation for $\mathrm{SE}(d)$ *synchronization).*

$$p_{\mathrm{SDP}}^* = \min_{0 \preceq Z \in \mathrm{Sym}(dn)} \mathrm{tr}(\tilde{Q}Z)$$
$$\text{s.t.} \quad \mathrm{BlockDiag}_{d \times d}(Z) = \mathrm{Diag}(I_d, \dots, I_d) \tag{11}$$

At this point it is instructive to compare the semidefinite relaxation (11) with the simplified maximum-likelihood estimation (8). For any $R \in \mathrm{SO}(d)^n$, the product $Z = R^\mathsf{T}R$ is positive semidefinite and has identity matrices along its $(d \times d)$-block-diagonal, and so is a feasible point of (11); in other words, (11) can be regarded as a relaxation of the maximum-likelihood estimation obtained by *expanding* (8)*'s feasible set*. This immediately implies that $p_{\mathrm{SDP}}^* \leq p_{\mathrm{MLE}}^*$. Furthermore, if it happens that a minimizer Z^* of Problem 3 admits a decomposition of the form $Z^* = R^{*\mathsf{T}} R^*$ for some $R^* \in \mathrm{SO}(d)^n$, then it is straightforward to verify that this R^* is also a minimizer of Problem 2, and so provides a *globally* optimal solution of the maximum-likelihood estimation Problem 1 via (9).

The crucial fact that justifies our interest in Problem 3 is that (as demonstrated empirically in our prior work [14] and analyzed in a simpler setting in [5]) this problem has a *unique* minimizer of just this form so long as the noise corrupting the available measurements $\tilde{x}_{ij}$ is not too large. More precisely, we prove:[6]

Proposition 1. *Let Q be the matrix of the form* (8b) *constructed using the true* (*latent*) *relative transforms $\underline{x}_{ij}$ in* (1). *There exists a constant $\beta \triangleq \beta(Q) > 0$ (depending upon Q) such that, if $\|\tilde{Q} - Q\|_2 < \beta$, then:*

(i) The semidefinite relaxation Problem 3 has a unique solution Z^*, and
(ii) $Z^* = R^{*\mathsf{T}}R^*$, where $R^* \in \mathrm{SO}(d)^n$ is a minimizer of the maximum-likelihood estimation Problem 2.

4   The SE-Sync algorithm

4.1   Solving the semidefinite relaxation

As a semidefinite program, Problem 3 can in principle be solved in polynomial-time using interior-point methods [41, 44]. In practice, however, the high computational cost of general-purpose semidefinite programming algorithms prevents these methods from scaling effectively to problems in which the dimension of the decision variable Z is greater than a few thousand [41]. Unfortunately, the instances of Problem 3 that arise in robotics and computer vision applications are typically one to two orders of magnitude larger than this maximum effective problem size, and are therefore well beyond the reach of these general-purpose methods. Consequently, in this subsection we develop a specialized optimization procedure for solving large-scale instances of Problem 3 efficiently.

Simplifying Problem 3   The dominant computational cost when applying general-purpose semidefinite programming methods to solve Problem 3 is the need to store and manipulate expressions involving the (large, dense) matrix variable Z. However, in the case that exactness holds, we know that the actual *solution* Z^* of Problem 3 that we seek has a very concise description in the factored form $Z^* = R^{*\mathsf{T}}R^*$ for $R^* \in \mathrm{SO}(d)^n$. More generally, even in those cases where exactness fails, minimizers Z^* of Problem 3 typically have a rank r not much greater than d, and therefore admit a symmetric rank decomposition $Z^* = Y^{*\mathsf{T}}Y^*$ for $Y^* \in \mathbb{R}^{r \times dn}$ with $r \ll dn$.

In a pair of papers, Burer and Monteiro [10, 11] proposed an elegant general approach to exploit the fact that large-scale semidefinite programs often admit such low-rank solutions: simply replace every instance of the decision variable Z with a rank-r product of the form $Y^\mathsf{T}Y$ to produce a *rank-restricted* version of the original problem. This substitution has the two-fold effect of (i) dramatically reducing the size of the search space and (ii) rendering the positive semidefiniteness constraint *redundant*, since $Y^\mathsf{T}Y \succeq 0$ for *any* choice of Y. The resulting

---

[6] Please see the extended version of this paper [38] for the proof.

rank-restricted form of the problem is thus a low-dimensional *nonlinear* program, rather than a *semidefinite* program.

Furthermore, following Boumal [6] we observe that after replacing Z in Problem 3 with $Y^\mathsf{T}Y$ for $Y = \left(Y_1 \ \cdots \ Y_n\right) \in \mathbb{R}^{r \times dn}$, the block-diagonal constraints in our specific problem of interest (11) are equivalent to $Y_i^\mathsf{T}Y_i = I_d$, for all $i \in [n]$, i.e., the columns of each block $Y_i \in \mathbb{R}^{r \times d}$ form an *orthonormal frame*. In general, the set of all orthonormal k-frames in $\mathbb{R}^p$ ($k \leq p$):

$$\mathrm{St}(k,p) \triangleq \left\{ Y \in \mathbb{R}^{p \times k} \mid Y^\mathsf{T}Y = I_k \right\} \tag{12}$$

forms a smooth compact matrix manifold, called the *Stiefel manifold*, which can be equipped with a Riemannian metric induced by its embedding into the ambient space $\mathbb{R}^{p \times k}$ [2, Sec. 3.3.2]. Together, these observations enable us to reduce Problem 3 to an equivalent *unconstrained* Riemannian optimization problem defined on a product of Stiefel manifolds:

Problem 4 (Rank-restricted semidefinite relaxation, Riemannian optimization form).

$$p^*_{\mathrm{SDPLR}} = \min_{Y \in \mathrm{St}(d,r)^n} \mathrm{tr}(\tilde{Q}Y^\mathsf{T}Y). \tag{13}$$

This is the optimization problem that we will actually solve in practice.

Ensuring optimality While the reduction from Problem 3 to Problem 4 dramatically reduces the size of the optimization problem that needs to be solved, it comes at the expense of (re)introducing the quadratic orthonormality constraints (12). It may therefore not be clear whether anything has really been gained by relaxing Problem 2 to Problem 4, since it appears that we may have simply replaced one difficult nonconvex optimization problem with another. The following remarkable result (adapted from Boumal et al. [9]) justifies this approach:

Proposition 2 (A sufficient condition for optimality in Problem 4). *If $Y \in \mathrm{St}(d,r)^n$ is a (row) rank-deficient second-order critical point[7] of Problem 4, then Y is a global minimizer of Problem 4 and $Z^* = Y^\mathsf{T}Y$ is a solution of the semidefinite relaxation Problem 3.*

Proposition 2 immediately suggests a procedure for obtaining solutions Z^* of Problem 3 by applying a Riemannian optimization method to search successively higher levels of the rank-restricted hierarchy of relaxations (13) until a *rank-deficient* second-order critical point is obtained.[8] This algorithm is referred to as the *Riemannian Staircase* [6, 9].

---

[7] That is, a point satisfying grad $F(Y) = 0$ and Hess $F(Y) \succeq 0$ (cf. (14)–(17)).

[8] Note that since *every* $Y \in \mathrm{St}(d,r)^n$ is row rank-deficient for $r > dn$, this procedure is guaranteed to recover an optimal solution after searching at most $dn + 1$ levels of the hierarchy (13).

A Riemannian optimization method for Problem 4 Proposition 2 provides a means of obtaining *global* minimizers of Problem 3 by *locally* searching for second-order critical points of Problem 4. In this subsection, we design a Riemannian optimization method that will enable us to rapidly identify these critical points in practice.

Equations (8b) and (10) provide an efficient means of computing *products* with $\tilde{Q}$ *without* the need to form $\tilde{Q}$ explicitly by performing a sequence of sparse matrix multiplications and sparse triangular solves. In turn, this operation is sufficient to evaluate the objective appearing in Problem 4, as well as its corresponding gradient and Hessian-vector products when considered as a function on the ambient Euclidean space $\mathbb{R}^{r \times dn}$:

$$F(Y) \triangleq \text{tr}(\tilde{Q}Y^\mathsf{T}Y), \qquad \nabla F(Y) = 2Y\tilde{Q}, \qquad \nabla^2 F(Y)[\dot{Y}] = 2\dot{Y}\tilde{Q}. \qquad (14)$$

Furthermore, there are simple relations between the ambient Euclidean gradient and Hessian-vector products in (14) and their corresponding Riemannian counterparts when $F(\cdot)$ is viewed as a function restricted to $\text{St}(d,r)^n \subset \mathbb{R}^{r \times dn}$. With reference to the orthogonal projection operator [19, eq. (2.3)]:

$$\text{Proj}_Y \colon T_Y\left(\mathbb{R}^{r \times dn}\right) \to T_Y\left(\text{St}(d,r)^n\right)$$
$$\text{Proj}_Y(X) = X - Y\,\text{SymBlockDiag}_{d \times d}(Y^\mathsf{T}X) \qquad (15)$$

the Riemannian gradient $\text{grad}\, F(Y)$ is simply the orthogonal projection of the ambient Euclidean gradient $\nabla F(Y)$ (cf. [2, eq. (3.37)]):

$$\text{grad}\, F(Y) = \text{Proj}_Y \nabla F(Y). \qquad (16)$$

Similarly, the Riemannian Hessian-vector product $\text{Hess}\, F(Y)[\dot{Y}]$ can be obtained as the orthogonal projection of the ambient directional derivative of the gradient vector field $\text{grad}\, F(Y)$ in the direction of $\dot{Y}$ (cf. [2, eq. (5.15)]). A straightforward computation shows that this is given by:

$$\text{Hess}\, F(Y)[\dot{Y}] = \text{Proj}_Y(\nabla^2 F(Y)[\dot{Y}] - \dot{Y}\,\text{SymBlockDiag}_{d \times d}(Y^\mathsf{T}\nabla F(Y))). \qquad (17)$$

Together, equations (8b), (10), and (14)–(17) provide an efficient means of computing $F(Y)$, $\text{grad}\, F(Y)$, and $\text{Hess}\, F(Y)[\dot{Y}]$. Consequently, we propose to employ the truncated-Newton *Riemannian Trust-Region* (RTR) method [1, 8] to efficiently compute high-precision estimates of second-order critical points of Problem 4 in practice.

4.2   Rounding the solution

In the previous subsection, we described an efficient algorithmic approach for computing minimizers $Y^* \in \text{St}(d,r)^n$ of Problem 4 that correspond to solutions $Z^* = Y^{*\mathsf{T}}Y^*$ of Problem 3. However, our ultimate goal is to extract an optimal solution of Problem 2 from Z^* whenever exactness holds, and a *feasible approximate solution* otherwise. In this subsection, we develop a rounding procedure satisfying these criteria.

Algorithm 1 Rounding procedure for solutions of Problem 4

Input: An optimal solution $Y^* \in \mathrm{St}(d, r)^n$ of Problem 4.
Output: A feasible point $\hat{R} \in \mathrm{SO}(d)^n$.
 1: **function** ROUNDSOLUTION(Y^*)
 2:     Compute a rank-d truncated singular value decomposition $U_d \Xi_d V_d^\mathsf{T}$ for Y^*
        and assign $\hat{R} \leftarrow \Xi_d V_d^\mathsf{T}$.
 3:     Set $N_+ \leftarrow |\{\hat{R}_i \mid \det(\hat{R}_i) > 0\}|$.
 4:     **if** $N_+ < \lceil \frac{n}{2} \rceil$ **then**
 5:         $\hat{R} \leftarrow \mathrm{Diag}(1, \ldots, 1, -1)\hat{R}$.
 6:     **end if**
 7:     **for** $i = 1, \ldots, n$ **do**
 8:         Set $\hat{R}_i \leftarrow$ NEARESTROTATION($\hat{R}_i$).   ▷ See e.g. [43]
 9:     **end for**
10:     **return** $\hat{R}$
11: **end function**

To begin, let us consider the case in which exactness obtains; here $Y^{*\mathsf{T}}Y^* = Z^* = R^{*\mathsf{T}}R^*$ for some optimal solution $R^* \in \mathrm{SO}(d)^n$ of Problem 2. Since $\mathrm{rank}(R^*) = d$, this implies that $\mathrm{rank}(Y^*) = d$ as well. Consequently, letting $Y^* = U_d \Xi_d V_d^\mathsf{T}$ denote a (rank-d) thin singular value decomposition of Y^*, and defining $\bar{Y} \triangleq \Xi_d V_d^\mathsf{T} \in \mathbb{R}^{d \times dn}$, it follows that $\bar{Y}^\mathsf{T}\bar{Y} = Z^* = R^{*\mathsf{T}}R^*$. This last equality implies that the $d \times d$ block-diagonal of $\bar{Y}^\mathsf{T}\bar{Y}$ satisfies $\bar{Y}_i^\mathsf{T}\bar{Y}_i = I_d$ for all $i \in [n]$, i.e. $\bar{Y} \in \mathrm{O}(d)^n$. Similarly, comparing the elements of the first block rows of $\bar{Y}^\mathsf{T}\bar{Y}$ and $R^{*\mathsf{T}}R^*$ shows that $\bar{Y}_1^\mathsf{T}\bar{Y}_j = R_1^*R_j^*$ for all $j \in [n]$. Left-multiplying this set of equalities by $\bar{Y}_1$ and letting $A = \bar{Y}_1 R_1^*$ shows $\bar{Y} = AR^*$ for some $A \in \mathrm{O}(d)$. Since any product of the form AR^* with $A \in \mathrm{SO}(d)$ is *also* an optimal solution of Problem 2 (by gauge symmetry), this shows that $\bar{Y}$ is optimal provided that $\bar{Y} \in \mathrm{SO}(d)$ specifically. Furthermore, if this is not the case, we can always make it so by left-multiplying $\bar{Y}$ by any orientation-reversing element of $\mathrm{O}(d)$, for example $\mathrm{Diag}(1, \ldots, 1, -1)$. This argument provides a means of recovering an optimal solution of Problem 2 from Y^* whenever exactness holds. Moreover, this procedure straightforwardly generalizes to the case that exactness fails, thereby producing a convenient rounding scheme (Algorithm 1).

4.3   The complete algorithm

Combining the efficient optimization approach of Section 4.1 with the rounding procedure of Section 4.2 produces *SE-Sync* (Algorithm 2), our proposed algorithm for synchronization over the special Euclidean group. When applied to an instance of $\mathrm{SE}(d)$ synchronization, SE-Sync returns a feasible point $\hat{x} \in \mathrm{SE}(d)^n$ for the maximum-likelihood estimation Problem 1 together with the lower bound $p^*_{\mathrm{SDP}} \leq p^*_{\mathrm{MLE}}$ for Problem 1's optimal value. Furthermore, in the case that Problem 3 is exact, the returned estimate $\hat{x} = (\hat{t}, \hat{R})$ *attains* this lower bound (i.e. $F(\tilde{Q}\hat{R}^\mathsf{T}\hat{R}) = p^*_{\mathrm{SDP}}$), which thus serves as a *computational certificate* of $\hat{x}$'s correctness. SE-Sync is thus a *certifiably correct* algorithm for $\mathrm{SE}(d)$ synchronization, as claimed.

Algorithm 2 The SE-Sync algorithm

Input: An initial point $Y \in \mathrm{St}(d, r_0)^n$, $r_0 \geq d + 1$.
Output: A feasible estimate $\hat{x} \in \mathrm{SE}(d)^n$ for the maximum-likelihood estimation Problem 1, and a lower bound $p^*_{\mathrm{SDP}} \leq p^*_{\mathrm{MLE}}$ for Problem 1's optimal value.
1: **function** SE-SYNC(Y)
2:     Set $Y^* \leftarrow$ RIEMANNIANSTAIRCASE(Y).   ▷ Solve Problems 3 & 4
3:     Set $p^*_{\mathrm{SDP}} \leftarrow F(\tilde{Q} Y^{*\mathsf{T}} Y^*)$.   ▷ $Z^* = Y^{*\mathsf{T}} Y^*$
4:     Set $\hat{R} \leftarrow$ ROUNDSOLUTION(Y^*).
5:     Recover the optimal translational estimates $\hat{t}$ corresponding to $\hat{R}$ via (9).
6:     Set $\hat{x} \leftarrow (\hat{t}, \hat{R})$.
7:     **return** $\{\hat{x}, p^*_{\mathrm{SDP}}\}$
8: **end function**

---

5   Experimental results

In this section, we evaluate SE-Sync's performance on a variety of simulated and real-world 3D pose-graph SLAM datasets. We consider two versions of the algorithm that differ in their approach to evaluating products with Π: the first (SE-Sync-Chol) employs equation (10) using the cached Cholesky factor L, while the second (SE-Sync-QR) computes the image of the orthogonal projection using a QR decomposition (see Appendix B.3 of [38] for details). As a basis for comparison, we also evaluate the performance of the Gauss-Newton method (GN), the *de facto* standard for solving pose-graph SLAM in robotics [24, 28, 29].

Our implementations of SE-Sync[9] and the Gauss-Newton method are written in MATLAB, and the former takes advantage of the truncated-Newton RTR method [1, 8] supplied by the Manopt toolbox [7]. The Gauss-Newton method is initialized using the state-of-the-art *chordal initialization* [15, 32], and we set $r_0 = 5$ in the Riemannian Staircase. Finally, while SE-Sync does not require a high-quality initialization in order to reach a globally optimal solution, it can still benefit (in terms of reduced computation time) from being supplied with one. Consequently, in the following experiments we employ two initialization procedures in conjunction with each version of SE-Sync: the first (rand) simply samples a point uniformly randomly from $\mathrm{St}(d, r_0)^n$, while the second (chord) supplies the same chordal initialization that the Gauss-Newton method receives, in order to enable a fair comparison of the algorithms' computational speeds.

5.1   Cube experiments

In this first set of experiments, we are interested in investigating how the performance of SE-Sync is affected by factors such as measurement noise, measurement density, and problem size. We simulate a robot traveling along a 3D grid world consisting of s^3 poses arranged in a cubical lattice. Odometric measurements are available between sequential poses, and loop closures between nearby nonsequential poses are available with probability p_{LC}. We fix default values of $\kappa = 16.67$ (corresponding to an expected RMS error of 10 degrees for $\tilde{R}_{ij}$), $\tau = 75$ (corresponding to an expected RMS error of .20 m for $\tilde{t}_{ij}$), $p_{LC} = .1$, and $s = 10$, and consider the effect of varying each of these individually; our complete dataset

---

[9] Available online: `https://github.com/david-m-rosen/SE-Sync`.

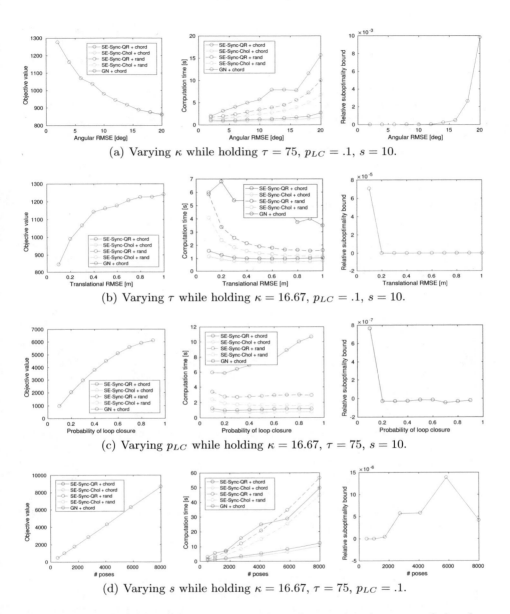

(a) Varying κ while holding $\tau = 75$, $p_{LC} = .1$, $s = 10$.

(b) Varying τ while holding $\kappa = 16.67$, $p_{LC} = .1$, $s = 10$.

(c) Varying p_{LC} while holding $\kappa = 16.67$, $\tau = 75$, $s = 10$.

(d) Varying s while holding $\kappa = 16.67$, $\tau = 75$, $p_{LC} = .1$.

Fig. 1. Results for the cube experiments. These figures plot the mean of the objective values (left column) and elapsed computation times (center column) attained by the Gauss-Newton and SE-Sync algorithms, as well as the upper bound $(F(\tilde{Q}\hat{R}^{\mathsf{T}}\hat{R}) - p^*_{\mathrm{SDP}})/p^*_{\mathrm{SDP}}$ for the relative suboptimality of the solution recovered by SE-Sync (right column), for 50 realizations of the cube dataset as functions of the measurement precisions κ (first row) and τ (second row), the loop closure probability p_{LC} (third row), and the problem size (fourth row).

consists of 50 realizations sampled for *each* joint setting of these parameters. Results for these experiments are shown in Fig. 1.

Consistent with our previous findings [14], these results suggest that the exactness of the relaxation (11) depends primarily upon the level of noise in the rotational observations $\tilde{R}_{ij}$ in (1). In these examples, exactness holds for RMS angular errors up to about 15 degrees, which is roughly an order of magnitude greater than the level of noise affecting sensors typically deployed in robotics and computer vision applications. Furthermore, in addition to its ability to recover certifiably optimal solutions, examining the center column of Fig. 1 shows that SE-Sync is consistently many-fold faster than Gauss-Newton, and that this speed differential increases superlinearly with the problem size (Fig. 1(d)).

5.2   SLAM benchmark datasets

Next, we evaluate SE-Sync on a suite of standard SLAM benchmark datasets containing both simulated and large-scale real-world examples. For the purpose of these experiments, we restrict attention to the version of SE-Sync employing QR factorization and chordal initialization. Results are summarized in Table 1.

	# Poses	# Edges	Gauss-Newton Objective value	Time [s]	SE-Sync Objective value	Time [s]	Max. suboptimality
sphere	2500	4949	1.687×10^3	14.98	1.687×10^3	2.81	1.410×10^{-11}
torus	5000	9048	2.423×10^4	31.94	2.423×10^4	5.67	7.276×10^{-12}
grid	8000	22236	8.432×10^4	130.35	8.432×10^4	22.37	4.366×10^{-11}
garage	1661	6275	1.263×10^0	17.81	1.263×10^0	5.33	2.097×10^{-11}
cubicle	5750	16869	7.171×10^2	136.86	7.171×10^2	13.08	1.603×10^{-11}
rim	10195	29743	5.461×10^3	575.42	5.461×10^3	36.66	5.639×10^{-11}

Table 1. Results for the SLAM benchmark datasets.

On each of these examples, both SE-Sync and Gauss-Newton converged to the same (globally optimal) solution. However, as in Section 5.1, SE-Sync did so considerably faster, outperforming Gauss-Newton in terms of computational speed by a factor of between 3.3 (on garage, the smallest dataset) and 15.7 (on rim, the largest). These results further support our claim that SE-Sync provides an effective means of recovering globally optimal pose-graph SLAM solutions under real-world operating conditions.

6   Conclusion

In this paper we developed SE-Sync, a certifiably correct algorithm for synchronization over the special Euclidean group. Our method enables the efficient recovery of *certifiably optimal* solutions of the SE(d) synchronization problem within a non-adversarial but operationally-relevant noise regime. Experimental evaluation on a variety of simulated and real-world pose-graph SLAM datasets shows that SE-Sync is capable of recovering globally optimal solutions when the available measurements are corrupted by noise up to an order of magnitude greater than that typically encountered in robotics and computer vision applications, and does so more than an order of magnitude faster than the Gauss-Newton-based approach that forms the basis of current state-of-the-art techniques.

Bibliography

[1] P.-A. Absil, C.G. Baker, and K.A. Gallivan. Trust-region methods on Riemannian manifolds. *Found. Comput. Math.*, 7(3):303–330, July 2007.

[2] P.-A. Absil, R. Mahony, and R. Sepulchre. *Optimization Algorithms on Matrix Manifolds*. Princeton University Press, 2008.

[3] R. Aragues, L. Carlone, G. Calafiore, and C. Sagues. Multi-agent localization from noisy relative pose measurements. In *IEEE Intl. Conf. on Robotics and Automation (ICRA)*, pages 364–369, Shanghai, China, May 2011.

[4] A.S. Bandeira. A note on probably certifiably correct algorithms. *Comptes Rendus Mathematique*, 354(3):329–333, 2016.

[5] A.S. Bandeira, N. Boumal, and A. Singer. Tightness of the maximum likelihood semidefinite relaxation for angular synchronization. *Math. Program.*, 2016.

[6] N. Boumal. A Riemannian low-rank method for optimization over semidefinite matrices with block-diagonal constraints. arXiv preprint: arXiv:1506.00575v2, 2015.

[7] N. Boumal, B. Mishra, P.-A. Absil, and R. Sepulchre. Manopt, a MATLAB toolbox for optimization on manifolds. *Journal of Machine Learning Research*, 15(1):1455–1459, 2014.

[8] N. Boumal, P.-A. Absil, and C. Cartis. Global rates of convergence for nonconvex optimization on manifolds. arXiv preprint: arXiv:1605.08101v1, 2016.

[9] N. Boumal, V. Voroninski, and A.S. Bandeira. The non-convex Burer-Monteiro approach works on smooth semidefinite programs. arXiv preprint arXiv:1606.04970v1, June 2016.

[10] S. Burer and R.D.C. Monteiro. A nonlinear programming algorithm for solving semidefinite programs via low-rank factorization. *Math. Program.*, 95:329–357, 2003.

[11] S. Burer and R.D.C. Monteiro. Local minima and convergence in low-rank semidefinite programming. *Math. Program.*, 103:427–444, 2005.

[12] L. Carlone and A. Censi. From angular manifolds to the integer lattice: Guaranteed orientation estimation with application to pose graph optimization. *IEEE Trans. on Robotics*, 30(2):475–492, April 2014.

[13] L. Carlone and F. Dellaert. Duality-based verification techniques for 2D SLAM. In *IEEE Intl. Conf. on Robotics and Automation (ICRA)*, pages 4589–4596, Seattle, WA, May 2015.

[14] L. Carlone, D.M. Rosen, G. Calafiore, J.J. Leonard, and F. Dellaert. Lagrangian duality in 3D SLAM: Verification techniques and optimal solutions. In *IEEE/RSJ Intl. Conf. on Intelligent Robots and Systems (IROS)*, Hamburg, Germany, September 2015.

[15] L. Carlone, R. Tron, K. Daniilidis, and F. Dellaert. Initialization techniques for 3D SLAM: A survey on rotation estimation and its use in pose graph

optimization. In *IEEE Intl. Conf. on Robotics and Automation (ICRA)*, pages 4597–4605, Seattle, WA, May 2015.

[16] L. Carlone, G.C. Calafiore, C. Tommolillo, and F. Dellaert. Planar pose graph optimization: Duality, optimal solutions, and verification. *IEEE Trans. on Robotics*, 32(3):545–565, June 2016.

[17] A. Chiuso, G. Picci, and S. Soatto. Wide-sense estimation on the special orthogonal group. *Communications in Information and Systems*, 8(3):185–200, 2008.

[18] F. Dellaert and M. Kaess. Square Root SAM: Simultaneous localization and mapping via square root information smoothing. *Intl. J. of Robotics Research*, 25(12):1181–1203, December 2006.

[19] A. Edelman, T.A. Arias, and S.T. Smith. The geometry of algorithms with orthogonality constraints. *SIAM J. Matrix Anal. Appl.*, 20(2):303–353, October 1998.

[20] T.S. Ferguson. *A Course in Large Sample Theory*. Chapman & Hall/CRC, Boca Raton, FL, 1996.

[21] J. Gallier. The Schur complement and symmetric positive semidefinite (and definite) matrices. unpublished note, available online: http://www.cis.upenn.edu/~jean/schur-comp.pdf, December 2010.

[22] G.H. Golub and C.F. Van Loan. *Matrix Computations*. Johns Hopkins University Press, Baltimore, MD, 3rd edition, 1996.

[23] G. Grisetti, C. Stachniss, and W. Burgard. Nonlinear constraint network optimization for efficient map learning. *IEEE Trans. on Intelligent Transportation Systems*, 10(3):428–439, September 2009.

[24] G. Grisetti, R. Kümmerle, C. Stachniss, and W. Burgard. A tutorial on graph-based SLAM. *IEEE Intelligent Transportation Systems Magazine*, 2 (4):31–43, 2010.

[25] R. Hartley, J. Trumpf, Y. Dai, and H. Li. Rotation averaging. *Intl. J. of Computer Vision*, 103(3):267–305, July 2013.

[26] S. Huang, Y. Lai, U. Frese, and G. Dissanayake. How far is SLAM from a linear least squares problem? In *IEEE/RSJ Intl. Conf. on Intelligent Robots and Systems (IROS)*, pages 3011–3016, Taipei, Taiwan, October 2010.

[27] S. Huang, H. Wang, U. Frese, and G. Dissanayake. On the number of local minima to the point feature based SLAM problem. In *IEEE Intl. Conf. on Robotics and Automation (ICRA)*, pages 2074–2079, St. Paul, MN, May 2012.

[28] M. Kaess, H. Johannsson, R. Roberts, V. Ila, J.J. Leonard, and F. Dellaert. iSAM2: Incremental smoothing and mapping using the Bayes tree. *Intl. J. of Robotics Research*, 31(2):216–235, February 2012.

[29] R. Kümmerle, G. Grisetti, H. Strasdat, K. Konolige, and W. Burgard. g2o: A general framework for graph optimization. In *IEEE Intl. Conf. on Robotics and Automation (ICRA)*, pages 3607–3613, Shanghai, China, May 2011.

[30] F. Lu and E. Milios. Globally consistent range scan alignment for environmental mapping. *Autonomous Robots*, 4:333–349, April 1997.

[31] Z.-Q. Luo, W.-K. Ma, A. So, Y. Ye, and S. Zhang. Semidefinite relaxation of quadratic optimization problems. *IEEE Signal Processing Magazine*, 27 (3):20–34, May 2010.

[32] D. Martinec and T. Pajdla. Robust rotation and translation estimation in multiview reconstruction. In *IEEE Intl. Conf. on Computer Vision and Pattern Recognition (CVPR)*, pages 1–8, Minneapolis, MN, June 2007.

[33] J. Nocedal and S.J. Wright. *Numerical Optimization*. Springer Science+Business Media, New York, 2nd edition, 2006.

[34] E. Olson, J.J. Leonard, and S. Teller. Fast iterative alignment of pose graphs with poor initial estimates. In *IEEE Intl. Conf. on Robotics and Automation (ICRA)*, pages 2262–2269, Orlando, FL, May 2006.

[35] J.R. Peters, D. Borra, B.E. Paden, and F. Bullo. Sensor network localization on the group of three-dimensional displacements. *SIAM J. Control Optim.*, 53(6):3534–3561, 2015.

[36] D.M. Rosen, M. Kaess, and J.J. Leonard. RISE: An incremental trust-region method for robust online sparse least-squares estimation. *IEEE Trans. on Robotics*, 30(5):1091–1108, October 2014.

[37] D.M. Rosen, C. DuHadway, and J.J. Leonard. A convex relaxation for approximate global optimization in simultaneous localization and mapping. In *IEEE Intl. Conf. on Robotics and Automation (ICRA)*, pages 5822–5829, Seattle, WA, May 2015.

[38] D.M. Rosen, L. Carlone, A.S. Bandeira, and J.J. Leonard. SE-Sync: A certifiably correct algorithm for synchronization over the special Euclidean group. Technical Report MIT-CSAIL-TR-2017-002, Computer Science and Artificial Intelligence Laboratory, Massachusetts Institute of Technology, Cambridge, MA 02139, USA, February 2017.

[39] A. Singer and Y. Shkolnisky. Three-dimensional structure determination from common lines in cryo-EM by eigenvectors and semidefinite programming. *SIAM J. Imaging Sci.*, 4(2):543–572, 2011.

[40] A. Singer and H.-T. Wu. Vector diffusion maps and the connection Laplacian. *Comm. Pure Appl. Math.*, 65:1067–1144, 2012.

[41] M.J. Todd. Semidefinite optimization. *Acta Numerica*, 10:515–560, 2001.

[42] R. Tron and R. Vidal. Distributed 3-D localization of camera sensor networks from 2-D image measurements. *IEEE Trans. on Automatic Control*, 59(12):3325–3340, December 2014.

[43] S. Umeyama. Least-squares estimation of transformation parameters between two point patterns. *IEEE Trans. on Pattern Analysis and Machine Intelligence*, 13(4):376–380, 1991.

[44] V. Vandenberghe and S. Boyd. Semidefinite programming. *SIAM Review*, 38(1):49–95, March 1996.

[45] H. Wang, G. Hu, S. Huang, and G. Dissanayake. On the structure of nonlinearities in pose graph SLAM. In *Robotics: Science and Systems (RSS)*, Sydney, Australia, July 2012.

[46] L. Wang and A. Singer. Exact and stable recovery of rotations for robust synchronization. *Information and Inference*, September 2013.

Autonomous Visual Rendering using Physical Motion

Ahalya Prabhakar, Anastasia Mavrommati, Jarvis Schultz, and Todd D. Murphey

Department of Mechanical Engineering, Northwestern University,
2145 Sheridan Road, Evanston, IL 60208, USA
`ahalyaprabhakar2013@u.northwestern.edu;stacymav@u.northwestern.`
`edu;jschultz@northwestern.edu;t-murphey@northwestern.edu`

Abstract. *This paper addresses the problem of enabling a robot to represent and recreate visual information through physical motion, focusing on drawing using pens, brushes, or other tools. This work uses ergodicity as a control objective that translates planar visual input to physical motion without preprocessing (e.g., image processing, motion primitives). We achieve comparable results to existing drawing methods, while reducing the algorithmic complexity of the software. We demonstrate that optimal ergodic control algorithms with different time-horizon characteristics (infinitesimal, finite, and receding horizon) can generate qualitatively and stylistically different motions that render a wide range of visual information (e.g., letters, portraits, landscapes). In addition, we show that ergodic control enables the same software design to apply to multiple robotic systems by incorporating their particular dynamics, thereby reducing the dependence on task-specific robots. Finally, we demonstrate physical drawings with the Baxter robot.*

Keywords: Robot art, Motion control, Automation

1 Introduction

An increasing amount of research is focused on using control theory as a generator of artistic expressions for robotics applications [8]. There is a large interest in enabling robots to create art, such as drawing [5], dancing [7], or writing. However, the computational tools available in the standard software repertoire are generally insufficient for enabling these tasks in a natural and interpretable manner. This paper focuses on enabling robots to draw and write by translating raw visual input into physical actions.

Drawing is a task that does not lend itself to analysis in terms of trajectory error. Being at a particular state at a particular time does not improve a drawing, and failing to do so does not make it worse. Instead, drawing is a process where the success or failure is determined after the entire time history of motion has been synthesized into a final product. How should "error" be defined for purpose of quantitative engineering decisions and software automation? Similarly, motion

© Springer Nature Switzerland AG 2020
K. Goldberg et al. (Eds.): *Algorithmic Foundations of Robotics XII*, SPAR 13, pp. 80–95, 2020.
https://doi.org/10.1007/978-3-030-43089-4_6

primitives can be an important foundation for tasks such as drawing (e.g., hatch marks to represent shading), but where should these primitives come from and what should be done if a robot cannot physically execute them? Questions such as these often lead to robots and their software being co-designed with the task in mind, leading to task-specific software enabling a task-specific robot to complete the task. How can we enable drawing-like tasks in robots as they *are* rather than as we would *like them to be*? And how can we do so in a manner that minimizes tuning (e.g., in the case of drawing the same parameters can be used for both faces and landscapes) while also minimizing software complexity? *In this paper we find that the use of ergodic metrics—and the resulting ergodic control— reduces the dependence on task-specific robots (e.g., robots mechanically designed with drawing in mind), reduces the algorithmic complexity of the software that enables the task (e.g., the number of independent processes involved in drawing decreases), and enables the same software solution to apply to multiple robotic instantiations.*

Moreover, this paper touches on a fundamental issue for many modern robotic systems—the need to communicate through motion. Symbolic representations of information are the currency of communication, physically transmitted through whatever communication channels are available (electrical signals, light, body language, written language and related symbolic artifacts such as drawings). The internal representation of a symbol must both be *perceivable* given a sensor suite (voltage readings, cameras, tactile sensors) and *actionable* given an actuator suite (signal generators, motors). *Insofar as all systems can execute ergodic control, we hypothesize that ergodic metrics provide a nearly-universal, actionable measure of spatially-defined symbolic information.* Specifically, in this paper we see that both letters (represented in a font) and photographs can be rendered by a robotic system working within its own particular physical capabilities. For instance, a hand-writing-like rendering of the letter N (seen later in Figure 2) is seen to be a consequence of putting a premium on efficiency for a first-order dynamical system rendering the letter. Moreover, in the context of drawing photographs (of people and landscapes), we see a) that other drawing algorithms implicitly optimize (or at least improve) ergodicity, and b) using ergodic control, multiple dynamical systems approach rendering in dramatically different manners with similar levels of success.

We begin by introducing ergodicity in Section 2.1, including a discussion of its characteristics. Section 2.2 includes an overview and comparison of the ergodic control methods used in this paper. We present some examples in Section 3, including comparisons of the results of the different ergodic control schemes introduced in the previous section, comparisons with existing robot drawing methods, and experiments using the Baxter robot.

2 Methods

2.1 Ergodicity

Ergodicity compares the spatial statistics of a trajectory to the spatial statistics of a desired spatial distribution. A trajectory is *ergodic* with respect to a spatial distribution if the time spent in a region is proportional to the density of the spatial distribution. In previous work, the ergodic metrics have been employed for other applications including search [15, 14] or coverage [18]. When used for search-based applications, ergodicity encodes the idea that the higher the information density of a region in the distribution, the more time spent in that region, shown in Figure 1.

The spatial distributions used in this paper represent the spatial distribution of the symbol or image being recreated through motion, introduced in [17]. The more intense the color in the image, the higher the value of the spatial distribution. Thus, ergodicity encodes the idea that the trajectory represents the path of a tool (e.g., marker, paintbrush, etc.), where the longer the tool spends drawing in the region the greater the intensity of the color in that region.

To evaluate the ergodicity, we define the ergodic metric to be the *distance from ergodicity* ε of the time-averaged trajectory from the spatial distribution $\phi(x)$. The ergodicity of the trajectory is computed as the sum of the weighted squared distance between the Fourier coefficients of the spatial distribution ϕ_k and the distribution representing the time-averaged trajectory c_k, defined below:

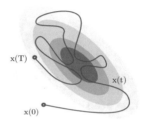

Fig. 1. An illustration of ergodic trajectories. Ergodic trajectories spend time in the workspace proportional to the spatial distribution.

$$\varepsilon = \sum_{k_1=0}^{K} \cdots \sum_{k_n=0}^{K} \Lambda_k |c_k - \phi_k|^2, \tag{1}$$

where K is the number of coefficients calculated along each of the n dimensions, and k is a multi-index $k = (k_1, ..., k_n)$. The coefficients Λ_k weight the lower frequency information higher and are defined as $\Lambda_k = \frac{1}{(1+||k||^2)^s}$, where $s = \frac{n+1}{2}$.

The Fourier basis functions are determined as below:

$$F_k(x) = \frac{1}{h_k} \prod_{i=1}^{n} \cos\left(\frac{k_i \pi}{L_i} x_i\right), \tag{2}$$

where h_k is a normalizing factor as defined in [13]. The spatial Fourier coefficients are computed from the inner product

$$\phi_k = \int_X \phi(x) F_k(x) dx, \tag{3}$$

and the Fourier coefficients of the trajectory $x(\cdot)$ are evaluated as

$$c_k = \frac{1}{T} \int_0^T F_k(x(t))dt, \tag{4}$$

where T is the final time of the trajectory [13].

2.2 Ergodic Control Algorithms

To demonstrate the different styles of resulting motions obtained from different methods, we compare the results of three ergodic control algorithms. All three algorithms generate trajectories that reduce the ergodic cost in (1), but each exhibits different time-horizon characteristics.

The algorithm with an infinitesimally small time horizon is a closed-form ergodic control (CFEC) method derived in [13]. At each time step, the feedback control is calculated as the closed-form solution to the optimal control problem with ergodic cost in the limit as the receding horizon goes to zero. The optimal solution is obtained by minimizing a Hamiltonian [6]. Due to its receding-horizon origins, the resulting control trajectories are piecewise continuous. The method can only be applied to linear first-order and second-order dynamics, with constant speed and forcing respectively. Thus, CFEC is an ergodic control algorithm that optimizes ergodicity along an infinitesimal time horizon at every time step in order to calculate the next control action.

The algorithm with a non-zero receding time horizon is Ergodic iterative-Sequential Action Control (E-iSAC), based on Sequential Action Control (SAC) [1, 21]. At each time step, E-iSAC uses hybrid control theory to calculate the control action that optimally improves ergodicity over a non-zero receding time horizon. Like CFEC, resulting controls are piecewise continuous. The method can generate ergodic trajectories for both linear and nonlinear dynamics, with saturated controls.

Finally, the method with a non-receding, finite time horizon is the ergodic Projection-based Trajectory Optimization (PTO) method derived in [14, 15]. Unlike the previous two approaches, it is an infinite-dimensional gradient-descent algorithm that outputs continuous control trajectories. Like E-iSAC, it can take into account the linear/nonlinear dynamics of the robotic system, but it calculates the control trajectory over the entire time duration that most efficiently minimizes the ergodic metric rather than simply the next time step. It also has a weight on control in its objective function that balances the ergodic metric, to achieve a dynamically-efficient trajectory that minimizes the ergodic metric.

Both CFEC and E-iSAC are efficient to compute, whereas PTO is computationally expensive as it requires numerical integration of several complex differential equations for the entire finite time horizon during each iteration of the algorithm. Next, we investigate the application of these techniques to examples including both letters and photographs.

Features	CFEC	E-iSAC	PTO
Finite Time Horizon			•
Closed Loop Control	•	•	
Nonlinear Dynamics		•	•
Control Saturation	•	•	
Receding Horizon		•	
Continuous Control			•
Weight on Control		•	•
Efficient Computation	•	•	

Table 1. Comparison of features for the different ergodic control methods used in the examples.

3   Examples

3.1   Writing Symbols

In the first example, we investigate how a robot can use ergodic control to recreate a structured visual input, such as a letter, presented as an image. In addition, because the input image is not merely of artistic interest but also corresponds to a recognizable symbol (in this case, the letter "N"), we show how a robot can render meaningful visual cues, without the prior knowledge of a dictionary or library of symbols. To do this, we represented the image of the letter as a spatial distribution as described in Section 2.1. We then determined the trajectories for the three different methods (CFEC, E-iSAC, and PTO) described in Section 2.2 for systems with first order dynamics and second order dynamics. We ran all the simulations for 60 seconds total with the same number of Fourier coefficients to represent the image spatially.

Figure 2 shows the resulting ergodic motions for each control algorithm with the drawing dynamics represented as a single integrator. From Figure 2, we can see that while all three methods produce results that are recognizable as the letter "N", the trajectories generated to achieve this objective are drastically different. The velocity-control characteristic of the single integrator system leads to the sharp, choppy turns evident in both discrete-time CFEC and E-iSAC methods. The infinitesimally small time horizon of the CFEC method, in contrast to the non-zero receding horizon of the E-iSAC method, results in the large, aggressive motions of the CFEC result compared to the E-iSAC result. Finally, the weight on the control cost and the continuous-time characteristics of the PTO method lead to a rendering that most closely resembles typical human penmanship.

For the double integrator system shown in Figure 3, the controls are accelerations rather than the velocities of the system. Because of this, the trajectories produced by the discrete controls for the CFEC and E-iSAC method are much smoother, without the sharp turns seen in Figure 2. Even though the CFEC result is smoother, its trajectory is more abstract and messy than the single integrator trajectory. The receding horizon E-iSAC produces much better results for systems with complicated or free dynamics (e.g., drift), including the

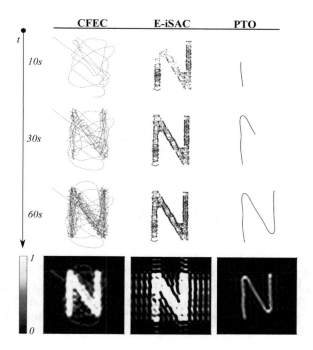

Fig. 2. Trajectories generated by the three different methods with single integrator dynamics for the letter "N" at 10 seconds, 30 seconds and 60 seconds with the spatial reconstructions of the time-averaged trajectories generated by each of the methods at the final time. The different methods lead to stylistic differences with an abstract representation from the CFEC method to a natural, human-penmanship motion from the PTO method.

double integrator system. The trajectory produced executes an "N" motion reminiscent of human penmanship and continues to draw similarly smooth motions over the time horizon. While PTO leads to a similarly smooth result compared to the single integrator system, it leads to a result that less resembles human penmanship.

From both examples, we can see how ergodicity can be used as a representation of symbolic spatial information (i.e., the letter "N") and ergodic control algorithms can be used to determine the actions needed to sufficiently render the information while incorporating the physical capabilities of the robot.

Figure 4a-4c shows the ergodic metric, or the difference between the sum of the trajectory Fourier coefficients and spatial Fourier coefficients, for the different methods over time. We can see that for both dynamic systems, all three methods produce trajectories that are similarly ergodic with respect to the letter by the end of the time horizon. Compared to E-iSAC, CFEC converges more slowly when more complex dynamics (i.e., double order dynamics) are introduced. PTO

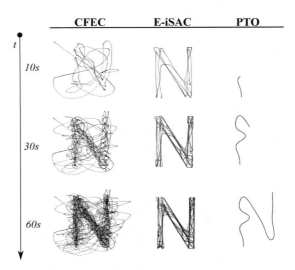

Fig. 3. The trajectories generated by the three different methods with double integrator dynamics for the letter "N" at 10 seconds, 30 seconds and 60 seconds. The double-order dynamics lead to much smoother motion from the discrete methods (CFEC and E-iSAC) and more ergodic results from the E-iSAC methods due to its receding-horizon characteristics.

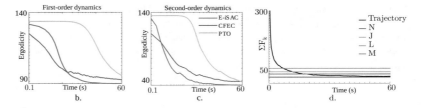

Fig. 4. a)-c) Time evolution of the ergodic metric, or the normalized difference of the squared sum of the trajectory Fourier coefficients and the spatial Fourier coefficients, for the three different methods with first-order and second-order dynamics for the letter "N" on logarithmic scale. Note that because ergodicity can only be calculated over a state trajectory of finite time duration (see Eq. 1), we start measuring the ergodicity values once 0.1 seconds of simulation have passed; hence the three approaches start at different ergodic metrics. All three methods result in similarly ergodic trajectories by the end of the time horizon, with differences due to their time-horizon characteristics. d) Sum of the Fourier coefficients for the trajectory over time compared to the spatial Fourier coefficients of different letters (N, J, L, and M). The trajectory coefficients converge to the spatial coefficients of the letter "N" that is being drawn, quantitatively representing the process of discriminating the specific symbol being drawn over time.

exhibits a lower rate of cost reduction because it performs finite-time horizon optimization. Note that E-iSAC always reaches the lowest ergodic cost value by the end of the 60 second simulation.

Figure 4d compares the sum of the absolute value of the Fourier coefficients trajectory generated by the CFEC method for the single integrator to the spatial Fourier coefficients of different letters. Initially, the difference between the trajectory coefficients and spatial Fourier coefficients is large, representing the ambiguity of the symbol being drawn. As the symbol becomes more clear, the Fourier coefficients of the trajectory converge to the coefficients of "N", representing the discrimination of the letter being drawn from the other letters. Moreover, letters that are more visually similar to the letter "N" have Fourier coefficients that are quantitatively closer to the the spatial coefficients of "N" and thus take longer to distinguish which symbol is being drawn. The ergodic control metric allows for representation of a symbol from visual cues and discrimination of that symbol from others without requiring prior knowledge of the symbols.

3.2   Drawing Images

In the second example, we consider drawing a picture from a photograph as opposed to drawing a symbol. Previously, we were concerned with the representation of the structured symbol. In this example, we move to representing a more abstract image for purely artistic expression. Here, we are drawing a portrait of Abraham Lincoln[1] with all three methods for single-order and double-order dynamics. We also render the portrait with a lightly damped spring system. The simulations are performed for the same 60-second time horizon and number of coefficients as the previous example.

Figure 5a compares the trajectories resulting from the different ergodic control methods for the single-integrator system. The weight on control and continuous-time characteristics of the PTO method that were desirable for the structured symbol example are disadvantageous in this case. While it reduces the ergodic cost, its susceptibility to local minima and its weight on energy lead to a far less ergodic result compared to the other methods.

Instead, the discrete nature of the other two methods produce trajectories that are more clearly portraits of Lincoln and are more ergodic with respect to the photo. The trajectory produced by the CFEC method initially covers much of the area of the face, but returns to the regions of high interest such that the final image produced matches the original image closely in shading. The E-iSAC method produces a trajectory that is much cleaner and does not cover the regions that are not shaded in (i.e., the forehead, cheeks). The velocity control of the single-order system leads to a disjointed trajectory similar to the results from Fig. 2.

Figure 5b compares the resulting trajectories for the double integrator system. As discussed previously, the PTO method produces a trajectory that is far

---

[1] The image was obtained from `https://commons.wikimedia.org/wiki/File:Abraham_Lincoln_head_on_shoulders_photo_portrait.jpg`.

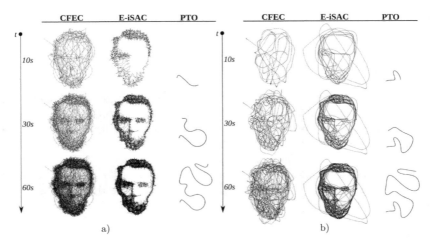

Fig. 5. Trajectories generated by the three different methods for drawing the portrait of Abraham Lincoln at 10 seconds, 30 seconds and 60 seconds. a) Single Integrator Dynamics: CFEC and E-iSAC result in distinguishable results with stylistic differences, whereas PTO is inadequate for this purpose. b) Double Integrator Dynamics: E-iSAC results in a smooth, clear portrait of Abraham Lincoln with a trajectory that naturally draws the face as a person would sketch one, without any preprogramming or library of motion primitives.

less ergodic with respect to the photo. The control on acceleration significantly impacts the stylistic rendering of the CFEC rendering. The E-iSAC method produced better results than the other methods for the double integrator system, due to its longer time horizon. The resulting trajectory is smoother and more natural compared to the single integrator results. Interestingly, this method creates a trajectory that naturally draws the contours of the face— the oval shape and the lines for the brows and nose— before filling in the details, similar to the way that humans sketch a face [11].

Fig. 6 compares the results of ergodicity with respect to time for the different methods. Similar to the symbolic example in Section 3.1, the E-iSAC trajectory creates a more ergodic image by the end for both cases. While the CFEC method results in a less ergodic trajectory for both systems, the ergodic cost for the single-order dynamics decreases faster for much of the time horizon. The CFEC method performs significantly worse for the double-order dynamics system and has a higher ergodic cost than the E-iSAC method throughout the entire time horizon. The inadequacy of the PTO method for this image is demonstrated here, having significantly higher ergodic costs at the final time for both systems.

Figure 7a compares the renderings of six different images with different content— portraits and monuments using the E-iSAC method with double-order dynamics. We show that E-iSAC successfully renders different images using a single set of parameters.

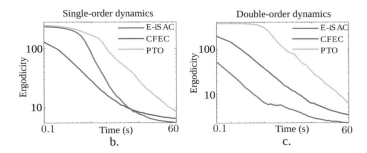

Fig. 6. Time evolution of the ergodic metric, or the normalized difference of the squared sum of the trajectory Fourier coefficients and the spatial Fourier coefficients for the three different methods with single-order and double-order dynamics for the Lincoln image on logarithmic scale. E-iSAC results in a more optimally ergodic trajectory for both systems. PTO performs poorly for both systems and CFEC performs significantly worse with the double-order dynamics.

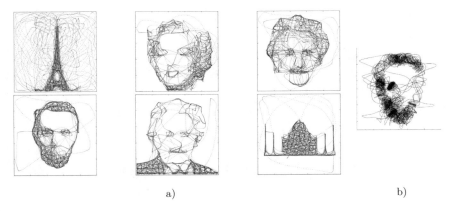

Fig. 7. a) Renderings of different images (Eiffel tower, Marilyn Monroe, Einstein's face, Lincoln's face, Einstein with suit, and Taj Mahal) from the double-order dynamical system using the E-iSAC method with the same set of parameters. E-iSAC is able to successfully render different images with different content (faces and monuments) with the identical parameters. b) Trajectory generated by the E-iSAC method for the Lincoln portrait image with damped spring dynamics. E-iSAC is able to produce a trajectory that reproduces the image while satisfying the dynamics of a system.

Finally, we demonstrate the ability of the E-iSAC method to take into account more complicated dynamics of the system. In Figure 7b, we simulate a system where the drawing mass is connected via a lightly damped spring to the controlled mass. The resulting Lincoln trajectory is harder to distinguish than the renderings from the single-order and double-order dynamical systems. However, the system is successfully able to optimize with respect to the dynamics and draw the main features of the face—the hair, the beard, and the eyes—within the time horizon, reducing the ergodic cost by 31.5% in 60 seconds.

3.3 Comparisons with Existing Robot Drawing Techniques

Existing drawing robots employ a multistage process, using preprocessing (e.g., edge detection and/or other image processing techniques) and postprocessing (e.g., motion primitives, shade rendering, path planning) to render the image [3, 4, 5, 12, 20]. To accomplish this, most robots and their software are co-designed specifically with drawing in mind, with most specializing in recreating scenes of specific structures, such as portraits of human faces. Similar multi-stage methods are commonly used for robot writing. They typically use image preprocessing, segmentation, waypoints, or a library of motion primitives to plan the trajectory and execute the trajectory using different motion control and trajectory tracking methods [9, 16, 19, 23].

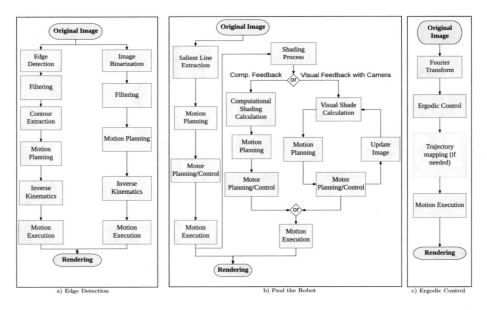

Fig. 8. Comparison of Drawing Methods. a) Edge detection and Binarization method used in [3] b) Method used by Paul the Drawing Robot from [20] and c) Ergodic Control. Ergodic Control is able to achieve comparable results with an algorithm requiring fewer independent processes.

Recently, some approaches using motion-driven machine learning have been used to enable robots to learn and mimic human motion [10, 19, 22]. These methods can be difficult and computationally costly and efforts to make them tractable (i.e., predefined dictionary of symbols) can be limiting in scope [9, 23].

Furthermore, they do not consider the robot's physical capabilities and can thus generate motions that are difficult for the robot to execute.

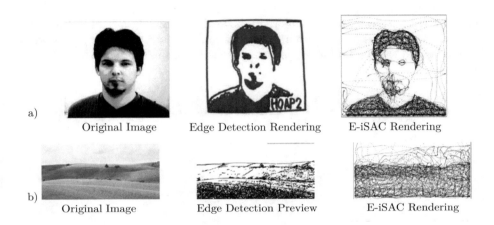

Fig. 9. Comparison of the method in [3] and the E-iSAC method. a) The original image and edge detection rendering using the HOAP2 robotic system come directly from [3]. While the edge detection method is successful, producing a tractable trajectory requires morphological filters to extract the facial features and predetermined drawing primitives to render the shading. b) Comparison for a landscape image, with a preview of the overlaid edge detection and binarization results compared to the E-iSAC trajectory. Producing a motion trajectory from the extracted contours of the preview would be computationally difficult (over 22000 contours result from the edge detection) or needs content-specific postprocessing, and requires a precise drawing system. The E-iSAC rendering results in a highly tractable result.

For comparison, we contrast ergodic control to the method employed in [3], a multi-stage process described in Figure 8a, with a preliminary stage to render the outlines of the important features using edge detction and a secondary stage to render shading using image binarization. Figure 9a shows the results of the method from [3] compared to the trajectory created using the E-iSAC method for the double-order system. While the edge detection method from [3] renders a successful recreation, obtaining a tractable trajectory requires parameter tuning and filtering to extract the most important features from the drawing, and predetermined drawing primitives and postprocessing to formulate the planar trajectory needed for rendering. Furthermore, the E-iSAC method is able to capture different levels of shading as opposed to the method in [3] that only renders a binary image (black and white).

In addition, because the processing (e.g., filtering, parameter tuning) used by [3] is tailored to drawing human portraits, the method is not robust to other content. To show this, we compare the results for rendering a landscape in Figure 9b. While the simulated preview of the rendering appears successful, the image binarization to render shading fails as it is tuned specifically for human portraits. Instead, the quality of the image comes entirely from the edge detection step. However, processing and rendering the contours would be difficult (over 22,000 contours are generated to render the image), and the filtering implemented to make the results tractable are tailored specifically for facial portraits. While the E-iSAC method results in a more abstract image, the trajectory produced is tractable and the method is robust to a variety of subjects (as shown in Figure 7a).

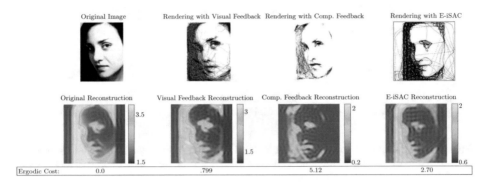

Fig. 10. Comparison of drawings rendered with Paul the Robot [20] and the drawing rendered with E-iSAC and the ergodic Fourier reconstructions of these results. The original image and the images of the renderings from feedback come directly from [20]. The E-iSAC rendering is able to perform comparably to the visual-feedback rendering (a closed-loop algorithm) and better than the computational-feedback rendering (an open-loop algorithm) with a simpler, open-loop algorithm. The reconstructions of the Fourier coefficients representing the different renderings with the respective ergodic costs show how ergodicity can be used as a quantitative metric for assessment of results.

Another drawing method is performed by Paul the robot [20], which uses a complicated multi-stage process (shown in Figure 8b) to render portraits. The first stage involves a salient-line extraction to draw the important features, and then performs a shading method using either visual feedback or computational feedback. The visual-feedback shading process uses an external camera to update the belief in real-time, while the computational-feedback shading process is based on the simulation of the line-extraction stage and is an open-loop process, similar to the E-iSAC method. Figure 10 compares the results of Paul the robot [20] with the E-iSAC method, and shows the reconstructions of the Fourier coefficients representing each rendering. While the robot successfully renders the image, the E-iSAC method performs comparably well with a much simpler, open-loop algorithm and could be improved with the integration of an external camera setup to update the drawing in real-time. Furthermore, the method from [20] relies on a highly engineered system that moves precisely and cannot take into account a change in the robotic system, unlike E-iSAC. In addition, Figure 10 shows the reconstructions of the Fourier representing each rendering, demonstrating how ergodicity can be used as a quantitative metric of assessment for the results, even if ergodic control is not used to determine the trajectories.

3.4   Baxter Experiments

We performed experiments executing the motion trajectories generated with the Baxter robot to demonstrate it physically rendering the portrait of Abraham Lincoln generated using the E-iSAC method from Section 3.2. The Baxter robot is able to successfully complete the trajectory and render a recognizable portrait of Lincoln. The main stylistic differences between the simulation and the experimental results derive primarily from the assumption of an infinitesimally small marker point in the simulation and the board's inability to render shading. Improving the algorithm to enable encoding characteristics of the rendering tool (e.g., marker size, paintbrush stroke) and integration of an external camera to update the belief of current drawing would improve the rendering capabilities of the robotic system in the future.

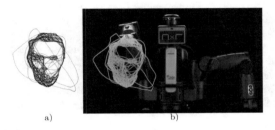

a)   b)

Fig. 11. Experimental results with the Baxter robot. a) Trajectory generated by the E-iSAC method b) Rendering executed by the Baxter robot using the Lightboard [2]. The Baxter robot is successfully able to render the portrait of Abraham Lincoln using the motion trajectory generated.

4   Conclusion

This paper presented an autonomous process for translating visual information to physical motion. We demonstrate how ergodic metrics can be used as an actionable measure of symbolic spatial information, and explore the use of ergodicity as a measure that enables the robot to actively distinguish among different symbols with no prior knowledge of letter structure, other than the associated ergodic value. In addition, ergodic control provides the robot the ability to naturally represent and recreate a wide range of visual inputs (from letters to portraits), while incorporating the robot's physical capabilities (e.g., dynamic drawing mechanics). Moreover, in the context of drawing, we see other drawing algorithms improve ergodicity, suggesting the use of ergodicity as a quantitative measure of assessment. Finally, we demonstrate experiments with the Baxter robot rendering these trajectories, and note that as optimal ergodic control can run in real-time, it can be ideal for developing interactive rendering behaviors in robots. In the future, we plan to adapt the algorithm to encode rendering characteristics of the system into the model and to integrate a visual feedback system to update the representation of the drawing in real-time.

5   Acknowledgements

This material is based upon work supported by the National Science Foundation under grants CMMI 1334609 and IIS 1426961. Any opinions, findings, and conclusions or recommendations expressed in this material are those of the author(s) and do not necessarily reflect the views of the National Science Foundation.

References

1. Ansari, A.R., Murphey, T.D.: Sequential action control: Closed-form optimal control for nonlinear and nonsmooth systems. IEEE Transactions on Robotics 32(5), 1196–1214 (Oct 2016)
2. Birdwell, J.A., Peshkin, M.: Capturing technical lectures on lightboard. In: 2015 ASEE Annual Conference & Exposition. ASEE Conferences, Seattle, Washington (June 2015), http://lightboard.info/
3. Calinon, S., Epiney, J., Billard, A.: A humanoid robot drawing human portraits. In: 5th IEEE-RAS International Conference on Humanoid Robots, 2005. pp. 161–166. IEEE (2005)
4. Deussen, O., Lindemeier, T., Pirk, S., Tautzenberger, M.: Feedback-guided stroke placement for a painting machine. In: Proceedings of the Eighth Annual Symposium on Computational Aesthetics in Graphics, Visualization, and Imaging. pp. 25–33. Eurographics Association (2012)
5. Jean-Pierre, G., Saïd, Z.: The artist robot: a robot drawing like a human artist. In: Industrial Technology (ICIT), 2012 IEEE International Conference on. pp. 486–491. IEEE (2012)
6. Kirk, D.E.: Optimal control theory: an introduction. Courier Corporation (2012)
7. LaViers, A., Chen, Y., Belta, C., Egerstedt, M.: Automatic sequencing of ballet poses. IEEE Robotics & Automation Magazine 18(3), 87–95 (2011)
8. Laviers, A., Egerstedt, M.: Controls and Art: Inquiries at the Intersection of the Subjective and the Objective. Springer Science & Business Media (2014)
9. Li, B., Zheng, Y.F., Hemami, H., Che, D.: Human-like robotic handwriting and drawing. In: Robotics and Automation (ICRA), 2013 IEEE International Conference on. pp. 4942–4947. IEEE (2013)
10. Liang, P., Yang, C., Li, Z., Li, R.: Writing skills transfer from human to robot using stiffness extracted from semg. In: Cyber Technology in Automation, Control, and Intelligent Systems (CYBER), 2015 IEEE International Conference on. pp. 19–24 (June 2015)
11. Loomis, A.: Drawing the Head and Hands. Titan Books Limited (2011), https://books.google.com/books?id=Cm1XuQAACAAJ

12. Lu, Y., Lam, J.H., Yam, Y.: Preliminary study on vision-based pen-and-ink drawing by a robotic manipulator. In: 2009 IEEE/ASME International Conference on Advanced Intelligent Mechatronics. pp. 578–583. IEEE (2009)
13. Mathew, G., Mezić, I.: Metrics for ergodicity and design of ergodic dynamics for multi-agent systems. Physica D: Nonlinear Phenomena 240(4–5), 432 – 442 (2011)
14. Miller, L.M., Murphey, T.D.: Trajectory optimization for continuous ergodic exploration. In: 2013 American Control Conference. pp. 4196–4201. IEEE (2013)
15. Miller, L.M., Silverman, Y., MacIver, M.A., Murphey, T.D.: Ergodic exploration of distributed information. IEEE Transactions on Robotics 32(1), 36–52 (2016)
16. Pérez-Gaspar, L.A., Trujillo-Romero, F., Caballero-Morales, S.O., Ramrez-Leyva, F.H.: Curve fitting using polygonal approximation for a robotic writing task. In: Electronics, Communications and Computers (CONIELECOMP), 2015 International Conference on. pp. 184–189 (Feb 2015)
17. Sahai, T., Mathew, G., Surana, A.: A chaotic dynamical system that paints. arXiv preprint arXiv:1504.02010 (2015)
18. Shell, D.A., Matarić, M.J.: Ergodic dynamics for large-scale distributed robot systems. In: International Conference on Unconventional Computation. pp. 254–266. Springer (2006)
19. Syamlan, A.T., Nurhadi, H., Pramujati, B.: Character recognition for writing robot control using anfis. In: 2015 International Conference on Advanced Mechatronics, Intelligent Manufacture, and Industrial Automation (ICAMIMIA). pp. 46–48 (Oct 2015)
20. Tresset, P., Leymarie, F.F.: Portrait drawing by paul the robot. Computers & Graphics 37(5), 348–363 (2013)
21. Tzorakoleftherakis, E., Ansari, A., Wilson, A., Schultz, J., Murphey, T.D.: Model-based reactive control for hybrid and high-dimensional robotic systems. IEEE Robotics and Automation Letters 1(1), 431–438 (Jan 2016)
22. Yin, H., Paiva, A., Billard, A.: Learning cost function and trajectory for robotic writing motion. In: 2014 IEEE-RAS International Conference on Humanoid Robots. pp. 608–615. IEEE (2014)
23. Yussof, S., Anuar, A., Fernandez, K.: Algorithm for robot writing using character segmentation. In: Third International Conference on Information Technology and Applications (ICITA'05). vol. 2, pp. 21–24. IEEE (2005)

Cloud-based Motion Plan Computation for Power-Constrained Robots

Jeffrey Ichnowski, Jan Prins, and Ron Alterovitz

University of North Carolina at Chapel Hill, USA
{jeffi,prins,ron}@cs.unc.edu

Abstract. We introduce a method for splitting the computation of a robot's motion plan between the robot's low-power embedded computer, and a high-performance cloud-based compute service. To meet the requirements of an interactive and dynamic scenario, robot motion planning may need more computing power than is available on robots designed for reduced weight and power consumption (e.g., battery powered mobile robots). In our method, the robot communicates its configuration, its goals, and the obstacles to the cloud-based service. The cloud-based service takes into account the latency and bandwidth of the connection between it and the robot and computes and returns a motion plan within the time frame necessary for the robot to meet requirements of a dynamic and interactive scenario. The cloud-based service parallelizes construction of a roadmap, and returns a sparse subset of the roadmap giving the robot the ability to adapt to changes between updates from the server. In our results, we show that with typical latency and bandwidth limitations, our method gains significant improvement in the responsiveness and quality of motion plans in interactive scenarios.

1 Introduction

Cloud-based computing offers a vast amount of low-cost computation power on-demand. It offers the ability to quickly scale up and down compute resources so that you can have more computing when you need it, and not pay for it when you do not. To place in context the price of cloud computation power, the July 2016 prices for 1 second of 360-cores of computation can be less than $0.0047 [1]. This implies that with an embarrassingly parallel algorithm [2], a 5-minute computation can be cut to less than 1 second. And because you pay for the resources that you use, the same computation would require $0.0047 whether using 1-core for 360 seconds, or 360-cores for 1 second. To access these immense computing resources, the only thing that is required is a connection to the internet.

Mobile robots are often designed and built to keep weight and power consumption as low as possible to achieve an acceptable duration of autonomy before requiring recharging. This design concern naturally dictates that the computation power on such a robot is limited—for example, to a low-power single-core processor. Motion planning is a computationally intensive process [3], and as

© Springer Nature Switzerland AG 2020
K. Goldberg et al. (Eds.): *Algorithmic Foundations of Robotics XII*, SPAR 13, pp. 96–111, 2020.
https://doi.org/10.1007/978-3-030-43089-4_7

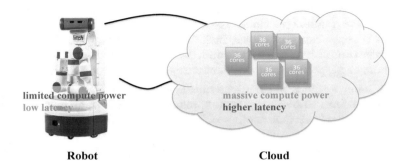

Fig. 1. Comparison of robot only and cloud computing for robot motion planning. The robot has limited computing power in order to reduce weight and increase battery life, however it has low latency access to its sensors and actuators. The cloud-computing has vast amounts of on-demand computing power available, but has a higher latency access to the robot and the information it sends.

such, if the mobile robot has more than a few degrees of freedom, its computational demands for motion planning can quickly exceed its available computational power.

In a static environment, the robot can compute its motion plan a priori and execute it. If the robot has no demands on when it needs to compute the motion plan, it can sit motionless while it computes the motion plan locally. On the other hand, if it needs a motion plan quickly, it can use cloud computing resources to greatly decrease the time to compute a motion plan, and start executing sooner.

In a dynamic environment, however, the robot must not only compute a complete motion plan, but it must also sense changes in the task's goal and the robot's environment and update its motion plan accordingly. As in a static environment, the robot can use a cloud-based computation to rapidly produce an initial motion plan. However, the network complicates matters when it comes to updating the plan due to changes in the environment since the network has limited bandwidth and introduces a network latency-based delay. The delay due to network latency and bandwidth may introduce enough of a lag that the mobile robot relying solely on cloud-based motion planning would not be able to respond to changes in its environment quickly enough to avoid a collision.

In this paper we propose a method for a mobile robot to compute and execute a motion plan by offloading much of the computational cost of motion planning to the cloud, while remaining reactive enough to respond to a dynamic environment and avoid obstacles.

2   Related Work

The NIST definition of cloud computing [4], provides a good high-level overview of the capabilities of the cloud. Broadly, cloud computing encompasses a "ubiquitous, convenient, on-demand network access to a shared pool of computing resources that can be rapidly provisioned and released...". *Cloud-robotics and automation* are a subset of cloud-based computing related to robotics—it encompasses a broad range of topics, including access to big-data libraries, high-performance computing, collective robot learning, and remote human interaction. Kehoe, et al. provides an excellent survey of cloud-robotics in [5].

In this paper we focus on cloud-computing as an on-demand high-performance computing platform to accelerate motion planning. Bekris et al. [6] use the cloud to precompute manipulation roadmaps. The robot uses the roadmap to compute the shortest collision-free path, lazily determining if edges on the roadmap are blocked as determined by the latest sensor data. They observe that a dense precomputed roadmap, while covering more space and capable of producing shorter paths between configurations, has the negative effect of increasing bandwidth requirements to transfer the roadmap and taking more time to perform a search. They thus use techniques such as SPARS and IRS (described below) to reduce the roadmap size and evaluate the tradeoffs. Our approach follows from that observations, but instead computes and updates the roadmap at an interactive rate.

In [7], Kehoe et al. use a cloud-based data service to facilitate recognition of objects for grasping. The approach uses a custom Google image recognition service that is trained to recognize objects and estimate grasp points. In a subsequent related paper [8], Kehoe et al. use cloud-based computation to massively accelerate through parallel computation, a Monte Carlo sampling-based grasp analysis and planning. The paper demonstrates the cloud's ability to scale to 500 compute nodes and achieve a $445\times$ speedup.

Parallel processing has been successfully used to accelerate motion planning computations. In [2], Amato et al. demonstrate that probabilistic roadmap generation is *embarrassingly parallel*—meaning that little effort is needed to separate the sample generation into multiple parallel processes. The method described in [9] uses lock-free synchronization to parallelize multi-core shared-memory sampling-based motion planning algorithms with minimal overhead and observe linear and super-linear speedup. Carpin et al. describes an OR-parallel RRT method [10] that allows for distributed generation of sampling-based motion plans among independent servers—the algorithm chooses the best plan generated from the servers participating, and the result is a probabilistically better plan. Otte et al.'s C-FOREST [11] algorithm improves upon OR-parallel RRT by exchanging information between computers about the best path found, resulting in speedup in the motion planning on all parallel threads.

Robots are increasingly integrated into networks of computers. With the advent of ROS [12] and similar systems, network connected robots are becoming the norm. ROS's network stack is designed for a high-bandwidth, low-latency, local private/protected network to facilitate unified access to the robot's sensor, actuators, and embedded systems. Cloud-based computing, on the other hand, has lower bandwidth, higher latency, and is generally publicly accessible (except, for example, when using a VPN), and thus requires additional consideration above the network stack.

The probabilistic roadmap method (PRM) [13] generates a connected graph of robotic configurations in a precomputing offline phase. The robot later uses the roadmap to find a path from an initial configuration to a goal configuration by following along the edges of the graph. The k-PRM* [14] method improves upon PRM by defining a connectivity level (k) needed to guarantee asymptotic optimality.

Sparse roadmaps and roadmap spanners such as SPARS [15] are an effective technique in reducing the complexity of motion planning roadmaps. They can produce asymptotically near-optimal roadmaps, which maintain reachability of the non-sparse graph, while limiting the size of the graph to thresholds needed for lower-end computing platforms. In our method we adopt and parallelize the incremental roadmap spanner (IRS) of [16] to reduce the roadmap size for transmission over the internet.

Once the robot has a roadmap, whether sparse or not, it needs a path finding algorithm to navigate its structure. Shortest-path finding algorithms such as Dijkstra's algorithm and A* search find optimal paths, but can suffer from a slow compute time that makes them inappropriate for reactive path planning. D* and D* Lite algorithms perform a search from goal to start and track information in the graph that allows them to be incrementally updated when changes to the roadmap (e.g., from moving obstacles) occur—this provides a performance benefit in that only a partial graph search is needed anytime there is a change in the roadmap. The Anytime Repairing A* [17] and Anytime D* Lite [18] algorithms use an inadmissible heuristic in A* to find a path quickly, then incrementally improve the plan in subsequent iterations.

3   Problem Definition

Let $\mathcal{C}$ be the *configuration space* for the robot—the k-dimensional space of all possible configurations the robot take. Let $\mathcal{C}_{\text{free}} \subseteq \mathcal{C}$ be the subset of configurations that are collision free. Let $\mathbf{q} \in \mathcal{C}$ be the k-dimensional complete specification of a single robotic configuration (e.g., the joint angles of an articulated robot). Let $\mathbf{Q}_{\text{goal}} \subseteq \mathcal{C}_{\text{free}}$ be the set of goal configurations. Given a starting configuration $\mathbf{q}_0$, the objective of motion planning in a static environment is to compute a path $\boldsymbol{\tau} = (\mathbf{q}_0, \mathbf{q}_1, \ldots, \mathbf{q}_n)$, such that the path between $\mathbf{q}_i$ and $\mathbf{q}_{i+1}$ is in $\mathcal{C}_{\text{free}}$ as traversed by a local planner, and $\mathbf{q}_n \in \mathbf{Q}_{\text{goal}}$

When the robot operates in a dynamic environment, $\mathcal{C}_{\text{free}}$ changes over time. Let $\mathcal{C}_{\text{free}}(t) \subseteq \mathcal{C}$ be the obstacle-free configuration space at time t, and let $\mathbf{Q}_{\text{goal}}(t) \subseteq \mathcal{C}_{\text{free}}(t)$ be the goal at time t. Given the robot starting configuration $\mathbf{q}_0$ at time t_0, the objective of motion planning in a dynamic environment is to compute a path $\boldsymbol{\tau} = \left(\left[\mathbf{q}_0^{\mathsf{T}}, t_0\right]^{\mathsf{T}}, \left[\mathbf{q}_1^{\mathsf{T}}, t_1\right]^{\mathsf{T}}, \ldots, \left[\mathbf{q}_n^{\mathsf{T}}, t_n\right]^{\mathsf{T}} \right)$, such that the path between $\mathbf{q}_i$ and $\mathbf{q}_{i+1}$ is in $\mathcal{C}_{\text{free}}(\cdot)$ as traversed by a local planner from time t_i to t_{i+1}, and $\mathbf{q}_n \in \mathbf{Q}_{\text{goal}}(t_n)$.

In a dynamic environment $\mathcal{C}_{\text{free}}(t)$ may only be known at time t, and within the sensing capabilities of the robot. We consider obstacles in the environment that fall into the following categories: (1) *known static* obstacles that do not change over the course of the task (e.g., a wall), (2) *unknown static* obstacles that are static, but are not known until sensed by the robot, and (3) *dynamic* obstacles that are moving through the environment and whose motion is unknown in advance.

The robot, being in its environment, has fast access to the input from its sensors, and is able to incorporate them into its planning to avoid moving obstacles. The cloud computing service does not have sensors relevant to the robot's scenario and thus only has access to the sensed environment via what the robot communicates to it.

Motion planning computation is split between two computing resources: (1) the robot's embedded *local computer*, and (2) the remote *cloud computer(s)*. Without loss of generality, we assume the robot's computer is a low-power single-core processor with some percentage of compute time dedicated to motion computations. The cloud-computing servers are fast multi-core computers.

The two computing resources communicate via a network with quantifiable bandwidth and latency. Bandwidth (R) is measured in bits per second, and is much lower than the bandwidth achievable between CPU and RAM. Latency (t_L) is measured as the time between when a bit is sent and when it is received. The bandwidth is low enough that sending a complete roadmap from client to server would hamper the robot's ability to adapt quickly to changing environment. The latency is high enough that the planning process must compensate for it in it requests updates to the motion plan.

4   Method

We introduce a new set of algorithms to effectively split motion plan computation between a robot and a cloud-based compute service based upon the strengths of each system. The robot is in the environment and has fast access to sensors, but it has a low-power processor—it is thus responsible for sensing the environment (i.e., detecting obstacles and estimating current state), reacting to dynamic obstacles, and executing collision-free motions. The cloud-based compute service is connected to the robot by a possibly high-latency low-bandwidth network, but has fast on-demand computing power—it is thus responsible for rapidly computing and sending to the robot a motion planning roadmap that encodes feasible collision-free motions.

When the robot starts a new task, it initiates a cloud-planning session by sending a request with the task and environment description to the cloud-based computing service. The cloud computer receives the request, starts a new cloud-based motion planning session, and computes a motion plan. Once the motion plan is of sufficient quality (as determined by the task), the cloud-based service sends the motion plan to the robot so that the robot can begin execution of the task.

The cloud-based service operates as a request-response service; each request the client makes results in a single response from the service. In the algorithms presented, the request-response communication is *asynchronous* unless otherwise

---

Algorithm 1 Robot Computation

Input: the initial configuration $\mathbf{q}_{robot}$, goal region $\mathbf{Q}_{goal}$, known static obstacles $\mathcal{W}$
 1: $\mathbf{G} = (\mathbf{V}, \mathbf{E}) \leftarrow (\varnothing, \varnothing)$ {roadmap is initially empty}
 2: $\tau \leftarrow \varnothing$ {path is initially empty}
 3: **send** plan_req$(t_0, \mathbf{q}_{robot}, \mathcal{W}, \mathbf{Q}_{goal}) \Rightarrow$ cloud
 4: **while** $\mathbf{q}_{robot} \notin \mathbf{Q}_{goal}$ **do**
 5:     $(\mathcal{W}, \mathcal{D}) \leftarrow$ (sensed static obstacles, tracked dynamic obstacles) {sense}
 6:     **if recv** roadmap_update $\Leftarrow$ cloud **then**
 7:         Incorporate update into robot's roadmap $\mathbf{G}$
 8:         $t_{req} \leftarrow$ (current time) $+ t_{step}$
 9:         $\mathbf{q}_{req} \leftarrow$ compute where robot will be at t_{req}
10:         **send** plan_req$(t_{req}, \mathbf{q}_{req}, \mathcal{W}, \mathbf{Q}_{goal}) \Rightarrow$ cloud
11:     **if** changes in $(\mathbf{G}, \mathcal{W}, \mathcal{D})$ **or** (Anytime D*'s ϵ) > 1 **then**
12:         $\tau \leftarrow$ compute/improve path using Anytime D*
13:     $\mathbf{q}_{robot} \leftarrow$ follow edges of shortest path τ {move}

---

stated. Within a planning session, the service retains state from one request-response cycle to the next so that it does not start from scratch at each point in the process.

4.1 Roadmap-Based Robot Computation

The robot's algorithm is shown in Alg. 1. It initializes the process and starts the cloud-planning session in lines 1 to 3. As part of initialization it creates an empty graph for the roadmap and sends an initial planning request. It then starts a sense-plan-move loop (line 4) in which it will remain until it reaches a goal.

The sensing process at the start of each loop iteration is responsible for processing sensor input to construct a model of the static obstacles in the environment ($\mathcal{W}$), and to track the movement of dynamic obstacles ($\mathcal{D}$). Since the static environment changes infrequently (e.g., as the robot rounds a corner to discover construction blocking its path), an implementation can save bandwidth by only sending changes to the static environment as it discovers them.

In the planning part of the loop, the robot incorporates new data from the cloud service, computes a local path around dynamic obstacles, and requests plan updates as it needs them. The robot internally represents its estimate of $\mathcal{C}_{free}$ using a roadmap encoded as a graph $\mathbf{G} = (\mathbf{V}, \mathbf{E})$, where $\mathbf{V}$ are configurations (the vertices) of the graph, and $\mathbf{E}$ are the collision-free motions (edges) between configurations in $\mathbf{V}$. On line 6, the robot checks if the cloud service has responded to the robot's most recent request with an update to the roadmap. When the robot receives the cloud's roadmap, it incorporates the new data into the robot's roadmap, and initiates a new cloud planning request with the latest information from the environment.

Alg. 1 requests updates as frequently as possible, however if excessive network utilization shortens battery life in an implementation, requests can be made less frequently, for example, only when the robot has moved sufficiently out of its available roadmap. To send a request, the robot computes where it will be at time

t_{step} in the future following its current plan. The value of t_{step} is a parameter of the system, and accounts for the network round-trip and cloud processing time to compute the update.

If the robot has encountered a change to the graph, or any of the static or dynamic obstacles, or its current path (τ) can be refined further, it computes or improves the path using an Anytime D* planner [18], with a time component as described in [19]. Anytime D* defines and uses a runtime value in ϵ (line 11) to incrementally refine the robot's path. It starts by setting the value of $\epsilon > 1$ which it uses to modify the A* heuristic to find a sub-optimal solution quickly. As the algorithm iterates, it decreases ϵ and correspondingly refines the path with the new heuristic, resulting in an improved plan. When $\epsilon = 1$, the solution is optimal. As the last part of the loop, Alg. 1 moves the robot along the shortest path it computed.

When the robot computes its local path it saves computation time by only considering collisions between paths on the roadmap and the dynamic obstacles. The robot does not need to recompute self-collision avoidance, collisions with static obstacles, or other motion constraints, as this information is incorporated into the roadmap that the cloud service computes.

4.2   Roadmap-Based Cloud Computation

Cloud-based computation in our algorithm computes a roadmap for a robot to use when navigating through an environment and around obstacles. Because this algorithm runs on the cloud-based compute service, it has access to immense computational resources, enabling computation of a large, detailed roadmap. When building a roadmap, the cloud-based computation only considers the obstacles in the environment that are sent to the cloud from the robot—since the robot only sends static obstacles, the roadmap does not include avoidance of dynamic obstacles.

The robot starts a cloud planning *session* with an initial request for a roadmap. A session corresponds to a single robotic task and cloud-computing process that spans multiple requests from the robot. At the start, both the cloud and the robot have an empty graph as a roadmap. The cloud computes an initial roadmap and sends the relevant portion of the roadmap to the robot to begin execution of the task. As the robot needs additional areas of the roadmap, it sends additional requests to the server, and the server responds with updates to the roadmap. Optionally, in parallel, cloud process optimizes and extends the roadmap between request/response cycles.

Alg. 2 shows the cloud computing process for a single cloud-based motion planning session. The session starts with an empty graph $\mathbf{G} = (\mathbf{V}, \mathbf{E}) = (\varnothing, \varnothing)$. The algorithm builds the graph (Sec. 4.3) by generating vertices ($\mathbf{V}$) and *dense* edges ($\mathbf{E}$); and selects and maintains a *sparse* subset of edges $\mathbf{E}_s \in \mathbf{E}$. The sparse edges retain graph connectivity and are used to reduce the transfer size, while the dense edges give the robot more options to react to dynamic obstacles. Alg. 2 also maintains a subgraph $\mathbf{G}_{\text{robot}} = (\mathbf{V}_{\text{robot}} \subseteq \mathbf{V}, \mathbf{E}_{\text{robot}} \subseteq \mathbf{E})$ that tracks the portion of the $\mathbf{G}$ sent to the robot.

Algorithm 2 Cloud Planning Session

1: $\mathbf{G} = (\mathbf{V}, \mathbf{E}_s \subseteq \mathbf{E}) \leftarrow (\varnothing, \varnothing)$
2: $\mathbf{G}_{\text{robot}} = (\mathbf{V}_{\text{robot}}, \mathbf{E}_{\text{robot}}) \leftarrow (\varnothing, \varnothing)$
3: $\mathcal{W} \leftarrow \varnothing$ {static obstacles}
4: **loop**
5:     **recv** `plan_req`$(t_{\text{req}}, \mathbf{q}_{\text{req}}, \mathcal{W}, \mathbf{Q}_{\text{goal}}) \Leftarrow$ robot {blocking wait for next request}
6:     $\mathbf{V} \leftarrow \{\mathbf{q} \in \mathbf{V} \cup \{\mathbf{q}_{\text{req}}\} \mid \forall w \in \mathcal{W} : \texttt{clear}(\mathbf{q} \mid w)\}$
7:     $\mathbf{E} \leftarrow \{(\mathbf{q}_a, \mathbf{q}_b) \in \mathbf{E} \mid \forall w \in \mathcal{W} : \texttt{link}(\mathbf{q}_a, \mathbf{q}_b \mid w)\}$
8:     **while** $t_{\text{now}} < t_{\text{req}} - t_{\text{res}}$ **and not** satisfactory solution **do**
9:         update $\mathbf{G}$ and $\mathbf{q}_{\text{goal}}$ using k-PRM*+IRS on $\mathcal{W}$ and $\mathbf{Q}_{\text{goal}}$
10:     $(\mathbf{V}'_{\text{robot}}, \mathbf{E}'_{\text{robot}}) \leftarrow \texttt{serialize\_graph}(\mathbf{G}, \mathbf{q}_{\text{req}}, \mathbf{Q}_{\text{goal}}, \mathbf{G}_{\text{robot}})$
11:     **send** `plan_res`$(\mathbf{V}'_{\text{robot}} \setminus \mathbf{V}_{\text{robot}}, \mathbf{E}'_{\text{robot}} \setminus \mathbf{E}_{\text{robot}}) \Rightarrow$ robot
12:     $(\mathbf{V}_{\text{robot}}, \mathbf{E}_{\text{robot}}) \leftarrow (\mathbf{V}'_{\text{robot}}, \mathbf{E}'_{\text{robot}})$

The cloud planning session starts when it receives a `plan_req` (plan request) from the robot (line 5). This request corresponds to the `plan_req` sent by the robot in Alg. 1 line 3. The cloud computer adds the requested configuration $\mathbf{q}_{\text{req}}$ to the graph and updates the existing graph for any new static obstacles that are added $\mathcal{W}$ (lines 6 and 7). This step makes use of two application-specific functions to produce a valid roadmap: $\texttt{clear}(\mathbf{q})$ computes whether or not $\mathbf{q} \in \mathcal{C}_{\text{free}}$ (e.g., via collision detection algorithms); and $\texttt{link}(\mathbf{q}_a, \mathbf{q}_b)$ checks if the path between $\mathbf{q}_a$ and $\mathbf{q}_b$ is in $\mathcal{C}_{\text{free}}$ as traversed by the robot's local planner. It then builds the roadmap until it runs out of time or it has a solution of satisfactory quality (lines 8 and 9). The compute time limit is the target completion time t_{req} minus the amount of time for the robot to receive the response t_{res}. Thus t_{res} is computed as the sum of graph serialization time and total network transfer time. The graph is then serialized using the method described in section 4.4, and the new vertices and edges selected for serialization are sent back to the robot as a `plan_res` (plan response) in line 11. Optionally, at the end of the loop the cloud computer may continue to update the roadmap in the background until it receives another `plan_req` from the robot.

4.3 Lock-free Parallel k-PRM* with a Roadmap Spanner

The cloud-based service computes a roadmap using k-PRM* [14] with the Incremental Roadmap Spanner (IRS) [16], accelerated by a lock-free parallelization construction we introduce in this section. k-PRM* is an asymptotically optimal sampling-based method that generates a roadmap. IRS selects an asymptotically near-optimal sparse subset of the edges generated by k-PRM* and results in a graph with significantly fewer edges as compared to k-PRM*. The edges from k-PRM* are the *dense* graph edges ($\mathbf{E}$). The edges selected by IRS are the *sparse* graph edges ($\mathbf{E}_s \subseteq \mathbf{E}$).

The server computes k-PRM*+IRS using a parallel lock-free algorithm in which all provisioned cores run Alg. 3 simultaneously to generate and add random samples to a graph in shared memory. The main portion of the algorithm proceeds similarly to the non-parallel version, with the key differences being that:

Algorithm 3 Lock-free Parallel k-PRM* IRS Thread

Input: $G = (V, E)$ is an initialized graph shared between threads, $\exists v \in V : \texttt{is\_goal}(v)$

1: **while not** done **do**
2:     $v_{\mathrm{rand}} \leftarrow$ new vertex with random sample and connected component C_{rand}
3:     $C_{\mathrm{rand}}.\mathrm{goal} \leftarrow \texttt{is\_goal}(v_{\mathrm{rand}})$
4:     **if** $\texttt{clear}(v_{\mathrm{rand}})$ **then**
5:         **for all** $v_{\mathrm{near}} \in \texttt{k\_nearest}(V, v_{\mathrm{rand}}, \{k =\}\lceil \log(|V| + 1) * k_{\mathrm{RRG}} \rceil)$ **do**
6:             **if** $\texttt{link}(v_{\mathrm{rand}}, v_{\mathrm{near}})$ **then**
7:                 sparse $\leftarrow \texttt{shortest\_path\_dist}(v_{\mathrm{rand}}, v_{\mathrm{near}}) < w_{\mathrm{stretch}} * \texttt{dist}(v_{\mathrm{rand}}, v_{\mathrm{near}})$
8:                 $\texttt{add\_edge}(v_{\mathrm{rand}}, v_{\mathrm{near}}, \text{sparse})$
9:                 $\texttt{add\_edge}(v_{\mathrm{near}}, v_{\mathrm{rand}}, \text{sparse})$
10:                solved $\leftarrow$ solved **or** $\texttt{merge\_components}(v_{\mathrm{rand}}.\mathrm{cc}, v_{\mathrm{near}}.\mathrm{cc})$
11:        $V \leftarrow V \cup v_{\mathrm{rand}}$

---

(1) nearest neighbor searching is fast and non-blocking due to the use the lock-free kd-tree described in [9], (2) graph edges are stored in lock-free linked lists (Alg. 4), and (3) progress towards a solution is tracked via connected components that are stored in lock-free linked trees (Alg. 5). As with k-PRM*, in each iteration this algorithm generates a random robot configuration and searches for its k-nearest neighbors using k from [14]. The algorithm checks if the path to each neighbor is obstacle-free (line 6), and if so, adds edges to the PRM graph (lines 8 and 9). As the algorithm builds the graph, it adds *dense* edges consistent with k-PRM*. When the shortest path distance between two vertices in the graph is shorter than a stretch weighted (w_{stretch}) straight-line distance, it adds *sparse* edges consistent with IRS.

Algorithm 4 $\texttt{add\_edge}(v_{\mathrm{from}}, v_{\mathrm{to}}, \text{sparse})$

1: $e_{\mathrm{dense}} \leftarrow$ new edge to v_{to} with $e_{\mathrm{dense}}.\mathrm{next} = v_{\mathrm{from}}.\mathrm{dense\_list\_head}$
2: **while not** $\texttt{CAS}(v_{\mathrm{from}}.\mathrm{dense\_list\_head}, e_{\mathrm{dense}}.\mathrm{next}, e_{\mathrm{dense}})$ **do**
3:     $e_{\mathrm{dense}}.\mathrm{next} \leftarrow v_{\mathrm{from}}.\mathrm{dense\_list\_head}$
4: **if** sparse **then**
5:     add edge to v_{to} to sparse list of edges with CAS loop similar to one for dense list
6: **while** $v_{\mathrm{from}}.\mathrm{cc}.\mathrm{parent} \neq \texttt{nil}$ **do**
7:     $v_{\mathrm{from}}.\mathrm{cc} \leftarrow v_{\mathrm{from}}.\mathrm{cc}.\mathrm{parent}$ {Lazy update of vertex's connected component}

The algorithm adds edges to the graph using Alg. 4. Each vertex in the graph has a reference to the head of two linked lists: one for $\mathbf{E}$, and one for $\mathbf{E_s}$. Updating the list makes use of a "compare-and-swap" (CAS) operation available on modern multi-core CPU architectures. $\texttt{CAS}(mem, old, new)$, in one atomic action, compares the value in mem to an expected old value, and if they match, updates mem to the new value. CAS, combined with the loop in line 2, updates the lists correctly even in the presence of competing concurrent updates.

The algorithm tracks progress towards a solution by maintaining information on each connected component ("cc" in Alg. 4) in the roadmap. When it adds an

---

Algorithm 5 merge_components(C_a, C_b)

1: **repeat**
2:   **while** C_a.parent $\neq$ nil **do** $C_a \leftarrow C_a$.parent
3:   **while** C_b.parent $\neq$ nil **do** $C_b \leftarrow C_b$.parent
4: **until** CAS(C_a.parent, nil, C_b)
5: **repeat**
6:   **while** C_b.parent $\neq$ nil **do** $C_b \leftarrow C_b$.parent
7:   $C_{\text{merged}} \leftarrow$ new component
8:   (C_{merged}.start, C_{merged}.goal) $\leftarrow$ (C_a.start **or** C_b.start, C_a.goal **or** C_b.goal)
9: **until** CAS(C_b.parent, nil, C_{merged})
10: **return** C_{merged}.start **and** C_{merged}.goal

Algorithm 6 serialize_graph($\mathbf{G}, \mathbf{q}_{\text{req}}, \mathbf{Q}_{\text{goal}}, \mathbf{G}_{\text{robot}}$)

Input: $\mathbf{G} = (\mathbf{V}, \mathbf{E}_s \subseteq \mathbf{E})$, $\mathbf{G}_{\text{robot}} = (\mathbf{V}_{\text{robot}}, \mathbf{E}_{\text{robot}})$, s.t. $\mathbf{V}_{\text{robot}} \subseteq \mathbf{V}$, $\mathbf{E}_{\text{robot}} \subseteq \mathbf{E}$
1: $(\mathbf{V}'_{\text{robot}}, \mathbf{E}'_{\text{robot}}) \leftarrow (\mathbf{V}_{\text{robot}}, \mathbf{E}_{\text{robot}})$
2: $\mathbf{V}_{\text{frontier}} = $ forward_frontier($\mathbf{q}_{\text{req}}, \mathbf{E}$)
3: $p(\cdot) \leftarrow$ path_to_frontier($\mathbf{Q}_{\text{goal}}, \mathbf{V}_{\text{frontier}}, \mathbf{E}$)
4: $\mathbf{V}'_{\text{robot}} \leftarrow \mathbf{V}'_{\text{robot}} \cup \mathbf{V}_{\text{frontier}}$
5: $\mathbf{Q} \leftarrow \{\mathbf{q} \in \mathbf{V}_{\text{frontier}}\}$ {populate FIFO queue}
6: **while** $|\mathbf{Q}| > 0$ **do**
7:   $\mathbf{q}_i \leftarrow$ remove head from $\mathbf{Q}$
8:   **for all** $(\mathbf{q}_i, \mathbf{q}_s) \in \mathbf{E}_s : \mathbf{q}_s \notin \mathbf{V}'_{\text{robot}}$ **do**
9:     $(\mathbf{V}'_{\text{robot}}, \mathbf{E}'_{\text{robot}}) \leftarrow (\mathbf{V}'_{\text{robot}} \cup \{\mathbf{q}_s\}, \mathbf{E}'_{\text{robot}} \cup \{(\mathbf{q}_i, \mathbf{q}_s)\})$
10:     append $\mathbf{q}_s$ to $\mathbf{Q}$
11:   **if** $p(\mathbf{q}_i) \neq$ nil **and** $(\mathbf{q}_i, p(\mathbf{q}_i)) \notin \mathbf{E}'_{\text{robot}}$ **then**
12:     **if** $p(\mathbf{q}_i) \notin \mathbf{V}_{\text{robot}}$ **then**
13:       append $p(\mathbf{q})$ to $\mathbf{Q}$
14:     $\mathbf{V}'_{\text{robot}} \leftarrow \mathbf{V}'_{\text{robot}} \cup p(\mathbf{q}_i)$
15:     $\mathbf{E}'_{\text{robot}} \leftarrow \mathbf{E}'_{\text{robot}} \cup (\mathbf{q}_i, p(\mathbf{q}_i))$
16: **return** $(\mathbf{V}'_{\text{robot}}, \mathbf{E}'_{\text{robot}})$

---

edge between two vertices, it also merges the connected components associated with the vertices (Alg. 5). This is done by maintaining a "parent" link from the pre-merged component to the post-merged component. The most recently merged component is thus found by repeatedly following parent links to the root of the connected components. Each connected component also maintains booleans tracking whether or not the component contains a vertex at the goal and/or start. Once a connected component is found that includes both a start and goal vertex, the graph contains a path between the two.

4.4 Roadmap Subset for Serialization

The roadmap serialization process selects a compact, relevant subset of a roadmap and converts it into a serial (linear) structure suitable for transmission over a network. Alg. 6 selects which vertices and edges of the graph to serialize. The process of converting the selected vertices and edges to sequence of bytes is left an implementation detail. Since bandwidth is limited, the process selects a

---

Algorithm 7 path_to_frontier($\mathbf{q}_{\text{goal}}, \mathbf{V}_{\text{frontier}}, \mathbf{E}$)

1: $g(\mathbf{q}_{\text{goal}}) \leftarrow 0$ {cost to goal}
2: $p(\mathbf{q}_{\text{goal}}) \leftarrow$ nil {forward pointers}
3: $\mathbf{U} \leftarrow \{\mathbf{q}_{\text{goal}}\}$ {priority queued ordered by $g(\cdot)$}
4: **while** $|\mathbf{V}_{\text{frontier}}| > 0$ **do**
5:     $\mathbf{q}_{\text{min}} \leftarrow$ remove (min $\mathbf{U}$) from $\mathbf{U}$
6:     $\mathbf{V}_{\text{frontier}} \leftarrow \mathbf{V}_{\text{frontier}} \setminus \{\mathbf{q}_{\text{min}}\}$
7:     **for all** $(\mathbf{q}_{\text{from}}, \mathbf{q}_{\text{min}}) \in \mathbf{E}$ **do**
8:         $d \leftarrow g(\mathbf{q}_{\text{min}}) + \text{cost}(\mathbf{q}_{\text{from}}, \mathbf{q}_{\text{min}})$
9:         **if** $\mathbf{q}_{\text{from}} \notin \mathbf{U}$ **or** $d < g(\mathbf{q}_{\text{from}})$ **then**
10:            $g(\mathbf{q}_{\text{from}}) \leftarrow d$
11:            insert/update $\mathbf{q}_{\text{from}}$ in $\mathbf{U}$
12:            $p(\mathbf{q}_{\text{from}}) \leftarrow \mathbf{q}_{\text{min}}$
13: **return** $p(\cdot)$

---

small subset of the configurations in the roadmap to send to the robot. To allow the robot to navigate around dynamic obstacles in its immediate vicinity, as well as find the best route to goal, the cloud selects a subset of configurations that includes ones reachable from $\mathbf{q}_{\text{req}}$ within a time bound t_{max}, as well as the path to goal for each such vertex.

Serialization selection begins by finding the frontier between the vertices reachable from $\mathbf{q}_{\text{req}}$ within the time bound t_{max}, and vertices not reachable (line 2). The forward_frontier algorithm is a modified Dijkstra's algorithm that terminates once it finds paths longer than t_{max}. Since Dijkstra's expands paths in increasing path length, this will terminate once it has found all paths reachable within t_{max}. It returns all vertices $\mathbf{V}_{\text{frontier}}$ reachable within the frontier. The selection process then computes the shortest path from all goals to the vertices in $\mathbf{V}_{\text{frontier}}$ (line 3). This process, shown in Alg. 7, is a modified Dijkstra's algorithm that terminates once it has found a path to all vertices in $\mathbf{V}_{\text{frontier}}$.

In the last step in Alg. 6, the vertices from the frontier set are appended to $\mathbf{V}'_{\text{robot}}$ along with all configurations along their shortest paths to goal and reachable by the sparse edges. Line 5 populates the queue from $\mathbf{V}_{\text{frontier}}$. The loop starting on line 6 iterates through each configuration in the queue, adding sparse neighbors and steps along the shortest path to goal as it encounters them. By checking the graph before appending to the queue, the algorithm ensures that vertices are queued at most once. When the loop completes, the new graph subset is ready for sending to the robot. Then the cloud service sends only the changes in the graph from one response to the next (Alg. 6 line 11).

5 Results

We evaluate our algorithm on a Fetch robot [20] by giving it an 8 degree-of-freedom task in an environment with a dynamic moving obstacle. Our cloud-compute server runs on a system with four Intel x7550 2.0-GHz 8-core Nehalem-EX processors for a total 32-cores. The cloud-computing process makes use of all 32-cores. The cloud-compute server is physically located approximately 6 km

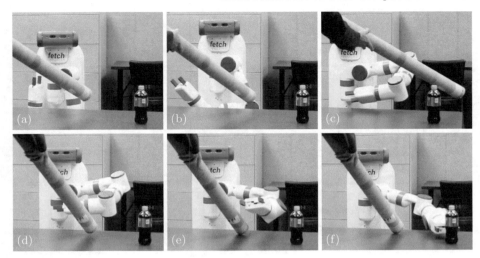

Fig. 2. The Fetch robot using our cloud-based motion planning for the task of grasping the bottle resting on the table while avoiding both the static obstacles (e.g., table) and the dynamic obstacle (a tube sensed via an RGBD camera). In frame (a) after the Fetch approaches the table with its arm in its standard rest configuration and it initiates the cloud-computation process. The Fetch's embedded CPU is tasked with sensing and avoiding dynamic obstacles, while a cloud-computer simultaneously generates and refines its roadmap. In frame (b), the Fetch begins its motion, only to be blocked in frame (c) by a new placement of the obstacle. The Fetch is again blocked in frame (d), moves again around the obstacle in frame (e), and reaches the goal in frame (f).

away from the robot, and the network connection between the server and robot supported a bandwidth in excess of 100 Mbps with a latency less than 20 ms. To model the impact of slower network connections, in our experiments we deliberately slowed packet transmission to model a fixed maximum bandwidth of R_{sim} and a fixed minimum round-trip latency of $t_{L_{\text{sim}}}$ subject to noise sampled from a Gaussian distribution with standard deviation of $0.16\,t_{L_{\text{sim}}}$.

We implemented our algorithm as a web-service accessible via HTTP [21]. The robot initiates a request by sending an HTTP POST to the server, and the server responds with an HTTP response code appropriate to the situation (e.g., "200 OK" for a successful plan, "503 Service Unavailable" when the server cannot acquire sufficient computing resources). Requests and responses are sent in a serialized binary form. To minimize overhead associated with establishing connections, both the cloud server and the robot use HTTP keep-alive to reuse TCP/IP connections between updates, and are configured to have a connection timeout that far exceeds expected plan computation time.

The Fetch robot has a 7 degree of freedom arm, a prismatic torso lift joint, and a mobile base. In our scenarios, prior to the cloud-based computation task, the Fetch robot navigates to the workspace using its mobile base without using the cloud service. This process introduces noise to the robot's base position and orientation. Once at the workspace, we give the Fetch robot the task of moving from a standard rest configuration (Fig. 2(a)) to a pre-grasp configuration over

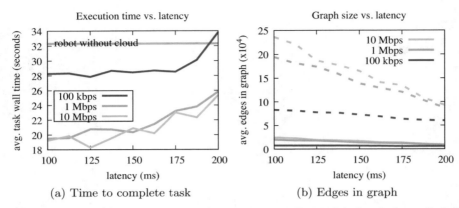

(a) Time to complete task   (b) Edges in graph

Fig. 3. Effect of different values for R_{sim} and $t_{L_{\text{sim}}}$. Graph (a) shows the wall-clock time for the Fetch robot to complete its pre-grasp motion task, where the orange line is the time for the robot to complete the task without the cloud service. Graph (b) compares the number of edges generated by the cloud computer (dashed lines) and the number of edges sent to the robot (solid lines) for the varying network conditions. The simulated network latency affects the amount of compute time that the cloud has for each update. Longer latencies lead to less time for available for computation, and thus leads to slower task completion time and fewer edges on the roadmap.

a table (Fig. 2(f)), requiring it to plan a motion using 8 degrees of freedom (i.e., the arm and prismatic torso lift joint). In this setting, the static obstacles are the table, floor, and surrounding office space. We also include a dynamic obstacle: a cylindrical tube that moves through the environment.

The sequence in Fig. 2 shows the full integrated system running, with the Fetch robot successfully moving its arm around the obstacles. At the beginning of a task, the Fetch communicates its position and orientation in the workspace to the cloud service and requests a roadmap for its task. The software uses custom tracking software and the Fetch's built in RGBD camera to determine the location of dynamic obstacles. When it computes a change in trajectory (e.g., to avoid a dynamic obstacle, or in response to a refined roadmap from the cloud), it sends the trajectory to the controller via a ROS/moveit interface.

We also ran our method in simulation to evaluate performance under varying networking conditions. We simulated the tube dynamic obstacle sweeping periodically over the table at a rate of 0.25 Hz (approximately 1 m/s). While the dynamic obstacle has a predictable motion consistent through all runs, the simulated sensors only sense the tube's position and orientation and do not predict its motion. As the tube obstacle is considered dynamic, the robot does not send information about it to the cloud computer, and it must avoid the tube by computing a path along the roadmap using its local graph. The robot and cloud are not given any pre-computation time; once given the task, the robot must begin and complete its motion as soon as it is able. We measure this as the "wall clock time to complete task."

The Fetch robot has a 2.9 GHz Intel i5-4570S processor with 4 cores. For our scenario, we limit our client-side planner to fully utilize a single core, under the assumption that in a typical scenario the remaining cores would leave sufficient compute power to run other necessary tasks, such as sensor processing.

As a baseline for comparison, we have the robot's computer generate a k-PRM* using a separate thread. This thread updates the graph used by the reactive planner at a period of 250 ms. The k-PRM* planner considers only the static environment and self-collision avoidance as the constraints on the roadmap generation, and generates a fully dense roadmap (no sparse edges). The reactive planner uses the roadmap to search for a path to the goal. While searching the roadmap, the robot lazily checks for collisions with the dynamic obstacle. In 50 runs, the robot completes the task with an average of 32.3 seconds.

We run the scenario using our method and simulate and vary the latency and bandwidth of the network between the robot and the 32-core cloud-computer. To maintain reactivity, the robot requests an update as soon as it receives the response to the previous request. Since the requested solve time (t_{req}) is set to 250 ms, an update is requested and received every 250 ms. The latency means that only a portion of the 250 ms can be used to compute a roadmap. The results in Fig. 3(a), averaging over 100 runs, show that the robot assisted by the cloud computation outperforms robot-only computation in almost all simulated cases. As we might expect, the slowest bandwidth and highest latency cause the performance benefit of using the cloud-based service to disappear. At the lowest latencies, the cloud-based solution outperforms the robot-only computation by 1.7×, reducing the task completion time to 19.0 seconds.

In Fig. 3(b), we show the savings that result from using the roadmap spanner and our serialization method. When latency is low, the cloud computer can spend more time computing, producing a roadmap that has on average 232649 edges. IRS and serialization reduce it to an average of 24236, a savings of close to 90%.

Fig. 4 shows the effect of roadmap serialization parameter t_{max} on our cloud-based motion planning. A smaller t_{max} implies less of the roadmap is sent to the robot, which results in reduced bandwidth usage but at a cost to the quality of the roadmap. As the robot executes its task, a proportionately higher portion of the server's dense roadmap is sent to the robot (see Fig. 4 (a)). From Fig. 4 (b), we see that if t_{max} is too small, the robot is slower to find a collision-free path past the dynamic obstacle. Conversely, there is little gain for increasing t_{max} beyond a certain threshold since unnecessary portions of the graph are sent to the robot, essentially wasting network bandwidth, leading to diminished performance.

6 Conclusion

Cloud computing offers access to vast amounts of computing power on demand. We introduce a method for power-constrained robots to accelerate their motion planning by splitting the motion planning computation between the robot and a high-performance cloud computing service. Our method rapidly computes an initial roadmap and then sends a mixed sparse/dense subgraph to the robot.

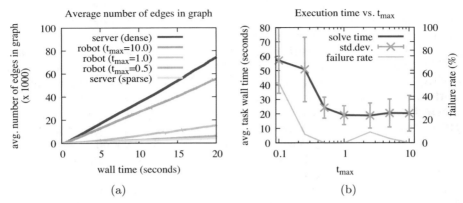

Fig. 4. The serialization parameter t_{max} affects the size of the graph on the robot and the robot's task completion wall time. In these graphs the simulated network is fixed at $R_{sim} = 1$ Mbps and $T_{L_{sim}} = 200$ ms and the server solve time is 250 ms. Graph (a) shows that larger values of t_{max} result in more of the dense edges of the graph being serialized and sent to the robot. In (b), we see that having t_{max} be too small results in a high failure rate (where failure means not reaching goal after 2 minutes), while having it too large increases the variance of the execution time.

The sparse portions of the graph retain connectivity and reduced transfer size, while the dense portions give the robot the ability to react to obstacles in its immediate vicinity. As the robot executes the plan, it periodically gets updates from the cloud to retain its reactive ability.

In our experiments, we applied our method to a Fetch robot, giving it an 8 degree of freedom task with a simulated dynamic obstacle. With our method, the split cloud/robot computation allows the robot to react to dynamic obstacles in the environment while attaining a more dense roadmap than possible with computation on the robot's embedded processor alone. The scenario requires a minimal amount of pre-computation time (less than a second) before the robot starts to execute its task. As a result, the task time-to-completion is significantly improved over the alternative without cloud computing.

In future work we will incorporate mobility into the planning for the Fetch robot to make use of its wheeled base. We will also implement the proposed framework using a commercial cloud-based service and investigate approaches that efficiently allocate cloud-computing resources, including for the short intervals of computation needed for single tasks. We will also evaluate this method on different scenarios and different robot types.

References

1. Amazon: EC2 instance pricing. https://aws.amazon.com/ec2/pricing/ Accessed: 2016-07.

2. Amato, N.M., Dale, L.K.: Probabilistic roadmap methods are embarrassingly parallel. In: Proc. IEEE Int. Conf. Robotics and Automation (ICRA). (May 1999) 688–694
3. Reif, J.H.: Complexity of the Mover's Problem and Generalizations. In: 20th Annual IEEE Symp. on Foundations of Computer Science. (1979) 421–427
4. Mell, P., Grance, T.: The NIST definition of cloud computing. http://dx.doi.org/10.6028/NIST.SP.800-145 (2011)
5. Kehoe, B., Patil, S., Abbeel, P., Goldberg, K.: A survey of research on cloud robotics and automation. IEEE Transactions on Automation Science and Engineering **12**(2) (2015) 398–409
6. Bekris, K., Shome, R., Krontiris, A., Dobson, A.: Cloud automation: Precomputing roadmaps for flexible manipulation. IEEE Robotics & Automation Magazine **22**(2) (2015) 41–50
7. Kehoe, B., Matsukawa, A., Candido, S., Kuffner, J., Goldberg, K.: Cloud-based robot grasping with the google object recognition engine. In: Robotics and Automation (ICRA), 2013 IEEE International Conference on, IEEE (2013) 4263–4270
8. Kehoe, B., Warrier, D., Patil, S., Goldberg, K.: Cloud-based grasp analysis and planning for toleranced parts using parallelized monte carlo sampling. IEEE Transactions on Automation Science and Engineering **12**(2) (2015) 455–470
9. Ichnowski, J., Alterovitz, R.: Scalable multicore motion planning using lock-free concurrency. IEEE Transactions on Robotics **30**(5) (2014) 1123–1136
10. Carpin, S., Pagello, E.: On parallel RRTs for multi-robot systems. In: Proc. 8th Conf. Italian Association for Artificial Intelligence. (2002) 834–841
11. Otte, M., Correll, N.: C-FOREST: Parallel shortest path planning with superlinear speedup. IEEE Trans. Robotics **29**(3) (2013) 798–806
12. ROS.org: Robot Operating System (ROS). http://ros.org (2012)
13. Kavraki, L.E., Svestka, P., Latombe, J.C., Overmars, M.: Probabilistic roadmaps for path planning in high dimensional configuration spaces. IEEE Trans. Robotics and Automation **12**(4) (1996) 566–580
14. Karaman, S., Frazzoli, E.: Sampling-based algorithms for optimal motion planning. Int. J. Robotics Research **30**(7) (June 2011) 846–894
15. Dobson, A., Bekris, K.E.: Sparse roadmap spanners for asymptotically near-optimal motion planning. The International Journal of Robotics Research **33**(1) (2014) 18–47
16. Marble, J.D., Bekris, K.E.: Asymptotically near-optimal planning with probabilistic roadmap spanners. IEEE Transactions on Robotics **29**(2) (2013) 432–444
17. Likhachev, M., Gordon, G.J., Thrun, S.: ARA*: Anytime A* with provable bounds on sub-optimality. In: Advances in Neural Information Processing Systems. (2003)
18. Likhachev, M., Ferguson, D.I., Gordon, G.J., Stentz, A., Thrun, S.: Anytime dynamic A*: An anytime, replanning algorithm. In: ICAPS. (2005) 262–271
19. Van Den Berg, J., Ferguson, D., Kuffner, J.: Anytime path planning and replanning in dynamic environments. In: Proc. IEEE Int. Conf. Robotics and Automation (ICRA). (2006) 2366–2371
20. Fetch Robotics: Fetch research robot. http://fetchrobotics.com/research/
21. Fielding, R., Gettys, J., Mogul, J., Frystyk, H., Masinter, L., Leach, P., Berners-Lee, T.: Hypertext transfer protocol–http/1.1. Technical report (1999)

Combining System Design and Path Planning

Laurent Denarie[1], Kevin Molloy[1], Marc Vaisset[1],
Thierry Siméon[1], and Juan Cortés[1]

LAAS-CNRS, Université de Toulouse, CNRS, Toulouse, France
`juan.cortes@laas.fr`

Abstract. This paper addresses the simultaneous design and path planning problem, in which features associated to the bodies of a mobile system have to be selected to find the best design that optimizes its motion between two given configurations. Solving individual path planning problems for all possible designs and selecting the best result would be a straightforward approach for very simple cases. We propose a more efficient approach that combines discrete (design) and continuous (path) optimization in a single stage. It builds on an extension of a sampling-based algorithm, which simultaneously explores the configuration-space costmap of all possible designs aiming to find the best path-design pair. The algorithm filters out unsuitable designs during the path search, which breaks down the combinatorial explosion. Illustrative results are presented for relatively simple (academic) examples. While our work is currently motivated by problems in computational biology, several applications in robotics can also be envisioned.

Keywords: robot motion planning, sampling-based algorithms, computational biology, protein design

1 Introduction

System design and path planning problems are usually treated independently. In robotics, criteria such as workspace volume, workload, accuracy, robustness, stiffness, and other performance indexes are treated as part of the system design [1, 2]. Path planning algorithms are typically applied to systems with completely fixed geometric and kinematic features. In this work, we propose an extension of the path planning problem, in which some features of the mobile system are not fixed *a priori*. The goal is to find the best design (i.e. values for the variable features) to optimize the motion between given configurations.

A brute-force approach to solve this problem would consist of individually solving motion planning problems for all possible designs, and then selecting the design providing the best result for the (path-dependent) objective function. However, because of the combinatorial explosion, only simple problems involving a small number of variable design features can be treated using this naive approach. We propose a more sophisticated approach that simultaneously considers system design and path planning. A related problem is the optimization of geometric and kinematic parameters of a robot to achieve a given end-effector

© Springer Nature Switzerland AG 2020
K. Goldberg et al. (Eds.): *Algorithmic Foundations of Robotics XII*, SPAR 13, pp. 112–127, 2020.
https://doi.org/10.1007/978-3-030-43089-4_8

trajectory, usually referred to as kinematic synthesis [3]. Nonetheless, the problem we address in this work (see Section 2.1 for details) is significantly different, since we assume that all kinematic parameters and part of the geometry of the mobile system are provided as input. The design concerns a discrete set of features that can be associated to the bodies of the mobile system, such as shape or electrostatic charge, aiming to find the best possible path between two given configurations provided a path cost function. Very few works have considered such a hybrid design and path planning problem. One of the rare examples is a recently proposed method for UAVs path planning [4] where the optimal path planning algorithm considers several possible flying speeds and wing reference areas to minimize path risk and time. Since the considered configuration space is two-dimensional, the proposed solution is based on an extension of Dijkstra's algorithm working on a discrete representation of the search-space. This type of approach cannot be applied in practice to higher-dimensional problems, as the ones we address.

Sampling-based algorithms have been developed since the late 90s for path planning in high-dimensional spaces [5,6], which are out of reach for deterministic, complete algorithms. Our work builds on this family of algorithms, which we extend to treat a combinatorial component in the search-space, associated to the systems design, while searching for the solution path.

Our approach presents some similarities with methods that extend sampling-based path planning algorithms to solve more complex problems such as manipulation planning [7] or minimum constraint removal (MCR) planning [8], which also involve search-spaces with hybrid structure. As in these other works, the proposed algorithm simultaneously explores multiple sub-spaces aiming to find solutions more efficiently. Nevertheless, the hybrid design problem addressed here is different.

This paper presents the Simultaneous Design And Path-planning algorithm (SDAP), which is based on the T-RRT algorithm [9]. As explained in Section 2.2, the choice of T-RRT as a baseline is guided by the type of cost function we apply for the evaluation of path quality. Nevertheless, other sampling-based algorithms can be extended following a similar approach.

We demonstrate the good performance of the method on relatively simple, academic examples (Section 4). These simple examples allow us to apply the naive exhaustive method, whose results can be used as a reference to evaluate the performance and the quality of the solutions produced by the SDAP algorithm. Results show that SDAP is able to find the best path-design pairs, requiring much less computing time than the naive method. This advantage increases with the complexity of the problem.

Although the application of the proposed approach to problems of practical interest is out of the scope of this paper, we note that our motivation comes from problems in computational biology. For more than a decade, robotics-inspired algorithms have been applied to this area [10–13]. The method presented in this paper aspires to solve problems related to computational protein design [14, 15], which aims to create or modify proteins to exhibit some desired properties.

Progress in this field promises great advances in pharmacology, biotechnology, and nanotechnology. Protein design is extremely challenging, and although there have been some considerable strides in the last years [16, 17], the problem remains largely open. Current approaches focus on a static picture (i.e. search the amino-acid sequence that stabilizes a given structure), whereas dynamic aspects related to protein function are rarely considered. Our goal behind this work is to develop new methods to optimize functional protein motions. In addition to computational protein design, applications of the proposed approach in robotics can be envisioned, as briefly mentioned in the conclusion.

2   Problem Formulation and Approach

This section defines the problem addressed in this work, along with some notation, and presents an overview of the proposed approach.

2.1   Problem Definition

Let us consider an articulated linkage $\mathcal{A}$ consisting of n rigid bodies, $A_1..A_n$. The kinematic parameters of $\mathcal{A}$ are static and are supplied as input. The geometry of the the rigid bodies A_i can admit some variability, as well as other physical properties (mass, electrostatic charge, ...). More precisely, a discrete set of m design features, $f_1..f_m$, is defined and each body $A_i \in \mathcal{A}$ is assigned a design feature $f_j \in \mathcal{F}$. We denote d as a vector of length n that represents the design features assigned to all the rigid bodies in $\mathcal{A}$, i.e. d defines a particular design. $\mathcal{D}$ denotes the set of possible combinations of assignments of features for $\mathcal{A}$, i.e. $\mathcal{D}$ defines all possible designs. $\mathcal{D}$ is referred to as the design space, which is a discrete space containing m^n elements. A given configuration of $\mathcal{A}$ is denoted by q. Let $\mathcal{C}$ denote the configuration space. Note that for each $q \in \mathcal{C}$, only a subset of the possible designs $\mathcal{D}$ can be assigned, since some designs are not compatible with some configurations due to self-constraints or environment constraints.

The workspace of $\mathcal{A}$ is constrained by a set of obstacles $O_i \in \mathcal{O}$. $\mathcal{C}^d_{free}$ denotes all valid, collision-free configurations of $\mathcal{A}$ for a given vector d of design features. A path P connecting two configurations q_{init} and q_{goal} of $\mathcal{A}$ with design d is defined as a continuous function $P : [0,1] \rightarrow \mathcal{C}$, such that $P(0) = q_{\text{init}}$ and $P(1) = q_{\text{goal}}$. The path is said to be collision-free if $\forall t \in [0,1], P(t) \in \mathcal{C}^d_{free}$. $\mathcal{C}_{free}$ is the union of all individual $\mathcal{C}^d_{free}$:

$$\mathcal{C}_{free} = \bigcup_{d \in \mathcal{D}} \mathcal{C}^d_{free} \ .$$

$\mathcal{P}_{free}$ denotes the set of all feasible, collision-free paths connecting q_{init} to q_{goal}, considering all possible designs ($\forall d \in \mathcal{D}$).

A cost function $c : \mathcal{C}_{free} \times \mathcal{D} \rightarrow \mathbb{R}_+$ associates to each pair (q, d) a positive cost value, $\forall q \in \mathcal{C}_{free}$ and $\forall d \in \mathcal{D}$. Another cost function $c_P : \mathcal{P}_{free} \times \mathcal{D} \rightarrow \mathbb{R}_+$ is also defined to evaluate the quality of paths. In this work, the path cost function

c_P is itself a function of the configuration cost function c, i.e. c_P is a functional. More precisely, we consider the *mechanical work* criterion as defined in [9, 18] to evaluate paths, which aims to minimize the variation of the configuration cost c along the path. This criterion is a suitable choice to evaluate path quality in many situations [9], and is particularly relevant in the context of molecular modeling. Nevertheless, other cost functions can be considered, such as the integral of c along the path. A discrete approximation, with constant step size $\delta = 1/l$, of the mechanical work (MW) cost of a path P for a system design d can be defined as:

$$c_P(P,d) = \sum_{k=1}^{l} \max \left\{ 0, \ c\left(P\left(\frac{k}{l}\right), d\right) - c\left(P\left(\frac{k-1}{l}\right), d\right) \right\} . \quad (1)$$

The goal of our method is to find the best pair (P^*, d^*) such that:

$$c_P(P^*, d^*) = \min\{c_P(P,d) \mid P \in \mathcal{P}_{free}, d \in \mathcal{D}\} . \quad (2)$$

2.2 Approach

A naive approach to solve the problem would be to compute the optimal cost path for each design $d \in \mathcal{D}$, and then choose the optimal design d^* that minimizes c_P. Such a brute-force approach can be applied in practice to simple problems involving a small number n of variable bodies and/or a few m design features (recall that the design space is size m^n). The method proposed below aims to solve the problem much more efficiently by combining both the discrete (design) and continuous (path) optimization in a single stage.

We assume that, for most problems of interest, the configuration space $\mathcal{C}$ is high-dimensional, so that exact/complete algorithms cannot be applied in practice to solve the path-planning part of the problem. For this, we build of sampling-based algorithms [19, 5], which have been very successful in the robotics community since the late 90s, and which have also been applied in other areas such as computational biology [10–13]. We also assume that the cardinality of the design space $\mathcal{D}$ is moderately high, such that a relatively simple combinatorial approach can be applied to treat the design part of the problem.

The idea is to explore $\mathcal{C}_{free}$ to find paths between q_{init} and q_{goal} simultaneously considering all possible deigns $d \in \mathcal{D}$. To reduce the number of configuration-design pairs (q, d) to be evaluated during the exploration, it is important to apply an effective filtering strategy. The choice of the particular sampling-based path planning algorithm and filtering strategy mainly depend on the type of objective function c_P being considered. The approach described below has been developed to find good-quality solutions with respect to the MW path evaluation criterion (1). In this work, we extend the T-RRT algorithm [9], which finds paths that tend to minimize cost variation by filtering during the exploration tree nodes that would produce a steep cost increase. Following a similar approach, alternative algorithms and the associated filtering strategies could be developed to optimize other path cost functions. For instance, variants of RRT* [20] or FMT* [21] could be considered for optimizing other types of monotonically increasing cost functions.

3   Algorithm

This section presents the SDAP algorithm, building upon a single-tree version of T-RRT. However, the approach is directly applicable to multi-tree variants [22]. First, the basic algorithm is introduced, followed by additional explanation on the tree extension strategy and a brief theoretical analysis.

3.1   SDAP Algorithm

The SDAP pseudo-code is shown in Algorithm 1. A search tree, $\mathcal{T}$, is created with q_{init} as the root node. The tree is grown in configuration space through a series of expansion operations. Each node s in $\mathcal{T}$ encodes a configuration q and a set of designs $D \subseteq \mathcal{D}$ for which the configuration is valid. Each node's set of designs D is a subset of its parent's designs, i.e. $Designs(s) \subseteq Designs(Parent(s))$.

During each iteration, a random configuration q_{rand} is generated (line 3). In T-RRT, a new node q_{new} is created by expanding the nearest node in $\mathcal{T}$ q_{near} in the direction of q_{rand} for a distance δ. q_{new} is then conditionally added to $\mathcal{T}$ based on a transition test. A common heuristic for this transition test is the Metropolis criterion, traditionally applied in Monte Carlo methods [23]. Transitions to lower cost nodes are always accepted and moves to higher costs nodes are probabilistically accepted. The probability to transition to a higher cost node is controlled by a temperature variable T. T-RRT dynamically controls T, as explained below.

SDAP modifies the expansion and transition test functions of the standard T-RRT algorithm in order to address the design and path planning problems simultaneously. At each iteration, SDAP attempts to expand at least one node per design in $\mathcal{D}$. The process is shown in Figure 1, where each design $d \in \mathcal{D}$ is encoded as a color on each node of the tree. During each iteration, SDAP expands a set of nodes that covers all designs $\mathcal{D}$. In other words, the NearestNeighbors function (line 4) returns a set, *Neighbors*, containing the closest node to q_{rand} for each design. In Figure 1, q_{rand} is shown in black and the 3 nodes in the set *Neighbors* are circled in red. Each node in *Neighbors* is extended towards q_{rand} (lines 6 - 10) creating new candidate nodes which are labeled s_1, s_2, and s_3 in Figure 1. All 3 designs in s_1 fail the transition test, so the new node is not added to $\mathcal{T}$. For s_2 the blue design passes the transition test and the node is added to $\mathcal{T}$. Finally for s_3, 1 of the 3 designs (yellow) fails the transition test, resulting in a node with 2 designs being added to $\mathcal{T}$.

3.2   Controlling Tree Expansion

The transition test is governed by the temperature parameter T. T-RRT automatically adjusts T during the exploration and has been shown highly effective in balancing tree exploration and tree refinement [9]. At each iteration, T-RRT adjusts T by monitoring the acceptance rate of new nodes. SDAP extends this idea by maintaining a separate temperature variable $T(d)$ for each design $d \in \mathcal{D}$. A given design d can appear in multiple nodes in *Neighbors*. For each design

---

Algorithm 1: SDAP Algorithm

---

input : the configuration space $\mathcal{C}$; the design space $\mathcal{D}$; the cost function c
the start state q_{init}; the goal state q_{goal}
number of iterations $MaxIter$

output: the tree $\mathcal{T}$

1 $\mathcal{T} \leftarrow$ InitTree($q_{\text{init}}, \mathcal{D}$)
2 **while** *not* StoppingCriterion ($\mathcal{T}, q_{\text{goal}}, MaxIter$) **do**
3 $\quad$ $q_{\text{rand}} \leftarrow$ Sample($\mathcal{C}$)
4 $\quad$ $Neighbors \leftarrow$ NearestNeighbors($\mathcal{T}, q_{\text{rand}}, \mathcal{D}$)
5 $\quad$ TransitionTest.Init()
6 $\quad$ **for** $s_{\text{near}} \in Neighbors$ **do**
7 $\quad\quad$ $q_{\text{new}} \leftarrow$ Extend($q_{\text{rand}}, s_{\text{near}}$)
8 $\quad\quad$ $D \leftarrow$ TransitionTest($\mathcal{T}, s_{\text{near}}, q_{\text{new}}, c$)
9 $\quad\quad$ **if** NotEmpty(D) **then**
10 $\quad\quad\quad$ AddNode($\mathcal{T}, s_{\text{near}}, q_{\text{new}}, D$)

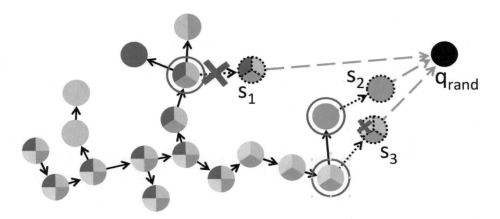

Fig. 1. An expansion operation for SDAP. Designs are encoded as colors within each node. The nodes being expanded are circled in red. The expansion towards node s_1 fails for all 3 designs, The expansion to s_2 succeeds, and the expansion to s_3 succeeds for 2 of the 3 designs.

d, the node in *Neighbors* closest to q_{rand} is identified. The temperature $T(d)$ is adjusted based on the success or failure of the extension operation from this node for design d.

The pseudocode for the transition test function is shown in Algorithm 2. The *Neighbors* set is processed in ascending order of the distance of each node from q_{rand} (line 6 of Algorithm 1). For the node being expanded (s_{near}), each design d has its cost evaluated (line 4). Transitions to lower cost nodes are always accepted (line 8). Transitions to higher cost nodes are subjected to probabilistic acceptance (line 10). The set V (lines 12 and 18) tracks designs which have had

---

Algorithm 2: TransitionTest$(\mathcal{T}, s_{\text{near}}, q_{\text{new}}, c)$

---

input   : the input tree $\mathcal{T}$; vector of temperatures T
             parent node s_{near}; new node q_{new} ; the cost function c
             temperature adjustment rate T_{rate}; Boltzmann constant K
internal: set of designs V with adjusted temperatures in this iteration
output : vector of designs D that pass the transition test

1 S $\leftarrow \phi$
2 **for** $d \in Designs(s_{\text{near}})$ **do**
3   **if** CollisionTest$(q_{\text{new}}, d) == \textit{False}$ **then**
4          $c_{\text{near}} = c(\text{Config}(s_{\text{near}}), d)$; $c_{\text{new}} = c(q_{\text{new}}, d)$
5   success $\leftarrow$ false
6          $\Delta c = c_{\text{new}}$ - c_{near}
7   **if** $\Delta c < 0$ **then**
8   success $\leftarrow$ true
9   **else**
10   **if** $exp(-\Delta c\,/\,(K \cdot T(d))) >$ UniformRand$()$ **then**
11   success $\leftarrow$ true
12   **if** $d \notin V$ **then**
13   **if** *success* **then**
14                    $T(d) \leftarrow T(d)\,/\,2^{(\Delta c\,/\,\text{energyRange}(\mathcal{T},d))}$
15   **else**
16                    $T(d) \leftarrow T(d) \cdot 2^{T_{\text{rate}}}$
17   **if** *success* **then** $D \leftarrow D \cup d$
18          $V \leftarrow V \cup d$

19 **return** (D)

---

their temperature adjusted during this iteration. The function returns the set D of designs that pass the transition test.

3.3   Theoretical Analysis

In this section we provide some theoretical analysis of SDAP algorithm's completeness and path optimality. A theoretical analysis of the complexity of SDAP with respect to the brute-force approach is difficult, since both are stochastic processes. In this work, we instead provide empirical results in Section 4 that clearly show SDAP's efficiency versus an exhaustive search of paths for all possible designs.

Probabilistic Completeness: SDAP's probabilistic completeness directly derives from that of RRT [19], which is inherited by T-RRT under the condition to guarantee a strictly positive probability of passing the transition test as explained in [9]. Since SDAP maintains this property by incorporating temperatures in the transition test for each given design $d \in \mathcal{D}$, it also ensures the positive transition probability and that each $\mathcal{C}^d_{free}$ will be completely sampled, thus maintaining the probabilistic completeness of the algorithm.

Path Optimality: The current SDAP implementation is based on T-RRT, which has been empirically shown to compute paths that tend to minimize cost with respect to the MW criterion [9], but without theoretical guarantee of optimality. Using anytime variants of T-RRT (AT-RRT or T-RRT*) [18] would provide asymptotic convergence guarantee. Implementing these within SDAP remains as future work.

4   Empirical Analysis and Results

As a proof of concept, we apply SDAP to a set of academic problems. SDAP is implemented as an adaptation of the Multi-T-RRT algorithm [22], with two trees growing from the initial and goal configurations. The search stops when the algorithm is able to join the two trees. For each problem, SDAP is compared against a naive approach consisting of multiple independent runs of Multi-T-RRT on each designs $d \in \mathcal{D}$.

4.1   Test System Description

The test system is a 2D articulated mechanism with a fixed geometry surrounded with fixed obstacles. The bodies $A_1..A_n$ are circles with radius R. The first body A_1 is a fixed base. The other bodies $A_2..A_n$ are articulated by a rotational joint centered on the previous rigid body that can move in the interval $[0, 2\pi)$. A configuration q is described by a vector of $n-1$ angles corresponding to the value of each rotational joint. The features $f_1..f_n$ assigned to each body are electrostatic charges in $\mathcal{F} = \{-1, 0, 1\}$ (i.e. $m = 3$). The design vector d contains n charges $f_1..f_n$ associated to each rigid body $A_1..A_n$ of the mechanism. In the following, d will be written as a string, with each charge $(-1, 0, 1)$ corresponding to N, U, and P respectively. For example, the design NPUN corresponds to the vector $d = (-1, 1, 0, -1)$. Obstacles $O_1..O_k$ are circles of radius R and have electrostatic charges with predefined values $g_i \in \mathcal{F}$.

The cost function is inspired by a simple expression of the potential energy of a molecular system. It contains two terms, one corresponding to the Lennard-Jones potential and the other to the electrostatic potential. It is defined as:

$$c(q, d) = LJ(q, d) + ES(q, d) \tag{3}$$

with:

$$LJ(q,d) = \sum_{i=1}^{|\mathcal{A}|-2} \left[\sum_{j=i+2}^{|\mathcal{A}|} \left(\frac{2 \cdot R}{\|A_i A_j\|} \right)^{12} - \left(\frac{2 \cdot R}{\|A_i A_j\|} \right)^6 \right]$$

$$+ \sum_{i=1}^{|\mathcal{A}|} \left[\sum_{j=1}^{|\mathcal{O}|} \left(\frac{2 \cdot R}{\|A_i O_j\|} \right)^{12} - \left(\frac{2 \cdot R}{\|A_i O_j\|} \right)^6 \right] \quad (4)$$

$$ES(q,d) = \sum_{i=1}^{|\mathcal{A}|-2} \left[\sum_{j=i+2}^{|\mathcal{A}|} \left(\frac{f_i \cdot f_j}{\|A_i A_j\|} \right) \right] + \sum_{i=1}^{|\mathcal{A}|} \left[\sum_{j=1}^{|\mathcal{O}|} \left(\frac{f_i \cdot g_j}{\|A_i O_j\|} \right) \right] \quad (5)$$

where $\|X_i X_j\|$ represents the Euclidean distance between the centers of the bodies/obstacles X_i and X_j.

SDAP is empirically tested using a 4 body and a 10 body scenarios described below. The objective is to find the path-design pair (P^*, d^*) that minimizes c_P.

Small 4 Body System: The first system consists of four bodies and five obstacles as shown in Figure 2. q_{init} and q_{goal} correspond to fully stretched configurations, to the left and to the right, represented with solid and dashed outlines respectively. The design space consists of $3^4 = 81$ possible combinations and the configuration space is 3 dimensional. This scenario favors designs with a negatively charged end-effector. The uncharged obstacles at the top and bottom of the workspace create a narrow passage that all solutions must pass through. Figure 3 shows a projection of the configuration-space costmap for two designs along with a solution path. One design has a negatively charged end-effector (UUUN) and one has a positively charged end-effector (UUUP). Angles 1 and 2 are projected onto the x and y axis respectively, with angle 3 being set to minimize the cost function. For the UUUN design, the costmap is highly favorable to the desired motion, starting at a high cost and proceeding downhill to a low cost area. The costmap associated with the UUUP design shows a non-favorable motion between the two states.

Larger 10 Body System: A larger system with 10 bodies and six obstacles is shown in Figure 2. The design space $\mathcal{D}$ contains $3^{10} = 59049$ possibilities, which cannot be exhaustively explored within a reasonable computing time, and is also challenging for SDAP because of memory issues (see discussions in Section 5). For that reason, two simplified versions of this scenario are constructed. The first one fixes the design for the first seven bodies $A_1..A_7$ as UUUUUUU. The remaining bodies (A_8, A_9, and A_{10}) can be designed, resulting in a design space of $3^3 = 27$ designs. The second version expands the design space to the last 4 bodies ($A_7..A_{10}$), resulting in a design space of $3^4 = 81$ designs. It both cases, the configuration space is 9 dimensional. Both versions of the 10-body system are constrained with the same obstacles. They were chosen so that designs with strongly positively or negatively charged end-effectors will be trapped at local minima resulting from attractive or repulsive forces generated by the bottom obstacles.

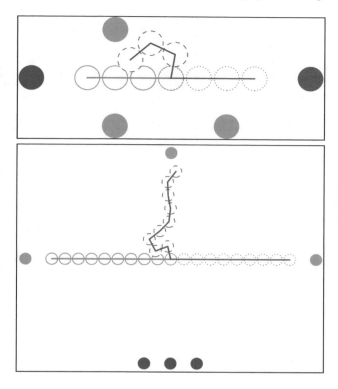

Fig. 2. A 4-body (top) and 10-body (bottom) scenario. Obstacles with positive charges are shown in solid red, negative in solid blue, neutral in gray. The initial state is shown in green with a solid line, a transition state shown in blue, and the goal state is shown in red with a dashed line.

4.2   Benchmark Results

We compare SDAP to a naive approach (solving individual problems for each design $d \in \mathcal{D}$) using the same Multi-T-RRT implementation. In other words, we compare one run of the SDAP algorithm against $|\mathcal{D}|$ runs of a single-design path search. Multiple runs are performed (100 for the 4-body scenario, 50 for the 10-body scenario with 3 designed bodies, and 20 for the 10-body scenario with 4 designed bodies) to avoid statistical variance inherent with stochastic methods. The single-design explorations can spend time trying to escape local minima associated with the costmaps of unfavourable designs, causing very long execution times and high-cost paths. As we are not interested in finding a solution path for every possible design but only for the designs with low-cost paths, a timeout is enforced for the single-design explorations of 300 seconds for the 4-body scenario and 1,200 seconds for the 10-body scenarios. The SDAP algorithm was considered unsuccessful if it did not find a solution within 2,400 seconds. In all the cases, the T-RRT parameter T_{rate} was set to 0.1, being the initial temperature $T = 1$ and $K = 0.00198721$ (note that T and K do not have much physical meaning in the present context). During the exploration, the maximal values reached by T were around 100, being slightly lower for SDAP compared to the naive approach.

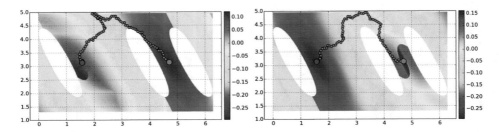

Fig. 3. Configuration-space costmap of the 4-body system projected onto the first two DOFs of the system expressed in radians. The initial configuration is indicated by the red dot on the left, and the goal by the green dot on the right. The plot on the left is for the UUUP design, the one to the right for UUUN. Each cell's cost is computed by finding the value of the third angle that minimizes the cost.

All the runs were performed in a single threaded process on a Intel(R) Xeon(R) CPU E5-2650 0 @ 2.00GHz processor with 32GB of memory.

Small Scenario Results: The runtimes for the small scenario are shown at the top in Figure 4. For the single-design approach, the sum of the 81 runs to cover $\mathcal{D}$ are plotted versus the SDAP runtime. The figure shows that SDAP is twice as fast as the single-design approach. Although the variance in execution time for SDAP seems much larger than for the naive approach, recall that each complete run of the latter involves 81 runs of the Multi-T-RRT algorithm, which attenuates the overall variance. However, the computing time variance for a specific design can be much larger. Figure 5 (top) compares the solutions found by the two methods. SDAP successfully identifies the designs corresponding to the lowest-cost paths. Recall that the current implementation of SDAP terminates when one valid path is found. Asymptotic convergence to the global optimum could be guaranteed by implementing an anytime variant of the algorithm such as AT-RRT [18] (this remains as future work). The high density of nodes created by the SDAP algorithm (19,784 nodes on average compared to 698 for each single-design search) could be well exploited to improve the path cost by incrementally adding cycles.

Larger scenario results: The runtimes for both versions of the larger scenario are shown in Figure 4 (middle and bottom). In both cases, the difference in computing time between the two approaches increases significantly compared to the 4-body scenario. In the 3-designed-body version, SDAP is 26 times faster than the single-design search on average. In the 4-designed-body version, SDAP is 46 times faster. Note that, while the cardinality of design space $\mathcal{D}$ is multiplied by 3 between the two versions of the large scenario, the execution time of SDAP is only multiplied by 2.5 on average. The variance of the execution times is now lower for SDAP compared to the naive approach. The reason is that the performance of Multi-T-RRT highly depends on the roughness of the

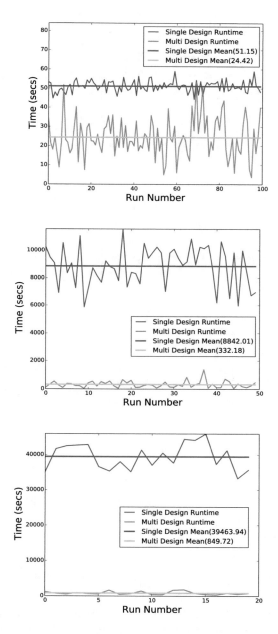

Fig. 4. Run times comparisons for the 4-body scenario (top), 10-body scenario with 3 designed bodies (middle) and 4 designed bodies (bottom). SDAP (green line) is compared against an exhaustive search using single-design explorations (blue line).

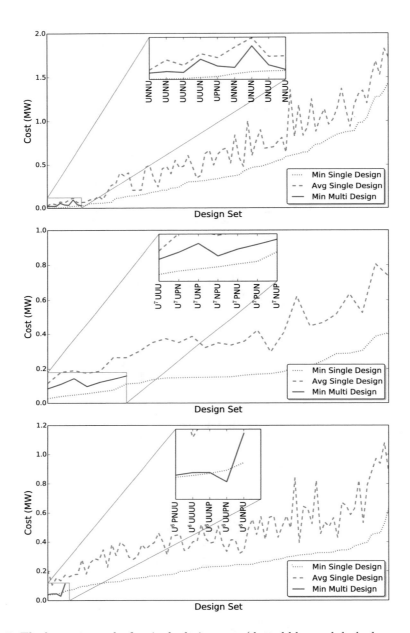

Fig. 5. The best cost paths for single-design runs (dotted blue and dashed green lines) and SDAP (red line) for the small scenario (top) and larger 3-designed-body (middle) and 4-designed-body (bottom). SDAP solutions shown only for discovered paths (does not exhaustively search). SDAP discovers the same low-cost designs as the exhaustive single-design searches.

configuration-space costmap. In a smooth costmap, Multi-T-RRT will be quite fast with a low variance, whereas the time required to find a solution in a rugged costmap will be higher and will have a larger variance. SDAP's computing time is only dependent on the difficulty to find the best designs, which typically have a smoother costmap, whereas the single-design search has to find a solution for every design, including those with a very rugged costmap. The 4-body scenarios is relatively simple, and thus even for very bad designs, a solution was found in close-to-constant time. But for the 10-body scenario, the problem is more complex, and the 27 (resp. 81) runs are not enough to attenuate a very high variance.

The single-design search reached the timeout 4 times over the 27 runs on average for the 3-designed-body version of the 10-body scenario, and 19 times over the 81 runs on average for the 4-designed-body version of the problem. The SDAP algorithm always found a solution before the timeout.

Figure 5 (middle and bottom) compares the solutions found by the two searches. Once again, SDAP successfully identifies the designs that yield the best path cost.

5   Conclusion

In this paper, we have presented an original formulation of a challenging problem combining system design and path planning, and have proposed a new approach to solve it building on sampling-based algorithms. The current implementation of SDAP is still preliminary, but it already shows significant gains in efficiency and accuracy compared to a brute-force approach.

The actual motivation of this work concerns computational biology. We are currently applying SDAP to help understand the effect of mutations in antibodies, which is a preamble to protein design (publication in preparation).

Several applications of SDAP in robotics can also be envisioned. In addition to the design of some robot's features to optimize its motion in a given workspace, it would also be possible to apply the proposed method to optimize the workspace layout for a given robot. One can also imagine applications for helping to the design of modular self-reconfigurable robots. We aim to implement SDAP within robot motion planning software in order to investigate these potential applications.

For future work, in addition to the implementation of a variant of SDAP with asymptotic optimality guarantees based on AT-RRT [18], we aim to introduce further improvements. The exploration of very-high-dimensional configuration and design spaces implies computer memory issues (the resulting tree/graphs are very large). A solution to this problem would be to introduce pruning stages during the exploration, as is done in the SST* algorithm [24]. Larger design spaces will also require SDAP to employ more sophisticated filters and heuristics, such as those that incorporate statistical learning of the structure of the space, to explore it more efficiently and control the size of the search tree.

Acknowledgment

This work has been partially supported by the French National Research Agency (ANR) under project ProtiCAD (project number ANR-12-MONU-0015), and by the European Union's Horizon 2020 research and innovation programme under grant agreement No 644271 AEROARMS.

References

1. C. Gosselin and J. Angeles, "A global performance index for the kinematic optimization of robotic manipulators," *Journal of Mechanical Design*, vol. 113, no. 3, pp. 220–226, 1991.
2. J.-P. Merlet, "Optimal design of robots," in *Robotics: Science and Systems*, 2005.
3. J. M. McCarthy and L. Joskowitz, "Kinematic synthesis," in *Formal Engineering Design Synthesis*, J. Cagan and E. Antonson, Eds. Cambridge Univ. Press., 2001.
4. E. S. Rudnick-Cohen, S. Azarm, and J. Herrmann, "Multi-objective design and path planning optimization of unmanned aerial vehicles," in *Proc. 16th AIAA/ISSMO Multidisciplinary Analysis and Optimization Conference, AIAA Aviation*, 2015.
5. L. E. Kavraki, P. Svestka, J.-C. Latombe, and M. Overmars, "Probabilistic roadmaps for path planning in high dimensional configuration spaces," *IEEE Transactions on Robotics and Automation*, vol. 12, no. 4, pp. 566–580, 1996.
6. S. M. LaValle, *Planning Algorithms*. New York: Cambridge University Press, 2006.
7. T. Siméon, J.-P. Laumond, J. Cortés, and A. Sahbani, "Manipulation planning with probabilistic roadmaps," *Int. J. Robot. Res.*, vol. 23(7), pp. 729–746, 2004.
8. K. Hauser, "The minimum constraint removal problem with three robotics applications," *Int. J. Robot. Res.*, vol. 33, no. 1, pp. 5–17, 2014.
9. L. Jaillet, J. Cortés, and T. Siméon, "Sampling-based path planning on configuration-space costmaps," *IEEE Trans. Robotics*, vol. 26, no. 4, pp. 635–46, 2010.
10. M. Moll, D. Schwarz, and L. E. Kavraki, *Roadmap Methods for Protein Folding*. Humana Press, 2007.
11. I. Al-Bluwi, T. Siméon, and J. Cortés, "Motion planning algorithms for molecular simulations: A survey," *Comput. Sci. Rev.*, vol. 6, no. 4, pp. 125–43, 2012.
12. B. Gipson, D. Hsu, L. Kavraki, and J.-C. Latombe, "Computational models of protein kinematics and dynamics: Beyond simulation," *Ann. Rev. Analyt. Chem.*, vol. 5, pp. 273–91, 2012.
13. A. Shehu, "Probabilistic search and optimization for protein energy landscapes," in *Handbook of Computational Molecular Biology*, S. Aluru and A. Singh, Eds. Chapman & Hall/CRC Computer & Information Science Series, 2013.
14. A. E. Keating, *Methods in protein design*, ser. Methods in enzymology. Amsterdam: Academic Press/Elsevier, 2013, vol. 523.
15. B. R. Donald, *Algorithms in Structural Molecular Biology*. The MIT Press, 2011.
16. C. E. Tinberg, S. D. Khare, J. Dou, L. Doyle, J. W. Nelson, A. Schena, W. Jankowski, C. G. Kalodimos, K. Johnsson, B. L. Stoddard, and D. Baker, "Computational design of ligand-binding proteins with high affinity and selectivity," *Nature*, vol. 501, pp. 212–6, 2013.

17. B. E. Correia, J. T. Bates, R. J. Loomis, G. Baneyx, C. Carrico, J. G. Jardine, P. Rupert, C. Correnti, O. Kalyuzhniy, V. Vittal, M. J. Connell, E. Stevens, A. Schroeter, M. Chen, S. MacPherson, A. M. Serra, Y. Adachi, M. A. Holmes, Y. Li, R. E. Klevit, B. S. Graham, R. T. Wyatt, D. Baker, R. K. Strong, J. E. Crowe, P. R. Johnson, and W. R. Schief, "Proof of principle for epitope-focused vaccine design," *Nature*, 2014.

18. D. Devaurs, , T. Siméon, and J. Cortés, "Optimal path planning in complex cost spaces with sampling-based algorithms," *IEEE Transactions on Automation Science and Engineering*, vol. 13, no. 2, pp. 415–424, 2015.

19. S. LaValle and J. Kuffner, "Rapidly-exploring random trees: progress and prospects," in *Algorithmic and Computational Robotics: New Directions*, 2001, pp. 293–308.

20. S. Karaman and E. Frazzoli, "Sampling-based Algorithms for Optimal Motion Planning," *Int. J. Rob. Res.*, vol. 30, no. 7, pp. 846–894, Jun. 2011.

21. L. Janson, E. Schmerling, A. Clark, and M. Pavone, "Fast marching tree," *Int. J. Rob. Res.*, vol. 34, no. 7, pp. 883–921, 2015.

22. D. Devaurs, T. Siméon, and J. Cortés, "A multi-tree extension of the transition-based RRT: Application to ordering-and-pathfinding problems in continuous cost spaces," in *IEEE/RSJ International Conference on Intelligent Robots and Systems*, 2014, pp. 2991–2996.

23. N. Metropolis, A. Rosenbluth, M. Rosenbluth, A. Teller, and E.Teller, "Equation of state calculations by fast computing machines," *Journal of Chemical Physics*, vol. 21, pp. 1087–1092, 1953.

24. Y. Li, Z. Littlefield, and K. E. Bekris, "Asymptotically optimal sampling-based kinodynamic planning," *Int. J. Robot. Res.*, vol. 35, pp. 528–564, 2016.

Language-Guided Sampling-based Planning using Temporal Relaxation

Francisco Penedo[1], Cristian-Ioan Vasile[2], and Calin Belta[1]

[1] Boston University, Boston, MA, U.S.A.
{franp, cbelta}@bu.edu
[2] Massachusetts Institute of Technology, Cambridge, MA, U.S.A.
cvasile@mit.edu

Abstract. In this paper, we focus on robot motion planning from timed temporal logic specifications. We propose a sampling-based algorithm and an associated language-guided biasing scheme. We leverage the notion of temporal relaxation of time-window temporal logic formulae (TWTL) to reformulate the temporal logic synthesis problem into an optimization problem. Our algorithm exhibits an exploration-exploitation structure, but retains probabilistic completeness. Moreover, if the problem does not have a solution due to time constraints, the algorithm returns a candidate path that satisfies a minimally relaxed version of the specification. The path may inform operators about timing problems with the specification or the system. We provide simulations to highlight the performance of the proposed algorithm.

1 Introduction

Motion planning is a fundamental problem in robotics. The objective is to generate control policies for a robot to drive it from an initial state to a goal region in its state space under kino-dynamic constraints [22]. Even without considering dynamics, the problem becomes increasingly difficult in high dimensions, and it has been shown to be PSPACE-complete [7]. Cell decomposition, potential fields and navigation functions [9] are some of the most used techniques to solve the problem. However, they scale poorly with the dimension of the state space and number of obstacles. To overcome these limitations, a class of algorithms was developed relying on randomly sampling the configuration space of the robot and planning local motions between these samples. Probabilistic Roadmaps [18] and Rapidly Exploring Random Trees [22] are among the most widely known examples, along with their asymptotically optimal variants, PRM* and RRT* [16].

Robots are increasingly required to perform complex tasks, where correctness guarantees, such as safety and liveness in human-robot teams and autonomous driving, are critical. One approach is to encode the tasks as temporal logic specifications and leverage formal methods techniques to generate control policies that are correct by construction [4]. As opposed to traditional methods restricted to reach-avoid setups, these frameworks are able to express more complex tasks such as sequencing (e.g., "Reach A, then B"), convergence ("Go to A and stay

© Springer Nature Switzerland AG 2020
K. Goldberg et al. (Eds.): *Algorithmic Foundations of Robotics XII*, SPAR 13, pp. 128–143, 2020.
https://doi.org/10.1007/978-3-030-43089-4_9

there forever"), persistent surveillance ("Visit A, B, and C infinitely often"), and more complex combinations of the above. Temporal logics, such as Linear Temporal Logic (LTL), Computational Tree Logic (CTL), and μ-calculus, and their probabilistic versions (PLTL, PCTL), have been shown to be useful as formal languages for motion planning [20, 38, 6, 15, 10]. Model-checking and automata game techniques were adapted [20, 8] to generate control policies for finite models of robot motion. These models were obtained through an abstraction process that partitions the robot configuration space and captures the ability of the robot to steer between regions in the partition [5]. As a result, these algorithms suffer from the same scalability issues as the cell-based decomposition methods.

Some applications additionally require time constraints [30, 27, 12]. For example, consider the following task: "Visit A, B, and C in this order. Perform action a for 2 time units at A within 10 time units. Then, perform b for 3 time units at B within 6 time units. Finally, in the time window [3, 9] after b is finished, perform c for 1 time unit at C. All three actions must be finished within 15 time units." Tasks with explicit time constraints may be expressed using bounded linear temporal logic (BLTL) [31, 14], metric temporal logic (MTL) [19], signal temporal logic (STL) [24], and time-window temporal logic (TWTL) [35, 1, 36].

A natural approach to generate control strategies from rich task specifications for robots with large state spaces is to combine sampling-based techniques with automata-based synthesis methods. The existing works in this area show that synthesis algorithms from specifications given in μ-calculus [15, 17] and LTL [34] can be adapted to scale incrementally with the graph constructed during the sampling process. In all of these, sampling is performed independently of the specification. Hierarchical planning frameworks were proposed in [28, 25], where a higher level planner performs a discrete search in a partition of the workspace of the robot that guides a lower level sampling-based planner in the configuration space. The idea of language-guided synthesis was also explored in [2], where an iterative partition-refinement algorithm is proposed together with cell-to-cell feedback controllers, and in [37], where candidate discrete plans are enumerated, and then motion plans are generated using an optimization-based procedure.

In this paper, we propose a language-guided sampling-based method to generate robot control policies satisfying tasks expressed as TWTL formulae. We leverage the notion of temporal relaxation [35] to reformulate a temporal logic motion problem as an optimization problem, where the objective is to minimize the temporal relaxation of the specification. This approach has two advantages: (a) the growth of the sampling graph is biased towards satisfaction of a relaxed version of the specification without initially taking into account deadlines, and (b) after an initial policy has been found, sampling can be focused on the parts of the plan that need to be improved. The two stages can be thought of as exploration-exploitation phases, where initially candidate solutions are found, and then focused local sampling is performed on the parts that need to be improved in order to satisfy the specification, i.e. time constraints. As a byproduct, a satisfying policy with respect to a minimally relaxed version of the specification may be returned when the original problem does not have a solution. Such a policy may inform operators about timing problems in the specification or system (i.e., robot dynamics and/or environment). The algorithm uses annotated finite state automata [35, 36, 1] to represent all possible temporal relaxations of a TWTL formulae. Lastly, we prove that our solution is probabilistically complete.

As opposed to [35], we do not assume a finite model of the system, and propose a sampling-based approach. Although the synthesis algorithm in [35] is more general (w.r.t. the range of specifications it can handle), it is not incremental and thus not suitable for use with sampling-based techniques. Moreover, we define a new temporal relaxation measure over τ-relaxations of TWTL formulae called *linear ramp temporal relaxation* that is better suited for incremental computation. Other studies have investigated minimal violations of LTL fragments [29, 33, 32, 21, 26, 13]. Our approach differs from [29, 33, 32] because we consider explicit time constraints in the specification, and a different semantics for relaxation of a specification. The minimum violation policies strive to minimize the duration that the specification is violated, i.e., satisfaction is preempted. Temporal relaxation on the other hand minimizes the deviation from the deadlines in the specification, and does not allow task interruption. In [13], the specifications were relaxed by minimally revising symbols associated with the transitions of the Büchi automata. Both [21, 26] modify LTL to accomodate partial satisfaction without considering explicit time bounds. Due to the fragment of TWTL we allow, our work is related to the scheduling literature, such as [23, 11]. However, these works do not allow partial satisfaction. In [3], explicit time constraints, as well as first order quantifiers, are allowed in the specification, but no partial completion is considered. A different approach is taken by [39] using the concept of resources to impose soft constraints in the specification.

Notation: Given $\mathbf{x}, \mathbf{x}' \in \mathbb{R}^n$, $n \geq 2$, the relationship $\mathbf{x} \sim \mathbf{x}'$, where $\sim \in \{<, \leq, >, \geq\}$, is true if it holds pairwise for all components. $\mathbf{x} \sim a$ denotes $\mathbf{x} \sim a\mathbf{1}_n$, where $a \in \mathbb{R}$ and $\mathbf{1}_n$ is the n-dimensional vector of all ones. Let S be a finite set. We denote the cardinality and the power set of S by $|S|$ and 2^S, respectively.

2   Preliminaries

2.1   Time-Window Temporal Logic (TWTL)

In this paper, we use Time-Window Temporal Logic (TWTL) as a specification language for temporal properties with time constraints. For details see [35] and [36, 1] for applications to robotics. TWTL is a linear-time logic encoding sets of discrete-time sequences with values in a finite alphabet. The syntax of TWTL formulae defined over a set of atomic propositions Π is

$$\phi ::= \mathcal{H}^d s \mid \mathcal{H}^d \neg s \mid \phi_1 \wedge \phi_2 \mid \phi_1 \vee \phi_2 \mid \neg \phi_1 \mid \phi_1 \cdot \phi_2 \mid [\phi_1]^{[a,b]}$$

where s is either the "true" constant $\top$ or an atomic proposition in Π; $\wedge$, $\vee$, and $\neg$ are the conjunction, disjunction, and negation Boolean operators, respectively; $\cdot$ is the concatenation operator; $\mathcal{H}^d$ with $d \in \mathbb{Z}_{\geq 0}$ is the *hold* operator; and $[\,]^{[a,b]}$ with $0 \leq a \leq b$ is the *within* operator. The semantics is defined with respect to finite (output) words $\mathbf{o} = o_0 o_1 ... o_k$ over the set 2^Π. The Boolean operators retain their usual semantics. The *hold* operator $\mathcal{H}^d s$ specifies that an atomic proposition $s \in \Pi$ should be serviced (satisfied) for d time units (i.e., $\mathbf{o} \models \mathcal{H}^d s$ if $o_t = s \; \forall t \in [0, d]$). For convenience, if $d = 0$ we simply write s and $\neg s$ instead of $\mathcal{H}^0 s$ and $\mathcal{H}^0 \neg s$, respectively. The *within* operator $[\phi]^{[a,b]}$ bounds the satisfaction of ϕ within $[a, b]$ time window (i.e., $\mathbf{o} \models [\phi]^{[a,b]}$ if $\exists k \in [0, b-a]$ s.t. $\mathbf{o}' \models \phi$ where $\mathbf{o}' = o_{a+k} \ldots o_b$). Lastly, the concatenation of ϕ_i and ϕ_j (i.e., $\phi_i \cdot \phi_j$) specifies that first ϕ_i must be satisfied and then immediately ϕ_j must be satisfied. The satisfaction of a TWTL formula by a word can be decided within bounded time.

The notion of *temporal relaxation* of a TWTL formula was introduced in [35, 1]. To illustrate the main ideas, consider the following TWTL formula:

$$\phi_1 = [\mathcal{H}^1 A]^{[0,2]} \cdot [\mathcal{H}^3 B \wedge [\mathcal{H}^2 C]^{[0,4]}]^{[1,8]}, \tag{1}$$

which reads as "Perform task A of duration 1 within 2 time units. Then, within the time interval $[1, 8]$ perform tasks B and C of durations 3 and 2, respectively. Furthermore, C must be finished within 4 time units from the start of B." If ϕ_1 cannot be satisfied, one way to relax ϕ_1 is to extend the deadlines for the time windows captured by the *within* operators:

$$\phi_1(\boldsymbol{\tau}) = [\mathcal{H}^1 A]^{[0,2+\tau_1]} \cdot [\mathcal{H}^3 B \wedge [\mathcal{H}^2 C]^{[0,4+\tau_2]}]^{[1,8+\tau_3]}, \tag{2}$$

where $\boldsymbol{\tau} = (\tau_1, \tau_2, \tau_3) \in \mathbb{Z}^3$. However, the choice of $\boldsymbol{\tau}$ must preserve the feasibility of the formula, i.e., the following must hold for $\phi_1(\boldsymbol{\tau})$: (i) $2+\tau_1 \geq 1$, (ii) $4+\tau_2 \geq 2$, and (iii) $7+\tau_3 \geq \max\{3, 4+\tau_2\}$. Note that $\boldsymbol{\tau}$ may be non-positive. In such cases, $\phi_1(\boldsymbol{\tau})$ becomes a stronger specification than ϕ_1, which implies that the sub-tasks are performed earlier than their actual deadlines.

Definition 1 (Feasible TWTL formula). *A TWTL formula ϕ is called feasible if the time window corresponding to each* within *operator is greater than the duration of the corresponding enclosed task expressed via the* hold *operators.*

Let ϕ be a TWTL formula. Then, a τ−relaxation of ϕ is defined as follows:

Definition 2 (τ−Relaxation of ϕ). *Let $\boldsymbol{\tau} \in \mathbb{Z}^m$, where m is the number of* within *operators contained in ϕ. A τ-relaxation of ϕ is a feasible TWTL formula $\phi(\boldsymbol{\tau})$, where each subformula of the form $[\phi_i]^{[a_i, b_i]}$ is replaced by $[\phi_i]^{[a_i, b_i + \tau_i]}$.*

Clearly, for any ϕ, we have $\phi(\mathbf{0}) = \phi$. Moreover, let $\boldsymbol{\tau}', \boldsymbol{\tau}'' \in \mathbb{Z}^m$ such that $\phi(\boldsymbol{\tau}')$ and $\phi(\boldsymbol{\tau}'')$ are feasible relaxations, where m is the number of *within* operators in ϕ. Note that if $\boldsymbol{\tau}' \leq \boldsymbol{\tau}''$, then $\mathbf{o} \models \phi(\boldsymbol{\tau}') \Rightarrow \mathbf{o} \models \phi(\boldsymbol{\tau}'')$.

Let ϕ be a TWTL formula and $\phi(\boldsymbol{\tau})$ its τ-relaxation, where $\boldsymbol{\tau} \in \mathbb{Z}^m$ and m is the number of *within* operators contained in ϕ. We denote by $\mathfrak{I}_\tau(\phi) = (\tau_1, \ldots, \tau_m)$ the ordered set of deadline deviations.

Definition 3. *Given an output word $\mathbf{o}$, we say that $\mathbf{o}$ satisfies $\phi(\infty)$, i.e., $\mathbf{o} \models \phi(\infty)$, if and only if $\exists \boldsymbol{\tau}' < \infty$ s.t. $\mathbf{o} \models \phi(\boldsymbol{\tau}')$.*

Similarly, if $\boldsymbol{\tau} < \infty$, then $\mathbf{o} \models \phi(\boldsymbol{\tau}) \Rightarrow \mathbf{o} \models \phi(\infty), \; \forall \; \boldsymbol{\tau}$.

Definition 4 (Concatenation Form). *A TWTL formula ϕ is in concatenation form (CF) if and only if $\phi = \phi_1 \cdot \phi_2 \cdot \ldots \cdot \phi_n$, where ϕ_i are TWTL formulae that do not contain concatenation operators.*

In the following, we use ϕ_i to refer to the ith subformula of a formula in CF in the same way as in the previous definition. We also denote as $\phi^j = \phi_1 \cdot \ldots \cdot \phi_j$ the subformula of ϕ that includes all subformulae $\phi_1, ..., \phi_j$, which is also in CF.

2.2   Specification Automaton and System Abstraction

In this section we provide a short presentation of the mathematical objects defined in [35] that we will be making use of.

Definition 5 (Deterministic Finite State Automaton). *A deterministic finite state automaton (DFA) is a tuple $\mathcal{A} = (S_\mathcal{A}, s_0, \mathfrak{A}, \delta_\mathcal{A}, F_\mathcal{A})$, where $S_\mathcal{A}$ is a finite set of states, $s_0 \in S_\mathcal{A}$ is the initial state, $\mathfrak{A}$ is the input alphabet, $\delta_\mathcal{A} : S_\mathcal{A} \times \mathfrak{A} \to S_\mathcal{A}$ is the transition function, and $F_\mathcal{A} \subseteq S_\mathcal{A}$ is the accepting set.*

A trajectory of the DFA $\mathbf{s} = s_0 s_1 \ldots s_{n+1}$ is generated by a sequence of symbols $\sigma = \sigma_0 \sigma_1 \ldots \sigma_n$ if $s_0 \in S_\mathcal{A}$ is the initial state of $\mathcal{A}$ and $s_{k+1} = \delta_\mathcal{A}(s_k, \sigma_k)$ for all $0 \leq k \leq n$. The function $\delta_\mathcal{A}^* : \mathfrak{A}^* \to S_\mathcal{A}$ is defined such that $s_{n+1} = \delta_\mathcal{A}^*(\sigma)$. An input word σ is accepted by a DFA $\mathcal{A}$ if $\delta_\mathcal{A}^*(\sigma) \in F_\mathcal{A}$.

Definition 6 (Transition System). *A transition system (TS) is a tuple $\mathcal{T} = (V, x_0, E, \Pi, h)$, where V is a finite set of states, $x_0 \in V$ is the initial state, $E \subseteq V \times V$ is a set of transitions, Π is a set of properties (atomic propositions), and $h : V \to 2^\Pi$ is a labeling function.*

A trajectory of the system is a finite or infinite sequence of states $\mathbf{x} = x_0 x_1 \ldots$ such that $(x_k, x_{k+1}) \in E$ for all $k \geq 0$. A state trajectory $\mathbf{x}$ generates an *output trajectory* (or word) $\mathbf{o} = o_0 o_1 \ldots$, where $o_k = h(x_k)$ for all $k \geq 0$.

Definition 7 (Product Automaton). *Given $\mathcal{T} = (V, x_0, E, \Pi, h)$ and $\mathcal{A} = (S_\mathcal{A}, s_0, \mathfrak{A}, \delta_\mathcal{A}, F_\mathcal{A})$, their product automaton, denoted by $\mathcal{P} = \mathcal{T} \times \mathcal{A}$, is a tuple $\mathcal{P} = (S_\mathcal{P}, p_0, \Delta_\mathcal{P}, F_\mathcal{P})$, where $S_\mathcal{P} = V \times S_\mathcal{A}$ is the set of states, $p_0 = (x_0, s_0)$ is the initial state, $\Delta_\mathcal{P} = \{((x, s), (x', s')) \mid (x, x') \in E \wedge s' = \delta_\mathcal{A}(s, h(x'))\}$ is the set of transitions, and $F_\mathcal{P} = V \times F_\mathcal{A}$ is the set of accepting states of $\mathcal{P}$.*

A trajectory of $\mathcal{P}$ is a sequence $\mathbf{p} = p_0 \ldots p_{n+1}$ such that $(p_k, p_{k+1}) \in \Delta_\mathcal{P}$ for all $0 \leq k \leq n$. A trajectory $\mathbf{p} = (x_0, s_0)(x_1, s_1) \ldots$ of $\mathcal{P}$ is accepted if and only if $s_0 s_1 \ldots$ is accepted by $\mathcal{A}$. A trajectory of $\mathcal{T}$ obtained from an accepting trajectory of $\mathcal{P}$ satisfies the given specification encoded by $\mathcal{A}$.

3   Problem Formulation

Consider a dynamical system $\Sigma(x_0) : \quad x_{k+1} = f(x_k, u_k)$, where $x_k \in \mathbb{R}^n$ and $u_k \in U \subset \mathbb{R}^m$ are the state and control input at time k, respectively, U is the control space, and x_0 is the initial state. Let $W \subset \mathbb{R}^n$ be a convex region denoting the workspace, $O \subset W$ the obstacle set, $W_{free} = W \setminus O$ the free space, $\Pi = \{\pi_i | i = 1, ..., p\}$ a set of atomic propositions, and $\mathcal{L} : W \rightarrow 2^\Pi$ the state labeling function. Let $\mathbf{x} = x_0 x_1 \ldots$ be a trajectory of $\Sigma(x_0)$. We say that $\mathbf{x}$ is collision-free if for all $k \geq 0$ and $\lambda \in [0, 1]$, $\lambda x_k + (1 - \lambda)x_{k+1} \in W_{free}$. The output word generated by $\mathbf{x}$ is $\mathbf{o} = o_0 o_1 \ldots$, with $o_k = \mathcal{L}(x_k)$. System $\Sigma(x_0)$ under control signal $\mathbf{u} = u_0 u_1 \ldots$ is said to satisfy a TWTL formula ϕ if $\mathbf{o} \models \phi$.

Problem 1. Given a TWTL formula ϕ over Π and an initial state $x_0 \in W$, find a control policy $\mathbf{u}^*$ with control inputs in U such that the trajectory of the closed-loop system $\Sigma(x_0)$ under policy $\mathbf{u}^*$ is collision-free and satisfies ϕ.

We can formulate an optimization problem equivalent to Problem 1 by taking advantage of temporal relaxations of TWTL formulae in the following way:

Definition 8. *The linear ramp temporal relaxation (LRTR) of a τ-relaxed formula $\phi(\tau)$ is defined as:*

$$|\phi(\tau)|_{LRTR} = \sum_{j \in \mathcal{I}_\tau(\phi)} \max\{0, \tau_j\}. \tag{3}$$

Problem 2. Let ϕ be a TWTL formula over Π, and $\phi(\tau)$ its τ-relaxation. Consider the optimization problem

$$\min |\phi(\tau)|_{LRTR} \quad \text{s.t.} \quad \exists \mathbf{u}^* : \Sigma(x_0) \text{ under } \mathbf{u}^* \text{ satisfies } \phi(\tau). \tag{4}$$

If the minimum obtained from (4) is equal to 0, find a corresponding policy $\mathbf{u}^*$.

Note that this notion of temporal relaxation, LRTR, is different from the one defined in [35], where it is denoted by $|\cdot|_{TR}$. Intuitively, LRTR measures the accumulation of positive deviations from the deadlines, as opposed to the worst deviation in TR. In both, when the temporal relaxation is 0, the unrelaxed formula is satisfied. The almost linear structure and monotonicity of LRTR allows the algorithm proposed in this paper to be incremental.

Example. Consider the workspace in the bottom right of Fig. 1, with areas of interest A, B and C, and several obstacles. The specification is "Perform tasks at A, B, and C, in this order, within time intervals $[0, 2]$, $[0, 3]$, and $[0, 4]$ from the end of the previous task, respectively." The corresponding TWTL formula is $\phi_{simple} = [A]^{[0,2]} \cdot [B]^{[0,3]} \cdot [C]^{[0,4]}$. The dynamics are given by $x_{k+1} = x_k + u_k$, with the state space equal to the workspace, $||u_k|| \leq 0.75$, and $x_0 = (1, 3)$.

4   Solution

We first present an overview of our approach to solve Prob. 1. Consider the algorithm in [35], which requires a discretization of W. First, a finite TS associated with the system Σ and the discretization of W is computed. Then, the formula ϕ is translated into an annotated DFA and the product automaton of the TS and the DFA is constructed. Finally, a path with minimum temporal relaxation is found in the product automaton using a shortest path algorithm. If ϕ is satisfied by this path, the associated control sequence is a solution to the problem.

In our approach, we assume that it is not possible to obtain a discretization of W. Instead, we incrementally construct a finite transition system, $\mathcal{T} = (V, x_0, E, \Pi, h)$, where $V \subset W_{free}$, $h = \mathcal{L}$ and (V, E) is a tree, using a sampling-based algorithm similar to RRT. Specifically, at each iteration a random sample $x \in W_{free}$ is added to V. Then, a feasible transition (according to Σ) from a state in V to x is included in E. The transition is selected so that a cost function for the path from x_0 to x is minimized. At the same time, the product automaton $\mathcal{P} = \mathcal{T} \times \mathcal{A}_\infty$ is also computed incrementally, where $\mathcal{A}_\infty$ is the annotated DFA obtained from ϕ. The path with minimum temporal relaxation in $\mathcal{P}$ is updated at each iteration by comparing the current best with the (single) path added.

In the following, we assume the following about ϕ: (1) negation operators appear only in front of atomic propositions; (2) all sub-formulae of ϕ correspond to unambiguous languages (no proper subset of the language is a prefix language of the difference). Both are required to translate the formula into a DFA, but are not overly restrictive (see [35] for details). Additionally, we require (3) ϕ to be in concatenation form. This last assumption allows us to divide the specification in tasks and induces a measure of progress towards an objective. It forms the basis of the language-guided sampling that is presented in detail in Sec. 4.3. The main limitation imposed by (3) is, intuitively, to constrain the specification to only describe "tasks" without "subtasks". For example, $\phi = [\mathcal{H}^3 A]^{[0,10]} \cdot [\mathcal{H}^2 B]^{[0,5]}$ (two high level hold tasks) satisfies (3), but $\phi = [[\mathcal{H}^3 A]^{[0,10]} \cdot [\mathcal{H}^2 B]^{[0,5]}]^{[0,20]}$ (a high level task with two hold substasks) does not.

Example (cont). The DFA corresponding to the specification ϕ_{simple} is shown in the upper left corner of Fig. 1. In the center of the figure, we show the grown TS (represented by the dots and arrows) as well as the product with the DFA.

4.1   Algorithm description

In Alg. 1 we present the procedure to incrementally generate a TS from a description of the workspace and a specification. The algorithm takes as inputs the TWTL formula ϕ, and the initial state $x_0 \in W_{free}$. The description of the workspace is implicit in the functions that use it.

The procedure starts by initializing the needed data structures in line 1: the formula is translated to the DFA $\mathcal{A}_\infty$ and the TS $\mathcal{T}$ is constructed with just the initial state. The TS is then grown for a number of iterations (line 2).

A sample pair of a state to expand in the TS and a state in the workspace is generated in line 3. The state that will be considered to be added to the TS, x_{new}, is computed by the steering function $Steer$ in line 4. If the path to the new state is feasible (line 5), it is added to the TS (line 8). In order to select the best node to connect x_{new} to, we look at the set of nodes within steering distance in the same state of the DFA as the node to expand (line 6). Then, a node with best cost and a feasible path to x_{new} is chosen from that set (line 7). After adding the new node, other nodes within sterring distance that can be reached from x_{new}

Algorithm 1: Algorithm

Input: ϕ – TWTL formula in concatenation form, x_0 – initial point
Output: $\mathcal{T}$ – transition system

1 $\mathcal{A}_\infty \leftarrow Translate(\phi)$; $\mathcal{T} = (V, x_0, E, \Pi, h) \leftarrow (\{x_0\}, x_0, \emptyset, \Pi, \mathcal{L})$
2 **for** $i = 1, ..., n$ **do**
3 $\quad$ $x_{exp}, x_{ran} \leftarrow Sample()$
4 $\quad$ $x_{new} \leftarrow Steer(x_{exp}, x_{ran})$
5 $\quad$ **if** $ColFree(x_{exp}, x_{new})$ **then**
6 $\quad\quad$ $V_{near} \leftarrow \{x \in V : \delta^*_{\mathcal{A}_\infty}(\vec{o}) = \delta^*_{\mathcal{A}_\infty}(\vec{o}_{exp})), \|x - x_{new}\| \le d_{steer}\}$
7 $\quad\quad$ $x_{min} \leftarrow \mathrm{argmin}_{\{x \in V_{near}:ColFree(x, x_{new})\}}\{Cost_{\mathcal{A}_\infty}(\vec{x})\}$
8 $\quad\quad$ $V \leftarrow V \cup \{x_{new}\}$; $E \leftarrow E \cup \{(x_{min}, x_{new})\}$
9 $\quad\quad$ $V_{next} \leftarrow \{x \in V : \delta^*_{\mathcal{A}_\infty}(\vec{o}) \in \delta_{\mathcal{A}}(\delta^*_{\mathcal{A}}(\vec{o}_{exp}), o_{new}), \|x - x_{new}\| \le d_{steer}\}$
10 $\quad\quad$ **foreach** $x_{next} \in V_{next}$ **do**
11 $\quad\quad\quad$ **if** $Cost_{\mathcal{A}_\infty}(\vec{x}_{next}) > Cost_{\mathcal{A}_\infty}(\vec{x}_{new} \oplus x_{next}) \wedge ColFree(x_{new}, x_{next})$
$\quad\quad\quad$ **then**
12 $\quad\quad\quad\quad$ $E \leftarrow (E \setminus \{(Parent(x_{next}), x_{next})\}) \cup \{(x_{new}, x_{next})\}$

13 **return** $\mathcal{T}$

(regarding both state consistency in the DFA and path feasibility) are considered for rewiring. If the path through x_{new} has better cost than their current one, the tree is rewired (lines 9-12, see more details on rewiring in Sec. 4.2).

Some primitive functions are assumed to be available in Alg. 1. *Sample* will be discussed in Sec. 4.3. *Steer* computes a state close to x_{ran} within steering distance of x_{exp} and *ColFree* checks if the path between two points is free of obstacles; a more in depth discussion of both can be found in [22]. Let $x \in V$ be a state in $\mathcal{T}$. We denote by $\vec{x}$ and $\vec{o}$ the (unique) path in $\mathcal{T}$ from the initial state x_0 to x and its corresponding output word, respectively. The function $Parent : V \rightarrow V \cup \{\bowtie\}$ returns the parent of x in the TS, with $\bowtie = Parent(x_0)$. We denote by $\vec{x} \oplus x_{new}$ the path resulting from appending the node x_{new} to $\vec{x}$. The cost associated with $\vec{x}$ is given by $Cost_{\mathcal{A}_\infty}(\vec{x})$. See Sec. 4.2 for details.

Note that even though we describe the algorithm as building the incremental TS $\mathcal{T}$, it is easy to see that the product automaton $\mathcal{P}$ is also being incrementally built: it is only necessary to consider each state x of the TS as augmented with the corresponding DFA state, $\delta^*_{\mathcal{A}_\infty}(\vec{o})$. From an implementation point of view, computing the product automaton does not incur in a performance penalty, since the number of states is equal to that of the TS. In fact, the function $\delta^*_{\mathcal{A}_\infty}$ is now immediately available for all states of $\mathcal{T}$ and only the transition function $\delta_{\mathcal{A}_\infty}$ needs to be computed for each new sample. In the following, we assume the algorithm incrementally builds the product automaton $\mathcal{P}$ explicitly.

4.2   Cost function

Each path $\vec{x}_{L-1} = \{x_i\}_{i=0}^{L-1}$ in the TS has a cost represented by the function $Cost_{\mathcal{A}_\infty}$. We first look for the largest j such that $\vec{x}_{L-1} \models \phi^j(\tau)$, for some τ.

Let τ^* be the tightest τ-relaxation of ϕ^j that $\vec{x}_{L-1}$ satisfies, i.e., the one that minimizes $\left|\phi^j(\tau)\right|_{LRTR}$. Then, we obtain the shortest subpath of $\vec{x}_{L-1}$ that satisfies $\phi^j(\tau^*)$, $\vec{x}_{S-1} = \{x_i\}_{i=0}^{S-1}$. Finally, we can define the cost as follows:

$$Cost_{\mathcal{A}_\infty}(\vec{x}_{L-1}) = L - S + \left|\phi^j(\tau^*)\right|_{LRTR}. \tag{5}$$

It is immediate to see that finding a path that satisfies $\phi(\tau)$ with minimum cost provides a solution for Prob. 2. Note that the cost increases by 1 with each node added to the path that does not render a longer subformula ϕ^{j+1} true. Moreover, when a node is added such that ϕ^{j+1} is satisfied, we only need to consider the subpath associated with ϕ_{j+1} in order to obtain the new τ^*. This leads to the following definition of $Cost_{\mathcal{A}_\infty}$, equivalent to the previous one:

$$
\begin{aligned}
Cost_{\mathcal{A}_\infty}(\vec{x}_{L-1}) = \\
\begin{cases}
Cost_{\mathcal{A}_\infty}(\vec{x}_{S-1}) + \max\{Cost_{\mathcal{A}_\infty}(\vec{x}_{L-2}) - Cost_{\mathcal{A}_\infty}(\vec{x}_{S-1}) + 1 - b_j, 0\}, & \text{if } \mathfrak{P}(\vec{x}) \\
Cost_{\mathcal{A}_\infty}(\vec{x}_{L-2}) + 1, & \text{otherwise}
\end{cases} \\
\mathfrak{P}(\vec{x}) = \vec{x}_{L-1} \models \phi^j(\tau) \wedge \vec{x}_{L-2} \not\models \phi^j(\tau) \wedge \vec{x}_{S-1} \models \phi^{j-1}(\tau) \wedge \vec{x}_{S-2} \not\models \phi^{j-1}(\tau)
\end{aligned} \tag{6}
$$

In the equation above, b_j is the upper bound of the time window for subformula ϕ_j. This cost function is easy to implement by storing the cost of $\vec{x}$ in the node x and obtaining the cost of new nodes recursively using Eq. (6).

Alg. 1 tries to keep minimum cost paths between the nodes of the tree. In lines 9-12, a rewiring process is executed after the tree has been extended. This rewiring is structurally very similar to the one performed by the RRT* algorithm [16]. However, while RRT* considers nodes nearby the last added node for rewiring, we add a consistency condition so that only nodes corresponding to a successor state in the product automaton will be considered for rewiring (see line 9). Moreover, the cost function deviates from the usual continuous additive cost functions considered in RRT*. In Sec. 4.4, we discuss the implications of rewiring from a (probabilistic) completeness and optimality point of view.

4.3   Sampling

The *Sample* function generates a random state to add to the TS and selects the candidate node to connect it to. The usual sampling function for RRT-type of algorithms requires the samples to be from the free space, W_{free}. However, as was pointed out when describing the algorithm, not only the TS $\mathcal{T}$, which has states in W_{free}, is constructed, but also the product automaton $\mathcal{P} = \mathcal{T} \times \mathcal{A}_\infty$. Therefore, we also need to sample a state from the DFA's set of states $S_{\mathcal{A}_\infty}$. In Fig. 1 we explicitly show the product automaton in layers corresponding to the different states of the DFA to illustrate this idea.

The straightforward approach to sampling would proceed as follows: obtain a random sample $p_{ran} = (x_{ran}, s_{ran})$ from $W_{free} \times S_{\mathcal{A}_\infty}$. Then, find the nearest state p_{exp} in $\mathcal{P}$ that can be connected to the sample p_{ran} (see below). However, in order to simplify the computation, first we sample from $S_{\mathcal{A}_\infty}$ a DFA state

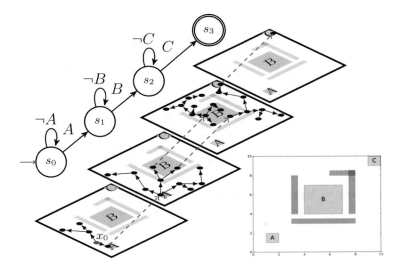

Fig. 1. A schematic representation of Alg. 1 applied to the example problem. Regions A, B and C are green, obstacles are red and the initial point x_0 is blue. The DFA corresponding to ϕ_{simple} is plotted in the upper left part. In each layer we plot in black the nodes of the product automaton that correspond to the accompanying DFA state. Black arrows represent transitions that do not change the DFA state, while magenta arrows indicate that the next node is in a new DFA state.

s_{exp} that restricts the product states we need to consider for extension. Then, a sample x_{ran} from W_{free} is obtained. Finally, p_{exp} is selected as before (with the DFA part now fixed) and s_{ran} is determined by $\delta_{\mathcal{A}_\infty}(\delta^*_{\mathcal{A}_\infty}(\overrightarrow{o}_{exp}), o_{ran})$. This procedure is sound in the sense that it does not generate "naive" random trees (i.e., biased towards the initial state, see [22]), since the DFA states are finite.

Algorithm 2: Sample

Input: $\mathcal{T} = (V, x_0, E, \Pi, h)$ – Current TS, ϕ – TWTL formula in concatenation form, τ – Best temporal relaxation found so far, $\mathcal{A}_\infty$ – The DFA associated with ϕ

Output: A state in $\mathcal{T}$ to extend from, and a random state to extend to

1 **if** $\text{Unif}([0,1]) < p_{bias}$ **then**
2 $\quad$ $k \leftarrow \text{argmax}_{i \in \mathfrak{I}_{\mathcal{A}_\infty}(\phi(\tau))}\{\tau_i\}$
3 $\quad$ $s_{ran} \leftarrow \text{Unif}(States_{\mathcal{A}_\infty}(\phi_k))$
4 **else**
5 $\quad$ $s_{ran} \leftarrow \text{Unif}(States_{\mathcal{A}_\infty}(\phi))$
6 $x_{ran} \leftarrow \text{Unif}(W_{free})$
7 $x_{exp} \leftarrow Nearest(\{x \in V : \delta^*_{\mathcal{A}_\infty}(\overrightarrow{o}) = s_{ran}\}, x_{ran})$
8 **return** (x_{exp}, x_{ran})

Intuitively, sampling the DFA state has the effect of choosing which layer, as shown in Fig. 1, we are exploring next. This observation allows us to develop heuristics that bias the sampling towards states that need to be better explored. In particular, we propose a sampling method in which the states associated with the subformula ϕ_i with largest τ_i are sampled with greater probability.

Our sampling procedure is presented in Alg. 2. The biased sampling described above is performed over a uniform sampling with probability p_{bias}. The $States_{\mathcal{A}_\infty}$ function returns the states in the DFA associated with a subformula; the Unif function chooses uniformly from a given set; and the $Nearest$ function returns the nearest point from a set to another point. The input parameter τ represents the best temporal relaxation (LRTR) for a path satisfying $\phi(\tau)$ so far, and it is important to keep it stored and updated throughout the execution of the algorithm in order to avoid unnecessary computation.

4.4   Completeness and optimality

We assume for our analysis an additional constraint for Prob. 1: the set U is discrete. Furthermore, the steering function is modified such that $Steer(x, y)$ returns the result of applying a random control input from U for one timestep from x. We start with an observation about the rewiring process:

Theorem 1. *Consider the graph $G = (V_G, E_G)$ obtained by running Alg. 1 for some number of iterations with line 12 deleted, as well as the tree $T = (V_T, E_T)$ obtained by running the unmodified algorithm. Then, $V_G = V_T$ and $\forall v \in V_G, Cost_{\mathcal{A}_\infty}(\overrightarrow{v}^G) \geq Cost_{\mathcal{A}_\infty}(\overrightarrow{v}^T)$, where $\overrightarrow{v}^T$ is the path from x_0 to v in T and $\overrightarrow{v}^G$ is any path from x_0 to v in G.*

Proof. Since the rewiring process only modifies the set of edges, the set of nodes remains the same. For the second property, note that line 11 ensures that the edges resulting in paths with a worse cost are removed.

Theorem 2 (Probabilistic Completeness). *If Prob. 1 with the discrete control input assumption has a solution, then Alg. 1 finds it with probability 1 as the number of iterations go to infinity.*

Proof. We sketch a proof closely following that in [22]. First, consider the algorithm without line 12. Suppose a solution path is $\mathbf{x} = \{x_k\}_{k=0}^{L}$. By induction on k, assume the graph contains a path until x_k for some k. Since the states of $\mathcal{A}_\infty$ are finite, with non-zero probability the state associated with the path from x_0 to x_k, $\delta^*_{\mathcal{A}_\infty}(\overrightarrow{\sigma}_k)$, will be selected for extension. Consider now the Voronoi diagram associated with the nodes in the selected state. With non-zero probability, a sample in the Voronoi cell associated with x_k will be obtained. Finally, with non-zero probability the correct control input that steers the system towards x_{k+1} will be selected. Therefore, the next point in the solution path will be constructed with probability tending to one. This process continues until the solution path is constructed in G. Since the path has cost 0, the path to its last node in the tree resulting from enabling rewiring has cost less than or equal to 0 by Theorem 1. Therefore, it is a solution path.

We now turn our attention to the behavior of the algorithm when faced with a scenario in which only a relaxation of the specification can be satisfied:

Theorem 3. *Suppose Prob. 1 has a solution for a τ-relaxation of the specification ϕ, $\phi(\tau)$, with $|\phi(\tau)|_{LRTR} = c$, and no τ-relaxation $\phi(\tau')$, with $|\phi(\tau')|_{LRTR} = c' < c$, yields a satisfiable specification. Then, if Alg. 1 is run for the original specification ϕ, it will find a path satisfying $\phi(\tau'')$, with $|\phi(\tau'')|_{LRTR} = c$, for some τ-relaxation $\phi(\tau'')$, with probability 1.*

Proof. Suppose the algorithm is run for the specification $\psi = \phi(\tau)$ and the rewiring process is disabled. The resulting graph has the same nodes as it would have if run for the specification ϕ. Moreover, a path with cost 0 with respect to ψ, or, equivalently, c with respect to ϕ, exists in the graph with probability 1, by Theorem 2. When rewiring is applied, the path to the final node will have less or equal cost by Theorem 1. However, by hypothesis, no path satisfying a relaxed formula exists with cost less than c. Therefore, a path with cost c is found with probability 1.

4.5 Complexity

Given the similarity of Alg. 1 with RRT*, it is not surprising that the computational complexity is similar. We analyze the number of calls to *ColFree* per iteration as well as the cost of some of the operations described in Sec. 4.1.

Consider an arbitrary ϕ, which gets translated into a DFA with set of states $\{s_i\}_{i=1}^{S}$. Note that *ColFree* is restricted to nodes in particular DFA states, given by the construction of the V_{near} and V_{next} sets, which is related to the sampled DFA state, s_{ran}. The number of considered states depends on certain features of the formula (such as the number of disjunction operands) that are not necessarily related to S, so we assume it fixed. Let p_i be the probability of a node being in one of those DFA states when s_i is sampled (which in general decreases when S increases, although its distribution depends on the structure of the formula and the workspace) and let P_S be its expected value. Then, the number of calls to *ColFree* in iteration n is $\mathcal{O}(P_S n)$.

We can proceed in a similar way in order to analyze the primitive procedures. Regarding the *ColFree* function itself, it can be executed in $\mathcal{O}(\log^d m)$, where d is the dimension of the space and m is the number of obstacles . The sets V_{near} and V_{next} computed in line 6 and line 9 respectively are also referred to as *range search problems* and can be approximately solved in $\mathcal{O}(\log P_S n + (1/\epsilon)^{d-1})$, where ϵ is a parameter controlling the precision. Finally, the optimization problem solved in line 7 to find x_{min} is an instance of the nearest neighbor search problem, which can be approximately solved in $\mathcal{O}(c_{d,\epsilon} \log P_S n)$, where $c_{d,\epsilon} \leq d\lceil 1 + 6d/\epsilon \rceil^d$. See [16] for a discussion on these results.

5   Case Studies

We implemented the algorithm in Python2.7 and we ran all examples on an Intel(R) Core(TM) i5-4690K CPU @ 3.50GHz with 8GB RAM. We compared

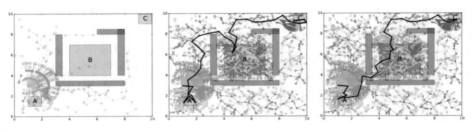

(a) TS after a few iterations (b) Path with $\tau^* = (0,3,0)$ (c) Satisfying path

Fig. 2. Evolution of Alg. 1 for case study 1. The regions of interest are colored in green and obstacles in red. The color of each node in the TS represents its associated DFA state. The current best path satisfying the relaxed formula is highlighted in bold black.

execution times with [35] using a grid with 20 divisions per dimension for the state space partition and skipping the computation of transitions between cells.

5.1 Case Study 1: Example revisited

We consider the following more complicated specification for the example: "Perform tasks at A, B, and C, in this order, of duration 2, 3, and 2 time units, within time intervals [3, 10], [0, 15], and [0, 15] from the end of the previous task, respectively." The corresponding TWTL formula is $\phi = [\mathcal{H}^2 A]^{[3,10]} \cdot [\mathcal{H}^3 B]^{[0,15]} \cdot [\mathcal{H}^2 C]^{[0,15]}$. The parameters of the algorithm are $d_{steer} = 0.75$ and $p_{bias} = 0.5$.

We show in Fig. 2 several snapshots of the state of the algorithm after some iterations. Note that the layers associated with each state in the DFA shown schematically in Fig. 1 can be identified in the figure by the different colors in nodes and edges. In Fig. 2a, we show the state of the algorithm when it has yet to find a path that satisfies any τ-relaxation of ϕ. At this point, the algorithm is exploring the state space in search of a candidate path. The cluster of nodes near x_0 is due to the sampling of states associated with delaying until the lower bound of the first time interval. The next snapshot, Fig. 2b, highlights a candidate path. After this iteration, the exploitation phase starts and the algorithm biases the sampling towards the subpath with worst deadline deviation (in this case, the subpath from A to B). If the algorithm is stopped at this iteration, it would return the highlighted path, which is a partial solution that violates the specification. We quantify the violation with the temporal relaxation. In Fig. 2c the candidate path was refined enough to finally satisfy ϕ. The two predominant colors in the figure, cyan and light magenta, correspond to the initial states of the second and third subformula respectively. The maximum, minimum and average times it took to solve the problem over 20 executions were 250, 7 and 61 seconds, respectively. We repeated the simulation with $p_{bias} = 0$ obtaining maximum, minimum and average times of 1766, 17 and 289 seconds respectively, which shows a performance decrease when language-guided sampling is

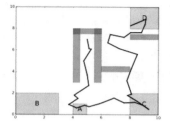

Fig. 3. Final path returned by the algorithm for the second case study, as seen when projected onto its first two components.

disabled. In comparison, the algorithm proposed in [35] is able to obtain a path in an average of 1 second.

5.2 Case Study 2

We consider a workspace in $\mathbb{R}^{10}$ with five obstacles and three regions of interest, A, B and C. The system dynamics are the same as in the previous example, with $||u_k|| \leq 2$ and initial point $x_0 = (5, 7, 3, \ldots, 3)$. The specification in this case is $\phi = [\mathcal{H}^1 A]^{[0,25]} \cdot ([\mathcal{H}^1 B]^{[0,15]} \vee [\mathcal{H}^1 C]^{[0,15]}) \cdot [\mathcal{H}^1 D]^{[0,35]}$, and we set $d_{steer} = 2$ and $p_{bias} = 0.5$. We show in Fig. 3 the satisfying path found by the algorithm. The execution time needed to solve an instance of the problem was 3834 seconds. In this case, the algorithm in [35] required too much memory to run.

6 Conclusion

In this paper, we introduced a sampling-based algorithm for solving motion planning problems under temporal logic specifications given as TWTL formulae. The algorithm initially finds a path that satisfies a temporally relaxed version of the specification. Then, sampling is biased towards the subpath that needs more improvement in order to satisfy the time bounds of the specification.

Our algorithm relies on the translation of TWTL formulae to annotated Deterministic Finite State Automata, a process recently developed in [35]. The design of the algorithm is inspired from RRT*, but differs in two major aspects. First, we incrementally construct the product between a Transition System, with states in the workspace, and the DFA. This allows us to not only grow a random tree in a similar way as RRT*, but to also keep track of our progress towards satisfying the specification. Second, we make use of a cost function related to the satisfaction of the formula that deviates from the usual metrics used by RRT*, like path length. We showed that for this cost function, not only our algorithm is probabilistically complete, but it can also obtain a "minimally violating" path in those cases were only a temporally relaxed version of the specification can be satisfied, again in a probabilistically complete way.

We implemented the algorithm in Python and we tested it for high dimensional problems. We obtained correct results at a moderately high computational cost, partially due to our naive implementation lacking the best known algorithms for solving nearest neighbors and range search problems.

As future work, we plan to extend the fragment of TWTL that we accept in order to allow specifications with "subtasks", which could be solved by recursive calls to the proposed algorithm. We also want to assess the performance when better algorithms for primitive operations are used.

Acknowledgments. This work was partially supported by the NSF under grant NRI-1426907 at Boston University.

References

[1] Aksaray, D., Vasile, C.I., Belta, C.: Dynamic Routing of Energy-Aware Vehicles with Temporal Logic Constraints. In: IEEE International Conference on Robotics and Automation. pp. 3141–3146. Stockholm, Sweden (May 2016)

[2] Aydin Gol, E., Lazar, M., Belta, C.: Language-Guided Controller Synthesis for Linear Systems. IEEE Trans. on Automatic Control 59(5), 1163–1176 (2014)

[3] Bacchus, F., Kabanza, F.: Planning for temporally extended goals. Annals of Mathematics and Artificial Intelligence 22(1-2), 5–27

[4] Baier, C., Katoen, J.P.: Principles of Model Checking. MIT Press (2008)

[5] Belta, C., Isler, V., Pappas, G.J.: Discrete abstractions for robot planning and control in polygonal environments. IEEE Trans. on Robotics 21(5), 864–874 (2005)

[6] Bhatia, A., Kavraki, L., Vardi, M.: Sampling-based motion planning with temporal goals. In: IEEE International Conference on Robotics and Automation (2010)

[7] Canny, J.F.: The Complexity of Robot Motion Planning. MIT Press, USA (1988)

[8] Chen, Y., Tumova, J., Belta, C.: LTL Robot Motion Control based on Automata Learning of Environmental Dynamics. In: IEEE International Conference on Robotics and Automation. Saint Paul, MN, USA (2012)

[9] Choset, H., Lynch, K., et al.: Principles of Robot Motion: Theory, Algorithms, and Implementations. MIT Press, Boston, MA (2005)

[10] Ding, X.C., Kloetzer, M., et al.: Formal Methods for Automatic Deployment of Robotic Teams. IEEE Robotics and Automation Magazine 18, 75–86 (2011)

[11] Doherty, P., Kvarnström, J., Heintz, F.: A temporal logic-based planning and execution monitoring framework for unmanned aircraft systems. Autonomous Agents and Multi-Agent Systems 19(3), 332–377 (Feb 2009)

[12] Gol, E.A., Belta, C.: Time-Constrained Temporal Logic Control of Multi-Affine Systems. Nonlinear Analysis: Hybrid Systems 10, 21–23 (2013)

[13] Guo, M., Dimarogonas, D.V.: Multi-agent plan reconfiguration under local LTL specifications. International Journal of Robotics Research 34(2), 218–235 (2015)

[14] Jha, S.K., Clarke, E.M., et al.: A bayesian approach to model checking biological systems. In: Computational Methods in Systems Biology. Springer-Verlag (2009)

[15] Karaman, S., Frazzoli, E.: Sampling-based Motion Planning with Deterministic μ-Calculus Specifications. In: IEEE Conference on Decision and Control (2009)

[16] Karaman, S., Frazzoli, E.: Sampling-based Algorithms for Optimal Motion Planning. International Journal of Robotics Research 30(7), 846–894 (June 2011)

[17] Karaman, S., Frazzoli, E.: Sampling-based Optimal Motion Planning with Deterministic μ-Calculus Specifications. In: American Control Conference (2012)

[18] Kavraki, L., Svestka, P., et al.: Probabilistic roadmaps for path planning in high-dimensional configuration spaces. IEEE Transactions on Robotics and Automation 12(4), 566–580 (1996)

[19] Koymans, R.: Specifying real-time properties with metric temporal logic. Real-time systems 2(4), 255–299 (1990)

[20] Kress-Gazit, H., Fainekos, G.E., Pappas, G.J.: Where's Waldo? Sensor-based temporal logic motion planning. In: IEEE International Conference on Robotics and Automation. pp. 3116–3121 (2007)

[21] Lahijanian, M., Almagor, S., et al.: This Time the Robot Settles for a Cost: A Quantitative Approach to Temporal Logic Planning with Partial Satisfaction. In: AAAI Conference on Artificial Intelligence. pp. 3664–3671. Austin, Texas (2015)

[22] LaValle, S.M., Kuffner, J.J.: Randomized Kinodynamic Planning. International Journal of Robotics Research 20(5), 378–400

[23] Luo, R., Valenzano, R.A., et al.: Using Metric Temporal Logic to Specify Scheduling Problems. In: Principles of Knowledge Representation and Reasoning (2016)

[24] Maler, O., Nickovic, D.: Monitoring temporal properties of continuous signals. In: Formal Techniques, Modelling and Analysis of Timed and Fault-Tolerant Systems, pp. 152–166. Springer (2004)

[25] Maly, M., Lahijanian, M., et al.: Iterative Temporal Motion Planning for Hybrid Systems in Partially Unknown Environments. In: Int. Conference on Hybrid Systems: Computation and Control (2013)

[26] Nakagawa, S., Hasuo, I.: Near-Optimal Scheduling for LTL with Future Discounting. In: Ganty, P., Loreti, M. (eds.) Trustworthy Global Computing, pp. 112–130. No. 9533 in Lecture Notes in Computer Science, Springer (2015)

[27] Pavone, M., Bisnik, N., et al.: A stochastic and dynamic vehicle routing problem with time windows and customer impatience. Mobile Networks and Applications 14(3), 350–364 (2009)

[28] Plaku, E., Kavraki, L.E., Vardi, M.Y.: Motion Planning with Dynamics by a Synergistic Combination of Layers of Planning. IEEE Trans. on Robotics 26(3), 469–482 (2010)

[29] Reyes Castro, L., Chaudhari, P., et al.: Incremental sampling-based algorithm for minimum-violation motion planning. In: IEEE Conference on Decision and Control. pp. 3217–3224 (2013)

[30] Solomon, M.M.: Algorithms for the vehicle routing and scheduling problems with time window constraints. Operations research 35(2), 254–265 (1987)

[31] Tkachev, I., Abate, A.: Formula-free Finite Abstractions for Linear Temporal Verification of Stochastic Hybrid Systems. In: Int. Conference on Hybrid Systems: Computation and Control. Philadelphia, PA (2013)

[32] Tumova, J., Marzinotto, A., et al.: Maximally satisfying LTL action planning. In: IEEE/RSJ Int. Conf. on Intelligent Robots and Systems. pp. 1503–1510 (2014)

[33] Tumova, J., Hall, G.C., et al.: Least-violating Control Strategy Synthesis with Safety Rules. In: Hybrid Systems: Computation and Control. pp. 1–10 (2013)

[34] Vasile, C., Belta, C.: Sampling-Based Temporal Logic Path Planning. In: IEEE/RSJ International Conference on Intelligent Robots and Systems (2013)

[35] Vasile, C.I., Aksaray, D., Belta, C.: Time Window Temporal Logic. Theoretical Computer Science p. (submitted), http://arxiv.org/abs/1602.04294

[36] Vasile, C.I., Belta, C.: An Automata-Theoretic Approach to the Vehicle Routing Problem. In: Robotics: Science and Systems Conference. USA (2014)

[37] Wolff, E.M., Topcu, U., Murray, R.M.: Automaton-guided controller synthesis for nonlinear systems with temporal logic. In: IEEE/RSJ International Conference on Intelligent Robots and Systems. pp. 4332–4339 (2013)

[38] Wongpiromsarn, T., Topcu, U., Murray, R.M.: Receding Horizon Temporal Logic Planning for Dynamical Systems. In: Conference on Decision and Control (2009)

[39] Yoo, C., Fitch, R., Sukkarieh, S.: Probabilistic temporal logic for motion planning with resource threshold constraints (2012)

Generating Plans that Predict Themselves

Jaime F. Fisac[1]⋆, Chang Liu[2]⋆, Jessica B. Hamrick[3]⋆,
Shankar Sastry[1], J. Karl Hedrick[2], Thomas L. Griffiths[3], Anca D. Dragan[1]

[1] Department of Electrical Engineering and Computer Sciences
[2] Department of Mechanical Engineering
[3] Department of Psychology
University of California, Berkeley, CA 94720 U.S.A.
{jfisac,changliu,jhamrick,shankar_sastry,tom_griffiths,anca}
@berkeley.edu

Abstract. Collaboration requires coordination, and we coordinate by anticipating our teammates' future actions and adapting to their plan. In some cases, our teammates' actions early on can give us a clear idea of what the remainder of their plan is, i.e. what action sequence we should expect. In others, they might leave us less confident, or even lead us to the wrong conclusion. Our goal is for robot actions to fall in the first category: we want to enable robots to select their actions in such a way that human collaborators can easily use them to correctly anticipate what will follow. While previous work has focused on finding initial plans that convey a set goal, here we focus on finding two portions of a plan such that the initial portion conveys the final one. We introduce t-predictability: a measure that quantifies the accuracy and confidence with which human observers can predict the remaining robot plan from the overall task goal and the observed initial t actions in the plan. We contribute a method for generating t-predictable plans: we search for a full plan that accomplishes the task, but in which the first t actions make it as easy as possible to infer the remaining ones. The result is often different from the most efficient plan, in which the initial actions might leave a lot of ambiguity as to how the task will be completed. Through an online experiment and an in-person user study with physical robots, we find that our approach outperforms a traditional efficiency-based planner in objective and subjective collaboration metrics.

1 Introduction

With robots stepping out of industrial cages and into mixed workspaces shared with human beings, human-robot collaboration is becoming more and more important [4, 8, 18]. There is unfortunately a history of serious and sometimes tragic failures in human-automation systems due to inadequate interaction between machines and their operators [12, 16, 17]. The most common reasons for this are mode confusion and "automation surprises", i.e. *misalignments* between what the automated agent is planning to do and what the human believes it is planning to do.

⋆ These authors contributed equally.

© Springer Nature Switzerland AG 2020
K. Goldberg et al. (Eds.): *Algorithmic Foundations of Robotics XII*, SPAR 13, pp. 144–159, 2020.
https://doi.org/10.1007/978-3-030-43089-4_10

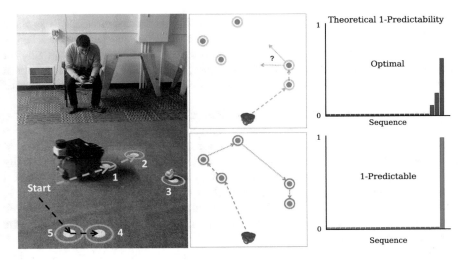

Fig. 1. Left: A participant controls the toy robot to clear the target that he believes will be third in the autonomous robot's sequence. Center: Schematic of the trajectory followed by the optimal (top) and 1-predictable (bottom) planner. Right: Theoretical 1-predictability of candidate sequences after the robot has visited a single target.

Our goal is to eliminate such misalignments: we want humans to be able to infer what a robot is planning to do during a collaborative task. This is important even beyond safety reasons [1], because it enables the human to adapt their own actions to the robot and more effectively achieve common goals [14, 21, 22].

Traditionally, human-robot collaboration work has focused on inferring human plans and adapting the robot's plan in response [11, 13]. In contrast, here we are interested in the *opposite*—making sure that humans can make these inferences about the robot. We envision that ultimately these two approaches will need to work in conjunction to achieve fluent collaboration. Though it is possible for robots to explicitly communicate their plans via language or text, we focus here on what the beginning of the plan itself implies because 1) people make inferences based on actions [2], and 2) certain scenarios such as the outdoors or busy factories might render explicit channels undesirable.

We introduce a property of a robot plan that we call *t-predictability*: a plan is *t*-predictable if a human can infer the robot's remaining actions in a task from having observed only the first t actions and from knowing the robot's overall goal in the task. We make the following contributions based on this property:

An algorithm for generating t-predictable plans. To generate robot plans that are t-predictable, we introduce a model for how humans might infer future plans, building on models for action interpretation and plan recognition [2, 3]. We then propose a planning algorithm that optimizes for t-predictability, i.e. it optimizes plans for how easy it will be for a human to infer the last actions from the initial ones.

Prior work has focused on legibility: generating trajectories that communicate a *set goal* via the initial robot *motion* [6, 9, 19, 20]. This is less useful in task planning situations, where the human already knows the task goal. Instead, t-predictability is about communicating the *sequence of future actions* that the robot will take given a *known goal*. The important difference is that these actions are not set *a priori*: optimizing for t-predictability means *changing not just the initial, but also the final part* of a plan. It is about finding a final part that is easy to communicate, along with an initial part that communicates it.

Our insight is that initial actions can be used to clarify what future actions will be. We find that in many situations, the robot can select initial actions that might seem somewhat surprising at first, but that make the remaining sequence of actions trivial to anticipate (or "auto-complete"). Fig. 1 shows an example. If the robot acts optimally for the shortest path through all targets, it will go to target 5 first, making target 4 an obvious candidate as a next step, but leaving the future ordering of targets 1, 2, 3 somewhat ambiguous (high probability on multiple sequences). On the other hand, if the robot instead chooses target 1, users can with high probability predict that the remaining sequence will be 2-3-4-5.

An online user study testing that we can generate t-predictable plans. We conduct an online user study in which participants observe a robot's partial plan, and anticipate the order of the remaining actions. We find that participants are significantly better at anticipating the correct order when the robot is planning for t-predictability. We also find a $r = 0.87$ correlation between our model's prediction of the probability of success, and the participants' actual success rate.

An in-person user study testing the implications on human-robot collaboration. Armed with evidence that we can make plans t-predictable, we move on to an in-person study that puts participants in a collaborative task with the robot, and study the advantages of t-predictability on objective and subjective collaboration metrics. We find that participants were more effective at the task, and prefer to work with a t-predictable robot than with an optimal robot.

2   Defining and Optimizing for t-Predictability

t-**Predictability.** We consider a task planning problem from a starting state $S \in \mathcal{S}$ with an overall goal $G \in \mathcal{G}$ that can be achieved through a series of actions, called a *plan*, within a finite horizon T. Let $\mathcal{A}$ denote the space of all feasible plans of length (up to) T that achieve the goal.

Definition 1. *The t-predictability $\mathcal{P}_t$ of a feasible plan* $\mathbf{a} = [a_1, a_2, ..., a_T]$ *that achieves an overall goal G is the probability of an observer correctly inferring* $[a_{t+1}, ..., a_T]$ *after observing* $[a_1, ..., a_t]$, *and knowing the overall goal G. Specifically, this is given by* $\mathcal{P}_t(\mathbf{a}) = P(a_{t+1}, .., a_T | S, G, a_1, .., a_t)$.

Definition 2. *A t-predictable planner generates the plan that maximizes t-predictability out of all those that achieve the overall goal G. That is, a t-predictable planner generates the action series* $\mathbf{a}^*$ *such that* $\mathbf{a}^* = \arg\max_{\mathbf{a} \in \mathcal{A}} \mathcal{P}_t(\mathbf{a})$.

This is equivalent, by the general product rule, to:

$$\mathbf{a}^* = \arg\max_{\mathbf{a} \in \mathcal{A}} \frac{P(a_1, ..., a_T | S, G)}{\sum_{[\tilde{a}_{t+1}, ..., \tilde{a}_T]} P(a_1, ..., a_t, \tilde{a}_{t+1}, ..., \tilde{a}_T | S, G)}. \tag{1}$$

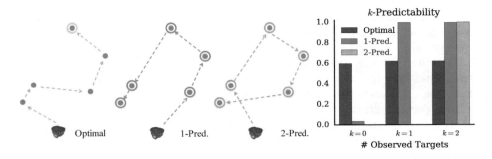

Fig. 2. Theoretical t-predictability. Left: sequences generated by the three planners for a typical task layout. Right: theoretical k-predictability of the same three sequences under different numbers k of observed targets. In all cases, the highest value corresponds to $t = k$.

Illustrative Example. Fig. 2 shows the outcome of optimizing for t-predictability in a Traveling Salesman context, with $t = 0$, 1, and 2 targets, and considers the theoretical k-predictability for each plan, with k the number of *actually* observed targets (which may be different from the t assumed by the planner). The 0-predictable plan (gray, left) is the best when the observer sees no actions, since it is the optimal plan. However, it is no longer the best plan when the observer gets to see the first action: whereas there are multiple low-cost remaining sequences after the first action in the 0-predictable plan, there is only one low-cost remaining sequence after the first action in the 1-predictable plan (blue, center). The first action in the 2-predictable (orange, right) seems irrational, but this plan is optimized for when the observer gets to see the first two actions: indeed, after the first two actions, the remaining plan is maximally clear.

Relation to Predictability. t-predictability generalizes predictability [5]. For $t = 0$, the t-predictability of a plan simply becomes its predictability, that is, the ease with which the entire sequence of actions can be inferred with knowledge of the overall goal G, i.e. $\mathcal{P}_0 = P(a_1, .., a_T | S, G)$.

Relation to Legibility. Legibility [6] as applied to task planning would maximize the probability of the goal given the beginning of the plan, i.e. $P(G | S, a_1, .., a_t)$. In contrast, with t-predictability the robot is given a high-level goal describing some state that the world needs to be brought into (for example, clearing all objects from a table), and the observer is *already* aware of this goal. Instead of communicating the goal, the robot conveys the remainder of the plan using the first few elements, maximizing $P(a_{t+1}, .., a_T | S, G, a_1, .., a_t)$.

One important implication is that for t-predictability, unlike for legibility, *there is no a-priori set entity to be conveyed*. The algorithm searches for *both* a beginning and a remainder of a plan such that, by observing the former, the observer can correctly guess the latter.

Furthermore, legibility and t-predictability entail a different kind of information encoding: in legibility, the robot uses a partial trajectory or action sequence to indicate a single goal state, whereas in t-predictability the robot uses a partial action sequence to indicate *another* action sequence. Therefore, one entails a mapping from a large space to a small set of possibilities (the finite candidate goal states), whereas the other entails a mapping between spaces of equivalent size.

The distinction between task-level legibility and t-predictability is crucially important, particularly in collaborative settings. If you are cooking dinner with your household robot, it is important for the robot to act legibly so you can infer *what* goal it has when it grabs a knife (e.g., to slice vegetables). But, it is equally important for the robot to act in a t-predictable manner so that you can predict *how* it will accomplish that goal (e.g., the order in which it will cut the vegetables).

Relation to Explicability. Explicability [23] has been recently introduced to measure whether the observer could assign labels to a plan. In this context, explicability would measure the existence of any remainder of a plan that achieves the goal, as opposed to optimizing the probability that the observer will infer the robot's plan.

Boltzmann Noisy Rationality. Computing t-predictability entails computing the conditional probability from (1). We model the human as expecting the robot to be noisily optimal, taking approximately the optimal sequence of actions to achieve G. Boltzmann probabilistic models of such noisy optimality (also known as the Luce-Shepard choice rule in cognitive science) have been used in the context of *goal* inference through inverse action planning [2]. We adopt an analogous approach for modeling the inference of *task plans*.

We define optimality via some cost function $c : \mathcal{A} \times \mathcal{S} \times \mathcal{G} \rightarrow \mathbb{R}^+$, mapping each feasible plan, from a starting state and for a particular goal, to a scalar cost. In our experiment, for instance, we use path length (travel distance) for c. Applying a Bolzmann policy [2] based on c, we get:

$$P(\mathbf{a}|S, G) = \frac{e^{-\beta c(\mathbf{a}, S, G)}}{\sum_{\tilde{\mathbf{a}} \in \mathcal{A}} e^{-\beta c(\tilde{\mathbf{a}}, S, G)}}. \tag{2}$$

Here $\beta > 0$ is termed the *rationality coefficient*. As $\beta \rightarrow \infty$ the probability distribution converges to one for the optimal sequence and zero elsewhere; that is, the human models the agent as rational. As $\beta \rightarrow 0$, the probability distribution becomes uniform over all possible sequences $\mathbf{a}$ and the human models the agent as indifferent.

t-**Predictability Optimization.** We make the assumption that cost is linearly separable, i.e. $c(\mathbf{a}, S, G) = \sum c(a_t, S_{\mathbf{a}}^t, G)$. Incorporating (2), (1) becomes:

$$\mathbf{a}^* = \arg \max_{\mathbf{a} \in \mathcal{A}} \frac{\exp\left(-\beta c(\mathbf{a}_{t+1:T}, S_{\mathbf{a}}^t, G)\right)}{\sum_{\tilde{\mathbf{a}}_{t+1:T} \in \mathcal{A}_{\mathbf{a}}^t} \exp\left(-\beta c(\tilde{\mathbf{a}}_{t+1:T}, S_{\mathbf{a}}^t, G)\right)}, \tag{3}$$

with $S_{\mathbf{a}}^t$ denoting the state reached by executing the first t steps of plan $\mathbf{a}$, and $\mathcal{A}_{\mathbf{a}}^t$ denoting the set of all feasible plans that achieve G from state $S_{\mathbf{a}}^t$ in $T - t$ steps or less.

Approximate Algorithm for Large-Scale Optimization. The challenge with the optimization in (3) is the denominator: it requires summing over all possible plan remainders. Motivated by the fact that plans with higher costs contribute exponentially less to the sum, we propose to approximate the denominator by only summing over the lowest-cost l plan remainders.

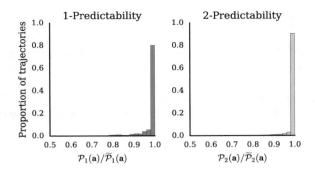

Fig. 3. Approximate t-predictability. Each subplot shows a histogram of ratios between exact t-predictability ($\mathcal{P}_t$) and approximate t-predictability ($\widetilde{\mathcal{P}}_t$) of all possible sequences across 270 unique layouts. The subplots show histograms of ratios for 1- and 2-predictability, respectively. In the majority of cases, the exact and approximate t-predictabilities are nearly identical.

Many tasks have the structure of Traveling Salesman Problems (TSP), where there are a number of subgoals whose order is not constrained but influences total cost. Van der Poort et al. [15] showed how to efficiently compute the l best solutions to the standard (cyclic tour) TSP using a branch-and-bound algorithm. The key mechanism is successively dividing the set of feasible plans into smaller subsets for which a lower bound on the cost can be computed by some heuristic. When a subset of solutions has a lower bound higher than the smallest l costs evaluated so far, it is discarded entirely, while the remaining subsets continue to be broken up. The process continues until only l feasible plans remain. This method is guaranteed to produce the l solutions with the least cost and can significantly reduce time complexity over exhaustive enumeration. In particular, it has been shown that for the standard TSP, computation is in $O(l(T - t)^3 2^{T-t})$ [15], while exhaustive enumeration requires computation in $O((T - t)!)$. While heuristics are domain-specific, we expect that this method can be widely applicable to robot task planning problems. Further, we expect $T - t$ to be a small number in realistic applications, limited by people's ability to reason about long sequences of actions.

To empirically evaluate the consequences of this approximation of t-predictability, we computed the exact and approximate (using $l = 2$) t-predictability for all possible plans in 270 randomly generated unique scenes (Fig. 3). If we choose the maximally t-predictable sequences for each scene using both the exact and approximate calculations of t-predictability, we find that these sequences agree in 242 (out of 270) scenes for 1-predictability and in 263 for 2-predictability[4]. For the sequences that disagree, the exact t-predictability of the sequence chosen using the approximate method is 89.5% of the optimal t-predictability in the worst case, and 99% of the optimal t-predictability on average. This shows that the proposed approximation is highly effective at producing t-predictable plans.

---

[4] For 0-predictability, the denominator is the same in all plans, so all 270 scenes agree trivially.

3   Online Experiment

We set up an experiment to test that our t-predictable planner is in fact t-predictable. We designed a web-based virtual human-robot collaboration experiment in a TSP setting, where the human had to predict the behavior of three robot avatars using different planners. Participants watched the robots move to a number of targets (either zero, one, or two) and had to predict the sequence of remaining targets the robot would complete.

3.1   Independent Variables

We manipulated two variables: the t-predictable planner (for $t \in \{0, 1, 2\}$) and the number of observed targets k (for $k \in \{0, 1, 2\}$).

Planner. We used three different planners which differed in their optimization criterion: the number t of targets assumed known to the observer. Each participant interacted with three robot avatars, each using one of the following three planners:

Optimal (0-predictable): This robot chooses the shortest path from the initial location visiting all target locations once; that is, the "traditional" solution to the open TSP. This robot solves (3) for $t = 0$.

1-predictable: This robot solves (3) for $t = 1$; the sequence might make an inefficient choice for the first target in order to make the sequence of remaining targets clear.

2-predictable: This robot solves (3) for $t = 2$; the sequence might make an inefficient choice for the first *two* targets in order to make the sequence of remaining targets clear.

Number of observed targets. Each subject was shown the first $k \in \{0, 1, 2\}$ targets of the robot's chosen sequence in each trial and was asked to predict the remainder of the sequence. This variable was manipulated between participants; thus, a given participant always saw the same number k of initial targets on all trials.

3.2   Procedure

The experiment was divided into two phases: a training phase to familiarize participants with TSPs and how to solve them, and an experimental phase. We additionally asked participants to fill out a survey at the end of the experiment.

In the training phase, subjects controlled a human avatar. They were instructed to click on targets in the order that they believed would result in the quickest path for the human avatar to visit all of them. The human avatar moved in a straight line after each click and "captured" the selected target, which was then removed from the display.

For the second phase of the experiment, participants saw a robot avatar move to either $k = 0$, $k = 1$, or $k = 2$ targets. After moving to these targets, the robot paused so that participants could predict the remaining sequence of targets by clicking on the targets in the order in which they believed the robot would complete them. Afterwards, participants were presented with an animation showing the robot moving to the rest of the targets in the sequence determined by the corresponding planner.

Stimuli. Each target layout displayed a square domain with five or six targets. There were a total of 60 trials, consisting of four repetitions of 15 unique target layouts in random order: one for the training phase, in addition to the three experimental conditions.

The trials were grouped so that each participant observed the same robot for three trials in a row before switching to a different robot. In the training trials, the avatar was a gender-neutral cartoon of a person on a scooter, and the robot avatars were images of the same robot in different poses and colors (either red, blue, or yellow).

Layout Generation. The layouts for the 15 trials were based from an initial database of 270 randomly generated layouts. This number was reduced down to 176 in which the chosen sequence was different between all three planners so that the stimuli were distinguishable. We also discarded some scenarios in which the robot's trajectory approached a target without capturing it, to avoid confounds. Out of these valid layouts, we chose the ones with the highest theoretical gain in 1-predictability to 2-predictability, to avoid scenarios where the information gain was marginal.

Attention Checks. After reading the instructions, participants were given an attention check in the form of two questions asking them the color of the targets and the color of the robot that they *would not* be evaluating.

Controlling for Confounds. We controlled for confounds by counterbalancing the colors of the robots for each planner; by using a human avatar in the practice trials; by randomizing the trial order; and by including attention checks.

3.3   Dependent Measures

Objective measures. We recorded the proportion of correct predictions of the robot's sequence of targets out of all 15 trials for each planner, resulting in a measure of *error rate*. We additionally computed the *Levenshtein distance* between predicted and actual sequences of targets. This is a more precise measure of how similar participants' predictions were to the actual sequences produced by the planner.

Subjective measures. After every ninth trial of the experiment, we asked participants to indicate which robot they preferred working with. At the end the experiment, each participant was also asked to complete a questionnaire to evaluate their perceived performance of three robots. An informal analysis of this questionnaire suggested similar results as those obtained from our other measures (see Section 3.6). Thus, because of space constraints, we have omitted specifics of the survey in this paper.

3.4   Hypotheses

H1 - Comparison with Optimal. *When showing 1 target, the 1-predictable robot will result in lower error than the optimal baseline. When showing 2 targets, the 2-predictable robot will result in lower error than the optimal baseline.*

H2 - Generalization. *The error rate will be lowest when $t = k$: the number of targets shown, k, equals the number of targets assumed by the t-predictable planner, t.*

H3 - Preference. *The perceived performance of the robots will be highest when $t = k$.*

3.5   Participants

We recruited a total of 242 participants from Amazon's Mechanical Turk using the psiTurk experimental framework [10]. We excluded 42 participants from analysis for

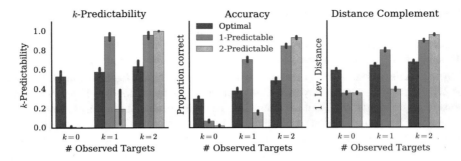

Fig. 4. Predictability, error rate and edit distance. Left: theoretical k-predictability of sequences generated by different t-predictable planners under different numbers k of observed targets, averaged over all task layouts used in the online experiment. In all cases, the highest value corresponds to $t = k$. Center: empirical proportion of correct predictions with different t-predictable planners for different numbers k of observed targets. Right: complement of the average empirical Levenshtein distance between predicted and correct sequences. The lowest experimental error rates under both metrics occur when $t = k$.

failing the attention checks, leaving a net total of $N = 200$ participants. All participants were treated in accordance with local IRB standards and were paid $1.80 USD for an average of 22 minutes of work, plus an average performance-based bonus of $0.47.

3.6 Results

Model validity. We first looked at the validity of our model of t-predictability with respect to people's performance in the experiment. We computed the theoretical k-predictability (probability of correctly predicting the robot's sequence from the k targets the user saw) for each task layout under each planner and number of targets the users observed. We also computed people's actual prediction accuracy on each of these layouts under each condition, averaged across participants.

We computed the Pearson correlation between k-predictability and participant accuracy, finding a correlation of $r = 0.87$, 95% CI $[0.81, 0.91]$; the confidence interval around the median was computed using 10,000 bootstrap samples (with replacement). *This high correlation suggests that our model of how people predict action sequences of other agents is a good predictor of their actual behavior.*

Accuracy. To determine how similar people's predictions of the robots' sequences were to the actual sequences, we used two objective measures of accuracy: first, overall error rate (whether they predicted the correct sequence or not), as well as the Levenshtein distance between the predicted and correct sequences (Fig. 4).

As the two measures have qualitatively similar patterns of results, and the Levenshtein distance is a more fine-grained measure of accuracy, we performed quantitative analysis only on the Levenshtein distance. We constructed a linear mixed-effects model with the number of observed targets k (k from 0 to 2) and the planner for t-predictability (t from 0 to 2) as fixed effects, and trial layout as random effects.

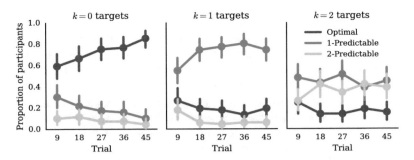

Fig. 5. Preferences over time. Participants prefer the 0-predictable (optimal) robot for $k = 0$ and the 1-predictable robot for $k = 1$, as well as $k = 2$ (despite performing better with the 2-predictable robot, subjects often report being confused or frustrated by its first 2 actions.)

This model revealed significant main effects of the number of observed targets ($F(2, 10299) = 1894.75$, $p < 0.001$) and planner ($F(2, 42) = 6.59$, $p < 0.01$) as well as an interaction between the two ($F(4, 10299) = 554.00$, $p < 0.001$). We ran post-hoc comparisons using the multivariate t adjustment. Comparing the planners across the same number of targets, we found that in the 0-targets condition the optimal (or 0-predictable) robot was better than the other two robots; in the 1-target condition, the 1-predictable robot was better than the other two; in the 2-target prediction, the 2-predictable robot was better than the optimal and 1-predictable robots. All differences were significant with $p < 0.001$ except the difference between the 2-predictable robot and the 1-predictable robot in the 2-target condition ($t(50) = 1.85$, $p = 0.56$). Comparing the performance of a planner across number of targets, we found significant differences in all contrasts, with one exception: the accuracy when using the optimal planner was not significantly different when seeing 1 target vs 2 targets ($t(10299) = 2.65$, $p = 0.11$). Overall, these results support our hypotheses **H1** and **H2**, that *accuracy is highest when t used in the planner equals k, the number of observed targets*.

Preferences over time. Fig. 5 shows the proportion of participants choosing each robot planner at each trial. We constructed a logistic mixed-effects model for binary preferences (where 1 meant the robot was chosen) with planner, number of observed targets, and trial as fixed effects and participants as random effects. The planner and number of observed targets were categorical variables, while trial was a numeric variable.

Using Wald's tests, we found a significant main effect of the planner ($\chi^2(2) = 13.66$, $p < 0.01$) and trial ($\chi^2(1) = 9.30$, $p < 0.01$). We detected only a marginal effect of number of targets ($\chi^2(2) = 4.67$, $p = 0.10$). However, there was a significant interaction between planner and number of targets ($\chi^2(4) = 20.26$, $p < 0.001$). We also found interactions between planner and trial ($\chi^2(2) = 24.68$, $p < 0.001$) and between number of targets and trial ($\chi^2(2) = 16.07$, $p < 0.001$), as well as a three-way interaction ($\chi^2(4) = 39.43$, $p < 0.001$). Post-hoc comparisons with the multivariate t adjustment for p-values indicated that for the 0-targets condition, the optimal robot was preferred over the 1-predictable robot ($z = -13.22$, $p < 0.001$) and the 2-predictable robot ($z = -14.56$, $p < 0.001$). For the 1-target condition, the 1-predictable robot was preferred over the optimal robot ($z = 12.97$, $p < 0.001$) and the 2-predictable robot ($z = 14.00$, $p < 0.001$). In the two-task condition, we did not detect a difference between the two 1-predictable and 2-predictable robots ($z = 2.26$, $p = 0.29$), though both were preferred over the optimal robot ($z = 7.44$, $p < 0.001$ for the 1-predictable robot and $z = 5.40$, $p < 0.001$ for the 2-predictable robot).

Overall, these results are in line with our hypothesis **H3** that *the perceived performance is highest when t used in the planner equals k, the number of observed targets.* This is the case for $k = 0$ and $k = 1$, but not $k = 2$: even though users tended to perform better with the 2-predictable robot, its suboptimal actions in the beginning seemed to confuse and frustrate users (see Qualtitative feedback results for details).

Final rankings. The final rankings of "best robot" and "worst robot" are shown in Fig. 6. For each participant, we assigned each robot a score based on their final rankings. The best robot received a score of 1; the worst robot received a score of 2; and the remaining robot received a score of 1.5. We constructed a logistic mixed-effects model for these scores, with planner and number of observed targets as fixed effects, and participants as random effects; we then used Walds tests to check for effects.

We found significant main effects of planner ($\chi^2(2) = 41.38, p < 0.001$) and number of targets ($\chi^2(2) = 12.97, p < 0.01$), as well as an interaction between them ($\chi^2(4) = 88.52, p < 0.001$). We again performed post-hoc comparisons using the multivariate t adjustment. These comparisons indicated that in the 1-target condition, people preferred the optimal robot over the 1-predictable robot ($z = 3.46$, $p < 0.01$) and the 2-predictable robot ($z = 5.60$, $p < 0.001$). In the 1-target condition, there was a preference for the 1-predictable robot over the optimal robot, however this difference was not significant ($z = -2.18$, $p = 0.27$). The 1-predictable robot was preferred to the 2-predictable robot ($z = -6.54$, $p < 0.001$). In the 2-target condition, both the 1-predictable and 2-predictable robots were preferred over the optimal robot ($z = -4.85$, $p < 0.001$ for the 1-predictable robot, and $z = -3.85$, $p < 0.01$ for the 2-predictable robot), though we did not detect a difference between the the 1-predictable and 2-predictable robots themselves ($z = -1.33$, $p = 0.84$). Overall, these rankings are in line with the preferences over time.

Qualitative feedback. At the end of the experiment, we asked participants to briefly comment on each robot. For $k = 0$, responses typically favored the optimal robot, often described as "efficient" and "logical", although they also showed some reservations: *"close to what I would do but just a little bit of weird choices tossed in"*. Conversely, for $k > 0$, the optimal robot was likened to "a dysfunctional computer", and described as "ineffective" or "very robotic": *"I feel like maybe I'm a dumb human and the [optimal] robot might be the most efficient, because I have no idea. It frustrated me."*

The 2-predictable robot had mixed reviews for $k = 2$: for some it was "easy to predict", others found it "misleading" or noted its "weird starting points". For $k < 2$, it was reported as "useless", "all over the place", and *"terribly unintuitive with an abysmal sense of planning"*; one participant wrote it *"almost seemed like it was trying to trip me up on purpose"* and another one declared *"I want to beat this robot against a wall."*

The 1-predictable robot seemed to receive the best evaluations overall: though for $k = 0$ many users found it "random", "frustrating" and "confusing", for $k > 0$ it almost invariably had positive reviews ("sensible", "reasonable", "dependable", "smart", "on

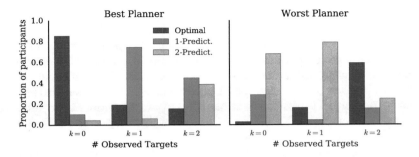

Fig. 6. Final rankings. Participants ranked the planners differently depending on how many targets were observed. For $k = 0$, people preferred the optimal planner; for $k = 1$ and $k = 2$, they preferred the 1-predictable planner.

top of it"), being likened to "a logical and rational human" and even eliciting positive emotions: *"You're my boy, Blue!"*, *"I like the little guy, he thinks like me"*, or *"It was my favorite. I started thinking 'Don't betray me, Yellow!' as it completed its sequence."*

Summary. Our t-predictability planner worked as expected, with the t-predictable robots leading to the highest user prediction accuracy given the first t targets. However, focusing on just 2-predictability at the expense of 0-predictability frustrated our users. Overall, we believe t-predictability will be important in a task for all ts, and hypothesize that optimizing for a weighted combination will perform best in practice.

We note that β is problem-specific and can be expected to decay as the difficulty of the task increases; in each setting, it can be estimated from participant data. Although β was chosen ahead of time in our experiment to be $\beta = 1$, our results are validated by the $r = 0.87$ correlation between expected and observed human error rates.

The optimal choice of t is also a subject for further investigation and is likely context-specific. Depending on the particular task, there should be a different trade-off between predictability of later actions and that of earlier actions.

4   User Study

Having tested our ability to produce t-predictable sequences, we next ran an in person study to test their implications. Participants used a smartphone to operate a remote-controlled Sphero BB-8 robot, and had to predict and adapt to the actions of an autonomous Pioneer P3DX robot in a collaboration scenario (Fig. 1).

4.1   Independent Variables

We manipulated one single variable, *planner*, as a within-subjects factor. Having confirmed the expected effects of the different planners in the previous experiment, and given the good overall performance of the 1-predictable planner across different conditions, we decided to omit the 2-predictable agent and focus on testing the implications of 1-predictable with respect to optimal in a more immersive collaborative context.

4.2 Procedure

At the beginning of the experiment, participants were told that they and their two robot friends were on a secret mission to deactivate an artifact. In each of 4 trials, the autonomous P3DX navigated to the 5 power sources and deactivated them in sequence; however, security sensors activated at each power source after 3 or more had been powered down. The subject's mission was to use BB-8 to jam the sensors at the third, fourth and fifth power sources before the P3DX arrived at them, by steering BB-8 into the corresponding sensor for a short period of time.

After an initial practice phase in which participants had a chance to familiarize themselves with the objective and rules of the task, as well as with the BB-8 teleoperation interface, there were two blocks of 4 trials whose order was counterbalanced across participants. In each block, the subject collaborated with the P3DX under a different task planner which we referred to as different robot "personalities".

Stimuli. Each of the 5 power sources (targets) in each trial was projected onto the floor as a yellow circle, using an overhead projector (Fig 1). Each circle was initially surrounded by a projected blue ring representing a dormant sensor. When the P3DX reached a target, both the circle and the ring were eliminated, except when the P3DX reached the third target, in which case the blue circles turned red symbolizing their switch into active state. Whenever BB-8 entered a ring, the ring turned green for 2 seconds and then disappeared, indicating successful jamming. If the P3DX moved over a red ring, a large red rectangle was projected, symbolizing capture and the trial ended in failure. Conversely, if the P3DX completed all 5 targets without entering a red ring, a green rectangle indicated successful completion of the trial.

Layout Generation. The 4 layouts used were taken from the larger pool of 15 layouts in the online experiment. There was a balance between layouts where online participants had been more accurate with the optimal planner, more accurate with the 1-predictable planner, or similarly accurate.

Controlling for Confounds. We controlled for confounds by counterbalancing the order of the planners; by using a practice layout; and by randomizing the trial order.

4.3 Dependent Measures

Objective measures. We recorded the number of successful trials for each subject and robot planner, as well as the number of trials where participants jammed targets in the correct sequence.

Subjective measures. After every block of the experiment, each participant was also asked to complete a questionnaire (adapted from [7]) to evaluate their perceived performance of the P3DX robot. At the end of the experiment, we asked participants to indicate which robot (planner) they preferred working with.

4.4 Hypotheses

H4 - Comparison with Optimal. *The 1-predictable robot will result in more successful trials than the optimal baseline.*

H5 - Preference. *Users will prefer working with the 1-predictable robot.*

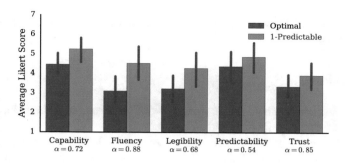

Fig. 7. Perceptions of the collaboration. Over all measures, participants ranked the 1-predictable planner as being preferable to the optimal planner.

4.5 Participants

We recruited 14 participants from the UC Berkeley community, who were treated in accordance with local IRB standards and paid $10 USD. The study took about 30 min.

4.6 Results

Successful completions. We first looked at how often participants were able to complete the task with each robot. We constructed a logistic mixed-effects model for completion success with planner type as a fixed effect and participant and task layout as random effects. We found a significant effect of planner type ($\chi^2(1) = 11.17, p < 0.001$), with the 1-predictable robot yielding more successful completions than the optimal robot ($z = 3.34$, $p < 0.001$). This supports **H4**.

Prediction accuracy. We also looked at how accurate participants were at predicting the robots' sequence of tasks, based on the order in which participants jammed tasks. We constructed a logistic mixed-effects model for prediction accuracy with planner type as a fixed effect and participant and task layout as random effects. We found a significant effect of planner type ($\chi^2(1) = 9.49, p < 0.01$), with the 1-predictable robot being more predictable than the optimal robot ($z = 3.08$, $p < 0.01$).

Robot preferences. We asked participants to pick the robot they preferred to collaborate with. We found that 86% ($N = 12$) of participants preferred the predictable robot, while the rest ($N = 2$) preferred the optimal robot. This result is significantly different from chance ($\chi^2(1) = 7.14, p < 0.01$). This supports **H5**.

Perceptions of the collaboration. We analyzed participants' perceptions of the robots' behavior (Fig. 7) by averaging each participant's responses to the individual questions for each robot and measure, resulting in a single score per participant, per measure, per robot. We constructed a linear mixed-effects model for the survey responses with planner and measure type as fixed effects, and with participants as random effects. We found a main effect of planner ($F(1, 117) = 16.42$, $p < 0.001$) and measure ($F(4, 117) = 5.45$, $p < 0.001$). Post-hoc comparisons using the multivariate t method for p-value adjustment indicated that participants preferred the predictable robot over the optimal robot ($t(117) = 4.05$, $p < 0.001$) by an average of 0.87 ± 0.21 SE points on the Likert scale.

5   Discussion

In what remains, we summarize our work and discuss future directions, including the application of t-predictability beyond task planning.

Summary. We enable robots to generate t-predictable plans, for which a human observing the first t actions can confidently infer the rest. We tested the ability to make plans t-predictable in a large-scale online experiment, in which subjects' predictions of the robot's action sequence significantly improved. In an in-person study, we found that t-predictability can lead to significant objective and perceived improvements in human-robot collaboration compared to traditional optimal planning.

t-predictability for Motion. Even though t-predictability is motivated by a task planning need, it does have an equivalent in motion planning: find an initial trajectory $\xi_{0:t}$, such that the remainder $\xi_{t:T}$ can be inferred by a human observer with knowledge of both the start state $\xi_0 = S$ and the goal state $\xi_T = G$. Modeling this conditional probability with a Boltzmann model yields

$$\xi^* = \arg \max_{\xi \in \Xi} P(\xi_{t:T} | G, \xi_{0:t}) = \arg \max_{\xi \in \Xi} \frac{e^{-\beta c(\xi_{t:T})}}{\int e^{-\beta c(\hat{\xi}_{t:T})} d\hat{\xi}_{t:T}},$$

where Ξ is the set of feasible trajectories from S to G. Using a second order expansion of the cost $c(\hat{\xi}_{t:T})$ about the optimal remaining trajectory, we get:

$$\max_{\xi} P(\xi_{t:T}|G, \xi_{0:t}) \approx \max_{\xi_{0:t}} \frac{e^{-\beta c(\xi_{t:T}^*)}}{e^{-\beta c(\xi_{t:T}^*)} \int e^{-\frac{1}{2}(\hat{\xi}_{t:T} - \xi_{t:T}^*)^T H(\xi_{t:T}^*)(\hat{\xi}_{t:T} - \xi_{t:T}^*)} d\hat{\xi}_{t:T}}$$

$$\equiv \max_{\xi_t} \det(H(\xi_{t:T}^*)), \tag{4}$$

where H denotes the Hessian. This implies that generating a t-predictable trajectory means finding a configuration for time t such that the optimal trajectory from that configuration to the goal is in a *steep*, high-curvature minimum: other trajectories from ξ_t to the goal would be significantly higher cost. For instance, if the robot had the option between two passages, it would choose the more *narrow* passage because that enables the observer to more accurately predict the remainder of its trajectory.

Limitations and Future Work. Our work is limited by the focus in our experiments on TSP scenarios (though we emphasize that t-predictability as it is formulated in Section 2 is not inherently limited to TSPs). This work is also limited by the choice of a user study that involved a tele-operated avatar to mediate the human's physical collaboration with the robot. Applications to other scenarios that involve direct physical collaboration and preconditions would be interesting topics to investigate. Additionally, while our work showcases the utility of t-predictability, a main remaining challenge is determining what t or combination of ts to use for arbitrary tasks. This decision requires more sophisticated human behavior models, which are the topic of our ongoing work.

References

[1] R Alami et al. "Safe and Dependable Physical Human-Robot Interaction in Anthropic Domains : State of the Art and Challenges". *Society* 6.1 (2006).

[2] C. L. Baker, R. Saxe, and J. B. Tenenbaum. "Action understanding as inverse planning". *Cognition* 113.3 (2009).

[3] E. Charniak and R. P. Goldman. "A Bayesian Model of Plan Recognition". *Artificial Intelligence* 64.1 (1993).

[4] M. B. Dias et al. "Sliding autonomy for peer-to-peer human-robot teams". *Intelligent Conference on Intelligent Autonomous Systems (IAS)* (2008).

[5] A. D. Dragan, K. C. T. Lee, and S. S. Srinivasa. "Legibility and predictability of robot motion". *International Conference on Human-Robot Interaction (HRI)* (2013).

[6] A. D. Dragan and S. Srinivasa. "Integrating human observer inferences into robot motion planning". *Autonomous Robots* (2014).

[7] A. D. Dragan et al. "Effects of Robot Motion on Human-Robot Collaboration". *International Conference on Human-Robot Interaction (HRI)* (2015).

[8] T. Fong et al. "The peer-to-peer human-robot interaction project". *AIAA Space* (2005).

[9] M. J. Gielniak and A. L. Thomaz. "Generating anticipation in robot motion". *International Workshop on Robot and Human Interactive Communication* (2011).

[10] T. M. Gureckis et al. "psiTurk: An open-source framework for conducting replicable behavioral experiments online". *Behavioral Research Methods* (2015).

[11] C. Liu, J. B. Hamrick, J. F. Fisac, et al. "Goal Inference Improves Objective and Perceived Performance in Human-Robot Collaboration". *International Conference on Autonomous Agents and Multiagent Systems (AAMAS)* (2016).

[12] D. T. McRuer. "Pilot-Induced Oscillations and Human Dynamic Behavior" (1995).

[13] S. Nikolaidis and J. Shah. "Human-robot cross-training: Computational formulation, modeling and evaluation of a human team training strategy". *International Conference on Human-Robot Interaction (HRI)* (2013).

[14] G. Pezzulo, F. Donnarumma, and H. Dindo. "Human sensorimotor communication: A theory of signaling in online social interactions". *PLoS One* 8.11 (2013).

[15] E. S. Van der Poort et al. "Solving the k-best traveling salesman problem". *Computers & operations research* 26.4 (1999).

[16] M. Saffarian, J. C. F. de Winter, and R. Happee. "Automated Driving: Human-Factors Issues and Design Solutions". *Proceedings of the Human Factors and Ergonomics Society Annual Meeting* 56.1 (2012).

[17] N. B. Sarter and D. D. Woods. "Team play with a powerful and independent agent: operational experiences and automation surprises on the Airbus A-320." *Human factors* 39.4 (1997).

[18] J. Shah and C. Breazeal. "An empirical analysis of team coordination behaviors and action planning with application to human-robot teaming." *Human factors* 52.2 (2010).

[19] D. Szafir, B. Mutlu, and T. Fong. "Communication of Intent in Assistive Free Flyers". *International Conference on Human-Robot Interaction (HRI)* (2014).

[20] L. Takayama, D. Dooley, and W. Ju. "Expressing thought: improving robot readability with animation principles". *International Conference on Human-Robot Interaction (HRI)* (2011).

[21] M. Tomasello et al. "Understanding and sharing intentions: the origins of cultural cognition". *Behavioral and brain sciences* 28.05 (2005).

[22] C. Vesper et al. "A minimal architecture for joint action". *Neural Networks* 23.8 (2010).

[23] Y. Zhang et al. "Plan Explicability for Robot Task Planning". *RSS Workshop on Planning for Human-Robot Interaction: Shared Autonomy and Collaborative Robotics* (2016).

Sensor-Based Reactive Navigation
in Unknown Convex Sphere Worlds

Omur Arslan and Daniel E. Koditschek

University of Pennsylvania, Philadelphia, PA 19103, USA

{omur,kod}@seas.upenn.edu

Abstract. We construct a sensor-based feedback law that provably solves the real-time collision-free robot navigation problem in a compact convex Euclidean subset cluttered with unknown but sufficiently separated and strongly convex obstacles. Our algorithm introduces a novel use of separating hyperplanes for identifying the robot's local obstacle-free convex neighborhood, affording a reactive (online-computed) piecewise smooth and continuous closed-loop vector field whose smooth flow brings almost all configurations in the robot's free space to a designated goal location, with the guarantee of no collisions along the way. We further extend these provable properties to practically motivated limited range sensing models.

Keywords: motion planning · collision avoidance · sensor-based planning

1 Introduction

Agile navigation in dense human crowds [1,2], or in natural forests, such as now negotiated by rapid flying [3,4] and legged [5,6] robots, strongly motivates the development of sensor-based reactive motion planners. By the term *reactive* [7, 8] we mean that motion is generated by a vector field arising from some closed-loop feedback policy issuing online force or velocity commands in real time as a function of instantaneous robot state. By the term *sensor-based* we mean that information about the location of the environmental clutter to be avoided is limited to structure perceived within some local neighborhood of the robot's instantaneous position — its sensor footprint.

In this paper, we propose a new reactive motion planner taking the form of a feedback law for a first-order (velocity-controlled), perfectly and relatively (to a fixed goal location) sensed and actuated disk robot, that can be computed using only information about the robot's instantaneous position and structure within its sensor footprint. We assume the a priori unknown environment is a static topological sphere world [9], whose obstacles are convex and have smooth boundaries whose curvature is "reasonably" high relative to their mutual separation. Under these assumptions, the proposed vector field planner is guaranteed to bring all but a measure zero set of initial conditions to the desired goal. To the best of our knowledge, this is the first time a sensor-based reactive motion planner has been shown to be provably correct w.r.t. a general class of environments.

1.1 Motivation and Prior Literature on Vector Field Planners

The simple, computationally efficient artificial potential field approach to real-time obstacle avoidance [10] incurs topologically necessary critical points [11], which, in practice, with no further remediation often include (topologically unnecessary) spurious local minima. In general, such local obstacle avoidance strategies [12–15] yield safe

© Springer Nature Switzerland AG 2020

K. Goldberg et al. (Eds.): *Algorithmic Foundations of Robotics XII*, SPAR 13, pp. 160–175, 2020.

https://doi.org/10.1007/978-3-030-43089-4_11

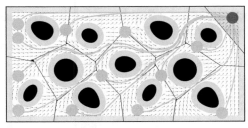

Fig. 1. Exact navigation of a disk-shaped robot using separating hyperplanes of the robot body (red at the goal) and convex obstacles (black solid shapes). Separating hyperplanes between the robot and obstacles define an obstacle-free convex neighborhood (the yellow region when the robot is at the goal) of the robot, and the continuous feedback motion towards the metric projection of a given goal (red) onto this convex set asymptotically steers almost all robot configurations (green) to the goal without collisions along the way. The grey regions represent the augmented workspace boundary and obstacles, and the arrows show the direction of the resulting vector field.

robot navigation algorithms but offer no assurance of (global) convergence to a designated goal location. Even in topologically simple settings such as the sphere worlds addressed here, constructions that eliminate these spurious attractors — e.g., navigation functions [16] — have largely come at the price of complete prior information.

Extensions to navigation functions partially overcoming the necessity of global prior knowledge of (and consequent parameter tuning for) a topologically and metrically simple environment have appeared in the last decade [17, 18]. Sequential composition [19] has been used to cover complicated environments with cellular local potential decompositions [20], but still necessitating prior global knowledge of the environment.

1.2 Contributions and Organization of the Paper

This paper abandons the smooth potential field approach to reactive planning, achieving an algorithm that is "doubly reactive" in the sense that not merely the integrated robot trajectory, but also its generating vector field can be constructed on the fly in real time using only local knowledge of the environment. Our piecewise smooth vector field combines some of the ideas of sensor-based exploration [21] with those of hybrid reactive control [20]. We use separating hyperplanes of convex bodies [22] to identify an obstacle-free convex neighborhood of a robot configuration, and build our safe robot navigation field by control action towards the metric projection of the designated point destination onto this convex set.

Our construction requires no parameter tuning and requires only local knowledge of the environment in the sense that the robot needs only locate those proximal obstacles determining its collision-free convex neighborhood. When the obstacles are sufficiently separated (Assumption 1 stipulates that the robot must be able to pass in between them) and sufficiently strongly convex at their "antipode" (Assumption 2 stipulates that they curve away from the enclosing sphere centered at the destination which just touches their boundary at the most distant point), the proposed vector field generates a smooth flow with a unique attractor at the specified goal along with (the topologically necessary number of) saddles — at least one associated with each obstacle. Since all of its critical points are nondegenerate, our vector field is guaranteed to steer almost all robot configurations to the goal, while avoiding collisions along the way, as illustrated in Fig. 1.

It proves most convenient to develop the theoretical properties of this construction under the assumption that the robot can identify and locate those nearby obstacles whose associated separating hyperplanes define the robot's obstacle-free convex neighborhood (a capability termed *Voronoi-adjacent*[9] *obstacle sensing* in Section 3.2), no matter how physically distant they may be. Thus, to accommodate more physically realistic sensors, we adapt the initial construction (and the proof) to the case of two different limited range sensing modalities, while extending the same formal guarantees as in the erstwhile (local but unbounded range) idealized sensor model.

In prior work [23], we propose a different construction based on power diagrams [24] for navigating among spherical obstacles using knowledge of Voronoi-adjacent[9] obstacles to construct the robot's local workspace [23, Eqn. (9)]. This paper introduces a new construction for that set in (7) based on separating hyperplanes, permitting an extension of the navigable obstacles to the broader class of convex bodies specified by Assumption 2, while providing the same guarantee of almost global asymptotic convergence (Theorem 3) to a given goal location. From the view of applications, the new appeal to separating hyperplanes permits the central advance of a purely reactive construction from limited range sensors (22), e.g., in the planar case from immediate lineof-sight appearance (27), with the same global guarantees.

This paper is organized as follows. Section 2 continues with a formal statement of the problem at hand. Section 3 briefly summarizes a separating hyperplane theorem of convex bodies, and introduces its use for identifying collision-free robot configurations. Section 4, comprising the central contribution of the paper, constructs and analyzes the reactive vector field planner for safe robot navigation in a convex sphere world, and provides its more practical extensions. Section 5 illustrates the qualitative properties of the proposed vector field planner using numerical simulations. Section 6 concludes with a summary of our contributions and a brief discussion of future work.

2   Problem Formulation

Consider a disk-shaped robot, of radius $r \in \mathbb{R}_{>0}$ centered at $x \in \mathcal{W}$, operating in a closed compact convex environment $\mathcal{W}$ in the n-dimensional Euclidean space $\mathbb{R}^n$, where $n \geq 2$, punctured with $m \in \mathbb{N}$ open convex sets $\mathcal{O} := \{O_1, O_2, \ldots, O_m\}$ with twice differentiable boundaries, representing obstacles.[1] Hence, the free space $\mathcal{F}$ of the robot is given by

$$\mathcal{F} := \left\{ x \in \mathcal{W} \mid \overline{B(x,r)} \subseteq \mathcal{W} \setminus \bigcup_{i=1}^m O_i \right\}, \tag{1}$$

where $B(x,r) := \{ q \in \mathbb{R}^n \mid \|q - x\| < r \}$ is the open ball centered at x with radius r, and $\overline{B(x,r)}$ denotes its closure, and $\|.\|$ denotes the standard Euclidean norm.

To maintain the local convexity of obstacle boundaries in the free space $\mathcal{F}$, we assume that our disk-shaped robot can freely fit in between (and thus freely circumnavigate) any of the obstacles throughout the workspace $\mathcal{W}$: [2]

---

[1] Here, $\mathbb{N}$ is the set of all natural numbers; $\mathbb{R}$ and $\mathbb{R}_{>0}$ ($\mathbb{R}_{\geq 0}$) denote the set of real and positive (nonnegative) real numbers, respectively.

[2] Assumption 1 is equivalent to the "isolated" obstacles assumption of [16].

Assumption 1. *Obstacles are separated from each other by clearance of at least* $d(O_i, O_j) > 2r$ *for all* $i \neq j$, *and from the boundary* $\partial \mathcal{W}$ *of the workspace* $\mathcal{W}$ *as* $d(O_i, \partial \mathcal{W}) > 2r$ *for all* $i = 1 \dots m$, *where* $d(A, B) := \inf\{\|a - b\| \mid a \in A, b \in B\}$.

Before formally stating our navigation problem, it is useful to recall a known topological limitation of reactive planners: if a continuous vector field planner on a generalized sphere world has a unique attractor, then it must have at least as many saddles as obstacles [9]. In consequence, the robot navigation problem that we seek to solve is:

Reactive Navigation Problem. *Assuming the first-order (completely actuated single-integrator) robot dynamics,*

$$\dot{x} = u(x) \,, \tag{2}$$

find a Lipschitz continuous vector field controller, $u : \mathcal{F} \to \mathbb{R}^n$, *that leaves the robot's free space* $\mathcal{F}$ *positively invariant and asymptotically steers almost all robot configurations in* $\mathcal{F}$ *to any given goal location* $x^* \in \mathcal{F}$.

3   Encoding Collisions via Separating Hyperplanes

3.1   Separating Hyperplane Theorem

A fundamental theorem of convex sets states that any two nonintersecting convex sets can be separated by a hyperplane such that they lie on opposite sides of this hyperplane:

Theorem 1 ([22, 25]). *For any two disjoint convex sets* $A, B \in \mathbb{R}^n$ *(i.e.,* $A \cap B = \varnothing$), *there exists* $a \in \mathbb{R}^n$ *and* $b \in \mathbb{R}$ *such that* $a^T x \geq b$ *for all* $x \in A$ *and* $a^T x \leq b$ *for all* $x \in B$.

For example, a usual choice of such a hyperplane is [22]:

Definition 1. *The* maximum margin separating hyperplane *of any two disjoint convex sets* $A, B \subset \mathbb{R}^n$, *with* $d(A, B) > 0$, *is defined to be*

$$H(A, B) := \left\{ x \in \mathbb{R}^n \,\middle|\, \|x - a\| = \|x - b\|, \|a - b\| = d(A, B), a \in \overline{A}, b \in \overline{B} \right\}, \tag{3}$$

where $d(x, H(A, B)) \geq \frac{d(A,B)}{2}$ *for all* $x \in A \cup B$.

Another useful tool for finding separating hyperplanes is metric projection:

Theorem 2 ([25]). *Let* $A \subset \mathbb{R}^n$ *be a closed convex set and* $x \in \mathbb{R}^n$. *Then there exists a unique point* $a^* \in A$ *such that*

$$a^* = \Pi_A(x) := \arg\min_{a \in A} \|a - x\|, \tag{4}$$

and one has $(x - \Pi_A(x))^T (\Pi_A(x) - a) \geq 0$ *for all* $a \in A$.
The map $\Pi_A(x)$ *is called the* metric projection *of* x *onto set* A.

Lemma 1. *The maximum margin separating hyperplane of a convex set* $A \subset \mathbb{R}^n$ *and the ball* $B(x, r)$ *of radius* $r \in \mathbb{R}_{>0}$ *centered at* $x \in \mathbb{R}^n$, *satisfying* $d(x, A) \geq r$, *is given by*

$$H(A, B(x, r)) = \left\{ y \in \mathbb{R}^n \,\middle|\, \left\| y - (\Pi_{\overline{B(x,r)}} \circ \Pi_{\overline{A}})(x) \right\| = \|y - \Pi_{\overline{A}}(x)\| \right\}, \tag{5}$$

where $(\Pi_{\overline{B(x,r)}} \circ \Pi_{\overline{A}})(x) = x - r \frac{x - \Pi_{\overline{A}}(x)}{\|x - \Pi_{\overline{A}}(x)\|}$.

Proof. See Appendix I-A in the Technical Report [26].   ∎

A common application of separating hyperplanes of a set of convex bodies is to discover their organizational structure. For instance, to model its topological structure, we define the generalized Voronoi diagrams $\mathcal{V} = \{V_1, V_2, \ldots, V_m\}$ of a convex environment $\mathcal{W}$ in $\mathbb{R}^n$ populated with disjoint convex obstacles $\mathcal{O} = \{O_1, O_2, \ldots, O_m\}$ (i.e., $d(O_i, O_j) > 0$ for all $i \neq j$), based on maximum margin separating hyperplanes, to be[3][4]

$$V_i := \left\{ q \in \mathcal{W} \middle| \|q - p_i\| \leq \|q - p_j\|, \|p_i - p_j\| = d(O_i, O_j), p_i \in \overline{O}_i, p_j \in \overline{O}_j \;\; \forall j \neq i \right\}, \quad (6)$$

which yields a convex cell decomposition of a subset of $\mathcal{W}$ such that, by construction, each obstacle is contained in its Voronoi cell, i.e., $O_i \subset V_i$, see Fig. 2. Note that for point obstacles, say $O_i = \{p_i\}$ for some $p_i \in \mathbb{R}^n$, the generalized Voronoi diagram of $\mathcal{W}$ in (6) simplifies back to the standard Voronoi diagram of $\mathcal{W}$, generated by points $\{p_1, \ldots, p_m\}$, i.e., $V_i = \{q \in \mathcal{W} | \|q - p_i\| \leq \|q - p_j\|, \;\; \forall j \neq i\}$ [32].

3.2 The Safe Neighborhood of a Disk-Shaped Robot

Throughout the sequel, we consider a disk-shaped robot, centered at $x \in \mathcal{W}$ with radius $r \in \mathbb{R}_{>0}$, moving in a closed compact convex environment $\mathcal{W} \subseteq \mathbb{R}^n$ populated with open convex obstacles, $\mathcal{O} = \{O_1, O_2, \ldots, O_m\}$, satisfying Assumption 1. Since the workspace, obstacles, and the robot radius are fixed, we suppress all mention of the associated terms wherever convenient, in order to simplify the notation.

Using the robot body and obstacles as generators of a generalized Voronoi diagram of $\mathcal{W}$, we define the robot's *local workspace*, $\mathcal{LW}(x)$, illustrated in Fig. 2(left), as,[5]

$$\mathcal{LW}(x) := \left\{ q \in \mathcal{W} \middle| \left\| q - x + r\frac{x - \Pi_{\overline{O}_i}(x)}{\|x - \Pi_{\overline{O}_i}(x)\|} \right\| \leq \|q - \Pi_{\overline{O}_i}(x)\|, \;\; \forall i \right\}. \quad (7)$$

Note that we here take the advantage of having a disk-shaped robot and construct the maximum margin separating hyperplane between the robot and each obstacle using the robot's centroid (Lemma 1).

A critical property of the local workspace $\mathcal{LW}$ is:

Proposition 1. *A robot placement* $x \in \mathcal{W} \setminus \bigcup_{i=1}^{m} O_i$ *is collision free, i.e.,* $x \in \mathcal{F}$, *if and only if the robot body is contained in its local workspace* $\mathcal{LW}(x)$, *i.e.,*[6]

$$x \in \mathcal{F} \iff \overline{B(x, r)} \subseteq \mathcal{LW}(x). \quad (8)$$

Proof. See Appendix I-B in the Technical Report [26].   ∎

---

[3] Generalized Voronoi diagrams and cell decomposition methods are traditionally encountered in the design of roadmap methods [8,21,27]. A major distinction between our construction and these roadmap algorithms is that the latter typically seek a global, one-dimensional graphical representation of a robot's environment (independent of any specific configuration), whereas our approach uses the local open interior cells of the robot-centric Voronoi diagram to determine a locally safe neighborhood of a given free configuration.

[4] It seems worth noting that our use of generalized Voronoi diagrams is motivated by application of Voronoi diagrams in robotics for coverage control of distributed sensor networks [28–31].

[5] Here, to solve the indeterminacy, we set $\frac{x}{\|x\|} = 0$ whenever $x = 0$.

[6] Note that $\mathcal{F} \subsetneq \mathcal{W} \setminus \bigcup_{i=1}^{m} O_i$ for a disk-shaped robot of radius $r > 0$.

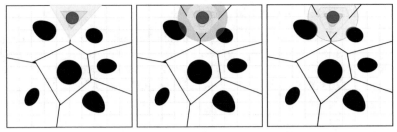

Fig. 2. Local workspace $\mathcal{LW}$ (yellow) and local free space $\mathcal{LF}$ (green) of a disk-shaped robot (blue) for different sensing modalities: (left) Voronoi-adjacent[9] obstacle sensing, (middle) a fixed radius sensory footprint (red), (right) a limited range line-of-sight sensor (red). The boundary of each generalized Voronoi cell is defined by the maximum margin separating hyperplanes of the robot body (blue) and obstacles (black).

Accordingly, we define the robot's *local free space*, $\mathcal{LF}(x)$, by eroding $\mathcal{LW}(x)$, removing the volume swept along its boundary, $\partial\mathcal{LW}(x)$, by the robot body radius, illustrated on the left in Fig. 2, as [33] [7]

$$\mathcal{LF}(x) := \mathcal{LW}(x) \setminus \left(\partial\mathcal{LW}(x) \oplus B(\mathbf{0}, r)\right) = \left\{ q \in \mathcal{LW}(x) \middle| \overline{B(q, r)} \subseteq \mathcal{LW}(x) \right\}. \quad (9)$$

Note that, for any $x \in \mathcal{F}$, $\mathcal{LF}(x)$ is a nonempty closed convex set, because $x \in \mathcal{LF}(x)$ and the erosion of a closed convex set by an open ball is a closed convex set.[8]

An immediate consequence of Proposition 1 is:

Corollary 1. *Any robot placement in the local free space $\mathcal{LF}(x)$ of a collision free robot location $x \in \mathcal{F}$ is also collision free, i.e., $\mathcal{LF}(x) \subseteq \mathcal{F}$ for all $x \in \mathcal{F}$.*

Finally, it is useful to emphasize that to construct its local workspace, the robot requires only local knowledge of the environment in the sense that the robot only needs to locate proximal obstacles — those whose Voronoi cells are adjacent[9] to the robot's (local workspace). This can be achieved by assuming an adjustable radius sensory footprint and gradually increasing its sensing range until the set of obstacles in the sensing range satisfies a certain geometric criterion guaranteeing that the detected obstacles exactly define the robot's local workspace [28]. We will refer to this sensing model as *Voronoi-adjacent obstacle sensing*.

4   Robot Navigation via Separating Hyperplanes

In this section, first assuming Voronoi-adjacent obstacle sensing, we introduce a new provably correct vector field controller for safe robot navigation in a convex sphere world, and list its important qualitative properties. Then we present its extensions for two more realistic sensor models (illustrated, respectively, in the middle and the right panels of Fig. 2): a fixed radius sensory footprint and a limited range line-of-sight sensor.

---

[7] Here, $\mathbf{0}$ is a vector of all zeros with the appropriate size, and $A \oplus B$ denotes the Minkowski sum of sets A and B defined as $A \oplus B = \{a + b \mid a \in A, b \in B\}$.

[8] The erosion of a closed half-space by an open ball is a closed half-space, and a closed convex set can be defined as (possibly infinite) intersection of closed half-spaces [22]. Thus, since the erosion operation is distributed over set intersection [33], and an arbitrary intersection of closed sets is closed [34], the erosion of a closed convex set by an open ball is a closed convex set.

[9] A pair of Voronoi cells in $\mathbb{R}^n$ is said to be *adjacent* if they share a $n - 1$ dimensional face.

4.1   Feedback Robot Motion Planner

Assuming the fully-actuated single-integrator robot dynamics in (2), for a choice of a desired goal location $x^* \in \mathcal{F}$, we propose a robot navigation strategy, called the "*move-to-projected-goal*" law, $u : \mathcal{F} \to \mathbb{R}^n$ that steers the robot at location $x \in \mathcal{F}$ towards the global goal x^* through the "*projected goal*", $\Pi_{\mathcal{LF}(x)}(x^*)$, as follows: [10]

$$u(x) = -k\left(x - \Pi_{\mathcal{LF}(x)}(x^*)\right) , \qquad (10)$$

where $k \in \mathbb{R}_{>0}$ is a fixed control gain and Π_A (4) is the metric projection onto a closed convex set $A \subset \mathbb{R}^n$, and $\mathcal{LF}(x)$ is continuously updated using the Voronoi-adjacent obstacle sensing and its relation with $\mathcal{LW}(x)$ in (9).

4.2   Qualitative Properties

Proposition 2. *The "move-to-projected-goal" law in (10) is piecewise continuously differentiable.*

Proof. An important property of generalized Voronoi diagrams in (6) inherited from the standard Voronoi diagrams of point generators is that the boundary of each Voronoi cell is a piecewise continuously differentiable function of generator locations [36, 37]. In particular, for any $x \in \mathcal{F}$, the boundary of the robot's local workspace $\mathcal{LW}(x)$ is piecewise continuously differentiable since it is defined by the boundary of the workspace and separating hyperplanes between the robot and obstacles, parametrized by x and $\Pi_{\overline{O}_i}(x)$, and metric projections onto convex cells are piecewise continuously differentiable [38]. Hence, the boundary of the local free space $\mathcal{LF}(x)$ is also piecewise continuously differentiable, because $\mathcal{LF}(x)$ is the nonempty erosion of $\mathcal{LW}(x)$ by a fixed open ball. Therefore, one can conclude using the sensitivity analysis of metric projections onto moving convex sets [39, 40] that the "move-to-projected-goal" law is Lipschitz continuous and piecewise continuously differentiable.   ∎

Proposition 3. *The robot's free space $\mathcal{F}$ in (1) is positively invariant under the "move-to-projected" law (10).*

Proof. Since x and $\Pi_{\mathcal{LF}(x)}(x^*)$ are both in $\mathcal{LF}(x)$ for any $x \in \mathcal{F}$, and $\mathcal{LF}(x)$ is an obstacle free convex neighborhood of x (Corollary 1), the line segment joining x and $\Pi_{\mathcal{LF}(x)}(x^*)$ is free of collisions. Hence, at the boundary of $\mathcal{F}$, the robot under the "move-to-projected-goal" law either stays on the boundary or moves towards the interior of $\mathcal{F}$, but never crosses the boundary, and so the result follows.   ∎

Proposition 4. *For any initial $x \in \mathcal{F}$, the "move-to-projected-goal" law (10) has a unique continuously differentiable flow in $\mathcal{F}$ (1) defined for all future time.*

---

[10] In general, the metric projection of a point onto a convex set can be efficiently computed using an off-the-shelf convex programming solver [22]. If $\mathcal{W}$ is a convex polytope, then the robot's local free space $\mathcal{LF}(x)$ is also a convex polytope and can be written as a finite intersection of half-spaces. Thus, the metric projection onto a convex polytope can be recast as quadratic programming and can be solved in polynomial time [35]. In the case of a convex polygonal environment, $\mathcal{LF}(x)$ is a convex polygon and the metric projection onto it can be solved analytically since the solution lies on one of its edges, unless the input point is inside $\mathcal{LF}(x)$.

Proof. The existence, uniqueness and continuous differentiability of its flow follow from the Lipschitz continuity of the "move-to-projected-goal" law in its compact domain $\mathcal{F}$, because a piecewise continuously differentiable function is locally Lipschitz on its domain [41], and a locally Lipschitz function on a compact set is globally Lipschitz on that set [42]. ∎

Proposition 5. *The set of stationary points of the "move-to-projected-goal" law (10) is* $\{x^*\} \cup \bigcup_{i=1}^{m} \mathfrak{S}_i$, *where*

$$\mathfrak{S}_i := \left\{ x \in \mathcal{F} \,\middle|\, d(x, O_i) = r, \frac{(x - \Pi_{\overline{O}_i}(x))^\mathrm{T}(x - x^*)}{\|x - \Pi_{\overline{O}_i}(x)\|\|x - x^*\|} = 1 \right\}. \tag{11}$$

Proof. It follows from (4) and $x^* \in \mathcal{LF}(x^*)$ that the goal x^* is a stationary point of (10). In fact, for any $x \in \mathcal{F}$, one has $\Pi_{\mathcal{LF}(x)}(x^*) = x^*$ whenever $x^* \in \mathcal{LF}(x)$. Hence, in the sequel of the proof, we only consider the set of robot locations satisfying $x^* \notin \mathcal{LF}(x)$.

Let $x \in \mathcal{F}$ such that $x^* \notin \mathcal{LF}(x)$. Recall from (7) and (9) that $\mathcal{LW}(x)$ is determined by the maximum margin separating hyperplanes of the robot body and obstacles, and $\mathcal{LF}(x)$ is obtained by eroding $\mathcal{LW}(x)$ by an open ball of radius r. Hence, x lies in the interior of $\mathcal{LF}(x)$ if and only if $d(x, O_i) > r$ for all i. As a result, since $x^* \notin \mathcal{LF}(x)$, one has $x = \Pi_{\mathcal{LF}(x)}(x^*)$ only if $d(x, O_i) = r$ for some i.

Note that if $d(x, O_i) = r$, then, since $d(O_i, O_j) > 2r$ (Assumption 1), $d(x, O_j) > r$ for all $j \neq i$. Therefore, there can be only one obstacle index i such that $x = \Pi_{\mathcal{LF}(x)}(x^*)$ and $d(x, O_i) = r$. Further, given $d(x, O_i) = r$, since $\Pi_{\mathcal{LF}(x)}(x^*)$ is the unique closest point of the closed convex set $\mathcal{LF}(x)$ to the goal x^* (Theorem 2), its optimality [22] implies that one has $x = \Pi_{\mathcal{LW}(x)}(x^*)$ if and only if the maximum margin separating hyperplane between the robot and obstacle O_i is tangent to the level curve of the squared Euclidean distance to the goal, $\|x - x^*\|^2$, at $\Pi_{\overline{O}_i}(x)$, and separates x and x^*, i.e.,

$$\frac{(x - \Pi_{\overline{O}_i}(x))^\mathrm{T}(x - x^*)}{\|x - \Pi_{\overline{O}_i}(x)\|\|x - x^*\|} = 1. \tag{12}$$

Thus, one can locate the stationary points of the "move-to-projected-goal" law in (10) associated with obstacle O_i as in (11), and so the result follows. ∎

Note that, for any equilibrium point $s_i \in \mathfrak{S}_i$ associated with obstacle O_i, one has that the equilibrium s_i, its projection $\Pi_{\overline{O}_i}(s_i)$ and the goal x^* are all collinear.

Lemma 2. *The "move-to-projected-goal" law (10) in a small neighborhood of the goal* x^* *is given by*

$$u(x) = -k(x - x^*), \quad \forall x \in B(x^*, \epsilon), \tag{13}$$

for some $\epsilon > 0$; *and around any stationary point* $s_i \in \mathfrak{S}_i$ *(11), associated with obstacle* O_i, *it is given by*

$$u(x) = -k\left(x - x^* + \frac{(x - \Pi_{\overline{O}_i}(x))^\mathrm{T}(x^* - h_i)}{\|x - \Pi_{\overline{O}_i}(x)\|^2}(x - \Pi_{\overline{O}_i}(x))\right), \tag{14}$$

for all $x \in B(s_i, \varepsilon)$ *and some* $\varepsilon > 0$, *where*

$$h_i := \frac{x + \Pi_{\overline{O}_i}(x)}{2} + \frac{r}{2}\frac{x - \Pi_{\overline{O}_i}(x)}{\|x - \Pi_{\overline{O}_i}(x)\|}. \tag{15}$$

Proof. See Appendix I-C in the Technical Report [26].   ∎

Since our "move-to-projected-goal" law strictly decreases the Euclidean distance to the goal x^* away from its stationary points (Proposition 7), to guarantee the existence of a unique stable attractor at x^*, we require the following assumption[11], whose geometric interpretation is discussed in detail in Appendix II in the Technical Report [26].

Assumption 2. *(Curvature Condition) The Jacobian matrix* $\mathbf{J}_{\Pi_{\overline{O}_i}}(s_i)$ *of the metric projection of any stationary point* $s_i \in \mathfrak{S}_i$ *onto the associated obstacle* O_i *satisfies*[12]

$$\mathbf{J}_{\Pi_{\overline{O}_i}}(s_i) \prec \frac{\left\| x^* - \Pi_{\overline{O}_i}(s_i) \right\|}{r + \left\| x^* - \Pi_{\overline{O}_i}(s_i) \right\|} \mathbf{I} \qquad \forall i, \tag{16}$$

where $\mathbf{I}$ *is the identity matrix of appropriate size.*

Proposition 6. *If Assumption 2 holds for the goal* x^* *and for all obstacles, then* x^* *is the only locally stable equilibrium of the "move-to-projected-goal" law (10), and all stationary points,* $s_i \in \mathfrak{S}_i$ *(11), associated with obstacles,* O_i, *are nondegenerate saddles.*

Proof. It follows from (13) that the goal x^* is a locally stable point of the "move-to-projected-goal" law, because its Jacobian matrix, $\mathbf{J}_u(x^*)$, at x^* is equal to $-k\,\mathbf{I}$.

To determine the type of any stationary point $s_i \in \mathfrak{S}_i$ associated with obstacle O_i, define

$$g(x) := \frac{\left(x^* - \Pi_{\overline{O}_i}(x)\right)^{\mathrm{T}}\left(x - \Pi_{\overline{O}_i}(x)\right)}{\left\| x - \Pi_{\overline{O}_i}(x) \right\|^2} - \frac{r}{2\left\| x - \Pi_{\overline{O}_i}(x) \right\|} - \frac{1}{2}, \tag{17}$$

and so the "move-to-projected-goal" law in a small neighborhood of s_i in (14) can be rewritten as

$$u(x) = -k\left(x - x^* + g(x)\left(x - \Pi_{\overline{O}_i}(x)\right)\right). \tag{18}$$

Hence, using $\left\| s_i - \Pi_{\overline{O}_i}(s_i) \right\| = r$, one can verify that its Jacobian matrix at s_i is given by

$$\mathbf{J}_u(s_i) = -kg(s_i)\left(\frac{\left\| x^* - \Pi_{\overline{O}_i}(s_i) \right\|}{r + \left\| x^* - \Pi_{\overline{O}_i}(s_i) \right\|}\mathbf{Q} - \mathbf{J}_{\Pi_{\overline{O}_i}}(s_i)\right) - \frac{k}{2}(\mathbf{I} - \mathbf{Q}), \tag{19}$$

where $g(s_i) = -\dfrac{\left\| x^* - \Pi_{\overline{O}_i}(s_i) \right\|}{r} - 1 < -2$, and

$$\mathbf{Q} = \mathbf{I} - \frac{\left(s_i - \Pi_{\overline{O}_i}(s_i)\right)\left(s_i - \Pi_{\overline{O}_i}(s_i)\right)^{\mathrm{T}}}{\left\| s_i - \Pi_{\overline{O}_i}(s_i) \right\|^2}. \tag{20}$$

Note that $\mathbf{J}_{\Pi_{\overline{O}_i}}(x)\left(x - \Pi_{\overline{O}_i}(x)\right) = 0$ for all $x \in \mathbb{R}^n \setminus \overline{O}_i$ [44, 45]. Hence, if Assumption 2 holds, then one can conclude from $g(s_i) < -2$ and (19) that the only negative eigenvalue of $\mathbf{J}_u(s_i)$ and the associated eigenvector are $-\frac{k}{2}$ and $\left(s_i - \Pi_{\overline{O}_i}(s_i)\right)$, respectively; and all other eigenvalues of $\mathbf{J}_u(s_i)$ are positive. Thus, s_i is a nondegenerate saddle point of the "move-to-projected-goal" law associated with O_i.   ∎

---

[11] A similar obstacle curvature condition is necessarily made in the design of navigation functions for spaces with convex obstacles in [43].

[12] For any two symmetric matrices $\mathbf{A}, \mathbf{B} \in \mathbb{R}^{N \times N}$, $\mathbf{A} \prec \mathbf{B}$ (and $\mathbf{A} \preccurlyeq \mathbf{B}$) means that $\mathbf{B} - \mathbf{A}$ is positive definite (positive semidefinite, respectively).

Proposition 7. *Given that the goal location* x^* *and all obstacles satisfy Assumption 2, the goal* x^* *is an asymptotically stable equilibrium of the "move-to-projected-goal" law (10), whose basin of attraction includes* $\mathcal{F}$, *except a set of measure zero.*

Proof. Consider the squared Euclidean distance to the goal as a smooth Lyapunov function candidate, i.e., $V(x) := \|x - x^*\|^2$, and it follows from (4) and (10) that

$$\dot{V}(x) = -k \underbrace{2(x - x^*)^{\mathrm{T}}\left(x - \Pi_{\mathcal{LF}(x)}(x^*)\right)}_{\substack{\geq \|x - \Pi_{\mathcal{LF}(x)}(x^*)\|^2 \\ \text{since } x \in \mathcal{LF}(x) \text{ and } \|x - x^*\|^2 \geq \|\Pi_{\mathcal{LF}(x)}(x^*) - x^*\|^2}} \leq -k\|x - \Pi_{\mathcal{LF}(x)}(x^*)\|^2 \leq 0 , \quad (21)$$

which is zero iff x is a stationary point. Hence, we have from LaSalle's Invariance Principle [42] that all robot configurations in $\mathcal{F}$ asymptotically reach the set of equilibria of (10). Therefore, the result follows from Proposition 2 and Proposition 6, because, under Assumption 2, x^* is the only stable stationary point of the piecewise continuous "move-to-projected-goal" law (10), and all other stationary points are nondegenerate saddles whose stable manifolds have empty interiors [46]. ∎

Finally, we find it useful to summarize important qualitative properties of the "move-to-projected-goal" law as:

Theorem 3. *The piecewise continuously differentiable "move-to-projected-goal" law in (10) leaves the robot's free space* $\mathcal{F}$ *(1) positively invariant; and if Assumption 2 holds, then its unique continuously differentiable flow, starting at almost any configuration* $x \in \mathcal{F}$, *asymptotically reaches the goal location* x^*, *while strictly decreasing the squared Euclidean distance to the goal,* $\|x - x^*\|^2$, *along the way.*

4.3   Extensions for Limited Range Sensing Modalities

Navigation using a Fixed Radius Sensory Footprint. A crucial property of the "move-to-projected-goal" law (10) is that it only requires the knowledge of the robot's Voronoi-adjacent[9] obstacles to determine the robot's local workspace and so the robot's local free space. We now exploit that property to relax our construction so that it can be put to practical use with commonly available sensors that have bounded radius footprint.[13] We will present two specific instances, pointing out along the way how they nevertheless preserve the sufficient conditions for the qualitative properties listed in Section 4.2.

Suppose the robot is equipped with a sensor with a fixed sensing range, $R \in \mathbb{R}_{>0}$, whose sensory output, denoted by $\mathcal{S}_R(x) := \{S_1, S_2, \ldots, S_m\}$, at a location, $x \in \mathcal{W}$, returns some computationally effective dense representation of the perceptible portion, $S_i := O_i \cap B(x, R)$, of each obstacle, O_i, in its sensory footprint, $B(x, R)$. Note that S_i is always open and might possibly be empty (if O_i is outside the robot's sensing range), see Fig. 2(middle); and we assume that the robot's sensing range is greater than the robot body radius, i.e., $R > r$.

---

[13] This extension results from the construction of the robot's local workspace (7) in terms of the maximum margin separating hyperplanes of convex sets. In consequence, because the intersection of convex sets is a convex set [22], perceived obstacles in the robot's (convex) sensory footprint are, in turn, themselves always convex.

As in (7), using the maximum margin separating hyperplanes of the robot and sensed obstacles, we define the robot's *sensed local workspace*, see Fig. 2(middle), as,

$$\mathcal{LW}_{\mathcal{S}}(x) := \left\{ q \in \mathcal{W} \cap \overline{B\left(x, \tfrac{r+R}{2}\right)} \middle| \left\| q - x + r \frac{x - \Pi_{\overline{S}_i}(x)}{\|x - \Pi_{\overline{S}_i}(x)\|} \right\| \le \left\| q - \Pi_{\overline{S}_i}(x) \right\|, \forall i \text{ s.t. } S_i \ne \varnothing \right\}. \tag{22}$$

Note that $\overline{B\left(x, \tfrac{r+R}{2}\right)}$ is equal to the intersection of the closed half spaces containing the robot body and defined by the maximum margin separating hyperplanes of the robot body, $\overline{B(x, r)}$, and all individual points, $q \in \mathbb{R}^n \setminus B(x, R)$, outside its sensory footprint.

An important observation revealing a critical connection between the robot's local workspace $\mathcal{LW}$ in (7) and its sensed local workspace $\mathcal{LW}_{\mathcal{S}}$ in (22) is:

Proposition 8. $\mathcal{LW}_{\mathcal{S}}(x) = \mathcal{LW}(x) \cap \overline{B\left(x, \tfrac{r+R}{2}\right)}$ *for all* $x \in \mathcal{W}$.

Proof. See Appendix I-D in the Technical Report [26].   ∎

In accordance with its local free space $\mathcal{LF}(x)$ in (9), we define the robot's *sensed local free space* $\mathcal{LF}_{\mathcal{S}}(x)$ by eroding $\mathcal{LW}_{\mathcal{S}}(x)$ by the robot body, illustrated in Fig. 2(middle), as,

$$\mathcal{LF}_{\mathcal{S}}(x) := \left\{ q \in \mathcal{LW}_{\mathcal{S}}(x) \middle| \overline{B(q, r)} \subseteq \mathcal{LW}_{\mathcal{S}}(x) \right\} = \mathcal{LF}(x) \cap \overline{B\left(x, \tfrac{R-r}{2}\right)}, \tag{23}$$

where the latter follows from Proposition 8 and that the erosion operation is distributed over set intersection [33]. Note that, for any $x \in \mathcal{F}$, $\mathcal{LF}_{\mathcal{S}}(x)$ is a nonempty closed convex set containing x as is $\mathcal{LF}(x)$.

To safely steer a single-integrator disk-shaped robot towards a given goal location $x^* \in \mathcal{F}$ using a fixed radius sensory footprint, we propose the following "move-to-projected-goal" law,

$$u(x) = -k\left(x - \Pi_{\mathcal{LF}_{\mathcal{S}}(x)}(x^*)\right), \tag{24}$$

where $k > 0$ is a fixed control gain, and $\Pi_{\mathcal{LF}_{\mathcal{S}}(x)}$ (4) is the metric projection onto the robot's sensed local free space $\mathcal{LF}_{\mathcal{S}}(x)$, which is assumed to be continuously updated.

Due to the nice relations between the robot's different local neighborhoods in Proposition 8 and (23), the revised "move-to-projected-goal" law for a fixed radius sensory footprint inherits all qualitative properties of the original one presented in Section 4.2.

Proposition 9. *The "move-to-projected-goal" law of a disk-shaped robot equipped with a fixed radius sensory footprint in (24) is piecewise continuously differentiable; and if Assumption 2 holds, then its unique continuously differentiable flow asymptotically steers almost all configurations in its positively invariant domain $\mathcal{F}$ towards any given goal location $x^* \in \mathcal{F}$, while strictly decreasing the (squared) Euclidean distance to the goal along the way.*

Proof. See the Technical Report [26].   ∎

Navigation using a 2D LIDAR Range Scanner. We now present another practical extension of the "move-to-projected-goal" law for safe robot navigation using a 2D LIDAR range scanner in an unknown convex planar environment $\mathcal{W} \subseteq \mathbb{R}^2$ populated with convex obstacles $\mathcal{O} = \{O_1, O_2, \ldots, O_m\}$, satisfying Assumption 1. Assuming an angular scanning range of 360 degrees and a fixed radial range of $R \in \mathbb{R}_{>0}$, we model the sensory measurement of the LIDAR scanner at location $x \in \mathcal{W}$ by a polar curve [47] $\rho_x : (-\pi, \pi] \to [0, R]$, defined as,

$$\rho_x(\theta) := \min \begin{pmatrix} R, \\ \min\{\|p - x\| \,\big|\, p \in \partial\mathcal{W}, \text{atan2}(p - x) = \theta\}, \\ \min_i \{\|p - x\| \,\big|\, p \in O_i, \text{atan2}(p - x) = \theta\} \end{pmatrix}. \quad (25)$$

Here, the LIDAR sensing range R is asummed to be greater than the robot body radius r.

Suppose $\rho_i : (\theta_{l_i}, \theta_{u_i}) \to [0, R]$ is a convex curve segment of the LIDAR scan ρ_x (25) at location $x \in \mathcal{W}$ (please refer to Appendix V in the Technical Report [26] for the notion of convexity in polar coordinates which we use to identify convex polar curve segments in a LIDAR scan, corresponding to the obstacle and workspace boundary), then we define the associated *line-of-sight obstacle* as the open epigraph of ρ_i whose pole is located at x [47],[7][14]

$$L_i := \{x\} \oplus \overset{\circ}{\text{epi}}\rho_i = \{x\} \oplus \left\{ (\varrho \cos\theta, \varrho \sin\theta) \,\big|\, \theta \in (\theta_{l_i}, \theta_{u_i}), \varrho > \rho_i(\theta) \right\}, \quad (26)$$

which is an open convex set. Accordingly, we assume the availability of a sensor model $\mathcal{L}_R(x) := \{L_1, L_2, \ldots, L_t\}$ that returns the list of convex line-of-sight obstacles detected by the LIDAR scanner at location x, where t denotes the number of detected obstacles and changes as a function of robot location.

Following the lines of (7) and (9), we define the robot's *line-of-sight local workspace* and *line-of-sight local free space*, illustrated in Fig. 2(right), respectively, as

$$\mathcal{LW}_{\mathcal{L}}(x) := \left\{ q \in L_{ft}(x) \cap \overline{B\left(x, \tfrac{r+R}{2}\right)} \,\Big|\, \left\| q - x + r\frac{x - \Pi_{\overline{L_i}}(x)}{\|x - \Pi_{\overline{L_i}}(x)\|} \right\| \leq \|q - \Pi_{\overline{L_i}}(x)\|, \forall i \right\}. (27)$$

$$\mathcal{LF}_{\mathcal{L}}(x) := \left\{ q \in \mathcal{LW}_{\mathcal{L}}(x) \,\big|\, \overline{B(q, r)} \subseteq \mathcal{LW}_{\mathcal{L}}(x) \right\}, \quad (28)$$

where $L_{ft}(x)$ denotes the LIDAR sensory footprint at x, given by the hypograph of the LIDAR scan ρ_x (25) at x, i.e.,

$$L_{ft}(x) := \{x\} \oplus \text{hyp}\rho_x = \{x\} \oplus \left\{ (\varrho \cos\theta, \varrho \sin\theta) \,\big|\, \theta \in (-\pi, \pi], 0 \leq \varrho \leq \rho_x(\theta) \right\}. (29)$$

Similar to Proposition 1 and Corollary 1, we have:

Proposition 10. *For any* $x \in \mathcal{F}$, $\mathcal{LW}_{\mathcal{L}}(x)$ *is an obstacle free closed convex subset of* $\mathcal{W}$ *and contains the robot body* $B(x, r)$. *Therefore,* $\mathcal{LF}_{\mathcal{L}}(x)$ *is a nonempty closed convex subset of* $\mathcal{F}$ *and contains* x.

Proof. See Appendix I-E in the Technical Report [26]. ∎

Accordingly, to navigate a fully-actuated single-integrator robot using a LIDAR scanner towards a desired goal location $x^* \in \mathcal{F}$, with the guarantee of no collisions along the way, we propose the following "move-to-projected-goal" law

$$u(x) = -k\big(x - \Pi_{\mathcal{LF}_{\mathcal{L}}(x)}(x^*)\big), \quad (30)$$

where $k > 0$ is fixed, and $\Pi_{\mathcal{LF}_{\mathcal{L}}(x)}$ (4) is the metric projection onto the robot's line-of-sight free space $\mathcal{LF}_{\mathcal{L}}(x)$ (28), which is assumed to be continuously updated.

---

[14] Here, $\overset{\circ}{A}$ denotes the interior of a set A.

We summarize important properties of the "move-to-projected-goal" law for navigation using a LIDAR scanner as:

Proposition 11. *The "move-to-projected-goal" law of a LIDAR-equipped disk-shaped robot in (30) leaves the robot's free space $\mathcal{F}$ (1) positively invariant; and if Assumption 2 holds, then its unique, continuous and piecewise differentiable flow asymptotically brings all but a mesure zero set of initial configurations in $\mathcal{F}$ to any designated goal location $x^* \in \mathcal{F}$, while strictly decreasing the (squared) Euclidean distance to the goal along the way.*

Proof. See Appendix I-F in the Technical Report [26].   ∎

As a final remark, it is useful to note that the "move-to-projected-goal" law in (30) might have discontinuities because of possible occlusions between obstacles. If there is no occlusion between obstacles in the LIDAR's sensing range, then the LIDAR scanner provides exactly the same information about obstacles as does the fixed radius sensory footprint of Section 4.3, and so the "move-to-projected-goal" law in (30) is piecewise continuously differentiable as is its version in (24). In this regard, one can avoid occlusions between obstacles by properly selecting the LIDAR's sensing range: for example, since $d(x, O_i) \geq r$ for any $x \in \mathcal{F}$ and $d(O_i, O_j) > 2r$ for any $i \neq j$ (Assumption 1), a conservative choice of R that prevents occlusions between obstacles is $r < R \leq 3r$.

5   Numerical Simulations

To demonstrate the motion pattern generated by our "move-to-projected-goal" law around and far away from the goal, we consider a 10×10 and a 50×10 environment cluttered with convex obstacles and a desired goal located at around the upper right corner, as illustrated in Fig. 3 and Fig. 4, respectively.[15] We present in these figures example navigation trajectories of the "move-to-projected-goal" law for different sensing modalities. We observe a significant consistency between the resulting trajectories of the "move-to-projected-goal" law and the boundary of the Voronoi diagram of the environment, where the robot balances its distance to all proximal obstacles while navigating towards its destination — a desired autonomous behaviour for many practical settings instead of following the obstacle boundary tightly. In our simulations, we avoid occlusions between obstacles by properly selecting the LIDAR's sensing range, and in so doing both limited range sensing models provide the same information about the environment away from the workspace boundary and the associated "move-to-projected-goal" laws yield almost the same navigation paths. It is also useful to note that the "move-to-projected-goal" law decreases not only the Euclidean distance, $\|x - x^*\|$, to the goal, but also the Euclidean distance, $\left\| \Pi_{\mathcal{L}\mathcal{F}(x)}(x^*) - x^* \right\|$, between the projected goal, $\Pi_{\mathcal{L}\mathcal{F}(x)}(x^*)$, and the global goal, x^*, illustrated in Fig. 3(a).

---

[15] For all simulations we set $r = 0.5$, $R = 2$ and $k = 1$, and all simulations are obtained through numerical integration of the associated "move-to-projected-goal" law using the `ode45` function of MATLAB. Please refer to Appendix VII in the Technical Report [26] and see the accompanying video for additional figures illustrating the navigation pattern far away from the goal for different sensing and actuation models.

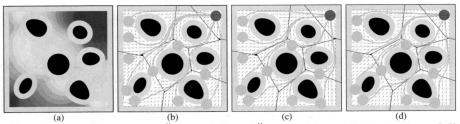

| (a) | (b) | (c) | (d) |

Fig. 3. (a) The Euclidean distance, $\left\| \Pi_{\mathcal{LF}(x)}(x^*) - x^* \right\|$, between the projected goal, $\Pi_{\mathcal{LF}(x)}(x^*)$, and the global goal, x^*, for Voronoi-adjacent[9] obstacle sensing. (b-d) Example navigation trajectories of the "move-to-projected-goal" law starting at a set of initial configurations (green) towards a designated point goal (red) for different sensing models: (c) Voronoi-adjacent[9] obstacle sensing, (d) a fixed radius sensory footprint, (e) a limited range LIDAR sensor.

6   Conclusions

In this paper we construct a sensor-based feedback law that solves the real-time collision-free robot navigation problem in a domain cluttered with unknown but sufficiently separated and strongly convex obstacles. Our algorithm introduces a novel use of separating hyperplanes for identifying the robot's local obstacle free convex neighborhood, affording a piecewise smooth velocity command instantaneously pointing toward the metric projection of the designated goal location onto this convex set. Given sufficiently separated (Assumption 1) and appropriately "strongly" convex (Assumption 2) obstacles, we show that the resulting vector field has a smooth flow with a unique attractor at the goal location (along with the topologically inevitable saddles — at least one for each obstacle). Since all of its critical points are nondegenerate, our vector field asymptotically steers almost all configurations in the robot's free space to the goal, with the guarantee of no collisions along the way. We also present its practical extensions for two limited range sensing models. We illustrate the effectiveness of the proposed navigation algorithm in numerical simulations.

Work now in progress targets a fully smoothed version of the move-to-projected-goal law (by recourse to reference governors [48]), permitting its lift to more complicated dynamical models such as force-controlled (second-order) and underactuated systems [49]. This will enable its empirical demonstration for safe, high-speed navigation in a forest-like environments [50] and in human crowds. We are also investigating the extension of these ideas for coordinated, decentralized feedback control of multirobot swarms. More generally, we seek to identify fundamental limits on navigable environments for a memoryless greedy robotic agent with a limited range sensing capability.

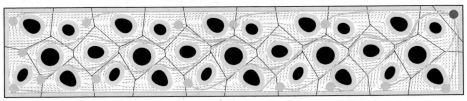

Fig. 4. Example navigation trajectories of the "move-to-projected-goal" law in (10) starting at a set of initial positions (green) far away from the goal (red).[15]

Acknowlegdment. This work was supported by AFOSR CHASE MURI FA9550-10-1-0567.

References

1. Trautman, P., Ma, J., Murray, R.M., Krause, A.: Robot navigation in dense human crowds: Statistical models and experimental studies of humanrobot cooperation. The International Journal of Robotics Research **34**(3) (2015) 335–356
2. Henry, P., Vollmer, C., Ferris, B., Fox, D.: Learning to navigate through crowded environments. In: Robotics and Automation, IEEE International Conference on. (2010) 981–986
3. Karaman, S., Frazzoli, E.: High-speed flight in an ergodic forest. In: Robotics and Automation (ICRA), IEEE International Conference on. (2012) 2899–2906
4. Paranjape, A.A., Meier, K.C., Shi, X., Chung, S.J., Hutchinson, S.: Motion primitives and 3D path planning for fast flight through a forest. The Int. J. Robot. Res. **34**(3) (2015) 357–377
5. Wooden, D., Malchano, M., Blankespoor, K., Howardy, A., Rizzi, A.A., Raibert, M.: Autonomous navigation for BigDog. In: IEEE Int. Conf. Robot. Autom. (2010) 4736–4741
6. Johnson, A.M., Hale, M.T., Haynes, G.C., Koditschek, D.E.: Autonomous legged hill and stairwell ascent. In: IEEE Int. Symp. on Safety, Security, Rescue Robotics. (2011) 134–142
7. Choset, H., Lynch, K.M., Hutchinson, S., Kantor, G.A., Burgard, W., Kavraki, L.E., Thrun, S.: Principles of Robot Motion: Theory, Algorithms, and Implementations. MIT Press (2005)
8. LaValle, S.M.: Planning Algorithms. Cambridge University Press, Cambridge, U.K. (2006)
9. Koditschek, D.E., Rimon, E.: Robot navigation functions on manifolds with boundary. Advances in Applied Mathematics **11**(4) (1990) 412 – 442
10. Khatib, O.: Real-time obstacle avoidance for manipulators and mobile robots. The International Journal of Robotics Research **5**(1) (1986) 90–98
11. Koditschek, D.E.: Exact robot navigation by means of potential functions: Some topological considerations. In: Robotics and Automation, IEEE International Conference on. (1987) 1–6
12. Borenstein, J., Koren, Y.: The vector field histogram-fast obstacle avoidance for mobile robots. IEEE Transactions on Robotics and Automation **7**(3) (1991) 278–288
13. Simmons, R.: The curvature-velocity method for local obstacle avoidance. In: Robotics and Automation (ICRA), IEEE International Conference on. (1996) 3375–3382
14. Fox, D., Burgard, W., Thrun, S.: The dynamic window approach to collision avoidance. IEEE Robotics Automation Magazine **4**(1) (1997) 23–33
15. Fiorini, P., Shiller, Z.: Motion planning in dynamic environments using velocity obstacles. The International Journal of Robotics Research **17**(7) (1998) 760–772
16. Rimon, E., Koditschek, D.E.: Exact robot navigation using artificial potential functions. Robotics and Automation, IEEE Transactions on **8**(5) (1992) 501–518
17. Lionis, G., Papageorgiou, X., Kyriakopoulos, K.J.: Locally computable navigation functions for sphere worlds. In: Robotics and Automation, IEEE Int. Conf. on. (2007) 1998–2003
18. Filippidis, I., Kyriakopoulos, K.J.: Adjustable navigation functions for unknown sphere worlds. In: IEEE Decision and Control and European Control Conf. (2011) 4276–4281
19. Burridge, R.R., Rizzi, A.A., Koditschek, D.E.: Sequential composition of dynamically dexterous robot behaviors. The International Journal of Robotics Research **18**(6) (1999) 535–555
20. Conner, D.C., Choset, H., Rizzi, A.A.: Flow-through policies for hybrid controller synthesis applied to fully actuated systems. Robotics, IEEE Transactions on **25**(1) (2009) 136–146
21. Choset, H., Burdick, J.: Sensor-based exploration: The hierarchical generalized Voronoi graph. The International Journal of Robotics Research **19**(2) (2000) 96–125
22. Boyd, S., Vandenberghe, L.: Convex Optimization. Cambridge University Press (2004)
23. Arslan, O., Koditschek, D.E.: Exact robot navigation using power diagrams. In: Robotics and Automation (ICRA), IEEE International Conference on. (2016) 1–8
24. Aurenhammer, F.: Power diagrams: Properties, algorithms and applications. SIAM Journal on Computing **16**(1) (1987) 78–96

25. Webster, R.: Convexity. Oxford University Press (1995)
26. Arslan, O., Koditschek, D.E.: Sensor-based reactive navigation in unknown convex sphere worlds. Technical report, University of Pennsylvania (2016) Online available at: `http://kodlab.seas.upenn.edu/Omur/WAFR2016`.
27. Ó'Dúnlaing, C., Yap, C.K.: A retraction method for planning the motion of a disc. Journal of Algorithms **6**(1) (1985) 104 – 111
28. Cortés, J., Martınez, S., Karatas, T., Bullo, F.: Coverage control for mobile sensing networks. Robotics and Automation, IEEE Transactions on **20**(2) (2004) 243–255
29. Kwok, A., Martnez, S.: Deployment algorithms for a power-constrained mobile sensor network. International Journal of Robust and Nonlinear Control **20**(7) (2010) 745–763
30. Pimenta, L.C., Kumar, V., Mesquita, R.C., Pereira, G.A.: Sensing and coverage for a network of heterogeneous robots. In: Decision and Control, IEEE Conference on. (2008) 3947–3952
31. Arslan, O., Koditschek, D.E.: Voronoi-based coverage control of heterogeneous disk-shaped robots. In: Robotics and Automation, IEEE International Conference on. (2016) 4259–4266
32. Okabe, A., Boots, B., Sugihara, K., Chiu, S.N.: Spatial Tessellations: Concepts and Applications of Voronoi Diagrams. 2nd edn. Volume 501. John Wiley & Sons (2000)
33. Haralick, R.M., Sternberg, S.R., Zhuang, X.: Image analysis using mathematical morphology. Pattern Analysis and Machine Intelligence, IEEE Transactions on **9**(4) (1987) 532–550
34. Munkres, J.: Topology. 2nd edn. Pearson (2000)
35. Kozlov, M., Tarasov, S., Khachiyan, L.: The polynomial solvability of convex quadratic programming. USSR Comp. Mathematics and Mathematical Physics **20**(5) (1980) 223–228
36. Bullo, F., Cortés, J., Martinez, S.: Distributed Control of Robotic Networks: A Mathematical Approach to Motion Coordination Algorithms. Princeton University Press (2009)
37. Rockafellar, R.: Lipschitzian properties of multifunctions. Nonlinear Analysis: Theory, Methods & Applications **9**(8) (1985) 867–885
38. Kuntz, L., Scholtes, S.: Structural analysis of nonsmooth mappings, inverse functions, and metric projections. Journal of Math. Analysis and Applications **188**(2) (1994) 346–386
39. Shapiro, A.: Sensitivity analysis of nonlinear programs and differentiability properties of metric projections. SIAM Journal on Control and Optimization **26**(3) (1988) 628–645
40. Liu, J.: Sensitivity analysis in nonlinear programs and variational inequalities via continuous selections. SIAM Journal on Control and Optimization **33**(4) (1995) 1040–1060
41. Chaney, R.W.: Piecewise C^k functions in nonsmooth analysis. Nonlinear Analysis: Theory, Methods & Applications **15**(7) (1990) 649 – 660
42. Khalil, H.K.: Nonlinear Systems. 3rd edn. Prentice Hall (2001)
43. Paternain, S., Koditschek, D.E., Ribeiro, A.: Navigation functions for convex potentials in a space with convex obstacles. IEEE Transactions on Automatic Control (submitted)
44. Holmes, R.B.: Smoothness of certain metric projections on Hilbert space. Transactions of the American Mathematical Society **184** (1973) 87–100
45. Fitzpatrick, S., Phelps, R.R.: Differentiability of the metric projection in Hilbert space. Transactions of the American Mathematical Society **270**(2) (1982) 483–501
46. Hirsch, M.W., Smale, S., Devaney, R.L.: Differential Equations, Dynamical Systems, and an Introduction to Chaos. 2nd edn. Academic Press (2003)
47. Stewart, J.: Calculus: Early Transcendentals. 7th edn. Cengage Learning (2012)
48. Kolmanovsky, I., Garone, E., Cairano, S.D.: Reference and command governors: A tutorial on their theory and automotive applications. In: American Control Conf. (2014) 226–241
49. Arslan, O., Koditschek, D.E.: Smooth extensions of feedback motion planners via reference governors. In: (accepted) Robotics and Automation (ICRA), IEEE Int. Conf. on. (2017)
50. Vasilopoulos, V., Arslan, O., De, A., Koditschek, D.E.: Minitaur bounds home through a locally sensed artificial forest. In: (submitted) IEEE/RSJ Int. Conf. Intel. Robot. Sys. (2017)

You can't save all the pandas: impossibility results for privacy-preserving tracking

Yulin Zhang and Dylan A. Shell

Department of Computer Science and Engineering,
Texas A&M University,
College Station, TX 77840, USA
yulinzhang|dshell@tamu.edu

Abstract. We consider the problem of target tracking whilst simultaneously preserving the target's privacy as epitomized by the panda tracking scenario introduced by O'Kane at *WAFR'08*. The present paper reconsiders his formulation, with its elegant illustration of the utility of ignorance, and the tracking strategy he proposed, along with its completeness. We explore how the capabilities of the robot and panda affect the feasibility of tracking with a privacy stipulation, uncovering intrinsic limits, no matter the strategy employed. This paper begins with a one-dimensional setting and, putting the trivially infeasible problems aside, analyzes the strategy space as a function of problem parameters. We show that it is not possible to actively track the target as well as protect its privacy for every nontrivial pair of tracking and privacy stipulations. Secondly, feasibility is sensitive, in several cases, to the information available to the robot at the start — conditions we call initial I-state dependent cases. Quite naturally in the one-dimensional model, one may quantify sensing power by the number of perceptual (or output) classes available to the robot. The number of initial I-state dependent conditions does not decrease as the robot gains more sensing power and, further, the robot's power to achieve privacy-preserving tracking is bounded, converging asymptotically with increasing sensing power. Finally, to relate some of the impossibility results in one dimension to their higher-dimensional counterparts, including the planar panda tracking problem studied by O'Kane, we establish a connection between tracking dimensionality and the sensing power of a one-dimensional robot.

1 Introduction

Most roboticists see uncertainty as something which should be minimized or even eliminated if possible. But, as robots become widespread, it is likely that there will be a shift in thinking— robots that know too much are also problematic in their own way. A robot operating in your home, benignly monitoring your activities and your daily routine, for example to schedule vacuuming at unobtrusive times, possesses information that is valuable. There are certainly those who could derive profit from it. A tension exists between information necessary for a robot to be useful and information which could be sensitive if mistakenly disclosed or stolen. But the problem of establishing the minimal information required to perform a particular task and the problem of analyzing the trade-off between information and performance, despite both being fundamental challenges, remain largely uncharted territory — notwithstanding [1] and [2].

© Springer Nature Switzerland AG 2020
K. Goldberg et al. (Eds.): *Algorithmic Foundations of Robotics XII*, SPAR 13, pp. 176–191, 2020.
https://doi.org/10.1007/978-3-030-43089-4_12

The present paper's focus is on the problem of tracking a target whilst preserving the target's privacy. Though fairly narrow, this is a crisply formulated instance of the broader dilemma of balancing the information a robot possesses: the robot must maintain some estimate of the target's pose, but information that is too precise is an unwanted intrusion and potential hazard if leaked. The setting we examine, the *panda tracker problem*, is due to O'Kane [3] who expressed the idea of uncertainty being valuable and aloofness deserving respect. The following, quoted verbatim from [3, p. 1], describes the scenario:

> "A giant panda moves unpredictably through a wilderness preserve. A mobile robot tracks the panda's movements, periodically sensing partial information about the panda's whereabouts and transmitting its findings to a central base station. At the same time, poachers attempt to exploit the presence of the tracking robot—either by eavesdropping on its communications or by directly compromising the robot itself—to locate the panda. We assume, in the worst case, that the poachers have access to any information collected by the tracking robot, but they cannot control its motions. The problem is to design the tracking robot so that the base station can record coarse-grained information about the panda's movements, without allowing the poachers to obtain the fine-grained position information they need to harm the panda."

Note that it is not sufficient for the robot to simply forget or to degrade sensor data via post-processing because the adversary may have compromised these operations, simply writing the information to separate storage.

Before formalizing our approach to the problem, which we do in detail in Section 2, we give an overview of the questions of interest to us. We also give an outline of the structure of the paper and summarize the reported results.

One can view the informational constraints as bounds: (1) A maximal- or upper-bound specifies how coarse the tracking information can be. The robot is not helpful in assuring the panda's well-being when this bound is exceeded. (2) A second constraint, a lower-bound, stipulates that if the information is more fine-grained than some threshold, a poacher may succeed in having his wicked way. The problem is clearly infeasible when the lower-bound exceeds the upper-bound. What of other circumstances? Is it always possible to ensure that one will satisfy both bounds indefinitely? In the original paper, O'Kane [3] proposed a tracking strategy for a robot equipped with a two-bit quadrant sensor, which certainly succeeds in several circumstances. As no claim of completeness was made, will the strategy work for all feasible bounds? And how are the strategies affected by improvements in the sensing capabilities of the robot?

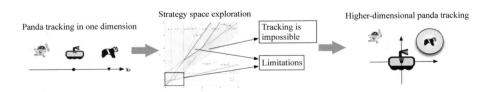

Fig. 1: An overview of the approach taken in the paper.

Our examination of these questions follows the outline shown in Figure 1. We start with a one-dimensional panda tracking problem, analyzing it in detail in Section 3, and find that it is impossible for the robot to achieve privacy-preserving tracking for all nontrivial tracking and privacy stipulations. In addition, there are instances where privacy-preserving tracking is limited, depending crucially on the initial information available to the robot. Then, in Section 4, we show that the impossibility conclusion holds for O'Kane's original planar panda tracking problem too.

2   Problem description: one-dimensional panda tracking

The original problem was posed in two dimensions with the robot and panda inhabiting a plane that is assumed to be free of obstacles. They move in discrete time-steps, interleaving their motions. A powerful adversary, who is interested in computing the possible locations of the panda, is assumed to have access to the full history of information. Any information is presumed to be used optimally by the adversary in reconstruction of possible locations of the panda — by which we mean that the region that results is both sound (that is, consistently accounted for by the information) but is also tight (no larger than necessary). The problem is formulated without needing to appeal to probabilities by considering only worst-case reasoning and by using motion- and sensor-models characterized by regions and applying geometric operations.

Information stipulation: The tracking and privacy requirements were specified as two disks. The robot is constrained to ensure that the region describing possible locations of the panda always fits inside the *tracking disk*, which has the larger diameter of the two. The *privacy disk* imposes the requirement that it always be possible to place the smaller disk wholly inside the set of possible locations.

Sensor model: As reflected in the title of his paper, O'Kane considered an unconventional sensor that outputs only two bits of information per measurement. With origin centered on the robot, the sensor outputs the quadrant containing the panda.

Target motion model: The panda moves unpredictably with bounded velocity. Newly feasible locations can be modeled as the convolution of the previous time-step's region with a disk, sized appropriately for the time-step interval.

Now, by way of simplification, consider pandas and robots that inhabit obstacle-free one-dimensional worlds, each only moving left or right along a line.

Information stipulation: Using the obvious 1-dimensional analogue, now the robot tracker has to bound its belief about the panda's potential locations to an interval of minimum size r_p (p for privacy) and maximum size r_t (t for tracking).

Sensor model: Most simply, the quadrant sensor corresponds to a one-bit sensor indicating whether the panda is on the robot's left- or right-hand side. When the robot is at u_1, the sensor indicates whether the panda is within $(-\infty, u_1]$ or (u_1, ∞).

We have sought a way to explore how modifying the robot's sensing capabilities alters its possible behavior. Our approach is to give the robot m set-points $u_1 < u_2 < \cdots < u_m$, each within the robot's control, so that it can determine which of the $m + 1$ non-overlapping intervals contains the panda. With m set-points one can model any sensor that fully divides the state space and produces at most $m + 1$ observations.* The case with $m = 1$ is the straightforward analogue of the quadrant sensor. Increasing m yields a robot with greater sensing power, since the robot has a wider choice of how observations should inform itself of the panda's location.

Target motion model: The convolution becomes a trivial operation on intervals.

Figure 2 is a visual example with $m = 2$. The panda is sensed by the robot as falling within one of the following intervals: $(-\infty, u_1]$, $(u_1, u_2]$, (u_2, ∞), where $u_1, u_2 \in \mathbb{R}$ and $u_1 < u_2$. And these three intervals are represented by observation values: 0, 1 and 2. For simplicity, no constraints are imposed on the robot's motion and we assume that each time-step, the robot can pick positions of $u_1 < \cdots < u_m$ as it likes.

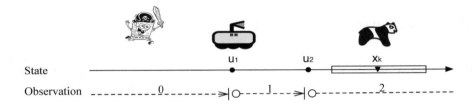

Fig. 2: Panda tracking in one dimension. (Inimitable artwork for the robot and panda adapted from the original in [3, p. 2].)

Notation and model:

The 1-dim. problem is formulated as follows. The panda's location is represented as a single coordinate indexed by discrete time. At stage k the location of the panda is $x_k \in \mathbb{R}$. Incorporating accumulated knowledge about the panda's possible location, sensor readings, and the movement model permits the robot (and adversary) to maintain knowledge of the panda's possible location after it moves and between sensing steps. The set of conceivable locations of the panda (a subset of $\mathbb{R}$) is a geometric realization of an information-state or I-state, as formalized and treated in detail by LaValle [4]. In this paper, we take the I-state as an interval.

The movement of the panda per time-step is bounded by length $\frac{\delta}{2}$, meaning that the panda can move at most $\frac{\delta}{2}$ in either direction. We use η_k to denote the robot's knowledge of the panda after the observation taken at time k. In evolving from η_k to η_{k+1} the robot's I-state first transits to an intermediate I-state, which we write η_{k+1}^- representing the state after adding the uncertainty arising from the panda's movement, but before observation $k + 1$. Since this update occurs before the sensing information is incorporated, we refer to η_{k+1}^- as the *prior* I-state for time $k + 1$. Updating I-state η_k involves mapping every $x_k \in \eta_k$ to $[x_k - \frac{\delta}{2}, x_k + \frac{\delta}{2}]$, the resultant I-state, η_{k+1}^-, being the union of the results.

---

*This is not a model of any physical sensor of which we are aware. The reader, finding this too contrived, may find merit in the choice later, e.g., for n-dimensional tracking (see Lemma 6).

Sensor reading updates to the I-state depend on the values of $u_1(k), u_2(k), \ldots, u_m(k)$, which are under the control of the robot. The sensor reports the panda's location to within one of the $m+1$ non-empty intervals: $(-\infty, u_1(k)], (u_1(k), u_2(k)], (u_2(k), u_3(k)], \ldots, (u_m(k), \infty)$. If we represent the observation at time k as a non-empty interval $y(k)$ then the *posterior* I-state η_k is updated as $\eta_k = \eta_k^- \cap y(k)$.

For every stage k, the robot chooses a sensing vector $\mathbf{v_k} = [u_1(k), u_2(k), \ldots, u_m(k)]$, $u_i(k) < u_j(k)$ if $i < j$, $u_i(k) \in \mathbb{R}$, so as to achieve the following conditions:

1. *Privacy Preserving Condition (PPC)*: The size of any I-state $\eta_k = [a, b]$ should be at least r_p. That is, for every stage k, $|\eta_k| = b - a \geq r_p$.
2. *Target Tracking Condition (TTC)*: The size of any I-state $\eta_k = [a, b]$ should be at most r_t. That is, for every stage k, $|\eta_k| = b - a \leq r_t$.

3   Privacy-preserving tracking

Given specific problem parameters, we are interested in whether there is always some $\mathbf{v_k}$ that a robot can select to track the panda while satisfying the preceding conditions.

Definition 1. *A 1-dim. panda tracking problem is a tuple $P_1 = (\eta_0, r_p, r_t, \delta, m)$, in which*
1) the initial I-state η_0 describes all the possible initial locations of the panda;
2) the privacy bound r_p gives a lower bound on the I-state size;
3) the tracking bound r_t gives a upper bound on the I-state size;
4) parameter δ describes the panda's (fastest) motion;
5) the sensor capabilities are given by the number m.

Definition 2. *The 1-dim. panda tracking problem $P_1 = (\eta_0, r_p, r_t, \delta, m)$ is privacy preserv-able, written as predicate $\mathbf{PP}(P_1)$, if starting with $|\eta_0| \in [r_p, r_t]$, there exists some strategy π to determine a $\mathbf{v_k}$ at each time-step, such that the Privacy Preserving Condition holds forever. Otherwise, the problem P_1 is not privacy preservable: $\neg \mathbf{PP}(P_1)$.*

Definition 3. *The 1-dim. panda tracking problem $P_1 = (\eta_0, r_p, r_t, \delta, m)$ is target track-able, $\mathbf{TT}(P_1)$, if starting with $|\eta_0| \in [r_p, r_t]$, there exists some strategy π to determine a $\mathbf{v_k}$ at each time-step, such that the Target Tracking Condition holds forever. Otherwise, the problem P_1 is not target trackable: $\neg \mathbf{TT}(P_1)$.*

To save space, we say a problem P_1 and also its strategy π are **PP** if $\mathbf{PP}(P_1)$. Similarly, both P_1 and its strategy π will be called **TT** if $\mathbf{TT}(P_1)$. Putting aside trivially infeasible P_1 where $r_p > r_t$, we wish to know which problems are both **PP** and **TT**. Next, we explore the parameter space to classify the various classes of problem instances by investigating the existence of strategies.

3.1   Roadmap of technical results

The results follow from several lemmas. The roadmap in Figure 3 provides a sense of how the pieces fit together to help the reader keep an eye on the broader picture.

Our approach begins, first, by dividing the strategy space into 'teeth' and 'gaps' according to the speed (or size) of the panda's motion. **PP** and **TT** tracking strategies

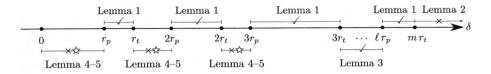

Fig. 3: Roadmap of results for 1-dim. panda tracking with m sensing parameters.

are shown to exist in teeth regions and the last gap region for all initial I-states that satisfy $|\eta_0| \in [r_p, r_t]$. The remaining regions differ: their feasibility depends not only on the panda's motion but also on the size of initial I-state. Dealing with the remaining gap regions is simplified considerably by focusing on $\mathbf{v_k}$ that divides η_k^- evenly. We use $s(i)$ to represent the choice of evenly dividing the prior I-state interval into i parts (the mnemonic is s for *split*). In practice, we will usually compare choices $s(a)$ and $s(a+1)$, for positive integer constant a determined from the speed of the panda's motion ($\frac{\delta}{2}$).

The action of evenly dividing the I-state interval η_k^- via $s(\cdot)$ is useful because, after the sensor reading has been processed, one knows the degree of the uncertainty involved, i.e., the size of the I-state interval η_k. An $s(i)$ action results in greater post-sensing uncertainty than an $s(i+1)$ does. Strategies consisting only of $s(i)$ and $s(i+1)$ actions enable examination of the evolution of interval sizes. If the I-state resulting from an $s(i)$ violates the tracking constraint, other actions dividing the I-state into at most i parts, though they be uneven parts, will also violate the tracking bound because we must guarantee a solution no matter which interval the panda happens to be in. Analogously, the $s(i+1)$ case is often useful in analyzing violation of the privacy bound.

We have found it helpful to think of the problem as playing a multi-stage game against suicidal pandas who strategically choose the movement that tends to violate the bounds. The robot's job is to save all of them. According to whether an $s(i)$ or $s(i+1)$ may be performed in the I-state or not, we are able to divide the remaining strategy space into four cases where two of them are non-**PP** or non-**TT**, one is **PP** and **TT** regardless of the initial I-state, and the other is I-state dependent.

3.2 Main results

In this section, we follow the roadmap in Figure 3.

Lemma 1. *For any 1-dim. panda tracking problem $P_1 = (\eta_0, r_p, r_t, \delta, m)$, if $\delta \in [ar_p, ar_t]$, where $a \in \mathbb{Z}^+$, $a \le m$, then $\mathbf{PP}(P_1) \wedge \mathbf{TT}(P_1)$.*

Proof. A **PP** and **TT** strategy is given in this proof. For any $|\eta_k| \in [r_p, r_t]$ the prior I-state has size $|\eta_{k+1}^-| = |\eta_k| + \delta$. Since $\delta \in [ar_p, ar_t](a \leq m)$, $|\eta_{k+1}^-| \in [ar_p + r_p, ar_t + r_t]$. By taking action $s(a+1)$, which is possible since $a \leq m$, we get $|\eta_{k+1}| = \frac{1}{a+1}|\eta_{k+1}^-| \in [r_p, r_t]$. That is, if $|\eta_0| \in [r_p, r_t]$ and we take action $s(a+1)$, then $|\eta_k| \in [r_p, r_t]$. Therefore, there exists a strategy (always take action $s(a+1)$) for P_1, so that the privacy-preserving tracking conditions *PPC* and *TTC* are always both satisfied when $\eta_0 \in [r_p, r_t]$. $\quad\square$

Lemma 2. *For any 1-dim. panda tracking problem* $P_1 = (\eta_0, r_p, r_t, \delta, m)$, *if* $\delta \in (mr_t, \infty)$, *then* $\neg TT(P_1)$.

Proof. The tracking stipulation is proved to be violated eventually. Given the constraint of m sensing parameters, the prior I-state $|\eta_{k+1}^-| = |\eta_k| + \delta$ can be divided into at most $m + 1$ parts. Among these $m + 1$ posterior I-states, if m of them reach the maximum size r_t, the size of the remaining I-state is $|\eta_k| + \delta - mr_t$. If none of the resulting I-states violate the tracking bound, then the size of the smallest resulting I-state is $|\eta_k| + \delta - mr_t$. Since $\delta > mr_t$, the size of the smallest resulting I-state must increase by some positive constant $\delta - mr_t$. After $\lceil \frac{r_t - r_p}{\delta - mr_t} \rceil$ stages, the I-state will exceed r_t. So it is impossible to ensure that the tracking bound will not eventually be violated. $\quad\square$

Lemma 3. *For 1-dim. panda tracking problem* $P_1 = (\eta_0, r_p, r_t, \delta, m)$, *if* $\delta \in (ar_t - r_t, ar_p)$, *where* $a \in \mathbb{Z}^+$, $a \leq m$ *and* $ar_t \geq ar_p + r_p$, *then* $PP(P_1) \wedge TT(P_1)$.

Proof. A **PP** and **TT** strategy is given in this proof. Since $\delta \in (ar_t - r_t, ar_p)$ and $ar_t > ar_p + r_p$, $|\eta_{k+1}^-| = |\eta_k| + \delta \in (ar_p, ar_t + r_t) \subset L_1 \cup L_2$, where $L_1 = [ar_p, ar_t]$ and $L_2 = [ar_p + r_p, ar_t + r_t]$. If action $s(a)$ is performed when $|\eta_{k+1}^-| \in L_1$ and $s(a+1)$ is performed when $|\eta_{k+1}^-| \in L_2$, the resulting I-state satisfies $|\eta_{k+1}| \in [r_p, r_t]$. Hence, there is a strategy consisting of $s(a)$ and $s(a+1)$, for the problem P_1 such that the *PPC* and *TTC* are always both satisfied when $\eta_0 \in [r_p, r_t]$. $\quad\square$

Lemma 4. *For any 1-dim. panda tracking problem* $P_1 = (\eta_0, r_p, r_t, \delta, m)$, *if* $\delta \in (ar_t - r_t, ar_p)$, *where* $a \in \mathbb{Z}^+$, $a \leq m$, $ar_t < ar_p + r_p$, *then* $\neg PP(P_1) \vee \neg TT(P_1)$ *when either:* *(i)* $r_p > ar_t - \delta$ *or* *(ii)* $r_t < (a+1)r_p - \delta$.

Proof. In case (i), $r_p > ar_t - \delta$, so the size of the prior I-state satisfies $|\eta_{k+1}^-| = |\eta_k| + \delta > r_p + \delta > ar_t$. That is, if η_{k+1}^- is divided into at most a parts, the largest posterior I-state η_{k+1} will violate the tracking stipulation at the next time-step. Thus, the prior I-state η_{k+1}^- must be divided into at least $a + 1$ parts. But by dividing $|\eta_{k+1}^-|$ into at least $a + 1$ parts, among all the resulting posterior I-states, it can be shown that the smallest I-state will eventually violate the privacy stipulation. The smallest size of the resulting posterior I-state $|\eta_{k+1}|_{smallest}$ is no greater than the average size $\frac{|\eta_k| + \delta}{a+1}$, when dividing η_{k+1}^- into $a + 1$ parts. For the smallest posterior I-state, the decrease in size is $\Delta_- = |\eta_k| - |\eta_{k+1}|_{smallest} \geq |\eta_k| - \frac{|\eta_k| + \delta}{a+1} \geq \frac{ar_p - \delta}{a+1} > 0$. Hence, eventually after $\lceil \frac{(a+1)(r_t - r_p)}{ar_p - \delta} \rceil$ steps, the smallest I-state will violate the privacy stipulation and put the panda in danger. Similarly, if the prior I-state is divided into less than $a + 1$ parts, the average size will become larger and the largest posterior I-state will violate the tracking stipulation, as it must increase at least as much as the average.

The same conclusion is reached for (ii), when $r_t < (a+1)r_p - \delta$, along similar lines. $\quad\square$

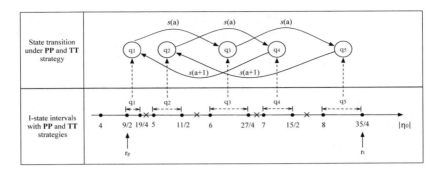

Fig. 4: A problem instance with $r_p = \frac{9}{2}, r_t = \frac{35}{4}, \delta = 2, a = 1$, where there exists a **PP** and **TT** tracking strategy for $|\eta_0| \in q_1 \cup q_2 \cup q_3 \cup q_4 \cup q_5$.

Lemma 5. *The tracking and privacy-preserving properties of a 1-dim. panda tracking problem $P_1 = (\eta_0, r_p, r_t, \delta, m)$ depend on the size of the initial I-state $|\eta_0|$, if $\delta \in (ar_t - r_t, ar_p)$, where $a \in \mathbb{Z}^+$, $a \leq m$ and $r_p \leq ar_t - \delta < (a+1)r_p - \delta \leq r_t$.*

Proof. The proof proceeds by showing, firstly, that there are certain initial I-states for which the problem is not both **PP** and **TT**. Next, we show that there exist other initial I-states where there are strategies which satisfy the conditions for the problem to be **PP** and **TT**.

First, for $|\eta_0| \in (ar_t - \delta, (a+1)r_p - \delta)$, if the prior I-state is divided into at most a parts, the size of the largest posterior I-state at the next time-step ($t = 1$) is $|\eta_1| = \frac{|\eta_1^-| + \delta}{a} > \frac{ar_t - \delta + \delta}{a} = r_t$, which will violate the tracking stipulation. If the prior I-state is divided into at least $a + 1$ parts, the size of posterior I-state at the next step ($t = 1$) is $|\eta_1| = \frac{|\eta_1^-| + \delta}{a+1} < \frac{(a+1)r_p - \delta}{a+1} = r_p$, which will violate the privacy stipulation. Hence, there are no strategies, which are both **PP** and **TT** if $|\eta_0| \in (ar_t - \delta, (a+1)r_p - \delta)$.

Second, we provide an example strategy that serves as an existence proof for **PP** and **TT** instances with $|\eta_0| \in [r_p, ar_t - \delta) \cup ((a+1)r_p - \delta, r_t]$. Consider the instance $r_p = \frac{9}{2}, r_t = \frac{35}{4}, \delta = 2, a = 1$, where there are **PP** and **TT** strategies for $\eta_0 \in q_1 \cup q_2 \cup q_3 \cup q_4 \cup q_5$, where $q_1 = [\frac{9}{2}, \frac{19}{4}], q_2 = [5, \frac{11}{2}], q_3 = [6, \frac{27}{4}], q_4 = [7, \frac{15}{2}], q_5 = [8, \frac{35}{4}]$. By always taking $s(a)$ and $s(a+1)$, the resulting I-state remains within $q_1, q_2, \ldots, q_5$ as shown in Figure 4. That is, there are strategies for those I-states in $q_1 \cup q_2 \cup \cdots \cup q_5$ and the problem is **PP** and **TT** under such circumstances. $\square$

Next, we collect the low-level results presented thus far to clarify their interplay.

3.3   Aggregation and summary of results

To have a clearer sense of how these pieces fit together, we found it helpful to plot the space of problem parameters and examine how the preceding theorems relate visually. Figure 5 contains subfigures for increasingly powerful robots (in terms of sensing) with $m = 1, 2, 3, 4$. The white regions represent the trivial $r_p > r_t$ instances; otherwise the whole space is categorized into the following subregions: **PP** and **TT** strategy space (colored green), not both **PP** and **TT** space (colored gray), and regions dependent on the initial I-states (colored pink — some caution is advised as particular values of η_0 are not visible in this projection). When summarized in this way, the results permit examination of how sensor power affects the existence of solutions.

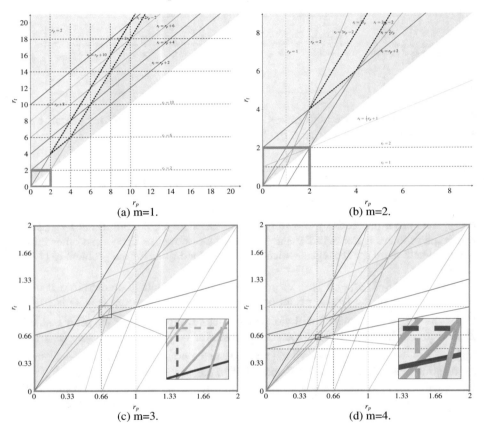

(a) m=1.

(b) m=2.

(c) m=3.

(d) m=4.

Fig. 5: The problem parameter space and the existence of strategies for robots with differing m. The green region depicts **PP** and **TT** conditions, where a suitable strategy exists no matter the initial I-state. The gray region represents the conditions that violate either **PP** or **TT** (or both). The pink region depicts conditions known to be I-state dependent. The white region represents trivially infeasible problems. The orange rectangles emphasize the different magnification-levels and highlight conditions where the differences in sensing power come into play. Both r_p and r_t are expressed in units of $\frac{\delta}{2}$.

Preserving the privacy of a target certainly makes some unusual demands of a sensor. O'Kane's quadrant sensor has pre-images for each output class that are infinite subsets of the plane, making it possible for his robot to increase its uncertainty if it must do so. But it remains far from clear how one might tackle the same problem with a different sensor. The privacy requirement makes it difficult to reason about the relationship between two similar sensors. For example, an octant sensor appears to be at least as useful as the quadrant sensor, but it makes preserving privacy rather trickier. Since octants meet at the origin at 45°, it is difficult to position the robot so that it does not discover too much. An advantage of the one-dimensional model is that the parameter m allows for a natural modification of sensor capabilities. This leads to three closely related results, each of which helps clarify how certain limitations persist even when m is increased.

Theorem 1.A (*More sensing won't grant omnipotence*) *The 1-dim. robot is not always able to achieve privacy-preserving tracking, regardless of its sensing power.*

Proof. The negative results in Lemmas 2 and 4 show that there are circumstances where it is impossible to find a tracking strategy satisfying both *PPC* and *TTC*. Though these regions depend on m, no finite value of m causes these regions to be empty.   □

Turning cases that depend on the initial I-state,[†] one might think that sensitivity to η_0 is a consequence of uncertainty that compounds because the sensors used are too feeble. But this explanation is actually erroneous. Observe that the I-state dependent region is more complicated than other regions within the strategy space: in Figure 5, the green region is contiguous, whereas the regions marked pink are not. (Bear in mind that the figure does not depict the η_0 itself, merely regions where it is known that η_0 affects the existence of a strategy.) The specific I-state dependent region for Lemma 5 under condition $a = 1$, visible clearly as chisel shape in Figures 5a and 5b, is invariant with respect to m, so remains I-state dependent in all circumstances (though outside the visible region in Figures 5c and 5d, it is present). As m increases, what happens is that the regions formerly marked as not both **PP** and **TT** are claimed as **PP** and **TT**, or become I-state dependent. The following expresses the fact that additional sensing power fails to reduce the number of I-state dependent strategy regions.

Theorem 1.B (*Information State Invariance*) *The number of initial I-state dependent strategy regions does not decrease by using more powerful sensors.*

Proof. We focus on the I-state dependent areas within the square between $(0,0)$ and $(2,2)$ as the parts outside this (orange) square do not change as m increases. According to Lemma 5, the I-state dependent conditions for any specific a are bounded by the following linear inequalities:

$$r_t < \frac{2}{a-1}, \qquad (1)$$

$$r_p > \frac{2}{a}, \qquad (2)$$

$$r_t \geq \frac{r_p}{a} + \frac{2}{a}, \qquad (3)$$

$$r_t \geq (a+1)r_p - 2, \qquad (4)$$

$$r_t < \frac{(a+1)}{a}r_p. \qquad (5)$$

[†]For brevity sometimes we will call such instances "I-state dependent" though it is strictly $|\eta_0|$, the size of the initial I-state, on which they depend.

Combining (1)–(3) gives both the bound for r_t as $r_t \in [\frac{2(a+1)}{a^2}, \frac{2}{a-1})$, and the bound for r_p as $r_p \in [\frac{2}{a}, \frac{2a}{a^2-1}]$. The I-state dependent condition, thus, is a bounded region.

Next, we show that (1) and (2) are dominated by (3)–(5). According to (4) and (5), we have $r_p < \frac{2a}{a^2-1}$. Applying this result to (5) produces (1). Similarly, combining (3) and (5) together yields (2). Hence, the I-state dependent conditions are fully determined by inequalities (3)–(5). (Figure 6 provides a visual example.)

To form a bounded region with three linear inequalities, the I-state region has to be a triangle. The three points of the triangle can be obtained by intersecting pairs of (3)–(5): $(\frac{2}{a}, \frac{2a+2}{a^2})$, $(\frac{2a}{a^2-1}, \frac{2}{a-1})$, $(\frac{2(a+1)}{a^2+a-1}, \frac{2(a+1)^2}{a^2+a-1} - 2)$. Since $a \in \{2,3,\cdots,m\}$, the triangle region will not be empty. Let $\Delta(a)$ denote the triangle with parameter a. Then the smallest y coordinate for $\Delta(a)$ is $minY(\Delta(a)) = \frac{2a+2}{a^2}$. And the largest y coordinate for $\Delta(a)$ is $maxY(\Delta(a)) = \frac{2}{a-1}$. For adjacent triangles $\Delta(a)$ and $\Delta(a+1)$, we have $minY(\Delta(a)) > maxY(\Delta(a+1))$. Hence, the triangles for different values of a do not overlap. $\qquad\square$

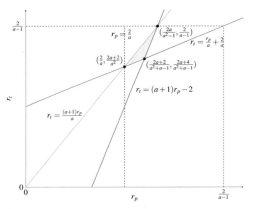

Fig. 6: Relationships of the linear inequalities in Lemma 5.

The preceding discussion showed that the number of I-state dependent regions increases with m and that, along with the chisel shaped region, there are $m-1$ triangles of decreasing size. This motivates our introduction of a quantitative measure of the robot's power as a function of m.

Definition 4. *A measure of tracking power, $p(m)$, for the robot with m sensing parameters should satisfy the following two properties: $p(m) > 0$, $\forall m \in \mathbb{Z}^+$ (positivity), and $p(a) > p(b)$, if $a,b \in \mathbb{Z}^+$ and $a > b$ (monotonicity).*

The plots in Figure 5 suggest that one way to quantify change in these regions is to measure changes in the various areas of the parameter space as m increases. As a specific measure of power in the one-dimensional setting, we might consider the proportion of cases (in the r_p vs. r_t plane) that are **PP** and **TT** (green) and I-state dependent (pink). Though the green and pink areas are unbounded in the full plane, the only changes that occur as m increases are in the square between $(0,0)$ and $(2,2)$. Thus, we take $p(m)$ to equal to the total volume of green and pink regions filling within the region $0 \leq r_p \leq r_t$ and $0 \leq r_t \leq 2$. This area satisfies the properties in Definition 4 and is indicative of the power of the robot as, intuitively, it can be interpreted as an upper-bound of the solvable cases.

Corollary 1 *(Asymptotic tracking power) The power $p(m)$ of a robot with m sensing parameters to achieve privacy-preserving tracking in the 1-dim. problem is bounded and $\lim_{m \to \infty} p(m) = L$, with $1.5 < L < 1.6$*

Proof. Inequalities (3)–(5) in the proof of Theorem 1.B give $m - 1$ triangles, one for each $a \in \{2, 3, \cdots, m\}$. An analytic expression gives the area of each of these triangles and the series describing the cumulative pink volume $p_{pink}(m)$ within $0 \leq r_p \leq r_t$ and $0 \leq r_t \leq 2$ can be shown (by the comparison test) to converge as $m \to \infty$. Similarly, the cumulative green volume $p_{green}(m)$ within $0 \leq r_p \leq r_t$ and $0 \leq r_t \leq 2$ converges. Numerical evaluation gives the value of the limit $\approx 1.54\overline{5}$ □

4 Beyond one-dimensional tracking

The inspiration for this work was the 2-dimensional case. This section lifts the impossibility result to higher dimensions.

4.1 Mapping from high dimension to one dimension

In the n-dimensional privacy-preserving tracking problem, the state for the panda becomes a point in $\mathbb{R}^n$. The panda can move with a maximum distance of $\frac{\delta}{2}$ in any direction in $\mathbb{R}^n$ within a single time-step, so that the panda's actions fill an n-dimensional ball. The privacy and tracking bound are also generalized from an interval of size r_p and r_t, to an n-dimensional ball of diameter r_p and r_t respectively. That is, the I-state should contain a ball of diameter r_p and be contained in a ball of diameter r_t, so as to achieve privacy-preserving tracking. The robot inhabits the n-dimensional space as well, and attention must be paid to its orientation too.

It is unclear what would form the appropriate higher dimensional analogue of parameter m, so we only consider n-dimensional tracking problems for robots equipped with a generalization of the quadrant sensor. The sensor's orientation is determined by that of the robot and it indicates which of the 2^n possible orthogonal cells the panda might be in. Adopting notation and definitions analogous to those earlier, we use a tuple for n-dimensional tracking problems—a subscript makes the intended dimensionality clear.

The following lemma shows that there is a mapping which preserves the tracking property from n-dimensional problem to 1-dimensional problems.

Lemma 6. *Given some 1-dim. panda tracking problem $P_1 = (\eta_0, r_p, r_t, n)$, there exists an n-dim. panda tracking problem $P_n = (\theta_0, r_p, r_t, \delta, 1^n)$ where, if $TT(P_n)$, then $TT(P_1)$.*

Proof. The approach to this proof has elements of a strategy stealing argument and simulation of one system by another. The robot faced with a 1-dim. problem constructs an n-dim. problem and uses the (hypothetical) strategy for this latter problem to select actions. The crux of the proof is that the 1-dim. robot can report back observations that are apposite for the n-dim. case. (Figure 7, below, gives a visual overview.)

For some $P_1 = (\eta_0, r_p, r_t, \delta, m)$, with $m = n$, we construct $P_n = (\theta_0, r_p, r_t, \delta, 1^n)$ as follows. Without sacrifice of generality, assume that in P_1 the initial I-state $\eta_0 = \{x \mid \eta_0^{min} \le x \le \eta_0^{max}\}$ is centered at the origin, so $\eta_0^{min} = -\eta_0^{max}$. (This simplifies the argument and a suitable translation of coordinate system rectifies the situation otherwise.) Then we choose θ_0 as the closed ball at the origin with radius η_0^{max}.

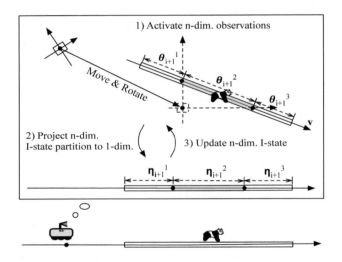

Fig. 7: Constructing a 1-dim. strategy π_1 from some n-dim. strategy π_n.

We show how, given some π_n on $P_n = (\theta_0, r_p, r_t, \delta, 1^n)$, we can use it to define a π_1 for use by the 1-dim. robot. The robot forms θ_0 and also has η_0. It picks an arbitrary unit-length vector $\hat{\mathbf{v}} = v_1\mathbf{e_1} + v_2\mathbf{e_2} + \cdots + v_n\mathbf{e_n}$, unknown to the source of π_n, which is the subspace that the 1-dim. panda lives in. For subsequent steps, the robot maintains $\theta_1^-, \theta_1, \theta_2^-, \theta_2, \ldots, \theta_k^-, \theta_k, \theta_{k+1}^-, \ldots$ along with the I-states in the original 1-dim. problem $\eta_1^-, \eta_1, \eta_2^-, \eta_2, \ldots, \eta_k^-, \eta_k, \eta_{k+1}^-, \ldots$. For any step k, the η_k can be seen as measured along $\hat{\mathbf{v}}$ within the higher dimensional space. Given θ_{k-1}, θ_k^- is constructed using Minkowski sum operations as before, though now in higher dimension. Given θ_k^-, strategy π_n determines a new pose for the n-dim. robot and, on the basis of this location and orientation, the n sensing planes slice through θ_k^-. Though the planes demarcate 2^n cells, the line along $\hat{\mathbf{v}}$ is cut into no more than $n+1$ pieces as the line can pierce each plane at most once (with any planes containing $\hat{\mathbf{v}}$ being ignored). Since the 1-dim. robot has $m = n$, it picks the $u_1, \ldots, u_n$ by measuring the locations that the sensing planes intersect the line $\mathbf{x} = \alpha\hat{\mathbf{v}}$, $\alpha \in \mathbb{R}$. (If fewer than n intersections occur, owing to planes containing the line, the extra u_i's are simply placed outside the range of the

I-state.) After the 1-dim. panda's location is determined, the appropriate orthogonal cell is reported as the n-dim. observation, and θ_k^- leads to θ_k via the intersection operation. This process comprises π_1. It continues indefinitely because θ_k must always entirely contain η_k along the line through $\hat{v}$ because, after all, a cantankerous n-dim. panda is free to choose to always limit its movements to that line.

If π_n is **TT**, then so too is the resulting strategy π_1 since the transformation relating $\theta_k \cap \{\alpha\hat{v} : \alpha \in \mathbb{R}\}$ with η_k preserves length and, thus, θ_k fitting within a ball of diameter r_t implies that $|\eta_k| < r_t$. □

4.2   Impossibility in high-dimensional privacy-preserving tracking

Now we are ready to connect the pieces together for the main result:

Theorem 2 *(Impossibility) It is not possible to achieve privacy-preserving panda tracking in n dimensions for every problem with $r_p < r_t$.*

Proof. To extend the lemmas that have shown this result for $n = 1$ to cases for $n > 1$, suppose such a solution existed for $P_n = (\theta_0, r_p, r_t, \delta, 1^n)$. Then according to Lemma 6, every 1-dim. panda tracking problem $P_1 = (\eta_0, r_p, r_t, \delta, n)$ is **TT**, since they can be mapped to an n-dim. **TT** panda tracking problem. But this contradicts the non-**TT** instances in Lemma 2, so no such strategy can exist for every non-trivial problem ($r_p < r_t$) in two, three, and higher dimensions. □

5   Related work

Information processing is a critical part of what most robots do—without estimating properties of the world, their usefulness is hampered, sometimes severely so. Granting our computational devices access to too much information involves some risk, and privacy conscious users may balk and simply opt to forgo the technology. Multiple models have been proposed to think about this tension in setting of data processing more generally, *cf.* formulations in [5,6]. Existing work has also explored privacy for networks, along with routing protocols proposed to increase anonymity [7], and includes wireless sensor problems where distinct notions of spatial and temporal privacy [8,9] have been identified.

Some recent work has begun to investigate privacy in settings more directly applicable to robots. Specifically three lines of work propose techniques to help automate the development of controllers that respect some limits on the information they divulge. O'Kane and Shell [10] formulated a version of the problem in the setting of combinatorial filters where the designer provides additional stipulations describing which pieces of information should always be plausibly indistinguishable and which must never be conflated. The paper describes hardness results and an algorithm for ascertaining whether a design satisfying such a stipulation is feasible. This determination of feasibility for a given set of privacy and utility choices is close in spirit to what this paper has explored for the particular case of privacy-preserving tracking: here those choices become quantities r_p and r_t. The second important line is that of Wu et al. who explore how to disguise some behavior in their control system's plant model, and show how to protect this secret behaviour by obfuscating the system outputs [11]. More recent work expresses both utility and privacy requirements in an automata framework and proposes algorithms to synthesize controllers satisfying these two constraints [12]. In the third line, Prorok and Kumar [13] adopt the differential privacy model to characterize the privacy of heterogeneous robot swarms, so that any individual robot cannot be determined to be of a particular type from macroscopic observations of the swarm.

Finally, we note that while we have considered panda tracking as a cute realization of this broader informationally constrained problem, wildlife monitoring and protection is an area in which serious prior robotics work exists (e.g., see [14]) and for which there is substantial and growing interest [15].

6   Conclusion and future work

In this paper we have reexamined the panda tracking scenario introduced by O'Kane [3], focusing on how various parameters the specify a problem instance (such as the capabilities of the robot and the panda) affect the existence of solutions. Our approach has been to study nontrivial instances of the problem in one dimension. This allows for an analysis of strategies by examining whether the sensing operations involved at each step increase or decrease the degree of uncertainty in a directly quantifiable way. Only if this uncertainty can be precisely controlled forever, can we deem the problem instance solved. We use a particular set of sensing choices, basically division of the region evenly, which we think of as a *split* operation. This operation is useful because the worst resulting I-state in other choices, those that are not evenly split, is weaker (in terms of satisfying the tracking and privacy constraints) than even divisions are. Thus, the split operation acts as a kind of basis: if the problem has no solution with these choices, the panda cannot be tracked with other choices either.

In examining the space of tracking and privacy stipulations, the existence of strategies is shown to be a function of the robot's initial belief and panda's movement. There exist regions without any solution, where it is impossible for the robot to actively track the panda as well as protect its privacy for certain nontrivial tracking and privacy bounds. Additionally, we have uncovered regions where solution feasibility is sensitive to the robot's initial belief, which we have called I-state dependent cases (or conditions). The simple one-dimensional setting also permits exploration of how circumstances change as the robot's sensing power increases. Perhaps surprisingly, the number of these I-state dependent strategy conditions does not decrease as the robot's sensing becomes more powerful. Finally, we connect the impossibility result back to O'Kane's setting by mapping between high-dimensional and one-dimensional versions, proving that the 2D planar panda tracking problem does not have any privacy-preserving tracking strategy for every non-trivial tracking and privacy stipulation.

The results presented in this paper reveal some properties of a particular —and it must be said somewhat narrow— instance of an informationally constrained problem. Future work could explore what sensor properties permit such a task to be achieved, perhaps identifying families of sensors that suffice more broadly. Also, probabilistic models impose different belief representations and observation functions and it is worth exploring analogous notions in those settings. Another thread is to weaken the poacher's capability, allowing the privacy bound to be relaxed somewhat. For example, in considering some latency in the poacher's reaction to information that has been leaked, it may be acceptable for the privacy bound to be relaxed so that it need not apply from every time-step to the next.

Acknowledgements

This work was supported by the NSF through awards IIS-1527436 and IIS-1453652. We thank the anonymous reviewers for their time and extraordinarily valuable comments, and are grateful to the conference committee for a student travel award.

References

1. B. R. Donald, "On information invariants in robotics," *Artificial Intelligence — Special Volume on Computational Research on Interaction and Agency, Part 1*, vol. 72, no. 1–2, pp. 217–304, 1995.
2. M. T. Mason, "Kicking the sensing habit," *AI Magazine*, vol. 14, no. 1, pp. 58–59, 1993.
3. J. M. O'Kane, "On the value of ignorance: Balancing tracking and privacy using a two'bit sensor," in *Proc. Int. Workshop on the Algorithmic Foundations of Robotics (WAFR'08)*, Guanajuato, Mexico, 2008.
4. S. M. LaValle, "Sensing and filtering: A fresh perspective based on preimages and information spaces," *Foundations and Trends in Robotics*, vol. 1, no. 4, pp. 253–372, 2010.
5. L. Sweeney, "*k*-anonymity: A model for protecting privacy," *Int. Journal on Uncertainty, Fuzziness and Knowledge-based Systems*, vol. 10, no. 5, pp. 557–570, 2002.
6. C. Dwork, "Differential privacy: A survey of results," in *Proc. Int. Conference on Theory and Applications of Models of Computation*. Springer, 2008, pp. 1–19.
7. I. Clarke, O. Sandberg, B. Wiley, and T. W. Hong, "Freenet: A distributed anonymous information storage and retrieval system," in *Int. Workshop on Designing Privacy Enhancing Technologies: Design Issues in Anonymity and Unobservability*, 2001, pp. 46–66.
8. P. Kamat, Y. Zhang, W. Trappe, and C. Ozturk, "Enhancing source-location privacy in sensor network routing," in *Proc. IEEE Int. Conference on Distributed Computing Systems (ICDCS'05)*, Columbus, Ohio, USA, 2005, pp. 599–608.
9. P. Kamat, W. Xu, W. Trappe, and Y. Zhang, "Temporal privacy in wireless sensor networks," in *Proc. Int. Conference on Distributed Computing Systems (ICDCS'07)*, Toronto, Ontario, Canada, 2007, pp. 23–23.
10. J. O'Kane and D. Shell, "Automatic design of discreet discrete filters," in *Proc. IEEE Int. Conference on Robotics and Automation (ICRA)*, Seattle, WA, USA, 2015, pp. 353–360.
11. Y.-C. Wu and S. Lafortune, "Synthesis of insertion functions for enforcement of opacity security properties," *Automatica*, vol. 50, no. 5, pp. 1336–1348, 2014.
12. Y.-C. Wu, V. Raman, S. Lafortune, and S. A. Seshia, "Obfuscator synthesis for privacy and utility," in *NASA Formal Methods Symposium*. Springer, 2016, pp. 133–149.
13. A. Prorok and V. Kumar, "A macroscopic privacy model for heterogeneous robot swarms," in *Proc. Int. Conference on Swarm Intelligence (ICSI)*. Springer, 2016, pp. 15–27.
14. D. Bhadauria, V. Isler, A. Studenski, and P. Tokekar, "A robotic sensor network for monitoring carp in minnesota lakes," in *Proc. IEEE Int. Conference on Robotics and Automation (ICRA)*, Anchorage, AK, USA, 2010, pp. 3837–3842.
15. F. Fang, P. Stone, and M. Tambe, "When security games go green: Designing defender strategies to prevent poaching and illegal fishing," in *Proc. Int. Joint Conference on Artificial Intelligence (IJCAI)*, Buenos Aires, Argentina, 2015.

Approximation Algorithms for Tours of Height-varying View Cones

Patrick A. Plonski and Volkan Isler

University of Minnesota[*]

Abstract. We introduce a novel coverage problem which arises in aerial surveying applications. The goal is to compute a shortest path which visits a given set of cones. The apex of each cone is restricted to lie on the ground plane. The common angle α of the cones represent the field of view of the onboard camera. The cone heights, which can be varying, correspond with the desired observation quality (e.g. resolution). This problem is a novel variant of the Traveling Salesperson Problem with Neighborhoods (TSPN). We name it *Cone-TSPN*.

Our main contribution is a polynomial time approximation algorithm for Cone-TPSN. We analyze its theoretical performance and show that it returns a solution whose length is at most $O((1 + \log(h_{\max}/h_{\min}))(1 + \tan \alpha))$ times the length of the optimal solution where $h_{\max}$ and $h_{\min}$ are the heights of the tallest and shortest input cones, respectively. We demonstrate the use of our algorithm in a representative precision agriculture application. We further study its performance in simulation using randomly generated cone sets. Our results indicate that the performance of our algorithm is superior to standard solutions.

1   Introduction

Imagine an aerial robot charged with observing n objects on the ground. It has a downward-facing camera with fixed field of view 2α. When the robot takes a picture, the number of pixels in the image corresponding to the object is proportional to the height of the camera when the image is taken. Therefore, in order to capture an image of an object with a desired resolution, the robot must enter an inverted cone positioned on the object with slope $\pi/2 - \alpha$ and a given height (Figure 1). Motivated by such visual coverage problems, we study the problem of computing the shortest path which visits a given set of cones.

The Traveling Salesperson Problem (TSP) is that of finding a shortest length tour which visits a given set of cities [1]. It is a classical optimization problem. In TSP with Neighborhoods (TSPN), the input is a collection of neighborhoods (e.g. disks). It suffices to visit any point in the neighborhood. In this

---

[*] P. A. Plonski and V. Isler are with the Department of Computer Science and Engineering, University of Minnesota, 200 Union Street SE, Minneapolis MN 55455 USA email: {`plonski, isler`}`@cs.umn.edu`. This material is based upon work supported by the National Science Foundation under grant numbers 1111638 and 1317788.

K. Goldberg et al. (Eds.): *Algorithmic Foundations of Robotics XII*, SPAR 13, pp. 192–207, 2020.
https://doi.org/10.1007/978-3-030-43089-4_13

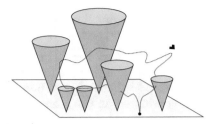

Fig. 1. Each cone represents a point which must be viewed from a given height and angle. We study the problem of computing the shortest path to visit all cones.

paper, we introduce a novel geometric TSPN variant which we call Cone-TSPN. Our main contribution is an approximation algorithm which runs in polynomial time and returns a solution that is guaranteed to be within a factor of $O((1 + \log(h_{max}/h_{min}))(1 + \tan \alpha))$ of the optimal solution where h_{max} and h_{min} are the respective heights of the tallest and shortest input cones, and α is the angle of the cones. In addition to proving the theoretical worst case approximation factor, we also validate our algorithm by executing it on Cone-TSPN instances generated from satellite imagery and by a random process.

Cone-TSP arises in many emerging applications of Unmanned Aerial Vehicles (UAVs). Small UAVs often have short flight times, yet they can still cover large areas because as the UAV flies higher, the image footprint gets bigger. This, of course, comes at the expense of resolution loss. The problem we study can be used to trade-off flight-time and coverage.

For example, consider a precision agriculture application in which aerial images are used to infer nitrogen or moisture levels or spread of diseases. When a farm is monitored periodically, we can use satellite imagery or partial data collected to form a prior estimate and determine areas of interest from that prior. A height value can be associated with each point based on the uncertainty levels. See e.g. [2]. In Section 5 we present simulations which emulate this scenario. Similar coverage problems arise in surveillance, search-and-rescue, sea floor exploration, environmental monitoring, and numerous other applications. Before formalizing Cone-TSPN, we start with an overview of related work.

1.1 Background and Related Work

Our goal of planning an informative trajectory for a UAV has received recent research attention. Sadat et al. [3] performed coverage with a UAV at varying height levels in response to online observations (interesting terrain was revisited at lower levels). The authors provide no guarantees on the length of their tour. Tokekar et al. [2] selected points of interest to visit to minimize misclassification error, and visited them with an aerial vehicle or ground vehicle. In contrast with this work, the measurement height was fixed a priori. Schwager et al. [4] designed control policies to minimize an error function based on area covered per pixel.

Their cost function pushed the set of mobile cameras to view the area of interest from as close as possible, while leaving none of it uncovered.

A similar TSPN formulation to ours has been used for data muling operations, where the goal is to compute a trajectory for a data mule to travel so that it passes close enough to sensors so that data can be wirelessly transferred [5,6]. These problems were only considered on the plane.

The Traveling Salesman Problem (TSP) is a widely studied optimization problem which is known to be NP-hard. As TSPN generalizes TSP, it too is NP-hard, even in a Euclidean metric space. It was shown by Arora [7] and Mitchell [8] that Euclidean TSP admits a Polynomial Time Approximation Scheme (PTAS) in $\mathbb{R}^d$ for any d. However, Euclidean TSPN is known to be APX-hard [9, 10], even if the neighborhoods are line segments of similar length [11].

If the structure of the neighborhoods is highly restricted, Euclidean TSPN can admit a PTAS using similar approaches as the known PTASs for Euclidean TSP. Dumitrescu and Mitchell [12] extended Mitchell's m-guillotine PTAS method to provide a PTAS for equal disjoint disks in a plane, or, more generally, nearly-disjoint "fat" polygons (polygons with lower-bounded volume compared to their width). The same work also provides constant factor approximations for arbitrary connected neighborhoods of similar diameter, using a technique similar to that described by Arkin and Hassin [13]. Bodlaender et al. [14] similarly extended Arora's shifted dissection PTAS method to provide a PTAS for disjoint fat polygons of similar size, for the related minimum corridor connection problem. Their algorithm applies in higher dimensions, and they claim that the same algorithm can work for the TSPN problem. The authors also provide a constant-factor approximation when the disjoint fat rooms are varying size. The remaining case known to admit a PTAS is the case of arbitrary fat disjoint neighborhoods on the plane, of possibly varying size, solved again by Mitchell [15].

The problem is more challenging when the neighborhoods are allowed to intersect. If the neighborhoods are fat and "weakly disjoint", Chan and Elbassioni [16] have provided a quasipolynomial-time approximation scheme, which works for any $\mathbb{R}^d$. This "α-fat weakly disjoint" requirement, in Euclidean space, is a requirement for disjoint balls contained entirely within the neighborhoods, with radius proportional to the neighborhoods' diameter.

For some classes of fat neighborhoods that may intersect, a constant factor approximation is known. On the plane, the work in [12] and [11] provided a constant factor approximation for neighborhoods of similar diameter. Recently, the approximation was improved by Dumitrescu and Tóth [17] for the case of disjoint unit disks, and a new constant factor approximation algorithm was presented for possibly-intersecting unit balls in $\mathbb{R}^3$. If polygons on a plane are not fat and not similarly sized, Elbassioni et al. [18] provided a $O(\log n)$ approximation algorithm. Otherwise, Mata and Mitchell [19] found an $O(\log n)$ approximation for arbitrary connected simple polygons in the plane.

Specifically for our proposed problem Cone-TSPN, the following is known:

- If viewing cones are disjoint, fat, similarly sized, there is a PTAS [14].
- If similarly sized cones are fat weakly-disjoint (this means they are allowed to intersect above a certain height), there is a QPTAS [16].

In this work we will show that there exists a strategy for Cone-TSPN with polynomial complexity which has an approximation factor that is independent of n but depends on α and the variability in cone heights. Our dependence on α is not the same as requiring fatness: our strategy has poor performance guarantees when $\tan \alpha$ is large, but it performs well when α is small; a cone is only considered fat for intermediate values of $\tan \alpha$.

2   Problem Statement

In this section we will formally state the Cone-TSPN problem and detail our approach for solving it.

Given: a set of n cones $\mathcal{C}$ with heights $\mathcal{H}$ and boundaries defined by the union of all lines that pass through an apex on the ground plane with angle α from the normal vector, and a takeoff position x_0 on the ground plane. Cone c_i has height h_i, and its upper disk, or *cap*, has radius $h_i \tan \alpha$.

Objective: find a minimal length tour T that intersects all cones. This problem is NP-Hard, so in this work we aim to find a tour in polynomial time that has guaranteed approximation factor between the length L of our tour T and the length $L^\star$ of the optimal tour $T^\star$.

Our approach is based on the following idea: When the height difference between cones is small, we can consider the problem in the intersection of the cones with a horizontal plane, where the problem can be solved more easily (because it reduces to TSPN with equal disks). By selecting a single plane (as shown in Section 4.1), we obtain an approximation factor which is $O\left(\left(\frac{\max(\mathcal{H})}{\text{mean}(\mathcal{H})} \right)^2 (1 + \tan \alpha) \right)$.

We can reduce the dependence on cone height variation from a quadratic to a logarithmic factor if we carefully select multiple planes with responsibility for visiting cones of different heights. This strategy (presented in Section 4.2) obtains an approximation factor which is $O\left(\left(1 + \log \frac{\max(\mathcal{H})}{\min(\mathcal{H})} \right) (1 + \tan \alpha) \right)$.

The $\tan \alpha$ terms are not usually significant in practical situations. A GoPro filming in medium FOV has 127°, which corresponds with $\tan \alpha \approx 2$. The field of view for planning purposes might be smaller than the FOV of the camera, because a certain pixel radius might be necessary to properly observe a target. Furthermore, when FOV is large, objects might appear distorted if they are near the boundary of the image. Therefore, it is often necessary to set α at less than the maximum pixel angle of the camera.

3   Optimal Maximum Height Attained

First we consider the problem of finding how high off the ground the optimal tour can travel, given that it must start on the ground. Let $h^\star$ be the maximum height attained by the optimal tour $T^\star$. If the cones are tall compared with the tour length, $h^\star$ might be shorter than the shortest cone. In this section we provide bounds on $h^\star$ and a method to search for a $\hat{h}$ with bounded distance from $h^\star$. Later we will use $\hat{h}$ to determine how high our tour can travel. Let T_0 be the optimal tour that lies entirely on the ground plane (this is the optimal TSP tour of the cone apexes). Define L_0 as the length of this tour.

Lemma 1. *If $h^\star \leq \min(\mathcal{H})$ and $h^\star \leq \frac{L_0}{2n\tan\alpha}$, $h_\ell \leq h^\star \leq h_u$ where*

$$h_\ell = \frac{L_0}{2} \frac{n\tan\alpha - \sin\alpha\sqrt{n^2-1}}{n^2\tan^2\alpha + 1} \quad and \quad h_u = \frac{L_0}{2} \frac{n\tan\alpha + \sin\alpha\sqrt{n^2-1}}{n^2\tan^2\alpha + 1}.$$

Proof. We start with bounds on $L^\star$ compared with $h^\star$. To upper bound $L^\star$, we construct a tour T_p from T_0 which is guaranteed to have length less than L_0 while still visiting all neighborhoods. Start with T_0 and project every point on it along the vector tilted α away from the normal vector, towards x_0. The distance to project is $\sin\alpha$ times the distance between the point and x_0; this ensures that the entire projected tour T_p lies on a virtual cone with angle $\pi/2 - \alpha$, with apex at x_0. This cone is shown in Figure 2. Every point on T_0 which visits the apex of a neighborhood is projected to a new point on the boundary of the same neighborhood: therefore, since T_0 visits all neighborhoods, so must T_p. If a robot executes T_0 at unit speed, the speed of its projection is $\cos\alpha$. Since the length of T_0 is L_0, the length of T_p is $L_0\cos\alpha$. Therefore,

$$L^\star \leq L_0\cos\alpha. \tag{1}$$

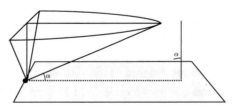

Fig. 2. Demonstration of our argument on the upper bound of $L^\star$ compared with L_0. By projecting every point on L_0 to the cone with angle $\pi/2 - \alpha$ and apex at x_0, we obtain a tour T_p which visits every cone that L_0 visits.

The lower bound on $L^\star$ comes from the following: Consider the projection of $T^\star$ to the ground plane. Denote it $T_0^\star$. Denote its length as $L_0^\star$. This tour $T_0^\star$ is no longer guaranteed to visit each neighborhood, but it is guaranteed to pass underneath each neighborhood. For each neighborhood c_i, we can add to $T_0^\star$ a detour of length at most $2h^\star\tan\alpha$ to ensure that c_i is visited at its apex. However, we also know that $L_0^\star \geq L_0$. Thus, $L_0^\star + 2nh^\star\tan\alpha \geq L_0$ (because T_0 is the optimal tour that visits each neighborhood at its apex). We know that $T^\star$ travels total horizontal distance $L_0^\star$, and it must travel total vertical distance not less than $2h^\star$ (enough to rise to $h^\star$ and return back to the ground). Therefore, $(L^\star)^2 \geq (2h^\star)^2 + (L_0^\star)^2$. We also know that $L_0^\star \geq 0$. Combining the above properties together, we achieve:

$$(L^\star)^2 \geq (2h^\star)^2 + \max(0, L_0 - 2nh^\star \tan \alpha)^2. \tag{2}$$

When $h^\star \leq \frac{L_0}{2n \tan \alpha}$, we can ignore the case where $0 > L_0 - 2nh^\star \tan \alpha$. Combining the bounds on $L^\star$, we achieve the following equation which is quadratic in $h^\star$:

$$L_0^2 \cos^2 \alpha \geq (2h^\star)^2 + (L_0 - 2nh^\star \tan \alpha)^2 \tag{3}$$

$$0 \geq (h^\star)^2 (n \tan^2 \alpha + 1) - h^\star (nL_0 \tan \alpha) + \frac{L_0^2 (1 - \cos^2 \alpha)}{4}. \tag{4}$$

Since the squared term is positive, the inequality is satisfied between the two roots h_ℓ and h_u.

Lemma 2. *If $h^\star \leq \min(\mathcal{H})$, $h^\star \leq \dfrac{L_0 \cos \alpha}{2}$.*

Proof. This follows by combining Equation 1 with a looser lower bound on $L^\star$:

$$L^\star \cos \alpha \geq L^\star \geq 2h^\star. \tag{5}$$

Lemma 3. $h^\star \in [h_\ell, h_u] \cup \left[\dfrac{L_0}{2n \tan \alpha}, \dfrac{L_0 \cos \alpha}{2} \right] \cup [\min(\mathcal{H}), max(\mathcal{H})].$

Proof. This directly follows from Lemmas 1 and 2, as well as the fact that the optimal tour can never travel higher than the tallest neighborhood.

Now we have three connected ranges where $h^\star$ might reside. To find a height estimate $\hat{h}$ which is close to $h^\star$ by a multiplicative factor, we can start at the lower bound of each range and multiply by a constant factor until the upper bound is reached.

Lemma 4. *It is possible to select $O\left(\log(n) + \log\left(\frac{\max(\mathcal{H})}{\min(\mathcal{H})} \right) \right)$ height guesses such that one of them $\hat{h}$ satisfies $\hat{h}/2 \leq h^\star \leq \hat{h} \leq 2h^\star$.*

Proof. By following a doubling strategy, we can search the range $[\min(\mathcal{H}), \max(\mathcal{H})]$ with $\log\left(\frac{\max(\mathcal{H})}{\min(\mathcal{H})} \right)$ guesses. For the other ranges, we will show that the quotient of their maximum value to their minimum value is polynomial in n, so they can be searched with $O(\log n)$ guesses. Consider the range $[h_\ell, h_u]$. The quotient Q_1 of the maximum to the minimum is described by:

$$Q_1 = \frac{h_u}{h_\ell} = \frac{n \sec(\alpha) + \sqrt{n^2 - 1}}{n \sec(\alpha) - \sqrt{n^2 - 1}} \leq \frac{n + \sqrt{n^2 - 1}}{n - \sqrt{n^2 - 1}} = \left(\sqrt{n^2 - 1} + n \right)^2. \tag{6}$$

Finally, consider the range $\left[\frac{L_0}{2n \tan \alpha}, \frac{L_0 \cos \alpha}{2} \right]$. The quotient Q_2 of the maximum to the minimum is described by:

$$Q_2 = n \cos(\alpha) \tan(\alpha) = n \sin \alpha \tag{7}$$

In the following sections we use the fact that by computing $O(\log n)$ tours with different height guesses, we ensure one of them will have been computed with a guess $\hat{h}$ that satisfies $\hat{h}/2 \leq h^\star \leq \hat{h}$. We truncate the tops of the cones so that $\max(\mathcal{H}) = \hat{h}$, without affecting the optimal tour.

4   Algorithm

Now that we have a method to bound $h^\star$, we can bound the effectiveness of visiting view cones at the same height. First, we will present the strategy for when the neighborhoods are disjoint; later we will find a tour for intersecting neighborhoods by first computing a Maximal Independent Set (MIS). Finally, we will show how we can obtain shorter tours by using multiple height slices.

4.1   Problem 1: Disjoint Cones of Similar Height

Define a function $f(T_x, h_x)$ which describes the sweep volume given by the Minkowski sum of: T_x projected to the ground plane, and a cone with angle α and input height h_x.

Lemma 5. *For any h, $h \geq \max(\mathcal{H})$, $h \leq 2h^\star$,*

$$f(T^\star, h) \leq \frac{1}{2}L^\star h^2 \tan\alpha + \frac{\pi}{3}h^3 \tan^2\alpha \leq L^\star h^2 \tan\alpha \left(\frac{1}{2} + \frac{\pi \tan\alpha}{3}\right).$$

Proof. The right hand side is the formula of the volume of a cone swept out along a path of length $L^\star$. Since our tour must return to the start, it is not a tight upper bound. We also use the property that $h \leq 2h^\star \leq L^\star$.

Lemma 6. *For any $h \geq \max(\mathcal{H})$, there exists a constant C_v such that*

$$f(T^\star, h) \geq C_v \tan^2\alpha \sum_{h_i \in \mathcal{H}} h_i^3.$$

Proof. If a tour T visits a cone with height h_i, the points swept out by its Minkowski sum include a portion which is the intersection of two cones of equal height h_i, which are offset by distance $\tan\alpha$. The volume of this intersection is proportional to the volume of the cone itself. These points are in cone i, and cones are disjoint, and all cones are visited.

Proposed strategy SLICE-VISIT: Truncate all cones to not be taller than the estimated max height $\hat{h}$. Intersect all cones with a plane at height h_t, where $h_t = \min(\mathcal{H})$. Find a TSPN tour T_h that visits the circular cross sections of the cones in this plane (with a PTAS or approximation) with x_0 projected to the plane as the starting point. Connect this tour with x_0 using a double line segment. An example execution of this strategy is shown in Figure 3. It is important that h_t not be greater than $\min(\mathcal{H})$ so that a tour computed on the plane at h_t will intersect all cones.

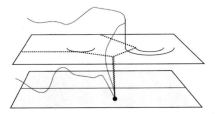

Fig. 3. Demonstration of SLICE-VISIT. If the optimal tour $T^\star$ visits a cone at a height lower than the height of a slice h_t, this doesn't lengthen the tour in the slice T_h. However, if $T^\star$ visits a cone at a greater height, T_h must lengthen in proportion to the height difference times $\tan \alpha$.

Lemma 7. *If T is computed with SLICE-VISIT using a $(1 + \epsilon)$ approximation, where ϵ is negligible,*

$$L - L^\star \leq 2h_t + 2\tan(\alpha) \sum_{h_i \in \mathcal{H}} (h_i - h_t) \leq 2h_t + 2n(\text{mean}(\mathcal{H}) - h_t)\tan \alpha.$$

Proof. First, project $T^\star$ to lie entirely at h_t. This can't make it longer. Now, for every cone i that was visited by $T^\star$ at a height higher than h_t, visit the cone in the plane by adding two line segments at height h_t, each with length less than $(h_i - h_t)\tan \alpha$. Now we have a tour which lies in the plane h_t and visits every cone, as well as passing directly above x_0 (because $T^\star$ must visit x_0). To connect with x_0, we need $2h_t$ additional length.

Theorem 1. *For any set of disjoint input cones, and given $\hat{h}$ such that $h^\star \leq \hat{h} \leq 2h^\star$ and $\hat{h} \geq \max(\mathcal{H})$, the strategy SLICE-VISIT solves Cone-TSPN with approximation factor*

$$3 + \frac{1}{C_v} \left(\frac{\hat{h}}{\text{mean}(\mathcal{H})} \right)^2 \left(1 + \frac{2\pi \tan \alpha}{3} \right).$$

Proof. First, combine the bounds on $f(T^\star, \hat{h})$:

$$L^\star \hat{h}^2 \tan \alpha \left(\frac{1}{2} + \frac{\pi \tan \alpha}{3} \right) \geq f(T^\star, \hat{h}) \geq C_v \tan^2 \alpha \sum_{h_i \in \mathcal{H}} h_i^3 \qquad (8)$$

$$L^\star \geq \frac{C_v \tan \alpha \sum_{h_i \in \mathcal{H}} h_i^3}{\left(\frac{1}{2} + \frac{\pi \tan \alpha}{3} \right) \hat{h}^2} \qquad (9)$$

$$n \tan \alpha \leq \frac{L^\star \left(\frac{1}{2} + \frac{\pi \tan \alpha}{3} \right) \hat{h}^2}{C_v \, \text{mean}(\mathcal{H})^3}. \qquad (10)$$

We use the property that $\sum h_i^3 \geq n \, \text{mean}(\mathcal{H})^3$.

Next, substitute Equation 10 into Lemma 7:

$$L - L^\star \leq 2h_t + (\text{mean}(\mathcal{H}) - h_t)\frac{L^\star \left(1 + \frac{2\pi \tan \alpha}{3}\right)\hat{h}^2}{C_v \, \text{mean}(\mathcal{H})^3} \leq L^\star \left(2 + \frac{\left(1 + \frac{2\pi \tan \alpha}{3}\right)\hat{h}^2}{C_v \, \text{mean}(\mathcal{H})^2}\right) \tag{11}$$

Here we use the property that since $h_t \leq \hat{h}$, it follows that $h_t \leq 2h^\star \leq L^\star$.

4.2 Problem 2: Disjoint Cones of Differing Heights

When there is high variability in cone height, SLICE-VISIT performs poorly on its own, because it cannot move higher than $\min(\mathcal{H})$. We address this by classifying cones according to their height and performing SLICE-VISIT separately for each each height class.

Proposed strategy HEIGHT-VISIT: Truncate all cones to not be taller than the estimated max height $\hat{h}$. Find $h_\epsilon = \min(\mathcal{H})$. Classify all cones in bins $[h_\epsilon, 2h_\epsilon), [2h_\epsilon, 4h_\epsilon), \ldots, [2^i h_\epsilon, 2^{i+1} h_\epsilon)$ until a bin is reached which is entirely taller than $\max(\mathcal{H})$. For each height bin of cones, construct a subtour by executing SLICE-VISIT. All subtours are connected along the line pointing up from x_0.

Theorem 2. *For any set of disjoint input cones, and given $\hat{h}$ such that $h^\star \leq \hat{h} \leq 2h^\star$ and $\hat{h} \geq \max(\mathcal{H})$, the strategy* HEIGHT-VISIT *solves Cone-TSPN with approximation factor*

$$\left(3 + \frac{16}{27C_v}\left(1 + \frac{2\pi \tan \alpha}{3}\right)\right)\left\lfloor 1 + \log_2 \frac{\hat{h}}{h_\epsilon}\right\rfloor.$$

Proof. For a slice at height h_t, We can compute the approximation factor of the subtour responsible for visiting the slice using Equation 11. For the subtour, we use the worst case $\hat{h} = 2h_t$ and $\text{mean}(\mathcal{H}) = (3/2)h_t$. There are $\left\lfloor 1 + \log_2 \frac{\hat{h}}{h_\epsilon}\right\rfloor$ subtours, joined together on the line straight up from x_0 at no increased cost.

4.3 Problem 3: Non-Disjoint Cones

Once we have a method to visit disjoint cones at heights comparable to their maximum heights, we can use the common trick [12,17,18] of selecting a Maximal Independent Set (MIS) of neighborhoods, finding a tour that visits the set, and modifying the tour to guarantee that it also visits all unselected neighborhoods.

First we construct our MIS. Proposed strategy HEIGHT-SELECT: This strategy converts a set of possibly intersecting cones into a disjoint set of cones that together intersect every input cone. Sort the cones by height from shortest to tallest. Select the shortest, add it to the output list and remove it and every cone that intersects it from the sorted list. Repeat until the sorted list is empty. Our selected set is the cones in the output list. An example execution of HEIGHT-SELECT is shown in Figure 4(a).

Lemma 8. *When a cone set is selected from $\mathcal{H}$ using* HEIGHT-SELECT, *the caps of the selected cones intersect every cone in $\mathcal{H}$.*

Proof. If a cone c_i is selected, it intersects its own cap. If c_i is not selected, it was removed from the sorted list before it was the shortest cone in the list. This means a shorter cone c_j intersects c_i. Therefore c_j's cap intersects c_i.

Proposed extension to HEIGHT-VISIT, for an input set of non-disjoint cones: Select the MIS to use with HEIGHT-SELECT. Execute HEIGHT-VISIT to find a tour that visits the MIS. Whenever a selected cone is visited at height h_t, add a circumference path at radius $h_t \tan \alpha$, and another circumference path at radius $3h_t \tan \alpha$. An example of this extension is shown in Figure 4(b).

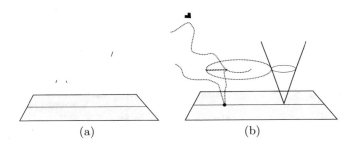

(a)   (b)

Fig. 4. (a) Demonstration of HEIGHT-SELECT. A MIS (shaded) is selected starting from the shortest cones. (b) Demonstration of extension to HEIGHT-VISIT. Two concentric circular paths at half the height of the selected cone x_i ensure that all cones that intersect the cap of x_i are visited by the tour. These paths need to be connected with the rest of the tour at cost proportional to radius.

Lemma 9. *The extension to* HEIGHT-VISIT *adds length of at most $h_t(8\pi + 6) \tan \alpha$ for each cone in the MIS visited at height h_t.*

Proof. The added circle paths have length $2\pi h_t \tan \alpha$ and $6\pi h_t \tan \alpha$. Since the original tour intersects the cone at height h_t, the circle paths can be added to the main tour using two line segments of length at most $3h_t \tan \alpha$.

Lemma 10. *If the MIS is selected using* HEIGHT-SELECT, *the added circle paths ensure that every cone is visited.*

Proof. We ensure that every cone that intersects the cap of a selected cone is visited by our added circle paths. If c_i is in the MIS and c_j intersects the disk at the top of c_i, its apex must not be farther than $2h_i \tan \alpha$ from the apex of c_i. This is not more than $4h_t \tan \alpha$, because $h_t \geq (1/2)h_i$. The outer circle intersects any cone that has distance from the apex of c_i in the range $[2h_t \tan \alpha, 4h_t \tan \alpha]$ and the inner circle intersects any cone in the range $[0, 2h_t \tan \alpha]$.

Theorem 3. *For any set of input cones, and given $\hat{h}$ such that $h^\star \leq \hat{h} \leq 2h^\star$ and $\hat{h} \geq \max(\mathcal{H})$, the strategy* HEIGHT-VISIT *solves Cone-TSPN with approximation factor*

$$\left(3 + \frac{4(4\pi + 3)}{C_v} \left(1 + \frac{2\pi \tan\alpha}{3} \right) \right) \left\lfloor 1 + \log_2 \frac{\hat{h}}{h_\epsilon} \right\rfloor .$$

Proof. Since $L^\star$ must visit our MIS, we can bound its length just as we did for Theorem 1. We look back at Lemma 7, and add the cost of added coverage to the length L of a single subtour:

$$L - L^\star \leq 2h_t + 2n(\text{mean}(\mathcal{H}) - h_t)\tan\alpha + nh_t(8\pi + 6)\tan\alpha \qquad (12)$$

$$\leq 2h_t + (\text{mean}(\mathcal{H}) + (4\pi + 2)h_t)2n\tan\alpha \qquad (13)$$

Now substituting Equation 10, and using the worst case $\text{mean}(\mathcal{H}) = h_t$ and $\hat{h} = 2h_t$, we obtain:

$$L - L^\star \leq 2h_t + (\text{mean}(\mathcal{H}) + (4\pi + 2)h_t)\frac{L^\star \left(1 + \frac{2\pi\tan\alpha}{3}\right)\hat{h}^2}{C_v \, \text{mean}(\mathcal{H})^3} \qquad (14)$$

$$\leq 2L^\star + L^\star\frac{4(4\pi + 3)}{C_v}\left(1 + \frac{2\pi\tan\alpha}{3}\right) \qquad (15)$$

And as before there are $\left\lfloor 1 + \log_2 \frac{\hat{h}}{h_\epsilon} \right\rfloor$ subtours.

5   Experiments

We have shown that HEIGHT-VISIT has bounded worst-case performance, but the question remains: is it a practical algorithm for real-world situations? In this section we will argue that there exist realistic Cone-TSPN instances where HEIGHT-VISIT computes a good (short) tour. We will evaluate the quality of the tour by comparing it with:

- The cone apex tour T_0. This represents an upper bound length of what any TSPN algorithm should be able to accomplish.
- The SLICE-VISIT tour. This tour is computed in a similar fashion to HEIGHT-VISIT, but it remains at the height of the shortest cone. The comparison with SLICE-VISIT demonstrates the value obtained by planning a tour embedded in more than one slice.

First we will consider a representative application of surveying a field of row crops, where cone heights are generated from satellite imagery with heights that vary depending on importance. Next, we will consider Cone-TSPN instances with cones drawn randomly from normally distributed positions and heights.

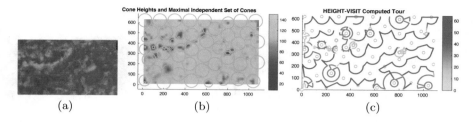

(a)   (b)   (c)

Fig. 5. (a) Satellite imagery showing a field of row crops. (b) Discrete maximum view heights chosen on a 20m grid. Heights were chosen linear to twice the green channel minus the sum of the remaining channels, with the minimum and maximum fixed at 8m and 150m respectively. Also shown are the outer boundaries of the MIS of view cones chosen by HEIGHT-SELECT at $\hat{h} = 64$m with $\alpha = \pi/4$. (c) Tour computed by HEIGHT-VISIT using (0,0) as the takeoff point x_0. The length of this tour is 13612m, the length of the tour computed by SLICE-VISIT is 27214m, and $L_0 = 33533$m. This demonstrates the utility of a varying height tour

5.1 Implementation

We implemented our algorithms using heuristics which improved practical performance without hurting the performance guarantees. We relaxed the requirement that the slice tours be performed in series: Instead, we compiled a set of points to visit in $\mathbb{R}^3$ for each slice and optimized a TSP tour of the union of all slice points using the Linkern module from the Concorde TSP solver[1].

The set of points to visit, $\mathcal{X}$, is chosen at height h_t for each cone in the MIS $c_i \in \hat{\mathcal{C}}$ so that every cone that intersects c_i is intersected by at least one of the points. Recall that h_t for cone i is greater than $(1/2)h_i$, after the cone heights are truncated based on $\hat{h}$. Denote the set of cones which intersect c_i and have apex distances in the range $(0, 2h_t \tan \alpha]$ as $\mathcal{N}_-$. Denote the set of cones which intersect c_i and have apex distances in the range $(2h_t \tan \alpha, 4h_t \tan \alpha]$ as $\mathcal{N}_+$.

- If both $\mathcal{N}_-$ and $\mathcal{N}_+$ are empty, add to $\mathcal{X}$ the center point of c_i at h_t. Mark this point as *adjustable*.
- If $\mathcal{N}_- \neq \varnothing$, compute a coverage radius r_c which is the minimum radius of a circle centered on the axis of c_i, contained in height h_t, which intersects all cones in $\mathcal{N}_-$. For each $c_j \in \mathcal{N}_-$, add a point to $\mathcal{X}$ that is at height h_t, in a vertical plane connecting the apexes of c_i and c_j, and at distance r_c from the center axis of c_i. In the worst case, $r_c = h_t \tan \alpha$, as in our previous specification of HEIGHT-VISIT, but when it is smaller this allows tighter coverage circles.
- If $\mathcal{N}_+ \neq \varnothing$, compute a coverage radius r_c which is the minimum radius of a circle centered on the axis of c_i, contained in height h_t, which intersects all cones in $\mathcal{N}_+$. For each $c_j \in \mathcal{N}_+$, add a point to $\mathcal{X}$ that is at height h_t, in a vertical plane connecting the apexes of c_i and c_j, and at distance r_c from the center axis of c_i.

---

[1] http://www.math.uwaterloo.ca/tsp/concorde.html

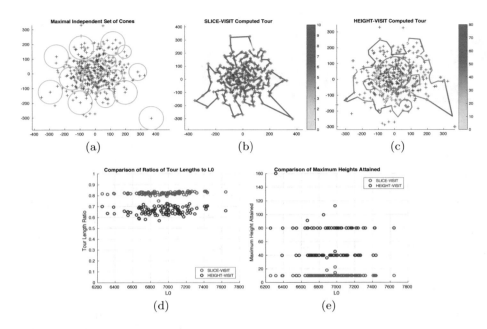

Fig. 6. (a) The apexes of a randomly generated Cone-TSPN instance, together with the outer boundaries of the MIS chosen by `HEIGHT-SELECT`. (b) The `SLICE-VISIT` Cone-TSPN solution. (c) The `HEIGHT-VISIT` Cone-TSPN solution. (d) The ratios of tour lengths to L_0, for 100 Cone-TSPN instances and both strategies. (e) The maximum heights achieved by both strategies, for 100 Cone-TSPN instances. `SLICE-VISIT` must remain at the level of the shortest cone, but `HEIGHT-VISIT` can travel higher. The most slices `HEIGHT-VISIT` ever used was 4. Neither the number of slices nor the length ratios significantly depended on L_0.

Finally, after the optimal TSP tour is computed, it is shortened through an iterative process. Each vertex of the tour which was marked *adjustable* is moved towards its previous vertex, so that it lies on the boundary of its cone instead of directly above the apex. If the input cones are disjoint, all the points used in the TSP optimization will be the centers of cone caps, subject to this iterative shortening. Even after adjusting the tour to bounce off the circle boundaries, the computed tour can be longer than the optimal TSPN tour of uniform circles by a constant factor, because it is computed from the tour of centers (see e.g. [12]). This hurts the proven performance guarantees by a constant factor, since they assume a PTAS with negligable ϵ is available.

In addition to the above procedure for `HEIGHT-VISIT`, we also executed a similar procedure for `SLICE-VISIT`. The difference was that cone heights were truncated to the height of the shortest cone. This provided a performance comparison with an algorithm that treated the input cones as uniform disks at a fixed height. The tour of cone apexes, T_0, demonstrated the performance of an algorithm that is unaware of the view cone properties.

5.2   Real Environment

We used a satellite image of a field or row crops to generate a realistic Cone-TSPN instance that we could execute our algorithms on. We generated view cones on a 20m grid for a rectangular field. Since a frequent aim of crop surveillance is to detect stressed plants, we required parts of the field with less green be imaged from closer. Specifically, we took twice the green channel and subtracted the other two channels to generate a simple approximation of crop stress. Then we selected cone heights linearly based on this stress metric: highest stress corresponded with 8m, lowest stress corresponded with 150m. The average cone height was 111.22m. We used the next grid square southwest from from the southwest grid square as the takeoff point x_0 (there was a road in this square). We used $\alpha = \pi/4$.

Using Concorde TSP solver we found the optimal tour T_0 of the cone apexes, and we executed both SLICE-VISIT and HEIGHT-VISIT. The generated cone heights and computed tours are shown in Figure 5. Since the input was a uniform grid, both T_0 and the SLICE-VISIT tour were boustrophedon-like coverage pattern, respectively at heights 0m and 8m (since 8m was the height of the shortest cone). Concorde found $L_0 = 33533$m, so the $\hat{h}$ search began from $h_\ell = 2.9336$m. The best $\hat{h}$ found for HEIGHT-VISIT was 64m, with 4 slices. The length of the SLICE-VISIT tour was 27214m, and the length of the HEIGHT-VISIT tour was 13612m. This was a cost reduction of 18.84% for SLICE-VISIT and 59.41% for HEIGHT-VISIT, over the baseline optimal tour of cone apexes.

These results show that HEIGHT-VISIT performs well when planning a tour of dense view cones of varying height.

5.3   Simulated Environments

To further validate our algorithms, we also executed them on randomly generated inputs. We generated 100 sets of 400 view cones which had normally distributed heights and positions. The position distribution had zero mean and independent variance of 100^2 in each direction. The takeoff point x_0 was chosen from the same distribution. The height distribution had 250 mean and a variance of 100^2, with a minimum of 10. We used $\alpha = \pi/4$. The tour lengths and tour heights attained are shown in Figure 6, as well as an example cone distribution. The average height attained by SLICE-VISIT was the average of the shortest cone in the input, which was 10.249. The average maximum height attained by HEIGHT-VISIT was 59.049. The average value for L_0 was 6934.9, while the average SLICE-VISIT tour was 0.8211 as long (17.89% shorter), and the average HEIGHT-VISIT tour was 0.6608 as long (33.92% shorter).

These results further demonstrate that HEIGHT-VISIT performs well when planning a tour of realistic view cones of varying height, concentrated in a particular area of interest.

6   Conclusion

In this work we have formulated a new optimization problem Cone-TSPN which is a form of TSPN in $\mathbb{R}^3$. Cone-TSPN is designed to formalize the problem of planning a minimum length tour for a UAV that must image specified points on the ground, where each point has a maximum distance that it can be imaged from while resolving sufficient detail.

We have presented new approximation algorithms for Cone-TSPN: SLICE-VISIT and HEIGHT-VISIT, which are the first to obtain approximation factors independent of the number of input cones n. The SLICE-VISIT algorithm remains at a constant height and obtains an approximation factor which is $O\left((h_{\max}/h_\mu)^2 (1 + \tan\alpha)\right)$, where $h_{\max}$ is the height of the tallest view cone, h_μ is the height of the average view cone, and α is the maximum field of view of the camera on the UAV. By executing SLICE-VISIT $O\left(1 + \log\left(h_{\max}/h_{\min}\right)\right)$ times at different heights, where $h_{\min}$ is the height of the shortest view cone, the HEIGHT-VISIT algorithm obtains an approximation factor which is $O\left((1 + \log\left(h_{\max}/h_{\min}\right))(1 + \tan\alpha)\right)$.

In addition to presenting proofs of the approximation factors, we have also validated the practical performance of the HEIGHT-VISIT algorithm through simulation. We generated Cone-TSPN instances from satellite imagery and randomly, and showed that HEIGHT-VISIT computed significantly shorter tours than those computed by SLICE-VISIT or by running a TSP solver on the positions of the cone apexes.

Here are some areas of future work:

- Obtaining improved approximation factors. It is likely that an approximation can be found that does not depend on the viewing angle α. It is also possible that a constant factor approximation algorithm exists for Cone-TSPN.
- The online version of Cone-TSPN. Particularly, what if the cone apex positions are known, but not the heights? An intelligent UAV might want to perform a loose, high-altitude coverage pattern, then drop down lower to image points of interest from closer up. For this situation SLICE-VISIT might be useful as a single-step subroutine.
- Implementation on real UAVs. Are the paths feasible to execute? Do they result in useful images?
- Multirobot planning. What if there are multiple robots, perhaps with varying capabilities?

References

1. W. Cook, *In pursuit of the traveling salesman: mathematics at the limits of computation.* Princeton University Press, 2012.
2. P. Tokekar, J. Vander Hook, D. Mulla, and V. Isler, "Sensor planning for a symbiotic uav and ugv system for precision agriculture," in *Intelligent Robots and Systems (IROS), 2013 IEEE/RSJ International Conference on.* IEEE, 2013, pp. 5321–5326.

3. S. A. Sadat, J. Wawerla, and R. T. Vaughan, "Recursive non-uniform coverage of unknown terrains for uavs," in *Intelligent Robots and Systems (IROS 2014), 2014 IEEE/RSJ International Conference on.* IEEE, 2014, pp. 1742–1747.

4. M. Schwager, B. J. Julian, M. Angermann, and D. Rus, "Eyes in the sky: Decentralized control for the deployment of robotic camera networks," *Proceedings of the IEEE*, vol. 99, no. 9, pp. 1541–1561, 2011.

5. D. Bhadauria, O. Tekdas, and V. Isler, "Robotic data mules for collecting data over sparse sensor fields," *Journal of Field Robotics*, vol. 28, no. 3, pp. 388–404, 2011.

6. O. Tekdas, D. Bhadauria, and V. Isler, "Efficient data collection from wireless nodes under the two-ring communication model," *The International Journal of Robotics Research*, p. 0278364912439429, 2012.

7. S. Arora, "Polynomial time approximation schemes for euclidean traveling salesman and other geometric problems," *Journal of the ACM (JACM)*, vol. 45, no. 5, pp. 753–782, 1998.

8. J. S. Mitchell, "Guillotine subdivisions approximate polygonal subdivisions: A simple polynomial-time approximation scheme for geometric tsp, k-mst, and related problems," *SIAM Journal on Computing*, vol. 28, no. 4, pp. 1298–1309, 1999.

9. M. de Berg, J. Gudmundsson, M. J. Katz, C. Levcopoulos, M. H. Overmars, and A. F. van der Stappen, "Tsp with neighborhoods of varying size," *Journal of Algorithms*, vol. 57, no. 1, pp. 22–36, 2005.

10. S. Safra and O. Schwartz, "On the complexity of approximating tsp with neighborhoods and related problems," *computational complexity*, vol. 14, no. 4, pp. 281–307, 2006.

11. K. Elbassioni, A. V. Fishkin, and R. Sitters, "Approximation algorithms for the euclidean traveling salesman problem with discrete and continuous neighborhoods," *International Journal of Computational Geometry & Applications*, vol. 19, no. 02, pp. 173–193, 2009.

12. A. Dumitrescu and J. S. Mitchell, "Approximation algorithms for tsp with neighborhoods in the plane," *Journal of Algorithms*, vol. 48, no. 1, pp. 135–159, 2003.

13. E. M. Arkin and R. Hassin, "Approximation algorithms for the geometric covering salesman problem," *Discrete Applied Mathematics*, vol. 55, no. 3, pp. 197–218, 1994.

14. H. L. Bodlaender, C. Feremans, A. Grigoriev, E. Penninkx, R. Sitters, and T. Wolle, "On the minimum corridor connection problem and other generalized geometric problems," *Computational Geometry*, vol. 42, no. 9, pp. 939–951, 2009.

15. J. S. Mitchell, "A ptas for tsp with neighborhoods among fat regions in the plane," in *Proceedings of the eighteenth annual ACM-SIAM symposium on Discrete algorithms.* Society for Industrial and Applied Mathematics, 2007, pp. 11–18.

16. T.-H. H. Chan and K. Elbassioni, "A qptas for tsp with fat weakly disjoint neighborhoods in doubling metrics," *Discrete & Computational Geometry*, vol. 46, no. 4, pp. 704–723, 2011.

17. A. Dumitrescu and C. D. Tóth, "The traveling salesman problem for lines, balls and planes," in *Proceedings of the Twenty-Fourth Annual ACM-SIAM Symposium on Discrete Algorithms.* SIAM, 2013, pp. 828–843.

18. K. Elbassioni, A. V. Fishkin, and R. Sitters, "On approximating the tsp with intersecting neighborhoods," in *Algorithms and Computation.* Springer, 2006, pp. 213–222.

19. C. S. Mata and J. S. Mitchell, "Approximation algorithms for geometric tour and network design problems," in *Proceedings of the eleventh annual symposium on Computational geometry.* ACM, 1995, pp. 360–369.

Competitive Two Team Target Search Game with Communication Symmetry and Asymmetry

Michael Otte[1,3], Michael Kuhlman[2], and Donald Sofge[1]

[1] U.S. Naval Research Lab, Washington D.C. 20375
[2] University of Maryland, College Park, MD 20742
[3] National Research Council RAP Postdoctoral Associate at NRL.

Abstract. We study a search game in which two multiagent teams compete to find a stationary target at an unknown location. Each team plays a mixed strategy over the set of search sweep-patterns allowed from its respective random starting locations. Assuming that communication enables cooperation we find closed-form expressions for the probability of winning the game as a function of team sizes, and vs. the existence or absence of communication within each team. Assuming the target is distributed uniformly at random, an optimal mixed strategy equalizes the expected first-visit time to all points within the search space. The benefits of communication enabled cooperation increase with team size. Simulations and experiments agree well with analytical results.

Keywords: Multiagent System, Competitive Search, Search and Rescue, Search Game.

1 Introduction

We consider the problem of team-vs.-team competitive search, in which two teams of autonomous agents compete to find a stationary target at an unknown location. The game is won by the team of the first agent to locate the target. We are particularly interested in how coordination within each team affects the outcome of the game. We assume that intra-team communication is a prerequisite for coordination, and examine how the expected outcome of the game changes if one or both of the teams lack the ability to communicate—and thus coordinate.

This game models, e.g., an adversarial scenario in which we are searching for a pilot that has crashed in disputed territory, and we want to find the pilot before the adversary does (see Figure 1). Both we and the adversary have multiple autonomous aircraft randomly located throughout the environment to aid in our respective searches (e.g., that were performing unrelated missions prior to the crash), but neither agents nor adversaries have formulated a plan *a priori*. In this paper we answer the questions: How does team size affect game outcome? How beneficial is communication? What is an optimal search strategy?

In Section 4 we derive a closed-form expression for the expected outcome of an "ideal game" in which both teams search at the maximum rate for the entire game. A mixed Nash equilibrium exists at the point that each team randomizes

© Springer Nature Switzerland AG 2020
K. Goldberg et al. (Eds.): *Algorithmic Foundations of Robotics XII*, SPAR 13, pp. 208–223, 2020.
https://doi.org/10.1007/978-3-030-43089-4_14

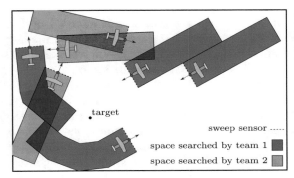

Fig. 1: Four agents (dark blue) compete against three adversaries (light red) to locate a target (black point). Communication enables members of the same team to cooperate (e.g., blue team), lack of communication prohibits cooperation (e.g., red team). The sweep sensors are, by definition, infinitesimally thin in the direction of travel.

target

sweep sensor ·····

space searched by team 1 ■

space searched by team 2 ▨

their distributed searches such that all points are swept at the same expected time. An ideal game is impossible to realize in many environments due to a number of boundary conditions; however, it provides a useful model that allows us to evaluate how coordination affects game outcome. In Section 4.5 we extend these results by bounding performance in non-ideal cases, and find that non-ideal games become asymptotically close to ideal as the size of the environment increases toward infinity. Related work is discussed in Section 2. Nomenclature, a formal statement of assumptions, and the formal game definition appear in Section 3. Supporting simulations and experimental results appear in Section 5; discussion and conclusions appear in Sections 6 and 7, respectively.

2 Related Work

The target search problem was formalized at least as early as 1957 by [19], who studied aircraft detection of naval vessels in a probabilistic framework. Variations of the problem have been studied in many different communities, resulting in a vast body of related work. Indeed, even the subset of related work involving multiagent teams is too large to cover here. Extensive surveys of different formulations and approaches can be found in [6, 31, 8]. Previous work on target search ranges from the purely theoretical (differential equations [21], graph theory [29], game theory [26], etc.) to the applied (numerical methods [2, 12], control theory [11, 15], heuristic search [23], etc.) and borrows ideas from fields as diverse as economics [4, 9] and biology [18, 28].

One difference between the current paper and previous work is the scenario that we consider, in which two teams compete to locate a target first. In *cooperative search* a single team of agents attempts to locate one or more targets [30, 1] that may be stationary [15] or moving [18], and a key assumption is that all searchers cooperate. In contrast, we assume an adversarial relationship exists between two different teams of searchers.

Closely related *pursuit-evasion* games assume that one agent/team actively tries to avoid capture by another agent/team [13, 22], leading to an adversarial relationship between the searchers(s) and the target(s). *Capture the flag* [17] assumes that one team is attempting to steal a target that is guarded by the other team. Our scenario differs from both pursuit-evasion and capture the flag in that the adversarial relationship is between two different teams of searchers, each individually performing cooperative search for the same target.

Our work shares similarities with 1-dimensional *linear search* [7], and *cow path problems* [32, 24]. Differences include our extensions to higher dimensional spaces, which themselves build on coverage methods that use lawn-mower sweep patterns [5]. Our world model shares many of the same assumptions as [5]. In particular, an initial uniform prior distribution over target location and perfect sensors. Using sweep patterns for single agent coverage is studied by [5], while [30] extends these ideas to a single multiagent team searching for a moving and possibly evading target. Spires and Goldsmith use the idea of space filling curves to reduce the 2D search problem to a 1D problem [25].

Our work explicitly considers how each team's ability to communicate affects the expected outcome of the search game. This allows us to analyze scenarios in which teams have asymmetric communication abilities. A number of previous methods have considered limited communication, but have done so in different ways than those explored here. For example, robots were required to move such that a communication link could be maintained [1], and/or the ability to communicate between agents was assumed to be dependent on distance [27, 15], limited by bandwidth [11], adversaries [3], other constraints [14], or impossible [10].

3   Preliminaries

The search space is denoted X. The multiagent team is denoted G, the adversary team is denoted A, and an arbitrary team is denoted T, i.e., $T \in \{G, A\}$. There are $n = |G|$ agents in the multiagent team, and $m = |A|$ adversaries in the adversary team. The i-th agent is denoted g_i and the j-th adversary a_j, where $i \in \{1, \ldots, n\}$ and $j \in \{1, \ldots, m\}$. Both teams search for the same target q. Agents, adversaries, and target are idealized as points, and we abuse our notation by allowing them to indicate their locations in the search space, $g_i, a_j, q \in X$.

The term 'actor' is used to describe a member of the set $G \cup A$. The state space $S = X \times \Theta$ of a single actor includes position X and directional heading Θ. Let S_{g_i} represent the state space of the i-th agent. The product state space of the team is $\mathcal{S}_G = S_1 \times \ldots \times S_n$. A particular configuration of the team is denoted $\mathbf{s}_G$, where $\mathbf{s}_G \in \mathcal{S}_G$. Similarly, for the adversary $\mathbf{s}_A \in \mathcal{S}_A = S_1 \times \ldots \times S_m$. It is convenient to define the product space of *locations* for each team. Formally, $\mathbf{g} = (g_1, \ldots, g_n) \in \mathcal{X}_G = X_1 \times \ldots \times X_n$ and $\mathbf{a} = (a_1, \ldots, a_m) \in \mathcal{X}_A = X_1 \times \ldots \times X_m$ where we continue our abuse of notation that actors denote their own locations.

We use the subscript '0' to denote a starting value. For example, the starting location of g_i is $g_{i,0}$ and the starting configuration of the team is $\mathbf{g}_0$.

3.1   Assumptions

We consider search spaces embedded in D-dimensional Euclidean space, $X \subset \mathbb{R}^D$. We assume X is "well behaved" such that X is bounded, convex, and has a

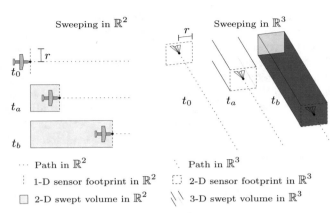

Fig. 2: Examples of sweep sensing in $\mathbb{R}^2$ (Left) and $\mathbb{R}^3$ (Right) for three different times $t_0 < t_a < t_b$. In $\mathbb{R}^2$ the sweep sensor footprint is a 1-D line segment oriented perpendicular to the direction of travel, in $\mathbb{R}^3$ it is a 2-D patch oriented perpendicular to the direction of travel. Swept volume increases as the robot moves forward.

boundary ∂X that can be decomposed into a finite number of locally Lipschitz continuous pieces. Our general formulation assumes X is a continuous[4] space.

The target is assumed to be stationary. Let $\mathcal{D}_X$ be the probability density function for a uniform distribution over X. Agents, adversaries, and targets are idealized as points, and have independent and identically distributed (i.i.d.) initial locations drawn from according to $\mathcal{D}_X$. The Lebesgue measure in $\mathbb{R}^D$ is denoted $\mathscr{L}_D(\cdot)$. Let Ω_X and Ω_S be the smallest σ-algebras over X and S, respectively. The (extension of) the Lebesgue measure in Ω_S is $\mathscr{L}_{\Omega_S}(\cdot)$. Given our assumptions, $\mathbb{P}(\hat{X}) = \int_{\hat{X}} \mathcal{D}_X(x) = \frac{\mathscr{L}_D(\hat{X})}{\mathscr{L}_D(X)}$ and $\mathbb{P}(\hat{S}) = \frac{\mathscr{L}_{\Omega_S}(\hat{S})}{\mathscr{L}_{\Omega_S}(S)}$ for all measurable subspaces $\hat{X} \subset X$ and $\hat{S} \subset S$, respectively, where $\mathbb{P}(\cdot)$ denotes the probability measure, and the integrals are Lebesgue. The probability spaces over starting locations and starting states are defined $(X, \Omega_X, \mathbb{P})$ and $(S, \Omega_S, \mathbb{P})$, respectively.

We assume agents and adversaries use *sweep sensors* with perfect accuracy (see Figure 2). A sweep sensor in $\mathbb{R}^D$ has an infinitesimally thin footprint defined by a subset of a $(D-1)$-dimensional hyperplane oriented perpendicular to the the direction of travel. We denote the sensor footprint B_r, where r refers to the radius of the smallest $(D-1)$-ball that contains the footprint, see Figure 2. Although $\mathscr{L}_D(B_r) = 0$ (e.g., the volume of a 2-dimensional disc is 0 in $\mathbb{R}^3$), the target is detected as the sensor footprint sweeps past it. We assume X is "large" in the sense that the minimum diameter of X is much greater than ("$\gg$") r. In "large" search spaces the sweep sensor provides a reasonable idealization of any sensor with finite observation volume[5].

---

[4]A discrete formulation can readily be obtained by replacing Lebesgue integrals over continuous spaces with summations over discrete sets, and reasoning about the probability of events directly instead of via probability density.

[5]Note that even for sensors with positive measure footprints, $0 < \mathscr{L}_D(B_r) < \infty$ (e.g., a D-ball instead of a $(D-1)$-ball) nearly all space is searched as the forward boundary of a sensor volume sweeps over it (in contrast to the space that is searched instantaneously at startup due to being within some agent's sensor volume).

We assume that agents and adversaries cannot detect the opposite team or physically interact with each other. This is a reasonable model when the environment is large such that chance encounters are unlikely. Actors are ignorant of their own teammates' locations *a priori* (for example, as if all actors are performing their own individual missions when the search scenario unexpectedly develops). Actors are assumed to sweep at a constant forward velocity v, where $0 < v < \infty$, and to have infinite rotational acceleration so that they are able to change direction instantaneously. Each actor may only change direction at most a countably infinite number of times[6]. We assume that each team is rational and will eventually sweep the entire space.

3.2 Paths, Multi-Paths, and Spaces

Let ρ denote a single actor's search path, $\rho \in S$. Let $B(s)$ denote the set of points in X swept by that actor's sensor when the actor is at a particular point $s \in \rho$. The set of points swept by an actor traversing path ρ is therefore $\bigcup_{s \in \rho} B(s)$. A *space covering* path is denoted $\hat{\rho}$ and has the property that its traversal will cause all points in the search space to be swept, i.e., $X \subset \bigcup_{s \in \hat{\rho}} B(s)$.

A search multi-path ψ_G is a set of paths containing one path ρ_i per agent in the team G, i.e., $\psi_G = \bigcup_{g_i \in G} \{\rho_i\}$. Let $\hat{\psi}_G$ denote a space covering search multi-path. One traversal of $\hat{\psi}_G$ by the members of G sweeps all points in the search space, $\bigcup_{s \in \mathbf{s}_G \in \hat{\psi}_G} B(s) \subset X$. Similarly quantities are defined for the adversary, $\psi_A = \bigcup_{a_j \in A} \{\rho_j\}$ and $\bigcup_{s \in \mathbf{s}_A \in \hat{\psi}_A} B(s) \subset X$. Let Ψ_G (and Ψ_A) denote the space of all possible search multi-paths given a team's state space $\mathcal{S}_G$ (and adversary's state space $\mathcal{S}_A$). Formally, $\Psi_G = \bigcup \{\psi \mid \mathcal{S}_G\}$ and $\Psi_A = \bigcup \{\psi \mid \mathcal{S}_A\}$.

3.3 Communication and Coordination Models

The function $\mathcal{C} : \{G, A\} \to \{0, 1\}$ denotes the communication ability of a team. Communication within a particular team T is either assumed to be perfect $\mathcal{C}(T) = 1$ or nonexistent $\mathcal{C}(T) = 0$. That is, team members can either communicate always or never. Communication enables coordination, which allows the team to find a target more quickly in expectation. When $\mathcal{C}(T) = 1$, the members of T *attempt* to equally divide the effort of searching X such that each $x \in X$ is swept by exactly one agent and each agent travels an equal distance. We investigate the 2 by 2 space of game scenarios this allows, $\mathcal{C}(G) \times \mathcal{C}(A) = \{0, 1\} \times \{0, 1\}$.

3.4 Game Formulation

Given our assumptions, the first team to sweep the target's location wins the game. The family of *competitive team target search games* we consider is defined:

---

[6]This prevents "cheating" where an agent that continuously rotates through an uncountably infinite number of points is able to use its zero-measure sweep sensor as if it were a volumetric sensor of non-zero-measure (the measure of a countably infinite union of sweep footprints is still 0).

Random team target search games: *Given a search space X, a stationary target $q \in X$, a multiagent team $G = \{g_1, \ldots, g_n\}$, and an adversary team $A = \{a_1, \ldots, a_m\}$; with initial locations drawn i.i.d. from $\mathcal{D}_X(q)$, $\mathcal{D}_X(g_{i,0})$, and $\mathcal{D}_X(g_{i,0})$, respectively; communication $\mathcal{C}(G), \mathcal{C}(A) \in \{0, 1\}$; and chosen movement along multi-paths $\psi_G \subset X$ and $\psi_A \subset X$; then team G wins iff $q \in B(g_i)$ for some $g_i \in \mathbf{s}_G \in \psi_G$ before $q \in B(a_j)$ for some $a_j \in \mathbf{s}_A \in \psi_a$.*

3.5   Game Outcomes, Multi-Path Spaces, and Strategies

Let $\Omega_{\text{outcome}} = \{\omega_{\text{lose}}, \omega_{\text{win}}, \omega_{\text{tie}}\}$ denote the space of game outcomes, where ω_{win} is the event that a member of team G finds the target first, ω_{lose} denotes the event that an adversary finds the target first, and ω_{tie} denotes a tie. Given our formulation within a continuous space, ties are a measure 0 set, $\mathbb{P}(\omega_{\text{tie}}) = 0$, that can be ignored for the purposes of analyzing expected performance. In discrete space one could break ties in a number of ways, e.g., by randomly selecting the actor that finds the target first.

Strategies are equivalent to multipaths—any valid multi-path ψ_G that starts at $\mathbf{g}_0$ is a particular search strategy for team G. Let Ψ denote the space of all strategies. Let Ψ_G be a function that maps starting configurations $\mathbf{g}_0$ to the subset of all valid strategies for G that begin at $\mathbf{g}_0$. Let Ω denote the (smallest) σ-algebra over Ψ. Formally, $\Psi_G : \mathcal{S}_G \to \Omega$. The subset of all valid strategies available to G given $\mathbf{g}_0$ is thus denoted $\Psi_G(\mathbf{g}_0)$, where $\Psi_G(\mathbf{g}_0) \subset \Psi$.

A *conditional mixed strategy* is both: (1) conditioned on the event that team G starts at a particular $\mathbf{g}_0$, and (2) mixed such that the particular strategy $\psi_G \in \Psi_G(\mathbf{g}_0)$ used by team G is drawn at random from $\Psi_G(\mathbf{g}_0)$ according to a chosen probability density $\mathcal{D}_{\mathbf{g}_0}(\psi_G)$. By designing $\mathcal{D}_{\mathbf{g}_0}(\psi_G)$ appropriately, it is possible for team G to play any valid conditional mixed strategy given $\mathbf{g}_0$.

Given $\mathcal{D}_{\mathbf{g}_0}(\psi_G)$, a probability measure function $\mathbb{P}_{\mathbf{g}_0}$ can be constructed such that $\int_{\Psi_G(\mathbf{g}_0)} \mathcal{D}_{\mathbf{g}_0}(\psi_G) = 1$ and such that for all subsets $\hat{\Psi} \subset \Psi_G(\mathbf{g}_0)$ we have $\mathbb{P}_{\mathbf{g}_0}(\psi_G \in \hat{\Psi}) = \int_{\hat{\Psi}} \mathcal{D}_{\mathbf{g}_0}(\psi_G)$. A particular conditional mixed strategy (conditioned on $\mathbf{g}_0 \in \mathcal{S}_G$) is thus a probability space that can be represented by the triple $(\Psi_G(\mathbf{g}_0), \Omega_G(\mathbf{g}_0), \mathbb{P}_{\mathbf{g}_0})$, where $\Omega_G(\mathbf{g}_0)$ is the (smallest) σ-algebra over $\Psi_G(\mathbf{g}_0)$.

A *mixed strategy* $(\Psi_G, \Omega_G, \mathbb{P}_G)$ is the set of conditional mixed strategies over all $\mathbf{g}_0 \in \mathcal{S}_G$, where $\Omega_G = \bigcup_{\mathbf{g}_0 \in \mathcal{S}_G} \Omega_G(\mathbf{g}_0)$ and $\mathbb{P}_G(\psi \,|\, \mathbf{g}_0) = \mathbb{P}_{\mathbf{g}_0}(\psi)$ for all $\mathbf{g}_0 \in \mathcal{S}_G$. Note that a mixed strategy triple *is not a probability space*, per se, because it does not include the probability measure of the starting configurations $\mathbf{g}_0$. That said, when a mixed strategy is combined with such a measure, e.g., the measure implied by $\mathcal{D}_X$, then a probability space is the result. Analogous quantities, $(\Psi_A, \Omega_A, \mathbb{P}_A)$ and $(\Psi_A(\mathbf{a}_0), \Omega_A(\mathbf{a}_0), \mathbb{P}_{\mathbf{a}_0})$, are defined for the adversary.

Given our assumption that the two teams cannot detect each other, one team's mixed strategy is necessarily independent of the other team's starting location. Let $t(\psi, x)$ denote the earliest time at which a team following ψ sweeps location $x \in X$. Given ψ_G and ψ_A, and a target at q (with location unknown to either team), team G wins if and only if $t(\psi_G, q) < t(\psi_A, q)$. Let $X_{\text{win}}(\psi_G, \psi_A) \subset X$

denote the subset of the search space where $t(\psi_G, x) < t(\psi_A, x)$.

$$X_{\text{win}}(\psi_G, \psi_A) = \{x \in X \mid t(\psi_G, x) < t(\psi_A, x)\}.$$

Team G wins if and only if $q \in X_{\text{win}}$. When G plays ψ_G and A plays ψ_A, we get:

Proposition 1. *Assuming the target is located uniformly at random in X, the probability team G wins is equal to the ratio of search space it sweeps before the adversary,* $\mathbb{P}\left(\omega_{\text{win}} | \psi_G, \psi_A\right) = \frac{\mathscr{L}_D(X_{\text{win}}(\psi_G, \psi_A))}{\mathscr{L}_D(X)}.$

The probability team G wins in a particular search space while playing a particular adversary is calculated by integrating $\mathcal{D}_{\mathbf{g}_0}(\psi_G)$ over $\Psi_G(\mathbf{g}_0)$ for all $\mathbf{g}_0$ and $\mathcal{D}_{\mathbf{a}_0}(\psi_A)$ over $\Psi_A(\mathbf{a}_0)$ for all $\mathbf{a}_0$. Assuming the target and teams are distributed uniformly at random, this is calculated:

$$\mathbb{P}(\omega_{\text{win}}) = \frac{1}{\mathscr{L}_{\Omega_{S_G}}(S_G)} \int_{S_G} \frac{1}{\mathscr{L}_{\Omega_{S_A}}(S_A)} \int_{S_A} \int_{\Psi_G(\mathbf{g}_0)} \mathcal{D}_{\mathbf{g}_0}(\psi_G) \int_{\Psi_A(\mathbf{a}_0)} \mathcal{D}_{\mathbf{a}_0}(\psi_A) \frac{\mathscr{L}_D(X_{\text{win}}(\psi_G, \psi_A))}{\mathscr{L}_D(X)} \quad (1)$$

where the Lebesgue integrals are respectively over all $\mathbf{g}_0 \in S_G$, all $\mathbf{a}_0 \in S_A$, all $\psi_G \in \Psi_G(\mathbf{g}_0)$ and all $\psi_A \in \Psi_A(\mathbf{a}_0)$.

We use "$*$" to denote quantities related to optimality. An optimal mixed strategy is defined: $(\Psi_G^*, \Omega_G, \mathbb{P}_G^*) = \arg\max_{((\Psi_G, \Omega_G, \mathbb{P}_G))} \mathbb{P}(\omega_{\text{win}})$.

4   Optimal Strategies for Ideal Games

Let X_{swept} denote the space team G has swept (X_{swept} is different from X_{win} in that X_{swept} may include space that has also been swept by the adversary). The instantaneous rate team G sweeps new space is given by: $\frac{d}{dt}[\mathscr{L}_D(X_{\text{swept}})]$. The optimal instantaneous rate at which an agent sweeps new space can be expressed as the agent's velocity multiplied by the $(D-1)$-dimensional hypervolume of the sensor footprint: $v\mathscr{L}_{D-1}(B_r)$. Given our assumptions, we have the following:

Proposition 2. *The optimal instantaneous normalized rate that a single agent sweeps new space is:* $c^* = v \frac{\mathscr{L}_{D-1}(B_r)}{\mathscr{L}_D(X)}.$

4.1   Both Teams Can Communicate (ideal case)

The optimal instantaneous normalized rate (c^*) occurs when there is no sensor overlap between agents. Building on Proposition 2 we get:

Corollary 1. *The optimal instantaneous normalized rate that n agents can cooperatively sweep new space is:* $\frac{d^*}{dt}[\frac{\mathscr{L}_D(X_{\text{swept}})}{\mathscr{L}_D(X)}] = nv\frac{\mathscr{L}_{D-1}(B_r)}{\mathscr{L}_D(X)} = nc^*.$

In an "ideal" cooperative search we assume that the team can maintain the optimal rate of sweep for the entire duration of search. The time required for an ideal search with n agents is $t_{n,\text{sweep}} = 1/(nc^*)$. The game is guaranteed to end by time $t_{\text{final}} = \min(t_{n,\text{sweep}}, t_{m,\text{sweep}})$.

We observe that any bias or predictability by a particular team (e.g., a mixed strategy that leads to a subset of the environment being swept sooner or later in expectation, over all possible starting locations) could be exploited by the opposing team. This observation leads to the following proposition.

Proposition 3. *A mixed strategy that causes some portion of the environment to be swept sooner or later, in expectation, over the set of all strategies and distributions of agent and adversary starting locations is a suboptimal strategy.*

As a corollary of proposition 3 we have the following:

Corollary 2. *If an optimal ideal mixed strategy $(\Psi_G^*, \Omega_G, \mathbb{P}_G^*)$ exists for a team G, then in that strategy the first sweep time for any point $x \in X$ is distributed uniformly at random between 0 and $t_{n,\text{sweep}}$ (over the space of all possible starting configurations).*

In an *ideal game* each team plays an optimal mixed strategy over a set of ideal search strategies. The following is true by the definition of a Nash equilibrium:

Proposition 4. *Assuming optimal ideal strategies exist for both teams, a mixed strategy Nash equilibrium exists when both teams play an optimal mixed strategy.*

At such a Nash equilibrium, the first sweep time of any point x by one team is completely decorrelated from the first sweep time of x by the other team (over the space of all possible actor starting locations).

Let $X_{\text{new}}(t)$ be the space that has not yet been swept by either team by time t, and $\frac{d}{dt}[\frac{\mathscr{L}_D(X_{\text{new}}(t))}{\mathscr{L}_D(X)}]$ be the instantaneous normalized rate team G sweeps this unswept space at time t. We note that, given a particular ψ_G and ψ_A,

$$\frac{\mathscr{L}_D(X_{\text{win}}(\psi_G, \psi_A))}{\mathscr{L}_D(X)} = \int_0^{t_{\text{final}}} \frac{d}{dt}\left[\frac{\mathscr{L}_D(X_{\text{new}}(t))}{\mathscr{L}_D(X)}\right] dt,$$

where $t_{\text{final}} = \min(\frac{1}{nc^*}, \frac{1}{mc^*})$. Thus, Equation 1 can be reformulated for the Nash equilibrium of an ideal game with cooperation within both teams as:

$$\mathbb{P}\left(\omega_{\text{win}}^*\right) = \frac{1}{\mathscr{L}_{\Omega_{S_G}}(\mathcal{S}_G)} \int_{\mathcal{S}_G} \frac{1}{\mathscr{L}_{\Omega_{S_A}}(\mathcal{S}_A)} \int_{\mathcal{S}_A} \int_{\Psi_G^*(\mathbf{g}_0)} \int_{\Psi_A^*(\mathbf{g}_0)} \mathcal{D}_{\mathbf{g}_0}(\psi_G)\mathcal{D}_{\mathbf{a}_0}(\psi_A) \int_0^{t_{\text{final}}} \frac{d}{dt}\left[\frac{\mathscr{L}_D(X_{\text{new}}(t))}{\mathscr{L}_D(X)}\right] dt$$

where integrals are Lebesgue. Using the independence of the two team's optimal mixed strategies, i.e., $\mathcal{D}_{\mathbf{g}_0}(\psi_G)$ and $\mathcal{D}_{\mathbf{a}_0}(\psi_A)$ for all $\mathbf{g}_0$ and $\mathbf{a}_0$ yields:

$$\mathbb{P}\left(\omega_{\text{win}}^*\right) = \int_0^{t_{\text{final}}} \frac{1}{\mathscr{L}_{\Omega_{S_G}}(\mathcal{S}_G)\mathscr{L}_{\Omega_{S_A}}(\mathcal{S}_A)} \int_{\mathcal{S}_G} \int_{\mathcal{S}_A} \int_{\Psi_G^*(\mathbf{g}_0)} \int_{\Psi_A^*(\mathbf{g}_0)} \mathcal{D}_{\mathbf{g}_0}(\psi_G)\mathcal{D}_{\mathbf{a}_0}(\psi_A) \frac{d}{dt}\left[\frac{\mathscr{L}_D(X_{\text{new}}(t))}{\mathscr{L}_D(X)}\right] dt$$

We observe that the quantity inside the outermost integral describes the expected value of $\frac{d}{dt}[\frac{\mathscr{L}_D(X_{\text{new}}(t))}{\mathscr{L}_D(X)}]$ over all $\mathcal{S}_G$, $\mathcal{S}_A$, Ψ_G^*, and Ψ_A^*. For brevity we denote the expected value of '·' over all $\mathcal{S}_G$, $\mathcal{S}_A$, Ψ_G^*, and Ψ_A^* as $\mathbb{E}^*[\cdot]$, i.e., $\mathbb{E}^*[\cdot] \equiv \mathbb{E}_{\mathcal{S}_G, \mathcal{S}_A, \Psi_G^*, \Psi_A^*}[\cdot]$. Thus, formally,

$$\mathbb{E}^*\left[\frac{d}{dt}\mathscr{L}_D(X_{\text{new}}(t))\right] = \frac{1}{\mathscr{L}_{\Omega_{S_G}}(\mathcal{S}_G)\mathscr{L}_{\Omega_{S_A}}(\mathcal{S}_A)} \int_{\mathcal{S}_G} \int_{\mathcal{S}_A} \int_{\Psi_G(\mathbf{g}_0)} \int_{\Psi_A(\mathbf{a}_0)} \mathcal{D}_{\mathbf{g}_0,}(\psi_G)\mathcal{D}_{\mathbf{a}_0}(\psi_A) \frac{d}{dt}\left[\frac{\mathscr{L}_D(X_{\text{new}}(t))}{\mathscr{L}_D(X)}\right]$$

Lemma 1. *Assuming optimal ideal mixed strategies exist and both teams play an optimal ideal mixed strategy,*

$$\mathbb{E}^*\left[\frac{d}{dt}\mathscr{L}_D(X_{\text{new}}(t))\right] = (1 - tmc^*)\, nc^* \tag{2}$$

Proof. At time t the adversary (operating according to its own ideal optimal strategy) has swept $tmv\frac{\mathscr{L}_D(B_r)}{\mathscr{L}_D(X)}$ portion of the entire search space. The interplay between the mixed ideal optimal strategies for each team forces the expected instantaneous overlap between teams to be uncorrelated. Thus, for all $t \in [0, t_{\text{final}}]$, the instantaneous expected rate team G sweeps $\mathscr{L}_D(X_{\text{new}})$ is discounted by a factor of $1 - tmv\frac{\mathscr{L}_D(B_r)}{\mathscr{L}_D(X)}$ vs. $\frac{d^*}{dt}\mathscr{L}_D(X_{\text{swept}})$.

$$\mathbb{E}^*\left[\frac{d}{dt}\mathscr{L}_D(X_{\text{new}}(t))\right] = \left(1 - tmv\frac{\mathscr{L}_D(B_r)}{\mathscr{L}_D(X)}\right)\frac{d^*}{dt}\left[\frac{\mathscr{L}_D(X_{\text{swept}})}{\mathscr{L}_D(X)}\right].$$

Substitution with Proposition 2 and Corollary 1 yields the desired result. $\qquad\square$

In other words, team G covers new territory at a rate that decreases, in expectation, proportionally to the proportion of space the adversary has covered up to time t. Substituting this result back into the previous equations and noting that $t_{\text{final}} = 1/(c^* \max(n, m))$:

$$\mathbb{P}\left(\omega_{\text{win}}^*\right) = \int_0^{1/(c^* \max(n,m))} (1 - tmc^*)\, nc^*\, dt.$$

Solving this equation yields the following theorem.

Corollary 3. *The probability team G wins an ideal game assuming both G and A are able to communicate and play optimal ideal mixed strategies is*

$$\mathbb{P}\left(\omega_{\text{win}}^*\right) = \begin{cases} n/(2m) & \text{when } n \le m \\ 1 - m/(2n) & \text{when } n \ge m \end{cases}.$$

4.2   Case 2: multiagent team G cannot communicate but the adversary team A can (ideal case)

Given that the starting location of each agent on team G is sampled i.i.d. a game in which team G cannot communicate is equivalent to the situation in which the adversary team A plays n sub-games, one vs. each member $\{g_i\} \subset G$, and A wins the overall game if and only if it wins all n sub-games. Because the target location is identical for each of these n sub-games, the n games are not indipendent (as was erroniously assumed in a preliminary version of this work). To win, the adversary must sweep the target before the particular agent of G that happens to sweep the target first among all members of G. This can be calculated by reformulating Equation 1 to integrate over the distribution of the smallest of n first sweep times:

$$\mathbb{P}\left(\omega_{\text{lose}}^*\right) = \frac{m - (m-1)^{n+1}m^{-n}}{n+1}$$

Combining $\mathbb{P}\left(\omega_{\text{tie}}^*\right) = 0$ with Corollary 3 we get:

Corollary 4. *The probability team G wins an ideal game, assuming team G cannot communicate but the adversary team A can, and the adversary team A plays optimal ideal mixed strategies, while each $\{g_i\} \subset G$ individually plays an optimal ideal mixed strategy, is: $\mathbb{P}\left(\omega_{\text{win}}^*\right) = 1 - \mathbb{P}\left(\omega_{\text{lose}}^*\right).$*

4.3   Case 3: Team G can communicate but the adversary's team A cannot (ideal case)

This case is complementary to the previous one, due to symmetry and the fact that $\mathbb{P}\left(\omega_{\text{tie}}^*\right) = 0$. We swap n and m and also ω_{lose} and ω_{win} from the results in the previous section to get:

Corollary 5. *The probability team G wins an ideal game, assuming team G can communicate but the adversary team A cannot, and team G plays an optimal ideal mixed strategy, while each of the adversary team's individual uncoordinated sub-teams $\{a_j\} \subset A$ for $1 \leq j \leq m$ plays an optimal ideal mixed strategy, is:*

$$\mathbb{P}\left(\omega_{\text{win}}^*\right) = \frac{n - (n-1)^{m+1} n^{-m}}{m+1}$$

We also note that $\mathbb{P}\left(\omega_{\text{lose}}^*\right) = 1 - \mathbb{P}\left(\omega_{\text{win}}^*\right)$.

4.4   Case 4: Neither team can communicate (ideal case)

The case when neither team has communication must be analyzed separately, but is somewhat trivial.

Theorem 1. *The probability team G wins an ideal game, assuming no team can communicate but all actors individually play an optimal ideal mixed strategy, is:*

$$\mathbb{P}\left(\omega_{\text{win}}^*\right) = \frac{n}{n+m} \qquad \mathbb{P}\left(\omega_{\text{lose}}^*\right) = \frac{m}{n+m}.$$

Proof. Our assumption of uniformly random i.i.d. starting locations of actors and target, combined with the fact that optimal mixed strategies decorrelate the expected sweep time of any particular point x, means that, in expectation, each actor has a $1/(n+m)$ chance of being the agent with the least amount of travel (i.e., time) required to sweep q. The probability that team G finds the target before A can be calculated as the ratio of agents to total actors.   □

4.5   Extensions to non-ideal games

The realization of an ideal game requires that an optimal mixed strategy exists such that Equation 2 holds. In practice, this idealization is often broken by both the startup locations of the actors and the boundary of the search space (see Figure 3). However, it is possible to modify the equations for an ideal game to obtain bounds on $\mathbb{P}\left(\omega_{\text{win}}\right)$. This is accomplished by breaking the multiagent search into two mutually exclusive phases: (1) a phase containing all portions of the search wherein the agents of G *are not* able to perform an ideal search and (2) a separate phase containing all other (ideal) portions of the search. Let t_{startup} denote the time required for the non-ideal portion of search. We assume that the adversaries in A are able to maintain an ideal search rate for the entire game, which produces a lower bound on $\mathbb{P}\left(\omega_{\text{win}}\right)$. Reversing the roles of adversaries

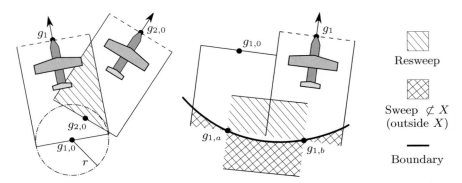

Fig. 3: Left: Two agents start within r of each other, causing some area to be swept by both of them (hashed). Right: a single robot turns near the boundary of the search space as it moves from $g_{1,0} \to g_{1,a} \to g_{1,b} \to g_1$; this causes some space to be swept multiple times (hashed) and space outside the search space to be swept (criss-crossed).

and agents provides an upper bound on $\mathbb{P}\left(\omega_{\text{win}}\right)$; let $\hat{t}_{\text{startup}}$ be the time it takes A to perform non-ideal search in this case. Recall that $c* = v\frac{\mathscr{L}_D(B_r)}{\mathscr{L}_D(X)}$. Also let $\hat{t}_{\text{final}} = \min(t_{\text{startup}} + \frac{1}{nc^*}, \frac{1}{mc^*})$ and $\tilde{t}_{\text{final}} = \min(\tilde{t}_{\text{startup}} + \frac{1}{mc^*}, \frac{1}{nc^*})$.

Theorem 2. *Assuming that both teams can communicate, and an optimal mixed strategy exists for both teams, and that both teams play an optimal mixed strategy, and that the game is ideal in every sense except for starting locations and boundary effects, the probability team G wins is bounded as follows:*

$$\left[\int_{t_{\text{startup}}}^{\hat{t}_{\text{final}}} (1 - tmc^*)\, nc^*\, dt\right] - t_{\text{startup}}mc^* \leq \mathbb{P}\left(\omega_{\text{win}}\right) \leq 1 - \left[\int_{\tilde{t}_{\text{startup}}}^{\tilde{t}_{\text{final}}} (1 - tnc^*)\, mc^*\, dt\right] - \tilde{t}_{\text{startup}}nc^*.$$

Proof. (Sketch) Non-ideal effects become increasingly detrimental to G's probability of winning the game as they occur earlier and earlier in the game. Thus, it is possible to construct a scenario that is even worse than a worst-case non-ideal search (in terms of team G's probability of winning the game) by: (1) assuming that all negative ramifications of a non-ideal search happen at the beginning of the search for team G, instead of whenever they actually occur, and (2) assuming the adversary team A is allowed to realize an ideal sweep rate for the entire game. The length of the non-ideal startup phase for the worse-than-worst case can be bounded as follows: $t_{\text{startup}} < c_1 r \mathscr{L}_{D-1}(\partial X)$, where c_1 is a dimensionally dependent constant, r is sweep radius, and $\mathscr{L}_{D-1}(\partial X)$ is the surface area of the search space boundary. $\square$

Theorem 2 shows that the proportion of time spent dealing with non-ideal startup and boundary approaches zero as environments get larger vs. sensor range, $\lim_{r \to 0} \frac{t_{\text{startup}}}{t_{\text{final}}} = 0$. In other words, the ideal equations model the non-ideal case more-and-more accurately as the size of the environment increases.

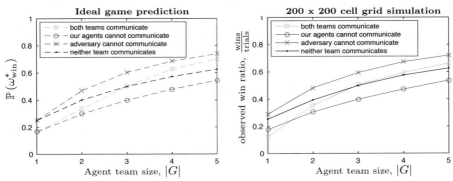

Fig. 4: Comparison of results predicted by analysis of the ideal case (left) vs. those from a simulation experiment (right) for games with 1 to 5 Agents vs. 3 Adversaries. Vertical axes measure the probability team G wins vs. its observed win ratio in repeated random trials performed in simulation, respectively. Colors denote which teams may or may not communicate. Each datapoint represents the mean result over 10^4 trials.

5   Simulations and Experiments

We compare the results derived in the previous section to repeated trials of search and rescue in contested environments performed both in simulation and on a mixed platform of real and virtual agents. For these simulations and experiments we assume a discrete grid environment where movement is allowed along the cardinal directions (note that this contrasts with the more general continuous space formulation assumed in previous sections).

Multipaths are selected from a library of predefined sweep patterns, such that each pattern forms a cycle, sweeps the entire space, and are designed to minimize sweep overlap between different parts of the search. If an agent/adversary cannot communicate with its team then it moves to the nearest point on a randomly selected cycle and then follows it. If an agent (resp. adversary) can communicate with its team then all team members agree on a cycle, divide the path into n (resp. m) contiguous sub-paths, and then allocate one sub-path per team member such that the cumulative distance traveled by the team to their start locations is minimized. Next, each agent/adversary moves to its start point and then searches its allotted sub-path. Unlike the ideal case, the probability of a tie is non-zero; ties are broken by a coin toss weighted proportionally to the number of {team members vs. all actors} that simultaneously discover the target.

Simulations are run in the Julia language using a 200 by 200 meter search space composed of 1 by 1 meter grid cells. B is an L-∞ ball of radius 5 meters. Locations of actors and target are determined uniformly at random over the set of grid cells. Selected results comparing predictions based on the ideal case vs. the average results from simulation of 4×10^5 trials (10^4 trials per datapoint) are

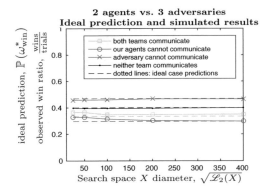

Fig. 5: The ideal case (dashed lines) more accurately predicts the mean result of Monte Carlo simulations (solid lines) as $\mathscr{L}_2(X)$ increases. These particular results are for games with 2 agents vs. 3 adversaries over various $\sqrt{\mathscr{L}_2(X)}$. Colors represent which teams may or may not communicate. Each datapoint represents the mean result over 10^4 trials.

presented in Figures 4 and 5, each datapoint in the simulation curves represents the mean results over 10^4 repeated trials.

The mixed platform combines Asctec Pelican quadrotor UAVs that have on-board Odroid single board computers with simulated agents that run on a laptop (Ubuntu 14.04). The quadrotors receive position measurements from a Vicon motion capture system and runs ETH-Zurich modular sensor fusion framework for state estimation [20]. Robot Operating System (ROS) is used on all computers for local interprocess communications and NRL's *Puppeteer* framework is used for coordination of all vehicles, which uses Lightweight Communications and Marshalling [16] for intervehicle communications. Grid cells are 2 by 2 meters, and the contested search space is 12 by 12 meters. We use a virtual target sensor such that an actor discovers the target if their locations are closer than 1 meter. All actors fly at an altitude of 2 meters, which corresponds to a field of view of approximately $60°$ when searching for ground targets with a downward facing camera. A random number generator is used to determine start location of the real actors as well as the virtual actors and the target. We perform repeated trials for a two agent team (consisting of one Asctec Pelican and one virtual agent) vs. an adversary (Asctec Pelican). We perform 10 successful trials: 5 trials for the case where team G can communicate and 5 for the case where it cannot. Results from experiments with the mixed platform appear in Figure 6.

6   Discussion

Our analysis, simulations, and experiments show that using the ideal case to predict $\mathbb{P}(\omega_{\text{win}})$ works reasonably well, and provides a more accurate prediction as the size of the environment increases. We also show that the relative effects of non-ideal startup locations and boundary conditions vanish, in the limit, as the size of the environment increases.

With respect to communication symmetry vs. asymmetry, our results verify the intuition that team G benefits from a situation in which G can communicate and team A cannot. More interesting is the result that moving from a scenario where both G and A can communicate to a scenario where neither G nor A can

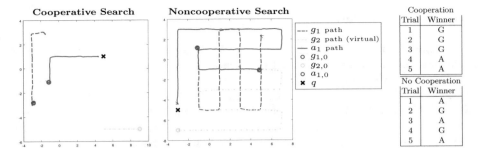

Fig. 6: Mixed platform experiments. A two agent team (consisting of a Pelican Quadrotor, blue, and a simulated agent, light-blue) vs. an adversary (Pelican Quadrotor, red) to find a target (black). Left and Center: Examples of paths when agents do and do not collaborate, respectively. Right: Game outcomes of repeated trials.

communicate benefits G only if $n < m$. The advantages of performing a coordinated search vs. uncoordinated search increase vs. team size. Uncommunicating larger teams will outperform uncommunicating smaller teams, in general.

7   Summary and Conclusions

We study the effects of cooperation on multiagent two-team competitive search games, a class of games in which two multiagent teams compete to locate a stationary target placed at an unknown location. Given an assumption that communication is required for coordination, this enables us to analyze how communication symmetry and asymmetry between teams affects the outcome of the game. For the case involving perfect finite sweep sensors, random initial placement of actors/target, and non-observability of the other team's movements, we find closed-form solutions for the probability of winning an "ideal game" in which transient boundary effects are ignored.

A team maximizes its chances of winning by playing a mixed strategy such that all points are eventually swept, the expected time a point is (first) swept is identical for all points, and there is as little search overlap as possible. A Nash equilibrium exists for an ideal game.

The chances of winning the search game increase vs. team size, and also increase if the team is able to communicate. Moving from a situation in which both teams can communicate to a situation where neither team can communicate will benefit the smaller team and hinder the larger team (this effect becomes stronger as the difference between the two teams' sizes increases).

Monte Carlo simulations over random start locations and experimental results on a platform with AscTec Pelican quadrotor UAVs validate that the observed outcomes of non-ideal games are predicted reasonably well by equations derived for the ideal case, and that these predictions become more accurate as the size of the search space increases.

Acknowledgments

We would like to thank Colin Ward, Corbin Wilhelmi, and Cyrus Vorwald for their help in facilitating the mixed platform experiments.

This work was performed at the Naval Research Laboratory and was funded by the Office of Naval Research under grant numbers N0001416WX01271 and N0001416WX01272. The views, positions and conclusions expressed herein reflect only the authors opinions and expressly do not reflect those of the Office of Naval Research, nor those of the Naval Research Laboratory.

References

1. Beard, R.W., McLain, T.W.: Multiple uav cooperative search under collision avoidance and limited range communication constraints. In: Decision and Control, 2003. Proceedings. 42nd IEEE Conference on. vol. 1, pp. 25–30 Vol.1 (Dec 2003)
2. Bertuccelli, L.F., How, J.P.: Robust UAV search for environments with imprecise probability maps. In: IEEE Conf. on Decision and Control and the European Control Conferenc. pp. 5680–5685 (Dec 2005)
3. Bhattacharya, S., Khanafer, A., Başar, T.: A Double-Sided Jamming Game with Resource Constraints, pp. 209–227. Springer International Publishing (2016)
4. Chandler, P., Pachter, M.: Hierarchical control for autonomous teams. In: Proceedings of the AIAA Guidance, Navigation, and Control Conference. pp. 632–642 (2001)
5. Choset, H., Pignon, P.: Coverage path planning: The boustrophedon cellular decomposition. In: Field and Service Robotics. pp. 203–209. Springer (1998)
6. Chung, T.H., Hollinger, G.A., Isler, V.: Search and pursuit-evasion in mobile robotics. Auton. Robots 31(4), 299–316 (2011)
7. Demaine, E.D., Fekete, S.P., Gal, S.: Online searching with turn cost. Theoretical Computer Science 361(2), 342–355 (2006)
8. Dias, M.B., Zlot, R., Kalra, N., Stentz, A.: Market-based multirobot coordination: A survey and analysis. Proceedings of the IEEE 94(7), 1257–1270 (July 2006)
9. Dias, M.B.: Traderbots: A new paradigm for robust and efficient multirobot coordination in dynamic environments. Ph.D. thesis, Carnegie Mellon University Pittsburgh (2004)
10. Feinerman, O., Korman, A., Lotker, Z., Sereni, J.S.: Collaborative search on the plane without communication. In: Proceedings of the 2012 ACM Symposium on Principles of Distributed Computing. pp. 77–86. PODC '12, ACM, New York, NY, USA (2012), http://doi.acm.org/10.1145/2332432.2332444
11. Flint, M., Polycarpou, M., Fernandez-Gaucherand, E.: Cooperative control for multiple autonomous uav's searching for targets. In: Decision and Control, 2002, Proceedings of the 41st IEEE Conference on. vol. 3, pp. 2823–2828 vol.3 (Dec 2002)
12. Forsmo, E.J., Grotli, E.I., Fossen, T.I., Johansen, T.A.: Optimal search mission with unmanned aerial vehicles using mixed integer linear programming. In: Unmanned Aircraft Systems (ICUAS), 2013 International Conference on. pp. 253–259 (May 2013)
13. Gerkey, B.P., Thrun, S., Gordon, G.: Parallel stochastic hill-climbing with small teams. In: Multi-Robot Systems. From Swarms to Intelligent Automata Volume III, pp. 65–77. Springer (2005)

14. Hollinger, G.A., Yerramalli, S., Singh, S., Mitra, U., Sukhatme, G.S.: Distributed data fusion for multirobot search 31(1), 55–66 (2015)
15. Hu, J., Xie, L., Lum, K.Y., Xu, J.: Multiagent information fusion and cooperative control in target search 21(4), 1223–1235 (July 2013)
16. Huang, A.S., Olson, E., Moore, D.C.: Lcm: Lightweight communications and marshalling. In: Intelligent robots and systems (IROS), 2010 IEEE/RSJ international conference on. pp. 4057–4062. IEEE (2010)
17. Huang, H., Ding, J., Zhang, W., Tomlin, C.J.: Automation-assisted capture-the-flag: A differential game approach. IEEE Transactions on Control Systems Technology 23(3), 1014–1028 (2015)
18. Kim, M.H., Baik, H., Lee, S.: Response threshold model based uav search planning and task allocation. Journal of Intelligent & Robotic Systems 75(3), 625–640 (2013)
19. Koopman, B.: The theory of search. ii. target detection. Operations Research 4(5), 503–531 (1956)
20. Lynen, S., Achtelik, M.W., Weiss, S., Chli, M., Siegwart, R.: A robust and modular multi-sensor fusion approach applied to mav navigation. In: 2013 IEEE/RSJ International Conference on Intelligent Robots and Systems. pp. 3923–3929 (Nov 2013)
21. Mangel, M.: Marcel Dekker, New York (1989)
22. Noori, N., Isler, V.: Lion and man with visibility in monotone polygons. The International Journal of Robotics Research p. 0278364913498291 (2013)
23. Sato, H., Royset, J.O.: Path optimization for the resource-constrained searcher. Naval Research Logistics 57(5), 422–440 (2010)
24. Spieser, K., Frazzoli, E.: The cow-path game: A competitive vehicle routing problem. In: Decision and Control (CDC), 2012 IEEE 51st Annual Conference on. pp. 6513–6520 (Dec 2012)
25. Spires, S.V., Goldsmith, S.Y.: Exhaustive geographic search with mobile robots along space-filling curves. In: Collective robotics, pp. 1–12. Springer (1998)
26. Sujit, P.B., Ghose, D.: Multiple agent search of an unknown environment using game theoretical models. In: American Control Conf. vol. 6, pp. 5564–5569 vol.6 (June 2004)
27. Sujit, P.B., Ghose, D.: Negotiation schemes for multi-agent cooperative search. Proc. of the Institution of Mech. Engineers, Part G: J. of Aero. Eng. 223(6), 791–813 (2009)
28. Sydney, N., Paley, D.A., Sofge, D.: Physics-inspired motion planning for information-theoretic target detection using multiple aerial robots. Auton. Robots pp. 1–11 (2015)
29. Trummel, K., Weisinger, J.: Technical notethe complexity of the optimal searcher path problem. Operations Research 34(2), 324–327 (1986)
30. Vincent, P., Rubin, I.: A framework and analysis for cooperative search using UAV swarms. In: ACM Symposium on Applied Computing. pp. 79–86. SAC '04, ACM, New York, NY, USA (2004), http://doi.acm.org/10.1145/967900.967919
31. Waharte, S., Trigoni, N.: Supporting search and rescue operations with UAVs. In: Int. Conf. on Emerging Security Technologies. pp. 142–147 (Sept 2010)
32. Zhu, M., Frazzoli, E.: On competitive search games for multiple vehicles. In: Decision and Control (CDC), 2012 IEEE 51st Annual Conference on. pp. 5798–5803 (Dec 2012)

Beyond the planning potpourri: reasoning about label transformations on procrustean graphs

Shervin Ghasemlou[1], Fatemeh Zahra Saberifar[2],
Jason M. O'Kane[1], and Dylan A. Shell[3]

[1] University of South Carolina, Columbia SC, USA,
[2] Amirkabir University of Technology, Tehran, Iran,
[3] Texas A&M University, College Station TX, USA

Abstract. We address problems underlying the algorithmic question of automating the co-design of robot hardware in tandem with its apposite software. Specifically, we consider the impact that degradations of a robot's sensor and actuation suites may have on the ability of that robot to complete its tasks. Expanding upon prior work that addresses similar questions in the context of filtering, we introduce a new formal structure that generalizes and consolidates a variety of well known structures including many forms of plans, planning problems, and filters, into a single data structure called a procrustean graph. We describe a collection of operations on procrustean graphs (both semantics-preserving and semantics-mutating), and show how a family of questions about the destructiveness of a change to the robot hardware can be answered by applying these operations. We also highlight the connections between this new approach and existing threads of research, including combinatorial filtering, Erdmann's strategy complexes, and hybrid automata.

1 Introduction

The process of designing effective autonomous robots—spanning the selection of sensors, actuators, and computational resources along with software to govern that hardware—is a messy endeavor. There appears to be little hope of fully automating this process, at least in the short term. There would, however, be significant value in *design tools* for roboticists that can manipulate partial or tentative designs, in interaction with a human co-designer.

To that end, this paper lays a formal foundation for answering questions about the relationship between a robot's hardware, specifically its sensors and actuators, and its ability to complete a given task. Interesting questions arise when one considers how modifications to a given robot's capabilities alter the planning and estimation efforts that robot must undertake. This paper develops theoretical tools that we believe to be helpful for thinking about such aspects.

This material is based upon work supported by the National Science Foundation under Grants IIS-1527436, IIS-1526862, IIS-0953503, IIS-1453652.

K. Goldberg et al. (Eds.): *Algorithmic Foundations of Robotics XII*, SPAR 13, pp. 224–239, 2020.
https://doi.org/10.1007/978-3-030-43089-4_15

Prior work by the current authors [10] made some preliminary progress in this direction by considering a limited form of *sensor map*, which describes a coarsification of a sensor model. This paper strengthens and extends those results in several ways.

1. We contribute, in Section 2, a new general representation called a *procrustean graph*[4] that unifies several previously distinct conceptual classes of object. This representation is constructive, in that it can be used to instantiate a data-structure from which various questions can be posed and addressed concretely. We detail, in Section 3, how this representation can be used to reason about planning problems and their solutions.

2. We extend, in Section 4, the existing notion of a sensor map to model modifications to both sensors and actuators, including modifications that introduce uncertainty directly. This generalized map is called *label map*. We show how to decide whether a label map is *destructive* in the sense of preventing the achievement of a previously-attainable goal. (Prior results used a much more restrictive notion of destructiveness, in which a map was considered destructive if it engendered any change to the robot's behavior; our new results deem non-destructive any map under which the robot can still reach its goal, even if its strategy for doing so is forced to change.)

The dénouement of the paper includes a review of related work interleaved with a discussion of the outlook for continued progress (Section 5) and some concluding remarks (Section 6).

2 Procrustean graphs

2.1 Basic definitions

The work in this paper is connected with a variety of design-time concerns that can be represented with a single formal construct: a graph with set-labelled transitions.

Definition 1 (p-graph). *A* procrustean graph (p-graph) *is an edge-labelled bipartite directed graph in which*

1. *the finite vertex set, of which each member is called a* state, *can be partitioned into two disjoint parts, called the* action vertices V_{u} *and the* observation vertices V_{y}, *with $V = V_{\mathrm{u}} \cup V_{\mathrm{y}}$,*
2. *each edge e originating at an action vertex is labeled with a set of actions $U(e)$ and leads to an observation vertex,*
3. *each edge e originating at an observation vertex is labeled with a set of observations $Y(e)$ and leads to an action vertex, and*

---

[4] Named for Procrustes (Προχρούστες), son of Poseidon, who, according to myth, took the one-size-fits-all concept to extremes.

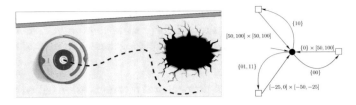

Fig. 1. [left] A differential drive robot with sensors for obstacles, both positive (walls) and negative (holes). [right] An example p-graph that models behavior in which the robot follows a wall while avoiding negative obstacles. This graph, and those that follow, have solid circles to represent elements of V_u, and empty squares for V_y. The arcs are labelled with sets; those that leave the central vertex have two digits, the first digit is '1' iff the wall is detected by the IR sensor on the left-hand side; the second digit is '1' iff the downward pointing IR sensor detects a cliff. The actions, on the edges leaving squares, represent sets of left and right wheel velocities, respectively.

4. *a non-empty set of states V_o are designated as* initial states, *which may be either exclusively action states ($V_o \subseteq V_u$) or exclusively observation states ($V_o \subseteq V_y$).*

The general intuition is to encode an interaction between an agent or robot (which selects actions) and its environment (which dictates the observations made by the robot). The definition is intentionally ecumenical in regard to the nature of that interaction, because we intend this definition to serve as a starting point for more specific structures which, once specific context and semantics are added, lead to special cases that represent particular (and familiar) objects involved planning, estimation, and the like.

Example 2 (wheels, walls, and wells). *A small example p-graph, intended to illustrate the basic intuition, appears in Figure 1. It models a Roomba-like robot that uses single-bit wall and cliff sensors to navigate through an environment. Action states are shown as unshaded squares; observation states are shaded circles. Action labels are subsets of $[0, 500] \times [0, 500]$, of which each element specifies velocities for the robot's left and right drive wheels, expressed in mm/s. Observations are bit strings of length 2, in which the left bit is the output of the wall sensor, and the right bit is the output of the cliff sensor.* ◇

Note that, to keep the model amenable to direct algorithmic manipulation, we require that a p-graph consist of only finitely many states. The labels for each edge, either $U(e)$ or $Y(e)$, need not be finite sets. We instead rely on the availability of some simple operations on labels such as unions, intersection, and membership tests; full details appear in [10].

2.2   Some things you know about are actually secretly p-graphs

Several well-known kinds of objects can be recognizably expressed as p-graphs.

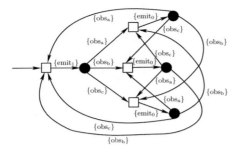

Fig. 2. The 'agents together' filter devised by Tovar *et al.* [12] expressed as a p-graph. The $emit_0$ action indicates that the agents are separated by a beam, and $emit_1$ indicates that the agents are together.

Example 3 (Combinatorial filters). *As formalized by LaValle [7], combinatorial filters are discrete expressions of estimation problems. Associated with each state in such a filter is a specific output. Filters can be cast as p-graphs by having observations and observation transitions exactly as in the filter, but with action vertices having only a single out-edge which has U as a singleton set bearing the output (which we label $emit_x$ for various outputs x). Figure 2 shows a canonical example in which the property of interest is whether two agents in an annulus-shaped environment with three beam sensors are apart or not.* ◇

Example 4 (Schoppers's universal plans). *For observable domains, a universal plan [11] describes an appropriate action for each circumstance that a robot might find itself in. As such, they can be cast as p-graphs in a straightforward way: The p-graph has a single observation vertex, with one uniquely-labeled out-edge corresponding to each world state, and one action state for each of the distinct available actions.* ◇

Example 5 (Erdmann-Mason-Goldberg-Taylor plans). *Several classic papers [5,6,8] find policies for manipulating objects in sensorless (or near sensorless) conditions. The problems are usually posed in terms of a polygonal description of a part; while the solutions to such problems are sequences of actions. Such plans can be expressed as p-graphs in which actions (e.g., a squeeze-grasp or a tray tilt at a particular orientation) or ranges of acceptable actions are interleaved with a special ε which constitutes the sole element in all $Y(e)$. Figure 3 shows an example of such a plan. Of particular note is the fact that that plan exhibits an unexpected dimension of nondeterminism: it indicates sets of allowable actions, rather than a single predetermined one, at each step. Also of note is that the graphs of knowledge states generally searched to produce such plans are p-graphs themselves.* ◇

Example 6 (Nondeterministic graphs). *Recent work by Erdmann [3,4] encodes planning problems using finite sets of states, along with nondeterministic actions represented as collections of edges 'tied' together into single actions. One might convert such a graph to a p-graph by replacing each group of action edges with an observation node, with an outgoing observation edge for each edge constituting the original action.* ◇

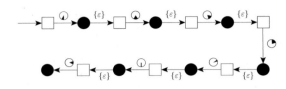

Fig. 3. A plan for orienting an Allen wrench via tray tilting, expressed as a p-graph. Action edges are labeled with sets of azimuth angles for the tray. There is a single dummy observation, ε. This plan is shown as Fig. 2 in Erdmann and Mason [5].

The intent in these examples is to illustrate that p-graphs form a general class that unifies, in a relatively natural way, a number of different kinds of objects that have been studied over a long period of time. The particular constraints applied in each case impose certain kinds of structure that proved useful in the original context. Our objective in this paper is to treat p-graphs, in a general sense, as first-class objects, suitable for manipulation by automated means.

2.3   Properties of p-graphs

At the most general level, we can view a p-graph as an implicit definition of a *language* of strings in which actions and observations alternate. The following definitions make this precise.

Definition 7 (event). *An* event *is an action or an observation.*

Definition 8 (transitions to). *For a given p-graph G and two states $v, w \in V(G)$, an event sequence $e_1 \cdots e_k$ transitions in G from v to w if there exists a sequence of states $v_1, \ldots, v_{k+1}$, such that $v_1 = v$, $v_{k+1} = w$, and for each $i = 1, \ldots, k$, there exists an edge $v_k \xrightarrow{E_k} v_{k+1}$ for which $e_k \in E_k$.*

Note that v and w need not be distinct: for every v, the empty sequence transitions in G from v to v. Longer cycles may result in non-empty sequences of states that start at some v and return.

Definition 9 (valid). *For a given p-graph G and a state $v \in V(G)$, an event sequence $e_1 \cdots e_k$ is* valid *from v if there exists some $w \in V(G)$ for which $e_1 \cdots e_k$ transitions from v to w.*

Observe that the empty sequence is valid from all states in any p-graph.

Definition 10 (execution). *An* execution *on a p-graph G is an event sequence valid from some start state in $V_0(G)$.*

The preceding definitions prescribe when a sequence is valid on a p-graph, placing few restrictions on the sets involved. There are several instances of 'choices' recognizable as forms of non-determinism: (i) there may be multiple elements in V_0; (ii) from any $v \in V_u$ some u may be an element in sets on multiple outgoing action edges; (iii) similarly, from any $w \in V_y$ some y may qualify for multiple outgoing observation edges.

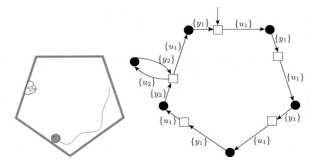

Fig. 4. [left] A robot wanders around a pentagonal environment; the segment with the lightning-bolt contains a battery charger. [right] A p-graph model of this world.

Example 11 (Pentagonal world). *Figure 4 presents concrete realizations of several of the preceding definitions in a single scenario. A robot moves in a pentagonal environment. Information—at least at a certain level of abstraction— describing the structure of the environment, operation of the robot's sensors, its actuators, and their inter-relationships is represented in the p-graph associated with the scenario. Both filtering and planning can be posed as problems on the p-graph representation in terms of valid event sequences.*   ◇

Thus far, other than the potentially infinite action- and observation-alphabets and their specific alternating structure, the definitions are close to classic formal language theory. The set of executions on G are taken to comprise $L(G)$, the language induced by p-graph G. Also, since $V_0(G) \subseteq V(G)$, the language of every p-graph includes the empty sequence.

2.4   Properties of pairs of p-graphs

In Section 3, we model both planning problems and plans themselves as p-graphs. The next definitions will be helpful for formalizing the relationships between those two p-graphs.

Definition 12 (joint-execution). *An event sequence $e_1 \cdots e_k$ is a joint-execution on a pair of p-graphs G and W if it is an execution in both G and W.*

These are the executions that make use of labels and transitions in both p-graphs, and the joint-executions make up the intersection of their respective languages.

Definition 13 (finite on). *A p-graph G is finite on p-graph W if there exists an integer k that bounds the length of every joint-execution of G and W.*

There are two types of p-graph: those that start by producing an action, and those that start by receiving an observation.

Definition 14 (akin). *Two p-graphs are* akin *if both have initial states that are action states, or have initial states that are observation states.*

A stronger notion of the relationship between two p-graphs is safety, which has no simple analogy to a property on the languages involved.

Definition 15 (safe). *P-graph G is* safe *on p-graph W if G is akin to W and if, for every joint-execution $e_1 \cdots e_k$ on G and W, the following property holds: For every state $v \in V(G)$ reached by $e_1 \cdots e_k$ in G, and every state $w \in V(W)$ reached by $e_1 \cdots e_k$ in W, from every possible initial state, we have*

1. *if v is an action state, then for every action u associated with an edge in G originating at v, there exists an edge e in W originating at w, for which $u \in U(e)$, and*
2. *if v is an observation state, then for every observation y associated with an edge in W originating at w, there exists an edge e in G originating at v, for which $y \in Y(e)$.*

The intuition is that if P is safe on Q, then P never executes any action that is not allowed by Q, and is always prepared to respond to any observation that may arrive if chosen by Q.

2.5   Basic constructions of new p-graphs from old

We have shown that several existing structures can be described by p-graphs but the question remains as to why they ought to be expressed as such. We give two examples of constructive operations, applicable to the p-graph structure, which produce new p-graphs as output.

Definition 16 (union of p-graphs). *The* union *of two p-graphs U and W, each akin to the other, denoted by $U \uplus W$, is the p-graph constructed by including both sets of vertices, both sets of edges, and with initial states equal to $V_0(U) \cup V_0(W)$.*

The intuition is to form a graph that allows, via the nondeterministic selection of the start state, executions that belong to either U or W.

Definition 17 (state-determined). *A p-graph P is in a* state-determined *presentation if $|V_0(P)| = 1$ and from every action vertex $u \in V_u$, the edges $e_u^1, e_u^2, \ldots, e_u^\ell$ originating at u bear disjoint labels: $U(e_u^i) \cap U(e_u^j) = \varnothing, i \neq j$, and from every observation vertex $y \in V_y$, the edges $e_y^1, e_y^2, \ldots, e_y^m$ originating at y bear disjoint labels: $Y(e_y^i) \cap Y(e_y^j) = \varnothing, i \neq j$.*

The intuition is that in a p-graph in a state-determined presentation it is easy to determine whether an event sequence is an execution: one starts at the unique initial state and always has an unambiguous edge to follow. We note, however, that the p-graph with a state-determined presentation for some set of executions need not be unique.

Given any p-graph it is possible to construct a new p-graph that has the same set of executions on it, but which is in a state-determined presentation. We only sketch the procedure as it is a generalization of the observation-only algorithm presented in detail in [10]. The basic idea is a forward search that performs a powerset construction on the input p-graph. We begin by constructing a single state to represent the "superposition" of all initial states, and push that onto a empty queue. While the queue has elements, remove a vertex and examine the edges leaving the set of vertices associated with it in the original input p-graph. The labels on those edges are *refined* by constructing a partition of the set spanned by the union of the labels in a way that the subsequent sets of states in the input p-graph is clear. Edges are formed with the refined sets connecting to their target vertices, constructing new ones as necessary, and placing these in the queue.

3   Plans and planning problems

While a p-graph induces a structured state space, further enrichment is needed in order to talk meaningfully about plans and planning problems.

Definition 18 (planning problem). *A* planning problem *is a p-graph G equipped with a* goal region $V_{\text{goal}} \subseteq V(G)$.

The idea is that for a pair that make up the planning problem, the p-graph describes the setting and form in which decisions must be made, while the V_{goal} characterizes what must be achieved.

Definition 19 (plan). *A* plan *is a p-graph P equipped with a* termination region $V_{\text{term}} \subseteq V(P)$.

The intuition is that the out-edges of each action state of the plan show one or more actions that may be taken from that point—if there is more than one such action, the robot selects one nondeterministically—and the out-edges of each observation state show how the robot should respond to the observations received from the environment. If the robot reaches a state in its termination region, it may decide to terminate there and declare success, or it may decide to continue on normally. We can now establish the core relationship between planning problems and plans.

Definition 20 (solves). *A plan (P, P_{term}) solves the planning problem (W, V_{goal}) if P is finite and safe on W, and every joint-execution $e_1 \cdots e_k$ of P on W either reaches a vertex in P_{term}, or is a prefix of some execution that reaches P_{term} and, moreover, all the $e_1 \cdots e_k$ that reach a vertex $v \in V(P)$ with $v \in P_{\text{term}}$, reach a vertex $w \in V(W)$ with $w \in V_{\text{goal}}$.*

Example 21 (Charging around and in the pentagonal world). *We can construct a planning problem from the p-graph of Figure 4, along with a goal region consisting of only the fully-charged state reached by action u_2. Figure 5*

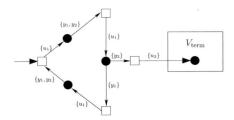

Fig. 5. A plan that directs the robot of Figure 4 to its charging station, along a hyperkinetic (that is, exhibiting more motion than is strictly necessary) path.

shows a plan that solves this problem. However, that plan, a cycle of three actions, is a bit surprising since it will take the robot along three full laps around its environment before terminating. The existence of such bizarre plans motivates our consideration of homomorphic plans, which behave rather more sensibly, in Section 3.1.   ◇

Given a plan (P, P_{term}) and a planning problem (W, V_{goal}), we can decide whether (P, P_{term}) solves (W, V_{goal}) in a relatively straightforward way. First, we convert both P and W into state-determined presentations, using the technique described in Section 2.5. Then, the algorithm conducts a forward search using a queue of ordered pairs (v, w), in which $v \in V(P)$ and $w \in V(W)$, beginning from the (unique, due to Definition 17) start states of each. For each state pair (v, w) reached by the search, we can test each of the properties required by Definition 20:

- If P and W are not akin, return false.
- If (v, w) has been visited by the search before, then we have detected the possibility of returning to the same situation multiple times in a single execution. This indicates that P is not finite on W. Return false.
- If v and w fail the conditions of Definition 15 (that is, if v is missing an observation that appears in w, or w omits an action that appears in v) then P is not safe on W. Return false.
- If v is a sink state not in P_{term}, or w is a sink state not in V_{goal}, then we have detected an execution that does not achieve the goal. Return false.
- If $v \in P_{\text{term}}$ and $w \notin V_{\text{goal}}$, then the plan might terminate outside the goal region. Return false.

If none of these conditions hold, then we continue the forward search, adding to the queue each state pair (v', w') reached by a single event from (v, w). Finally, if the queue is exhausted, then—knowing that no other state pair can be reached by any execution—we can correctly conclude that (P, P_{term}) does solve (W, V_{goal}).

It may perhaps be surprising that both planning problems and plans are defined by giving a p-graph, along with a set of states at which executions should end. We view this symmetry as a feature—not a bug—in the sense that it clearly illuminates the duality between the robot and the environment with which it interacts. Observations can be viewed as merely "actions taken by nature" and vice versa. At an extreme, the planning problem and the plan may be identical:

Lemma 22 (self-solving plans). *If P is a p-graph which is acyclic and the set of its sink nodes is V_{sink}, then (P, V_{sink}) is both a planning problem and a plan. Moreover, (P, V_{sink}) solves (P, V_{sink}).*

Proof: The plan is obviously finite and safe on itself. As joint-executions are essentially just executions, the result follows from the fact that every execution on P either reaches an element of V_{sink}, or is the prefix of one that does.   □

We have described, in Definitions 16–17, operations to construct new p-graphs out of old ones. We can extend these in natural ways to apply to plans.[5]

Definition 23 (∪-product of plans). *The ∪-product of plans (U, V_{goal}) and (W, V'_{goal}), with U and W akin, is a plan $(U \uplus W, V_{goal} \cup V'_{goal})$.*

Theorem 24 (state-determined ∪-products). *Given two plans (P, P_{term}) and (Q, Q_{term}), with P and Q akin, construct a new plan whose p-graph, denoted R, is the expansion of $P \uplus Q$ into a state-determined presentation. Recall that the expansion means that every state $s \in V(R)$ corresponds to sets $P_s \subseteq V(P)$ and $Q_s \subseteq V(Q)$ of states in the original p-graphs (either possibly empty, but not both). Define a termination region R_{term} as follows:*

$$R_{term} := \{ s \in V(R) \mid (P_s \neq \varnothing \wedge P_s \backslash P_{term} = \varnothing) \vee (Q_s \neq \varnothing \wedge Q_s \backslash Q_{term} = \varnothing) \}.$$

Then (R, R_{term}) is equivalent to $(P \uplus Q, P_{term} \cup Q_{term})$, in the sense that they have identical sets of executions on them, and moreover that any problem solved by the former is also solved by the latter.

Proof: The result follows directly from the executions that underlie the state-determined expansion, and the definition of the ∪-product.   □

This result illustrates how the state-determined expansion is useful — it permits a construction that captures the desired behavioral properties and, by working from a standardized presentation, can do this directly by examining states rather than posing questions quantified over the set of executions.

3.1   Homomorphic solutions

The following are a subclass of all solutions to a planning problem.

Definition 25 (homomorphic solution). *For a plan (P, V_{term}) that solves planning problem (W, V_{goal}), consider the relation $R \subseteq V(P) \times V(W)$, in which $(v, w) \in R$ if and only if there exists a joint execution on P and W that can end at v in P and in w in W. A plan for which this relation is a function is called an homomorphic solution.*

The name for this class of solutions comes via analogy to the homomorphisms —that is, structure-preserving maps— which arise in algebra. In this context, an homomorphic solution is one for which each state in the plan corresponds to exactly one state in the planning problem.

---

[5] ... and—via the symmetry between Definitions 18 and 19—in the same stroke, to planning problems, though in this paper we'll use these operations only on plans.

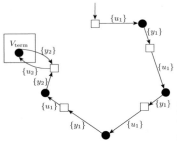

Fig. 6. An alternative, more direct plan that solves the problem of navigating Figure 4's robot to its charger. This plan is an homomorphic solution.

Example 26. *Recall Example 21, which shows a cyclic solution that involves tracing around the cyclic planning problem multiple times, until the least common multiple of their cycle lengths is found, in this case a series of 30 states in each graph. This plan is not an homomorphic solution, because each plan state corresponds to multiple problem states. However, a simpler plan, depicted in Figure 6, can be formed in which each plan state maps to only one problem state. This solution is therefore an homomorphic one.*   ◇

The preceding example is a particular instance of a more general pattern.

Theorem 27. *If there exists a plan to solve a planning problem, then there exists an homomorphic solution.*

Proof: Suppose (P, P_{term}) is a solution to (W, V_{goal}). If every joint-execution arriving at v in P arrives at the same w in W, then $(v, w) \in R$ is a function, so (P, P_{term}) is an homomorphic solution. Thus, consider the cases for which there are elements $(v, w) \in R$ and $(v, y) \in R$, with $w \neq y$. Let $R_{\text{last}} \subset R$ be the relation where $(v_p, v_w) \in R_{\text{last}}$ iff there is a joint-execution $e_1 \cdots e_k$ arriving at v_p on P and v_w on W, and there are no joint-executions which extend the execution (e.g., $e_1 \cdots e_k \cdots e_m$, $m > k$) that arrive at v_w again. Then construct a new plan (Q, Q_{term}) with $V(Q) = V(W)$ and $V_0(Q) = V_0(W)$. For all edges departing $v \in P$ associated with $w \in Q$ where $(v, w) \in R_{\text{last}}$, we collect the label sets by unioning them to form V_e. Then edges departing w are included in Q by carrying over edges from W, intersecting V_e with all the labels of edges departing w, and dropping those for which the result is empty. Finally, an element w is included in Q_{term} if there is a $v \in P_{\text{term}}$ with $(v, w) \in R_{\text{last}}$. Then (Q, Q_{term}) is a solution to (W, V_{goal}) because, though (P, P_{term}) and (Q, Q_{term}) have different sets of executions, every execution on P that reaches P_{term} is transformed into another on Q reaching Q_{term} (and V_{goal}). Moreover, this ensures that R is a bijection, so that (Q, Q_{term}) is an homomorphic solution to (W, V_{goal}).   □

4   Label maps and the damage they inflict

Since p-graphs are capable of representing several structures of interest, the next question is how they might enable a roboticist to evaluate tentative designs and to better understand solution space trade-offs. One class of interesting design-time questions arises when one considers how modifications to a given robot's capabilities alter the planning and estimation efforts that the robot must undertake.

4.1   Label maps

We express modification of capabilities through maps that mutate the labels attached to the edges of a p-graph.

Definition 28 (action, observation, and label maps). *An* action map *is a function* $h_u : U \to 2^{U'}\backslash\{\varnothing\}$ *mapping from an action space U to a non-empty set of actions in a different action space U'. Likewise, an* observation map *is a function* $h_y : Y \to 2^{Y'}\backslash\{\varnothing\}$ *mapping from an observation space Y to a non-empty set of observations in a different observation space Y'. A* label map *combines an action map h_u and a sensor map h_y:*

$$h(a) = \begin{cases} h_u(a) & \text{if } a \in U \\ h_y(a) & \text{if } a \in Y \end{cases}.$$

Definition 29 (label maps on sets and p-graphs). *Given a label map h, its extension to sets is a function that applies the map to a set of labels:*

$$h(E) = \bigcup_{e \in E} h(e).$$

The extension to p-graphs is a function that mutates p-graphs by replacing each edge label E with $h(E)$. We will write $h(P)$ for application of h to p-graph P.

Example 30 (label maps on intervals). *If the action or observation space is $\mathbb{R}$, then we can implicitly represent some subsets of those events as a finite union of intervals [10]. To represent a label map on such an event space, we might, for example, take bounding polynomials $p_1(x)$ and $p_2(x)$, and define*

$$h(x) = \{x' \mid p_1(x) \leq x' \leq p_2(x)\}.$$

Given a finite-union-of-intervals label $\ell \subset \mathbb{R}$, we can evaluate this kind of h by decomposing h into monotone sections, selecting the minimal and maximal values of p_1 and p_2 within that range, and computing the union of the results across all of the monotone sections. Figure 7 shows an example.   ◇

Label maps allow one to express weakening of capabilities as follows. If multiple elements in the domain of $h(\cdot)$ map to sets that are not disjoint, this expresses a conflation of two elements that formerly were distinct.   When they are observations, this directly models a sensor undergoing a reduction in fidelity since the sensor loses the ability to distinguish elements. When they are actions, this models circumstances where uncertainty increases because a single

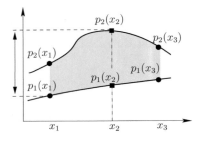

Fig. 7. A label map from $\mathbb{R}$ to $2^{\mathbb{R}}$ may be described by functions p_1 and p_2 as lower and upper bounds, respectively. The marked vertical interval, spanning $p_1(x_1)$ to $p_2(x_2)$ illustrates the image of h across the monotone segment from x_1 to x_2. Values for other monotone segments would be computed similarly.

action can now potentially produce multiple outcomes, and the precise outcome is unknown until after its execution.

Further, when the image of element E is a set with multiple constituents, this also expresses the fact that planning becomes more challenging. For observations, it means that several observations may result from the same state and, as observations are non-deterministic, this increases the onus for joint-executions to maintain safety (for example, plans must account for more choices).

For actions, while there is a seemingly larger choice of actions, this increase does not represent an increase in control authority because several actions behave identically. In both action and observation instances, the map may become detrimental when the outputs of $h(E)$ intersect for multiple Es and thus 'bleed' into each other. Broadly, one would expect that this is more likely when the output sets from $h(\cdot)$ are larger.

4.2   Destructive or not?

If a label map can express a change in a p-graph, the question is whether this change matters. One can pose this question meaningfully for planning problems as the added ingredients provide semantics that yield the notion of solubility.

Definition 31 (destructive and non-destructive). *A label map h is destructive on a set of solutions S to planning problem (G, V_{goal}) if, for every plan $(P, V_{\text{term}}) \in S$, $(h(P), V_{\text{term}})$ cannot solve $(h(G), V_{\text{goal}})$. We say that h is non-destructive on S if for every plan $(P, V_{\text{term}}) \in S$, $(h(P), V_{\text{term}})$ does solve $(h(G), V_{\text{goal}})$.*

Intuitively, destructiveness requires that the label map break all existing solutions; non-destructiveness requires that the label map break none of them.

Example 32 (single plans). *If $S = \{s\}$ is a singleton set, then we can determine whether h is destructive on S by applying the label map h —recall Definition 28— to compute $h(s)$ and $h(G)$, and then testing whether $h(s)$ solves $h(G)$ —recall the algorithm described in Section 3. If $h(s)$ solves $h(G)$, then h is nondestructive on S; otherwise, h is destructive on S. In this singleton case, we say simply that h is (non-)destructive on s.*   ◇

Definition 31 depends on a selection of some class of solutions. Of particular interest is the maximal case, in which every solution is part of the class.

Definition 33 (strongly destructive and strongly non-destructive). *A label map h is strongly (non-)destructive on a planning problem (G, V_{goal}) if it is (non-)destructive on the set of all solutions to (G, V_{goal}).*

Note that, while strong destructiveness may be decided by attempting to generate a plan for $h(G)$ (perhaps by backchaining from V_{goal}), strong non-destructiveness may be quite difficult to verify in general, if only due to the sheer variety of extant solutions. (Recall Example 21, which solves its problem in an unexpected way.) The next results, while not sufficient in general to decide whether a map is strongly non-destructive, do perhaps shed some light on how that might be accomplished.

Lemma 34 (label maps preserve safety). *If P is safe on G, then for any label map h, $h(P)$ is safe on $h(G)$.*

Proof: Consider each pair of states (v, w), with $v \in V(P)$ and $w \in V(G)$ reached by some joint-execution on P and G. Suppose for simplicity that v is an action state. (The opposite case is similar.) Let E_1 denote the union of all labels for edges outgoing from v, and likewise E_2 for labels of edges outgoing from w. Since P is safe on G, we have $E_1 \subseteq E_2$. Then, in $h(P)$ and $h(G)$, observe that

$$h(E_1) = \bigcup_{e \in E_1} h(e) \subseteq \bigcup_{e \in E_2} h(e) = h(E_2),$$

and conclude that $h(P)$ is safe on $h(G)$. □

Lemma 35 (label maps never introduce homomorphism). *If (P, P_{term}) is a non-homomorphic solution to (G, V_{goal}) then no label map h results in $(h(P), P_{term})$ being an homomorphic solution to $(h(G), V_{goal})$.*

Proof: Since (P, P_{term}) is a non-homomorphic solution to (G, V_{goal}), there exist two joint-executions $e_1 \cdots e_k$ and $e'_1 \cdots e'_m$ on P and G such that both arrive at $v \in V(P)$ in P, but on G, the former arrives at $w \in V(G)$ and the latter arrives at $w' \in V(G)$ with $w \neq w'$. Now, given any $h(\cdot)$, pick any particular sequence $(h_1 \in h(e_1)) \cdots (h_k \in h(e_k))$, and $(h'_1 \in h(e'_1)) \cdots (h'_m \in h(e'_m))$, making choices arbitrarily. These are joint-executions on $h(P)$ and $h(G)$. Application of the label map means there is a way of tracing both $(h_1 \in h(e_1)) \cdots (h_k \in h(e_k))$ and $(h'_1 \in h(e'_1)) \cdots (h'_m \in h(e'_m))$ on $h(P)$ to arrive at v, while there is a way of tracing the former on $h(G)$ to arrive at w, and the latter at w'. So $(h(P), P_{term})$ cannot be an homomorphic solution to $(h(G), V_{goal})$. □

Theorem 36 (extensive destructiveness). *For a planning problem (G, V_{goal}), let $\mathcal{H}$ denote the set of homomorphic solutions that problem. Then any label map that is destructive on $\mathcal{H}$ is strongly destructive.*

Proof: Since h is destructive on $\mathcal{H}$, we know that $(h(G), V_{goal})$ can only have homomorphic solutions if some formerly non-homomorphic solution can become an homomorphic one under h, but Lemma 35 precludes that eventuality. This implies, via Theorem 27, that *no* plan solves $h(G)$. Therefore h is strongly destructive on (G, V_{goal}). □

The interesting thing here is that Theorem 36 shows that the class of homomorphic solutions play a special role in the space of all plans: By examining the behavior of h on $\mathcal{H}$, we can gain some insight into its behavior on the space of all plans. Informally, $\mathcal{H}$ seems to function as a 'kernel' of the space of all plans.

5   Related work and outlook

The examples presented early in the paper illustrate how p-graphs allow for a uniformity in treating multiple existing formal objects; the assortment of constructs is surely indicative of the ongoing search for foundational forms.

Combinatorial filters: This paper builds on our prior work [9, 10], which is strongly influenced by the combinatorial filtering perspective, with its use of simple, discrete objects that generalize beyond the methods used in traditional estimation theory, which has a strong reliance on probabilistic models. In both LaValle [7], which provides a tutorial introduction and overview to the approach, and the substantial paper on the topic [12], more work is needed to extend the theory, which only provides for inference, to express aspects of feedback-aware control for achieving tasks. The present paper makes some progress in extending the approach to deal with active (rather than merely passive) systems.

Strategy complexes: One formulation that emphasizes action from the outset is Erdmann's more recent work on strategy complexes [3]. He uses tools from classic and computational topology to relate plans, formulated broadly to include sources of non-determinism, to high-dimensional objects—his loopback complexes—whose homotopy type provides information about whether the planning problem can be solved. We speculate that preservation of plan existence under label maps might be productively studied across planning problems by examining the map's operation on loopback complexes: classes of maps that can be shown to preserve the homotopy type of such complexes (perhaps over restricted classes of planning problems) can be declared non-destructive.

Sensor maps on filters: The maps we have introduced here generalize that of our prior work [10] in three crucial ways: (1) These definitions consider modification of actuators and related resources involved in generating actions. (2) By mapping each element of some label set to another set and then taking the union on graph edges, one may express the notion of loss of information by having E grow under the action of h. This idea was expressed as a valuable feature for mutations in [10], but is not correctly achievable with the definitions therein. (3) The notion of non-destructiveness in that work is stronger than Definition 31, not only requiring that plan $(h(P), V_{\text{term}})$ solve $(h(G), V_{\text{goal}})$ but also that it solve it in the same way as (P, V_{term}) solves (G, V_{goal}).

Co-design: As we have already argued, the right formalism should aim to provide the representational basis for objects that can manipulated by algorithms in order to guide the design process. We mention here, perhaps related mostly by broader spirit, the recent work of Censi [2] wherein he poses and solves what he terms co-design problems where, given a network of monotone constraints, the selection of components is a process that can be automated. Part of the interest in studying labels maps is that they can model aspects of different components; forging connections with the co-design approach is interesting.

Hybrid automata: More immediately related, and also adopting an algorithmic stance on the design process, are methods based on hybrid automata (HA), several of which leverage powerful synthesis and verification techniques [1]. Despite some similarities, including extensive use of non-determinism, the relationship between p-graphs and HA is somewhat involved: guard expressions in a rich logical specification language have structure missing from the label sets we study; nor are actions labels intended to model continuous dynamics.

6 Conclusion

We believe that the most crucial intellectual contributions of the present work are in achieving a degree of abstraction of prior ideas in two ways: (1) We separate those entities which have been formalized in robotics because they have some interpretation that is useful (e.g., the idea of a plan, a filter), from their representation. The p-graph, in and of itself, lacks an obvious interpretation. Its definition does not include semantics belying a single anticipated use, rather context and any specific interpretation are only added for the special subclasses. (2) Even if the p-graph is a unifying representation, it is not a *canonical form*. This paper represents an important mental shift in lifting most of the notions of equivalence up to sets of executions (languages), rather than depending on operations on some specific graph. The present work continues to separate the notion of behavior from presentation.

References

1. Belta, C., Bicci, A., Egerstedt, M., Frazzoli, E., Klavins, E., Pappas, G.J.: Symbolic Control and Planning of Robotic Motion. IEEE Transactions on Robotics and Automation 14(1), 51–70 (Mar 2007)
2. Censi, A.: A Class of Co-Design Problems With Cyclic Constraints and Their Solution. IEEE Robotics and Automation Letters 2(1), 96–103 (Jan 2017)
3. Erdmann, M.: On the topology of discrete strategies. International Journal of Robotics Research 29(7), 855–896 (2010)
4. Erdmann, M.: On the topology of discrete planning with uncertainty. in advances in applied and computational topology. In: Zomorodian, A. (ed.) Proc. Symposia in Applied Mathematics. vol. 70. American Mathematical Society (2012)
5. Erdmann, M., Mason, M.T.: An Exploration of Sensorless Manipulation. IEEE Transactions on Robotics and Automation 4(4), 369–379 (Aug 1988)
6. Goldberg, K.Y.: Orienting Polygonal Parts without Sensors. Algorithmica 10(3), 201–225 (1993)
7. LaValle, S.M.: Sensing and Filtering: A Fresh Perspective Based on Preimages and Information Spaces. Foundations and Trends in Robotics 1(4), 253–372 (Apr 2012)
8. M. T. Mason and K. Y. Goldberg and R. H. Taylor: Planning Sequences of Squeeze-Grasps to Orient and Grasp Polygonal Objects. In: Seventh CISM-IFToMM Symposium on Theory and Practice of Robots and Manipulators (1988)
9. O'Kane, J.M., Shell, D.: Concise planning and filtering: Hardness and algorithms. IEEE Transactions on Automation Science and Engineering (2017), to appear.
10. Saberifar, F.Z., Ghasemlou, S., O'Kane, J.M., Shell, D.: Set-labelled filters and sensor transformations. In: Proc. Robotics: Science and Systems (2016)
11. Schoppers, M.J.: Universal Plans for Reactive Robots in Unpredictable Environments. In: Proc. International Joint Conference on AI. pp. 1039–1046 (1987)
12. Tovar, B., Cohen, F., Bobadilla, L., Czarnowski, J., LaValle, S.M.: Combinatorial filters: Sensor beams, obstacles, and possible paths. ACM Transactions on Sensor Networks 10(3) (2014)

Importance Sampling for Online Planning under Uncertainty

Yuanfu Luo, Haoyu Bai, David Hsu, and Wee Sun Lee

National University of Singapore, Singapore 117417, Singapore

Abstract. The partially observable Markov decision process (POMDP) provides a principled general framework for robot planning under uncertainty. Leveraging the idea of Monte Carlo sampling, recent POMDP planning algorithms have scaled up to various challenging robotic tasks, including, e.g., real-time online planning for autonomous vehicles. To further improve online planning performance, this paper presents IS-DESPOT, which introduces *importance sampling* to DESPOT, a state-of-the-art sampling-based POMDP algorithm for planning under uncertainty. Importance sampling improves the planning performance when there are critical, but rare events, which are difficult to sample. We prove that IS-DESPOT retains the theoretical guarantee of DESPOT. We present a general method for learning the importance sampling distribution and demonstrate empirically that importance sampling significantly improves the performance of online POMDP planning for suitable tasks.

1 Introduction

Uncertainty in robot control and sensing presents significant barriers to reliable robot operation. The partially observable Markov decision process (POMDP) provides a principled general framework for robot decision making and planning under uncertainty [20]. While POMDP planning is computationally intractable in the worst case, approximate POMDP planning algorithms have scaled up to a wide variety of challenging tasks in robotics and beyond, e.g., autonomous driving [1], grasping [8], manipulation [13], disaster rescue management [25], and intelligent tutoring systems [4]. Many of these recent advances leverage probabilistic sampling for computational efficiency. This work investigates efficient sampling distributions for planning under uncertainty under the POMDP framework.

A general idea for planning under uncertainty is to sample a finite set of "scenarios" that capture uncertainty approximately and compute an optimal or near-optimal plan under these sampled scenarios on the average. Theoretical analysis reveals that a small sampled set guarantees near-optimal online planning, provided there exists an optimal plan with a compact representation [22]. In practice, the sampling distribution may have significant effect on the planning performance. Consider a de-mining robot navigating in a mine field. Hitting a mine may have low probability, but severe consequence. Failing to sample these rare, but critical events often results in sub-optimal plans.

Importance sampling [10] provides a well-established tool to address this challenge. We have developed IS-DESPOT, which applies importance sampling to DESPOT [22], a state-of-the-art sampling-based online POMDP algorithm (Section 3). The idea is to sample probabilistic events according to their "importance" instead of their natural probability of occurrence and then reweight the samples when computing the plan. We

© Springer Nature Switzerland AG 2020
K. Goldberg et al. (Eds.): *Algorithmic Foundations of Robotics XII*, SPAR 13, pp. 240–255, 2020.
https://doi.org/10.1007/978-3-030-43089-4_16

prove that IS-DESPOT retains the theoretical guarantee of DESPOT. We also present a general method for learning the importance sampling distribution (Section 4). Finally, we present experimental results showing that importance sampling significantly improves the performance of online POMDP planning for suitable tasks (Section 5).

2   Background

2.1   POMDP Preliminaries

A POMDP models an agent acting in a partially observable stochastic environment. It is defined formally as a tuple (S, A, Z, T, O, R, b_0), where S, A and Z are the state space, the action space, and the observation space, respectively. The function $T(s, a, s') = p(s'|s, a)$ defines the probabilistic state transition from $s \in S$ to $s' \in S$, when the agent takes an action $a \in A$. It can model imperfect robot control and environment changes. The function $O(s, a, z) = p(z|s, a)$ defines a probabilistic observation model, which can capture robot sensor noise. The function $R(s, a)$ defines a real-valued reward for the agent when it takes action $a \in A$ in state $s \in S$.

Because of imperfect sensing, the agent's state is not known exactly. Instead, the agent maintains a *belief*, which is a probability distribution over S. The agent starts with an initial belief b_0. At time t, it infers a new belief, according to Bayes' rule, by incorporating information from the action a_t taken and the observation z_t received:

$$b_t(s') = \tau(b_{t-1}, a_t, z_t) = \eta O(s', a_t, z_t) \sum_{s \in S} T(s, a_t, s') b_{t-1}(s), \tag{1}$$

where η is a normalizing constant.

A POMDP *policy* maps a belief to an action. The goal of POMDP planning is to choose a policy π that maximizes its *value*, i.e., the expected total discounted reward, with initial belief b_0:

$$V_\pi(b_0) = \mathbb{E}\left(\sum_{t=0}^{\infty} \gamma^t R(s_t, a_{t+1}) \,\middle|\, b_0, \pi \right) \tag{2}$$

where s_t is the state at time t, $a_{t+1} = \pi(b_t)$ is the action taken at time t to reach s_{t+1}, and $\gamma \in (0, 1)$ is a discount factor. The expectation is taken over the sequence of uncertain state transitions and observations over time.

A key idea in POMDP planning is the *belief tree* (Fig. 1a). Each node of a belief tree corresponds to a belief b. At each node, the tree branches on every action $a \in A$ and every observation $z \in Z$. If a node b has a child node b', then $b' = \tau(b, a, z)$. To find an optimal plan, one way is to traverse the tree from the bottom up and and compute an optimal action recursively at each node using the Bellman's equation:

$$V^*(b) = \max_{a \in A} \left\{ \sum_{s \in S} b(s) R(s, a) + \gamma \sum_{z \in Z} p(z|b, a) V^*\big(\tau(b, a, z)\big) \right\}. \tag{3}$$

2.2 Importance Sampling

We want to calculate the expectation $\mu = \mathbb{E}(f(s)) = \int f(s)p(s)\,\mathrm{d}s$ for a random variable s distributed according to p, but the function $f(s)$ is not integrable. One idea is to estimate μ by Monte Carlo sampling: $\hat{\mu} = (1/n)\sum_{i=1}^{n} f(s_i)$, where $s_i \sim p$ for all samples $s_i, i = 1, 2, \ldots, n$. The estimator $\hat{\mu}$ is unbiased, with variance $\mathrm{Var}(\hat{\mu}) = \sigma^2/n$, where $\sigma^2 = \int (f(s) - \mu)^2 p(s)\,\mathrm{d}s$.

Importance sampling reduces the variance of the estimator by carefully choosing an *importance distribution* q for sampling instead of using p directly:

$$\hat{\mu}_{\text{UIS}} = \frac{1}{n}\sum_{i=1}^{n} \frac{f(s_i)p(s_i)}{q(s_i)} = \frac{1}{n}\sum_{i=1}^{n} f(s_i)w(s_i), \quad s_i \sim q, \tag{4}$$

where $w(s_i) = p(s_i)/q(s_i)$ is defined as the *importance weight* of the sample s_i and $q(s) \neq 0$ whenever $f(s)p(s) \neq 0$. The estimator $\hat{\mu}_{\text{UIS}}$ is also unbiased, with variance $\mathrm{Var}(\hat{\mu}) = \sigma^2/n$, where $\sigma^2 = \int (\frac{f(s)p(s)}{q(s)} - \mu)^2 q(s)\,\mathrm{d}s$. Clearly the choice of q affects the estimator's variance. The optimal importance distribution $q^*(s) = |f(s)|\,p(s)/\mathbb{E}_p(|f(s)|)$ gives the lowest variance [16].

When either p or q is unnormalized, an alternative estimator normalizes the importance weights:

$$\hat{\mu}_{\text{NIS}} = \frac{\sum_{i=1}^{n} f(s_i)w(s_i)}{\sum_{i=1}^{n} w(s_i)}, \quad s_i \sim q, \tag{5}$$

which requires $q(s) \neq 0$ whenever $p(s) \neq 0$. This estimator is biased, but is asymptotically unbiased as the number of samples increases [16]. The performance of $\hat{\mu}_{\text{NIS}}$ versus that of $\hat{\mu}_{\text{UIS}}$ is problem-dependent. While $\hat{\mu}_{\text{NIS}}$ is biased, it often has lower variance than $\hat{\mu}_{\text{UIS}}$ in practice, and the reduction in variance often outweighs the bias [12].

2.3 Related Work

There are two general approaches to planning under uncertainty: offline and online. Under the POMDP framework, the offline approach computes beforehand a policy contingent on all possible future events. Once computed, the policy can be executed online very efficiently. While offline POMDP algorithms have made dramatic progress in the last decade [15,17,21,23], they are inherently limited in scalability, because the number of future events grows exponentially with the planning horizon. In contrast, the online approach [18] interleaves planning and plan execution. It avoids computing a policy for all future events beforehand. At each time step, it searches for a single best action for the current belief only, executes the action, and updates the belief. The process then repeats at the new belief. The online approach is much more scalable than the offline approach, but its performance is limited by the amount of online planning time available at each time step. The online and offline approaches are complementary and can be combined in various ways to further improve planning performance [5,7]

Our work focuses on online planning. POMCP [19] and DESPOT [22] are among the fastest online POMDP algorithms available today, and DESPOT has found applications in a range of robotics tasks, including autonomous driving in a crowd [1], robot de-mining in Humanitarian Robotics and Automation Technology Challenge 2015 [27],

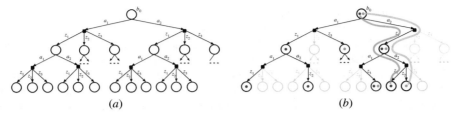

(a) $\qquad\qquad\qquad\qquad\qquad$ (b)

Fig. 1: Online POMDP planning performs lookahead search on a tree. (a) A standard belief tree of height $D = 2$. Each belief tree node represents a belief. At each node, the tree branches on every action and observation. (b) A DESPOT (black), obtained under 2 sampled scenarios marked with blue and orange dots, is overlaid on the standard belief tree. A DESPOT contains all actions branches, but only sampled observation branches.

and push manipulation [14]. One key idea underlying both POMCP and DESPOT is to use Monte Carlo simulation to sample future events and evaluate the quality of candidate policies. A similar idea has been used in offline POMDP planning [3]. It is, however, well known that standard Monte Carlo sampling may miss rare, but critical events, resulting in sub-optimal actions. Importance sampling is one way to alleviate this difficulty and improve planning performance.

We are not aware of prior use of importance sampling for robot planning under uncertainty. However, importance sampling is a well-established probabilistic sampling technique and has applications in many fields, e.g., Monte Carlo integration [10], ray tracing for computer graphic rendering [24], and option pricing in finance [6].

3   IS-DESPOT

3.1   Overview

Online POMDP planning interleaves planning and action execution. At each time step, the robot computes a near-optimal action a^* at the current belief b by searching a belief tree with the root node b and applies (3) at each tree node encountered during the search. The robot executes the action a^* and receives a new observation z. It updates the belief with a^* and z, through Bayesian filtering (1). The process then repeats. A belief tree of height D contains $\mathcal{O}(|A|^D|Z|^D)$ nodes. The exponential growth of the tree size is a major challenge for online planning when a POMDP has large action space, large observation space, or long planning horizon.

The *DEterminized Sparse Partially Observable Tree* (DESPOT) is a sparse approximation of the belief tree, under K sampled "scenarios" (Fig. 1b). The belief associated with each node of a DESPOT is approximated as a set of sampled states. A DESPOT contains all the action branches of a belief tree, but only the sampled observation branches. It has size $\mathcal{O}(|A|^D K)$, while a corresponding full belief tree has size $\mathcal{O}(|A|^D|Z|^D)$. Interestingly, K is often much smaller than $|Z|^D$ for a DESPOT to approximate a belief tree well, under suitable conditions. Now, to find a near-optimal action, the robot searches a DESPOT instead of a full belief tree. The reduced tree size leads to much faster online planning.

Clearly the sampled scenarios have major effect on the optimality of the chosen action. The original DESPOT algorithm samples scenarios with their natural probability of occurrence, according to the POMDP model. It may miss those scenarios that incur large reward or penalty, but happen with low probability. To address this issue, IS-DESPOT samples from an *importance distribution* and reweights the samples, using (4) or (5). We show in this section that IS-DESPOT retains the theoretical guarantee of DESPOT for all reasonable choices of the importance distribution. We also demonstrate empirically that IS-DESPOT significantly improves the planning performance for good choices of the importance distribution (see Section 5).

3.2 DESPOT with Importance Sampling

We define the DESPOT constructively by applying a *deterministic simulative model* to all possible action sequences under K sampled scenarios. Formally, a *scenario* $\phi_b = (s_0, \varphi_1, \varphi_2, \ldots)$ for a belief b consists of a state s_0 sampled from b and a sequence of random numbers $\varphi_1, \varphi_2, \ldots$ sampled independently and uniformly over the range $[0, 1]$. The deterministic simulative model is a function $\mathcal{G} \colon S \times A \times \mathrm{R} \mapsto S \times Z$, such that if a random number φ is distributed uniformly over $[0, 1]$, then $(s', z') = \mathcal{G}(s, a, \varphi)$ is distributed according to $p(s', z'|s, a) = T(s, a, s')O(s', a, z')$. Intuitively, $\mathcal{G}$ performs one-step simulation of the POMDP model. It is deterministic simulation of a probabilistic model, because the outcome is fixed by the input random number φ. To simulate a sequence of actions $(a_1, a_2, \ldots)$ under a scenario $\phi_b = (s_0, \varphi_1, \varphi_2, \ldots)$, we start at s_0 and apply the deterministic simulative model $\mathcal{G}$ at each time step. The resulting simulation sequence $\zeta = (s_0, a_1, s_1, z_1, a_2, s_2, z_2, \ldots)$ traverses a path $(a_1, z_1, a_2, z_2, \ldots)$ in the belief tree, starting at its root (Fig. 1b). The nodes and edges along the path are added to the DESPOT. Further, each belief node contains a set of sampled states, commonly called a *particle set*, which approximates the corresponding belief. If ζ passes through the node b at time step t, the state s_t is added to the particle set for b. Repeating this process for all possible action sequences under all K sampled scenarios completes the construction of the DESPOT. Clearly, the size of a DESPOT with height D is $\mathcal{O}(|A|^D K)$.

A DESPOT policy π can be represented as a policy tree derived from a DESPOT $\mathcal{T}$. The policy tree contains the same root as $\mathcal{T}$, but it contains at each internal node b only one action branch determined by $a = \pi(b)$. We define the size of such a policy, $|\pi|$, as the number of internal policy tree nodes. A singleton policy tree thus has size 0.

Given an initial belief b, the value of a policy π can be approximated by integrating over $\mathcal{Z}$, the space of all possible D-step simulation sequences under π:

$$V_\pi(b) \approx \int_{\zeta \in \mathcal{Z}} V_\zeta\, p(\zeta|b, \pi)\, \mathrm{d}\zeta,$$

where $p(\zeta|b, \pi) = b(s_0) \prod_{t=0}^{D-1} p(s_{t+1}, z_{t+1}|s_t, a_{t+1})$ is the probability of ζ and $V_\zeta = \sum_{t=0}^{D-1} \gamma^t R(s_t, a_{t+1})$ is the total discounted reward of ζ. To estimate $V_\pi(b)$ by unnormalized importance sampling (4), IS-DESPOT samples a subset $\mathcal{Z}' \subset \mathcal{Z}$ according to an importance distribution

$$q(\zeta|b, \pi) = q(s_0) \prod_{t=0}^{D-1} q(s_{t+1}, z_{t+1}|s_t, a_{t+1}), \tag{6}$$

where $q(s_0)$ is the distribution for sampling the initial state and $q(s_{t+1}, z_{t+1}|s_t, a_{t+1})$ is the distribution for sampling the state transitions and observations. Then,

$$\hat{V}_\pi(b) = \frac{1}{|\mathcal{Z}'|} \sum_{\zeta \in \mathcal{Z}'} w(\zeta) V_\zeta = \frac{1}{|\mathcal{Z}'|} \sum_{\zeta \in \mathcal{Z}'} \sum_{t=0}^{D-1} w(\zeta_{0:t}) \gamma^t R(s_t, a_{t+1}), \qquad (7)$$

where $w(\zeta) = p(\zeta|b, \pi)/q(\zeta|b, \pi)$ is the importance weight of ζ, $\zeta_{0:t}$ is a subsequence of ζ over the time steps $0, 1, \ldots, t$, and $w(\zeta_{0:t})$ is the importance weight of $\zeta_{0:t}$.

To avoid over-fitting to the sampled scenarios, IS-DESPOT optimizes a regularized objective function:

$$\max_{\pi \in \Pi_\mathcal{T}} \left\{ \hat{V}_\pi(b) - \lambda|\pi| \right\}, \qquad (8)$$

where $\Pi_\mathcal{T}$ is the set of all policy trees derived from a DESPOT $\mathcal{T}$ and $\lambda \geq 0$ is regularization constant. More details on the benefits of regularization are available in [26].

3.3 Online Planning

IS-DESPOT is an online POMDP planning algorithm. At each time step, IS-DESPOT searches a DESPOT $\mathcal{T}$ rooted at the current belief b_0. It obtains a policy π that optimizes (8) at b_0 and chooses the action $a = \pi(b_0)$ for execution.

To optimize (8), we substitute (7) into (8) and define the *regularized weighted discounted utility* (RWDU) of a policy π at each DESPOT node b:

$$\nu_\pi(b) = \frac{1}{|\mathcal{Z}'|} \sum_{\zeta \in \mathcal{Z}'_b} \sum_{t=\Delta(b)}^{D-1} w(\zeta_{0:t}) \gamma^t R(s_t, a_{t+1}) - \lambda|\pi_b|, \qquad (9)$$

where $\mathcal{Z}'_b \subset \mathcal{Z}'$ contains all simulation sequences traversing paths in $\mathcal{T}$ through the node b; $\Delta(b)$ is the depth of b in $\mathcal{T}$; π_b is the subtree of π with the root b. Given a policy π, there is one-to-one correspondence between scenarios and simulation sequences. So, $|\mathcal{Z}'| = K$. We optimize $\nu_\pi(b_0)$ over $\Pi_\mathcal{T}$ by performing a tree search on $\mathcal{T}$ and applying Bellman's equation recursively at every internal node of $\mathcal{T}$, similar to (3):

$$\nu^*(b) = \max\left\{ \frac{\gamma^{\Delta(b)}}{K} \sum_{\zeta \in \mathcal{Z}'_b} w(\zeta_{0:\Delta(b)}) V_{\pi_0, s_{\zeta, \Delta(b)}}, \max_{a \in A}\left\{ \rho(b,a) + \sum_{z \in Z_{b,a}} \nu^*(\tau(b,a,z)) \right\} \right\} \qquad (10)$$

where

$$\rho(b, a) = \frac{1}{K} \sum_{\zeta \in \mathcal{Z}'_b} \gamma^{\Delta(b)} w(\zeta_{0:\Delta(b)}) R(s_{\zeta, \Delta(b)}, a) - \lambda.$$

In (10), π_0 denotes a given default policy; $s_{\zeta, \Delta(b)}$ denotes the state in ζ at time step $\Delta(b)$; $V_{\pi_0, s}$ is the value of π_0 starting from a state s. At each node b, we may choose to follow a default policy π_0 or one of the action branches. The out maximization in (10) chooses between these two options, while the inner maximization chooses the specific action branch. The maximizer at b_0, the root of $\mathcal{T}$, gives the optimal action.

There are many tree search algorithms. One is to traverse $\mathcal{T}$ from the bottom up. At each leaf node b of $\mathcal{T}$, the algorithm sets $\nu^*(b)$ as the value of the default policy π_0

and then applies (10) at each internal node until reaching the root of $\mathcal{T}$. The bottom-up traversal is conceptually simple, but $\mathcal{T}$ must be constructed fully in advance. For very large POMDPs, the required number of scenarios, K, may be huge, and constructing the full DESPOT is not practical.

To scale up, an alternative is to perform anytime heuristic search. To guide the heuristic search, the algorithm maintains at each node b of $\mathcal{T}$ a lower bound and an upper bound on $\nu^*(b)$. It constructs and searches $\mathcal{T}$ incrementally, using K sampled scenarios. Initially, $\mathcal{T}$ contains only a single root node with belief b_0. The algorithm makes a series of explorations to expand $\mathcal{T}$ and reduces the gap between the upper and lower bounds at the root node b_0 of $\mathcal{T}$. Each exploration follows the heuristic and traverses a promising path from b_0 to add new nodes at the end of the path. The algorithm then traces the path back to b_0 and applies (10) to both the lower and upper bounds at each node along the way. The explorations continue, until the gap between the upper and lower bounds reaches a target level or the allocated online planning time runs out.

Online planning for IS-DESPOT is very similar to that for DESPOT. We refer the reader to [26] for details.

3.4 Analysis

We now show that IS-DESPOT retains the theoretical guarantee of DESPOT. The two theorems below generalize the earlier results [22] to the case of importance sampling. To simplify the presentation, this analysis assumes, without loss of generality, $R(s, a) \in [0, R_{\max}]$ for all states and actions. All proofs are available in the appendix.

Theorem 1 shows that with high probability, importance sampling produces an accurate estimate of the value of a policy.

Theorem 1. *Let b_0 be a given belief. Let $\Pi_\mathcal{T}$ be the set of all IS-DESPOT policies derived from a DESPOT $\mathcal{T}$ and $\Pi_{b_0, D, K} = \bigcup_\mathcal{T} \Pi_\mathcal{T}$ be the union over all DESPOTs with root node b_0, with height D, and constructed with all possible K importance-sampled scenarios. For any $\tau, \alpha \in (0, 1)$, every IS-DESPOT policy $\pi \in \Pi_{b_0, D, K}$ satisfies*

$$V_\pi(b_0) \geq \frac{1-\alpha}{1+\alpha}\hat{V}_\pi(b_0) - \frac{R_{\max}W_{\max}}{(1+\alpha)(1-\gamma)} \cdot \frac{\ln(4/\tau) + |\pi|\ln(KD|A||Z|)}{\alpha K} \quad (11)$$

with probability at least $1 - \tau$, where

$$W_{\max} = \left\{ \max_{\substack{s,s'\in S \\ a\in A, z\in Z}} \frac{p(s, z|s', a)}{q(s, z|s', a)} \right\}^D$$

is the maximum importance weight.

The estimation error bound in (11) holds for all policies in $\Pi_{b_0, D, K}$ simultaneously. It also holds for any constant $\alpha \in (0, 1)$, which is a parameter that can be tuned to tighten the bound. The additive error on the RHS of (11) depends on the size of policy π. It also grows logarithmically with $|A|$ and $|Z|$, indicating that IS-DESPOT scales up well for POMDP with very large action and observation spaces.

Theorem 2 shows that we can find a near-optimal policy $\hat{\pi}$ by maximizing the RHS of (11).

Theorem 2. *Let π^* be an optimal policy at a belief b_0. Let $\Pi_{\mathcal{T}}$ be the set of policies derived from a DESPOT $\mathcal{T}$ that has height D and is constructed with K importance-sampled scenarios for b_0. For any $\tau, \alpha \in (0,1)$, if*

$$\hat{\pi} = \arg\max_{\pi \in \Pi_{\mathcal{T}}} \left\{ \frac{1-\alpha}{1+\alpha} \hat{V}_\pi(b_0) - \frac{R_{\max} W_{\max}}{(1+\alpha)(1-\gamma)} \cdot \frac{|\pi| \ln(KD|A||Z|)}{\alpha K} \right\},$$

then with probability at least $1 - \tau$,

$$V_{\hat{\pi}}(b_0) \geq \frac{1-\alpha}{1+\alpha} V_{\pi^*}(b_0) - \frac{R_{\max} W_{\max}}{(1+\alpha)(1-\gamma)}$$
$$\times \left(\frac{\ln(8/\tau) + |\pi^*| \ln(KD|A||Z|)}{\alpha K} + (1-\alpha)\left(\sqrt{\frac{2\ln(2/\tau)}{K}} + \gamma^D \right) \right).$$

$$(12)$$

The estimation errors in both theorems depend on the choice of the importance distribution. By setting the importance sampling distribution to the natural probability of occurrence, we recover exactly the same results for the original DESPOT algorithm [22].

3.5 Normalized Importance Sampling

As we discussed in Section 2.2, normalized importance sampling often reduces the variance of an estimator in practice. The algorithm remains basically the same, other than normalizing the importance weights. The analysis is also similar, but is more involved. Our experiments show that IS-DESPOT with normalized importance weights produces better performance in some cases (see Section 5).

4 Importance Distributions

We now derive the optimal importance distribution for IS-DESPOT and then present a general method for learning it. We focus on the importance distribution for sampling state transitions and leave that for sampling observations as future work.

4.1 Optimal Importance Distribution

The value of a policy π at a belief b, $V_\pi(b)$, is the expected total discounted reward of executing π, starting at a state s distributed according to b. It can be obtained by integrating over all possible starting states:

$$V_\pi(b) = \int_{s \in S} \mathbb{E}(v|s, \pi) b(s) \, \mathrm{d}s,$$

where v is a random variable representing the total discounted reward of executing π starting from s and $\mathbb{E}(v|s, \pi)$ is its expectation. Compared with standard importance sampling (Section 2.2), IS-DESPOT estimates $f(s) = \mathbb{E}(v|s, \pi)$ by Monte Carlo simulation of π rather than evaluates $f(s)$ deterministically. Thus the importance distribution must account for not only the mean $\mathbb{E}(v|s, \pi)$ but also the variance $\mathrm{Var}(v|s, \pi)$ resulting from Monte Carlo Sampling, and give a state s increased importance when either is large. The theorem below formalizes this idea.

Theorem 3. *Given a policy π, let v be a random variable representing the total discounted reward of executing π, starting from state s, and let $V_\pi(b)$ be the expected total discounted reward of π at a belief b. To estimate $V_\pi(b)$ using unnormalized importance sampling, the optimal importance distribution is $q_\pi^*(s) = b(s)/w_\pi(s)$, where*

$$w_\pi(s) = \frac{\mathbb{E}_b\left(\sqrt{\left[\mathbb{E}(v|s, \pi)\right]^2 + \mathrm{Var}(v|s, \pi)}\right)}{\sqrt{\left[\mathbb{E}(v|s, \pi)\right]^2 + \mathrm{Var}(v|s, \pi)}}. \tag{13}$$

Theorem 3 specifies the optimal importance distribution for a given policy π, but IS-DESPOT must evaluate many policies within a set $\Pi_\mathcal{T}$, for some DESPOT $\mathcal{T}$. Our next result suggests that we can use the optimal importance distribution for a policy π to estimate the value of another "similar" policy π'. This allows us to use a single importance distribution for IS-DESPOT.

Theorem 4. *Let $\hat{V}_{\pi,q}(b)$ be the estimated value of a policy π at a belief b, obtained by K independent samples with importance distribution q. Let v be the total discounted reward of executing a policy π, starting from a state $s \in S$. If two policies π and π' satisfy $\frac{\mathrm{Var}(v|s,\pi')}{\mathrm{Var}(v|s,\pi)} \leq 1 + \epsilon$ and $\frac{\left[\mathbb{E}(v|s,\pi')\right]^2}{\left[\mathbb{E}(v|s,\pi)\right]^2} \leq 1 + \epsilon$ for all $s \in S$ and some $\epsilon > 0$, then*

$$\mathrm{Var}\left(\hat{V}_{\pi',q_\pi^*}(b)\right) \leq (1 + \epsilon)\left(\mathrm{Var}(\hat{V}_{\pi,q_\pi^*}(b)) + \frac{1}{K}V^*(b)^2\right), \tag{14}$$

where q_π^* is an optimal importance distribution for estimating the value of π and $V^*(b)$ is the value of an optimal policy at b.

4.2 Learning Importance Distributions

According to Theorems 3 and 4, we can construct the importance distribution for IS-DESPOT, if we know the mean $\mathbb{E}(v|s, \pi)$ and the variance $\mathrm{Var}(v|s, \pi)$ for all $s \in S$ under a suitable policy $\pi \in \Pi_\mathcal{T}$. Two issues remain. First, we need to identify π. Second, we need to represent $\mathbb{E}(v|s, \pi)$ and $\mathrm{Var}(v|s, \pi)$ compactly, as the state space S may be very large or even continuous. For the first issue, we use the policy generated by the DESPOT algorithm, without importance sampling. Theorem 4 helps to justify this choice. For the second issue, we use offline learning, based on discrete feature mapping. Specifically, we learn the function

$$\xi(s) = \sqrt{\left[\mathbb{E}(v|s, \pi)\right]^2 + \mathrm{Var}(v|s, \pi)}, \tag{15}$$

which is proportional to the inverse importance weight $1/w_\pi(s)$. We then set the importance distribution to $q(s) = \eta p(s)\,\xi(s)$, where η is the normalization constant.

To learn $\xi(s)$, we first generate data by running DESPOT many times offline in simulation without importance sampling. Each run starts at a state s_0 sampled from the initial belief b_0 and generates a sequence $\{s_0, a_1, r_1, s_1, a_2, r_2, s_2, \ldots, a_D, r_D, s_D\}$, where a_t is the action that DESPOT chooses, s_t is a state sampled from the model $p(s_t|s_{t-1}, a_t) = T(s_{t-1}, a_t, s_t)$, and $r_t = R(s_{t-1}, a_t)$ is the reward at time t for $t = 1, 2, \ldots$. Let $v_t = \sum_{i=t}^{D} \gamma^{(i-t)} r_i$ be the total discounted reward from time t. We collect all pairs (s_{t-1}, v_t) for $t = 1, 2, \ldots$ from each run. Next, we manually construct a set of features over the state space S so that each state $s \in S$ maps to its feature value $\mathcal{F}(s)$. Finally, We discretize the feature values into a finite set of bins of equal size and insert each collected data pair (s, v) into the bin that contains $\mathcal{F}(s)$. We calculate the mean $\mathbb{E}(v)$ and the variance $\text{Var}(v)$ for each bin. For any state $s \in S$, $\xi(s)$ can then be approximated from the mean and the variance of the bin that contains $\mathcal{F}(s)$.

This learning method scales up to high-dimensional and continuous state spaces, provided a small feature set can be constructed. It introduces approximation error, as a result of discretization. However, the importance distribution is a heuristic that guides Monte Carlo sampling, and IS-DESPOT is overall robust against approximation error in the importance distribution. The experimental results in the next section show that the importance distribution captured by a small feature set effectively improves online planning for large POMDPs.

5 Experiments

We evaluated IS-DESPOT on a suite of five tasks (Table 1). The first three tasks are known to contain *critical* states, states which are not encountered frequently, but may lead to significant consequences. The other two, RockSample [21] and Pocman [19], are established benchmark tests for evaluating the scalability of POMDP algorithms.

We compared two versions of IS-DESPOT, unnormalized and normalized, with both DESPOT and POMCP. See Table 1 for the results. For all algorithms, we set the maximum online planning time to one second per step. For DESPOT and IS-DESPOT, we set the number of sampled scenarios to 500 and used an uninformative upper bound and fixed-action lower bound policy without domain-specific knowledge [26] in Asymmetric Tiger, CollisionAvoidance, and Demining. Following earlier work [26], we set the number of sampled scenarios to 500 and 100, respectively, in RockSample and Pocman, and used domain-specific heuristics. For fair comparison, we tuned the exploration constant of POMCP to the best possible and used a rollout policy with the same domain knowledge as that for DESPOT and IS-DESPOT in each task.

Overall, Table 1 shows that both unnormalized and normalized IS-DESPOT substantially outperform DESPOT and POMCP in most tasks, including, in particular, the two large-scale tasks, Demining and Pocman. Normalized IS-DESPOT performs better than unnormalized IS-DESPOT in some tasks, though not all.

AsymmetricTiger Tiger is a classic POMDP [9]. A tiger hides behind one of two doors, denoted by states s_L and s_R, with equal probabilities. An agent must decide which door to open, receiving a reward of $+10$ for opening the door without the tiger and receiving -100 otherwise. At each step, the agent may choose to open a door or to listen. Listening incurs a cost of -1 and provides the correct information with probability 0.85. The task resets once the door is opened. The only uncertainty here is the tiger's location. Tiger is a toy problem, but it captures the key trade-off between information

Table 1: Performance comparison. The table shows the average total discounted rewards (95% confidence interval) of four algorithms on five tasks. UIS-DESPOT and NIS-DESPOT refer to unnormalized and normalized IS-DESPOT, respectively.

	AsymmetricTiger	CollisionAvoidance	Demining	RockSample(15,15)	Pocman
$|S|$	2	9,720	$\sim 10^{49}$	7,372,800	$\sim 10^{56}$
$|A|$	3	3	9	20	4
$|Z|$	2	18	16	3	1,024
POMCP	-5.60 ± 1.51	-4.19 ± 0.07	-17.11 ± 2.74	15.32 ± 0.28	294.16 ± 4.06
DESPOT	-2.20 ± 1.78	-1.05 ± 0.27	-11.09 ± 2.45	18.37 ± 0.28	317.90 ± 4.17
UIS-DESPOT	3.70 ± 0.49	-0.87 ± 0.19	-3.45 ± 1.76	18.30 ± 0.32	315.13 ± 4.92
NIS-DESPOT	3.75 ± 0.47	-0.44 ± 0.12	-3.69 ± 1.80	18.68 ± 0.28	326.92 ± 3.89

state	$\mathbb{E}(v|s)$	$\mathrm{Var}(v|s)$	$p(s)$	$q(s)$
s_L	4.36	7.34	0.99	0.40
s_R	-65.84	566978.9	0.01	0.60

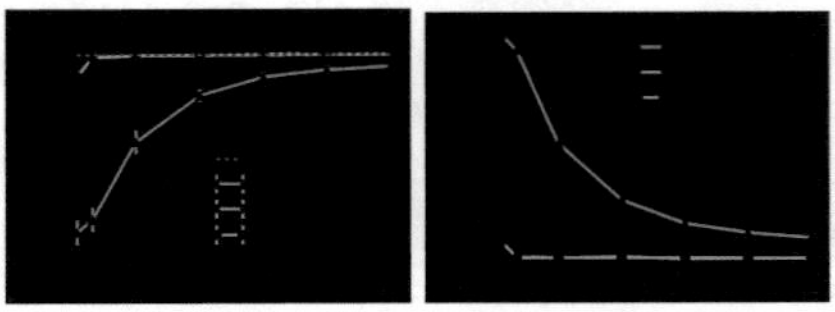

(a)   (b)   (c)

Fig. 2: AsymmetricTiger. (a) Importance distribution q. DESPOT samples according to p. IS-DESPOT samples according to q. (b) Average total discounted reward versus K, the number of sampled scenarios. (c) The rate of opening the door with the tiger behind versus K.

gathering and information exploitation prevalent in robot planning under uncertainty. We use it to gain understanding on some important properties of IS-DESPOT.

We first modify the original Tiger POMDP. Instead of hiding behind the two doors with equal probabilities, the tiger hides behind the right door with much smaller probability 0.01. However, if the agent opens the right door with the tiger hiding there, the penalty increases to $-10,000$. Thus the state s_R occurs rarely, but has a significant consequence. Sampling s_R is crucial.

Since this simple task only has two states, we learn the weight for each state and use them to construct the importance distribution for IS-DESPOT (Fig. 2a).

Table 1 shows clearly that both versions of IS-DESPOT significantly outperform DESPOT and POMCP. This is not surprising. DESPOT and POMCP sample the two states s_L and s_R according to their natural probabilities of occurrence and fail to account for their *significance*, in this case, the high penalty of s_R. As a result, they rarely sample s_R. In contrast, the importance distribution enables IS-DSPOT to sample s_R much more frequently (Fig. 2a). Unnormalized and normalized IS-DESPOT are similar in performance, as normalization has little effect on this small toy problem.

We conducted additional experiments to understand how K, the number of sampled scenarios, affects performance (Fig. 2b, c). To calibrate the performance, we ran an offline POMDP algorithm SARSOP [15] to compute an optimal policy. When K is small, the performance gap between DESPOT and IS-DESPOT is significant. As K increases, the gap narrows. When K is 32, both versions of IS-DESPOT are near-optimal, while

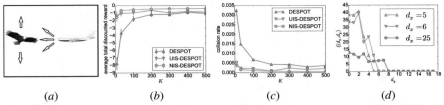

(a)   (b)   (c)   (d)

Fig. 3: CollisionAvoidance. (a) An aircraft changes the direction to avoid collision. (b) Average total discounted reward versus K, the number of scenarios. (c) Collision rate versus K. (d) $\xi(d_x, d_y)$ learned from data.

DESPOT does not reach a comparable performance level even at $K = 500$. All these confirm the value of importance sampling for small sample size.

One may ask how DESPOT and IS-DESPOT compare on the original Tiger task. The original Tiger has two symmetric states. The optimal importance distribution is flat. Thus DESPOT and IS-DESPOT have the same sampling distribution, and IS-DESPOT has no performance advantage over DESPOT. Importance sampling is beneficial only if the planning task involves significant events of low probability.

CollisionAvoidance This is an adaptation of the aircraft collision avoidance task [2]. The agent starts moving from a random position in the right-most column of a 18×30 grid map (Fig. 3a). An obstacle randomly moves in the left-most column of this map: moves up with probability 0.25, down with 0.25, and stay put with 0.50. The probabilities become 0, 0.25, and 0.75 respectively when the obstacle is at the top-most row, and become 0.25, 0, and 0.75 respectively when it is at the bottom-most row. Each time, the agent can choose to move upper-left, lower-left or left, with a cost of $-1, -1$, and 0 respectively. If the agent collides with the obstacle, it receives a penalty of $-1, 000$. The task finishes when the agent reaches the left-most column. The agent knows its own position exactly, but observes the obstacle's position with a Gaussian noise $\mathcal{N}(0, 1)$. The probability of the agent colliding with the obstacle is small, but the penalty of collision is high.

To construct the importance distribution for IS-DESPOT, we map a state s to a feature vector (d_x, d_y) and learn $\xi(d_x, d_y)$. The features d_x and d_y are the horizontal and vertical distances between the obstacle and the agent respectively. The horizontal distance d_x is the number of steps remaining for the agent to move, and $(d_x, d_y) = (0, 0)$ represents a collision. These features capture the changes in the mean and variance of policy values for different states, because when the agent is closer to the obstacle, the probability of collision is higher, which leads to worse mean and larger variance.

Fig. 3d plots $\xi(d_x, d_y)$ for a few chosen horizontal distance d_x: $\xi(d_x = 5, d_y)$, $\xi(d_x = 6, d_y)$ and $\xi(d_x = 25, d_y)$. Overall, $\xi(d_x, d_y)$ is higher when the obstacle is closer to the agent. At $d_x = 5$, $\xi(d_x, d_y)$ is close to zero for all $d_y >= 4$, and $\xi(d_x, d_y) = 0$ when $d_y >= 10$. This indicates that the DESPOT policy rarely causes collisions when the vertical distance is larger than 4, because the collision would require the agent and obstacle move towards each other for several steps in the remaining $d_x = 5$ steps, which is unlikely to happen since the DESPOT policy is actively avoiding the collision. Furthermore, collision is not possible at all when $d_y >= 10$ because it

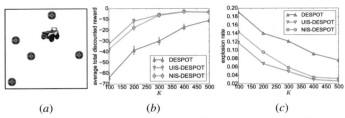

Fig. 4: Demining. (*a*) A robot moves in a 10×10 map to detect and report landmines. (*b*) Average total discounted reward versus K, the number of scenarios. (*c*) The rate of hitting mines versus K.

requires at least 6 steps for the agent and the obstacle to meet. With this importance distribution, IS-DESPOT can ignore the states that do not cause collision and focus on sampling states that likely lead to collisions. This could significantly improve the convergence rate, as shown in Fig. 3*b* and Fig. 3*c* : IS-DESPOT performs better than DESPOT with small K. NIS-DESPOT improves the performance further. It also performs well with small K, while DESPOT cannot reach the same level of performance even with a large K.

Demining This is an adapted version of the robotic de-mining task in Humanitarian Robotics and Automation Technology Challenge (HRATC) 2015 [27]. It requires an agent to sweep a 10×10 grid field to detect and report landmines. Each grid cell has a mine with probability 0.05. At each time step, the agent can move in four directions, and then observe the four adjacent cells with 0.9 accuracy. DESPOT is a core component of the system winning HRATC 2015. We show that IS-DESPOT handles critical states in the task better than DESPOT, resulting in better performance overall. If the agent reports the mine correctly, it receives reward $+10$, and -10 otherwise. Stepping over the mine causes high penalty $-1,000$ and task termination, which is a critical state.

To construct the importance distribution for IS-DESPOT, we map a state s to a feature vector (n, d) and learn $\xi(n, d)$, where n is the number of unreported mines and d is the Manhattan distance between the agent and the nearest mine. These two features capture the changes in the mean and variance of the rollout values for different states, because n determines the maximum total discounted rewards that the agent can get, and d affects the probability of stepping over mines.

Fig. 4*b* shows the performance comparison between IS-DESPOT and DESPOT for different number of scenarios K. IS-DESPOT converges faster than DESPOT, and significantly outperforms DESPOT even for relatively large number of scenarios ($K = 500$). Fig. 4*c* shows that the performance improvement results from a decrease in the number of explosions. For this task, the size of the state space is about 10^{49}, and IS-DESPOT clearly outperforms DESPOT and POMCP (Table 1), affirming the scalability of IS-DESPOT.

RockSample RockSample is a standard POMDP benchmark problem [21]. In the problem $RockSample(n, k)$, the agent moves on an $n \times n$ grid map which has k rocks. Each of the rocks can either be good or bad. The agent knows each rock's position but does not know its state (good or bad). At each time step, the agent can choose to observe the states, move, or sample. The agent can sample a rock it steps on, and get a reward of

Fig. 5: RockSample. A robot senses rocks to identify "good" ones and samples them. Upon completion, it exits the east boundary.

$+10$ if the rock is good, -10 otherwise. A rock turns bad after being sampled. Moving or sensing has no cost. The agent can sense the state of the rock with accuracy decreasing exponentially with respect to the distance to the rock. When the agent exits from the east boundary of the map, it gets a reward of $+10$ and the task terminates.

To construct the importance distribution for IS-DESPOT, we map a state s to a feature vector (n, d_+, d_-), where n is the number of remaining good rocks, d_+ and d_- are the Manhattan distances to the nearest good rock and the nearest bad rock respectively. This task does not contain critical states, nevertheless, IS-DESPOT maintains the same level of performance as DESPOT (Table 1).

Pocman Pocman [19] is a partially observable variant of the popular Pacman game (Fig. 6a). An agent and four ghosts move in a 17×19 maze populated with food pallets. The agent can move from a cell in the maze to an adjacent one if there is no wall in between. Each move incurs a cost of -1. A cell contains the food pallet with probability 0.5. Eating a food pallet gives the agent a reward of $+10$. Getting caught by ghosts incurs a penalty of -100. There are four power pills. After eating a power pill, the agents retains it for the next 15 steps and acquires the power to kill ghosts. Killing a ghost gives a reward of $+25$. Let d be the Manhattan distance between the agent and a ghost. When $d \leq 5$, the ghost chases the agent with probability 0.75 if the agent does not possess the power pill; the ghost runs away, but slips with probability 0.25 if the agent possesses the power pill. When $d > 5$, the ghost moves uniformly at random to feasible adjacent cells. In contrast with that in the original Pacman game, the Pocman agent does not know the exact locations of ghosts, but sees approaching ghosts when they are in a direct line of sight and hears them when $d \leq 2$.

To construct the importance distribution for IS-DESPOT, we map a state s to a feature vector $(n, d_{\min})$, where $d_{\min}$ is the Manhattan distance to the nearest ghost and n is the number of remaining steps for which the agent retains a power pill.

Table 1 shows that unnormalized IS-DESPOT does not bring in improvement over DESPOT. However, normalized IS-DESPOT does, because normalization of importance weight reduces variance in this case.

To understand the benefits of learning the importance distribution, we tried to construct an importance distribution q_M manually. The states, in which a ghost catches the agent, are critical. They do not occur very often, but incur high penalty. An effective importance distribution must sample such states with increased probability. One crucial aspect of the ghost behaviour is the decision of chasing the agent, governed by

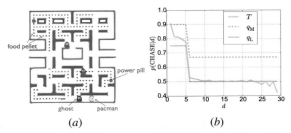

(a)   (b)

Fig. 6: Pocman. (a) The Pacman/Pocman game. (b) Compare the natural state-transition distribution T, the manually constructed importance distribution q_M, and the learned importance distribution q_L.

the distribution $p(\text{CHASE} \mid d)$. To obtain q_M, we shift $p(\text{CHASE} \mid d)$ upward by a constant amount, reasoning that having the ghost chase the agent more often increases the sampling of critical states in which a ghost catches an agent. Unfortunately q_M, when used in normalized IS-DESPOT, achieved an average total discounted reward of 317.02 ± 4.21, weaker than that of the learned importance distribution q_L. To understand why, let us compare q_M with q_L (Fig. 6b). Although q_L does have the ghost chase the agent more often as q_M does, it does not increase $p(\text{CHASE}|d)$ uniformly. The increase depends on the distance d. It is not realistic to handcraft such an importance distribution without the help of learning from data.

6   Conclusion

This paper introduces importance sampling to sampling-based online POMDP planning. Specifically, IS-DESPOT retains the theoretical guarantee of the original DESPOT algorithm, and it outperforms two state-of-the-art online POMDP algorithms on a test suite of five distinct tasks.

There are multiple directions to extend this work. First, our current method for learning the importance distribution focuses on critical states, but observations are equally important for planning under uncertainty. Extending IS-DESPOT to handle critical observations is straightforward. Second, we want to fully automate the importance distribution construction through feature learning. Finally, the idea of importance sampling is general and can be applied to other MDP and POMDP planning algorithms (e.g., [3,11,19]).

Acknowledgments. This work is supported in part by NUS AcRF Tier 1 grant R-252-000-587-112, A*STAR grant R-252-506-001-305, and US Air Force Research Laboratory under agreement number FA2386-15-1-4010.

References

1. Bai, H., Cai, S., Ye, N., Hsu, D., Lee, W.S.: Intention-aware online POMDP planning for autonomous driving in a crowd. In: Proc. IEEE Int. Conf. on Robotics & Automation (2015)
2. Bai, H., Hsu, D., Kochenderfer, M.J., Lee, W.S.: Unmanned aircraft collision avoidance using continuous-state POMDPs. Robotics: Science and Systems VII 1 (2012)

3. Bai, H., Hsu, D., Lee, W.S., Ngo, V.A.: Monte Carlo value iteration for continuous-state POMDPs. In: Algorithmic Foundations of Robotics IX (2010)
4. Folsom-Kovarik, J., Sukthankar, G., Schatz, S.: Tractable POMDP representations for intelligent tutoring systems. ACM Trans. on Intelligent Systems & Technology 4(2) (2013)
5. Gelly, S., Silver, D.: Combining online and offline knowledge in UCT. In: Proc. Int. Conf. on Machine Learning (2007)
6. Glasserman, P.: Monte Carlo methods in financial engineering, vol. 53. Springer Science & Business Media (2003)
7. He, R., Brunskill, E., Roy, N.: Efficient planning under uncertainty with macro-actions. J. Artificial Intelligence Research 40(1) (2011)
8. Hsiao, K., Kaelbling, L.P., Lozano-Perez, T.: Grasping POMDPs. In: Proc. IEEE Int. Conf. on Robotics & Automation (2007)
9. Kaelbling, L.P., Littman, M.L., Cassandra, A.R.: Planning and acting in partially observable stochastic domains. Artificial Intelligence 101(1) (1998)
10. Kalos, M., Whitlock, P.: Monte Carlo Methods, vol. 1. John Wiley & Sons, New York (1986)
11. Kearns, M., Mansour, Y., Ng, A.Y.: A sparse sampling algorithm for near-optimal planning in large Markov decision processes. Machine Learning 49(2-3) (2002)
12. Koller, D., Friedman, N.: Probabilistic graphical models: principles and techniques. MIT press (2009)
13. Koval, M., Pollard, N., Srinivasa, S.: Pre- and post-contact policy decomposition for planar contact manipulation under uncertainty. Int. J. Robotics Research 35(1–3) (2016)
14. Koval, M., Hsu, D., Pollard, N., Srinivasa, S.: Configuration lattices for planar contact manipulation under uncertainty. In: Algorithmic Foundations of Robotics XII—Proc. Int. Workshop on the Algorithmic Foundations of Robotics (WAFR) (2016)
15. Kurniawati, H., Hsu, D., Lee, W.S.: SARSOP: efficient point-based POMDP planning by approximating optimally reachable belief spaces. In: Proc. Robotics: Science & Systems (2008)
16. Owen, A.B.: Monte Carlo theory, methods and examples (2013)
17. Pineau, J., Gordon, G., Thrun, S.: Point-based value iteration: An anytime algorithm for POMDPs. In: Proc. Int. Jnt. Conf. on Artificial Intelligence (2003)
18. Ross, S., Pineau, J., Paquet, S., Chaib-Draa, B.: Online planning algorithms for POMDPs. J. Artificial Intelligence Research 32 (2008)
19. Silver, D., Veness, J.: Monte-Carlo planning in large POMDPs. In: Advances in Neural Information Processing Systems (NIPS) (2010)
20. Smallwood, R., Sondik, E.: The optimal control of partially observable Markov processes over a finite horizon. Operations Research 21 (1973)
21. Smith, T., Simmons, R.: Heuristic search value iteration for POMDPs. In: Proc. Uncertainty in Artificial Intelligence (2004)
22. Somani, A., Ye, N., Hsu, D., Lee, W.S.: DESPOT: Online POMDP planning with regularization. In: Advances in Neural Information Processing Systems (NIPS) (2013)
23. Spaan, M., Vlassis, N.: Perseus: Randomized point-based value iteration for POMDPs. J. Artificial Intelligence Research 24 (2005)
24. Veach, E.: Robust Monte Carlo methods for light transport simulation. Ph.D. thesis, Stanford University (1997)
25. Wu, K., Lee, W.S., Hsu, D.: POMDP to the rescue: Boosting performance for RoboCup rescue. In: Proc. IEEE/RSJ Int. Conf. on Intelligent Robots & Systems (2015)
26. Ye, N., Somani, A., Hsu, D., Lee, W.S.: DESPOT: Online POMDP planning with regularization. J. Artificial Intelligence Research (to appear)
27. Humanitarian robotics and automation technology challenge (HRATC) 2015, http://www.isr.uc.pt/HRATC2015

Reactive Motion Planning in Uncertain Environments via Mutual Information Policies

Ryan A. MacDonald and Stephen L. Smith

Department of Electrical and Computer Engineering
University of Waterloo, Waterloo ON, Canada
{ryan.macdonald,stephen.smith}@uwaterloo.ca

Abstract. This paper addresses path planning with real-time reaction to environmental uncertainty. The environment is represented as a graph and is uncertain in that the edges of the graph are unknown to the robot a priori. Instead, the robots prior information consists of a distribution over candidate edge sets. At each vertex, the robot can take a measurement to determine the presence or absence of an edge. Within this model, the Reactive Planning Problem (RPP) provides the robot with a start location and a goal location and asks it to compute a policy that minimizes the expected travel and observation cost. In contrast to computing paths that maximize the probability of success, we focus on complete policies (i.e., policies that produce a path, or determine no such path exists). We prove that the RPP is NP-Hard and provide a suboptimal, but computationally efficient, solution. This solution, based on mutual information, returns a complete policy and a bound on the gap between the policy's expected cost and the optimal. Finally, simulations are run on a flexible factory scenario to demonstrate the scalability of the proposed approach.

Keywords: Motion and Path Planning, Planning under Uncertainty, Mutual Information

1   Introduction

Robot motion planning under uncertainty is typically concerned with uncertainty in the robot's state within an environment and/or uncertainty in the outcome of a selected action on the robot's state [1–5]. In this work, we consider motion planning with uncertainty in the set of *motion* actions that a robot has access to at a given state. This problem arises in scenarios where the robot is given a set of possible locations for obstacles in an environment. The obstacles restrict the set of motions available to the robot at each point in the environment. By taking sensor measurements, the robot can narrow down the set of feasible obstacle locations and thus the motion actions it has available. Our goal is to compute motion and sensing policies prior to robot deployment that enable the robot to efficiently navigate in such environments. In this paper, we focus on the task of moving from a start location to a goal location while minimizing the expected action cost. The challenge in this problem is that future costs (for obtaining information and moving between locations) are dependent on the information the

© Springer Nature Switzerland AG 2020
K. Goldberg et al. (Eds.): *Algorithmic Foundations of Robotics XII*, SPAR 13, pp. 256–271, 2020.
https://doi.org/10.1007/978-3-030-43089-4_17

robot has obtained thus far. We present conditions where exploration is no longer helpful. When these conditions are met, the robot should exploit the known motion action set to reach the goal. We also develop a policy that provides constant time lookup for the next action given the outcomes of prior observations. This allows for implementation on robots where on-board computational resources are limited at deployment, or in which high-speed motion is required.

Related Work: In robotics, there are several effective methods for dealing with uncertainty. Point-to-point motion is addressed in [6] using persistent paths, which maximize the probability of success. However, if the computed path is obstructed, the robot ends in failure. To avoid failure, Partially Observable Markov Decision Processes (POMDPs) can be used to compute reactive motion policies. A POMDP selects actions based on partially observed states, but the computation of policies is in general a PSPACE-Complete problem [7]. In our work, we are interested in cases where the environment has a very large state space; for these cases, the POMDP's scalability becomes a barrier to use [8]. To avoid the computational complexity, algorithms like A* and D* lite allow for replanning during execution [9, 10]. These re-computations may present a bottleneck in real-time performance. Informative path planning is studied in several works [1–4, 11], all of which provide methods for real-time reaction to information within the environment. Research in this area focuses on tasks ranging from underwater inspection [3] to maximizing information from start to goal [1]. These works plan prior to deployment of the robot and react to new information collected by the robot, but their possible actions are known prior. In contrast, we consider cases where information may not be attainable until the robot has explored parts of the environment, which is not captured in this prior work.

In operations research, a closely related problem is planning with recourse and the Canadian Travelers Problem (CTP). Planning with recourse in [12] provides possible obstacle locations, but assumes obstacles will never make movement from the start to goal impossible. The CTP is PSPACE-Complete [13] and does not include prior information on obstacles. To make the problem more tractable, [14] provides possible obstacle locations in their problem R-SSPPR and seeks to minimize the expected cost between start and goal. In contrast to this work, R-SSPPR uses move actions, which can always be taken (i.e., a path to goal always exists), that have costs sensed for no cost. In transportation research, [15] presents an integer linear program to solve a route blocked problem, but they select a primary path and only switch to a secondary path when the primary fails. Our work wishes to minimize the expected cost to get to the goal or realize the goal is unreachable.

Our work leverages the concept of mutual information within discrete environments. Mutual Information is widely used to develop efficient sub-optimal solutions for gaining information in planning [5,11,16,17]. In particular, a mutual information gradient controller is presented in [5], where multiple robots search for targets and avoid hazards. The authors present a discretized probabilistic model where targets and hazards may exist and focus on positions of failure. In contrast, our work considers unknown environments where locations may not be reachable until the robot acquires a certain level of environmental awareness.

Contributions: The contributions of this paper are fourfold. First, we introduce the Reactive Planning Problem (RPP) and prove it is NP-Hard. Second, we provide properties that allow for a compact representation of a RPP policy. Third, we present an efficient algorithm for a sub-optimal solution to RPP that utilizes mutual information to guide exploration and uses an estimation of the cost-to-go for exploitation. Fourth, we provide a method to bound the gap between the expected cost of our policy and that of the optimal.

Organization: Section 2 introduces background terminology from graph theory and the Informative Path Planning problem. The Reactive Planning Problem is defined in Section 3 along with the environmental and robotic models. Section 4 provides problem properties as well as proof of computational complexity. Several properties from Section 4 are then expanded as a base for scalable policy generation in Section 5. Section 6 provides simulation results.

2 Background

2.1 Graph Terminology

A directed graph G is defined by the pair $G = (\mathcal{V}, \mathcal{E})$ and a cost function $c : \mathcal{E} \to \mathbb{R}$. The set $\mathcal{V}$ with $n = |\mathcal{V}|$ is the set of vertices that are connected by the set of edges $\mathcal{E}$, and $c(e)$ gives the cost of traversing an edge $e \in \mathcal{E}$. A path P in a graph is defined by a sequence of vertices $v_1, \ldots, v_k$ that satisfies $(v_i, v_{i+1}) \in \mathcal{E}$ for all $i \in \{1, \ldots, k-1\}$ with cost of traversal defined by $c(P) = \sum_{i=1}^{k-1} c(v_i, v_{i+1})$. With some abuse of notation for $v, w \in \mathcal{V}$, $c(v, w)$ refers to the minimum cost of a path from v to w. Given a graph $G = (\mathcal{V}, \mathcal{E})$, the subgraph $G_E = (V, E)$ is induced by $E \subseteq \mathcal{E}$ with $V \subseteq \mathcal{V}$ given by the endpoints of E.

An edge $e = (v, u) \in E$ is said to be *incident* with vertices v and u. As the graph is directed, e is outgoing at v and incoming at u. Therefore, e is *incident-in* to u and is *incident-out* to v with the set of edges *incident-out* to v, $I_v \subseteq \mathcal{E}$.

2.2 Informative Path Planning Review

In [11], the Informative Path Planning (IPP) problem under noiseless observation is defined as a tuple (X, d, H, ρ, O, Z, r). A robot starts at r and can visit the set of sensing locations X. The cost of travel between these locations is $d(x, y)$ for $x, y \in X$. There is a finite set of hypotheses H, which has a probability mass function ρ, and a set of observations O, which are sensed with $Z(x, h, o)$ for $x \in X$, $h \in H$ and $o \in O$. The function Z returns 1 when o agrees with h and 0 otherwise. The problem then asks to minimize the expected cost of identifying the correct hypothesis. An optimal policy can be encoded as a binary tree where nodes contain sensing information and the outgoing edges are selected via the sensing outcome. From [11], IPP is NP-hard as it contains the optimal decision tree problem [18] as a special case. We will use IPP to prove our problem is NP-Hard (decision form NP-Complete).

3   Problem Definition

We consider a single robot in a discrete environment. The robot and environment models are defined using a directed and doubly weighted graph $G = (\mathcal{V}, \mathcal{E}, c, \mu)$ where $c(e) \in \mathbb{R}_{\geq 0}$ defines the robot's cost for traversing $e \in \mathcal{E}$ and $\mu(e) \in \mathbb{R}_{\geq 0}$ defines robot's cost for sensing if edge e is obstructed. The robot knows the vertex it occupies, but does not know which edges leaving that vertex are free to traverse (that is, which edges are obstructed by obstacles). If the robot is unsure of an edge's state, it must first inspect the edge and incur the sensing cost.

3.1   Environmental Model

The unknown environment is one of m subgraphs of G, denoted $G_1, \ldots, G_m$, and we refer to the indices of these subgraphs as *environmental states* with *environmental state space* $\mathbb{N}_m = \{1, \ldots, m\}$. Each subgraph G_i is induced by a subset $E_i \subseteq \mathcal{E}$ for $i \in \mathbb{N}_m$. The robot is given the set of possible edge subsets $\mathcal{S} = \{E_1, \ldots, E_m\}$ along with a probability mass function capturing the likelihood of each subgraph. We encode the probability as a random variable X that takes values from $\mathbb{N}_m$. Given a random draw x from X, the edge subset E_x induces the *realization* $G_x = (V_x, E_x, c, \mu)$; the robot must operate in G_x without knowing x.

Note that if every edge subset is possible, $m = 2^{|\mathcal{E}|}$, then the absence or presence of an edge cannot imply the absence or presence of other edges. In this paper, we focus on cases where $m \ll 2^{|\mathcal{E}|}$, and thus observing one edge allows the robot to infer the state of other edges. This is motivated in Section 4.2 by the space complexity required for a control policy.

Example 1. To illustrate the problem, consider Fig. 1 as a simplified model of a building. A robot is tasked with delivering a package to room Q. The robot starts outside the building at S, which has two main entrances A and B. Both A and B will be locked in an emergency. From entrance A, there are two paths to Q. However, at times, one of these paths is obstructed. From entrance B, there is a single path. For this problem, we use $m = 3$ *environmental states* since they directly affect the completion of the robot's task. In state 1, no dashed edges exist; in state 2, all edges except (A, Q) exist; and in state 3, all edges exist.

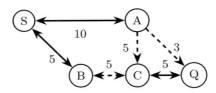

Fig. 1. Building model. Travel costs are shown on edges. Dashed edges are uncertain.

Remark 1. Given G_i for any $i \in \mathbb{N}_m$, the connected component containing the start must be strongly connected.

3.2   Robot Model

When the robot is located at vertex v, it can gather information on the environment by inspecting *incident-out* edges at v. This process is defined by the following model.

Definition 1 (Robot Observations). *Given a graph G and a vertex $v \in \mathcal{V}$ with $e \in I_v$, a robot observation is a mapping $\gamma : \mathcal{E} \to \{0, 1\}$ where $\gamma(e) = 1$ if $e \in E_x$ and $\gamma(e) = 0$ otherwise. The robot must occupy v to attain $\gamma(e)$.*

When we say the robot travels to an *observation* e, we mean it travels to v such that $(v, u) = e \in \mathcal{E}$. Given the robot is at v, the set I_v represents all possible *observations*.

Observations allow the robot to rule out *environmental states*. If an *observation* indicates an edge e does not exist in G_x, the robot knows that all edge subsets that contain e cannot be correct.

Definition 2 (Consistent). *Given a set of observed edges O, an edge subset E is consistent with O if and only if for every $e \in O$, $\gamma(e) = 1$ for $e \in E$ and $\gamma(e) = 0$ for $e \notin E$.*

Given a set of *observed* edges O, we define $Y \subseteq \mathbb{N}_m$ to be the set of *environmental states consistent* with O. To avoid collisions with an obstacle in the environment, we impose the restriction that an edge e can be traversed only if $\mathbb{P}(e|Y) = (\sum_{j \in Y} \mathbb{P}(X = j))^{-1} \sum_{i \in Y \mid e \in E_i} \mathbb{P}(X = i) = 1$. If an edge's existence is uncertain then the robot must first *observe* the edge and incur the corresponding *observation* cost, before proceeding across the edge.

3.3   Policy Space

The robot state is characterized by the set of *environmental states* Y that are *consistent* with its *observations* and the vertex v it occupies. Thus, the robot state space is $(2^{\mathbb{N}_m}, \mathcal{V})$. At each state (Y, v), the robot selects an action $a = (e, d)$ from $A_v = I_v \times \{0, 1\}$ to *observe* an edge ($d = 0$) or to move over an edge ($d = 1$) unless it terminates denoted by $\emptyset$. A policy maps the robot state space to the set of actions, $\pi : (2^{\mathbb{N}_m}, \mathcal{V}) \to (\cup_{v \in \mathcal{V}} A_v) \cup \emptyset$. Our safety constraint allows $d = 1$ for $e \in I_v$ only if $\mathbb{P}(e|Y) = 1$.

Given a start and goal $s, g \in \mathcal{V}$, the *environmental state space* $\mathbb{N}_m$ is partitioned into $Y_{\text{goal}} = \{i \in \mathbb{N}_m | c(s, g)$ calculated on G_i is finite$\}$ and $Y_{\text{no goal}}$ otherwise. We restrict policies to satisfy the following definition:

Definition 3 (Complete Policy). *A policy π is complete if for any realization it produces a sequence of actions that reach the goal (i.e., a state (Y, g) with $Y \subseteq Y_{\text{goal}}$) or that determine no path exists (i.e., a state (Z, v) with $Z \subseteq Y_{\text{no goal}}$).*

Consider the function $f : ((2^{\mathbb{N}_m}, \mathcal{V}), (\cup_{v \in \mathcal{V}} A_v) \cup \emptyset) \to (2^{\mathbb{N}_m}, \mathcal{V})$ where f updates Y after $d = 0$ and updates v after $d = 1$. Given $i \in \mathbb{N}_m$, a *complete* policy emits a sequence of states and actions $r_1, a_1, r_2, a_2, \ldots, a_T$ where $a_T = \emptyset$. The cost of this sequence is the sum of movement costs ($d = 1$) and *observation* costs ($d = 0$) defined by $\text{cost}(\pi|X = i) = \sum_{j=1}^{T-1} \mu(e_j)(1 - d_j) + c(e_j)d_j$.

Remark 2. Note that the domain of the policy has $\mathcal{O}(n2^m)$ states. In the following section we will derive properties that enable a more compact representation.

3.4   The Reactive Planning Problem

The expected cost of a *complete* policy π is found by taking the expectation over the *environmental states,*

$$\mathbb{E}_X(\pi) = \sum_{i \in \mathbb{N}_m} \text{cost}(\pi | X = i) \mathbb{P}(X = i) . \tag{1}$$

Problem 1 (Reactive Planning Problem, RPP). Given a graph G, start and goal vertices $s, g \in \mathcal{V}$ and a set of edge subsets $\mathcal{S}$ with corresponding random variable X that has a known probability mass function, find a *complete* policy π that minimizes $\mathbb{E}_X(\pi)$ over induced subgraph G_x for random draw x from X.

4   Properties and Complexity of Reactive Planning

In this section, we establish several properties of robot actions that enable us to efficiently represent *complete* policies along with the complexity of the RPP.

4.1   Action Properties

As the robot moves along a path P in G_x, it gathers *observations* $O_v \subseteq I_v \cup \emptyset$ for all $v \in P$, where $\emptyset$ is used to denote that no observation is taken at v. We define this sequence of actions to be an *observed path.*

Definition 4 (Observed Path). *Given a path* $P = v_1, \ldots, v_k$ *with observations* O_v *for all* $v \in P$, *the observed path is the sequence* $O_P = O_{v_1}, \ldots, O_{v_k}$.

The cost of an *observed path* can be found as the sum of travel costs and *observation* costs along the path: $\text{cost}(O_P) = c(P) + \mu(O_P)$. The robot's understanding of G_x, namely Y, is based on the *observed path* beginning at a starting vertex s. Two important subgraphs can be formed within this understanding.

Definition 5 (Known Subgraph). *Given a set of environmental states* Y, *the graph* $\overline{G} = (\overline{V}, \overline{E}, c, \mu)$ *induced by* $\overline{E} = \{e | \mathbb{P}(e|Y) = 1\}$ *is the known subgraph.*

Definition 6 (Consistent Subgraph). *Given a set of environmental states* Y, *the graph* $\underline{G} = (\underline{V}, \underline{E}, c, \mu)$ *induced by* $\underline{E} = \{e | \mathbb{P}(e|Y) > 0\}$ *is the consistent subgraph.*

The *known subgraph* includes only edges that are sure to exist, while the *consistent subgraph* includes all edges that may still exist. If an *observation* provides new information, then it partitions Y as follows.

Definition 7 (Constructive Observation). *Given environmental states* Y, *a constructive observation* o *updates* Y *to* $Y_1 = \{i \in Y \mid o \in E_i \text{ for } \gamma(o) = 1\}$, *and* $Y_0 = Y \setminus Y_1$ *such that* $Y_0, Y_1 \neq \emptyset$.

An *observed path* can be broken into smaller sections called *legs* that start at one *constructive observation* and end at the next *constructive observation*.

Definition 8 (Leg). *Given an observed path* O_P, *a leg is a subpath of* P, *namely* $v_1', v_2', \ldots, v_y'$ *such that* $v_{i+j} = v_j'$ *for* $1 \leq j \leq y$ *and* $0 \leq i \leq k - y$, *where the only non-empty observations are* $O_{v_1'}$ *and* $O_{v_y'}$.

A *leg* can be thought of as a meta-edge between *constructive observations*. Since the robot can move only on edges which are known to exist, a *leg* is composed only of edges which are understood to exist after the *leg's* first *observation* set $O_{v_1'}$. Therefore, a *leg* is a sequence of move actions that join *construction observation* actions.

The order in which *observations* can be visited depends on *observations* to date. The following definition provides a property of an optimal *complete* policy that can react to the environment without re-computation of that policy.

Definition 9 (Reachable). *Given a known subgraph* $\overline{G}$ *and a vertex* v, *an observation* o *is reachable from* v *if there exists a path from* v *to* o *in* $\overline{G}$.

The following result ties the notion of *reachability* to that of *legs* between *constructive observation*.

Lemma 1. *Consider two consecutive constructive observations* o_1 *and* o_2 *on a path* P. *Let* Y *be the environmental state after* o_1. *Then, in the known subgraph* $\overline{G}$ *defined by* Y, *observation* o_2 *is reachable from* o_1.

Proof. After $o_1 = (v_1, u_1)$ the understanding of the environment Y is fixed until it gains new information at $o_2 = (v_2, u_2)$. The robot can only select move actions for edges that are known to exist. $\overline{G}$, defined by Y, contains only edges that do not need to be *observed* before traversal; therefore, the robot can only reach o_2 if there exists a path from v_1 to v_2 in $\overline{G}$. $\qquad\square$

4.2   Control Policy Properties

We now show how a *complete* policy can be efficiently represented by a binary tree $\pi = (N, L)$. The nodes N are tuples (Y, o) where Y corresponds to the possible *environmental states* prior to *constructive observation* o. The edges are defined by *legs* L between *constructive observations*. Every non-leaf node $(Y, o) \in N$ must have one *leg incident-in* and two *legs incident-out*. The two *legs incident-out* to (Y, o) are *incident-in* to $(Y_0, o_0), (Y_1, o_1) \in N$ corresponding to $\gamma(o) = 0$ and $\gamma(o) = 1$ respectively where $Y = Y_0 \sqcup Y_1$ and $Y_0, Y_1 \neq \emptyset$. This allows real-time reaction in every possible *environmental state* by Lemma 1.

Lemma 2. *A complete policy can be encoded as a binary tree using* $\mathcal{O}(nm + m^2)$ *space with* n *the number of* G's *vertices and* m *the number of edge subsets.*

Proof. A *constructive observation* has a worst-case partition of Y_0 and Y_1 where one *environmental state* is ruled out (i.e., $|Y_0|$ or $|Y_1|$ is 1). In this case, the robot will need to make $m - 1$ *constructive observations* nodes of size $\mathcal{O}(m)$. We know each *leg* connecting these *observations* will visit at most n vertices. Therefore, the policy can be stored as a lookup table of size $\mathcal{O}(nm + m^2)$. $\qquad\square$

Remark 3. Note that the policy size scales linearly with m which motivates $m \ll 2^{|\mathcal{E}|}$. A POMDP with nm states and a MDP with $n2^m$ states can be encoded for the RPP, but for our cases this is still very large.

Continuing Example 1, suppose the robot travels to B and *observes* the edge from B to C (o_1) to find $\gamma(o_1) = 0$. The robot knows the *environment state* is 1, and thus it has reached the no goal terminal state. Alternatively, if $\gamma(o_1) = 1$, the robot then travels from B to C to Q and delivers the package. Fig. 2 displays this policy. Let *observation* costs be 0. We can specialize Eq. 1 for *observed paths* labelled $O_{P|X=i}$ for each $i \in \mathbb{N}_m$. The expected cost of a policy π is

$$\mathbb{E}_X(\pi) = \sum_{i \in \mathbb{N}_m} \mathbb{P}(X = i)\mathrm{cost}(O_{P|X=i}) \,. \tag{2}$$

If the PMF of X is $\{0.05, 0.5, 0.45\}$, then the policy in Fig. 2 using Eq. 2 renders expected cost of 14.5. Note: Q is reached without fully knowing x.

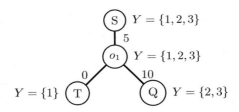

Fig. 2. Weights are the costs of L. T indicates no goal terminal state.

Remark 4. The policy in Fig. 2 satisfies both the *reachability* condition in Lemma 1 and the *constructive observation* property.

4.3   Computational Complexity

Consider this slight variation to the Reactive Planning Problem:

Problem 2 (Probable World Problem, PWP). Given a graph G, a start vertex $s \in \mathcal{V}$ and a set of edge subsets $\mathcal{S}$ with corresponding random variable X that has a known probability mass function, find a policy π that minimizes $\mathbb{E}_X(\pi)$ and identifies induced subgraph G_x for a random draw x from X.

Proposition 1. *The Probable World Problem is NP-Hard.*

Proof. Consider the tuple (X, d, H, ρ, O, Z, r) that defines an instance of IPP from Section 2.2. We will reduce IPP to PWP. Create a graph of vertices $\mathcal{V} = A \cup B$ where A mirrors X and B mirrors O. Let $r = s$. Create an edge subset E_h for every $h \in H$. In every E_h, connect A with edges of cost defined by d. For each $a \in A$ and $b \in B$, add an edge from a to b for subset E_h only if $Z(a, h, b) = 1$. Let random variable X's PMF be in line with ρ. Set $\mu((a, b)) = 0$ for all *observations*. Consider a solution S for PWP. Change each visited vertex of A to X and each *constructive observation* to respective elements of O for a solution S'. The *legs* of S contain no verticies of B as B has no path to *constructive observations*. Given S identifies random draw x, S' identifies true h. Given IPP (perfect sensing) is NP-Hard [11], PWP must be NP-Hard.   □

Theorem 1. *The Reactive Planning Problem is NP-Hard.*

Proof. We will prove this result by reducing PWP to RPP. Consider an instance of PWP. Given the graph for PWP, add a set of vertices Q with $|Q| = m$, an intermediary vertex h and a goal vertex g. Connect every $v \in \mathcal{V}$ to h with 0 cost for all $E \in \mathcal{S}$. Let α be the maximum of all traversal and *observation* costs. We can upper bound the expected cost of any optimal policy with $\alpha(mn + m^2)$ by Lemma 2. Connect h bidirectionally with each $q \in Q$ with traversal cost of U for all $E \in \mathcal{S}$ such that $(1 - \mathbb{P}(X = y))U \gg \alpha(mn + m^2)$ where $E_y \in \mathcal{S}$ is the most probable edge subset. Add an edge to $E_i \in \mathcal{S}$ from $q_i \in Q$ to g with cost of 0. In other words, there will only ever be one edge from Q to g, and it is always different for each subset. This new problem is in the form of RPP.

Suppose, by way of contradiction, there existed a solution to this RPP without solving the original PWP. This would imply there were at least two *environmental states Y consistent* with the *observations* of an *observed path* (starting at s) of the policy before attempting to reach g. This policy would move the robot to $q_i \in Q$ and *observe* the edge from q_i to g for $i \in Y$ (only exists in E_i). The policy must react to $\gamma((q_i, g)) = 0$. The resulting expected cost is at least $p_i U + (1 - p_i)2U$. Given $(1 - p_i)U \gg \alpha(mn + m^2)$, there exists a policy that can do better as $\alpha(mn + m^2)$ upper bounds the expected cost of an optimal policy providing a contradiction. This shows RPP solves PWP. Given PWP is NP-Hard by Proposition 1, RPP is NP-Hard.   □

Remark 5. In the decision version of RPP we are given a budget B and asked to find a *complete* policy with expected cost less than or equal to B. From Lemma 2, it is straightforward to see that the decision version is in NP, and thus is NP-Complete. [14] provides a similar result for R-PPSSR.

5   Scalable Policy Generation

The Reactive Planning Problem seeks information to reach the goal. The robot explores until it is beneficial to exploit the *observed* information and move to the goal. To address this, we maximize weighted mutual information (explore) and establish a condition to prune *observations* (exploit).

5.1   Exploration

Consider RPP. By Lemma 1, information can only be collected at the set of *reachable observations*. To select which *constructive observation* is benificial, we maximize mutual information extended from [11, 16, 17].

Let X_Y encode the probability distribution over environments given a set of *consistent environmental states* Y: The pmf of X_Y is $\mathbb{P}(X_Y = i) = \frac{\mathbb{P}(X=i)}{\sum_{j \in Y} \mathbb{P}(X=j)}$ for each $i \in Y$. Mutual information is the difference between entropy of X_Y and conditional entropy of X_Y given *observation* $o \in R_v$ where R_v is the set of *reachable constructive observations* from (Y, v). Formally,

$$MI(X_Y, o) = H(X_Y) - H(X_Y | o) . \tag{3}$$

The entropy $H(X_Y)$ does not depend on o; therefore, this problem can be reduced to minimization of conditional entropy,

$$H(X_Y | o) = -\sum_{j=0}^{1} \mathbb{P}(\gamma(o) = j) \sum_{i \in Y} \mathbb{P}(X_Y = i | \gamma(o) = j) \log \left(\mathbb{P}(X_Y = i | \gamma(o) = j) \right) .$$

5.2   Exploitation

The robot must be able to decide when it has collected enough information. We begin with the following inequality from the principle of optimality,

$$c_G(v, g) \le c_G(v, u) + \mu(o) + c_G(u, g) \quad (u, w) = o \in R_v , \tag{4}$$

where the subscript on the cost function c indicates the realization of the environment in which the cost is calculated.

Intuitively, making a measurement and going to the goal is at least as expensive as going straight to the goal in G. The cost calculated in G often performs poorly as an under-estimator for G_x. To address this, a new cost-to-go function is calculated as an expectation over the possible *environmental states* Y. The expected cost-to-go,

$$\mathbb{C}_Y(u, g) = \sum_{i \in Y} c_{G_i}(u, g) \mathbb{P}(X_Y = i) , \tag{5}$$

is found for every vertex $u \in \mathcal{V}$. To calculate $c_{G_i}(u, g)$, the edges are flipped in each G_i and a shortest path algorithm is run from g to all other $u \in V_i$. If $c_i(u, g)$ is infinite, we set such costs to zero as the robot will not travel any further (i.e., no goal terminal state).

Eq. 4 is augmented to include the current environmental understanding and the expected cost-to-go. The pruning condition can be written as

$$c_{\overline{G}}(v, g) \le c_{\overline{G}}(v, u) + \mu(o) + \mathbb{C}_Y(u, g) \quad (u, w) = o \in R_v . \tag{6}$$

If going straight to the goal is less expensive than gaining information and going to the goal, the information should not be collected.

Lemma 3. *Given a robot state r with constructive observations R_v, if all $o \in R_v$ satisfy Eq. 6, the robot should move to the goal.*

Proof. Consider $c_{\overline{G}}(v, g) = \infty$. This implies there is no known path to goal. No *observation* satisfies Eq. 6, so this trivially holds. Now, consider the case where enough information has been gathered to $r = (Y, v)$ for $c_{\overline{G}}(v, g) < \infty$. If all $o \in R_v$ satisfy Eq. 6, the known cost of making any *observation* and the expected cost-to-go is more than the known cost to complete the task. Thus, the robot should move to the goal.   □

Lemma 4. *The expected cost-to-go from the start, $\mathbb{C}_{\mathbb{N}_m}(s, g)$, forms a lower bound on the expected cost of any policy π.*

Proof. Consider any two *environmental states* $i, j \in \mathbb{N}_m$. If G_i and/or G_j do not have paths to the goal, the robot must identify the *environmental state* and return no goal terminal state. To do this, the robot uses an *observed path* to gain the information. The cost of such a path is at least 0. The expected cost-to-go for these cases is always 0. Suppose G_i and G_j can both reach the goal. There is at least one *leg* the robot must travel for both G_i and G_j. The expected cost-to-go selects the optimal paths independently. Therefore, the expected cost of the *observed paths* from π for i and for j can never be less than the expected cost-to-go, even if the robot acts optimally otherwise.   □

Remark 6. We find this bound performs well in practice (See Section 6). Although, the bound performs poorly when $\mathbb{P}(X \in Y_{\text{no goal}})$ and *observation* costs are large relative to $\mathbb{P}(X \in Y_{\text{goal}})$ and travel costs respectively since they are unaccounted for in $\mathbb{C}_{\mathbb{N}_m}(s, g)$.

5.3   Combining Exploration and Exploitation

Information gain and motion to goal can be combined as a function of the exploration metric $H(X_Y|o)$ and the exploitation metric $c_{\overline{G}}(v, u) + \mu(o) + \mathbb{C}_Y(u, g)$.

Weighted conditional entropy [19] is a well-studied method for combining entropy with a second metric, and thus in the following presentation and simulations we take the product of the two metrics and select *observations* satisfying

$$o_{\min} = \underset{(u,w)=o \in R_v}{\arg\min} \ (c_{\overline{G}}(v, u) + \mu(o) + \mathbb{C}_Y(u, g))H(X_Y|o) . \tag{7}$$

The product of the exploitation and exploration terms is non-negative. A value of 0 is achieved when o fully determines Y, or when traversal, *observation*, and the expected cost-to-go all have 0 cost.

5.4   Algorithm

Algorithm 1 minimizes every *leg* based on Eq. 7. It calls $Reachable(G, \mathcal{S}, (Y, v))$ which computes the minimum path lengths $d[u]$ from v to all other vertices u in the *known subgraph* $\overline{G}$. The set R_v is formed from edges that may or may not exist which render finite path cost from v, i.e. $R_v = \{(u, w) = e | \exists i, j \in Y \text{ s.t. } e \in E_i, e \notin E_j, d[u] \neq \infty\}$ with their corresponding distances D_v.

Algorithm 1: RPP Minimization of Conditional Entropy Policy

Data: Graph G, edge subsets $\mathcal{S}$, vertices s & g, states $\mathbb{N}_m$, probabilities p

Result: policy π for RPP and expected cost lower bound L

1   Compute $c_{G_i}(g, v)$ for all $v \in V$ and $i \in \mathbb{N}_m$;

2   Let Q contain only $(\mathbb{N}_m, s)$;

3   **while** Q *not empty* **do**

4 | Remove (Y, v) from Q;

5 | d = dijkstra($\underline{G}, v$);

6 | **if** $d[g] = \infty$ **then**

7 | | Mark π, at v for Y, no goal terminal state;

8 | **else**

9 | | Compute $(R_v, D_v) = Reachable(G, \mathcal{S}, (Y, v))$;

10 | | Remove elements of R_v that satisfy Eq. 6;

11 | | **if** $|R_v| = 0$ **then**

12 | | | Add *leg* from v to g, marked goal terminal state, to π;

13 | | **else**

14 | | | Let $o = (u', w')$ be the minimum of Eq. 7;

15 | | | Add *leg* from v to u' and node (Y, o) to π;

16 | | | Add $(Y_{\gamma(o)=0}, u')$ and $(Y_{\gamma(o)=1}, u')$ to Q;

17   Let $L = \mathbb{C}_{\mathbb{N}_m}(s, g)$;

18   Return π and L;

Remark 7. The runtime is dominated by the m calls to Dijkstra's Algorithm, which gives $\mathcal{O}(m(|\mathcal{V}|+|\mathcal{E}|) \log |\mathcal{V}|)$ (priority queue implemented as a binary heap).

The biased cost from Eq. 7 and the pruning condition from Eq. 6 complement each other to provide incentive toward the goal. The biased cost encourages *observation* selection closer to the goal. Once enough information is gained, the pruning condition removes information that is not important for the task. When the pruning condition removes all *observations* from R_v, the robot makes its way to the goal.

Remark 8. The policy π, returned by Algorithm 1, is independent of the order states are removed from Q in line 4 (potential for parallel computation).

To avoid learning the *environmental state* when it is impossible to reach the goal, the *consistent subgraph* $\underline{G}$ is checked for a path to goal. If the distance from the current location to g is infinite, there is no point continuing.

Theorem 2. *Algorithm 1 returns a complete policy.*

Proof. Suppose by contradiction, Algorithm 1 did not return a *complete* policy. This would imply either it terminates at (Y, g) for $Y \subseteq Y_{\text{no goal}}$ (false positive) or it terminates a (Z, v) for $v \neq g$ and $\exists z \in Z$ s.t. $z \in Y_{\text{goal}}$ (false negative).

    False positive: Algorithm 1 must have directed the robot to travel on an edge that does not exist because $Y \subseteq Y_{\text{no goal}}$. Since the environment does not have a path to goal, $\underline{G}$ for $Y \subseteq Y_{\text{no goal}}$ will not have a path to goal. Line 6 directly catches this case.

False negative: Algorithm 1 would not be able to find a path in $\underline{G}$, but since the *environmental state* z is still possible, $\underline{G}$ for Z will have a path to goal. This contradicts the fact that π is marked no goal terminal state in Line 7.   □

6   Simulation Results

In this section we provide simulation results on a large scale practical example and on randomly generated environments. Tests where run on a single Intel Core i7-6700 at 3.4GHz.

6.1   Flexible Factory

Flexible factories often spend considerable downtime between contracts due to changes in infrastructure and machinery. Consider Fig. 3 as a simple flexible factory that produces D items per hour. We are interested in knowing if the robot can move this volume. The dashed vertices indicate areas that require heavy use. For clarity, in Table 1 the column labelled "Vertex Obstructions" indicates the properties of the environment obstruction. For instance, in region 0 (vertices labelled 0) up to two vertices may be missing from the graph. Regions 1 and 2 each contain one forklift obstruction (which corresponds to removing the two adjacent vertices it occupies). When regions 5 and 6 are obstructed, all other vertices exist.

We cast this as a Reactive Planning Problem by enumerating all combinations of the obstructed vertices and removing their *incident* edges. This generates $m = 34561$ edge subsets each with a corresponding probability. We compute policies from S to A, from S to B, from A to S, and from B to S. The robot can move faster when not loaded, so the movement costs of A to S and B to S are decreased by a factor of 2.

We implemented a mixed integer linear program for RPP, but thus far have not been able to scale to problems of this size (in our formulation, the number of variables scales with m^3). Instead, we will allow re-planning during the online phase for comparison purposes (note these results do not have constant action lookup time). We compare against A* and maximum probability of success ($\mathbb{P}_s$). Both approaches generate a path, which we follow until it is obstructed. Then we take the edges of the path as *observations* and use this new information to re-plan. This is completed for every realization $x \in X$. The cost of the corrected paths from s to a terminal state and X's PMF are used to calculate the expected cost found in Table 2. Note that the proposed algorithm provides significantly lower expected cost than the two comparison policies, and in three of the four cases, it also provides a lower variance. In addition, for each task the expected cost of Algorithm 1 is within 30% of the lower bound (L) on the optimal cost.

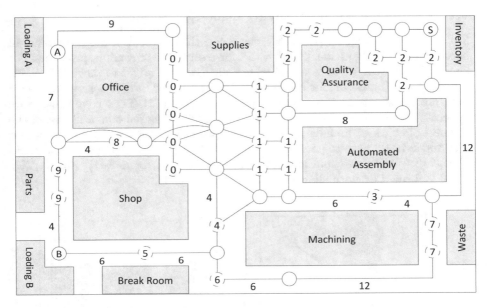

Fig. 3. Flexible Factory model: Dashed vertices may be obstructed. Edges with cost of 2 or 3 are not labelled for simplicity. The curved edges cost 2 more for waiting.

Table 1. Flexible Factory model parameters used in simulations.

Regions	Vertex Obstructions	Obstruction Probability	*Observation* Cost
0	≤ 2 independent	uniform over combinations	0.5
1,2	2 adjacent vertices	uniform over combinations	0.25
3,4	1	0.3	0.25
5,6	both or none	0.02	0.5
7,9	≤ 1 independent	0.1	0.5
8	1	0.4	0.5

Table 2. Flexible Factory simulation results. $|N|$ gives the number of nodes in the binary tree policy, $\mathbb{V}_X$ denotes the variance, Time indicates duration to compute the policy, and L denotes the lower bound.

	Algorithm 1					A*		Max $\mathbb{P}_s$	
Task	$|N|$	L	$\mathbb{E}_X(\pi)$	$\mathbb{V}_X(\pi)$	Time (s)	$\mathbb{E}_X$	$\mathbb{V}_X$	$\mathbb{E}_X$	$\mathbb{V}_X$
S→A	114	39.0	**42.5**	**12.6**	461	47.7	74.5	85.3	659
S→B	6869	41.4	**50.0**	**164.2**	739	50.1	282.0	61.7	206
A→S	84	19.5	**23.8**	**6.6**	510	30.2	81.9	43.6	135
B→S	175	20.7	**26.1**	51.3	497	28.3	**30.3**	33.9	135

6.2   Testing via Random Environment Generation

Since A* significantly outperformed the policy that maximizes the probability of success, we test Algorithm 1 against A* for a sequence of random problems. For simplicity the test graph is a grid where vertices have edges to move left, right, up and down (unless on the boundary). Travel costs are 1 and *observation* costs are drawn from the uniform random variable W on $[0, 1]$. Then edge subsets are generated by removing edges randomly. To do this, we incrementally relax the removal of edges on edge subset until a path to goal exists for 950 cases and vice versa for no path to goal for another 50 cases ($m = 1000$). The robot starts at cell $(2, 2)$ with the goal located in cell (width $-$ 2, depth $-$ 1). Finally, X's PMF is formed by normalizing m random draws of W.

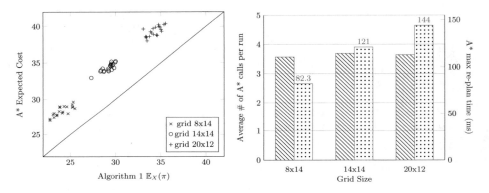

Fig. 4. Results of random environment generation.

The results in Fig. 4 show Algorithm 1 consistently outperforms A*. We also show the number of times A* re-plans and the maximum amount of time spent in a re-plan. The average runtimes, in seconds, of Algorithm 1 were 161, 302 and 569 for 8x14, 14x14 and 20x12 respectively. The average gaps between Algorithm 1's expected cost and L were 6.3, 5.9 and 6.0, meaning Algorithm 1 typically provides solutions within 25% of the optimal. We ran one 40x40 grid to test timing in a larger environment. The max re-plan time for A* was over 1(s) which is not viable for many real-time applications (Algorithm 1: 3349(s)).

7   Conclusion

A reactionary *complete* policy based on environmental *observations* was presented that finishes a task or identifies it is impossible to finish. For future work, we wish to remove the requirement of prior environmental knowledge so the robot may learn environmental trends during repetitive tasks. We are also interested in allowing multiple termination conditions (including multiple goals), faulty sensor models and extending Algorithm 1 to parallel computation.

References

1. Binney, J., Sukhatme, G.S.: Branch and bound for informative path planning. In: ICRA, Citeseer (2012) 2147–2154
2. Yu, J., Schwager, M., Rus, D.: Correlated orienteering problem and its application to informative path planning for persistent monitoring tasks. In: International Conference on Intelligent Robots and Systems, IEEE (2014) 342–349
3. Hollinger, G.A., Englot, B., Hover, F.S., Mitra, U., Sukhatme, G.S.: Active planning for underwater inspection and the benefit of adaptivity. The International Journal of Robotics Research (2012) 3–18
4. Javdani, S., Chen, Y., Karbasi, A., Krause, A., Bagnell, D., Srinivasa, S.S.: Near optimal bayesian active learning for decision making. In: AISTATS. (2014) 430–438
5. Dames, P., Schwager, M., Kumar, V., Rus, D.: A decentralized control policy for adaptive information gathering in hazardous environments. In: 51st IEEE Conference on Decision and Control (CDC), IEEE (2012) 2807–2813
6. Bhattacharya, S., Ghrist, R., Kumar, V.: Persistent homology for path planning in uncertain environments. IEEE Transactions on Robotics **31**(3) (2015) 578–590
7. Papadimitriou, C.H., Tsitsiklis, J.N.: The complexity of markov decision processes. Mathematics of operations research **12**(3) (1987) 441–450
8. LaValle, S.M.: Planning algorithms. Cambridge university press (2006)
9. Hart, P.E., Nilsson, N.J., Raphael, B.: A formal basis for the heuristic determination of minimum cost paths. IEEE transactions on Systems Science and Cybernetics **4**(2) (1968) 100–107
10. Koenig, S., Likhachev, M.: Fast replanning for navigation in unknown terrain. IEEE Transactions on Robotics **21**(3) (2005) 354–363
11. Lim, Z.W., Hsu, D., Lee, W.S.: Adaptive informative path planning in metric spaces. The International Journal of Robotics Research **35**(5) (2015) 585–598
12. Andreatta, G., Romeo, L.: Stochastic shortest paths with recourse. Networks **18**(3) (1988) 193–204
13. Papadimitriou, C.H., Yannakakis, M.: Shortest paths without a map. Theoretical Computer Science **84**(1) (1991) 127–150
14. Polychronopoulos, G.H., Tsitsiklis, J.N.: Stochastic shortest path problems with recourse. Networks **27**(2) (1996) 133–143
15. Issac, P., Campbell, A.M.: Shortest path problem with arc failure scenarios. EURO Journal on Transportation and Logistics (2015) 1–25
16. Dames, P.M., Schwager, M., Rus, D., Kumar, V.: Active magnetic anomaly detection using multiple micro aerial vehicles. IEEE Robotics and Automation Letters **1**(1) (2016) 153–160
17. Charrow, B., Kumar, V., Michael, N.: Approximate representations for multi-robot control policies that maximize mutual information. Autonomous Robots **37**(4) (2014) 383–400
18. Chakaravarthy, V.T., Pandit, V., Roy, S., Awasthi, P., Mohania, M.: Decision trees for entity identification: approximation algorithms and hardness results. In: Proceedings ACM Symp. on Principles of Database Systems, ACM (2007) 53–62
19. Suhov, Y., Stuhl, I., Sekeh, S.Y., Kelbert, M.: Basic inequalities for weighted entropies. Aequationes mathematicae (2015) 1–32

Linearization in Motion Planning under Uncertainty

Marcus Hoerger[1,2], Hanna Kurniawati[1], Tirthankar Bandyopadhyay[2], and Alberto Elfes[2]

[1] The University of Queensland, St Lucia, Brisbane, QLD, 4072, Australia
{m.hoerger, hannakur}@uq.edu.au

[2] CSIRO, Pullenvale, Brisbane, QLD 4069, Australia

Abstract. Motion planning under uncertainty is essential to autonomous robots. Over the past decade, the scalability of such planners have advanced substantially. Despite these advances, the problem remains difficult for systems with non-linear dynamics. Most successful methods for planning perform forward search that relies heavily on a large number of simulation runs. Each simulation run generally requires more costly integration for systems with non-linear dynamics. Therefore, for such problems, the entire planning process remains relatively slow. Not surprisingly, linearization-based methods for planning under uncertainty have been proposed. However, it is not clear how linearization affects the quality of the generated motion strategy, and more importantly where to and where not to use such a simplification. This paper presents our preliminary work towards answering such questions. In particular, we propose a measure, called *Statistical-distance-based Non-linearity Measure (SNM)*, to identify where linearization can and where it should not be performed. The measure is based on the distance between the distributions that represent the original motion-sensing models and their linearized version. We show that when the planning problem is framed as the Partially Observable Markov Decision Process (POMDP), the difference between the value of the optimal strategy generated if we plan using the original model and if we plan using the linearized model, can be upper bounded by a function linear in SNM. We test the applicability of this measure in simulation via two venues. First, we compare SNM with a negentropy-based Measure of Non-Gaussianity (MoNG) —a measure that has recently been shown to be a suitable measure of non-linearity for stochastic systems [1]. We compare their performance in measuring the difference between a general POMDP solver [2] that computes motion strategies using the original model and a solver that uses the linearized model (adapted from [3]) on various scenarios. Our results indicate that SNM is more suitable in taking into account the effect that obstacles have on the effectiveness of linearization. In the second set of tests, we use a local estimate of SNM to develop a simple on-line planner that switches between using the original and the linearized model. Simulation results on a car-like robot with second order dynamics and a 4-DOFs and 6-DOFs manipulator with torque control indicate that our simple planner appropriately decides if and when linearization should be used.

Keywords: Motion planning, Motion planning under uncertainty, POMDP

1 Introduction

An autonomous robot must be able to compute reliable motion strategies, despite various errors in actuation and prediction on its effect on the robot and its environment, and despite various errors in sensing and its interpretation. Computing such robust strategies is computationally hard even for a 3 DOFs point robot [4,5]. Conceptually, this problem can be solved in a systematic and principled manner when framed as the Partially

© Springer Nature Switzerland AG 2020
K. Goldberg et al. (Eds.): *Algorithmic Foundations of Robotics XII*, SPAR 13, pp. 272–287, 2020.
https://doi.org/10.1007/978-3-030-43089-4_18

Observable Markov Decision Process (POMDP) [6]. POMDP represents the aforementioned errors as probability distributions, and estimates the system's states (following a sequence of actions and observations that the robot has performed and perceived) as a probability distributions called *beliefs*. It then computes the best motion strategy with respect to beliefs rather than with respect to single states, because the actual state is uncertain due to errors in the system's dynamics and sensing. Although the concept of POMDP was proposed since early 1960 [7], only in recent years that POMDP starts to become practical for robotics problems (e.g., [8,9]). This advancement is achieved by trading optimality with approximate optimality for speed and memory. But even then, in general, computing close to optimal strategies for systems with complex non-linear dynamics remains relatively slow.

Several general POMDP solvers —one that does not restrict the type of dynamics and sensing model of the system, nor the type of distributions used to represent uncertainty— can now compute good motion strategies on-line with 1-10Hz update rate for a number of robotics problems [2,10,11,12]. However, their speed degrades when the robot has complex non-linear dynamics. To find a good strategy, these methods simulate the effect of many sequences of actions from different beliefs. A simulation run generally invokes many numerical integrations, and more complex dynamics tends to increase the cost of each numerical integration, which in turn significantly increases the total planning cost of these methods. Of course, this cost will increase even more for problems that require more or longer simulation runs, such as in problems with long planning horizon.

Many linearization based methods have been proposed [3,13,14,15,16]. These methods evaluate the effect of many sequences of actions from different beliefs too, but uses a linearized model of the dynamics and sensing for simulation, so as to reduce the cost of each simulation run. Together with linearization, many of these methods assume that each reachable belief must be a Gaussian distribution. This assumption improves the speed of simulation further, because the subsequent belief after an action is performed and an observation is perceived can be computed by propagating only the mean and co-variance matrix. In contrast, the aforementioned general solvers use particle representation and therefore must compute the subsequent belief by propagating each particle. As a result, the linearization-based planners require less time to simulate the effect of performing a sequence of actions from a belief, and therefore can *potentially* find a good strategy faster than the general method. However, it is known that linearization in control and estimation performs well only when the system's non-linearity is "weak" [17]. The question is, what constitute "weak" non-linearity in motion planning under uncertainty? Where will it be useful and where will it be damaging to use linearization (and Gaussian) simplifications?

This paper presents our preliminary work to answer the above questions. We propose a measure of non-linearity, called *Statistical-distance-based Non-linearity Measure (SNM)*, to help identify the suitability of linearization in a given problem of motion planning under uncertainty. SNM is based on the total variation distance between the original dynamics and sensing models, and their corresponding linearized models. It is general enough to be applied to any type of motion and sensing errors, and any linearization technique, regardless of the type of approximation to the true beliefs (e.g., with and without Gaussian simplification). We showed that the difference between the value of the optimal strategy generated if we plan using the original model and if we plan using the linearized model, can be upper bounded by a function linear in SNM. Furthermore, our experimental results (Section 6) indicate that compared to recent state-of-the-art methods of non-linearity measures for stochastic systems, SNM is more sensitive to the effect that obstacles have on the effectiveness of linearization, which is critical for motion planning.

To further test the applicability of SNM in motion planning, we develop a simple on-liner planner that uses a local estimate of SNM to automatically switch between a general planner[2] that uses the original POMDP model and a linearization-based planner (adapted from [3]) that uses the linearized model. Experimental results on a car-like robot with acceleration control, and a 4-DOFs and 6-DOFs manipulators with torque control indicate that this simple planner can appropriately decide if and when linearization should be used and therefore computes better strategies faster than each of the component planner.

2   Related work

Linearization is a common practice in solving non-linear control and estimation problems. It is known that linearization performs well only when the system's non-linearity is "weak" [17]. To identify the effectiveness of linearization in solving non-linear problems, many non-linearity measure have been proposed in the control and information fusion community.

Many non-linearity measures (e.g., [18,19,20]) have been designed for deterministic systems. For instance, [18] proposed a measure derived from the curvature of the non-linear function. The work in [19,20] compute the measure based on the distance between the non-linear function and its nearest linearization. A brief survey of non-linearity measures for deterministic systems is available in [17].

Only recently more work on non-linearity measures for stochastic systems started to flourish. For instance, [17] extends the measures in [19,20] to be based on the average distance between the non-linear function that models the motion and sensing of the system, and the set of all possible linearizations of the function.

Very recently, [1] proposes a different class of measures, which is based on the distance between distribution over states and its Gaussian approximation, called Measure of Non-Gaussianity (MoNG), rather than based on the non-linear function itself. They assume a passive stochastic systems, and computes the negentropy of the non-linear function of the transformed belief — that is, the non-linearity measure of a belief is computed as the negentropy between the subsequent beliefs and their Gaussian approximations. Their results indicate that this non-Gaussianity measure is more suitable to measure the non-linearity of stochastic systems, as it takes into account the effect that non-linear transformations have on the shape of the transformed beliefs. This advancement is encouraging and we will use this measure as a comparator of SNM. However, for this purpose, this measure must be modified because our system is not passive, and in fact, eventually, we would like to have a measure that can be used to decide what strategy to use (e.g., to use a linearized or a general planner). The exact modifications we made can be found in the supplementary material[3].

Despite the various non-linearity measures that have been proposed, most are not designed to take into account the effect of obstacles to the non-linearity of the robotic system. Except for MoNG, all of the aforementioned non-linearity measures will have difficulties in taking into account the effect of obstacles, even when these effects are embedded in the motion and sensing models. For instance, curvature-based measures requires the non-linear function to be twice continuously differentiable, but the presence of obstacles is very likely to break the differentiability of the motion model. Furthermore, the effect of obstacles is likely to violate the additive Gaussian error, required for instance by [17]. Although MoNG can potentially take into account the effect of obstacles, it is not designed to. The measure is based on the Gaussian approximation to the subsequent belief. In the presence of obstacles this subsequent belief would have support only in the valid region of the state space, and therefore computing the difference between this subsequent belief and its Gaussian approximation is likely to underestimate the effect of obstacles to the effectiveness of linearization. This is exactly the problem we try to alleviate in our proposed non-linearity measure SNM.

Instead of adopting existing approaches in non-linearity measure, SNM adopts the approach commonly used for sensitivity analysis[21,22] of Markov Decision Processes (MDP) —a special class of POMDP where uncertainty is only in the effect of performing actions. It is based on the statistical distance measure between the true transition dynamics and its perturbed versions. Linearized dynamics can be viewed as a special case of perturbed dynamics, and hence this statistical distance measure can be applied as a non-linearity measure too. We do need to extend these analysis, as they are generally defined for discrete state space and are defined with respect to only the dynamics models (MDP assumes the state of the system is fully observable). Nevertheless, such extensions are feasible and the generality of this measure could help decide which linearization method to use.

3   Problem modelling

In this paper, we consider motion planning problems, in which a robot must move from a given initial state to a state in the goal region while avoiding obstacles. The robot operates inside deterministic, bounded, and perfectly known 2D or 3D environments populated by static obstacles.

The robot's dynamics and sensing are uncertain and are defined as follows. Let $S \subset \mathbb{R}^n$ be the bounded n-dimensional state space, $A \subset \mathbb{R}^d$ the bounded d-dimensional control space and $O \subset \mathbb{R}^l$ the bounded l-dimensional observation space of the robot. The robot evolves according to a discrete-time non-linear stochastic system, which we model in the general form $s_{t+1} = f(s_t, a_t, v_t)$ where f is a known non-linear stochastic dynamic system, $s_t \in S$ is the state of the robot at time t, $a_t \in A$ the control input at time t, and $v_t \in \mathbb{R}^d$ is a random control error. At each time step t, the robot receives imperfect information regarding its current state according to a non-linear stochastic function of the form $o_t = h(s_t, w_t)$ where $o_t \in O$ is the observation at time t and $w_t \in \mathbb{R}^d$ is a random observation error.

---

[3]The paper with supplementary materials is available in http://robotics.itee.uq.edu.au/~hannakur/dokuwiki/papers/wafr16_linearization.pdf

We now define the motion planning under uncertainty problem for the above system as a Partially Observable Markov Decision Process (POMDP) problem.

Formally, a POMDP is a tuple $\langle S, A, O, T, Z, R, b_0, \gamma \rangle$. The notations S, A and O are the state, action, and observation spaces. The notation T is a conditional probability function $p(s' | s, a)$ (where $s, s' \in S$ and $a \in A$) that represents uncertainty in the effect of actions, while Z is a conditional probability function $p(o | s, a)$ that represents uncertainty on sensing. The notation R is the reward function, which depends on the state–action pair and acts as an objective function. The notations b_0 and $\gamma \in (0, 1)$ are the initial belief and discount factor.

At each time-step, a POMDP agent is at a state $s \in S$, takes an action $a \in A$, perceives an observation $o \in O$, receives a reward based on the reward function $R(s, a)$, and moves to the next state. Now due to uncertainty in the results of action and sensing, the agent never knows its exact state and therefore, estimates its state as a probability distribution, called belief. The solution to the POMDP problem is an optimal policy (denoted as π^*), which is a mapping $\pi^* : \mathbb{B} \to A$ from beliefs ($\mathbb{B}$ denotes the set of all beliefs, which is called the belief space) to actions that maximizes the expected total reward the robot receives, i.e., $V^*(b_0) = \max_{a \in A} \left(R(b, a) + \gamma \int_{o \in O} p(o | b, a) V^*(\tau(b, a, o)) \, do \right)$, where $\tau(b, a, o)$ computes the updated belief estimate after the robot performs action $a \in A$ and perceived $o \in O$ from belief b, and is defined as:

$$b'(s') = \tau(b, a, o)(s') = \eta \, Z(s', a, o) \int_{s \in S} T(s, a, s') b(s) \, ds \tag{1}$$

For our motion planning problem, S, A, and O of the POMDP problem is the same as those of the robotic system (for simplicity, we use the same notation). The transition T represents the dynamics model f, while Z represents the sensing model h. The reward function represents the task' objective, for example, high reward for goal states and low negative reward for states that cause the robot to collide with the obstacles. The initial belief b_0 represents uncertainty on the starting state of the robot.

4   Statistical-distance-based Non-linearity Measure (SNM)

Intuitively, our proposed measure SNM is based on the total variation distance between the effect of performing an action and perceiving an observation under the true dynamics and sensing model, and the effect under the linearized dynamic and sensing model. The total variation distance $dist_{TV}$ between two probability functions θ and θ' over a measurable space Ω is defined as $dist_{TV}(\theta, \theta') = sup_{E \in \Omega} |\theta(E) - \theta'(E)|$. More formally, SNM is defined as:

Definition 1. *Let $P = \langle S, A, O, T, Z, R, b_0, \gamma \rangle$ be the POMDP model of the system and $\widehat{P} = \langle S, A, O, \widehat{T}, \widehat{Z}, R, b_0, \gamma \rangle$ be a linearization of P, where $\widehat{T}$ is a linearization of the transition function T and $\widehat{Z}$ is a linearization of the observation function Z of P, while all other components of P and $\widehat{P}$ are the same. Then, the SNM (denoted as Ψ) between P and $\widehat{P}$ is $\Psi(P, \widehat{P}) = \Psi_T(P, \widehat{P}) + \Psi_Z(P, \widehat{P})$, where*

$$\Psi_T(P, \widehat{P}) = \sup_{s \in S, a \in A} \left| dist_{TV}(T(s, a, s'), \widehat{T}(s, a, s')) \right| = \sup_{s, s' \in S, a \in A} \left| T(s, a, s') - \widehat{T}(s, a, s') \right|$$

$$\Psi_Z(P, \widehat{P}) = \sup_{s \in S, a \in A} \left| dist_{TV}(Z(s, a, o), \widehat{Z}(s, a, o)) \right| = \sup_{s \in S, a \in A, o \in O} \left| Z(s, a, o) - \widehat{Z}(s, a, o) \right|$$

Note that SNM can be applied as both a global and a local measure. For local measure, the supremum over the state s can be restricted to a subset of S, rather than the entire state space. Also, SNM is general enough for any approximation to the true dynamics and sensing model, which means that it can be applied to any type of linearization and belief approximation techniques, including those that assume and those that do not assume Gaussian belief simplifications.

We want to use the measure $\Psi(P,\widehat{P})$ to bound the difference between the expected total reward received if the system were to run the optimal policy of the true model P and if it were to run the optimal policy of the linearized model $\widehat{P}$. Note that since our interest is in the actual reward received, the values of these policies are evaluated with respect to the original model P (we assume P is a faithful model of the system). More precisely, we want to show that:

Theorem 1. *Let π^* be the optimal policy of POMDP problem $P = \langle S,A,O,T,Z,R,b_0,\gamma \rangle$ and $\widehat{\pi}$ be the optimal policy of its linearized version $\widehat{P} = \langle S,A,O,\widehat{T},\widehat{Z},R,b_0,\gamma \rangle$, where $\widehat{T}$ is a linearization of the transition function T, $\widehat{Z}$ is a linearization of the observation function Z of P. Suppose the spaces S, A, and O for both models are the same and all represented as $[0,1]^n$, $[0,1]^d$, and $[0,1]^l$, respectively, with n, d, and l are the dimensions of the respective spaces. And, the reward function is Lipschitz continuous in S with Lipschitz constant K. Then,*

$$V_{\pi^*}(b_0) - V_{\widehat{\pi}}(b_0) \leq 2\gamma \left(\frac{R_{\max}}{(1-\gamma)^2} + \frac{KC\sqrt{n}}{1-\gamma} \right) \Psi(P,\widehat{P})$$

where $V_\pi(b) = R(b,\pi(b)) + \gamma \int_{o \in O} Z(b,a,o) V_\pi(\tau(b,a,o)) do$ and $\tau(b,a,o)$ is the belief transition function as defined in (1) for the true model P. The notation C is a constant, defined as $C = \frac{1+\eta_1}{\eta_1 \eta_2}$, where η_1 and η_2 are the normalization constants in the computation of $\tau(b,a,o)$ and its linearized version, respectively.

To prove this theorem, we first need to compute the upper bounds of two other difference functions. First is the total variation distance between the beliefs propagated under the true model P and under its linearized version $\widehat{P}$. More precisely,

Lemma 1. *Suppose $b' = \tau(b,a,o)$ and $\widehat{b'} = \widehat{\tau}(b,a,o)$ are the belief transition functions (as defined in (1)) for P and $\widehat{P}$, respectively. Then for any $a \in A$ and any $o \in O$,*

$$dist_{TV}(b',\widehat{b'}) = \sup_{s \in S} |b'(s) - \widehat{b'}(s)| \leq C\Psi(P,\widehat{P})$$

where $C = \frac{1+\eta_1}{\eta_1 \eta_2}$, and η_1 and η_2 are the normalization constants in the computation of $\tau(b,a,o)$ and $\widehat{\tau}(b,a,o)$ respectively.

The second function is the difference between the observation function and its linearized version, given a belief and an action. Slightly abusing the notation Z, we use $Z(b,a,o)$ and $\widehat{Z}(b,a,o)$ to denote the conditional probability density (mass) function of perceiving observation $o \in O$ after action $a \in A$ is performed from belief b, under the model P and its linearized version $\widehat{P}$. We want to show that:

Lemma 2. *For any $b \in B$, $a \in A$, and $o \in O$, $Z(b,a,o) - \widehat{Z}(b,a,o) \leq \Psi(P,\widehat{P})$.*

In addition, we need to show that the optimal value function is Lipschitz continuous under the total variation distance, i.e.:

Lemma 3. *Let* $P = \langle S, A, O, T, Z, R, b_0, \gamma \rangle$ *be a POMDP problem where* S *is a metric space with dimension* n *and* $|R(s, a) - R(s', a)| \leq K\,dist(s, s')$ *for any* $s \in S$ *and* $a \in A$. *Then,* $|V^*(b) - V^*(b')| \leq K\sqrt{n}\,dist_{TV}(b, b')$.

The proofs for the above lemmas are available in the supplementary material[3].

Now, we can show Theorem 1. For this purpose, we first divide the difference in the expected total reward into two components:

$$V_{\pi^*}(b_0) - V_{\widehat{\pi}}(b_0) = \left(V_{\pi^*}(b_0) - \widehat{V}(b_0) \right) + \left(\widehat{V}(b_0) - V_{\widehat{\pi}}(b_0) \right) \tag{2}$$

where $\widehat{V}(b_0)$ is the optimal value of the linearized model $\widehat{P}$ when the belief is at b_0.

We will start by computing an upper bound for the first component. Note that the value V_{π^*} is the same as the optimal value V^* of the POMDP problem P, and therefore we can compute the bound of the first component as:

$$\left(V_{\pi^*}(b_0) - \widehat{V}(b_0) \right) = \max_{a \in A} \left(R(b_0, a) + \gamma \int_{o \in O} Z(b_0, a, o) V^*(b_1) do \right) -$$

$$\max_{a \in A} \left(R(b_0, a) + \gamma \int_{o \in O} \widehat{Z}(b_0, a, o) \widehat{V}(\widehat{b_1}) do \right)$$

$$\leq \gamma \max_{a \in A} \left(\int_{o \in O} Z(b_0, a, o) V^*(b_1) - \widehat{Z}(b_0, a, o) \widehat{V}(\widehat{b_1}) do \right)$$

$$= \gamma \max_{a \in A} \left(\int_{o \in O} \left(Z(b_0, a, o) V^*(b_1) - Z(b_0, a, o) \widehat{V}(\widehat{b_1}) \right) + \right.$$

$$\left. \left(Z(b_0, a, o) \widehat{V}(\widehat{b_1}) - \widehat{Z}(b_0, a, o) \widehat{V}(\widehat{b_1}) \right) do \right)$$

$$= \gamma \max_{a \in A} \left(\int_{o \in O} \left(Z(b_0, a, o) \left[V^*(b_1) - \widehat{V}(\widehat{b_1}) \right] + \right.\right.$$

$$\left.\left. \left[Z(b_0, a, o) - \widehat{Z}(b_0, a, o) \right] \widehat{V}(\widehat{b_1}) \right) do \right) \tag{3}$$

Replacing the difference between Z and $\widehat{Z}$ in the last term of (3) with the upper bound in Lemma 2 and assuming the volume of O is one, in addition to dividing the difference $[V^*(b_1) - \widehat{V}(\widehat{b_1})]$ into two components $(V^*(b_1) - \widehat{V}(b_1)) + (\widehat{V}(b_1) - \widehat{V}(\widehat{b_1}))$, allows us to rewrite (3) as:

$$\left(V_{\pi^*}(b_0) - \widehat{V}(b_0) \right) \leq \gamma \Psi(P, \widehat{P}) \frac{R_{max}}{1 - \gamma} + \gamma \max_{a \in A} \left(\int_{o \in O} Z(b_0, a, o) \left[V^*(b_1) - \widehat{V}(b_1) \right] do \right.$$

$$\left. + \int_{o \in O} Z(b_0, a, o) \left| \widehat{V}(b_1) - \widehat{V}(\widehat{b_1}) \right| do \right) \tag{4}$$

We bound the last term of the right-hand-side of (4) using Lemma 3 and rewrite (4) as:

$$\left(V_{\pi^*}(b_0) - \widehat{V}(b_0) \right) \leq \gamma \left(\Psi(P, \widehat{P}) \frac{R_{max}}{1 - \gamma} + KC\sqrt{n}\,\Psi(P, \widehat{P}) \right) +$$

$$\gamma \max_{a \in A} \left(\int_{o \in O} Z(b_0, a, o) \left[V^*(b_1) - \widehat{V}(b_1) \right] do \right)$$

Since V^* is equivalent to V_{π^*} for any beliefs, the last term in the above equation is essentially a recursion. Solving this recursion completes the upper-bound for the first component of (2), i.e.:

$$\left(V_{\pi^*}(b_0) - \widehat{V}(b_0)\right) \leq \gamma \left(\frac{R_{\max}}{(1-\gamma)^2} + \frac{KC\sqrt{n}}{1-\gamma}\right) \Psi(P,\widehat{P}) \tag{5}$$

Now, we compute the upper bound for the second component of (2) in a similar manner:

$$\widehat{V}(b_0) - V_{\widehat{\pi}}(b_0) \leq \left(R(b_0, \widehat{\pi}(b_0)) + \gamma \int_{o\in O} \widehat{Z}(b_0, \widehat{\pi}(b_0), o)\widehat{V}(\widehat{b}_1)do\right) -$$

$$\left(R(b_0, \widehat{\pi}(b_0)) + \gamma \int_{o\in O} Z(b_0, \widehat{\pi}(b_0), o)V_{\widehat{\pi}}(b_1)do\right)$$

$$= \gamma \left(\int_{o\in O} \widehat{Z}(b_0, \widehat{\pi}(b_0), o)\widehat{V}(\widehat{b}_1) - Z(b_0, \widehat{\pi}(b_0), o)V_{\widehat{\pi}}(b_1)do\right)$$

$$= \gamma \left(\int_{o\in O} \left(\widehat{Z}(b_0, \widehat{\pi}(b_0), o)\left[\widehat{V}(\widehat{b}_1) - V_{\widehat{\pi}}(b_1)\right] + \right.\right.$$

$$\left.\left. \left[\widehat{Z}(b_0, \widehat{\pi}(b_0), o) - Z(b_0, \widehat{\pi}(b_0), o)\right] V_{\widehat{\pi}}(b_1)\right) do\right) \tag{6}$$

By dividing $\left[\widehat{V}(\widehat{b}_1) - V_{\widehat{\pi}}(b_1)\right]$ into two components $\left(\widehat{V}(\widehat{b}_1) - \widehat{V}(b_1)\right), \left(\widehat{V}(b_1) - V_{\widehat{\pi}}(b_1)\right)$ and using similar arguments we made in (3)–(5), we can bound the second component of (2) in the same way we did for its first component, i.e.:

$$\left(\widehat{V}(b_0) - V_{\widehat{\pi}}(b_0)\right) \leq \gamma \left(\frac{R_{\max}}{(1-\gamma)^2} + \frac{KC\sqrt{n}}{1-\gamma}\right) \Psi(P,\widehat{P}) \tag{7}$$

The sum of the right-hand-side of (5) and (7) yields the upper bound in Theorem 1.

The upper bound in Theorem 1 is relatively loose. However, the results in Section 6 indicate that this bound can be used as a sufficient condition to identify where linearization should and should not be applied.

5   SNM-Planner: An Application of SNM for Planning

SNM-Planner is an on-line planner that uses SNM as a heuristic to decide whether a general POMDP solver or a linearization-based motion planner should be used. The general solver used is Adaptive Belief Tree (ABT)[2], while the linearization-based method called Modified High Frequency Replanning (MHFR), which is an adaptation of HFR[3]. HFR is designed for chance-constraint POMDPs, i.e., it explicitly minimizes the collision probability, while MHFR is a POMDP solver where the objective is to maximize the expected total reward. During run-time, at each step, SNM-Planner approximates the local value of SNM around the current belief b. This value and a given threshold will then be used to decide whether to use ABT or MHFR to decide what action to take from b. An overview of the algorithm is in Algorithm 1.

5.1   Approximating SNM

Given the current belief b_i, SNM-Planner approximates the local value of SNM around b_i by approximating each component of SNM, i.e., Ψ_T and Ψ_Z, separately, using a simple Monte-Carlo approach. To approximate Ψ_T, SNM-Planner uses a Monte Carlo approach to construct a histogram representation of $T(s,a,s')$ in the support set of b_i. For

---

Algorithm 1 SNM-Planner (initial belief b_0, threshold μ, max planning time t, max time to approximate SNM t_m, #steps N, goal region $\mathbb{G}$)

1: $T_{ABT} \leftarrow InitializeABT(P)$
2: $T_{MHFR} \leftarrow InitializeMHFR(P)$
3: $t_p \leftarrow t - t_m, i \leftarrow 0$
4: **while** $s_i \notin G$ and collided = False and $i < N$ **do**
5:    ▷ s_i is the actual state at step-i. The system never knows s_i, but it knows if it is at a goal state.
6:        $\Psi \leftarrow approximatePsi(t_m, b_i)$
7:        **if** $\Psi < \mu$ **then**
8:            $a \leftarrow MHFR(T_{MHFR}, t_p, b_i)$
9:        **else**
10:            $a \leftarrow ABT(T_{ABT}, t_p, b_i)$
11:        **end if**
12:        executeAction(a)
13:        $o \leftarrow getObservation(b_i, a)$
14:        $b_{i+1} \leftarrow \tau(b_i, a, o)$ ▷ We use Sequential Importance Resampling [23]
15:        $i \leftarrow i + 1$
16: **end while**

---

this purpose, SNM-Planner starts by sampling a set of state-action pairs (s, a), denoted as U, where each s is a state sampled from b_i and a is sampled uniformly at random from the action space A. For each pair (s, a), SNM-Planner samples a set of L possible next states according to $T(s, a, s')$ (denoted as $X_{(s,a)}$) and another set of L possible next states according to $\widehat{T}(s, a, s')$ (denoted as $\hat{X}_{(s,a)}$, where L is a given constant. It then constructs a histogram for each set of possible next states. Both histograms are constructed with respect to the same discretization of the state space S. Suppose K is the number of bins in this discretization, the approximate local value $\widehat{\Psi_T}(b_i)$ can be approximated as:

$$\widehat{\Psi_T}(b_i) \approx \max_{(s,a) \in U} \max_{k \in [1,K]} \frac{1}{L} \left| n_{s,a}^k - \hat{n}_{s,a}^k \right| \tag{8}$$

where $n_{s,a}^k$ is the number of elements in $X_{(s,a)}$ that lies in bin-k, while $\hat{n}_{s,a}^k$ is the number of elements in $\hat{X}_{(s,a)}$ that lies in bin-k.

To approximate Ψ_Z around b_i, the same procedure is performed. However, here, SNM-Planner samples the observation from $Z(s, a, o)$ and $\widehat{Z}(s, a, o)$. The histogram is then constructed by discretizing the observation space O.

At a first glance, the above method seems rather inefficient, due to the large number of histogram bins for high-dimensional state and observation spaces. However, three problem properties significantly reduce the computation cost. First, often a large portion of the histogram bins are empty and do not need to be considered, allowing for more efficient data structures, such as associative maps. Second, since SNM-Planner is a threshold-based method, as soon as the local approximation of SNM hits the threshold, the computation can be stopped. In our experiments we have seen that when the robot operates near obstacles where the local SNM is high, only a few state-action samples are needed to exceed the threshold. Last, calculating $\widehat{\Psi_T}^{s,a}(b_i)$ and $\widehat{\Psi_Z}(b_i)$ for different states and actions is trivially parallelizable.

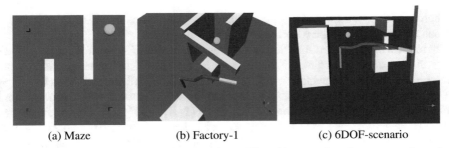

(a) Maze	(b) Factory-1	(c) 6DOF-scenario

Fig. 1: Test scenarios for the different robots. The objects colored grey are obstacles, while the green sphere is the goal region. (a) The car-like robot scenario. The purple square represents the beacons, while the red square at the bottom left represents the initial state. (b) The 4DOF-manipulator scenario. (c) The 6DOF-manipulator scenario. For (b-c), the robot is shown in red color with yellow end-effector.

6   Experimental Results

6.1   Experimental Setup

Our experiment is two-fold: To test SNM and to test the planner as proposed in Section 5. For our first objective, we compare SNM with a modified version of the measure of non-Gaussianity (MoNG) [1]. Details on these modifications are presented in the supplementary materials[3]. We use ABT as the general POMDP solver and MHFR as the linearization-based POMDP solver.

All algorithms are implemented in C++, while all experiments are conducted on Intel Xeon E5-2650 CPUs with 16GB RAM. For the parallel construction of the RRTs in MHFR, we utilize 8 CPU cores throughout the experiments. All parameters are set based on preliminary runs over the possible parameter space, the parameters that generate the best results are then chosen to generate the experimental results. For the comparison between SNM and MoNG, we use a car-like robot with 2nd order control and a 4-DOFs manipulator with torque control. To test the proposed planner, we use these two scenarios plus a scenario involving a 6-DOFs manipulator with torque control. Details of these scenarios are as follows.

Car-like robot with 2nd Order Control. A nonholonomic car-like robot of size (0.12×0.07) drives on a flat xy-plane inside a 3D environment populated by obstacles (Figure 1(a)). The robot must drive from a known start state to a position inside the goal region (marked as a green sphere) without colliding with any of the obstacles. The state of the robot at time t is defined as a 4D vector $s_t = (x_t, y_t, \theta_t, \upsilon_t) \in \mathbb{R}^4$ where $x_t \in [-1, 1]$ and $y_t \in [-1, 1]$ are the position of the center of the robot on the xy-plane, $\theta_t \in [-3.14rad, 3.14rad]$ is the orientation, and $\upsilon_t \in [0, 0.2]$ is the linear velocity of the robot. The initial state of the robot is $(-0.7, -0.7, 1.57rad, 0)$, while goal region is centered at $(0.7, 0.7)$ with radius 0.1. The control input at time t, $a_t = (\alpha_t, \phi_t)$ is a 2D real vector consisting of the acceleration $\alpha \in [0, 1]$ and the steering wheel angle $\phi \in [-1rad, 1rad]$. The robot's dynamics is subject to control noise $v_t = (\tilde{\alpha}_t, \tilde{\phi}_t)^T \sim N(0, \Sigma_v)$. The robot's transition model is

$$s_{t+1} = \left[x_t + \Delta t \upsilon cos\theta_t \; ; \; y_t + \Delta t \upsilon sin\theta_t \; ; \; \theta + \Delta t \; tan(\phi_t + \tilde{\phi})/0.11 \; ; \; \upsilon + \Delta t(\alpha_t + \tilde{\alpha})\right]$$

where Δt is the duration of a time step and the value 0.1 is the distance between the front and rear axles of the wheels.

The robot will enter a terminal state and receive a penalty of -500 if it hits an obstacle. It will also enter a terminal state, but with a reward of $1{,}000$ after reaching a goal region. All other actions incur a cost of -1. The robot localizes itself with the help of a velocity sensor mounted on the car and two beacons (marked with purple square in Figure 1(a)). Suppose the beacons are located at $(\hat{x}_1, \hat{y}_1)$ and $(\hat{x}_2, \hat{y}_2)$. In our experiment, the first beacon is at $(-0.7, 0.7)$ and the second beacon is at $(0.7, -0.7)$ Then, the signals the robot receives from these two beacons is a function of the distance to them, with additive Gaussian noise w_t. More formally, the robot's observation model is:

$$z_t = \left[1/((x_t - \hat{x}_1)^2 + (y_t - \hat{y}_1)^2 + 1); 1/((x_t - \hat{x}_2)^2 + (y_t - \hat{y}_2)^2 + 1); \upsilon_t\right] + w_t$$

4-DOFs and 6-DOFs Manipulator with torque control. We describe these robotic systems in a general manner for a k-DOFs robot. This robot has k rotational joints, with limits at each of their joint angles and velocities, mounted on a static base. The manipulator operates in an environment populated by obstacles, and must move from the initial state to a state where the end-effector lies inside the goal region, without colliding with any of the obstacles. The environment scenarios for the 4-DOFs and 6-DOFs are in Figure 1(b) and Figure 1(c), respectively.

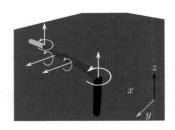

Fig. 2: The configuration of the 4DOFs-manipulator.

A state of the manipulator is defined as $s = (\theta, \dot{\theta}) \in \mathbb{R}^{2k}$, where $\theta \in [-3.14rad, 3.14rad]$ is the vector of joint angles and $\dot{\theta} \in [-3rad, 3rad]$ is the vector of rotational joint velocities. The rotational axes for the 4-DOFs manipulator are presented in Figure 2. For the 6-DOFs manipulator, the rotational axis of the first 4 joints are exactly the same as those of the 4-DOFs manipulator, and the additional two joints rotates around the Z axis. The mass of each link is 0.8kg. The control input $a \in A \subset \mathbb{R}^k$ is the joint torques. The torque limits for the 4-DOFs manipulator are $(\pm 20Nm/s, \pm 20Nm/s, \pm 10Nm/s, \pm 10Nm/s)$, while the torque limits for the 6-DOFs manipulator are $(\pm 20Nm/s, \pm 20Nm/s, \pm 20Nm/s, \pm 10Nm/s, \pm 10Nm/s, \pm 10Nm/s)$. The motion of the robot is disturbed by a k-dimensional error vector $v \sim N(0, \Sigma_v)$. The dynamics of the manipulator are modelled using the well-known Euler-Lagrangian formalism [24]. Note that although the error is Normally distributed, due to the non-linearity of the dynamics, the resulting belief estimate will generally not be Normally distributed. The initial state for both the 4DOFs and the 6DOFs manipulator is a state where all joint angles and joint velocities are zero. When the robot collides with an obstacle or with itself, it will move to a terminal state and receive a penalty of -500. When its end-effector reaches a goal region, the robot will move to a terminal state too, but it will receive a reward of 1000. All other actions incur a cost of -1. The robot is equipped with two types of sensors. The first sensor measures the position of the end-effector in the robot's workspace. The second sensor measures the joint velocities.

Empty environment				Maze scenario			
$e_T = e_Z$	SNM	MoNG	$\frac{V_{ABT}(b_0) - V_{MHFR}(b_0)}{V_{ABT}(b_0)}$	$e_T = e_Z$	SNM	MoNG	$\frac{V_{ABT}(b_0) - V_{MHFR}(b_0)}{V_{ABT}(b_0)}$
1.25	0.205	0.597	0.220	1.25	0.206	0.556	-0.660
2.50	0.200	0.611	0.262	2.50	0.213	0.675	-0.567
3.75	0.278	0.638	0.157	3.75	0.302	0.705	0.369
5.00	0.326	0.679	0.130	5.00	0.393	0.719	8.598

Table 1: The measure computed using SNM and MoNG, and the relative value difference between ABT and MHFR for the car-like robot in an empty environment, and in the maze scenario for increasing e_T and e_Z.

Suppose $g : \mathbb{R}^{2k} \mapsto \mathbb{R}^3$ is a function that maps the state of the robot to an end-effector position in the workspace and $w_t \sim N(0, \Sigma_w)$ is the error vector, then the observation model is defined as $z_t = [g(s_t), \dot{\theta}_t] + w_t$.

6.2 Testing SNM

In this set of experiments, we want to understand the performance of SNM compared to existing non-linearity measures for stochastic systems in various scenarios. In particular, we are interested in the effect that motion and sensing errors have and the effect that obstacles have on the effectiveness of SNM, compared to MoNG.

To this end, we perform experiments on the car-like robot and the 4-DOFs manipulator, with increasing the motion and sensing error of these robotic systems, operating in empty environments and environments populated by obstacles.

For experiments with increasing motion and sensing error, recall Section 6.1 that the control and sensing errors are drawn from zero-mean multivariate Gaussian distributions with covariance matrices Σ_v and diagonal entries $(\sigma_1, ..., \sigma_n)$. We then define relative control error (denoted as e_T) to be the percentage of the value range of the control inputs for each control dimension respectively. The square of the resulting values are then the diagonal entries of Σ_v. The observation error (denoted as e_Z) is defined in a similar fashion.

To investigate how SNM and MoNG perform as the control and sensing errors increase, we run experiments with multiple relative control and observation errors, ranging between 1.25% and 5.0%. To investigate the effect of obstacles to each measure, we ran each robotic system in an empty environment and in the environments as presented in Figure 1(a)-(b). For each scenario and each control-sensing error value (we set $e_T = e_Z$), we ran 100 simulation runs using ABT and MHFR, respectively. Since both ABT and MHFR are on-line planners, in each simulation run, each planner was allowed a planning time of 1s per planning step for the car-like robot, and 2s per planning step for the 4-DOFs manipulator. The average measures and value differences between ABT and MHFR are presented in Table 1 and Table 2.

The results indicate that in all scenarios, both SNM and MoNG are sensitive to an increase in the relative motion and sensing error. This increase generally resonates well with the increase in the difference between the average total discounted reward received if ABT were used and if MHFR were use, except for the case of the car-like robot operating in an empty environment. In this particular scenario, the relative difference between the general and the linearized solver decreases as the motion and sensing errors

Empty environment				Factory-1 scenario			
$e_T = e_Z$	SNM	MoNG	$\frac{V_{ABT}(b_0) - V_{MHFR}(b_0)}{V_{ABT}(b_0)}$	$e_T = e_Z$	SNM	MoNG	$\frac{V_{ABT}(b_0) - V_{MHFR}(b_0)}{V_{ABT}(b_0)}$
1.25	0.099	0.979	0.018	1.25	0.556	0.968	0.389
2.50	0.113	1.026	0.005	2.50	0.529	1.071	0.882
3.75	0.127	1.017	0.082	3.75	0.604	1.094	0.889
5.00	0.193	1.049	0.172	5.00	0.638	1.108	1.239

Table 2: The measure computed using SNM and MoNG, and the relative value difference between ABT and MHFR for the 4-DOFs manipulator in an empty environment, and in the factory-1 scenario for increasing e_T and e_Z.

increase. The reason is the performance of ABT decreases as the errors increase (similar as in the other three scenarios), but the performance of MHFR is almost unaffected. Figure 3(left) presents the plot of the average total reward of ABT and MHFR for the car-like robot operating in an empty environment. The performance of ABT decreases because as the motion and sensing errors increases, more particles are needed to represents the stochastic uncertainty well, which means if planning time per step does not increase, the quality of the generated strategy will decrease. The performance of MHFR is almost unaffected because in terms of computation time, MHFR is almost unaffected, it remains to use only the mean and covariance of the Gaussian distribution for planning. Furthermore, in this scenario, the penalty of making a wrong estimate will only be a longer route. Since the cost of a single action is -1 and due to the discount factor, the increase in the path length have an almost negligible effect on the total discounted reward.

Now, one may question why then the difference in value increases in the case of a 4-DOFs manipulator operating in an empty environment? As the plot in Figure 3(right) indicated, in this scenario, the performance of MHFR degrades

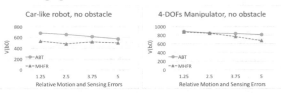

Fig. 3: Mean total discounted reward when no obstacle is present.

as the relative motion and sensing errors increases. The reason is although the environment is empty, a 4-DOFs manipulator may have self-collision, and the increase in the motion and sensing errors causes the robot to be more susceptible to self-collision. Therefore, this scenario produces a similar trend to the test scenarios where obstacles are present (i.e., the maze and factory-1 scenarios).

In terms of sensitivity on the effect of obstacles to the effectiveness of linearization, both Table 1 and Table 2 indicate that SNM is more sensitive than MoNG. Overall, obstacles significantly increase the difference between the average total discounted reward if ABT were run and if MHFR were run. Similar to this trend, SNM shows significant increase in its non-linearity measure when obstacles are introduced. However, the measures computed using MoNG are unaffected by the introduction of obstacles.

6.3   Testing SNM-Planner

In this set of experiments, we want to test the performance of SNM-Planner. To this end, we tested our planner against the two component planners ABT and MHFR for three different scenarios: The maze problem for the car-like robot, a 4-DOFs manipulator, and a 6-DOFs manipulator (as shown in Figure 1). We fixed both e_T and e_Z to 2.5% in these experiments. The SNM-threshold that is being used throughout these experiments is 0.3 for the car-like robot and 0.4 for both manipulator scenarios. The planning time that is being used for each algorithm is 1 sec per step for the car-like robot, 2 sec for the 4-DOFs manipulator and 7 sec for the 6-DOFs manipulator. Note that for SNM-Planner, the planning time per step consists of the time to approximate SNM and the planning time for the individual component planners. We allow a maximum of 20% of the total planning time per step for the approximation of SNM, while the other 80% is used for the component planners.

Planner	Car-like robot	4-DOFs manipulator	6-DOFs manipulator
ABT	128.59 ± 43.59	213.62 ± 64.68	737.23 ± 37.41
MHFR	141.39 ± 75.08	13.41 ± 116.76	265.39 ± 109.95
SNM-Planner	230.30 ± 70.56	382.62 ± 102.79	652.86 ± 69.48

Table 3: Mean total discounted reward +/- 95 % confidence interval over 100 simulation runs. The proportion of using ABT in the car-like robot, 4-DOFs and 6-DOFs manipulator scenarios are 29.53%, 36.13%, and 51.64% of the planning steps, respectively.

The results in Table 3 indicate that SNM-Planner can appropriately identify where to run linearization and where to not run linearization with a small enough cost, such that it can be used to generate motion strategies that are at least comparable to the suitable method.

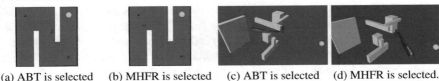

(a) ABT is selected   (b) MHFR is selected   (c) ABT is selected   (d) MHFR is selected.

Fig. 4: Typical situations for the car-like robot and the 6DOFs-manipulator robot for which the SNM is above and below the threshold

It is interesting to note that ABT is often selected when the robot operates in the vicinity of the obstacles. Figure 4 illustrates typical beliefs where ABT is selected (in the car-like robot and 6-DOFs manipulator scenarios). When the robot operates in the vicinity of an obstacle, SNM is usually larger than in free areas. In these situations, where careful planning is mandatory in order to avoid collisions, SNM-Planner prefers to use ABT, while in free areas where a more coarse planning would suffice, SNM-Planner prefers to use MHFR.

A critical aspect in SNM-Planner is how well can we approximate SNM in a limited time (i.e., 0.2s, 0.4s, and 1.4s for the car-like robot, 4-DOFs manipulator, and 6-DOFs manipulator, respectively). To understand this issue better, we tested the convergence rate of SNM in the car-like robot and the 4-DOFs manipulator scenario with various motion and sensing errors. For this purpose, we generate a trajectory for the maze and factory-1 scenario, and for each belief in the trajectory, we perform 100 independent Monte Carlo runs for different time limits to estimate the local value of SNM around the belief. The average of these estimates when we use the various time limits are presented in Figure 5.

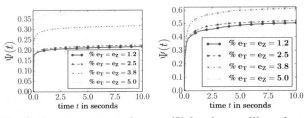

Fig. 5: Convergence to the true Ψ for the car-like robot (left) and 4DOFs-manipulator (right).

As expected, the results indicate that the size of the state, action, and observation space significantly influence the convergence rate. The car-like robot can converge to a good estimate faster than the 4-DOFs robot. It is interesting to note that the motion and sensing errors have little to no effect on the convergence of our method. These results also indicate that the estimate we use for testing SNM-Planner would have been a reasonable estimate, though not perfect, even though it takes very little time.

7   Summary and Future Work

This paper presents our preliminary work in identifying the suitability of linearization for motion planning under uncertainty. To this end, we present a general measure of non-linearity, called Statistical-distance-based Non-linearity Measure (SNM), which is based on the distance between the distributions that represent the system's motion–sensing model and its linearized version. Comparison studies with one of state-of-the-art methods for non-linearity measure indicate that SNM is more suitable in taking into account obstacles in measuring the effectiveness of linearization.

We also propose a simple on-line planner that uses a local estimate of SNM to select whether to use a general POMDP solver or a linearization-based solver for robot motion planning under uncertainty. Experimental results indicate that our simple planner can appropriately decide where linearization should be used and generates motion strategies that are comparable or better than each of the component planner.

Future work abounds. For instance, the question for a better measure remains. Total variation distance relies on computing a maximization, which is often difficult to estimate. Statistical distance function that relies on expectation exists and can be computed faster. How suitable are these functions as a non-linearity measure? Furthermore, our upper bound result is relatively loose and can only be applied as a sufficient condition to identify if linearization will perform well. It would be useful to find a tighter bound that remains general enough for the various linearization and distribution approximation methods in robotics.

References

1. Duník, J., Straka, O., Šimandl, M.: Nonlinearity and non-gaussianity measures for stochastic dynamic systems. In: Information Fusion (FUSION), IEEE (2013) 204–211
2. Kurniawati, H., Yadav, V.: An online POMDP solver for uncertainty planning in dynamic environment. In: ISRR. (2013)
3. Sun, W., Patil, S., Alterovitz, R.: High-frequency replanning under uncertainty using parallel sampling-based motion planning. IEEE Transactions on Robotics **31**(1) (2015) 104–116
4. Canny, J., Reif, J.: New lower bound techniques for robot motion planning problems. In: Foundations of Computer Science, 1987., 28th Annual Symposium on, IEEE (1987) 49–60
5. Natarajan, B.: The complexity of fine motion planning. The International journal of robotics research **7**(2) (1988) 36–42

6. Kaelbling, L., Littman, M., Cassandra, A.: Planning and acting in partially observable stochastic domains. AI **101** (1998) 99–134
7. Drake, A.W.: Observation of a Markov process through a noisy channel. PhD thesis, Massachusetts Institute of Technology (1962)
8. Horowitz, M., Burdick, J.: Interactive Non-Prehensile Manipulation for Grasping Via POMDPs. In: ICRA. (2013)
9. Temizer, S., Kochenderfer, M., Kaelbling, L., Lozano-Pérez, T., Kuchar, J.: Unmanned aircraft collision avoidance using partially observable markov decision processes. Project Report ATC-356, MIT Lincoln Laboratory, Advanced Concepts Program, Lexington, Massachusetts, USA (September 2009)
10. Silver, D., Veness, J.: Monte-Carlo Planning in Large POMDPs. In: NIPS. (2010)
11. Somani, A., Ye, N., Hsu, D., Lee, W.S.: DESPOT: Online POMDP planning with regularization. In: NIPS. (2013) 1772–1780
12. Seiler, K., Kurniawati, H., Singh, S.: An online and approximate solver for pomdps with continuous action space. In: ICRA. (2015)
13. Agha-Mohammadi, A.A., Chakravorty, S., Amato, N.M.: Firm: Sampling-based feedback motion planning under motion uncertainty and imperfect measurements. IJRR (2013)
14. Berg, J., Abbeel, P., Goldberg, K.: LQG-MP: Optimized Path Planning for Robots with Motion Uncertainty and Imperfect State Information. In: RSS. (2010)
15. Berg, J., Wilkie, D., Guy, S., Niethammer, M., Manocha, D.: LQG-Obstacles: Feedback Control with Collision Avoidance for Mobile Robots with Motion and Sensing Uncertainty. In: ICRA. (2012)
16. Prentice, S., Roy, N.: The belief roadmap: Efficient planning in linear pomdps by factoring the covariance. In: Robotics Research. Springer (2010) 293–305
17. Li, X.R.: Measure of nonlinearity for stochastic systems. In: Information Fusion (FUSION), 2012 15th International Conference on, IEEE (2012) 1073–1080
18. Bates, D.M., Watts, D.G.: Relative curvature measures of nonlinearity. Journal of the Royal Statistical Society. Series B (Methodological) (1980) 1–25
19. Beale, E.: Confidence regions in non-linear estimation. Journal of the Royal Statistical Society. Series B (Methodological) (1960) 41–88
20. Emancipator, K., Kroll, M.H.: A quantitative measure of nonlinearity. Clinical chemistry **39**(5) (1993) 766–772
21. Mastin, A., Jaillet, P.: Loss bounds for uncertain transition probabilities in markov decision processes. In: CDC, IEEE (2012) 6708–6715
22. Müller, A.: How does the value function of a markov decision process depend on the transition probabilities? Mathematics of Operations Research **22**(4) (1997) 872–885
23. Arulampalam, M.S., Maskell, S., Gordon, N., Clapp, T.: A tutorial on particle filters for online nonlinear/non-gaussian bayesian tracking. IEEE Transactions on signal processing **50**(2) (2002) 174–188
24. Spong, M.W., Hutchinson, S., Vidyasagar, M.: Robot Modeling and Control. Volume 3. Wiley New York (2006)
25. Kurniawati, H., Patrikalakis, N.: Point-Based Policy Transformation: Adapting Policy to Changing POMDP Models. In: WAFR. (2012)
26. Gibbs, A.L., Su, F.E.: On choosing and bounding probability metrics. International statistical review **70**(3) (2002) 419–435
27. Lavalle, S.M., Kuffner Jr, J.J.: Rapidly-exploring random trees: Progress and prospects. In: Algorithmic and Computational Robotics: New Directions, Citeseer (2000)

Motion Planning for Active Data Association and Localization in Non-Gaussian Belief Spaces

Saurav Agarwal, Amirhossein Tamjidi, and Suman Chakravorty

Dept. of Aerospace Engineering,
Texas A&M University,
sauravag,ahtamjidi,schakrav@tamu.edu

Abstract. This paper presents a method for motion planning under uncertainty to resolve situations where ambiguous data associations result in a multimodal hypothesis on the robot state. Simultaneous localization and planning for a lost (or kidnapped) robot requires that given little to no a priori pose information, a planner should generate actions such that future observations allow the localization algorithm to recover the correct pose of a mobile robot with respect to a global reference frame. We present a Receding Horizon approach, to plan actions that sequentially disambiguate a multimodal belief to achieve tight localization on the correct pose in finite time. In our method, disambiguation is achieved through active data associations by picking target states in the map which allow distinctive information to be observed for each belief mode and creating local feedback controllers to visit the targets. Experimental results are presented for a kidnapped physical ground robot operating in an artificial maze-like environment.

1   Introduction

In practical mobile robot motion planning problems, situations may arise where data association between what is observed and the robot's map leads to a multimodal hypothesis on the state, for example a kidnapped robot with no a priori information or a mobile robot operating in a symmetric environment (see Fig. 1). Figure 1 depicts a problem wherein belief (the probability distribution over all possible robot states) modes are widely separated in an environment with symmetry. In such cases if a robot begins with an equal likelihood on all hypothesis, it is difficult to ascertain the true hypothesis as local sensing may result in identical information for all belief modes. Thus in practice a robot often has to seek information that helps to disambiguate its belief.

Simply relying on randomized actions to correctly recover robot pose is known to be unreliable and inefficient in practice [1]. Further, existing methods to disambiguate multimodal hypothesis [1–3] rely on heuristics-based strategies (e.g., picking random targets, wall following etc.) to seek disambiguating information. As opposed to [1–3], our approach disambiguates, i.e., rejects incorrect hypothesis in a multimodal belief by actively seeking maximally disambiguating information in the map for each mode, and recovers the robot pose with a higher certainty threshold than current state-of-the-art.

© Springer Nature Switzerland AG 2020
K. Goldberg et al. (Eds.): *Algorithmic Foundations of Robotics XII*, SPAR 13, pp. 288–303, 2020.
https://doi.org/10.1007/978-3-030-43089-4_19

Our Multi-Modal Motion Planner (M3P) achieves disambiguation in a multimodal belief by first finding a neighboring location (referred to as target state) for each belief mode and then creating a candidate action to guide the belief mode to its target state such that these actions lead to information gathering behavior. The target states are chosen such that different modes of the robot's belief are expected to observe distinctive information at the target locations, thus accepting or rejecting hypotheses in the belief. We represent a multimodal hypothesis with a Gaussian Mixture Model (GMM) and use an Extended Kalman filter (EKF) based Multi-Hypothesis Tracking (MHT) approach to propagate the belief [2, 4, 5]. The main contributions of this work can be summarized as follows; (i) we develop a novel method for picking target states and creating candidate trajectories for a multimodal belief, our method then chooses a candidate trajectory such that maximum disambiguating information is observed which helps in rejecting incorrect hypotheses, (ii) we prove that under certain realistic assumptions, through a process of iterative hypothesis elimination, our method can localize to the true robot pose, (iii) we demonstrate an application in which a kidnapped ground robot is tasked to recover its pose.

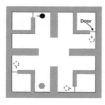

Fig. 1: A scenario depicting a multi-hypothesis localization problem with widely separated modes in a world with 4 rooms with identical doors. The true hypothesis is depicted by the solid black disk, whereas others are depicted by dashed circles. As the robot cannot distinguish between the doors, all hypotheses are equally likely.

We motivate this work with the kidnapped robot scenario since it is one of the hardest localization problems and measures the ability of an algorithm to recover from global localization failures [6]. However, the method proposed is general and can be extended to any planning situation where a multimodal belief arises in the robot state due to ambiguous data associations (a common practical issue in robot localization [6]). In the proceeding section, we present relevant related work, and discuss how our approach compares with them. In Section 3 we state some preliminaries followed by the problem description. In Section 4 we present our method followed by experimental results in Section 5.

2   Related Work

Recent work in sampling-based methods for belief space planning has shown promising results. Gaussian (unimodal) belief space planning methods such as [7–11] provide solutions that depend on the initial belief. Recent developments in [12, 13] extend Gaussian belief space planning to multi-query settings (cases where multiple planning requests are made sequentially) by creating a belief space variant of a Probabilistic RoadMap (PRM) [14]. We note that the aforementioned methods assume that data associations between observations and information sources (e.g., landmarks) are known and unambiguous. In contrast, we do not assume that data associations are unambiguous or that belief is unimodal. In our problem ambiguous data associations lead to a multimodal belief where the modes are widely separated (see Fig. 1), this violates the underlying Gaussian unimodal belief assumption in previously mentioned methods.

Recent work in [15, 16] extends belief space planning to non-Gaussian beliefs where the belief modes are not widely separated. The authors investigate a grasping problem with a multimodal hypothesis on the gripper's state. Their method picks the most-likely hypothesis and a fixed number of samples from the belief distribution, then using an RHC approach, belief space trajectories are found that maximize the observation gap between the most-likely hypothesis and the drawn samples, which helps to accept or reject the most-likely hypothesis. The method in [17] builds upon the work in [15] wherein the author transposes the non-convex trajectory planning problem in belief space to a convex problem. Among other recent works, [18] reduces the computational complexity of planning for a non-Gaussian hypothesis but also assumes distributions without widely separated modes. Compared to [15–18], our method is better suited to deal with more severe cases of non-Gaussian belief space planning such as the kidnapped robot scenario. Such scenarios may not be possible to address using the trajectory optimization based techniques of [15–18] in their current form, due to the difficulty of generating an initial feasible plan for the widely separated modes in the presence of obstacles (as shown in Fig. 1).

To the extent of our knowledge, a limited number of methods approach the problem of recovering global robot pose for a mobile robot with an initial multimodal hypothesis. The analysis in [19] showed that finding the optimal (shortest) plan to re-localize a robot with multiple hypotheses in a deterministic setting (no sensing or motion uncertainty) is NP-hard. At best a greedy localization strategy can be developed whose plan length is upper bounded by a factor of the optimal plan. In the localization strategy of [19] reference points are chosen in the environment at which observations may lead to disambiguation, the robot is then driven the minimum distance over all active hypothesis-reference point combinations to make a perfect range-scan observation with infinite range (not available in practice). Section 5 presents a discussion on differences of [19] from M3P that highlight why [19] may not work well for a physical robot. In a symmetric environment, [20] showed that for a robot equipped with only perfect odometery, no sequence of actions can disambiguate a pair of symmetric configurations. In [1], the authors develop an active localization method in a grid based scheme for a known map. Their planning method considers arbitrary targets in the robot's local coordinate frame as atomic actions (e.g., move 1m right and 4m forward). The optimal candidate action is selected based on the path cost and the expected decrease in entropy at the target. Compared to [1], our target selection methodology (Section 4.2) is active, i.e., M3P uses the a priori map information to select targets such that by visiting them, belief modes expect to see maximally disambiguating information (e.g., seeing a landmark with a distinctive appearance can immediately confirm or reject a hypothesis, see Fig. 2). In [2], the authors present a greedy heuristic-based planning strategy to disambiguate a multimodal hypothesis for a kidnapped robot. The method of [3] plans safe trajectories by picking a point in the vicinity of obstacles to disambiguate the hypothesis. Compared to [2, 3], we present a planning approach that explicitly reasons about belief evolution as a result of actions in the planning stage and picks the optimal policy among a set of candidates.

3   Preliminaries and Problem

Let C be the configuration space and $C_{free} \subset C$ be the set of collision free configurations. Let $x_k \in \mathbb{X}$, $u_k \in \mathbb{U}$, and $z_k \in \mathbb{Z}$ represent the system state, control input, and observation at time step k respectively. $\mathbb{X}$, $\mathbb{U}$, and $\mathbb{Z}$ denote the state, control, and observation spaces respectively. It should be noted that in our work, the state x_k refers to the state of the mobile robot, i.e., we do not model the environment and obstacles in it as part of the state. The non-linear state evolution model f and measurement model h are denoted as $x_{k+1} = f(x_k, u_k, w_k)$ and $z_k = h(x_k, v_k)$, where $w_k \sim \mathcal{N}(0, Q_k)$ and $v_k \sim \mathcal{N}(0, R_k)$ are zero-mean Gaussian process and measurement noise, respectively. The belief b_k at time t_k can be represented by a Gaussian Mixture Model (GMM) as a weighted linear summation over Gaussian densities. Let $w_{i,k}$, $\mu_{i,k}$ and $\Sigma_{i,k}$ be the weight, mean vector, and covariance matrix associated to the ith Gaussian $m_{i,k}$ respectively at time t_k, then $b_k = \sum_{i=1}^{M_k} w_{i,k} m_{i,k}$, $m_{i,k} \sim \mathcal{N}(\mu_{i,k}, \Sigma_{i,k})$, where M_k is the number of modes at time t_k. We state our problem as follows:

Given an a priori map, system dynamics and observation models, construct a belief space planner $G(b_k)$ such that under the planner G, given an initial multimodal belief b_0, the sequence of future observations allow a robot to localize about its true pose.

Note that there may be degenerate cases, where the map may not allow actions that lead to hypothesis elimination such that the belief converges to a unimodal distribution (e.g., in a map with two identical closed rooms, if a robot is kidnapped and placed in either room, it cannot distinguish which room it is in). In such cases, M3P attempts to minimize the number of modes M_k (by design), but it is not possible to pre-compute what this minimum value of M_k is without explicit knowledge of the true hypothesis [19] in a multimodal belief.

4   Methodology

We begin by defining certain key concepts used in the M3P planner.

Uniqueness Graph: A graph U_g, whose nodes are states sampled from the collision free space C_{free} and whose edges relate the similarity of information observed at the sampled locations.

Target State: A target state $v_i^{tt} \in U_g$ for mode m_i is a node of the uniqueness graph which belongs to some neighborhood of radius R of the mode's mean μ_i such that if each mode were to visit its target, the observations at the target would lead to disambiguation in the belief.

Candidate Policy: A candidate policy π_i for mode m_i is a local feedback controller that guides the mode to its target v_i^{tt}.

The M3P methodology has two phases, an offline phase in which we generate U_g and an online phase in which we use the offline computations and plan in a receding horizon manner to disambiguate the belief.

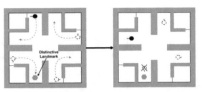 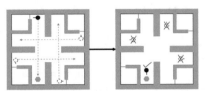

(a) Candidate A leads to negative information for the mode in lower left corner. It expects to see the distinctive landmark which robot doesn't observe, and is thus rejected.

(b) Candidate B leads the true hypothesis to be confirmed as the robot sees the distinctive landmark.

Fig. 2: Extending the example in Fig. 1, we depict 2 candidate trajectories; candidates A & B and the effect of their execution. Candidate A leads to one disambiguation and candidate B results in complete disambiguation. Candidate B is a better choice, however the difficulty of picking B lies in the fact that robot does not know its true hypothesis a priori.

4.1   Computing the Uniqueness Graph: Offline Phase

The uniqueness graph U_g is constructed by uniformly sampling the configuration space and adding these samples as nodes of U_g. Once a node is added, we simulate the observation for the state represented by that node. Let v_α be one such node and z^{v_α} be the observation if the robot were to be in state v_α. We add an edge $E_{\alpha\beta}$ (undirected) between two nodes v_α and v_β if the simulated observations at both nodes are similar. Further, the edges are weighted and the weight is dependent on the similarity in information observed, i.e., for edge $E_{\alpha\beta}$ the weight $\omega_{\alpha\beta} = \tau(z^{v_\alpha}, z^{v_\beta})$ where $\tau : \mathbb{Z} \times \mathbb{Z} \to \mathbb{R}$ computes a measure of similarity between two observations. Note that the form of τ is general and can be changed to suit the problem domain (perception model). Figure 3 explains this concept visually for a landmark based observation model, where each landmark has some discrete signature (identifier) that a robot can detect. In Fig. 3 state v_α observes z^{v_α} with signatures $^s z^{v_\alpha} = \{s_1, s_2, s_3\}$, i.e., the landmarks with signature s_1, s_2 and s_3 and at v_β observes $^s z^{v_\beta} = \{s_1, s_2, s_4\}$, the edge weight $\omega_{\alpha\beta}$ for edge $E_{\alpha\beta}$ is $\omega_{\alpha\beta} = \tau(z^{v_\alpha}, z^{v_\beta}) = |^s z^{v_\alpha} \cap {}^s z^{v_\beta}| = |\{s_1, s_2\}| = 2$. A higher edge weight signifies that the states represented by the vertices of that edge are more likely to observe similar information. The lack of an edge between two nodes means that if a robot were to make an observation at those two states, it would see distinctly different information.

The complexity for the construction of U_g is $\mathcal{O}(n^2)$ (where n is the number of samples) as each sample (node) is checked with every other for information overlap. Due to its random nature, sampling may often occur in regions of low information density (e.g., regions where there are few or no landmarks). One can often circumvent this issue by increasing the number of samples. As U_g is computed offline, the online performance is not significantly affected. Recent work in [21] suggests a localization aware sampling strategy which may be explored in future work.

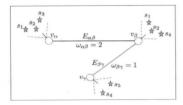

Fig. 3: Simple example of a uniquess graph with 3 nodes $\{v_\alpha, v_\beta, v_\gamma\}$ and 2 edges $\{E_{\alpha\beta}, E_{\beta\gamma}\}$. The nodes v_α and v_γ do not see any similar landmark hence there is no edge between them. Here $\tau(z^{v_i}, z^{v_j}) = |{}^s z^{v_i} \cap {}^s z^{v_j}|$ for $i, j \in \{\alpha, \beta, \gamma\}$.

4.2 RHC based Planning: Online Phase

In a multimodal scenario, we claim that the best action to take is one that guides a robot without collision through a path that results in information gain such that a disambiguation occurs (one or more hypotheses are rejected, see Fig. 2). Algorithm 1 describes the online planning process. In step 3, the planner picks target states for each belief mode such that visiting a target can either prove or disprove the hypothesis. In step 4, the planner generates a set of candidate policies to drive each mode to its target. In step 5, the expected information gain for each policy is computed and we pick the best one, and in step 7, the multimodal belief is propagated according to the action and observations. We proceed to describe steps 3, 4, 5 and 7 of Algorithm 1 below.

Picking the target state for a mode Algorithm 2 describes in detail the steps involved to pick a target state for a belief mode. Let us pick a mode $m_{i,k} \sim \mathcal{N}(\mu_{i,k}, \Sigma_{i,k})$ from the belief. To find the target $v_{i,k}^{tt}$ for $m_{i,k}$, we first choose the set of nodes $N_{i,k} \in U_g$ (Section 4.1) which belong to the neighborhood of the mean $\mu_{i,k}$ at time t_k (steps 3 and 4, Alg. 2). Then, we find the target node $v_{i,k}^{tt} \in N_{i,k}$ which observes information that is least similar in appearance to that observed by nodes in the neighborhoods $N_{j,k}$ of all other modes $m_{j,k}$ where $j \neq i$. To do this, after computing $N_{i,k}$, we calculate the total weight of the outgoing edges from every node $v_{i,k} \in N_{i,k}$ to nodes in all other neighborhoods $N_{j,k}$ where $j \neq i$ (steps 7-13, Alg. 2). The node which has the smallest outgoing edge weight (steps 14-16, Alg. 2), is the target candidate $v_{i,k}^{tt}$ for $m_{i,k}$ as the observation $z^{v_{i,k}^{tt}}$

Algorithm 1: M3P: MultiModal Motion Planner

---

1 Input: b
2 **while** $b \neq \mathcal{N}(\mu, \Sigma)$ **do**
3     $\{v^{tt}\} \leftarrow$ Pick target states for belief modes (see Alg. 2);
4     $\Pi \leftarrow$ Generate candidate policies to connect each mode to its target;
5     $\pi^* \leftarrow$ Pick optimal policy from Π;
6   **forall** $u \in \pi^*$ **do**
7        $b \leftarrow$ Apply action u and update belief (see Alg. 3 for weight update calculation);
8   **if** *Change in number of modes or Expect a belief mode to violate constraints* **then**
9   break;

10 **return** b;

---

would be least similar to the information observed in the neighborhood of all other modes m_j where $j \neq i$.

Algorithm 2: Finding the target for i-th mode

---

1 Input: b_k, i , U_g
2 Output: $v_{i,k}^{tt}$
3 **forall** $l \in [1, M_k]$ **do**
4     $N_{l,k} \leftarrow$ Find neighborhood nodes for $\mu_{l,k}$ in U_g;

5 $minWeight \leftarrow$ Arbitrarily large value;
6 $v_{i,k}^{tt} \leftarrow -1$;
7 **forall** $v \in N_{i,k}$ **do**
8     $w \leftarrow 0$;
9   **for** $N_{j,k} \in \{N_{1,k}, \ldots, N_{M_k,k}\} \setminus N_{i,k}$ **do**
10   **forall** $e \in Edges$ *connected to* v **do**
11   **forall** $p \in N_{j,k}$ **do**
12   **if** *p is a target of edge e* **then**
13                $w \leftarrow w + \texttt{edgeWeight}(e)$;

14   **if** $w < minWeight$ **then**
15        $minWeight \leftarrow w$;
16        $v_{i,k}^{tt} \leftarrow v$;

17 **return** $v_{i,k}^{tt}$;

Generating candidate policies for belief modes Once the targets corresponding to each mode have been picked, we need to find the control action that can take a mode from its current state to the target state. We generate the candidate trajectory that takes each mode to its target using the RRT* planner [22]. Once an open loop trajectory is computed, we generate a local policy π_i (feedback controller) for the i-th mode, which drives the i-th mode along this trajectory. Let Π be the set of all such policies for the different modes.

Picking the Optimal Policy After generating the set Π of candidate policies, we evaluate the expected information gain ΔI_i for each policy π_i and pick the optimal policy π^* that maximizes this information gain. We model this information gain as the discrete change in the number of modes. To compute the expected change in the number of belief modes, we simulate the most-likely belief trajectory, i.e., approximating noisy observations and actions with their most-likely values [11, 23–25]. The steps to calculate the expected information gain for a candidate policy $\pi_i \in \Pi$ are as follows:

1. For every belief mode $m_{j,k} \in b_k$.
   (a) Assume that robot is at $m_{j,k}$.
   (b) Simulate π_i and propagate all the modes.
   (c) Compute information gain $\Delta I_{i,m_{j,k}}$ for π_i.
2. Compute the weighted information gain $\Delta I_i = \sum_{j=1}^{M_k} w_{j,k} \Delta I_{i,m_{j,k}}$.

After computing the expected information gain for each policy, we pick the gain maximizing policy. The computational complexity of this step is $\mathcal{O}(M_k^3 L_{max})$ (where M_k is the number of belief modes and L_{max} is the maximum candidate trajectory length). This is due to the fact that each policy is simulated for each mode for the length of policy, where at every step of policy execution, there are M_k filter updates.

Belief Propagation Using GMM We first discuss our decision to use EKF based MHT over a particle filtering approach. In practical localization problems, a relatively small number of Gaussian hypotheses are sufficient for maintaining the posterior over the robot state, secondly the filtering complexity grows linearly in the number of hypotheses and finally due to the computational complexity of picking the optimal policy (see previous section), the number of samples required for a particle filter would make re-planning significantly harder.

Now, we proceed to describe the belief update step which propagates each mode's mean and covariance and determines how likely each mode is in the belief. The mean and covariance of each belief mode is updated with its own Kalman filter, i.e., a parallel bank of Kalman filters is used. Each filter undergoes an identical prediction update and the measurement update is a function of the data association between robot's observation and most-likely observation for belief mode. Note that during belief propagation our estimate remains conservative in that covariance is not truncated when passing through obstacles. In this regard [26] provides a direction for future enhancements; the method of [26] truncates estimation covariance by first applying an affine transformation to the 2D Gaussian and then truncating the 1D distribution according to given linear inequality constraints. The weights $w_{i,k}$'s are updated based on the measurement likelihood function as

$$w_{i,k+1} = w_{i,k}e^{-\frac{1}{2}D^2_{i,k+1}}, \tag{1}$$

where $D_{i,k+1}$ is the Mahalanobis distance between the sensor observation and most-likely observation for mode m_i such that

$$D^2_{i,k+1} = (z_{k+1} - h(\mu_{i,k+1},0))^T R_k^{-1}(z_{k+1} - h(\mu_{i,k+1},0)). \tag{2}$$

The weights are normalized such that $\sum_{i=1}^{M_k} w_{i,k+1} = 1$.

A known issue with EKF-based MHT is that a naive implementation is unable to process negative information [6]. Negative information refers to the lack of information which one may expect to see and can certainly help in disproving a hypothesis (see Fig. 2(a)). We now proceed to describe how negative information is factored into the weight update.

Factoring Negative Information: Depending on the state of the robot, individual hypotheses and data association results, we might have several cases. We discuss this issue in the context of a landmark based measurement model. At time t_{k+1}, let $n_{z_{k+1}}$ be the number of landmarks observed by the robot and $n_{z^p_{i,k+1}}$ be the number of landmarks that we predict to see for m_i where $z^p_{i,k+1} = h(\mu_{i,k+1},0)$ is the predicted observation. Then $n_{z_{k+1}} = n_{z^p_{i,k+1}}$ means that the i-th mode expected to see as many landmarks as the robot observed; $n_{z_{k+1}} > n_{z^p_{i,k+1}}$ implies the robot observes more landmarks than predicted for the mode; $n_{z_{k+1}} < n_{z^p_{i,k+1}}$ implies the robot observes less landmarks than predicted for the mode. Also, we can have the number of data associations to be less than the number of predicted or measured observations or both. This means that we may not be able to make a unique association between each predicted and observed landmark. At time t_{k+1}, we estimate the Mahalanobis distance $D_{i,k+1}$ (Eq. 2) for mode m_i between the predicted and observed landmarks that are matched by the data association module and update weight according to Eq. 1. Then we multiply the updated weight by a factor γ, which models the effect of duration $\beta_{i,k+1}$ for which the robot observes different landmarks than the i-th mode's prediction; and the discrepancy α in the number of data associations. When a belief mode is initialized, we set $\beta_{i,0} = 0$. The weight update procedure is described in Algorithm 3. After each weight update step, we remove modes with negligible contribution to the belief, i.e., when $w_{i,k+1} \leq \delta_w$ where δ_w is user defined.

4.3   Analysis

In this section, we show that under certain assumptions on the structure of the environment, the receding horizon planner M3P can guarantee that an initial multimodal belief is driven into a unimodal belief in finite time. We now proceed to state our assumptions.

Assumption 1 *For every mode m_i, the environment allows for the existence of some target state v_i^{tt} and some homotopy class of paths through which the robot can visit v_i^{tt} if the robot is actually at mode m_i.*

Algorithm 3: GMM Weight Update

---

1 Input: $w_{i,k}, \mu_{i,k+1}, \beta_{i,k}, \delta t$

2 Output: $w_{i,k+1}, \beta_{i,k+1}$

3 $z_{k+1}, n_{z_{k+1}} \leftarrow$ Get sensor observations;

4 $z^p_{i,k+1}, n_{z^p_{i,k+1}} \leftarrow$ Get predicted observations for $\mu_{i,k+1}$;

5 $n_{z_{k+1} \cap z^p_{i,k+1}} \leftarrow$ Do data association;

6 $w'_{i,k+1} \leftarrow$ Update and normalize weight according to likelihood function;

7 $\gamma \leftarrow 1$;

8 **if** $n_{z^p_{i,k+1}} \neq n_{z_{k+1}}$ or $n_{z^p_{i,k+1}} \neq n_{z_{k+1} \cap z^p_{i,k+1}}$ **then**

9 $\quad \alpha \leftarrow max(1 + n_{z_{k+1}} - n_{z_{k+1} \cap z^p_{i,k+1}}, 1 + n_{z^p_{i,k+1}} - n_{z_{k+1} \cap z^p_{i,k+1}})$;

10 $\quad \beta_{i,k+1} \leftarrow \beta_{i,k} + \delta t$;

11 $\quad \gamma \leftarrow e^{-\alpha \beta_{i,k+1} 10^{-4}}$;

12 **else**

13 $\quad \beta_{i,k+1} \leftarrow 0$;

14 $w_{i,k+1} \leftarrow w'_{i,k+1} \gamma$;

15 **return** $w_{i,k+1}, \beta_{i,k+1}$;

Assumption 2 *If the robot is actually at mode m_i, and its associated target state is v^{tt}_i, let $B_r(v^{tt}_i)$ to be a neighborhood of radius $r > 0$ centered at the target v^{tt}_i such that if robot state $x \in B_r(v^{tt}_i)$ (robot in vicinity of target), exteroceptive observations can confirm that m_i is the true hypothesis.*

Due to the uncertain nature of the actuation and sensing process, the existence of a path to visit a target location does not guarantee that a robot can drive its belief along this path or that on reaching neighborhood $B_r(v^{tt}_i)$, localization uncertainty will be sufficiently low so as to make a disambiguating data association. Let the true belief be mode m_i. Let $F \subset C \setminus C_{free}$ be the set of failure states, and let L be the finite stopping time for policy π_i defined as the time at which collision occurs or the belief mean μ_i reaches the neighborhood $B_r(v^{tt}_i)$. Denote $P^{\pi_i}(x_L \in F | m_i)$ as the probability that policy π_i drives the underlying state x into a collision given the initial belief is m_i.

Assumption 3 *Given Assumption 1, let mode $m_i \sim \mathcal{N}(\mu_i, \Sigma_i)$, with $||\Sigma_i|| < \bar{P} < \infty$ (initial covariance is bounded) be the true hypothesis. We assume that under the feedback policy π_i, the failure probability $P^{\pi_i}(x_L \in F | m_i)$ is sufficiently low such that we can drive the robot state x into the neighborhood $B_r(v^{tt}_i)$ with a high probability $\int_{B_r(v^{tt}_i)} p^{\pi_i}(x_L | m_i, \neg F) dx > 1 - \delta$ for any $\delta > 0$ where $p^{\pi_i}(x_L | m_i, \neg F)$ is the terminal pdf on the state under policy π_i when the robot does not collide.*

Assumption 4 *The environment (world) in which the robot operates is static.*

Proposition 1. *Under Assumptions 1, 2, 3 and 4, given any initial multimodal belief $b_0 = \sum_i w_{i,0} m_{i,0}$, the receding horizon planner M3P drives the belief process into a unimodal belief $b_T = m_T \approx \mathcal{N}(\mu_T, \Sigma_T)$ in some finite time T.*

Proof. Given an initial belief b_0, let π_{i^*}, i.e., candidate policy for mode m_{i^*}, be the one that results in most information gain as required by M3P. We have only two possibilities; (i) Case 1: Mode m_{i^*} is the true hypothesis, or (ii) Case 2: Mode m_{i^*} is *not* the true hypothesis. If case 1 is true, due to Assumptions 1, 2 and 3, M3P can confirm that m_{i^*} is the true hypothesis by visiting the target location and rejecting all other hypotheses in the process (see Fig. 2(b)). If case 2 is true then the robot is at some other mode m_j where $j \neq i^*$. In case 2, as policy π_{i^*} is executed, two situations can arise, either (i) π_{i^*} is executed fully in which case m_{i^*} will expect to see distinctive information at its target location which the robot will not observe, leading to a disambiguation immediately due to negative information (see Fig. 2(a)) or (ii) the policy π_{i^*} becomes unfeasible at some point of its execution in which case we immediately know that the robot is not at mode i^* since we know that the map did not change during the execution of π_{i^*} (Assumption 4) and thus, there is a disambiguation whereby mode i^* is discarded. Thus we see that either π_{i^*} confirms the true hypothesis or the number of modes is reduced by at least one. After this disambiguation, we restart the process as before and we are assured that at least one of the modes is going to be disambiguated and so on. Thus, it follows given that we had a finite number of modes to start with, the belief eventually converges to a unimodal belief. Further, since each of the disambiguation epochs takes finite time, a finite number of such epochs also takes a finite time, thereby proving the result.

Remarks: The above result shows that the M3P algorithm will stabilize the belief process to a unimodal belief under Assumptions 1, 2, 3 and 4. In the case that Assumption 1 is violated we are either (i) unable to find a target which allows the robot to observe distinctive information (e.g., trivial case of a robot operating in a world with identical infinite corridors) or (ii) we may find such a target but the environment geometry does not allow for any path to visit it (e.g., robot stuck in one of many identical rooms and the doors are closed). These violations refer to degenerate cases that rarely occur in practical motion planning problems. Assumptions 2 and 3 can be violated when all candidate trajectories pass through regions lacking enough information, either because the region is unknown or featureless. In such a case the localization uncertainty on each mode may grow so high that we cannot make data associations at the target location to disambiguate the multimodal belief. Thus these two assumptions imply that the known map has enough information sources (see Fig. 4). Handling the issue of maps that are either unknown, partially known or sparse in information sources is beyond the scope of this paper and presents an important direction for future research. Assumption 4 (static world) is common in localization literature, though it may be violated in certain scenarios. In such cases, if the map is not changing rapidly, one may use sensory observations to incorporate new constraints into the map and trigger replanning.

5   Experimental Results

We present experimental results for two motion planning scenarios wherein the robot is placed randomly at a location in an environment which is identical to

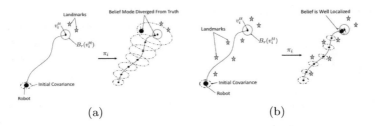

(a)   (b)

Fig. 4: Evolution of the true belief mode in environments with and without sufficient information. (a) No landmarks present along the candidate trajectory,
leading to high uncertainty at the end. The belief mode has diverged from the
robot pose and it is no more possible to make an accurate data association for the
landmarks at the target. (b) Sufficient information along the candidate trajectory leads the belief mode to be well localized at the end, allowing unambiguous
data association for the landmarks at the target.

other locations in appearance[1]. Thus the initial belief is multimodal, the goal of
the experiment is to use the non-Gaussian planner M3P described in Section 4
to localize the robot pose. We first describe the system setup to motivate the
experiment followed by the results.

5.1   System Description

We used a low-cost Arduino based differential drive robot equipped with an
Odroid U3 computer running ROS on Ubuntu 14.04 and an off-the-shelf Logitech
C-310 webcam for sensing. The onboard computer uses a wifi link to communicate with the ground control station (laptop running ROS on Ubuntu 14.04).
The ground station runs the planner and image processing algorithms while communicating with the robot via wifi. The kinematics of the robot are represented
by a standard unicycle motion model. The observation model is a vision-based
range bearing sensor augmented with appearance information (see [6] Sec. 6.6.2)
such that $z_k = h(x_k, v_k) = [(r_1, \phi_1, s_1)^T, (r_2, \phi_2, s_2)^T, \ldots]$ where r_l, ϕ_l, s_l are
the range, bearing and signature for the l-th observed landmark. The signature
is an integer value and identical landmarks have the same signature[2]. For this
observation model, the function τ (compute information overlap between two observations, see Sec. 4.1) is identical to that described in Fig. 3. In the real world,
landmark appearances may change due to environmental conditions (e.g., light-
ing), perspective etc., which may adversely affect detection, such issues require
more complex perception models and map representations which are outside the
scope of this work.

---

[1] Due to paucity of space we only present one experiment here, a supplementary video
[27] is provided that clearly depicts every stage of both our experiments.

[2] A detailed description of the motion and observation model parameters is omitted in
the interest of space, we refer the reader to our pre-print version [28] of this paper.

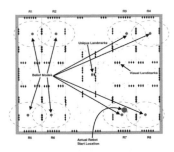

Fig. 5: Environment with 8 rooms marked R1-R8 and belief at the start of first run. Robot is placed in room R7 (blue disk), initial sampling leads to 8 belief modes, one in each room. The black diamonds mark the locations of augmented reality markers in the environment. Unique landmarks are placed inside the narrow passage, such that if robot enters the passage from either side, it sees distinctive information.

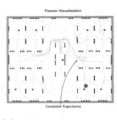 

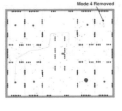 

(a) The planner visualization showing the candidate trajectories (green). The top right image shows the view from the onboard camera, with the detected marker information overlaid. The bottom-right image shows the top-view of the maze in which the robot is run.

(b) Robot observes landmark ID 55 on the door of the opposite room causing the weights of modes $m_1, m_3, m_4, m_5, m_6, m_8$ to gradually decrease which leads to these modes being removed from the belief.

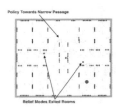 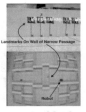

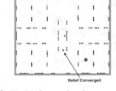 

(c) Robot has exited the room and looks at the outside wall of the narrow passage. Modes m_2 and m_7 are symmetrically located due to information in the map observed by the robot.

(d) Belief mode has converged to the the true belief as the robot enters the narrow passage and observes the unique landmark (ID 39).

Fig. 6: Snapshots of first run of the experiment at different times.

5.2 Scenario

We constructed a symmetrical maze that has 8 identical rooms (R1-8) as shown in Fig. 5. Augmented reality (AR) markers were placed on the walls which act as the landmarks detected by the vision-based sensing system of the robot [29].

When the robot sees a landmark, it can detect the range, bearing as well as its signature. To create ambiguity in the data association, we placed multiple AR markers with the same signature in different parts of the environment. For example, one of the symmetries in our experiment is the inside of each room. Each room in the maze appears identical to the robot as markers with the same appearance are placed on each room's walls with an identical layout. Thus, if the robot is placed in a location with markers similar to another part of the environment, the data associations lead the robot to believe it could be in one of these many locations, which leads to a multimodal belief on the state. We also place four unique markers in a narrow passage in the center of maze as marked in Fig. 5. To successfully localize, the robot must visit this location in order to converge to its true belief.

The robot is initially placed in room R7 and not given any prior information about its state. To estimate initial belief b_0, we discretize the 2D environment with a grid cell size of 0.04m and at each grid point place four hypothesis separated in orientation by $90°$ starting at $0°$. The initial covariance for each hypothesis is $\Sigma_0 = diag([0.04, 0.04, 0.04])$. For our maze with dimensions 4.56m $\times$ 3.81m, it results in 42408 initial hypothesis wherein only collision-free states are preserved and the rest are discarded. While the initial number of hypothesis appears to be large, it is important to have sufficient number of hypothesis in order to guarantee that at least one hypothesis' pdf sufficiently captures true robot state. After this, the robot remains stationary and the sensory measurements are used to update the belief state and remove the unlikely modes with weight $w \leq \delta_w = 0.01$. This process of elimination continues until we converge to a fixed number of modes. Figure 5 shows the initial belief. The robot plans its first set of candidate actions as shown in Fig. 6(a). After the candidates are evaluated, the policy based on mode m_5 in room R5 is chosen and executed. As the robot turns, it sees a landmark on the wall outside the room (shown in Fig. 6(b)). This causes mode m_4 to be deleted. Immediately, replanning is triggered and a new set of candidate trajectories is created. In successive steps, we see that first modes m_3 and m_5 are deleted and then after the next two replanning steps, modes m_8, m_1 and m_6 are deleted. We notice that the robot does not move till only the 2 most-likely modes are remaining. The reason for this is that seeing the marker on the outside wall has the effect of successively lowering the weights of the unlikely modes. As the mode weights fall below the threshold, they are deleted, which triggers the replanning condition. Once the belief has converged to the two most-likely modes m_2, m_7 (as expected by the symmetry) a new set of candidate policies is created and the policy based on mode m_2 is chosen. This policy leads the modes out of the rooms, and towards the narrow passage. Figure 6(c) shows both belief modes executing the policy based on mode m_2. While executing this policy, replanning is triggered as the robot exceeds maximum horizon (60 secs) for policy execution. The final policy drives the robot into the narrow passage and the unique landmarks are observed (Fig. 6(d)) which leads the belief to converge to the robot pose.

5.3   Discussion

Our approach results in a behavior which guides the robot to seek disambiguating information. Candidate trajectories are regenerated every time a belief mode is rejected or a constraint violation is foreseen and time to re-plan reduces drastically as number of modes reduce. Thus, first few actions are the hardest which is to be expected as we start off with a large number of hypotheses. Finally, the planner is able to localize the robot safely.

In the method of [19], once an observation is made, the simulated robot retraces its path to reset odometery-based estimation error; this may be inefficient in practical scenarios due to limited on-board power. Further, as opposed to [19] our robot is not equipped with a global heading sensor. In case a global orientation sensor is present, fewer hypothesis may be required due to the additional constraint of known heading. In [1], the authors showed in simulation that random motion is inefficient and generally incapable of localizing a robot within reasonable time horizons especially in cases with symmetry (e.g., office environments with long corridors and similar rooms). In [2] the authors consider a physical robot localized in their experiments when one mode attains a weight ≥ 0.8, in contrast our approach is more conservative in that we only consider the robot localized when a mode has weight ≥ 0.99. We can afford to be more conservative as our localization strategy actively seeks disambiguating information using prior map knowledge as opposed to a heuristic based strategy. While our experiment acts as a proof of concept, there are certain phenomenon such as cases where the belief modes split into child modes, or dynamic environments which were not covered and will be addressed in future work.

6   Conclusion

In this work, we studied the problem of mobile robot motion planning for active data association in order to correctly localize a robot when the initial underlying belief is multimodal (non-Gaussian). Our main contribution in this work is a planner M3P that generates a sequentially disambiguating policy through active data association, which leads the belief to converge to the true hypothesis. We are able to show in practice that the robot is able to recover from a kidnapped state and localize in environments that present ambiguous data associations such that the underlying belief modes are widely separated. Compared to previous works, we take a non-heuristic approach to candidate policy generation and selection, while remaining conservative in accepting the true hypothesis. A current limitation may be the computational cost for the policy selection step in large maps which lead to a high number of hypotheses. Future work will look at reducing this cost and experiments will be extended to larger problems (e.g., symmetric office environments), with more complex perception models and drastic localization failures (e.g., sequential kidnappings). Finally, there may be tasks which are feasible with a multimodal distribution on the belief. Such cases present an interesting area for future motion planning research.

References

[1] Fox, D., Burgard, W., Thrun, S.: Active markov localization for mobile robots. Robotics and Autonomous Systems **25**(34) (1998) 195 – 207 Autonomous Mobile Robots.

[2] Jensfelt, P., Kristensen, S.: Active global localization for a mobile robot using multiple hypothesis tracking. Robotics and Automation, IEEE Transactions on **17**(5) (Oct 2001) 748–760

[3] Gasparri, A., Panzieri, S., Pascucci, F., Ulivi, G.: A hybrid active global localisation algorithm for mobile robots. In: Robotics and Automation, 2007 IEEE International Conference on. (April 2007) 3148–3153

[4] Reuter, J.: Mobile robot self-localization using pdab. In: Robotics and Automation, 2000. Proceedings. ICRA '00. IEEE International Conference on. Volume 4. (2000) 3512–3518 vol.4

[5] Roumeliotis, S.I., Bekey, G.A.: Bayesian estimation and kalman filtering: A unified framework for mobile robot localization. In: Robotics and Automation, 2000. Proceedings. ICRA'00. IEEE International Conference on. Volume 3., IEEE (2000) 2985–2992

[6] Thrun, S., Burgard, W., Fox, D.: Probabilistic robotics. MIT press (2005)

[7] Prentice, S., Roy, N.: The belief roadmap: Efficient planning in belief space by factoring the covariance. International Journal of Robotics Research **28**(11-12) (October 2009)

[8] Bry, A., Roy, N.: Rapidly-exploring random belief trees for motion planning under uncertainty. In: ICRA. (2011) 723–730

[9] van den Berg, J., Abbeel, P., Goldberg, K.: LQG-MP: Optimized path planning for robots with motion uncertainty and imperfect state information. In: Proceedings of Robotics: Science and Systems (RSS). (June 2010)

[10] Kurniawati, H., Bandyopadhyay, T., Patrikalakis, N.: Global motion planning under uncertain motion, sensing, and environment map. Autonomous Robots **33**(3) (2012) 255–272

[11] Platt, R., Tedrake, R., Kaelbling, L., Lozano-Perez, T.: Belief space planning assuming maximum likelihood observatoins. In: Proceedings of Robotics: Science and Systems (RSS). (June 2010)

[12] Agha-mohammadi, A., Chakravorty, S., Amato, N.: FIRM: Sampling-based feedback motion planning under motion uncertainty and imperfect measurements. International Journal of Robotics Research **33**(2) (2014) 268–304

[13] Agha-mohammadi, A., Agarwal, S., Mahadevan, A., Chakravorty, S., Tomkins, D., Denny, J., Amato, N.: Robust online belief space planning in changing environments: Application to physical mobile robots. In: IEEE Int. Conf. Robot. Autom. (ICRA), Hong Kong, China (2014)

[14] Kavraki, L., Svestka, P., Latombe, J., Overmars, M.: Probabilistic roadmaps for path planning in high-dimensional configuration spaces. IEEE Transactions on Robotics and Automation **12**(4) (1996) 566–580

[15] Platt, R., Kaelbling, L., Lozano-Perez, T., , Tedrake, R.: Efficient planning in non-Gaussian belief spaces and its application to robot grasping. In: Proc. of International Symposium of Robotics Research, (ISRR). (2011)

[16] Platt, R., Kaelbling, L., Lozano-Perez, T., Tedrake, R.: Non-gaussian belief space planning: Correctness and complexity. In: IEEE International Conference on Robotics and Automation (ICRA). (2012)

[17] Platt, R.: Convex receding horizon control in non-Gaussian belief space. In: Workshop on the Algorithmic Foundations of Robotics (WAFR). (2012)

[18] Rafieisakhaei, M., Tamjidi, A., Chakravorty, S., Kumar, P.: Feedback motion planning under non-gaussian uncertainty and non-convex state constraints. In: 2016 IEEE International Conference on Robotics and Automation (ICRA), IEEE (2016) 4238–4244

[19] Dudek, G., Romanik, K., Whitesides, S.: Localizing a robot with minimum travel. SIAM Journal on Computing **27**(2) (1998) 583–604

[20] O'Kane, J.M., LaValle, S.M.: Localization with limited sensing. IEEE Transactions on Robotics **23**(4) (Aug 2007) 704–716

[21] Pilania, V., Gupta, K.: A localization aware sampling strategy for motion planning under uncertainty. In: Intelligent Robots and Systems (IROS), 2015 IEEE/RSJ International Conference on. (Sept 2015) 6093–6099

[22] Karaman, S., Frazzoli, E.: Sampling-based algorithms for optimal motion planning. International Journal of Robotics Research **30**(7) (June 2011) 846–894

[23] Bertsekas, D.: Dynamic Programming and Optimal Control: 3rd Ed. Athena Scientific (2007)

[24] Chakravorty, S., Erwin, R.S.: Information space receding horizon control. In: IEEE Symposium on Adaptive Dynamic Programming And Reinforcement Learning (ADPRL). (April 2011)

[25] He, R., Brunskill, E., Roy, N.: Efficient planning under uncertainty with macro-actions. Journal of Artificial Intelligence Research **40** (February 2011) 523–570

[26] Patil, S., van den Berg, J., Alterovitz, R.: Estimating probability of collision for safe motion planning under gaussian motion and sensing uncertainty. In: Robotics and Automation (ICRA), 2012 IEEE International Conference on. (May 2012) 3238–3244

[27] Agarwal, S., Tamjidi, A., Chakravorty, S.: Video of M3P physical experiments. https://www.youtube.com/watch?v=ufZlrGlzhxI

[28] Agarwal, S., Tamjidi, A., Chakravorty, S.: Motion planning for global localization in non-gaussian belief spaces. (2015) arXiv:1511.04634 [cs.RO].

[29] Garrido-Jurado, S., Muoz-Salinas, R., Madrid-Cuevas, F., Marn-Jimnez, M.: Automatic generation and detection of highly reliable fiducial markers under occlusion. Pattern Recognition **47**(6) (2014) 2280 – 2292

Integrated Perception and Control at High Speed: Evaluating Collision Avoidance Maneuvers Without Maps

Pete Florence[1], John Carter[1], and Russ Tedrake[1]

MIT Computer Science and Artificial Intelligence Laboratory, Cambridge, MA
{peteflo,jcarter,russt}@csail.mit.edu

Abstract. We present a method for robust high-speed quadrotor flight through unknown cluttered environments using integrated perception and control. Motivated by experiments in which the difficulty of accurate state estimation was a primary limitation on speed, our method forgoes maintaining a map in favor of using only instantaneous depth information in the local frame. This provides robustness in the presence of significant state estimate uncertainty. Additionally, we present approximation methods augmented with spatial partitioning data structures that enable low-latency, real-time reactive control. The probabilistic formulation provides a natural way to integrate reactive obstacle avoidance with arbitrary navigation objectives. We validate the method using a simulated quadrotor race through a forest at high speeds in the presence of increasing state estimate noise. We pair our method with a motion primitive library and compare with a global path-generation and path-following approach.

1 Introduction

A primary challenge in improving robot performance is to increase robustness in regimes of fast motion, proximity to obstacles, and significant difficulty of estimating state. A robotics platform that is at the center of all of these challenges is a UAV navigating quickly through unknown, cluttered environments. Although compelling progress has been made [1–4], the goal of autonomous, robust, agile flight into unknown environments remains an open problem.

In this paper, we present an integrated approach for perception and control, which we apply to the high-speed collision avoidance problem. Our approach departs from the paradigm of building maps, optimizing trajectories, and tracking trajectories. Central to the approach is considering routes to achieve control objectives (fly fast, and don't crash into obstacles) and taking advantage of model-based state-space without relying on full-state feedback.

Our approach is directly motivated by the success of reactive control that is "straight from sensors to control input" but uses tools from more rigorous state space control. We show that in order to get the performance of a motion planning system, the robot doesn't need to build a map, doesn't need precise estimates of its full state, and doesn't need to heavily optimize trajectories.

© Springer Nature Switzerland AG 2020
K. Goldberg et al. (Eds.): *Algorithmic Foundations of Robotics XII*, SPAR 13, pp. 304–319, 2020.
https://doi.org/10.1007/978-3-030-43089-4_20

A key insight we explore in this paper is that we can both estimate the probability of collision for any action without building a locally consistent map, and execute that action without the use of position-control feedback. The basics steps of our method are: evaluate maneuvers probabilistically for collision and impose field of view constraints, choose a maneuver based on an unconstrained objective combining collision avoidance and navigation, and execute this at high rate with a model-predictive control type approach. This method offers a mapless collision avoidance approach that does not depend on position, rigorously considers robustness, is amenable to low-latency implementations, and integrates seamlessly with arbitrary navigation objectives. We note, however, that the mapless method cannot escape dead-ends by itself without a layered global planner.

Our primary contribution is the novel synthesis of our approach combining typically separate perception, control, and state estimation considerations. This synthesis is implemented for robustness at speed by a combination of: local frame estimation of path collision probabilities that considers field of view (FOV) constraints, motion primitives defined in the local frame, acceleration by spatial partitioning, and high-rate robust model-predictive control that doesn't depend on trajectory-tracking. This is also the first paper known to the authors to describe stochastic receding horizon control with depth sensor data for a UAV. Additionally, we present simulation experiments in which a benchmark approach cannot provide robust collision avoidance at high speeds, while our method enables the quadrotor to navigate a simulated forest environment at 12 m/s even in the presence of significant state estimate noise.

2   Related Work

The close integration of perception and control, where the realities of perceptual information inform the control approach, is a concept of active interest in robotics. Visual servoing methods for robotic manipulation [5], for example, are an application where control is designed to work with partial information (relative positions in image space) rather than full-state feedback.

In the application area of UAV navigation in unknown environments, the predominant approach is to instead impose the separation principle between perception and control, and separately build a map, plan an optimal trajectory in that map, and execute trajectory-tracking feedback control along the nominal plan. In this map-plan-track paradigm, the goal is to produce a map as close as possible to full obstacle knowledge and produce highly accurate estimates of full state. These methods work well in regimes of good information, such as motion capture rooms with pre-prescribed obstacle locations. They are particularly fragile, however, when exposed to significant state estimate uncertainty, causing mapping and tracking to fail. Planning-heavy approaches also tend towards high latency, although offline-computed libraries enable low-latency response [1,3].

A different approach to UAV navigation is offered by reactive control, which has achieved some of the most impressive obstacle avoidance results demonstrated to date [6–8]. Three primary types of reactive approaches have shown success: optic flow methods [7,9,10], artificial potential fields [6,11], and imitation learning [8]. Reactive methods by definition do not fit into the map-plan-track paradigm since they do not plan a time-sequence of states into the future, but are also generally characterized by not performing full-state feedback.

In that our method neither builds a map nor executes trajectory-tracking control, it departs from the map-plan-track paradigm. In that it does not perform position-control feedback, it is more similar to the mentioned reactive methods, yet it does plan states in the local frame into the future and reason about state-space uncertainty, which does not fit the definition of a reactive method.

The theory of motion planning under uncertainty has been well studied, at least in the domain of full obstacle knowledge. One approach is that of chance-constrained optimization [12–15], in which the probability of collision at any time is upper-bounded as a constraint in an optimization. In the planning portion of our approach we use a variant where collision avoidance is included in the objective, not as a constraint, and we estimate collision probabilities for entire paths, then choose among a finite library. An important component of this approach requires path collision probability estimation, which has been well studied [16].

Several other works are notably related to various components of our integrated approach. One related method for online stochastic receding-horizon control is that of "funnel" computation and sequential composition [17–19], which notably can handle nonlinear models. The focus of those works, however, is not on integrated perception and control considerations, as ours is here. A somewhat related work is by Matthies et al. [20] since it presents field-of-view-limited planning with depth image information for collision avoidance, but their approach is a map-plan-track approach, and doesn't consider uncertainty. Probabilistic collision detection in point clouds has been studied [21] and integrated with sampling-based motion-planners [22], but not to our knowledge has been applied to the collision avoidance problem with field-of-view constraints. Another complementary approach aims to learn, through supervised training in simulation, collision probabilities outside of conservative field of view approximations [23].

3   Generalized Formulation for Collision Avoidance

First, we consider the problem of estimating the probability of collision for a time-varying distribution of configurations using only instantaneous depth information. We then present approximation methods that enable fast computation for collision avoidance at high speeds. Additionally, we discuss the use of spatial partitioning data structures and the incorporation of global navigation objectives. This section is generalized to allow for application to an arbitrary robot. In the next section, a particular implementation for a quadrotor is presented.

3.1   Evaluating Collision Probabilities from Instantaneous Depth Information

We wish to evaluate the probability of collision for:

$$P\big(\text{Collision during } t \in [0, t_f] \mid \mathbf{D}, p_t(\mathbf{q})\big) \tag{1}$$

where $p_t(\mathbf{q})$ is the time-varying distribution of configuration, t_f is the final time, and $\mathbf{D}$ is a vector of depth sensor returns $[\mathbf{d}_0, ..., \mathbf{d}_n]$. This probability cannot be calculated with certainty, due to the large amount of unknown space $\mathcal{U} \subset \mathbb{R}^3$ caused by occlusions and the finite FOV (field of view) of the depth sensor. Each depth return corresponds to an occupied frustum $\mathcal{F}_{\mathbf{d}_j} \subset \mathbb{R}^3$ whose volume is defined by the image resolution, depth return distance, and sensor discretization. Together these occupied frustums comprise the known occupied subset of space, $\mathcal{O}_{known} = \bigcup_j \mathcal{F}_{\mathbf{d}_j}, \mathcal{O}_{known} \subset \mathbb{R}^3$. Each depth return also creates a portion of unknown space $\mathcal{F}_{(occluded\ by\ \mathbf{d}_j)} \subset \mathcal{U}$ which is a frustum that joins the unknown space at the sensor horizon. For handling the FOV constraints, the conservative route is to make the assumption that all unknown space $\mathcal{U}$ is occupied ($\mathcal{U} \cup \mathcal{O}_{known} = \mathcal{O}$), which provides a mapping from $\mathbf{D} \to \mathcal{O}$ that is strictly conservative.

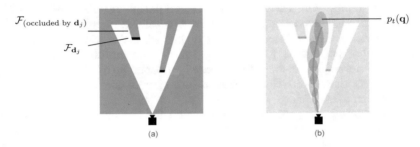

Fig. 1: Depictions of (a) depth measurements (black) and conservative assumption of unknown space as occupied (blue), and (b) time-varying distribution of configuration (purple).

At any given point in time and given the distribution $p_t(\cdot)$ over robot configuration $\mathbf{q}$, the probability of collision is obtained by the probability that the robot is in collision with any of the sensor returns or occupies any unknown space:

$$P\big(\text{Collision}, p_{t_i}(\mathbf{q})|\ \mathcal{O}_{known}, \mathcal{U}\big) = P\big(\mathbf{q}(t_i) \in \{\mathcal{O}_{known}\text{ or }\mathcal{U}\}\big) \qquad (2)$$

Note that the probabilities are not disjoint, since for any non-zero-volume robot, a given configuration can be in collision with multiple frustums, occupied or unknown. To evaluate this probability, an integral over all possible configurations must be integrated. Even given a solution to this integral, however, this only provides an evaluation of one possible distribution of configuration at some future time, and hence the probability of collision for the time-varying distribution of configuration is still difficult to evaluate, given that all future positions in time are dependent on previous positions in time. One route to estimating this probability is through Monte Carlo simulation, but approximations offer computationally efficient routes. Although the literature does not typically account for FOV constraints, a good review of available options for estimating path collision probabilities with full obstacle knowledge is included in a recent paper by Janson et al. [16].

Additionally, even with the conservative assumption, the form of $\mathcal{U}$ (large subsets of space) is of a different form than $\mathcal{O}_{known}$ (small frustums). Our current formulation addresses this by converting $\mathcal{O}_{known}$ into a point cloud and evaluating the probability distribution $p_t(\mathbf{q})$ at these points, whereas for $\mathcal{U}$ we perform a binary evaluation of the mean of $p_t(\mathbf{q})$ entering $\mathcal{U}$. Future work could evaluate both of these probabilities more rigorously by integrating the probability distribution $p_t(\mathbf{q})$ over the volumes of both $\mathcal{O}_{known}$ and $\mathcal{U}$, at additional computational cost.

3.2   Fast Approximation of Maneuver Collision Probabilities

Given the goal to evaluate collision probabilities in real time for the purpose of collision avoidance, some approximations are in order. Although these are significantly simplifying assumptions, the simulation results presented in this paper suggest that even these approximations offer a significant improvement over deterministically collision-checking trajectories. We consider maneuvers of the form: $\mathcal{M} = \{\mathbf{u}(t), p_t(\mathbf{q})\}$, i.e. control inputs as a function of time $\mathbf{u}(t)$ that produce a time-varying distribution of configurations $p_t(\mathbf{q})$. Our choice of open-loop maneuvers is a choice that represents our control decision to not depend on position-control feedback.

For estimating the probability of collision for the entire maneuver we use an independence approximation. Future positions are sampled in time, and the maneuver's probability of collision is approximated as the subtraction from unity of the product of the no-collision probabilities at each sampled time t_i:

$$P\big(\text{Collision}, p_t(\mathbf{q})\big) \approx 1 - \prod_{i=1}^{n_t} \big[1 - P(\text{Collision}, p_{t_i}(\mathbf{q}))\big] \qquad (3)$$

For the evaluation of the no-collision probabilites at each time t_i, we assign a no-collision probability of 0 (definite collision) if the mean of $p_{t_i}(\mathbf{q})$ is in $\mathcal{U}$, and otherwise evaluate the probability of collision with the point cloud. Evaluating only the mean in $\mathcal{U}$ is a large oversimplification, but avoids integrating over many small occluded frustums:

$$\big[1 - P(\text{Collision}, p_{t_i}(\mathbf{q}))\big] = \begin{cases} 0, & \text{if } \mu(p_{t_i}) \in \mathcal{U} \\ \prod_{j=1}^{n_d} \big[1 - P(\text{Collision}, p_{t_i}(\mathbf{q}), \mathbf{d}_j)\big], & \text{otherwise} \end{cases}$$
$$(4)$$

Checking if $\mu(p_{t_i}) \in \mathcal{U}$ can be done by a projective transform into depth image space, and checking if the projection is either out of bounds of the depth image (outside FOV), or a depth return at that pixel has less depth (occluded). If not in $\mathcal{U}$, the probability of collision with $\mathcal{O}_{known}$ is approximated by an additional independence approximation: each collision with all n_d depth returns is assumed an independent probability event. To evaluate each event $P(\text{Collision}, p_{t_i}(\mathbf{q}), \mathbf{d}_j)$ above, we must choose a dynamic model with uncertainty. Thus far, the discussion has been generalizable to any model. In Section 4 we describe how we evaluate this term for a simplified model with Gaussian noise.

Naively, the complexity of the computation above is $O(n_{\mathcal{M}} \times n_t \times n_{\mathbf{d}})$. Even for a "low-resolution" depth image, the number of depth points can be high, for example a 160 x 120 image is $n_{\mathbf{d},\text{Total}} = 19,200$ points. Only the closest depth returns to the mean of the robot's distribution, however, will have the highest probability of impact, and this additionally offers a route to lower computational complexity. Thus, rather than evaluate Equation 4 for all depth returns, we query only the closest $n_d < n_{\mathbf{d},\text{Total}}$ points with a k-d-tree.

In contrast to deterministic collision checking, collision probability approximation significantly benefits from three-dimensional spatial partitioning as opposed to operating directly on the depth image. This is because with the probabilistic collision checking, we care about "long-tails" of the robot position distribution, rather than just deterministically collision-checking the mean. To deterministically collision check, there is no faster way than using the raw depth image [20], but in order to consider long-tail positions in the direct depth image method, a large block of pixels needs to be checked. The depth image structure provides information about proximity in two dimensions (neighboring pixels), but not the third (depth). We also note, however, that since the direct depth image method requires no building of a new data structure, highly parallelized implementations may tip computational time in its favor (as opposed to sequentially building a k-d-tree, then searching it).

Briefly, we analyze the limitations of the approximation accuracy. In the context of full obstacle knowledge, the independence approximation over time has been shown to provide overly conservative estimates of collision probability [16]. Additionally, the independence approximation between depth returns contributes to more overestimation, and picking only one point from each cluster has been recommended to reduce this overestimation [21]. In our method, the FOV constraints contribute even more to over-conservatism, but there is not available information to improve this approximation without adding risk going into the unknown. Learned priors, however, can intelligently minimize this risk [23]. We note that with our unconstrained formulation, it is the relative differences between maneuver collision probabilities (see Figure 4b), not their absolute scale, that impacts control decisions.

At additional computational cost, additional accuracy could be achieved through Monte Carlo (MC) evaluation, whereby randomly sampled trajectories are deterministically collision-checked and the proportion of collision-free trajectories is the collision probability. In the limit of infinite samples the probability is exact, but the computational cost is approximately $n_{MC} \times T_D$, where T_D is the time to deterministically collision-check, and n_{MC} is the number of samples. As we show in Table 1 (Section 6), deterministic collision-checking takes approximately the same amount of time as our independence approximation evaluation. Hence, naive MC evaluation is slower than our method by approximately the factor n_{MC}. Smart MC sampling strategies have been demonstrated to enable path collision probability approximations on the order of seconds for reasonable models [16], but our requirement is a few orders of magnitude faster (milliseconds) to replan at the rate of depth image information (30-150 hz).

3.3   Integrating Reactive and Navigation Objectives

A benefit of the probabilistic maneuver evaluation approach is that it naturally offers a mathematical formulation that integrates reactive-type obstacle avoidance with arbitrary navigation objectives. Whereas other "layered" formulations might involve designed weightings of reactive and planning objectives, the probabilistic formulation composes the expectation of the reward, $\mathbb{E}[R]$. Given some global navigation function that is capable of evaluating a reward $R_{nav}(\mathcal{M}_i)$ for a given maneuver, the expected reward is:

$$\mathbb{E}[R(\mathcal{M}_i)] = P(\text{No Collision}, \mathcal{M}_i)R_{nav}(\mathcal{M}_i) + P(\text{Collision}, \mathcal{M}_i)R_{collision} \quad (5)$$

As we show in the simulation experiments, $R_{nav}(\mathcal{M}_i)$ may not even need to consider obstacles, and collision avoidance can still be achieved. The global navigation function can be, for example, just Euclidean progress to the global goal for environments with only convex obstacles, or for environments with dead-ends could for example be a cost-to-go using Dijkstra's algorithm (Figure 3a). A key point is that with the instantaneous mapless approach handling collision avoidance, $R_{nav}(\mathcal{M}_i)$ can be naive, and/or slow, although a good $R_{nav}(\mathcal{M}_i)$ is only a benefit. One parameter that must be chosen, and can be tuned up/down for less/more aggressive movement around obstacles, is the cost (negative reward) of collision, $R_{collision}$,

Given a library of maneuvers, the optimal maneuver $\mathcal{M}^*$ is then chosen as:

$$\mathcal{M}^* = \operatorname*{argmax}_i \ \mathbb{E}[R(\mathcal{M}_i)] \quad (6)$$

4   Implementation for High Speed Quadrotor Flight

The formulation presented above is generalizable for different robot models and for evaluating different types of discrete action libraries. In this section we present a specific implementation for high-speed quadrotor control.

4.1   High-Rate Replanning with a Motion Primitive Library

We use an approach similar to a traditional trajectory library, except our library is generated online based on a simplified dynamical model. In the sense that a model is used for real-time control, and we use no trajectory-tracking controller, this is MPC (Model Predictive Control), but since we perform no continuous optimization but rather just select from a discrete library, this is a motion primitive library approach. This high-rate replanning with no trajectory-tracking controller offers a route to controlling collision avoidance without a position estimate. Since the uncertainty of the maneuvers is considered open-loop, this can be categorized as OLRHC (open-loop receding horizon control). Another control approach is to "shrink" the future uncertainty with a feedback controller [17, 18, 24], but this assumes that a reasonable position estimate will be available. It is crucial to our method that we do not shrink the uncertainty in this way, since this enables sensible avoidance decisions and control without ever needing a position estimate.

4.2   Dynamical Model and Propagating Uncertainty

To build intuition of our simple quadrotor model, we first describe the basic version of a constant-input double-integrator (constant-acceleration point-mass) modeled around the attitude controller. This version approximates the quadrotor as a point-mass capable of instantaneously producing an acceleration vector of magnitude $||\mathbf{a}|| \leq a_{max}$ in any direction. Together with gravitational acceleration, this defines the achievable linear accelerations. This model is applied with the inner-loop attitude and thrust controller in feedback, as depicted in Figure 2. Given a desired acceleration $\mathbf{a}_i$, geometry defines the mapping to {roll, pitch thrust} required to produce such an acceleration, given any yaw.

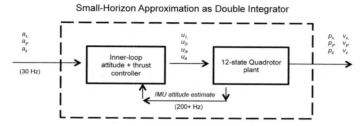

Fig. 2: Dynamics approximation considered: the quadrotor is modeled in feedback with the inner loop attitude and thrust controller.

A motivating factor for this model is that the overwhelmingly ubiquitous implementation for quadrotor control involves a high-rate ($\sim$200+ Hz) inner-loop attitude and thrust controller. The desirability of quickly closing a PID or similar loop around the IMU makes this an attractive control design choice.

The only source of uncertainty we consider is the state estimate. In particular, since the quadrotor's initial position is by definition the origin in the local frame, we only consider uncertainty in the velocity estimate. We use Gaussian noise for the initial linear velocity estimate $\mathbf{v}_0 \sim \mathcal{N}(\mathbf{v}_{0,\mu}, \Sigma_{v_0})$ which gets propagated through the linear model. We use the notation $\mathbf{p} \in \mathbb{R}^3$ to refer to the configuration since it is just position (point-mass is rotation-invariant). Accordingly we have:

$$\mathbf{p}_i(t) \sim \mathcal{N}\left(\frac{1}{2}\mathbf{a}_i t^2 + \mathbf{v}_{0,\mu}t,\ t^2\Sigma_{v_0}\right) \qquad (7)$$

for maneuver $\mathcal{M}_i = \{\mathbf{a}_i, \mathbf{p}_i(t)\}, t \in [0, t_f]$

where $\mathbf{p}_i(t)$ is a random variable defining the distribution referred to as $p_t(\mathbf{q})$ in Section 3. The chosen acceleration $\mathbf{a}_i$ defines the maneuver $\mathcal{M}_i$.

Extension to Piecewise Triple-Double Integrator Model The limitations of the constant-acceleration model are clear, however: it does not consider attitude dynamics, even though they are fast ($\sim$100-200 ms to switch between extremes of roll/pitch) compared to linear dynamics. It is preferable to have a model that does include attitude dynamics: for example, the initial roll of the vehicle should affect "turn-left-or-right" obstacle-dodging decisions.

Accordingly, we use a triple integrator for the first segment, and a double integrator for the remaining ("triple-double" integrator for short). Each maneuver $\mathcal{M}_i$ is still defined uniquely by $\mathbf{a}_i$, but during $t \in [0, t_{jf}]$, we use a jerk $\mathbf{j}_i$ that linearly interpolates from the initial acceleration $\mathbf{a}_0$ to the desired acceleration:

$$\mathbf{j}_i = \frac{\mathbf{a}_i - \mathbf{a}_0}{t_{jf}} \tag{8}$$

During the initial constant-jerk $t \in [0, t_{jf}]$ period, this gives

$$\mathbf{p}_i(t) \sim \mathcal{N}\left(\frac{1}{6}\mathbf{j}_i t^3 + \frac{1}{2}\mathbf{a}_0 t^2 + \mathbf{v}_{0,\mu}t, t^2 \Sigma_{v_0}\right) \quad \forall t \in [0, t_{jf}] \tag{9}$$

and for $t \in [t_{jf}, t_f]$ the double integrator model (Equation 7) is used with the appropriate forward-propagation of position and velocity. Note that for the constant-jerk portion, an initial acceleration estimate, $\mathbf{a}_0$ is required. We assume this to be a deterministic estimate. Since roll, pitch, and thrust are more easily estimated than linear velocities, this is a reasonable assumption.

The maneuvers produced by this piecewise triple-double integrator retain the properties of being closed-form for any future $t \in [0, t_f]$, of being linear with Gaussian noise, and cheap to evaluate. Although the actual attitude dynamics are nonlinear, a linear approximation of the acceleration dynamics during the constant-jerk period is an improved model over the constant-acceleration-only model. We approximate t_{jf} as 200 ms for our quadrotor.

4.3   Maneuver Library and Attitude-Thrust Setpoint Control

We use a finite maneuver library (Figure 3b), where the maneuvers are determined by a set of desired accelerations $\mathbf{a}_i$ for the piecewise triple-double integrator. Our method is compatible for a 3D library, but for the purposes of the simulation comparison against a global-planning 2D method in the next section, we use a library constrained to a single altitude plane. To build a suitable discrete set of maneuvers, we approximate the maximum horizontal acceleration and sample over possible horizontal accelerations around a circle in the horizontal plane. The max horizontal acceleration is approximated as the maximum thrust vector (T_{max}) angled just enough to compensate for gravity: $a_{max} = \frac{\sqrt{T_{max}^2 + (mg)^2}}{m}$. By sampling both over horizontal accelerations with just a few discretizations (for example, $[a_{max}, 0.6a_{max}, 0.3*a_{max}]$) and just 8 evenly spaced θ over $[0, 2\pi]$, this yields a useful set in the horizontal plane. We also add a $[0,0,0]$ acceleration option, for 25 maneuvers total in the plane, and use $t_f = 1.0$ seconds.

Executing the chosen maneuver is achieved by commanding a desired roll and pitch to the attitude controller. For this 2D-plane implementation, a PID loop on z-position maintains desired altitude by regulating thrust. We allow for slow yawing at 90 degrees per second towards the direction $\mathbf{p}(t_f) - \mathbf{p}_0$, which in practice has little effect on the linear model and allows for slow yawing around trees.

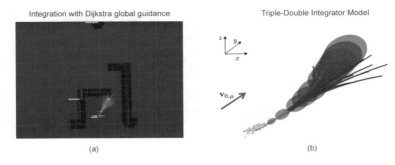

Fig. 3: (a) Visualization of integrating Dijkstra global guidance, where $\mathcal{R}_{nav}$ is the cost-to-go (blue is lower, purple is higher) of the final maneuver position. (b) Visualization of the piecewise triple-double integrator maneuver library. The library of maneuvers is shown with a positive x, positive y initial velocity $\mathbf{v}_{\mu,0}$, and the 1-σ of the Gaussian distribution is shown for one of the maneuvers. The $t_{jf} = 200$ ms constant-jerk period shown in orange. Note that due to the initial roll-left of the vehicle, it can more easily turn left than right.

4.4 Evaluation of Collision Probability and Global Navigation

Each maneuver is sampled at n_t positions (we use $n_t = 20$), for a total of 500 positions to be evaluated in our $n_{\mathcal{M}} = 25$ library. To allow for speeds past 10 m/s, given $t_f = 1.0$ s, we do not consider positions beyond our simulated depth image horizon of 10 meters to be in collision. All mean robot positions are evaluated for n_d nearest neighbors in the k-d-tree. In practice we have found success with $n_d = 1$, although larger n_d is still fast enough for online computation, as shown in Table 1 in Section 6.

For each robot position mean $\mathbf{p}_{i,\mu}$ evaluated, we use a small-volume approximation of the probability that a depth return point $\mathbf{d}_j$ and the robot are in collision, by multiplying the point Gaussian probability density by the volume V_r of the robot's sphere:

$$P(\text{Collision}, \mathbf{p}_i(t)) \approx V_r \times \frac{1}{\sqrt{\det(2\pi\Sigma_p)}} \exp\left[-\frac{1}{2}(\mathbf{p}_{i,\mu} - \mathbf{d}_j)^T \Sigma_p^{-1}(\mathbf{p}_{i,\mu} - \mathbf{d}_j)\right]$$
(10)

where Σ_p is the covariance of the robot position as described by the model. This small-volume spherical approximation has been used in the chance-constrained programming literature [12]. If the above equation evaluates to > 1 (possible with the approximation), we saturate it to 1. A key implementation note is that using a diagonal covariance approximation enables the evaluation of Equation 10 approximately an order of magnitude faster than a dense 3×3 covariance. Rather than use online-estimated covariances of velocity, we choose linear velocity standard deviations $\sigma_{v\{x,y,z\}}$ that scale with linear velocity.

For our quadrotor race through the forest, since the obstacles are all convex and so navigating out of dead-ends is not a concern, we use a simple Euclidean progress metric as our navigation function R_{nav}, plus a cost on terminal speed $v_f = ||\mathbf{v}_i(t_f)||_2$ if it is above the target max speed, v_{target}:

$$R_{nav}(\mathcal{M}_i) = ||\mathbf{p}_0 - \mathbf{p}_{goal}|| - ||\mathbf{p}_i(t_f) - \mathbf{p}_{goal}|| + R_v(v_f) \tag{11}$$

$$R_v(v_f) = \{0 \text{ if } v_f < v_{target}, \ kv_f \text{ if } v_f \geq v_{target}\} \tag{12}$$

Where we used $k = 10$, and $R_{collision} = -10,000$.

5   Simulation Experimental Setup

5.1   Simulator Description

To facilitate the comparison study, simulation software was developed to closely mimic the capabilities of our hardware platform for the Draper-MIT DARPA FLA (Fast Lightweight Autonomy) research team. The sensor configuration includes a depth sensor that provides dense depth information at 160x120 resolution out to a range of 10 meters, with a FOV (field of view) limited to 58 degrees horizontally, 45 degrees vertically. A simulated 2D scanning lidar provides range measurements to 30 meters. Both sensors are simulated at 30 Hz.

Drake [25] was used to simulate vehicle dynamics using a common 12-state nonlinear quadrotor model [26] while the Unity game engine provides high fidelity simulated perceptual data that includes GPU-based depth images and raycasted 2D laser scans. The flight controller uses a version of the Pixhawk [27] firmware running in the loop (SITL) that utilizes an EKF over noisy simulated inertial measurements to estimate attitude and attitude rates of the vehicle.

(a)   (b)

Fig. 4: (a) Screenshot from our race-through-forest simulation environment in Unity. (b) Screenshot from Rviz which shows the evaluation of the 25-maneuver real-time-generated motion library. The chosen maneuver and the 1-σ of the Gaussian distribution over time are visualized. The small sphere at the end of each maneuver indicates approximated collision probabilities from low to high (green to red).

5.2   Experimental Setup

The experiments were carried out in a virtual environment that consists of an artificial forest valley that is 50 meters wide and 160 meters long. The corridor is filled with 53 randomly placed trees whose trunks are roughly 1 meter in diameter. A timer is started when the vehicle crosses the 5 meter mark and stopped either when a collision occurs or when the 155 meter mark is reached. If the vehicle is able to navigate the forest without colliding with any of the trees or terrain in under a predetermined amount of time, the trial is considered a success. Collisions and time-outs are considered failures.

The experiments were repeated for each algorithm at various target velocities $v_{target} = \{$ 3, 5, 8, 12$\}$ meters per second and with increasing levels of state estimate noise for $x, \dot{x}, y, \dot{y}$. We do not simulate noise in the altitude or in the orientations since these are more easily measurable quantities. To simulate noise that causes position to drift over time, we take the true difference in x, y over a timestep, $\Delta\mathbf{p}_{x,y}$, and add zero-mean Gaussian noise which is scaled linearly with the velocity vector. The three noise levels we use are $\sigma = \{0, 0.1, 1\}$ which is scaled by $\frac{\sigma}{10}\mathbf{v_{true}}$. This linearly increases noise with higher speed. We also add true-mean Gaussian noise to $\dot{x}$ and $\dot{y}$, with standard deviations that are the same as for position noise. Accordingly we have:

$$\mathbf{p}_{noisy}[i+1] \sim \mathcal{N}(\mathbf{p}_{true}[i+1] - \mathbf{p}_{true}[i], \frac{\sigma}{10}\mathbf{v_{true}}) \tag{13}$$

$$\mathbf{v}_{noisy}[i] \sim \mathcal{N}(\mathbf{v}_{true}[i], \frac{\sigma}{10}\mathbf{v_{true}}) \tag{14}$$

The total time taken and the trial outcome was recorded for 10 trials at each noise and speed setting, for a total of 360 simulation trials.

5.3   Dijkstra's Algorithm with Pure Pursuit Description

We compare our method to a typical map-based robotics navigation solution that consists of a global path planner that is paired with a path following algorithm. The particular implementation we chose functions by maintaining a global probabilistic occupancy grid (Octomap [28]) with a 0.2 meter voxel size. At a specified rate, a horizontal slice of the map is extracted and a globally optimal path is computed using Dijkstra's algorithm. The path planning includes a soft cost on proximity to obstacles. We then use a pure pursuit algorithm to command a vehicle velocity along the resulting path to the goal. This approach has been heavily tested on our hardware, and shown considerable success in complex environments in the range of 2.0 to 5.5 m/s with little state estimate noise.

6   Simulation Results and Discussion

The key metric for our comparison of the three methods is the no-collison success rate of reaching the finish line, and is presented in Figure 5. Additional data is presented in Figure 6: average time to goal for successful trials, and example paths at various noise levels.

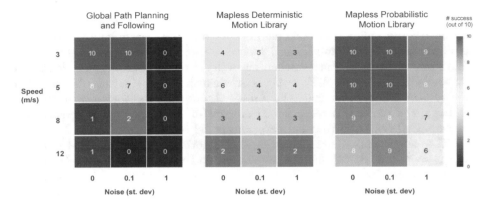

Fig. 5: Comparison summary of number of successful collision-free trials for the different approaches tested in in our simulated quadrotor race through the forest. Ten trials were run for each of the three approaches, for four different speeds {3, 5, 8, 12} meters per seconds, and for three different levels of 2-dimensional state estimate noise as described in Section 5.2.

The results for the global path planning and following approach show both the limitations on handling higher speed, and on handling higher state estimate noise. The approach was not able to handle any of the severe noise ($\sigma = 1$) for any of the speeds and was only able to reliably reach the goal at 5 m/s and below, with zero or little state estimate noise. These limits on speed and state estimate noise match well our experimental results in hardware. Primary inhibiting factors for this approach's success are (i) dependence on a global position estimate, (ii) latency incurred by processing sensor data into a global map (up to ~50 ms), (iii) latency incurred by path planning on the local map (up to ~200 ms), and (iv) neglect of vehicle dynamics, which are increasingly important for obstacle avoidance at higher speeds.

For comparison, we also compare with the approach of deterministically collision-checking our motion primitive library. For this deterministic method, the average time to goal on a successful run was faster than the probabilistic method by approximately 14%. The deterministic nature of the collision checking, however, causes the method to leave little margin for error while navigating around obstacles. Thus, small inaccuracies in the linear planning model (which approximates the nonlinear model used for simulation) or in the state estimate can lead to fatal collisions.

The results for the probabilistic method demonstrate a marked increase in robustness at higher speeds and with noise levels an order of magnitude higher than was manageable by the path following approach. The sacrifice in average time to goal compared to the deterministic method is outweighed by the gains in robustness.

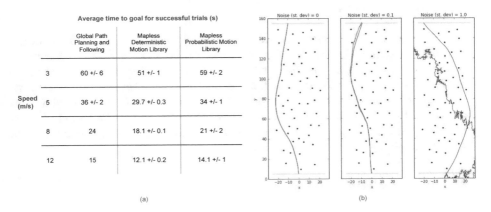

(a)   (b)

Fig. 6: (a) Comparison summary of the average time to goal for successful trials for $\sigma = 0$, which all methods were at least able to get 1 trial across the finish line. (b) Visualization of the different noise levels $\sigma = \{0, 0.1, 1.0\}$ and representative paths for the probabilistic motion library navigating successfully through the forest at 12 m/s. The path of the noisy x, y state estimates (red) are plotted together with the ground truth path (blue). The brown circles represent the tree obstacles at the flying altitude of 1.8 m.

Additionally, an important practical consideration is that, given our fast collision probability approximations, the total computation times of the probabilistic and deterministic methods are nearly identical ($\sim$3-4 ms total), as is displayed in Table 1. This is a strong argument for replacing deterministic collision checking with fast collision probability approximation in a wide number of scenarios. We also emphasize that these approximate computation times are achievable on our actual flight vehicle hardware, which uses an Intel i7 NUC.

Subprocess	Deterministic, N=1		Probabilistic, N=1		Probabilistic, N=10	
	Average time (μs)	Percentage time (%)	Average time (μs)	Percentage time (%)	Average time (μs)	Percentage time (%)
Building kd-tree	1900 +/- 700	50.5	2000 +/- 500	57.8	1900 +/- 400	42.6
Evaluating future positions from real-time generated 25-maneuver motion library	40 +/- 10	1.0	40 +/- 10	1.1	40 +/- 10	0.9
Evaluating collision probabilities with N-nearest neighbor search on kd-tree	1800 +/- 800	47.9	1400 +/- 600	40.5	2500 +/- 1000	56.1
Evaluating expected reward, given R_{nav}	2 +/- 1	0.1	2 +/- 1	0.1	2 +/- 1	0.0
Calculating attitude setpoint for attitude controller	17 +/- 5	0.5	17 +/- 5	0.5	17 +/- 5	0.4

Table 1: Measured averages and standard deviations of subprocess latencies, from one representative run each. Implementation on single-thread Intel i7.

7 Future Work

There are several components to this line of work that we would like to extend. For one, we plan to present validation experiments of the method in hardware. Additionally, the highly parallel nature of the fast collision probability approximation algorithm is amenable to data-parallel implementations on a GPU. We also plan to expand on the motion primitive library, including true 3D flight, increased variety of maneuvers, and analysis of the accuracy of the model. We also plan to characterize the performance of the collision probability approximation with more elaborate global navigation functions.

8 Acknowledgements

This work was supported by the DARPA Fast Lightweight Autonomy (FLA) program, HR0011-15-C-0110. We also thank Brett Lopez and William Nicholas Greene for fruitful discussions on modeling, control, and depth images.

References

1. Barry, A.J.: High-speed autonomous obstacle avoidance with pushbroom stereo, PhD Thesis, MIT (2016)
2. Barry, A.J., Tedrake, R.: Pushbroom stereo for high-speed navigation in cluttered environments. In: 2015 IEEE International Conference on Robotics and Automation (ICRA), IEEE (2015) 3046–3052
3. Daftry, S., Zeng, S., Khan, A., Dey, D., Melik-Barkhudarov, N., Bagnell, J.A., Hebert, M.: Robust monocular flight in cluttered outdoor environments. arXiv preprint arXiv:1604.04779 (2016)
4. Liu, S., Watterson, M., Tang, S., Kumar, V.: High speed navigation for quadrotors with limited onboard sensing. In: 2016 IEEE International Conference on Robotics and Automation (ICRA), IEEE (2016) 1484–1491
5. Bateux, Q., Marchand, E.: Histograms-based visual servoing. IEEE Robotics and Automation Letters **2**(1) (2017) 80–87
6. Scherer, S., Singh, S., Chamberlain, L., Elgersma, M.: Flying fast and low among obstacles: Methodology and experiments. The International Journal of Robotics Research **27**(5) (2008) 549–574
7. Beyeler, A., Zufferey, J.C., Floreano, D.: Vision-based control of near-obstacle flight. Autonomous robots **27**(3) (2009) 201–219
8. Ross, S., Melik-Barkhudarov, N., Shankar, K.S., Wendel, A., Dey, D., Bagnell, J.A., Hebert, M.: Learning monocular reactive uav control in cluttered natural environments. In: Robotics and Automation (ICRA), 2013 IEEE International Conference on, IEEE (2013) 1765–1772
9. Conroy, J., Gremillion, G., Ranganathan, B., Humbert, J.S.: Implementation of wide-field integration of optic flow for autonomous quadrotor navigation. Autonomous robots **27**(3) (2009) 189–198
10. Hyslop, A.M., Humbert, J.S.: Autonomous navigation in three-dimensional urban environments using wide-field integration of optic flow. Journal of guidance, control, and dynamics **33**(1) (2010) 147–159

11. Nieuwenhuisen, M., Behnke, S.: Hierarchical planning with 3d local multiresolution obstacle avoidance for micro aerial vehicles. In: ISR/Robotik 2014; 41st International Symposium on Robotics; Proceedings of, VDE (2014) 1–7

12. Du Toit, N.E., Burdick, J.W.: Probabilistic collision checking with chance constraints. IEEE Transactions on Robotics **27**(4) (2011) 809–815

13. Blackmore, L., Ono, M., Williams, B.C.: Chance-constrained optimal path planning with obstacles. IEEE Transactions on Robotics **27**(6) (2011) 1080–1094

14. Ono, M., Pavone, M., Kuwata, Y., Balaram, J.: Chance-constrained dynamic programming with application to risk-aware robotic space exploration. Autonomous Robots **39**(4) (2015) 555–571

15. Luders, B., Kothari, M., How, J.P.: Chance constrained rrt for probabilistic robustness to environmental uncertainty. In: AIAA Guidance, Navigation, and Control Conference, Guidance, Navigation, and Control. (2010)

16. Janson, L., Schmerling, E., Pavone, M.: Monte carlo motion planning for robot trajectory optimization under uncertainty. International Symposium on Robotics Research (ISRR) (2015)

17. Majumdar, A., Tedrake, R.: Funnel libraries for real-time robust feedback motion planning. CoRR **abs/1601.04037** (2016)

18. Majumdar, A., Tedrake, R.: Robust online motion planning with regions of finite time invariance. In: Algorithmic Foundations of Robotics X. Springer (2013) 543–558

19. Tedrake, R., Manchester, I.R., Tobenkin, M., Roberts, J.W.: Lqr-trees: Feedback motion planning via sums-of-squares verification. The International Journal of Robotics Research (2010)

20. Matthies, L., Brockers, R., Kuwata, Y., Weiss, S.: Stereo vision-based obstacle avoidance for micro air vehicles using disparity space. In: 2014 IEEE International Conference on Robotics and Automation (ICRA), IEEE (2014) 3242–3249

21. Pan, J., Chitta, S., Manocha, D.: Probabilistic collision detection between noisy point clouds using robust classification. In: International Symposium on Robotics Research (ISRR). (2011) 1–16

22. Pan, J., Manocha, D.: Fast probabilistic collision checking for sampling-based motion planning using locality-sensitive hashing. The International Journal of Robotics Research (2016) 0278364916640908

23. Richter, C., Vega-Brown, W., Roy, N.: Bayesian learning for safe high-speed navigation in unknown environments. In: Proceedings of the International Symposium on Robotics Research (ISRR 2015), Sestri Levante, Italy (2015)

24. Van Den Berg, J., Abbeel, P., Goldberg, K.: Lqg-mp: Optimized path planning for robots with motion uncertainty and imperfect state information. The International Journal of Robotics Research **30**(7) (2011) 895–913

25. Tedrake, R.: Drake: A planning, control, and analysis toolbox for nonlinear dynamical systems (2014)

26. Mellinger, D., Michael, N., Kumar, V.: Trajectory generation and control for precise aggressive maneuvers with quadrotors. Int. J. Rob. Res. **31**(5) (April 2012) 664–674

27. Meier, L., Tanskanen, P., Heng, L., Lee, G.H., Fraundorfer, F., Pollefeys, M.: Pixhawk: A micro aerial vehicle design for autonomous flight using onboard computer vision. Auton. Robots **33**(1-2) (August 2012) 21–39

28. Hornung, A., Wurm, K.M., Bennewitz, M., Stachniss, C., Burgard, W.: Octomap: An efficient probabilistic 3d mapping framework based on octrees. Auton. Robots **34**(3) (April 2013) 189–206

Risk-Aware Graph Search with Dynamic Edge Cost Discovery

Ryan Skeele, Jen Jen Chung, Geoffrey A. Hollinger

Oregon State University, Corvallis, OR, USA
{skeeler,jenjen.chung,geoff.hollinger}@oregonstate.edu

Abstract. In this paper we introduce a novel algorithm for incorporating uncertainty into lookahead planning. Our algorithm searches through connected graphs with uncertain edge costs represented by known probability distributions. As a robot moves through the graph, the true edge costs of adjacent edges are revealed to the planner prior to traversal. This locally revealed information allows the planner to improve performance by predicting the benefit of edge costs revealed in the future and updating the plan accordingly in an online manner. Our proposed algorithm, Risk-Aware Graph Search (RAGS), selects paths with high probability of yielding low costs based on the probability distributions of individual edge traversal costs. We analyze RAGS for its correctness and computational complexity and provide a bounding strategy to reduce its complexity. We then present results in an example search domain and report improved performance compared to traditional heuristic search techniques. Lastly, we implement the algorithm on satellite imagery to show the benefits of risk-aware planning through uncertain terrain.

Keywords: Risk-Aware Planning, Graph Search, Planning Under Uncertainty

1   Introduction

When planning in unstructured environments there is a greater need for fast, reliable path planning methods capable of operating under uncertainty. Planning under uncertainty allows for robustness when faced with unknown and partially known environments. We introduce a method, Risk-Aware Graph Search (RAGS), for finding paths through graphs with uncertain edge costs (similar to Stochastic Shortest Path Problems with Recourse [1]). In the domain of interest, a robot moves through an environment represented by a graph, and the true costs for edges adjacent to the robot's location are dynamically revealed. Our method accounts for both the uncertainty in the edge costs and the dynamic revealing of these costs. Thus, we bridge the gap between traditional search methods [2,3] and risk-aware planning under uncertainty [4,5].

Traditionally, graphs are composed of nodes and edges, with a node representing some state and an edge representing the transition between states. Effectively searching through a graph with known edges has been extensively researched and applies across disciplines in robotics, computer science, and optimization. We aim to expand the capabilities of graph search algorithms by allowing for uncertainty in traversal costs and dynamic discovery of those costs,

© Springer Nature Switzerland AG 2020
K. Goldberg et al. (Eds.): *Algorithmic Foundations of Robotics XII*, SPAR 13, pp. 320–335, 2020.
https://doi.org/10.1007/978-3-030-43089-4_21

while avoiding the blowup in computation from more expressive frameworks (e.g., POMDPs). Our novel approach searches over uncertain edge costs with known distributions and properly adjusts as new information about the edge costs becomes available. This formulation allows for computationally efficient methods of reducing the risk of traversing paths with high cost.

The main novelty of this work is the introduction of a non-myopic graph search algorithm for risk-aware planning with uncertain edge costs and dynamic local edge discovery. RAGS is an online planning mechanism that incorporates live feedback for deciding when to be conservative and when to be aggressive. With edge costs modeled as probability distributions, we can derive a principled way of leveraging information further down a path. This leads to a tradeoff between revealing the true cost of a large number of edges (exploration) and traversing uncertain edges with low mean cost (exploitation). RAGS addresses this tradeoff in a principled way, and the result is a low probability of executing a path with high cost.

We compare our method to A^*, sampled A^* [5], and a greedy approach on a large number of random connected graphs. The results show that RAGS reduces the number of high cost runs compared to all other tested methods. In addition to testing over random graphs, we validate RAGS using satellite imagery. By applying a series of filters to satellite data, we are able to extract potential obstacles that may impede a robot moving through the map. To deal with the imprecise nature of obstacle extraction, we can utilize the benefits of RAGS to find paths that are less likely to be substantially delayed. We show examples of different cases in which RAGS finds preferable paths. The algorithm is run over 100 satellite images taken from a broad set of landscapes. The resulting path costs over the image database confirms the benefit of the RAGS algorithm in a real-world planning domain.

The remainder of this paper is organized as follows. First we review related work and research in probabilistic planners (Section 2). We then introduce the path search problem with uncertain traversal costs and dynamic edge cost discovery (Section 3). Next we derive a method for quantifying path risk (Section 4.1) and present a way to reduce the search space by removing dominated paths (Section 4.2). In Section 4.3 we present the RAGS algorithm and show comparisons to existing search algorithms in Section 5. We then highlight the utility of RAGS by demonstrating its effectiveness for planning through various terrain captured from satellite imagery (Section 6). Finally, we draw conclusions and propose future directions for this line of research (Section 7).

2   Related Work

Planning under uncertainty is a challenging problem in robotics. An underlying assumption in many existing planning algorithms is that the search space is not stochastic. Under this assumption, researchers have developed many powerful algorithms like A^* [2] and RRT^* [6] that perform efficient point to point planning over expected costs (see [7] and [8] for surveys). These algorithms are efficient for problems with deterministic actions and a well-defined search space, but are often not well suited to planning under uncertainty. Recent work has explored ways of incorporating uncertainty into similar planners in field robotics applications [4,9].

Reasoning probabilistically in robotic planning allows performance to degrade gracefully when encountering the unexpected. Notable work has been done on incorporating uncertainty from sensors into the state estimation of the system, [10,11], or in the path planning itself [12,13]. However, uncertainty can lie in both the state and the world model, so we must address both sources of uncertainty to plan effectively.

Prior work in uncertain traversability of graph edges has focused on a binary status of the edge. This family of problems is a variant of the shortest path problem known as the Canadian Traveler Problem (CTP) [14]. The CTP is inspired by drivers in northern Canada who sometimes have to deal with snow blocking roads and causing delays. The focus of CTP, and variations like it is to plan paths/policies when graph connections are uncertain. A slight variation, known as Stochastic Shortest Path with Recourse (SSPR) [1], adds random arc costs. Our problem formulation, similar to the CTP and SSPR, provides local information during traversal. This is also consistent with the algorithm PAO^* for domains where there is a hidden state in the graph [15]. The hidden state relates this work to Partially Observable Markov Decision Process (POMDPs) [16], which provides an expressive framework for uncertain states, actions, and observations. In our case, we assume there is only uncertainty in the transition costs, which avoids computational blowup often found in large POMDPs.

Planning over the expected cumulative cost is another relevant variant of the shortest path problem [17]. Risk-Sensitive Markov Decision Process are one approach to such stochastic planning problems. These solutions address cases where large deviations from the expected behavior can have detrimental effects [18]. Previous approaches have used a parameter for risk aversion to solve Risk-Sensitive MDPs [19]. Like these techniques, we aim to avoid large deviations from the expected value while planning for low costs; however we also look to exploit local information available during execution. We build on work in risk-aware planning (e.g., Risk-Sensitive MDPs), which deal with probability distributions over outcomes. Other risk-aware planning techniques in the literature use bounding of likelihood [20] by minimizing the path length with respect to a lower bound on the probability of success. This is similar to work in [21], which instead bounds the average risk. These algorithms define reasonable ways of assigning a value for trading off between risk and a primary search objective like distance, but they do not incorporate dynamically revealed information as part of the search.

The stochastic edge cost problem has been approached using an iterative sampling method when dealing with uncertainty in terrain classification [5]. In this prior work, the edge values between landmarks are sampled from modeled cost distributions, and A^* is used over each sampling to generate a list of paths. The paths most frequently taken are considered the most likely to return low cost paths, which yields a method (*sampled A**) that we test our algorithm against. In our work, we derive a formula for reducing the risk of a path based on the uncertainty of the traversal costs themselves, which allows for a more expressive framework than heuristic searches and reduced computational requirements than sampled approaches.

3   Problem Formulation

In this paper, we consider the problem of planning and executing a risk-aware path through an uncertain environment where knowledge of the true traversal costs are revealed only as we arrive within some proximity of an area. This planning scenario can be described over a graph with edge costs initially represented by some set of distributions, with the true edge costs identified as we arrive at the parent node of each edge during the path execution. Although we focus on a path cost minimization objective in this paper, we note that this formulation could represent informative path planning objectives by considering the equivalent maximization problem [22]. We now formulate the path search problem with uncertain edge costs and introduce notation that will be used throughout the paper.

Consider a graph $\mathcal{G} = (\mathcal{V}, \mathcal{E})$ where the cost of traversing edge $E \in \mathcal{E}$ is drawn from a normal distribution $\mathcal{N}\left(\mu_E, \sigma_E^2\right)$. The cost of a path $\mathcal{P} \subset \mathcal{E}$ is the sum of the edge traversal costs, each of which are normally distributed. Thus the total path cost is drawn from $\mathcal{N}\left(\mu_{\mathcal{P}}, \sigma_{\mathcal{P}}^2\right)$, where $\mu_{\mathcal{P}} = \sum_{\mathcal{P}} \mu_E$ and $\sigma_{\mathcal{P}}^2 = \sum_{\mathcal{P}} \sigma_E^2$. This formulation is similar to that of the random network used in [1].

Assumption 1 *The true edge costs are drawn as i.i.d. samples from the respective cost distributions when queried.*

An important implication of Assumption 1 is that although multiple paths may share a subset of edges, all paths can be treated as having independent cost distributions. This is analogous to a problem where traversal costs are not fixed, but instead vary over time, and edge costs are queried upon arrival at the parent node. This assumption simplifies the computation of risk and dominance (shown later in Section 4) without significantly changing the nature of the problem.

Given any pair of start and goal vertices in the graph, $V_s, V_g \in \mathcal{V}$, the task is to traverse the graph along the *acyclic* path of least risk. More precisely, since each edge transition prunes the available set of paths to the goal, the path of least risk retains the highest probability of traversing the overall least-cost path as each transition is executed from V_s to V_g.

The decision at each vertex can be formulated by considering the next available transitions and their associated path sets. For example, say edge connections exist between the current vertex V_t and each of the vertices in the set $\mathcal{V}_{t+1} = \{A, B, C, \cdots\}$; furthermore, m acyclic paths exist from vertex A to V_g, while n acyclic paths exist from vertex B to V_g, etc. Let $\mathcal{A}$ be the set of random variables A_i, $i = \{1, \cdots, m\}$, where $A_i \sim \mathcal{N}\left(\mu_{A_i}, \sigma_{A_i}^2\right)$ represents the cost distribution of path i from vertex A to V_g (see Figure 1). Let c_{A_i} be a sample of

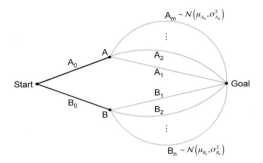

Fig. 1. Setup of pairwise comparison for sequential lookahead planning. Given the path cost distributions, we can directly compute the probability that traveling from Start to Goal through A will ultimately yield a cheaper path than traveling via B.

the random variable A_i and let c_{A_0} be the known cost of transitioning between V_t and A, then the lowest-cost path from V_t to A and onwards to V_g has cost:

$$c_{A_{min}} = c_{A_0} + \min_{A_i \in \mathcal{A}} c_{A_i}. \tag{1}$$

Similar statements can be made for path sets $\mathcal{B}, \mathcal{C}, \cdots$. To traverse the path of least-risk, each edge transition must select the next vertex $V \in \mathcal{V}_{t+1}$ such that the following probability holds,

$$P\left(c_{V_{min}} < c_{V'_{min}}\right) \geq 0.5, \quad \forall V' \in \mathcal{V}_{t+1} \setminus V. \tag{2}$$

That is, transitioning to vertex V results in a higher probability of executing a lower cost path than transitioning to any other neighboring vertex V'.

The pairwise comparisons between the available path sets obey a total ordering of preference. That is, given

$$P\left(c_{A_{min}} < c_{B_{min}}\right) = \gamma,$$
$$P\left(c_{A_{min}} < c_{C_{min}}\right) = \epsilon,$$

then,

$$\gamma \leq \epsilon \Leftrightarrow P\left(c_{B_{min}} < c_{C_{min}}\right) \geq 0.5,$$

with equality iff $\gamma = \epsilon$. Thus $|\mathcal{V}_{t+1}| - 1$ pairwise comparisons are needed to solve for the next vertex V using (2).

4   Risk-Aware Graph Search (RAGS)

4.1   Quantifying Path Risk

The pairwise comparison $P\left(c_{A_{min}} < c_{B_{min}}\right)$ describes the probability that the lowest-cost path in the set $\mathcal{A}$ is cheaper than the lowest-cost path in the set $\mathcal{B}$.

Given Assumption 1, this probability can be expanded to,

$$P\left(c_{A_{min}} < c_{B_{min}}\right) = \int_{-\infty}^{\infty} P\left(c_{B_{min}} = x\right) \cdot P\left(c_{A_{min}} < x\right) dx. \tag{3}$$

We can now express each term in the integral according to the path cost distributions of each respective set. Let,

$$f(x, \mathcal{A}) = P\left(c_{A_{min}} < x\right),$$
$$g(x, \mathcal{B}) = P\left(c_{B_{min}} = x\right).$$

Since each of the path costs are drawn from normal distributions, then

$$f(x, \mathcal{A}) = 1 - \prod_{i=1}^{m} \tfrac{1}{2}\mathrm{erfc}\left(d(A_i)\right), \tag{4}$$

$$g(x, \mathcal{B}) = \sum_{j=1}^{n} \left[\frac{1}{\sqrt{2\pi}\sigma_{B_j}} \exp\left(-d(B_j)^2\right) \cdot \prod_{\substack{k=1 \\ k \neq j}}^{n} \tfrac{1}{2}\mathrm{erfc}\left(d(B_k)\right) \right], \tag{5}$$

where $d\left(\zeta_i\right) = \frac{x - c_{\zeta_0} - \mu_{\zeta_i}}{\sqrt{2}\sigma_{\zeta_i}}$.

As an aside, note that $f'(x, \cdot) = g(x, \cdot)$. Thus, using integration by parts, we can show that

$$P\left(c_{A_{min}} < c_{B_{min}}\right) = 1 - P\left(c_{B_{min}} < c_{A_{min}}\right),$$

confirming that these two events are indeed complementary.

Equations (4) and (5) give some intuitive insight into the calculation of path risk. This formulation performs a tradeoff between the number of available paths in each set and the quality of the paths in each set, the latter represented by the means and variances of the path cost distributions. For example, $f(x, \mathcal{A})$ calculates the probability that the best path in $\mathcal{A}$ has a path cost less than x (the plot of $f(x, \mathcal{A})$ is shown in Figure 2 for $\mu_{\mathcal{A}_0} = \{20\}$ and $\sigma_{\mathcal{A}_0} = \{5\}$ as well as six other path set variants). From (4), we note that as $m \to \infty$, $f(x, \mathcal{A}) \to 1$, $\forall x \in (-\infty, \infty)$ and this is shown in the plots of $f(x, \mathcal{A}_{1,2})$. Path set $\mathcal{A}_1$ has twice as many paths (of equal means and variances) as $\mathcal{A}_0$, while $\mathcal{A}_2$ has three times as many. The curves show a trend towards an earlier and more rapid transition to 1 as m increases; however it is also apparent that adding more paths results in diminishing returns in terms of driving $f(x, \mathcal{A}) \to 1$, $\forall x \in (-\infty, \infty)$.

Other trends can be observed when adding paths of varying cost distributions to the set. $\mathcal{A}_3$ includes a second path with lower mean and equivalent variance to $\mathcal{A}_0$. The corresponding $f(x, \mathcal{A}_3)$ is shifted significantly to the left of $f(x, \mathcal{A}_0)$, causing the transition towards 1 to occur much earlier. On the other hand, the addition of a second path with higher mean and equivalent variance results in almost no change to the curve, as shown by the overlap between $f(x, \mathcal{A}_4)$ and

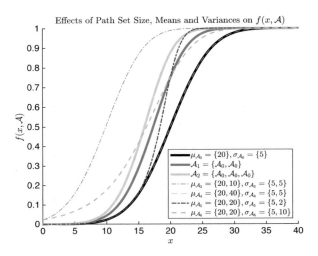

Fig. 2. The plot of $f(x, \mathcal{A})$ for seven path sets, $\mathcal{A}_0$ has a single path with cost drawn from $\mathcal{N}\left(20, 5^2\right)$. This plot shows the effects of including additional paths to the set whose costs are drawn from varying distributions.

$f(x, \mathcal{A}_0)$. Adding a path with equivalent mean but a lower variance results in the plot of $f(x, \mathcal{A}_5)$, which shows a much more rapid transition to 1 compared to either $f(x, \mathcal{A}_0)$ or $f(x, \mathcal{A}_1)$. However, the addition of a path with equal mean and higher variance increases the overall uncertainty associated with the path set, as shown by the shallower gradation in the probability curve. Furthermore, this means that although the transition towards 1 begins at lower values of x, as $x \to \infty$, $f(x, \mathcal{A}_6) \to 1$ more slowly than for $f(x, \mathcal{A}_1)$ or $f(x, \mathcal{A}_5)$.

This analysis motivates a bounding approach that compares the values of path cost means and variances to retain only the most relevant paths for the calculation of (3). Bounding the path set is especially desirable since the computation of each pairwise comparison has complexity $\mathcal{O}\left(n^2 m\right)$, where n and m are the sizes of the path sets under consideration.

4.2 Non-Dominated Path Set

Existing graph search algorithms, such as A*, use a total ordering of vertices based on the calculated cost-to-arrive to find a single optimal path between defined start and goal vertices. In contrast, an implementation of RAGS requires two sweeps of the graph, since an initial pass is required to gather cost-to-go information at each node for quantifying the path risk at execution. If all possible acyclic paths between start and goal are to be considered, then this initial pass is exponential in the average branching factor of the search tree. To reduce this computation, we introduce a partial ordering condition based on the path cost means and variances to sort the priority queue and terminate the search. This bounds the set of resultant paths to be considered during execution by accepting only those that exhibit desirable path cost mean and variance characteristics.

As discussed in Section 4.1, the addition of paths with higher mean costs to the existing path set results in little improvement in the overall risk of committing to that set. Similarly, adding paths with higher variances on the path cost results in a slower convergence of $f(x, \mathcal{A})$ to 1 as well as greater uncertainty regarding the true path cost outcomes for the set. Thus, an intuitive bounding condition is to only accept paths with lower path cost means and/or lower path cost variances. Furthermore, the same condition can be applied to the cost-to-arrive distributions to provide a partial ordering for the node expansions.

In practice, the partial ordering considers whether a path is *dominated* by an existing path, either in the open or closed set. For any two paths, A and B,

$$A \succ B \Leftrightarrow (\mu_A < \mu_B) \wedge \left(\sigma_A^2 < \sigma_B^2\right). \tag{6}$$

That is, if both the mean and variance of the cost of path B are greater than those of path A, then path A is said to *dominate* path B. This is a similar method to the one described in [13]. Using (6) to sort the priority queue guarantees that all non-dominated nodes are expanded before any dominated nodes are considered and only non-dominated paths to the goal vertex are accepted. This also allows the search to terminate once all paths in the open set are dominated by paths in the closed set. Thus the set of accepted paths from the initial search through the graph is referred to as the *non-dominated path set*. It is worth noting that the non-dominated path set is not guaranteed to contain the true optimal path; however it is guaranteed to contain the A* path calculated on the mean.

Proposition 1. *The non-dominated path set includes the A* path calculated on the mean of the edge cost distributions.*

Proof. The A* path calculated on the mean is the path with the lowest total mean cost. This path cannot be dominated by another path since the first inequality on the right hand side of (6) will never be true.   □

4.3   RAGS Dynamic Execution

Given the non-dominated path set, path execution can occur by conducting edge transitions at each node to select the path set of least risk according to (2). The true edge transition costs, which become available for all neighboring edges to the current node, are included by directly substituting the known value of c_{V_0} into (1).[1] The pseudo code for the complete RAGS algorithm is provided in Algorithm 1; an initial sweep of the graph is conducted to search for the non-dominated path set while a second sweep is conducted during execution to incorporate edge cost information as it is received.

---

[1] Note that knowledge of the true neighboring edge costs is not required for RAGS to formulate a path. The immediate transition costs c_{A_0} and c_{B_0} can be included as distributions in (3) to determine a path from start to goal. Dynamic replanning is only necessary if new edge cost information is discovered.

Algorithm 1 RISK-AWARE GRAPH SEARCH

```
    // INITIAL SWEEP
 1: Initialize open, closed, N ← ∅        ▷ Initialize open and closed sets, and node data
 2: V_s ← start, V_g ← goal
 3: N.append(V_s)
 4: open.push(N)                          ▷ Place the start node in the open set
 5: while open ≠ ∅ do
 6:     N_0 ← open.pop()                                      ▷ Current search node
 7:     V_{t+1} ← getNeighbors(N_0, G)
 8:     for V in V_{t+1} do                         ▷ Assess all neighboring vertices
 9:         N ← N_0.append(V)                               ▷ Compute child node
10:         if notAncestor(V, N_0) ∧ nonDom(N, closed) then
11:             if V = V_g then
12:                 closed.push(N)
13:             else
14:                 open.push(N)              ▷ Open set sorted according to dominance
15:         if ¬nonDom(open.top(), closed) then   ▷ End search if all nodes in the open
16:             break                            set are dominated by the closed set

    // PATH EXECUTION
17: G_{ND} ← closed             ▷ Directed graph formed by non-dominated path set
18: N ← ∅
19: V_0 ← V_s
20: while V_0 ≠ V_g do
21:     N.append(V_0)
22:     V_{t+1} ← getNeighbors(N, G_{ND})
23:     V_{ordered} ← ComparePathSets(V_{t+1}, G_{ND})  ▷ Total ordering of vertices from (3)
24:     V_0 = V_{ordered}.pop()                         ▷ Execute edge traversal
```

---

5 Comparison to Existing Search Algorithms

We compared RAGS against a *naïve A** implementation, a *sampled A** method, and a *greedy* approach. *Naïve A** finds and executes the lowest-cost path based on the mean edge costs and does not perform any replanning. The *greedy* search is performed over the set of non-dominated paths and selects the cheapest edge to traverse at each step. The *sampled A** method searches over graphs constructed by sampling over the edge cost distributions and executes the path which is most frequently found. To provide a fair comparison, the planning time of *sampled A** (related to the number of sampled graphs it plans over before selecting the most frequent path) is limited to the time RAGS needed to find a path. We chose *A** to compare against as a simplified solution to the problem and *sampled A** because the method was previously introduced in a similar domain. The *greedy* approach provides a baseline for comparison to a purely reactive implementation.

The search algorithms were tested on a set of graphs generated with a uniform random distribution of 100 vertices over a space 100×100 in size and connected according to the PRM* radius [6]. Edge costs were represented by normal distributions with mean equal to the Euclidean distance between vertices plus an additional cost drawn from a uniform random distribution over $[0, 100]$. The variance of each distribution was drawn from a uniform distribution over $[0, \sigma_{max}^2]$, where $\sigma_{max}^2 = \{5, 10, 35\}$ for the three separate sets of experiments. Note that a minimum cost of the Euclidean distance was enforced in the fol-

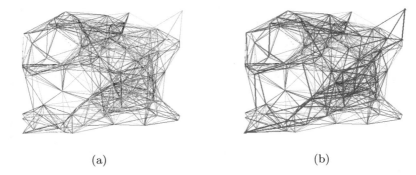

(a)   (b)

Fig. 3. A sample search graph is shown in (a). The mean cost is the sum of the Euclidean distance plus a random additional cost. Edge variances are represented using greyscale on the graph. The darker the edges, the less variance there is on the cost. The non-dominated path set (red) and the executed RAGS path (blue) shown in (b) demonstrates the algorithm's ability to account for both the path cost distributions as well as the available path options to goal. Not only does it favor traversing edges with balanced mean costs and variances, it also maintains a high number of path options towards the goal.

lowing experiments. The start vertex was defined at $V_s = (0,0)$, with the goal at $V_g = (100, 100)$. Figure 3a gives an example of a randomly generated graph, with edge variance shown in greyscale (darker lines having a smaller variance). In Figure 3b the non-dominated path set is shown in red, and the RAGS path is shown in blue.

In Figure 4 we show the results for 100 trials, where each trial drew new edge costs from the same distribution as described above. *Naïve A** performs well in the mean but is prone to outliers of more expensive paths, especially as the edge uncertainty increases due to edges with high variance along the path. RAGS is able to mitigate against such outliers by choosing safer routes with lower variances and more path options, demonstrating the benefit of risk-aware planning. Sampling all edges in the graph and planning iteratively, as in the *sampled A** approach, can account for variability in costs along the edges as long as it can sample enough paths. However this method is similarly prone to risky paths yielding high-cost outliers because it doesn't account for local information like RAGS. The performance of *greedy* search is prone to more variability in final costs as the edge cost variances increase. This trend can be attributed to the fact that the *greedy* algorithm performs search over the non-dominated path set, which becomes less representative of the true optimal path as the variances in the edge costs increase. At low cost variances, the non-dominated path set is a tighter representation of the optimal path set, and this bound becomes looser as cost variances increase. Due to the myopic nature of its planning, *greedy* is fallible to arriving at vertices with few path alternatives that all turn out to have high cost. RAGS does not suffer from the same performance decay since it accounts for the full path-to-goal cost distribution at each decision instance.

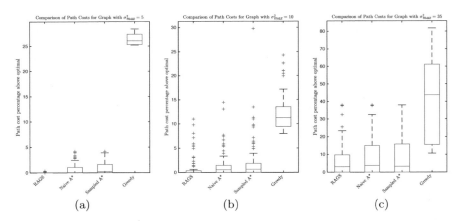

Fig. 4. Path search results over 100 samples of three graphs with uncertain edge costs; edge cost variances are drawn uniformly between 0 and $\{5, 10, 35\}$ for the three graphs (a)-(c), respectively. The difference in cost of the executed path and the true optimal path as a percentage of the optimal path cost is shown. Note that the true optimal path cost can only be calculated in hindsight.

The *greedy* results suggest that similar replanning strategies based only on the immediate edge cost updates may suffer from similar one-step lookahead myopia if downstream risks are not incorporated into the planning.

Table 1. Path Search and Execution Time (s)

σ^2_{max}	RAGS	A*	Greedy	Sampled A*
5	0.895 ± 0.150	0.102 ± 0.007	0.071 ± 0.007	109 Samples
10	$0.998 \pm .245$	$.111 \pm 0.008$	0.066 ± 0.006	111 Samples
20	$1.287 \pm .228$	$.108 \pm 0.008$	0.072 ± 0.006	130 Samples

(Sampled A* times are matched with RAGS)

The computation time averaged over 100 samples of 20 different graphs is presented in Table 1. Graph search and execution was calculated on a 3.2GHz Intel Core i7-4702HQ laptop. As expected, both $A*$ and *greedy* planning execute significantly faster than either RAGS or *sampled A**. The initial $A*$ search is faster than any of the other tested algorithms since it returns only a single path which it then executes. Similarly, the *greedy* execution-time decisions require only the comparison of immediately neighboring edge costs. The main increase in computation time over $A*$ is due to the initial sweep for the non-dominated path set. Even so, this overhead is comparatively small when considering the size of the non-dominated set. On average, this set contained 69 paths and took 0.04s longer to compute than the $A*$ path.

As discussed in Section 4.1, the computation time of RAGS is heavily influenced by the pairwise risk comparison at each execution step and thus is sensitive to the branching factor of the non-dominated path set. Although this set is pruned as edge transitions are executed, causing path traversal decisions to increasingly become faster to compute, the RAGS execution decision (3) actively attempts to maintain a large set of future path options. Despite this, these computation times suggest that real-time implementation on board a platform is feasible, and future work will investigate methods for improving the computational speeds of the algorithm.

6   Satellite Data Experiments

We also applied RAGS to a real world domain using satellite data. In robotic path planning there is often prior information available on the environment, but this information is not necessarily reliable. An example of this is overhead satellite imagery. In these trials, we used available satellite images, along with some filtering, to extract potential obstacles for a ground robot or a low-flying UAV. To convert the imagery into a useful mapping of obstacles, we performed a series of filters to provide a correspondence between obstacle density and pixel intensity. The satellite images were first converted to greyscale and were then blurred using a Gaussian filter. We then increased the contrast of the image. Finally, we eroded and then reconstructed the image to better identify trees and obstacles.

After the filtering process, the images provided a rough estimate of obstacles that could force the vehicle to slow down or fly around. In Figures 5-7, the brighter the pixel, the more likely there is to be an obstacle. The satellite information is too pixelated to provide fully reliable information, but we can use the imagery as an estimate of the obstacles in the environment. To do this we calculated the mean and variance of the pixel intensity values over an edge and used these to characterize the edge cost distributions. Similar to the previous comparison trials, the mean intensity was added to the Euclidean distance to provide a spatial scaling. Using the same method as before, we randomly sampled the space to generate a connected graph. The edges were assigned distributions, and then RAGS was used to search through the graph for a path from the top left start vertex to the bottom right goal vertex. Actual values of the edge costs were drawn from the distribution as the simulated robot moved through the terrain.

6.1   Results

We compared the performance of RAGS to the three other planning algorithms across 100 satellite images. The images are of fields with trees of varying tree densities and may also contain houses or other built structures. Images were captured at different resolutions as well as at different altitudes. The majority of the data have tree clusters scattered around the image to provide interesting path planning dilemmas. Three distinct environments are shown in Figures 5, 6, and 7. To avoid visual clutter, only the RAGS, A^* and hindsight optimal paths are shown here, compiled results for all four planning algorithms are presented in Figure 8.

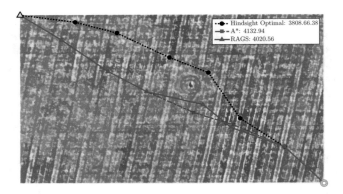

Fig. 5. The RAGS, A^*, and the global optimal (known only in hindsight) paths are shown in the figure with the final path costs. The goal is to traverse from the top left start vertex to the bottom right goal vertex. The empty field is a test case showing that both algorithms plan direct paths as expected in an obstacle-free environment.

Fig. 6. Paths are planned through a dense cluster of trees surrounding the goal. Here we can see RAGS plans around the cluster, where there are more low cost paths, before traversing through a less cluttered area. RAGS takes advantage of the wide open region instead of searching for narrow tracks within the tree cluster.

In Figure 5 the paths through an empty field are straight from start (yellow triangle) to goal (orange circles). As expected there is little variance in edge costs, and the trajectories for RAGS and A^* are quite similar. Analyzing the paths found in Figure 6 is more interesting. Here we see the benefit of RAGS in obstacle-dense environments. The path from start to goal is blocked by a large cluster of semi-permeable forest. A^* executes a path through the center of the cluster that has a low cost in the mean but does not allow for easy deviations if the path is found to be untraversable. On the other hand, the path executed by RAGS demonstrates the nonmyopic nature of this algorithm. RAGS selects

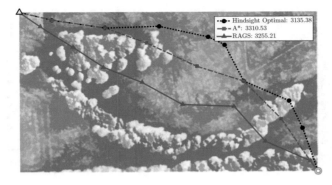

Fig. 7. The RAGS path is shown traveling straight through the initial cluster of trees to take full advantage of the direct route through the open field. A^* finds a path that does not navigate completely around the trees and so incurs additional traversal costs on top of taking a less direct route. The costs after execution show that RAGS is actually a cheaper path due to balancing the risk of finding a path through the initial tree cluster that connects to a more direct path to goal. This demonstrates the benefit of RAGS, knowing when to take risks and when to act conservatively.

a path that travels part way around the cluster to minimize the portion of the path within the dense section of the forest. Values are drawn from the edge distributions to calculate what would have been the optimal path in hindsight. The optimal path (known only after execution) is shown in black, and we can see that it follows a similar trajectory as RAGS.

The final test case can be seen in Figure 7, where a small cluster of trees stands between a direct path from the start to the goal. In blue, RAGS plans a path through the trees that is able to take advantage of the clearing in the center. In this example, RAGS is able to assess the risk of taking the shorter, more direct route through the trees and compare this to the expected cost of traveling around the cluster. In comparison, the A^* solution finds a less direct route to the goal; however it does not navigate completely around the trees and so incurs additional traversal costs on top of taking a longer path.

The compiled results for all tested algorithms are shown in a box plot of percent above the hindsight optimal in Figure 8. From the comparison on the three randomly generated graphs in Section 4.3, we showed that the relative performance of RAGS increases as edge variance increases. This is accounted for by the fact that the other planning algorithms are merely searching over the single heuristic of mean traversal cost. If variance is low then this can be enough to solve for a path that is close to the optimal solution. However, Figure 8 reveals that real world data sets can contain significant noise, and it is valuable to account for that variability during planning. This is especially evident in the *sampled* A^* result, which shows an overall lower mean cost but contains significant outliers (over twice as expensive) compared to the worst-case RAGS path.

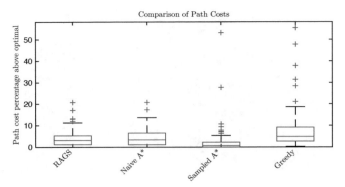

Fig. 8. Results from satellite data experiment, using 100 images. Box plots represent the path cost percent above what would have been in hindsight the optimal path. The same trends in performance are seen as with the simulated graphs. RAGS accounts for uncertainty in traversing the world, balancing risk of travel time (cost) to reduce expensive outliers seen in the comparative algorithms.

7   Conclusion

In this paper we have introduced a novel approach to incorporating and address-ing uncertainty in planning problems. The proposed RAGS algorithm combines traditional deterministic search techniques with risk-aware planning. RAGS is able to trade off the number of future path options, as well as the mean and variance of the associated path cost distributions to make online edge traversal decisions that minimize the risk of executing a high-cost path. The algorithm was compared against existing graph search techniques on a set of graphs with randomly assigned edge costs, as well as over a set of graphs with traversability costs generated from satellite imagery data. In all cases, RAGS was shown to reduce the probability of executing high-cost paths over A^*, *sampled A^** and a *greedy* planning approach.

Our next step will be to implement the RAGS algorithm in an informa-tion optimization domain. In this case, edge cost probability distributions will represent environmental information, and paths will be generated based on the amount of information the vehicle may collect in different parts of the map. Additionally, we will investigate methods to reduce the computational complex-ity and memory requirements of the algorithm for implementation on a robotic platform.

Acknowledgements

This research has been funded in part by NASA grant NNX14AI10G and Office of Naval Research grant N00014-14-1-0509.

References

1. Polychronopoulos, G.H., Tsitsiklist, J.N.: Stochastic shortest path problems with recourse. Networks **27** (1996) 133–143

2. Hart, P.E., Nilsson, N.J., Raphael, B.: A formal basis for the heuristic determination of minimum cost paths. IEEE Transactions on Systems Science and Cybernetics **4**(2) (1968) 100–107
3. Koenig, S., Likhachev, M.: Fast replanning for navigation in unknown terrain. IEEE Transactions on Robotics **21**(3) (2005) 354–363
4. Hollinger, G.A., Pereira, A.A., Binney, J., Somers, T., Sukhatme, G.S.: Learning uncertainty in ocean current predictions for safe and reliable navigation of underwater vehicles. Journal of Field Robotics **33**(1) (2016) 47–66
5. Murphy, L., Newman, P.: Risky planning on probabilistic costmaps for path planning in outdoor environments. IEEE Transactions on Robotics **29**(2) (2013) 445–457
6. Karaman, S., Frazzoli, E.: Sampling-based algorithms for optimal motion planning. International Journal of Robotics Research **30**(7) (2011) 846–894
7. Latombe, J.C.: Robot Motion Planning. Springer Science & Business Media (2012)
8. LaValle, S.M.: Planning Algorithms. Cambridge University Press (2006)
9. Bohren, J., Rusu, R.B., Jones, E.G., Marder-Eppstein, E., Pantofaru, C., Wise, M., Mösenlechner, L., Meeussen, W., Holzer, S.: Towards autonomous robotic butlers: Lessons learned with the PR2. In: Proc. IEEE International Conference on Robotics and Automation. (2011) 5568–5575
10. Kalman, R.E.: A new approach to linear filtering and prediction problems. Journal of Fluids Engineering **82**(1) (1960) 35–45
11. Kurniawati, H., Hsu, D., Lee, W.S.: SARSOP: Efficient point-based POMDP planning by approximating optimally reachable belief spaces. In: Robotics: Science and Systems. (2008)
12. Chaves, S.M., Walls, J.M., Galceran, E., Eustice, R.M.: Risk aversion in belief-space planning under measurement acquisition uncertainty. In: Proc. IEEE/RSJ International Conference on Intelligent Robots and Systems. (2015) 2079–2086
13. Bry, A., Roy, N.: Rapidly-exploring random belief trees for motion planning under uncertainty. In: Proc. IEEE International Conference on Robotics and Automation. (2011) 723–730
14. Papadimitriou, C.H., Yannakakis, M.: Shortest paths without a map. Theoretical Computer Science **84**(1) (1991) 127–150
15. Ferguson, D., Stentz, A., Thrun, S.: PAO for planning with hidden state. In: Proc. IEEE International Conference on Robotics and Automation. Volume 3. (2004) 2840–2847
16. Monahan, G.E.: State of the Art—-A survey of partially observable Markov decision processes: Theory, models, and algorithms. Management Science **28**(1) (1982) 1–16
17. Hou, P., Yeoh, W., Varakantham, P.R.: Revisiting risk-sensitive MDPs: New algorithms and results. In: Proc. International Conference on Automated Planning and Scheduling. (2014)
18. Carpin, S., Chow, Y.L., Pavone, M.: Risk aversion in finite Markov Decision Processes using total cost criteria and average value at risk. In: Proc. IEEE International Conference on Robotics and Automation. (2016) 335–342
19. Marcus, S.I., Fernández-Gaucherand, E., Hernández-Hernandez, D., Coraluppi, S., Fard, P.: Risk sensitive Markov decision processes. In: Systems and Control in the Twenty-First Century. Springer (1997) 263–279
20. Sun, W., Patil, S., Alterovitz, R.: High-frequency replanning under uncertainty using parallel sampling-based motion planning. IEEE Transactions on Robotics **31**(1) (2015) 104–116
21. Feyzabadi, S., Carpin, S.: Risk-aware path planning using hierarchical constrained Markov Decision Processes. In: Proc. IEEE International Conference on Automation Science and Engineering. (2014) 297–303
22. Meliou, A., Krause, A., Guestrin, C., Hellerstein, J.M.: Nonmyopic informative path planning in spatio-temporal models. In: Proc. AAAI Conference. (2007) 602–607

Symbolic Computation of Dynamics on Smooth Manifolds

Brian Bittner and Koushil Sreenath

Dept. of Mechanical Engineering, Carnegie Mellon University, Pittsburgh, USA.

Abstract. Computing dynamical equations of motion for systems that evolve on complex nonlinear manifolds in a coordinate-free manner is challenging. Current methods of deriving these dynamical models is only through cumbersome hand computations, requiring expert knowledge of the properties of the configuration manifold. Here, we present a symbolic toolbox that captures the dynamic properties of the configuration manifold, and procedurally generates the dynamical equations of motion for a great variety of systems that evolve on manifolds. Many automation techniques exist to compute equations of motion once the configuration manifold is parametrized in terms of local coordinates, however these methods produce equations of motion that are not globally valid and contain singularities. On the other hand, coordinate-free methods that explicitly employ variations on manifolds result in compact, singularity-free, and globally-valid equations of motion. Traditional symbolic tools are incapable of automating these symbolic computations, as they are predominantly based on scalar symbolic variables. Our approach uses Scala, a functional programming language, to capture scalar, vector, and matrix symbolic variables, as well as the associated mathematical rules and identities that define them. We present our algorithm, along with its performance, for computing the symbolic equations of motion for several systems whose dynamics evolve on manifolds such as $\mathbb{R}$, $\mathbb{R}^3$, S^2, $SO(3)$, and their product spaces.

1 Introduction

Computing dynamics of systems directly on nonlinear manifolds involves laborious manual computation and expert knowledge of the properties of the configuration space. The motivation for such computation lies in the end result of obtaining a compact, globally valid, and singularity free dynamical model. These traits enable the use of geometric controllers to obtain almost global stability properties. While such dynamical models for certain popular, modestly complex mechanical systems are readily available (through hand computations), there exists no procedural or automatic way to compute the symbolic dynamics on manifolds. The primary reason is that we do not have a way to capture non-scalar symbolic variables and perform non-scalar symbolic arithmetic. The contribution of this paper addresses this gap.

© Springer Nature Switzerland AG 2020
K. Goldberg et al. (Eds.): *Algorithmic Foundations of Robotics XII*, SPAR 13, pp. 336–351, 2020.
https://doi.org/10.1007/978-3-030-43089-4_22

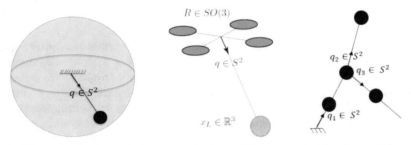

Fig. 1: The spherical pendulum, quadrotor with suspended load, and three-link walker systems that evolve on complex nonlinear manifolds. The dynamics for each are symbolically computed using the presented framework, see codebase at https://gitlab.com/HybridRobotics/dynamics_on_manifolds.

The paper can be summarized by the following: For systems that evolve on manifolds (see Figure 1), traditional ways to obtain dynamical models involve taking a local parametrization of the configuration space, such as Euler angles, and then solving for the dynamics. However, obtaining the dynamics on manifolds, without local parametrization, results in a dynamical model that is condensed to a compact expression, is globally valid, and is in a singularity free form. But, there exist no tools that can automate solving for the symbolic dynamics on manifolds. We provide a symbolic algebraic framework to capture geometric axioms and identities that govern the **scalar, vector, and matrix** elements of dynamical systems. An algorithm is shown to automate computation of the dynamics of simple mechanical systems on cartesian products of $\mathbb{C}^n$, $\mathbb{R}^n$, S^2, and $SO(3)$, as well as including computation of dynamics of a set of complex robotic systems.

The rest of the introduction places our proposed algorithm amongst prior work involving dynamics solvers and computer algebra systems, followed by an outline of the paper. With respect to dynamics solvers, Euler-Angle parametrization has long been the standard for system development. Introducing these local parametrizations of manifolds enables the use of well-established analytical methods and corresponding software algorithms for deriving the equations of motion. For instance, the Newton-Euler method is implemented in Neweul [16] and SD/FAST [14]. The Recursive Newton-Euler Algorithm has been used to efficiently compute dynamics of serial chain manipulators, [7,9]. The Lagrangian method has been implemented on various multibody systems [22]. Additionally, the Piogram Method [5], the Composite Rigid Body Algorithm [11], and the Articulated Body Algorithm [11], all provide efficient algorithmic approaches for solving the dynamical equations of long kinematic chains. Commercial systems such as ADAMS [24], SimMechanics, MapleSim, LMS Dads [27], etc., not only solve the equations of motion, but also offer efficient simulations. Symoro [15] allows for symbolic modeling and simulation for high dimensional kinematic tree-structures and closed kinematic chains. Although there are several efficient methods and corresponding software tools that automate the computation of equations of motion, every single one of them is based on local parametrizations that employ *scalar* symbolic variables.

Automating symbolic computation on nonlinear manifolds requires treating vectors and matrices as single-entity symbolic variables with operations defined directly on them, which existing computing tools do not support. We achieve this by creating symbolic variables that represent either scalars, vectors, or matrices, with additional properties such as constants, unit vectors, symmetric matrices, etc. Moreover, using a functional programming approach, we enable a programmatic understanding of mathematical axioms and identities between these variables.

We looked at the field's best reputed computer algebra systems to encapsulate our dynamics framework. Commonly used Computer Algebra Systems include Mathematica [3] and Matlab's Symbolic Math Toolbox [2], (see [1] for full list of CASs). These only support scalar symbolic variables (vector and matrix symbolic variables exist only indirectly through scalars.) ATLAS 2 [10], an extension of Mathematica, enables high quality visualizing of n-dimensional manifolds, but is not applicable for our computation since it also requires local frame parametrizations. Our solution is to use Scala, a functional programming language with a vast set of object oriented programming capabilities [25]. This enables the design of strong expression classes and an intuitive functional programming environment for capturing mathematical identities. This coupled with existence of an extendable ScalaMathDSL [13] package made Scala preferable to the frameworks SymPy [30] and Axiom [8]. We extend prior work on scalar expressions in ScalaMathDSL [13] to handle vector and matrix classes, resulting in a powerful framework to implement the dynamics algorithm. Note that our modeling tool does not provide any type of numerical computation.

The rest of the paper is structured as follows: Section II, covers the mathematical background. Section III details the dynamics framework and walks through the algorithm to automate computation of the symbolic dynamics. Section IV tabulates results of applying the dynamics solver on a set of mechanical systems. Finally Sections V and VI provide a discussion and concluding remarks.

2   Mathematical Background

Many robotic systems evolve on nonlinear manifolds. However, we typically compute the dynamics of these systems through local parametrizations, such as Euler angles. Robotic systems are hindered by local parametrizations that induce singularities, model complexity, and unwanted features such as the unwinding phenomena [4] during control. The Lagrange-d'Alembert principle (see [20], [21], [18] for more details.), detailed in Section 2.1, allows the computation of dynamics of systems evolving on complex, nonlinear manifolds to arrive at singularity-free, globally valid, and compact equations of motion.

2.1   Lagrange d'Alambert Principle for Computing Dynamics

Let the degrees of freedom for a dynamical system exist on the configuration manifold Q. Then, the Lagrangian of the system, $\mathcal{L} : TQ \to \mathbb{R}$ is defined as $\mathcal{L} = \mathcal{T} - \mathcal{U}$, where $\mathcal{T} : TQ \to \mathbb{R}$ and $\mathcal{U} : Q \to \mathbb{R}$ are the kinetic and potential energies respectively. Here TQ is the tangent bundle of the configuration space. The virtual work $\mathcal{W} : Q \to \mathbb{R}$ captures energy input to the system via actuation. The dynamical equations of motion can be computed by minimization of the action integral S through the Least Action Principle [12], where the action integral is defined as:

$$S = \int_0^t \mathcal{L} \, \mathrm{d}t + \int_0^t \mathcal{W} \mathrm{d}t. \tag{1}$$

The minimization is done by setting the variation of the action integral to zero, i.e., $\delta S = 0$.

2.2   Computation on S^2 Manifold

To concretely illustrate the procedure, we carry out the following computational steps to derive the equations of motion of the spherical pendulum on S^2 [19]. (To observe the computation on $SO(3)$ the reader is directed to [6].)

Step 1: Supply the Lagrangian ($\mathcal{L}$) and virtual work ($\mathcal{W}$), providing the components of the action integral (S) for the spherical pendulum, as given below,

$$\mathcal{L} = \frac{1}{2} m l^2 (\dot{q} \cdot \dot{q}) - mgl(q \cdot e_3), \qquad \mathcal{W} = q \cdot \tau,$$

$$S = \int_0^t \frac{1}{2} m l^2 (\dot{q} \cdot \dot{q}) - mgl(q \cdot e_3) \, \mathrm{d}t + \int_0^t q \cdot \tau \, \mathrm{d}t.$$

Here unit vector $q \in S^2 \subset \mathbb{R}^3$ has the property $q \cdot q = 1$ and kinematic constraint $q \cdot \dot{q} = 0$. The constant scalars $(m, g, l) \in \mathbb{R}$ correspond to the mass of the pendulum, magnitude of acceleration due to gravity, and length of the pendulum respectively. Constant $-e_3 \in \mathbb{R}^3$ provides the direction of the gravitational field, while $\tau \in \mathbb{R}^3$ defines the torque via virtual work at the base of the pendulum. Generalized vector $\xi \in \mathbb{R}^3$ is orthogonal to the evolution of q such that the variation of q is given as, $\delta q = \xi \times q$.

Step 2: Take the variation of the action integral and set to zero:

$$\delta S = \int_0^t m l^2 (\delta \dot{q} \cdot \dot{q}) - mgl(\delta q \cdot e_3) \, \mathrm{d}t + \int_0^t \delta q \cdot \tau \, \mathrm{d}t = 0.$$

Step 3: Replace variations δq and $\delta \dot{q}$ with the generalized vector ξ:

$$\int_0^t m l^2 ((\dot{\xi} \times q + \xi \times \dot{q}) \cdot \dot{q}) - mgl(\xi \times q) \cdot e_3) \, \mathrm{d}t + \int_0^t (\xi \times q) \cdot \tau \, \mathrm{d}t = 0.$$

Step 4: Group equations with respect to $\dot{\xi}$ and ξ and apply simplifications (e.g.: $\dot{q} \times \dot{q} = 0$):

$$\int_0^t \dot{\xi} \cdot (m l^2 q \times \dot{q}) + \xi \cdot (m l^2 \dot{q} \times \dot{q} - mglq \times e_3 + q \times \tau) \, \mathrm{d}t = 0.$$

Step 5: Perform integration by parts on $\dot{\xi}$:

$$\int_0^t -\xi \cdot (ml^2(\dot{q} \times \dot{q} + q \times \ddot{q})) + \xi \cdot (-mglq \times e_3 + q \times \tau) \, \mathrm{d}t = 0.$$

Step 6: Collect for ξ and simplify (e.g.: $\dot{q} \times \dot{q} = 0$):

$$\int_0^t \xi \cdot (-ml^2 q \times \ddot{q} - mglq \times e_3 + q \times \tau) \, \mathrm{d}t = 0.$$

Step 7: These equations of motion can be extracted to describe the dynamics:

$$\dot{q} = \omega \times q,$$

$$q \times (ml^2 \ddot{q} + mgle_3) = q \times \tau,$$

where the first identity is a kinematic identity (derived from q's action on S^2) with ω being the angular velocity, and the second identity obtained from Step 6 by setting the term that is dotted with ξ as zero. This variation-based computation returns a set of compact, globally-valid equations of motion that are singularity-free. Having discussed the mathematical background, we will next describe the computational framework of our dynamics solver that automates this method.

3   Computational Design of Dynamics Solver

This section will provide a description of the data structures used to represent mathematical objects and cover the algorithmic techniques that operate on them. Critical computational steps in the algorithm are covered, followed by a user-level code snippet that implements the symbolic computation of dynamics on a spherical pendulum.

3.1   Class Hierarchies for Mathematical Expressions

In Section II we used symbolic scalars (m, l, g) and symbolic vectors (q, $\dot{q}$, e_3) to compute the dynamics for the spherical pendulum. These two expression types along with symbolic matrices make up the *base classes* used to implement the dynamics solver. The class hierarchy consists of each base class (ScalarExp, VectorExp, MatrixExp) and its children, which are comprised of a set of case classes. As we will see, *case classes* enable pattern matching to encode mathematical rules (e.g.: the chain rule for differentiation, scalar and vector triple product rules, etc.) Each case class can inherit *traits* to possess additional properties (e.g.: vectors having unit magnitude or matrices having skew symmetric properties.) The unit vector trait allows dynamics to be computed on the S^2 configuration space. The matrix traits constant, skew, and symmetric enable distinction of skew, skew symmetric, and rotation matrices. This enables direct and intelligent computation of dynamics on $SO(3)$. Each case class serves as a unique mathematical entity in the symbolic evaluator, see Figure 2 for the class hierarchy. Next we will discuss how these case classes interact to create a meaningful data structure for symbolic computation.

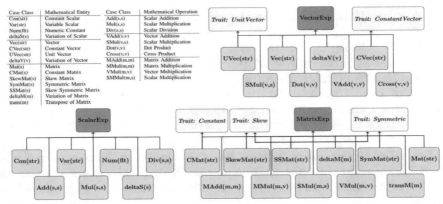

Fig. 2: All mathematical expressions inherit from a parent class corresponding to its identity as a scalar, vector, or matrix. Class traits are inherited to add additional properties to symbolic expressions.

3.2 Abstract Syntax Trees

Our mathematical expressions are stored as abstract syntax trees (ASTs). Mathematical operations, such as addition and multiplication, are represented as parent nodes which contain a specified number of leaves appropriate for their respective functionality. Leaves on the other hand represent symbolic scalars, vectors, and matrices. Mathematical algorithms are then written as functional programs with the ASTs serving as the input and output data structures. Figure 3 provides an AST of the Lagrangian for the spherical pendulum. As detailed next, ASTs are coupled with pattern matching functions to encode mathematical rules.

3.3 Embedding Mathematical Axioms

Writing the framework in Scala, a functional and object oriented programming language, we are able to succinctly embed important mathematical axioms governing the properties of the configuration manifolds. Because the elements of an AST are each designated a case class with inherited mathematical traits, this structure can be quickly parsed to enforce mathematical axioms and identities, which are used to simplify the dynamics and perform calculus operations.

3.4 Mathematical Operations

Each mathematical step in the computational pipeline, is achieved through specialized pattern matching functions. These functions perform a mathematical evaluation or restructuring of the AST. Pattern matching via the extensive class hierarchy for elements of manifold dynamics enables brief, yet powerful functions that are intuitive to write and understand.

To illustrate this, we will walk through the various functions implemented to execute each step of the variation-based method. Accompanied by illustrations of tree manipulation for a spherical pendulum, we discuss the procedural nature of each step of the algorithm (steps enumerated 1-7). We will talk about trees of n nodes containing p parents (operands) and e literals (mathematical elements).

Tree Conditioning: For the steps below, the tree conditioning is defined as a full expansion of the expression, (i.e. $a*(b+c)$ is converted to $a*b+a*c$). Isolating the dynamical terms enable succinct, procedural code to be written for tree operations. Expanding the tree requires traversing the tree (covered next) and invokes 3 parent operations per expansion ($a*b+a*c$). A maximum of $e-1$ expansions can take place during traversal. Table 1 provides a condensed layout of the computational cost for the algorithm.

Tree Traversal: Each algorithm requires traversal of the tree, from the top down. The general run time for this case is $O(n)$. If we balance the tree and parallelize the operations (allowed by the structured commutativity above), we could theoretically assert an $O(log(n))$ traversal time. The algorithm enables fast computation of the dynamics (see Results Tables 2 and 3), with most systems requiring computation time on the order of milliseconds.

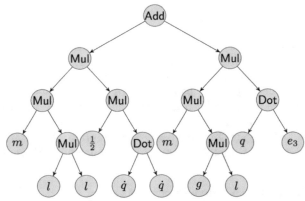

Fig. 3: **Step 1:** Abstract Syntax Tree representation of the Lagrangian of a spherical pendulum, given by $\mathcal{L} = \frac{1}{2}ml^2(\dot{q} \cdot \dot{q}) - mgl(q \cdot e_3)$. ASTs are the data structure used to represent all symbolic expressions.

Step 1 requires initialization of the Lagrangian AST. This involves supplying the information of the mechanical configuration and energy of the system, resulting in the input AST shown in Figure 3. The Lagrangian AST of e elements is constructed in $e-1$ operations.

Step 2 takes the variation of the Lagrangian. To respect space constraints, we display variation of only the kinetic energy component of the Lagrangian AST in Figure 4a. Taking the variation of an element changes its string identifier and possibly its case class. For example, δq loses q's trait as a unit vector. This function also applies the chain rule, $\delta(u*v) = \delta u*v + u*\delta v$. The chain rule invokes 3 parent operations and can be applied $(e-1)$ times. The variation function is run on each element once. Taking the time derivative has equivalent structure to this algorithm, but generates elements of different properties.

Step 3 shown in Figure 4b, applies constraints through substitution. The tree is searched for a node containing the expression candidate for replacement. In this case, when $\delta\dot{q}$ is found, it is replaced with $\dot{\xi} \times q + \xi \times \dot{q}$. Substitution allows for the variations and kinematic constraints of the system to be enforced, along with simplifying expressions such as $\dot{q} \times \dot{q} = 0$. Each node is checked once. The tree is then **reconditioned** to fully expanded form.

Step 4 collects an expression with respect to a vector or scalar. Figure 4c illustrates collecting for the generalized vector $\dot{\xi}$. For each isolated energy term, each element is checked for a match to the collected expression. A collected scalar is set as the leftmost multiplied element, allowed by commutativity. A collected vector is placed on the left hand side of the dot product by vector algebraic axioms such as the scalar triple product. Each collection requires 2 parent operations. Examples collecting for a include $b * (a * c) = a * (b * c)$ for scalars and $b \cdot (a \times c) = a \cdot (b \times c)$ for vectors.

Step 5 performs integration by parts on the dot product containing the collected generalized vectors from the previous step. In Figure 4d, the mathematical operation is enforced by negating and integrating $\dot{\xi}$ while differentiating the right hand side of the dot product. The result leaves the expression entirely in terms of generalized vector ξ. Since integration by parts requires two expressions, the negation, integration, and differentiation involved can operate on up to $e - 1$ elements. The tree is **reconditioned** again. This allows for the expression to be **collected with respect to ξ which is done in Step 6.** This allows for the right hand side of the dot product to be set to zero, resulting in the dynamical model obtained in **Step 7, where the dynamics can be extracted line by line,** one group of terms set to zero for each generalized vector. One parent operation is required to remove a term from the tree. This can occur e times.

Process	Traversal	Parent Operations	Literal Operations
Step 1:	–	$e - 1$	e
Expand:	n	$3(e - 1)$	–
Step 2:	n	$3(e - 1)$	e
Step 3:	n	–	e
Expand:	n	$3(e - 1)$	–
Step 4:	n	$2(e - 1)$	–
Step 5:	n	$2(e - 1)$	$3(e - 1)$
Expand:	n	$3(e - 1)$	–
Step 6:	n	$2(e - 1)$	–
Step 7:	n	e	–
Run Time:	$O(n)$	$O(e)$	$O(e)$

Table 1: This table presents the run time of critical steps in the computation.

Linear run time convergence enables broad inspection of high degree of freedom dynamics, the laborious computational nature of which made model extraction previously intractable. The coupled simplicity and effectiveness of the algorithm is enabled by computing on coordinate free configuration spaces. There is no trigonometric book-keeping, which grows rapidly for systems described on S^1 and $SO(2)$.

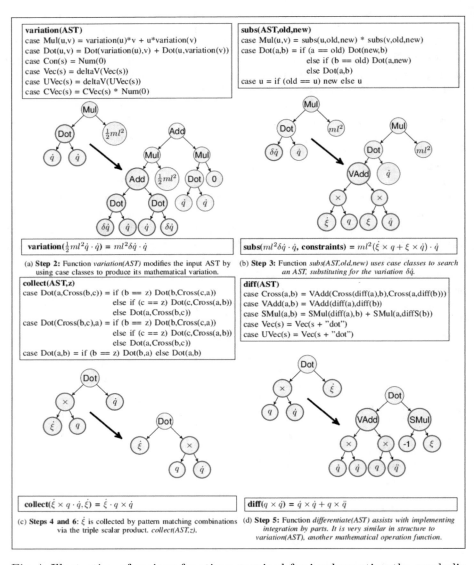

Fig. 4: Illustration of various functions required for implementing the symbolic dynamics solver. Shown for each function are the input and output mathematical expressions, the corresponding input and output ASTs (with changes highlighted in red), and the corresponding code-snippets that invoke the tree manipulation.

3.5 Example with Spherical Pendulum

Listing 1 illustrates the user code for deriving the dynamics of a spherical pendulum. The user provides the Lagrangian and specifies the actuated variables of the system. These are input to the *computeDynamics* function (which implements the algorithm detailed in Section 3.4 to derive the dynamics for the system. Specification of configuration variables and manifold based constraints is not required, since the constraint information is encapsulated within mathematical type declaration of each variable. This simple example code will enable users to effortlessly set up problems for solving the dynamics of new robotic systems. Constraint manifolds are duduced from configuration variable types, i.e. $q \in S^2$ has kinematic constraint $\delta q = \xi \times q$. The codebase is available at `https://gitlab.com/HybridRobotics/dynamics_on_manifolds`.

```
object SphericalPendulum {
    def main() {
        // define constant scalars
        val g = Cons("g")   // gravitational constant
        val m = Cons("m")   // point mass
        val l = Cons("l")   // length at which point mass hangs

        // define vectors
        val e3  = CVec(  "e3")   // orientation of gravity
        val q   = UVec(  "q")    // point mass acts on S^2
        val u   = Vec(  "u")     // virtual work done on system

        // set configuration variables
        // (scalars, vectors, matrices)
        val configVars = Tuple3(List(), List(q), List())

        // define lagrangian
        val KE = Num(0.5) * m * l * l * (diffV(q) dot diffV(q))
        val PE = m * g * l * (q dot e3)
        val L = KE - PE

        // specify infinitesimal virtual work of system
        val infWork = Dot(deltaV(q),u)

        var eoms = computeDynamics(L, W, configVars)}
```

Listing 1: User code to compute dynamics for a spherical pendulum.

4 Results of Computing Symbolic Models

Having presented an overview of our computational framework for the symbolic dynamics solver, we next test this across many robotic systems with dynamics evolving on several different manifolds. Table 1 provides results on simple pendulum systems with dynamics on manifolds S^2 and $SO(3)$. Table 2, on the other hand, introduces real robotic systems evolving on highly complex manifolds. The user supplies the Lagrangian, Virtual Work, and configuration variables. The table lists the runtime and equations produced for each mechanical system. Next, we briefly describe the systems tabulated in Table 1, for which our proposed symbolic algorithm computes the dynamics on manifolds.

4.1   Simple Mechanical Systems

A **Spherical Pendulum** [6] is the first system considered, its parametrization is described in section 2.2.
Configuration Manifold: S^2
Kinematic Constraints: $\dot{q} \cdot q = 0$
Variations: $\delta q = \xi \times q$, $\dot{\delta q} = \dot{\xi} \times q + \xi \times \dot{q}$, $\xi \in \mathbb{R}^3$
Energy Terms:

$$T = \frac{1}{2}ml^2(\dot{q} \cdot \dot{q}), \qquad U = mgl(q \cdot e_3).$$

The **3D Pendulum** [28] provides a system which acts on the $SO(3)$ manifold. The pendulum is represented as a rigid body of mass m and symmetric inertia matrix $J \in \mathbb{R}^{3 \times 3}$. It is pivoted at the point from which vector $\rho \in \mathbb{R}^3$ extends toward the center of mass. The orientation of the 3D pendulum is specified by rotation matrix $R \in SO(3)$ with $\Omega, M \in \mathbb{R}^3$ being the angular velocity and moment input of the 3D pendulum respectively. The generalized vector $\eta \in \mathbb{R}^3$ is utilized to describe the variation of R on $SO(3)$.
Configuration Manifold: $SO(3)$
Kinematic Constraints: $\dot{R} = R\hat{\Omega}$
Variations: $\delta R = R\hat{\eta}$, $\delta \Omega = \hat{\Omega}\eta + \dot{\eta}$, $\eta \in \mathbb{R}^3$
Energy Terms:

$$T = \frac{1}{2}\Omega \cdot J\Omega, \qquad U = mgR\rho \cdot e_3.$$

A **Double Spherical Pendulum** [20] is an extension of the singular Spherical Pendulum. A second pendulum with point mass m_2 is pivoted from the end of the first pendulum containing point mass m_1. Each pendulum has a respective length l_i from its pivot to point mass. The unit vector $q_i \in S^2$ points along each l_i with its origin at the respective pivots. The torque at the base of each pendulum is given by $\tau_i \in \mathbb{R}^3$.
Configuration Manifold: $S^2 \times S^2$
Kinematic Constraints: $\dot{q}_i \cdot q_i = 0$
Variations: $\delta q_i = \xi_i \times q_i$, $\dot{\delta q_i} = \dot{\xi}_i \times q_i + \xi_i \times \dot{q}_i$, $\xi_i \in \mathbb{R}^3$
Energy Terms:

$$T = \frac{1}{2}(m_1 + m_2)l_1^2\dot{q}_1 \cdot \dot{q}_1 + m_1 l_1 l_2 \dot{q}_1 \cdot \dot{q}_2 + \frac{1}{2}m_2 l_2^2 \dot{q}_2 \cdot \dot{q}_2$$
$$U = (m_1 + m_2)gl_1 q_1 \cdot e_3 + m_2 gl_2 q_2 \cdot e_3$$

Similarly, the **Double 3D Pendulum** [17] is an extension of the Singular 3D Pendulum. The rigid body of mass m_2 is pivoted from the point that vector $\lambda \in \mathbb{R}^3$ points to. The rigid bodies' orientations at point masses m_i are specified

System	Config. Manifold	DOF	Time to Compute	Equations of Motion Produced by our Symbolic Dynamics Solver
 Spherical Pendulum [6]	S^2	2	0.019 sec.	$\dot{q} = \omega \times q$ $q \times (ml^2\ddot{q} + mgle_3) = q \times \tau$
 3D Pendulum [28]	$SO(3)$	3	0.015 sec.	$\dot{R} = R\hat{\Omega}$ $J\dot{\Omega} + \Omega \times J\Omega = mg\rho \times R^T e_3 + M$
 Double Spherical Pendulum [19]	$S^2 \times S^2$	4	0.032 sec.	$\dot{q}_1 = \omega_1 \times q_1$ $q_1 \times ((m_1 + m_2)l_1^2\ddot{q}_1 + m_2l_1l_2\ddot{q}_2 +$ $(m_1 + m_2)gl_1e_3) = q_1 \times \tau_1$ $\dot{q}_2 = \omega_2 \times q_2$ $q_2 \times (m_2l_1l_2\ddot{q}_1 + m_2l_2^2\ddot{q}_2 + m_2gl_2e_3) =$ $q_2 \times \tau_2$
 Double 3D Pendulum [17]	$SO(3) \times SO(3)$	6	0.033 sec.	$(J_1 - m_1\hat{\lambda}^2 - m_2\hat{\rho}_2^2)\dot{\Omega}_1 +$ $m_2\hat{\rho}_1 R_1^T R_2\hat{\rho}_2\dot{\Omega}_2 + \hat{\Omega}_1(J_1 - m_1\hat{\lambda}^2 -$ $m_2\hat{\rho}_1^2)\Omega_1 + m_1 g\lambda R_1^T e_3 +$ $m_2 g\hat{l}_1 R_1^T e_3 + m_2\hat{\rho}_1 R_1^T R_2\hat{\Omega}_2^2\lambda = M_1$ $(J_2 - m_2\hat{\rho}_2^2)\dot{\Omega}_2 - m_2\hat{\rho}_2 R_2^T R_1\hat{\rho}_1\dot{\Omega}_1 +$ $\hat{\Omega}_2(J_2 - m_2\hat{\rho}_2^2)\Omega_2 + m_2 g\hat{\rho}_2 R_2^T e_3 +$ $m_2\hat{\rho}_2 R_2^T R_1\hat{\Omega}_1^2\rho_1 = M_2$

Table 2: The equations of motion were computed on these simple systems using the proposed symbolic computation tool with the kinematic constraints, variations, and Lagrangian (T-U) provided for each system.

by rotation matrices R_i.

Configuration Manifold: $SO(3) \times SO(3)$

Kinematic Constraints: $\dot{R} = R\hat{\Omega}$

Variations: $\delta R_i = R_i \hat{\eta}_i$, $\delta \Omega_i = \hat{\Omega}_i \eta_i + \dot{\eta}_i$, $\eta_i \in \mathbb{R}^3$

Energy Terms:

$$T = \frac{1}{2}\Omega_1 \cdot J_1\Omega_1 + \frac{1}{2}\Omega_2 \cdot J_2\Omega_2 + \frac{1}{2}m_2\dot{x}_2 \cdot \dot{x}_2$$
$$U = m_1 g R_1 \rho_1 \cdot e_3 + m_2 g R_2 \rho_2 \cdot e_3$$

4.2 Robotic Systems

Having seen the algorithm work for simple mechanical systems, we next consider robotic systems with more complex configuration spaces, shown in Table 3.

The **Three Link Walker** [23] represents an idealized mechanical configuration for bipedal walkers. Masses m, m_H, and m_T, are dynamically actuated by torques τ_2 and τ_3. Unit vectors q_1, q_2, and q_3 specify the motion of the three links. The **Stance Dynamics** are as shown below. The **Flight Dynamics** exist on $S^2 \times S^2 \times S^2 \times \mathbb{R}^3$, and account for the walkers evolution in Euclidean Space.

Configuration Manifold: $S^2 \times S^2 \times S^2$

Kinematic Constraints: $\dot{q}_i \cdot q_i = 0$

Variations: $\delta q_i = \xi_i \times q_i$, $\dot{\delta q}_i = \dot{\xi}_i \times q_i + \xi_i \times \dot{q}_i$

Energy Terms:

$$T = (\frac{5m}{8} + \frac{m_H}{2} + \frac{m_T}{2})l^2\dot{q}_1 \cdot \dot{q}_1 + l^2(\frac{m}{2}\dot{q}_1 + \frac{m}{8}\dot{q}_2) \cdot \dot{q}_2 + \frac{m_T}{2}(lL\dot{q}_1 + L^2\dot{q}_2) \cdot \dot{q}_3$$
$$U = (\frac{3m}{2}gl + m_H l + m_T gl)q_1 \cdot e_3 \frac{m}{2}glq_2 \cdot e_3 + m_T gLq_3 \cdot e_3$$

The **Quadrotor with Tethered Load** [29] and **Reaction Mass Pendulum** [26] can be referred to at the respective sources for information on configuration variables and composition of the Lagrangian.

Manifolds $\mathbb{C}^n$ and $\mathbb{R}^n$ are directly computed by the same steps shown in Section III, with the exception of Step 3, which introduces constraint manifolds for S^2 and $SO(3)$. This step is not required for variables that act on $\mathbb{C}^n$ and $\mathbb{R}^n$. The computational results for each system were validated by a published derivation and hand calculation. These results establish the validity and efficiency of computing dynamics on complex manifolds using the symbolic evaluator.

5 Discussion

It should be noted that manually computing and validating the dynamics for several of the systems in Table 3 would have taken a novice one to several days. Our proposed method of automating the computation reduces this to fractions of seconds. This will enable a broader study of novel robotic systems whose dynamics evolve on complex manifolds, systems that were previously not broadly

System	Config. Manifold	DOF	Time to Compute	Equations of Motion Produced by our Symbolic Dynamics Solver
3-Link Walker Stance Dynamics [23]	$S^2 \times S^2 \times S^2$	6	0.032 sec.	$\dot{q}_1 = \omega_1 \times q_1$ $q_1 \times ((\frac{5}{4}m + m_H + m_T)l^2\ddot{q}_1 + \frac{1}{2}ml^2\ddot{q}_2 + m_T lL\ddot{q}_3 + (\frac{3}{2}m + m_H + m_T)gle_3) = 0$ $\dot{q}_2 = \omega_2 \times q_2$ $q_2 \times (\frac{1}{2}ml^2\ddot{q}_1 + \frac{1}{4}ml^2\ddot{q}_2 + \frac{1}{2}mgle_3 = q_2 \times \tau_2$ $\dot{q}_3 = \omega_3 \times q_3$ $q_3 \times (m_T lL\ddot{q}_1 + m_T L^2\ddot{q}_3 + m_T gLe_3 = q_3 \times \tau_3$
Quadrotor with Suspended Load [29]	$SO(3) \times S^2 \times \mathbb{R}^3$	8	0.024 sec.	$\dot{x}_L = v_L$ $(m_Q + m_L)\dot{V}_L + m_Q l(\dot{q} \cdot \dot{q})q + (m_Q + m_L)ge_3 = fRe_3 + p \times (p \times fRe_3)$ $\dot{q} = \omega \times q$ $q \times (ml^2\ddot{q} + mgle_3) = q \times T$ $\dot{R} = R\hat{\Omega}$ $J\dot{\Omega} + \Omega \times J\Omega = M$
3 Link Walker Flight Dynamics [23]	$S^2 \times S^2 \times S^2 \times \mathbb{R}^3$	10	0.064 sec.	$\dot{q}_1 = \omega_1 \times q_1$ $q_1 \times ((\frac{5}{4}m + m_h + m_t)l^2\ddot{q}_1 + \frac{1}{2}ml^2\ddot{q}_2 + m_t lL\ddot{q}_3 + (\frac{7}{4}m + m_h + 2m_t)l\ddot{x} + (\frac{3}{2}m + m_h + m_t)lge_3) = 0$ $\dot{q}_2 = \omega_2 \times q_2$ $q_2 \times (\frac{1}{2}ml^2\ddot{q}_1 + \frac{1}{4}ml^2\ddot{q}_2 + \frac{3}{4}ml\ddot{x} + \frac{1}{2}mlge_3) = q_2 \times \tau_2$ $\dot{q}_3 = \omega_3 \times q_3$ $q_3 \times (m_t lL\ddot{q}_1 + m_t L^2\ddot{q}_3 + 2m_t L\ddot{x} + m_t Lge_3) = q_3 \times \tau_3$ $(\frac{7}{4}lm + lmh + 2lmt)\ddot{q}_1 + (\frac{3}{4}ml + \frac{1}{2}m_t L)\ddot{q}_2 + \frac{3}{2}m_t L\ddot{q}_3 + (\frac{5}{2}m + mh + 4m_t)\ddot{x} + (2m + 1mh + 2mt)ge_3 = 0$
Reaction Mass Pendulum [26]	$\mathbb{C} \times SO(3) \times SO(3) \times S$	11	0.073 sec.	$m\ddot{\rho} = -m\rho\Omega_L^T\hat{e}_3^2\Omega_L + mge_3^T R_L^T e_3$ $\dot{R}_L = R_L\hat{\Omega}_L$ $J_L\dot{\Omega}_L = -\Omega_L \times J_L\Omega_L + 2m\rho\dot{\rho}\hat{e}_3^2\Omega_L + mg\rho\hat{e}_3 R_L^T e_3 + \tau_L - R_L^T R_P \tau_D$ $\dot{R}_P = R_P\hat{\Omega}_P$ $J_P\dot{\Omega}_P = -\Omega_P \times J_P\Omega_P + 4\sum m_p s_i\dot{s}_i\hat{e}_i^2\Omega_P + \tau_D$ $2m_P\ddot{s}_i = -2m_P s_i\Omega_P^T\hat{e}_i^2\Omega_P + U_i$

Table 3: The equations of motion were computed on these robotic systems using the proposed symbolic computation tool with the kinematic constraints, variations, and Lagrangian (T-U) provided for each system.

approachable by the robotics community. The compact models will also bring new insight into the dynamics of existing robotic systems, enabling novel control designs for achieving highly dynamical maneuvers. A limitation of this method is that it cannot evaluate systems with nonholonomic constraints. Dynamics can only be computed on Cartesian products of $\mathbb{C}^n$, $\mathbb{R}^n$, S^2, and $SO(3)$.

6   Conclusion

We have presented an algorithm which automates the computation of dynamical equations for systems evolving on manifolds. By utilizing pattern matching within the Scala framework, we are able to capture the geometric axioms and identities of scalar, vector, and matrix elements of a dynamical system. Using the framework we implement a generalized algorithm that works across a wide variety of manifolds. The time efficient computation allows for the software to provide near-instantaneous output of dynamical equations. The dynamics generated are compact, globally-valid, and free of singularities, as they are described directly on the configuration manifold. This tool is released publicly and will enable broader inspection of systems that act on complex, nonlinear manifolds.

7   Acknowledgements

B. Bittner would like to thank Carnegie Mellon's URO, who supported him with a Summer Undergraduate Research Fellowship. K. Sreenath would like to thank R. Ravindran for discussions that led to the use of Scala. This work is supported by NSF grants IIS-1464337 and CMMI-1538869.

References

1. List of computer algebra systems. http://en.wikipedia.org/wiki/List_of_computer_algebra_systems.
2. *Symbolic Math Toolbox.*   Mathworks Inc, 1993.
3. *Mathematica Version 10.0.*   Wolfram Research Inc., 2014.
4. S. P. Bhat and D. S. Bernstein, "A topological obstruction to continuous global stabilization of rotational motion and the unwinding phenomenon," *Systems & Control Letters*, vol. 39, no. 1, pp. 63–70, 2000.
5. P.-y. Cheng, C.-i. Weng, and C.-k. Chen, "Symbolic derivation of dynamic equations of motion for robot manipulators using piogram symbolic method," *IEEE Journal of Robotics and Automation*, vol. 4, no. 6, pp. 599–609, 1988.
6. L. Consolini and M. Tosques, "On the exact tracking of the spherical inverted pendulum via an homotopy method," *Systems & Control Letters*, vol. 58, no. 1, pp. 1–6, 2009.
7. P. I. Corke, "An automated symbolic and numeric procedure for manipulator rigid-body dynamic significance analysis and simplification," in *IEEE Conf. on Robotics and Automation*, 1996, pp. 1018–1023.
8. T. Daly, D. V. Chudnovsky, and G. V. Chudnovsky, *Axiom The 30 Year Horizon.* The Axiom Foundation, 2007.

9. E. Dean-leon, S. Nair, and A. Knoll, "User friendly matlab-toolbox for symbolic robot dynamic modeling used for control design," in *IEEE Conf. on Robotics and Biomimetics*, 2012, pp. 2181–2188.

10. I. DigiArea, *Atlas 2 for Mathematica*, 2012. [Online]. Available: http://www.digi-area.com/Mathematica/atlas/guide/Atlas.php

11. R. Featherstone and D. Orin, "Robot dynamics: equations and algorithms," in *IEEE Conf. on Robotics and Automation*, 2000, pp. 826–834.

12. R. P. Feynman, "The principle of least action in quantum mechanics," Ph.D. dissertation, Princeton University Princeton, New Jersey, 1942.

13. T. Flaherty, "Scala Math DSL," https://github.com/axiom6/ScalaMathDSL, 2013.

14. M. G. Hollars, D. E. Rosenthal, and M. A. Sherman, *SD/FAST users manual*. Symbolic Dynamics Inc, 1991.

15. W. Khalil, F. Bennis, C. Chevallereau, and J. Kleinfinger, "Symoro: A software package for the symbolic modelling of robots," in *Proc. of the 20th ISIR*, 1989.

16. E. Kreuzer and W. Schiehlen, "Neweul software for the generation of symbolical equations of motion," in *Multibody Systems Handbook*. Springer, 1990.

17. T. Lee, "Computational geometric mechanics and control of rigid bodies," Ph.D. dissertation, University of Michigan, Ann Arbor, 2008.

18. T. Lee, M. Leok, and N. H. McClamroch, "Lagrangian mechanics and variational integrators on two-spheres," *International Journal for Numerical Methods in Engineering*, vol. 79, no. 9, pp. 1147–1174, 2009.

19. ——, "Lagrangian mechanics and variational integrators on two-spheres," *J. for Numerical Methods in Engineering*, vol. 79, pp. 1147–1174, 2009.

20. ——, "Stable manifolds of saddle equilibria for pendulum dynamics on S2 and SO(3)," in *IEEE Conf. on Decision and Control*, 2011, pp. 3915–3921.

21. ——, "Dynamics and control of a chain pendulum on a cart," in *IEEE Conf. on Decision and Control*, 2012, pp. 2502–2508.

22. R. Lot and M. D. A. Lio, "A symbolic approach for automatic generation of the equations of motion of multibody systems," *Multibody System Dynamics*, vol. 12, pp. 147–172, 2004.

23. P. X. Miranda, L. Hera, A. S. Shiriaev, B. Freidovich, S. Member, U. Mettin, and S. V. Gusev, "Stable walking gaits for a three-link planar biped robot with one actuator," *IEEE Trans. on Robotics*, vol. 29, no. 3, pp. 589–601, 2013.

24. G. Nonlinearity, "Adams 2014," pp. 1–5, 2014.

25. M. Odersky and Al., "An Overview of the Scala Programming Language," EPFL, Lausanne, Switzerland, Tech. Rep. IC/2004/64, 2004.

26. A. K. Sanyal and A. Goswami, "Dynamics and balance control of the reaction mass pendulum: A three-dimensional multibody pendulum with variable body inertia," *J. Dyn. Sys., Meas., Control*, vol. 136, no. 2, p. 021002, Nov. 2013.

27. D. Services, "LMS Virtual . Lab The Unified Environment for Functional Performance Engineering," pp. 1–16, 2007.

28. J. Shen, A. K. Sanyal, N. A. Chaturvedi, D. Bernstein, and H. McClamroch, "Dynamics and control of a 3d pendulum," in *IEEE Conf. on Decision and Control*, 2004, pp. 323–328.

29. K. Sreenath, T. Lee, and V. Kumar, "Geometric control and differential flatness of a quadrotor UAV with a cable-suspended load," in *IEEE Conf. on Decision and Control*, 2013, pp. 2269–2274.

30. SymPy Development Team, *SymPy: Python library for symbolic mathematics*, 2014. [Online]. Available: http://www.sympy.org

A Linear-Time Variational Integrator for Multibody Systems

Jeongseok Lee[1], C. Karen Liu[2], Frank C. Park[3], and Siddhartha S. Srinivasa[1]

[1] Carnegie Mellon University, Pittsburgh, PA, USA 15213
{jeongsel, ss5}@andrew.cmu.edu
[2] Georgia Institute of Technology, Atlanta, GA, USA 30332
karenliu@cc.gatech.edu
[3] Seoul National University, Seoul 151-742, Korea
fcp@snu.ac.kr

Abstract. We present an efficient variational integrator for simulating multibody systems. Variational integrators reformulate the equations of motion for multibody systems as discrete Euler-Lagrange (DEL) equation, transforming forward integration into a root-finding problem for the DEL equation. Variational integrators have been shown to be more robust and accurate in preserving fundamental properties of systems, such as momentum and energy, than many frequently used numerical integrators. However, state-of-the-art algorithms suffer from $O(n^3)$ complexity, which is prohibitive for articulated multibody systems with a large number of degrees of freedom, n, in generalized coordinates. Our key contribution is to derive a quasi-Newton algorithm that solves the root-finding problem for the DEL equation in $O(n)$, which scales up well for complex multibody systems such as humanoid robots. Our key insight is that the evaluation of DEL equation can be cast into a *discrete inverse dynamic* problem while the approximation of inverse Jacobian can be cast into a *continuous forward dynamic* problem. Inspired by Recursive Newton-Euler Algorithm (RNEA) and Articulated Body Algorithm (ABA), we formulate the DEL equation individually for each body rather than for the entire system, such that both inverse and forward dynamic problems can be solved efficiently in $O(n)$. We demonstrate scalability and efficiency of the variational integrator through several case studies.

Keywords: variational integrator · discrete mechanics · multibody systems · dynamics · computer animation & simulation

1 Introduction

We address the problem of accurately and efficiently simulating the dynamics of complex multibody systems, often referred to as the forward dynamics problem. Existing state-of-the-art approaches use the Lagrangian formalism, expressing the difference between kinetic and potential energy (the Lagrangian) in generalized coordinates, and derive the Euler-Lagrange second-order differential equations from them via the principle of least action. The state of the system

© Springer Nature Switzerland AG 2020
K. Goldberg et al. (Eds.): *Algorithmic Foundations of Robotics XII*, SPAR 13, pp. 352–367, 2020.
https://doi.org/10.1007/978-3-030-43089-4_23

at any time t is then obtained by integrating these differential equations from initial conditions.

However, the long-term conservation of conserved quantities like energy and momentum of the system remains a key open challenge. In particular, discrete-time simulations, even with advanced algorithms for solving differential equations, eventually produce alarming and physically implausible behaviors, even for simple dynamical systems like N-link pendulums, due to the accumulation of numerical errors.

To address this problem, Marsden and West [1] introduced the discrete Lagrangian, which approximates the integral of the Lagrangian over a small time interval. They then derived its variation via the principle of least action, creating the discrete Euler-Lagrange (DEL) equations. They also showed that variational integrators based on the DEL formulation were symplectic (energy-conserving) and crucially decoupled energy behavior from step size [1,2].

Unfortunately, despite their benefits for stability, variational integrators suffer from computational complexity. Variational integrators transform the integration of the equations of motion into a root-finding problem for the DEL equation. This introduces complexity in three places as most nonlinear root-finding algorithms require: (1) the evaluation of the DEL equation, (2) computation of their gradient (Jacobian), and (3) the inversion of the gradient. Although there exist efficient algorithms for evaluating the DEL equation, they *do not* use generalized coordinates but instead treat each link as a free-body and apply constraint forces to enforce joints [3,4,5]. This becomes especially complicated with branching multi-body systems and joint constraints.

Recently Johnson and Murphey [6] proposed a scalable variational integrator that represents the DEL equation in generalized coordinates. By representing the multibody system as a tree structure in generalized coordinates, they showed that the DEL equation, as well as the gradient and Hessian of the Lagrangian, can be calculated recursively. However, the complexity of their algorithm is $O(n^2)$ for evaluating the DEL equation, and $O(n^3)$ for computing the Jacobian. When coupled with traditional root-finders, *e.g.*, Newton's method, that require the inverse of the Jacobian, this adds an approximately $O(n^3)$ complexity for matrix inversion.

In this paper, we introduce a new variational integrator for multibody dynamic systems. The primary contribution is an $O(n)$ algorithm which solves the root-finding problem for the DEL equation. Our key insight is that the evaluation of DEL can be cast into a discrete inverse dynamics problem [7,8] while the root updating can be cast into a continuous forward dynamics problem. Both inverse and forward dynamics problems can be solved efficiently in $O(n)$ using a recursive Lie group formulation of the dynamics [9,10,11,12].

Inspired by Recursive Newton-Euler Algorithm (RNEA) and Articulated Body Algorithm (ABA), we formulate the DEL equation individually for each body rather than for the entire system. By taking advantage of the recursive relations between body links, it becomes possible to evaluate the DEL function using a discrete inverse dynamics algorithm in linear-time. The same recursive representation is applied to update the root using an impulse-based forward dynamics algorithm. Together with these two algorithms, we propose an $O(n)$ quasi-Newton method specialized for finding the root of DEL equation, resulting in a *Linear-Time Variational Integrator*.

We compare our method with the state-of-the-art variational integrator in generalized coordinates [6]. The results show that, for the same computation method of root updating, the performance of our recursive evaluation of the DEL equation (linear-time DEL algorithm) is 15 times faster for a system with 10 degrees of freedom (DOFs) and 32 times faster for 100 DOFs. For the same evaluation method of the DEL equation (*i.e.*, linear-time DEL algorithm), our results show that the performance of our new quasi-Newton method is 3.8 times faster for a system with 10 DOFs, and 53 times faster for 100 DOFs. Further analysis shows that for higher DOF systems, the impulse-based Jacobian approximation becomes increasingly more effective compared to our linear-time DEL algorithm.

2   Background

Our work is built on the concepts of discrete mechanics and variational integrators. In this section, we will briefly describe the standard formulation of discrete mechanics [6], followed by a reformulation using the Lie group representation for the Special Euclidean group $\mathsf{SE}(3)$ of rigid body motions [11].

2.1   Variational Integrators in Generalized Coordinates

We begin with the definition of Lagrangian, $L(\mathbf{q}, \dot{\mathbf{q}}) \in \mathbb{R}$, the difference between the total kinetic energy and the total potential energy of a system characterized by generalized coordinates $\mathbf{q} \in \mathbb{R}^n$ where n denotes the degrees of freedom of the system. For continuous-time systems, the principle of least action states that the system will follow the trajectory that minimizes the *action integral* $\int_{t_1}^{t_2} L(\mathbf{q}(t), \dot{\mathbf{q}}(t))dt$.

However, when we simulate the mechanical system on a computer, the mechanical system takes *discrete time steps* rather than following the continuous trajectory. Loosely speaking, the idea of discrete mechanics is that the system will follow the *discretized trajectory* that minimizes the approximated action integral defined on the discretized trajectory. If we discretize a continuous trajectory $\mathbf{q}(t)$ into a sequence of configurations $\mathbf{q}^0, \mathbf{q}^1, \cdots, \mathbf{q}^N$, we can define a discrete Lagrangian that approximates the integral of $L(\mathbf{q}(t), \dot{\mathbf{q}}(t))$ over a short interval Δt:

$$L_d(\mathbf{q}^k, \mathbf{q}^{k+1}) \approx \int_{k\Delta t}^{(k+1)\Delta t} L(\mathbf{q}(t), \dot{\mathbf{q}}(t))dt. \tag{1}$$

Using the discrete Lagrangian, we can define the *action sum* $\sum_{k=0}^{N-1} L_d(\mathbf{q}^k, \mathbf{q}^{k+1})$ as an approximation of the action integral. Minimizing the action sum with respect to $\{\mathbf{q}^k\}$ ($k = 1, 2, \cdots, N-1$), we arrive at the discrete Euler-Lagrange

(DEL) equation:

$$D_2 L_d(\mathbf{q}^{k-1}, \mathbf{q}^k) + D_1 L_d(\mathbf{q}^k, \mathbf{q}^{k+1}) = 0, \tag{2}$$

where $D_i : \mathbb{R} \to \mathbb{R}^n$ denotes differential operator with respect to the i-th parameter of the function, and the differentials of L_d can be analytically computed [6]. Note that the boundary configurations q^0 and q^N are not varied.

Instead of numerically integrating the Euler-Lagrange equation to simulate the trajectory, discrete mechanics solves a *root-finding problem* to obtain the next configuration. Specifically, given two previous configurations $\mathbf{q}^{k-1}$ and $\mathbf{q}^k$, we solve the next configuration $\mathbf{q}^{k+1}$ by finding the root of the following function:

$$f(\mathbf{q}^{k+1}) = D_2 L_d(\mathbf{q}^{k-1}, \mathbf{q}^k) + D_1 L_d(\mathbf{q}^k, \mathbf{q}^{k+1}) = 0. \tag{3}$$

The superior energy behavior of variational integrators compared to the traditional integrators like Euler and Runge-Kutta methods have been shown using a discrete version of Noether's theorem [1]. One geometric interpretation of variational integrators is that the DEL equation plays the role of constraints, enforcing the discrete system to evolve on the constraint manifold such that $f(\mathbf{q}^{k+1}) = 0$, *i.e.*, satisfying the least action principle on the approximated action. In that sense, the process of root-finding can be seen as a feedback controller to find the physically correct configuration for the next time step, with the DEL equation being used by the feedback law to indicate how far away the given configuration is from the manifold. Traditional integrators do not have such indicators, only account for the rate of change based on the current state, which leads to the numerical error accumulation.

This nonlinear, high-dimensional, continuous root-finding problem can be solved efficiently by Newton's method, provided that the partial derivatives of f, $J_f(\mathbf{q})$ (*i.e.*, the Jacobian matrix), can be evaluated:

Algorithm 1 Newton's Method for Solving DEL Equation

1: **Initial Guess q_0**
2: **do**
3:     Evaluate $f(\mathbf{q}^{k+1})$ ▷ $O(n^2)$ time
4:     **if** $\|f(\mathbf{q}^{k+1}) < \epsilon\|$ **return** $\mathbf{q}^{k+1}$
5:     Update $\mathbf{q}^{k+1} \leftarrow \mathbf{q}^{k+1} - \left[J_f(\mathbf{q}^{k+1})\right]^{-1} f(\mathbf{q}^{k+1})$ ▷ $O(n^3)$ time
6: **while** num_iteration < max_iteration

To avoid the computation of the Jacobian and its inversion, various quasi-Newton methods can be applied to approximate $\left[J_f(\mathbf{q}^{k+1})\right]^{-1}$. In Section 3.2, we introduce a linear-time algorithm to approximate the product of $\left[J_f(\mathbf{q}^{k+1})\right]^{-1}$ and $f(\mathbf{q}^{k+1})$ for finding the root of DEL equation.

2.2 Variational Integrators in SE(3)

The linear-time root-finding algorithm we will introduce in the next section leverages the idea of reformulating DEL equation for each rigid body rather than for the entire system. We begin with the expression of the DEL equation in SE(3) for a single rigid body.

The configuration of the rigid body can be represented by matrices of the form:

$$T = \begin{bmatrix} R & p \\ 0 & 1 \end{bmatrix} \in \mathsf{SE}(3), \tag{4}$$

where $R \in \mathsf{SO}(3)$ is a 3×3 rotation matrix, and $p \in \mathbb{R}^3$ is a position vector. The spatial velocity of the rigid body $V = (w, v) \in \mathfrak{se}(3)$ or twist can be represented in six-dimensional vector or 4×4 matrix form:

$$V = \begin{pmatrix} w \\ v \end{pmatrix}, \quad [V] = \begin{bmatrix} \hat{w} & v \\ 0 & 0 \end{bmatrix}, \tag{5}$$

where $w \in \mathfrak{so}(3)$ and $v \in \mathbb{R}^3$ denote the angular velocity and linear velocity, respectively, and $\hat{w}$ is the 3×3 skew symmetric matrix for w such that $\hat{w}^T = -\hat{w}$. In this paper, we use brackets $[\cdot]$ to denote matrix representations.

The Lagrangian of a rigid body can be compactly expressed using the Lie group representation ([9,13]) in the space of SE(3):

$$L(T, V) = \frac{1}{2} V^T G V - P(T), \tag{6}$$

where $P : \mathsf{SE}(3) \to \mathbb{R}$ is the potential energy. G is the spatial inertia matrix that has the following structure:

$$G = \begin{bmatrix} \mathcal{I} & 0 \\ 0 & mI \end{bmatrix} \in \mathbb{R}^{6 \times 6}, \tag{7}$$

where $\mathcal{I}$ is the inertia matrix, m is the mass, and I is 3×3 identity matrix when the center of mass is at the origin of the body frame.

Analogous to Equation (1), the discrete Lagrangian for a single rigid body can be expressed as

$$L_d(T^k, T^{k+1}) \approx \int_{k\Delta t}^{(k+1)\Delta t} L(T, V) dt. \tag{8}$$

In this paper, we use the trapezoidal quadrature approximation for the discrete Lagrangian of the single body system as

$$L_d(T^k, T^{k+1}) \triangleq \frac{\Delta t}{2} L(T^k, V^k) + \frac{\Delta t}{2} L(T^{k+1}, V^k), \tag{9}$$

where the *average velocity* V^k can be defined as

$$V^k = \frac{1}{\Delta t} \log(\Delta T^k), \tag{10}$$

with the *log map* log : $\mathsf{SE}(3) \to \mathfrak{se}(3)$, the inverse of the *exponential map* exp : $\mathfrak{se}(3) \to \mathsf{SE}(3)$ [11,13], and $\Delta T^k = T^{k^{-1}} T^{k+1}$, the displacement of the rigid body's configuration during the discrete times of t_k and t_{k+1}.

To derive the DEL equation for a single rigid body in $\mathsf{SE}(3)$, we need to take the variational calculus on V^k with respect to T^k and T^{k+1}. This requires the derivative of log map defined as

$$\left(\frac{\partial}{\partial T} \log(T)\right)[W] = d\log_V \left([W]\exp(-[V])\right), \tag{11}$$

where $V = \log(T)$, and $W \in \mathfrak{se}(3)$ is an arbitrary twist, and $d\log_V : \mathfrak{se}(3) \to \mathfrak{se}(3)$ is the inverse of the right trivialized tangent $d\exp_V : \mathfrak{se}(3) \to \mathfrak{se}(3)$ as an linear operator [11,14]:

$$d\log_V(W) = \sum_{j=0}^{\infty} \frac{B_j}{j!} \mathrm{ad}_V^j(W). \tag{12}$$

The Lie bracket operator $\mathrm{ad}_V : \mathfrak{se}(3) \to \mathfrak{se}(3)$ is defined as $\mathrm{ad}_V(W) = [V][W] - [W][V]$. $d\log_V$ can be alternatively represented in matrix form as

$$[d\log_V] = \sum_{j=0}^{\infty} \frac{B_j}{j!} [\mathrm{ad}_V]^j, \quad [\mathrm{ad}_V] = \begin{bmatrix} \hat{w} & 0 \\ \hat{v} & \hat{w} \end{bmatrix}, \tag{13}$$

where B_j are the Bernoulli numbers ($B_0 = 1, B_1 = -1/2, B_2 = 1/6, B_3 = 0, \dots$) [15].

Using Equation (10) and (11), we can now express the variation of V^k as

$$\delta V^k = \frac{1}{\Delta t} d\log_{\Delta t V^k} \left(-T^{k^{-1}} \delta T^k + \mathrm{Ad}_{\exp(\Delta t[V^k])} \left(T^{k+1^{-1}} \delta T^{k+1}\right)\right), \tag{14}$$

where δT^k and δT^{k+1} are variations, and $\mathrm{Ad}_T : \mathfrak{se}(3) \to \mathfrak{se}(3)$ is the adjoint action of $T \in \mathsf{SE}(3)$ on $V \in \mathfrak{se}(3)$ defined as $\mathrm{Ad}_T V = T[V]T^{-1}$. The adjoint action can be regarded as an linear operator in the 6×6 matrix form of:

$$[\mathrm{Ad}_T] = \begin{bmatrix} R & 0 \\ \hat{p}R & R \end{bmatrix}. \tag{15}$$

By the least action principle with Equation (9), (10), and (14), we can derive the DEL equation for a single rigid body in $\mathsf{SE}(3)$, which is the well known *discrete reduced Euler-Poincaré* equations [11,16]:

$$D_2 L_d(T^{k-1}, T^k) + D_1 L_d(T^k, T^{k+1}) = 0 \quad \in \mathbb{R}^6, \tag{16a}$$

where

$$D_2 L_d(T^{k-1}, T^k) = -[\mathrm{Ad}_{\exp(\Delta t[V^{k-1}])}]^T [d\log_{\Delta t V^{k-1}}]^T GV^{k-1} + \frac{\Delta t}{2} T^{k^*} P(T^k) \tag{16b}$$

$$D_1 L_d(T^k, T^{k+1}) = [d\log_{\Delta t V^k}]^T GV^k + \frac{\Delta t}{2} T^{k^*} P(T^k). \tag{16c}$$

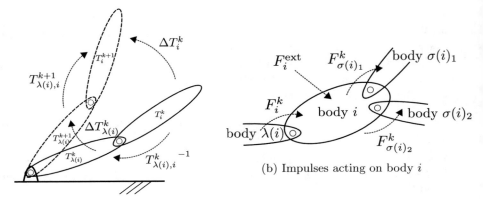

(a) Displacement of body i's configuration

(b) Impulses acting on body i

Fig. 1: Recurrence relationships of configuration displacement and impulses

By Lagrange-d'Alembert principle, Equation (16a) can be straightforwardly extended to a forced system [11]:

$$D_2 L_d(T^{k-1}, T^k) + D_1 L_d(T^k, T^{k+1}) + F^k = 0, \qquad (17)$$

where $F^k \in \mathfrak{se}^*(3)$ is the integral of the virtual work performed by the force over the time interval Δt.

3   Linear-Time Variational Integrator

We introduce a new linear-time variational integrator which, at each time instance t_k, solves for the root of Equation (3). Our variational integrator consists of two linear-time algorithms for evaluating the DEL equation and updating the root, which, as shown in Algorithm 1, determine the time complexity of the root-finding algorithm. We first derive the DEL equation for multibody systems in a recursive manner, resulting a linear-time procedure to evaluate the function $f(\mathbf{q})$. Next, we introduce an impulse-based dynamics algorithm, which is also linear-time, to estimate the next configuration. Replacing Line 3 and Line 5 in Algorithm 1 with these two algorithms, we present a new linear-time quasi-Newton root-finding method for finding the root of DEL equation.

3.1   Linear-Time Evaluation of the DEL Equation

If we view the function $f(\mathbf{q}) = \mathbf{0}$ as a dynamic constraint that enforces the equation of motion, any nonzero value of $f(\mathbf{q})$ indicates the residual impulse that violates the equation of motion. As such, evaluating $f(\mathbf{q})$ can be considered a discrete inverse dynamics problem which solves the residual impulse of the system given $\mathbf{q}^{k-1}$, $\mathbf{q}^k$, and $\mathbf{q}^{k+1}$. We derive a recursive DEL equation using similar formulation as recursive Newton-Euler algorithm (RNEA) [7,8], which solves the inverse dynamics for continuous systems in linear time with respect to the degrees of freedom of the system.

Assuming that the multibody system can be represented as a tree-structure where each body has at most one parent and an arbitrary number of children, connected by joints, our goal is to expand Equation (17) to account for the dynamics of entire tree-structure.

We begin with the recursive definition for a rigid body's configuration and the displacement of the configuration. Let us denote $\{0\}$ as an inertial frame which is stationary in the space, $\{i\}$ as body frame of i-th body in the tree structured system, and $\{\lambda(i)\}$ as a body frame of the parent of the i-th body. The configuration of a body in the system can be represented as

$$T_i^k = T_{\lambda(i)}^k T_{\lambda(i),i}^k, \tag{18}$$

where T_i^k and $T_{\lambda(i)}^k$ denote the transformations from the inertial frame to $\{i\}$ and $\{\lambda(i)\}$, respectively, while $T_{\lambda(i),i}^k$ denotes the relative transformation from $\{\lambda(i)\}$ to $\{i\}$ represented as a function of the i-th joint configuration $\mathbf{q}_i^{k+1}$. From Equation (18), the configuration displacement of a rigid body can be written as

$$\Delta T_i^k = {T_{\lambda(i),i}^k}^{-1} \Delta T_{\lambda(i)}^k T_{\lambda(i),i}^{k+1}. \tag{19}$$

Fig. 1 (a) gives a geometric interpretation of the recurrence relationship of the configuration displacements between ΔT_i^k and $\Delta T_{\lambda(i)}^k$.

Plugging Equation (19) into Equation (10), we can obtain the average velocity of i-th rigid body as $V_i^k = \frac{1}{\Delta t}\log(\Delta T_i^k)$. Unlike the continuous velocity of i-th body $V_i = S_i \dot{\mathbf{q}}_i$ where S_i is the joint Jacobian [13], the equation for the average velocity is implicit with respect to $\mathbf{q}^{k+1}$ due to the log map. The use of log map, with $d\log_V$, is the key reason that makes the DEL equation implicit with respect to $\mathbf{q}^{k+1}$.

For a rigid body in a multibody system, the impulse term F^k in Equation (17) includes the impulse transmitted from the parent link F_i^k, impulses transmitting to the child links F_c^k, and other external impulses $F_i^{\text{ext},k}$ applied by the environment as (Fig. 1 (b)):

$$F^k = F_i^k - \sum_{c \in \sigma(i)} \mathrm{Ad}^*_{T_{i,c}^k}{}^{-1} F_c^k + F_i^{\text{ext},k}. \tag{20}$$

Note that F_i^k is expressed in the i-body coordinates so the coordinate frame transformation is required for F_c^k as $\left[\mathrm{Ad}_{T_{i,c}^k}{}^{-1}\right]^T F_c^k$.

Plugging these forces into Equation (17) and using the definitions in Equation (16b) and (16c), we express the equations of motion for the i-th body as

$$F_i^k = \mu_i^k - \left[\mathrm{Ad}_{\exp(\Delta t[V_i^{k-1}])}\right]^T \mu_i^{k-1} + \sum_{c \in \sigma(i)} \left[\mathrm{Ad}_{T_{i,c}^k}{}^{-1}\right]^T F_c^k - F_i^{\text{ext},k} \tag{21a}$$

$$\mu_i^k = \left[d\log_{\Delta t V_i^k}\right]^T G_i V_i^k, \tag{21b}$$

where μ_i^k is the discrete momentum of body i and $\sigma(i)$ denotes the set of child bodies to body i. The required generalized impulse of joint i to achieve the motion q^{k+1} is simply the projection of F_i^k onto the joint Jacobian as $S_i^T F_i^k$ where $S_i \in \mathbb{R}^{6 \times n_i}$ is the i-th joint Jacobian [13]. The residual impulse then can be obtained by subtracting the joint impulses, Q_i^k, such as joint actuation or joint friction, from the required impulse:

$$f_i = S_i^T F_i^k - Q_i^k \quad \in \mathbb{R}^{n_i}. \tag{22}$$

Algorithm 2 summarizes the recursive procedure, which we call discrete recursive Newton-Euler algorithm (DRNEA). DRNEA consists a forward pass from the root of the tree structure to the leaf nodes and a backward pass in the reverse order. The forward pass computes the velocity of each body while the backward pass computes force transmitted between joints. By exploiting the recursive relationship between a parent body and its child bodies, the computation for each pass is $O(n)$, where n is the number of rigid body links in the system assuming the degree of freedom of each joint is one.

Algorithm 2 Discrete recursive Newton-Euler algorithm (DRNEA)

1: **for** $i = 1 \to n$ **do**
2:      $T_{\lambda(i),i}^{k+1} = $ function of q_i^{k+1}
3:      $\Delta T_i^k = {T_{\lambda(i),i}^k}^{-1} \Delta T_{\lambda(i)}^k T_{\lambda(i),i}^{k+1}$
4:      $V_i^k = \frac{1}{\Delta t} \log\left(\Delta T_i^k\right)$
5: **end for**
6: **for** $i = n \to 1$ **do**
7:      $\mu_i^k = \left[d\log_{\Delta t V_i^k}\right]^T G_i V_i^k$
8:      $F_i^k = \mu_i^k - \left[\mathrm{Ad}_{\exp(\Delta t[V_i^{k-1}])}\right]^T \mu_i^{k-1} - F_i^{\mathrm{ext},k} + \sum_{c \in \sigma(i)} \left[\mathrm{Ad}_{{T_{i,c}^k}^{-1}}\right]^T F_c^k$
9:      $f_i = S_i^T F_i^k - Q_i^k$
10: **end for**

For clarity, the mathematical symbols used in DRNEA are listed below.

- i: index of the i-th body.
- $\lambda(i)$: index of the parent body of the i-th body.
- $\sigma(i)$: set of indices of the child bodies of the i-th body.
- $q_i^k \in \mathbb{R}^{n_i}$: generalized coordinates of the i-th joint which connects the i-th body with its parent body where n_i denotes the dimension of the coordinates.
- $Q_i \in \mathbb{R}^{n_i}$: generalized force exerted by the i-th joint.
- $T_{\lambda(i),i} \in \mathsf{SE}(3)$: relative transformation matrix from the $\{\lambda(i)\}$ to $\{i\}$.
- $V_i^k \in \mathfrak{se}(3)$: the spatial average velocity of the i-th body, expressed in $\{i\}$ at time step k
- $S_i^k \in \mathbb{R}^{6 \times n_i}$: Jacobian of $T_{\lambda(i),i}$ expressed in $\{i\}$.
- $G_i \in \mathbb{R}^{6 \times 6}$: the spatial inertia of the i-th body, expressed in $\{i\}$.

- $F_i^k \in \mathfrak{se}^*(3)$: the spatial impulse transmitted to the i-th body from its parent through the connecting joint, expressed in $\{i\}$.
- $F_i^{\text{ext},k} \in \mathfrak{se}^*(3)$: the spatial impulse acting on the i-th body, expressed in $\{i\}$.

3.2 Linear-Time Root Updating

Besides function evaluation, Newton-like methods also require the update of Jacobian to estimate the root, which is usually the computation bottleneck in each iteration. Here we describe a recursive impulse-based method to efficiently update the root in linear-time.

Let us denote the current iteration in Newton's method as l and the current estimate of the configuration at next time step as $\mathbf{q}_{(l)}^{k+1}$. Evaluating the forced DEL equation (17) gives the residual impulse, $f(\mathbf{q}_{(l)}^{k+1}) = \mathbf{e}_{(l)}$, in the system. If the magnitude of $\mathbf{e}_{(l)}$ is zero or less than the tolerance, $\mathbf{q}_{(l)}^{k+1}$ is the next configuration that satisfies the forced DEL equation. Otherwise, $\mathbf{e}_{(l)}$ can be regarded as the residual impulse needed to result in $\mathbf{q}_{(l)}^{k+1}$ at the next time step. If we apply the negative residual force, $-\mathbf{e}_{(l)}/\Delta t$, to the system, we should arrive at a configuration closer to the root of $f(\mathbf{q}^{k+1})$. Applying such a force to the system can be done by continuous forward dynamics in linear-time [8].

Given the approximation of $\dot{\mathbf{q}}^k$ as $\frac{1}{\Delta t}\left(\mathbf{q}^k - \mathbf{q}^{k-1}\right)$, the continuous forward dynamics equation can be used to evaluate the generalized acceleration:

$$\ddot{\mathbf{q}}^k = M^{-1}(\mathbf{q}^k)\left(-C(\mathbf{q}^k, \dot{\mathbf{q}}^k)\dot{\mathbf{q}}^k + Q\right), \tag{23}$$

where $M(\mathbf{q}^k)$ is the mass matrix and $C(\mathbf{q}^k, \dot{\mathbf{q}}^k)$ is the Coriolis force in generalized coordinates. Q indicates the sum of other external and internal forces applied to the system in generalized coordinates.

Using the 2nd order central difference to approximate $\mathbf{q}^{k+1} = \Delta t^2 \ddot{\mathbf{q}}^k + 2\mathbf{q}^k - \mathbf{q}^{k-1}$, we can apply the negative residual force to improve the estimate of root:

$$\mathbf{q}_{(l+1)}^{k+1} = \Delta t^2 M^{-1}(\mathbf{q}^k)\left(-C(\mathbf{q}^k, \dot{\mathbf{q}}^k)\dot{\mathbf{q}}^k + Q - \sum_{m=0}^{l}\frac{\mathbf{e}_{(m)}}{\Delta t}\right) + 2\mathbf{q}^k - \mathbf{q}^{k-1}. \tag{24}$$

Consolidating the quantities on the RHS of Equation (24) gives the update rule for $\mathbf{q}^{k+1}$:

$$\mathbf{q}_{(l+1)}^{k+1} = \mathbf{q}_{(l)}^{k+1} - \Delta t M^{-1}(\mathbf{q}_{(l)}^k)\mathbf{e}_{(l)}, \tag{25}$$

where $\Delta t M^{-1}(\mathbf{q}_{(l)}^k)\mathbf{e}_{(l)}$ can be evaluated in $O(n)$ using recursive impulse-based dynamics (ABI algorithm: articulated body inertia algorithm) introduced by Featherstone [8]. Specifically, ABI is a forward dynamics algorithm which computes Equation (23). If we set $\dot{\mathbf{q}} \equiv \mathbf{0}$ (to eliminate the Coriolis force) and $Q \equiv \Delta t \mathbf{e}_{(l)}$, ABI will return exactly $\Delta t M^{-1}(\mathbf{q}_{(l)}^k)\mathbf{e}_{(l)}$.

Comparing to the Newton's method in Algorithm 1, the inverse of Jacobian matrix is approximated by the inverse mass matrix multiplied by Δt. We name this algorithm RIQN (Recursive Impulse-based Quasi-Newton method) and summarize it in Algorithm 3.

Algorithm 3 Recursive Impulse-based Quasi-Newton method (RIQN)

1: **Initial Guess** $\mathbf{q}_0^{k+1}$
2: **do**
3:      Use DRNEA to evaluate $\mathbf{e} \leftarrow f(\mathbf{q}^{k+1})$ ▷ $O(n)$ time
4:      **if** $\|\mathbf{e} < \epsilon\|$ **return** $\mathbf{q}^{k+1}$
5:      Use ABI to compute $\Delta t M^{-1}(\mathbf{q}^k)\mathbf{e}$ ▷ $O(n)$ time
6:      Update $\mathbf{q}^{k+1} \leftarrow \mathbf{q}^{k+1} - \Delta t M^{-1}(\mathbf{q}^k)\mathbf{e}$
7: **while** num_iteration < max_iteration

3.3   Initial Guess

Similar to other Newton-like methods, our algorithm requires the initial guess to be sufficiently close to the solution. We propose three different ways to produce an initial guess for RIQN.

- IG1: Directly use the current configuration as the initial guess of the next configuration: $\mathbf{q}_{(0)}^{k+1} = \mathbf{q}^k$.
- IG2: Apply explicit Euler integration, $\mathbf{q}_{(0)}^{k+1} = \mathbf{q}^k + \Delta t\ \dot{\mathbf{q}}^k$, where $\dot{\mathbf{q}}^k$ is approximated by $\frac{1}{\Delta t}\left(\mathbf{q}^k - \mathbf{q}^{k-1}\right)$.
- IG3: Compute the acceleration via the equations of motion, $\ddot{\mathbf{q}}^k = M^{-1}\left(-C + Q\right))$, and apply semi-implicit Euler integration to integrate velocity, $\dot{\mathbf{q}}^{k+1} = \dot{\mathbf{q}}^k + \Delta t\ \ddot{\mathbf{q}}^k$, followed by position, $\mathbf{q}_{(0)}^{k+1} = \mathbf{q}^k + \Delta t\ \dot{\mathbf{q}}^{k+1}$.

4   Experimental Results

In this section, we describe the implementation of the proposed algorithms, RIQN and DRNEA, and verify the algorithms in terms of efficiency and scalability by comparing them to the state-of-are algorithms through case studies. We used fixed time step of 1 millisecond for all the experiments.

4.1   Implementation

The algorithms introduced by this paper and several state-of-art algorithms were implemented on top of DART [17,18], which is an C++ open source dynamics library for multibody systems. All of the simulations were performed on a Intel Core i7-4970K @ 4.00 GHz desktop computer.

All the source code of the implementations is available at https://github.com/jslee02/wafr2016.

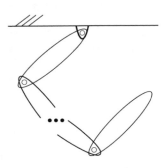
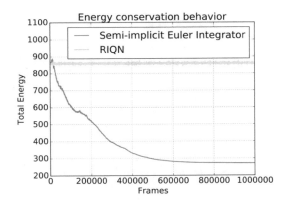

Fig. 2: (a) Serial chain of N-bodies connected by revolute joints, (b) Energy conservation behavior over simulation frames

4.2 Energy Conservation

We first show that our linear-time variational integrator inherits the energy conservation property, which is one of the important features of variational integrators. We simulate a serial chain that consists of N-bodies connected by revolute joints (Fig. 2 (a)) with RIQN (variational integrator) and semi-implicit Euler method, which is an easy-to-implement standard method. In this experiment, we use a 10-body serial chain with no joint actuation nor external forces except for the gravity. The total energy (kinetic energy + potential energy) of this passive system should remain constant.

Fig. 2 (b) shows the energy evolution of the serial chain over simulation frames for both integration methods. RIQN does not artificially dissipate the energy while the Euler method does.

4.3 Performance Comparisons

The major factors that affect on the computational time of variational integrator are (1) evaluation of DEL equation and (2) the evaluation of Jacobian inverse. We consider various of the root-finding algorithm that are combination of methods for (1) and (2).

For (1), we compare our DRNEA to the scalable variational integrator (SVI) [6]. For (2), we compare the proposed RIQN to Newton's method and Broyden method (quasi-Newton method) [19].

Newton's method requires the (exact) Jacobian of the DEL equation. When combining with DRNEA, for a fair comparison we also derive a recursive algorithm to evaluate the derivatives of the DEL equation with respect to $\mathbf{q}^{k+1}$. Please see the Appendix for the algorithm.

For all the root-finding methods, we measure computation time of serial chain forward dynamics simulations for 10k frames. To reveal the scalability of

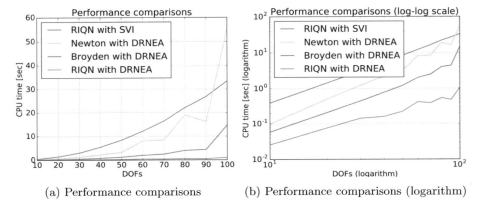

(a) Performance comparisons   (b) Performance comparisons (logarithm)

Fig. 3: Absolute computation time versus DOFs for the various root-finding methods.

the methods, we vary the number of bodies of the serial chain (Fig. 3). RIQN method with DRNEA shows the best performance. We also noticed that, for the same method for (2), DRNEA shows better performance than SVI. Further analyses show that the impulse-based Jacobian approximation contributes more than our linear-time DEL algorithm for the higher DOFs systems.

4.4   Convergence

We consider the convergent rate of RIQN comparing to Newtons method. We inspect the convergence of error $f(\mathbf{q}_{(l)}^{k+1}) = \mathbf{e}$ during the iterations in solving the DEL equation for one simulation time step. For quantitatively visible convergence, we use the zero configurations as the initial guess $\mathbf{q}_0^{k+1} = 0$ instead of the proposed initial guesses in Section 3.3.

Fig. 4a shows that under the tolerance RIQN converges more slowly than Newton's method. This observation is expected because Newton's method has a quadratic convergence rate which is in theory faster than that of Quasi-Newton methods. However, in Section 4.3, we observed that the absolute computation time of the proposed method (DRNEA+RIQN) showed the best performance.

Fig. 4b shows the average iteration numbers per each simulation step in the root-finding process. As expected, Newton's method requires less iteration numbers than RIQN.

5   Conclusion

We introduced a novel linear-time variational integrator for simulating multi-body dynamic systems. At each simulation time step, the integrator solves a root-finding problem for the DEL equation using our quasi-Newton algorithm, *RIQN*, which consists of two primary contributions:

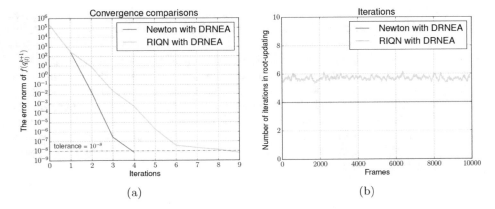

Fig. 4: (a) Convergence rate comparison for Newton's method and RIQN (b) Iteration numbers over simulation frames. Newton's method: mean = 4, $\sigma = 0.0$. RIQN: mean = 5.69, $\sigma = 1.16$

- **DRNEA:** Based on the variational integrator on Lie group and inspired by RNEA, we derived an $O(n)$ recursive algorithm that evaluates DEL equations of tree-structured multibody systems. Unlike the previous work, which formulates and solves the DEL equation for the entire system, in our approach the DEL equation for each body is solved recursively.
- **Root updating:** By leveraging existing forward dynamic algorithm for multibody systems, we introduced an $O(n)$ impulse-based dynamic algorithm to estimate the configuration at next time step.

We evaluated our linear-time variational integrator on a n-DOF open chain system and compared the results with existing state-of-art algorithms. The results show that, for the same computation method of root updating, the performance of our recursive evaluation of the DEL equation (linear-time DEL algorithm) is 15 times faster for a system with 10 degrees of freedom (DOFs) and 32 times faster for 100 DOFs. For the same evaluation method of the DEL equation (*i.e.*, linear-time DEL algorithm), our results show that the performance of our new quasi-Newton method is 3.8 times faster for a system with 10 DOFs, and 53 times faster for 100 DOFs. Further analysis shows that for higher DOF systems, the impulse-based Jacobian approximation becomes increasingly more effective compared to our linear-time DEL algorithm.

One of the future directions is to apply the linear-time variational integrator on constrained dynamic systems. This paper demonstrates the performance gain on multibody systems with joint constraints, but does not address other types of constrains, such as contacts or closed-loop chains. The standard way to handle constraints in a dynamic system is to solve the DEL equations and constraints simultaneously using Lagrangian multipliers [1,2]. To preserve the performance gain achieved by RIQN, one possible extension to constrained systems is to solve constraint force using the similar idea of impulse-based forward dynamics [8,20].

Our current implementation of RIQN can be improved by using variable time step size. Although the variational integrator allows for larger time step size than other numerical integrators for the same accuracy, the variable time step size can still be exploited to achieve further stability and time performance. However, naively changing the time step size can have negative impact on the qualitative behavior of a simulation [15,21]. Previous work has shown that additional constraints are needed when using the scheme of variable time step size. Integrating this line of work to our linear-time variational integrator can be a fruitful future research direction.

Acknowledgments

This work was (partially) funded by the National Science Foundation IIS (#1409003), Toyota Motor Engineering & Manufacturing (TEMA), and the Office of Naval Research.

References

1. Marsden, J.E., West, M.: Discrete mechanics and variational integrators. Acta Numerica 2001 **10** (2001) 357–514
2. West, M.: Variational integrators. PhD thesis, California Institute of Technology (2004)
3. Betsch, P., Leyendecker, S.: The discrete null space method for the energy consistent integration of constrained mechanical systems. part ii: Multibody dynamics. International journal for numerical methods in engineering **67**(4) (2006) 499–552
4. Leyendecker, S., Marsden, J.E., Ortiz, M.: Variational integrators for constrained dynamical systems. ZAMM-Journal of Applied Mathematics and Mechanics/Zeitschrift für Angewandte Mathematik und Mechanik **88**(9) (2008) 677–708
5. Leyendecker, S., Ober-Blöbaum, S., Marsden, J.E., Ortiz, M.: Discrete mechanics and optimal control for constrained systems. Optimal Control Applications and Methods **31**(6) (2010) 505–528
6. Johnson, E.R., Murphey, T.D.: Scalable variational integrators for constrained mechanical systems in generalized coordinates. Robotics, IEEE Transactions on **25**(6) (2009) 1249–1261
7. Luh, J.Y., Walker, M.W., Paul, R.P.: On-line computational scheme for mechanical manipulators. Journal of Dynamic Systems, Measurement, and Control **102**(2) (1980) 69–76
8. Featherstone, R.: Rigid body dynamics algorithms. Springer (2014)
9. Park, F.C., Bobrow, J.E., Ploen, S.R.: A lie group formulation of robot dynamics. The International Journal of Robotics Research **14**(6) (1995) 609–618
10. Lee, T.: Computational geometric mechanics and control of rigid bodies. ProQuest (2008)
11. Kobilarov, M.B., Marsden, J.E.: Discrete geometric optimal control on lie groups. Robotics, IEEE Transactions on **27**(4) (2011) 641–655

12. Kobilarov, M., Crane, K., Desbrun, M.: Lie group integrators for animation and control of vehicles. ACM Transactions on Graphics (TOG) **28**(2) (2009) 16

13. Murray, R.M., Li, Z., Sastry, S.S.: A mathematical introduction to robotic manipulation. CRC press (1994)

14. Bou-Rabee, N., Marsden, J.E.: Hamilton–pontryagin integrators on lie groups part i: Introduction and structure-preserving properties. Foundations of Computational Mathematics **9**(2) (2009) 197–219

15. Hairer, E., Lubich, C., Wanner, G.: Geometric numerical integration: structure-preserving algorithms for ordinary differential equations. Volume 31. Springer Science & Business Media (2006)

16. Fan, T., Murphey, T.: Structured linearization of discrete mechanical systems on lie groups: A synthesis of analysis and control. In: 2015 54th IEEE Conference on Decision and Control (CDC), IEEE (2015) 1092–1099

17. Liu, C.K., Stillman, M., Lee, J., Grey, M.X.: DART - Dynamic Animation and Robotics Toolkit. (2011 (accessed October 30, 2016)) http://dartsim.github.io.

18. Liu, C.K., Jain, S.: A short tutorial on multibody dynamics. Technical Report GIT-GVU-15-01-1, Georgia Institute of Technology, School of Interactive Computing (08 2012)

19. Broyden, C.G.: A class of methods for solving nonlinear simultaneous equations. Mathematics of computation **19**(92) (1965) 577–593

20. Mirtich, B., Canny, J.: Impulse-based simulation of rigid bodies. In: Proceedings of the 1995 symposium on Interactive 3D graphics, ACM (1995) 181–ff

21. Kharevych, L.: Geometric interpretation of physical systems for improved elasticity simulations. PhD thesis, Citeseer (2009)

Appendix: Derivative of DRNEA

Algorithm 4 Derivative of DRNEA for computing $\frac{\partial f(\mathbf{q}^{k+1})}{\partial \mathbf{q}^{k+1}} \in \mathbb{R}^{n \times n}$

1: **for** $j = 1 \to n$ **do**

2: **for** $i = 1 \to n$ **do**

3:   $\frac{\partial T_{\lambda(i),i}^{k+1}}{\partial q_j^{k+1}} = T_{\lambda(i),i}^{k+1}[S_i]\delta_{ij}$ $\qquad \triangleright \delta_{ij} = \begin{cases} 1, & \text{if } i = j \\ 0, & \text{otherwise} \end{cases}$

4:   $\frac{\partial \Delta T_i^k}{\partial q_j^{k+1}} = T_{\lambda(i),i}^{k}{}^{-1} \frac{\partial \Delta T_{\lambda(i)}^k}{\partial q_j^{k+1}} T_{\lambda(i),i}^{k+1} + \Delta T_i^k[S_i]\delta_{ij}$

5:   $\left[\frac{\partial V_i^k}{\partial q_j^{k+1}}\right] = \frac{1}{\Delta t}d\log_{\Delta t V_i^k}\left(\frac{\partial \Delta T_i^k}{\partial q_j^{k+1}}\exp\left(-\Delta t[V_i^k]\right)\right)$

6: **end for**

7: **for** $i = n \to 1$ **do**

8:   $\frac{\partial \mu_i^k}{\partial q_j^{k+1}} = \frac{\partial}{\partial q_j^{k+1}}\left[d\log_{\Delta t V_i^k}\right]^T G_i V_i^k + \left[d\log_{\Delta t V_i^k}\right]^T G_i \frac{\partial V_i^k}{\partial q_j^{k+1}}$

9:   $\frac{\partial F_i^k}{\partial q_j^{k+1}} = \frac{\partial \mu_i^k}{\partial q_j^{k+1}} + \sum_{c \in \sigma(i)}\left[\text{Ad}_{(T_{i,c}^k)^{-1}}\right]^T \frac{\partial F_c^k}{\partial q_j^{k+1}} - \frac{\partial F_i^{\text{ext},k}}{\partial q_j^{k+1}}$

10:   $\frac{\partial f(\mathbf{q}^{k+1})}{\partial q_j^{k+1}} = S_i^T \frac{\partial F_i^k}{\partial q_j^{k+1}} - \frac{\partial Q_i^k}{\partial q_j^{k+1}}$ $\qquad \triangleright$ j-th column of $\frac{\partial f(\mathbf{q}^{k+1})}{\partial \mathbf{q}^{k+1}}$

11: **end for**

12: **end for**

A General Algorithm for Time-Optimal Trajectory Generation Subject to Minimum and Maximum Constraints

Stephen D. Butler, Mark Moll, and Lydia E. Kavraki

Rice University, Department of Computer Science, Houston, TX 77005, USA

Abstract. This paper presents a new algorithm which generates time-optimal trajectories given a path as input. The algorithm improves on previous approaches by generically handling a broader class of constraints on the dynamics. It eliminates the need for heuristics to select trajectory segments that are part of the optimal trajectory through an exhaustive, but efficient search. We also present an algorithm for computing all achievable velocities at the end of a path given an initial range of velocities. This algorithm effectively computes bundles of feasible trajectories for a given path and is a first step toward a new generation of more efficient kinodynamic motion planning algorithms. We present results for both algorithms using a simulated WAM arm with a Barrett hand subject to dynamics constraints on joint torque, joint velocity, momentum, and end effector velocity. The new algorithms are compared with a state-of-the-art alternative approach.

1 Introduction

A path is a continuous curve describing the joint positions desired for a robot to move through. It is often represented by a sequence of waypoints with positions between them obtained through interpolation. A trajectory is a re-parameterization of a path as a function of time. The trajectory generation process is subject to the dynamics constraints of the underlying system. Finding time-optimal trajectories for robots with many degrees of freedom (DOFs) subject to complex dynamics and geometric constraints is a challenging problem that arises naturally in many robotics applications ranging from welding and painting to humanoid robots [1] and even spacecraft [2]. In practice, conservative safety constraints (e.g., by assuming quasistatic dynamics) are often used to simplify the trajectory generation problem, but this leads to either sub-optimal performance or more powerful hardware requirements than strictly necessary. If a task cannot be accomplished quasistatically, the task may still be dynamically feasible, in which case there are necessarily *minimum* velocity constraints which have generally, previously been ignored.

Many in the control community, such as [3], have made progress in solving planning problems which are constrained by dynamics, e.g., through constraints on torque, velocity, and momentum. Likewise, those in the path planning community, typically using sampling-based planning [4,5], have made progress in cluttered environments where collision and geometric constraints arise, e.g., on pose, orientation, and end effector contact. Nevertheless, motion planning problems for high DOF systems subject to both complex geometric and dynamics constraints still pose a formidable challenge. The typical motion planning pipeline for high DOF systems often consists of first performing sampling-based planning to generate a path which is then fed into a trajectory generator or controller. In this case, any sense of optimality or feasibility for the dynamics is lost as

© Springer Nature Switzerland AG 2020
K. Goldberg et al. (Eds.): *Algorithmic Foundations of Robotics XII*, SPAR 13, pp. 368–383, 2020.
https://doi.org/10.1007/978-3-030-43089-4_24

the path is generated without the dynamics. Kinodynamic motion planners that consider both the dynamics and geometric constraints do exist, e.g. [6], but the size of the state space of the problem still increases exponentially for high-DOF systems. Our work aspires to push the envelop on solving high-DOF problems with dynamics efficiently.

Admissible velocity propagation (AVP) [7] is a promising new method which combines sampling-based planning and time-optimal trajectory generation to make the kinodynamic planning problem *significantly* more tractable. AVP builds on tree-based motion planning algorithms (e.g., [8,9]), which iteratively construct a tree of configurations connected by path segments. In AVP, a tree-based planner performs an additional step for each new path segment: it computes a bundle of all dynamically feasible trajectories. In this manner, the planner finds the interval of all admissible velocities (AVI) that can be reached at each sampled configuration. By computing the path segment-AVI pairs the search space is significantly shrunk since while position is sampled as a point, the velocity component of state is sampled as a volume. By sampling entire volumes of the state space, kinodynamic motion planning times are significantly improved. Unfortunately, the classical algorithms [10,11] which compute the AVI needed by AVP planners have been criticized as difficult to implement, not very robust, and difficult to extend to other dynamics constraints such as observed in [12,13]. The classical methods only work for those specific constraints for which they were written. No proof of completeness exists for these algorithms despite their determinism.

This paper presents two algorithms. The first algorithm, which we will refer to as Traj_Alg hereafter, generates a time-optimal trajectory over a path given some starting and ending velocity. That is, given a path, a robot description, and a set of additional constraint functions, the trajectory returned is of the shortest time possible. The second algorithm, which we will refer to as AVI_Alg hereafter, computes the AVI over a path instead of a single trajectory like Traj_Alg.

Both algorithms we present are efficient, robust, and generalize to a large class of constraints. While previous algorithms would require considerable extension or indeed a complete rework to reason over constraints other than torque or velocity, our algorithms require the user only input a parameterization of their novel constraint function, which is the bear minimum to even define a problem. So long as the user's constraint adheres to the class of constraints which we detail in 6, our algorithm will either return the solution or report that no solution exists. This class includes all those constraints presented previously in the phase plane literature and a considerable number more for which we present a few examples.

In addition to a considerable expansion in generality, we show in the results section that our algorithms outperform the state-of-the-art methods as implemented in TOPP [14] in terms of compute time. Additionally, our algorithms never fail to find a solution despite the stressful test cases which cause TOPP to not find solutions. Our algorithms challenge the fundamental assumptions of the classical methods through an exhaustive, but efficient search process. We plan as future work to present a proof of completeness.

2   Related Work

Work on the path-constrained time-optimal problem began in the early 1980s by [10,11]. These works attempt to encode dynamical constraints into a phase plane delineated by inadmissible regions and proved that time-optimal trajectories result from finding profiles in the phase plane which maximize the integral of phase velocity. [10,11] are both specific only to torque constraints, fail at undifferentiable points, and cannot guarantee that they will find a solution even if one exists.

Recently, convex optimization [13,15] is another approach which has been used to solve the path parameterization problem. Convex optimization is similarly efficient in terms of computation time to phase plane navigation and has the added benefit of being able to optimize for quantities other than time. However, introducing new constraints requires that convexity is maintained and no extension has been presented so far to efficiently solve the AVI problem.

Phase plane navigation, however, can solve the AVI problem as efficiently as it solves the path parameterization problem. Much like the move from the workspace to the configuration space for geometric robot motion planning, the phase plane takes the state of the robot and transforms it into a point and trajectories into curves in the phase plane. This gives us a new space in which to solve the optimal-trajectory generation problem. Unfortunately, the early algorithms of [10,11] are difficult to extend to new constraints beyond joint torque and joint velocity constraints [16].

Recent work in phase plane navigation by [12], which builds on [11,16], enforces a specific type of path to simplify the problem and works only over velocity and acceleration constraints. Specifically, [12] uses a type of "circular blend" at each way-point which prevents a general path which does not have this feature from being input to the algorithm. There is no clear way to extend this method to torque or other constraints.

In the case of torque constraints, which the classical method, e.g. [10,11], was designed for, zero inertia points along the path which cause the maximum and minimum phase acceleration to be undefined were overlooked. These points were identified by [17], but continued to be mishandled or ignored leading to poor, erratic looking solution trajectories. An approach for identifying and determining the slope at zero inertia switching states for rigid body robots was developed by [18].

The TOPP library [14] is the most recent implementation of the algorithms originated by [10,11] and incorporates the extensions and improvements by the authors mentioned previously, e.g., [16,17,18], and we use it as our point of comparison.

Through all of the mentioned related works, the focus has been on specific constraints, typically torque and velocity. This paper generalizes to a much broader class of constraints, as defined in 6, and eliminates the need for a case-by-case analysis, thereby making it trivial to provide novel constraints as input to our algorithms. Additionally, these previous works operate by seeking only specifically those switching states and corresponding profiles which form the solution. We abandon this tenet and perform an exhaustive search over all possible switching states.

3   Problem Definitions

Traj_Alg computes a time-optimal trajectory from the following inputs:

A path. A path is specified by C^2 curves for each DOF, $\mathbf{q}(s)$. Without C^2 continuity the acceleration is undefined where the second derivative is discontinuous which could lead to undefined behavior in the navigation policy at these points.

System-control dynamics. Specifically, the algorithm requires functions that compute the maximum acceleration $\ddot{s}_{max}(s, \dot{s})$ and minimum acceleration $\ddot{s}_{min}(s, \dot{s})$ in the phase plane. The phase plane itself is described in more detail in section 4.

Path start and end states. The initial position and velocity $(s_o, \dot{s}_o)$ and the final position and velocity $(s_f, \dot{s}_f)$ along the path.

Dynamics constraints. Inequalities of the form $\dot{s}_{min}(s) \leq \dot{s}(s) \leq \dot{s}_{max}(s)$ which determine the maximum or minimum phase velocity for a point s along the path. We will show in detail how one parameterizes constraints for the phase plane in section 6.

The **objective function** for minimization is time: $T = \int_{t_o}^{t_f} dt$. The **output** of our algorithm is then a trajectory, $\mathbf{q}(t)$ and $\dot{\mathbf{q}}(t)$ $(t_o \leq t \leq t_f)$, where $\mathbf{q}$ is a full configuration of the system.

AVI_Alg extends Traj_Alg and takes as input in place of discrete starting and ending states a *range* of starting velocities $[\dot{s}_{min,o}, \dot{s}_{max,o}]$, which we will refer to as the input AVI. The output correspondingly is the AVI at s_f: $[\dot{s}_{min,f}, \dot{s}_{max,f}]$. For any admissible output velocity we can efficiently extract the corresponding time-optimal trajectory.

4   Classical Phase Plane Navigation Overview

Before we present our algorithms, it is important to understand classical phase plane navigation. The phase plane is defined by the path velocity $\dot{s}$ and the path position s. The scalar s indicates the position along the path $\mathbf{q}(s)$. The velocity and acceleration for all joints in the configuration space can be derived from $\mathbf{q}(s(t))$ using the chain rule.

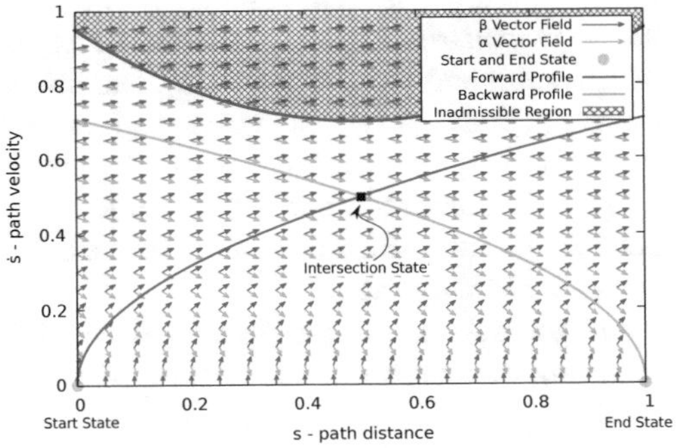

Fig. 1. Example phase plane

$$\dot{\mathbf{q}} = \frac{d\mathbf{q}}{ds}\dot{s} \qquad \ddot{\mathbf{q}} = \frac{d\mathbf{q}}{ds}\ddot{s} + \frac{d^2\mathbf{q}}{ds^2}\dot{s}^2 \qquad (1)$$

The transformation of constraints to the phase plane is useful because it creates a lower-dimensional space in which the time-optimal trajectory can be computed more efficiently. A *profile*, such as the blue and gold curves in figure 1, is defined as a curve in the phase plane. A *solution trajectory* is any profile or concatenation of profiles from the starting to ending state which do no cross into the inadmissible region. If a solution trajectory maximizes the area under its curve, then it is the *time optimal-trajectory* [10]. Hence, maximizing $\dot{s}$ over a path is equivalent to minimizing the total time for its corresponding trajectory.

For the simple example in figure 1, the solution trajectory can be found by following the β and α vector fields *forward* and *backward* from the start and end states, respectively. The β vector field is defined by the maximum acceleration, $\beta = \ddot{s}_{max}(s, \dot{s})$, which the system is capable of at a given state and α is defined likewise as the minimum acceleration, $\alpha = \ddot{s}_{min}(s, \dot{s})$. We refer to profiles which progress monotonically left to right, and thus try to follow the β vector field as forward profiles and having the forward directional type. Backward profiles are the opposite, flowing right to left along α.

Clearly, there are no valid profiles originating at the start or end state which can rise to values of $\dot{s}$ above those shown in figure 1 since they were created by applying extremal control inputs, i.e., following the β/α fields. Thus, the trajectory resulting from these profiles is time-optimal.

In figure 1, the intersection between the forward and backward profile is denoted by an intersection state. In prior work this intersection would not be computed explicitly, intersections were assumed when it was detected that one profile was above another at some value s. For our methods, intersections between profiles are discovered using curve intersection methods. These intersections are valuable to our algorithms since we need them to develop a graph of profiles in the phase plane.

To construct profiles we *navigate*. We refer to their construction as *navigation*, instead of merely integration, because a *navigation policy* is employed to step-by-step generate profiles. For example, to construct the profile originating from the starting state, the algorithm asks the navigation policy what phase acceleration $\ddot{s}$ it should use to integrate the equations of motion based on its current $[s, \dot{s}]$ state. The phase acceleration which is returned by the policy is then integrated to arrive at some new state. Usually, the acceleration returned by the policy will be equivalent to the β or α vector fields. This "bang-bang control" ensures that profiles are always maximizing $\dot{s}$ with each step.

However, the navigation policy may deviate to some value between β and α when it hits an *inadmissible region*. Inadmissible regions are induced by the dynamics constraints of a system and are delineated by maximum or minimum velocity curves (maxVC) (minVC). Since, the states in these regions violate constraints they cannot be entered by a valid profile. In the instance where a profile has intersected the maxVC, the navigation policy will attempt to guide the profile along the maxVC and so may return a value between β and α. Since no valid states exists above the maxVC, following the maxVC constitutes maximizing $\dot{s}$.

In general, the inadmissible regions and β/α vector field interactions are complex, requiring many profiles and a *phase plane navigation algorithm* is needed. At a high level, all phase plane navigation algorithms perform the following steps, which will be explained in more detail below:

1. Compute maximum and minimum velocity curves (maxVC/minVC), respectively.
2. Compute constraint switching states.
3. Construct maximum limiting profiles (MLP) from constraint switching states.
4. Connect start to end state (Traj_Alg) or connect start to end AVI (AVI_Alg)

Computing the minVC and maxVC is usually a straightforward substitution of the terms in equation 1 into the functions which define specific system constraints. We show how this is done for several constraints in section 6.

If a profile originating at the start or end state intersects the maxVC and has no viable direction of navigation, then it is necessary to identify new states in the space to connect to. These states are referred to as *switching states*. In terms of the control, they are points where the control input switches from following α to β or vice versa. A priori, it is unknown which switching states will be needed to traverse the plane. Any point along the maxVC could potentially be a switching state. In [14], the identification of these necessary switching states follows from several heuristics designed for different types of constraints.

The MLP is formed from a set of profiles which originate from switching states along the maxVC. MLP construction follows by navigating profiles forward and backward from the constraint switching states, which we will refer to as *expansion*. In [14], profiles are expanded without means of accounting for connectivity between profiles or continuity to states such as the start or end states. This lack of information leads to many problems, both in determining if the resulting profiles form a solution trajectory and in determining if the resulting velocity intervals are in fact feasible. That is, the prior methods may report a feasible trajectory infeasible, i.e., a false negative, or report an infeasible trajectory as feasible, i.e., a false positive. Our algorithms do not have this problem thanks to a key contribution in the form a graph data structure which allow us to efficiently extract the time-optimal trajectory if it exists and report definitively that no solution trajectory exists if it does not.

For trajectory generation, the classical method attempts to connect the starting and ending states either directly together as we saw in figure 1 or through the MLP. For AVI, the algorithm presented in [14] attempts a bisection search. Profiles are iteratively expanded from the start range to find the maximum final $\dot{s}$. It then proceeds to perform the same search at the end trying to find the minimum final $\dot{s}$. This process is described in more detail later as we deviate from the classical method in AVI computation.

5   Traj_Alg and AVI_Alg, Algorithms for Phase Plane Navigation

This section elaborates the key contributions of our work in phase plane navigation. Our algorithms follow the same general flow as what is presented in the previous section. However, we replace the methods used in each step from section 4 with our own to resolve the issues we identified and improve the generality to new novel dynamics constraints. For example, rather than trying to only find the subset of switching states which forms the solution, we exhaustively find all possible switching states as outlined in 5.1. To manage the exhaustive set of profiles and intersection states, we introduce a data structure that can be searched for the time-optimal trajectory or used in service of finding the AVI as outlined in 5.2. The following subsections apply to both Traj_Alg and AVI_Alg until section 5.3 where they split.

5.1   Identification of Constraint Switching States

We refer to a state as *navigable* if there exists an infinitesimal expansion to the left and to the right in the phase plane which does not enter the inadmissible region. In figure 2, we can see that point marked 1 on the maxVC is navigable. If the slope of the maxVC is between the slopes formed by the β and α vectors, which we will refer to as the *navigation cone*, then the maxVC can be followed. The navigation cone for the point marked 2 is directed into the inadmissible region and thus this point is un-navigable.

We do not seek switching states along the minVC. If there exists a state above the minVC at some s, then this state is necessarily better than the one on the minVC, since it has a higher $\dot{s}$. However, if the maxVC and minVC coincide, i.e., no state exists above the minVC, then the path is infeasible as there is no way to traverse the plane without entering the inadmissible region.

General Switching States In general, any state along the maxVC which is navigable is a potential switching state. Also, all states within a navigable range are equivalent. That is, any state on the maxVC is equivalent to its navigable neighbors because expansion of said state

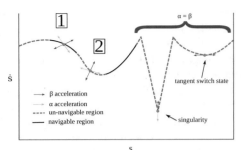

Fig. 2. Phase plane plot illustrating navigability along the maxVC. The α and β vectors are shown only the points marked in the figure.

will generate a profile which necessarily passes through its neighbors. The navigation policy always directs a profile in the direction which maximizes $\dot{s}$, which in this case is along the maxVC since no direction can result in a higher $\dot{s}$ in the next step. Thus, finding all navigable regions and picking at least one point in the range is equivalent to finding all possible constraint switching states. We define a function $navigable(s,\dot{s})$ which evaluates whether the slope of the maxVC bisects the navigation cone. This function is evaluated at finely discretized points along the path to find the valid regions and hence all general switching states. We assume $navigable(s,\dot{s})$ is Lipschitz continuous and that the integration step size is chosen such that the function is monotonic between any two consecutive discretized points. There are two degenerate cases: tangent switching states between discrete states and zero inertia switching states.

Tangent Switching States When the β vector is tangent to the slope of the maxVC we have a point of navigability. If the β vector is above the maxVC at a discrete state and below the maxVC at an adjacent discrete state, then we can use bisection to find the tangent switch state in the interval between them.

Zero Inertia Points As we will show in section 6, a change in sign of the inertial term causes a divide-by-zero scenario. The zero inertia points (ZIPs) corresponding to such an event are easy to identify. Similar to [14], we find these switching states via checking each discretized neighboring state along the maxVC for whether there is a change in sign of the denominator term in the equation which defines $\ddot{s}_{max}$ and $\ddot{s}_{min}$. The ZIP is then located through interpolation between these two states.

Once located, a simple method to determine the slope of best fit for a ZIP is to project segments to the left and right of the ZIP. For each segment a search can be performed to find the angle which minimizes the difference in slope between the segment and the β or α vector at the segments end point. Typically, the segment's s component should be held equal to the discretization length of the path to ensure that that the segment's projected endpoint is sufficiently far from the ZIP which is being avoided.

5.2   Constructing the Maximum Limiting Profiles

With all switching states found, the order of expansion does not matter and, indeed, they can be generated in parallel. Additionally, all switching states are expanded in the same

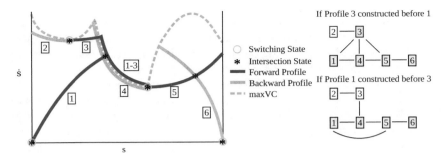

Fig. 3. Phase plane plot with numbered profiles and corresponding continuity graphs. Shaded nodes have continuity to the start and end states.

way except for those which occur at ZIPs which we discussed in section 5.1. This makes our algorithms very easy to parallelize.

In the event that any expansion would lead to a profile intersecting the minVC the algorithm will report that the path is infeasible and halt. Since the profile which intersected the minVC was following the β or α vector field, there exists no control which would avoid the inadmissible region. Also, the profile in question must have originated from the maxVC since it was a candidate MLP profile. Since, it originates from the maxVC, it is then clear that no profile can span the phase plane without also intersecting this profile and hence leading to the minVC. Thus, in this event, the path is known to be infeasible.

We will construct a graph, called the *continuity data structure* or *continuity graph*, with profiles and their intersections corresponding to vertices and edges, respectively. However, an encoding of profiles as edges and intersections as vertices is also viable. Our algorithms will search this graph to find the profiles and intersections leading from start to end state. If such a path from start to end exists, this path can be used to extract the time-optimal trajectory, or else the path is reported infeasible.

One who is familiar with [7] might wonder why such a graph is necessary. First, lemma 2 from [7] does not hold. By contradiction, this lemma states that there must be switching points from the left and right originating from the maxVC which would other wise fill a continuity hole in the CLC, an analogue to the MLP, with α and β profiles. However, if an α profile intersects the maxVC short of s_o there may not be a navigable switching state on the maxVC between the intersection and s_o, which frequently occurs in practice. By symmetry the same can occur with a β profile and s_f. The MLP, can still be made continuous to s_o by iteratively expanding states at s_o while maximizing $\dot{s}$ and seeking to intersect the MLP.

Second, since we have abandoned the heuristics that [7] uses to identify and expand switch points, the profile expanded from s_o might intersect several profiles some of which are not part of what we redefine in section 5.3 as the MLP. Or, if several profiles already intersect s_o, the continuity graph tells us which connected profiles actually form the MLP. Thus, the continuity graph is a necessary instrument for us to exhaustively expand all switching states and abandon selection heuristics.

Since switching states in our algorithm can be expanded in any order, there are many possible graphs which can be constructed for the same problem. One might then be concerned that this could lead to multiple solution trajectories. We illustrate in figure 3 two possible graphs, which result depending on the order of construction of profiles 1 and 3. However, the nodes and edges in our graph all have spatial meaning. Regardless of which graph is constructed, the spatial relationship between elements is the same. For example, paths [1-3-4-5-6] and [1-4-5-6] are equivalent as when we evaluate the paths monotonically from left to right, the exact same curve in the phase plane is constructed. Hence, the first path returned by any search method over the graph is the time-optimal solution. The algorithms cannot generate multiple spatially-distinct paths between states in the phase plane. The proof is a subject of future work but some of the intuition is apparent by the properties of profiles. For example, as was proved in [14], no α and β profile can intersect more than once and like profiles terminate after a single intersection.

5.3   Solution Extraction

Traj_Alg With all of the hard work finished, i.e., constructing the MLP, all that Traj_Alg needs to do is expand the starting and ending states. The resulting continuity graph as described in the previous section can then be searched for the solution trajectory.

AVI_Alg Computing admissible velocity intervals requires a bit more work. There is not a single $\dot{s}$ value to start from and so there is not a single start state to expand. Different $\dot{s}_o$ values can result in different $\dot{s}_{f,max}$ and $\dot{s}_{f,min}$. We illustrate the various steps of AVI_Alg in figure 4 needed to find $\dot{s}_{f,max}$ and $\dot{s}_{f,min}$.

First, AVI_Alg will find the $\dot{s}_{f,max}$. In the ideal situation, plot 2 in figure 4 or step 1(b) in AVI_Alg 1, the MLP already intersects the starting AVI in which case we can

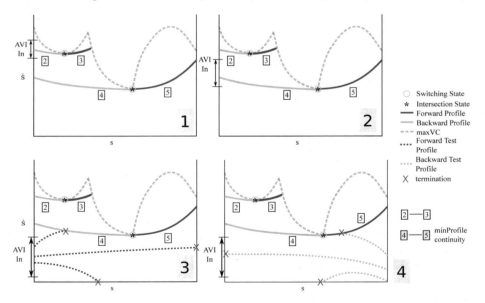

Fig. 4. Plot 1 illustrates condition 1a. Plot 2 illustrates condition 1b and 1c. Plot 3 illustrates condition 1d. Plot 4 illustrates both conditions 1(e) and 2(a) in algorithm 1.

Algorithm 1: AVI_Alg: algorithm to find the maximum and minimum reachable $\dot{s}$ at the end of the path.

---

1. Find the $\dot{s}_{max,f}$:
   (a) If the MLP (the set of profiles with continuity to the minProfile) reaches s_o and is below the starting AVI, return no solution exists, else go to step 1(b).
   (b) If the MLP intersects the starting AVI, go to step 1(c), else go to step 1(d).
   (c) If the MLP reaches s_f, then $\dot{s}_{max,f}$ is equivalent to this intersection and go to step 2, else go to step 1(e).
   (d) Bisection search, using forward *test profiles*, from start AVI to MLP or to highest $\dot{s}_f$. If MLP is intersected go to step 1(c), else if s_f reached then $\dot{s}_{max,f}$ has been found, else return no solution.
   (e) Bisection search, using backward *test profiles*, for highest $\dot{s}$ at $\dot{s}_f$ which intersects the MLP or reaches the start AVI. If MLP intersected or start AVI reached, $\dot{s}_{max,f}$ is found, else return no solution exists.
2. Find the $\dot{s}_{min,f}$:
   (a) Bisection search, using backward *test profiles*, to find $\dot{s}_{min,f}$ between the minVC(s_f) as a lower bound and $\dot{s}_{max,f}$ found in part 1 as the upper bound.

---

proceed to step 1(c). However, a set of profiles, such as 1 and 2, expanded during MLP construction may be disconnected from the profiles which actually span the plane such as in plot 1. To deal with this situation we redefine the MLP.

We will refer to the profile, out of all profiles constructed for the MLP, with the lowest value of $\dot{s}$ along its entire length as the *minProfile*. Clearly, if the MLP is to span the plane, this profile must be part of the solution. This profile like all those generated during MLP construction originates on the maxVC. No other profile can span the s space of the minProfile without also passing through the origin switching state of the minProfile. AVI_Alg can then use the continuity data structure to find all those profiles with continuity to the minProfile which now constitute the new MLP.

If the start AVI is above the MLP or below the minVC, steps 1(a) and 1(b) in algorithm 1, then the path velocity is too fast or too slow and so AVI_Alg returns no solution exists.

If we reach step 1(d), i.e., the MLP either reaches s_o and is above the AVI input or does not reach s_o, then AVI_Alg will perform a bisection search. The bisection search is conducted over the starting AVI, i.e., at s_o, from $\dot{s}_{o,min}$ to $\dot{s}_{o,max}$ as illustrated in plot 3. Over this interval of starting states, forward *test profiles* are expanded. The nature of the termination condition for a test profile determines if the search moves up or down the $\dot{s}_o$ interval until the test $\dot{s}$ converges on some ε. If the test profile terminates on the minVC, then the test $\dot{s}$ increases as higher $\dot{s}$ are less likely to hit the minVC. Likewise, if the test profile terminates on the maxVC then $\dot{s}$ is decreased. If the test profile terminates at s_f, then the $\dot{s}$ is increased. If the MLP is intersected at any point, then AVI_Alg proceeds to step 1(c). These are all the features in the space which can cause termination and thus the search must converge with one of these final conditions. The resulting conditions and the branch they cause in the algorithm are enumerated in step 1(d).

If step 1(e) is reached, AVI_Alg has not found $s_{f,max}$ but has connected the start AVI to the MLP and the MLP does not reach s_f. Thus we perform a similar bisection search as 1(d) except at s_f. Clearly, backward test profiles are used in the place of forward test profiles. The $\dot{s}_f$ interval used for bisection spans from the $maxVC(s_f)$ to the $minVC(s_f)$.

With $\dot{s}_{f,max}$ found, the next step, 2(a), is to find $\dot{s}_{f,min}$. To do this, perform the same search as in 1(e), except instead of biasing search up after terminating at a profile which has continuity to the start, we bias our search down as the goal is to find the smallest $\dot{s}$ possible. The $\dot{s}_f$ interval used for bisection spans from the $\dot{s}_{f,max}$ to the $minVC(s_f)$.

6   Phase Plane Parameterizations

The constraints which Traj_Alg and AVI_Alg can handle is defined as those constraints which form maximum and/or minimum curves which do not lead to more than one homotopy class of solution profiles. Our algorithms currently do not handle those constraints which [11] described as "inadmissible islands," so called because all of the space around these regions is admissible isolating them from other regions of inadmissibility. All of the constraints presented here fall into the class which Traj_Alg and AVI_Alg can handle. Determining if a constraint can create "inadmissible islands" is usually straight forward as the minimum and maximum curves defined by such a constraint form an *or relationship*, i.e. a state is valid if it is above the minimum or below the maximum, instead of an *and relationship*, i.e. a state is valid if it is both above the minimum and below the maximum.

Torque Constraints For torque constraints, consider rigid body robots whose dynamics can be described as:

$$\tau_{\mathbf{min}} \leq \mathbf{M(q)\ddot{q}} + \mathbf{\dot{q}}^T \mathbf{C(q)\dot{q}} + \mathbf{G(q)} \leq \tau_{\mathbf{max}} \tag{2}$$

We substitue the terms from 1 into 2 and alias the terms as a, b, c as seen in [14]. We have also changed our notation so that evaluation results in a scalar value of torque on the axis of some joint i.

$$a_i(s) = \mathbf{M}_i(\mathbf{q}(s))\frac{d\mathbf{q}}{ds} \quad b_i(s) = \mathbf{M}_i(\mathbf{q}(s))\frac{d^2\mathbf{q}}{ds^2} + \frac{d\mathbf{q}}{ds}^T \mathbf{C}_i(\mathbf{q}(s))\frac{d\mathbf{q}}{ds} \quad c_i(s) = \mathbf{G}_i(s) \tag{3}$$

Substituting 3 back into 2 leaves us with the simplified phase plane parameterized equations of motion for a rigid body robot.

$$\tau_{i,min} \leq a_i\ddot{s} + b_i\dot{s}^2 + c_i \leq \tau_{i,max} \tag{4}$$

Note that the a_i term is the inertial component, b_i is the Coriolis component, and c_i incorporates the external force terms. Rearranging terms from 4 we get

$$\frac{\tau_{i,min}-b_i\dot{s}^2-c_i}{a_i} \leq \ddot{s}_i \leq \frac{\tau_{i,max}-b_i\dot{s}^2-c_i}{a_i} \tag{5}$$

taking care to note that the sign of the inertial term a will determine whether the resulting $\ddot{s}$ is that joints maximum, $\ddot{s}_{i,max}$, or minimum, $\ddot{s}_{i,min}$, since the sign of a can flip the direction of the inequalities. With the system parameterized for the phase plane, we see that it is very easy to solve for the maximum acceleration for a given state $[s, \dot{s}]$ from 5.

To compute the maxVC and minVC we must solve for when $\alpha(s,\dot{s}) \geq \beta(s,\dot{s})$. Take a maximum constraint, i, and a minimum constraint, j, as defined in 5 to get:

$$\frac{-b_i\dot{s}^2-c_i}{a_i} \leq \frac{-b_j\dot{s}^2-c_j}{a_j} \quad \Rightarrow \quad \frac{-b_i\dot{s}^2}{a_i} + \frac{b_j\dot{s}^2}{a_j} \leq \frac{-c_j}{a_j} + \frac{c_i}{a_i} \tag{6}$$

Note that we have pulled the constant τ term into the constant c term. Isolating the $\dot{s}$ terms, we see that whether $\dot{s}$ is a solution to the maximum or minimum velocity curve at s is dependent upon the sign of the denominator in 7.

$$\dot{s}(s) \leq \sqrt{\left(\frac{-c_j}{a_j} + \frac{c_i}{a_i}\right) / \left(\frac{-b_i}{a_i} + \frac{b_j}{a_j}\right)} \tag{7}$$

Thus we can evaluate 7 with every combination of joints i, and joints j to find the minVC and maxVC.

Velocity Constraints Unlike torque constraints, which involve considering the full body dynamics of the robot, velocity constraints are much more straightforward. Here we repeat the constraint as seen in prior work [16]:

$$\dot{s}_{max}(s) = \min_i \dot{q}_{i,max} / \frac{dq}{ds}_i(s) \tag{8}$$

Momentum Constraints While the torque constraint parameterization appears intimidating, many constraints are very easy to parameterize. A common constraint in safety critical environments where robots are in close proximity to people or fragile hardware is the momentum constraint. Here we present a parameterization to prevent the robot from exceeding some maximum linear momentum for a given link in the system. The angular momentum is parameterized similarly. Linear momentum in a rigid body robot, for a link i, can be computed from the Jacobian and center of mass as follows

$$\mathbf{h}_i = \mathbf{v}_i m_i = \mathbf{J}_i(\mathbf{q})\dot{\mathbf{q}} m_i \tag{9}$$

Substituting 1 into 9 and solving for $\dot{s}$ gives us

$$\dot{s}_{i,max}(s) = \frac{\mathbf{h}_{i,max}}{m_i}\left(\mathbf{J}_i(\mathbf{q}(s))\frac{dq}{ds}\right)^{-1} \tag{10}$$

which we can compute for each link. As usual, we take the lowest $\dot{s}$ as the maxVC.

Workspace Proximity Constraints In the early work of [10,11], the phase plane constraints were parameterized in the context of the end effector for workspace planning and control instead of joint constraints. We extend this to a new, interesting constraint wherein a user might wish to slow the end effector or some other link when it is in proximity to some tool or object they wish it to interact with. When controlling an end effector to grasp an object, we might wish to add a threshold or some function which lowers the velocity limit as the end effector gets close to an object to improve accuracy. A hard threshold on end effector velocity when within a certain distance of an object can create jumps in the maxVC which our results show are handled without fail. We can find the maxVC for an arbitrary point on the robot by removing mass from equation 10:

$$\dot{s}_{max}(s) = \mathbf{v}_{max}(s)\left(\mathbf{J}(\mathbf{q}(s))\frac{dq}{ds}\right)^{-1} \tag{11}$$

7   Results

To evaluate our algorithms, we set up a series of simulation experiments for a 7-DOF WAM arm with the attached 4-DOF Barret hand using OpenRAVE [19] (see Fig. 5). To generate paths for the experiments, we randomly sample four configurations in the C-space of the Barret-WAM arm. A C^2 curve is interpolated through these four configurations via piece-wise quintic spline interpolation. In this fashion, we generate a test set of 1,000 random paths. The s length of each path segment is set to one resulting in paths of length three and discretized by $\varepsilon = 0.001$. Every state along these curves is reachable quasi-statically by the Barret-WAM arm. The input AVI are also set for each curve to span from the minVC to the maxVC. Thus, if an algorithm reports the path infeasible, it is a failure of the algorithm, i.e., a false negative. To check for false positives we evaluate the solution profiles for continuity and constraint violation. Only those cases which pass both tests are counted in the success rate.

In our experiments we compare the TOPP library [14] with our algorithms. TOPP supports both joint torque and velocity constraints and so we compare our algorithms using both of these types of constraints. These limits are included in the OpenRAVE supplied model. Our algorithms generalize to a broad class of constraints. We show below results for momentum and end effector velocity constraints which are within this class. Prior methods (like TOPP) cannot handle these types of constraints, so no direct comparison is possible. For momentum, we impose linear momentum limits of $[0.75, 0.75, 0.75, 0.65, 0.65, 0.75, 0.75, 0.75]$ (kg·m/s) to links wam0 through wam7, respectively. For end effector velocity we impose a limit of 0.5 m/s on wam7's center of mass.

Fig. 5. A path for the Barrett-WAM arm between four random waypoints, used as input for TOPP and our algorithms.

Traj_Alg The objective of this work is to improve the success rate and generality of time-optimal trajectory generation; however, the algorithms presented also outperform the state-of-the-art method in computation time (see table 1). One would think that growing the number of potential switching states, profiles, and introducing intersection routines would decrease performance. However, the prior method, such as implemented in TOPP, expends considerable effort vetting switching states before expansion as including an unnecessary or missing a necessary switching state can result in failure. In table 1 we can see that this effect results in an over 65% improvement in the time it takes to generate the profiles which store the solution. Our graph structure also greatly speeds up the time it takes to extract the trajectory by several orders of magnitude.

Comparing the success rates, TOPP does fairly well. However, considering that these are deterministic algorithms we should expect 100% success rates. For Traj_Alg we see that the time-optimal trajectory was found for every test case.

AVI_Alg In this section we evaluate the performance of TOPP and AVI_Alg. We measure success rate, generation time, and the height of the velocity intervals found. The velocity intervals found should be fairly similar for both algorithms given the same constraints, but one would expect some loss in TOPP due to the heuristics employed.

For TOPP, we see a notable drop in the success rate for the harder AVI problem which is the focus of this work. This is mostly due to the fact that some of the heuristics

Table 1. Success rate and trajectory computation time for TOPP and Traj_Alg. The trajectory generation time can be broken down into generation time (T_g) and extraction time (T_e). Results are an average over 1000 random test cases.

Algorithm	Constraints	Success (%)	T_g (s)	T_e (s)	$T_g + T_e$ (s)
TOPP	torque	98.0	0.248	0.174	0.422
TOPP	torque + velocity	99.6	0.278	0.324	0.601
Our algorithm	torque	100	0.184	0.004	0.188
Our algorithm	torque + velocity	100	0.184	0.003	0.187
Our algorithm	torque + momentum	100	0.191	0.003	0.194
Our algorithm	torque + momentum + end eff. vel.	100	0.192	0.002	0.194

Table 2. Success rate and AVI generation time for TOPP and AVI_Alg. Results are an average over 1000 random test cases. †: 3 successful cases are not enough to produce statistically significant times or ranges for comparison.

Algorithm	Constraints	Success (%)	AVI Generation Time (s)	AVI Size ($\dot{s}$ units)
TOPP	torque	90.8	0.226	0.757
TOPP	torque + velocity	0.3	†	†
Our algorithm	torque	100	0.182	0.761
Our algorithm	torque + velocity	100	0.181	0.399
Our algorithm	torque + momentum	100	0.188	0.316
Our algorithm	torque + momentum + end eff. vel.	100	0.190	0.272

used by TOPP for the point-to-point problem cannot be easily generalized to the AVI problem. Comparing tables 1 and 2, we see a drop from 98.0% to 90.8% for the cases with joint torque constraints and complete failure when joint velocity constraints are added. Ignoring the complete failure case, losing 10% of cases would be catastrophic for a kinodynamic motion planning algorithm relying on these algorithms for determining feasibility. This data suggests that the algorithm implemented in [14] is not complete.

Other arms The performance differences between TOPP and our algorithms vary based on the kinematics and dynamics of the robot, but they can be quite large. For example, for a 5-DOF arm with unit-length and unit-mass links the success rate from TOPP drops to 84.3% for trajectory generation and 52.7% for AVI computation under torque constraints of 20 units for each joint. In contrast, the success rate of our algorithms remains 100%.

Minimum Velocity Constraints: A Case Study In this section we illustrate a fully dynamic planning problem representative of the painting/welding application mentioned in the introduction. In addition to all the previous constraints, we now also impose a minimum velocity limit on the end effector of 0.05 m/s. The end effector velocity limits make sure that neither too little nor too much paint is deposited. The momentum limits ensure some measure of safety for human co-workers. The joint torque and velocity limits simply follow from the robot's dynamics constraints.

Finding a time-optimal trajectory for a given path subject to all these constraints is extremely challenging. Standard workspace time-optimal control methods that solve boundary value problems cannot be applied as we wish to follow a *given* path. Algorithms

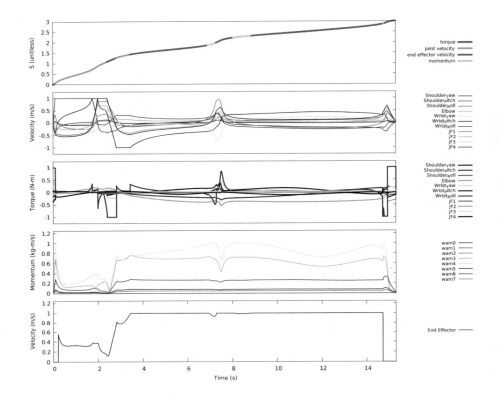

Fig. 6. Phase plane plot along with the resulting trajectories which correctly saturate the constraints. The colored time vs. *s* plot indicates which constraint is saturated.

like TOPP do not support minimum constraints, momentum constraints, or end effector velocity constraints. In addition, they tend to perform poorly on non-quasistatic paths. In contrast, by merely inputing the parameterizations for these constraints (discussed in Section 6) into our algorithm we are able to find time-optimal trajectories.

Figure 6 shows the trajectory generated for a given path. The plots below it show the torques and velocities for all joints, the momentum for each link, and the end effector velocity, respectively. Each curve has been normalized based on its constraint limit. A value above 1 or below −1 would indicate that a constraint is being violated. The first plot is color coded based on the active constraint at each point in time. For example, the first purple segment indicates that joint velocity is the active constraint. We can trace down to the velocity vs. time plot and see that indeed at least one joint's limit is saturated. This is indicated by the plateau at the normalized value of 1. Walking along the first plot we can see that at every time point some constraint is saturated.

8   Conclusion

We have presented novel algorithms for time-optimal trajectory generation and the computation of AVI. We have shown that they outperform classical methods in terms of success rate and performance, and generalize to a broader class of constraints than prior work. A proof of completeness and further theoretical analysis of the algorithms are future work.

Acknowledgments Work on this paper has been supported in part by NSF 1139011, NSF 1317849, and NSF 1514372.

References

1. Pham, Q.C., Nakamura, Y.: Time-optimal path parameterization for critically dynamic motions of humanoid robots. In: IEEE-RAS Intl. Conf. on Humanoid Robots. (2012) 165–170
2. Nguyen, H., Pham, Q.: Time-optimal path parameterization of rigid-body motions: Applications to spacecraft reorientation. J. of Guidance Control and Dynamics **39**(7) (2016) 1667–1671
3. Ansari, A.R., Murphey, T.D.: Sequential action control: Closed-form optimal control for nonlinear systems. IEEE Trans. on Robotics **32**(5) (October 2016) 1196–1214
4. Berenson, D., Srinivasa, S.S., Ferguson, D., Kuffner, J.J.: Manipulation planning on constraint manifolds. In: IEEE Intl. Conf. on Robotics and Automation. (2009) 625–632
5. Jaillet, L., Porta, J.: Path planning under kinematic constraints by rapidly exploring manifolds. IEEE Trans. on Robotics **29**(1) (2013) 105–117
6. Elbanhawi, M., Simic, M.: Sampling-based robot motion planning: A review. IEEE Access **2** (2014) 56–77
7. Pham, Q.C., Caron, S., Nakamura, Y.: Kinodynamic planning in the configuration space via admissible velocity propagation. In: Robotics: Science and Systems, Berlin, Germany (2013)
8. Hsu, D., Latombe, J.C., Motwani, R.: Path planning in expansive configuration spaces. Intl. J. of Computational Geometry and Applications **9**(4-5) (1999) 495–512
9. Kuffner, J., LaValle, S.M.: RRT-Connect: An efficient approach to single-query path planning. In: IEEE Intl. Conf. on Robotics and Automation, San Francisco, CA (2000) 995–1001
10. Bobrow, J., Dubowsky, S., Gibson, J.: Time-optimal control of robotic manipulators along specified paths. The Intl. Journal of Robotics Research **4**(3) (1985) 3–17
11. Shin, K., McKay, N.: Minimum-time control of robotic manipulators with geometric path constraints. IEEE Trans. on Automatic Control **30**(6) (1985) 531–541
12. Kunz, T., Stilman, M.: Time-optimal trajectory generation for path following with bounded acceleration and velocity. In: Robotics: Science and Systems, Sydney, Australia (2012)
13. Hauser, K.: Fast interpolation and time-optimization on implicit contact submanifolds. In: Robotics: Science and Systems, Berlin, Germany (2013)
14. Pham, Q.C.: A general, fast, and robust implementation of the time-optimal path parameterization algorithm. IEEE Trans. on Robotics **30**(6) (2014) 1533–1540
15. Verscheure, D., Demeulenaere, B., Swevers, J., Schutter, J.D., Diehl, M.: Time-optimal path tracking for robots: A convex optimization approach. IEEE Trans. on Automatic Control **54**(10) (2009) 2318–2327
16. Žlajpah, L.: On time optimal path control of manipulators with bounded joint velocities and torques. In: IEEE Intl. Conf. on Robotics and Automation. (1996) 1572–1577
17. Shiller, Z.: On singular time-optimal control along specified paths. IEEE Trans. on Robotics and Automation **10**(4) (1994) 561–566
18. Pham, Q.C.: Characterizing and addressing dynamic singularities in the time-optimal path parameterization algorithm. In: IEEE/RSJ Intl. Conf. on Intelligent Robots and Systems. (2013) 2357–2363
19. Diankov, R.: Automated Construction of Robotic Manipulation Programs. PhD thesis, Carnegie Mellon University, Robotics Institute (August 2010)

Dynamic Walking on Stepping Stones with Gait Library and Control Barrier Functions

Quan Nguyen[1], Xingye Da[2], J. W. Grizzle[3], Koushil Sreenath[1]

[1]Dept. of Mechanical Engineering, Carnegie Mellon University, Pittsburgh, PA 15213
[2]Dept. of Mechanical Engineering, University of Michigan, Ann Arbor, MI 48109
[3]Dept. of Electrical Engineering and Computer Science, University of Michigan, Ann Arbor, MI 48109

Abstract. Dynamical bipedal walking subject to precise footstep placements is crucial for navigating real world terrain with discrete footholds such as stepping stones, especially as the spacing between the stone locations significantly vary with each step. Here, we present a novel methodology that combines a gait library approach along with control barrier functions to enforce strict constraints on footstep placement. We numerically validate our proposed method on a planar dynamical walking model of MARLO, an underactuated bipedal robot. We show successful single-step transitions from a periodic walking gait with a step length of 10 (cm) to a stepping stone with a 100 (cm) separation (10x step length change), while simultaneously enforcing motor torque saturation and ground contact force constraints. The efficacy of our method is further demonstrated through dynamic walking over a randomly generated set of stepping stones requiring single-step step length changes in the range of [10:100] (cm) with a foot placement precision of 2 (cm).

1   Introduction

An important advantage of robotic systems employing legged locomotion is the ability to traverse terrain with discrete footholds, such as "stepping stones." Current approaches to handling this form of terrain primarily rely on simplistic methods, both at the level of models of bipedal robots (e.g., linear inverted pendulum) and control (e.g., ZMP) to achieve the desired foot placements. The overarching goal of this work is to create a formal framework that will enable bipedal humanoid robots to achieve dynamic and rapid locomotion over a randomly placed, widely varying, set of stepping stones.

Footstep placement control for fully actuated legged robots initially relied on quasi-static walking and resulted in slow walking speeds [13], [14], [5]. Impressive results in footstep planning and placements in obstacle filled environments with vision-based sensing have been carried out in [15], [4]. The DARPA Robotics Challenge inspired several new methods, some based on mixed-integer quadratic

© Springer Nature Switzerland AG 2020
K. Goldberg et al. (Eds.): *Algorithmic Foundations of Robotics XII*, SPAR 13, pp. 384–399, 2020.
https://doi.org/10.1007/978-3-030-43089-4_25

programs [7]. However, as mentioned in [8, Chap. 4], mixed-integer-based foot-step planning does not offer dynamic feasibility even on a simplified model. These methods therefore are not applicable for dynamic walking with faster walking gaits. The approach developed in [24] allows aperiodic gaits with varying step lengths designed on a complete dynamical model, but requires the *a priori* design of controllers that realize precise transitions between each pair of elements of the gait library, resulting in exponential (factorial) growth in the number of pre-designed controllers.

Instead of relying on kinematics of quasi-static motion planning of simplified dynamical models, such as a linear inverted pendulum with massless legs [9], [22], this paper presents a novel control strategy based on the full nonlinear hybrid dynamic model of the robot to achieve precise foot placement with guarantees on stability and constraint enforcement. We do this by combining a pre-computed library of walking gaits [6] with control barrier function-based quadratic programs (CBF-QPs) for enforcing stepping stone constraints [17], [2]. The gait library is populated with a small number of asymptotically stable periodic walking gaits with pre-determined fixed step lengths, satisfying torque limits, ground reaction forces and other key constraints. Instead of pre-computing transition gaits between discrete elements of the gait library, the gait library is linearly interpolated online to obtain a nominal gait with a desired step length in steady state. To ensure precise foot placement during transients associated with varying distances between stepping stones, the CBF-QP based controller relaxes the tracking behavior of the nominal gait and strictly enforces a set of state-dependent safety constraints that guide the swing foot trajectory to the discrete footholds. Our method enables dealing with a continuum of widely varying desired foothold separations, while achieving foot placement on small footholds. This work builds off our recent work on gait libraries in [6] and precise footstep placement using CBFs in [17]. In this paper, we will use exponential control barrier functions (ECBFs) [18] to handle safety constraints. In comparison to our prior work, this paper makes the following additional contributions:

- We present gait optimization and a gait-library-interpolation approach for achieving underactuated dynamic bipedal walking with a continuum of desired step lengths in steady state.
- We incorporate exponential control barrier functions and control Lyapunov functions to achieve precise transient footstep placement.
- We significantly enlarge the range of variation on step length that can be handled.
- We provide a way to handle sustained step length perturbations.
- Through our QP-based real-time controller, we address simultaneously footstep placement, foot scuffing avoidance, friction constraints and input saturation.

The remainder of the paper is organized as follows. Section 2 presents the hybrid dynamical model of 2D MARLO, an underactuated planar bipedal robot. Section 3 presents gait optimization and a gait library interpolation strategy.

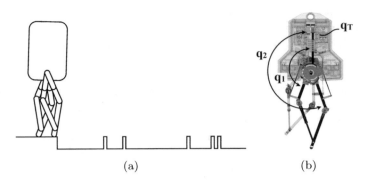

Fig. 1: (a) The problem of dynamically walking over a randomly generated set of discrete footholds. Simulation video: https://youtu.be/udpxZUXBi_s. (b) The coordinates of the biped are illustrated with the torso world frame pitch angle denoted by q_T, and the body coordinates denoted by (q_1, q_2).

Section 4 presents the proposed ECBF-CLF-QP based feedback controller for enforcing precise footstep placement for dynamic walking. Section 5 presents numerical validation of the controller on MARLO. Finally, Section 6 provides concluding remarks.

2   Dynamical Model for Walking

The bipedal robot shown in Fig. 1b is a planar representation of MARLO. Its total mass is 63 kg, with approximately 50% of the mass in the hips, 40% in the torso, and with light legs formed by a four-bar linkage. The robot is approximately left-right symmetric.

The configuration variables for the system can be defined as $q := (q_T, q_{1R}, q_{2R}, q_{1L}, q_{2L}) \in \mathbb{R}^5$. The variable q_T corresponds to the world frame pitch angle of the torso, while the variables $(q_{1R}, q_{2R}, q_{1L}, q_{2L})$ refer to the local coordinates for the linkages. Each of the four linkages are actuated by a DC motor behind a 50:1 gear ratio harmonic drive, with the robot having one degree of under-actuation. The four-bar linkage mechanism comprising of the leg coordinates (q_1, q_2) map to the leg angle and knee angle (q_{LA}, q_{KA}), as $q_{LA} := \frac{1}{2}(q_1 + q_2)$ and $q_{KA} := q_2 - q_1$. With the state $x = (q, \dot{q})$ denoting the generalized positions and velocities of the robot and u denoting the joint torques, a hybrid model of walking can be expressed as

$$\begin{cases} \dot{x} = f(x) + g(x)\, u, & x^- \notin \mathcal{S} \\ x^+ = \Delta(x^-), & x^- \in \mathcal{S}, \end{cases} \tag{1}$$

where $\mathcal{S}$ is the impact surface and Δ is the reset or impact map. A more complete description of the robot and a derivation of its model is given in [21].

3   Optimization and Gait library

Having described the dynamical model of MARLO, we will now present a model-based approach for designing a continuum of stable periodic walking gaits that satisfy physical constraints arising from the robot and its environment. The method combines virtual constraints, parameter optimization, and an interpolation strategy for creating a continuum of gaits from a finite library of gaits.

3.1   Gait Design Using Virtual Constraints

The nominal feedback controller is based on the virtual constraints framework presented in [23]. Virtual constraints are kinematic relations that synchronize the evolution of the robot's coordinates via continuous-time feedback control. One virtual constraint in the form of a parametrized spline can be imposed for each (independent) actuator. Parameter optimization is used to find the spline coefficients so as to create a periodic orbit satisfying a desired step length, while respecting physical constraints on torque, motor velocity, and friction cone. The optimization method used here is the direct collocation code from [12], although other methods, such as [11] or `fmincon` can be used as well.

The virtual constraints are expressed as an output vector

$$y = y_0(q) - y_d(s(q), \alpha), \tag{2}$$

to be asymptotically zeroed by a feedback controller. Here, $y_0(q)$ specifies the quantities to be controlled

$$y_0(q) = \begin{bmatrix} q_{LA}^{st} \\ q_{KA}^{st} \\ q_{LA}^{sw} \\ q_{KA}^{sw} \end{bmatrix}, \tag{3}$$

where st and sw designate the stance and swing legs, respectively, and $y_d(s, \alpha)$ is a 4-vector of Beziér polynomials in the parameters α specifying the desired evolution of $y_0(q)$, where s is a gait phasing variable defined as

$$s := \frac{\theta - \theta_{init}}{\theta_{final} - \theta_{init}}, \tag{4}$$

with $\theta = q_T + q_{LA}^{st}$ being the absolute stance leg angle and $\theta_{init}, \theta_{final}$ being the values of θ at the beginning and end of the gait respectively.

The cost function and constraints for the optimization are formulated as in [23] [Chap. 6.6.2], with the optimization constraints given in Table 1 and the cost taken as integral of squared torques over step length:

$$J = \frac{1}{L_{step}} \int_0^T ||u(t)||_2^2 \, dt. \tag{5}$$

Having presented an optimization approach to create an individual walking gait, we will next discuss the design of a finite set of gaits and a means to create from it a continuum of gaits, called the gait library.

Table 1: Optimization constraints

Motor Toque	$|u| \leq 5$ Nm
Impact Impulse	$F_e \leq 15$ Ns
Friction Cone	$\mu_f \leq 0.4$
Vertical Ground Reaction Force	$F_{st}^v \geq 200$ N
Mid-step Swing Foot Clearance	$h_f|_{s=0.5} \geq 0.1$ m
Dynamic Constraints	Eq. (1)

3.2   Gait Library and Interpolation

The optimization problem posed in the previous section is used to generate five gaits having step lengths $L_{step} = \{0.08, 0.24, 0.40, 0.56, 0.72\}$ meters[1]. For values of step length between the discrete values, $L_{step,i}$, $1 \leq i \leq 5$, define the Beziér coefficients α in (2) by linear interpolation of the coefficients α_i for the five nominal step lengths. In particular, define,

$$\zeta(L_{step}) = \frac{L_{step} - L_{step,i}}{L_{step,i+1} - L_{step,i}}, \quad 1 \leq i \leq 4 \tag{6}$$

$$\alpha(L_{step}) = (1 - \zeta(L_{step}))\alpha_i + \zeta(L_{step})\alpha_{i+1}. \tag{7}$$

For step lengths longer than 0.72, linear extrapolation is used. As in [6, Eqn. (8,9)], this defines a continuum of gaits, called the gait library

$$\mathcal{A} = \{\alpha(L_{step}) \mid 0.08 \leq L_{step} \leq 0.72\}. \tag{8}$$

The update resets the periodic orbit to adapt the step length, while respecting the physical constraints and approximately optimizing the cost on the periodic orbit. During steady-state, there is no theoretical guarantee that the interpolated gait results in exactly the desired step length. However, due to continuity of each pre-defined gait in the library, the interpolated gait will result in a step length that is close enough to the desired step length. On the other hand, during a transient following a change in commanded step length, the footstep placement and optimization constraints shown in Table.1 are not guaranteed to be satisfied. In the next Section, we will introduce the method of control barrier functions to handle transients in the form of real-time constraints on footstep placement, scuffing avoidance, friction cone, and input saturation.

---

[1] The number of gaits is arbitrary. A finer grid did not change the results. A coarser grid was not tried.

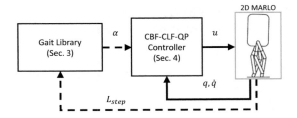

Fig. 2: Diagram of the controller structure integrating the gait library and CBF based controller. Solid lines indicate signals in continuous time representing a within-stride controller; dashed lines indicate signals in discrete time representing a stride-to-stride controller.

4 Control Barrier Function based controller for stepping stones

Having presented the creation of a library of gaits for a small set of step lengths and an associated switching controller, we now discuss the low-level continuous-time controller design that uses control Lyapunov functions for driving the outputs in (2) to zero and control barrier functions for strictly enforcing foot placement constraints. We will incorporate both features through a quadratic program that will also be used to enforce torque saturation, ground contact force and friction cone constraints. The control diagram for the combination of gait library and CBF based controller is shown in Fig.2.

4.1 Control Lyapunov Function based Quadratic Programs Revisited

In this section we will review recent innovations in control Lyapunov functions for hybrid systems and control Lyapunov function based quadratic programs, introduced in [3] and [10] respectively.

Input-output linearization Consider the control output vector $y(q)$ defined in (2) with vector relative degree 2, then the second derivative takes the form

$$\ddot{y} = L_f^2 y(x) + L_g L_f y(x) \, u. \tag{9}$$

We can then apply the following pre-control law

$$u(x) = u^*(x) + (L_g L_f y(x))^{-1} \, \mu, \tag{10}$$

where

$$u^*(x) := -(L_g L_f y(x))^{-1} L_f^2 y(x), \tag{11}$$

and μ is a stabilizing control to be chosen. Defining transverse variables $\eta = [y, \dot{y}]^T$, and using the IO linearization controller above with the pre-control law (10), we have,

$$\begin{bmatrix} \dot{y} \\ \ddot{y} \end{bmatrix} = \dot{\eta} = \bar{f}(x) + \bar{g}(x)\mu, \tag{12}$$

where

$$\bar{f}(x) = F\eta, \quad \bar{g}(x) = G, \quad \text{with} \quad F = \begin{bmatrix} 0 & I \\ 0 & 0 \end{bmatrix}, \quad G = \begin{bmatrix} 0 \\ I \end{bmatrix}. \tag{13}$$

CLF-based Quadratic Programs A control approach based on control Lyapunov functions, introduced in [3], provides guarantees of exponential stability for the traverse variables η. In particular, a function $V(\eta)$ is an *exponentially stabilizing control Lyapunov function (ES-CLF)* for the system (12) if there exist positive constants $c_1, c_2, \lambda > 0$ such that

$$c_1\|\eta\|^2 \le V(\eta) \le c_2\|\eta\|^2, \tag{14}$$
$$\dot{V}(x, \mu) + \lambda V(\eta) \le 0. \tag{15}$$

In our problem, we chose a CLF candidate as follows

$$V(\eta) = \eta^T P \eta, \tag{16}$$

where P is the solution of the Lyapunov equation $A^T P + PA = -Q$ (with A being a Hurwitz matrix such that $\dot{\eta} = A\eta$ is exponentially stable, and Q being any symmetric positive-definite matrix). The time derivative of the CLF (16) is computed as

$$\dot{V}(x, \mu) = L_{\bar{f}}V(x) + L_{\bar{g}}V(x)\mu, \tag{17}$$

where

$$L_{\bar{f}}V(x) = \eta^T(F^T P + PF)\eta; \qquad L_{\bar{g}}V(x) = 2\eta^T PG. \tag{18}$$

The CLF condition in (15) then takes the form

$$L_{\bar{f}}V(x) + L_{\bar{g}}V(x)\mu + \lambda V(\eta) \le 0. \tag{19}$$

If this inequality holds, then it implies that the output η will be exponentially driven to zero by the controller. The following CLF-QP based controller, initially presented in [10], takes the form:

___

CLF-QP:

$$\begin{aligned} \mu^*(x) =\underset{\mu, d_1}{\text{argmin}} \quad & \mu^T\mu + p_1\, d_1^2 \\ \text{s.t.} \quad & \dot{V}(x, \mu) + \lambda V(\eta) \le d_1, \qquad \textbf{(CLF)} \\ & A_{AC}(x)\, \mu \le b_{AC}(x), \quad \textbf{(Constraints)} \end{aligned} \tag{20}$$

where p_1 is a large positive number that represents the penalty of relaxing the CLF condition (15) and A_{AC}, b_{AC} represent additional constraints such as torque constraints, contact force constraints, friction constraints and joint limit constraints. This formulation opened a novel method to guarantee stability of the nonlinear system with respect to additional constraints, such as torque saturation in [10] and L_1 adaptive control in [16].

Having presented control Lyapunov function based quadratic programs, we will next introduce control barrier functions and control barrier function based quadratic programs.

4.2 Exponential Control Barrier Function based Quadratic Programs

Consider the affine control system shown in the continuous dynamics of (1), with the goal to design a controller to keep the state x in the set

$$\mathcal{C} = \{x \in \mathbb{R}^n : h(x) \geq 0\}, \tag{21}$$

where $h : \mathbb{R}^n \to \mathbb{R}$ is a continuously differentiable function.

In order to systematically design safety-critical controllers for higher order relative degree constraints, we will use "Exponential Control Barrier Functions" (ECBFs), introduced in [18].

With application to precise footstep placement, our constraints will be position based, $h(q) \geq 0$, which has relative degree 2. For this problem, we can design an Exponential CBF as follows:

$$B(x) = \dot{h}(x) + \gamma_1 h(q), \tag{22}$$

and the Exponential CBF condition will be simply defined as:

$$\dot{B}(x, u) + \gamma B(x) \geq 0, \tag{23}$$

where $\gamma_1 > 0, \gamma > 0$ play the role of pole locations for the constraint dynamics $\ddot{h}(x, u)$ (see [18]). Enforcing (23) will then enforce $B(x) \geq 0$ for all time, provided $B(x(0)) \geq 0$. It then follows that $h(x) \geq 0$ for all time, provided $h(q(0)) \geq 0$.

Combination of ECBF and CLF-QP We have the exponential CBF constraint $B(x)$ as a real-valued function with relative degree one, i.e,

$$\dot{B}(x, u) = L_f B(x) + L_g B(x) \, u, \tag{24}$$

where $L_g B \neq 0$. Substituting for the pre-control law (10), we can rewrite the above in terms of the control input μ, i.e., $\dot{B}(x, \mu)$. We then have the following QP based controller:

ECBF-CLF-QP:

$$\mu^*(x) = \underset{\mu, d_1}{\text{argmin}} \quad \mu^T \mu + p_1 \, d_1^2$$

$$\text{s.t.} \qquad \dot{V}(x, \mu) + \lambda V(\eta) \leq d_1, \qquad\qquad\qquad \textbf{(CLF)}$$

$$\dot{B}(x, \mu) + \gamma B(x) \geq 0, \qquad\qquad\qquad \textbf{(ECBF)}$$

$$u_{min} \leq u(x, \mu) \leq u_{max}, \quad \textbf{(Input Saturation)}$$

(25)

---

where $B(x)$ is constructed based on the safety constraint $h(x)$ in (22).

Having revisited control barrier function based quadratic programs, we will now formulate our controller to achieve dynamic walking with precise footstep placements.

4.3   Safety-Critical Control for Dynamical Bipedal Walking with Precise Footstep Placement

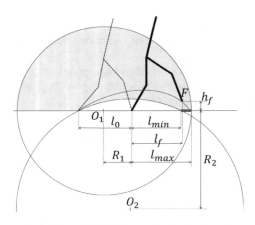

Fig. 3: Geometric explanation of CBF constraints for the problem of bipedal walking over discrete footholds. If we can guarantee the trajectory of the swing foot F (the red line) to be limited in the blue domain, we will force our robot to step onto a discrete foothold position (thick red range on the ground). This approach therefore also provides a safety guarantee against foot scuffing or swing foot being always above the ground prior to contact.

Constraints on Footstep Placement If we want to force the robot to step onto a specific position (see Fig. 1a), we need to guarantee that the step length when the robot swing foot hits the ground is bounded within a given range $[l_{min}; l_{max}]$. Let $h_f(q)$ be the height of the swing foot to the ground and $l_f(q)$

be the horizontal distance between the stance and swing feet. We define the step length at impact as,

$$l_s := l_f(q)|_{h_f(q)=0, \dot{h}_f(q,\dot{q})<0}. \tag{26}$$

The discrete foothold constraint to be enforced then becomes,

$$l_{min} \le l_s \le l_{max}. \tag{27}$$

However, in order to guarantee this final impact-time constraint, we construct a state-based constraint for the evolution of the swing foot during the whole step, so that at impact the swing foot satisfies the discrete foothold constraint (27). We now offer a solution for this issue. The geometric explanation for this is presented in Fig. 3. If we can guarantee the trajectory of the swing foot, F, to be bounded between the domain of the two circles O_1 and O_2, it will imply that the step length when the swing foot hits the ground is bounded within $[l_{min}; l_{max}]$. These two constraints can be represented as:

$$O_1 F \le R_1 + l_{max}; \quad O_2 F \ge \sqrt{R_2^2 + \left(\frac{l_{min} + l_0}{2}\right)^2}.$$

When the swing foot hits the ground at the end of the step, the step length is l_s (see (26)), implying the discrete foothold constraint (27).

Remark 1. The gait library provides gaits that enforce constraints during steady-state but not during transients that occur when the gait is switched. The CBF controller guarantees constraints during transients, thereby preventing a combinatorial explosion of pre-computed transient gaits. The combination of the gait library and the CBF controller is then able to handle constraints during both steady-state and transient phases. The typical failure mode of the gait library is foot scuffing of the swing foot during gait switches. Using a CBF to maintain the swing foot position to be outside the circle O_2, guarantees both the lower bound constraint on step length ($l_s \ge l_{min}$) and foot scuffing avoidance simultaneously. The choice of R_2 in Fig. 3 can be designed based on the desired mid-step swing foot clearance.

We now define the two barrier constraints based on this approach, through the position constraints

$$h_1(q) = R_1 + l_{max} - O_1 F \ge 0,$$

$$h_2(q) = O_2 F - \sqrt{R_2^2 + (\frac{l_{min} + l_0}{2})^2} \ge 0. \tag{28}$$

We can then apply the ECBF-CLF-QP based controller (25) for the above constraints. This involves creating two barriers B_1, B_2 for the corresponding position functions h_1, h_2 respectively.

Constraints on Friction Cone In bipedal robotic walking, contact force constraints are very important for the problem of robotic walking. Any violation of these constraints will result in the leg slipping and the robot potentially falling. Although walking gait optimization is usually designed to respect these constraints, we cannot guarantee these constraints when switching between different walking gaits. In particular, we consider, $F_{st}^h(x, u)$ and $F_{st}^v(x, u)$, the horizontal and vertical contact force between the stance foot and the ground (or the friction force and the normal ground reaction force). Then, the constraints to avoid slipping during walking are,

$$F_{st}^v(x, u) \geq \delta_N > 0,$$
$$\frac{|F_{st}^h(x, u)|}{|F_{st}^v(x, u)|} \leq k_f. \tag{29}$$

where δ_N is a positive threshold for the vertical contact force, and k_f is the friction coefficient. We enforce the above ground contact constraints with $\delta_N = 150(N), k_f = 0.6$.

Remark 2. Note that since the gait optimization is performed offline, we enforce stricter constraints (ground reaction force $F_{st}^v \geq 200(N)$ and friction cone $\mu \leq 0.4$) (see Table 1), allowing for a margin of safety. These constraints hold only for the gaits in the gait library and not for the transient steps generated by the gait library controller. Our ECBF-CLF-QP controller enforces the constraints $F_{st}^v \geq 150(N)$ and friction cone $\mu \leq 0.6$ in real-time for the transient steps.

We then have the following ECBF-CLF-QP based controller that can handle simultaneously footstep placement, scuffing avoidance, friction constraint and input saturation:

$$
\begin{aligned}
\mu^*(x) = \underset{\mu, d_1}{\text{argmin}} \quad & \mu^T \mu + p_1\, d_1^2 \\
\text{s.t.} \quad & \dot{V}(x, \mu) + \lambda V(\eta) \leq d_1 && \textbf{(CLF)} \\
& \dot{B}_1(x, \mu) + \gamma B_1(x) \geq 0 && \textbf{(ECBF on } l_s \leq l_{max}) \\
& \dot{B}_2(x, \mu) + \gamma B_2(x) \geq 0 && \textbf{(ECBF on } l_s \geq l_{min} \\
& && \textbf{\& Foot Scuffing)} \\
& F_{st}^v(x, u(x, \mu)) \geq \delta_N > 0 && \textbf{(Normal Force)} \\
& \frac{|F_{st}^h(x, u(x, \mu))|}{|F_{st}^v(x, u(x, \mu))|} \leq k_f && \textbf{(Friction Cone)} \\
& u_{min} \leq u(x, \mu) \leq u_{max} && \textbf{(Input Saturation)}
\end{aligned}
\tag{30}
$$

Remark 3. Note that all the constraints are affine in μ and thus the above optimization problem is still a quadratic program that can be solved in real-time.

Remark 4. The gait library approach offers a switching strategy under a wide range of step lengths. Based on the desired step length, the interpolation between different gaits in the library will result in a new walking gait for the next step. If the system state is on or close enough to the periodic orbit, it will converge to the desired step length while maintaining physical constraints mentioned in Table.1. However, in our problem, we want the robot to be able to switch between two gaits with very different step lengths, the initial condition is basically very far from the periodic orbit of the next step. Therefore, the transition to the new gait is not guaranteed to satisfy constraints such as friction constraints as well as scuffing avoidance. In the simulation, these two main reasons make the gait library approach fail almost all the time.

Note that the CBF-CLF-QP controller in [17] is only based on one nominal gait and tries to adjust the control inputs so as to enforce the footstep placement constraint, friction constraints and input saturation while following the nominal gait. Due to the limitation of having only one walking gait, the working range of step length is therefore limited.

In this paper, we attempt to combine the advantages of each method and develop the ECBF-CLF-QP controller with foot scuffing constraints and combine it with the gait library approach (see Fig.2). Given a desired step length, the gait library assigns an interpolated gait for the next walking step and the ECBF-CLF-QP controller tracks the outputs corresponding to this gait by solving a quadratic program in real-time to find the control input that follows this new gait while maintaining all above constraints (footstep placement, friction constraints, scuffing avoidance and input saturation).

In the next Section, we present numerical validation of our proposed controller on the dynamical model of the bipedal robot MARLO.

5   Numerical Validation

In this Section, we will demonstrate the effectiveness of the proposed method by conducting numerical simulations on the model of MARLO. We validate the performance of our proposed approach through dynamic bipedal walking on MARLO, while simultaneously enforcing foot placement, scuffing avoidance, ground contact force constraints and input saturation. Furthermore, in order to demonstrate the effectiveness of the method, we compare three controllers on different ranges of desired step lengths:

$$\begin{cases} \text{I: Gait Library} \\ \text{II: CBF (with nominal step length of 56 cm)} \\ \text{III: CBF \& Gait Library} \end{cases} \tag{31}$$

For each range of step length (see Table 2), we randomly generated 100 problem sets, where each set has 10 randomly placed "stepping stones" with a stone size of 5 (cm) (see Fig. 1a). The controller is considered successful for a trial run if the bipedal robot is able to walk over this terrain without violation of foot placement, ground contact, friction, and input constraints. The percentage

Table 2: **(Main Result)** Percentage of successful tests of three controllers (see (31)) with different ranges of desired step length.

Step Length Range (cm)	Gait Library	CBF	Gait Library & CBF
[50:60]	6%	100%	100%
[40:70]	1%	44%	100%
[30:80]	1%	17%	100%
[25:85]	1%	12%	100%
[20:90]	1%	3%	97%
[15:95]	1%	0%	92%
[10:100]	0%	0%	78%

of successful tests for each of the three controllers is tabulated in Table 2 for various ranges of step lengths. The approach based on the combination of CBF and Gait Library outperforms the approaches that rely on only the CBF or only the Gait Library. For example, with the step length range of [20:90] (cm), the percentage of successful tests on controller III (CBF and Gait Library) is 97 %, while that of controller II (CBF only) and controller I (Gait Library only) are just 3% and 1% respectively. Thus the proposed controller not only achieves dynamic walking over discrete footholds, it also dramatically increases the range of step lengths that are handled compared to our prior work in [17].

We show here one simulation of MARLO walking over 20 stepping stones with desired step lengths randomly generated in the range of $[10 : 100]$ (cm), where the stone size is smaller, i.e., $l_{max} - l_{min} = 2$ (cm). Fig. 4 shows the satisfaction of foot step placement constraints as well as CBF constraints, without a violation of the friction cone or input saturation (see Fig. 5). Furthermore, in order to illustrate how aggressively our proposed method can traverse a set of stepping stones, Fig. 6 shows a simulation where the robot has to switch between very a large step length (95 cm) and a very small step length (15 cm).

Remark 5. Potential reasons of the gait library and CBF controller not reaching 100% success in the last three cases include (a) the QP (30) becoming feasible; and (b) the gait library being limited to a pre-computed gait at 72 (cm) (see Section 3.2), which is extrapolated by 25-39% to reach step lengths of 90-100 (cm). Including a pre-computed gait with a step length of 100 (cm) could potentially improve performance.

6   Conclusion

We have presented a model-based control framework that allows transition among widely and randomly varying stepping stones, without an exponential explosion in the number of pre-computed motion primitives. The control design begins with model-based optimization producing a small number of periodic walking gaits that meet desired physical constraints and span a range of step lengths.

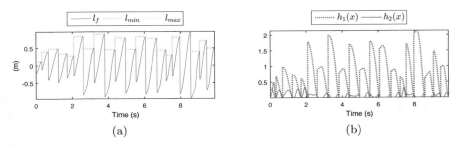

Fig. 4: (a) Footstep placement constraint: $l_{min} \leq l_s \leq l_{max}$, where the step length l_s is the value of the distance between swing and stance feet l_f at impact (see (26)). (b) CBF constraints: $h_1(x) \geq 0, h_2(x) \geq 0$ (see (28)).

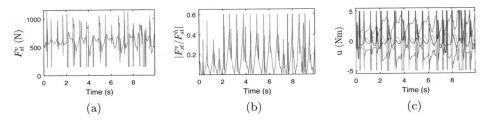

Fig. 5: (a) Ground reaction force: $F_{st}^v \geq 150(N)$. (b) Friction cone: $|F_{st}^h / F_{st}^v| \leq 0.6$. (c) Control inputs are saturated at 5 (Nm) ($|u| \leq 5$); recall the 50:1 gear ratio from the motors to the links.

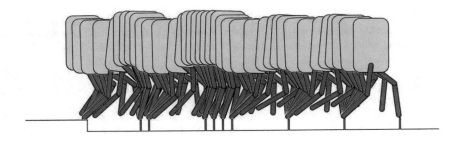

Fig. 6: Simulation of MARLO walking over stepping stones with desired step lengths of $\{95, 15, 95, 15, 15, 15, 95, 95, 95, 15, 15, 15\}(cm)$ and stone size of 2 (cm). For clarity of visualization, the rear links of the 4-bar are suppressed. Simulation video: https://youtu.be/udpxZUXBi_s.

In an outer-loop, a gait library is formed by interpolating this set of walking gaits to provide controllers that realize a continuum of step lengths. In an inner-loop, a quadratic program mediates safety, interpreted as landing the swing foot on a stepping stone, and performance, which involves joint-level tracking commands, friction cone, scuffing avoidance and torque bounds. The resulting controller achieved dynamic walking while enforcing strict constraints on foot step placement at impact, resulting in dynamic walking over stepping stones. Numerical illustration of the proposed method on MARLO, an underactuated bipedal robot, included the robot handling random step length variations that are between $[10:100]$ (cm) with a foot placement precision of 2 (cm).

The proposed method, however, still possesses some disadvantages. Since, we swap between different gaits passively, we usually start with an initial condition that is far away from the orbit or the desired gait, requiring the robot to have high torque right after the impact (see Fig. 5c) to be able to converge back to the desired trajectory. This aggressive behavior may cause infeasibility for the QP. Having a better switching policy to handle transitions would be an interesting future direction.

In the future, the proposed method using CBFs and gait library will be extended to 3D robots so that the stepping-stone course in the W-Prize, [1], can be attempted. In addition to the challenges of 3D locomotion, the heights of the stepping stones vary over the course and the stepping stones could topple over. Our preliminary results towards this include CBF-based controller for 3D dynamic walking on stepping stones, [20], and robust CBF-based controllers to handle constraints under model uncertainty [19].

Acknowledgements
This work is supported by NSF grants IIS-1526515, IIS-1525006, and CNS-1239037.

References

1. "The W-Prize on stepping stones," *http://www.wprize.org/SteppingStones.html.*
2. A. D. Ames, J. Grizzle, and P. Tabuada, "Control barrier function based quadratic programs with application to adaptive cruise control," in *IEEE Conference on Decision and Control*, 2014, pp. 6271–6278.
3. A. D. Ames, K. Galloway, J. W. Grizzle, and K. Sreenath, "Rapidly Exponentially Stabilizing Control Lyapunov Functions and Hybrid Zero Dynamics," *IEEE Trans. Automatic Control*, vol. 59, no. 4, pp. 876–891, 2014.
4. J. Chestnutt, J. Kuffner, K. Nishiwaki, and S. Kagami, "Planning biped navigation strategies in complex environments," in *IEEE International Conference on Humanoid Robotics*, 2003, pp. 117–123.
5. J. Chestnutt, M. Lau, G. Cheung, J. Kuffner, J. Hodgins, and T. Kanade, "Footstep planning for the honda asimo humanoid," *IEEE International Conference on Robotics and Automation.*, pp. 629 – 634, 2005.
6. X. Da, O. Harib, R. Hartley, B. Griffin, and J. Grizzle, "From 2d design of underactuated bipedal gaits to 3d implementation: Walking with speed tracking," *IEEE Access*, vol. PP, no. 99, pp. 1–1, 2016.

7. R. Deits and R. Tedrake, "Footstep planning on uneven terrain with mixed-integer convex optimization." *IEEE/RAS International Conference on Humanoid Robots*, pp. 279–286, 2014.

8. R. L. H. Deits, "Convex segmentation and mixed-integer footstep planning for a walking robot," Master's thesis, Massachusetts Institute of Technology, 2014.

9. R. Desai and H. Geyer, "Robust swing leg placement under large disturbances," *IEEE International Conference on Robotics and Biomimetics*, pp. 265–270, 2012.

10. K. Galloway, K. Sreenath, A. D. Ames, and J. W. Grizzle, "Torque saturation in bipedal robotic walking through control lyapunov function based quadratic programs," *IEEE Access*, vol. PP, no. 99, p. 1, April 2015.

11. A. Hereid, C. M. Hubicki, E. A. Cousineau, J. W. Hurst, and A. D. Ames, "Hybrid zero dynamics based multiple shooting optimization with applications to robotic walking," in *IEEE International Conference on Robotics and Automation*, 2015, pp. 5734–5740.

12. M. S. Jones, "Optimal control of an underactuated bipedal robot," Master's thesis, Oregon State University, ScholarsArchive@OSU, 2014.

13. S. Kajita, F. Kanehiro, K. Kaneko, K. Fujiwara, K. Harada, K. Yokoi, and H. Hirukawa, "Biped walking pattern generation by using preview control of zero-moment point," *the IEEE International Conference on Robotics and Automation*, vol. 2, pp. 1620 – 1626, 2003.

14. J. J. Kuffner, K. Nishiwaki, S. Kagami, M. Inaba, and H. Inoue, "Footstep planning among obstacles for biped robots," *the IEEE/RSJ International Conference on Intelligent Robots and Systems*, vol. 1, pp. 500 – 505, 2001.

15. P. Michel, J. Chestnutt, J. Kuffner, and T. Kanade, "Vision-guided humanoid footstep planning for dynamic environments," in *Humanoids*, 2005, pp. 13–18.

16. Q. Nguyen and K. Sreenath, "L1 adaptive control for bipedal robots withcontrol lyapunov function based quadratic programs," in *American Control Conference*, 2015, pp. 862–867.

17. ——, "Safety-critical control for dynamical bipedal walking with precise footstep placement," in *The IFAC Conference on Analysis and Design of Hybrid Systems*, 2015, pp. 147–154.

18. ——, "Exponential control barrier functions for enforcing high relative-degree safety-critical constraints," in *American Control Conference*, 2016, pp. 322–328.

19. ——, "Optimal robust control for constrained nonlinear hybrid systems with application to bipedal locomotion," in *American Control Conference*, 2016.

20. ——, "3d dynamic walking on stepping stones with control barrier functions," in *2016 IEEE 55th Conference on Decision and Control*, To Appear, 2016.

21. A. Ramezani, J. W. Hurst, K. Akbari Hamed, and J. W. Grizzle, "Performance Analysis and Feedback Control of ATRIAS, A Three-Dimensional Bipedal Robot," *Journal of Dynamic Systems, Measurement, and Control*, vol. 136, no. 2, 2014.

22. M. Rutschmann, B. Satzinger, M. Byl, and K. Byl, "Nonlinear model predictive control for rough-terrain robot hopping," *IEEE/RSJ International Conference on Intelligent Robots and Systems*, pp. 1859–1864, 2012.

23. E. R. Westervelt, J. W. Grizzle, C. Chevallereau, J. Choi, and B. Morris, *Feedback Control of Dynamic Bipedal Robot Locomotion*, ser. Control and Automation. Boca Raton, FL: CRC, June 2007.

24. T. Yang, E. Westervelt, A. Serrani, and J. P. Schmiedeler, "A framework for the control of stable aperiodic walking in underactuated planar bipeds," *Autonomous Robots*, vol. 27, no. 3, pp. 277–290, 2009.

Algorithmic Foundations of Realizing Multi-Contact Locomotion on the Humanoid Robot DURUS

Jacob P. Reher, Ayonga Hereid, Shishir Kolathaya, Christian M. Hubicki and
Aaron D. Ames

Georgia Institute of Technology, Atlanta, GA, 30332

Abstract. This paper presents the meta-algorithmic approach used to
realize multi-contact walking on the humanoid robot, DURUS. This systematic methodology begins by decomposing human walking into a sequence of distinct events (e.g. heel-strike, toe-strike, and toe push-off).
These events are converted into an alternating sequence of domains and
guards, resulting in a hybrid system model of the locomotion. Through
the use of a direct collocation based optimization framework, a walking gait is generated for the hybrid system model emulating human-like
multi-contact walking behaviors – additional constraints are iteratively
added and shaped from experimental evaluation to reflect the machine's
practical limitations. The synthesized gait is analyzed directly on hardware wherein feedback regulators are introduced which stabilize the walking gait, e.g., modulating foot placement. The end result is an energy-
optimized walking gait that is physically implementable on hardware.
The novelty of this work lies in the creation of a systematic approach
for developing dynamic walking gaits on 3D humanoid robots: from formulating the hybrid system model to gait optimization to experimental
validation refined to produce multi-contact 3D walking in experiment.

1 Introduction

Biological bipeds, such as humans, demonstrate walking patterns which are efficient, agile, fast, and robust to a degree not yet attainable by robotic systems.
While humans and other biological bipeds can perform these motions with relative ease, translation of these capabilities to 3D humanoid systems is fraught
with complexities in the form of nonlinearities, modeling errors, and high degrees of freedom which must be coordinated. With the goal bridging this gap in
natural and efficient locomotion on robots, it is advantageous develop algorithmic approaches capable of exploiting the natural dynamics of the robot. While
some researchers argue robotics is currently limited by physical hardware capabilities, a lack of fundamental knowledge in the area has yet to be bridged as
well. Robotic walking presents a wide range of mathematical and algorithmic
challenges that provides an fertile proving ground for addressing these gaps.

Many of the approaches currently employed revolve around the use of reduced order inverted pendulum models. Perhaps the most prevalent approach

© Springer Nature Switzerland AG 2020
K. Goldberg et al. (Eds.): *Algorithmic Foundations of Robotics XII*, SPAR 13, pp. 400–415, 2020.
https://doi.org/10.1007/978-3-030-43089-4_26

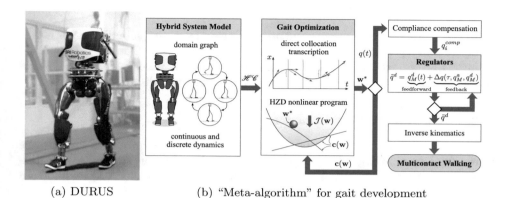

(a) DURUS   (b) "Meta-algorithm" for gait development

Fig. 1: (a) The humanoid robot, DURUS, walking heel-to-toe. (b) The "meta-algorithm" followed to achieve walking control, with decision points (diamonds) representing iteration based on the results of robot experiments.

uses a framework known as the Zero Moment Point (ZMP) criterion [19]. The resulting walking motions are typically flat footed and quasi-static. Human walking patterns, on the other hand, consist of multiple phases (or domains) with changes in contact conditions, impacts, and underactuation [18]. In an attempt to generate more human-like walking patterns, multi-contact methods have been implemented which allowed for longer walking strides and increased energy efficiency through heel and toe contact conditions [7, 15]. One difficulty with this approach is that its inherent assumptions prevent it from utilizing the natural forward momentum of the robot in a manner similar to humans.

A method which has been used to generate dynamic walking motions with stability guarantees through underactuated domains is termed Hybrid Zero Dynamics (HZD) [3, 9, 21]. The stability of these methods on bipeds has been validated experimentally [17], and it has also been shown that HZD methods can be extended to 3D robots [16]. However, to the authors knowledge, HZD methods have not been utilized to obtain multi-contact walking behaviors on a 3D humanoid robot with both heel-toe contact motions and underactuation.

The goal of this paper is to provide a foundation upon which HZD based multi-contact walking behaviors can be formally generated and then experimentally realized on humanoid robots. With this goal in mind, we begin with a discussion of human walking patterns and their relation to the domain structure hybrid model of humanoid walking in Sec. 2. The optimization method, including the cost function and constraints necessary to arrive at an experimentally realizable gait, are presented in Sec. 3. The experimental methods and results along with the discrete feedback compensation algorithms used for experimental stabilization are presented in Sec. 4, in which a mean cost of transport over 200 steps is shown as 1.02, the lowest electrical cost of transport yet reported on a 3D humanoid robot. Finally, an analysis of the overall methodology performance for multi-contact gait generation is presented in Sec. 5.

2   Robot Walking Model

In this section, we will discuss human walking and the corresponding partitioning of the walking behavior into the domains heel strike, toe strike, toe lift, and heel lift. This four-domain structure is incorporated for the robotic model as a hybrid system with an alternating sequence of continuous and discrete events.

2.1   Human Walking Domains

In studies of human locomotion, multi-contact behaviors have been found to be essential in reducing joint torques and increasing walking speeds [11]. In this work, a walking gait for a 3D humanoid robot is designed with a hybrid domain breakdown matching that of the temporal domain pattern observed in natural human walking motions [4]. From Fig. 2 it can be observed that human walking has four distinct phases: heel strike (*hs*) when the swinging foot strikes the ground, toe strike (*ts*) when the toe of the foot goes down and the legs switch, toe lift (*tl*) when the other foot takes off the ground and becomes the swinging foot and finally heel lift (*hl*) when the stance heel goes up with the stance toe being the only contact point with ground. The behavior goes from fully actuated (hs) to overactuated (ts) to fully actuated (tl) to underactuated (hl) phases of motion in a sequence in a repeated manner where a domain is considered underactuated or overactuated if the humanoid has less or more actuators and contact constraints than degrees of freedom in the system.

What determines the phase in which the robot is currently operating is the set of contact points $\mathcal{C} = \{swh, swt, sth, stt\}$ (swing heel, swing toe, stance heel, stance toe). The switch of each leg from stance to swing and vice versa occurs after toe strike (see Fig. 2). Recent work by the authors [22] detailed the implementation of multi-contact locomotion on two 2D robots. In this work,

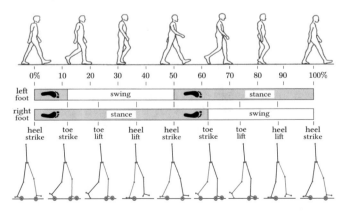

Fig. 2: Multi-contact locomotion diagram of a typical human gait cycle [2] (top) and multi-contact domain breakdown of two steps of one subject based on the changes of heel and toe contact condition (bottom). Blue circles represent one specific point in contact with the walking surface.

three domains were used to represent locomotion, with the removal of the domain corresponding to the toe liftoff before swing (ts). While this domain is relatively short in comparison to the overall gait cycle, the inclusion of this phase allows for walking which is more closely aligned with human locomotion and permits longer steps. This four domain representation of human walking is next framed in the context of constructing a hybrid system model.

2.2 Multi-Domain Hybrid System

Bipedal robots display both continuous and discrete behaviors, lending themselves naturally to *hybrid systems* models. Continuous evolution of the biped dynamics occurs when there is a fixed number of contact points with the environment, in which we say the robot is on a *vertex*. There is a discrete change in the biped dynamics when the number of contact points with the environment changes, on which we say the robot has reached an *edge*. The study of human walking data shows how a periodic human-like walking gait can be described by a directed cycle based on contact conditions. The directed cycle is given by a graph $\Gamma = (V, E)$ consisting of four vertices and edges:

$$V = \{ts, tl, hl, hs\}, \ \ E = \{ts \rightarrow tl, tl \rightarrow hl, hl \rightarrow hs, hs \rightarrow ts\}, \tag{1}$$

where each vertex represents a continuous domain and each edge corresponds to a transition between these domains, as shown in Fig. 3. For example, ts denotes the toe strike and the corresponding vertex denotes all possible states where the toe is in contact with the ground. Motivated by this breakdown, we adapt this human like multi-domain formulation with four vertices and edges. Interested readers are referred to [4, 22] for a full definition of the multi-domain hybrid system. For the n-DOF robot of configuration $q \in \mathcal{Q} \subset \mathbb{R}^n$, tangent bundle with (local) coordinates $(q, \dot{q}) \in T\mathcal{Q} \subset \mathbb{R}^{2n}$, the multi-domain hybrid control system is defined to be a tuple:

$$\mathscr{HC} = (\Gamma, \mathcal{D}, \mathcal{U}, S, \Delta, FG). \tag{2}$$

 - Γ is the directed cycle specified by (1).
 - $\mathcal{U}$ is the set of admissible control inputs.
 - $\mathcal{D} = \{\mathcal{D}_v\}_{v \in V}$ is the set of domains of admissibility. Each domain $\mathcal{D}_v \subseteq T\mathcal{Q} \times \mathcal{U}$ which can be interpreted as the set of possible states the robot can assume given the constraints on the feet for the corresponding domain. For example,

$$\mathcal{D}_{ts} = \{(q, \dot{q}, u) \in T\mathcal{Q} \times \mathcal{U} | \left[h_{stt}, h_{sth}, h_{swt}\right]^T = 0, h_{swh} \geq 0\}, \tag{3}$$

where h_{swh}, h_{swt}, h_{sth}, and h_{stt} are the vertical positions of the foot contact points. An alternative definition of the domain can also be obtained by using the holonomic constraints $h_v : \mathcal{Q} \rightarrow \mathbb{R}^l$ wherein the position and orientation of the contact points $\mathcal{C}$ are fixed.

– $S = \{S_{ts \to tl}, S_{tl \to hl}, S_{hl \to hs}, S_{hs \to ts}\}$ is the set of guards which form the transition point from one domain to another. For example,

$$S_{hl \to hs} = \{(q, \dot{q}) \in \mathbb{R}^{2n} | \left[h_{stt}, h_{swh}\right]^T = 0, \dot{h}_{swh} \le 0\}. \quad (4)$$

– $\Delta = \{\Delta_{ts \to tl}, \Delta_{tl \to hl}, \Delta_{hl \to hs}, \Delta_{hs \to ts}\}$ is the set of reset maps from one domain to the next domain. In the presence of an impact, the reset map emits the post impact state of the robot. Each reset map $\Delta_e : S_e \to \mathcal{D}_{v_{\text{target}}}$, with $e = \{v_{\text{source}} \to v_{\text{target}}\} \in E$ is computed by assuming that the impacts are plastic and instantaneous [10].

– FG provides the set of vector fields given by the equation: $\dot{x} = f_v(x) + g_v(x)u$, where $x = (q, \dot{q})$, $u \in \mathcal{U}$. f_v, g_v are defined in each domain by the Euler-Lagrangian dynamics. More details on the dynamics are given below.

Given the state $(q, \dot{q}) \in TQ$, the dynamics of the system with foot contact constraints for each domain are given by:

$$D(q)\ddot{q} + H(q, \dot{q}) - Bu - J_v^T(q)\lambda_v = 0,$$
$$J_v(q)\ddot{q} + \dot{J}_v(q, \dot{q})\dot{q} = 0 \quad (5)$$

where D, H have the usual meaning from the Euler-Lagrangian dynamics, B is the mapping of torques to the joints, λ_v is the set of ground reaction forces and J_v is the Jacobian for the contact points where λ_v is applied.

3   Direct Collocation Based HZD Optimization

Using the hybrid system model, a gait optimization framework is now introduced which is used to generate walking gaits for the robot with a set of parameters to yield a hybrid invariant periodic orbit. This guarantees that at least for simulation, the bipedal walking is stable. This comprises the next element

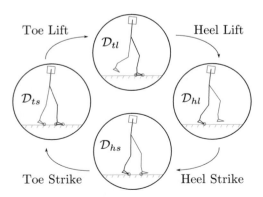

Fig. 3: The directed cycle of four domains for 3D multi-contact walking. The red circles represent foot contact points.

Domain	Human	DURUS
$\mathcal{D}_{ts}$	6%	4.6%
$\mathcal{D}_{tl}$	59%	52.4%
$\mathcal{D}_{hl}$	18%	29.6%
$\mathcal{D}_{hs}$	17%	13.4%

Table 1: A comparison of domain durations which has been shown to be consistent in human locomotion versus the walking gait implemented on DURUS through optimization.

in the "meta-algorithm", in which these walking gaits are obtained, recorded, and tested on hardware. In the authors' previous work, a direct collocation formulation of HZD gait optimization has been successfully applied to DURUS to generate flat-footed walking gaits [8]. In this paper, we extend this methodology to the multi-contact case. Similar methods have been proposed utilizing constrained dynamical systems with open loop controllers to achieve multi-contact walking gaits in simulation [13].

Feedback Linearization. The feedback linearizing controller introduced in this section allows for the formulation of a set of stability criteria used throughout the optimization. Specifically, to lend itself to formal analysis, we must set up a problem in which we can theoretically drive the actual robot configuration to the desired outputs of the system. We have the set of actual outputs of the robot as $y^a : TQ \to \mathbb{R}^{2n}$, and the desired outputs as $y^d : \mathbb{R}^+ \to \mathbb{R}^m$. y^d is usually modulated by a phase (or time) variable $\tau : Q \to \mathbb{R}^+$ (or $\tau : \mathbb{R}^+ \to \mathbb{R}^+$ for time based). By adapting a feedback linearizing controller, we can drive the relative degree one outputs $y_{1,v}(q, \dot{q}) = y^a_{1,v}(q, \dot{q}) - y^d_{1,v}(\tau, \alpha_v)$ and relative two outputs $y_{2,v}(q) = y^a_{2,v}(q) - y^d_{2,v}(\tau, \alpha_v)$ to zero, with v denoting the domain, α denoting the parameters of the desired trajectory. These outputs are generally called *virtual constraints* [20]. Note that any effective tracking controller will theoretically suffice (the experimental implementation uses PD control [9]). The feedback linearizing controller [4] that drives $y_{1,v} \to 0$, $y_{2,v} \to 0$ is given by:

$$u_v = \begin{bmatrix} L_g y_{1,v} \\ L_g L_f y_{2,v} \end{bmatrix}^{-1} \left(- \begin{bmatrix} L_f y_{1,v} \\ L_f^2 y_{2,v} \end{bmatrix} + \mu_v \right), \qquad (6)$$

where L_g, L_f denote the Lie derivatives and μ_v denotes the auxiliary input applied after the feedback linearization. Applying u_v in (6) yields linear dynamics:

$$\dot{\eta}_v(q, \dot{q}) = \mu_v, \qquad (7)$$

where $\dot{\eta}_v := [\dot{y}_{1,v}, \ddot{y}_{2,v}]^T$, so that μ_v can be carefully chosen to stabilize the output dynamics.

Partial Hybrid Zero Dynamics. When the control objective is met such that $y_{1,v}, y_{2,v} = 0$ for all time then the system is said to be operating on the *partial zero dynamics surface* [3] (full zero dynamics for purely relative degree two outputs):

$$\mathbb{Z}_v = \{(q, \dot{q}) \in \mathcal{D}_v | y_{2,v} = 0, L_f y_{2,v} = 0\}, \qquad (8)$$

for the domain $\mathcal{D}_v$. The controller u_v, being domain specific, guarantees partial zero dynamics only in the continuous dynamics. Therefore, for a multi-domain hybrid system, *partial hybrid zero dynamics* can be guaranteed if and only if the discrete maps Δ_e are invariant of the partial zero dynamics in each domain. As a result, the parameters α_v of the outputs must be chosen in a way which renders the surface invariant through impacts. can be mathematically formulated as:

$$\Delta_e(\mathbb{Z}_{v_{\text{source}}} \cap S_e) \subset \mathbb{Z}_{v_{\text{target}}}, \quad e = \{v_{\text{source}} \to v_{\text{target}}\} \in E. \qquad (9)$$

The best way to ensure hybrid invariance under a discrete transition is by a careful selection of the desired trajectories (desired gait) via the parameterization: $y_v^d(\tau, \alpha_v)$. Hence if the desired trajectories are a function of Bézier polynomials, the parameters α_v are the coefficients. These coefficients are chosen by using a direct collocation based walking gait optimization problem explained in the following section.

Collocation Algorithm. Here, we simply introduce the main idea of the direct collocation optimization [8]. In particular, the solution of each domain, $\mathcal{D}_v$, is discretized based on the time discretization $0 = t_0 < t_1 < t_2 < \cdots < t_{N_v} = T_{I,v}$, assuming $T_{I,v} > 0$ is the time at which the system reaches the guard associated with a given domain. Let x^i and $\dot{x}^i$ be the approximated states and first order derivatives at node i, the defect constraints are defined at each odd node as:

$$\zeta^i := \dot{x}^i - 3(x^{i+1} - x^{i-1})/2\Delta t_v^i + (\dot{x}^{i-1} + \dot{x}^{i+1})/4 = 0, \tag{10}$$

$$\delta^i := x^i - (x^{i+1} + x^{i-1})/2 - \Delta t_v^i(\dot{x}^{i-1} - \dot{x}^{i+1})/8 = 0, \tag{11}$$

where $\Delta t_v^i = t_{i+1} - t_{i-1}$ is the time interval. Moreover, the first order derivatives must satisfy the system dynamics, i.e.,the restricted partial zero dynamics in the context of HZD.

Constrained Hybrid Dynamics. Recall that reduced dimensional restricted dynamics, i.e., the zero dynamics, is determined via the full order dynamics (5) subject to the holonomic constraints and virtual constraints being zero. Thus, the restricted dynamics can be described as the differential algebraic equations:

$$F_v(q, \dot{q}, \ddot{q}, u, \lambda_v, \alpha_v) := \begin{bmatrix} D(q)\ddot{q} + H(q, \dot{q}) - Bu - J_v^T(q)\lambda_v \\ J_v(q)\ddot{q} + \dot{J}_v(q, \dot{q})\dot{q} \\ \dot{\eta}_v(q, \dot{q}, \ddot{q}, \alpha_v) - \mu_v \end{bmatrix} = 0, \tag{12}$$

subject to the initial value conditions:

$$h_v(q^+) = \bar{h}_v, \quad J_v(q^+)\dot{q}^+ = 0, \tag{13}$$

$$y_{2,v}(q^+) = 0, \quad \dot{y}_{2,v}(q^+, \dot{q}^+) = 0, \tag{14}$$

where $\bar{h}_v$ is a vector of constants, and $(q^+, \dot{q}^+)$ are the initial state values. This system can be considered as an implicit form that is equivalent to the zero dynamics equation by its definition.

Moreover, the trajectories of the system states of two neighboring domains are connected via the discrete dynamics captured in the reset maps. Specifically, suppose that $(q^+, \dot{q}^+)$ are the post-impact states of a particular domain and $(q^-, \dot{q}^-)$ are the pre-impact states of the previous domain, then they must satisfy

$$(q^+, \dot{q}^+) - \Delta_e(q^-, \dot{q}^-) = 0, \tag{15}$$

where $e \in E$ corresponds to the transition between these two domains. This constraint together with the initial value constraint in (14) guarantee that the hybrid invariant conditions are satisfied, therefore, the solution to the optimization lies on the *partial hybrid zero dynamics* manifold given in (8).

General Formulation. Let $\mathbf{w}$ be a vector containing all optimization variables and $\mathbf{c}(\mathbf{w})$ be a vector of constraint functions given in (10) – (15), we then state the optimization problem as,

$$\mathbf{w}^* = \underset{\mathbf{w}}{\operatorname{argmin}} \ \mathcal{J}(\mathbf{w}) \tag{16}$$

$$\text{s.t} \quad \mathbf{c}^{\min} \leq \mathbf{c}(\mathbf{w}) \leq \mathbf{c}^{\max},$$

$$\mathbf{w}^{\min} \leq \mathbf{w} \leq \mathbf{w}^{\max}.$$

Cost Function. Despite the goal of human-like walking, we do not impose any human specific constraints in the optimization. Instead, we define the cost function of our gait optimization as minimizing the total mechanical cost of transport of the gait:

$$\mathcal{J} := \frac{1}{mgl_{step}} \sum_{v \in V} \int_{t_v^+}^{t_v^-} P(\dot{q}(t), u(t)) dt \tag{17}$$

where t_v^+ and t_v^- is the initial and final time of a domain v, mg is the weight of the robot and l_{step} is the step length of one gait cycle, respectively, and $P(\dot{q}(t), u(t))$ is the 2-norm sum of the mechanical power, given as

$$P(\dot{q}(t), u(t)) := \sum_{j=1}^{m} \|u_j(t) \cdot \dot{q}_j(t)\| \tag{18}$$

where $u_j(t)$ and $\dot{q}_j(t)$ is the computed torque and joint velocity of each actuated joint j. In the context of the direct collocation optimization, the numerical integration in (17) is computed with the discrete state and control variables using quadrature rules [8].

Physical Constraints. In addition to the constraints defined in (10) – (15), other physical constraints can be easily added into $\mathbf{c}$ to ensure the resulting gaits are feasible on hardware. For example, torque bounds, joint velocity limits and angle limits, etc., can be imposed directly as the boundary value of corresponding optimization variables in $\mathbf{w}^{\min}$ or $\mathbf{w}^{\max}$. Hence, the method lends itself naturally to the addition of physical constraints based on actual hardware considerations for the physical hardware. Using this approach, the following constraints are added to the gait optimization and are configured specifically to provide favorable conditions for experimental walking.

• *Torso Movement.* The optimization tends to find energetically minimal walking gaits in which the torso inertia is used similar to arm-swing to counter moments generated by the swinging legs. When implemented experimentally, gaits with particularly large torso swing tend to worsen unwanted contact conditions, such as loss of foot contacts or early striking. This can be prevented by constraining the torso movement in the gait design. Let $\phi_{tor}(q) : \mathcal{Q} \to \mathbb{R}^3$ be the three dimensional orientations of the upper torso link, we restrict them within a small range $[\phi_{tor}^{\min}, \phi_{tor}^{\max}]$, i.e.,

$$\phi_{tor}^{\min} \leq \phi_{tor}(q) \leq \phi_{tor}^{\max}. \tag{19}$$

• *Impact Velocity.* If the swing foot impacts the ground too hard, it can destabilize the robot. Therefore, we constrain the impact velocities of the heel to be within a reasonable range. Let $v_x^{\max}, v_y^{\max}, v_z^{\max} > 0$ be the maximum allowable impact velocities in x, y, and z direction respectively, then the swing heel velocities $\dot{h}_{swh}(q^-, \dot{q}^-)$ should satisfy

$$|\dot{h}_{swh}^x(q^-, \dot{q}^-)| \leq v_x^{\max}, \quad |\dot{h}_{swh}^y(q^-, \dot{q}^-)| \leq v_y^{\max}, \quad |\dot{h}_{swh}^z(q^-, \dot{q}^-)| \leq v_z^{\max}, \qquad (20)$$

where $(q^-, \dot{q}^-) \in \mathcal{D}_{hl} \cap S_{hl \to hs}$.

• *Swing Leg Roll.* Due to the existence of unmeasured compliance in the mechanical system, the swing leg can strike the stance leg if they are not separated enough. The separation of legs can be expressed as the difference between stance and swing hip roll angles. Assuming the right leg is the stance leg, then the following constraint should be enforced:

$$\phi^{\min} \leq \phi_{rh} - \phi_{lh} \leq \phi^{\max}, \qquad (21)$$

where ϕ_{rh} and ϕ_{lh} are the right and left hip roll angles, and $\phi^{\max} > \phi^{\min} \geq 0$ are the maximum and minimum allowable leg separation angles.

• *Ground Reaction Wrench Constraints.* In Sec. 2, we model the ground-foot contact as holonomic constraints. However, these constraints are unilateral in essence. Thus the ground reaction wrenches resulting from the contact conditions cannot be infinitely large. The limitations of ground reaction wrenches are often described as the Zero Moment Point (ZMP) constraints, which are discussed thoroughly in [6]. In particular, we enforce the ZMP constraints only during the single support domain $\mathcal{D}_{tl}$ when the stance foot is flat on the ground. In addition, we also constrain the yaw reaction moment of the stance foot, λ_{sf}^{mz}, to be reasonably small:

$$\|\lambda_{sf}^{mz}\| \leq \lambda^{\max} \qquad (22)$$

where $\lambda^{\max}$ is the maximum acceptable yaw reaction moment.

4   Experimental Validation

The main goal of this work is to allow for the formal generation of HZD based multi-contact walking gaits which can then be rapidly prototyped on hardware to ensure suitability for the application of more advanced control methods. This section details the implementation method used to validate the suitability of gaits generated via the methods proposed in Sec. 3 on hardware. Desired motor trajectories are interpolated through the time based trajectory generated using the formal methods detailed in Sec. 3 and played back as a *feedforward* element. The joint configuration of the robot is then adjusted as:

$$\tilde{q}^d = \underbrace{q_M^d(t, \alpha_v)}_{\text{feedforward}} + \underbrace{\Delta q(\tau, \alpha_v, x_M^a)}_{\text{feedback}}, \qquad (23)$$

wherein three compensators are used as *feedback*.

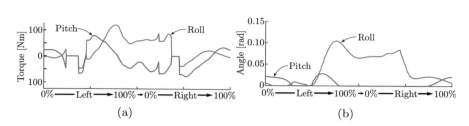

Fig. 4: (a) The anticipated torque computed on the robot using the time-based trajectories over two steps. (b) Angle compensation applied directly to the hip roll and pitch joints over the corresponding steps.

4.1 Experimental Methods

In order to ensure the robot is able to compensate for small modeling uncertainties during gait evaluation, we introduce three compensators which comprise the experimental implementation aspect of the "meta-algorithmic" approach. These feedback actions manifest as small modifications to the joint angle commands correcting for unmeasured compliance, lateral lean, and heel strike orientation.

Compliance Compensation. A prevalent problem with humanoid hardware is unmeasured compliance in the system. Throughout the walking behaviors presented in this work, DURUS exhibited unmeasured compliance in its hips with deflections reaching over 0.1 radians. This has also been cited as an issue on ATLAS [12], in which a linear compliance assumption is introduced to augment the measured angles fed to a fullbody estimator. A similar approach is used here with the primary difference that the compensator directly adjusts the desired joint configuration via a position command:

$$q_i^{comp} = q_{M,i}^d + u_i/K_i, \tag{24}$$

where $q_i \in \mathcal{Q}$ are the corresponding joints, q_i^{comp} is the preprocessed joint angle to be passed to the controller, $q_{M,i}^d$ is the feedforward joint angle, u_i is the feedback linearizing torque computed at each time step, and K_i is the stiffness coefficient which has been measured for each joint. The compliance parameters, measured with a force gauge and caliper in units of Nm/rad, for the worst case joints on DURUS were found to be $K_{lhp} = 1284$, $K_{lhr} = 900$, $K_{rhp} = 1124$, and $K_{rhr} = 815$. The anticipated torque at the joint is computed online and the values q_i^{comp} are displayed in Fig. 4.

Hip Regulators. To account for differences between the physical robot and the ideal model and to stabilize the robot to minor perturbations in the lateral plane, a regulator structure is introduced expanding on previous regulation approaches used on DURUS for flat-footed walking [14]. Discrete logic is used to handle a smooth blending factor (s) increasing linearly through the toe-lift domain $\mathcal{D}_{tl}$ according to the change in the normalized phase variable, Λ. Throughout all other domains $\{\mathcal{D}_{hs}, \mathcal{D}_{ts}, \mathcal{D}_{hl}\}$, the blending factor is held constant in order to

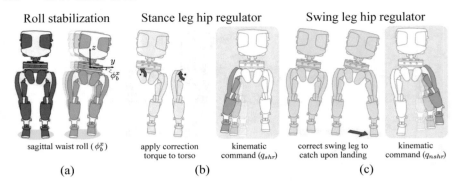

Roll stabilization   Stance leg hip regulator   Swing leg hip regulator

sagittal waist roll (ϕ_b^x)   apply correction torque to torso / kinematic command (q_{shr})   correct swing leg to catch upon landing / kinematic command (q_{nshr})

(a)   (b)   (c)

Fig. 5: Pictured is an illustration of the regulator response to (a) an excessive frontal waist roll. During stance, (b) a counter-rotating torque on the torso is desired to correct the torso roll (left), so a kinematic command is given to the stance leg to adjust the abduction/adduction angle (right). During swing (c) the regulator widens the strike stance between the two legs (left) by kinematically adjusting the swing leg abduction/adduction angle (right).

prevent opposing motions of the legs while both legs are in contact with the ground.

Regulation is provided for two scenarios in which it can compensate for rolling to the outside of the stance leg or towards the swing leg, pictured in Fig. 5. Each regulator performs motion in one direction; adduction for the stance leg and abduction for the swing. Each leg is assigned both a stance and nonstance blending factor for the stance and nonstance hips $s_{i,nsh}$ and $s_{i,sh}$ where $i \in \{stance, nonstance\}$. The stance hip blending factor is increased for the hip currently in stance ($s_{s,sh}$) and decreased for the swing hip ($s_{ns,sh}$). The converse is true for the swing blending factor. The regulator action for each leg is then:

$$\Delta q_{i,nshr}^d = -s_{i,nsh}K_{nsh}(y^a - y^d), \tag{25}$$

$$\Delta q_{i,shr}^d = -s_{i,sh}K_{sh}(y^a - y^d), \tag{26}$$

where $\Delta q_{i,hr}^d$ and $\Delta q_{i,nshr}^d$ are the angle abduction and adduction angles added to the trajectories as regulation, $y^a := \phi_b^{x,a}$, and $y^d := \phi_b^{x,d}$ are the measured waist roll and time based waist roll recorded in simulation, and $K_{nsh,sh}$ are the tunable nonstance hip and stance hip proportional gains.

Ankle Inverse Kinematics. While performing multi-contact walking the heel holonomic constraint is very important, particularly as it impacts the ground and transitions between the domains $\mathcal{D}_{hl}$ and $\mathcal{D}_{hs}$. If the robot strikes the ground with a foot configuration which does not have the heel parallel to the ground, then it will be thrown off balance. To ensure the holonomic constraint at this transition is satisfied, we implement an inverse kinematics solver. Angles for the motors controlling the push-rod transmission in the swing foot are solved with a Newton-Raphson method and ensure the swing heel strikes the ground evenly.

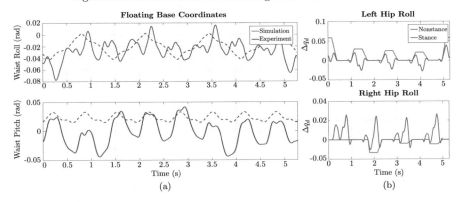

Fig. 6: A plot of the floating base coordinates (a) measured by IMU (solid) versus the simulated walking data (dashed) and the corresponding stance and swing regulator actions (b) over six steps.

4.2   Experimental Results

This section discusses the experimental results which were obtained through the implementation of the walking gait found through the multi-domain optimization problem[1]. The limit cycles achieved experimentally on DURUS shown in Fig. 7 exhibit a closed behavior, indicating the multi-contact walking behaviors are stable. It is clear from the hip roll limit cycle that this the joint most heavily augmented by the feedback regulators. These regulators react to the floating base coordinate, pictured in Fig. 6. We can see that while the system has a limit cycle on the floating base coordinates there is a mismatch between the experiment and simulation which is particularly evident in the pitch direction. This mismatch is a key point that the authors hope to address in future work in which more advanced controllers can be applied to the multi-contact gait.

The experimentally implemented walking gait ambulated with a forward velocity of 0.60 m/s and a stride length of 0.39, for which the specific cost of electrical transport (CoT) $c_{et,i}$ for each step is computed according to [5] as:

$$c_{et,i} = \frac{1}{mgd_i} \int_{t_i^+}^{t_i^-} P_{el} + \sum_{j=1}^{15} I_j(t)V_j(t)dt, \qquad (27)$$

where $P_{el} = 86.4$W is logic power consumed by the on-board computer and motor controllers, d_i is the x-position traveled by the non-stance foot of the robot through the ith step, and $I_j(t)$ and $V_j(t)$ are the currents and voltage recorded for the jth motor. The mean total power consumed over all 15 actuators for 200 steps, along with the cost of transport per step can be seen in Fig. 8. These results indicate that the mean CoT for DURUS during steady-state multi-contact locomotion is $\bar{c}_{et} = 1.02$, which is 37% more efficient than experimental results obtained on DURUS for flat-footed walking [14], which was previously the lowest recorded CoT on a humanoid robot.

---

[1] See [1] for a video of DURUS walking with the multi-contact behaviors.

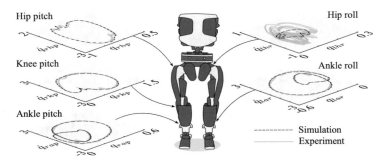

Fig. 7: Pictured is the trace of continuous walking limit cycles over 10 steps (solid) compared to the nominal time based trajectory from simulation (dashed). (Units: *rad* and *rad/s*)

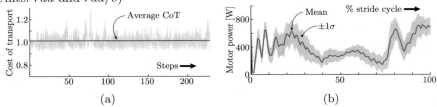

Fig. 8: (a) The specific cost of transport over 200 steps of continuous walking. (b) Mean motor power consumed per step over 200 step interval.

5   Methodology Performance

This section provides some metrics and performance discussion on the overall algorithm. As the primary component in designing these walking gaits, an efficient implementation of the optimization approach proposed in Sec. 3 is crucial to the rapid design of multi-contact walking gaits. The optimization is implemented in MATLAB using the software package IPOPT[2] with linear solver ma57 on a laptop computer with an Intel Core i7-3820QM processor (2.7 GHz) and 12 GB of RAM. The number of cardinal nodes are picked as 10, 15, 20, and 12 for the toe-strike, toe-lift, heel-strike, and toestrike domains, respectively. Based on the formulation presented in Sec. 3 we arrive at an NLP with $21,309$ variables, $22,721$ constraints, and a Jacobian sparsity of 0.05%. Typical computation times for the multi-contact behavior in this work is 647 seconds over 418 iterations.

Since this methodology begins with analysis and utilization of a domain sequencing mirroring that of humans, we would like to better determine whether the approach generates behaviors in line with that of nominal human walking. To provide a more quantitative measure of "human-likeness" of DURUS' walking, the *human-based cost* of eight human subjects[3] and DURUS are computed with respect to the nominal human domain cycle presented in Table 1. We define a *walking cycle* as a pair (γ, l) with $l = (V, E)$ the graph presented in Fig. 3 and

---

[2] https://projects.coin-or.org/Ipopt

[3] The human walking cycles analyzed are derived from the dataset presented in [4].

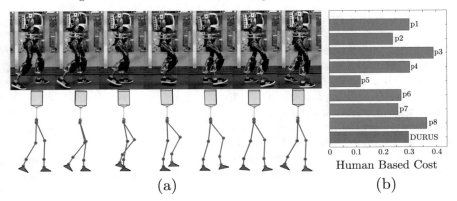

(a)   (b)

Fig. 9: (a) Tiles of DURUS walking seen at an angle from front. (b) Human-based cost of DURUS compared to eight healthy human subjects.

$\gamma : l \to \mathbb{R}^{|V|}$ is a function such that $\gamma(v) \geq 0$ and $\sum_{v \in V} \gamma(v) = 1$. The human-based cost for the multi-contact walking gaits can be found as the *cut distance* between the the nominal human cycle (γ^*, l^*) and the optimized cycle (γ_r, l_r). We then view γ^* and γ_r as functions on $V^* \cup V_r$ by letting $\gamma^*(i) \equiv 0$ if $i \in V_r \backslash V^*$ and $\gamma_r(j) \equiv 0$ if $j \in V^* \backslash V_r$. The cut distance is then computed as in [4] by:

$$d(\gamma^*, l^*, \gamma_r, l_r) = \max_{I, J \subset V^* \cup V_r} \left| \sum_{i \in I, j \in J} (\gamma^*(i)\gamma^*(j)\beta^*(i, j) - \gamma_r(i)\gamma_r(j)\beta_r(i, j)) \right|$$

$$+ \sum_{k \in V^* \cup V_r} |\gamma^*(k) - \gamma_r(k)|,$$

where $\beta^*(i, j) = 1$ for all edges $(i, j) \in E^*$ and $\beta_r(i, j) = 1$ for all edges $(i, j) \in E_r$. The eight healthy subjects feature human-based costs ranging from 0.12 to 0.36 in which DURUS has a cost of 0.30. A comparison of the human-based costs for each of these subjects and DURUS is pictured in Fig. 9(b).

The end result of the approach presented in this paper is a stable multi-contact dynamic walking gait which lends itself to experimental implementation on humanoid robots. Additionally, we have shown that the duration percentages of each discrete domain of the optimal gait match very closely to human walking [4], that is, we have recovered human-like behavior without explicit human reference as a consequence of the natural dynamics of the robot. Walking tiles from the experiment and simulation are synchronized and shown in Fig. 9 (a). While this work primarily focused on the development of periodic multi-contact walking motions on flat terrain it is extensible to the prototyping of other behavior cycles which can be represented within in the framework presented in Sec. 2. This could include alternating step lengths or speeds, stairs, and ramps to name a few. Future work will attempt to apply more rigorous controllers to the walking behaviors generated using this behavior generation methodology.

6   Conclusions

This paper presented a complete algorithmic approach, from modeling to hardware implementation, which was used to obtain multi-contact locomotion on the 3D humanoid robot, DURUS. Through a hybrid model mirroring closely that of a naturally walking human, an optimization problem was formulated for gait generation. Additional constraints allowed for the adjustment of parameters in the gaits generated by the optimizer are introduced and iteratively tuned until a gait is produced which performs well on hardware as a feedforward trajectory playback term. To stabilize the robot when the behaviors are implemented on hardware, three compensation terms are used: a compliance compensator, hip roll proportional regulators, and an inverse kinematics solver which ensures that the swing heel strikes parallel to the ground, thus satisfying the holonomic constraint at the beginning of the following domain. The end result is dynamic and efficient locomotion on the physical hardware utilizing the full-order dynamics of the 3D humanoid robot. This walking was shown to be both stable and energy efficient, demonstrating an even lower electrical cost of transport than previously attained on DURUS. Therefore, this paper presented an algorithmic framework producing the most efficient walking realized on a humanoid robot that is notably human-like when compared to the breakdown of multi-contact foot behavior present in human locomotion.

Acknowledgment

The authors would like to SRI for the design and support of DURUS and Eric Ambrose for designing the custom foot used on DURUS in this work. This work is supported by the National Science Foundation through NRI-1526519.

References

1. DURUS walks like a human. `https://youtu.be/1fC7b2LjVW4`.
2. M. Ackermann. *Dynamics and Energetics of Walking with Prostheses*. PhD thesis, University of Stuttgart, Stuttgart, 2007.
3. Aaron D Ames. Human-inspired control of bipedal walking robots. *IEEE Transactions on Automatic Control*, 59(5):1115–1130, 2014.
4. Aaron D Ames, Ramanarayan Vasudevan, and Ruzena Bajcsy. Human-data based cost of bipedal robotic walking. In *Proceedings of the 14th international conference on Hybrid systems: computation and control*, pages 153–162. ACM, 2011.
5. Steve Collins, Andy Ruina, Russ Tedrake, and Martijn Wisse. Efficient bipedal robots based on passive-dynamic walkers. *Science*, 307(5712):1082–1085, 2005.
6. Jessy W Grizzle, Christine Chevallereau, Ryan W Sinnet, and Aaron D Ames. Models, feedback control, and open problems of 3D bipedal robotic walking. *Automatica*, 50(8):1955–1988, 2014.
7. Nandha Handharu, Jungwon Yoon, and Gabsoon Kim. Gait pattern generation with knee stretch motion for biped robot using toe and heel joints. In *Humanoids 2008-8th IEEE-RAS International Conference on Humanoid Robots*, pages 265–270. IEEE, 2008.

8. Ayonga Hereid, Eric A. Cousineau, Christian M. Hubicki, and Aaron D. Ames. 3D dynamic walking with underactuated humanoid robots: A direct collocation framework for optimizing hybrid zero dynamics. In *IEEE International Conference on Robotics and Automation (ICRA)*. IEEE, 2016.

9. Ayonga Hereid, Shishir Kolathaya, Mikhail S Jones, Johnathan Van Why, Jonathan W Hurst, and Aaron D Ames. Dynamic multi-domain bipedal walking with ATRIAS through SLIP based human-inspired control. In *Proceedings of the 17th international conference on Hybrid systems: computation and control*, pages 263–272. ACM, 2014.

10. Y. Hurmuzlu. Dynamics of bipedal gait; part i: Objective functions and the contact event of a planar five-link biped. *ASME Journal of Applied Mechanics*, 60(2):331–336, 1993.

11. Verne T Inman. Human locomotion. *Canadian Medical Association Journal*, 94(20):1047, 1966.

12. Matthew Johnson, Brandon Shrewsbury, Sylvain Bertrand, Tingfan Wu, Daniel Duran, Marshall Floyd, Peter Abeles, Douglas Stephen, Nathan Mertins, Alex Lesman, et al. Team IHMC's lessons learned from the DARPA robotics challenge trials. *Journal of Field Robotics*, 32(2):192–208, 2015.

13. M. Posa, S. Kuindersma, and R. Tedrake. Optimization and stabilization of trajectories for constrained dynamical systems. In *2016 IEEE International Conference on Robotics and Automation (ICRA)*, pages 1366–1373, May 2016.

14. Jacob Reher, Eric A. Cousineau, Ayonga Hereid, Christian M. Hubicki, and Aaron D. Ames. Realizing dynamic and efficient bipedal locomotion on the humanoid robot DURUS. In *IEEE International Conference on Robotics and Automation (ICRA)*. IEEE, 2016.

15. Ramzi Sellaouti, Olivier Stasse, Shuuji Kajita, Kazuhito Yokoi, and Abderrahmane Kheddar. Faster and smoother walking of humanoid HRP-2 with passive toe joints. In *2006 IEEE/RSJ International Conference on Intelligent Robots and Systems*, pages 4909–4914. IEEE, 2006.

16. Ching-Long Shih, JW Grizzle, and Christine Chevallereau. From stable walking to steering of a 3D bipedal robot with passive point feet. *Robotica*, 30(07):1119–1130, 2012.

17. Koushil Sreenath, Hae-Won Park, Ioannis Poulakakis, and Jessy W. Grizzle. Compliant hybrid zero dynamics controller for achieving stable, efficient and fast bipedal walking on MABEL. *International Journal of Robotics Research*, 30(9):1170–1193, August 2011.

18. DH Sutherland, KR Kaufman, and JR Moitoza. Kinematics of normal human walking. *Human walking*, 2:23–44, 1994.

19. Miomir Vukobratović and Branislav Borovac. Zero-moment point thirty five years of its life. *International Journal of Humanoid Robotics*, 1(01):157–173, 2004.

20. E. Westervelt, J.W. Grizzle, and D.E. Koditschek. Hybrid zero dynamics of planar biped walkers. *IEEE Transactions on Automatic Control*, 48(1):42–56, January 2003.

21. E. R. Westervelt, J. W. Grizzle, C. Chevallereau, J.-H. Choi, and B. Morris. *Feedback Control of Dynamic Bipedal Robot Locomotion*. Control and Automation. Boca Raton, FL, June 2007.

22. Huihua Zhao, Ayonga Hereid, Wen-loong Ma, and Aaron D Ames. Multi-contact bipedal robotic locomotion. *Robotica*, pages 1–35, 2015.

Synthesis of Energy-Bounded Planar Caging Grasps using Persistent Homology

Jeffrey Mahler[1,*], Florian T. Pokorny[1,2,*], Sherdil Niyaz[1], Ken Goldberg[1]

[1] AUTOLAB and Berkeley Artificial Intelligence Research Laboratory
Dept. of IEOR and EECS, University of California, Berkeley, USA
{jmahler, sniyaz, goldberg}@berkeley.edu
[2] CAS/CVAP, KTH Royal Institute of Technology, Sweden
fpokorny@kth.se
* These authors contributed equally

Abstract. Caging grasps restrict object motion without requiring complete immobilization, providing a robust alternative to force- and form-closure grasps. Energy-bounded cages are a new class of caging grasps that relax the requirement of complete caging in the presence of external forces such as gravity. In this paper, we address the problem of synthesizing energy-bounded cages by identifying optimal gripper and force-direction configurations that require the largest increases in potential energy for the object to escape. We present Energy-Bounded-Cage-Synthesis-2D (EBCS-2D), a sampling-based algorithm that uses persistent homology, a recently-developed multiscale approach for topological analysis, to efficiently compute candidate rigid configurations of obstacles that form energy-bounded cages of an object from an α-shape approximation to the configuration space. We also show that constant velocity pushing in the horizontal plane generates an energy field analogous to gravity in the vertical plane that can be analyzed with our algorithm. EBCS-2D runs in $O(s^3 + sn^2)$ time where s is the number of samples and n is the total number of object and obstacle vertices, where typically $n << s$. We observe runtimes closer to $O(s)$ for fixed n. We implement EBCS-2D using the Persistent Homology Algorithms Toolbox (PHAT) and study performance on a set of seven planar objects and four gripper types. Experiments suggest that EBCS-2D takes 2-3 minutes on a 6 core processor with 200,000 pose samples. We also confirm that an RRT* motion planner is unable to find escape paths with lower energy. Physical experiments suggest that push grasps synthesized by EBCS-2D are robust to perturbations. Additional proofs, data, and code are available at http://berkeleyautomation.github.io/caging/.

1 Introduction

In manufacturing and logistics, there are many applications where parts must be reliably grasped and moved without precise constaints on object pose (as required for example in assembly). Caging configurations, in which an object's mobility is bounded by a set of obstacles such as a gripper and / or an energy

© Springer Nature Switzerland AG 2020
K. Goldberg et al. (Eds.): *Algorithmic Foundations of Robotics XII*, SPAR 13, pp. 416–431, 2020.
https://doi.org/10.1007/978-3-030-43089-4_27

field such as gravity, are a promising alternative to form- and force-closure as they provide robustness to perturbations in object pose.

The standard model of caging (complete caging) considers whether a set of obstacles can be placed in a configuration such that the object cannot escape because its mobility is restricted to a bounded set in the free configuration space $\mathcal{F}$ [35, 41] as illustrated in the left part of Fig. 1. When an energy-potential $U : \mathcal{C} \times \mathcal{C} \rightarrow \mathbb{R}$ specifying an energy field such as gravity is defined on the configuration space $\mathcal{C}$, the notion of caging can be generalized to *energy-bounded caging* [24], where the object is constrained to a bounded path-component of the subset of the free configuration space $\mathcal{F}$ with energy less than some threshold u. This arises, for example, when a constant force-field such as gravity acts on the object as illustrated in the middle and right of the figure. We show that energy-bounded cages also occur in the context of planar pushing.

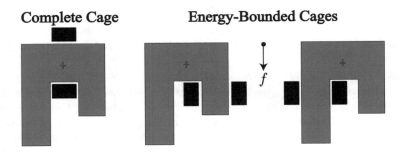

Fig. 1: Complete and energy-bounded cages. Left: a complete cage. The blue object is constrained to a bounded component of the free configuration space by the rigid arrangement of the two gripper fingers (black). Middle and right figure: Two energy-bounded cages with respect to a downward vertical gravity force direction f. The blue object can only escape from its initial configuration when lifted against the gravity field. Note that the rightmost configuration requires more energy to escape.

This paper presents Energy-Bounded-Cage-Synthesis-2D (EBCS-2D), a sampling based algorithm for synthesis of energy-bounded cages given a polygonal object and a rigid configuration of polygonal obstacles under a concave energy field defined over object translations, such as gravity or linear pushing. EBCS-2D synthesizes an ordered list of energy-bounded cages with nonzero minimum escape energy using persistent homology, a tool from computational topology that efficiently computes representatives for bounded components of the free configuration over varying escape energy thresholds. EBCS-2D constructs a weighted α-shape from samples of object poses and a lower-bound on their penetration depth [24], finds a set of candidate energy bounded cages using persistence, and prunes the candidates based on collisions and energy level. The escape energies returned by EBCS-2D provably lower bound the true minimum escape energy for each returned cage. If the returned escape energy is infinite then the object is completely caged.

We implement EBCS-2D using the Persistent Homology Algorithms Toolbox (PHAT) [3] to efficiently identify the most robust energy bounded cages. We evaluate EBCS-2D on a set of seven polygonal parts with parallel-jaw grippers using a push energy field and use it to synthesize optimal push directions. In each case, RRT* optimal path planning was unable to find an escape path with lower energy than the estimated lower bound within 120 seconds. We also apply EBCS-2D to the problem of planar pushing on a Zymark Zymate robot and find that configurations synthesized by EBCS-2D successfully push objects on a planar worksurface.

2   Related Work

Complete caging vs energy-bounded caging: The standard concept of caging, which we refer to as "complete" caging, was introduced by Kuperberg [19] in 1990 and extended by Rimon and Blake [34]. Caging is distinct from complete immobilization of an object by means of form or force closure grasps [29]. Unlike approaches that depend on the local contact geometry, a complete cage of an object causes the object to be constrained to a bounded subset of its free configuration space and requires reasoning about global properties of the configuration space.

Early research on caging studied the caging condition for n points in the plane caging a planar object [19, 35]. Rimon and Blake [34] described the space of cages for a two-finger gripper with one degree of freedom. Sudsang and Ponce [38, 39] proposed caging-based methods for manipulating polygonal objects by means of disc-shaped robots in the plane. Recently, Allen, Burdick and Rimon [2] proposed an algorithm to find all two-finger cage formations of planar polygonal objects by two point-fingers. Vahedi and van der Stappen [41] studied the computation of two and three-finger cages on polygons and used a classification into squeezing and stretching cages. Rodriguez and Mason [36] established and studied caging as a pre-stage to force-closure grasping. Diankov et al. [9] demonstrated that caging grasps can be used to manipulate articulated objects such as door handles.

Recent research has focused on computing cages for specific object families or approximate algorithms due to the difficulty of computing the configuration space for complex gripper and object geometries. These lines of research have primarily focused on synthesizing caging grasps from features in the object surface [33, 20] (e.g. handles) using features of the object surface to rank potential caging configurations [25]. Other research has focused on cell-based approximations of the configuration space based on sampling [42]. Mahler et al. [24] defined energy-bounded caging and presented EBCA-2D, an analysis algorithm that can provably lower bound the minimum escape energy to verify energy-bounded cages for a fixed object and obstacle configuration. The present paper proposes a synthesis algorithm, Energy-Bounded-Cage-Synthesis-2D (EBCS-2D) and considers energy-bounded cages in the context of planar pushing.

Pushing for manipulation: Mason introduced the study of planar pushing to robotics [27] and studied mechanics and planning problems for pushing operations [26]. Constant-velocity quasi-static planar pushing in the horizontal plane can be modeled by an energy potential. Peshkin and Sanderson [30] gave a method to find the locus of the centers of rotation of a planar object for all possible pressure distributions of the object on a planar worksurface. Planar pushing can reduce grasp uncertainty using mechanical compliance and can be used to orient parts [1]. Goldberg [17] gave the first complete algorithm for synthesizing a sequence of pushes to orient polygonal parts without sensory feedback. Lynch and Mason investigated controllability of planar pushing, to determine whether an object can be moved between two configurations purely by pushing actions using point and line contacts [23]. Dogar et al. [10] used a physics-based analysis of two-dimensional contact wrenches to compute push-grasps in clutter and proposed a combinatorial search method to plan push-grasps in [11]. Koval et al. [18] decomposed grasping policies into a pre- and post-contact strategy to reduce uncertainty during pushing actions preceding a grasp using a POMDP planner.

Representations and algorithmic approach: We utilize sampling and a discrete representation of the collision space using α-shapes to reason about cages, building on previous work on motion planning and computational topology. Semi-algebraic functions can be used to prove path non-existence [22], but in practice can be prohibitively expensive to compute. Zhang et al. [44] utilized a rectangular cell-decomposition of the configuration space to prove path non-existence for motion planning by assigning cells to the collision space based on penetration depth. McCarthy et al. [28] use (weighted) α-shapes, a simplicial complex construction defined by Edelsbrunner [13], to represent the collision space from pose samples, and present an algorithm that can prove path non-existence. We build on our previous work [24], which showed that an α-shape-based approximation to the configuration space could be used to analyze a given object and obstacle configuration to check it it is a complete or energy-bounded cage. The present paper also builds on recent advances in topological data analysis [7] and the concept of *persistent homology* [12] to identify "voids" corresponding to cages. Other applications of persistent homology in robotics include methods for clustering trajectories [31] and for motion planning [32, 4].

3   Definitions and Problem Statement

Given a rigid polygonal object $\mathcal{O}$, a rigid configuration of obstacles $\mathcal{G}$ on a planar worksurface, and a potential function P, we consider the problem of finding the set of energy-bounded cages of $\mathcal{O}$ by $\mathcal{G}$ with nonzero minimum escape energy.

Complete Caging and Energy-Bounded Caging

We consider a planar configuration space $\mathcal{C} \subseteq SE(2)$ of a compact polygonal planar object $\mathcal{O} \subset \mathbb{R}^2$ placed in a planar workspace with obstacles defined by fixed positions of a set of k polygons $\mathcal{G} = \mathcal{P}_1 \cup \ldots \cup \mathcal{P}_k \subset \mathbb{R}^2$, such as the jaws of a robotic gripper. We assume the center of mass is known for both the object and obstacles. We denote the object polygon in pose $\mathbf{q} = (x, y, \theta) \in SE(2) = \mathbb{R}^2 \times \mathbb{S}^1$ relative to a reference pose $\mathbf{q}_0$ by $\mathcal{O}(\mathbf{q})$. We define the *collision space* of $\mathcal{O}$ relative to $\mathcal{G}$ by $\mathcal{Z} = \{\mathbf{q} \in SE(2) : int(\mathcal{O}(\mathbf{q})) \cap \mathcal{G} \neq \emptyset\}$ and denote by $\mathcal{F} = SE(2) - \mathcal{Z}$ the *free configuration space*.

We define the energy required to move the object between poses by an *energy function* $U : SE(2) \times SE(2) \to \mathbb{R}$ satisfying $U(\mathbf{q}, \mathbf{q}) = 0, \forall \mathbf{q} \in SE(2)$. This is consistent with [24], in which the reference pose was implicit in the energy function. For a fixed threshold $u \in \mathbb{R}$ and reference $\mathbf{q}_0 \in SE(2)$ define the *u-energy forbidden space* by $\mathcal{Z}_u(\mathbf{q}_0) = \mathcal{Z} \cup \{\mathbf{q} \in \mathcal{C} : U(\mathbf{q}, \mathbf{q}_0) > u\}$ and the *u-energy admissible space* $\mathcal{F}_u(\mathbf{q}_0) = SE(2) - \mathcal{Z}_u(\mathbf{q}_0)$. In this work we use the following definitions of caging as in [24] (see Fig. 2):

Definition 1 (Complete and Energy-bounded caging). *A configuration* $\mathbf{q}_0 \in \mathcal{F}$ *is completely caged if* $\mathbf{q}_0$ *lies in a compact path-component of* $\mathcal{F}$. *We call* $\mathbf{q}_0$ *a u-energy-bounded cage of* $\mathcal{O}$ *with respect to* U *if* $\mathbf{q}_0$ *lies in a compact path-component of* $\mathcal{F}_u(\mathbf{q}_0)$. *Furthermore, the* minimum escape energy, u^*, *for an object* $\mathcal{O}$ *and obstacle configuration* $\mathcal{G}$, *is the infimum over values of* u *such that* q *is not a u-energy-bounded cage of* $\mathcal{O}$, *if a finite such* u^* *exists. Otherwise, we define* $u^* = \infty$.

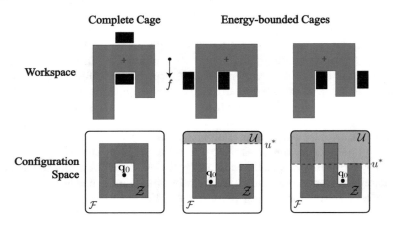

Fig. 2: The top row depicts gripper jaws $\mathcal{G}$ (in black) and an object $\mathcal{O}$ (in blue) in three configurations. The bottom row illustrates conceptually the corresponding point $\mathbf{q}_0 \in SE(2)$ in configuration space. While a complete cage corresponds to an initial pose $\mathbf{q}_0$ completely enclosed by forbidden space $\mathcal{Z}$, the energy-bounded cage on the right instead correpsonds to a case where $\mathbf{q}_0$ is enclosed by $\mathcal{Z}_u = \mathcal{Z} \cup \mathcal{U}(\mathbf{q}_0, u)$ where $\mathcal{U}(\mathbf{q}_0, u) = \{\mathbf{q} \in \mathcal{C} : U(\mathbf{q}, \mathbf{q}_0) > u\}$ for U that is strictly increasing with increasing vertical coordinate. The smallest value of u such that $\mathbf{q}_0$ is not enclosed is called the minimum escape energy, u^*.

While energy-bounded cages can be defined for any energy function U, finding bounded components of $\mathcal{C}$ for all possible pairs of poses in the energy function may be computationally expensive. Thus, for synthesis, we require that the energy function can be derived from a univariate *potential function* $P : SE(2) \to \mathbb{R}$: $U(\mathbf{q}_i, \mathbf{q}_j) = P(\mathbf{q}_i) - P(\mathbf{q}_j)$. In this work, we further assume that P depends only on the translation component $\mathbb{R}^2$ of $SE(2)$ and that P is concave on that space, which guarantees that the point of minumum potential within any convex set is on the boundary of the set. Given such an energy field U, the objective is to synthesize all energy-bounded cages $\mathbf{q}_i \in SE(2)$ with nonzero minimum escape energy.

Energy Functions

We now derive energy functions for gravity in the vertical plane and constant force pushing in the horizontal plane. We develop such functions based on the energy (mechanical work) that wrenches must exert to transport the object between two poses under a nominal wrench resulting from pushing or gravity.

Gravity in the Vertical Plane. Let m denote the mass of the object. Then the energy required to move the object from a reference configuration $\mathbf{q}_i$ to configuration $\mathbf{q}_j$ is $U(\mathbf{q}_j, \mathbf{q}_i) = mg(y_j - y_i)$, where $g = 9.81 m/s^2$ is the acceleration due to gravity in the y-direction [16, 24]. This corresponds to the potential $P(\mathbf{q}) = mgy$.

Constant-Velocity Linear Pushing in the Horizonal Plane. Consider an object being pushed along a fixed direction $\hat{\mathbf{v}}$ by a gripper with a constant velocity on a horizonal worksurface under quasi-static conditions and Coloumb friction with uniform coefficient of friction μ [27, 26, 30]. Then the energy function $U(\mathbf{q}_j, \mathbf{q}_i) = F_p(\mathcal{O}, \mathcal{G}, \mu)\hat{\mathbf{v}} \cdot (x_j - x_i, y_j - y_i) - \kappa(\mathcal{O}, \mathcal{G}, \mu)$ is a lower bound on the energy required to move the object from pose $\mathbf{q}_i$ to $\mathbf{q}_j$ relative to $\mathcal{G}$, where $F_p(\mathcal{O}, \mathcal{G}, \mu) \in \mathbb{R}$ is a bound on the possible resultant force due to contact between the object and gripper and $\kappa(\mathcal{O}, \mathcal{G}, \mu) \in \mathbb{R}$ is a bound on the possible contact torques and frictional wrenches. A justification is given in the supplemental file at http://berkeleyautomation.github.io/caging/. Therefore we propose to use the linear potential $P(\mathbf{q}) = F_p\hat{\mathbf{v}} \cdot (x, y)$ to lower bound the minimum energy required for the object to escape under the nominal push wrench.

Configuration Spaces and α-Complexes

As in [24], we utilize a family of simplicial complexes called α-complexes [13] to approximate the collision space $\mathcal{Z}$ and u-energy forbidden space. For this purpose, we first uniformly sample a collection of s poses $Q = \{\mathbf{q}_1, \ldots, \mathbf{q}_s\}$, $\mathbf{q}_i = (x_i, y_i, \theta_i)$ in $\mathcal{Z}$ and determine the radius $r(\mathbf{q}_i) > 0$ for each $\mathbf{q}_i$, such that the metric ball $\mathbb{B}(\mathbf{q}_i) = \{\mathbf{q} \in SE(2) : d(\mathbf{q}, \mathbf{q}_i) \leqslant r(\mathbf{q}_i)\}$ is completely contained in $\mathcal{Z}$. These radii are computed using a penetration depth solver and the standard metric d on $SE(2)$; details can be found in [24]. The union of these balls $B(Q) = \cup_{i=1}^s \mathbb{B}(\mathbf{q}_i)$ forms a subset of the collision space that approximates $\mathcal{Z}$. See the left part of Fig. 3 for a conceptual illustration.

We can construct a cell-based approximation to $\mathcal{Z}$ using weighted α-shapes to guarantee that the cells are a subset of $\mathcal{Z}$. First, we follow the approach of [24] to lift samples from Q to a set $X \subset \mathbb{R}^3$ for computational reasons (see [24] for details). We then construct a weighted α-shape representation of $B(X)$ [13, 14] since the shape of the union of balls is difficult to analyze computationally.

Weighted α-shapes are a type of *simplicial complex*. A geometric k-simplex $\sigma = [\mathbf{v}_0, \ldots, \mathbf{v}_k]$ in $\mathbb{R}^d$ is a convex hull of $k + 1$ ordered affinely independent elements $\mathbf{v}_0, \ldots, \mathbf{v}_k \in \mathbb{R}^d$ and a convex hull of an ordered subset of these elements

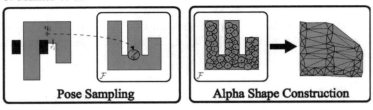

Fig. 3: (Left) We sample a set of poses Q and their penetration depth. (Right) An approximation of the forbidden space $\mathcal{Z} \subset SE(2)$ from Fig. 2 by unions of balls around sampled points Q results in an α-shape simplicial complex $A(X)$ (gray triangles) that is a subset of $\mathcal{Z}$. The triangles of the weighted Delaunay triangulation $D(X)$ that are not in $A(X)$ approximate the free space (red triangles).

is called a face τ of σ, indicated by $\tau \leqslant \sigma$. A finite simplicial complex $\mathcal{K}$ is a non-empty set of simplices such that if $\sigma \in \mathcal{K}$ and $\tau \leqslant \sigma$, then $\tau \in \mathcal{K}$ and if $\sigma, \sigma' \in \mathcal{K}$ then $\sigma \cap \sigma'$ is empty or an element of $\mathcal{K}$. In dimension 3, a simplicial complex $\mathcal{K}$ is a union of points, line-segments, triangles and tetrahedra whose intersections are either empty or another simplex in $\mathcal{K}$, thus generalizing the idea of both a graph and a triangulation in $\mathbb{R}^3$. The α-shape simplicial complex $A(X)$ corresponding to $B(X)$ lies strictly inside $B(X)$ and is homotopy-equivalent to $B(X)$, meaning that topological properties of $B(X)$ can be computed directly from $A(X)$ [13]. Additionally, all simplices in $A(X)$ are contained in $D(X)$, the weighted Delaunay triangulation of X, a data structure that triangulates the convex hull of X. Fig. 3 provides a conceptual illustration.

Persistent Second Homology

Persistent Homology [12] studies the topological features (e.g. holes, voids) that are created and destroyed over one parameter families of simplicial complexes called *filtrations*. Fig. 4 provides a conceptual visualization of 2D slices of "voids" found by persistence for a 3D filtration and a qualitative persistence diagram. A simplex-wise filtration of a simplicial complex $K = \cup_{i=1}^{n}\sigma_i$ is a collection of simplicial complexes K_i such that $K_i = \cup_{i=1}^{i}\sigma_i$, so that K_{i+1} is the result of adding a single simplex σ_{i+1} to K_i. We call i the filtration index. Such a filtration can arise naturally when a function $f : K \to \mathbb{R}$ is defined on the set simplices of K and simplices are ordered in decreasing values of f: $f(\sigma_i) \geqslant f(\sigma_j)$ for all $i \leqslant j$. Thus persistence finds the topological features that emerge as the simplices are added in order of decreasing f. Here, $f(\sigma_i)$ is called the filtration value corresponding to filtration index i. The j-th persistence diagram measures the dimension of the j-th homology group $H_j(K_i)$ that corresponds to a vector space (with finite field coefficients). The dimension of each of these spaces is a topological invariant that does not vary under continuous deformations of the underlying simplicial complex $H(K_i)$. In this work, we are interested in sub-complexes K_i of the weighted Delaunay triangulation $D(X) \subset \mathbb{R}^3$ and the *second homology group* $H_2(K_i)$. Intuitively, the dimension of the second homology group corresponds to "voids" in K_i which are completely enclosed by the complex K_i. These voids in K_i can appear as we add new simplices with increasing i, or they can disappear as voids are filled in. The persistent second homology diagram enables us to visualize these topological changes. Each point (x, y) in the diagram corresponds to a pair of filtration indices (i, j) recording the fact that a void has "appeared" at index i and disappeared at index j. For a geometric simplicial complex, these index pairs (i, j) correspond to simplices (σ_i, σ_j) where σ_j is a *tetrahedron* (a 3-simplex) which destroys or "fills in" a void, while σ_i corresponds to a *triangle* (2-simplex) that corresponds to the last complex needed to first create a fully enclosed void. The set of (i, j) pairs can be displayed in the (index)-persistence diagram, or alternatively, when the filtration arises from a function f, we may display the set of points $(f(\sigma_i), f(\sigma_j))$. By considering the vertical distance $|f(\sigma_i) - f(\sigma_j)|$ from the diagonal, we can read off the parameter range of f during which a void exists in the evolution of the filtration.

4   The EBCS-2D Algorithm

EBCS-2D (Algorithm 1) takes as input a polygonal object $\mathcal{O}$, obstacle configuration $\mathcal{G}$, and potential function P, and outputs a set of energy-bounded cages that require nonzero energy to escape.

Using uniform sampling, the algorithm first generates s object poses in collision $Q = \{\mathbf{q}_1, ..., \mathbf{q}_s\}$ and their corresponding penetration depths $R = \{r_1, ..., r_s\}$. We lift the poses to $\mathbb{R}^3$ and construct an α-shape approximation to the configuration space as described in Section 3. Next, we construct a filtration by sorting all simplices in the free space in order of decreasing energy level and use persistent homology to identify path components fully surrounded by u-energy forbidden space for all u thresholds. Finally, we examine the simplices within each bounded component in order of increasing energy to check for a collision-free object pose, and return the poses extracted from each component. Fig. 4 illustrates the use of persistence in our algorithm.

Filtrations and Persistence from Energy Functions: To synthesize energy-bounded cages with persistence, we first order the simplices of the α-shape approximation by decreasing energy level. We assumed that the potential $P : SE(2) \rightarrow \mathbb{R}^3$ depends only on the translational component $\mathbb{R}^2$ of $SE(2)$ and is concave on that space, which implies that . In this case, for any k-simplex $\sigma = \text{Conv}(\mathbf{v}_0, \ldots, \mathbf{v}_k) \in D(X) - A(X)$ the maximum principle of convex optimization [5] implies that the minimum occurs at one of the vertices:

$$\min_{\mathbf{x} \in \sigma} P(\pi(\mathbf{x})) = \min\{P(\pi(\mathbf{v}_0)), \ldots, P(\pi(\mathbf{v}_k))\}$$

where $\pi : \mathbb{R}^3 \rightarrow SE(2)$ denotes the projection to $SE(2)$. Using this fact, we construct a function $D(X) \rightarrow \mathbb{R}$:

$$f(\sigma) = \begin{cases} \min_{\mathbf{x} \in \sigma} P(\pi(\mathbf{x})) & \sigma \in D(X) - A(X) \\ \infty & \sigma \in A(X) \end{cases}$$

This gives rise to a filtration $K = K(X, U) : \emptyset = K_0 \subset K_1 \subset \ldots \subset K_n \subset D(X)$ of simplices in $D(X)$ with respect to P as described in Section 3, which we can use to find bounded path components corresponding to energy-bounded cages.

EBCS-2D finds pairs of simplices σ_i, σ_j corresponding to the birth and death, respectively, of a bounded path component $C(X) \subset D(X)$ in the free configuration space using persistent homology. All collision-free configurations within the bounded path component are energy-bounded cages by definition. Therefore, EBCS-2D next searches for the configuration $\mathbf{q} \in C(X) \cap \mathcal{F}$ with the highest minimum escape energy by iterating over the set of centroids of simplices in $C(X)$. While the set of simplex centroids only approximates $C(X) \cap \mathcal{F}$, in practice the centroids cover the space well due to the large number of samples used to construct the configuration space. The algorithm runs in $O(s^3 + sn^2)$ time where s is the number of samples and n is the total number of object and obstacle vertices, since α-shape construction is $O(s^2 + sn^2)$ [24, 28] and the boundary matrix reduction used in persistent homology is $O(s^3)$ in the worst case [8].

---

1 **Input:** Polygonal robot gripper $\mathcal{G}$, Polygonal object $\mathcal{O}$, Potential function P,
   Number of pose samples s
   Result: $\hat{Q}$, set of energy-bounded cages with estimated escape energies
   `// Sample poses in collision`
2 $Q = \varnothing, R = \varnothing, \ell = diam(\mathcal{G}) + diam(\mathcal{O})$;
3 $\mathcal{W} = [-\ell, \ell] \times [-\ell, \ell] \times [0, 2\pi)$;
4 **for** $i \in \{1, ..., s\}$ **do**
5     $\mathbf{q}_i = \text{RejectionSample}(\mathcal{W})$;
6     $r_i = \text{LowerBoundPenDepth}(\mathbf{q}_i, \mathcal{O}, \mathcal{G})$;
7   **if** $r_i > 0$ **then**
8       $Q = Q \cup \{\mathbf{q}_i\}, R = R \cup \{r_i\}$;
9 **end**
10 $X = \text{ConvertToEuclideanSpace}(Q)$;
   `// Create alpha shape`
11 $D(X, R) = \text{WeightedDelaunayTriangulation}(X, R)$;
12 $A(X, R) = \text{WeightedAlphaShape}(D(X, R), \alpha = 0)$;
   `// Run Persistent Homology`
13 $K = \text{Filtration}(D(X, R), A(X, R))$;
14 $\Delta = \text{ComputeSecondHomologyPersistencePairs}(K)$;
   `// Find Energy-Bounded Cages`
15 **for** $(i, j) \in \Delta$ **do**
16     $C(K_i, \sigma_j) = \text{PathComponent}(\sigma_j, K_i)$;
17   **for** $\sigma \in Sorted(C(K_i, \sigma_j)), P)$ **do**
18       $\mathbf{q} = \text{Centroid}(\sigma)$;
19       $u = P(\sigma_i) - P(\mathbf{q})$;
20   **if** *CollisionFree(*$\mathbf{q}$*)* and $u > 0$ **then**
21         $\hat{Q} = \hat{Q} \cup \{(\mathbf{q}, u)\}$;
22   **end**
23   **end**
24 **end**
25 **return** $\hat{Q}$;

Algorithm 1: Energy-Bounded-Cage-Synthesis-2D

Correctness: EBCS-2D returns energy-bounded cages with a provable lower bound on the minimum escape energy:

Theorem 1. *Let* $\hat{Q} = \{(\hat{\mathbf{q}}_1, \hat{u}_1), ...(\hat{\mathbf{q}}_n, \hat{u}_n)\}$ *denote the energy bounded cages returned by EBCS-2D. For each* $(\hat{\mathbf{q}}_i, \hat{u}_i) \in \hat{Q}$, $\hat{\mathbf{q}}$ *is a* $\hat{u}$*-energy bounded cage of* $\mathcal{O}$ *with respect to* U.

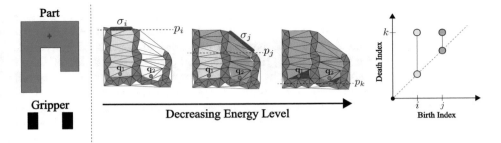

Fig. 4: Persistence diagram for ranking energy-bounded cages. Left: polygonal part and gripper polygons serve as input. We sample object poses X in collision and generate an α-shape representation (shown in gray in the three middle figures). Given an energy potential, we insert simplices in $D(X) - A(X)$ in decreasing order of energy potential, creating a filtration of simplicial complexes. Voids (yellow and orange) are born with the addition of edges σ_i and σ_j (red) at threshold potential levels p_i and p_j respectively, and die with the additions of the last triangle in each void at potential p_k (red). The associated second persistence diagram (right figure) reveals voids with large persistence corresponding to energy-bounded cages. In particular, configuration $\mathbf{q}_1$ is identified as more persistent than configuration $\mathbf{q}_2$. There escape energy fof each configuration is equal to the difference in potentials: $u_1 = p_k - p_i$ and $u_2 = p_k - p_j$, and by the filtration ordering this implies that $\mathbf{q}_1$ has higher escape energy than $\mathbf{q}_2$.

See the supplemental file at http://berkeleyautomation.github.io/caging/ for a proof.

Extension to Pushing EBCS-2D can be applied to push grasping in the horizontal plane. We use it to find push directions that yield robust energy-bounded cages by running EBCS-2D for a set of sampled push directions using the constant velocity linear push energy of Section 3. The extension runs EBCS-2D using P push angles uniformly sampled from $[\frac{\pi}{2} - \varphi, \frac{\pi}{2} + \varphi]$ and returns a ranked list of push directions and energy-bounded cages that can be reached by a linear, collision-free path along the push direction. While the potential changes for each such push direction, the simplices only need to be re-sorted and therefore the sampling and α-complex construction only need to be performed once.

5   Experiments

We implemented EBCS-2D in C++ and evaluated its performance on a set of polygonal objects under both gravitational and pushing energy fields. We used CGAL [40] to compute α-shapes, the GJK-EPA algorithm of libccd [15] to compute penetration depth, and the twist reduction algorithm implemented in PHAT [3] to compute the second persistence diagram. Our dataset consisted of seven polygonal parts created by triangulating the projections of models from YCB [6] and 3DNet [43] onto a plane. All experiments ran on an Intel Core i7-4770K 350GHz processor with 6 cores.

Energy Bounded Cages Under Linear Push Energy We consider a linear push energy field with a push force bound of $F_p = 1.0$ for the set of parts with four grippers: rectangular parallel jaws, an overhead projection of a Zymark Zymate gripper with parallel jaws [21], an overhead projection of a Barrett hand with a pregrasping shape inspired by [11], and a four finger disc gripper inspired by [37]. We ran the pushing extension to EBCS-2D for the rectangular parallel jaws, Zymark gripper, and Barrett hand with $s = 200,000$ samples, an angle limit of $\varphi = \frac{\pi}{4}$, and $P = 5$ push directions to sweep from $-\frac{\pi}{4}$ to $\frac{\pi}{4}$ in intervals of $\frac{\pi}{8}$, and pruned all pushes with $\hat{u} < 0.5$ to ensure that our set of pushes was robust. For the four finger gripper, we ran EBCS-2D with a fixed vertical push direction to illustrate the ability of our algorithm to prove complete cages. EBCS-2D took approximately 170 seconds to run on average for a single push direction. Fig. 5 illustrates configurations synthesized by EBCS-2D with the estimated minimum escape energy $\hat{u}$, which is the distance against the linear push energy that the object must travel to escape. To evaluate the lower bound of Theorem 1, we also used RRT* to attempt to plan an object escape path over the set of collision-free poses with energy less than $\hat{u}$, which was not able to find an escape path with energy less than $\hat{u}$ in 120 seconds of planning [24].

Sample and Time Complexity We also studied the sensitivity of the estimated escape energy for the highest energy configurations synthesized by EBCS-2D for a fixed push direction and the algorithm runtime to the number of pose samples s. The left panel of Fig. 6 shows the ratio of $\hat{u}$ for $s \in \{12.5, 25, 50, 100, 200, 400\} \times 10^3$ pose samples to $\hat{u}$ at $s = 400,000$ pose samples for each of the displayed objects and parallel jaw griipers configurations. We averaged the ratios over 5 independent trials per value of s. Object A is only within 80% of the value at $s = 400,000$ after $s = 200,000$ samples, possibly because of the long thin portion of the configuration space as observed in [24]. Objects B and C both converge to within 95% after about $s = 200,000$ samples. This is comparable to the sample complexity for analysis of a single, fixed configuration with EBCA-2D. The right panel of Fig. 6 shows the relationship between the runtime of EBCS-2D in seconds versus the number of pose samples s over 5 independent runs of the algorithm for the same objects. We broke down the run time by the section of the algorithm: sampling poses, constructing the α-shape to aproximate $\mathcal{C}$, sorting the simplices for the filtration, and computing and pruning candidate energy bounded cages with persistence. The runtime is approximately linear in the number of pose samples, and the largest portion of runtime is the time to sample poses and compute penetration depth. This suggests that the runtime is considerably below the worst case s^3 scaling in practice. The persistence diagram computation in particular has been observed to commonly exhibit sub-quadratic runtime [8] despite its worst-case cubic complexity.

Physical Experiments We evaluated the pushes synthesized by EBCS-2D for the three object configurations with the Zymark gripper illustrated in Fig. 5 on a set of extruded fiberboard polygonal parts [21] using a Zymark Zymate robot to push the objects at a constant velocity on a planar worksurface. Fig. 7 illustrates

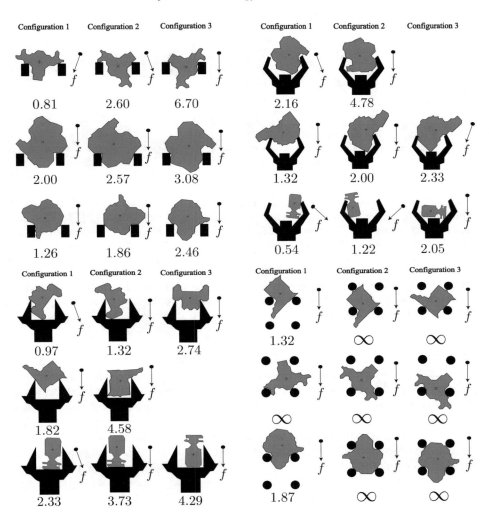

Fig. 5: Illustration of highest energy configurations and push directions synthesized using EBCS-2D ranked from left to right for seven example polygonal objects (blue) and grippers (black) under a linear planar pushing energy field with a push force bound of $F_p = 1.0$. Displayed are three objects for each of the following grippers: (left-to-right, top-to-bottom) parallel-jaw grippers with rectangular jaws, a Barrett hand with fixed preshape, a Zymark Zymate gripper with fixed opening width, and a four finger disc gripper. Below each object the escape energy $\hat{u}$ estimated by EBCS-2D using $s = 200,000$ pose samples, which is the distance the object would have to travel against the pushing direction, and to the right is the synthesized push energy direction f. For each test case we searched over 5 energy directions from $-\frac{\pi}{4}$ to $\frac{\pi}{4}$ and checked push reachability as described in Section 4 except for the four finger gripper, for which we ran only EBCS-2D to illustrate complete cages. The energy of the synthesized configurations is not always directly related to the depth of the part within the object, such as the first row of results for the parallel jaw and Zymark gripper configurations. EBCA-2D also synthesizes several complete cages for the four finger gripper.

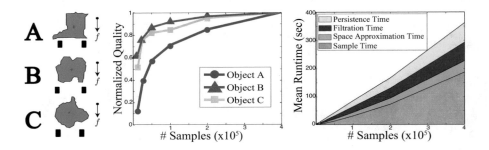

Fig. 6: (Middle) The sample complexity of EBCS-2D. Plotted is the ratio of the highest minimum escape energy out of the energy-bounded cages synthesized by EBCS-2D, $\hat{u}^*$, for the number of pose samples $s = \{12.5, 25, 50, 100, 200, 400\} \times 10^3$ on the object and parallel-jaw displayed on the left. (Right) The mean runtime of EBCS-2D in seconds is broken down by component of the algorithm for varying numbers of pose samples $s = \{12.5, 25, 50, 100, 200, 400\} \times 10^3$. Each datapoint is averaged over five independent runs for each of the object and gripper configurations on the left. Despite the theoretical worst case s^3 runtime, the algorithm runtime is approximately linear in s, and is dominated by sampling time.

our experiments. For each configuration, the object was placed in the center of a turntable, rotated to align the push direction with the arm's major axis, and pushed forward while the turntable oscillated with an amplitude of ± 0.1 radians to simulate external wrenches on the object. To test robustness we added zero-mean Gaussian noise with standard deviation of 5mm to the gripper translation and 0.04 radians to the gripper rotation in the plane. We then evaluated whether or not the object was captured and remained within the gripper jaws after being pushed 150mm. Pushes planned by EBCS-2D had a success rate of 100% versus 41% for a baseline of pushes planned by choosing gripper poses uniformly at random from (x, y) in the object bounding box and θ in $[0, 2\pi)$.

6   Discussion and Future Work

We present EBCS-2D, an algorithm to synthesize energy-bounded cages for polygonal objects and rigid configurations of objects under a 2D energy field. We also extend EBCS-2D to synthesize push directions under a linear constant velocity push energy field. In future work, we will perform additional experiments and model caging as a pre-stage to force-closure grasping and stretching cages [36, 41]. We also plan to study additional energy-functions to model task-specific caging, and to explore extensions of our algorithms to caging in 3D.

Acknowledgements: This research was performed at the AUTOLAB at UC Berkeley in affiliation with the AMP Lab, BAIR, and the CITRIS "People and Robots" (CPAR) Initiative: http://robotics.citris-uc.org. The authors were supported in part by the U.S.

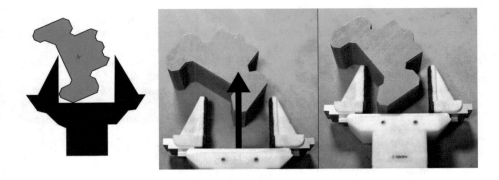

Fig. 7: Illustration of execution of an energy-bounded cage synthesized with EBCS-2D using a Zymark Zymate robot. (Left) The synthesized configuration. (Middle) The planned push direction. (Right) The object remains in the gripper as it is pushed.

National Science Foundation under NRI Award IIS-1227536: Multilateral Manipulation by Human-Robot Collaborative Systems, the Department of Defense (DoD) through the National Defense Science & Engineering Graduate Fellowship (NDSEG) Program, Google, UC Berkeley's Algorithms, Machines, and People Lab, and the Knut and Alice Wallenberg Foundation. We thank the anonymous WAFR reviewers for valuable feedback. We also thank our colleagues who provided helpful suggestions, in particular Subhrajit Bhattacharya, Animesh Garg, David Gealy, Sanjay Krishnan, Michael Laskey, Jacky Liang, Zoe McCarthy, Stephen McKinley, Lauren Miller, and A. Frank van der Stappen.

References

1. S. Akella and M. T. Mason, "Parts orienting by push-aligning," in *Robotics and Automation, 1995. Proceedings., 1995 IEEE International Conference on*, vol. 1. IEEE, 1995, pp. 414–420.
2. T. F. Allen, J. W. Burdick, and E. Rimon, "Two-finger caging of polygonal objects using contact space search," *IEEE Trans. Robotics*, vol. 31, pp. 1164–1179, 2015.
3. U. Bauer, M. Kerber, and J. Reininghaus, "Phat - persistent homology algorithm toolbox," 2013. [Online]. Available: https://code.google.com/p/phat/
4. S. Bhattacharya, R. Ghrist, and V. Kumar, "Persistent homology for path planning in uncertain environments," *IEEE Transactions on Robotics (T-RO)*, March 2015.
5. S. Boyd and L. Vandenberghe, *Convex optimization*. Cambridge university press, 2004.
6. B. Calli, A. Walsman, A. Singh, S. Srinivasa, P. Abbeel, and A. M. Dollar, "Benchmarking in manipulation research: The ycb object and model set and benchmarking protocols," *arXiv preprint arXiv:1502.03143*, 2015.
7. G. Carlsson, "Topology and data," *Bull. Amer. Math. Soc. (N.S.)*, vol. 46, no. 2, pp. 255–308, 2009.
8. C. Chen and M. Kerber, "Persistent homology computation with a twist," in *Proceedings 27th European Workshop on Computational Geometry*, vol. 11, 2011.

9. R. Diankov, S. S. Srinivasa, D. Ferguson, and J. Kuffner, "Manipulation planning with caging grasps," in *Humanoid Robots, 2008. Humanoids 2008. 8th IEEE-RAS International Conference on.* IEEE, 2008, pp. 285–292.

10. M. Dogar, K. Hsiao, M. Ciocarlie, and S. Srinivasa, "Physics-based grasp planning through clutter," in *Robotics: Science and Systems VIII*, 2012.

11. M. Dogar and S. Srinivasa, "A framework for push-grasping in clutter," *Robotics: Science and systems VII*, vol. 1, 2011.

12. H. Edelsbrunner and J. Harer, "Persistent homology-a survey," *Contemporary mathematics*, vol. 453, pp. 257–282, 2008.

13. H. Edelsbrunner, *Weighted alpha shapes.* University of Illinois at Urbana-Champaign, Department of Computer Science, 1992.

14. H. Edelsbrunner and J. Harer, *Computational topology: an introduction.* American Mathematical Soc., 2010.

15. D. Fiser, "libccd - collision detection between convex shapes," http://libccd.danfis.cz/.

16. D. C. Giancoli, *Physics: principles with applications.* Pearson Education, 2005.

17. K. Y. Goldberg, "Orienting polygonal parts without sensors," *Algorithmica*, vol. 10, no. 2-4, pp. 201–225, 1993.

18. M. C. Koval, N. S. Pollard, and S. S. Srinivasa, "Pre-and post-contact policy decomposition for planar contact manipulation under uncertainty," *The International Journal of Robotics Research*, vol. 35, no. 1-3, pp. 244–264, 2016.

19. W. Kuperberg, "Problems on polytopes and convex sets," in *DIMACS Workshop on polytopes*, 1990, pp. 584–589.

20. T. H. Kwok, W. Wan, J. Pan, C. C. Wang, J. Yuan, K. Harada, and Y. Chen, "Rope caging and grasping," in *Proc. IEEE Int. Conf. Robotics and Automation (ICRA)*, 2016.

21. M. Laskey, J. Lee, C. Chuck, D. Gealy, W. Hsieh, F. T. Pokorny, A. D. Dragan, and K. Goldberg, "Robot grasping in clutter: Using a hierarchy of supervisors for learning from demonstrations," in *Proc. IEEE Conf. on Automation Science and Engineering (CASE).* IEEE, 2016.

22. S. M. LaValle, *Planning algorithms.* Cambridge university press, 2006.

23. K. M. Lynch and M. T. Mason, "Controllability of pushing," in *Robotics and Automation, 1995. Proceedings., 1995 IEEE International Conference on*, vol. 1. IEEE, 1995, pp. 112–119.

24. J. Mahler, F. T. Pokorny, A. F. van der Stappen, and K. Goldberg, "Energy-bounded caging: Formal definition and 2d lower bound algorithm based on weighted alpha shapes," in *IEEE Robotics & Automation Letters.* IEEE, 2016.

25. T. Makapunyo, T. Phoka, P. Pipattanasomporn, N. Niparnan, and A. Sudsang, "Measurement framework of partial cage quality based on probabilistic motion planning," in *IEEE Int. Conf. on Robotics and Automation (ICRA)*, 2013, pp. 1574–1579.

26. M. T. Mason, "Mechanics and planning of manipulator pushing operations," *The International Journal of Robotics Research*, vol. 5, no. 3, pp. 53–71, 1986.

27. ——, *Mechanics of Robotic Manipulation.* Cambridge, MA, USA: MIT Press, 2001.

28. Z. McCarthy, T. Bretl, and S. Hutchinson, "Proving path non-existence using sampling and alpha shapes," in *Robotics and Automation (ICRA), 2012 IEEE International Conference on.* IEEE, 2012, pp. 2563–2569.

29. R. M. Murray, Z. Li, and S. S. Sastry, *A mathematical introduction to robotic manipulation.* CRC press, 1994.

30. M. A. Peshkin and A. C. Sanderson, "The motion of a pushed, sliding workpiece," *IEEE Journal on Robotics and Automation*, vol. 4, no. 6, pp. 569–598, 1988.

31. F. T. Pokorny, M. Hawasly, and S. Ramamoorthy, "Multiscale topological trajectory classification with persistent homology," in *Proceedings of Robotics: Science and Systems*, July 2014.

32. F. T. Pokorny and D. Kragic, "Data-driven topological motion planning with persistent cohomology," in *Proceedings of Robotics: Science and Systems*, Rome, Italy, July 2015.

33. F. T. Pokorny, J. A. Stork, and D. Kragic, "Grasping objects with holes: A topological approach," in *Proc. of the IEEE International Conference on Robotics and Automation (ICRA)*, Karlsruhe, Germany, 2013.

34. E. Rimon and A. Blake, "Caging 2d bodies by 1-parameter two-fingered gripping systems," in *Proc. IEEE Int. Conf. Robotics and Automation (ICRA)*, 1996, pp. 1458–1464.

35. E. Rimon and J. W. Burdick, "Mobility of bodies in contact. i. A 2nd-order mobility index for multiple-finger grasps," vol. 14, no. 5, 1998, pp. 696–708.

36. A. Rodriguez, M. T. Mason, and S. Ferry, "From caging to grasping," *Int. J. Robotics Research (IJRR)*, pp. 1–15, 2012.

37. J. Su, H. Qiao, Z. Ou, and Z.-Y. Liu, "Vision-based caging grasps of polyhedron-like workpieces with a binary industrial gripper," *IEEE Transactions on Automation Science and Engineering*, vol. 12, no. 3, pp. 1033–1046, 2015.

38. A. Sudsang and J. Ponce, "On grasping and manipulating polygonal objects with disc-shaped robots in the plane," in *Proc. IEEE Int. Conf. Robotics and Automation (ICRA)*, 1998, pp. 2740–2746.

39. ——, "A new approach to motion planning for disc-shaped robots manipulating a polygonal object in the plane," in *Robotics and Automation, 2000. Proceedings. ICRA'00. IEEE International Conference on*, vol. 2. IEEE, 2000, pp. 1068–1075.

40. The CGAL Project, *CGAL User and Reference Manual*, 4th ed. CGAL Editorial Board, 2015.

41. M. Vahedi and A. F. van der Stappen, "Caging polygons with two and three fingers," *The International Journal of Robotics Research*, vol. 27, no. 11-12, pp. 1308–1324, 2008.

42. W. Wan, R. Fukui, M. Shimosaka, T. Sato, and Y. Kuniyoshi, "A new grasping by caging solution by using eigen-shapes and space mapping," in *Proc. IEEE Int. Conf. Robotics and Automation (ICRA)*. IEEE, 2013, pp. 1566–1573.

43. W. Wohlkinger, A. Aldoma, R. B. Rusu, and M. Vincze, "3dnet: Large-scale object class recognition from cad models," in *Proc. IEEE Int. Conf. Robotics and Automation (ICRA)*, 2012, pp. 5384–5391.

44. L. Zhang, Y. J. Kim, and D. Manocha, "Efficient cell labelling and path nonexistence computation using c-obstacle query," *The International Journal of Robotics Research*, vol. 27, no. 11-12, pp. 1246–1257, 2008.

Equilateral Three-Finger Caging of Polygonal Objects Using Contact Space Search

Hallel A. Bunis, Elon D. Rimon[1]*, Thomas F. Allen[2], and Joel W. Burdick[3]

[1] Dept. of ME, Technion - Israel Institute of Technology
[2] Pneubotics, San Francisco
[3] Dept. of ME, California Institute of Technology

Abstract. Multi-finger caging offers a robust object grasping approach. However, while efficient computation of two-finger caging grasps is well developed, the computation of three-finger caging grasps has remained a challenge. This paper considers the caging of polygonal objects with three-finger hands which maintain an *equilateral triangle* formation during the grasping process. While the c-space of such hands is four dimensional, their *contact space* which represents all two and three finger contacts along the grasped object's boundary forms a two-dimensional stratified manifold. The paper describes a *caging graph* that can be constructed in the hand's relatively simple contact space. Starting from a desired immobilizing grasp of the polygonal object, the caging graph can be readily searched for the largest finger opening that maintains a three-finger cage about the object to be grasped. Any equilateral finger placement within the corresponding *caging regions* guarantees robust object grasping.

1 INTRODUCTION

Multi-finger caging offers a robust object grasping approach by robot hands. In a *cage*, the object to be grasped is surrounded by fingers such that the object may have local freedom to move but cannot escape from the cage formed by the fingers. To securely grasp an object, the robot places the fingers in a cage formation around the object, then closes the cage until the object is securely immobilized by the fingers. The *caging regions* that surround a desired immobilizing grasp allow initial placement of the robot fingers under huge position uncertainty, while subsequent closing of the cage allows the hand to grasp the object without any need for accurate positioning of the fingers relative to the object.

Since robotic caging was introduced by Rimon and Blake [1, 2], the synthesis of caging grasps initially focused on two-finger hands. Notable examples are the comprehensive two-finger caging algorithms of Vahedi and Van der Stappen [3] and Pipattanasomporn and Sudsang [4], which compute all two-finger cage formations directly in the hand's four-dimensional configuration space. Allen, Rimon and Burdick [5] recently showed how these cage formations can be alternatively computed in the hand's two-dimensional *contact space*, which offers

---

* This work was supported by grant number 1253/14 of the Israel Science Foundation.

K. Goldberg et al. (Eds.): *Algorithmic Foundations of Robotics XII*, SPAR 13, pp. 432–447, 2020.
https://doi.org/10.1007/978-3-030-43089-4_28

implementation simplicity and geometric verifiability along the grasped object's boundary.

While two point or disc fingers can immobilize polygonal objects which possess concavities, three point or disc fingers can immobilize and hence securely grasp almost all polygonal objects. This insight motivated several three-finger caging algorithms. Erickson et al. [6] attempted to generate small caging regions localized around three particular edges of the grasped polygon in $O(n^6)$ time, where n is the number of the polygon's edges. Subsequent work by Vahedi and Van der Stappen [3] was able to compute all three-finger caging grasps in $O(n^6 \log^2 n)$ time, provided that two fingers are held fixed as base fingers. Having three as well as higher number of fingers in mind, Pipattanasomporn and Sudsang [7] proposed to use convex functions of the finger locations, called *dispersion functions*, to compute the maximal finger dispersion that maintains a cage formation as a convex optimization problem. Wan considered the efficient numerical computation of three-finger cage formations with fixed base fingers [8, 9]. Wan recently proposed the topological enumeration of the hand's free c-space boundary components, a technique that can efficiently report when a candidate three or higher number of fingers formation forms a cage about a given object [10].

The literature survey would be incomplete without mentioning advances in caging theory made by Rodriguez [11] and Allen [12] in the context of three and higher number of fingers cage formations, and the inclusion of gravity in partial cage formations [13]. Caging has also been considered for robust object manipulation [14, 15], as well as the robust transfer of warehouse objects by mobile robots [16, 17].

This paper extends the contact space approach to three-finger hands which maintain an *equilateral triangle* finger formation, whose size is determined by a single scalar parameter. The contact space of single parameter three-finger hands has been used by Davidson and Blake [18] to compute the critical grasp at which the cage surrounding an object is broken. However, they used a numerical method to compute and classify the entire set of critical grasps, which is often costly. Next, they used a numerical gradient descent to identify which of the pre-computed critical grasps is associated with a desired immobilizing grasp. This paper offers an alternative method. It describes a *caging graph* that is constructed analytically in the hand's contact space and does not require any critical point classification. Starting from a desired immobilizing grasp of the polygonal object, a simple search on the caging graph gives the caging regions surrounding the immobilizing grasp. The caging graph method can be performed in $O(n^4 \log n)$ time, which is more comparable to those of [3, 6, 7]. In addition, it can be intuitively observed as the motion of the fingers along the object's boundary.

The three-finger hands considered in this paper are motivated by hand mechanism design. As discussed by Dollar et al. [19, 20], roboticists seek general purpose *minimalistic hand designs* that will posses the smallest number of fingers as well as the smallest number of actuators. Three-finger hands form an attractive minimalistic design, since three point or disc fingers can immobilize every polygonal object which does not possess opposing parallel edges [21]. Moreover, by limiting the three-finger hand to a particular one-parameter family of formations, a *single actuator* can simultaneously close the three fingers, thus allowing for extremely simple hand designs.

While the current paper considers equilateral finger formations, we plan to extend the contact space approach to other one-parameter triangular formations. In addition, we plan to extend the work done by Allen et al. on two-finger caging of 3D polyhedra using contact space [22], to a higher number of fingers.

The paper is structured as follows. Section 2 introduces the three-finger caging problem. Section 3 provides background on the representation of cage formations in the hand's four dimensional configuration space. Section 4 describes the hand's *contact space*. Section 5 describes the *caging graph* construction in the hand's contact space. Section 6 presents the *caging graph search algorithm* which is used to compute the critical grasp which defines the caging regions that surround an immobilizing grasp. The conclusion discusses using contact space for three-finger hands that maintain more general triangular finger formations.

2   The Three-Finger Caging Problem

Consider the caging of polygonal objects using a robot hand made of a rigid *palm* and three finger mechanisms attached to the palm (Fig. 1(a)). The hand is modeled by a *palm frame*, $\mathcal{F}_B$, that can freely translate and rotate in $\mathbb{R}^2$, and three point or disc fingers that form an equilateral triangle fixed to the hand's palm frame (Fig. 1(b)). The distance between each pair of fingers (and hence the size of the equilateral triangle) is specified by the scalar parameter $\sigma \geq 0$. The hand's configuration is thus specified by the pair (q, σ), where $q = (d, \theta) \in \mathbb{R}^2 \times \mathbb{S}$ is the position and orientation of the palm frame with respect to a fixed world frame, while σ determines the hand's *size*.

The object to be caged, denoted $\mathcal{B}$, is a rigid polygon that can freely translate and rotate in $\mathbb{R}^2$. In this paper we assume that the only hand-object interactions occur through frictionless point contacts. The object is said to be *immobilized* by the fingers, when no motion of $\mathcal{B}$ is possible relative to the fingers (Fig. 1(a)). In general there are two types of immobilizing grasps: *squeezing grasps* where the hand is closed around the grasped object, and *stretching grasps* where the hand is opened from within a cavity of the grasped object. This paper focuses on squeezing grasps. The object is said to be *caged* by the fingers when the inter-finger distance σ is held fixed, and the hand cannot move to infinity without one of its fingers penetrating the stationary object.

Every immobilizing grasp of $\mathcal{B}$ is associated with *caging regions*. When the fingers are placed in these caging regions such that the inter-finger distance is kept under some critical value, $\sigma < \sigma_{max}$, the object will be caged by the fingers. Starting at such a cage and monotonically decreasing σ while maintaining the hand's equilateral triangle shape will lead to an immobilizing grasp of $\mathcal{B}$, while the object remains caged throughout this process. Thus, given a geometric

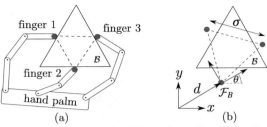

Fig. 1. (a) A triangular object immobilized by a three-finger hand which maintains an equilateral triangle formation. (b) The three-finger hand is modeled by a palm frame and three point or disc fingers, which maintain an equilateral triangle formation fixed to the palm frame.

description of $\mathcal{B}$ and a desired two or three-finger immobilizing grasp of this object, the *three-finger caging problem* seeks to compute the critical inter-finger distance, σ_{max}, associated with the immobilizing grasp of $\mathcal{B}$. The value of σ_{max} can then be used to determine the caging regions which form the output of the three-finger caging problem.

3   Representation of Cages in the Hand's Configuration Space

3.1   The Free Configuration Space Boundary

Let the object $\mathcal{B}$ lie stationary in $\mathbb{R}^2$, so that it forms an obstacle from the hand's perspective. The hand's configuration space (*c-space*), denoted $\mathcal{C}$, can be thought of as $(q, \sigma) \in \mathbb{R}^4$. Since the fingers cannot penetrate the stationary object, the object induces three *c-obstacles* in the hand's c-space (Fig. 2). The union of the c-space obstacles is denoted $\mathcal{CB}$. The hand's *free c-space*, $\mathcal{F}$, is the complement of $\mathcal{CB}$'s interior: $\mathcal{F} = \mathcal{C} - \text{int}(\mathcal{CB})$, where int denotes set interior. The boundary of $\mathcal{F}$ (which is the same as the boundary of $\mathcal{CB}$) consists of all hand configurations at which one or more fingers touch the boundary of the stationary object $\mathcal{B}$. The boundary consists of three types of submanifolds: 3D manifolds formed by single-finger contacts, 2D manifolds formed by two-finger contacts, and 1D curves formed by three-finger contacts.

The union of the two-finger contact manifolds together with their intersection curves along the three-finger contact manifolds defines the *contact submanifold*, which will be the basis for the contact space approach for computing cage formations.

Definition 1 *The three-finger hand's **contact submanifold**, denoted $\mathcal{S}$, is the union of the submanifolds of the free c-space $\mathcal{F}$ associated with all hand configurations at which at least two fingers contact the stationary object.*

3.2   C-Space Representation of Cage Formations

Let us define the *equilibrium grasps*, which play a key role in the c-space representation of cage formations. Let x_i denote the i'th finger contact along the object's

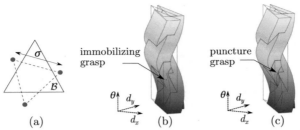

Fig. 2. (a) An object caged by an equilateral triangle finger formation. (b) The c-space slice through $\sigma = \sigma_0$, where the object is immobilized by the fingers and the c-obstacle slices completely surround the immobilizing configuration point. (c) The c-space slice through $\sigma = \sigma_1$, where a puncture point appears on the free c-space cavity wall.

boundary, expressed in a fixed world frame. Let f_i denote the i'th finger force applied on $\mathcal{B}$ along the object's inward normal at x_i, expressed in the fixed world frame. We do allow finger contacts at the vertices of $\mathcal{B}$. In such cases the point fingers are modeled as small discs, and their applied force lies within the object's *generalized contact normal* at the object's vertex. When the hand's palm frame coincides with the fixed world frame, the *wrench* (i.e. force and torque) generated by a finger force f_i acting on $\mathcal{B}$ at x_i is given by $\boldsymbol{w}_i = (f_i, x_i \times f_i) \in I\!R^3$, where the torque component is given by the scalar $x_i \times f_i = x_i^T J f_i$, such that $J = \begin{bmatrix} 0 & 1 \\ -1 & 0 \end{bmatrix}$. The object $\mathcal{B}$ is held in a *feasible equilibrium grasp* when the finger contact forces can apply a zero net wrench on $\mathcal{B}$, as stated in the following definition.

Definition 2 *A rigid object $\mathcal{B}$ is held via $k \geq 2$ finger contacts in a **feasible equilibrium grasp** (in the absence of external influences such as gravity) when there exist finger forces satisfying the condition:*

$$\lambda_1 \begin{pmatrix} f_1 \\ x_1 \times f_1 \end{pmatrix} + \cdots + \lambda_k \begin{pmatrix} f_k \\ x_k \times f_k \end{pmatrix} = \boldsymbol{0} \tag{1}$$

such that $\lambda_1, \ldots, \lambda_k \geq 0$ and are not all zero.

An *immobilizing grasp* of $\mathcal{B}$ is a feasible equilibrium grasp which occurs at a hand configuration, $(q_0, \sigma_0) \in \mathcal{S}$, at which the point q_0 is completely surrounded by the c-obstacles in the σ_0 slice of $\mathcal{C}$ (Fig. 2(b)). A small increase of σ above σ_0 will cause the c-obstacle slices to move away from each other, forming a small *bounded cavity* in $\mathcal{F}$. This cavity allows the fixed-σ hand to locally move in a bounded neighborhood of the immobilizing grasp, while the object is kept stationary. Further increasing σ would cause the cavity to expand until eventually a *puncture point* will appear on its boundary. At this instant, the puncture point might connect the cavity to an adjacent cavity associated with a different immobilizing grasp of $\mathcal{B}$, thus being an *intermediate puncture point*, or can connect to infinity, in which case it represents the *escape puncture point* with critical value σ_{max} (Fig. 2(c)). The latter puncture point will be denoted (q_1, σ_1), where $\sigma_1 = \sigma_{max}$. Importantly, every puncture point in $\mathcal{F}$ corresponds to an equilibrium grasp of $\mathcal{B}$ which involves two or three finger contacts [2].

Starting at an immobilizing grasp (q_0, σ_0) and increasing σ up to σ_1 allows the hand's palm frame to move in the largest bounded area in $\mathbb{R}^2$, while maintaining a cage around $\mathcal{B}$. The three fingers move accordingly in three bounded regions which form the *caging regions* in $I\!\!R^2$. To complete the picture in terms of caging theory, the union of fixed-σ cavities in $\mathcal{F}$ for $\sigma_0 \leq \sigma \leq \sigma_1$ is termed the *caging set* in $\mathcal{F}$. Given a desired immobilizing configuration (q_0, σ_0), our objective is to compute the critical value of $\sigma_1 = \sigma_{max}$ as well as the corresponding escape puncture point at (q_1, σ_1), then use this information to determine the caging regions that surround the immobilizing grasp.

4   The Three-Finger Hand Contact Space

The contact submanifold $\mathcal{S}$ consists of the two-finger contact manifolds, together with their intersection along the three-finger contact curves in $\mathcal{F}$. The hand's *contact space* is a parametrization of $\mathcal{S}$ in terms of the finger contacts along the object's boundary. Contact space will be used to find an *escape path* which starts at (q_0, σ_0), passes through (q_1, σ_1), and ends at a three-finger *pinching configuration* within contact space, thus reducing the caging problem from a search in $\mathbb{R}^4$ to a search in a space which can be thought of as $\mathbb{R}^2$.

4.1   The Two-Finger Contact Spaces

Let us assume that the immobilizing grasp is specified along the object's outer boundary, in case the object contains holes. This boundary will be parametrized by arclength in counterclockwise direction using the scalar parameter $s \in [0, L]$, where L is the object's perimeter (Fig. 3(a)). Let the three fingers be labeled in counterclockwise order, and let fingers i and $i+1$ contact the object's boundary at the points $p(s_i)$ and $p(s_{i+1})$, where $i=1,2,3$ mod 3 (Fig. 3(b)). A formal definition of the two-finger contact space follows.

Definition 3 *Let a polygonal object be contacted by fingers i and $i+1$ at $p(s_i)$ and $p(s_{i+1})$. The **two-finger contact space** is the parameterization of all two-finger contacts along the object's boundary, given by the set $\mathcal{U}_{i,i+1} = [0, L] \times [0, L]$ in the (s_i, s_{i+1}) plane.*

A three-finger hand has *three* two-finger contact spaces: $\mathcal{U}_{12}$, $\mathcal{U}_{23}$, and $\mathcal{U}_{31}$. Each of these spaces is partitioned into *rectangles*, each associated with a particular pair of object edges, including same-edge pairs. The rectangles that partition a two-finger contact space $\mathcal{U}_{i,i+1}$ will be denoted $\mathcal{R}_{jk}$, meaning that fingers i and $i+1$ are contacting edges j and k of $\mathcal{B}$. The two-finger *diagonal*, $\Delta_{i,i+1} = \{(s_i, s_{i+1}) \in \mathcal{U}_{i,i+1} : s_i = s_{i+1}\}$, represents all two-finger pinching configurations along the object's boundary. Next we define the inter-finger distance function on the individual two-finger contact spaces.

Definition 4 *Let $\mathcal{U}_{i,i+1}$ be the two-finger contact space associated with fingers i and $i+1$. The inter-finger distance function is the scalar valued function $d : \mathcal{U}_{i,i+1} \to \mathbb{R}$ given by $d(s_i, s_{i+1}) = \|p(s_i) - p(s_{i+1})\|$.*

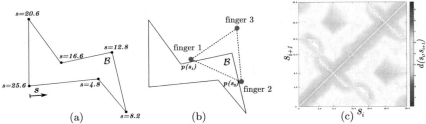

Fig. 3. (a) The s parameterization of the object's boundary. (b) When fingers 1 and 2 contact the object at $p(s_1)$ and $p(s_2)$, the position of finger 3 is uniquely determined under counterclockwise ordering of the finger indices. (c) The two-finger contact space $\mathcal{U}_{i,i+1}$, overlaid with the contours of $d(s_i, s_{i+1})$ induced by the object shown in (a). Note that $\mathcal{U}_{i,i+1}$ is partitioned into rectangles corresponding to the object's edge pairs.

Consider the projection function $\pi(q, \sigma) = \sigma$ which maps the hand's configuration to the inter-finger distance σ. The inter-finger distance function $d(s_i, s_{i+1})$ is the restriction of $\pi(q, \sigma)$ to the contact submanifold $\mathcal{S}$, using contact space parametrization. The contours of $d(s_i, s_{i+1})$ in $\mathcal{U}_{i,i+1}$ are depicted in Fig. 3(c). Note that $d(s_i, s_{i+1})$ is non-negative and continuous on $\mathcal{U}_{i,i+1}$, and attains a global minimum of zero along the diagonal $\Delta_{i,i+1}$. Moreover, $d(s_i, s_{i+1})$ forms a *convex function* in the individual contact space rectangles. Also note that the contours of $d(s_i, s_{i+1})$ have an identical layout in the three two-finger contact spaces $\mathcal{U}_{12}$, $\mathcal{U}_{23}$ and $\mathcal{U}_{31}$.

4.2   The Third-Finger Contact Space Obstacles

When fingers i and $i+1$ move along the object's boundary, the third finger moves in a way that maintains the hand's equilateral triangle formation. Since the third finger may not penetrate the stationary object $\mathcal{B}$ during this motion, *contact space obstacles* are induced in $\mathcal{U}_{i,i+1}$. A graphical technique for constructing these obstacles is next described. The position of finger i along a particular edge of $\mathcal{B}$ is given by:

$$p(s_i) = p_i^0 + \left(s_i - s_i^0 \right) t_i \quad s_i \in \left[s_i^0, s_i^0 + L_i \right]$$

where $p_i^0 = p(s_i^0)$ is the edge's initial vertex, t_i is the edge's *unit tangent*, and L_i is the edge's length. The position of finger $i+1$ along some other edge of $\mathcal{B}$ is similarly described by $p(s_{i+1})$. Let R_ϕ denote a $60°$ counterclockwise rotation matrix in $\mathbb{R}^2$. Since the hand maintains an equilateral triangle formation (and the finger indices maintain a counterclockwise order), the position of the third finger, denoted X_{i+2}, is given by:

$$X_{i+2}(s_i, s_{i+1}) = p(s_i) + R_\phi \left(p(s_{i+1}) - p(s_i) \right)$$

Substituting for $p(s_i)$ and $p(s_{i+1})$ gives:

$$X_{i+2}(s_i, s_{i+1}) = u + \left(s_i - s_i^0 \right) v + \left(s_{i+1} - s_{i+1}^0 \right) w$$

where $u = p_i^0 + R_\phi \left(p_{i+1}^0 - p_i^0 \right)$, $v = (I - R_\phi) t_i$, and $w = R_\phi t_{i+1}$ are constant vectors. From this last expression it is clear that $X_{i+2}(s_i, s_{i+1})$ is a parametrization of a *parallelogram* in $\mathbb{R}^2$, with vertices located at u, $u + L_i v$, $u + L_{i+1} w$, and $u + L_i v + L_{i+1} w$. The parallelogram construction is illustrated in Fig. 4.

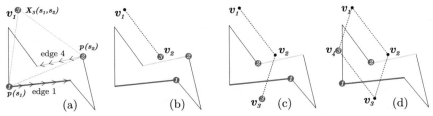

Fig. 4. The third-finger parallelogram construction procedure: (a)-(d) The vertices of a parallelogram, v_1, v_2, v_3 and v_4, are drawn at the position of the third finger $i+2$, when fingers i and $i+1$ contact the endpoints of two edges of $\mathcal{B}$.

Note that each parallelogram is associated with a particular contact space rectangle $\mathcal{R}_{jk}$. The parallelograms are next used to determine the following contact space obstacles.

Definition 5 *When fingers i and $i+1$ contact the object $\mathcal{B}$, the **contact space obstacles**, denoted $\mathcal{CB}_{i,i+1}$, consist of all points $(s_i, s_{i+1}) \in \mathcal{U}_{i,i+1}$ for which $X_{i+2}(s_i, s_{i+1})$ lies within $\mathcal{B}$.*

The intersection between the object and the area enclosed by the parallelogram determines the contact space obstacles in $\mathcal{R}_{jk}$ (Fig. 5(a)). The third finger parallelogram, $X_{i+2}(s_i, s_{i+1})$, can be thought of as the result of a linear mapping applied to the contact space rectangle $\mathcal{R}_{jk}$. By applying the *inverse* transformation on the intersection between the object and the area enclosed by the parallelogram, one can obtain the contact space obstacles in $\mathcal{R}_{jk}$. Denote by T the 2×2 matrix which transforms the contact space basis vectors $(1, 0)$ and $(0, 1)$ to v and w. The inverse transformation matrix is given by $T^{-1} = \frac{1}{w^T J v}\left[w \ -v\right]^T J$, where $J = \begin{bmatrix} 0 & 1 \\ -1 & 0 \end{bmatrix}$.

The inverse mapping of a parallelogram onto $\mathcal{R}_{jk}$ is illustrated in Fig. 5(b). Figure 5(c) shows the entire collection of contact space obstacles depicted as black regions in $\mathcal{U}_{i,i+1}$, with the obstacles layout being identical in all three two-finger contact spaces. The contact space obstacles form polygonal regions as stated in the following lemma.

Lemma 4.1 *The contact space obstacles of an equilateral three-finger hand form **polygonal regions** in the two-finger contact spaces $\mathcal{U}_{i,i+1}$, where $i = 1, 2, 3$.*

Note that every *corner point* along a contact space obstacle boundary corresponds to a hand configuration at which at least one finger is contacting a vertex of $\mathcal{B}$. This fact can be asserted by inspecting Fig. 5.

Finally consider how the two-finger contact spaces relate to each other. The boundary curves of the contact space obstacles correspond to configurations at which all three fingers contact the object, and can be considered as *gluing seams* which connect the three two-finger contact spaces to each other. A formal definition of the hand's contact space follows.

Definition 6 *The **three-finger hand contact space**, denoted $\mathcal{U}$, is given by the gluing of the individual two-finger contact spaces $\mathcal{U}_{12}$, $\mathcal{U}_{23}$, and $\mathcal{U}_{31}$ along the contact space obstacle boundaries:*

$$\mathcal{U} = \mathcal{U}_{12} \cup \mathcal{U}_{23} \cup \mathcal{U}_{31} / \mathsf{bdy}\,(\mathcal{CB}_{i,i+1}) \sim \mathsf{bdy}\,(\mathcal{CB}_{i+1,i+2})$$

where $\mathsf{bdy}\,(\mathcal{CB}_{i,i+1})$ is the contact space obstacle boundary, and the quotient identifies the corresponding point triplets on $\mathsf{bdy}\,(\mathcal{CB}_{12})$, $\mathsf{bdy}\,(\mathcal{CB}_{23})$ and $\mathsf{bdy}(\mathcal{CB}_{31})$.

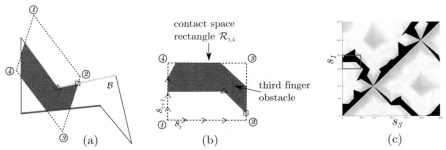

Fig. 5. (a) A parallelogram with non-feasible third finger positions marked by a red region. (b) Applying T^{-1} on the red region gives the contact space obstacle in the contact space rectangle $\mathcal{R}_{1,4}$. The brown triangle and blue box mark *corner points* of the contact space obstacle, which lie within $\mathcal{R}_{1,4}$ and on one of its bounding lines. (c) The contact space $\mathcal{U}_{31}$ overlaid with the contact space obstacles induced by finger 2. The contact space rectangle $\mathcal{R}_{1,4}$ marked with red corresponds to edges 1 and 4 of $\mathcal{B}$, and is the same as the one obtained in (b).

The following proposition formally states that the hand's contact space $\mathcal{U}$ is topologically equivalent to the contact submanifold $\mathcal{S}$.

Proposition 4.2 *The three-finger hand's contact space $\mathcal{U}$ is **topologically equivalent** (homeomorphic) to the contact submanifold $\mathcal{S}$ in the hand's free c-space $\mathcal{F}$.*

A search for an escape path in $\mathcal{S}$ can therefore be equivalently performed in the hand's contact space $\mathcal{U}$.

4.3 Contact Space Representation of Immobilizing, Puncture and Pinching Grasps

Points in $\mathcal{U}$ corresponding to immobilizing, puncture, and pinching grasps will form nodes of the caging graph. The following proposition characterizes these grasps in contact space [23].

Proposition 4.3 *The immobilizing and puncture grasps of $\mathcal{B}$ appear as **local minima** and **saddle points** of the inter-finger distance function $d\,(s_i, s_{i+1})$ in contact space $\mathcal{U}$.*

To identify the extremum points of $d\,(s_i, s_{i+1})$ in $\mathcal{U}$, recall that immobilizing and puncture grasps are equilibrium grasps. The two-finger equilibrium grasps occur in three cases (Figs. 6(a)-(c)): (1) One finger contacts a vertex and the other contacts an opposing edge of $\mathcal{B}$. (2) The two fingers contact opposing vertices of $\mathcal{B}$. (3) The two fingers contact opposing parallel edges of $\mathcal{B}$. This leads to the following lemma.

Lemma 4.4 *A two-finger equilibrium grasp of $\mathcal{B}$ is an extremum point of $d(s_i, s_{i+1})$ which lies at a **corner point** of $\mathcal{R}_{jk}$, or at an **interior point** of a bounding line of $\mathcal{R}_{jk}$ in $\mathcal{U}$.*

The three-finger equilibrium grasps correspond to extremum points of $d(s_i, s_{i+1})$ along the *boundary* of the contact space obstacles. Since $d(s_i, s_{i+1})$ is convex in each contact space rectangle $\mathcal{R}_{jk}$, it can have at most one extremum point in the interior of each line segment along a contact space obstacle boundary, leading to the following lemma.

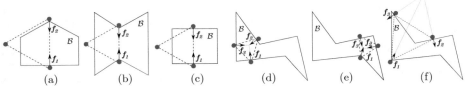

Fig. 6. Equilibrium grasps of $\mathcal{B}$: (a) A grasp at a vertex and an opposing edge. (b) A grasp at two opposing vertices. (c) A grasp at parallel edges. (d) A grasp at interior points of edges of $\mathcal{B}$. (e) A grasp where one finger contacts a vertex of $\mathcal{B}$. (f) A three-finger puncture grasp.

Lemma 4.5 *A three-finger equilibrium grasp of $\mathcal{B}$ is an extremum point of $d(s_i, s_{i+1})$ which lies at a **corner point** of a contact space obstacle boundary, or at a single point in the interior of each obstacle boundary line segment in $\mathcal{U}$.*

Three-finger equilibrium grasps are illustrated in Figs. 6(d)-(f), which depict two immobilizing grasps and one puncture grasp. Note that all of the extremum points of $d(s_i, s_{i+1})$ in $\mathcal{U}$ will become nodes of the caging graph.

Let us finally define the escape points and *escape nodes* in $\mathcal{U}$. When a three-finger hand maintains an equilateral triangle formation, every two-finger pinching configuration automatically forms a three-finger pinching configuration, leading to the following definition of escape points in $\mathcal{U}$.

Definition 7 *The **escape points** in $\mathcal{U}$ are all points on the three-finger diagonal Δ, which correspond to three-finger pinching configurations along the object's boundary.*

All corner points of the contact space rectangles which lie on the two-finger diagonals $\Delta_{i,i+1}$ will form the *escape nodes* of the caging graph. These nodes correspond to hand configurations from which the hand can escape to infinity.

5   The Three-Finger Caging Graph

This section describes the *caging graph*, which allows a search for the escape puncture point directly in the hand's contact space $\mathcal{U}$. The caging graph is constructed in two stages. First the caging graph's nodes and edges which are embedded in $\mathcal{U}$ are computed. The caging graph is then augmented with *tunnel edges*, which ensure sublevel equivalence with the hand's free c-space $\mathcal{F}$.

5.1   The Naive Caging Graph

Let $G(V, E)$ denote the *caging graph* whose nodes and edges are defined as follows.

Definition 8 *The **nodes** of G are the corner points of each contact space rectangle $\mathcal{R}_{jk}$, the extremum points of $d(s_i, s_{i+1})$ along the rectangle's bounding lines, the corner points of the contact space obstacles in $\mathcal{R}_{jk}$, and all extremum points of $d(s_i, s_{i+1})$ along the contact space obstacles' boundaries in $\mathcal{U}$. The **edges** of G are all linear segments that connect pairs of nodes in each contact space rectangle $\mathcal{R}_{jk}$ without intersecting any contact space obstacle.*

By construction, all the extremum points of the inter-finger distance function $d(s_i, s_{i+1})$ are nodes of the caging graph. Such extremum points represent the collection of all feasible and non-feasible equilibrium grasps of the object $\mathcal{B}$.

Computation of the caging graph nodes: First consider the computation of the two-finger nodes. The nodes at the *corners* of a contact space rectangle $\mathcal{R}_{jk}$ correspond to hand configurations at which two fingers contact vertices of $\mathcal{B}$. The nodes at the *extremum points* of $d(s_i, s_{i+1})$ along the bounding lines of $\mathcal{R}_{jk}$ can be readily computed by evaluating the formula:

$$\nabla d\left(s_i, s_{i+1}\right) = \frac{1}{d\left(s_i, s_{i+1}\right)} \begin{pmatrix} (p_i - p_{i+1}) \cdot t_{i,j} \\ -(p_i - p_{i+1}) \cdot t_{i+1,k} \end{pmatrix} = \mathbf{0}$$

along each bounding line. A two-finger node is feasible only if the position of the non-contacting third finger, $X_{i+2}(s_i, s_{i+1})$, at this node lies *outside* the object $\mathcal{B}$. Using a simple $O(n)$ test if a given point lies in a polygon (where n is the number of edges of $\mathcal{B}$), the computation of all two-finger nodes can be performed in $O(n^3)$ steps.

Next consider the computation of the three-finger nodes, at which all three fingers are at contact with $\mathcal{B}$ while maintaining an equilateral formation. Nodes at the *corner points* of the contact space obstacles correspond to hand configurations at which at least one finger contacts a vertex of $\mathcal{B}$. Nodes at the *extremum points* of $d(s_i, s_{i+1})$ along the contact space obstacle boundaries correspond to three-finger equilibrium grasps of $\mathcal{B}$. Let the i'th finger wrench be represented by $\boldsymbol{w}_{i,j} = \lambda_i \left(n_{i,j}, p_i^T J n_{i,j}\right)$, where $n_{i,j}$ is the inward normal of the object edge j contacted by finger i, and $J = \begin{bmatrix} 0 & 1 \\ -1 & 0 \end{bmatrix}$. Then for a three-finger equilibrium grasp, the finger wrenches must satisfy $\boldsymbol{w}_{1,j} + \boldsymbol{w}_{2,k} + \boldsymbol{w}_{3,l} = \mathbf{0}$ for $\lambda_1, \lambda_2, \lambda_3$ which are all non-zero, where j, k and l denote the edges of $\mathcal{B}$ contacted by the three fingers. This linear dependency of the finger wrenches can be written as the condition $\det\left(\boldsymbol{w}_{1,j}, \boldsymbol{w}_{2,k}, \boldsymbol{w}_{3,l}\right) = 0$. The computation of three-finger nodes also requires $O(n^3)$ steps. A pseudo-code for computing the graph nodes is summarized in [23]. Figure 7 shows a two-finger contact space with the four types of caging graph nodes as well as the edges of G embedded in it.

The following lemma is the basis for the *node-connecting test*, which will be used to check whether two nodes can be connected by a feasible graph edge.

Lemma 5.1 *A* **line segment** *in the contact space rectangle* $\mathcal{R}_{jk}$ *in the two-finger contact space* $\mathcal{U}_{i,i+1}$ *corresponds to a* **linear path** *traversed by finger* $i+2$ *in* $\mathbb{R}^2$ *while fingers* i *and* $i+1$ *move along edges* j *and* k *of the stationary object* $\mathcal{B}$.

Lemma 5.1 can be asserted by restricting the algebraic expression of the parallelogram $X_{i+2}(s_i, s_{i+1})$ to a line segment in $\mathcal{R}_{jk}$. The node-connecting test is demonstrated in Fig. 8.

We next consider the computational complexity of the graph edges which are embedded in $\mathcal{U}$. We have already noted that the nodes of $G(V, E)$ can be computed in $O(n^3)$ steps, where n is the number of edges of $\mathcal{B}$. To compute the graph edges, the node-connecting test is performed on all node pairs located in the same contact space rectangle $\mathcal{R}_{jk}$. Since the corresponding parallelogram might contain all n vertices of $\mathcal{B}$, each $\mathcal{R}_{jk}$ might contain $O(n)$ nodes, and hence contain $O(n^2)$ possible node pairs. Thus, the intersection of the n edges of $\mathcal{B}$ with $O(n^2)$ line segments must be checked. Using a sweep line algorithm such as [24], it is possible to compute all intersections between k line segments in $O(k \log k)$ steps. Hence, the number of computations required to determine the graph edges in each $\mathcal{R}_{jk}$ can reach $O(n^2 \log n)$ steps, and for all contact space rectangles, it can reach $O(n^4 \log n)$ steps.

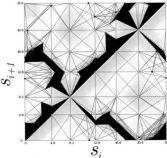

Fig. 7. The two-finger contact space $\mathcal{U}_{i,i+1}$ overlaid with the caging graph G, for $i=1,2,3$ mod 3.

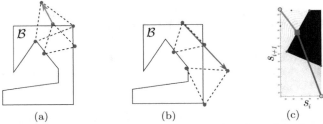

$$(a) \qquad\qquad\qquad (b) \qquad\qquad\qquad (c)$$

Fig. 8. Performing the node-connecting test on two pairs of graph nodes. (a) The green line connecting the positions of finger $i+2$ does *not* intersect the object $\mathcal{B}$, hence the corresponding graph nodes can be connected by an edge. (b) The red line connecting the positions of finger $i+2$ intersects the object $\mathcal{B}$, hence the corresponding graph nodes cannot be connected. (c) The green line between the first node pair does not intersect the contact space obstacle, while the red line between the second pair intersects the obstacle.

The caging graph $G\,(V,E)$ is next examined in relation to the topology of the inter-finger distance function $d(s_i, s_{i+1})$ in $\mathcal{U}$. Let us introduce the sublevel subgraphs of G. The notation $d\,(s_i\,(\boldsymbol{v})\,, s_{i+1}\,(\boldsymbol{v}))$ will denote the inter-finger distance at the point in $\mathcal{U}_{i,i+1}$, represented by the graph node $\boldsymbol{v}$.

Definition 9 *Let* $G\,(V,E)$ *be the caging graph over contact space* $\mathcal{U}$. *For each* $c \geq 0$, *the* **c-sublevel subgraph** *of* G, *denoted* $G_c\,(V_c, E_c)$, *is the subgraph of* G *whose nodes are given by* $V_c = \{\boldsymbol{v} \in V\,:\, d\,(s_i\,(\boldsymbol{v})\,, s_{i+1}\,(\boldsymbol{v})) \leq c\}$.

The following theorem asserts that the caging graph G topologically captures the sublevel structure of $\mathcal{U}$.

Theorem 1 *The caging graph* G **topologically captures the sublevel structure** *of* $\mathcal{U}$: *there exists a contact space path in* $\mathcal{U}$ *between two nodes of* G, $\boldsymbol{v_1}$ *and* $\boldsymbol{v_2}$, *lying entirely in a* c-sublevel set of $d\,(s_i, s_{i+1})$ *in* $\mathcal{U}$ *if and only if there exists a caging graph path between* $\boldsymbol{v_1}$ *and* $\boldsymbol{v_2}$ *lying entirely in a* c-sublevel subgraph of G.

Proof sketch: The proof is based on the convexity of $d(s_i, s_{i+1})$ in each $\mathcal{R}_{jk}$. Given a contact space path between two nodes of G, the idea is to partition it into path segments according to the contact space rectangles $\mathcal{R}_{jk}$ it passes through, and then show that each path segment has an equivalent piecewise linear path which passes between nodes of G and lies in the same sublevel set of $d(s_i, s_{i+1})$ in $\mathcal{U}$. Next, a caging graph path is considered between two nodes in a sublevel subgraph of G, and the contact space path corresponding to the graph edges in each $\mathcal{R}_{jk}$ is shown to lie in the same sublevel set of $d(s_i, s_{i+1})$ in $\mathcal{U}$, containing the sublevel subgraph. An extended and illustrated proof can be found in [23].

5.2   The Augmented Caging Graph

A search for an escape path in $\mathcal{U}$ is valid only if it yields the same result as a search in the hand's full free c-space $\mathcal{F}$. To ensure the same result, *sublevel equivalence* between $\mathcal{S}$ and $\mathcal{F}$ in terms of connectivity must be established. Let $\mathcal{F}_c = \{(q, \sigma) \in \mathcal{F} : \sigma \leq c\}$ denote the sublevel set of free configurations in the hand's free c-space $\mathcal{F}$, and let $\mathcal{S}_c$ denote the subset $\mathcal{F}_c \cap \mathcal{S}$. Sublevel equivalence between $\mathcal{S}$ and $\mathcal{F}$ is defined next.

Definition 10 *The contact submanifold $\mathcal{S}$ and the ambient free c-space $\mathcal{F}$ are **sublevel equivalent** in terms of connectivity, if in any connected component of $\mathcal{F}_c$ the subset $\mathcal{S}_c$ is also connected, for any $c > 0$.*

Recall the projection function $\pi(q, \sigma) = \sigma$. Squeezing immobilizing grasps are local minima of $\pi(q, \sigma)$ in $\mathcal{F}$ (whereas stretching immobilizing grasps are local maxima of $\pi(q, \sigma)$ in $\mathcal{F}$). Sublevel equivalence between $\mathcal{F}$ and $\mathcal{S}$ breaks when *local minima* of the restriction of $\pi(q, \sigma)$ to $\mathcal{S}$ are *not* local minima of $\pi(q, \sigma)$ in $\mathcal{F}$. These local minima correspond to grasps at which the hand can close without penetrating the object, termed *non-feasible* grasps. Since $\mathcal{S}$ and $\mathcal{U}$ are topologically equivalent, such configurations appear as *non-feasible local minima* of $d(s_i, s_{i+1})$ in $\mathcal{U}$. Hence, to ensure sublevel equivalence of $\mathcal{F}$ with $\mathcal{S}$, a graph node at a non-feasible local minimum of $d(s_i, s_{i+1})$ in $\mathcal{U}$ must be connected to a node in a lower sublevel subgraph of G by a *tunnel edge* which represents a σ-decreasing free space path termed a *tunnel curve* which connects two disjoint subsets of $\mathcal{S}$ in a connected component of $\mathcal{F}$. The caging graph G *augmented* with tunnel edges is denoted by $G(V, \overline{E})$. The procedure for augmenting G with tunnel edges is demonstrated in Fig. 9.

We next consider the computational complexity of the tunnel edges. First, note that only the nodes that represent configurations at which only two fingers contact the object should be considered, since non-feasible local minima of the restriction of $\pi(q, \sigma)$ to $\mathcal{S}$ generically involve only two contacting fingers. Next note that each contact space rectangle $\mathcal{R}_{jk}$ can contain at most one non-feasible local minimum of $d(s_i, s_{i+1})$ in $\mathcal{U}$. Hence, to compute the tunnel edges, the two-contact nodes in each $\mathcal{R}_{jk}$ must be sorted according to their inter-finger distance value, and only the node with the smallest value must be checked. For each of these nodes, a ray shooting method is then used to determine whether the hand can be closed without penetrating the object, and if so, which hand configuration will be reached when it closes. Finally, the node-connecting test which might require $O(n \log n)$ computational steps must be used to find a node which can be connected to the non-feasible local minimum node. Therefore, the augmentation of $G(V, E)$ with the tunnel edges might require $O(n^3 \log n)$ computational steps. Note that if the inter-finger distance function $d(s_i, s_{i+1})$ does *not* have any non-feasible local minima, then $\mathcal{F}$ and $\mathcal{S}$ are automatically sublevel equivalent, and the search for the escape path can be executed on the caging graph $G(V, E)$.

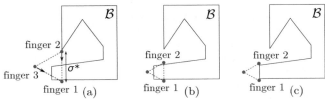

Fig. 9. Caging graph augmentation procedure: (a) A hand configuration represented by the non-feasible local minimum node v_1 in $\mathcal{U}$, with $\sigma(v_1) = \sigma^*$. (b) While one finger maintains contact and the hand orientation remains fixed, σ is decreased until a second finger contacts the object. This hand configuration corresponds to a point u in a contact space rectangle $\mathcal{R}_{jk}$. (c) A node v_2 in $\mathcal{R}_{jk}$ which can be connected to u by a line segment and satisfies $\sigma(v_2) \leq \sigma^*$ is found, and the tunnel edge (v_1, v_2) is added to $G(V, \overline{E})$.

6   The Caging Graph Search Algorithm

The caging graph $G(V, E)$ augmented with tunnel edges is sublevel equivalent to the free c-space $\mathcal{F}$. Hence, a search along the augmented caging graph $G(V, \overline{E})$ will give the same result as a search in $\mathcal{F}$. The caging problem, i.e. the search for an escape path, is thus reduced from a search in $\mathbb{R}^4$ ($\mathcal{F}$) to a search over discrete points in what can be considered as $\mathbb{R}^2$ ($\mathcal{U}$).

The search is based on the following intuitive notion. The inter-finger distance function $d(s_i, s_{i+1})$ can be considered as a potential energy function, where opening the hand while at least two fingers maintain contact with the object requires an increase of energy. The following search algorithm can be considered as a search for a path on the surface of $d(s_i, s_{i+1})$, which connects one of its local minima (the initial immobilizing grasp) to its global minimum set Δ (comprised of pinching configurations), while passing through the saddle point which requires the least increase in energy (the escape puncture grasp). The caging graph search algorithm follows.

Algorithm 1 Caging Graph Search Algorithm

    Data structures: open list $\mathcal{O}$, closed list $\mathcal{C}$.
    Input: $G(V, \overline{E})$, initial immobilizing grasp node v.
    Output: closed list $\mathcal{C}$.
    Initialize: $\mathcal{O} = \emptyset$, $\mathcal{C} = \emptyset$, current node $n = v$.
    Mark n as explored.
    while $\sigma(n) \neq 0$ **do**
        Add the unexplored neighbors of n to $\mathcal{O}$ and mark them as explored.
        Place n at the end of $\mathcal{C}$.
        Sort the nodes in $\mathcal{O}$ by ascending values of σ.
        Remove the node with minimal σ value from $\mathcal{O}$ and set it as n.
    end while   ▷ n is now an escape node
    Place n at the end of $\mathcal{C}$.
    Return: closed list $\mathcal{C}$.

Algorithm 1 uses two node lists: the open list $\mathcal{O}$ of explored nodes and the closed list $\mathcal{C}$ which contains the escape path nodes. The node in $\mathcal{C}$ with the maximal σ value, denoted σ_{max}, corresponds to the escape puncture grasp. Beyond σ_{max} the cage around the object is broken. A computational example of Algorithm 1, which includes a graphical rendering of the caging regions can be found in [23].

Computational complexity of the search algorithm: Let each node have two flags: explored and unexplored. During the search process, only unexplored nodes are added to $\mathcal{O}$, which is held sorted. Each node might have $O\left(n\right)$ neighbors, hence adding $O\left(n\right)$ neighbor nodes to $\mathcal{O}$ at each search iteration might require $O\left(n\log n\right)$ steps. If $O\left(n^3\right)$ iterations are required for the $O\left(n^3\right)$ graph nodes, then $O\left(n^4\log n\right)$ computational steps might be required to complete the search. Note that computing all the graph edges also requires $O\left(n^4\log n\right)$ steps.

7   Conclusion

The paper considered the caging of polygonal objects by three-finger hands which maintain equilateral triangle formations during the grasping process. While the hand's configuration space is four dimensional, the paper presented a caging algorithm that can be implemented in the hand's *contact space* $\mathcal{U}$, which can be thought of as $I\!\!R^2$. The hand's contact space contains obstacles which represent forbidden third-finger positions when two fingers contact the grasped object's boundary. The paper defined a *caging graph*, G, whose nodes include all the extremum points of the inter-finger distance function $d(s_i, s_{i+1})$ in $\mathcal{U}$, as well as all corner points of the contact space obstacles. The caging graph edges consist of line segments that connect adjacent nodes in $\mathcal{U}$, as well as *tunnel edges* which connect nodes at local minima of $d(s_i, s_{i+1})$ that represent nonfeasible equilibrium grasps, to lower sublevel subgraphs of G. The paper established that when the caging graph is augmented with tunnel edges, it is sublevel equivalent to the hand's free c-space $\mathcal{F}$. Starting from an immobilizing grasp, the caging graph can be searched for the puncture grasp through which the object can escape to infinity. The caging regions associated with the immobilizing grasp can then be rendered in $I\!\!R^2$.

We are currently considering the possible extension of the contact space approach to more general three-finger hand formations. While equilateral finger formations seem natural and can immobilize almost every polygonal object without opposing parallel edges, we identified special polygonal shapes that *cannot* be immobilized by equilateral finger formations. In contrast, it seems that isosceles triangle formations, for example, can immobilize and hence securely grasp every polygonal object without opposing parallel edges. Such hand mechanisms still possess the sought after mechanical simplicity, and their ability to securely grasp generic 2D objects justifies the development of efficient tools for synthesizing robust grasp sequences for such hands.

References

[1] Rimon, E., Blake, A.: Caging 2D bodies by 1-parameter two-fingered gripping systems. IEEE Int. Conf. on Robotics and Automation (1996) 1458–1464
[2] Rimon, E., Blake, A.: Caging planar bodies by 1-parameter two-fingered gripping systems. Int. Journal of Robotics Research **18**(3) (1999) 299–318
[3] Vahedi, M., van der Stappen, A.F.: Caging polygons with two and three fingers. Int. Journal of Robotics Research **27**(11-12) (2008) 1308–1324
[4] Pipattanasomporn, P., Sudsang, A.: Two-finger caging of nonconvex polytopes. IEEE Trans. on Robotics **27**(2) (2011) 324–333
[5] Allen, T.F., Rimon, E., Burdick, J.W.: Two-finger caging of polygonal objects using contact space search. IEEE Trans. on Robotics **31**(5) (2015) 1164–1179
[6] Erickson, J., Thite, S., Rothganger, F., Ponce, J.: Capturing a convex object with three discs. IEEE Trans. on Robotics **23**(6) (2007) 1133–1140
[7] Pipattanasomporn, P., Vongmasa, P., Sudsang, A.: Caging rigid polytopes via finger dispersion control. IEEE Int. Conf. on Robotics and Automation (2008) 1181–1186

[8] Wan, W., Fukui, R., Shimosaka, M., Sato, T., Kuniyoshi, Y.: Grasping by caging: A promising tool to deal with uncertainty. IEEE Int. Conf. on Robotics and Automation (2012) 5142–5149

[9] Wan, W., Fukui, R., Shimosaka, M., Sato, T., Kuniyoshi, Y.: A new grasping by caging solution using eigen-shapes and space mapping. IEEE Int. Conf. on Robotics and Automation (2013) 1566–1573

[10] Wan, W., Fukui, R.: Efficient planar caging test using space mapping. IEEE Trans. Autom. Sci. Eng. **PP**(99) (1-12, 2016)

[11] Rodriguez, A., Mason, M.T., Ferry, S.: From caging to grasping. Int. Journal of Robotics Research **31**(7) (2012) 886–900

[12] Allen, T.F., Rimon, E., Burdick, J.W.: Robust three finger three-parameter caging of convex polygons. IEEE Int. Conf. on Robotics and Automation (2015) 4318–4325

[13] Mahler, J., Pokorny, F.T., McCarthy, Z., van der Stappen, A.F., Goldberg, K.: Energy-bounded caging: Formal definition and 2D energy lower bound algorithm based on weighted alpha shapes. IEEE Robotics and Automation Letters **1**(1) (2016) 508–515

[14] Wang, Z., Kumar, V.: Object closure and manipulation by multiple cooperating mobile robots. IEEE Int. Conf. on Robotics and Automation (394-399, 2002)

[15] Sudsang, A., Rothganger, F., Ponce, J.: Motion planning for disc-shaped robots pushing a polygonal object in the plane. IEEE Trans. on Robotics and Automation **18**(4) (2002) 550–562

[16] Fink, J., Hsieh, M.A., Kumar, V.: Multi-robot manipulation via caging in environments with obstacles. IEEE Int. Conf. Robotics and Automation (2008) 1471–1476

[17] Pereira, G., Campos, M., Kumar, V.: Decentralized algorithms for multi-robot manipulation via caging. Int. Journal of Robotics Research **23**(7-8) (2004) 783–795

[18] Davidson, C., Blake, A.: Caging planar objects with a three-finger one-parameter gripper. IEEE Int. Conf. on Robotics and Automation **3** (1998) 2722–2727

[19] Backus, S.B., Dollar, A.M.: An adaptive three-fingered prismatic gripper with passive rotational joints. IEEE Robotics and Automation Letters **1**(2) (2016) 668–675

[20] Howe, R.D., Dollar, A.M., Claffee, M.: Robots get a grip. IEEE Spectrum (2014) 42–47

[21] Rimon, E., Burdick, J.W.: New bounds on the number of frictionless fingers required to immobilize planar objects. J. of Robotic Systems **12**(6) (1995) 433–451

[22] Allen, T.F., Rimon, E., Burdick, J.W.: Two-finger caging of 3D polyhedra using contact space search. IEEE Int. Conf. on Robotics and Automation (2014) 2005–2012

[23] Bunis, H.A., Rimon, E.D.: Equilateral three-finger caging of polygonal objects using contact space search. Tech. report, Dept. of ME, Technion, http://robots.net.technion.ac.il/publications/ (2016)

[24] Bartuschka, U., Mehlhorn, K., Naher, S.: A robust and efficient implementation of a sweep line algorithm for the straight line segment intersection problem. Workshop on Algorithm Engineering (1997) 124–135

On the Distinction between Active and Passive Reaction in Grasp Stability Analysis [*]

Maximilian Haas-Heger[1], Garud Iyengar[2], and Matei Ciocarlie[1]

[1] Department of Mechanical Engineering, Columbia University, New York, NY 10027, USA; m.haas@columbia.edu, matei.ciocarlie@columbia.edu
[2] Department of Industrial Engineering and Operations Research, Columbia University, New York, NY 10027, USA; garud@ieor.columbia.edu

Abstract. Stability analysis for multi-fingered robotic grasping is often formalized as the problem of answering the following questions: can the hand exert contact forces on the object either without a net resultant wrench (thus loading the contacts while creating purely internal forces) or in order to counterbalance an external disturbance (by applying an equal and opposite wrench). However, some of the most commonly used methods for performing this analysis do not distinguish between active torque generation and passive resistance at the joints (and, in fact, many commonly used methods disregard the actuation mechanism entirely). In this study, we introduce an analysis framework constructed to capture such differences, and present evidence showing that this is an important distinction to make when assessing the stability of a grasp employing some of the most commonly used actuation mechanisms.

1 Introduction

Stability analysis is one of the foundational problems for multi-fingered robotic manipulation. Grasp planning can be posed as a search over the space of possible grasps looking for instances that satisfy a measure of stability. The formulation and characteristics of the stability measure thus play a key role in this search, and, by extension, in any task that begins by planning and executing a grasp.

In turn, multi-fingered grasp stability relies on studying the net resultant wrench imparted by the hand to the grasped object. Ferrari and Canny [8] introduced a very efficient geometric method for determining the total space of possible resultant wrenches as long as each individual contact wrench obeys (linearized) friction constraints. This method answers the simplest form of what we refer to here as the existence problem: given a desired output, are there legal contact wrenches that achieve it, and, if so, how large is their needed magnitude in relation to the output? This approach has been at the foundation of numerous planning algorithms proposed since.

Consider a grasp that scores highly according to the quality metric described above. This means that any desired resultant can be produced by a computable

---

[*] This work was supported in part by the Office of Naval Research Young Investigator Program under award N00014-16-1-2026.

© Springer Nature Switzerland AG 2020
K. Goldberg et al. (Eds.): *Algorithmic Foundations of Robotics XII*, SPAR 13, pp. 448–463, 2020.
https://doi.org/10.1007/978-3-030-43089-4_29

set of contact wrenches (of bounded magnitude). In turn, the contact wrenches can be balanced by a set of joint torques, which can also be computed [13]. However, this approach is based on a string of assumptions:

- First, we have assumed that, at any given moment, the control mechanism knows what resultant wrench is needed on the object.
- Second, we have assumed that the joint torques needed to balance this resultant can be actively commanded by the motor outputs.
- Third and finally, we have assumed that the desired motor output torques can be obtained accurately.

In practice, these assumptions do not always hold. The external wrench in need of balancing is difficult to obtain: to account for gravity and inertia, one needs the exact mass properties and overall trajectory of the object, which are not always available; any additional external disturbance will be completely unknown, unless the hand is equipped with tactile sensors. Finally, many commonly used robot hands use highly geared motors unable to provide accurate torque sensing or regulation.

A much simpler approach, applicable to more types of hardware, is to simply select a set of motor commands that generate some level of *preload* on the grasp, and maintain that throughout the task. This method assumes that the chosen motor commands will not only lead to an adequate and stable preload for grasp creation, but also prove suitable for the remainder of the task. A key factor that allows this approach to succeed is *the ability of a grasp to absorb forces that would otherwise unbalance the system without requiring active change of the motor commands*.

Following the arguments above, we believe it is important to not only consider the wrenches the hand can apply actively by means of its actuators, but also the reactions that arise passively. Thus, in this study we are interested in the distinction between *active force generation*, directly resulting from forces applied by a motor, and *passive force resistance*, arising in response to forces external to the contacts or joints. Consider the family of highly geared, non-backdrivable motors: the torque applied at a joint can exceed the value provided by the motor, as long as it arises passively, in response to torques applied by other joints, or to external forces acting on the object. Put another way, joint torques are not always the same as output motor torques, even for direct-drive hands.

From a practical standpoint, a positive answer to the existence problem outlined above is not useful as long as the joint torques necessary for equilibrium will not be obtained given a particular set of commands sent to the motors. We focus on stability from an inverse perspective: given a set of commanded torques to be actively applied to the robot's joints, what is the net effect expected on the grasped object, accounting for passive reactions?

Overall, the main contribution of this paper is a quasi-static grasp stability analysis framework to determine the *passive* response of the hand-object system to applied joint torques and externally applied forces. This method was designed to account for actuation mechanisms such as non-backdrivable or position-controlled motors.

2   Related Work

The problem of force distribution between an actively controlled robotic hand and a rigid object has been considered by a number of authors [14, 1, 10, 16]. A great simplification to grasp analysis is the assumption that any contact force can be applied by commanding the joint motors accordingly. This assumption neglects the deficiency of the kinematics of many commonly used robotic hands in creating arbitrary contact forces. The idea that the analysis of a grasp must include not only the geometry of the grasp but also the kinematics of the hand is central to this paper.

Bicchi [2] showed that for a kinematically deficient hand only a subset of the internal grasping forces is actively controllable. Using a quasi-static model, the subspace of internal forces was decomposed into subspaces of active and passive internal forces. Making use of this decomposition Bicchi proposed a method to pose the problem of optimal distribution of contact forces with respect to power consumption and given an externally applied wrench as a quadratic program [3]. He proposed a definition of force-closure that makes further use of this decom-position and developed a quantitative grasp quality metric that reflects on how close a grasp is to losing force-closure under this definition [4].

There have been rigid body approaches to the analysis of active and passive grasp forces. Yoshikawa [15] studied the concept of active and passive closures and the conditions for these to hold. Melchiorri [11] decomposed contact forces into four subspaces using a rigid body approach. Burdick and Rimon [5] formally defined four subspaces of contact forces and gave physically meaningful interpre-tations. They analyzed active forces in terms of the injectivity of the transposed hand Jacobian matrix. They note that the rigid body modeling approach is a limitation, as a compliance model is required to draw conclusions on the stability of a grasp.

An important distinction between our work and that of the above authors lies in the definition of what qualifies as a "passive" contact force. In addition to contact forces that lie in the null space of the transposed hand Jacobian, we consider contact forces arising from joints being loaded passively (due to the non-backdrivability of highly geared motors) and not arising from the commanded joint torque as passive. Furthermore, we define preload forces as the internal forces that arise from selecting a set of motor commands that achieve a grasp in stable equilibrium, previous to the application of any external wrench.

3   Problem Statement

Consider a grasp establishing multiple contacts between the robot hand and the grasped object. We denote the vector of contact wrenches by c. The grasp map matrix G relates contact wrenches to the net resultant wrench w, while the transpose of the grasp Jacobian J relates contact wrenches to joint torques τ:

$$Gc = w \tag{1}$$
$$J^T c = \tau \tag{2}$$

The classical approach: In combination with a contact constraint model, the relatively simple system of Eqs. (1)&(2) expresses the static equilibrium constraints for the grasp. However, for rigid bodies, the problem of computing the exact force distribution across contacts in response to an applied wrench is statically indeterminate. Thus, previous studies such as those of Bicchi [2–4] make use of a linear compliance matrix that characterizes the elastic elements in a grasp and solves the indeterminacy. For a comprehensive study on how to compute such a compliance matrix see the work by Cutkosky and Kao [7]. A compliance matrix allows us to consider the force distribution across contacts as the sum of a particular and a homogeneous solution. The contact forces c_0 create purely internal forces and hold the object in equilibrium. This is the homogeneous solution and as noted in the Introduction, it can be of great importance to grasp stability. The contact forces c_p associated with the application of an external wrench w are considered the particular solution. Bicchi formulates a *force distribution problem* [3] given by $c = c_p + c_0 = G_K^R w + c_0$ where G_K^R is the K-weighted pseudoinverse of the grasp map matrix G. K is the stiffness matrix of the grasp and is given by the inverse of the grasp compliance matrix.

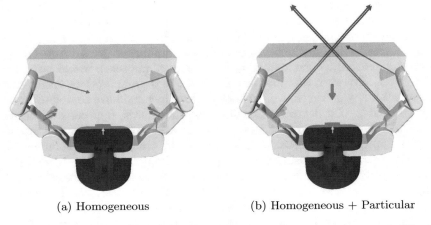

(a) Homogeneous   (b) Homogeneous + Particular

Fig. 1: Illustration of the shortcomings of a linear compliance model. The homogeneous solution was obtained using the algorithm presented in this paper. The particular solution was computed using the linear compliance approach [3]. Contacts have unity stiffness in the normal direction. The stiffness in the frictional direction was set equal to the coefficient of friction. The joints are assumed infinitely stiff. Friction cones are shown in red and corresponding contact forces are shown as blue arrows. The violet arrow denotes the applied force.

Given the subspace of controllable internal forces (see [2]), the particular solution computed in this way can be used to compute a homogeneous solution such that c satisfies all contact constraints. Using Eq. (2) the required equilibrium joint torques that satisfy this system τ_{eq} can then be calculated.

Limitations of this approach: The use of a linear compliance matrix is an important limitation, as it assumes a linear stiffness of the contacts and the joints. However, a contact force may only "push", it cannot "pull" and hence contact forces behave in a nonlinear fashion. Furthermore, a linear compliance model disregards the nonlinearity of frictional forces obeying Coulombs law of friction. We consider contacts of the *point contact with friction* type, which means that the contact force must lie within its friction cone. A linear compliance model has no notion of this friction constraint and thus cannot distribute forces accordingly once the frictional component of a contact force reaches its limit.

To illustrate this issue, consider Fig. 1. Fig. 1a shows a homogeneous solution to a force distribution problem. Fig. 1b shows the sum of the homogeneous and particular solutions when an external wrench pushing the object towards the palm of the robot is applied. Applying a downward force has caused the contact forces on the distal links to violate the friction constraint (they lie outside their respective cones), perhaps leading us to believe that we have to increase the internal forces in the grasp in order to resist the applied wrench. In reality, however, the contacts on the distal links will only apply as much frictional force as they may and more force will be distributed to the contacts on the proximal links. Indeed, experimental results indicate that this grasp withstands arbitrary downward forces applied to the object even in the absence of internal forces.

Furthermore, the compliance of the joints in a robotic hand may be non-linear. Consider the distinction between the equilibrium joint torques τ_{eq}, and those commanded by the motors τ_c. The simplest approach would be to simply set $\tau_c = \tau_{eq}$. However, this approach is subject to the assumptions outlined in the Introduction, requiring that w be known, that the hand have complete authority over all needed joint degrees of freedom (in other words, that one can obtain τ_{eq} by commanding τ_c), and that τ_c can be accurately produced by the motors. A simpler approach, and more commonly used in practice, is to command a τ_c, and rely on $\tau_{eq} \neq \tau_c$ arising through passive reactions. For the large family of hands powered by geared, non-backdrivable motors, at any joint i the resulting torque τ_i can exceed the commanded value, but only passively, in response to the torques τ_j, $j \neq i$ and the wrench w. We state our problem as follows: given commanded torques τ_c, can the system find quasi-static equilibrium for a net resultant object wrench w, assuming passive reaction effects at the joints? In the next section, we present our method for computing an answer, then apply it to the problem of assessing stability for a range of grasps and wrenches.

4   Grasp Model

Due to the above limitations of linear models, we propose a model that accounts for non-linear effects due to the behavior of contact forces and non-backdrivable actuators. To capture the passive behavior of the system in response to external disturbance, we (as others before [9, 2–4]) rely on computing virtual object movements in conjunction with virtual springs placed at the contacts between the rigid object and the hand mechanism. Unlike previous work however, we attempt to also capture effects that are non-linear w.r.t. virtual object movement.

In order to express (linearized) friction constraints at each contact, contact forces are expressed as linear combinations of the edges that define contact friction pyramids, and restricted to lie inside the pyramid:

$$\boldsymbol{D}\boldsymbol{\beta} = \boldsymbol{c} \tag{3}$$

$$\boldsymbol{F}\boldsymbol{\beta} \leq 0 \tag{4}$$

Details on the construction of the linear force expression matrix $\boldsymbol{D}$ and the friction matrix $\boldsymbol{F}$ can be found in the work of Miller and Christensen [12]. Note that while the friction model is linear, in contrast to the linear compliance model discussed in the previous section the frictional forces are not linearly related to virtual object movements. In fact, friction forces are not related to virtual object motion at all. Instead, we propose an algorithm that searches for equilibrium contact forces everywhere inside the friction cones. In this study we use the Point Contact With Friction model, however, the formulation is general enough for other linearized models, such as the Soft Finger Contact [6]. For our purposes, the vector $\boldsymbol{\beta}$, denoting force amplitudes along the edges of the friction pyramids, completely defines contact forces.

Assuming virtual springs placed at the contacts, the normal force at a contact i is determined by the virtual relative motion between the object and the robot hand at that contact in the direction of the contact normal. This can be expressed in terms of virtual object displacements $\boldsymbol{x}$ and virtual joint movements $\boldsymbol{q}$. A subscript n denotes a normal component of a contact force or relative motion. For simplicity, we choose unity stiffness for the virtual contact spring $(k = 1)$.

$$\beta_{i,n} = k(\boldsymbol{G}^T\boldsymbol{x} - \boldsymbol{J}\boldsymbol{q})_{i,n} \tag{5}$$

However, a contact may only apply positive force (it may only push, not pull). Hence, if the virtual object and joint movements are such that the virtual spring is extended from its rest position, the contact force must be zero.

$$\beta_{i,n} \geq 0 \tag{6}$$

$$\beta_{i,n} - k(\boldsymbol{G}^T\boldsymbol{x} - \boldsymbol{J}\boldsymbol{q})_{i,n} \geq 0 \tag{7}$$

$$\beta_{i,n} \cdot (\beta_{i,n} - k(\boldsymbol{G}^T\boldsymbol{x} - \boldsymbol{J}\boldsymbol{q})_{i,n}) = 0 \tag{8}$$

This is a non-convex quadratic constraint and as such not readily solvable. (Note that if re-posed as a Linear Complementarity Problem it produces a non positive-definite matrix relating the vectors of unknowns). However, the same problem can be posed as a set of linear inequality constraints instead, which can be solved by a mixed-integer programming solver.

$$\beta_{i,n} \geq 0 \tag{9}$$

$$\beta_{i,n} \leq k_1 \cdot y_i \tag{10}$$

$$\beta_{i,n} - k(\boldsymbol{G}^T\boldsymbol{x} - \boldsymbol{J}\boldsymbol{q})_{i,n} \geq 0 \tag{11}$$

$$\beta_{i,n} - k(\boldsymbol{G}^T\boldsymbol{x} - \boldsymbol{J}\boldsymbol{q})_{i,n} \leq k_2 \cdot (1 - y_i) \tag{12}$$

Each contact i is assigned a binary variable y_i determining if the normal force at that contact is equal to the force in the virtual spring (for positive spring forces) or zero. Constants k_1 and k_2 are virtual limits that have to be carefully chosen such that the magnitude of the expressions on the left-hand side never exceed them. However, they should not be chosen too large or the problem may become numerically ill-conditioned.

The mechanics of the hand place constraints on the virtual motion of the joints. To clarify this point, consider the equilibrium joint torque τ, at which the system settles, and which may differ from the commanded joint torque τ_c. At any joint j the torque may exceed the commanded value, but only passively. In non-backdrivable hands this means the torque at a joint may only exceed its commanded level if the gearing between the motor and the joint is absorbing the additional torque. In consequence, a joint at which the torque exceeds the commanded torque may not display virtual motion. Defining joint motion, which closes the hand on the object as positive, this constraint can be expressed as another linear complementarity. Similarly to the linear complementarity describing normal contact forces this constraint can be posed as a set of linear inequalities.

$$\tau_j \geq \tau_{c,j} \tag{13}$$

$$\tau_j \leq \tau_{c,j} + k_3 \cdot z_j \tag{14}$$

$$q_j \leq k_4 \cdot (1 - z_j) \tag{15}$$

Each joint is assigned a binary variable z_j that determines if the joint may move or is being passively loaded and hence stationary. Similarly to k_1 and k_2 the constants k_3 and k_4 are virtual limits and should be chosen with the same considerations in mind.

5   Solution Method

The computational price we pay for considering these non-linear effects is that virtual object movement is not directly determined by the compliance-weighted inverse of the grasp map matrix; rather, it becomes part of the complex mixed-integer problem we are trying to solve. In general, if a solution exists, there is an infinite number of solutions satisfying the constraints. The introduction of an optimization objective leads to a single solution. A physically well motivated choice of objective might be to minimize the energy stored in the virtual springs. We formulate a *passive response problem* (or PRP) as outlined in Algorithm 1.

In certain circumstances, this formulation proves to be insufficient. The rigid, passively loaded fingers allow an optimization formulation with unconstrained object movement to "wedge" the object between contacts creating large contact forces. This allows the grasp to withstand very large applied wrenches by performing unnatural virtual displacements (see Fig. 2a). To address this, we constrain the object movement such that motion is only allowed in the direction of the unbalanced wrench acting on the object: $x = sw$, $s \in \mathbb{R}_{\geq 0}$. We replace the objective of the optimization formulation such as to minimize the net resultant wrench $r = w + G^T \beta$ (applied wrench and contact forces). However, under

---

Algorithm 1

Input: $\boldsymbol{\tau}_c$ - commanded joint torques, $\boldsymbol{w}$ - applied wrench
Output: $\boldsymbol{\beta}$ - equilibrium contact forces
procedure PASSIVE RESPONSE PROBLEM$(\boldsymbol{\tau}_c, \boldsymbol{w})$
   minimize: $\boldsymbol{\beta}_n^T \boldsymbol{\beta}_n$ ▷ energy stored in virtual springs
   subject to:
      Eqs. (1)&(2) ▷ equilibrium
      Eq. (4) ▷ friction
      Eqs. (9) − (12) ▷ virtual springs complementarity
      Eqs. (13) − (15) ▷ joint complementarity
   return $\boldsymbol{\beta}$
end procedure

---

this constraint, the solver will generally not be able to completely balance out the wrench in a single step; even after the optimization, some level of unbalanced wrench may remain. To eliminate it, we call the same optimization procedure in an iterative fashion, where, at each step we allow additional object movement in the direction of the unbalanced wrench $\boldsymbol{r}$ remaining from the previous call:

$$\boldsymbol{x}_{next} = \boldsymbol{x} + s\boldsymbol{r}, \quad s \in \mathbb{R}_{\geq 0} \tag{16}$$

After each iteration, we check for convergence by comparing the incremental improvement to a threshold ϵ. If the objective has converged to a sufficiently small net wrench, we deem the grasp to be stable; otherwise, if the objective converges to a larger value, we deem the grasp unstable. Thus, we formulate a *movement constrained passive response problem* as outlined in Algorithm 2 to be solved iteratively as outlined in Algorithm 3.

The computation time of this process is directly related to the number of iterations required until convergence. A single iteration takes of the order of 10^{-2} to 10^{-1} seconds, depending on the complexity of the problem. Most problems

---

Algorithm 2

Input: $\boldsymbol{\tau}_c$ - commanded joint torques, $\boldsymbol{w}$ - applied wrench, $\boldsymbol{x}$ - previous object displacement, $\boldsymbol{r}$ - previous net wrench
Output: $\boldsymbol{\beta}$ - contact forces, $\boldsymbol{x}_{next}$ - next step object displacement, $\boldsymbol{r}_{next}$ - next step net wrench
procedure MOVEMENT CONSTRAINED PRP$(\boldsymbol{\tau}_c, \boldsymbol{w}, \boldsymbol{x}, \boldsymbol{r})$
   minimize: $\boldsymbol{r}_{next}^T \boldsymbol{r}_{next}$ ▷ net wrench
   subject to:
      Eq. (4) ▷ friction
      Eqs. (9) − (12) ▷ virtual springs complementarity
      Eqs. (13) − (15) ▷ joint complementarity
      Eq. (16) ▷ object movement
   return $\boldsymbol{\beta}, \boldsymbol{x}_{next}, \boldsymbol{r}_{next}$
end procedure

Algorithm 3

Input: τ_c - commanded joint torques, w - applied wrench
Output: β - contact forces, r - net resultant
procedure ITERATIVE PASSIVE RESPONSE PROBLEM(τ_c, w)
    $x = 0$
    $r = w$
    loop
        $(\beta, x_{next}, r_{next}) = $ **Movement Constrained PRP**(τ_c, w, x, r)
        if $norm(r - r_{next}) < \epsilon$ **then**   ▷ Check if system has converged
            break
        end if
        $x = x_{next}$
        $r = r_{next}$
    end loop
    return β, r
end procedure

---

converge within less than 50 iterations. All computations were performed on a commodity computer with a 2.80GHz Intel Core i7 processor.

We use this procedure to answer the question if a grasp, in which the joints are preloaded with a certain commanded torque can resist a given external wrench. In much of the analysis introduced in the next section we are interested in how the maximum external wrench, which a grasp can withstand depends on the direction of application. We approximate the maximum resistible wrench along a single direction using a binary search limited to 20 steps, which requires computation time of the order of tens of seconds. In general, investigating the magnitude of the maximum resistible wrench in every direction involves sampling wrenches in 6 dimensional space. Within our current framework this is prohibitively time consuming and hence we limit ourselves to sampling directions in 2 dimensions and then using the aforementioned binary search to find the maximum resistible wrench along those directions.

6   Analysis and Results

We illustrate the application of our method on two example grasps using the Barrett hand. We show force data collected by replicating the grasp on a real hand and testing resistance to external disturbances. We model the Barrett hand as having all non-backdrivable joints. Our qualitative experience indicates that the finger flexion joints never backdrive, while the spread angle joint backdrives under high load. For simplicity we also do not use the breakaway feature of the hand; our real instance of the hand also does not exhibit this feature. We model the joints as rigidly coupled for motion, and assume that all the torque supplied by each finger motor is applied to the proximal joint.

To measure the maximum force that a grasp can resist in a certain direction, we manually apply a load to the grasped object using a Spectra wire in series

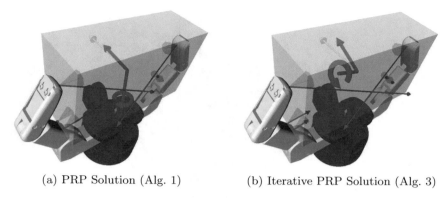

(a) PRP Solution (Alg. 1)   (b) Iterative PRP Solution (Alg. 3)

Fig. 2: Illustration of the shortcomings of directly solving the PRP problem defined above. A force normal to the closing plane of the fingers (illustrated by the green arrow) is applied to the object at its center of mass. The translational and rotational components of the resulting object movement are shown in violet. The PRP algorithm finds a way to wedge the object between the fingers by rotating the object in an unnatural fashion. This enables the solver to find ways to resist arbitrary wrenches. The iterative approach yields the natural finite resistance.

with a load cell (Futek, FSH00097). In order to apply a pure force, the wire is connected such that the load direction goes through the center of mass of the object. We increase the load until the object starts moving, and take the largest magnitude recorded by the load cell as the largest magnitude of the disturbance the grasp can resist in the given direction.

Case 1. We consider first the case illustrated in Fig. 3. Note that this is the same grasp we used in the Problem Statement section to explain the limitations of the linear compliance model. This grasp can be treated as a 2D problem, considering only forces in the grasp plane, simple enough to be studied analytically, but still complex enough to give rise to interesting interplay between the joints and contacts. Since our simulation and analysis framework is built for 3D problems, we can also study out-of plane forces and in-plane moments.

Consider first the problem of resisting an external force applied to the object CoM and oriented along the Y axis. This simple case already illustrates the difference between active and passive resistance. Resistance against a force oriented along positive Y requires active torque applied at the joints in order to load the contacts and generate friction. The force can be resisted only up to the limit provided by the preload, along with the friction coefficient. If the force is applied along negative Y, resistance happens passively, provided through the contacts on the proximal link. Furthermore, this resistant force does not require any kind of preload, and is infinite (up to the breaking limit of the hand mechanism, which does not fall within our scope here).

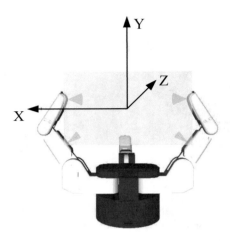

Fig. 3: Grasp example 1.

For an external force applied along the X axis, the problem is symmetric between the positive and negative directions. Again, the grasp can provide passive resistance, through a combination of forces on the proximal and distal links. For the more general case of forces applied in XY plane, we again see a combination of active and passive resistance effects. Intuitively, any force with a negative Y component will be fully resisted passively. However, forces with a positive Y component and non-zero X component can require both active and passive responses. Fig. 4 shows the forces that can be resisted in the XY plane, both predicted by our framework and observed by experiment. Note that our formulation predicts the distinction between finite and infinite resistance directions, in contrast to the results obtained using the linear compliance model.

For both real and predicted data, we normalize the force values by dividing with the magnitude of the force obtained along the positive direction of the Y axis (note thus that both predicted and experimental lines cross the Y axis at y=1.0). The plots should therefore be used to compare trends rather than absolute values. We use this normalization to account for the fact that the absolute torque levels that the hand can produce, and which are needed by our formulation in order to predict absolute force levels, can only be estimated and no accurate data is available from the manufacturer. The difficulty in obtaining accurate assessments of generated motor torque generally limits the assessments we can make based on absolute force values.

Moving outside of the grasp plane, Fig. 5 shows predicted and measured resistance to forces in the XZ plane. Again, we notice that some forces can be resisted up to arbitrary magnitudes thanks to passive effects, while others are limited by the actively applied preload.

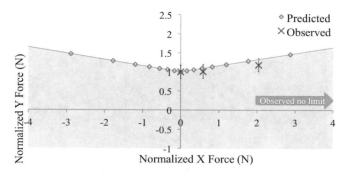

Fig. 4: Normalized forces in the XY plane that can be resisted by grasp example 1: observed by experiment (mean ± one standard deviation) and predicted by our framework (normalized as explained in the text). In all directions falling below the blue line, the prediction framework hit the upper limit of the binary search (arbitrarily set to 1.0e3 N). Hence we deem forces in the shaded area resistible. In the direction denoted by "Observed no limit", the grasp was not disturbed even when hitting the physical limit of the testing setup.

Case 2. One advantage of studying the effect of applied joint torques on grasp stability is that it allows us to observe differences between different ways of preloading the same grasp. For example, in the case of the Barrett hand, choosing at which finger(s) to apply preload torque can change the outcome of the grasp, even though there is no change in the distribution of contacts. We illustrate this approach on the case shown in Fig. 6. We compare the ability of the grasp to resist a disturbance applied along the X axis in the negative direction if either finger 1 or finger 2 apply a preload torque to the grasp. Our formulation predicts that by preloading finger 1 the grasp can resist a disturbance that is 2.37 times higher in magnitude than if preloading finger 2. Experimental data (detailed in Table 1) indicates a ratio for the same disturbance direction of 2.23. The variance in measurements again illustrates the difficulty of verifying such simulation results with experimental data. Nevertheless, experiments confirmed that preloading finger 1 is significantly better for this case.

This result can be explained by the fact that, somewhat counter-intuitively, preloading finger 1 leads to larger contact forces than preloading finger 2, even if the same torque is applied by each motor. Due to the orientation of finger 1, the contact force on finger 1 has a smaller moment arm around the finger flexion axes than is the case for finger 2. Thus, if the same flexion torque is applied in turn at each finger, the contact forces created by finger 1 will be higher. In turn, due to passive reaction, this will lead to higher contact forces on finger 2, even if finger 1 is the one being actively loaded. Finally, these results hold if the spread degree of freedom is rigid and does not backdrive; in fact, preloading finger 1 leads to a much larger passive (reaction) torque on the spread degree of freedom than when preloading finger 2.

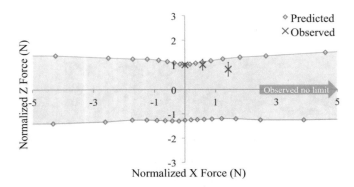

Fig. 5: Normalized forces in the XZ plane that can be resisted by grasp example 1: predicted by our framework, and observed by experiment. In all directions falling between the blue lines (shaded), the prediction framework hit the upper limit of the binary search (arbitrarily set to 1.0e3 N). In the direction denoted by "Observed no limit", the grasp was not disturbed even when hitting the physical limit of the testing setup.

	Measured resistance				Predicted	
	Values(N)	Avg.(N)	St. Dev.	Ratio	Value(N)	Ratio
F1 load	12.2, 10.8, 7.5, 7.9, 9.3	9.6	1.9	2.23	16.5	2.37
F2 load	3.7, 4.1, 5.0	4.3	0.7		7.0	

Table 1: Predicted and measured resistance to force applied along the negative X axis in the grasp problem in Fig. 6. Each row shows the results obtained if the preload is applied exclusively by finger 1 or finger 2 respectively. Experimental measurements were repeated 5 times for finger 1 (to account for the higher variance) and 3 times for finger 2. Predicted values are also shown both in absolute force, and ratio between the two preload cases.

7 Discussion

Limitations Our iterative approach allows us to constrain virtual object movement to the successive directions of unbalanced wrenches. However, such an iterative approach is not guaranteed to converge, or to converge to the physically meaningful state of the system. In a subset of cases, the solver reports large resistible wrenches relative to neighboring states; for example, in the grasp Case 2 from the previous section (Fig. 6), when computing resistance to disturbances sampled from the positive X half plane of the XY plane (Fig. 7), we obtain an outlier in the case of Finger 1 preload that does not follow the trend of the surrounding points. It is possible that the numerical setup of the problem for that particular direction allows the solver to "wedge" the object into the grasp, increasing resistance to the external wrench. These effects will require further

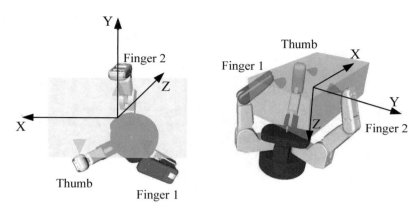

Fig. 6: Top and side views for grasp example 2 also indicating finger labels. Note that the spread angle degree of freedom of the Barrett hand changes the angle between finger 1 and finger 2; the thumb is only actuated in the flexion direction.

investigation. A promising avenue we are exploring is to also include energy constraints in the core solver, requiring that the energy used responding to an external wrench equal that provided by the wrench itself.

Furthermore, we would like to analyze the effect of uncertainties (e.g. in exact contact location) on our model. We believe exploring the sensitivity of the model to such uncertainties may yield many valuable insights.

Alternative Approaches As was described in the Problem Statement, a simpler alternative is to disregard non linear effects w.r.t. virtual object movement, i.e. assume that the joints are fixed (thus joint torque can both increase and decrease passively), and that friction forces also behave in spring-like fashion. The price for this simplicity is, that the results may not be physically sound.

At the other and of the spectrum, our iterative approach allowing successive virtual object movements in the direction of the net resultant wrench shares some of the features of a typical time-stepping dynamics engine. One could therefore forgo the quasi-static nature of our approach, assume that unbalanced wrenches produce object acceleration or impulses, and perform time integration to obtain new object poses. This approach can have additional advantages: even an unstable grasp can eventually transform into a stable one, as the object settles in the hand; a fully dynamic simulation can capture such effects. However, in highly constrained cases, such as grasps at or near equilibrium, any inaccuracy can lead to the violation of interpenetration or joint constraints, in turn requiring corrective penalty terms which add energy to the system. Our quasi-static approach only attempts to determine if an equilibrium can exist in the given state, and thus only reasons about virtual object movements, without dynamic effects.

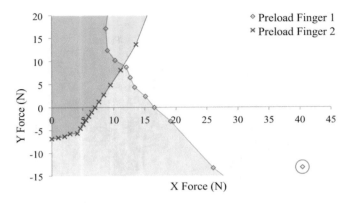

Fig. 7: Forces in an XY halfplane that can be resisted by grasp example 2 (shaded) as predicted by our framework, depending on which finger is preloaded. An outlier prediction is circled in orange.

8   Conclusions

In this paper, we have introduced an algorithm that aims to answer what we believe to be not only a meaningful theoretical question, but also one with important practical applications: *once a given joint preload has been achieved, can a grasp resist a given wrench passively, i.e. without any change in commanded joint torques?* In the inner loop of a binary search, the same algorithm allows us to determine the largest magnitude that can be resisted for a disturbance along a given direction.

In the examples above we show how the actively set joint preload combines with passive effects to provide resistance to external wrenches; our algorithm captures these effects. Furthermore, we can also compute how preloads set for some of the hand joints can cause the other joints to load as well, and the combined effects can exceed the intended or commanded torque levels. We can also study what subset of the joints is preferable to load with the purpose of resisting specific disturbances.

Our directional goal is to enable practitioners to choose grasps for a dexterous robotic hand knowing that all disturbances they expect to encounter will be resisted without further changes in the commands sent to the motors. Such a method would have wide applicability, to hands that are not equipped with tactile or proprioceptive sensors (and thus unable to sense external disturbances) and can not accurately control joint torques, but are still effective thanks to passive resistance effects.

In its current form, the algorithm introduced here can answer "point queries", for specific disturbances or disturbance directions. However, its computational demands do not allow a large number of such queries to be answered if a grasp is to be planned at human-like speeds; furthermore, the high dimensionality of the complete space of possible external wrenches generally prevents sampling approaches. GWS-based approaches efficiently compute a global measure of wrenches that can be resisted assuming perfect information and controllability of contact forces. We believe passive resistance has high practical importance for the types of hands mentioned above, but no method is currently available to efficiently distill passive resistance abilities into a single, global assessment of the grasp. We will continue to explore this problem in future work.

References

1. Aicardi, M., Casalino, G., Cannata, G.: Contact force canonical decomposition and the role of internal forces in robust grasp planning problems. International Journal of Robotics Research 15(4), 351–364 (1996)
2. Bicchi, A.: Force distribution in multiple whole-limb manipulation. In: Robotics and Automation, 1993. Proceedings., 1993 IEEE International Conference on. pp. 196–201. IEEE (1993)
3. Bicchi, A.: On the problem of decomposing grasp and manipulation forces in multiple whole-limb manipulation. Robotics and Autonomous Systems 13(2), 127–147 (1994)
4. Bicchi, A.: On the closure properties of robotic grasping. The International Journal of Robotics Research 14(4), 319–334 (1995)
5. Burdick, J., Rimon, E.: Wrench resistant multi-finger hand mechanisms. In: International Conference on Robotics and Automation (2016)
6. Ciocarlie, M., Lackner, C., Allen, P.: Soft finger model with adaptive contact geometry for grasping and manipulation tasks. In: Joint Eurohaptics Conference and IEEE Symp. on Haptic Interfaces. pp. 219–224 (2007)
7. Cutkosky, M.R., Kao, I.: Computing and controlling the compliance of a robotic hand. IEEE Transactions on Robotics and Automation 5(2) (1989)
8. Ferrari, C., Canny, J.: Planning optimal grasps. In: IEEE International Conference on Robotics and Automation. pp. 2290–2295 (1992)
9. Hanafusa, H., Asada, I.: Stable prehension by a robot hand with elastic fingers. In: Proc. of the 7th ISIR, Tokyo (1977)
10. Kerr, J., Roth, B.: Analysis of multifingered hands. International Journal of Robotics Research 4(4), 3–17 (1986)
11. Melchiorri, C.: Multiple whole-limb manipulation: An analysis in the force domain. Robotics and Autonomous Systems 20(1), 15–38 (1997)
12. Miller, A., Christensen, H.: Implementation of multi-rigid-body dynamics within a robotic grasping simulator. In: IEEE Intl. Conference on Robotics and Automation. pp. 2262–2268 (2003)
13. Prattichizzo, D., Trinkle, J.: Grasping. Springer Handbook of Robotics (2008)
14. Salisbury, J., Roth, B.: Kinematic and force analysis of articulated mechanical hands. ASME Journal of Mechanisms, Transmissions, and Automation in Design 105, 35–41 (1983)
15. Yoshikawa, T.: Passive and active closures by constraining mechanisms. In: IEEE International Conference on Robotics and Automation. vol. 2, pp. 1477–1484 (1996)
16. Yoshikawa, T., Nagai, K.: Manipulating and grasping forces in manipulation by multifingered robot hands. IEEE Transactions on Robotics and Automation 7(1), 67–77 (1991)

Robust Planar Dynamic Pivoting by Regulating Inertial and Grip Forces

Yifan Hou, Zhenzhong Jia, Aaron M. Johnson, and Matthew T. Mason

Robotics Institute, Carnegie Mellon University
5000 Forbes Ave., Pittsburgh, PA 15213, USA
yifanh@cmu.edu, zhenzjia@cmu.edu, amj1@cmu.edu, matt.mason@cs.cmu.edu
http://mlab.ri.cmu.edu

Abstract. In this paper, we investigate the planar *dynamic pivoting* problem, in which a pinched object is reoriented to a desired pose through wrist swing motion and grip force regulation. Traditional approaches based on friction compensation do not work well for this problem, as we observe the torsional friction at the contact has large uncertainties during pivoting. In addition, the discontinuities of friction and the lower bound constraint on the grip force all make dynamic pivoting a challenging task for robots. To address these problems, we propose a robust control strategy that directly uses friction as a key input for dynamic pivoting, and show that active friction control by regulating the grip force significantly improves system stability. In particular, we embed a Lyapunov-based control law into a quadratic programming framework, which also ensures real-time computational speed and the existence of a solution. The proposed algorithm has been validated on our dynamic pivoting robot that emulates human wrist-finger configuration and motion. The object orientation can quickly converge to the target even under considerable uncertainties from friction and object grasping position, where traditional methods fail.

Keywords: Robot, Manipulation, Pivoting, Friction, Robust Control

1 Introduction

Compared with humans, robots have limited dexterity. In particular, certain tasks that are simple for humans can be quite challenging for robots. Beyond the dexterity of human hand, one important advantage of human manipulation is the use of a richer set of force resources (*extrinsic dexterity* [1]), including gravity, inertial forces, contact forces and friction. One such example is regrasping - to shift an object from one grasp pose to another. For robots, a common solution is to put the object on the ground or in a fixture, then move the robot hand to the desired pose and grasp again. The human hand, however, often employs a more direct approach: in-hand regrasp through arm, wrist, and finger motions without breaking contact.

In this paper, we study *dynamic pivoting* – a common nonprehensile regrasping manipulation. One example is shown in Fig. 1: the hand holds a cellphone

© Springer Nature Switzerland AG 2020
K. Goldberg et al. (Eds.): *Algorithmic Foundations of Robotics XII*, SPAR 13, pp. 464–479, 2020.
https://doi.org/10.1007/978-3-030-43089-4_30

Fig. 1. Human pivots an in-hand object in the horizontal plane (top view).

between two finger tips, and rotates it to a desired angle relative to the finger by varying grip force and wrist rotation. Pivoting is a simple way to orient objects in the hand, and is faster than pick-and-place [2]. This operation is interesting because a human can modulate friction force and switch the contact mode between left sliding, right sliding and sticking, which is a hybrid control strategy.

This paper investigates techniques that enable a robot to perform human-like dynamic pivoting. The task is challenging for robots, because the contact brings three problems: [3–5].

The *Modeling Uncertainty.* On the one hand, we may not exactly know which area on the object is grasped, due to slipping or noise in the initial grasp. On the other hand, precise contact friction modeling is usually not realistic [6], so we have to tolerate some degree of uncertainty with practical friction models. Dynamic pivoting has even more frictional uncertainty due to noise in grip force control during fast motion. Recent work indicates that for certain robotic applications, a very detailed friction model is unnecessary [7–9]. However, a closed-loop strategy is preferred in these cases.

Positive lower bound constraint on grip force. We need a positive contact normal force to maintain grasping and reduce slip. In our preliminary simulations, however, a traditional nonlinear controller often commands large negative grip force when trying to pull the system back from error. If we truncate the grip force to satisfy the constraint, the traditional controller no longer works.

Friction discontinuity. We cannot assume a certain sliding direction, since the closed-loop system will need to fight against the position error in both directions during pivoting. As detailed in section 3.2, the stiction phenomenon introduces several different continuous modes, thereby directly introducing discontinuities into the gain matrix of the system.

In this work we propose a robust control strategy for robotic dynamic pivoting that addresses the above issues. A control law based on sliding mode control (SMC) calculates wrist torque and grip force within each continuous friction mode. The sliding condition, which leads to Lyapunov convergence, is imposed by a soft constraint and solved under control saturation as hard constraints. Following a hybrid system routine, when transition to multiple modes is possible, we add constraints to prevent undesired mode transitions when solving for a specific mode.

In particular, our algorithm:

- works with a discontinuous friction model;
- satisfies control saturation constraints including the positive lower bound;
- converges in experiments even when friction modeling are simplified and imprecise.

The paper is organized as follows. Section 2 describes the related work. Section 3 introduces the robotic pivoting system and modeling highlights. Section 4 presents the control strategy design. Section 5 presents the implementation and experimental results. Finally, Section 6 gives the summary and directions for future work.

2   Related Work

2.1   Pivoting

Rao *et al.* [10] were the first to use the term "pivoting" for the rotation of a grasped object relative to the contact point with fingers. Since there is no active joint at the contact, there must be an extrinsic source of actuation that drives the object, i.e. the *extrinsic dexterity* [1]. One such source is gravity [10, 2, 11, 9, 8, 12]. Brock calculated possible twists for an object in a multi-fingered hand under gravity by maximizing the virtual work [11]. Rao *et al.* [10], described how to choose a grasp so that after lifting up polyhedral objects from the ground, the object rotates to a desired stable pose under gravity. Holladay *et al.* [2], extended [10] by planning a whole trajectory for the gripper with the consideration of dynamics, and utilizing contact with the ground to discretize the final poses.

Gravity is a good source of actuation, for its perfectly precise direction and magnitude. The disadvantage is also obvious: its direction cannot change. We refer to those works as *gravity pivoting* in this paper.

Pivoting can be done by making contact with environment [1, 13]. We call them *contact pivoting*. Chavan-Dafle *et al.* implemented open-loop contact pivoting [13], where a firm grasp was maintained all the time during sliding. Contact pivoting is shown to be reliable in slow motion, as the object position can be inferred from the contact position.

We use inertial force as source of actuation, and call it *inertial pivoting*. Like gravity, inertial force does not rely on contact with the environment; however its direction is also controllable. Shi *et al.* [7], proposed an open-loop strategy for a three-DOF planar sliding problem, where an object grasped by a parallel gripper slides under inertial force and gravity. This strategy, though verified in simulation, showed notable error in experiments; the reason could be the lack of feedback, according to the authors. When the object is treated as an additional link of the robot with frictionless joint, pivoting reduces to a passive last joint manipulator problem, where partial feedback linearization is shown to be successful [14–18]. These approaches are extended in this paper in two ways. Firstly, we add a robust control term to the feedback linearization control law, so as to explicitly tackle the non-trivial, uncertain contact friction and some amount of slip during pivoting. Secondly, we utilize the grip force as an additional source of control, which brings notable stability improvement as well as new difficulties in controller design.

There are very few studies using contact normal force to control friction directly, except for vibration suppression [19]. Closely related to our work, Via *et al.* [8, 9] performed controlled *gravity pivoting* to a spoon by controlling the grip force, and closed the loop with vision feedback. Robust control [8], and adaptive control techniques [9] were used to ensure convergence under friction uncertainty. They also improved the performance by adopting a more precise soft finger model. The controller was verified on a parallel gripper, for which the grip force control was implemented by compressing soft finger with tactile feedback. Sintov *et al.* [12] used optimal control to solve the *gravity pivoting* control problem by linearizing the dynamics for one mode. They also designed a strategy to swing the object up above the desired angle. The main limitation for both work is the dependency on gravity, which makes it hard to recover from overshoot.

Pivoting is a typical example of nonprehensile manipulation (except [13]), where the object is manipulated without a firm grasp. Analysis of nonprehensile manipulation dates back to the 1980s. A more thorough list of nonprehensile manipulation can be found in [20, 21].

2.2   Friction Modeling

Friction determines the interaction between robots and grasped objects in pivoting [13, 7, 11]. The tribology community has extensive researches on precise friction modeling [22]. Static friction models treat friction as a memoryless function of contact normal force, contact sliding velocity and external force, [5]. More detailed static friction phenomena and modeling can be found in [5]. In the robotics community, Goyal analyzed the Coulomb friction in rigid body 2D planar sliding, described the relation between the wrench acting on the object and 3D twist of the object using *limit surface* concept. Zhou *et al.* proposed a polynomial approximation of the limit surface [23], and a method for fast identification from pushing experiments.

The discontinuity and lack of expressiveness of static friction models motivates dynamic friction modeling [5, 6, 4], which provides smooth friction behavior even during friction direction transitions. Dynamic friction models use one or more hidden state variables to describe microscopic asperities in contact [6, 4]. Complicated friction models provide a more precise description of friction phenomena. The cost is more effort and more data required for parameter estimation. In this work, we will stick to a static model while relying on hardware design to minimize unexpected dynamic frictional behavior.

3   Modeling

3.1   Robot Hardware

Fig. 2 shows our robot tailored for this task by emulating human operation. The wrist joint motor generates rotational torque and provides inertial force to the hand assembly. The top motor generate grip force through a lever mechanism. The bottom fingertip, which is mounted on a low-friction small-inertia shaft, contacts the object with high friction rubber and rotates with it. The rotational angles of the object is thus measured by an encoder on the shaft. The lower finger is installed on a loadcell to provide grip force feedback.

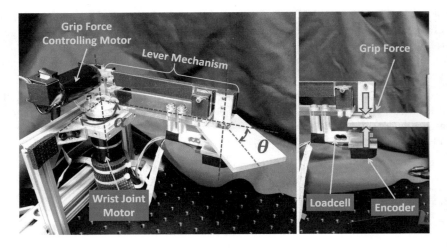

Fig. 2. The robot and gripper designed for dynamic pivoting.

The top finger provides frictional torque through a piece of hard fingertip. It is chosen to be hard and thin to minimize spring-like stiction behavior [5], which is reported as troublesome for pivoting in [9]. Large stiction makes it hard to predict when sliding will occur. The size of the fingertip is critical. A smaller contact area will result in a more flat limit surface [24], which makes rotation easier than slip. The fingertip should also be large enough to provide frictional torque in order to handle the object's momentum. It is also important to select the right material so that a non-trivial range of friction can be provided by the grip force. In our experiments, a 11mm-diameter round piece cut from a 0.8mm-thick Teflon sheet is used.

To drive the wrist joint, we choose a Maxon RE-40 DC motor (with a 4.3:1 gearbox) operated in current control mode with 0.8Nm maximum continuous torque output. The torsional friction in this joint is modeled and compensated as constant stiction plus viscous friction. The grip force is produced by a current controlled Maxon RE-36 DC motor. Through a lever mechanism, the motor can provide a maximum of 40N grip force, while staying close to the wrist axis and contributing a small moment of inertia. There is a significant hysteresis nonlinearity in motor stall torque; hence we use a 5kg loadcell to provide grip force feedback, then close the loop with a PI plus feed-forward controller. Typical response time of the grip force control is around 25ms, which is limited by the latency in loadcell reading.

3.2   Two Link Model

The following assumptions are made on the pivoting robot shown in Fig. 2:

1. Dimensional/inertial properties of the robot and the object are known.
2. The robot wrist joint axis is parallel to gravity. The object always stays within the horizontal plane during motion. Consequently, the gravity does not affect the rotation of the robot or the object.
3. The object is initially grasped at rest, but our knowledge of the grasping position may not be exact.
4. The object may have translational slip during motion, but will not slip off the gripper.

Denote α as the wrist joint angle. Instead of coping with a known object with uncertain position, we model the contact as pin joint (call it *pivoting joint*, joint variable denoted by θ), and treat the object as an additional link whose inertia properties have uncertainty. Denote $\mathbf{x} = [\theta, \alpha]^T$ as the joint state vector. The Lagrange dynamics of the whole 2-DOF system are:

$$M(\mathbf{x})\ddot{\mathbf{x}} + C(\dot{\mathbf{x}}, \mathbf{x})\dot{\mathbf{x}} + N(\mathbf{x}) = \begin{pmatrix} \tau_f(\dot{\mathbf{x}}, N_f) \\ \tau \end{pmatrix}, \tag{1}$$

where the joint torque vector consists of wrist torque τ and contact frictional torque τ_f. For friction modeling we use Coulomb friction plus stiction, which is a trade-off between accurancy and simplicity. Then the frictional torque can be related to the contact normal force N_f by [5]:

$$\tau_f = \begin{cases} -\mu N_f \mathrm{sgn}(\dot{\theta}) & \text{if } \dot{\theta} \neq 0 \\ -F_e & \text{if } \dot{\theta} = 0 \text{ and } |F_e| \leq \mu N_f \\ -\mu N_f \mathrm{sgn}(F_e) & \text{otherwise} \end{cases}, \tag{2}$$

where F_e is the external torque [5] acting on the contact. Note the normal force is subjected to unilateral constraint:

$$N_f \geq N_f^{(\mathrm{low})} > 0 \tag{3}$$

Denote control vector as $\mathbf{u} = [N_f, \tau]^T$. We can express the hybrid dynamics (1) with a compact form:

$$\ddot{\mathbf{x}} = F(\mathbf{x}, \dot{\mathbf{x}}) + B(\mathbf{x}, \dot{\mathbf{x}})\mathbf{u}. \tag{4}$$

We suppress the arguments $\mathbf{x}, \dot{\mathbf{x}}, \ddot{\mathbf{x}}$ in what follows for conciseness. For slipping mode $|\dot{\theta}| \neq 0$, we have:

$$F = -M^{-1}(C + N), \quad B = B_{\text{Slipping}} := M^{-1} \begin{bmatrix} -\mu\,\text{sgn}(\dot{\theta}) & 0 \\ 0 & 1 \end{bmatrix}. \tag{5}$$

The gain matrix B is of full rank, so feedback linearization is possible. Similarly, when $\dot{\theta} = 0$ and $|F_e| > \mu N_f$, the dynamics are still of the form in (4), but with:

$$B = B_{\text{ToSlip}} := M^{-1} \begin{bmatrix} -\mu\,\text{sgn}(F_e) & 0 \\ 0 & 1 \end{bmatrix}. \tag{6}$$

During sticking, however, the system reduces to 1D. Denote m_c as the momentum of inertia of the whole assembly, the dynamics satisfies:

$$F = \mathbf{0}, \quad B = B_{\text{Sticking}} := \begin{bmatrix} 0 & 0 \\ 0 & m_c^{-1} \end{bmatrix} \tag{7}$$

Here the zeros are vectors of suitable size. In practice, we replace the condition $\dot{\theta} = 0$ by a range $|\dot{\theta}| < \xi$, where ξ describes the noise level in angular velocity measurement.

3.3   Uncertainty Analysis

Our experiment shows that there is a considerable amount of uncertainty in friction, which is a compound result of a simple friction model, non-perfect friction parameter estimation, and the noise in grip force control. This uncertainty directly affects the gain matrix B in (4). Another source of uncertainty is the grasping position. In our pin joint model, this uncertainty will affect the inertia matrix M in the Lagrange dynamics (1), which will eventually affect both F and B in (4). The influence of all other uncertainty sources, including measurement noise, can be modeled as an uncertainty in F. Denoting by $\hat{F}$ and $\hat{B}$ our estimation of F and B, respectively, we can describe the bounded uncertainty as:

$$\begin{aligned} \hat{F} = F + \Delta_F & \qquad |\Delta_F| \prec \delta_F \\ B = \Delta_B \hat{B} & \qquad |\Delta_B - I| \prec \delta_B. \end{aligned} \tag{8}$$

where $|\cdot|$, $\prec$ denote element-wise absolute value and inequality. $\delta_F \succ \mathbf{0}, \delta_B \succ \mathbf{0}$ are estimated error bounds. With this notation, the true dynamics (4) can be expressed by the estimated model with bounded uncertainty as:

$$\ddot{\mathbf{x}} = \hat{F}(\mathbf{x}, \dot{\mathbf{x}}) - \Delta_F + \Delta_B \hat{B}(\mathbf{x}, \dot{\mathbf{x}})\mathbf{u}. \tag{9}$$

Note that if Δ_B has non-zero off-diagonal component we can decompose it into diagonal and off-diagonal terms, $\Delta_B = I_D + O_D$. Then, move the off-diagonal terms out of gain matrix:

$$\ddot{\mathbf{x}} = \hat{F} - (\Delta_F - O_D \hat{B}\mathbf{u}) + I_D \hat{B}\mathbf{u},$$

and treat the quantity inside the parentheses as the new $\varDelta_F$, which is still bounded. Consequently, we consider $\varDelta_B$ to be diagonal from now on:

$$\varDelta_B = \begin{bmatrix} d_{b1} & 0 \\ 0 & d_{b2} \end{bmatrix}, d_{b1}, d_{b2} > 0. \tag{10}$$

It is not trivial to estimate an error bound on $\hat{F}$ or $\hat{B}$, as the error is a compound result of multiple sources of uncertainty. Instead, we treat them as parameters and tune them according to experimental results.

4   Robust Controller Design

4.1   Robust Controller Design Within a Continuous Mode

Within any certain mode, the system is continuous with uncertainties described in Section 3.3. Before considering constraints, we can use the sliding mode control (SMC) [25] to solve the unconstrained problem, for its ability to converge under bounded uncertainty. Denote $\mathbf{x}_r(t)$ as a smooth reference state trajectory, the control task is to make the tracking error $\mathbf{x}(t) - \mathbf{x}_r(t)$ converge to zero. The 2-D sliding mode $\mathbf{s} = [s_1, s_2]^T$ is defined as:

$$\mathbf{s}(t) = G_D \dot{\tilde{\mathbf{x}}}(t) + G_P \tilde{\mathbf{x}}(t) + G_I \int_0^t \tilde{\mathbf{x}}(\tau) d\tau, \tag{11}$$

where $\tilde{\mathbf{x}}(t) = \mathbf{x}(t) - \mathbf{x}_r(t)$ is the state tracking error, and G_P, G_I, and G_D are diagonal positive definite coefficient matrices. When the system stays on the *sliding surface* $\mathbf{s} = \mathbf{0}$, equation (11) indicates that $\tilde{\mathbf{x}}(t)$ will converge to zero exponentially. Thus the control problem for the original system is equivalent to the problem of stabilizing $\mathbf{s}$, which is only a first-order system described by:

$$\dot{\mathbf{s}} = G_D F + \tilde{G} + G_D B \mathbf{u}, \quad \tilde{G} := -G_D \ddot{\mathbf{x}}_r + G_P \dot{\tilde{\mathbf{x}}} + G_I \tilde{\mathbf{x}} \tag{12}$$

Use the following Lyapunov function:

$$V = \frac{1}{2} \mathbf{s}^T \mathbf{s}, \tag{13}$$

And choose the controller structure to be a feedback linearization term plus a robust control term: (the measured $\hat{B}, \hat{F}$ are described in (8))

$$\mathbf{u} = (G_D \hat{B})^{-1} (-G_D \hat{F} - \tilde{G} - K \operatorname{sgn}(\mathbf{s})), \tag{14}$$

where K is the gain matrix: $K = \begin{bmatrix} k_1 & 0 \\ 0 & k_2 \end{bmatrix} \succ 0$, Now we can express $\dot{V}$ as:

$$\begin{aligned} \dot{V} &= \mathbf{s}^T \dot{\mathbf{s}} \\ &= \mathbf{s}^T \left(-G_D \varDelta_F + G_D (I - \varDelta_B) \hat{F} + (I - \varDelta_B) \tilde{G} \right) - \mathbf{s}^T \varDelta_B K \operatorname{sgn}(\mathbf{s}) \end{aligned} \tag{15}$$

The function V becomes a Robust Control Lyapunov Function (RCLF) and guarantees convergence if there exists $\mathbf{u}$ to make its derivative negative under all possible uncertainties:

$$\dot{V} < -\eta||\mathbf{s}||, \qquad \forall |\Delta_F| \prec \delta_F, |\Delta_B - I| \prec \delta_B. \tag{16}$$

This is called *the sliding condition* in sliding control literature [25], as it ensures $\mathbf{s}$ converges to sliding surface exponentially. The sliding condition is satisfied if

$$D_{\text{low}}\mathbf{k} > C_{\text{up}}. \tag{17}$$

where $\mathbf{k} = [k_1, k_2]^T$, $D_{\text{low}}, C_{\text{up}}$ are lower bound and upper bound of

$$D = \begin{bmatrix} |s_1||d_{b1}| \\ |s_2||d_{b2}| \end{bmatrix}^T,$$
$$C = \mathbf{s}^T \left(-G_D \Delta_F + G_D(I - \Delta_B)\hat{F} + (I - \Delta_B)\tilde{G} \right) + \eta||\mathbf{s}||. \tag{18}$$

Here we use the fact that G_D, Δ_B are diagonal to simplify the derivation. The bound can be obtained by linear programming over Δ_F and Δ_B. In traditional sliding control, we solve equation (17) for $\mathbf{k}$, calculate controls $\mathbf{u}$ from (14). However, in pivoting we also need to consider control saturation constraints:

$$N_f^{(\text{low})} < N_f < N_f^{(\text{high})},$$
$$\tau^{(\text{low})} < \tau < \tau^{(\text{high})}. \tag{19}$$

where the *positive lower bound* N_{low} is causing problem. In simulation the controller often produces negative grip force, and the control would fail if we truncate grip force to satisfy saturation constraints. Instead of direct truncation, we need to leverage wrist rotation more when grip force can not attain a desired value, i.e. find a solution to for both (17) and (19). Unfortunately, the two constraints together are infeasible, if we do not have a tight uncertainty bound in Δ_B, Δ_F. In practice we enforce (17) by soft constraint and solve a constrained optimization at each time step, similar with the optimization performed in [26]. We solve for the two-dimensional gain $\mathbf{k}$ by:

$$\min_{\mathbf{k}} \quad (D_{\text{low}}\mathbf{k} - C_{\text{up}})^2 + w\mathbf{k}^T\mathbf{k}, \tag{20}$$

with (19) as the only constraint. The second term is a regularization term. From (14), we know $\mathbf{u}$ is linear in gain vector $\mathbf{k}$. Hence, the saturation constraints are linear on $\mathbf{k}$. Therefore, we end up with a quadratic programming problem that can be efficiently solved by an off-the-shelf QP solver. The overall computation time, including solving the LP and QP, is less than 1ms for each control loop.

The optimization formulation above sacrifices robust convergence guarantee for feasibility. However, in experiments we still obtain good convergence, indicating the worst case guarantee is unnecessary in our case.

4.2   Control Strategy Among Different Modes

We design one controller for each mode. A such controller will not make sense if it drives the system to any other modes. This could happen when the pivoting velocity $\dot{\theta}$ equals zero, as shown in equation (2). Depending on the external torque F_e, the contact dynamics can end up in one of three possible modes:

- Rotating with $\ddot{\theta} > 0$, if $F_e > \mu N_f$;
- Rotating with $\ddot{\theta} < 0$, if $F_e < -\mu N_f$;
- Sticking, $\ddot{\theta} = 0$, if $|F_e| < \mu N_f$.

To resolve this ambiguity, we solve each of the three modes with the condition above as additional constraints. Then we just pick the solution with optimal cost. In hybrid systems theory, the additional condition is called *guard condition* [27]. Note F_e is linear in $\mathbf{u}$, thus the guard conditions are linear in $\mathbf{k}$, and the problem is still quadratic programming.

The last issue is when to stop the controller. We observe in experiments that the closed-loop system has small-amplitude oscillations around the goal. To stop the oscillation, we set the goal region to be $|\theta - \theta_{\text{goal}}| < \sigma$, and stop the controller as soon as the object stops within this region. The overall algorithm at each control time step is described as follows:

1. If $|\dot{\theta}| > \xi$, solve the corresponding mode for control.
2. While $|\dot{\theta}| <= \xi$,
   (a) If $|\theta - \theta_{\text{goal}}| < \sigma$, stop the control loop and apply the maximum grip force with zero wrist torque.
   (b) Otherwise, solve the three possible modes under guard condition respectively, and pick the one with the best cost value.

5   Experiments

The proposed algorithm is implemented on the hardware described in Section 3.1. The robust control loop runs at 50Hz. The object to be rotated is an acrylic board with a protective paper cover.

5.1   System Identification and Parameter Tuning

Robot mass	700g	Robot moment of inertia	$8.9 \times 10^{-3} \text{kgm}^2$
Object mass	44g	Object moment of inertia	$8.96 \times 10^{-5} \text{kgm}^2$
Length of wrist link	0.16m	Contact friction coefficient μ	$4.5 \times 10^{-4} \text{m}$
Grip force range	4N ~ 15N	Wrist joint torque range	$-0.5\text{Nm} \sim 0.5\text{Nm}$

Table 1. Physical properties of the robot and the object.

The inertia parameters of the robot are identified offline from torque-speed profile. To measure the friction coefficient between the object and the finger, we fix the wrist joint and apply a certain grip force on the object. Then we give the object an initial rotational velocity and record the deceleration curve. The friction coefficient estimated are shown in Fig. 3. An affine relation is fitted, with a rate representing the Coulomb friction coefficient. Physical parameters and actuation constraints are listed in table 5.1. Note all frictional coefficients are torsional.

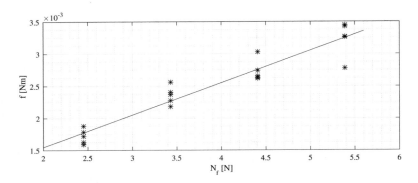

Fig. 3. Measured torsional friction f under different grip forces N_f.

Now we briefly explain how to tune parameters. δ_B can be estimated from friction measuring data, and here we use $\delta_B = \begin{bmatrix} 0.4 \\ & 0.1 \end{bmatrix}$. The performance is not sensitive to η, we pick $\eta = 5$. Next, start with $\delta_F = 0$, we firstly tune the gains G_P, G_I, G_D as if we are tuning a normal PID controller. When performance is peaked, go back to tune δ_F. Repeat the last two steps until satisfactory performance is obtained.

5.2   Experiment I: Pivoting under Grasping Position Uncertainty

In real-life manipulation, the object may not be grasped exactly at the expected position. The ability to endure grasping position uncertainty is crucial to pivoting control. Here we implement and compare our method with two baselines: The first baseline controller is based on partial feedback linearization (PFL), which is a typical approach in the passive manipulator literature [14–18]. In this controller, only wrist joint torque is used as control. The control is chosen such that the dynamics of the passive joint (i.e. the object) is linear and stable, while the stability of the wrist joint dynamics is determined by the zero dynamics of the system, which can be shown to be stable using center manifold theorem if we ignore friction [14]. In our simulation, the approach converges to the goal if there is no friction and no slip. However the inertia force generated on the object is usually very low. If we add friction, the inertia force is not able to overcome stiction, the controller would diverge by keeping accelerating wrist joint. The phenomenon is verified in experiments. The second baseline is a robust controller that only uses wrist joint torque as control. It extends the first controller

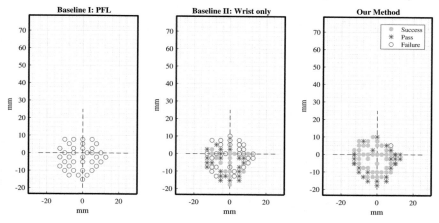

Fig. 4. Results of experiment I. Each dot represents the actual initial grasping position with respect to the object (the outer solid line frame) for one pivoting trail, while the cross of dotted lines denotes the nominal grasping position. A green solid circle denotes a success, a blue star denotes a pass, and a red circle denotes a failure.

in that friction (and its uncertainty) is explicitly handled. Here we set the target object rotation to be $\theta = 1$ rad, starting from zero initial condition $\theta = \alpha = 0$. The reference trajectory θ_{ref} in $\mathbf{x}_{ref}$ is generated by simulating PID control on an integrator, so we can tune its shape. α_{ref} in $\mathbf{x}_{ref}$ is simply set to zero, as we expect the wrist to stay close to the origin and move gently. The two baselines do not utilize grip force, thus we set grip force to the minimal possible value (4N) to make them work better.

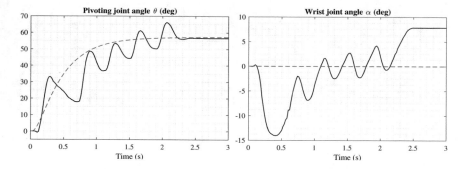

Fig. 5. The reference (dashed line) and response (solid line) trajectories of object orientation θ and robot wrist angle α of our method.

We perform the experiment multiple times with different initial positions offset unknown to the controller. The results are plotted in Fig. 4. A success is defined as converging and stopping in the target region [0.95, 1.05] within 4 seconds. A pass means the object motion is converging, but not fast enough to stop within 4s. Failure means the trail diverged. Our full controller almost converges all the time and outperforms the other two approaches. Fig. 5 shows a typical response for our full controller. Note that during pivoting, the θ converges to the desired region with a steady-state error less than 0.05rad, while the wrist joint stays close to the origin (within about 0.25rad).

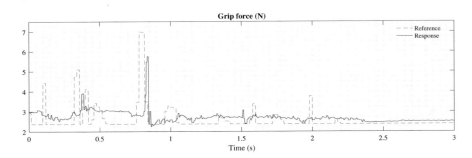

Fig. 6. The grip force control in pivoting.

The calculated grip force and the loadcell feedback are shown in Fig. 6, with notable tracking error. The main source of error is the noise generated from fast motion. In our grip force control test, the error grows immediately as the object rotates. Although the lower-level force control is not perfect, this source of control still appears to be crucial to the overall stability.

Fig. 7. Snapshots of consecutive pivoting experiments comparing wrist only controller (top) and full controller (bottom). The blue, green, red and yellow lines are hand orientation, object orientation, goal orientation and the next goal, respectively.

The ability to tolerate grasping position uncertainty becomes important in tasks where multiple pivots are performed consecutively, as the slips will accumulate. Starting from a precise initial grasping position (error < 0.3mm), the controller with grip force control is able to perform at least four pivots stably in a row, while the controller without grip force control will diverge before finishing the second pivot, as shown in Fig. 7. We omit partial feedback linearization controller to save space, as it never works. See attached video for more details.

5.3   Experiment II: Disturbance Recovery

Our closed-loop controller is able to recover from unexpected disturbances. To illustrate this, we run the controller with goal at the origin, and directly perturb the object through external intervention. The controller with both wrist torque and grip force control can always recover from disturbance and converge back to the origin, while the one without grip force control will diverge quickly. The results are shown in Fig. 8. Again we omit partial feedback linearization controller to save space, as it diverges immediately after perturbation. See the attached video for all three experiments.

Fig. 8. Snapshots of pivoting under disturbance, comparing wrist only controller (top) and full controller (bottom)

5.4   Discussion

An interesting phenomenon observed from our experiments is the small-amplitude oscillation shown in Fig. 5 and Fig. 6. The oscillation has a pattern sometimes observed in human pivoting. This is partly due to the fact that the robust controller, or a human, tries to move the object towards the goal even in the worst possible friction. This will make it hard to avoid overshoot within one control time step. The observation suggests a faster control loop with lower latency is likely to reduce the oscillation.

6   Conclusions and Future Work

We have performed robust dynamic pivoting driven by inertial force and grip force, and described the robust control algorithm being used. The work helps us better understand how to cope with friction in manipulation tasks. It is worth noting that a coarse grip force control can significantly improve the stability of the closed-loop pivoting system. The experiment results also help us better understand how humans cope with friction. We use a low control frequency (50Hz) on a highly dynamic motion. This agrees with the human case where control frequency is also low because of slow nerve conduction velocity. The auto-generated oscillation pattern is similar to human pivoting strategy, which may suggest the importance of incorporating feedback information and actively recover from error.

We believe a precise, less noisy grip force controller with lower latency is likely to improve the performance of pivoting. Our gripper only approximates a parallel gripper for objects of a certain thickness, which is another limitation to be resolved in the future. Our algorithm behaves greedily in terms of friction mode switching. A higher level planning on the sequence of modes may bring further performance improvement. Finally, we are also interested in exploring learning and adaptive control strategies that can handle more general problems, where knowledge of the object inertia property and friction coefficient is not fully available.

7   Acknowledgment

We appreciate Francisco E. Viña B., Yiannis Karayiannidis and François R. Hogan for insightful discussions and suggestions. This material is based upon work supported by the National Science Foundation under Grant No. 1662682.

References

1. N. C. Dafle, A. Rodriguez, R. Paolini, B. Tang, S. S. Srinivasa, M. Erdmann, M. T. Mason, I. Lundberg, H. Staab, and T. Fuhlbrigge, "Extrinsic dexterity: In-hand manipulation with external forces," in *2014 IEEE International Conference on Robotics and Automation (ICRA)*, 2014, pp. 1578–1585.
2. A. Holladay, R. Paolini, and M. T. Mason, "A general framework for open-loop pivoting," in *2015 IEEE International Conference on Robotics and Automation (ICRA)*, pp. 3675–3681.
3. C. Canudas, K. Astrom, and K. Braun, "Adaptive friction compensation in dc-motor drives," *IEEE Journal of Robotics and Automation*, vol. 3, no. 6, pp. 681–685, 1987.
4. C. C. De Wit, H. Olsson, K. J. Astrom, and P. Lischinsky, "A new model for control of systems with friction," *IEEE Transactions on Automatic Control*, vol. 40, no. 3, pp. 419–425, 1995.
5. H. Olsson, K. J. Åström, C. C. De Wit, M. Gäfvert, and P. Lischinsky, "Friction models and friction compensation," *European journal of control*, vol. 4, no. 3, pp. 176–195, 1998.
6. B. Armstrong-Hélouvry, P. Dupont, and C. C. De Wit, "A survey of models, analysis tools and compensation methods for the control of machines with friction," *Automatica*, vol. 30, no. 7, pp. 1083–1138, 1994.
7. J. Shi, J. Z. Woodruff, and K. M. Lynch, "Dynamic in-hand sliding manipulation," in *2015 IEEE/RSJ International Conference on Intelligent Robots and Systems*, 2015, pp. 870–877.
8. B. Vina, E. Francisco, Y. Karayiannidis, K. Pauwels, C. Smith, and D. Kragic, "In-hand manipulation using gravity and controlled slip," in *2015 IEEE/RSJ International Conference on Intelligent Robots and Systems*, pp. 5636–5641.

9. F. E. Vi, Y. Karayiannidis, C. Smith, D. Kragic *et al.*, "Adaptive control for pivoting with visual and tactile feedback," in *2016 IEEE International Conference on Robotics and Automation (ICRA)*, pp. 399–406.

10. A. Rao, D. J. Kriegman, and K. Y. Goldberg, "Complete algorithms for feeding polyhedral parts using pivot grasps," *IEEE Transactions on Robotics and Automation*, vol. 12, no. 2, pp. 331–342, 1996.

11. D. L. Brock, "Enhancing the dexterity of a robot hand using controlled slip," in *1988 IEEE International Conference on Robotics and Automation*, 1988, pp. 249–251.

12. A. Sintov, O. Tslil, and A. Shapiro, "Robotic Swing-Up regrasping manipulation based on the ImpulseMomentum approach and cLQR control," *Ieee T Robot*, vol. 32, no. 5, pp. 1079–1090, 2016.

13. N. Chavan-Dafle and A. Rodriguez, "Prehensile pushing: In-hand manipulation with push-primitives," 2015.

14. M. W. Spong, "The swing up control problem for the acrobot," *Control Systems, IEEE*, vol. 15, no. 1, pp. 49–55, 1995.

15. M. W. Spong and D. J. Block, "The Pendubot: a mechatronic system for control research and education," in *Proceedings of the 34th IEEE Conference on Decision and Control*, vol. 1, 1995, pp. 555–556 vol.1.

16. A. De Luca, R. Mattone, and G. Oriolo, "Stabilization of an underactuated planar 2r manipulator," *International Journal of Robust and Nonlinear Control*, vol. 10, no. 4, pp. 181–198, 2000.

17. Y. Nakamura, T. Suzuki, and M. Koinuma, "Nonlinear behavior and control of a nonholonomic free-joint manipulator," *IEEE Transactions on Robotics and Automation*, vol. 13, no. 6, pp. 853–862, 1997.

18. A. De Luca and G. Oriolo, "Trajectory planning and control for planar robots with passive last joint," *The International Journal of Robotics Research*, vol. 21, no. 5-6, pp. 575–590, 2002.

19. L. Gaul and R. Nitsche, "Friction control for vibration suppression," *Mechanical Systems and Signal Processing*, vol. 14, no. 2, pp. 139–150, 2000.

20. K. M. Lynch and M. T. Mason, "Stable pushing: Mechanics, controllability, and planning," *The International Journal of Robotics Research*, vol. 15, no. 6, pp. 533–556, 1996.

21. ——, "Dynamic nonprehensile manipulation: Controllability, planning, and experiments," *The International Journal of Robotics Research*, vol. 18, no. 1, pp. 64–92, 1999.

22. B. Armstrong-Hélouvry, P. Dupont, and C. Canudas de Wit, "A Survey of Models, Analysis Tools and Compensations Methods for the Control of Machines with Friction," *Automatica*, vol. 30, no. 7, pp. 1083–1138, 1994.

23. J. Zhou, R. Paolini, J. A. Bagnell, and M. T. Mason, "A convex polynomial force-motion model for planar sliding: Identification and application," in *2016 IEEE International Conference on Robotics and Automation (ICRA)*, 2016, pp. 372–377.

24. S. Goyal, A. Ruina, and J. Papadopoulos, "Planar sliding with dry friction part 1. limit surface and moment function," *Wear*, vol. 143, no. 2, pp. 307–330, 1991.

25. J.-J. E. Slotine, W. Li *et al.*, *Applied nonlinear control*.   Prentice-Hall Englewood Cliffs, NJ, 1991, vol. 199, no. 1.

26. M. Spong, J. Thorp, and J. Kleinwaks, "The control of robot manipulators with bounded input," *IEEE Transactions on Automatic Control*, vol. 31, no. 6, pp. 483–490, 1986.

27. R. Goebel, R. G. Sanfelice, and A. R. Teel, "Hybrid dynamical systems," *IEEE Control Systems*, vol. 29, no. 2, pp. 28–93, 2009.

Re-configuring knots to simplify manipulation

Weifu Wang and Devin Balkcom

Dartmouth College Computer Science Department

Abstract. Humans often change the geometry of flexible objects during manipulation so that the goal is easier to accomplish with either simple motions or simple controls. This paper explores how to change the geometry of a knot to allow simpler tying or untying. The paper presents algorithms that modify the knot configuration to allow the knot to be arranged into the correct topological structure or untangled by moving the tip of the string along a straight line, with only a few re-grasps. The paper also presents proof-of-concept physical experiments in which robot arms arrange and untangle several knots.

1 Introduction

Humans stretch clothes while getting dressed, bend string during knitting and weaving, and change the shape of knots while tying, perhaps using techniques learned from others or based on their own experience. How can robots automatically determine how to change the geometry of flexible objects for easy manipulation? This paper explores algorithms for systematically discovering tricks and shortcuts for tying and untangling new and different knots.

The wide variety of knots, the diverse and complex geometries, and the flexible nature of string all make tying knots with robots challenging. Researchers have successfully tied some simple knots with arms [12, 13, 27, 28, 29, 31]; most attempts explore motions that trace some particular geometry.

In this paper, we explore how to modify the geometry to make the tying or untying process simpler, mainly in the perspective of using fewer re-grasps. We present a general approach to changing knot geometry so as to allow tying and untangling using simple motions of a robot arm. For tying knots, the geometry is changed virtually – choosing a more convenient goal geometry than the input geometry. For detangling (untying loose knots), the re-configuration is physical, re-configuring the knot as a first step before pulling on a particular end.

We present an algorithm to change the geometry of a knot so that the knot can be arranged or untied by dragging the tip of the string along a straight line. Another algorithm shows that for any knot, there is a sufficient number of such straight line motions that can arrange or untangle the knot. Our algorithm takes the Gauss code, a text-based description of a knot, as input, and outputs the target geometry of the knot.

We also demonstrate a relation between re-grasps and the layout of the knots, and show why many re-grasps are usually necessary, unless the knot geometry is chosen carefully. We show that through manipulation of the knot geometry, we can reduce the use of re-grasps during knot arrangement. Re-grasping is a common practice in knot tying, especially when using a finite DOF robot arm mounted on a fixed base, because

© Springer Nature Switzerland AG 2020
K. Goldberg et al. (Eds.): *Algorithmic Foundations of Robotics XII*, SPAR 13, pp. 480–495, 2020.
https://doi.org/10.1007/978-3-030-43089-4_31

(a) Arranging part of the knot without re-grasping.

(b) Completing the arrangement of a knot 7_1 with one re-grasp.

Fig. 1: Arranging a 7_1 knot with a Da Vinci robot arm.

the arm needs to arrange string both over and under other segments of the laid-out string. However, re-grasping is difficult, since it may require precise information about and control over the environment and the object being manipulated.

In physical experiments, we have arranged and untangled several different knots, including the *double-coin knot* and a knot known as 7_1 in the *standard knot table* [10], using different robot arms to show that the motions needed to tie or untie the knots in this work are in fact simple.

The paper is organized as follows. We first introduce some fundamental concepts about knots and show the importance of geometry and topology in knot tying in Section 2. Then in Section 3, we show the simple arrangement of knots through the manipulation of the knot geometry, using few re-grasps by dragging the tip of the string along a straight line. In Section 4, we show the simple knot untangling approach using the same knot manipulation scheme. Experiments are conducted to show the success in the manipulation of the geometry of knots in arrangement and untangling.

1.1 Related work

Even though changes of geometry are quite common when manipulating flexible objects, large-scale deformation away from the given goal geometry is often considered an undesirable error, instead of a means to simply the manipulation process to achieve the goal. One notable study about changing geometry to simplify manipulation is the work by Demaine *et al.* [11] on *kirigami*, in which paper is first folded into a particular shape, so that a single cut can be made such that when the paper is unfolded, a desired pattern or scene is created.

As a special case of flexible object manipulation, knot tying has been studied by roboticists as far back as the early 1980s [13] where a robot arm was used to tie knots using sensor feedback. More recently, researchers have attempted to use pairs of arms to tie knots [27, 29, 28]; there is also a rich body of work on machine suturing [15, 16, 14, 18]. Hopcroft *et al.* developed a graph-based language to design knot-tying motions and tested the approach using a robot arm [12]. Wakamatsu, Arai, and Hirai's tree search planner finds sequences of motions (selected from four primitives) to tie or untie a knot [31]. Untying knots has also been recently studied as a vision and learning problem [21].

Recent work on knots also includes the authors' work on fixture-based knot tying [7, 8, 34, 35, 33] that separates the knot tying process into *arrangement* and *tightening*, and provides the first bounds on the complexity of knot tying [32]. Apart from knot tying, other examples have been studied as a gateway to understanding string manipulation. Elastic rods have been used to model wires and string for manipulation [30].

We use many terms from *knot theory* in this paper. A knot is described by its projection onto a plane, called its *knot diagram* [1, 2, 19, 26, 20]. *Physical knot theory* [17] studies the geometry of tight knots formed with thick string [3]. The tightness of a knot has been studied in applied mathematics and physics [24, 6, 22].

Friction plays an important role in tightening a knot, or in untying tight string. Analysis of the frictional forces is frequently a key component of analysis of rigid-body systems [5, 23, 9], and as string wraps around other segments of string with a certain thickness, the friction can be studied by applying the capstan equations [4]. In the present work, we consider only string that is loose enough that friction can be neglected from the analysis.

2 Knots and knot geometry

Knots are usually projected onto a plane for simple description and illustration. If no three points on the knot project to the same point, and no vertex projects to the same point as any other point on the knot [20], the projection is said to be *regular*, and the projected diagram is called a *knot diagram*.

On the drawing, broken lines are used to indicate where one part of the knot undercrosses the other part of the knot that is directly above the broken lines; such locations are called *crossings*. Each crossing is labeled with a unique number, indicating the order of appearance when tracing along the diagram. Figure 2a shows a shoelace knot diagram. In this work, when we say that we "remove" a crossing, we are referring to the result of manipulating one end of the string so that the projected diagram no longer contains the crossing.

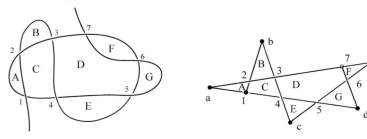

(a) A knot diagram of a shoelace knot, with numbered crossings, and cells labeled with letters.

(b) A polygonal shoelace knot diagram.

Fig. 2: Shoelace knot diagrams, with Gauss code $1^+, 2^-, 3^-, 4^+, 5^-, 6^-, 7^+, 3^+, 2^+, 1^-, 4^-, 5^+, 6^+, 7^-$.

We can also use the labels of the crossings to describe the knot. One such description is the *Gauss code*: a sequence of labels for crossings indicating a walk along the diagram from a given starting point. We use numbers to label crossings, and a superscript "+" or "-" to indicate over- or under-crossing. For example, an overhand knot with Gauss code $G = \{1^+, 2^-, 3^+, 1^-, 2^+, 3^-\}$ has 3 crossings, so the length of G is 6 ($|G| = 6$). The following paragraphs give definitions of a few terms from [32].

Each crossing appears twice on a Gauss code for any knot. A sequence of one or more curves connecting two adjacent labels in the Gauss code is called a *c-path* (crossing path). The projected knot diagram separates the plane into several disconnected closed *cells*, labeled by capital letters in Figure 2a. Call the cell that extends to infinity the *exterior cell*, and all other cells *interior cells*.

A c-path is called an *exterior c-path* if it contacts the exterior cell. For example, Figure 2b has c-paths $(1,2)$ through a contacting interior cell A, $(2,3)$ through b contacting interior cell B, $(3,7)$ contacting interior cell D, $(7,6)$ through e contacting interior cell F, $(6,5)$ through d contacting interior cell G, $(5,4)$ through c contacting interior cell E, and $(4,1)$ contacting interior cell C. All of these exterior c-paths contacting the exterior cell that is the complement of the polygonal shape.

Let an *exterior crossing* be a crossing that is the common endpoint of two adjacent exterior c-paths; all the other crossings are *interior crossings*. If a connection between two crossings is not an exterior c-path, it is called an *interior c-path*.

Sometimes the number and the order of crossings can be different even for the same knot; *Reidemeister moves* [25] can be used to transform the crossings without changing the topology of the knot. Determining whether two different Gauss codes represent the same knot is one of the most challenging and fundamental problems in knot theory; we exclude Reidemeister moves in this work.

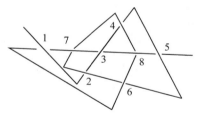

Fig. 3: The resulting configuration of a double-coin knot so that the last five crossings are on a straight line.

2.1   Knots and weaving

On a weaving loom, the warp is the set of strings that form the basic structure around which the weft (the string pulled by the shuttle) is woven. The approach we take in this work to knot tying is to find a simple substructure of the knot in such a way that the under-crossing always appears before its over-crossing so that this spiral-like structure is analogous to the warp. The rest of the knot, like the weft, is then arranged with respect to this warp to construct the more challenging crossings.

 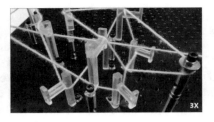

(a) Arranging the warp crossings of a double-coin knot.   (b) Completing the arrangement of a double-coin knot with one re-grasp.

Fig. 4: Arranging a double-coin with Da Vinci robot arm.

Let us consider the following example. For a double-coin knot with Gauss code $G = \{1^+, 2^-, 3^+, 4^-, 5^+, 6^-, 2^+, 7^-, 4^+, 8^-, 6^+, 1^-, 7^+, 3^-, 8^+, 5^-\}$, the last five crossings counting from the right open end are 5, 8, 3, 7, and 1. These five crossings can be formed by dragging the right open end along a straight line simulating the motion of a shuttle on a loom, provided that the other segments of string are arranged appropriately. An example of the rearranged polygonal configuration of a double-coin knot is shown in Figure 3. We implemented the knot arrangement using the proposed layout with a Da Vinci robot arm. Figure 4 shows the results of implementation.

3   Arranging knots

In this section, we introduce how to change the geometry of a knot to allow the division of knot arrangement into warp and weft stages. Let us refer to all the crossings formed by the warp stage as *warp crossings*, and all the crossings formed in the weft stage as *weft crossings*. Formally, a crossing i is a weft crossing if and only if after all the crossings to the right (left) of i^a have been removed, and the crossing to the left (right) of i^a has a different sign than a. Remember that each crossing appears twice on a Gauss code, once with a $^+$ superscript (over-crossing), and once with a $^-$ superscript (under-crossing).

We will show that for an arbitrary given knot, we can find patterns on its Gauss code, so that by arranging m alternating sequence of warp and weft crossings, the knot can be arranged with a upper bounded m re-grasps. Since re-grasping is not a trivial task, our approach gives a simpler knot arrangement method compared to knot arrangement methods where re-grasps have to be performed between the arrangement of over- and under-crossings. What is more, in our knot arrange method, weft crossings on the same sequence can be arranged by pulling string to follow a straight line.

3.1   Forming or removing crossings based on weaving sequence

We define a *minimal Gauss code* as a Gauss code that cannot be simplified by performing Reidemeister moves. Adding or removing of a crossing from a structure with minimal Gauss code through physical manipulation of the string can only be achieved if one of the two appearances of the crossing number is at one end of the Gauss code.

What happens if we remove crossings one-by-one from the open ends? Intuitively, we know that after a certain number of removals, the remaining crossing pattern is no longer knotted, because eventually the knot is untied if we remove all crossings. The knotting and unknotting process are symmetric, and it is easier to see the pattern when removing crossings from an existing sequence of crossings, so we choose the unknotting process for analysis.

We define a *weaving sequence* as a sequence of alternating over- and under-crossings that have to be formed or deleted in the given order indicated by the Gauss code. Consider the example of *unknotting* a double-coin knot, whose Gauss code is $G = \{1^+, 2^-, 3^+, 4^-, 5^+, 6^-, 2^+, 7^-, 4^+, 8^-, 6^+, 1^-, 7^+, 3^-, 8^+, 5^-\}$. Starting from the right end, crossing 5^- is adjacent to 8^+ in the Gauss code, and the crossings have *different* superscript signs. Therefore, we can identify these two crossings as part of a weaving sequence — a sequence of weft crossings, and remove crossing 5.

We continue to remove crossings that are part of the same weaving sequence from the right end, including crossings 8, 3, 7, and 1, in order. After we remove the last five crossings in the Gauss code, we have $G = \{2^-, 4^-, 6^-, 2^+, 4^+, 6^+\}$. Now, the next two crossings from the right have the same superscripts, so they are no longer part of a weaving sequence. We know that the five deleted crossings can be formed by a single weft (weaving) motion, dragging the string along a straight line. We continue searching for weaving sequences from right to left. In this example, there are none, and the remaining structure consists only of warp crossings.

The following algorithm, which takes the Gauss code of the knot as input, finds weaving sequences for an arbitrary knot. With a single pass through the Gauss code, the algorithm outputs a sufficient number of m weaving sequences that can be used to form the knot; m is also a sufficient number of re-grasps to tie the knot with a fixed-base arm.

Algorithm 1: WEAVE

1. Select either left or right end of the Gauss code.
2. Delete crossings from the selected end.
3. If the crossing to be removed has a different sign from the next crossing to removed, then the two crossings belong to the same weaving sequence. Register a new weaving sequence if the current crossing to be removed is not already on a weaving sequence. If the crossing to be removed has the same sign as the next crossing to be removed, then terminate the current weaving sequence; if the crossing to be removed has the same label as the next crossing to be removed (for example, $i^- j^- j^+$ where j^- and j^+ have the same label), then compare the sign to the first crossing with a different label (compare j^+ with i^- in the given example).
4. Repeat steps 2 and 3 until only one crossing is left, attach the last crossing to the on-going pattern.

For a Gauss code with k crossings where $|G| = 2k$, we can find $O(k^2)$ different sequence of labels that are the results of removing the crossings at the beginning or the end of the Gauss code. However, since a weaving sequence can only be formed by weaving with one end of the string, we only need to check the sequence of labels that are the results of removing crossings from solely the left or right end. The total length of such a sequence of labels is $2k$.

The algorithm only checks if the current crossing has a different sign from the adjacent one. This approach may overlook some structures that are unknotted but still contain adjacent crossings that have different signs, such as $\{1^+, 2^-, 3^+, 4^-, 4^+, 3^-, 2^+, 1^-\}$. This structure is unknotted, but still contains one weaving sequence.

Knots such as the overhand knot contains only one weaving sequence associated with the last two crossings, while the first crossing can be formed by a type I Reidemeister move. Similarly, the *figure eight knot* with Gauss code $G = \{1^+, 2^-, 3^+, 4^-, 2^+, 1^-, 4^+, 3^-\}$, also contains only one weaving sequence associated the last three crossings where the first crossing is achieved by a type I Reidemeister move. The double-coin knot shown earlier contains one weaving sequence, with the crossings 2, 4 and 6 forming the initial unknotted structure.

Lemma 1. *If a knot can be arranged by following a single weaving sequence, then a motion that removes all weft crossings unties the knot.*

Proof. If we remove all the crossings on the weaving sequence and there is only one weaving sequence, the remaining crossings are all warp crossings by definition. Then, if we continue to remove crossings, every crossing they remove will have the same sign as the the next crossing to remove. Without loss of generality, let us assume the first crossing we will remove is an over-crossing. Then, since all remaining crossings are warp crossings, whenever we are trying to remove a crossing i^a, a is $+$ until all crossings are removed. Then, all these warp crossings can be arranged on two layers. One plane contains only the over-crossings, while the other plane contains all the under-crossings, with finitely many vertical line segments connecting two planes. This structure has the topology of a circle when the ends of the string are connected to each other, an *unknot*.

The lemma shows that even though there are knots of many crossings, but they are in fact simple knots. A single motion can untie the knot. We believe that the number of crossings may not be the best way to illustrate how complex a knot is.

3.2   Aligning crossings on a straight line for simple manipulation

A straight-line motion is easy to achieve even for simple robotic devices. This section will show that weft crossings can always be aligned on a single straight line, without changing the knot topology.

Theorem 1. *In a weaving sequence, each crossing label appears only once.*

Proof. A weaving sequence contains an alternating over- and under-crossing pattern, and all the crossings on the weaving sequence are adjacent to each other in the Gauss code. If the same label j appears twice, let crossings i^- and k^- be the two crossings in the Gauss code adjacent to j^+, and let s^+ and t^+ be the two crossings adjacent to j^-. The crossings i, k, s and t have the corresponding signs because they are adjacent crossings to j, and they are on the same weaving sequence. Without loss of generality, let j^- be closer to the open end. After the deletion of the crossing j^-, crossings i and k are now adjacent in the Gauss code, and they have the same sign, so they cannot be on the same weaving sequence.

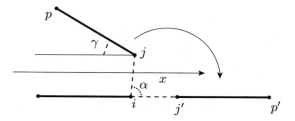

Fig. 5: Rotating extreme segments to align all weaving crossings on a straight line.

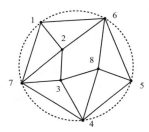
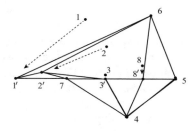

(a) A polygonal knot diagram for a double-coin knot, computed using methods proposed in [32].

(b) The rearranged configuration for the double-coin knot, by rotating extreme segments using the proposed method indicated in the Algorithm 2.

Fig. 6: Reconfiguration of a double-coin knot to align weft crossings onto a straight line.

Since no crossing appears twice on a weaving sequence, we can move the crossings so that they appear on a straight line. Therefore, a single straight line motion of the string can form multiple crossings at once. Define an *extreme segment* of the weaving sequence as the segment between two exterior crossings.

The algorithm below shows how to arrange extreme segments on a straight line if they are on the the the same weaving sequence. The input to the algorithm is the locations of all the crossings, which may be computed using the technique implied by the proof of Theorem 10 in [32]. The algorithm computes the geometry of the knot during arrangement, and computes the placement of the fixtures that are used in our experiments. Because the geometry computed are straight line configurations supported by the fixtures, no vision feedback is needed in our approach.

Algorithm 2: ALIGN

1. For each given extreme segment, connect a line between the two exterior crossings (or an interior to an exterior crossing), and move each of the crossings on the extreme segment to its projection on the connected line;
2. Let crossing i $((x_i, y_i))$ and crossing j $((x_j, y_j))$ be the two adjacent exterior crossings on two adjacent extreme segments, with k connections between them; without loss of generality, let $y_j < y_i$; let p be the other end point on the extreme segment with j as an end point;

3. Rotate all the points above extreme segment between p and j (because $y_j < y_i$) around j then around i in the stated order, so that (x'_j, y'_j) is the new location of crossing j where $y'_j = y_i$, as shown in Figure 5; The angle rotated around crossing i can be calculated as the acute angle α between the x axis and the vector ij, and the rotation angle around crossing j is $\beta = \pi - \alpha - \gamma$, where γ is the angle between pj and the x axis;

4. Along the line $x = (x_i + x'_i)/2$, find k points above (or under) the line $y = y_i$ if $x'_i \geq x_i$ ($x'_i < x_i$) with equal distance. For the endpoints of the k pairs of connection between two extreme segments, connect to k points along the line $x = (x_i + x'_i)/2$ in order based on the distance of the end point on segment pj to the exterior crossing j, such that no additional intersection is introduced;

5. Adjust the z coordinates of all crossings so that the crossings on the rearranged weaving sequence lie on a straight line in three dimensions;

The result of applying the process to a double-coin knot is shown in Figure 6.

In our previous work [32], we have shown that an arbitrary (polygonal) knot with k crossings can be laid out based on its Gauss code using no more than $3k - 2$ line segments. We use this projected configuration to compute where each crossing should be placed so the crossings on a weaving sequence is on a straight line in our experiments.

3.3   Re-grasping and weaving

This section analyzes the number of re-grasps needed to arrange each of the two types of crossings. If we do not choose the geometry of the final knot configuration carefully, the number of re-grasps needed to arrange a knot maybe be as large as the number of weft crossings, plus one re-grasp for each sequence of warp crossings.

We show that the output of Algorithm 1 also gives a sufficient bound for the number of re-grasps needed to arrange a given knot, if each sequence of weft crossings are aligned on a straight line following Algorithm 2.

We can use a fixed-base robot arm to lay out the warp crossings without re-grasping:

Lemma 2. *No re-grasp is needed to arrange a collection of warp crossings if for each warp crossing, its under-crossing appears before its over-crossing.*

Proof. The structure is unknotted and belongs to two layers. If for each layer, we trace along the configuration with a robot arm, then the two layers form the corresponding crossings. Therefore, no re-grasping is needed.

Sometimes, changing the geometry of the knot can reduce the number of degrees of freedom that a fixed-base arm must have to tie the knot using a particular number of re-grasps. Notice that an elephant-trunk arm with infinite degrees of freedom can arrange any knot without re-grasping.

The next lemma shows that changing the geometry of the knot is in fact sometimes necessary to optimize this tradeoff between degrees of freedom of the arm and the number of re-grasps required: some knot geometries are quite difficult for any finite-DOF arm. For example, a sequence of crossings of the form i^a, j^b, k^a, where a and b have opposite signs. A 4 DOF robot arm needs at least one re-grasp to arrange this sequence of the crossings, if j^a has already been arranged. Such sequence of crossings are the basic structures of the weaving sequence, so traditional knot tying approach of complex knot usually involves many re-grasps.

A weaving sequence contains a sequence of consecutive over- and under-crossings. However, if we imagine the robot end-effector as the shuttle on the loom, it does not need to re-grasp every time the string switches between over- and under-crossing, if all the crossings on this weaving sequence is on a straight line. However, for a weaving sequence, one re-grasp is still needed.

Lemma 3. *To arrange a weaving sequence with a fixed-base robot arm grasping the ends of the string, at least one re-grasp is needed.*

Proof. Let the two ends of a knot be S_1 and S_2, and let S_1 be fixed to the ground. Let the base of the robot arm be B, and the end effector be E. The robot arm needs to grasp S_2 with E during the arrangement of the knot. Let us assume no re-grasp is performed during the arrangement of a weaving sequence w. At the end of the arrangement of the weaving sequence, let the configuration of the string be fixed in space. Let curve c_1 be the current configuration of the robot arm, connecting from B to E. Let curve c_2 be the curve of a different robot arm configuration connecting from B to E, where the entire robot arm is outside the convex hull of the knot. In both configurations, the E is attached to S_2. Then, the two curves belong to two different homotopy classes with respect to the string. The curve connecting c_2 to S_2 and then to S_1 has the correct knot topology. Therefore, at least one re-grasp is needed to arrange a weaving sequence.

The number m output by Algorithm 2 gives a sufficient number of re-grasps needed to tie a given knot. For many knots, including the double-coin knot, the number is 1. Since these knots are in a different topology class from a topological loop when one of their end points are grasped by a robot arm and the other end point is attached to the ground, at least one re-grasp is needed, so this number is also a lower bound.

3.4   Knot weaving with Da Vinci

We conducted experiments with a Da Vinci surgical robot, which has two symmetric high precision arms. We computed knot layouts and built fixtures to support string arranged at various heights, allowing weaving to be implemented with a single translation. Based on the computed knot configuration, such as the example shown in Figure 3, the robot manipulator follows predefined paths without vision feedback.

In all the experiments conducted, during the layout of the warp crossings, the string is supported by fixtures. Each fixture is either straight rods or upside down "L" shape fixtures on top of rods. These fixtures are placed at various locations to support warp crossings to form the computed geometry, so that the weft crossings are all aligned. The location of the fixtures are also computed automatically by placing a fixture on the inside of each turn of the string, so that the string wraps around them follows the shortest path in the homotopy class [34]. Because the fixtures are simply rods placed at various computed locations, we consider the approach simple and general.

Figure 4a shows the layout of the warp crossings of a double-coin knot, arranged without re-grasps. After layout, the effector of the second arm grasps the tip of the string and uses a pure translation to complete the knot, as shown in Figure 4b and in the multimedia attachments. In the attached video, the author had to manually intervene to loosen the string wrap around the fixtures, due to high tension along the string. The need for human intervention is the combined results of the use of yarn, the manipulator of Da Vinci robot cannot fully close, and the high friction coefficient between yarn and the printed "L" shaped fixture. One of the future work is to study how to automatically use robots to loosen string during manipulation.

(a) Arranging warp crossings of a double-coin (b) Completing the arrangement of a double-
knot at the same height.   coin knot with a single re-grasp.

Fig. 7: Arranging a double-coin by laying out the warp crossings on the same height.

The preliminary arrangement of the string requires many support structures laid out in the workspace of the robot. For simplicity, we programmed the robot to just arrange the warp crossings of the string at the same heights around simple fixtures. Figures 1a and 1b show the arrangement of a 7_1 knot. Figures 7a and 7b shows the arrangement of a double-coin knot.

Our approach is able to arrange a double-coin knot with only a simple re-grasp. There has not been any other successful attempt to arrange a double-coin knot with a robot arm. However, by following the manipulation sequence suggested in previous work such as [28], five re-grasps are needed to arrange the double-coin knot. Even with state of the art robotic manipulation strategies, a re-grasp takes a long time to execute. Even though we did not compare the execution time or our approach to other knot tying approaches, we believe when a robot arm following two paths of similar lengths, the fewer re-grasps are performed, the shorter the execution time.

Even though weaving sequence can be found in many knots, different knots containing the same number of crossings can have different number weaving sequences. For example, the double coin knot contains 8 crossings, and is labeled 8_{18} on the standard knot table. This knot, as shown above, can be arranged with a single re-grasp. However, the knot labeled 8_4 contains two weaving sequences, even though it contains the same number of 8 crossings. Following our knot arrangement strategy, arranging the double-coin knot will be simpler compared to knot 8_4. We are not able to determine how the number of weaving sequences on a knot is related to the number of crossings.

3.5   Robot-human collaboration

When we arrange the segments of string that are not part of a weaving sequence at the same height, the robot arm weaves around arranged segments of string. Even though the

(a) Arranging warp crossings of a 8_{10} knot at the same height.

(b) Completing the arrangement of a 8_{10} knot by a human weaving the string.

Fig. 8: Arranging a 8_{10} knot by robot and human collaborating together.

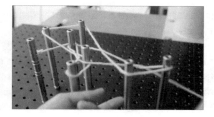

(a) Arranging warp crossings of a 9_{31} knot at the same height.

(b) Completing the arrangement of a 9_{31} knot by a human weaving the string.

Fig. 9: Arranging a 9_{31} knot by robot and human collaborating together.

locations of the string segments are known, the motion still may not be easy to perform for a robot. Humans, however, can arrange the weaving sequence easily with re-grasps.

For example, a double-coin knot can be tied using the robot to lay out the structure, and allowing the human to finish the weft crossings. We used this technique to tie figure-eight knots, and knots 7_1, 8_2, 8_5, 8_{10}, 9_{31}, 9_{32} from the standard knot table. Applying Algorithm 2, all the listed knots can be tied with one re-grasp. Figures 8, 9, and 10 show the examples of a human collaborating with an Adept Cobra industrial arm to tie knot 8_{10}, knot 9_{31} and knot 9_{32}.

4   Untangling knots

In the previous approach, we changed geometry of knots to simplify the knot arrangement, based on the identification of wrap and weft crossings on the knots, and we identify the weft crossings by processing the Gauss code.

The same change of geometry can also be used to untangle knots. In this work, we will focus on untangling knots from a loose configuration rather than untying knots from a tight configuration. We untangle the knot by changing the geometry of the knot and pulling the string several times along straight lines. With each pull of the string, we remove all consecutive weft crossings that extend to the current end of the string.

(a) Arranging warp crossings of a 9_{32} knot at the same height.

(b) Completing the arrangement of a 9_{32} knot by a human weaving the string.

Fig. 10: Arranging a 9_{32} knot by robot and human collaborating together.

However, non-consecutive weft crossings may not be able to be removed in a single motion.

Given two different sequences of weft crossings, in order for them to be aligned, other crossings need to be relocated. In order to change the knot configuration of the knot to align the second consecutive sequence of weft crossings on a straight line, the crossings on the first consecutive sequence of weft crossings need to be relocated, which may break the alignment. Therefore, without knowing which specific knot we are trying to untangle and detailed analysis of the specific knot, the best we can do with each grasp is to align a sequence of consecutive weft crossings, and remove them by moving the string along a straight line.

Even when a knot is in a loose configuration, the untangling of knots usually have to overcome friction. After we have identified all the weft crossings, and attempt to use a single motion to untangle them, friction may prevent the untangling, such as shown in Figure 11 where we attempt to remove the last five crossings by pulling the string direction without changing the geometry. Therefore, a pulling motion of string that involves least friction is desirable. It appears that pulling string along a straight line can keep friction relatively low.

The process of manipulating the geometry can be described as follows. We first choose one end of the string to untangle the knot. Along the chosen end of the string, we determine a side that is closer to the boundary, left or right. Starting from the chosen end of the string, we identify all the cells on the chosen side to the string in sequence, until the last consecutive weft crossing, and then we find the largest inscribed circle in each cell. Using the same algorithm we presented in the previous section, we will delete crossings from the chosen end, and record how the crossings change, either from under-crossing to over-crossing, or from over-crossing to under-crossing.

Fig. 11: Friction prevents the untying of a double-coin knot.

Given the x-y plane on which the knot diagram is projected, we place a vector parallel to the z axis at each center of the largest inscribed circle we have identified. The direction of the vector is positive if the crossings associated with the cell change from under-crossing to over-crossing, otherwise negative.

We then use a robot arm manipulating a rod to follow these vectors parallel to the z axis to the points above and below the $z = 0$ plane, and connect between these points by following linear motions with the end effector. After tracing all the vectors, we have aligned all the weft crossings. Figures 12 and 13 show the change of the geometry.

(a) The initial configuration of double-coin knot before untangling.

(b) Aligning several crossings on a straight line for untangling.

(c) Pulling string along straight line to untangle.

(d) Knot is fully untangled after removing the rod.

Fig. 12: Untangling a double-coin knot.

(a) The initial configuration of knot 7_1 before untangling.

(b) Aligning several crossings on a straight line for untangling.

(c) Pulling string along straight line to untangle.

(d) Knot is fully untangled after removing the rod.

Fig. 13: Untangling knot 7_1.

We then identify the last weft crossing we have aligned, and the warp crossing adjacent to it, and let the robot grasp any point between the two crossings. The robot arm then pulls the string along the direction parallel to the vector point along the rod we used to align the crossings. After all the consecutive weft crossings are removed and the rod is removed, the knot will be untangled. Even though we have only demonstrated the untangling of loose knots, the principle can be applied to tight knots, if we can identify the crossings and thread a needle through those enclosed cells.

5 Conclusions and future work

This work shows that some knots can be tied or untied with simple motions by changing the geometry of the knots. We discussed an algorithm for changing the geometry of the knots, and another algorithm for discovering different knot tying or untying phases using the Gauss code descriptor for a knot. We also showed practical implementations using simple robot arms, and also as a collaboration between a robot arm and a human.

For future work, we would like to better understand how motions can be designed to mechanically simplify tying knots and untying even tightened knots. We are also interested to know if by changing the geometry of the goal, we can manipulate other flexible objects, such as cloth, using simple motions.

We are particularly interested in knots like the shoelace and sheepshank; humans tie these knots by pulling loops through loops. We can identify these structures from the Gauss code, as Type II Reidemeister moves: for each adjacent appearance, crossings i and j have the same sign, and the sign is different from the crossings adjacent to the ij (or ji) sequence.

We would like to thank Dmitry Berenson and Gregory Fisher for letting us use the Da Vinci robot for some of the experiments. This work is supported by NSF grant IIS-1217447.

References

[1] C.C. Adams. *The Knot Book: An Elementary Introduction to the Mathematical Theory of Knots.* American Mathematical Society, 2004.

[2] J. W. Alexander. Topological invariants of knots and links. *Trans. Amer. Math. Soc.*, 20:275–306, 1923.

[3] Ted Ashton, Jason Cantarella, Michael Piatek, and Eric Rawdon. Knot tightening by constrained gradient descent. *Experimental Mathematics*, 20(1):57–90, 2011.

[4] Stephen W. Attaway. The mechanics of friction in rope rescue. *International Technical Rescue Symposium*, 1999.

[5] Devin J. Balkcom, Jeffrey C. Trinkle, and E. J. Gottlieb. Computing wrench cones for planar contact tasks. In *ICRA*, pages 869–875. IEEE, 2002.

[6] J. Baranska, S. Przybyl, and P. Pieranski. Curvature and torsion of the tight closed trefoil knot. *The European Physical Journal B - Condensed Matter and Complex Systems*, 66(4):547–556, 2008.

[7] Matthew P. Bell. Flexible Object Manipulation. Technical Report TR2010-663, Dartmouth College, Computer Science, Hanover, NH, February 2010.

[8] Matthew P. Bell, Weifu Wang (co-first author), Jordan Kunzika, and Devin Balkcom. Knot-tying with four-piece fixtures. *International Journal of Robotics Research (IJRR)*, vol 33, no. 11:1481–1489, Sep, 2014.

[9] Stephen Berard, Kevin Egan, and Jeffrey C. Trinkle. Contact modes and complementary cones. In *ICRA*, pages 5280–5286, 2004.

[10] G. Burde. Knoten. *Jahrbuch Ueberblicke Mathematik*, pages 131–147, 1978.

[11] Erik D. Demaine, Martin L. Demaine, Andrea Hawksley, Hiro Ito, Po-Ru Loh, Shelly Manber, and Omari Stephens. Making polygons by simple folds and one straight cut. In *Revised Papers from the China-Japan Joint Conference on Computational Geometry, Graphs and Applications (CGGA 2010)*, Lecture Notes in Computer Science, pages 27–43, Dalian, China, November 3–6 2010.

[12] John E. Hopcroft, Joseph K. Kearney, and Dean B. Krafft. A case study of flexible object manipulation. *International Journal of Robotic Research*, 10(1):41–50, 1991.

[13] H. Inoue and M. Inaba. Hand-eye coordination in rope handling. *Robotics Research: The first International Symposium*, pages 163–174, 1985.

[14] H Kang and J.T. Wen. Robotic knot tying in minimally invasive surgeries. In *IEEE/RSJ International Conference on Intelligent Robots and Systems (IROS)*, 2002.

[15] Hyosig Kang and John T. Wen. Endobot: a robotic assistant in minimally invasive surgeries. In *Proc. IEEE International Conference on Robotics and Automation*, volume 2, pages 2031–2036, 2001.

[16] Hyosig Kang and J.T. Wen. Robotic assistants aid surgeons during minimally invasive procedures. *Engineering in Medicine and Biology Magazine, IEEE*, 20(1):94–104, Jan 2001.

[17] L.H. Kauffman. *Knots and Physics*. K & E series on knots and everything. World Scientific, 1991.

[18] Makoto Kudo, Yasuo Nasu, Kazuhisa Mitobe, and Branislav Borovac. Multi-arm robot control system for manipulation of flexible materials in sewing operation. *Mechatronics*, 10(3):371 – 402, 2000.

[19] W.B.R. Lickorish. *An Introduction to Knot Theory*. Graduate Texts in Mathematics. Springer New York, 1997.

[20] Charles Livingston. *Knot theory*. The carus mathematical monographs, Volume Twenty Four. The mathematical Association of America, Washington D.C., 1993.

[21] Wen Hao Lui and Ashutosh Saxena. Tangled: Learning to untangle ropes with RGB-D perception. In *2013 IEEE/RSJ International Conference on Intelligent Robots and Systems, Tokyo, Japan, November 3-7, 2013*, pages 837–844, 2013.

[22] J.H.Maddocks M. Carlen, B. Laurie and J. Smutny. Biarcs, global radius of curvature, and the computation of ideal knot shapes. *Physical and numerical models in knot theory*, 36 of Ser. Knots Everything:75–108, 2005.

[23] Matthew T. Mason. *Mechanics of Robotic Manipulation*. MIT Press, Cambridge, MA, August 2001.

[24] Eric J. Rawdon. Approximating the thickness of a knot. *Ideal knots*, 19 of Ser. Knots Everything:143–150, 1998.

[25] Kurt Reidemeister. Elementare begrndung der knotentheorie. *Abhandlungen aus dem Mathematischen Seminar der Universitt Hamburg*, 5(1):24–32, 1927.

[26] D. Rolfsen. *Knots and Links*. AMS/Chelsea Publication Series. AMS Chelsea Pub., 1976.

[27] Mitul Saha and Pekka Isto. Motion planning for robotic manipulation of deformable linear objects. In *Proc. IEEE International Conference on Robotics and Automation*, pages 2478–2484, May 2006.

[28] Mitul Saha and Pekka Isto. Manipulation planning for deformable linear objects. *IEEE Transaction on Robotics*, 23(6):1141–1150, December 2007.

[29] Mitul Saha, Pekka Isto, and J.-C. Latombe. Motion planning for robotic knot tying. In *Proc. International Symposium on Experimental Robotics*, July 2006.

[30] Zoe McCarthy Timothy W Bretl. Quasi-static manipulation of a Kirchhoff elastic rod based on a geometric analysis of equilibrium configurations. *International Journal of Robotics Research (IJRR)*, June 2013.

[31] Hidefumi Wakamatsu, Eiji Arai, and Shinichi Hirai. Knotting/unknotting manipulation of deformable linear objects. *International Journal of Robotics Research*, 25:371–395, 2006.

[32] Weifu Wang and Devin Balkcom. Grasping and folding knots. In *IEEE International Conference on Intelligent Robots and Systems (ICRA)*, pages 3647–3654, 2016.

[33] Weifu Wang and Devin Balkcom. Towards tying knots precisely. In *IEEE International Conference on Intelligent Robots and Systems (ICRA)*, pages 3639–3646, 2016.

[34] Weifu Wang, Matthew P. Bell, and Devin J. Balkcom. Towards arranging and tightening knots and unknots with fixtures. In *International Workshop on the Algorithmic Foundations of Robotics, WAFR 2014*, pages 677–694, 2014.

[35] Weifu Wang, Matthew P. Bell, and Devin J. Balkcom. Towards arranging and tightening knots and unknots with fixtures. *IEEE T. Automation Science and Engineering*, 12(4):1318–1331, 2015.

Continuous Pseudoinversion of a Multivariate Function: Application to Global Redundancy Resolution

Kris Hauser

Department of Electrical and Computer Engineering, Duke University,
`kris.hauser@duke.edu`

Abstract. This paper seeks to generate a continuous pseudoinverse of a function that maps a higher dimensional compact set to a lower dimensional one. Continuity and smoothness should be attained if possible, but otherwise the volume of the discontinuity boundary should be minimized. A sampling-based approximation technique is presented that uses discretized roadmaps of both the domain and image, and minimizes discontinuities of the inverse function. The method is applied to kinematic redundancy resolution for redundant robots, which have more degrees of freedom than workspace dimensions. The output is a global redundancy resolution, which has the convenient property that whenever the robot returns to the same workspace point, it uses the same joint-space pose. If a global resolution cannot be found, then the method minimizes discontinuities and maps them in workspace. Results are demonstrated on toy problems with up to 20 DOF, and on several robot arms.

1 Introduction

Function inversion is frequently encountered in many fields of science and engineering, including robotics, computer graphics, control, and mechanical design. Past techniques typically consider calculating a point-wise inverse, i.e., find a point x such that $f(x) = y$. In contrast, this paper is interested in generating a functional map of the inverse across entire regions of space, i.e., find a function $f^{-1}(y)$ such that $f(f^{-1}(y)) = y$ across all y. This is often an underconstrained problem, because the preimage of a point in the range may contain multiple or an infinite number of points in the domain and some inverses may be wildly varying or highly discontinuous. This paper seeks a systematic way to obtain maximally continuous and smooth inverse functions in high-dimensional spaces.

We ground this problem in the terminology of forward and inverse kinematics (IK), which is likely to be more familiar for readers in robotics. For example, consider the IK problem for a 2R planar robot manipulator restricted by joint limits (Fig. 1). Here the forward map f maps joint angles to Cartesian end effector (workspace) points via forward kinematics. Points in the CCW extremes of the workspace are reachable only with "elbow-down" configurations, while points in CW extremes are reachable only with "elbow-up" configurations. In

© Springer Nature Switzerland AG 2020
K. Goldberg et al. (Eds.): *Algorithmic Foundations of Robotics XII*, SPAR 13, pp. 496–511, 2020.
https://doi.org/10.1007/978-3-030-43089-4_32

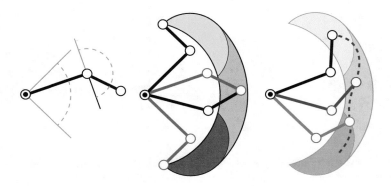

Fig. 1. A planar 2R manipulation with joint limits of $\pm 45°$ and $\pm 90°$, respectively. There are no joint space resolutions of Cartesian end effector paths (e.g., the dotted curve) in the interior of the reachable workspace from the upper region to the lower region.

between, both elbow-down and elbow-up configurations are valid. Observe that any workspace path from the CCW extreme to the CW extreme must cause the robot to at some point flip between elbow down and elbow up. In fact, there is no continuous solution for such a workspace path unless it touches the workspace boundary, causing the robot to pass through a singularity.

If this robot were to be changed to a kR manipulator with $k \geq 3$ (i.e., a redundant manipulator), would a continuous inverse function exist? In contrast to the 2R case, there are an continuous infinity of inverses at each point, and it may seem as though this leaves sufficient flexibility to choose an everwhere-continuous inverse map. Hence, it is perhaps surprising that this is not the case, and in fact no solution exists for many settings of joint limits and link lengths.

This paper describes the redundancy problem for general robots and work-spaces, and presents approximate algorithms for computing answers to questions of redundancy resolution existence and optimality. We draw a distinction between *pointwise*, *pathwise*, and *global* redundancy resolution, listed in order of increasing restrictiveness (Fig. 2). Pathwise redundancy resolution is the problem of generating a continuous configuration-space path for a given workspace path. While local methods may get stuck, we present a simple probabilistically complete technique that inspires our work in global redundancy resolution. Global redundancy resolution has the property that every workspace cycle causes the robot to return to the same configuration. This may be a useful property to make robots behave more predictably during Cartesian movement than either pathwise or pointwise resolution, in which the robot may adopt different poses depending on the workspace path taken, or may get stuck in local minima.

We present an approximate global resolution algorithm that generates a roadmap of sampled points in the workspace and associates each point with a set of configuration samples in its preimage. A configuration space roadmap is then generated along these samples, and a constraint satisfaction optimization (MAX-

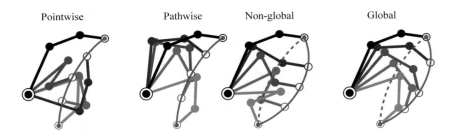

Fig. 2. Illustrating three types of redundancy resolution. Pointwise resolution gives a solution for each point, but does not guarantee continuity along a path. Pathwise resolution gives a continuous solution along a workspace path, but the robot may return to the starting point with a different configuration, and some paths may not have a valid resolution from a given starting configuration. Global resolution gives a continuous solution such that every cyclic path returns to the same configuration.

CSP) is applied to generate a pseudoinverse map that minimizes the number of discontinuous edges. The algorithm is applied to kinematic models of the Rethink Robotics Baxter, Boston Dynamics ATLAS, Kinova Jaco, NASA's Robonaut2, and JPL's Robosimian, showing that some robots are nearly globally-resolvable and others are not. Moreover, it calculates rich workspace maps that help visualize the boundaries of discontinuity in the optimized redundancy resolution.

2   Related work

Redundancy resolution is an old topic in robotics with many pointwise heuristics proposed to avoid singularities, joint limits, and obstacles [7]. Pathwise redundancy resolution has also long been studied, with several researchers identifying the problem of local minima of pointwise resolution methods [3,6,8,10]. Our algorithm is most closely related to [8], who use a randomized tree-growing approach to explore self-motion manifolds along a discretized workspace path.

Unlike planning a path, which is a 1-dimensional map to configuration space, the global redundancy resolution problem is concerned with computing an m-dimensional map to configuration space. The closest related works in global resolution are [2, 5]. The method of [5] constructs a topological network of connected components of self-motion manifolds across a discretization of the workspace, and was applied to planar problems. However it is difficult to analyze connected components for arbitrary robots with constraints. The method of [2] generates IK tables via local optimization and selection of similar IK solutions to yield a smooth mapping for each leg of the Robosimian robot. These tables were then used to simplify motion planning for legged locomotion trajectories in [9]. We use a similar optimization method, although in this case, using coordinate descent, to smooth our maps in postprocessing as described in Sec. 5.2.

Global redundancy resolution has several potential applications. They could be used to select positions for a mobile base for a robot to perform certain

Cartesian movements [15]. Qualitatively, we observe these discontinuities correspond to workspace regions that are frustrating for teleoperation [4]. Finally, these mappings may also be useful for reduced-dimensionality motion planning in Cartesian space [9].

3 Problem definition

We wish to compute a continuous pseudoinverse of a functional mapping between a higher dimensional compact, bounded set and a lower dimensional compact, bounded set. These concepts and their equivalents in kinematic redundancy resolution are defined here.

Function pseudoinverse. A pseudoinverse of a surjective function $f : A \to B$ is an injective function $f^+ : f[A] \to A$ such that each element of the image $f[A]$ is mapped to an element in its preimage, i.e., $f(f^+(y)) = y$ for all $y \in f[A]$.

A pseudoinverse always exists even though an inverse may not, and if f is a bijection, then the pseudoinverse is identically the inverse. We also note that when A is of higher dimension than B, the preimage of each element in $f[A]$ is, in general, an infinite set, and hence the number of pseudoinverses is infinite. Although it may be easy to compute a pseudoinverse *pointwise*, e.g., by Newton's method, pointwise pseudoinverses often do not satisfy desirable properties, such as continuity or smoothness.

Continuous pseudoinverse. A continuous pseudoinverse f^+ is a pseudoinverse that is continuous in the parameters y across all of $f[A]$.

Through the rest of the paper we will use robot kinematics terminology as follows. Here, $\mathcal{C}$ is a configuration space (C-space) consisting of the robot's degrees of freedom, $\mathcal{W}$ is the Cartesian workspace, and $f : \mathcal{C} \to \mathcal{W}$ is the forward kinematics mapping. The workspace typically contains either position or orientation of the end effector, or both. In redundant problems, we have $dim(\mathcal{C}) > dim(\mathcal{W})$. We will use q to denote configurations and y to denote workspace points. Define the free space $\mathcal{F} \subseteq \mathcal{C}$ as the subset of configurations that lie within joint bounds and are collision-free. Define the reachable workspace $\mathcal{W}_C \subseteq \mathcal{W}$ as the subset reachable by free space configurations $\mathcal{W}_C = f[\mathcal{F}]$.

Pointwise redundancy resolution. A function f^+ is a pointwise resolution if it is a pseudoinverse of f over $\mathcal{W}_C$.

Pathwise redundancy resolution. A function f^+ is a pathwise resolution over the workspace path $y : [0, 1] \to \mathcal{W}_C$ if it is a continuous pseudoinverse over the path. In other words, $f(f^+(y(t))) = y(t)$ and $\lim_{u \to t} f^+(y(u)) = f^+(y(t))$ for all $t \in [0, 1]$. In this case we call $q(t) \equiv f^+(y(t))$ the resolved path. We also speak of *endpoint constrained pathwise resolution* where boundary conditions $q_0 = f^+(y(0))$ and/or $q_1 = f^+(y(1))$ exist at one or both of the endpoints.

Global redundancy resolution. A function f^+ is a global resolution if it is a continuous pseudoinverse of f over $\mathcal{W}_C$.

We will say that a problem is (pointwise, pathwise, or globally) resolvable if a (pointwise, pathwise, or global) resolution exists. It is straightforward to observe that global $\implies$ pathwise $\implies$ pointwise resolvability. (In fact, pointwise

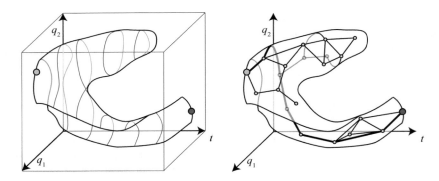

Fig. 3. Left: Illustrating the pathwise redundancy resolution problem. Right: a probabilistically complete roadmap-based solution.

resolvability holds true by definition.) As discussed above, the converse does not necessarily hold true.

4   Pathwise Redundancy Resolution

As a preliminary step, we will describe an algorithm to solve the simpler problem of pathwise redundancy resolution, illustrated in Fig. 3. At any workspace point, the kinematic constraint limits the set of valid points to a submanifold of C-space known as the self-motion manifold, and boundaries are introduced to the manifold due to feasibility constraints (e.g., joint limits). As the path parameter sweeps from 0 to 1, the self-motion manifolds sweeps out a manifold of one higher dimension in the Cartesian product $[0, 1] \times C$. The goal is to find a path along this manifold with a monotonically increasing path parameter.

First, let us define a few commonly used subroutines.

- Solve(y, q_{init}) solves a root-finding problem $f(q) = y$ numerically using q_{init} as the initial point. If it fails, it returns *nil*. It is assumed that the result q lies close to q_{init}.
- SampleF(y) first samples a random configuration $q_{rand} \in C$ and then uses Solve(y, q_{rand}). If the result is *nil* or infeasible, then *nil* is returned.
- Visible(y, t_s, t_g, q_s, q_g) is an incomplete, deterministic method for local path resolution of a path $y(t)$ over the interval $[t_s, t_g]$. We require that $f(q_s) = y(t_s)$ and $f(q_g) = y(t_g)$ be given as the endpoints of the interval. Pseudocode for Visible is given in Alg. 1, and is similar to the method of [14] except with collision handling.

Here $0.5 < c < 1$ is a parameter that controls the maximum amount of drift away from a straight line path. Without Line 5, the bisected path could grow without bound. Usually c is set close to 1. Examples in this paper use 0.9.

Using these primitives, we present a probabilistically complete path resolution method using a slightly modified probabilistic roadmap (PRM) algorithm.

Algorithm 1 Visible(y, t_s, t_g, q_s, q_g)

1: **if** $d(q_s, q_g) \leq \epsilon$ **then return** "true"
2: Let $y_m \leftarrow y((t_s + t_g)/2)$ and $q_m \leftarrow (q_s + q_g)/2$
3: Let $q \leftarrow Solve(y_m, q_m)$
4: **if** $q = nil$ or $q \notin \mathcal{F}$ **then return** "false"
5: **if** $max(d(q, q_s), d(q, q_g)) > c \cdot d(q_s, q_g)$ **then return** "false"
6: **if** Visible(y, t_s, t_m, q_s, q_m) and Visible(y, t_m, t_g, q_m, q_g) **then return** "true"
7: **return** "false"

Algorithm 2 PRM-Path-Resolution(y, N)

1: Initialize empty roadmap $\mathcal{R} = (V, E)$
2: **if** $q(0)$ and $q(1)$ are given **then**
3:     Add $(0, q(0))$ and $(1, q(1))$ to V
4: **else**
5:     Sample $O(N)$ start configurations using SampleF($y(0)$)
6:     Sample $O(N)$ goal configurations using SampleF($y(1)$)
7: **for** $i = 1, ..., N$ **do**
8:     Sample $t_{sample} \sim U([0, 1])$
9:     Sample $q_{sample} \leftarrow$ SampleF($y(t_{sample})$)
10:    **if** $q_{sample} \neq nil$ **then** add (t_{sample}, q) to V
11: **for** all nearby pairs of vertices $(t_u, q_u), (t_v, q_v)$ with $t_u < t_v$ **do**
12:    **if** Visible(y, t_u, t_v, q_u, q_v) **then**
13:        Add the (directed) edge to E
14: Search $\mathcal{R}$ for a path from $t = 0$ to $t = 1$

Here, the PRM is built in the space $[0, 1] \times \mathcal{F}$ of time-configuration pairs (t, q), subject to the manifold constraint $f(q) = y(t)$. The two necessary modifications to PRM are 1) maintaining the manifold constraint, and 2) restricting forward progress along the time domain by constructing a directed graph. Pseudocode is given in Alg. 2.

5 Approximate global redundancy resolution

Although planning paths on constraint manifolds has been well studied [1, 11], an algorithm for global resolution must, essentially, plan C-space motions for all workspace paths. We assume the workspace is discretized into a network of workspace paths, whose nodes are sufficiently dense to interpolate behavior across the entire workspace via standard function approximation techniques (Sec. 6). This section describes two methods for generating the resolution, one local method and one global sampling-based method. This latter method takes inspiration from Alg.2 in that we build C-space roadmaps, except the roadmap is built along all workspace paths, and rather than finding one path we seek a connected "sheet" that spans the projection to workspace.

Let $G_W = (V_W, E_W)$ be a workspace roadmap. This can be generated either in the form of a probabilistic roadmap via random sampling, or as a grid. The solution we seek is a mapping from the vertices to C-space $g : V_W \to \mathcal{F}$ such that for any two adjacent workspace points $(y, y') \in E_W$, the straight line path $\overline{yy'}$ is locally pathwise resolvable between $g[y]$ and $g[y']$. For notational convenience, let the local reachability indicator function $R(y, y', q, q')$ be 1 if $\text{Visible}(\overline{yy'}, 0, 1, q, q')$ yields "success" and 0 otherwise.

In the case that g is not a resolution, we propose the use of the following primary error metric that measures the number of unresolved edges:

$$U(g) = |E_W| - \sum_{(y,y') \in E_W} R(y, y', g[y], g[y']) \tag{1}$$

If the problem is not resolvable, we wish to find a g that minimizes $U(g)$.

Note that for a globally resolvable problem, there are also an infinite number of pseudoinverses, some of which are smoother than others. It is then a desirable secondary objective to maximize smoothness in the redundant dimensions. Distance is a good proxy for smoothness, so for a secondary error metric we use an edge cost metric that measures total C-space path length:

$$L(g) = \sum_{(y,y') \in E_W} d(g[y], g[y']) R(y, y', g[y], g[y']). \tag{2}$$

In each of our algorithms, we will generate a configuration roadmap $\mathcal{R}_C = (V_C, E_C)$ whose vertices and edges map (surjectively) to corresponding vertices and edges of G_W. The connection is given by $Y[q]$, which maps C-space vertices to associated workspace vertices, and its inverse $Q[y]$, which map workspace vertices to a set (possibly empty) of associated C-space vertices. We will maintain:

$$E_C = \{(q, q') \mid (Y[q], Y[q']) \in E_W \text{ and } R(Y[q], Y[q'], q, q') = 1\}. \tag{3}$$

Each algorithm has a different method of sampling $\mathcal{R}_C$ and selecting the elements of the map $g[y] \in Q[y]$.

5.1   Pointwise global resolution

We first present a fast and simple pointwise method that locally propagates pointwise solutions across the workspace roadmap. Here each set $Q[y]$ contains at most 1 configuration, and hence the extraction of g is straightforward. Pseudocode is given in Alg. 3.

Here $N(y)$ is the neighborhood of a vertex y in the workspace graph. Note that the bookkeeping associated with maintaining the associations in Y and Q is fairly straightforward, but since it is rather tedious to write it will be left implicit in the remainder of the pseudocode. Specifically, in lines 6 and 7, we implicitly assume that the maps Y and Q will be updated appropriately: $Y[q] \leftarrow y$ and $Q[y] \leftarrow Q[y] \cup \{q\}$.

Algorithm 3 Pointwise-Global-Resolution(G_W, N_q)

1: Initialize empty roadmap $\mathcal{R}_C = (V_C, E_C)$
2: **for** each $y \in V_W$ **do**
3:     Let $Q_{seed} \leftarrow \cup_{w \in N(y)} Q[w]$
4:     **for** each $q_s \in Q_{seed}$ **do**
5:         Run $q \leftarrow$Solve(y, q_s)
6:         **if** $q \neq nil$ **then** add q to V_C and go to Step 2, proceeding to the next y.
7:     Run SampleF(y) up to N_q times. If any sample q succeeds, add it to V_C.
8: **for** all edges $(y, y') \in E_W$ such that $|Q(y)| > 0$ and $|Q(y')| > 0$ **do**
9:     Let q be the only member of $Q(y)$ and q' the only member of $Q(y')$
10:     **if** R(y, y', q, q')=1 **then**
11:         Add (q, q') to E_C
    return $\mathcal{R}_C$

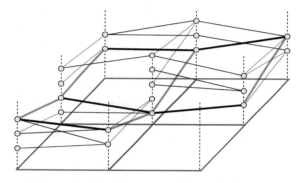

Fig. 4. Illustrating global resolution on a small problem. Each point on a workspace roadmap (3x3 grid on horizontal plane) corresponds to a certain self-motion manifold (vertical dashed lines). Feasible configuration space samples (circles) are sampled in each manifold, and edges are tested for feasibility along all workspace edges. Global resolution seeks to select a single configuration for each workspace point to form a maximally connected "sheet" in the configuration space roadmap.

5.2   Constraint satisfaction-based resolution

The pointwise method can yield poor results, leaving several edges unnecessarily unresolved. To improve performance, we present the following algorithm that samples many configurations in the preimage of each workspace point, connects them with feasible edges, and seeks a "sheet" in the C-space roadmap (Fig. 4). In order to perform this latter step, redundancy resolution is cast as a constraint satisfaction problem (CSP).

The algorithm proceeds by sampling many configurations per workspace point and adding them to $Q[y]$. Each pair of configurations along workspace edge is then tested for visibility and added to E_C. This leads to the formulation of a MAX-CSP in which a value $g[y] \in Q[y]$ is sought for each workspace point, such that for all neighboring points y', the C-space edge is visible, that is, $(g[y], g[y']) \in E_C$. If there is no such resolution, then MAX-CSP seeks an assignment g that minimizes the number of conflicts $U(g)$. We also maintain $L(g)$ as a secondary optimization criterion.

Since we typically have thousands of workspace points and often dozens or hundreds of configurations in each point's domain, backtracking CSP methods are usually prohibitively expensive, especially when the problem is not resolvable. For example, the open source Gecode library exhausted our machine's RAM before finding even a suboptimal solution on the smallest problems in our experiments [12]. The Sugar algorithm, which converts CSPs to boolean satisfaction (SAT) problems, produced intermediate SAT problems exceeding the limit of 4 Gb on even smaller problems with 500 workspace nodes [13]. As a result we employ a heuristic local search method. The algorithm proceeds in three phases.

Heuristic descent. Our search phase proceeds down a single path of a backtracking search in the dual graph CSP, and stops when no more workspace edges can be assigned. In this formulation, the edges (y, y') are the variables, values are the $g[y]$, $g[y']$ pairs at their endpoints, and the domains are the set of all edges $(q, q') \in E_C$ such that $Y[q] = y$ and $Y[q'] = y'$. Pairs of edges that meet at the same workspace point are constrained so a configuration at one endpoint must match the value of the adjacent edge. The advantage of this formulation is that maximizing the number of values assigned leads to a direct minimization of $U(q)$. Forward checking and domain consistency are used for constraint propagation, and variable ordering heuristics of most constrained variable and most constraining variable are applied. Value selection is done using the least-constraining value heuristic. ties are broken randomly. No backtracking is performed, and as many non-conflicting variables are assigned as possible.

Min-conflicts. The second phase does a more straightforward min-conflicts operation, this time on the primal graph where variables are nodes, values are assigned configurations $g[y]$, and domains are sampled configurations $Q[y]$. Here, the algorithm iterates from the assignment produced by phase 1 by choosing a conflicting variable at random, if one exists. If none exists, then we are done. The value in its domain that minimizes conflicts with its neighbors is then selected. This process repeats some number of iterations.

Length optimization via coordinate descent. At this point, the number of conflicts is usually close to a minimum. The final phase then attempts to optimize $L(g)$ while keeping $U(g)$ constant. Here, a coordinate descent algorithm is used. For each C-space node $g[y]$, we compute the average configuration q_{avg} of its neighbors in $\mathcal{R}_C$ amongst those whose endpoints match in g. It is then used as a target for optimization $\min_q d(q, q_{avg})$, under the constraint that $f(q) = y$ must be satisfied and that reachability of q from all neighbors is satisfied. To do this, we use bisection. First, $q \leftarrow Solve(y, q_{avg})$ is calculated. If q is not reachable from its neighbors, then q_{avg} is set to the midpoint of the line segment from $g[y]$ to q_{avg}. Solving and bisection repeats until a reachable configuration is found.

The overall algorithm is listed in Alg. 4. Experiments suggest that all three of these phases cooperate to produce high-quality results. Skipping the heuristic descent step leads to poor performance for more complex problems, because

---

Algorithm 4 CSP-Global-Resolution($G_W, N_q, N_{mc}, N_{opt}$)

1: Initialize $\mathcal{R}_C \leftarrow$ Pointwise-Global-Resolution(G_W, N_q)
2: **for** each $y \in V_W$ **do**
3:     Run SampleF(y) N_q times. Add all samples that succeed to V_C.
4: **for** all edges $(y, y') \in E_W$ such that $|Q(y)| > 0$ and $|Q(y')| > 0$ **do**
5:     **for** all $(q, q') \in Q(y) \times Q(y')$ **do**
6:         **if** R(y, y', q, q') = 1 **then**
7:             Add (q, q') to E_C
8: $g \leftarrow$ HeuristicDescentCSP($\mathcal{R}_C$)
9: $g \leftarrow$ MinConflictsCSP($\mathcal{R}_C, g, N_{mc}$)
10: CoordinateDescent($\mathcal{R}_C, g, N_{opt}$)
11: **return** g

---

pointwise assignment leaves the solution in a deep local minimum. Skipping min-conflicts fails to clean up some errors in heuristic descent, and skipping length optimization leads to maps that are much less smooth.

The parameter values are as follows:

- N_q: the number of C-space samples drawn per workspace node. More samples are typically needed for highly redundant systems. We use 50–100 in our experiments.
- N_{mc}: the number of min-conflicts iterations. A small number of passes, such as 10× the number of conflicts, usually works well.
- N_{opt}: the number of coordinate descent iterations. We have observed most problems converging to less than 0.1% change in objective function value in about 20 iterations.

5.3 Performance considerations

In our experiments, the algorithms above range in computation time from minutes to hours for thousands of workspace points. The limiting step is C-space roadmap construction (Steps 2–7), and visibility checking in particular (Step 6), which is significantly more expensive than any other primitive operation.

Scalability-wise, the pointwise assignment algorithm performs $O(N_q|V_W|)$ C-space samples (although in practice this is closer to $O(|V_W|)$), and $O(|E_W|)$ visibility checks. The CSP algorithm performs $O(N_q|V_W|)$ configuration samples and $O(N_q^2|E_W|)$ visibility checks in Lines 2–7. With efficient implementation of CSP updates and a priority queue for sorting edges, heuristic descent runs in time $O(d_W N_q^2|E_W|\log|E_W|)$, where d_W is the degree of G_W. The min-conflicts operation finds initial conflicts in $O(d_W N_q^2|V_W|)$ and each of the N_{mc} iterations takes time $O(d_W N_q^2)$ time. Finally, optimization takes $O(N_{opt}|V_W|)$ optimization steps and $O(N_{opt}|E_W|)$ visibility checks.

The other question is whether the algorithm produces an optimal or near-optimal resolution? There are two factors influencing this: whether the C-space roadmap contains a solution, and whether the constraint satisfaction solver finds

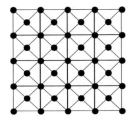

Fig. 5. Staggered grid network used to discretize the workspace.

it. For problems with narrow passages in free space, finer workspace and configuration roadmaps are needed to resolve them. As a result, it is expected that as $|G_W|$ and N_q grows, the probability that $\mathcal{R}_C$ contains a resolution will grow towards 1. In practice, it is often effective to employ a second sampling pass at the vertices of unresolved edges. As for the CSP solver, even though the algorithm performs well in practice, it is indeed a heuristic. Producing an efficient, optimal method seems challenging in light of the NP-hardness of CSPs and the poor performance of existing state-of-the-art backtracking solvers.

6   Interpolation between sample points

Once a resolution $g[y]$ over the workspace roadmap G_W is produced, we extend it to the continuous workspace $\mathcal{W}_C$ via function approximation. In particular, we yield $f^+(y) = Solve(g_{interp}(y), y)$ where $g_{interp}(y)$ is a weighted combination of resolved configurations at workspace points

$$g_{interp}(y) = \sum_i w_i(y)c_i(y)g[y_i] / \sum_j w_j(y)c_j(y) \qquad (4)$$

and the sums are taken over indices of workspace points $y_i \in V_W$. The terms w_i are distance-based weights that decrease as y grows more distant from y_i, while the c_i terms are indicator functions that prevent interpolation across discontinuity boundaries in the resolution. For a regular grid, barycentric coordinates in the simplex are used as weights. If G_W is a probabilistic roadmap it is possible to use scattered data interpolation techniques.

The term c_i is 1 if the neighborhood of y is globally resolved, but if there is a discontinuity, it is used only to average points on one side of the boundary. To compute it, we take the subgraph of the C-space roadmap $\mathcal{R}_C$ corresponding to the resolved points $g[y_i]$ for all workspace points y_i with nonzero weights. We then compute connected components of this subgraph, and for all nodes in the largest connected component c_i is set to 1. It is set to 0 for all other nodes.

7   Results

In all of the following experiments the workspace was the Cartesian coordinates of a point on the robot's end effector in 2- or 3-dimensional space. A staggered

Table 1. Summary of experiments.

Robot	$dim(\mathcal{C})$	$dim(\mathcal{W})$	Grid points	% disconnected	Distance ratio (rad/m)
Planar 3R	3	2	2,000	1.42	0.28
Planar 20R	20	2	2,000	0.96	0.21
ATLAS	7	3	16,000	1.61	3.03
Baxter	7	3	8,000	3.92	1.95
Jaco	6	3	8,000	0.062	3.81
Robonaut2	6	3	8,000	4.50	3.12
Robosimian	7	3	4,000	13.9	2.92

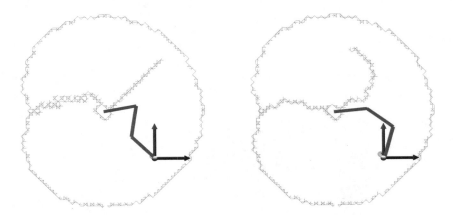

Fig. 6. Pseudoinverse discontinuity boundaries for a planar 3R arm. Left: pointwise resolution. Right: the optimized solution has slightly fewer discontinuities.

workspace grid was used in each example (Fig. 5). Statistics for these tests are given in Tab. 1. Most of these parameters are self-explanatory except for the final two columns, which give the percentage of disconnected workspace graph edges (% disconnected) and the overall ratio of configuration space distance to workspace distance summed along all edges on the roadmap (Distance ratio).

The first set of experiments are performed on planar kR robots. Fig. 6, left, shows the pointwise assignment boundaries for a 3R robot with joint limits of ± 2 rad and a workspace grid of 2,000 points. In this case, pointwise assignment works fairly well. Our method (Fig. 6, right) only reduces the number of disconnections by 24%, but reduces the total C-space path length by about 60%.

Results for a 20R robot with joint limits of ± 0.6 rad are shown in Fig. 7. In this case, the pointwise assignment causes a wide swath of disconnections in the upper part of the workspace (3.69% of reachable edges). Using the CSP method, the number of disconnections are greatly reduced (0.96% of reachable edges) and the total path length is reduced by 38%.

Our next set of experiments apply the method to robot arms with 6 or more DOFs in 3D Cartesian space. Arbitrary orientation of the end effector is permitted, and arm configurations are required to obey joint limits and avoid self-

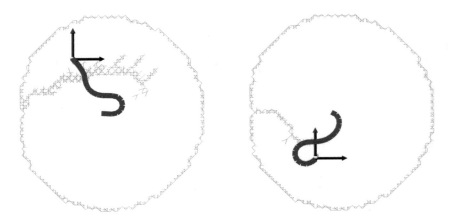

Fig. 7. A planar 20R arm. Left: pointwise resolution. Right: optimized solution.

Fig. 8. Results for the Jaco arm. Left: after pointwise resolution, the discontinuity boundary spans nearly the entire workspace. Right: optimized solution is resolved almost everywhere. (For clarity, the outer workspace boundary is not drawn in these or any other 3D figures.)

collision. In each case, pointwise assignment does worse than in 2D. Fig. 8, left, shows results for a Kinova Jaco arm. The purple region illustrates the surface of disconnections. The IK problem is to move the center of the hand to a specific position, and the workspace grid has 8,000 points. A huge number of disconnections are caused by pointwise resolution. After CSP resolution (Fig. 8, right), the number of disconnections drops to a minimal number (0.062% of reachable edges). This favorable result is likely because the Jaco has three continuous rotation joints.

The next four figures show optimized results for single arms of the Boston Dynamics ATLAS (Fig. 9), the Rethink Robotics Baxter (Fig. 10), the NASA Robonaut2 (Fig. 11), and a leg of the JPL Robosimian (Fig. 12). None of these robots has continuous rotation joints. The kinematic structure and limits of each robot is different, with the ATLAS, Baxter, and Robonaut2 having an

Fig. 9. Optimized results for the arm of the ATLAS humanoid. Two views of the discontinuity boundary are shown.

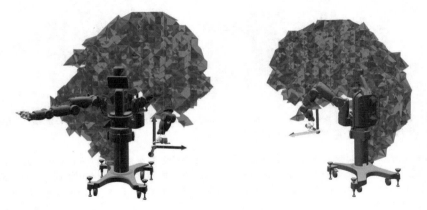

Fig. 10. Optimized results for the arm of the Baxter robot. Two views of the discontinuity boundary are shown.

anthropomorphic elbow and 3-axis shoulder, but with different axes and joint limits. The Robosimian's joints are arranged so that each pair of 2 joints have intersecting axes, and each joint can rotate 360°.

ATLAS has a discontinuous region near a singularity at the shoulder, as well as over the head where the upper arm comes close to colliding with the head. Baxter has a much larger region above the shoulder, likely because the arm naturally bends downward, and it approaches joint limits as the arm bends upward. As a result, to pass from front to back in that upward region, the hand must pass down and around. It also has discontinuities the opposite side of the torso near the rear, where the arm can reach points either around the front or around the back of the torso. Robonaut2 has essentially the opposite problem, with discontinuities along the underside of the arm. Joint limits in the shoulder mean that its hand must move up and around to pass from elbow-back to elbow-

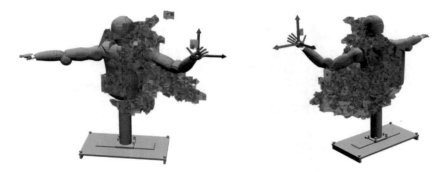

Fig. 11. Optimized results for the arm of the Robonaut2. Two views of the discontinuity boundary are shown.

Fig. 12. Optimized results for one limb of the Robosimian quadruped. Two views of the discontinuity boundary are shown.

front configurations. Other small discontinuities are scattered about the torso, which are likely free space narrow passages caused by self-collision. Perhaps a denser sampling of configurations or workspace would help reduce these artifacts.

Robosimian is an interesting case. Although its joints have a larger range of motion than each of the four other robots, its kinematics are quite different. Its workspace is quite large, as each limb can reach completely around the body in all directions. Although it is hard to see from the photos, most of the discontinuous boundaries lie close to edge of the workspace or near the robot's body, and a large cavity of resolved space lies below the robot. Unlike the manipulator arms, it is unable to rotate a "shoulder" about an arbitrary axis, and hence it needs to perform rather large joint space motions to reconfigure itself to move the end effector when close to singularities. Also, when close to self-collision, many end effector movements are likely to cause links near the body to come into collision.

8   Conclusion

This paper defined and presented an approximate method for computing maximally continuous pseudoinverses of multivariate functions, with the application of global redundancy resolution for Cartesian workspace movements of robot manipulators. The resulting maps may be useful for robot mechanism design, teleoperation, or reduced dimensionality motion planning. Performance of the CSP solver might be improved via connectivity analysis of self-motion manifolds to reduce the size of the constraint graph. Future work should also study application of global resolution to higher dimensional workspaces.

Acknowledgment

This work is partially supported by NSF CAREER Award #1253553 and NSF NRI #1527826.

References

1. D. Berenson, S. S. Srinivasa, D. Ferguson, and J. J. Kuffner. Manipulation planning on constraint manifolds. In *IEEE Int. Conf. Rob. Aut.*, 2009.
2. K. Byl, M. Byl, and B. Satzinger. Algorithmic optimization of inverse kinematics tables for high degree-of-freedom limbs. In *Proc. ASME 2014 Dynamic Systems and Control Conference*, Oct. 2014.
3. P. Freeman. *Minimum Jerk Trajectory Planning for Trajectory Constrained Redundant Robots*. PhD thesis, Washington University of Saint Louis, 2011.
4. M. Goel, A. A. Maciejewski, V. Balakrishnan, and R. W. Proctor. Failure tolerant teleoperation of a kinematically redundant manipulator: an experimental study. *IEEE Transactions on Systems, Man, and Cybernetics - Part A: Systems and Humans*, 33(6):758–765, Nov 2003.
5. C. L. Lück. Self-motion representation and global path planning optimization for redundant manipulators through topology-based discretization. *J. Intelligent and Robotic Systems*, 19:23–28, 1997.
6. R. V. Mayorga and A. K. C. Wong. A global approach for the optimal path generation of redundant robot manipulators. *J. Robotics Systems*, 7(1):107–128, 1990.
7. Y. Nakamura. *Advanced Robotics: Redundancy and Optimization*. Addison-Wesley, New York, 1991.
8. G. Oriolo and C. Mongillo. Motion planning for mobile manipulators along given end-effector paths. In *IEEE Int. Conf. Rob. Aut.*, 2005.
9. B. Satzinger, J. I. Reid, M. Bajracharya, P. Hebert, and K. Byl. More solutions means more problems: Resolving kinematic redundancy in robot locomotion on complex terrain. In *IEEE/RSJ Int. Conf. Intel. Rob. Sys.*, 2014.
10. S. Seereeram and J. T. Wen. A global approach to path planning for redundant manipulators. *IEEE T. Robotics and Automation*, 11(1):152–160, Feb. 1995.
11. M. Stilman. Global manipulation planning in robot joint space with task constraints. *IEEE Trans. Robotics*, 26:576–584, 2010.
12. G. Tack. *Constraint Propagation - Models, Techniques, Implementation*. PhD thesis, Saarland University, Germany, 2009.
13. N. Tamura, A. Taga, S. Kitagawa, and M. Banbara. Compiling finite linear csp into sat. *Constraints*, 14(2):254–272, June 2009.
14. R. H. Taylor. Planning and execution of straight line manipulator trajectories. *IBM J. Res. Dev.*, 23(4):424–436, July 1979.
15. F. Zacharias, C. Borst, M. Beetz, and G. Hirzinger. Positioning mobile manipulators to perform constrained linear trajectories. In *IEEE/RSJ Int. Conf. Intel. Rob. Sys.*, 2008.

A Generalized Label Correcting Method for Optimal Kinodynamic Motion Planning

Brian Paden and Emilio Frazzoli

Massachusetts Institute of Technology
bapaden@mit.edu,   frazzoli@mit.edu

Abstract. An optimal kinodynamic motion planning algorithm is presented and described as a generalized label correcting (GLC) method. Advantages of the technique include a simple implementation, convergence to the optimal cost with increasing resolution, and no requirement for a point-to-point local planning subroutine. The principal contribution of this paper is the construction and analysis of the GLC conditions which are the basis of the proposed algorithm.

1   Introduction

Motion planning is a challenging and fundamental problem in robotics research with numerous algorithms fine-tuned to variations of the problem. Among the most popular is the PRM [1], an algorithm for planning in relatively high dimensions, but requiring a point-to-point local planning subroutine or steering function to connect pairs of randomly sampled states. A variation, PRM* [2], converges to an optimal solution with respect to some objective provided the steering function is optimal.

The EST [3] and RRT [4] algorithms were developed to address suboptimal planning with differential constraints where a steering function is unavailable. The steering function in these algorithms is replaced by a subroutine which forward integrates the system dynamics with a selected control input.

Optimal planning under differential constraints can be addressed by the kinodynamic variant of the RRT* algorithm [5], but again requires an optimal steering function like PRM*. The effectiveness of RRT* has motivated several general approaches for optimal steering discussed in [6,7]. However, the computation of these solutions can slow down the iteration time of RRT* considerably.

This paper addresses optimal planning under differential constraints without the use of a steering function. The method is based on a discrete approximation of the problem as a graph together with a generalization of label correcting techniques such as Dijkstra's algorithm [8].

Label correcting algorithms compare the relative cost of paths in a graph terminating at the same vertex and discard paths with non-minimal cost. This is effective when there are multiple paths reaching each vertex from the root vertex. Without a steering function it is difficult to construct such a graph approximating motion planning solutions since multiple trajectories terminating at specified states must be generated. The intuitive generalization is to compare the cost of paths related to trajectories terminating "close enough" to one another. This concept first appeared in [9] as the Hybrid A*

© Springer Nature Switzerland AG 2020
K. Goldberg et al. (Eds.): *Algorithmic Foundations of Robotics XII*, SPAR 13, pp. 512–527, 2020.
https://doi.org/10.1007/978-3-030-43089-4_33

algorithm, but without conditions for resolution completeness. More recently, the SST algorithm [10] provided a more principled approximation of the problem and an algorithm converging asymptotically to an approximately optimal solution.

This paper refines this approach further with an algorithm producing approximate solutions in finite-time. The key difference allowing the algorithm to terminate in finite time is a slightly more conservative comparison between paths terminating in the same region of the state space. This comparison is described in Section 2. Section 3 provides examples illustrating convergence to optimal cost solutions with increasing resolution. The analysis of the algorithm is addressed in Section 4 and is the principal contribution of the paper.

2   GLC Methods

Label correcting methods: Shortest path algorithms are methods of searching over paths of a graph for a minimum-cost path between an origin or root vertex to a set of goal vertices. In a conventional label correcting method, the algorithm maintains a best known path terminating at each vertex of the graph. This path *labels* that vertex. At a particular iteration, if a path under consideration does not have lower cost than the path labeling the terminal vertex, the path under consideration is discarded. As a consequence, the subtree of paths originating from the discarded path will not be evaluated.

Generalizing the notion of a label: Observe that the label of a vertex in conventional label correcting algorithms is in fact a label for the paths terminating at that vertex. Then each vertex identifies an equivalence class of paths. Paths within each equivalence class are ordered by their cost, and the efficiency of label correcting methods comes from narrowing the search to minimum cost paths in their associated equivalence class. The generalization is to identify paths associated to trajectories terminating in the same region of the state space instead of the same state. However, this generalization prevents a direct comparison of cost between two related paths. Instead, the difference in cost must exceed a threshold described by the GLC conditions introduced in Section 2.3.

2.1   Problem Formulation

Consider a system whose configuration and relevant quantities are described by a state in $\mathbb{R}^n$. The decision variable of the problem is an input signal or continuous history of actions u from a *signal space* $\mathcal{U}$ affecting a state trajectory x in a *trajectory space* $\mathcal{X}_{x_0}$. The signal space is constructed from a set of admissible inputs $\Omega \subset \mathbb{R}^m$ bounded by u_{max}. The input signal space is defined

$$\mathcal{U} := \bigcup_{\tau > 0} \{u \in L_1([0, \tau]) : u(t) \in \Omega \ \forall t \in [0, \tau]\}. \tag{1}$$

The signal and trajectory are related through a model of the system dynamics described by a differential constraint,

$$\frac{d}{dt} x(t) = f(x(t), u(t)), \quad x(0) = x_0. \tag{2}$$

A *system map* $\varphi_{x_0} : \mathcal{U} \to \mathcal{X}_{x_0}$ is defined to relate signals to associated trajectories (the solution to (2) [11, cf. pg. 42]) with domain equal to the domain of the input signal. The initial state x_0 parametrizes the map and trajectory space. To simplify notation, $\tau(x)$ for a function x with domain $[t_1, t_2]$ denotes the maximum of the domain, t_2.

In addition to the differential constraint, feasible trajectories for a particular problem must satisfy point-wise constraints defined by a subset X_{free} of $\mathbb{R}^n$ and a specified initial state x_{ic}. The subset of feasible trajectories $\mathcal{X}_{\text{feas}}$ are defined

$$\mathcal{X}_{\text{feas}} := \{x \in \mathcal{X}_{x_{\text{ic}}} : \; x(t) \in X_{\text{free}} \; \forall t \in [0, \tau(x)]\}. \tag{3}$$

Similarly, the subset X_{goal} of $\mathbb{R}^n$ is used to encode a terminal constraint. The subset of $\mathcal{X}_{\text{feas}}$ consisting of trajectories x with $x(\tau(x)) \in X_{\text{goal}}$ defines $\mathcal{X}_{\text{goal}}$.

The decision variable for the problem is the input signal u which must be chosen such that the trajectory $\varphi_{x_{\text{ic}}}(u)$ is in $\mathcal{X}_{\text{goal}}$. Naturally, input signals mapping to trajectories in $\mathcal{X}_{\text{feas/goal}}$ are defined by the inverse relation $\mathcal{U}_{\text{feas/goal}} := \varphi_{x_{\text{ic}}}^{-1}(\mathcal{X}_{\text{feas/goal}})$.

A general cost functional which integrates a running-cost g of the state and input at each instant along a trajectory is used to compare the merit of one input signal over another. Restricted to solutions of (2), the cost functional depends only on the control and intitial state,

$$J_{x_0}(u) = \int_{[0,\tau(u)]} g([\varphi_{x_0}(u)](t), u(t)) \, d\mu(t). \tag{4}$$

The notation $[\varphi_{x_0}(u)](t)$ denotes the evaluation of the trajectory satisfying (2) with the input signal u and initial state x_0 at time t.

The domain of J is extended with the object NULL such that $J_{x_0}(\text{NULL}) = \infty$ for all $x_0 \in \mathbb{R}^n$. Further, since (4) may not admit a minimum, we address the following relaxed optimal kinodynamic motion planning problem:

Problem 1. Find a sequence $\{u_R\} \subset \mathcal{U}_{\text{goal}} \cup \text{NULL}$ such that

$$\lim_{R \to \infty} J_{x_{\text{ic}}}(u_R) = \inf_{u \in \mathcal{U}_{\text{goal}}} J_{x_{\text{ic}}}(u) := c^*. \tag{5}$$

With the convention that $\inf_{u \in \emptyset} J_{x_{\text{ic}}}(u) = \infty$, a solution sequence exists so the problem is well-posed. An algorithm parameterized by a resolution $R \in \mathbb{N}$ whose output for each R forms a sequence solving this problem will be called *resolution complete*.

Assumptions. The problem data are assumed to satisfy the following:

A-1 The sets X_{free} and X_{goal} are open with respect to the standard topology on $\mathbb{R}^n$.
A-2 There are known constants $L_f \geq 0$ and $M \geq 0$ such that $\|f(x_1, u) - f(x_2, u)\|_2 \leq L_f \|(x_1 - x_2)\|_2$, and $\|f(x_1, u)\|_2 \leq M$ for all $x_1, x_2 \in X_{\text{free}}$ and $u \in \Omega$.
A-3 There is a known constant $L_g \geq 0$ such that $\|g(x_1, u_1) - g(x_2, u_2)\|_2 \leq L_g \|(x_1 - x_2, u_1 - u_2)\|_2$ for all $x_1, x_2 \in X_{\text{free}}$ and $u_1, u_2 \in \Omega$.
A-4 $J_{x_{ic}}(u) > 0$ for all $u \in \mathcal{U}$.

Remark 1. We do not require the reachable set to have a nonempty interior as in the kinodynamic variant of RRT* [5] and SST [10].

2.2   Approximation of $\mathcal{U}$

The signal space $\mathcal{U}$ is approximated by a finite subset $\mathcal{U}_R$ where $R \in \mathbb{N}$ is a resolution parameter. To construct $\mathcal{U}_R$ it is assumed that we have access to a family of finite subsets $\Omega_R \subset \Omega$ of the input space, such that the dispersion in Ω converges to zero as $R \to \infty$. A family of such subsets exists and is often easily obtained with regular grids or random sampling for a given Ω.

The approximated signal space $\mathcal{U}_R$ consists of piecewise constant signals on time intervals $\left[\frac{c \cdot (i-1)}{R}, \frac{c \cdot i}{R} \right)$ for $c > 0, i \in \{1, ..., d\}$, and all values d less than a user specified function $h(R)$ (take the last interval to be closed). The constant values of the signal are inputs from Ω_R.

The function $h : \mathbb{N} \to \mathbb{N}$ defines a horizon limit and can be *any* function satisfying

$$\lim_{R \to \infty} R/h(R) = 0. \tag{6}$$

For example, $h(R) = R \log(R)$ is acceptable. This ensures that the horizon limit is unbounded in R so that any finite time domain can be approximated for sufficiently large R.

A *parent* of an input signal $w \in \mathcal{U}_R$ with domain $\left[0, \frac{c \cdot i}{R}\right]$ is defined as the input signal $u \in \mathcal{U}_R$ with domain $\left[0, \frac{c \cdot (i-1)}{R}\right]$ such that $w(t) = u(t)$ for all $t \in \left[0, \frac{c \cdot (i-1)}{R}\right)$. In this case, w is a *child* of u. Two signals are *siblings* if they have the same parent. A tree (graph) is defined with $\mathcal{U}_R$ as the vertex set, and edges defined by ordered pairs of signals (u, w) such that u is the parent of w. To serve as the root of the tree, $\mathcal{U}_R$ is augmented with the special input signal $Id_{\mathcal{U}}$ defined such that $J_{x_0}(Id_{\mathcal{U}}) = 0$ and $[\varphi_{x_0}(Id_{\mathcal{U}})](0) = x_0$. $Id_{\mathcal{U}}$ has no parent, but is the parent of signals with domain $[0, c/R]$.

The signal w is an *ancestor* of u if $\tau(w) \leq \tau(u)$ and $w(t) = u(t)$ for all $t \in [0, \tau(w))$. In this case u is a *descendant* of w. The *depth* of a control in $\mathcal{U}_R$ is the number of ancestors of that control.

2.3   Partitioning X_{free} and the GLC Conditions

A partition of X_{free} is used to define comparable signals. We say the partition has radius r if the partition elements are each contained in a neighborhood of radius r. Like the discretization of $\mathcal{U}$, the radius of the partition is parametrized by the resolution R.

For brevity we only consider hypercube partitions. A user specified function $\eta : \mathbb{N} \to \mathbb{R}_{>0}$ controls the side length of the hypercube. For states $p_1, p_2 \in \mathbb{R}^n$ we write $p_1 \overset{R}{\sim} p_2$ if $\lfloor \eta(R)p_1 \rfloor = \lfloor \eta(R)p_2 \rfloor$, where $\lfloor \cdot \rfloor$ is the coordinate-wise floor map (e.g. $\lfloor (2.9, 3.2) \rfloor = (2, 3)$). Then the equivalence classes of the $\overset{R}{\sim}$ relation define a simple hypercube partition of radius $\sqrt{n}/\eta(R)$. We extend this relation to control inputs by comparing the terminal state of the resulting trajectory. For $u_1, u_2 \in \mathcal{U}_R$ we write $u_1 \overset{\mathcal{U}_R}{\sim} u_2$ if the resulting trajectories terminate in the same hypercube. That is,

$$u_1 \overset{\mathcal{U}_R}{\sim} u_2 \Leftrightarrow [\varphi_{x_{\text{ic}}}(u_1)](\tau(u_1)) \overset{R}{\sim} [\varphi_{x_{\text{ic}}}(u_2)](\tau(u_2)). \tag{7}$$

Figure 1 illustrates the intuition behind this equivalence relation.

To compare signals we write $u_1 \prec_R u_2$ if the GLC conditions are satisfied:

GLC-1 $u_1 \overset{\mathcal{U}_R}{\sim} u_2$,

GLC-2 $\tau(u_1) \le \tau(u_2)$,

GLC-3 $J_{x_{\text{ic}}}(u_1) + \frac{\sqrt{n}}{\eta(R)} \frac{L_g}{L_f} \left(e^{\frac{L_f h(R)}{R}} - 1 \right) \le J_{x_{\text{ic}}}(u_2)$.

A signal u_1 is called *minimal* if there is no $u_2 \in \mathcal{U}_R$ such that $u_2 \prec_R u_1$. Such a signal can be thought of as being a good candidate for later expansion during the search. Otherwise, it can be discarded. In order for the GLC method to be a resolution complete algorithm, the scaling parameter η must satisfy

$$\lim_{R \to \infty} \frac{R}{L_f \eta(R)} \left(e^{\frac{L_f h(R)}{R}} - 1 \right) = 0. \tag{8}$$

Observe that (8) implies the cost threshold in **GLC-3** is in $O(1/R)$ and converges to zero. A condition yielding the same theoretical results, but asymptotically requiring more signals to be evaluated would be to replace the threshold with an arbitrarily small positive constant.

Additionally, (8) and **GLC-3** simplify in some cases. For kinematic problems $L_f = 0$. Taking the limit $L_f \to 0$, the constraint (8) becomes $h(R)/\eta(R) \to 0$. The second special case is minimum-time problems where $g(x, u) = 1$ in (4) so that $L_g = 0$. Then **GLC-3** simplifies to $J_{x_{\text{ic}}}(u_1) \le J_{x_{\text{ic}}}(u_2)$.

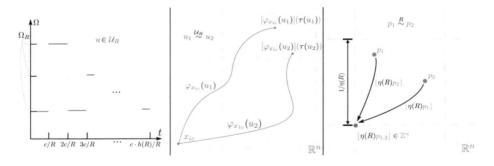

Fig. 1. Color coded depictions of: (left) a signal from $\mathcal{U}_R$ with signals represented in red, (middle) the mapping into the trajectory space $\mathcal{X}_{x_{\text{ic}}}$ in blue for two equivalent signals $u_1 \overset{\mathcal{U}_R}{\sim} u_2$, and (right) the mapping from terminal states in $\mathbb{R}^n$ shown in green into $\mathbb{Z}^n$ by the floor map.

2.4   Algorithm Description

Pseudocode for the GLC method is described in Algorithm 1 below. A set Q serves as a priority queue of candidate signals. A set Σ contains signals representing labels of $\overset{\mathcal{U}_R}{\sim}$ equivalence classes.

The method expand(u) returns the set of all children of u. The method pop(Q) deletes from Q, and returns an input signal $\hat{u}$ such that

$$\hat{u} \in \underset{u \in Q}{\operatorname{argmin}} \{J_{x_{\text{ic}}}(u)\}. \tag{9}$$

The addition of an admissible heuristic [12] in (9) can be used to guide the search without affecting the solution accuracy.

The method $\mathtt{find}(u, \Sigma)$ returns $w \in \Sigma$ such that $u \overset{R}{\sim} w$ or NULL if no such w is present in Σ. Problem specific collision and goal checking subroutines are used to evaluate $u \in \mathcal{U}_{\text{feas}}$ and $u \in \mathcal{U}_{\text{goal}}$. The method $\mathtt{depth}(u)$ returns the number of ancestors of u.

Algorithm 1 Generalized Label Correcting (GLC) Method

1: $Q \leftarrow \{Id_{\mathcal{U}}\}$, $\Sigma \leftarrow \emptyset$, $S \leftarrow \emptyset$
2: **while** $Q \neq \emptyset$
3:    $u \leftarrow \mathtt{pop}(Q)$
4:    **if** $u \in \mathcal{U}_{\text{goal}}$
5:       **return** $(J_{x_{ic}}(u), u)$
6:    $S \leftarrow \mathtt{expand}(u)$
7:    **for** $w \in S$
8:       $z \leftarrow \mathtt{find}(w, \Sigma)$
9:       **if** $(w \notin \mathcal{U}_{feas.} \vee (z \prec_R w) \vee \mathtt{depth}(w) \geq h(R))$
10:          $S \leftarrow S \setminus \{w\}$
11:       **else if** $J_{x_{ic}}(w) < J_{x_{ic}}(z)$
12:          $\Sigma \leftarrow (\Sigma \setminus \{z\}) \cup \{w\}$
13:    $Q \leftarrow Q \cup S$
14: **return** (∞, NULL)

The algorithm begins by adding the root $Id_{\mathcal{U}}$ to the queue (line 1), and then enters a loop which recursively removes and expands the top of the queue (line 3) adding children to S (line 4). If the queue is empty the algorithm terminates (line 2) returning NULL (line 14). Each signal removed from Q (line 5) is checked for membership in $\mathcal{U}_{\text{goal}}$ in which case the algorithm terminates returning a feasible solution with approximately the optimal cost. Otherwise, the signals are checked for infeasibility or suboptimality by the GLC conditions (line 9). Next, a relabeling condition for the associated equivalence classes (i.e. grid cells) of remaining signals is checked (line 11). Finally, remaining signals in S are added to the queue (line 13).

The main result of this paper, justified in Section 4, is the following:

Theorem 1. *Let w_R be the signal returned by the GLC method for resolution R. Then* $\lim_{R \to \infty} J_{x_{ic}}(w_R) = c^*$. *That is, the GLC method is a resolution complete algorithm for the optimal kinodynamic motion planning problem.*

This conclusion is independent of the order in which children of the current node are examined in line 7.

3 Numerical Experiments

The GLC method (Algorithm 1) was tested on five problems and compared, when applicable, to the implementation of RRT* from [13] and SST from [14]. The goal is to examine the performance of the GLC method on a wide variety of problems. The examples include under-actuated nonlinear systems, multiple cost objectives, and environments with/without obstacles. Note that adding obstacles effectively speeds up the GLC method since it reduces the size of the search tree.

Another focus of the examples is on real-time application. In each example the running time for GLC method to produce a (visually) acceptable trajectory is comparable to the execution time. Of course this will vary with problem data and computing hardware.

Implementation Details: The GLC method was implemented in C++ and run with a 3.70GHz Intel Xeon CPU. The set Q was implemented with an STL priority queue so that the pop(Q) method and insertion operations have logarithmic complexity. The set Σ was implemented with an STL set which uses a binary search tree so that find(w, Σ) has logarithmic complexity as well.

Sets X_{free} and X_{goal} are described by algebraic inequalities. The approximation of the input space Ω_R is constructed by uniform deterministic sampling of R^m controls from Ω. Recall m is the dimension of the input space.

Evaluation of $u \in \mathcal{U}_{\text{feas}}$ and $u \in \mathcal{U}_{\text{goal}}$ is approximated by first numerically approximating $\varphi_{x_{\text{ic}}}(u)$ with Euler integration (except for RRT* which uniformly samples along the local planning solution). The number of time-steps is given by $N = \lceil \tau(u)/\Delta \rceil$ with duration $\tau(u)/N$. Maximum time-steps Δ are 0.005 for the first problem, 0.1 for the second through fourth problem, and 0.02 for the last problem. Feasibility is then approximated by collision checking at each time-step along the trajectory.

Shortest path problem: A shortest path problem can be represented by the dynamics $f(x, u) = u$ with the running-cost $g(x, u) = 1$ and input space $\Omega = \{u \in \mathbb{R}^2 : \|u\|_2 = 1\}$. The free space and goal are illustrated in Figure 2. The parameters for the GLC method are $c = 10$, $\eta(R) = R^2/300$, and $h(R) = 100R\log(R)$ with resolutions $R \in \{20, 25, ..., 200\}$. The exact solution to this problem is known so we can compare the relative convergence rates of the GLC method, SST, and RRT*.

Torque limited pendulum swing-up: The system dynamics are $f(\theta, \omega, u) = (\omega, u - \sin(\theta))^T$ with the running-cost $g(\theta, \omega, u) = 1$ and input space $\Omega = [-0.2, 0.2]$. The free space is modeled as $X_{free} = \mathbb{R}^2$. The initial state is $x_{ic} = (0, 0)$ and the goal region is $X_{goal} = \{x \in \mathbb{R}^2 : \|(\theta \pm \pi, \omega)\|_2 \leq 0.1\}$. The parameters for the GLC method are $c = 6$, $\eta(R) = R^{2.5}/16$, and $h(R) = 100R\log(R)$ with resolutions $R \in \{4, 5, 6, 7, 8\}$. The optimal solution is unknown so only the running time vs. cost can be plotted. Without a local planning solution RRT* is not applicable. The same is true for the remaining examples.

Torque limited acrobot swing-up: The acrobot is a double link pendulum actuated at the middle joint. The expression for the four dimensional system dynamics are cumbersome to describe and we refer to [15] for the details. The model parameters, free space, and goal region are identical to those in [14] with the exception that the radius of the goal region is reduced to 0.5 from 2.0. The running-cost is $g(x, u) = 1$ and the input space is $\Omega = [-4.0, 4.0]$. The parameters for the GLC method are $c = 6$, $\eta(R) = R^2/16$, and $h(R) = 100R\log(R)$ with resolutions $R \in \{5, 6, ..., 10\}$.

Acceleration limited 3D point robot: To emulates the mobility of an agile aerial vehicle (e.g. a quadrotor with high bandwidth attitude control), the system dynamics are $f(x, v, u) = (v, 5.0u - 0.1v\|v\|_2)^T$ where x, v, and u are each elements of $\mathbb{R}^3$; there are a total of six states. The quadratic dissipative force antiparallel to the velocity v models aerodynamic drag during high speed flight. The running-cost is $g(x, v, u) = 1$ and the input space is $\Omega = \{u \in \mathbb{R}^3 : \|u\|_2 \leq 1\}$. The free space and goal are illustrated in Figure 2. The sphere in the rightmost room illustrates the goal configuration. The terminal velocity is left free. The velocity is initially zero, and the cylinder indicates the starting configuration. The parameters for the GLC method are $c = 10$, $\eta(R) = R^{3/2}/64$, and $h(R) = 100R \log(R)$ with resolutions $R \in \{8, 9, ..., 14\}$. A guided search is also considered with heuristic given by the distance to the goal divided by the maximum speed v of the robot (A maximum speed of $\sqrt{50}$ can be determined from the dynamics and input constraints).

Nonholonomic wheeled robot: The system dynamics emulating the mobility of a wheeled robot are $f(x, y, \theta, u) = (\cos(\theta), \sin(\theta), u)^T$. The running cost is $g(x, y, \theta, u) = 1 + 2u^2$, with input space $\Omega = [-1, 1]$. The parameters for the GLC method are $c = 10$, $\eta(R) = 15R^{5/\pi}$, and $h(R) = 5R \log(R)$ with resolutions $R \in \{4, 5, ..., 9\}$. Note the quadratic penalty on angular rate which is relevant to rider comfort specifications in driverless vehicle applications.

3.1   Observations and Discussion

The RRT* algorithm was only tested on the first example since a steering function is unavailable for the remaining four. The SST algorithm was tested in all examples but the last since the implementation in [14] only supports min-time objectives.

We observe the run-time vs. objective-cost curves for the GLC method is several orders of magnitude faster than the SST algorithm. In the shortest path problem, the steering function for RRT* is a line segment between two points. In this case RRT* outperforms both the GLC and SST methods and is the more appropriate algorithm. A more complex steering function can increase the running time of RRT* by a considerable constant factor making it a less competitive option.

In the 3D point robot example, we see the running time for the GLC method to produce a (visually) good quality trajectory is roughly equal to the execution time. This suggests six states is roughly the limit for real-time application without better heuristics.

In the wheeled robot example a minimum time cost function was compared to a cost function which also penalized lateral acceleration. The minimum time solution results in a path with abrupt changes in angular rate making it unsuitable for autonomous driving applications where passenger comfort is a consideration. Penalizing angular velocity resulted in a solution with more gradual changes in angular velocity.

Each data point in Figure 2 represents the running time and solution cost of a complete evaluation of the GLC method while RRT* and SST are incremental methods running until interrupted. Each run of Algorithm 1 operates on $\mathcal{U}_R$ for a fixed R. Since $\mathcal{U}_R \not\subset \mathcal{U}_{R+1}$ it is possible that an optimal signal in $\mathcal{U}_R$ may be better than any signal in $\mathcal{U}_{R+1}$ for some R which explains the non-monotonic convergence observed.

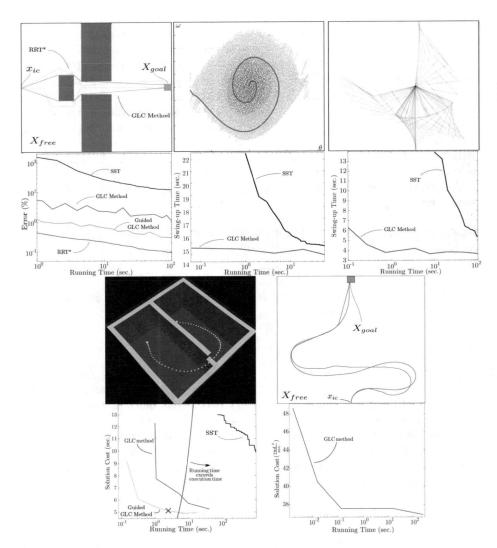

Fig. 2. Running times are based on 10 trial average for randomized planners and only reported if a solution was found at a particular time in all 10 trials. **Shortest path example (top-left)**; the opaque and solid paths illustrate first and last solutions obtained respectively. **Pendulum example (top-middle)**; the graphic illustrates the GLC solution for $R = 7$ with the colored markers indicating the cost of grid labels. **Acrobot example (top-right)**; The state space has four dimensions. Only the configuration is illustrated. **Acceleration limited point robot (bottom-left)**; The graphic shows the best GLC solution with a running time less than the execution time. The $\times$ indicates the running time of the solution in the graphic. The state space has six dimensions. Only the configuration is illustrated. **Wheeled robot (bottom-right)**; The blue path indicates a GLC solution with a quadratic angular rate penalty while the red path is a GLC solution with a shortest path objective.

4 Analysis of the GLC Condition

Section 4.1 begins by equipping $\mathcal{U}$ and $\mathcal{X}_{x_0}$ with metrics in order to discuss continuity of φ_{x_0} and J_{x_0}. Using these metrics we can also discuss in what sense $\mathcal{U}_R$ approximates $\mathcal{U}$ in Section 4.2. Finally, in Lemma 6 and Theorem 2, we derive a bound on the gap between the cost of the solution output by Algorithm 1 and the optimal cost for the problem and show that the gap converges to zero as the resolution is increased.

4.1 Metrics on $\mathcal{X}_{x_0}$ and $\mathcal{U}$, and Continuity of Relevant Maps

The metrics introduced by Yershov and Lavalle [16] will be used. Recall, the signal space $\mathcal{U}$ was defined in (1). A metric on this space is given by

$$d_{\mathcal{U}}(u_1, u_2) := \int_{[0,\min(\tau(u_1),\tau(u_2))]} \|u_1(t) - u_2(t)\|_2 \, d\mu(t) + u_{max}|\tau(u_1) - \tau(u_2)|. \quad (10)$$

The family of trajectory spaces are now more precisely defined as

$$\mathcal{X}_{x_0} := \bigcup_{\tau > 0} \left\{ x : [0,\tau] \to \mathbb{R}^n \; : \; x(0) = x_0, \; \left\| \frac{x(t_1) - x(t_2)}{|t_1 - t_2|} \right\|_2 \le M \; \forall t_{1,2} \in [0,\tau] \right\}, \quad (11)$$

where M is that of assumption A-2. This set is equipped with the metric

$$d_{\mathcal{X}}(x_1, x_2) \equiv \max_{t \in [0,\min\{\tau(x_1),\tau(x_2)\}]} \{\|x_1(t) - x_2(t)\|\} + M|\tau(x_1) - \tau(x_2)|. \quad (12)$$

Several known continuity properties of φ_{x_0} are reviewed here. Recall (e.g. [17, pg. 95]), that the distance between solutions to (2) with initial conditions x_0 and z_0 is bounded by

$$\|[\varphi_{x_0}(u)](t) - [\varphi_{z_0}(u)](t)\|_2 \le \|x_0 - z_0\|_2 e^{L_f t}. \quad (13)$$

In addition to continuous dependence on the initial condition parameter, the map φ_{x_0} is also continuous from $\mathcal{U}$ into $\mathcal{X}_{x_0}$ (cf. [16, Theorem 1]). It is a useful observation then that $\mathcal{X}_{feas}$ is open when assumption A-1 is satisfied (cf. [16, Theorem 2]) since it follows directly from the definition of continuity that $\mathcal{U}_{feas}$ and $\mathcal{U}_{goal}$ are open subsets of $\mathcal{U}$; recall that they are defined as the preimage of $\mathcal{X}_{feas}$ and $\mathcal{X}_{goal}$ under $\varphi_{x_{ic}}$.

Similar observations for the cost functional are developed below.

Lemma 1. $J_{x_0} : \mathcal{U} \to \mathbb{R}$ is continuous for any $x_0 \in \mathbb{R}^n$.

Proof. Let $u_1, u_2 \in \mathcal{U}$ and without loss of generality, let $\tau(u_1) \le \tau(u_2)$. Denote trajectories $\varphi_{x_0}(u_1)$ and $\varphi_{x_0}(u_2)$ by x_1 and x_2 respectively. The associated difference in cost is

$$|J_{x_0}(u_1) - J_{x_0}(u_2)| = \left| \int_{[0,\tau(u_1)]} g(x_1(t), u_1(t)) - g(x_2(t), u_2(t)) \, d\mu(t) \right.$$
$$\left. - \int_{[\tau(u_1),\tau(u_2)]} g(x_2(t), u_2(t)) \, d\mu(t) \right|. \quad (14)$$

Using the Lipschitz constant of g (cf. A-3), the definition of $d_{\mathcal{U}}$ in (10), and the definition of $d_{\mathcal{X}}$ in (12) the difference is bounded as follows:

$$
\begin{aligned}
|J_{x_0}(u_1) - J_{x_0}(u_2)| &\leq \left| \int_{[0,\tau(u_1)]} L_g \, \|x_1(t) - x_2(t)\|_2 + L_g \, \|u_1(t) - u_2(t)\|_2 \, d\mu(t) \right. \\
&\quad \left. - \int_{[\tau(u_1),\tau(u_2)]} g\,(x_2(t), u_2(t)) \, d\mu(t) \right| \\
&\leq L_g \tau(u_1) \|x_1 - x_2\|_{L_\infty[0,\tau(u_1)]} + L_g \|u_1 - u_2\|_{L_1[0,\tau(u_1)]} \\
&\quad + \int_{[\tau(u_1),\tau(u_2)]} |g\,(x_2(t), u_2(t))| \, d\mu(t) \\
&\leq L_g \tau(u_1) d_{\mathcal{X}}(x_1, x_2) + L_g d_{\mathcal{U}}(u_1, u_2) \\
&\quad + \int_{[\tau(u_1),\tau(u_2)]} |g\,(x_2(t), u_2(t))| \, d\mu(t).
\end{aligned}
\tag{15}
$$

Since x_2 is continuous, u_2 is bounded, and g is continuous, there exists a bound G on $g(x_2(t), u_2(t))$ for $t \in [\tau(u_1), \tau(u_2)]$. Thus, the difference in cost is further bounded by

$$
|J_{x_0}(u_1) - J_{x_0}(u_2)| \leq L_g \tau(u_1) d_{\mathcal{X}}(x_1, x_2) + L_g d_{\mathcal{U}}(u_1, u_2) + G|\tau(u_2) - \tau(u_1)|.
\tag{16}
$$

Sine φ_{x_0} is continuous, for $d_{\mathcal{U}}(u_1, u_2)$ sufficiently small the resulting trajectories will satisfy $L_g \tau(u_1) d_{\mathcal{X}}(x_1, x_2) < \varepsilon/3$. Additionally, for u_1, u_2 satisfying $d_{\mathcal{U}}(u_1, u_2) < \frac{\varepsilon u_{max}}{3G}$ and $d_{\mathcal{U}}(u_1, u_2) < \frac{\varepsilon}{3L_g}$ we have $G|\tau(u_2) - \tau(u_1)| < \varepsilon/3$ and $L_g d_{\mathcal{U}}(u_1, u_2) < \varepsilon/3$. For such a selection of u_1, u_2, we have $|J_{x_0}(u_1) - J_{x_0}(u_2)| < \varepsilon$. Thus, J_{x_0} is continuous.   □

Lemma 2. *For any $u \in \mathcal{U}_{\text{feas}}$ and $x_0, z_0 \in \mathbb{R}^n$,*

$$
|J_{x_0}(u) - J_{z_0}(u)| \leq \|x_0 - z_0\|_2 \cdot \frac{L_g}{L_f} \left(e^{L_f \tau(u)} - 1 \right)
\tag{17}
$$

Proof. The difference is bounded using the Lipschitz continuity of g. This is further bounded using (13). Denoting $x(t) = \varphi_{x_0}(u)$ and $z(t) = \varphi_{z_0}(u)$,

$$
\begin{aligned}
|J_{x_0}(u) - J_{z_0}(u)| &= \left| \int_{[0,\tau(u)]} g(x(t), u(t)) - g(z(t), u(t)) \, d\mu(t) \right| \\
&\leq \int_{[0,\tau(u)]} |g(x(t), u(t)) - g(z(t), u(t))| \, d\mu(t) \\
&\leq \int_{[0,\tau(u)]} L_g \, \|x(t) - z(t)\|_2 \, d\mu(t) \\
&\leq \int_{[0,\tau(u)]} \|x_0 - z_0\|_2 L_g e^{L_f t} \, d\mu(t) \\
&= \|x_0 - z_0\|_2 \frac{L_g}{L_f} \left(e^{L_f \tau(u)} - 1 \right).
\end{aligned}
\tag{18}
$$

□

4.2   Properties of the Approximation of $\mathcal{U}$ by $\mathcal{U}_R$

Note that Lemma 3 is not a statement about the dispersion of $\mathcal{U}_R$ in $\mathcal{U}$ which does not actually converge. For numerical approximations of function spaces the weaker statement that $\limsup_{R \to \infty} \mathcal{U}_R$ is dense in $\mathcal{U}$ will be sufficient. Equivalently,

Lemma 3. *For each $u \in \mathcal{U}$ and $\varepsilon > 0$, there exists $R^* > 0$ such that for any $R > R^*$ there exists $w \in \mathcal{U}_R$ such that $d_{\mathcal{U}}(u, w) < \varepsilon$.*

Proof. By Lusin's Theorem [18, pg. 41], there exists a continuous $v : [0, \tau(u)] \to \Omega$ such that

$$\mu(\{t \in [0, \tau(u)] : \; v(t) \neq u(t)\}) < \frac{\varepsilon}{3 u_{max}}. \tag{19}$$

The domain of v is compact so v is also uniformly continuous. Denote its modulus of continuity by $\delta(\epsilon)$,

$$|\sigma - \gamma| < \delta(\epsilon) \Rightarrow \|v(\sigma) - v(\gamma)\|_2 < \epsilon. \tag{20}$$

To construct an approximation of v by $w \in \mathcal{U}_R$ choose R sufficiently large so that (i) $h(R)/R > \tau(u)$, (ii) there exists an integer r such that $0 < \tau(u) - r/R < 1/R < \delta\left(\frac{\varepsilon}{6\tau(u)}\right)$, and (iii) the dispersion of Ω_R in Ω is less than $\frac{\varepsilon}{6\tau(u)}$.

Then for $t \in [(i-1)/R, i/R)$ $i = 1, ..., r$ there exists $v_i \in \Omega_R$ such that $\|v_i - v(t)\|_2 < \frac{\varepsilon}{3\tau(u)}$. Select $w \in \mathcal{U}_R$ which is equal to v_i on each interval. Combining (i)-(iii),

$$\begin{aligned}
d_{\mathcal{U}}(v, w) &= \int_{[0, r/R]} \|v(t) - w(t)\|_2, \, d\mu(t) + u_{max} |r/R - \tau(\omega)| \\
&< \int_{[0, r/R]} \|v(t) - w(t)\|_2, \, d\mu(t) + \frac{\varepsilon}{6} \\
&< \int_{[0, r/R]} \frac{\varepsilon}{3\tau(u)}, \, d\mu(t) + \frac{\varepsilon}{6} \\
&< \frac{\varepsilon}{2}.
\end{aligned} \tag{21}$$

Thus, by the triangle inequality

$$d_{\mathcal{U}}(u, w) \leq d_{\mathcal{U}}(u, v) + d_{\mathcal{U}}(v, w) < \varepsilon. \tag{22}$$

$\square$

We use $cl(\cdot)$ and $int(\cdot)$ to denote the closure and interior of subsets of $\mathcal{U}$.

Lemma 4. *For any $w \in cl\,(int\,(\mathcal{U}_{goal}))$ and $\varepsilon > 0$ there exists $R^* > 0$ such that for any $R > R^*$*

$$\min_{u \in \mathcal{U}_R \cap \mathcal{U}_{goal}} \{|J_{x_{ic}}(u) - J_{x_{ic}}(w)|\} < \varepsilon. \tag{23}$$

Proof. $\omega \in cl\,(int\,(\mathcal{U}_{goal}))$ implies that for all $\delta > 0$, $B_{\delta/2}(\omega) \cap int(\mathcal{U}_{goal}) \neq \emptyset$. Then each $v \in int(\mathcal{U}_{goal})$ has a neighbourhood $B_\rho(v) \subset int(\mathcal{U}_{goal})$ with $\rho < \frac{\delta}{2}$. Take $v \in B_{\delta/2}(\omega) \cap int(\mathcal{U}_{goal.})$. By Lemma 3, for sufficiently large R, there exists $u \in \mathcal{U}_R$ such that $d_{\mathcal{U}}(v, u) < \delta/2$ which implies $u \in B_\rho(v) \subset int(\mathcal{U}_{goal})$. Then $u \in \mathcal{U}_{goal}$ and $d_{\mathcal{U}}(\omega, u) < d_{\mathcal{U}}(\omega, v) + d_{\mathcal{U}}(v, u) < \delta$. Now by the continuity of $J_{x_{ic}}$ for δ sufficiently small, $d_{\mathcal{U}}(\omega, v) < \delta$ implies $|J_{x_{ic}}(u) - J_{x_{ic}}(w)| < \varepsilon$ from which the result follows. $\square$

A sufficient condition for every $\omega \in \mathcal{U}_{goal}$ to be contained in the closure of the interior of $\mathcal{U}_{goal}$ is that $\mathcal{U}_{goal}$ be open which is the case when Assumption A-1 is satisfied.

4.3 Pruning $\mathcal{U}_R$ with the GLC Condition

To describe trajectories remaining on the ε-interior of X_{free} at each instant and which terminate on the ε-interior of X_{goal} the following sets are defined,

$$\begin{aligned}
\mathcal{X}_{goal}^\varepsilon &:= \{x \in \mathcal{X}_{goal} : \; B_\varepsilon(x(t)) \subset X_{free} \, \forall t \in [0, \tau(x)], \, B_\varepsilon(x(\tau(x)) \subset X_{goal}\}, \\
\mathcal{U}_{goal}^\varepsilon &:= \{u \in \mathcal{U}_{goal} : \; \varphi_{x_{ic}}(u) \in \mathcal{X}_{goal}^\varepsilon\},
\end{aligned} \tag{24}$$

and similarly for c_R,

$$c_R^\varepsilon := \min_{u \in \mathcal{U}_R \cap \mathcal{U}_{\text{goal}}^\varepsilon} \{J_{x_{\text{ic}}}(u)\}. \tag{25}$$

Since $\mathcal{U}_{\text{goal}}^\varepsilon \subset \mathcal{U}_{\text{goal}}$ we have that $0 \leq c_R \leq c_R^\varepsilon$.

Lemma 5. *If* $\lim_{R \to \infty} \epsilon(R) = 0$ *then* $\lim_{R \to \infty} c_R^{\epsilon(R)} = c^*$.

Proof. By the definition of c^* in (5), for any $\varepsilon > 0$ there exists $\omega \in \mathcal{U}_{\text{goal}}$ such that $J_{x_{\text{ic}}}(\omega) - \varepsilon/2 < c^*$. Since $\mathcal{U}_{\text{goal}}$ is open and $\varphi_{x_{\text{ic}}}$ is continuous there exists $\tilde{r} > 0$ and $\rho > 0$ such that $B_{\tilde{r}}(\omega) \subset \mathcal{U}_{\text{goal}}$ and $\varphi_{x_{\text{ic}}}(B_{\tilde{r}}(\omega)) \subset B_\rho(\varphi_{x_{\text{ic}}}(\omega))$. Thus, $\omega \in \mathcal{U}_{\text{goal}}^{\tilde{r}}$. Similarly, there exists a positive $r < \tilde{r}$ such that $\varphi_{x_{\text{ic}}}(B_r(\omega)) \subset B_{\rho/2}(\varphi_{x_{\text{ic}}}(\omega))$ and $B_r(\omega) \subset \mathcal{U}_{\text{goal}}^{\rho/2}$. From the continuity of J in Lemma 1 there also exists a positive $\delta < r$ such that for any signal v with $d_\mathcal{U}(v, \omega) < \delta$ we have $|J_{x_{\text{ic}}}(\omega) - J_{x_{\text{ic}}}(v)| < \varepsilon/2$.

Next, choose R^* to be sufficiently large such that $R > R^*$ implies $\epsilon(R) < \rho/2$ and $B_\delta(\omega) \cap \mathcal{U}_R \neq \emptyset$. Such a resolution R^* exists by Lemma 4 and the assumption $\lim_{R \to \infty} \epsilon(R) = 0$. Now choose $u \in B_\delta(\omega) \cap \mathcal{U}_R$. Then $|J_{x_{\text{ic}}}(u) - J_{x_{\text{ic}}}(\omega)| < \varepsilon/2$ and $u \in \mathcal{U}_{\text{goal}}^{\rho/2} \subset \mathcal{U}_{\text{goal}}^{\epsilon(R)}$. Then by definition of $c_R^{\epsilon(R)}$, $u \in \mathcal{U}_{\text{goal}}^{\epsilon(R)}$ implies $c_R^{\epsilon(R)} \leq J_{x_{\text{ic}}}(u)$. Finally, by triangle inequality,

$$|J_{x_{\text{ic}}}(u) - c^*| < |J_{x_{\text{ic}}}(u) - J_{x_{\text{ic}}}(\omega)| + |J_{x_{\text{ic}}}(\omega) - c^*| < \varepsilon. \tag{26}$$

Rearranging the expression yields $J_{x_{\text{ic}}}(u) < c^* + \varepsilon$ and thus, $c_R^{\epsilon(R)} < c^* + \varepsilon$. The result follows since the choice of ε is arbitrary. $\qquad\square$

To simplify the notation in what follows, a concatenation operation on elements of $\mathcal{U}$ is defined. For $u_1, u_2 \in \mathcal{U}$, their concatenation $u_1 u_2$ is defined by

$$[u_1 u_2](t) := \begin{cases} u_1(t), & t \in [0, \tau(u_1)) \\ u_2(t - \tau(u_1)), & t \in [\tau(u_1), \tau(u_1) + \tau(u_2)] \end{cases}. \tag{27}$$

The concatenation operation will be useful together with the following equalities which are readily verified,

$$J_{x_0}(u_1 u_2) = J_{x_0}(u_1) + J_{[\varphi_{x_0}(u_1)](\tau(u_1))}(u_2) \tag{28}$$

$$\varphi_{x_0}(u_1 u_2) = \varphi_{x_0}(u_1) \varphi_{[\varphi_{x_0}(u_1)](\tau(u_1))}(u_2). \tag{29}$$

The concatenation operation on $\mathcal{X}_{x_0}$ in (29) is defined in the same way as in (27).

Lemma 6. *Let* $\delta_0 = \frac{\sqrt{n}}{\eta(R)} e^{\frac{L_f h(R)}{R}}$. *For* $\varepsilon \geq \delta_0$ *and* $u_i, u_j \in \mathcal{U}_R \cap \mathcal{U}_{\text{feas}}$, *if* $u_i \prec_R u_j$, *then for each descendant of* u_j *in* $\mathcal{U}_{\text{goal}}^{\varepsilon - \delta_0}$ *with cost* c_j, *there exists a descendant of* u_i *in* $\mathcal{U}_{\text{goal}}$ *with cost* $c_i \leq c_j$.

Proof. Suppose there is a $w \in \mathcal{U}_R$ such that $u_j w \in \mathcal{U}_R$ is a descendant of u_j and $u_j w \in \mathcal{U}_{\text{goal}}^\varepsilon$. Since $u_i \overset{R}{\sim} u_j$,

$$\|[\varphi_{x_{\text{ic}}}(u_i w)](\tau(u_i)) - [\varphi_{x_{\text{ic}}}(u_j w)](\tau(u_j))\|_2 \leq \frac{\sqrt{n}}{\eta(R)}. \tag{30}$$

Then, by equation (13), for all $t \in [0, \tau(w)] \subset [0, h(R)/R]$,

$$\|[\varphi_{x_{\mathrm{ic}}}(u_i w)](t + \tau(u_i)) - [\varphi_{x_{\mathrm{ic}}}(u_j w)](t + \tau(u_j))\| \leq \frac{\sqrt{n}}{\eta(R)} e^{\frac{L_f h(R)}{R}}. \tag{31}$$

Then for $\delta_0 = \frac{\sqrt{n}}{\eta(R)} e^{\frac{L_f h(R)}{R}}$ we have $u_i w \in \mathcal{U}_{\mathrm{goal}}^{\epsilon - \delta_0}$. As for the cost, from equation (28)

$$
\begin{aligned}
J_{x_{\mathrm{ic}}}(u_i w) &= J_{x_{\mathrm{ic}}}(u_i) + J_{[\varphi_{x_{\mathrm{ic}}}(u_i)](\tau(u_i))}(w) \\
&\leq J_{x_{\mathrm{ic}}}(u_j) + \frac{\sqrt{n}}{\eta(R)} \frac{L_g}{L_f} \left(e^{\frac{L_f h(R)}{R}} - 1 \right) + J_{[\varphi_{x_{\mathrm{ic}}}(u_j)](\tau(u_j))}(w) \\
&\leq J_{x_{\mathrm{ic}}}(u_j) + J_{[\varphi_{x_{\mathrm{ic}}}(u_j)](\tau(u_j))}(w) \\
&\leq J_{x_{\mathrm{ic}}}(u_j w)
\end{aligned}
\tag{32}
$$

The first step applies the conditions for $u_i \prec_R u_j$. The second step combines Lemma 2 and $u_i \overset{N}{\sim} u_j$. $\qquad \square$

In the above Lemma, the quantity δ_0 was constructed based on the radius of the partition of X_{free} and the sensitivity of solutions to initial conditions. Let δ_k be defined as a finite sum of related quantities, $\delta_k := \sum_{i=1}^{k} \frac{\sqrt{n}}{\eta(R)} e^{\frac{L_f (h(R)-i)}{R}}$. The following bound will be useful to state our main result in the next Theorem,

$$\delta_{h(R)} = \sum_{i=1}^{h(R)} \frac{\sqrt{n}}{\eta(R)} e^{\frac{L_f (h(R)-i)}{R}} \leq \frac{R\sqrt{n}}{L_f \eta(R)} \left(e^{L_f h(R)/R} - 1 \right). \tag{33}$$

Theorem 2. *Let* $\epsilon(R) = \frac{R\sqrt{n}}{L_f \eta(R)} \left(e^{\frac{L_f h(R)}{R}} - 1 \right)$. *Algorithm 1 terminates in finite time and returns a solution with cost less than or equal to* $c_R^{\epsilon(R)}$.

Proof. The queue is a subset of $\mathcal{U}_R \cup \{Id_{\mathcal{U}}\}$ and at line 3 in each iteration a lowest cost signal u is removed from the queue. In line 13, only children of the current signal u are added to the queue. Since $\mathcal{U}_R$ is organized as a tree and has no cycles, any signal u will enter the queue at most once. Therefore the queue must be empty after a finite number of iterations so the algorithm terminates.

Next, consider as a point of contradiction the hypothesis that the output has cost greater than $c_R^{\epsilon(R)}$. Then it is necessary that $c_R^{\epsilon(R)} < \infty$ and by the definition of $c_R^{\epsilon(R)}$ in (25), it is also necessary that $\mathcal{U}_R \cap \mathcal{U}_{\mathrm{goal}}^{\epsilon(R)}$ is non-empty.

Choose $u^* \in \mathcal{U}_R \cap \mathcal{U}_{\mathrm{goal}}^{\epsilon(R)}$ with cost $J_{x_{\mathrm{ic}}}(u^*) = c_R^{\epsilon(R)}$. It follows from the hypothesis that u^* does not enter the queue. Otherwise, by (9) it would be evaluated before any signal of cost greater than $c_R^{\epsilon(R)}$. If u^* does not enter the queue, then a signal u_0 must at some iteration be present in Σ which prunes an ancestor a_0 of u^* ($u_0 \prec_R a_0$ in line 9). This ancestor must satisfy $\mathrm{depth}(a_0) > 0$ since the ancestor with depth 0 is $Id_{\mathcal{U}}$ which enters queue in line 1. By Lemma 6, u_0 has a descendant of the form $u_0 d_0 \in \mathcal{U}_{\mathrm{goal}}^{\epsilon(R)-\delta_1}$ and $J_{x_{\mathrm{ic}}}(u_0 d_0) \leq c_R^{\epsilon(R)}$. Additionally, $\mathrm{depth}(a_0) > 0$ implies $\mathrm{depth}(d_0) \leq h(R) - 1$.

Having pruned u^* in line 9, the signal $u_0 \in \Sigma$, or a sibling which prunes u_0 (and by transitivity, prunes u^*) must at some point be present in the queue (cf. line 12-13).

Of these two, denote the one that ends up in the queue by $\tilde{u}_0$. Since $\tilde{u}_0$ is at some point present in the queue and $\tilde{u}_0 d_0 \in \mathcal{U}_{\text{goal}}^{\epsilon(R)-\delta_1}$, a signal $u_1 \in \Sigma$ must prune an ancestor a_1 of $\tilde{u}_0 d_0$ ($u_1 \prec_R a_1$ in line 9). Since $\tilde{u}_0$ is at some point present in the queue, the ancestor a_1 of $\tilde{u}_0 d_0$, must have greater depth than $\tilde{u}_0$. By Lemma 6, u_1 has a descendant of the form $u_1 d_1 \in \mathcal{U}_{\text{goal}}^{\epsilon(R)-\delta_2}$ and $J_{x_{\text{ic}}}(u_1 d_1) \leq c_R^{\epsilon(R)}$. Additionally, $\texttt{depth}(a_1) > \tilde{u}_0$ implies $\texttt{depth}(d_1) \leq \texttt{depth}(d_0) - 1 \leq h(R) - 2$.

Continuing this line of deduction leads to the observation that a signal $u_{h(R)-1}$, with a descendant of the form $u_{h(R)-1} d_{h(R)-1} \in \mathcal{U}_{\text{goal}}^{\epsilon(R)-\delta_{h(R)-1}}$ and $J_{x_{\text{ic}}}(u_{h(R)-1} d_{h(R)-1}) \leq c_R^{\epsilon(R)}$, will be present in the queue; and $\texttt{depth}(d_{h(R)-1}) \leq 1$. Since $u_{h(R)-1}$ is at some point present in the queue, a signal $u_{h(R)} \in \Sigma$ must prune an ancestor $a_{h(R)}$ of $u_{h(R)-1} d_{h(R)-1}$ ($u_{h(R)} \prec_R a_{h(R)}$ in line 9). Since $u_{h(R)-1}$ is at some point present in the queue, the ancestor $a_{h(R)}$ of $u_{h(R)-1} d_{h(R)-1}$, must have greater depth than $u_{h(R)-1}$, and therefore, is equal to $u_{h(R)-1} d_{h(R)-1}$. Thus, $u_{h(R)} \in \mathcal{U}_{\text{goal}}^{\epsilon(R)-\delta_{h(R)}}$ (note that $\epsilon(R) - \delta_{h(R)}) \geq 0$ by (33)) and $J(u_{h(R)}) \leq c_R^{\epsilon(R)}$. Then $u_{h(R)}$ or a sibling which prunes $u_{h(R)}$ will be added to the queue; a contradiction of the hypothesis since this signal will be removed from the queue and the algorithm will terminate, returning this signal in line 7.   □

The choice of $\epsilon(R)$ in Theorem 2 converges to zero by (8). Then by Lemma 5 we have $c_R^{\epsilon(R)} \to c^*$. An immediate corollary is that the GLC method is resolution complete which is the main contribution of this work stated in Theorem 1.

5   Conclusion

In this paper we described a simple grid-based approximation of the optimal kinodynamic motion planning problem and developed the appropriate generalization of label correcting methods to efficiently search the approximation. The advantage of the GLC method is that it does not require a point-to-point local planning subroutine. Moreover, numerical experiments demonstrate that the GLC method is considerably faster the related SST algorithm, and is more broadly applicable than the RRT* algorithm.

The focus of the paper, and its main contribution was the theoretical investigation showing that the cost of the feasible solutions returned by the GLC method converge to the optimal cost for the problem.

From a practical point of view the proposed algorithm is easy to implement and may be used directly for motion planning and trajectory optimization. Future investigations will include convergence rate analysis, improved partitioning schemes over the hypercube grid we presented, and the construction of admissible heuristics for kinodynamic motion planning to guide the search.

Acknowledgements

This research was funded in part by the Israeli Ministry of Defense.

We are also grateful to our colleagues Michal Čáp and Dmitry Yershov for their insightful comments.

References

1. Kavraki, L.E., Svestka, P., Latombe, J.C., Overmars, M.H.: Probabilistic Roadmaps for Path Planning in High-Dimensional Configuration Spaces. IEEE Transactions on Robotics and Automation 12(4) (1996) 566–580
2. Karaman, S., Frazzoli, E.: Sampling-Based Algorithms for Optimal Motion Planning. The International Journal of Robotics Research 30(7) (2011) 846–894
3. Hsu, D., Kindel, R., Latombe, J.C., Rock, S.: Randomized kinodynamic motion planning with moving obstacles. The International Journal of Robotics Research 21(3) (2002) 233–255
4. LaValle, S.M., Kuffner, J.J.: Randomized kinodynamic planning. The International Journal of Robotics Research 20(5) (2001) 378–400
5. Karaman, S., Frazzoli, E.: Optimal Kinodynamic Motion Planning Using Incremental Sampling-Based Methods. In: IEEE Conference on Decision and Control. (2010) 7681–7687
6. Perez, A., Platt Jr, R., Konidaris, G., Kaelbling, L., Lozano-Perez, T.: LQR-RRT*: Optimal Sampling-Based Motion Planning with Automatically Derived Extension Heuristics. In: International Conference on Robotics and Automation, IEEE (2012) 2537–2542
7. Xie, C., van den Berg, J., Patil, S., Abbeel, P.: Toward Asymptotically Optimal Motion Planning for Kinodynamic Systems Using a Two-Point Boundary Value Problem Solver. In: International Conference on Robotics and Automation, IEEE (2015) 4187–4194
8. Dijkstra, E.W.: A Note on Two Problems in Connexion with Graphs. Numerische mathematik 1(1) (1959) 269–271
9. Dolgov, D., Thrun, S., Montemerlo, M., Diebel, J.: Path planning for autonomous vehicles in unknown semi-structured environments. The International Journal of Robotics Research 29(5) (2010) 485–501
10. Li, Y., Littlefield, Z., Bekris, K.E.: Sparse Methods for Efficient Asymptotically Optimal Kinodynamic Planning. In: Algorithmic Foundations of Robotics XI. Springer (2015) 263–282
11. Coddington, E.A., Levinson, N.: Theory of Ordinary Differential Equations. McGraw-Hill Education (1955)
12. Hart, P.E., Nilsson, N.J., Raphael, B.: A Formal Basis for the Heuristic Determination of Minimum Cost Paths. Systems Science and Cybernetics 4(2) (1968) 100–107
13. Karaman, S.: RRT* Library. http://karaman.mit.edu/software.html
14. Li, Y., Littlefield, Z., Bekris, K.E.: Sparse RRT package. https://bitbucket.org/pracsys/sparse_rrt/ Accessed: Jan. 2016.
15. Spong, M.W.: The Swing Up Control Problem for the Acrobot. Control Systems, IEEE 15(1) (1995) 49–55
16. Yershov, D.S., LaValle, S.M.: Sufficient Conditions for the Existence of Resolution Complete Planning Algorithms. In: Algorithmic Foundations of Robotics IX. Springer (2011) 303–320
17. Khalil, H.K., Grizzle, J.: Nonlinear Systems. Volume 3. Prentice hall New Jersey (1996)
18. Kolmogorov, A.N., Fomin, S.V.: Elements of the Theory of Functions and Functional Analysis. Vol. 2, Measure. The Lebesgue Integral. Hilbert Spaces. Graylock Press (1961)

Asymptotically optimal planning under piecewise-analytic constraints

William Vega-Brown and Nicholas Roy

Computer Science and Artificial Intelligence Laboratory
Massachusetts Institute of Technology
Cambridge, Massachusetts 02139
{wrvb,nickroy}@csail.mit.edu

Abstract. We present the first asymptotically optimal algorithm for motion planning problems with piecewise-analytic differential constraints, like manipulation or rearrangement planning. This class of problems is characterized by the presence of differential constraints that are local in nature: a robot can only move an object once the object has been grasped. These constraints are not analytic and thus cannot be addressed by standard differentially constrained planning algorithms. We demonstrate that, given the ability to sample from the locally reachable subset of the configuration space with positive probability, we can construct random geometric graphs that contain optimal plans with probability one in the limit of infinite samples. This approach does not require a hand-coded symbolic abstraction. We demonstrate our approach in simulation on a simple manipulation planning problem, and show it generates lower-cost plans than a sequential task and motion planner.

1   Introduction

Consider a robot tasked with building a structure; a general-purpose planning algorithm for such a task must infer plans that respect both kinematic constraints, such as joint limits and collision between moving objects, as well as differential constraints, such as the limitation that the robot must be in contact with an object in order to affect its configuration. There exist provably asymptotically optimal algorithms for motion planning under analytic differential constraints. However, problems that involve discrete decisions, such as whether to grasp or release an object, cannot be described by analytic differential constraints.

This class of problems includes all contact-based manipulation. In the assembly task, for instance, the robot must decide where to place any object it sets down. This decision will affect all future decisions, as once the object has been set down, the robot cannot affect its position without grasping it again. While standard algorithms can decide *how* to place an object, they cannot make optimal decisions about *where* to place the object. See fig. 1 for a simple illustration of this problem.

There are two families of approaches to address this problem. One option is to break up the planning problem into a sequence of simpler planning problems in which the constraints are analytic. For example, the robot may first use a task

© Springer Nature Switzerland AG 2020
K. Goldberg et al. (Eds.): *Algorithmic Foundations of Robotics XII*, SPAR 13, pp. 528–543, 2020.
https://doi.org/10.1007/978-3-030-43089-4_34

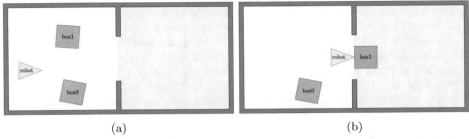

(a)   (b)

Fig. 1: a: In this block pushing problem, the triangular robot seeks to move both the red blocks into the shaded room. b: A locally-optimal decision early on in the plan can have dramatic consequences later. Because it placed the first block just over the threshold of the door to the target room, the robot cannot move the second block without grasping the first block again.

planner to choose a sequence of high-level actions, such as grasping or placing a component. A motion planner can then generate a detailed plan corresponding to each action.

Substantial research effort has gone into accounting for the interplay between the high-level task planner and the low-level geometric motion planner.For example, Kaelbling and Lozano-Pérez [10] use a hierarchy to guide high-level decision making, resolving low-level decisions arbitrarily and trusting in the reversibility of the system to ensure hierarchical completeness. Wolfe et al. [17] ensure hierarchical optimality by expanding high-level plans in a best-first way, using domain specific subtask irrelevance detection to increase efficiency. Krontiris and Bekris [13] use a backtracking search as a low-level planner in the context of a high-level task planning problem. While these approaches are often computationally efficient, the guarantees they provide are *hierarchical*: they will return the best feasible plan that can be expressed within the task hierarchy they search. These hierarchies are typically hand-coded and domain-specific; the quality of the solution and the reliability of the planner hinges on the existence of a concise symbolic description of the planning domain, which may be expensive or impossible to find.

An alternative strategy is to solve the problem in the joint configuration space of the robot and objects, modifying differentially constrained planning algorithms like RRT* [11] to account for non-analytic differential constraints. This approach avoids the need for hand-coded motion primitives or a symbolic representation of the domain and allows for guarantees of probabilistic completeness under various assumptions. For example, Van Den Berg et al. [16] present a probabilistically complete algorithm for planning in domains including moveable objects. Hauser and Latombe [5] and Berenson et al. [2] show how to plan when the differential constraints create a fixed, finite set of manifolds on which the dynamics are unconstrained, as when a single object may be lifted or dragged. The core idea of both algorithms involves building many separate planning graphs on subsets of the configuration space and connecting them to formulate a global planner. Others have expanded on this idea to choose which local planning graphs to construct in an automated fashion. For example, Jaillet and Porta [8] construct an atlas of local planning graphs and use that atlas to search for paths on manifolds. Their approach allows for efficient planning in situations where working in the ambient space would be prohibitively expensive. While their algorithm is asymptotically optimal, it cannot be directly applied to problems with non-analytic constraints. Algorithms that do apply to non-analytically constrained domains, such as Hauser and Ng-Thow-Hing's extension of their earlier work [6], are probabilistically complete but are not known to be asymptotically optimal.

In this work, we present two algorithms that are provably asymptotically optimal for problems involving differential constraints that are piecewise-analytic, a broad class that includes manipulation planning problems. Our first algorithm extends the result of Hauser and Ng-Thow-Hing to ensure asymptotic optimality, much as the PRM* algorithm [11] extends the PRM algorithm [7]. This extension leads to a provably optimal but prohibitively computationally expensive algorithm; our second algorithm mitigates this complexity by factoring the configuration space. Our key analytical result is that we can ensure asymptotic optimality by considering only a finite collection of subsets of the configuration space, each of which is subject only to analytic constraints. By building a graph on each of these subsets and connecting the resulting collection of graphs, we can construct a random graph that spans the configuration space; as the collection grows sufficiently large, it will contain a near-optimal plan with probability one and an optimal plan in the limit of infinite samples. We show experimentally that these algorithms obtain plans with lower cost than conventional sequential task and motion planning approaches.

2   Background and notation

Formally, we can define a differentially-constrained planning domain by a tuple $(f, \mathcal{X})$, where $\mathcal{X}$ is an N-dimensional Riemannian manifold with tangent bundle $T\mathcal{X}$ defining the configuration space, and $f : \mathcal{X} \times T\mathcal{X} \to \mathbb{R}^k$ is a set of k constraints on the allowable motions, $K < N$. We assume this manifold is specified by an embedding into a higher-dimensional Euclidean space, such that it can be represented as the zero level set of some function $g : \mathcal{X} \to \mathbb{R}^M$, $N < M$. Then a continuous function $\sigma : [0, T] \to \mathcal{X}$ is a *feasible path* if $f(\sigma(t), \dot{\sigma}(t)) = 0 \, \forall t \in [0, T]$. We denote the set of feasible paths in a given domain by $\Sigma_{\mathcal{X}}$. Note that this formulation can model kinematic constraints, such as collisions between objects or joint limits, and dynamic constraints.

We define a planning problem as a tuple $(x_0, \mathcal{X}_G, c)$, where $x_0 \in \mathcal{X}$ is an initial configuration, $\mathcal{X}_G$ is a set of goal configurations, and $c : \Sigma \to \mathbb{R}^+$ is a piecewise Lipschitz continuous cost function. We restrict our attention to problems where the cost of a path is independent of the time taken to traverse the path; accordingly, the cost of a plan is the line integral of the cost function over the path.

Definition 1. *A solution to a planning problem* $(x_0, \mathcal{X}_G, c)$ *in a domain* $(f, \mathcal{X})$ *is a continuously differentiable path* $\sigma^* : [0, T] \to \mathcal{X}$, *where*

$$\sigma^* = \arg\min_{\sigma \in \Sigma_\mathcal{X}} \int_0^T c(\sigma(t)) \|\dot{\sigma}(t)\| \, \mathrm{d}t \tag{1}$$

$$\text{s.t.} \quad f(\sigma(\tau), \dot{\sigma}(\tau)) \geq 0 \qquad \forall \tau \in [0, T] \qquad \text{(feasibility)}$$

$$\sigma(0) = x_s, \quad \sigma(1) \in X_g$$

We can solve problems of this form under a variety of conditions, by constructing and searching a *random geometric graph*. A random geometric graph $G_n^{\mathrm{RGG}} = (\mathcal{V}, \mathcal{E})$ is a graph whose vertices are a randomly chosen finite set of configurations $\mathcal{V} \subset \mathcal{X}$, and whose edges are paths between nearby configurations, with adjacency determined using simple geometric rules. There are many motion planning algorithms that exploit this underlying concept; see LaValle [14] for a survey.

Planning algorithms based on random geometric graphs are provably optimal under a variety of conditions. For holonomic systems, Karaman and Frazzoli [11] demonstrated that a random geometric graph with n vertices drawn uniformly at random from the configuration space $\mathcal{X}$ will contain an optimal path with probability one in the limit as $n \to \infty$. This result was proven for a graph with an edge between any pair of samples with geodesic distance less than $\gamma_\mathcal{X} (\frac{\log n}{n})^{\frac{1}{k}}$, provided the straight-line path between those vertices is collision free; $\gamma_\mathcal{X}$ is a constant depending only on the manifold $\mathcal{X}$ and k is the dimensionality of the configuration space $\mathcal{X}$. The same authors extended this result [12] to the broad class of non-holonomic real analytic dynamical systems. Note these approaches can be applied to planning on manifolds, assuming we know how to sample random configurations from the manifold and how to connect nearby configurations.

The class of systems addressed by these algorithms does not include many systems of practical interest in robotics. Consider a simple model of object manipulation. A holonomic robot is tasked with moving K boxes. When in contact with a box, the robot can rigidly grasp the box, so that the robot-box pair behave as a single rigid body as long as the box is grasped. Any box not grasped by the robot does not move. The robot can release a grasped box at any time. This model avoids much of the complexity of real contact dynamics; there is no consideration of momentum, or even of form or force closure. These constraints can be written explicitly; we define the indicator function $f_{\mathrm{contact}} : \mathrm{SE}(2) \times \mathrm{SE}(2) \to \{0, 1\}$ such that for any object o, $f_{\mathrm{contact}}(x_r, x_o) = 0$ if the robot can grasp the object when the robot pose is x_r and the object pose is x_o, and $f_{\mathrm{contact}}(x_r, x_o) = 1$ otherwise. Then the constraints on permissible motions can be expressed as

$$f_{\mathrm{contact}}(x_r, x_o) \dot{x}_o = 0 \quad \forall o. \tag{2}$$

Because the support of the function $f_{\mathrm{contact}}(x_r, x_o)$ is compact, the function is not analytic. Note that although f_{contact} is not smooth, a similar result holds for smooth constraints, provided they have compact support.

Non-analytic constraints cause motion planning algorithms to fail for a simple reason: arbitrary pairs of configurations can be quite close in geodesic distance but require very long trajectories to connect. Consider two world configurations where the position of each object is perturbed slightly from one configuration to the other; to move the world from the first configuration to the second, the robot must travel to each object in turn. Typical PRM* implementations use either straight-line connections or two-point boundary value solvers as local planners; in order to apply an algorithm like PRM* to this domain, we would need a more powerful local planner capable of computing the very long path between the original and perturbed configurations. Similarly, the rapid exploration of algorithms like the RRT* is dependent on a steer function that finds a plan that moves toward an arbitrary configuration; in manipulation planning problems, simple implementations of the steer function may not bring the world configuration closer to a perturbed configuration, leading to slow exploration. Note this does not constitute a proof of incompleteness of RRT* or PRM* in problems with non-analytic constraints; little is known about the completeness or optimality of sampling-based motion planning in settings where the geodesic distance does not accurately capture the complexity of moving between configurations.

3   Modes and orbits

Although the conditional constraints that are a fundamental characteristic of manipulation planning domains cannot be represented as analytic differential constraints, they can be represented using the more general class of *piecewise-analytic* constraints. We define a function f as piecewise-analytic if there exists a finite set $\mathcal{M}$ of $N_{\mathcal{X}}$ connected Riemannian manifolds $\{\mathcal{X}_i\}_{i=1}^{N_{\mathcal{X}}}$, each a subset of the configuration space $\mathcal{X}$, such that $\mathcal{X} = \bigcup_{i=1}^{N_{\mathcal{X}}} \mathcal{X}_i$ and such that the restriction of f to the interior of $\mathcal{X}_i$ is analytic for each i. Following Alami et al. [1] and Hauser and Latombe [5], we refer to the manifolds $\mathcal{X}_i$ as *modes*. In the simplified block-pushing problem with k blocks, there are $k+1$ modes: one describing the motion of the robot when not in contact with any blocks, and one for each block describing the evolution of the system when the robot grasps that block.

Recall that in our block-pushing problem, a block moves only if grasped by the robot. Consequently, it is impossible to move between arbitrary configurations on a mode without grasping and releasing blocks—that is, without switching modes. Consider the mode defined by which block the robot has grasped; the reachable configurations *within* this mode are defined by the locations of all the other blocks. We will refer to the collection of configurations reachable from an arbitrary configuration as an *orbit*.

Formally, an orbit $\mathbf{O}_{\mathcal{X}_i}(x)$ of a mode $\mathcal{X}_i$ through a configuration $x \in \mathcal{X}_i$ is the subset of $\mathcal{X}_i$ connected to x by a feasible path that lies wholly on the manifold $\mathcal{X}_i$. Because the constraints are analytic on each mode, the orbits are disjoint submanifolds of the same dimensionality [15]. This means that a planning problem where the start and goal states both lie on the same orbit is straightforward to solve using standard sampling-based planning techniques. The key to our approach is recognizing that we can solve arbitrary planning problems in a given domain by choosing a finite set of orbits, building a random geometric graph on each orbit, and connecting the resulting set of graphs. In the sections that follow, we will describe in detail how to construct and link these graphs in a way that preserves the optimality guarantees of the PRM* while also remaining computationally tractable.

Note that as with most sampling-based motion planning algorithms, we do not require an explicit geometric representation of the modes or orbits in order to plan; we require only the ability to sample from them. This may be non-trivial, as standard rejection-sampling approaches will fail if the modes or orbits have dimension less than that of the configuration space. However, for many problems of interest, it is straightforward to write subroutines that uniformly sample configurations from a given orbit. These subroutines are also a prerequisite for many other manipulation planning algorithms.

For the purposes of this work, we assume that the modes and orbits are exogenously given in a form that permits sampling. That is, we assume that given a configuration x, there exists a subroutine `modes(x)` that returns a list of identifiers of the modes that contain x, a subroutine `sample(x)` that generates a configuration chosen uniformly at random from orbits containing x, and a subroutine `sample_boundary(x)` that generates a configuration uniformly at random from the boundary of one of the orbits containing x, such that the probability of the generated sample belonging to any boundary manifold is greater than zero.

4   Algorithms

We first describe the implementation of our algorithms on the block-pushing domain, then describe the general formulation. The complexity of the algorithms involved leads to unwieldy pseudocode; instead, we provide an open source Python implementation[1] of the algorithms described here. In addition to the subroutines `sample` and `sample_boundary` mentioned in section 3, we assume we have a local planner available, which must satisfy several regularity conditions described in section 5.

Our algorithms both follow the same basic procedure: we construct an implicit random geometric graph by choosing a set of configurations (the *vertex set*) and specifying a subroutine to generate the neighboring configurations for a given vertex. In addition to a problem specification (x_0, X_G, c), we take as input an integer parameter n. As n increases, the size of the graph increases, which increases the computational resources required to search for a trajectory but also improves the quality of the path returned.

Two vertices x and x' are connected by an edge in our graph if the geodesic distance on some orbit including x and x' is less than a critical threshold value depending on n. We use the local planner to determine the cost of an edge; if the local planner cannot find a feasible path, perhaps due to the presence of an obstacle, the edge is assigned a cost of ∞. Together, the vertex set and the neighbor function specify an implicit random geometric graph. An eager PRM* implementation would explicitly evaluate the cost of every edge; instead, we lazily evaluate only those edges that may be part of an optimal path.

---

[1]   https://github.com/robustrobotics/forgg

We then search the implicit random geometric graph using the A* algorithm of Hart et al. [4]. Only when a vertex is expanded do we actually invoke the local planner to determine whether it can generate a collision-free feasible trajectory to any of the neighbors of the expanded vertex; if it can, the neighboring vertex is added to a priority queue with priority equal to the minimal cost to reach that vertex. Note that this idea of lazily searching a random geometric graph is not novel; the core idea was described by Bohlin and Kavraki [3], and recent work from Janson et al. [9], among others, suggest this approach can be significantly more efficient than RRT* or searching an explicit graph constructed with the PRM* algorithm. The innovation in our approach is the way in which we construct the graph, which ensures that the graphs constructed on each orbit are constructed in a way that ensures asymptotic optimality.

4.1   Orbital Bellman trees

To extend the asymptotic optimality of PRM* to problems with piecewise-analytic constraints, we must ensure that as the graph size n tends to infinity, the number of orbits on the connected component containing x_s tends to infinity and that the number of samples on each of those orbits tends to infinity. Building on the Random-MMP algorithm described by Hauser and Ng-Thow-Hing [6], we provide a graph construction and search algorithm that incrementally builds the graph such that as n increases, the graph contains an increasing number of samples from an increasing number of orbits yet remains connected with probability one as $n \to \infty$.

The algorithm has two free parameters, $\nu \in (0, 1)$ and $\eta \in (0, \infty)$. Increasing η increases the number of edges in the graph by inflating the radius defining whether two vertices are connected. ν represents a trade-off between exploring the interior of each orbit and exploring the relations between different orbits.

We initialize the vertex set to include the initial configuration x_0 and place the initial configuration in a priority queue with priority zero. We then repeatedly remove the lowest-cost vertex from the queue and perform two algorithmic operations. First, we check whether the removed vertex belongs to any orbits that do not contain samples and add samples from those orbits if so. Concretely, if we remove the vertex x from the priority queue and for some mode $\mathcal{M}$ containing x the orbit $\mathbf{O}_{\mathcal{M}}(x)$ does not contain any samples, we call the subroutine sample n times to sample n configurations from $\mathbf{O}_{\mathcal{M}}(x)$. Then for each mode $\mathcal{M}'$ adjacent to $\mathcal{M}$, we call the sample_boundary subroutine νn times to generate νn samples from the intersection of $\mathbf{O}_{\mathcal{M}}(x)$ and $\mathcal{M}'$. Finally, we build a search index such as a k-d tree or a cover tree from the n samples, to enable efficient lookup of the neighboring vertices.

Second, we perform a standard iteration of A*, considering the neighbors of x on each orbit to which it belongs. For each neighbor x', we evaluate the cost of the path returned by the local planner from x to x'; if that cost is finite, we add the neighbor x' to the priority queue as usual. The algorithm terminates when a vertex is expanded that lies in X_g, when the queue is empty, or when a predefined maximum number of samples have been removed from the priority queue.

As the search proceeds, this process grows a graph of interconnected orbits. We refer to this graph construction as a random orbital geometric graph (ORGG), as it is a random geometric graph that respects the structure of the orbits created by the constraints. As with any best-first search algorithm, the search produces a tree in which each vertex v is labelled with the minimal cost of a path through the ORGG from the start vertex to v, the parent of each vertex except the start vertex lies along that minimum cost path. We refer to this tree as an orbital Bellman tree, as the cost to reach each vertex satisfies Bellman's equations; we refer to the search algorithm as the orbital Bellman tree (OBT) algorithm.

The OBT algorithm is asymptotically optimal for any ν, η if the set of neighbors of a given vertex includes all configurations within a distance $r_{\mathbf{O}}(n)$, for each orbit $\mathbf{O}$ containing the configuration. The function $r_{\mathbf{O}}(n)$ is determined by ν, η, the dimensionality $d_{\mathbf{O}}$, and the Lebesgue measure $\text{vol}(\mathbf{O})$ of the orbit, where the measure is induced by the volume form of the manifold $\mathbf{O}$. As with other sampling-based motion planners, the Lebesgue measure of the orbit can be approximated using rejection sampling.

$$r_{\mathbf{O}}(n) = (1 + \eta)\gamma_{\mathbf{O}} \left(\frac{\log n}{n} \right)^{1/d_{\mathbf{O}}} \tag{3}$$

$$\gamma_{\mathbf{O}} = 4 \left((1 + \frac{1}{d_{\mathbf{O}}}) \frac{\text{vol}(\mathbf{O})}{\zeta_{d_{\mathbf{O}}}} \right)^{\frac{1}{d_{\mathbf{O}}}} \tag{4}$$

$$\zeta_{d_{\mathbf{O}}} = \frac{\pi^{d_{\mathbf{O}}}}{\Gamma(\frac{d_{\mathbf{O}}}{2} + 1)} \qquad \text{(volume of a } d_{\mathbf{O}}\text{-sphere)}$$

Note this connectivity radius is nearly identical to that presented by Karaman and Frazzoli; the only change is the factor of 4 in γ, which is needed to ensure optimality when the path obtained is on a different orbit from the optimal path.

OBT is different from Random-MMP in two important ways. First, OBT obtains provably asymptotically optimal paths across each orbit by constructing an optimal random geometric graph on each orbit. Random-MMP instead adds a single path across an orbit to a configuration on another mode; finding this path is sufficient to guarantee completeness but not optimality. Second, by choosing a fixed fraction of samples from each orbit to be from the intersection of the orbit and the adjacent modes, OBT ensures that enough orbits are considered to guarantee optimality.

4.2   Factored orbital Bellman trees

The OBT algorithm is extremely computationally intensive. Each new orbit considered requires sampling n new configurations and building a new search index; this takes $\mathcal{O}(n \log n)$ time per orbit. Consider the block-pushing problem; every grasp configuration is a part of two orbits, as the robot can either maintain the grasp and move with the object or immediately drop the object at that location and move only itself. This means that we must consider n new orbits whenever the robot grasps an object. Consequently, the number of samples we must store grows exponentially with the number of objects the shortest plan must grasp. There are K objects to grasp whenever the robot is not holding an object. Each time we consider a possible set of grasps and releases, we generate $\mathcal{O}(Kn)$ samples and take $\mathcal{O}(Kn \log n)$ time, and the OBT algorithm does so for each combination of grasps and releases, leading to $\mathcal{O}((Kn \log n)^d)$ time and $\mathcal{O}(K^d n^d)$ space to search through all possible combinations of objects to grasp and locations to release in a sequence of d grasps.

One avenue toward reducing this computational burden is to take advantage of the structure of the problem domain to reduce the number of samples and search indices we must generate. The block-pushing domain has a key feature that makes this possible: the constraints that define the modes and orbits *factor*, allowing us to consider different parts of the configuration space independently. The constraint that an object cannot move unless grasped is unary: it does not affect the permissible locations of the other objects. The constraint that the robot is grasping an object is binary: it affects the robot and the object, but has no effect on the configurations of the other objects.

We can encode this structure in a factor graph representing the uniform distribution over a mode. The vertices of this factor graph represent the configuration of each object, and the unnormalized factors are the constraints defining the orbits. We can generate samples from this uniform distribution over a mode by sampling from each connected component of this factor graph independently and then taking the Cartesian product of the resulting sample sets. We can exploit this factorization to generate samples from the full configuration space efficiently.

In the block-pushing domain, this factored sampling amounts to sampling a set of poses of each object and a set of grasping poses of the robot for each object. A graph whose vertices are the union of the products of these sets of poses is an orbital random geometric graph. Because the vertices of such a graph are generated by factoring the uniform distribution over a mode, we refer to this graph construction as a factored orbital random geometric graph (FORGG). This factorized sampling strategy can be generalized to arbitrary configuration spaces and sets of constraints, and will be beneficial if the factor graph encoding a uniform distribution over a given orbit has multiple connected components. Arbitrarily complex models of contact dynamics satisfy this requirement due to the local nature of contact dynamics.

The algorithm that constructs and searches a factored orbital random geometric graph closely resembles the OBT search algorithm, with the key distinction that the samples are generated from independent factors. Accordingly, we refer to this search algorithm as the factored orbital Bellman tree (FOBT). On the block pushing problem, this reduces the amount of space needed to store the orbital geometric graph from $\mathcal{O}((nk)^D)$ to $\mathcal{O}(nk)$ and the amount of time required from $\mathcal{O}((kn \log n)^D)$ to $\mathcal{O}(kn \log n)$, with no dependence on D, the number of grasps in the shortest solution. Note that we have not avoided the exponential cost of graph search; the search itself still takes $\mathcal{O}(b^d)$ time and space, where b is the graph branching factor and d is the depth of the shortest solution. Although this is only a constant factor improvement, it represents a significant practical advance.

5 Analysis

We now prove the asymptotic optimality of OBT and FOBT, subject to a regularity condition on the local planner used. For brevity, we present several propositions without proof; proofs of these propositions are available in our supplementary material.[2] In addition to the piecewise-analyticity of the constraints, we require an additional technical condition on the local planner π used to connect nearby states. First, there must exist a radius $r_0 > 0$ such that the planner will return a feasible path if invoked to connect two configurations that lie inside an open geodesic ball of free space with radius less than r_0. Second, for any $\epsilon > 0$ there must exist $r_\epsilon > 0$ such that for all $x, x' \in \mathcal{M} : d_\mathcal{M}(x, x') < r_\epsilon, L_\mathcal{M}(\pi(x, x')) < (1 + \epsilon)d_\mathcal{M}(x, x')$. If a local planner has these two properties, we say it is locally complete. Note that a local planner that connects trajectories with geodesic curves is locally complete.

Theorem 1 (Optimality of OBT). *Given a planning problem (x_s, X_g, c) in a domain (X, f), let c_n be the shortest path between x_s and X_g on an orbital random geometric graph G_n with n vertices, built using a locally complete local planner. Then $\mathbb{P}(\{\limsup\limits_{n \to \infty} c_n = c^*\}) = 1$.*

5.1 Construction of a sequence of paths

Let $\bar{\sigma}^*$ be an optimal solution to the planning problem. Decompose $\bar{\sigma}^*$ into a sequence of M paths $\{\sigma_m^*\}_{m \in [1, M]}$, each lying on a single manifold.

$$\bar{\sigma}^* = \bigoplus_{m=1}^{M} \sigma_m^* \tag{5}$$

Define $\varphi_m \in V_M$ as the mode on which the path σ_m^* lies. Let d_m be the dimensionality of the orbit containing σ_m^*. Let $\bar{d}$ be the maximum dimension of any intersection orbit expressed in the path: $\dim(\varphi_m \cap \varphi_{m+1}) \leq \bar{d} \forall m$.

Because the modes are analytic manifolds, there exists $\delta > 0$ such that each path σ_m^* is homotopy-equivalent to a path that lies in the union of the δ-interior of the mode φ_m, an open ball of radius δ whose closure contains $\sigma_m^*(0)$, and an open ball of radius δ whose closure contains $\sigma_m^*(1)$. Define the weakly monotonically decreasing sequence $\{\delta_n\}_{n \in \mathbb{Z}}$.

$$\delta_n = \min\left(\delta, n^{-\frac{1}{2d}}\right) \tag{6}$$

Clearly, this sequence satisfies $0 < \delta_n \leq \delta$ and $\lim_{n \to \infty} \delta_n = 0$; let $n_0 = \min\{n \in \mathbb{Z} : \delta_n < \delta\}$. Because the problem is δ-robust, there exists a sequence $\{\bar{\sigma}_n\}_{n \in \mathbb{N}}$ such that $\bar{\sigma}_n$ has δ_n-clearance. Decompose each path $\bar{\sigma}_n$ into a sequence of M paths $\sigma_{n,m}$, just as with $\bar{\sigma}^*$.

---

[2] The supplementary material is available at https://people.csail.mit.edu/wrvb/papers/vega-brown_wafr16_supplement.pdf

5.2   Construction of balls on the intersections between modes

Define $r_{\cap,n,m} = a_m n^{-\frac{1}{2d}}$, where a_m is recursively defined to ensure that if a leaf intersects $r_{\cap,n,m}$, it also intersects $r_{\cap,n,m+1}$.

$$a_M = \delta \tag{7}$$

$$a_m = \sup\{a > 0 : \forall y \in B(\sigma_{n,m}, a) \sup_{t\in(0,1)} \inf_{y'\in\mathcal{O}_m(y)} d(\sigma_{n,m}(t), y') < a_{m+1}\} \tag{8}$$

Note that $r_{\cap,n,m} \leq \delta_n \,\forall m$ for large n. For each path $\sigma_{n,m}$, define the region $B_{\cap,n,m}$ as the geodesic ball centered at $\sigma_{n,m}(0)$ on the manifold $\varphi_{m-1}\cap\varphi_m$ with radius $r_{\cap,m,n}$, Let $E_{\cap,n,m}$ be the event that the ball $B_{\cap,n,m}$ contains a sample: that is, $E_{\cap,n,m}$ occurs when the intersection of the vertex set $\mathcal{V}_n$ and the ball $B_{\cap,n,m}$ is nonempty. Let $A_{\cap,n} = \bigcap_m E_{\cap,n,m}$ be the event that each ball $B_{\cap,n,m}$ contains a sample.

5.3   Construction of balls on an arbitrary orbit

Fix $\theta \in (0,1)$ and $r > 0$; let $\sigma : [0,1] \to \mathcal{M}$ be a feasible path on a mode $\mathcal{M}$ such that there exist real numbers $t_-, t_+, 0 \leq t_- < t_+ \leq 1$ and the following conditons hold: for all $t \in (t_-, t_+)$, $\sigma(t) \in \mathrm{Int}_r(\mathcal{M})$; for all $t \in (0, t_-]$, $\sigma(t) \in B_r(\sigma(t_-))$; and for all $t \in [t_+, 1)$, $\sigma(t) \in B_r(\sigma(t_+))$. Then there exists a finite collection of configurations $Y(\sigma, y_0, r) = \{y_k\}$ drawn from the orbit containing y_0 with the property that if $z_k \in B(y_k, \frac{r}{4+\theta})$, $z_{k+1} \in B(y_{k+1}, \frac{r}{4+\theta})$ are two vertices in an orbital random geometric graph, the OBT algorithm will call the local planner for the pair (z_k, z_{k+1}), and the local planner will succeed. We provide a construction of Y in two steps. First, we consider the part of the path that lies in the r-interior of the manifold. Define a strictly monotonically increasing sequence (t_k) as follows.

$$\tau_0 = t_- \tag{9}$$

$$\tau_{k+1} = \sup_{\tau\in(\tau_k,t_+)} \{d_{\mathcal{M}}(\sigma(\tau_k), \sigma(\tau)) < \frac{\theta r}{4+\theta})\} \tag{10}$$

Let K be the smallest integer k such that $\tau_k = t_+$. Define $(x_k)_{k\in[K]}$ so that $x_k = \sigma(\tau_k)$. Define $(y_k)_{k\in[K]}$ so that $d(y_k, x_k) < \frac{r}{4+\theta}$; by the assumptions on the leaf, such a sequence must exist. Define the set of balls $B_{k\,k\in[K]}$, where $B_k = B(y_k, \frac{r}{4+\theta})$.

Let z_k be an arbitrary configuration in B_k.

$$d(z_k, x_k) \leq d(z_k, y_k) + d(y_k, x_k) \leq \frac{r}{4+\theta} + \frac{r}{4+\theta} \leq r \tag{11}$$

$$d(z_{k+1}, x_k) \leq d(z_{k+1}, y_{k+1}) + d(y_{k+1}, x_{k+1}) + d(x_{k+1}, x_k)$$
$$\leq \frac{\theta r}{4+\theta} + \frac{r}{4+\theta} + \frac{r}{4+\theta} \leq \frac{2+\theta}{4+\theta}r \leq r \tag{12}$$

$$d(z_k, z_{k+1}) \leq d(z_k, x_k) + d(x_k, z_{k+1}) \leq \frac{2}{4+\theta}r + \frac{2+\theta}{4+\theta}r \leq r \tag{13}$$

From eq. (13), if the set of vertices includes a configuration in each of the pair of balls B_k and B_{k+1}, the local planner will be invoked for the pair; from eq. (11) and eq. (12), both samples lie inside the ball $B(x_k, r)$, and therefore by the assumptions of the theorem the local planner will succeed if called. Note that with the exception of τ_K, sequential centers $\sigma(\tau_k)$ and $\sigma(\tau_{k+1})$ are separated by $\frac{\theta r}{4+\theta}$; if $L(\sigma)$ is the length of the path, it follows that

$$K \leq \left\lceil \frac{4+\theta}{\theta r} L(\sigma) \right\rceil + 1. \tag{14}$$

Next, we consider the part of the path that lies near the boundary. We will prove the result for $t \in (0, t_-)$, assuming $t_- \neq 0$; the proof for $t \in (t_+, 1)$ is similar. Fix an arbitrary $y_0 \in \partial M \cap L$ such that $d(\sigma(0), y_0) \leq \frac{r}{4+\theta}$. Define a chart $\phi_{y_0} : U \subset M \to S \times V \times W$ in *collar coordinates*, such that $S \subseteq \mathbb{R}_{\geq 0}, V \subseteq \mathbb{R}^{k-1}, W \subseteq \mathbb{R}^{n-k}$. Note that the coordinate s is equal to the minimum distance of a configuration to the boundary of the manifold. Note also that $S \times V$ is diffeomorphic to a subset of the leaf $\mathbf{O}$; any curve with for which the coordinates in W are constant will be a feasible path. For sufficiently small r, such a chart must exist [15]. Without loss of generality assume $\phi_{y_0}(y_0) = (0, 0, 0)$.

Assume r is small enough that $B_r(\sigma(t_-)) \subset U$. Choose $y_- \in \mathbf{O}$ such that $d(y_-, \sigma(t_-)) \leq \frac{r}{4+\theta}$, and define $\phi_{y_0}(y_-) = (s_-, v_-, 0)$. Let $y_k = \phi_{y_0}^{-1}((1 - \alpha_k)\frac{3r}{4+\theta} + \alpha_k s_-, \alpha_k v_-, 0)$, where $\alpha_1 = 0$ and the sequence (α_k) is defined recursively as follows.

$$\alpha_{k+1} = \sup_{\alpha \in (\alpha_k, 1)} \{ d(y_k, y_{k+1}) < \frac{2r}{4+\theta} \} \tag{15}$$

Note that the total distance from y_0 to y_- is upper-bounded.

$$d(y_0, y_-) \leq d(y_0, \sigma(0)) + d(\sigma(0), \sigma(t_-)) + d(\sigma(t_-), y_-) \tag{16}$$

$$\leq \frac{r}{4+\theta} + 2r + \frac{r}{4+\theta} = \frac{10 + 2\theta}{4+\theta} \tag{17}$$

Since with the exception of the first and last centers the distance between successive centers (y_k) is at least $\frac{2r}{4+\theta}$, this part of the construction adds at most $2 + (\frac{10+2\theta}{4+\theta}r)/(\frac{2}{4+\theta}r) = 7 + \theta \leq 8$ configurations to the set.

The following claim is proven as proposition 1 in our supplementary material; we omit the proof here. If the set of vertices includes configurations $z_k \in B(y_k, \frac{r}{4+\theta})$ and $z_{k+1} \in B(y_{k+1}, \frac{r}{4+\theta})$, these configurations will be connected by an edge. Similarly, if the set of vertices includes y_0 and a configuration $z_k \in B(y_k, \frac{r}{4+\theta})$, those vertices will be connected by an edge.

If $t_+ \neq 1$, we can apply a similar construction at the other end of the path. In total, we have constructed a set of at most

$$K_\sigma = \left\lceil \frac{4+\theta}{\theta r} L(\sigma) \right\rceil + 17 \tag{18}$$

balls, such that if each ball contains a sample, the resulting graph will contain a path with the desired properties. Then the cardinality of the set $Y(\sigma, y_0, r)$ is upper-bounded by $\lceil \frac{4+\theta}{\theta r} L(\sigma) \rceil + 17$.

We now apply the construction $Y(\sigma, y_0, r)$ to each mode. If the event $E_{\cap,n,m}$ occurs, there exists some $y_{n,m,0} \in B_{\cap,n,m} \cap V_n$. Let $r_{n,m} = \gamma_m (\frac{\log n}{n})^{\frac{1}{d_m}}$. Let $Y_{n,m} = Y(\sigma_{n,m}, y_{n,m,0}, r_{n,m})$. Let $K_{n,m} = \text{card}(Y_{n,m})$. Let $B_{n,m,k}$ be the geodesic ball centered at $y_{n,m,k}$ of radius $\frac{1}{4+\theta} r_{n,m}$. Let $E_{n,m,k}$ be the event that the intersection of the vertex set V_n and the ball $B_{n,m,k}$ is nonempty. Let A_n be the event that all balls on each mode contains a sample. Note that A_n occurs only if $A_{\cap,n}$ occurs, as A_n is meaningful only if there exists an $y_{n,m,0} \in B_{\cap,n,m}$ to define the orbit on which $Y_{n,m}$ is defined. By construction, if A_n occurs, then algorithm OBT will return a solution with finite cost.

5.4 Bounding the cost of the path returned

Fix an arbitrary $\beta \in (0,1)$, and assume there exists $x_m \in B_{\cap,n,m}$ $\forall m$. For each $y_{n,m,k}$, define a smaller ball $\tilde{B}_{n,m,k}$ with the same center and radius $\frac{\beta r_{n,m}}{4+\theta}$. Let $I_{n,m,k}$ be the indicator for the event that the intersection of the vertex set V_n and the ball $\tilde{B}_{n,m,k}$ is nonempty.

$$I_{n,m,k} = \begin{cases} 1 & \text{card}(\tilde{B}_{n,m,k} \cap V_n) > 0 \\ 0 & \text{else} \end{cases} \tag{19}$$

Let $S_{n,m,k} = \sum_{m=1}^{M} \sum_{k=1}^{K_{n,m}} I_{n,m,k}$ be the number of smaller balls $\{\tilde{B}_{n,m,k}\}$ containing a configuration, and let $K_n = \sum_{m=1}^{M} K_{n,m}$ be the total number of smaller balls. If the cost function c is Lipschitz continuous, then for any $\epsilon > 0, \theta > 0$ there exists $\alpha > 0, \beta > 0, n_0 > 0$ such that if $S_n \geq \alpha K_n$, then for all $n > n_0$, $c(\sigma_n) \leq (1+\epsilon)c^*$ (proposition 2 in the supplement). Let $A_{n,\alpha,\beta}$ be the event that $S_n \geq \alpha K_n$.

We can then upper-bound the probability that the path returned by OBT has cost more than $1 + \epsilon$ times the optimal cost in terms of the probabilities $\mathbb{P}(\overline{A_{\cap,n}})$, $\mathbb{P}(\overline{A_n}|A_{\cap,n})$, and $\mathbb{P}(\overline{A_{n,\alpha,\beta}}|A_{\cap,n})$ (proposition 3 in the supplement).

$$\mathbb{P}(c_n \geq (1+\epsilon)c^*) \leq \mathbb{P}(\overline{A_{\cap,n}}) + \mathbb{P}(\overline{A_n}|A_{\cap,n}) + \mathbb{P}(\overline{A_{n,\alpha,\beta}}|A_{\cap,n}) \tag{20}$$

Since each of the terms on the right side of eq. (20) is summable (proposition 4, proposition 5, and proposition 6 in the supplement) the term on the left side of eq. (20) is summable. Consequently, the term on the left side of eq. (20) is summable.

$$\sum_{n=1}^{\infty} \mathbb{P}(\overline{A_{\cap,n}}) < \infty, \qquad \sum_{n=1}^{\infty} \mathbb{P}(\overline{A_n}|A_{\cap,n}) < \infty, \qquad \sum_{n=1}^{\infty} \mathbb{P}(\overline{A_{n,\alpha,\beta}}|A_{\cap,n}) < \infty \tag{21}$$

$$\therefore \sum_{n=1}^{\infty} \mathbb{P}(c_n \geq (1+\epsilon)c^*) < \infty \tag{22}$$

Therefore, by the Borel-Cantelli lemma, the event that the algorithm returns a feasible path with cost less than $(1 + \epsilon)c^*$ occurs infinitely often as $n \to \infty$. The sequence c_n then converges almost surely to c^*. □

The proof of asymptotic optimality of FOBT employs the same geometric construction as for OBT. We need only modify the proof that the terms on the right side of eq. (20) are summable. The first two terms can be shown to be summable using identical logic to OBT, by noting that a ball in a product space contains a product of smaller balls in each component space and applying the union bound. The third term requires more effort to adapt, as the proof relies on the independence of the small balls. However, we can define a looser bound that does not require independence (proposition 7 in the supplment). Due to space constraints, we omit this construction here.

6 Computational experiments

We implemented FOBT for the simplified block-pushing problem (fig. 2a). The goal is to move the block labelled 'box1' into the region shaded red, past a move-able obstacle. The planner must either decide to go around or must choose where to place the moveable object to get it out of the way. For comparison, we considered a simple task and motion planning algorithm approach, labelled TAMP. The TAMP planner can invoke a motion planner as a subroutine to accomplish a set of tasks, such as grasping an object or moving a grasped object to one of a fixed, hand-coded set of regions. In practice, this amounts to evaluating both sensible plans and choosing the one with the lower-cost solution.

As expected, we find that while TAMP can often quickly find a solution, more computational time does not allow that solution to be improved. In contrast, FOBT continues to perform better as the available computational time increases. Note that TAMP is suboptimal because it can only consider a finite set of goal locations; this set does not grow as n increases. By injecting domain knowledge in the form of a better task hierarchy, it is likely the TAMP planner could find plans as good as FOBT; FOBT finds these plans without such domain knowledge.

The quantitative comparison in fig. 2c highlights the main deficiency of FOBT: it is computationally demanding. As the parameter n increased above 1000, the implementation exhausted available memory and the algorithm failed due to space constraints. Improving the computational efficiency of FOBT is an important avenue for future work. Augmenting our algorithm with intelligent heuristics or nonuniform sampling strategies derived from domain knowledge could greatly increase computational efficiency.

7 Conclusion

To our knowledge, these are the first algorithms for asymptotically optimal motion planning that are applicable to piecewise-analytically constrained problem domains like manipulation planning. We note that the ideas in these algorithms can likely be combined with the ideas in many other sampling-based motion planning and graph search algorithms. This would improve performance and

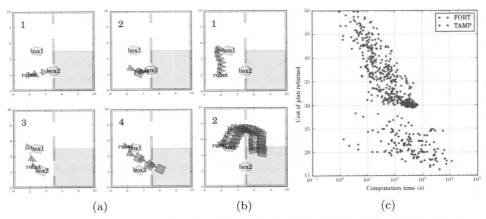

Fig. 2: Plans obtained with FOBT and TAMP for a simple block pushing problem. The goal is to move the block labeled 'box1' to the region shaded in red, but the most direct path is blocked by another box. (a): The four-step plan obtained by FOBT. (b) A simpler, but more expensive, plan returned by a sequential task and motion planning algorithm. (c) Quantitative comparison of the cost of the plan returned and the computational time used for various graph sizes with both methods.

extend our results to domains such as kinodynamic planning, planning under uncertainty, and adversarial planning. In particular, if we combine symbolic task hierarchies with asymptotically optimal motion planning algorithms like the ones presented here, we can perhaps create task and motion planning algorithms with strong asymptotic performance guarantees.

In addition, our analytical results provide a foundation for a rigorous evaluation of the performance gap between hierarchically optimal planners and asymptotically optimal planners like those presented here. In principle, we could utilize such a bound to learn better symbolic representations, rather than requiring they be hand-coded by domain experts. This presents a promising route toward linking recent advances in planning and unsupervised learning.

Bibliography

[1] Rachid Alami, Jean-Paul Laumond, and Thierry Siméon. Two manipulation planning algorithms. In *Proceedings of the workshop on Algorithmic foundations of robotics*, pages 109–125. AK Peters, Ltd., 1995.

[2] Dmitry Berenson, Siddhartha S Srinivasa, Dave Ferguson, and James J Kuffner. Manipulation planning on constraint manifolds. In *Robotics and Automation, 2009. ICRA'09. IEEE International Conference on*, pages 625–632. IEEE, 2009.

[3] Robert Bohlin and Lydia E Kavraki. Path planning using lazy prm. In *Robotics and Automation, 2000. Proceedings. ICRA'00. IEEE International Conference on*, volume 1, pages 521–528. IEEE, 2000.

[4] Peter E Hart, Nils J Nilsson, and Bertram Raphael. A formal basis for the heuristic determination of minimum cost paths. *Systems Science and Cybernetics, IEEE Transactions on*, 4(2):100–107, 1968.

[5] Kris Hauser and Jean-Claude Latombe. Multi-modal motion planning in non-expansive spaces. *The International Journal of Robotics Research*, 2009.

[6] Kris Hauser and Victor Ng-Thow-Hing. Randomized multi-modal motion planning for a humanoid robot manipulation task. *The International Journal of Robotics Research*, 30(6):678–698, 2011.

[7] David Hsu, Jean-Claude Latombe, and Rajeev Motwani. Path planning in expansive configuration spaces. In *Robotics and Automation, 1997. Proceedings., 1997 IEEE International Conference on*, volume 3, pages 2719–2726. IEEE, 1997.

[8] Léonard Jaillet and Josep Porta. Path planning under kinematic constraints by rapidly exploring manifolds. *IEEE Transactions on Robotics*, 29(1):105–117, 2013.

[9] Lucas Janson, Edward Schmerling, Ashley Clark, and Marco Pavone. Fast marching tree: A fast marching sampling-based method for optimal motion planning in many dimensions. *The International Journal of Robotics Research*, 2015.

[10] Leslie Pack Kaelbling and Tomás Lozano-Pérez. Hierarchical task and motion planning in the now. In *Proceedings of the International Conference on Robotics and Automation*, pages 1470–1477. IEEE, 2011.

[11] Sertac Karaman and Emilio Frazzoli. Sampling-based algorithms for optimal motion planning. *The International Journal of Robotics Research*, 30(7):846–894, 2011.

[12] Sertac Karaman and Emilio Frazzoli. Sampling-based optimal motion planning for non-holonomic dynamical systems. In *International Conference on Robotics and Automation (ICRA)*. IEEE, 2013.

[13] Athanasios Krontiris and Kostas E Bekris. Dealing with difficult instances of object rearrangement. In *Robotics: Science and Systems (RSS)*, 2015.

[14] Steven M LaValle. *Planning algorithms*. Cambridge university press, 2006.

[15] Peter W Michor. *Topics in differential geometry*, volume 93. American Mathematical Soc., 2008.

[16] Jur Van Den Berg, Mike Stilman, James Kuffner, Ming Lin, and Dinesh Manocha. Path planning among movable obstacles: a probabilistically complete approach. In *Algorithmic Foundation of Robotics VIII*, pages 599–614. Springer, 2009.

[17] Jason Wolfe, Bhaskara Marthi, and Stuart J Russell. Combined task and motion planning for mobile manipulation. In *the proceedings of the International Conference on Automated Planning and Scheduling*, pages 254–258, 2010.

Decidability of Semi-Holonomic Prehensile Task and Motion Planning

Ashwin Deshpande, Leslie Pack Kaelbling, and Tomás Lozano-Pérez

Massachusetts Institute of Technology, Cambridge, Massachusetts, USA
ashwind,lpk,tlp@csail.mit.edu

Abstract. In this paper, we define *semi-holonomic controllability* (SHC) and a general task and motion planning framework. We give a perturbation algorithm that can take a prehensile task and motion planning (PTAMP) domain and create a *jointly-controllable-open* (JC-open) variant with practically identical semantics. We then present a decomposition-based algorithm that computes the reachability set of a problem instance if a controllability criterion is met. Finally, by showing that JC-open domains satisfy the controllability criterion, we can conclude that PTAMP is decidable.

1   Introduction

The last few decades of robotic planning have been dominated by sample-based techniques. Sample-based techniques are very useful tools to quickly find solutions in many domains. However, they suffer from the notable drawback that they cannot prove that a solution does not exist for a particular problem.

The existence of a probabilistically complete algorithm for a planning problem does not settle the question of whether a complete decision procedure, an algorithm that indicates whether a solution does or does not exist for any problem instance, exists. For "classic" motion planning, a holonomic robot among static obstacles, we know that exact algorithms exist for the general case [1, 2]. However, for motion planning in the presence of movable objects, the results are much more limited.

The formal treatment of the problem of planning among movable objects was initiated by Wilfong [3]. When the number of placements and grasps is finite, the problem can be shown to be decidable by building a manipulation graph consisting of a finite number of transfer and transit paths [4]. Decidability for continuous grasps and placements, but involving a single movable object, was shown by Dacre-Wright et al. [5]. More recently, decidability was shown for planning with two objects under restrictive geometries and dynamics [6].

In this paper, we consider a much more general version of planning in the presence of movable obstacles. We allow an arbitrary dimensional world with an arbitrary number of robots, objects, and obstacles, all with semi-algebraic geometries. We also assume that each robot can be holonomically controlled, and each object can be holonomically manipulated. In this manner, we can account for various continuous polynomial dynamics including translations, rotations, stretching, twisting, and morphing. We do restrict our attention to prehensile manipulation, where objects are

© Springer Nature Switzerland AG 2020
K. Goldberg et al. (Eds.): *Algorithmic Foundations of Robotics XII*, SPAR 13, pp. 544–559, 2020.
https://doi.org/10.1007/978-3-030-43089-4_35

rigidly attached to appropriate robots during manipulation. We call the resulting class of problems "prehensile task and motion planning" (PTAMP).

In the first section, we define a general task and motion planning framework capable of representing a large variety of planning problems including PTAMP. At the core of the framework is the concept of semi-holonomic controllability (SHC), which accurately describes the intrinsic dynamics of many task and motion planning problems including PTAMP.

Next, we state the central result of the paper: *jointly-controllably-open* (JC-open) domains are decidable. We then give a perturbation algorithm and show that any PTAMP problem can be rewritten to be JC-open.

Finally, we give a constructive proof of the decidability of JC-open domains. Our algorithm is divided in four parts. First, we describe the decomposition algorithm which decomposes the configuration space into a finite number of manifolds with special properties. Next, we use techniques from differential geometry to calculate the internal controllability of each manifold. Afterwards, for every manifold, we calculate its stratified controllability, i.e. the controllability gained by leaving a manifold and utilizing the controllability of neighboring manifolds. To accomplish this step, we present the convergence condition, which we shows holds for JC-open domains. Finally, we execute a graph search to calculate the reachability set for our initial configuration and test for the existence of a solution.

2   Semi-holonomic task and motion planning framework

We consider an example PTAMP domain with one robot A and several movable objects $B_1,...,B_k$ as shown in figure 1a. The configuration space of the problem is the Cartesian product of individual configuration spaces for the robot and each movable object, i.e. $\mathcal{C}_A \times \mathcal{C}_{B_1} \times ... \times \mathcal{C}_{B_k}$. There are two types of operators: MOVEROBOT, in which A transits around the space, and MANIPULATE-B_i, in which A manipulates object B_i while remaining in contact. We are given an initial configuration and set of goal configurations.

Semantically, for MOVEROBOT, we need to be able to modify the dimensions $\mathcal{C}_A$ without affecting any other dimensions. We enforce that each operator exhibits *semi-holonomic controllability* (SHC) in that a subset of dimensions, F, are marked as "free dimensions" and can be holonomically modified by the operator; non-free dimensions must be held constant. In our example, $F_{\text{MOVEROBOT}}$ and $F_{\text{MANIPULATE-}B_i}$ are $\{\mathcal{C}_A\}$ and $\{\mathcal{C}_A, \mathcal{C}_{B_i}\}$ respectively.

For each operator, we also set a predicate R which encodes the set of configurations that can be in any valid execution trajectory of the operator. $R_{\text{MOVEROBOT}}$ includes the set of all configurations in which there are no collisions between the robot, the objects, and the walls. $R_{\text{MANIPULATE-}B_i}$ is similar to $R_{\text{MOVEROBOT}}$, but it also eliminates configurations in which A and B_i are not in contact. Note that while R is a subset of the configuration space, any specific operator instance is confined to a cross-section of R corresponding to the free dimensions F. In addition, R can only be tested against the current configuration and does not have a memory; therefore, if the robot starts MANIPULATE-B_i with one grasp, it may end the operation with another grasp as

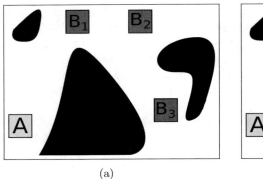 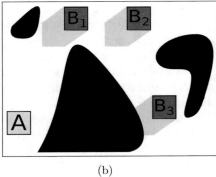

(a)   (b)

Fig. 1: (a) Basic PTAMP domain with one robot, three movable objects and obstacles. (b) Augmented PTAMP domain with shadows in green.

long as the object and robot remain in manipulation range. This "sliding grasp" phenomenon is corrected by replacing that trajectory with a sequence of transit and transfer operations, which is possible if the system allows the robot a small amount of leeway. So far, the predicates cover the typical geometric constraints of manipulation planning.

However, our framework affords us additional flexibility. For example, consider an extended scenario in which the robot is equipped with a solar panel and the movable objects cast shadows on the ground as shown in figure 1b. The robot can transit as before but cannot be in the shade when it manipulate objects as it requires additional power. In the updated scenario, $F_{\text{MOVEROBOT}}$, $F_{\text{MANIPULATE-B}_i}$ and $R_{\text{MOVEROBOT}}$ remain constant. However, $R_{\text{MANIPULATE-B}_i}$ must be updated to remove all areas that are in the shade. Therefore, unlike in typical manipulation planning, a shaded area can be traversable by one operator and impassable by another. This highlights the power of our representation as each operator can have its own unique dynamics.

A *domain* is a tuple $\mathcal{M} = (D, O)$ where:

- D is a set of n configuration dimensions of the entire domain, including robots and movable objects, each defined over $\mathbb{R}$. We assume D has been augmented with all the requisite dimensions utilized by dimension theory in embedding any elements of the configuration space that are typically expressed in alternate spaces to Euclidean space, e.g. angles in S^1 to $\mathbb{R}^2$ [2, 7].
- O is a set of operators. Each operator $o = (F, R)$, contains a set of free dimensions $F \subseteq D$ and a predicate R. Each operator is assumed to exhibit SHC (definition 2).
- R is a predicate such that every trajectory of its operator o must be contained within R. R is defined as a semi-algebraic subset of $\mathbb{R}^n$, i.e., there exists a finite set of finite-coefficient rational polynomials f of finite degree such that:

$$R = \left\{ x \in \mathbb{R}^n \mid \bigcup_i \bigcap_j f_{ij}(x) \otimes 0 \right\} \tag{1}$$

where each $\otimes$ is a binary relation in the set $\{>, <, \geq, \leq, =, \neq\}$. Let R^{all} be the set of R for all operators.

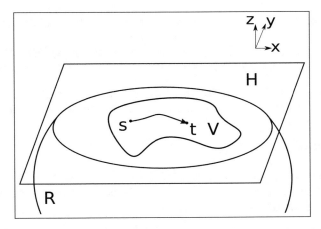

Fig. 2: Illustration of semi-holonomic controllability. In this example, the free dimensions of the operator are $\{x,y\}$, so H is a $\{x,y\}$-section. V shows a connected set on $H \bigcap R$. For any two points (s,t) inside V, there is a path between them without leaving V.

A *problem instance* is a tuple $I = (\mathcal{M}, x_0, G)$ where:

- $\mathcal{M}$ is a domain.
- $x_0 \in \mathbb{R}^n$ is the initial configuration.
- G is a semi-algebraic subset of $\mathbb{R}^n$ that represents the set of goal configurations.

An operator execution trajectory is a continuous trajectory in the configuration space. A sequence of operator execution trajectories is legal only if the terminus of each trajectory serves as the initial state of the subsequent trajectory. A sequence of operator execution trajectories is a solution to a problem instance, $\mathcal{I}$, if starting from x_0 and applying each execution trajectory sequentially, we end on a configuration contained within G. A decision procedure gives a proof that either such a solution does or does not exist in bounded time.

Before describing SHC, we first describe extrusion sets and sections, which extrude spatial subsets along some dimensions:

Definition 1. *Let $F \subseteq D$ be a subset of dimensions, $\overline{F} = (D \setminus F)$, and $V \subseteq \mathbb{R}^{|\overline{F}|}$ be a subset of the span of $\overline{F}$. For a configuration $x \in \mathbb{R}^n$, let $\mathrm{proj}_{\overline{F}}(x) : \mathbb{R}^n \to \mathbb{R}^{|\overline{F}|}$ drop the dimensions F from x. $P(F,V)$ is an F-**extrusion set** with respect to V if:*

$$P(F,V) = \{x \in \mathbb{R}^n \mid \mathrm{proj}_{\overline{F}}(x) \in V\}$$

*If V is a singleton set, then we describe the F-extrusion set $P(F,V)$ as an F-**section**.*

We can now define semi-holonomic controllability, which allows for holonomic behavior in a subset of the dimensions as shown in figure 2:

Definition 2. *Let* $o = (F, R)$ *be an operator,* $s \in R$ *be a configuration,* $H = P(F, \text{proj}_{\overline{F}}(s))$, *V be a connected subset of* $H \cap R$, *and* $t \in V$ *be another configuration. Then,* o *exhibits* **semi-holonomic controllability (SHC)** *if there exists a trajectory of* o *from* s *to* t *while staying inside* V *for all* V, s, *and* t.

A domain is SHC if all of its operators are SHC. Note that SHC is a form of factored holonomicity and cannot be used to model general non-holonomic problems.

3 Decidability result

In this section, we give our main decidability result: PTAMP is decidable. Unfortunately, not every problem in our framework is decidable, so we first define the concept of an *open* domain and state a primary result:

Definition 3. *An* **open** *domain is one in which for every operator* $o = (F,R)$, *R is an open set in* $\mathbb{R}^n$.

Theorem 1. *Every problem instance in an open, SHC domain is decidable.*

Our attempt to apply this theorem to PTAMP immediately fails as $R_{\text{MANIPULATEB}_i}$ may have codimension 1 and is semantically meaningful. Unfortunately, PTAMP is not open in general. Furthermore, we cannot merely drop all low dimensional predicates as some like $R_{\text{MANIPULATEB}_i}$ as it is semantically meaningful. However, we can perturb the domain in a manner similar to Canny to create an open domain with approximately the same dynamics [8]. This involves making every predicate full dimensional.

We start by taking the conjunctions of all combinations of predicates and storing the resulting 2^n sets of states as $\mathcal{R}$. As the operators are only dependent upon predicates, we can safely ignore their negations. For each $r \in \mathcal{R}$, we expand r on its closed borders and contract r on its open borders by some very small, positive amount ϵ. The expansion operation is accomplished by computing the Minkowski sum of closed border and an open ball of radius ϵ; the contraction operation is analogously computed with two additional complement operations. For a small enough ϵ, all perturbed $r \in \mathcal{R}$ would become full dimensional. Unlike sample-based approaches, our algorithm is agnostic to the magnitude of ϵ, so an arbitrarily small value can be chosen. As semi-algebraic geometry is closed under Minkowski sums and complementation, the resulting system is also semi-algebraic. The perturbation may affect the solvability of a domain. However, for real-world systems, the effect is negligible and the difference between the original and perturbed system cannot be physically measured for small enough ϵ.

However, a potentially significant effect of the perturbation involves identities like trigonometric functions. Typically, an angle θ in S^1 is embedded in $\mathbb{R}^2$ on a pair of real dimensions θ_S and θ_C, corresponding to $\sin(\theta)$ and $\cos(\theta)$, when dynamics are polynomial in terms of θ_S and θ_C. The extra degree of freedom is removed by the identity $\theta_S^2 + \theta_C^2 = 1$. We call this trigonometric identity a *semantic-invariant*, since perturbing the identity would fundamentally affect the domain semantics no matter how small the value of ϵ. We address this issue by relaxing the requirement that domains be open to that domains be *jointly-controllably-open* (JC-open). We first define a set of dimensions to be *jointly-controllable* when they can always be manipulated together:

Definition 4. *Let $B = \{B_1, B_2, ..., B_k\} \subseteq D$ be a subset of dimensions. B is **jointly-controllable** if:*

$$\forall o_i \in O. \forall j, k. B_j \in F_i \leftrightarrow B_k \in F_i$$

In our trigonometric example, θ_S is only ever modified when θ_C is modified and vice-versa, so those dimensions are jointly-controllable.

Definition 5. *Let v be an n-vector. The* nonzero *function returns a subset of D such that:*

$$\text{nonzero}(v) = \{d_i \in D \mid v[i] \neq 0\}$$

Definition 6. *Let $J = \{J_1, J_2, ...\}$ be a maximal partitioning of dimensions into jointly-controllable sets for a domain. A manifold M is **JC-Open** if for every point $x \in M$, there exists a set of basis vectors $V_x = \{v_x^1, v_x^2, ...\}$ that span the tangent space of M at x, $T_x M$, such that*

$$\forall v_x^i \in V_x. \exists J_k \in J. \text{nonzero}(v_x) \subseteq J_k$$

Definition 7. *A domain is JC-Open if all predicates in R^{all} are JC-Open.*

We can now broaden theorem 1:

Theorem 2. *Every problem instance in a JC-open, SHC domain is decidable.*

Note that every open domain is JC-open with singleton jointly-controllable sets; therefore, theorem 1 follows from theorem 2, and we only provide a proof for theorem 2 in the next section. We can perturb a domain to be JC-open while preserving semantic-invariants by not perturbing any predicate that is dependent on only the variables in a single jointly-controllable set of dimensions. Such perturbations are unneeded, as the system always exhibits either holonomic controllability or no controllability in those dimensions by construction.

In the general variant of PTAMP, we have may several robots and several movable objects. There are two types of operators, one to transit one or more robots and the other for one or more robots to manipulate one or more objects when the robots and objects are in contact. We make the assumption that any time a group of robots move, all the dimensions in the configuration spaces of the robots can be modified. Similarly, in manipulation operations, all the dimensions in the joint configuration space of the involved robots and objects can be modified. The configuration dimensions for each robot and object therefore constitute a maximal jointly-controllable set. We assume that the predicates defining these operators are semi-algebraic. We also make the assumption that any semantic-invariants arising from dynamics for each robot or object are only dependent upon the dimensions in the configuration space for that robot or object. The only elements of the problem that combine jointly-controllable sets are those that model physical collisions and robot-object contact. However, both constraints can be rewritten by perturbation, as they are not semantic-invariants as they model purely physical phenomenon. Finally, as PTAMP can be rewritten to be JC-open, we can use the Chow-Rashevsky theorem to approximate a sliding grasp trajectory to a sequence of transit and manipulation operations with standard fixed grasps. Within the scope of the stated conditions, PTAMP is decidable regardless of the dynamics, number of robots and objects, and complexity of the semi-algebraic geometries.

4   Decision procedure

In this section, we describe a decision procedure which determines the existence of a solution for any problem instance in a JC-open domain. The primary intuition behind the entire approach is that since all the operators are SHC, the dynamics of the operators are strongly tied to the Euclidean axes. Therefore, rather than taking the typical differential geometry view that any particular coordinate system is irrelevant, we decompose the configuration space relative to the Euclidean axes into a finite number of manifolds with properties that are subsequently defined. For each manifold, we calculate its controllabilities, the set of directions that can be traversed from a point within the manifold using sequences of operations. These controllabilities can be used to calculate orbits, a foliation of reachable sets within the manifold. We first calculate the internal and exterior controllability for each such manifold. Next, with the aid of a constraint, we calculate the stratified controllability for each manifold by examining which additional controllabilities can be achieved by leveraging the controllabilities of adjacent manifolds. Once the controllabilities of each manifold are computed, the initial configuration is used to calculate the reachability set, which is then intersected against the goal condition to test for satisfiability.

The following sections describe, give pseudocode, and prove key properties of the four phases: decomposition, internal/exterior controllability, stratified controllability, and reachability. The following pseudocode is a roadmap for decision procedure DP.

Algorithm   DP — Input: $\mathcal{I} = (\mathcal{M}, x_0, G) = ((D, O), x_0, G)$ — Output: Solvability

$\mathfrak{A} \leftarrow$ DECOMPOSITION(R^{all})
$\mathcal{E}, \mathcal{D} \leftarrow$ INTERNAL/EXTERIOR CONTROLLABILITY($\mathfrak{A}, O$)
$\mathcal{S} \leftarrow$ STRATIFIED CONTROLLABILITY($\mathfrak{A}, \mathcal{D}, \mathcal{E}$)
return REACHABILITY($\mathfrak{A}, x_0, \mathcal{D}, \mathcal{S}, G$)

4.1   Decomposition phase

The decomposition phase decomposes the configuration space along extrusion sets using algebraic geometry techniques. First, we give an overview of cylindrical algebraic decomposition (CAD), full cylindrical algebraic decomposition (FCAD), and the decomposition phase algorithm. Then, we define geometric correctness conditions for this phase and show that the algorithm satisfies them.

Cylindrical algebraic decomposition (CAD) is an algorithm to decompose semialgebraic sets into a stratification and was used to prove the decidability of motion planning [1]. We use the notation E_i^j to represent an ordering of a subset of dimensions, $[e_i, e_{i+1}, ..., e_{j-1}, e_j]$. CAD requires two inputs: a decomposition ordering E_1^n (a permutation of D) and the set of predicates Q, which are collectively defined by the polynomial set $Y(E_1^n)$. CAD iteratively processes dimensions backwards over E_1^n. On each projection iteration, CAD identifies two types of events: the intersection set of two polynomials (or a self-intersection) or the set on a polynomial in which the normal is orthogonal to the dimension being processed. On the i-th iteration, CAD extrudes these events along the processed dimension as the set of polynomials $Y(E_1^{n-i})$. This process is continued to create polynomials of the form $Y(E_1^i)$ for $1 \leq i \leq n$. We omit a detailed description of the projection operators as CAD is covered in depth in various other publications [9, 10].

While CAD has traditionally been used to decompose semi-algebraic sets, we are especially interested in CAD for some of the side effects it produces based on the decomposition ordering. The full cylindrical algebraic decomposition (FCAD) algorithm takes a set of predicates as its input, runs CAD over every decomposition ordering, and intersects the results as shown in figure 3. Although there are $O(n!)$ unique decomposition orderings, only the set of extruded dimensions is relevant (not their specific ordering), necessitating only 2^n total projection iterations. We run FCAD once with the predicate set R^{all}. We then run FCAD a second time with the predicate set $FCAD(R^{all})$ and let $\mathfrak{A}$ be the resulting set of manifolds.

Algorithm Decomposition — Input: R^{all} — Output: $\mathfrak{A}$

return $\mathfrak{A} \leftarrow FCAD(FCAD(R^{all}))$

Since the geometry of manifolds and the dynamics of operators are linked by their relation to the Euclidean axes, we define characteristic sets and dimensional sets to mathematically express these properties.

Definition 8. *A **characteristic set** is a set of subsets of D. A **dimensional set** is a subset of D. Let $\mathcal{C}$ be a characteristic set. As a shorthand, we define $\mathcal{C}^* = \bigcup \mathcal{C}$ to create a dimensional set from $\mathcal{C}$.*

Characteristic sets and dimensional sets can be used to express both the controllability of a manifold and its geometry. In the decomposition phase, we concern ourselves only with the latter. In particular, we are interested in when vectors are orthogonal to the Euclidean axes. The tangent characteristic set of a manifold is a boolean representation of the geometry of the tangent space at each point.

Definition 9. *The **tangent characteristic set**, $\mathcal{T}$, of a manifold M at a point x is defined as:*

$$\mathcal{T}_x M = \{B \subseteq D \mid \exists v_x \in T_x M . \text{nonzero}(v_x) = B\}$$

As shown in figures 5a and 5b, in general, the tangent characteristic set of a manifold can vary from point to point. However, for *aligned* manifolds as shown in figures 5c and 5d, the tangent characteristic set is constant. Aligned manifolds are useful as SHC operators have uniform controllability throughout the manifold.

Definition 10. *If $\mathcal{T}_x M = \mathcal{T}_y M$ for all $x, y \in M$, then M is **aligned**. We denote the tangent characteristic set of the entire aligned manifold M as $\mathcal{T} M$.*

Proposition 1. *For any manifold M, if $B_1 \in \mathcal{T} M$ and $B_2 \in \mathcal{T} M$, then $B_1 \bigcup B_2 \in \mathcal{T} M$.*

Proof Sketch. The tangent space of a manifold is a locally, finitely generated distribution, so the span of two components of the tangent space must also be in the tangent space. □

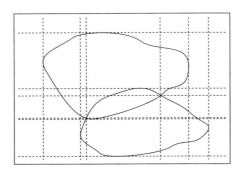

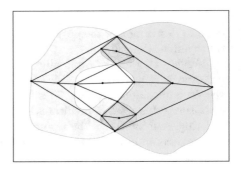

Fig. 3: Example FCAD decomposition. The resulting decomposition is the union of performing CAD with the decomposition orderings $\{x, y\}$ and $\{y,x\}$.

Fig. 4: Example adjacency graph of two predicates. Vertices are maximal, connected regions that are invariant to the polynomials constituting the predicates and edges indicate adjacent regions.

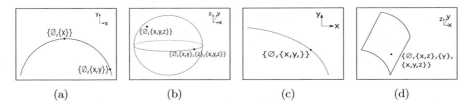

(a)   (b)   (c)   (d)

Fig. 5: Tangent characteristic set is marked at given points. (a) and (b) show non-aligned manifolds. (c) and (d) show aligned manifolds.

Next, we define *adjacency graphs* to characterize the adjacency of regions of the configuration space as shown in figure 4.

Definition 11. *Let $Q = \{Q_1, Q_2, ..., Q_k\}$ be a set of subsets of $\mathbb{R}^n$. Let T be the set of maximal, connected regions that are sign-invariant to the polynomials constituting Q. We define the **adjacency graph** of Q as the graph having nodes T and edges $(t_1, t_2) \in T^2$ whenever $t_1 \bigcup t_2$ is contiguous.*

Definition 12. *Let $B \subseteq D$ be a subset of dimensions. A manifold M is B-**uniform** to Q if M is a B-extrusion set and every B-section through M has the same adjacency graph of Q.*

We start our analysis by noting that CAD produces uniform extrusion sets.

Proposition 2. *Assume CAD is run on the set of predicates Q with decomposition ordering E_1^n. Let $W_i = \bigcup_{j=1}^{n-i-1} Y(E_1^j)$ be the set of polynomials created by CAD that are independent of E_{n-i}^n. W_i divides the space into a finite number of E_{n-i}^n-extrusion sets, each sign-invariant with respect to W_i. Each such E_{n-i}^n-extrusion set, M, is E_{n-i}^n-uniform to Q.*

Proof Sketch. Let J_1 and J_2 be two E_{n-i}^n-sections that intersect M. Let S_M, S_1 and S_2 be the projection of M, J_1 and J_2 respectively onto the dimensions E_1^{n-i-1}. Let T be a path from S_1 to S_2 such that $T \subseteq S_M$. For contradiction, we examine the two cases when the adjacency graphs of J_1 and J_2 might differ:

- **A vertex disappears** By the mean-value theorem, a sign-invariant manifold can only appear or disappear between J_1 and J_2 if for all T, there exists some value $S_3 \in T$ such that within the E_{n-i}^n extrusion of S_3, either two polynomials in W_i intersect or a polynomial in W_i is orthogonal to some dimension in E_1^{n-i-1}. The set of S_3 for all T divides S_1 and S_2 and its extrusion must be a member of W_i. Therefore, M is not sign-invariant to W_i, which is a contradiction.
- **An edge disappears** Similarly, by the mean-value theorem, for all T, there exists a set of E_{n-i}^n-sections at which the adjacency between the two predicate-invariant manifolds disappears. At that point, two polynomials forming Q must intersect. Following the same reasoning as before, M is not sign-invariant to W_i, which is a contradiction. $\qquad\square$

The decomposition computed by this procedure has the following properties for each manifold $M \in \mathfrak{A}$:

Proposition 3. *M is sign-invariant to R^{all}.*

Proof Sketch. CAD decomposes the configuration space into cells that are sign-invariant to the input predicate set regardless of decomposition ordering, and the intersection operator preserves sign-invariance. $\qquad\square$

Proposition 4. *M is aligned.*

Proof Sketch. If M is n-dimensional, then trivially it is aligned. Otherwise, assume for contradiction that there exists two configurations $x \in M$ and $y \in M$ and a vector $v_x \in T_x M$, such that there does not exist a vector $v_y \in T_y M$ where $\mathrm{nonzero}(v_x) = \mathrm{nonzero}(v_y)$. Since CAD produces regular manifolds, M must be regular and have a constant dimensionality. Therefore, at y there must exist some vector $z_y \in T_y M$ such that $\mathrm{nonzero}(z_y) \notin \mathcal{T}_x M$ and either $\mathrm{nonzero}(v_x) \subset \mathrm{nonzero}(z_y)$ or $\mathrm{nonzero}(v_x) \supset \mathrm{nonzero}(z_y)$. Therefore, either x or y must lie on a polynomial created by CAD, so x and y cannot be in the same cell that is sign-invariant to R^{all}, which is a contradiction. $\qquad\square$

Proposition 5. *For every subset of dimensions $B \subseteq D$, M is B-uniform to Q. Furthermore, any pair of B-sections intersecting M, isomorphic manifolds have the same alignment.*

Proof Sketch. As FCAD runs CAD in every decomposition ordering, by proposition 2, after one iteration of FCAD, every manifold in the resulting decomposition, L, is H-uniform for every $H \subseteq D$. Since propositions 3 and 4 apply after even a single application of FCAD, every manifold in L is aligned and sign-invariant to R^{all}. As the subsequent FCAD decomposition is run on the predicates L, by proposition 2, each resulting manifold in the second decomposition is B-uniform with respect to L. Therefore, isomorphic manifolds in adjacency graphs of $\mathfrak{A}$ must have the same alignment. $\qquad\square$

4.2 Internal/exterior controllability phase

For the next phase, we calculate both the internal and exterior controllability of every manifold $M \in \mathfrak{A}$. The exterior characteristic set is the set of directions that can be utilized to exit M. The internal characteristic set is the set of directions that can be traversed within M without leaving the manifold and define internal orbits of M.

Definition 13. *Let ϕ_M be the set of operators such that $\forall o_i \in \phi_M. M \subseteq R_i$.*

The exterior characteristic set, $\mathcal{E}$, is easily computed, because we can ignore interaction between operators. We take the union of all the directions in which each operator can individually utilize to leave the manifold.

Definition 14. *Let* Pow *be the powerset function. The **exterior characteristic set** of M, $\mathcal{E}M$ is:*

$$\mathcal{E}M = \bigcup_{o_i \in \phi_M} \mathrm{Pow}(F_i)$$

The internal characteristic set, $\mathcal{D}$, describes the set of directions that can be traversed within M without leaving the manifold. The calculation is slightly more complex since both the interactions of operators as well as the geometry of M need to be taken into account. First, we define the internal characteristic set of a single operator.

Definition 15. *Let M be an aligned manifold and $o_i \in \phi_M$ be an operator. The **internal characteristic set** of M with respect to o_i, $\mathcal{D}_{o_i} M$ is:*

$$\mathcal{D}_{o_i} M = \mathrm{Pow}(F_i) \bigcap \mathcal{T}M \tag{2}$$

The internal dimensional set of M, $\mathcal{D}^* M$, with respect to all operators is:

$$\mathcal{D}^* M = \bigcup_{o_i \in \phi_M} \mathcal{D}^*_{o_i} M \tag{3}$$

Therefore, every manifold M has internal orbits that are $\mathcal{D}^* M$-sections.

Algorithm Internal/Exterior Controllability — Input: $\mathfrak{A}, O$ — Output: $\mathcal{D}, \mathcal{E}$

---

for each $M \in \mathfrak{A}$ **do**
    $\mathcal{E}M = \bigcup_{o_i \in \phi_M} \mathrm{Pow}(F_i)$
    for each $o_i \in \phi_M$ **do**
        $\mathcal{D}_{o_i} M = \mathrm{Pow}(F_i) \bigcap \mathcal{T}M$
    $\mathcal{D}^* M = \bigcup_{o_i \in \phi_M} \mathcal{D}^*_{o_i} M$
return $\mathcal{E}, \mathcal{D}$

We prove that equation 3 follows from equation 2:

Proposition 6. *For any aligned manifold M, $\mathcal{D}^* M = \bigcup_{o_i \in \phi_M} \bigcup \mathcal{D}_{o_i} M$.*

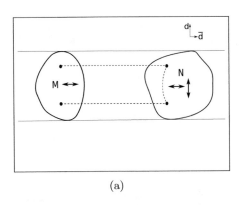 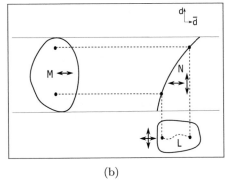

(a)   (b)

Fig. 6: (a) M utilizes the controllability of N to gain controllability in the d dimension. (b) Calculating the stratified controllability of M requires calculating the stratified controllability of N.

Proof Sketch. First, we show that $\bigcup \mathcal{D}_{o_i} M = \mathcal{D}_{o_i}^* M \in \mathcal{D}_{o_i} M$ and therefore orbits within M are $\mathcal{D}_{o_i}^* M$-sections. Let $B_1 \in \mathcal{D}_{o_i} M$ and $B_2 \in \mathcal{D}_{o_i} M$. By construction, $B_1 \bigcup B_2 \subseteq F_i$. By proposition 1, $B_1 \bigcup B_2 \subseteq \mathcal{T} M$. Therefore, $B_1 \bigcup B_2 \subseteq \mathcal{D}_{o_i} M$. $\mathcal{D}_{o_i}^* M$ is just the union of all $B \in \mathcal{D}_{o_i} M$ and so must also be a member of $\mathcal{D}_{o_i} M$.

Let $H = \bigcup_{o_i \in \phi_M} \mathcal{D}_{o_i}^* M$. $H \in \mathcal{T} M$ by proposition 1. The Lie closure of ϕ_M spans every non-empty H-section of M. Therefore, by the Chow-Rashevsky theorem [11], the intersection of those H-sections and M constitute the orbits of M.   □

4.3   Stratified controllability phase

In the stratified controllability phase, for each manifold M, we calculate the stratified controllability as the dimensional set $\mathcal{S}^* M$ that can be achieved by leaving M and utilizing the controllability of adjacent manifolds [12] as shown in figure 6a. For every dimension d, we attempt to gain stratified controllability d for M. We travel in a $\bar{d}$-section from M and look for a reachable manifold N that has the ability to travel in the d direction. We then travel in the d direction on N before retracing our steps back to M.

Determining when manifold N allows for movement in the d direction based on its internal controllability is straightforward, but is tricky when the stratified controllability of N itself needs to be calculated to compute the stratified controllability of M as shown in figure 6b. In order to prevent cyclic computations, we state a condition, which holds for JC-Open domains, that allows us to immediately ascertain the controllability of N in the d dimension.

Condition 3 (Convergence condition). *For every manifold $N \in \mathfrak{A}$:*

$$|\mathcal{E}^* N| \leq 1 \vee (\forall d \in D.d \in \mathcal{T}^* N \wedge d \in \mathcal{E}^* N \rightarrow d \in \mathcal{D}^* N \vee d \in \mathcal{S}^* N)$$

Roughly stated, the convergence condition ensures that for every manifold N whose operators allow for travel in more than one direction, for every dimension d, if some tangent vector of N contains a nonzero d component and an operator allows for movement in some direction with a nonzero d component, then N must be controllable in the d direction via either internal controllability or stratified controllability.

Algorithm   Stratified Controllability — Input: $\mathfrak{A}, \mathcal{D}, \mathcal{E}$ — Output: $\mathcal{S}$

---

for each manifold $M \in \mathfrak{A}$ **do**
    for each dimension $d \in D$ **do**
        for each manifold $N \in \mathfrak{A}$ that is reachable in some $\bar{d}$-section **do**
            if $d \in \mathcal{E}^* N$ **then**
                Add d to $S^* M$
return $\mathcal{S}$

First, we address when it is possible to traverse between a pair of adjacent manifolds. As CAD produces a stratification, two resulting manifolds can only be adjacent if one is in the stratum of the other.

Definition 16. *Let M_1 and M_2 be adjacent manifolds. Assume M_2 is in the stratum of M_1. Let $B \subseteq D$ be a subset of dimensions. M_1 and M_2 are B-**traversable** if for every point $x \in M_2$, there exists some vector v such that $\mathrm{nonzero}(v) = B$, $\mathrm{nonzero}(v) \in \mathcal{E}M_1$, $\mathrm{nonzero}(v) \in \mathcal{E}M_2$, and $x + \epsilon v \in M_1$ for small enough ϵ.*

Proposition 7. *If M_1 and M_2 are B-traversable, then for each configuration $x_1 \in M_1$, there exists a reachable configuration $x_2 \in M_2$ and vice-versa.*

Proof Sketch. Again, assume M_2 is contained in the closure of M_1. Since M_1 and M_2 are B-traversable, by definition 16, every point in M_2 can reach a point in M_1. Let I be the intersection of the closures of M_1 and M_2. The decomposition phase extrudes I in every subset of dimensions. Assume for contradiction that $\hat{x}_1 \in M_1$ cannot reach M_2. Then $\hat{x}_1 \notin P(B, \mathrm{proj}_{\overline{B}}(I))$, so M_1 is not invariant with respect to the polynomials created by FCAD, which is a contradiction. □

We use proposition 7 to construct paths from chains of traversable manifolds.

Proposition 8. *If M can reach manifold N on some d-section, then M can reach N on any d-section that intersects M.*

Proof. By proposition 5, the adjacency graphs and alignment of manifolds of every d-section that intersects M is the same. Therefore, proposition 7 must hold for every manifold-manifold transition in any d-section. □

Proposition 9. *If the convergence condition holds, then we compute the closure of stratified controllability.*

Proof Sketch. Let N be a manifold that might grant M the ability to move in the d dimension. Therefore, $d \in \mathcal{E}^* N$. If $|\mathcal{E}^* N| \leq 1$, then N is unreachable from any $\bar{d}$-section. If $d \notin \mathcal{T}^* N$, then by proposition 5, $d \notin \mathcal{T}^* M$, so M cannot travel in the d dimension with stratified controllability regardless of the controllability of N. Therefore, N must be able to travel in the d direction either internally or via stratified controllability via the convergence condition, which exhausts all cases. □

Proposition 10. *JC-open domains satisfy the convergence condition.*

Proof Sketch. Consider the case when $|\mathcal{E}^*N|>1$, $d\in\mathcal{T}^*N$, and $d\in\mathcal{E}^*N$ as it is the only scenario in which the convergence condition may be violated. Let d be a member of J, a maximal jointly-controllable set of dimensions. Since $d\in\mathcal{E}^*N$, $\forall j\in J.j\in\mathcal{E}^*N$. Since $d\in\mathcal{T}^*N$, by definition 7, we know there must exist a JC-open neighborhood around every point in N. Since at least one vector at every point has a d component and is in the tangent space of the open neighborhood, we can travel in the direction of that vector to effect movement in the d dimension before returning to N. Therefore, $d\in\mathcal{S}^*N$, which satisfies the convergence condition. $\qquad\square$

4.4 Reachability phase

In the first three phases, only the domain information is needed; for the final phase, the initial configuration and goal are also taken into account. From the initial configuration x_0, we calculate the reachability set $\mathfrak{R}$. By using a graph search across manifolds, we add the reachabilities of adjacent manifolds until all reachable manifolds are visited. Since we have already computed the internal and stratified controllability of each manifold, there is no need to ever revisit a manifold in order to gain additional reachability. Finally, after $\mathfrak{R}$ has been fully computed, we intersect $\mathfrak{R}$ with the goal condition and test for emptiness to determine the existence of a solution.

We calculate $\mathfrak{R}$ by a graph search on the graph $\mathfrak{G}$. Initially, the vertices in $\mathfrak{G}$ are $\mathfrak{A}$, and there are no edges in $\mathfrak{G}$. However, over the course of the algorithm, we add edges from visited manifolds to reachable manifolds. Through the computation, we let I_M be the initial reachability set for manifold M. Let M_0 be the manifold that contains the initial configuration x_0. We set I_{M_0} to x_0.

When visiting a manifold M with initial reachable set I_M, we construct H_M, the reachability set of M, by take the following steps:

- Initialize H_M to I_M.
- Extrude H_M according to the controllability of M. For any dimension d such that $d\in\mathcal{D}^*M$ or $d\in\mathcal{S}^*M$, we extrude H_M in the dimension d within the confines of M.
- Add edges from M to all reachable neighboring manifolds. For each adjacent manifold N, we test if N can be reached from the reachable region of M by testing if (M,N) is B-traversable for some B and $\text{Cl}(H_M)\bigcap\text{Cl}(N)\neq\emptyset$ where Cl is the closure operator. For each such manifold, we add the edge (M,N) and set $I_N=\text{Cl}(H_M)\bigcap\text{Cl}(N)$.
- Add H_M to $\mathfrak{R}$.

After the graph search terminates, we check for the existence of a solution by testing the emptiness of $\mathfrak{R}\bigcap G$.

Algorithm   Reachability — Input: $\mathfrak{A}, x_0, \mathcal{D}, \mathcal{S}, G$ — Output: Solvablility

---

$\mathfrak{G} \leftarrow (\mathfrak{A}, \emptyset)$
$I_{M_0} \leftarrow x_0$
while $M \leftarrow$ next visited node in graph search of $\mathfrak{G}$ **do**
$\quad H_M \leftarrow I_M$
$\quad$ **for** each $d \in D$ **do**
$\quad\quad$ **if** $d \in \mathcal{D}^* M$ or $d \in \mathcal{S}^* M$ **then**
$\quad\quad\quad$ Extrude H_M in dimension d
$\quad$ **for** each manifold N where (M, N) is traversable and $\mathrm{Cl}(H_M) \bigcap \mathrm{Cl}(N) \neq \emptyset$ **do**
$\quad\quad$ $I_N \leftarrow \mathrm{Cl}(H_M) \bigcap \mathrm{Cl}(N)$
$\quad\quad$ Add edge (M, N)
$\quad$ Add H_M to $\mathfrak{R}$
return $\mathfrak{R} \bigcap G \neq \emptyset$

Proposition 11. *Let H_M be the maximal reachable set within M. For an adjacent and reachable manifold N, if $I_N = \mathrm{Cl}(H_M) \bigcap \mathrm{Cl}(N)$, then H_N constitutes the maximal reachable set within N.*

Proof Sketch. Since H_M is the maximal reachable set within M, I_N constitutes the maximal region that can be reached on the border of M and N. The extrusion of H_N in all dimensions of $\mathcal{D}^* N$ and $\mathcal{S}^* N$ must be reachable and constitute the maximal reachability of of N since we have computed the maximal internal and stratified controllabilities.   □

Proposition 12. *The reachability set within any manifold N is independent of the order of the graph search algorithm.*

Proof Sketch. Let N be reached via two different paths from adjacent manifolds M_1 and M_2. As x_0 can be reached from the reachable set of M_1, the reachability set within N when entering from M_1 is a superset of the reachability set within N when entering from M_2. However, the argument can also be reversed. Therefore, the reachability set within N is independent of its predecessor.   □

5   Conclusion

In this paper, we defined SHC and JC-open domains. We described how the general PTAMP problem can be perturbed to be JC-open without significant semantic changes. We then gave a decision procedure, DP, for domains that satisfy the convergence condition. Last, we showed that JC-open domains satisfy the convergence condition. Therefore, the general PTAMP problem is decidable.

Acknowledgments We gratefully acknowledge support from NSF grants 1420927 and 1523767, from ONR grant N00014-14-1- 0486, and from ARO grant W911NF1410433. Any opinions, findings, and conclusions or recommendations expressed in this material are those of the authors and do not necessarily reflect the views of our sponsors.

References

1. Schwartz, J.T., Sharir, M.: On the piano movers problem. ii. general techniques for computing topological properties of real algebraic manifolds. Advances in applied Mathematics **4**(3) (1983) 298–351
2. Canny, J.: The complexity of robot motion planning. MIT press (1988)
3. Wilfong, G.: Motion planning in the presence of movable obstacles. In: Proceedings of the fourth annual symposium on Computational geometry, ACM (1988) 279–288
4. Alami, R., Laumond, J.P., Siméon, T.: Two manipulation planning algorithms. In: Proceedings of the Workshop on Algorithmic Foundations of Robotics. WAFR, Natick, MA, USA, A. K. Peters, Ltd. (1995) 109–125
5. Dacre-Wright, B., Laumond, J.P., Alami, R.: Motion planning for a robot and a movable object amidst polygonal obstacles. In: Robotics and Automation, 1992. Proceedings., 1992 IEEE International Conference on, IEEE (1992) 2474–2480
6. Vendittelli, M., Laumond, J.P., Mishra, B.: Decidability of robot manipulation planning: Three disks in the plane. In: WAFR. (2014)
7. Munkres, J.: Topology. Featured Titles for Topology Series. Prentice Hall, Incorporated (2000)
8. Canny, J.: Computing roadmaps of general semi-algebraic sets. The Computer Journal **36**(5) (1993) 504–514
9. Collins, G.E.: Quantifier elimination for real closed fields by cylindrical algebraic decompostion. In: Automata Theory and Formal Languages 2nd GI Conference Kaiserslautern, May 20–23, 1975, Springer (1975) 134–183
10. Hong, H.: An improvement of the projection operator in cylindrical algebraic decomposition. In: Proceedings of the international symposium on Symbolic and algebraic computation, ACM (1990) 261–264
11. Jurdjevic, V.: Geometric Control Theory. Cambridge Studies in Advanced Mathematics. Cambridge University Press (1997)
12. Sussman, H.: Lie brackets, real analyticity and geometric control. Differential Geometric Control Theory **27** (1982) 1–116

Approximation Algorithms for Time-Window TSP and Prize Collecting TSP Problems

Jie Gao[1], Su Jia[1], Joseph S. B. Mitchell[1], and Lu Zhao[1]

Stony Brook University, Stony Brook, NY 11794, USA.
{jie.gao, su.jia, joseph.mitchell, lu.zhao}@stonybrook.edu.

Abstract. We give new approximation algorithms for robot routing problems that are variants of the classical traveling salesperson problem (TSP). We are to find a path for a robot, moving at speed at most s, to visit a set $V = \{v_1, \ldots, v_n\}$ of sites, each having an associated time window of availability, $[r_i, d_i]$, between a release time r_i and a deadline d_i. In the *time-window prize collecting problem (TWPC)*, the objective is to maximize the number of sites visited within their time windows. In the *time-window TSP problem (TWTSP)*, the objective is to minimize the length of a path that visits *all* of the sites V within their respective time windows, if it is possible to do so within the speed bound s. For sites on a line, we give approximation algorithms for TWPC and TWTSP that produce paths that visit sites v_i at times within the relaxed time windows $[r_i - \varepsilon L_i, d_i + \varepsilon L_i]$, for fixed $\varepsilon > 0$, where $L_i = d_i - r_i$; the running time is $O((nL_{max})^{O(\frac{\log L_{max}}{\log(1+\varepsilon)})})$, where $L_{max} = \max_i L_i$. For TWPC, the computed path visits at least k^* (the cardinality of an optimal solution to TWPC) sites; for TWTSP, the computed path is of length at most λ^* (the length of an optimal TWTSP solution). For general instances of sites in a metric space, we give approximation algorithms that apply to instances with certain special structure of the time windows (that they are "dyadic" or that they are "elementary"), giving paths whose lengths are within a bounded factor of the optimal length, $\lambda^*(s)$, for the given speed s, while relaxing the speed to be a factor greater than s; for arbitrary time windows, we give an $O(\log n)$-approximation for TWTSP, assuming unbounded speed ($s = \infty$).

1   Introduction

Advances in mobile robotics and autonomous vehicles have given rise to a variety of new applications and research challenges. As the hardware and control systems have matured, a number of new strategic planning problems have emerged as robots and autonomous vehicles become deployed in our living spaces, urban areas, and roadways. Our work is motivated by the scheduling and motion planning problem when the tasks that need to be done are associated with both locations and with time windows. For example, an autonomous vehicle may be tasked with performing package pickup/delivery operations, each with a physical location and a window of time during which the pickup/delivery is expected to

© Springer Nature Switzerland AG 2020
K. Goldberg et al. (Eds.): *Algorithmic Foundations of Robotics XII*, SPAR 13, pp. 560–575, 2020.
https://doi.org/10.1007/978-3-030-43089-4_36

take place. Our goal, then, is to find an efficient path to pick up or deliver all of the packages (or as many as possible), subject to the given time windows.

The classical *travelling salesperson problem (TSP)* seeks a shortest path or cycle to visit a set $V = \{v_1, \ldots, v_n\}$ of n *sites* in a metric space (e.g., the Euclidean plane); the challenge is to determine the order in which to visit the sites. In this paper we study two variants of the TSP. In each variant, we assume that there is a single mobile robot, which can move at a maximum speed s. The robot is initially located (at time $t = 0$) at location v_0; the location v_0 might be given, as a fixed *depot*, or might be flexible, allowing us to determine the best choice for v_0. The input to our problems includes, for each site $v_i \in V$, a specified *time window*, $[r_i, d_i]$, with *release time* r_i and *deadline* d_i, during which the site v_i is to be visited. (A further generalization of our problems includes a *processing time* associated with each site v_i, indicating the amount of time that must be spent at v_i; we assume here that processing times are zero.)

In the *time-window prize collecting problem (TWPC)*, the objective is to determine a path P for the robot that maximizes the number of sites (or, more generally, the sum of "prizes" associated with sites) visited within their time windows; we let k^* denote the number of sites visited by an optimal TWPC path, P^*.

In the *time-window TSP problem (TWTSP)*, the objective is to determine a path P of minimum length that visits *all* of the sites V within their respective time windows, if it is possible to do so within the speed bound s; we let λ^* denote the length of an optimal TWTSP path P^*.

Related Work. Both the TWPC and TWTSP problems are NP-hard, in general, since they generalize the classic TSP. The TSP has been studied extensively (see, e.g., [11]), and polynomial-time approximation schemes are known for geometric instances (see, e.g., [2, 12, 13]). In 1D (i.e., for points on a line), the TSP is trivial. However, the TWTSP and TWPC problems are known to be strongly NP-complete even in 1D [15]. Bockenhauer *et al.* [7] showed that there is no polytime constant-factor approximation algorithm for TWTSP in metric spaces, unless P = NP.

Approximation algorithms for the time window prize collecting (TWPC) problem have been studied. Bar-Yehuda *et al.* [5] gave an $O(\log n)$-approximation algorithm for n sites on a line. For general metric spaces, Bansal *et al.* [4] gave an an $O(\log^2 n)$-approximation algorithm in general and an $O(\log n)$-approximation for the special case with release times $r_i = 0$. Chekuri *et al.* [8] gave an algorithm with approximation $O(\text{poly}(\log \frac{L_{max}}{L_{min}}))$, where L_{max} and L_{min} are the maximum and minimum lengths of the time windows. The online version was recently studied by Azar *et al.* [3]. The special case of the TWPC problem in which all release times are zero ($r_i = 0$) and all deadlines are the same ($d_i = d$, for all i) is known as the *orienteering problem*; the objective is to visit as many sites as possible with a path of length at most d/s. Approximation algorithms are known for orienteering, including polynomial-time approximation schemes for geometric instances (see, e.g., [1, 6, 9, 14]).

The TWTSP has also been studied in the operations research literature, using integer programming and branch-and-bound techniques; see [10] for a survey. These algorithms are not reviewed here, as our emphasis is on provable approximation algorithms that are polynomial-time (or potentially quasipolynomial-time).

Preliminaries. The input set of n sites $V = \{v_1, \ldots, v_n\}$ lie in a metric space; we let $\delta(v_i, v_j)$ denote the distance between v_i and v_j. We assume that time windows $[r_i, d_i]$ associated with the sites v_i have release times r_i and deadlines d_i, with $r_i, d_i \in [0, T]$ for time horizon T. We let $L_i = d_i - r_i$ denote the length of the time window associated with site v_i, and we let $L_{max} = \max_i L_i$ be the length of the largest time window.

A time window $[r_i, d_i]$ is *dyadic* if $L_i = 2^m$, for some integer $m \geq 0$, and r_i is an integer multiple of L_i. We say that an input is a *dyadic instance* if all time windows are dyadic. We say that an input is an *elementary instance* if (1) each time window is either of unit length (i.e., $d_i = r_i + 1$) or is of full length, with $[r_i, d_i] = [0, T]$, for integer T; and (2) for each integer $j \in [0, T-1]$, there exists at least one site having time window $[j, j+1]$.

Our Contributions. We provide a collection of new results for the TWPC and TWTSP, including:

1. For points V on a line (i.e., 1D), we provide dual approximation algorithms, running in time $O((nL_{max})^{O(\frac{\log L_{max}}{\log(1+\varepsilon)})})$ for both the TWPC and TWTSP problem for any fixed speed bound s. In other words, we find a path that performs as well as the optimal, but that allows each site v_i to be visited in the relaxed time window $[r_i - \varepsilon L_i, d_i + \varepsilon L_i]$, where $L_i = d_i - r_i$.
   Our method also provides an approximation for TWPC in 1D when only relaxation of the deadline is allowed, computing a path to visit $\geq k^*$ sites, allowing each site to be visited in the time window $[r_i, (1 + \varepsilon) \cdot d_i]$. This improves the results by Bansal *et al.* (in [4]), which found in polynomial time a path to visit $\Omega(\frac{k^*}{\log 1/\varepsilon})$ sites with the same relaxation on time windows.
2. As a byproduct of our method, we give new approximation algorithms for the Monotone TSP with Neighborhoods (TSPN) and Monotone Orienteering with Neighborhoods problem in 2D for arbitrary regions (arbitrary size, overlapping and fatness).
3. For TWTSP with finite speed s in a metric space, we present an (α, β) dual approximation algorithm, using speed $\leq \alpha \cdot s$ and travel distance $\leq \beta \cdot \lambda^*(s)$, where $\lambda^*(s)$ is the length of an optimal path subject to speed bound s, $\alpha, \beta = O(1)$ for an elementary instance, and $\alpha, \beta = O(\log L_{max})$ for a dyadic instance. For $s = \infty$, we give an $O(\log n)$-approximation for arbitrary time windows.

While the TWPC problem is well studied (for over 20 years), the best known factor for TWPC in 1D is still $O(\log L_{max})$ or $O(\log n)$ (as it is in metric spaces), if we strictly insist on visiting points in their time windows. Little attention has been given to dual approximations. Dual approximation schemes are natural approaches to addressing hard optimization problems. Further, relaxation of time windows is realistic, since the release times and deadlines often have some flexibility. Our results show that if we are allowed to relax the time windows even by a little bit, we can obtain much better approximation. Our results also reveal that the time window variants of TSP in 1D are significantly different from the problems in higher dimensions or in metric spaces.

For the TWTSP problem in a metric space, one may ask whether it is possible to apply the known method for TWPC. The difficulty is as follows. Recall the strategy in [4, 5]: first, classify the vertices into $g = O(\log n)$ or $O(\log^2 n)$ groups according to their time windows, such that the time windows in each group are roughly the same. Then, note that when all time windows are the same, the TWPC problem is just the ordinary prize collecting problem, so we can find a constant-factor approximation for each group. Among these g solutions, choose the one with the largest prize. A natural idea to apply to the TWTSP problem is to find a constant-factor approximation for each group and then to paste them together in an appropriate way. This strategy works for TWTSP with infinite speeds but fails when the speed is bounded, since it does not give any guaranteed bounds in speed, and consequently we may return a solution with speed much higher than s. We explain briefly the techniques we use in this paper.

Overview of Our Approach. We view the 1D problem in space-time, in the (t, x) plane, where t is the (horizontal) time coordinate and x is the (vertical) spatial coordinate along the line containing the sites V. Then, the sites $v_i \in V$ are points along the (vertical) x-axis, and the time windows $[r_i, d_i]$ are horizontal line segments σ_i in the (t, x) plane. A path P corresponds to a slope-bounded piecewise-linear function, $P(t)$, with absolute value of slope at most s. Visiting a site v_i within its time window corresponds to the path $P(t)$ visiting the corresponding (horizontal) line segment σ_i.

To get our results for 1D problems, we first look at special cases called *dyadic instances* when all time windows are *dyadic*. These dyadic intervals can be considered as intervals of a binary recursive partition. We can run dynamic programming to find the best path. Then we introduce the h-dyadic instance, which, intuitively can be solved by partitioning each dyadic interval at at most h places called partition points. Our method consists of the following steps: (1) we give an $O((nL_{max})^{O(h)})$-time algorithm for h-dyadic instances; (2) we show how arbitrary time windows can be expanded to have endpoints among a set of h carefully selected partition points placed within dyadic intervals, thereby transforming a general instance into an h-dyadic instance. We solve h-dyadic instances using a carefully designed dynamic programming algorithm, taking advantage of the fact that dyadic intervals have a hierarchical structure. Within each subinterval of a dyadic interval, the subproblems of our dynamic program keep track of the x-extent (min-x and max-x) of a solution path.

Step (2) may appear to be straightforward, but there are two challenges in assigning partition points. One is the tradeoff between the precision of our relaxation and the number of partition points: if the partition points are too dense, then we end up with a high running time; if the partition points are too sparse, then the relaxation of time windows is too coarse. The other difficulty is that the partitions for dyadic intervals are not independent; they are correlated, in the sense that the partition of any dyadic interval must inherit all the partition points from its parent.

For the dual approximation for TWTSP in metric spaces on elementary instances, we first group the sites so that the weight of the minimum spanning tree (MST) of V restricted to each group is $O(\log L)\lambda^*(s)$. Then, we solve a matching problem on a bipartite graph whose "red" nodes correspond to sites having unit-length time windows and whose "blue" nodes correspond to the groups, allowing us to assign the groups to red nodes in a balanced manner, thereby avoiding the need for the speed to be increased by more than a particular bound.

2   TWTSP and TWPC on a Line

We start with the case when the nodes are on a line (or on a curve). We first look at the special case when the robot may travel with infinite speed. For this case, the problem of visiting all sites within their time windows is always feasible. The goal therefore is to minimize the travel distance. We show a dual algorithm for this instance as below.

Theorem 1. *Given an instance for 1D TWTSP problem when the speed bound s for the robot is ∞, let L_{max} be the maximum length of the input segments. Then for any $\varepsilon > 0$, in $O(n^{O(\frac{\log L_{max}}{\log(1+\varepsilon)})} \log L_{max})$ time we can find a path P, such that*

1. *the length of P is at most OPT, the optimum of the problem,*
2. *each segment σ_i is visited in time window $[r_i - \varepsilon L_i, d_i + \varepsilon L_i]$, where $L_i = d_i - r_i$.*

To prove this theorem we first look at special cases called *dyadic instances* when all time windows are *dyadic*. That is, when the length of each time window is a power of two, and the release time is a nonnegative integer multiple of its length. For a dyadic interval I, let I_L and I_R be its the left and right child interval respectively, when we cut the interval I at its midpoint. In Subsection 2.1 we first give a polytime algorithm for dyadic instances using dynamic programing.

In Subsection 2.2 we generalize the above algorithm to an $O(n^{O(h)} \log L_{max})$ time algorithm for h-*dyadic instances*, which is defined below. Consider a dyadic interval $I = [a, b]$ of the t-axis, and let $S(I)$ be the segments fully contained in the slab $I \times [-\infty, \infty]$ and stabbed by the midline of this slab, i.e. $\{(a+b)/2\} \times [-\infty, \infty]$. We call an instance h-*dyadic*, if we can partition each dyadic interval I at interger values (called the partition points) into at most h pieces so that (1) for each segment s in $S(I)$, the two endpoints project to partition points on the t axis; and (2) for every dyadic interval $I = [a, b]$, every partition point for I is also a partition point for the children intervals of I, i.e. $[a, \frac{1}{2}(a + b)]$ and $[\frac{1}{2}(a + b), b]$, but not vice-versa.

Last we show that for any $\varepsilon > 0$, we can transform any instance $\mathcal{I}$ to an $O(\frac{\log L_{max}}{\log(1+\varepsilon)})$-dyadic instance $\mathcal{I}'$, such that each time window is stretched by at most $(1 + \varepsilon)$ times (Subsection 2.3), which completes the proof.

After we have a good understanding of the case when the robot can travel with infinite speed, we now handle the general case when the robot speed is bounded by s.

Theorem 2. *Given an instance for 1D TWPC problem with bounded velocity s, let L_{max} be the maximum length of the input segments, and assume the shortest time window has length ≥ 1. Then for any $\epsilon > 0$, in $O((nL_{max})^{O(\frac{\log L_{max}}{\log(1+\epsilon)})})$ time we can find a path P, such that*

1. *the number of segments that P visits is at least OPT,*
2. *each segment σ_i is visited in $[r_i - \epsilon L_i, d_i + \epsilon L_i]$, where $L_i = d_i - r_i$.*

Similar result holds for 1D TWTSP with finite speed.

2.1 Infinite Speed ($s = \infty$) for Dyadic Instance

Lemma 1. *For dyadic instances, the 1D TWTSP problem with infinite speed ($s = \infty$) can be solved in poly(n) time.*

Proof. The main idea is to use dynamic programming. For that, we now introduce the subproblems. Let $|P|$ be the length of a path P. Given a dyadic interval $I = [a, b]$, we use $P^* = OPT(I; \theta)$, where $\theta = (x_N, x_S, x_b, x_e)$, to denote the optimum of the following subproblem:

Minimize $|P|$, s.t.

1. P visits all segments that are fully contained in I;
2. the points with maximum and minimum x-coordinate that P visits in I are x_N and x_S respectively;
3. P starts and ends in x_b and x_e respectively.

If there is no feasible solution to the problem of $OPT(I; \theta)$, then we take its value to be $-\infty$. This happens when the parameters in θ are contradictory to each other. For example, $x_N < x_S$, or $x_b > x_N$ or $x_e < x_S$, we do not enumerate them here.

Now define $P^*|_{I_L}$ and $P^*|_{I_R}$ as the restriction of P on the two children intervals I_L, I_R (that is, I is partitioned in the middle and the left one is I_L and the right one is I_R). By an exchange argument, $P^*|_{I_L}$ is in fact the optimal of $OPT(I_L; \theta_L^*)$, where θ_L^* is the parameter induced by P_L^*: $\theta_L^* = (x_{N,L}, x_{S,L}, x_{b,L}, x_{e,L})$, where $x_{N,L}/x_{S,L}$ is the max/min x-coordinates position that P^* visits in I_L and $x_{b,L}/x_{e,L}$ is the starting point/ending point of P^* in I_L. Specifically, if there is another path P' on I_L having parameter θ_L^* whose length is shorter than $P^*|_{I_L}$, then the concatenation $P' \cup P^*|_{I_R}$ should be shorter than P^*. That is a contradiction. Similarly, we know $P^*|_{I_R} = OPT(I_R; \theta_R^*)$.

Given $\theta, \theta_L, \theta_R$, where $\theta_L = (x_{N,L}, x_{S,L}, x_{b,L}, x_{e,L})$, and $\theta_R = (x_{N,R}, x_{S,R}, x_{b,R}, x_{e,R})$, we say θ_L, θ_R are *compatible* with θ, if $\max\{x_{N,L}, x_{N,R}\} = x_N$, $\min\{x_{S,L}, x_{S,R}\} = x_S$, and $x_{e,L} = x_{b,R}$. Hence, we have the following recurrence: $OPT(I; \theta) = \min\{OPT(I_L; \theta_L) + OPT(I_R; \theta_R): \theta_L, \theta_R$ are compatible with $\theta\}$.

Therefore, if we know $OPT(J; \theta)$ for each Level(k) interval J and all θ, then we are able to compute $OPT(I; \theta')$ for all Level($k+1$) interval and all θ' in $O(n^{12}L_{max} \log L_{max})$ time. Specifically, for a fixed θ, we find the minimum value over $O(n^4) \cdot O(n^4) = O(n^8)$ choices of θ_L and θ_R. Since there are $O(n^4)$ different θ, it takes $O(n^{12})$ to find $OPT(I; \theta)$ for all θ. Since there are $O(\log L_{max})$ rows in our lookup table, each row with $O(L_{max})$ dyadic intervals, the total running time for computing all subproblems is $O(n^{12}L_{max} \log L_{max})$. Now by rescaling the t-axis L_{max} can be viewed as linear in n. So the running time is $O(n^{13} \log n)$.

2.2   Infinite Speed ($s = \infty$) for h-dyadic Instance

To generalize the dynamic programming idea used in the dyadic instance, we consider the notion of an h-*dyadic* instance by capturing the intuition that the solution for a larger interval can be solved by using solutions for h subintervals. To do that, we first explain the definitions of h-dyadic instances.

Given an interval $[a, b]$, let $\pi: a = t_0 \leq t_1 \leq \dots \leq t_{k-1} \leq t_k = b$ be a partition of $[a, b]$ into $k \leq h$ subintervals, we call such a partition an h-*partition*, and each t_i is called a partition point. Now we define the *Inheriting Property*. Given a collection of dyadic intervals each associated with a h-partition, $\pi(I)$ for the dyadic interval I, we say this family of partitions has the inheriting property, if for any two sibling dyadic intervals I_1, I_2 (i.e., whose union/parent I is also a dyadic interval), the union of the partitioning points of $\pi(I_1), \pi(I_2)$ is a superset of the partitioning points of $\pi(I)$.

An instance for TWTSP is called h-*dyadic*, if we can associate an h-partition $\pi(I)$ to each dyadic interval I, such that this family of partitions has the inheriting property, and for each input segment σ_i in the instance, both endpoints of σ_i are partition points of $\pi(W(\sigma_i))$, where $W(I)$ as the minimal dyadic interval containing I.

Now we modify the dynamic programming for dyadic instances to h-dyadic instances. We need the following subproblem definition (See Figure 2.2).

Definition 1. *(Subproblem for h-dyadic instance) Let $\mathcal{I}$ be an instance to 1D TWTSP problem with infinite speed bound. Let π be an h-partition of $I = [a, b]$, with partition points $\{t_i\}$. Define $OPT(I; \pi; \theta)$, where $\theta = (x_N^1, \dots, x_N^h; x_S^1, \dots, x_S^h; x_b, x_e)$, as the optimum of the following problem:*
Minimize $|P|$, s.t.

1. *P visits all segments that are fully contained in I;*
2. *the maximum and minimum x-coordinates that P visits in $[t_{j-1}, t_j]$ are x_S^j and x_N^j respectively, denoted as the vertical range $[x_S^j, x_N^j]$;*
3. *P starts and ends in x_b and x_e respectively.*

Let $t_i \leq t_j$, define $V[t_i, t_j]$ as the set of segments whose projection to the t-axis is *exactly* $[t_i, t_j]$. Here is a crucial observation (refer to Fig 2.2): given a dyadic interval $I = [a, b]$, associated with an h-partition $\pi : t_0 = a \leq t_1 \leq \dots \leq t_h = b$. Then, path P visits all the segments in $V[t_i, t_j]$ if and only if

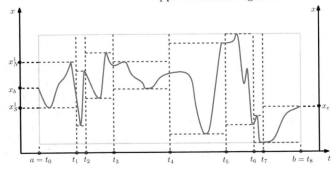

Fig. 1. Illustration of $OPT(I; \pi; \theta)$. The red dashed segments are the max/min x-coordinates P visits in each interval $[t_{i-1}, t_i]$. The points x_N^1, x_S^1, x_b, x_e are highlighted.

the maximum x-coordinates that P visits in the interval $[t_i, t_j]$ is "higher" than the "highest" in $V[t_i, t_j]$, and minimum x-coordinates that P visits in $[t_i, t_j]$ is "lower" than the "lowest" segment in $V[t_i, t_j]$.

Theorem 3. *The 1D TWTSP with $s = \infty$ for h-dyadic instances can be solved in $O(n^{O(h)} \log L_{max})$ time.*

Proof. The dynamic programming algorithm differs from the dyadic case in two ways. First, we have a more complicated parameter θ, which now encodes the vertical ranges in *every* subinterval, hence having $O(h)$ length. Second, we have a slightly different recursive structure: in addition to the requirement that θ_L, θ_R be compatible with θ, we also require that all non-dyadic segments in I be visited.

Let π, π_L, π_R be the h-partitions associated with dyadic intervals I, I_L, I_R respectively, where I is the parent of I_L, I_R. Given a segment σ parallel to t-axis, let $x(\sigma)$ be its x-coordinate. We have the following recursive relation:

$$OPT(I; \pi; \theta) = \min_{\theta_L, \theta_R} \{OPT(I_L; \pi_L; \theta_L) + OPT(I_R; \pi_R; \theta_R)\},$$

for any $j \leq h$,

$$x_N^j = \max\{x_N^i : t_{j-1} \leq t_{i-1}^L \leq t_i^L \leq t_j \text{ or } t_{j-1} \leq t_{i-1}^R \leq t_i^R \leq t_j\},$$

$$x_S^j = \min\{x_S^i : t_{j-1} \leq t_{i-1}^L \leq t_i^L \leq t_j \text{ or } t_{j-1} \leq t_{i-1}^R \leq t_i^R \leq t_j\},$$

and for any pair $i, j \leq h$,

$$\max\{x_{N,L}^i, ... x_{N,L}^h, x_{N,R}^1, ..., x_{N,R}^j\} \geq \max\{x(\sigma) : \sigma \in V[t_i, t_j]\},$$

$$\min\{x_{S,L}^i, ... x_{S,L}^h, x_{S,R}^1, ..., x_{S,R}^j\} \leq \min\{x(\sigma) : \sigma \in V[t_i, t_j]\}.$$

There are four constraints in the relation above. The first two say that if the max/minimum x-coordinates that P visits in $[t_i, t_j]$ is x, then P visits x in at least one subinterval of π_L or π_R contained in $[t_i, t_j]$. The last two constraints are saying that P must visit all segments in $\mathcal{S}(I)$. Indeed, for each i, j, the segments in $V[t_i, t_j]$ are all visited by P if *and only if* the union of vertical ranges of P in $[t_i, t_j]$ fully contains the vertical range of $V[t_i, t_j]$, i.e. $[S, N] \subset \cup_{i \leq k \leq j}[x_S^k, x_N^k]$, where $S = \min\{x(s) : s \in V[t_i, t_j]\}$, $N = \max\{x(s) : s \in V[t_i, t_j]\}$.

Since there are $O(n^{O(h)} \log L_{max})$ entries in the lookup table, the proof is complete.

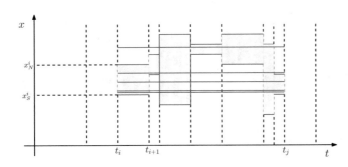

Fig. 2. Illustration of Theorem 3. Segments in $V[t_i, t_j]$ are all visited by P *if and only if* the union of vertical ranges of P in $[t_i, t_j]$ fully contains the vertical range of $V[t_i, t_j]$, For example, all segments in $V[t_i, t_j]$ (colored blue) are all visited by P.

2.3   Infinite Speed ($s = \infty$) for General Case and Generalizations

For a general instance, how do we approximate a general instance using an h-dyadic instance? Our idea is as follows: associate a partition to each dyadic interval, and then stretch the endpoints of each segment σ to some partition points of $\pi(W(\sigma))$, where $W(\sigma)$ is the minimal dyadic interval containing σ.

But how to find these partitions? The challenge is as follows: on one hand, we want the partition points to be sparse, so that we have fewer entries in our dynamic programming table and hence having a small running time; on the other hand, we want them to be dense, so that we can stretch each segment by a factor of at most $1 + \varepsilon$. Moreover, we wish this family of partitions to have a nice structure – the "inheriting property" – so that we can use dynamic programming. Below we define this partition formally.

Given an integer interval $I = [a, b]$, let $l = b - a$ and $c = \frac{1}{2}(a + b)$. We say a partition π is a *symmetric two-sided logarithmic ε-dense partition* (in short ε-*dense* partition) of I if (1) all partition points are integers, (2) this partition is symmetric with respect to the midpoint c, (3) for any integer $q \le \log_{1+\varepsilon} l$, the subinterval $[(1 + \varepsilon)^q + a, (1 + \varepsilon)^{q+1} + a]$ contains at least one partition point, unless it contains no integers at all.

The following shows that such a family of partitions exists. The proof is omitted in this version.

Lemma 2. (*ε-dense Partition with Inheriting Property*) Let $L_{max} > 0$ be a power of 2. We can assign a partition $\pi(I)$ to each dyadic interval $I \subseteq [0, L_{max}]$, such that

1. $\{\pi(I)\colon$ dyadic $I \subseteq [0, L_{max}]\}$ is a family of partitions with the inheriting property,
2. each $\pi(I)$ has at most $O(\frac{\log L_{max}}{\log(1+\varepsilon)})$ partition points,
3. $\pi(I)$ is ε-dense for each dyadic interval $I \subseteq [0, L_{max}]$.

Finally, we show our dual approximation for 1D TWTSP with infinite speed.

Proof (Theorem 1). First we construct a family of ε-dense partition Π. Then, transform our instance into an h-dyadic instance $\mathcal{I}'$, where $h = O(\frac{\log L_{max}}{\log(1+\varepsilon)})$, by stretching each segment σ_i so that both its endpoints are the partition points of $\pi(W(\sigma_i))$. Find the optimal solution P for $\mathcal{I}'$ using the dynamic programming. Clearly $|P| \leq OPT$. Note that we have stretched each segment by at most $1 + \varepsilon$ times, so P visits each segment σ_i in time interval $[r_i - \varepsilon L_i, d_i + \varepsilon L_i]$, and the proof is complete.

2.4   Bounded Speed ($s < \infty$)

The generalization of the above algorithm to a finite speed scenario is non-trivial. First the problem may not admit a feasible solution if the robot speed is too slow. Therefore we look at the prize-collecting problem (TWPC). Second, it is possible that we need to sacrifice some immediate interest in order to obtain more long-term interest. So the solution structure might completely change when the speed cap is placed. We will overcome this by carefully defining subproblems and encoding more parameters in our dynamic programming.

Without loss of generality, we consider rectilinear paths, travelling at most distance 1 in unit time. For simplicity, we assume the prize placed at each point is one, though the proof can be easily generalized to the case where arbitrary prize are allowed at each point.

Let S be a set of segments parallel to the t-axis, and $I = [a, b]$ be any integer interval. Let $S(I)$ be the set of segments whose projection on the t-axis is fully inside I. Given path P, let y_N and y_S denote the maximum and minimum coordinates among the segments P visits in $S([a, b])$, define $B(P, a, b)$ as the rectangle $[a, b] \times [y_S, y_N]$. For an axis-parallel rectangle/box B in the plane, denote the upper/lower boundary as $\partial^N(B)$ and $\partial^S(B)$ respectively. Given a t-monotone path P, let $P(\tau)$ denote the x-coordinate of P at time τ.

Definition 2. *Given a dyadic interval $I = [a, b]$, numbers $\tau_N^-, \tau_N^+, \tau_S^-, \tau_S^+$ and τ_b, τ_e, define $OPT(I; \theta)$, where $\theta = (y_N, y_S; x_N, x_S; x_b, x_e; \tau_b, \tau_e, \tau_N^-, \tau_N^+, \tau_S^-, \tau_S^+)$, as the optimal solution to the following problem.*
Maximize number of segments in $S(I)$ visited by P, s.t.

1. *the maximum and minimum x-coordinates that P visits in I are x_N and x_S respectively,*
2. *among the segments in $S(I)$ that P visits, the maximum and minimum x-coordinates are y_N and y_S respectively,*
3. *P starts/stops at location x_b and x_e respectively,*
4. *$\tau_b = \min\{\tau \in [a, b] : P(\tau) \in [y_S, y_N]\}$, $\tau_e = \max\{\tau \in [a, b] : P(\tau) \in [y_S, y_N]\}$,*
5. *P arrives at y_N at time τ_N^- and moves to x_N, and comes back to y_N at time τ_N^+; it also arrives at y_S at time τ_S^- and moves to x_S, and comes back to y_S at time τ_S^+, and finally arrives at x_e.*

To understand this, consider a path as a (monotone) rope, we hit 8 nails into the wall to fix the rope. The coordinates of the 8 nails are: (a, x_b), (τ_b, x_b), (τ_N^-, y_N), (τ_N^+, y_N), (τ_S^-, y_S), (τ_S^+, y_S), (τ_e, x_e), (b, x_e). You are allowed to change the shape of the rope, as long as the nails are fixed, and the maximum and minimum x-coordinate visited are x_N, x_S. Among all such ropes, find the one hitting maximum number of segments in $\mathcal{S}(I)$. See Fig 3.

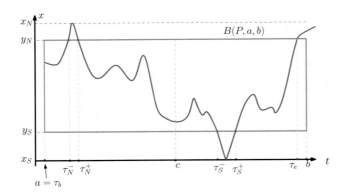

Fig. 3. Illustration of the subproblem $OPT(I; \theta)$. The big green box is $B = B(P, a, b)$.

Now we perform the dynamic programming. Compared with the one in the last section, we have more parameters for finding $OPT(I; \theta)$. We need the maximum and minimum x-coordinate that P visits in I, as well as the coordinates of the highest and lowest segment in $\mathcal{S}(I)$ that P visits, whose position are denoted by y_N and y_S, and the time that P leaves/enters $I \times [y_S, y_N]$.

Define $B := B(P, a, b)$, $B_L := B(P, a, c)$, $B_R := B(P, c, b)$. The following lemma says that a "reasonable" path P crosses the boundary of each of these boxes for $O(1)$ times. With space constraints we omit the proofs in this version.

Lemma 3. *(Polynomial Boundary Complexity of P^*)* Let $P^* = OPT(\mathcal{S}([a, b]); \theta)$, and τ_b, τ_e be defined as in Definition 2. Then, P^* leaves/enters each of $\partial^N B_L$, $\partial^S B_L$, $\partial^N B_R$, $\partial^S B_R$ at most once between τ_b, τ_e.

Corollary 1. *(Foundation of DP)* If $P^* = OPT(\mathcal{S}([a, b]); \theta)$, then there exists parameters θ_L and θ_R, such that $P^*|_{[a,c]} = OPT(\mathcal{S}([a, c]); \theta_L)$ and $P^*|_{[c,b]} = OPT(\mathcal{S}([c, b]); \theta_R)$.

Lemma 4. *The 1D TWPC problem with dyadic time windows can be solved in* $\mathrm{poly}(n, L_{max})$ *time.*

As in the case of TWTSP, this dynamic programming can be generalized to h-dyadic instances, and leads to the algorithm in Theorem 2.

2.5   Generalization

This section we discuss generalization of the results to several other scenarios.
TWTSP/TWPC for m-robots. The result for TWTSP and TWPC with infinite/finite speed can be easily generalized when m-robots are used, with running time $O((nL_{max})^{O(m\frac{\log L_{max}}{\log(1+\varepsilon)})})$.

Monotone TSPN. Our results can be used to show new results for traveling salesman problem with neighborhood when the path needs to be monotone. Given a set of regions $\{Q_i\}_{i=1,\ldots,n}$ in the plane, we wish to find a horizontally monotone TSPN path which minimizes the vertical distance travelled. Suppose that the width of any region at least 1, and let L_{max} be the largest width. Then for any $\varepsilon > 0$, in $O(n^{O(\frac{\log L_{max}}{\log(1+\varepsilon)})} \log L_{max})$ time we can find a path P, whose vertical distance travelled is at most λ^*, and $\delta(P, Q_i) \leq \varepsilon \cdot width(Q_i)$ for each Q_i. Note that this theorem holds for regions with arbitrary overlapping, shape, and size.

Improvement of Bansal *et al.*'s approximation ([4]) As a byproduct, we also improve Bansal *et al.*'s approximation ([4]) in 1D, but in $O((nL_{max})^{O(\frac{\log L_{max}}{\log(1+\varepsilon)})})$ time. They gave the following bi-criteria approximation for the TWPC problem in general metric space: in $poly(n)$ time find a path that visits $O(\frac{1}{\log\frac{1}{\varepsilon}} k^*)$ points in $[r_i, (1+\varepsilon)d_i]$.

Theorem 4. *For the 1D TWPC problem, in $O((nL_{max})^{O(\frac{\log L_{max}}{\log(1+\varepsilon)})})$ time, we can find a path P which visits at least k^* segments in $[r_i, (1+\varepsilon)d_i]$.*

To prove this theorem, we need 2 lemmas.

Lemma 5. *Given path P and point v on it, let $t_P(v)$ be the time that P visits v. Suppose $t_P(v) \in [\frac{1}{1+\varepsilon}r(v), (1+\varepsilon)d(v)]$ for each v visited by P for some $\varepsilon < 1$. Then, by slowing down P by $(1+\varepsilon)$, we obtain a path P' such that $t_{P'}(v) \in [r(v), (1+3\varepsilon)d(v)]$ for each v visited by P.*

Proof. Since $t_{P'}(v) = (1+\varepsilon)t_P(v)$, we know $t_{P'}(v) \in [r(v), (1+\varepsilon)^2 d(v)]$. Since $\varepsilon < 1$, we have $(1+\varepsilon)^2 \leq 1+3\varepsilon$.

Lemma 6. *For any $\varepsilon > 0$, there exists a family of partitions $\Pi = \{\pi(I): interval I \subseteq [0, L] is dyadic\}$ with the inheriting property, such that*

1. *for any dyadic I, and any integer q with $[(1+\varepsilon)^q, (1+\varepsilon)^{q+1}] \cap I \neq \emptyset$, there is a partition point of $\pi(I)$ in the interval $[(1+\varepsilon)^q, (1+\varepsilon)^{q+1}]$, and*
2. *each $\pi(I)$ has $O(\frac{\log L}{\log(1+\varepsilon)})$ partition points.*

Proof. Consider the sequence $Z = \{\lceil (1+\varepsilon)^q \rceil : q = 1, 2, \ldots \lceil \frac{\log L}{\log(1+\varepsilon)} \rceil\}$. For each dyadic interval $I = [a, b]$, let $\pi(I) = \{a, b\} \cup Z$. Clearly this family of partitions satisfies the inheriting property. By definition, it satisfies condition (1). To verify (2), we note that Z has $O(\frac{\log L}{\log(1+\varepsilon)})$ partition points, hence the number of partition points in $\pi(I)$ is at most $O(\frac{\log L}{\log(1+\varepsilon)}) + 2 = O(\frac{\log L}{\log(1+\varepsilon)})$.

Proof of Theorem 4. For each s_i, with time window $[r_i, d_i]$, stretch its left endpoint leftwards to the nearest partition point of $\pi(W(s_i))$, and stretch its right endpoint rightwards to the nearest partition point of $\pi(W(s_i))$. Let the new segment be s_i', with time window $[r_i', d_i']$. Then, $r_i' \geq \frac{1}{1+\varepsilon}r_i$ and $d_i' \leq (1+\varepsilon)d_i$. Since the new instance is $O(\frac{\log L_{max}}{\log(1+\varepsilon)})$-dyadic, we can find an optimal solution P^* in $O((nL_{max})^{O(\frac{\log L_{max}}{\log(1+\varepsilon)})})$ time. We complete the proof by slowing down P^* by a factor of $(1+\varepsilon)$.

3 General Metric Space

Given an instance of time window TSP problem in a metric space with metric $\delta(\cdot,\cdot)$. Let s_{min} be the smallest speed s such that there exists a feasible solution. For a given speed bound $s \geq s_{min}$, let $\lambda^*(s)$ be the minimum possible travel distance. For $s \geq s_{min}$, an algorithm is an (α, β) dual approximation if the robot moves with speed $\leq \alpha s$, and travels a total distance $\leq \beta\lambda^*(s)$. The key result for our dual-approximation is the following theorem for dyadic time windows.

Theorem 5. *Under the assumption above, if $s \geq s_{min}$, then there is an algorithm with $(O(\log L), O(\log L))$ dual approximation for the TWTSP problem for dyadic instance, with running time $O(nL_{max} \log n + n^{1.5}L_{max})$.*

When $s = \infty$, we can have the following stronger result.

Theorem 6. *For $s = \infty$ and arbitrary time windows, there is a polytime $O(\log n)$-approximation for the TWTSP problem in a metric space.*

We do not have space to present the proofs for the above cases. But we will present the results and proofs for the most important special case which leads to the results in the general case.

3.1 Proof for an Elementary Case

We begin with a special case called an elementary instance: All release times are integers; Each time window is either unit-length or $[0, L]$, for some fixed integer L; For each $i \leq L$, there is at least one node with time window $[i - 1, i]$.

Theorem 7. *For TWTSP problem in general metric space and $s \geq s_{min}$, there is a poly(n) time $(O(1), O(1))$ dual approximation for an elementary instance.*

We call the nodes with unit time window "red nodes", and those with time window $[0, L]$ "black nodes". By losing a constant factor in speed, we can assume the path takes the following form (See Fig 4): it starts from a red node with time window $[0, 1]$, say Red_1, then visits some black nodes (the black cycle in Fig 4), then returns to Red_1, and visits all other red nodes with time window $[0, 1]$ (the red cycle in Fig 4), and then goes to a red node with time window $[1, 2]$, say Red_2, and repeat. Hence, among all the nodes with time window $[i, i + 1]$, there is a "representative" red node which is attached with two cycles each of length at most s, one black and one red, denoted by $Cycle_B(Red_i)$ and $Cycle_R(Red_i)$ respectively.

We first find an MST of the sites and cut the tree into subtrees called "blocks" $\mathcal{T} = \{T_1, ..., T_k\}$, such that (1) all blocks have size between $C_1 s$ and $C_2 s$ except at most one block (called the "exceptional block") whose size is less than $C_1 s$, and (2) every vertex is contained in at most two blocks. For each $i \leq L$, pick an arbitrary red node with time window $[i - 1, i]$ as Red_i, build an unweighted bipartite graph H on the chosen red nodes $\{Red_i\}_{1 \leq i \leq L}$ and subtrees $\{T_j\}$: add an edge (Red_i, T_j) if $\delta(T_j, Red_i) \leq s$.

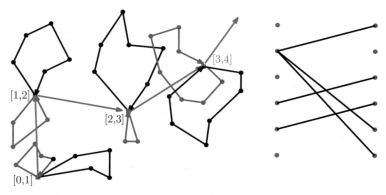

Fig. 4. (a) Left: Each black and red cycle is of length at most s. The distance between Red_i and Red_{i+1} is at most s. (b) Right: A perfect Δ-matching, $\Delta = 3$.

Find a perfect Δ-matching M in H. Here a Δ-matching is a subgraph such that each red node has at most Δ edges in the matching and each subtree has exactly one edge in the matching. See Fig 4(b). If $(Red_i, T_j) \in M$, then we say block T_j is "assigned" to red node Red_i.

For each $i \le L$, start from Red_i, traverse all the blocks assigned to it, then return to Red_i, then visit all other red nodes with time window $[i - 1, i]$ along an $O(1)$-approximate TSP tour on them, and finally add the edge between Red_i and Red_{i+1}.

The proof that the above algorithm achieves what we want comes from the following four lemmas. The proofs of some of them are omitted.

Lemma 7. *(Generalized Hall's Marriage Theorem) In a bipartite graph $G = (V_R \cup V_B, E)$, if for any subset X of V_B, the neighboring vertices set $\Gamma_G(X)$ satisfies $|\Gamma_G(X)| \ge \frac{1}{\Delta}|X|$, then there exists a perfect Δ-matching.*

Lemma 8. *Given a weighted undirected graph $G = (V, E)$, let $T_1, ... T_m$ be some node-disjoint subtrees of $MST(G)$. Let $V' = \cup_{1 \le i \le m} V(T_i)$ and G' be the restriction of G on V'. Then $MST(G') \ge \sum_i |T_i|$.*

Lemma 9. *Let $A = \{T_1, ... T_k\} \subseteq \mathcal{T}$, and $V(A)$ be the union of nodes in the blocks in A, i.e. $V(A) = \cup_{i \in A} V(T_i)$. Then the black cycles associated with the red nodes in $\Gamma(A)$ together cover $V(A)$, i.e. $V(A) \subseteq \cup_{r \in \Gamma(A)} BlackCycle(r)$.*

Lemma 10. *Suppose $s \ge s_{min}$, then for any set of blocks $A = \{T_1, ... T_k\} \subseteq \mathcal{T}$, we have $|\Gamma(A)| \ge \mu|A|$, where $\mu = \frac{C_1}{\alpha_{st}(C_2 + 3)}$.*

Proof. Define $V(A) = \cup_{T_i \in A} V(T_i)$, where $V(T_i)$ are the sites in T_i. The idea is to use $MST(V(A) \cup \Gamma(A))$ as an intermediate quantity to derive an inequality of $|A|$ and $|\Gamma(A)|$.

Assume that for any red node $r \in \Gamma(A)$, there is another red node $r' \in \Gamma(A)$, s.t. $\delta(r, r') \le (C_2 + 2)s$. We will show how to remove this assumption later.

First we derive an upper bound for $MST(V(A) \cup \Gamma(A))$. Since $s \ge s_{min}$, there is a path P visiting all vertices in their time windows. We will use P to construct a spanning tree for $V(A) \cup \Gamma(A)$. By adding $|\Gamma(A)| - 1$ edges, each with length $\leq s$, we can connect all the black cycles of the red nodes in $\Gamma(A)$. From Lemma 9, we know that this subgraph visits all nodes in $V(A) \cup \Gamma(A)$. Hence,

$$MST((V(A)) \cup \Gamma(A)) \leq |\Gamma(A)|s + (|\Gamma(A)| - 1) \cdot (C_2 + 2)s$$
$$\leq ((C_2 + 3) \cdot |\Gamma(A)| - 1)s.$$

On the other hand,

$$MST((V(A)) \cup \Gamma(A)) \geq \frac{1}{\alpha_{st}} MST(V(A))$$

$$\geq \frac{1}{\alpha_{st}} \sum_{i=1}^{|A|} |T_i| \geq \frac{C_1 \cdot s}{\alpha_{st}} |A|.$$

The second last inequality is due to Lemma 8. It follows that

$$\frac{|\Gamma(A)|}{|A|} \geq \mu = \frac{C_1}{\alpha_{st}(C_2 + 3)}.$$

Next we remove the assumption that for any set of red nodes $r \in \Gamma(A)$, there is another red node $r' \in \Gamma(A)$, s.t. $\delta(r, r') \leq (C_2 + 2)s$. Build the following auxiliary graph G': its nodes are all red nodes in $\Gamma(A)$, for each pair of nodes in G', say u, v, add an edge (u, v) if $\delta(u, v) \leq 2C_2 s$. Then, these red nodes are classified into several connected components, say $\{C^i\}_i$. For each i, we have $r_i \geq \mu a_i$, where $r_i = |C^i|$ and $a_i = |\Gamma(C^i)|$. It follows that $|\Gamma(A)| \geq \mu \sum_i a_i$. Since the size of T_j is at most $C_2 s$, for every $T_j \in A$, there is at most one C^i s.t. $\delta(C^i, T_j) \leq s$. Clearly, each $T_j \in A$ intersects at least one C^i, thus $\sum_i a_i = |A|$. The proof is complete by combing this with $|\Gamma(A)| \geq \mu \sum_i a_i$.

Now we are ready to prove Theorem 7.

Proof of Theorem 7. Fix constants C_1, C_2. Since $s \geq s_{min}$, by lemma 10, there exists a perfect Δ-matching in H, where $\Delta = \frac{\alpha_{st}(C_2+3)}{C_1}$. Since (i) each red node is assigned with at most Δ blocks, (ii) each block is no larger than $C_2 s$, and (iii) the distance between a block and the red node it is assigned to is at most s, it follows that the total distance we travel in each unit length interval is at most $O(\Delta)s = O(1)s$. Since the the blocks are edge-disjoint subtrees of the MST, the total distance travelled is at most $O(1)\lambda^*(s)$.

4   Conclusion and Future Work

In this paper we presented new results for time window TSP and prize-collecting problems. There are still rooms to improve or obtain good approximation ratios in the general case, or present new inapproximation results for these problems. We hope that these results will be useful in guiding scheduling of autonomous vehicles in practice.

Acknowledgements

J. Gao acknowledges support from AFOSR (FA9550-14-1-0193) and NSF (DMS-1418255, NSF CCF-1535900, CNS-1618391). J. Mitchell acknowledges support from NSF (CCF-1526406) and from Sandia National Labs.

References

1. E. M. Arkin, J. S. Mitchell, and G. Narasimhan. Resource-constrained geometric network optimization. In *Proceedings of the fourteenth annual symposium on Computational geometry*, pages 307–316. ACM, 1998.
2. S. Arora. Polynomial time approximation schemes for euclidean TSP and other geometric problems. In *Foundations of Computer Science, 1996. Proceedings., 37th Annual Symposium on*, pages 2–11. IEEE, 1996.
3. Y. Azar and A. Vardi. TSP with time windows and service time. *arXiv preprint arXiv:1501.06158*, 2015.
4. N. Bansal, A. Blum, S. Chawla, and A. Meyerson. Approximation algorithms for deadline-TSP and vehicle routing with time-windows. In *Proceedings of the Thirty-sixth Annual ACM Symposium on Theory of Computing*, STOC '04, pages 166–174, 2004.
5. R. Bar-Yehuda, G. Even, and S. M. Shahar. On approximating a geometric prize-collecting traveling salesman problem with time windows. *Journal of Algorithms*, 55(1):76–92, 2005.
6. A. Blum, S. Chawla, D. R. Karger, T. Lane, A. Meyerson, and M. Minkoff. Approximation algorithms for orienteering and discounted-reward TSP. *SIAM Journal on Computing*, 37(2):653–670, 2007.
7. H.-J. Bockenhauer, J. Hromkovic, J. Kneis, and J. Kupke. The parameterized approximability of TSP with deadlines. *Theory of Computing Systems*, 41(3):431–444, 2007.
8. C. Chekuri, N. Korula, and M. Pál. Improved algorithms for orienteering and related problems. *ACM Transactions on Algorithms (TALG)*, 8(3):23, 2012.
9. K. Chen and S. Har-Peled. The orienteering problem in the plane revisited. In *Proceedings of the twenty-second annual symposium on Computational geometry*, pages 247–254. ACM, 2006.
10. M. Desrochers, J. K. Lenstra, M. W. Savelsbergh, and F. Soumis. Vehicle routing with time windows: optimization and approximation. *Vehicle routing: Methods and studies*, 16:65–84, 1988.
11. G. Gutin and A. P. Punnen. *The Traveling Salesman Problem and Its Variations*. Springer, 2007.
12. J. S. Mitchell. Guillotine subdivisions approximate polygonal subdivisions: A simple polynomial-time approximation scheme for geometric TSP, k-MST, and related problems. *SIAM Journal on Computing*, 28(4):1298–1309, 1999.
13. J. S. B. Mitchell. *Encyclopedia of Algorithms*, chapter Approximation Schemes for Geometric Network Optimization Problems, pages 1–6. Springer Berlin Heidelberg, Berlin, Heidelberg, 2015.
14. V. Nagarajan and R. Ravi. Approximation algorithms for distance constrained vehicle routing problems. *Networks*, 59(2):209–214, 2012.
15. J. N. Tsitsiklis. Special cases of traveling salesman and repairman problems with time windows. *Networks*, 22(3):263–282, 1992.

Resolution-Exact Planner for Thick Non-Crossing 2-Link Robots*

Chee K. Yap, Zhongdi Luo, and Ching-Hsiang Hsu

Department of Computer Science
Courant Institute, NYU
New York, NY 10012, USA
{yap,zl562,chhsu}@cs.nyu.edu

Abstract. We consider the path planning problem for a 2-link robot amidst polygonal obstacles. Our robot is parametrizable by the lengths $\ell_1, \ell_2 > 0$ of its two links, the thickness $\tau \geq 0$ of the links, and an angle κ that constrains the angle between the 2 links to be strictly greater than κ. The case $\tau > 0$ and $\kappa \geq 0$ corresponds to "thick non-crossing" robots. This results in a novel 4DOF configuration space $\mathbb{R}^2 \times (\mathbb{T}^2 \setminus \Delta(\kappa))$ where $\mathbb{T}^2$ is the torus and $\Delta(\kappa)$ the diagonal band of width κ.

We design a resolution-exact planner for this robot using the framework of Soft Subdivision Search (SSS). First, we provide an analysis of the space of forbidden angles, leading to a soft predicate for classifying configuration boxes. We further exploit the T/R splitting technique which was previously introduced for self-crossing thin 2-link robots.

Our open-source implementation in Core Library achieves real-time performance for a suite of combinatorially non-trivial obstacle sets. Experimentally, our algorithm is significantly better than any of the state-of-art sampling algorithms we looked at, in timing and in success rate.

1   Introduction

Motion planning is one of the key topics of robotics [9,3]. The dominant approach to motion planning for the last two decades has been based on sampling, as represented by PRM [7] or RRT [8] and their many variants. An alternative (older) approach is based on subdivision [2,20,1]. Recently, we introduced the notion of **resolution-exactness** which might be regarded[1] as the well-known idea of "resolution completeness" but with a suitable converse [16,18]. Briefly, a planner is resolution-exact if in addition to the usual inputs of path planning, there is an input parameter $\varepsilon > 0$, and there exists a $K > 1$ such that the planner will output a path if there exists one with clearance $K\varepsilon$; it will output NO-PATH if there does not exist one with clearance K/ε. Note that its output is indeterminate if the optimal clearance lies between K/ε and $K\varepsilon$. This provides the theoretical basis for exploiting the concept of **soft predicates**, which

---

* This work is supported by NSF Grants CCF-1423228 and CCF-1564132.

[1] In the theory of computation, a computability concept that has no such converse (e.g., recursive enumerability) is said to be "partially complete".

K. Goldberg et al. (Eds.): *Algorithmic Foundations of Robotics XII*, SPAR 13, pp. 576–591, 2020.
https://doi.org/10.1007/978-3-030-43089-4_37

is roughly speaking the numerical approximation of exact predicates. Such predicates avoid the hard problem of deciding zero, leading to much more practical algorithms than exact algorithms. To support such algorithms, we introduce an algorithmic framework [18, 19] based on subdivision called **Soft Subdivision Search** (SSS). The present paper studies an SSS algorithm for a 2-link robot. Figure 1 shows a path found by this robot in a nontrivial environment.

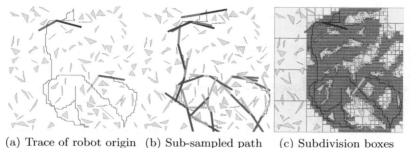

(a) Trace of robot origin   (b) Sub-sampled path   (c) Subdivision boxes

Fig. 1: 100 Random Triangles Environment: non-crossing path found ($\kappa = 115°$)

Link robots offer a compelling class of non-trivial robots for exploring path planning (see [6, chap. 7]). In the plane, the simplest example of a non-rigid robot is the **2-link robot**, $R_2 = R_2(\ell_1, \ell_2)$, with links of lengths $\ell_1, \ell_2 > 0$. The two links are connected through a rotational joint A_0 called the **robot origin**. The 2-link robot is in the intersection of two well-known families of link robots, **chain robots** and **spider robots**   (see [11]). One limitation of link robots is that links are unrealistically modeled by line segments. On the other hand, a model of mechanical links involving complex details may require algorithms that currently do not exist or have high computational complexity. As a compromise, we introduce **thick links** by forming the Minkowski sum of each link with a ball of radius $\tau > 0$ (**thin links** correspond to $\tau = 0$). To our knowledge, no exact algorithm for thick R_2 is known; for a single link R_1, an exact algorithm based on retraction follows from [14]. In this paper, we further parametrize R_2 by a "bandwidth" κ which constrains the angle between the 2 links to be greater than or equal to κ ("self-crossing" links is recovered by setting $\kappa < 0$). Thus, our full robot model

$$R_2(\ell_1, \ell_2, \tau, \kappa)$$

has four parameters; our algorithms are uniform in these parameters.

To illustrate the non-crossing constraint, we use the simple "T-room" environment in Figure 2. Suppose the robot has to move from the start configuration α to the goal configuration β as indicated in Figure 2(a). There is an obvious path from α to β as illustrated in Figure 2(b): the robot origin moves directly from its start to goal positions, while the link angles simultaneously adjust to their goal angles. However, such paths require the two links to cross each other. To achieve a "non-crossing" solution from α to β, we need a less obvious path such as found by our algorithm in Figure 2(c): the robot origin must first move

(a) α (above), β (below)   (b) Self-crossing path   (c) Non-crossing path

Fig. 2: Path from configurations α to β in T-Room Environment

away from the goal, towards the T-junction, in order to maneuver the 2 links into an appropriate relative order before moving toward the goal configuration.

We had chosen $\varepsilon = 2$ in Figure 2(b,c); also, κ is 7 for the non-crossing instance. But if we increase either ε to 3 or κ to 8, then the non-crossing instance would report NO-PATH. It is important to know that the NO-PATH output from resolution-exact algorithms is not never due to exhaustion ("time-out"). It is a principled answer, guaranteeing the non-existence of paths with clearance $> K \cdot \varepsilon$ (for some $K > 1$ depending on the algorithm). In our view, the narrow passage problem is, in the limit, just the **halting problem** for path planning: algorithms with narrow passage problems will also have non-halting issues when there is no path. Our experiments suggests that the "narrow passage problem" is barely an issue for our particular 4DOF robot, but it could be a severe issue for sampling approaches. But no amount of experimental data can truly express the conceptual gap between the various sampling heuristics and the *a priori* guaranteed methods such as ours.

Literature Review. Our theory of resolution-exactness and SSS algorithms apply to any robot system but we focus on algorithmic techniques to achieve the best algorithms for a 2-link robot. The main competition is from sampling approaches – see [3] for a survey. In our experiments, we compare against the well-known PRM [7] as well as variants such as Toggle PRM [4] and lazy version [5]. Another important family of sampling methods is the RRT with variants such as RRT-connect [8] and retraction-RRT [12]. The latter is comparable to Toggle PRM's exploitation of non-free configurations, except that retraction-RRT focuses on contact configurations. Salzman et al [13] introduce a "tiling technique" for handling non-crossing links in sampling algorithms that is quite different from our subdivision solution [11].

Overview of Paper. Section 2 describes our parametrization of the configuration space of R_2, and implicitly, its free space. Section 3 analyzes the forbidden angles of thick links. Section 4 shows our subdivision representation of the non-crossing configuration space. Section 5 describes our experimental results. Section 6 concludes the paper. Omitted proofs and additional experimental data are available as appendices in the full paper [17].

2   Configuration Space of Non-Crossing 2-Link Robot

The configuration space of R_2 is $C_{space} := \mathbb{R}^2 \times \mathbb{T}^2$ where $\mathbb{T}^2 = S^1 \times S^1$ is the torus and $S^1 = SO(2)$ is the unit circle. We represent S^1 by the interval $[0, 2\pi]$ with the identification $0 = 2\pi$. Closed angular intervals of S^1 are denoted by $[s, t]$ where $s, t \in [0, 2\pi]$ using the convention

$$[s, t] := \begin{cases} \{\theta : s \leq \theta \leq t\} & \text{if } s \leq t, \\ [s, 2\pi] \cup [0, t] & \text{if } s > t. \end{cases}$$

In particular, $[0, 2\pi] = S^1$ and $[2\pi, 0] = [0, 0]$. The standard Riemannian metric $d : S^1 \times S^1 \to \mathbb{R}_{\geq 0}$ on S^1 is given by $d(\theta, \theta') = \min \{|\theta - \theta'|, 2\pi - |\theta - \theta'|\}$. Thus $0 \leq d(\theta, \theta') \leq \pi$.

To represent the non-crossing configuration space, we must be more specific about interpreting the parameters in a configuration $(x, y, \theta_1, \theta_2) \in C_{space}$: there are two distinct interpretations, depending on whether R_2 is viewed as a chain robot or a spider   robot. We choose the latter view: then (x, y) is the **footprint** of the joint A_0 at the center of the spider and θ_1, θ_2 are the independent angles of the two links. This has some clear advantage over viewing R_2 as a chain robot, but we can conceive of other advantages for the chain robot view. That will be future research. In the terminology of [11], the robot R_2 has three named points A_0, A_1, A_2 whose **footprints** at configuration $\gamma = (x, y, \theta_1, \theta_2)$ are given by

$$A_0[\gamma] := (x, y), \quad A_1[\gamma] := (x, y) + \ell_1(\cos \theta_1, \sin \theta_1), \quad A_2[\gamma] := (x, y) + \ell_2(\cos \theta_2, \sin \theta_2).$$

The **thin footprint** of R_2 at γ, denoted $R_2[\gamma]$, is defined as the union of the line segments $[A_0[\gamma], A_1[\gamma]]$ and $[A_0[\gamma], A_2[\gamma]]$. The **thick footprint** of R_2 is given by $Fprint_\tau(\gamma) := D(\mathbf{0}, \tau) \oplus R_2[\gamma]$, the Minkowski sum $\oplus$ of the thin footprint with disc $D(\mathbf{0}, \tau)$ of radius τ centered at $\mathbf{0}$.

The **non-crossing configuration space** of bandwidth κ is defined to be

$$C_{space}(\kappa) := \mathbb{R}^2 \times (\mathbb{T}^2 \setminus \Delta(\kappa))$$

where $\Delta(\kappa)$ is the **diagonal band** $\Delta(\kappa) := \{(\theta, \theta') \in \mathbb{T}^2 : d(\theta, \theta') \leq \kappa\} \subseteq \mathbb{T}^2$. Note three special cases:

- If $\kappa < 0$ then $\Delta(\kappa)$ is the empty set.
- If $\kappa = 0$ then $\Delta(\kappa)$ is a closed curve in $\mathbb{T}^2$.
- If $\kappa \geq \pi$ then $\Delta(\kappa) = S^1$.

Configurations in $\mathbb{R}^2 \times \Delta(0)$ are said to be **self-crossing**; all other configurations are **non-crossing**. Here we focus on the case $\kappa \geq 0$. For our subdivision below, we will split $\mathbb{T}^2 \setminus \Delta(0)$ into two connected sets: $\mathbb{T}^2_< := \{(\theta, \theta') \in \mathbb{T}^2 : 0 \leq \theta < \theta' < 2\pi\}$ and $\mathbb{T}^2_> := \{(\theta, \theta') \in \mathbb{T}^2 : 0 \leq \theta' < \theta < 2\pi\}$. For $\kappa \geq 0$, the diagonal band $\Delta(\kappa)$ retracts to the closed curve $\Delta(0)$. In $\mathbb{R}^2$, if we omit such a set, we will get two connected components. In contrast, that $\mathbb{T}^2 \setminus \Delta(\kappa)$ remains connected. CLAIM: $\mathbb{T}^2 \setminus \Delta(\kappa)$ *is topologically a cylinder with two boundary components.* Thus, the non-crossing constraint has changed the topology of the configuration space. To see claim, consider the standard model of $\mathbb{T}^2$ represented by a square with opposite sides identified as in Figure 3(a) (we show the case $\kappa = 0$). By rearranging the two triangles $\mathbb{T}^2_<$ and $\mathbb{T}^2_>$ as in Figure 3(b), our claim is now visually obvious.

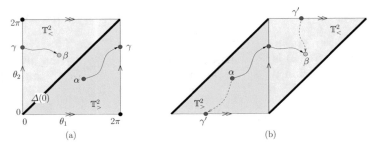

Fig. 3: Paths in $\mathbb{T}^2 \setminus \Delta(0)$ from $\alpha \in \mathbb{T}^2_>$ to $\beta \in \mathbb{T}^2_<$

3   Forbidden Angle Analysis of Thick Links

Towards the development of a soft-predicate for thick links, we must first extend our analysis in [11] which introduced the concept of forbidden angles for thin links. Let $L(\ell, \tau)$ be a single link robot of length $\ell > 0$ and thickness $\tau \geq 0$. Its configuration space is $SE(2) = \mathbb{R}^2 \times S^1$. Given a configuration $(b, \theta) \in SE(2)$, the **footprint** of $L(\ell, \tau)$ at (b, θ) is

$$Fprint_{\ell, \tau}(b, \theta) := L \oplus D(\mathbf{0}, \tau)$$

where $\oplus$ denotes Minkowski sum, L is the line segment $[b, b + \ell(\cos\theta, \sin\theta)]$ and $D(\mathbf{0}, \tau)$ is the disk as above. When ℓ, τ is understood, we simply write "$Fprint(b, \theta)$" instead of $Fprint_{\ell, \tau}(b, \theta)$.

Let $S, T \subseteq \mathbb{R}^2$ be closed sets. An angle θ is **forbidden** for (S, T) if there exists $s \in S$ such that $Fprint(s, \theta) \cap T$ is non-empty. If $t \in Fprint(s, \theta) \cap T$, then the pair $(s, t) \in S \times T$ is a **witness** for the forbidden-ness of θ for (S, T). The set of forbidden angles of (S, T) is called the **forbidden zone** of S, T and denoted $\mathrm{Forb}_{\ell, \tau}(S, T)$. Clearly, $\theta \in \mathrm{Forb}_{\ell, \tau}(S, T)$ iff there exists a witness pair $(s, t) \in S \times T$. Moreover, we call (s, t) a **minimum witness** of θ if the Euclidean norm $\|s - t\|$ is minimum among all witnesses of θ. If (s, t) is a minimum witness, then clearly $s \in \partial S$ and $t \in \partial T$.

Lemma 1. *For any sets $S, T \subseteq \mathbb{R}^2$, we have*

$$\mathrm{Forb}_{\ell, \tau}(S, T) = \pi + \mathrm{Forb}_{\ell, \tau}(T, S).$$

Proof. For any pair (s, t) and any angle α, we see that

$$t \in Fprint(s, \alpha) \text{ iff } s \in Fprint(t, \pi + \alpha).$$

Thus, there is a witness (s, t) for α in $\mathrm{Forb}_{\ell, \tau}(S, T)$ iff there is a witness (t, s) for $\pi + \alpha$ in $\mathrm{Forb}_{\ell, \tau}(T, S)$. The lemma follows.   **Q.E.D.**

¶1. The Forbidden Zone of two points Consider the forbidden zone $\mathrm{Forb}_{\ell, \tau}(V, C)$ defined by two points $V, C \in \mathbb{R}^2$ with $d = \|V - C\|$. (The notation

V suggests a vertex of a translational box B^t and C suggests a corner of the obstacle set.) In our previous paper [11] on thin links (i.e., $\tau = 0$), this case is not discussed for reasons of triviality. When $\tau > 0$, the set $\mathrm{Forb}_{\ell,\tau}(V, C)$ is more interesting. Clearly, $\mathrm{Forb}_{\ell,\tau}(V, C)$ is empty iff $d > \ell + \tau$ (and a singleton if $d = \ell + \tau$). Also $\mathrm{Forb}_{\ell,\tau}(V, C) = S^1$ iff $d \leq \tau$. Henceforth, we may assume

$$\tau < d < \ell + \tau. \tag{1}$$

The forbidden zone of V, C can be written in the form

$$\mathrm{Forb}_{\ell,\tau}(V, C) := [\nu - \delta, \nu + \delta]$$

for some ν, δ. Call ν the **nominal angle** and δ the **correction angle**. By the symmetry of the footprint, ν is equal to $\theta(V, C)$ (see Figure 4).

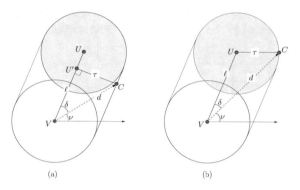

(a)   (b)

Fig. 4: $\mathrm{Forb}_{\ell,\tau}(V, C)$

It remains to determine δ. Consider the configuration $(V, \theta) \in SE(2)$ of our link $L(\ell, \tau)$ where link origin is at V and the link makes an angle θ with the positive x-axis. The angle δ is determined when the point C lies on the boundary of $Fprint(V, \theta)$. The two cases are illustrated in Figure 4 where $\theta = \nu + \delta$ and other endpoint of the link is U; thus $\|VU\| = \ell$ and $\|VC\| = d$, and $\delta = \angle(CVU)$. Under the constraint (1), there are two ranges for d:

(a) d is short: $d^2 \leq \tau^2 + \ell^2$. In this case, the point C lies on the straight portion of the boundary of the footprint, as in Figure 4(a). From the right-angle triangle $CU'V$, we see that $\delta = \arcsin(\tau/d)$.

(b) d is long: $d^2 > \tau^2 + \ell^2$. In this case, the point C lies on the circular portion of the boundary of the footprint, as in Figure 4(b). Consider the triangle CUV with side lengths of d, ℓ, τ. By the cosine law, $\tau^2 = d^2 + \ell^2 - 2d\ell \cos \delta$ and thus

$$\delta = \arccos\left(\frac{\ell^2 + d^2 - \tau^2}{2d\ell}\right).$$

This proves:

Lemma 2. *Assume* $\|VC\| = d$ *satisfies (1). Then*

$$\mathrm{Forb}_{\ell,\tau}(V,C) = [\nu - \delta, \nu + \delta]$$

where $\nu = \theta(V,C)$ *and*

$$\delta = \delta(V,C) = \begin{cases} \arcsin(\tau/d) & \text{if } d^2 \leq \tau^2 + \ell^2, \\ \arccos\left(\frac{\ell^2 + d^2 - \tau^2}{2d\ell}\right) & \text{if } d^2 > \tau^2 + \ell^2. \end{cases} \tag{2}$$

¶2. The Forbidden Zone of a Vertex and a Wall Recall that the boundary of a box B^t is divided into four **sides**, and two adjacent sides share a common endpoint which we call a **vertex**. We now determine $\mathrm{Forb}_{\ell,\tau}(V,W)$ where V is a vertex and W a wall feature. Choose the coordinate axes such that W lies on the x-axis, and $V = (0, -\sigma)$ lies on the negative y-axis, for some $\sigma > 0$. Let the two corners of W be C, C' with C' lying to the left of C. See Figure 5.

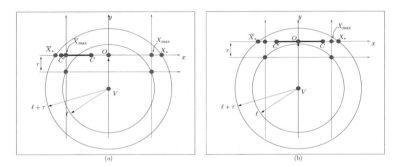

Fig. 5: Stop Analysis for $\mathrm{Forb}_{\ell,\tau}(V,W)$ (assuming $\sigma > \tau$)

We first show that the interesting case is when

$$\tau < \sigma < \ell + \tau. \tag{3}$$

If $\sigma \geq \ell + \tau$ then $\mathrm{Forb}_{\ell,\tau}(V,W)$ is either a singleton ($\sigma = \ell + \tau$) or else is empty ($\sigma > \ell + \tau$). Likewise, the following lemma shows that when $\sigma \leq \tau$, we are to point-point case of Lemma 2:

Lemma 3. *Assume* $\sigma \leq \tau$. *We have*

$$\mathrm{Forb}_{\ell,\tau}(V,W) = \begin{cases} S^1 & \text{if } D(V,\tau) \cap W \neq \emptyset, \\ \mathrm{Forb}_{\ell,\tau}(V,c) & \text{else} \end{cases}$$

where $c = C$ *or* C'.

Proof. Recall that we have chosen the coordinate system so that W lies on the x-axes and $V = (0, -\sigma)$. It is easy to see that $\text{Forb}_{\ell,\tau}(V, W) = S^1$ iff the disc $D(V, \tau)$ intersects W. So assume otherwise. In that case, the closest point in W to V is c, one of the two corners of W. The lemma is proved if we show that

$$\text{Forb}_{\ell,\tau}(V, W) = \text{Forb}_{\ell,\tau}(V, c).$$

It suffices to show $\text{Forb}_{\ell,\tau}(V, W) \subseteq \text{Forb}_{\ell,\tau}(V, c)$. Suppose $\theta \in \text{Forb}_{\ell,\tau}(V, W)$. So it has a witness (V, c') for some $c' \in W$. However, we see that the minimal witness for this case is (V, c). This proves that $\theta \in \text{Forb}_{\ell,\tau}(V, c)$.   **Q.E.D.**

In addition to (3), we may also assume the wall lies within the annulus of radii $(\tau, \tau + \ell)$ centered at V:

$$\|VC\|, \|VC'\| \in (\tau, \ell + \tau) \tag{4}$$

Using the fact that $V = (0, -\sigma)$ and W lies in the x-axis, we have:

Lemma 4. *Assume (3) and (4).*
Then $\text{Forb}_{\ell,\tau}(V, W)$ *is a non-empty connected interval of* S^1,

$$\text{Forb}_{\ell,\tau}(V, W) = [\alpha, \beta] \subseteq (0, \pi).$$

Our next goal is to determine the angles α, β in this lemma. Consider the footprints of the link at the extreme configurations $(V, \alpha), (V, \beta) \in SE(2)$. Clearly, W intersects the boundary (but not interior) of these footprints, $Fprint(V, \alpha)$ and $Fprint(V, \beta)$. Except for some special configurations, these intersections are singleton sets. Regardless, pick any $A \in W \cap Fprint(V, \alpha)$ and $B \in W \cap Fprint(V, \beta)$. Since α is an endpoint of $\text{Forb}_{\ell,\tau}(V, W)$, we see that $A \in (\partial W) \cap \partial(Fprint(V, \alpha))$. We call A a **left stop** for the pair (V, W) because[2] for any $\delta' > 0$ small enough, $A \in Fprint(V, \alpha + \delta')$ while $W \cap (V, \alpha - \delta') = \emptyset$. Similarly the point B is called a **right stop** for the pair (V, W). Clearly, we can write

$$\alpha = \theta(V, A) - \delta(V, A), \qquad \beta = \theta(V, B) + \delta(V, B)$$

where $\delta(V, \cdot)$ is given by Lemma 2. We have thus reduced the determination of angles α and β to the computation of the left A and right B stops.

We might initially guess that the left stop of (V, W) is C, and right stop of (V, W) is C'. But the truth is a bit more subtle. Define the following points $X_*, X_{\max}$ on the positive x-axis using the equation:

$$\|OX_*\| = \sqrt{(\ell + \tau)^2 - \sigma^2},$$
$$\|OX_{\max}\| = \sqrt{\ell^2 - (\sigma - \tau)^2}.$$

These two points are illustrated in Figure 5. Also, let $\overline{X}_*$ and $\overline{X}_{\max}$ be mirror reflections of X_* and $X_{\max}$ across the y-axis. The points $X_*, \overline{X}_*$ are the two

---

[2] Intuitively: At configuration (V, α), the single-link robot can rotate about V to the right, but if it tries to rotate to the left, it is "stopped" by A.

points at distance $\ell + \tau$ from V. The points $X_{\max}, \overline{X}_{\max}$ are the left and right stops in we replace W by the infinite line through W (i.e., the x-axis).

With the natural ordering of points on the x-axis, we can show that

$$\overline{X}_* < \overline{X}_{\max} < O < X_{\max} < X_*$$

where O is the origin. Since $\|VC\|$ and $\|VC'\|$ lie in $(\tau, \tau + \ell)$, it follows that

$$\overline{X}_* < C' < C < X_*.$$

Two situations are shown in Figure 5. The next lemma is essentially routine, once the points $X_{\max}, \overline{X}_{\max}$ have defined:

Lemma 5. *Assume (3) and (4).*
The left stop of (V, W) is

$$\begin{cases} C' & \text{if } X_{\max} \leq C' & (L1) \\ X_{\max} & \text{if } C' < X_{\max} < C & (L2) \\ C & \text{if } C \leq X_{\max} & (L3) \end{cases}$$

The right stop of (V, W) is

$$\begin{cases} C & \text{if } C \leq \overline{X}_{\max} & (R1) \\ \overline{X}_{\max} & \text{if } C' < \overline{X}_{\max} < C & (R2) \\ C' & \text{if } \overline{X}_{\max} \leq C' & (R3) \end{cases}$$

The cases (L1-3) and (R1-3) in this lemma suggests 9 combinations, but 3 are logically impossible: (L1-R1), (L1-R2), (L2-R1). The remaining 6 possibilities for left and right stops are summarized in the following table:

	(R1)	(R2)	(R3)
(L1)	*	*	(C', C')
(L2)	*	$(X_{\max}, \overline{X}_{\max})$	$(X_{\max}, C')$
(L3)	(C, C)	$(C, \overline{X}_{\max})$	(C, C')

Observe the extreme situations (L1-R3) or (L3-R1) where the the left and right stops are equal to the same corner, and we are reduced to the point-point analysis. Once we know the left and right stops for (V, W), then we can use Lemma 2 to calculate the angles α and β.

¶3. The Forbidden Zone of a Side and a Corner We now consider the forbidden zone $\text{Forb}_{\ell, \tau}(S, C)$ where S is a side and C a corner feature. Note that is complementary to the previous case of $\text{Forb}_{\ell, \tau}(V, W)$ since C and V are points and S and W are line segments. We can exploit the principle of reflection symmetry of Lemma 1:

$$\text{Forb}_{\ell, \tau}(S, C) = \pi + \text{Forb}_{\ell, \tau}(C, S)$$

where $\text{Forb}_{\ell, \tau}(C, S)$ is provided by previous Lemma (with C, S instead of V, W).

¶4. Cone Decomposition We have now provided formulas for computing sets of the form $\text{Forb}_{\ell,\tau}(V, W)$ or $\text{Forb}_{\ell,\tau}(S, C)$; such sets are called **cones**. We now address the problem of computing $\text{Forb}_{\ell,\tau}(B^t, W)$ where $B^t \subseteq \mathbb{R}^2$ is a (translational) box. We show that this set of forbidden angles can be written as the union of at most 3 cones, generalizes a similar result in [11]. Towards such a cone decomposition, we first classify the disposition of a wall W relative to a box B^t. There is a preliminary case: if W intersects $B^t \oplus D(0, \tau)$, then we have

$$\text{Forb}_{\ell}(B^t, W) = S^1.$$

Call this **Case (0)**. Assuming W does not intersect $B^t \oplus D(0, \tau)$, there are three other possibilities, **Cases (I-III)** illustrated Figure 6.

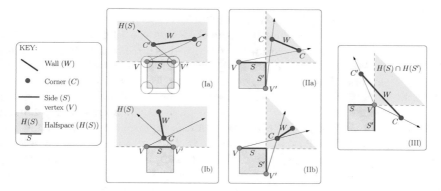

Fig. 6: Cases (I-III) of $\text{Forb}_{\ell,\tau}(B^t, W)$

We first need a notation: if $S \subseteq \partial(B^t)$ is a side of the box B^t, let $H(S)$ denote the open half-space which is disjoint from B^t and is bounded by the line through S. Then we have these three cases:

(I) $W \subseteq H(S)$ for some side s of box B^t.
(II) $W \subseteq H(S) \cap H(S')$ for two adjacent sides S, S' of box B^t.
(III) None of the above. This implies that $W \subseteq H(S) \cup H(S')$ for two adjacent sides S, S' of box B^t.

Theorem 1. $\text{Forb}_{\ell,\tau}(B^t, W)$ *is the union of at most three thick cones.*

Sketch proof: we try to reduce the argument to the case $\tau = 0$ which is given in [11]. In that case, we could write

$$\text{Forb}_{\ell}(B^t, W) = C_1 \cup C_2 \cup C_3$$

where each C_i is a thin cone or an empty set. In the non-empty case, the cone C_i has the form $\text{Forb}_{\ell}(S_i, T_i)$ where $S_i \subseteq \partial B^t, T_i \subseteq W$. The basic idea is that we now "transpose" $\mathrm{Forb}_\ell(S_i, T_i)$ to the thick version $C_i' := \mathrm{Forb}_{\ell,\tau}(S_i, T_i)$. In case C_i is empty, C_i' remains empty. Thus we would like to claim that

$$\mathrm{Forb}_\ell(B^t, W) = C_1' \cup C_2' \cup C_3'.$$

This is almost correct, except for one issue. It is possible that some C_i is empty, and yet its transpose C_i' is non empty. See proof in the full paper [17]. In case of thin cones, the C_i's are non-overlapping (i.e., they may only share endpoints). But for thick cone decomposition, the cones will in general overlap.

4   Subdivision for Thick Non-Crossing 2-Link Robot

A resolution-exact planner for a thin self-crossing 2-link robot was described in [11]. We now extend that planner to the thick non-crossing case.

   We will briefly review the ideas of the algorithm for the thin self-crossing 2-link robot. We begin with a box $B_0 \subseteq \mathbb{R}^2$ and it is in the subspace $B_0 \times \mathbb{T}^2 \subseteq C_{space}$ where our planning problem takes place. We are also given a polygonal obstacle set $\Omega \subseteq \mathbb{R}^2$; we may decompose its boundary $\partial\Omega$ into a disjoint union of corners (=points) and edges (=open line segments) which are called (boundary) **features**. Let $B \subseteq C_{space}$ be a box; there is an exact classification of B as $C(B) \in \{\texttt{FREE}, \texttt{STUCK}, \texttt{MIXED}\}$ relative to Ω. But we want a soft classification $\widetilde{C}(B)$ which is correct whenever $\widetilde{C}(B) \neq \texttt{MIXED}$, and which is equal to $C(B)$ when the width of B is small enough. Our method of computing $\widetilde{C}(B)$ is based on computing a set $\phi(B)$ of features that are relevant to B. A box $B \subseteq C_{space}$ may be written as a Cartesian product $B = B^t \times B^r$ of its translational subbox $B^t \subseteq \mathbb{R}^2$ and rotational subbox $B^r \subseteq \mathbb{T}^2$. In the T/R splitting method (simple version), we split B^t until the width of B^t is $\leq \varepsilon$. Then we do a single split of the rotational subbox B^r into all the subboxes obtained by removing all the forbidden angles determined by the walls and corners in $\widetilde{\phi}(B^t)$. This "rotational split" of B^r is determined by obstacles, unlike the "translational splits" of B^t.

   ¶5. **Boxes for Non-Crossing Robot.**   Our basic idea for representing boxes in the non-crossing configuration space $C_{space}(\kappa)$ is to write it as a pair $(B, \texttt{XT})$ where $\texttt{XT} \in \{\texttt{LT}, \texttt{GT}\}$, and $B \subseteq C_{space}$. The pair $(B, \texttt{XT})$ represents the set $B \cap (\mathbb{R}^2 \times \mathbb{T}^2_{\texttt{XT}})$ (with the identification $\mathbb{T}^2_{\texttt{LT}} = \mathbb{T}^2_<$ and $\mathbb{T}^2_{\texttt{GT}} = \mathbb{T}^2_>$). It is convenient to call $(B, \texttt{XT})$ an X-**box** since they are no longer "boxes" in the usual sense.

   An angular interval $\Theta \subseteq S^1$ that[3] contains a open neighborhood of $0 = 2\pi$ is said to be **wrapping**. Also, call $B^r = \Theta_1 \times \Theta_2$ wrapping if either Θ_1 or Θ_2 is wrapping. Given any B^r, we can decompose the set $B^r \cap (\mathbb{T}^2 \setminus \Delta(\kappa))$ into the union of two subsets $B^r_{\texttt{LT}}$ and $B^r_{\texttt{GT}}$, where $B^r_{\texttt{XT}}$ denote the set $B^r \cap \mathbb{T}^2_{\texttt{XT}}$. In case B^r is non-wrapping, this decomposition has the nice property that each subset $B^r_{\texttt{XT}}$ is connected. For this reason, we prefer to work with non-wrapping boxes. Initially, the box $B^r = \mathbb{T}^2$ is wrapping. The initial split of $\mathbb{T}^2$ should be done

---

[3] Wrapping intervals are either equal to S^1 or has the form $[s, t]$ where $2\pi > s > t > 0$.

in such a way that the children are all non-wrapping: the "natural" (quadtree-like) way to split $\mathbb{T}^2$ into four congruent children has[4] this property. Thereafter, subsequent splitting of these non-wrapping boxes will remain non-wrapping.

Of course, $B^r_{\mathtt{XT}}$ might be empty, and this is easily checked: say $\Theta_i = [s_i, t_i]$ $(i = 1, 2)$. Then $B^r_<$ is empty iff $t_2 \leq s_1$. and $B^r_>$ is empty iff $s_2 \geq t_1$. Moreover, these two conditions are mutually exclusive.

We now modify the algorithm of [11] as follows: as long as we are just splitting boxes in the translational dimensions, there is no difference. When we decide to split the rotational dimensions, we use the T/R splitting method of [11], but each child is further split into two X-boxes annotated by $\mathtt{LT}$ or $\mathtt{GT}$ (they are filtered out if empty). We build the connectivity graph G (see Appendix A) with these X-boxes as nodes. This ensures that we only find non-crossing paths. Our algorithm inherits resolution-exactness from the original self-crossing algorithm.

The predicate $\mathtt{isBoxEmpty}(B^r, \kappa, \mathtt{XT})$ which returns true iff $(B^r_{\mathtt{XT}}) \cap (\mathbb{T}^2 \backslash \Delta(\kappa))$ is empty is useful in implementation. It has a simple expression when restricted to non-wrapping translational box B^r:

Lemma 6.
Let $B^r = [a, b] \times [a', b']$ be a non-wrapping box.
(a) $\mathtt{isBoxEmpty}(B^r, \kappa, \mathtt{LT}) = \mathtt{true}$ *iff* $\kappa \geq b' - a$ *or* $2\pi - \kappa \leq a' - b$.
(b) $\mathtt{isBoxEmpty}(B^r, \kappa, \mathtt{GT}) = \mathtt{true}$ *iff* $\kappa \geq b - a'$ *or* $2\pi - \kappa \leq a - b'$.

5   Implementation and Experiments

We implemented our thick non-crossing 2-link planner in C++ and OpenGL on the Qt platform. A preliminary heuristic version appeared [11, 10]. Our code, data and experiments are distributed[5] with our open source **Core Library**. To evaluate our planner, we compare it with several sampling algorithms in the open source OMPL [15]. Besides, based on a referee's suggestion, we also implemented the 2-link (crossing and non-crossing) versions of Toggle PRM and Lazy Toggle PRM (in lieu of publicly available code). We benefited greatly from the advice of Prof. Denny in our best effort implementation. The machine we use is a MacBook Pro with 2.5 GHz Intel Core i7 and 16GB DDR3-1600 MHz RAM.

Tables 1 and 2 summarize the results of two groups of experiments, which we call **Narrow Passages** and **Easy Passages**. Each row in the tables represents an experiment, each column represents a planner. There are 8 planners: 3 versions of SSS, 3 versions of PRM and 2 versions of RRT. Only Table 1 is listed in the paper; Table 2 (as well as other experimental results) are relegated to an appendix. We extract two bar charts from Table 1 in Figure 7 for visualization to show the average times and success rates of the planners in Narrow Passages.

---

[4] This is not a vacuous remark – the quadtree-like split is determined by the choice of a "center" for splitting. To ensure non-wrapping children, this center is necessarily $(0, 0)$ or equivalently $(2\pi, 2\pi)$. Furthermore, our T/R splitting method (to be introduced) does not follow the conventional quadtree-like subdivision at all.

[5] http://cs.nyu.edu/exact/core/download/.

Timing is in milliseconds on a Log10 scale. E.g., the bar chart Figure 7(a) for Narrow Passages shows that the average time for the SSS(I) planner in the T-Room experiment is about 2.9. This represents $10^{2.9} \simeq 800$ milliseconds; indeed the actual value is 815.9 as seen in Table 1. On this bar chart, each unit represents a power of 10. E.g., if one bar is at least k units shorter than another bar, we say the former is "k orders of magnitude" faster (i.e., at least 10^k times faster). **Conclusion:** (1) SSS is at least an order of magnitude faster than each sampling method, and (2) success rates of RRT-connect and Toggle PRM are usually (but not always) best among sampling methods, but both are inferior to SSS.

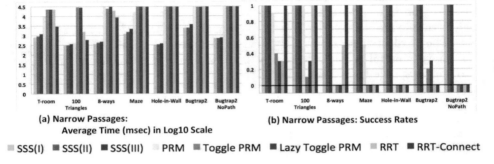

(a) Narrow Passages:
Average Time (msec) in Log10 Scale

(b) Narrow Passages: Success Rates

■ SSS(I)   ■ SSS(II)   ■ SSS(III)   PRM   ■ Toggle PRM   ■ Lazy Toggle PRM   RRT   ■ RRT-Connect

Fig. 7: Bar Charts of Average Times and Success Rates

Two general remarks are in order. First, as in our previous work, we implemented several search strategies in SSS. But for simplicity, we only use the Greedy Best First (GBF) strategy in all the SSS experiments; GBF is typically our best strategy. Next, OMPL does not natively support articulated robots such as R_2. So in the experiments of Tables 1 and 2, we artificially set $\ell_2 = 0$ for all the sampling algorithms (so that they are effectively one-link thick robots). This is a suboptimal experimental scenario, but it only reinforces any exhibited superiority of our SSS methods. In the SSS versions, we set $\ell_2 = 0$ for SSS(I) but SSS(II) and SSS(III) represent (resp.) crossing and non-crossing 2-link robots where ℓ_2 has the values shown in the column 4 header of Table 1.

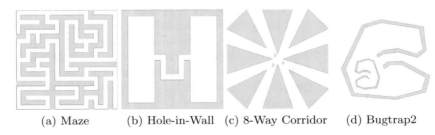

(a) Maze   (b) Hole-in-Wall   (c) 8-Way Corridor   (d) Bugtrap2

Fig. 8: More environments in our experiments

Reading the Tables: Each experiment (i.e., row) corresponds to a fixed environment, robot parameters, initial and goal configurations. Figure 8 depicts these environments (save for the T-Room and 100 triangles from the introduction). We name the experiments after the environment. E.g., column 1 for the Maze experiment, tells us that $\ell_1 = 16, \tau = 10$. The last two experiments use the "double bugtrap" environment, but the robot parameters for one of them ensures NO-PATH. For each experiment, we perform 40 runs of the following planners: SSS(I-III), PRM, RRT, RRT-connect (all from OMPL), Toggle PRM and Lazy Toggle PRM (our implementation). Each planner produces 4 statistics:

Average Time / Best Time / Standard Deviation / Success Rate,

abbreviated as **Avg/Best/STD/Success**, respectively. Success Rate is the fraction of the 40 runs for which the planner finds a path (assuming there is one) out of 40 runs. But if there is no path, our SSS planner will always discover this, so its **Success** is 1; simultaneously, the sampling methods will time out and hence their **Success** is 0. All timing is in milliseconds (msec). Column 2 contains the **Record Statistics**, i.e., the row optimum for these 4 statistics. E.g., the Record Statistics for the T-Room experiment is **815.9/743.6/21.9/1**. This tells us the row optimum for **Avg** is 815.9 ms, for **Best** is 743.6 ms, for **STD** is 21.9 ms, and for **Success** is 1. "Optimum" for the first three (resp., last) statistics means minimum (resp., maximum) value. The four optimal values may be achieved by different planners. In the rest part of the Table, we have one column for each Planner, showing the ratio of the planner's statistics relative to the Record Statistics. The best performance is always indicated by the ratio of 1. E.g., for T-Room experiment, the row maximum for **Success** is 1, and it is achieved by all SSS planners and RRT-Connect. The row minimum for **Avg**, **Best** and **STD** are achieved by SSS(I), RRT-Connect and SSS(II), resp. We regard the achievement of row optimum for **Success** and **Avg** (in that order) to be the main indicator of superiority. Table 1 (and Table 2) show that our planner is consistently superior to sampling planners. E.g., Table 1 shows that in the T-Room experiment, the record average time of 815.9 milliseconds is achieved by SSS(I). But SSS(III) is only 1.5 times slower, and the best sampling method is RRT-Connect which is 3.6 times slower. For the Maze experiment, again SSS(I) achieves the record average time of 1193.2 milliseconds, SSS(III) is 1.8 times slower, but none of the sampling methods succeeded in 40 trials.

In the Appendix, we have Table 2 which is basically the same as Table 1 except that we decrease the thickness τ in order to improve the success rates of the sampling methods. Our planner needs an ε parameter, which is set to 1 in Table 1 and 2 in Table 2 (this is reasonable in view of narrow passage demands). Sampling methods have many more tuning parameters; but we choose the defaults in OMPL because we saw no systematic improvements in our partial attempts to enhance their performance. In Toggle PRM, we use small k for k-nearest neighbors in the obstacle graph and the similar default k as in OMPL in the free graph. We set the time-out to be 30 seconds; with this cutoff, it takes 18 hours to produce the data of Table 1. In the appendix, we mention some experiments to allow the sampling algorithm up to 0.5 hour.

Environment (ℓ, τ)	Record Statistics (Avg./Best/STD/ Success)	Ratios Relative to Record Statistics							
		SSS (I) $(\ell_2: 0)$	SSS (II) $(\ell_2: 20, 10, 10, 10, 10, 5, 5)$	SSS (III) (non-crossing)	PRM	Toggle PRM	Lazy Toggle PRM	RRT	RRT-Connect
T-Room (120, 24)	815.9/743.6/21.9/1	1/1/1.3/1	1.1/1.1/1/1	1.5/1.5/1.4/1	11.6/1.4/414.7/0.9	27.6/3.6/468.1/0.4	27.4/1.8/539.1/0.3	28/1.1/571.9/0.3	3.6/1/213.5/1
100 Triangles (35, 20)	301.1/268.4/18.7/1	1/1/2/1	1/1.1/1/1	1.2/1.2/3.2/1	1.3/1.3/41.5/1	98.2/89.2/74.8/0.1	92.5/60.1/225.6/0.3	5.1/3.7/28.3/1	2/1.2/12.9/1
8-Ways (48, 11)	370.8/329.3/17.4/1	1/1/3.8/1	1.1/1.1/2.1/1	1.3/1.4/1/1	81.1/x/x/0	64.6/x/x/0	80.9/x/x/0	52.5/2.4/728.9/0.5	23/2.4/302.3/1
Maze (16, 10)	1193.2/1137.1/26.8/1	1/1/1/1	1.3/1.3/1.3/1	1.8/1.8/2.5/1	13.1/8.4/211.9/0	25.1/x/x/0	25.1/x/x/0	25.2/x/x/0	25.2/x/x/0
Hole-in-Wall (50, 12)	319.2/284.9/21.5/1	1/1/1.2/1	1.1/1.1/1/1	1.2/1.4/1.3/1	94.2/x/x/0.5	94/x/x/0	94/x/x/0	94.2/x/x/0	94.2/x/x/0
Bugtrap2 (36, 9)	2335/2213.2/75.2/1	1/1/1/1	1/1.1/1.2/1	1.6/1.5/4.9/1	12.9/x/x/0	12.5/9.5/27.9/0.2	12.5/11.2/22.9/0.3	12.9/x/x/0	12.9/x/x/0
Bugtrap2 NoPath (70, 8)	686.6/666.5/11.8/1	1/1/1.3/1	1/1.1/1/1	1.1/1.1/1.2/1	43.8/x/x/0	43.7/x/x/0	43.7/x/x/0	43.8/x/x/0	43.8/x/x/0

Table 1: Narrow Passages

6   Conclusion and Limitations

We have introduced a novel and efficient planner for thick non-crossing 2-link robots. Our work contributes to the development of practical and theoretically sound subdivision planners [16, 18]. It is reasonable to expect a tradeoff between the stronger guarantees of our resolution-exact approach versus a faster running time for sampling approaches. But our experiments suggest no such tradeoffs at all: *SSS is consistently superior to sampling*. We ought to say that although we have been unable to improve the sampling planners by tuning their parameters, it is possible that sampling experts might do a better job than us. But to actually exceed our performance, their improvement would have to be dramatic. SSS has no tuning, except in the choice of a search strategy (Greedy Best First), and a value for ε. But we do not view ε as a tuning parameter, but a value determined by the needs of the application.

Conventional wisdom maintains that subdivision will not scale to higher DOF's, and our current experience have been limited to at most 4DOF. We interpret this wisdom as telling us that new subdivision techniques (such as the T/R splitting idea) are needed to make higher DOF's robots perform in real-time. This is a worthy challenge for SSS which we plan to take up.

References

1. M. Barbehenn and S. Hutchinson. Toward an exact incremental geometric robot motion planner. In *Proc. IROS*, vol. 3, pp. 39–44, Pittsburgh, PA. 1995.
2. R. A. Brooks and T. Lozano-Perez. A subdivision algorithm in configuration space for findpath with rotation. In *8th IJCAI – Vol. 2*, pp. 799–806, San Francisco, CA, USA, 1983. Morgan Kaufmann Pub. Inc.

3. H. Choset, K. M. Lynch, S. Hutchinson, G. Kantor, W. Burgard, L. E. Kavraki, and S. Thrun. *Principles of Robot Motion: Theory, Algorithms, and Implementations.* MIT Press, Boston, 2005.

4. J. Denny and N. M. Amato. Toggle PRM: A coordinated mapping of C-free and C-obstacle in arbitrary dimension. In *WAFR 2012*, volume 86 of *Springer Tracts in Advanced Robotics*, pp. 297–312. MIT, Cambridge, USA. June 2012.

5. J. Denny, K. Shi, and N. M. Amato. Lazy Toggle PRM: a Single Query approach to motion planning. In *Proc. ICRA*, pp. 2407–2414. 2013.

6. S. L. Devadoss and J. O'Rourke. *Discrete and Computatational Geometry.* Princeton University Press, 2011.

7. L. Kavraki, P. Švestka, C. Latombe, and M. Overmars. Probabilistic roadmaps for path planning in high-dimensional configuration spaces. *IEEE Trans. Robotics and Automation*, 12(4):566–580, 1996.

8. J. J. Kuffner Jr and S. M. LaValle. RRT-connect: An efficient approach to single-query path planning. In *Proc. ICRA*, vol. 2, pp. 995–1001. 2000.

9. S. M. LaValle. *Planning Algorithms.* Cambridge Univ. Press, Cambridge, 2006.

10. Z. Luo. Resolution-exact Planner for a 2-link planar robot using Soft Predicates. Master thesis, New York University, Courant Institute, Jan. 2014.

11. Z. Luo, Y.-J. Chiang, J.-M. Lien, and C. Yap. Resolution exact algorithms for link robots. In *Proc. 11th Intl. Workshop on Algorithmic Foundations of Robotics (WAFR '14)*, vol. 107 of *Springer Tracts in Advanced Robotics* , pp. 353–370, 2015. 3-5 Aug 2014, Istanbul, Turkey.

12. J. Pan, L. Zhang, and D. Manocha. Retraction-based rrt planner for articulated models. In *Proc. ICRA*, pages 2529–2536, 2010.

13. O. Salzman, K. Solovey, and D. Halperin. Motion planning for multi-link robots by implicit configuration-space tiling, 2015. CoRR abs/1504.06631.

14. M. Sharir, C. O'D'únlaing, and C. Yap. Generalized Voronoi diagrams for moving a ladder II: efficient computation of the diagram. *Algorithmica*, 2:27–59, 1987.

15. I. Şucan, M. Moll, and L. Kavraki. The Open Motion Planning Library. *IEEE Robotics & Automation Magazine*, 19(4):72–82, 2012.

16. C. Wang, Y.-J. Chiang, and C. Yap. On Soft Predicates in Subdivision Motion Planning. In *29th ACM Symp. on Comp. Geom.*, pages 349–358, 2013. SoCG'13, Rio de Janeiro, Brazil, June 17-20, 2013. Journal version in Special Issue of *CGTA*.

17. C. Yap, Z. Luo, and C.-H. Hsu. Resolution-exact planner for thick non-crossing 2-link robots. *ArXiv e-prints*, 2017. Submitted. See also proceedings of 12th WAFR, 2016, and http://cs.nyu.edu/exact/ for proofs and more experimental data.

18. C. K. Yap. Soft Subdivision Search in Motion Planning. In A. Aladren et al., editor, *Proceedings, 1st Workshop on Robotics Challenge and Vision (RCV 2013)*, 2013. A Computing Community Consortium (CCC) **Best Paper Award**, Robotics Science and Systems Conference (RSS 2013), Berlin. In arXiv:1402.3213.

19. C. K. Yap. Soft Subdivision Search and Motion Planning, II: Axiomatics. In *Frontiers in Algorithmics*, volume 9130 of *Lecture Notes in Comp.Sci.*, pages 7–22. Springer, 2015. Plenary Talk at 9th FAW. Guilin, China. Aug 3-5, 2015.

20. J. Yu, C. Yap, Z. Du, S. Pion, and H. Bronnimann. Core 2: A library for Exact Numeric Computation in Geometry and Algebra. In *3rd Proc. Int'l Congress on Mathematical Software (ICMS)*, LNCS No. 6327, pages 121–141. Springer, 2010.

21. D. Zhu and J.-C. Latombe. New heuristic algorithms for efficient hierarchical path planning. *IEEE Transactions on Robotics and Automation*, 7:9–20, 1991.

Fast and Bounded Probabilistic Collision Detection for High-DOF Robots in Dynamic Environments

Chonhyon Park, Jae Sung Park, and Dinesh Manocha

University of North Carolina, Chapel Hill, NC 27599, USA,
{chpark,jaesungp,dm}@cs.unc.edu
http://gamma.cs.unc.edu/PCOLLISION

Abstract. We present a novel approach to performing probabilistic collision detection between a high-DOF robot and imperfect obstacle representations in dynamic, uncertain environments. These uncertainties are modeled using Gaussian distributions. We present an efficient algorithm for bounded collision probability approximation. We use our probabilistic collision algorithm for trajectory optimization in dynamic scenes for 7-DOF robots. We highlight its performance in challenging simulated and real-world environments with robot arms operating next to dynamically moving human obstacles.

1 Introduction

Robots are increasingly being used in living spaces, factories, and outdoor environments. In such environments, various elements of or parts of the robot tend to be in close proximity to humans or other moving objects. This proximity gives rise to two kinds of challenges in terms of motion planning. First, we have to predict the future actions and reactions of moving obstacles or agents in the environment to avoid collisions with the obstacles. Therefore, the collision avoidance algorithm needs to deal with uncertain and imperfect representations of obstacle motions. Second, the computed robot motion still needs to be reasonably efficient and all such collision computations have to be performed at almost realtime rates.

Various uncertainties arise from control errors, sensing errors, or environmental errors (i.e. imperfect environment representation) in the estimation and prediction of environment obstacles. Typically, these uncertainties are modeled using Gaussian distributions. In this paper, we limit ourselves to environmental errors. Current motion planning algorithms use probabilistic collision detection algorithms to compute appropriate trajectories with imperfect obstacle representations. Typically, these uncertainties are modeled using Gaussian distributions and stochastic algorithms are used to approximate the collision probability [4, 12, 16, 17]. However, it is almost impossible to guarantee a perfect collision-free trajectory as the Gaussian distributions corresponding to the obstacle positions have non-zero probabilities in the entire workspace.

© Springer Nature Switzerland AG 2020
K. Goldberg et al. (Eds.): *Algorithmic Foundations of Robotics XII*, SPAR 13, pp. 592–607, 2020.
https://doi.org/10.1007/978-3-030-43089-4_38

Many of the stochastic algorithms used to approximate the collision probability [4, 12] tend to be computationally expensive or limited to 2D workspaces. Most prior planning approaches for high-DOF robots perform exact collision checking with scaled objects that enclose the potential object volumes [3, 5, 13, 24]. Although these planning approaches can guarantee probabilistic safety bounds, they tend to overestimate the collision probability. This overestimation can either result in less optimal trajectories or may fail to compute a feasible trajectory in the limited planning time in dynamic scenes. Therefore, it is desirable to balance the safety and efficiency in terms of the planned trajectory.

Main Results: In this paper, we present a novel approach to perform probabilistic collision detection. In particular, we present an algorithm for fast approximation of collision probability between the high-DOF robot and obstacles. Our approach computes more accurate probabilities as compared to prior approaches that perform exact collision checking with enlarged obstacle shapes. Moreover, we can guarantee that our computed probability is an upper bound on the actual probability. Moreover, we present a trajectory optimization algorithm for high-DOF robots in dynamic, uncertain environments based on our probabilistic collision detection, and a practical belief space estimation algorithm that accounts for both spatial and temporal uncertainties in the position and motion of each obstacle in dynamic environments. We have evaluated our planner using 7-DOF robot arms operating in a simulation and a real workspace environment with high-resolution point cloud data corresponding to moving human obstacles, captured using a Kinect device[1].

The paper is organized as follows. Section 2 gives a brief overview of prior work on probabilistic collision detection and motion planning. We introduce the notation and describe our probabilistic collision detection algorithm in Section 3. We describe the trajectory planning algorithm in Section 4. We highlight planning performance in challenging human environment scenarios in Section 5.

2   Related Work

In this section, we give a brief overview of prior work on probabilistic collision detection and trajectory planning in dynamic environments.

2.1   Probabilistic Collision Detection

Collision detection is an integral part of any motion planning algorithm and most prior techniques assume an exact representation of the robot and obstacles. Given some uncertainty or imperfect representation of the obstacles, certain algorithms perform probabilistic collision detection. Typically, these uncertainties are modeled using Gaussian distributions, and stochastic techniques are used to approximate the collision probability [4, 9, 12]. In stochastic algorithms, a large number of sample evaluations is required to compute an accurate collision probability.

---

[1] https://developer.microsoft.com/en-us/windows/kinect

If it can be assumed that the sizes of the objects are small, the collision probability between objects can be approximated using the probability at a single configuration and corresponds to the mean of the probability distribution function (PDF) [6]. This approximation is fast, but the computed probability cannot provide a bound; i.e. it can be higher or lower than the actual collision probability, and the error increases as the object becomes larger.

For high-dimensional spaces, a common approach to check collisions for imperfect or noisy objects is to perform exact collision checking with a large volume that encloses the object poses [3, 19]. Prior approaches generally enlarge an object shape, which may correspond to a robot or an obstacle, to compute the space occupied by the object for a given standard deviation. This may correspond to a sphere [5] or a sigma hull [13]. These approaches tend to compute a bounding volume for the given confidence level. However, the computed volume overestimates the probability and can be much bigger than the actual volume corresponding to the confidence level. Therefore, these approaches can result in a failure to find existing feasible trajectories for motion planning.

Other approaches have been proposed to perform probabilistic collision detection on point cloud data. Bae et al. [1] presented a closed-form expression for the positional uncertainty of point clouds. Pan et al. [16] reformulated the probabilistic collision detection problem as a classification problem and computed per point collision probability. However, these approaches assume that the environment is mostly static. Other techniques are based on broad phase data structures that handle large point clouds for realtime collision detection [17].

2.2   Planning in Dynamic and Uncertain Environments

There is considerable literature on motion planning in dynamic scenes. In many scenarios, the future positions of the obstacles are not known. As a result, the planning problem is typically solved using replanning algorithms, which interleave planning with execution. These methods include sampling-based planners [22], grid searches [15], and trajectory optimization [19].

Applications that require high responsiveness use control-based approaches [11], which can compute trajectories in realtime. They compute the robot trajectory in the workspace of the robot, according to the sensor data. However, the mapping from the Cartesian trajectory to the trajectory in the configuration space of high-DOF robots can be problematic as there can be multiple configurations for a single pose defined in the Cartesian workspace. Furthermore, control-based approaches tend to compute less optimal robot trajectories as compared to the planning approaches that incorporate the estimation of the future obstacle poses. Planning algorithms can compute better robot trajectories in applications in which a good prediction about obstacle motions in a short horizon can be obtained.

The unknown future obstacle positions are one of the source of uncertainties. POMDPs (partially-observable Markov decision processes) provide a mathematically rigorous and general approach for planning under uncertainty [10]. They handle the uncertainty by reasoning over the *belief* space. However, the POMDP formulation is regarded as computationally intractable [18] for problems that are high-dimensional or have a large number of actions. Many efficient approximations use Gaussian belief spaces, which are estimated using Bayesian filters (e.g., Kalman filters) [14, 23]. Gaussian belief spaces have also been used for the motion planning of high-DOF robots [3, 24], but most planning algorithms do not account for environment uncertainty or imperfect obstacle information. Under the conditions of dynamic environments, planning with uncertainty algorithms are mainly limited to 2D spaces [2, 7].

3   Probabilistic Collision Detection for High-DOF Robots

In this section, we first introduce the notation and terminology used in the paper; then we present our probabilistic collision detection algorithm for detecting collisions between a high-DOF robot and the dynamic environment.

3.1   Notation and Assumptions

Our goal is to compute the collision probability between a high-DOF robot configuration and a given obstacle representation of dynamic environments, where the obstacle representation is a probability distribution that accounts for uncertainties in the obstacle motion.

For an articulated robot with D one-dimensional joints, we represent a single robot configuration as $\mathbf{q}$, which is a vector composed of the joint values. The D-dimensional vector space of $\mathbf{q}$ is the configuration space $\mathcal{C}$ of the robot. We denote the collision-free subset of $\mathcal{C}$ as $\mathcal{C}_{free}$, and the other configurations corresponding to collisions as $\mathcal{C}_{obs}$.

We assume that the robot consists of J links $R_1, ..., R_J$, where $J \leq D$. Furthermore, for each robot link R_j, we use a sequence of bounding volumes $B_{j1}, ..., B_{jK}$ to tightly enclose $R_j(\mathbf{q})$, which corresponds to a robot configuration $\mathbf{q}$, i.e.,

$$\forall j : R_j(\mathbf{q}) \subset \bigcup_{k=1}^{K} B_{jk}(\mathbf{q}) \text{ for } (1 \leq j \leq J). \tag{1}$$

We denote obstacles in the environment as O_l ($1 \leq l \leq L$). The configuration of these obstacles is specified based on their poses. As is the case for the robot, we use the bounding volumes $S_{l1}, ..., S_{lM}$ to enclose each obstacle O_l in the environment:

$$\forall l : O_l \subset \bigcup_{m=1}^{M} S_{lm} \text{ for } (1 \leq l \leq L). \tag{2}$$

For dynamic obstacles, we assume the predicted pose of a bounding volume S_{lm} at time t is estimated as a Gaussian distribution $\mathcal{N}(\mathbf{p}_{lm}, \mathbf{\Sigma}_{lm})$.

3.2   Fast and Bounded Collision Probability Approximation

The collision probability between a robot configuration $\mathbf{q}_i$ with the environment at time t_i, $P(\mathbf{q}_i \in \mathcal{C}_{obs}(t_i))$ can be evaluated by checking their bounding volumes for possible overlaps, which can be formulated as

$$P(\mathbf{q}_i \in \mathcal{C}_{obs}(t_i)) = P\left(\left(\bigcup_j \bigcup_k B_{jk}(\mathbf{q}_i)\right) \cap \left(\bigcup_l \bigcup_m S_{lm}(t_i)\right) \neq \emptyset\right). \quad (3)$$

We assume the robot links R_j and obstacles O_l are independent of each other, as their poses depend on corresponding joint values or obstacle states. Then (3) can be computed as

$$P(\mathbf{q}_i \in \mathcal{C}_{obs}(t_i)) = 1 - \prod_j \prod_l \overline{P_{col}(i,j,l)}, \quad (4)$$

where $P_{col}(i,j,l)$ is the collision probability between $R_j(\mathbf{q}_i)$ and $O_l(t_i)$. Because poses of bounding volumes B_{jk} and S_{lm} are determined by joint values or obstacle states of the corresponding robot link or obstacle, bounding volumes for the same object are dependent on each other, and $P_{col}(i,j,l)$ can be approximated as

$$P_{col}(i,j,l) \approx \max_{k,m} P_{col}(i,j,k,l,m) \quad (5)$$

$$P_{col}(i,j,k,l,m) = P(B_{jk}(\mathbf{q}_i) \cap S_{lm}(t_i) \neq \emptyset), \quad (6)$$

where $P_{col}(i,j,k,l,m)$ denotes the collision probability between $B_{jk}(\mathbf{q}_i)$ and $S_{lm}(t_i)$.

Fig. 1 illustrates how $P_{col}(i,j,k,l,m)$ can be computed for $S_{lm}(t_i) \sim \mathcal{N}(\mathbf{p}_{lm}, \mathbf{\Sigma}_{lm})$. We assume that the robot's bounding volume $B_{jk}(\mathbf{q}_i)$ is a sphere centered at $\mathbf{o}_{jk}(t_i)$, similar to the environment bounding volume S_{lm}, and denote the radii of B_{jk} and S_{lm} as r_1 and r_2, respectively. We assume radii r_1 and r_2 are constants. Then the exact probability of collision between them is given as:

$$P_{col}(i,j,k,l,m) = \int_{\mathbf{x}} I(\mathbf{x}, \mathbf{o}_{jk}(t_i)) p(\mathbf{x}, \mathbf{p}_{lm}, \mathbf{\Sigma}_{lm}) d\mathbf{x}, \quad (7)$$

where the indicator function $I(\mathbf{x}, \mathbf{o})$ and the obstacle function $p(\mathbf{x}, \mathbf{p}, \mathbf{\Sigma})$ are defined as,

$$I(\mathbf{x}, \mathbf{o}) = \begin{cases} 1 \text{ if } \|\mathbf{x} - \mathbf{o}\| \leq (r_1 + r_2) \\ 0 \qquad \text{otherwise} \end{cases} \text{ and} \quad (8)$$

$$p(\mathbf{x}, \mathbf{p}, \mathbf{\Sigma}) = \frac{e^{-0.5(\mathbf{x}-\mathbf{p})^T \mathbf{\Sigma}^{-1}(\mathbf{x}-\mathbf{p})}}{\sqrt{(2\pi)^3 \|\mathbf{\Sigma}\|}}, \quad (9)$$

respectively. It is known that there is no closed form solution for the integral given in (7).

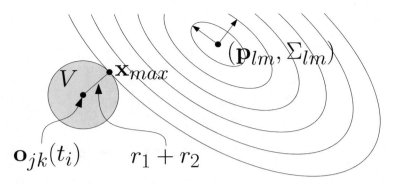

Fig. 1: Approximation of probabilistic collision detection between a sphere obstacle of radius r_2 with a probability distribution $\mathcal{N}(\mathbf{p}_{lm}, \mathbf{\Sigma}_{lm})$ and a rigid sphere robot bounding volume $B_{jk}(\mathbf{q}_i)$ centered at $\mathbf{o}_{jk}(t_i)$ with radius r_1 for a robot configuration $\mathbf{q}_i$. The collision probability is approximated as $V \cdot \mathbf{x}_{max}$, where V is the volume of the sphere with the radius computed as the sum of two radii, $V = \frac{4\pi}{3}(r_1 + r_2)^3$, and $\mathbf{x}_{max}$ is the position which has the maximum probability of $\mathcal{N}(\mathbf{p}_{lm}, \mathbf{\Sigma}_{lm})$.

Du Toit and Burdick approximate (7) as $V \cdot p(\mathbf{o}_{jk}(t_i), \mathbf{p}_{lm}, \mathbf{\Sigma}_{lm})$, where V is the volume of the sphere, i.e., $V = \frac{4\pi}{3}(r_1 + r_2)^3$ [6]. However, this approximated probability can be either smaller or larger than the exact probability. If the covariance $\mathbf{\Sigma}_{lm}$ is small, the approximated probability can be much smaller than the exact probability. Planners using this approximation may underestimate the collision probability and may compute unsafe robot motion.

In order to avoid this problem, we compute $\mathbf{x}_{max}$, the position that has the maximum probability of $\mathcal{N}(\mathbf{p}_{lm}, \mathbf{\Sigma}_{lm})$ in $\mathbf{B}_{jk}(\mathbf{q}_i)$, and compute the upper bound of $P_{col}(i, j, k, l, m)$ as

$$P_{approx}(i, j, k, l, m) = V \cdot p(\mathbf{x}_{max}, \mathbf{p}_{lm}, \mathbf{\Sigma}_{lm}). \tag{10}$$

Although $\mathbf{x}_{max}$ has no closed-form solution, it can be computed efficiently using numerical techniques.

Lemma 1. $\mathbf{x}_{max}$, the position has the maximum probability of $\mathcal{N}(\mathbf{p}_{lm}, \mathbf{\Sigma}_{lm})$ in $\mathbf{B}_{jk}(\mathbf{q}_i)$, is formulated as a one-dimensional search of a parameter λ,

$$\mathbf{x}_{max} = \{\mathbf{x} | \|\mathbf{x} - \mathbf{o}_{jk}(t_i)\| = (r_1 + r_2) \, and \, \mathbf{x} \in \mathbf{x}(\lambda)\}, \, where \tag{11}$$

$$\mathbf{x}(\lambda) = (\mathbf{\Sigma}_{lm}^{-1} + \lambda \mathbf{I})^{-1}(\mathbf{\Sigma}_{lm}^{-1}\mathbf{p}_{lm} + \lambda \mathbf{o}_{jk}(t_i)). \tag{12}$$

Proof. The problem of finding the position with the maximum probability in a convex region can be formulated as an optimization problem with a Lagrange multiplier λ [8],

$$\mathbf{x}_{max} = \arg \min_{\mathbf{x}} \left\{ (\mathbf{x} - \mathbf{p}_{lm})^T \mathbf{\Sigma}_{lm}^{-1}(\mathbf{x} - \mathbf{p}_{lm}) + \lambda(\mathbf{x} - \mathbf{o}_{jk})^2 \right\}. \tag{13}$$

The solution of (13) satisfies

$$\nabla \left\{ (\mathbf{x} - \mathbf{p}_{lm})^T \Sigma_{lm}^{-1} (\mathbf{x} - \mathbf{p}_{lm}) + \lambda (\mathbf{x} - \mathbf{o}_{jk})^2 \right\} = 0, \qquad (14)$$

and can be computed as

$$2\Sigma_{lm}^{-1}(\mathbf{x} - \mathbf{p}_{lm}) + 2\lambda(\mathbf{x} - \mathbf{o}_{jk}) = 0 \qquad (15)$$

$$\mathbf{x} = (\Sigma_{lm}^{-1} + \lambda \mathbf{I})^{-1})(\Sigma_{lm}^{-1}\mathbf{p}_{lm} + \lambda \mathbf{o}_{jk}). \qquad (16)$$

The approximated probability (10) is guaranteed as an upper bound of the exact collision probability (7).

Theorem 1. *The approximated probability $P_{approx}(i, j, k, l, m)$ (10) is always greater than or equal to the exact collision probability $P_{col}(i, j, k, l, m)$ (7).*

Proof. $p(\mathbf{x}_{max}, \mathbf{p}_{lm}, \Sigma_{lm}) \geq p(\mathbf{x}, \mathbf{p}_{lm}, \Sigma_{lm})$ for $\{\mathbf{x} | \|\mathbf{x} - \mathbf{o}_{jk}(t_i)\| \leq (r_1 + r_2)\}$ from Lemma 1. Therefore,

$$P_{approx}(i, j, k, l, m) = V \cdot p(\mathbf{x}_{max}, \mathbf{p}_{lm}, \Sigma_{lm}) \qquad (17)$$

$$= \int_{\mathbf{x}} I(\mathbf{x}, \mathbf{o}_{jk}(t_i))d\mathbf{x} \cdot p(\mathbf{x}_{max}, \mathbf{p}_{lm}, \Sigma_{lm}) \qquad (18)$$

$$= \int_{\mathbf{x}} I(\mathbf{x}, \mathbf{o}_{jk}(t_i)) \cdot p(\mathbf{x}_{max}, \mathbf{p}_{lm}, \Sigma_{lm})d\mathbf{x} \qquad (19)$$

$$\geq \int_{\mathbf{x}} I(\mathbf{x}, \mathbf{o}_{jk}(t_i)) \cdot p(\mathbf{x}, \mathbf{p}_{lm}, \Sigma_{lm})d\mathbf{x} \qquad (20)$$

$$= P_{col}(i, j, k, l, m). \qquad (21)$$

3.3   Comparisons with Other Algorithms

In Fig. 2, we illustrate two cases of the collision probability computation between a circle B (in gray), and a point (in black) with uncertainties, $\mathbf{x} \sim (\mathbf{p}, \Sigma)$, in 2D. We evaluate the exact collision probabilities using the numerical integration of the PDF. The collision probability of Case I is 0.09%, which is feasible with $\delta_{CL} = 0.99$, while the probability of Case II is 1.72%, which is infeasible. The contours in Fig. 2 represent the bounds for different confidence levels. Approaches that use the exact collision checking with enlarged bounding volumes [3, 19] for a given confidence level (e.g., the blue ellipse for $\delta_{CL} = 0.99$) determine both Case I and Case II have collisions and infeasible, i.e., the collision probability is 100%, while the collision probability for Case I is only 0.09%.

Du Toit and Burdick [6] use the probability of the center point (shown in green in Fig. 2) to approximate the collision probability, as described in Section 3.2. However, their approach cannot guarantee upper bounds, and the approximated probability can be significantly smaller than the exact probability if the covariance value is small. Case II in Fig. 2 shows that the approximated probability of their approach is 0.89%, and that satisfies the safety with $\delta_{CL} = 0.99$ and determines Case II as a feasible configuration, which is not true for the exact probability 1.72%.

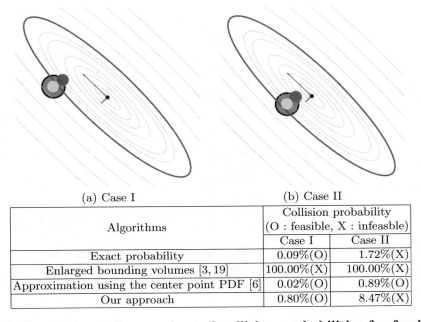

| | Collision probability (O : feasible, X : infeasble) ||
Algorithms	Case I	Case II
Exact probability	0.09%(O)	1.72%(X)
Enlarged bounding volumes [3, 19]	100.00%(X)	100.00%(X)
Approximation using the center point PDF [6]	0.02%(O)	0.89%(O)
Our approach	0.80%(O)	8.47%(X)

Fig. 2: **Comparison of approximated collision probabilities for feasible** $(P(\mathbf{x}) \leq 1 - \delta_{CL})$ **and infeasible** $(P(\mathbf{x}) > 1 - \delta_{CL})$ **scenarios for** $\delta_{CL} = 0.99$: We compare the exact collision probability (computed using numerical integration) with approximated probabilities of 1) enlarged bounding volumes (blue contour) [3, 19], 2) approximation using object center point (in green) [6], and 3) our approach that uses the maximum probability point (in red). Our approach guarantees that we do not underestimate the probability, while our approximated probability is close to the exact probability.

Unlike [6], we approximate the probability of the entire volume using the maximum probability value of a single point (shown in red in Fig. 2), as described in Section 3.2. Our approach guarantees computation of the upper bound of collision probability, while the approximated probability is closer to the exact probability than of the enlarged bounding volume approaches.

4 Trajectory Optimization using Probabilistic Collision Detection

In this section, we present our trajectory optimization approach which uses the probabilistic collision detection to avoid collision in the dynamic environment.

We define the time-space domain $\mathcal{X}$, which adds a time dimension to the configuration space, i.e., $\mathcal{X} = \mathcal{C} \times T$. The robot's trajectory, $\mathbf{q}(t)$, is represented as a function of time from the start configuration $\mathbf{q}_s$ to the goal configuration

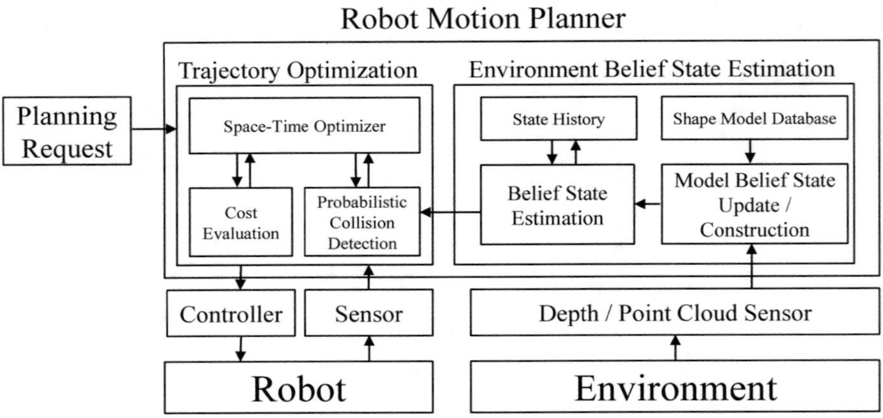

Robot Motion Planner

Fig. 3: **Trajectory Planning:** We highlight various components of our algorithm. These include belief space estimation of environment (described in [20]), probabilistic collision checking (described in Section 3), and trajectory optimization.

$\mathbf{q}_g$. It is represented using the matrix $\mathbf{Q}$,

$$\mathbf{Q} = \begin{bmatrix} \mathbf{q}_s \; \mathbf{q}_1 \; \cdots \; \mathbf{q}_{n-1} \; \mathbf{q}_g \\ t_0 \; t_1 \; \cdots \; t_{n-1} \; t_n \end{bmatrix}, \tag{22}$$

which corresponds to $n + 1$ configurations at discretized keyframes, $t_i = i\Delta_T$, which have a fixed interval Δ_T. We denote the i-th column of $\mathbf{Q}$ as $\mathbf{x}_i = \begin{bmatrix} \mathbf{q}_i^T \; t_i \end{bmatrix}^T$.

Fig. 3 highlights various components of our planning algorithm. The pseudocode description is given in Algorithm 1 for a single planning step. Our overall trajectory planning algorithm consists of two main components: environment belief state estimation and trajectory optimization.

We update the belief state of the environment $\mathbf{b} = (\mathbf{p}, \Sigma)$ using means and covariances $\mathbf{p}_{lm}$ and Σ_{lm} of the poses of the existing bounding volumes S_{lm}. That is, $\mathbf{p} = \begin{bmatrix} \mathbf{p}_{11}^T \; \cdots \; \mathbf{p}_{LM}^T \end{bmatrix}^T$ and $\Sigma = \mathrm{diag}(\Sigma_{11}, ..., \Sigma_{LM})$, where Σ is a block diagonal matrix of the covariances. Moreover, we use a Bayesian estimator to predict the future belief state of the environment, which is used for probabilistic collision detection.

We use incremental trajectory optimization, which repeatedly refines a motion trajectory using an optimization formulation [19]. The planner initializes the robot trajectory $\mathbf{Q}$ as a smooth trajectory of predefined length T between $\mathbf{q}_s$ and $\mathbf{q}_g$, and refines it in every planning step ΔT.

We define the collision avoidance constraint based on the following probability computation formulation:

$$\forall \mathbf{x}_i : P(\mathbf{q}_i \in \mathcal{C}_{obs}(t_i)) < 1 - \delta_{CL}. \tag{23}$$

Algorithm 1 $\mathbf{Q}^*$ =PlanWithEnvUncertainty($\mathbf{Q}$, $\{\mathbf{d}_k\}$, t_i)
: Compute the optimal robot trajectory $\mathbf{Q}^*$ during the planning step ΔT for the
environment point clouds $\{\mathbf{d}\}$ at time t_i

Input: initial trajectory $\mathbf{Q}$, environment point clouds $\{\mathbf{d}\}$, time t_i
Output: Optimal robot trajectory $\mathbf{Q}^*$ for time step ΔT
 1: $\mathbf{p}_i$ = EnvironmentStateComputation($\{\mathbf{d}\}$) // *compute the environment state*
 2: **for** $k \in \{i, ..., i + \Delta T\}$ **do**
 3:     $\mathbf{B}_k$ = BeliefStateEstimation($\mathbf{B}_0, ..., \mathbf{B}_{k-1}$, $\mathbf{p}_i$) //*estimate the current and future belief states*
 4: **end for**
 5: **while** elapsed time $< \Delta T$ **do**
 6:     P=ProbCollisionChecking($\mathbf{Q}$, $\{\mathbf{B}_i, ..., \mathbf{B}_{i+\Delta T}\}$) // *perform probabilistic collision detection*
 7:     $\mathbf{Q}^*$=Optimize($\mathbf{Q}, P$) // *compute the optimal trajectory for high-DOF robot*
 8: **end while**

We can compute $P(\mathbf{q}_i \in \mathcal{C}_{obs}(t_i))$ using (4) in Section 3. The computed trajectories that satisfy (23) guarantee that the probability of collision with the obstacles is bounded by the confidence level δ_{CL}, i.e. the probability that a computed trajectory has no collision is higher than δ_{CL}. Use of a higher confidence level computes safer, but more conservative trajectories. The use of a lower confidence level increases the success rate of planning, but also increases the probability of collision.

The objective function for trajectory optimization at time t_k can be expressed as the sum of trajectory smoothness cost, and collision constraint costs for dynamic uncertain obstacles and static known obstacles,

$$f(\mathbf{Q}) = \min_Q \sum_{i=k+m}^{n} \left(\|\mathbf{q}_{i-1} - 2\mathbf{q}_i + \mathbf{q}_{i+1}\|^2 + C_{static}(\mathbf{Q}_i) \right)$$
$$+ \sum_{i=k+m}^{k+2m} \max(P(\mathbf{q}_i \in \mathcal{C}_{obs}(\mathbf{x}_i)) - (1 - \delta_{CL}), 0), \tag{24}$$

where m is the number of time steps in a planning time step ΔT.

Unlike the previous optimization-based planning approaches [19, 25] which maintain and cannot change the predefined trajectory duration for the computed trajectory, we adjust the duration of trajectory T to avoid collisions with the dynamic obstacles. When the trajectory planning starts from $\mathbf{t}_i$ ($\mathbf{t}_i$ can be different from $\mathbf{t}_s$ due to replanning) and if the computed trajectory $\mathbf{Q}$ violates the collision probability constraint (23) at time t_j, i.e., $P(\mathbf{q}_j \in \mathcal{C}_{obs}(t_j)) \geq \delta_{CL}$, we repeatedly add a new time step $\mathbf{x}_{new}$ before $\mathbf{x}_j$ and rescale the trajectory from $[\mathbf{t}_i, ..., \mathbf{t}_{j-1}]$ to $[\mathbf{t}_i, ..., \mathbf{t}_{j-1}, \mathbf{t}_{new}]$, until $\mathbf{x}_{new}$ is collision-free. Then, the next planning step starts from $\mathbf{x}_{new}$. It allows the planner to slow the robot down when it cannot find a safe trajectory for the previous trajectory duration due to the dynamic obstacles. If the optimization algorithm converges, our algorithm computes the optimal trajectory,

$$\mathbf{Q}^* = \arg\min_{\mathbf{Q}} f(\mathbf{Q}), \tag{25}$$

which provides a collision-free guarantee for the given confidence level δ_{CL} in dynamic environments. Further details of the integration of probabilistic collision detection with trajectory optimization can be found in [20].

5   Results

In this section, we describe our implementation and highlight the performance of our probabilistic collision checking and trajectory planning algorithm on different benchmark scenarios. We measure the performance of our planning algorithm in simulated environments with difference benchmark scenarios and robot arm models, and validate our algorithm using experiments with a real 7-DOF Fetch robot arm. In our experiments, bounding spheres are automatically generated along the medial axis of each robot link. The environments have some complex static obstacles such as tools or furnitures in a room. The dynamic obstacle is a human, and we assume that the robot operates in close proximity to the human; however, the human does not intend to interact with the robot. We use a Kinect device as the depth sensor, which can represent a human as 30-35k point clouds. We compute the state of human obstacle model which has 60 DOFs. Details of the belief state estimation of dynamic obstacles is given in [20].

5.1   Experimental Results

Robot	Robot BV	Human BV	Prob. Col BV Pairs	Prob. Col Computation Time (ms)
IIWA	40	336	13440 (40x336)	0.147
UR5	56	336	18816 (56x336)	0.282
Fetch	76	336	25536 (76x336)	0.526

Table 1: **Performance of our probabilistic collision detection:** We measure the computation time of the probabilistic collision detection per single robot configuration.

Table 1 shows the computation time of the probabilistic collision detection per single robot configuration. We evaluate (10) in Section 3 for each bounding volume pair correspond to a robot and a human obstacle, and the computation time is linear to the number of pairs.

Table 2 describes the benchmark scenarios and the performance of the planning results for simulated environments. We set $\delta_{CL} = 0.95$, except the second benchmark scenarios where the confidence levels vary.

Benchmarks		Scenarios	Planning Results		
Name	Robot		Minimum Distance (m)	Trajectory Duration (sec)	Trajectory Length (m)
Bookshelf	UR5 (6 DOFs)	Stationary obstacle	0.29	3.7	1.29
		Moving obstacle	0.35	5.4	2.14
Tool	IIWA (7 DOFs)	$\delta_{CL} = 0.95$, $\mathbf{v}_t = \mathbf{0}$	0.06	6.0	1.60
		$\delta_{CL} = 0.95$, $\mathbf{v}_t = 0.005\mathbf{I}_{3\times3}$	0.30	6.9	1.92
		$\delta_{CL} = 0.95$, $\mathbf{v}_t = 0.05\mathbf{I}_{3\times3}$	0.32	7.1	2.01
		$\delta_{CL} = 0.99$, $\mathbf{v}_t = 0.05\mathbf{I}_{3\times3}$	0.38	8.3	2.43
Comparisons using Different Prob. Collision Computations	IIWA (7 DOFs)	Our Approach	0.32	7.1	2.01
		Enlarged bounding volumes [3, 19]	0.40	8.8	2.32
		Approximation using the center point PDF [6]	-0.05	3.4	1.38

Table 2: **Planning results in our benchmarks:** We measure the planning results of the computed trajectories: the minimum distance to the human obstacle, trajectory duration, and trajectory length, for different benchmark scenarios.

In our first benchmark, the planner computes a motion for 6-DOF UR5 robot to move an object on the table to a point on the bookshelf. When a human is dashing toward the robot at a fast speed, the robot is aware of the potential collision with the predicted future human position and changes its trajectory (Fig. 4(a)). However, if a standing human only stretches out an arm toward the robot, even if the velocity of the arm is fast, the model-based prediction prevents unnecessary reactive motions, which is different from the prediction models with constant velocity or acceleration extrapolations (Fig. 4(b)).

The second benchmark shows the difference in planning results due to the different confidence and noise levels, for the same recorded human motion. Fig. 4(c)-(e) shows a robot trajectory with different confidence levels and sensor noises. If the obstacle states are assumed as exact and have no noise, the robot can follow the shortest and smoothest trajectory that is close to the obstacle (Fig. 4(c)). However, as the noise of the environment state or expected confidence level becomes higher, the computed robot trajectories become longer and less smooth to avoid potential collision with the obstacles (Fig. 4(d)-(e)).

Fig. 5 shows a 7-DOF Fetch robot arm motion which is computed using our algorithm to avoid collisions with human motion captured in run-time.

5.2   Probabilistic Collision Checking and Trajectory Planning

In the next benchmark, we plan trajectories using the different probabilistic collision detection algorithms which discussed in Section 3.3. We measure the minimum distance between the robot and the human obstacle along the computed trajectory as a safety metric, and the duration and length of the end-effector trajectory as efficiency metrics. The results for the planners with three different probabilistic collision detection algorithms are shown in Table 2. The enlarged bounding volumes have the largest safety margins, but the durations and lengths

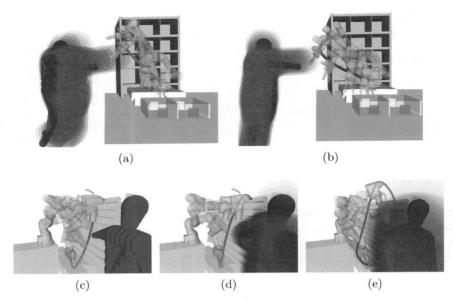

(a)   (b)

(c)   (d)   (e)

Fig. 4: **Robot Trajectory with Dynamic Human Obstacles:** Static obstacles are shown in green, the estimated current and future human bounding volumes are shown in blue and red, respectively. Our planner uses the probabilistic collision detection to compute the collision probability between the robot and the uncertain future human motion. (a) When a human is approaching the robot, our planner changes its trajectory to avoid potential future collisions. (b) When a standing human only stretches out an arm, our model-based prediction prevents unnecessary reactive motions, which results in a better robot trajectory than the prediction using simple extrapolations. (c)-(e) Robot trajectory with different confidence and noise levels: (c) A trajectory for zero-noise obstacles. (d) $\delta_{CL} = 0.95$ and $\mathbf{v}_t = 0.005 I_{3 \times 3}$. (e) $\delta_{CL} = 0.99$ and $\mathbf{v}_t = 0.05 I_{3 \times 3}$.

of the computed trajectories are longer than other approaches, since the overestimated collision probability makes the planner compute trajectories that are unnecessarily far from the obstacles. On the other hand, the approximating approach that uses the probability of the object center point underestimates the collision probability and causes several collisions in the planned trajectories, i.e., the minimum distance between the robot and human obstacle become negative. Our approach shows a similar level of safety with the approach using enlarged bounding volumes, while it also computes efficient trajectories that have shorter trajectory durations and lengths. These benchmarks demonstrate the benefits of our probabilistic collision checking on trajectory planning.

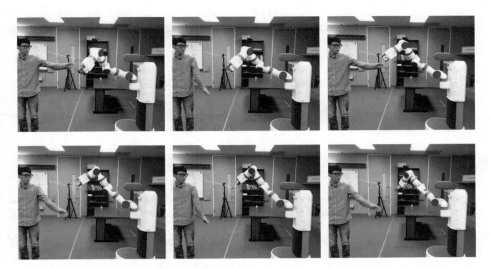

Fig. 5: **Real Robot Experiment:** We evaluate our planning algorithm on 7-DOF Fetch robot arm to compute collision-free robot motion. Our bounded probabilistic collision checking is used for computing safe trajectories.

6   Conclusions, Limitations and Future Work

We present a novel algorithm for collision probability approximation for high-DOF robots in dynamic, uncertain environments. Our approach is fast, and works well in our simulated and real robot results where it can compute efficient collision-free paths with a high confidence level. Our probabilistic collision detection computes tighter upper bounds of the collision probability as compared to prior approaches, and can be used with different planning algorithms. We highlight the performance of our planner on different benchmarks with human obstacles.

Our approach has some limitations. Some of the assumptions used in belief space estimation in terms of Gaussian distribution and Kalman filter may not hold. Moreover, our approach needs pre-defined shape representations of the obstacles. The trajectory optimization may get stuck at a local minima and may not converge to a global optimal solution. Furthermore, our approach assumes that the obstacles in the scene undergo rigid motion. There are many avenues for future work. Our approach only takes into account the imperfect information about the moving obstacles. Uncertainties from control errors or sensor errors, which are rather common with the controllers and sensors, need to be integrated in our approach. Finally, we would to integrate our approach with other robots and evaluate the performance in different scenarios.

Recently, we have extended our approach to general convex polytopes and proposed specialized algorithms for bounding shapes such as AABB, OBB and k-

DOPs [21]. Furthermore, we show that by using bounding volume hierarchies, we can improve the speed and the accuracy of the collision probability computation for non-convex cases.

7   Acknowledgments

This research is supported in part by ARO Contract W911NF-14-1-0437 and NSF award 1305286.

References

1. Bae, K.H., Belton, D., Lichti, D.D.: A closed-form expression of the positional uncertainty for 3d point clouds. Pattern Analysis and Machine Intelligence, IEEE Transactions on 31(4), 577–590 (2009)
2. Bai, H., Cai, S., Ye, N., Hsu, D., Lee, W.S.: Intention-aware online pomdp planning for autonomous driving in a crowd. In: Robotics and Automation (ICRA), 2015 IEEE International Conference on. pp. 454–460. IEEE (2015)
3. Van den Berg, J., Wilkie, D., Guy, S.J., Niethammer, M., Manocha, D.: LQG-Obstacles: Feedback control with collision avoidance for mobile robots with motion and sensing uncertainty. In: Robotics and Automation (ICRA), 2012 IEEE International Conference on. pp. 346–353. IEEE (2012)
4. Blackmore, L.: A probabilistic particle control approach to optimal, robust predictive control. In: Proceedings of the AIAA Guidance, Navigation and Control Conference. No. 10 (2006)
5. Bry, A., Roy, N.: Rapidly-exploring random belief trees for motion planning under uncertainty. In: Robotics and Automation (ICRA), 2011 IEEE International Conference on. pp. 723–730. IEEE (2011)
6. Du Toit, N.E., Burdick, J.W.: Probabilistic collision checking with chance constraints. Robotics, IEEE Transactions on 27(4), 809–815 (2011)
7. Du Toit, N.E., Burdick, J.W.: Robot motion planning in dynamic, uncertain environments. Robotics, IEEE Transactions on 28(1), 101–115 (2012)
8. Groetsch, C.W.: The theory of Tikhonov regularization for Fredholm equations of the first kind, vol. 105. Pitman Advanced Publishing Program (1984)
9. Guibas, L.J., Hsu, D., Kurniawati, H., Rehman, E.: Bounded uncertainty roadmaps for path planning. In: Algorithmic Foundation of Robotics VIII, pp. 199–215. Springer (2010)
10. Kaelbling, L.P., Littman, M.L., Cassandra, A.R.: Planning and acting in partially observable stochastic domains. Artificial intelligence 101(1), 99–134 (1998)
11. Kroger, T., Wahl, F.M.: Online trajectory generation: Basic concepts for instantaneous reactions to unforeseen events. Robotics, IEEE Transactions on 26(1), 94–111 (2010)
12. Lambert, A., Gruyer, D., Pierre, G.S.: A fast monte carlo algorithm for collision probability estimation. In: Control, Automation, Robotics and Vision, 2008. ICARCV 2008. 10th International Conference on. pp. 406–411. IEEE (2008)
13. Lee, A., Duan, Y., Patil, S., Schulman, J., McCarthy, Z., van den Berg, J., Goldberg, K., Abbeel, P.: Sigma hulls for gaussian belief space planning for imprecise articulated robots amid obstacles. In: Intelligent Robots and Systems (IROS), 2013 IEEE/RSJ International Conference on. pp. 5660–5667. IEEE (2013)

14. Leung, C., Huang, S., Kwok, N., Dissanayake, G.: Planning under uncertainty using model predictive control for information gathering. Robotics and Autonomous Systems 54(11), 898–910 (2006)
15. Likhachev, M., Ferguson, D., Gordon, G., Stentz, A., Thrun, S.: Anytime dynamic A*: An anytime, replanning algorithm. In: Proceedings of the International Conference on Automated Planning and Scheduling (2005)
16. Pan, J., Chitta, S., Manocha, D.: Probabilistic collision detection between noisy point clouds using robust classification. In: International Symposium on Robotics Research (ISRR) (2011)
17. Pan, J., Şucan, I.A., Chitta, S., Manocha, D.: Real-time collision detection and distance computation on point cloud sensor data. In: Robotics and Automation (ICRA), 2013 IEEE International Conference on. pp. 3593–3599. IEEE (2013)
18. Papadimitriou, C.H., Tsitsiklis, J.N.: The complexity of markov decision processes. Mathematics of operations research 12(3), 441–450 (1987)
19. Park, C., Pan, J., Manocha, D.: ITOMP: Incremental trajectory optimization for real-time replanning in dynamic environments. In: Proceedings of International Conference on Automated Planning and Scheduling (2012)
20. Park, C., Park, J.S., Manocha, D.: Fast and bounded probabilistic collision detection in dynamic environments for high-dof trajectory planning. CoRR abs/1607.04788 (2016), http://arxiv.org/abs/1607.04788
21. Park, J.S., Park, C., Manocha, D.: Efficient probabilistic collision detection for nonconvex shapes. CoRR abs/1610.03651 (2016), http://arxiv.org/abs/1610.03651
22. Petti, S., Fraichard, T.: Safe motion planning in dynamic environments. In: Proceedings of IEEE/RSJ International Conference on Intelligent Robots and Systems. pp. 2210–2215 (2005)
23. Platt Jr, R., Tedrake, R., Kaelbling, L., Lozano-Perez, T.: Belief space planning assuming maximum likelihood observations. In: Proceedings of Robotics: Science and Systems (2010)
24. Sun, W., van den Berg, J., Alterovitz, R.: Stochastic extended lqr: Optimization-based motion planning under uncertainty. In: Algorithmic Foundations of Robotics XI, pp. 609–626. Springer (2015)
25. Zucker, M., Ratliff, N., Dragan, A.D., Pivtoraiko, M., Klingensmith, M., Dellin, C.M., Bagnell, J.A., Srinivasa, S.S.: CHOMP: Covariant hamiltonian optimization for motion planning. International Journal of Robotics Research (2012)

Efficient Nearest-Neighbor Search for Dynamical Systems with Nonholonomic Constraints

Valerio Varricchio, Brian Paden, Dmitry Yershov, and Emilio Frazzoli

Massachusetts Institute of Technology

Abstract. Nearest-neighbor search dominates the asymptotic complexity of sampling-based motion planning algorithms and is often addressed with k-d tree data structures. While it is generally believed that the expected complexity of nearest-neighbor queries is $O(\log(N))$ in the size of the tree, this paper reveals that when a classic k-d tree approach is used with sub-Riemannian metrics, the expected query complexity is in fact $\Theta(N^p \log(N))$ for a number $p \in [0, 1)$ determined by the degree of nonholonomy of the system. These metrics arise naturally in nonholonomic mechanical systems, including classic wheeled robot models. To address this negative result, we propose novel k-d tree build and query strategies tailored to sub-Riemannian metrics and demonstrate significant improvements in the running time of nearest-neighbor search queries.

1 Introduction

Sampling-Based algorithms such as Probabilistic Roadmaps (PRM) [9], Rapidly exploring Random Trees (RRT) [11] and their asymptotically optimal variants (PRM*, RRT*) [7] are widely used in motion planning. These algorithms build a random graph of motions between points on the robot's configuration manifold.

During the graph expansion, nearest-neighbor search is used to limit the computation to regions of the graph close to the new configurations and it is shown to dominate the asymptotic complexity of randomized planners. The notion of closeness appropriate for motion planning is induced by the length of the shortest paths between configurations, or in general, by the minimum cost of controlling a system between states.

As compared to exhaustive linear search, efficient algorithms with reduced complexity have been studied in computational geometry and their use in motion planning has been highlighted as a factor of dramatic performance improvement [12]. Among a variety of approaches, k-d trees [1] are ideal due to their remarkable efficiency in low-dimensional spaces, typical of motion planning problems.

Classic k-d trees are shown to have logarithmic average case complexity for distance functions *strongly equivalent* to L-p metrics [4]. While this requirement is reasonable for many applications, it does not apply to distances induced by the shortest paths of nonholonomic systems. Therefore, identifying nearest neighbors in the sense of a generic control cost remains an important open problem in sampling-based motion planning. In both literature and practical implementations, when searching for neighbors, randomized planners resort to distance functions that only approximate the true control cost. Arguably, the most common choices are Euclidean distance or quadratic forms [12]. This ad-hoc approach

© Springer Nature Switzerland AG 2020
K. Goldberg et al. (Eds.): *Algorithmic Foundations of Robotics XII*, SPAR 13, pp. 608–623, 2020.
https://doi.org/10.1007/978-3-030-43089-4_39

can significantly slow down the convergence rate of sampling-based algorithms if an inappropriate metric is selected.

A number of heuristics have been proposed to resolve this issue, such as the *reachability* and *utility guided RRTs* [17,3] which bias the tree expansion towards promising regions of the configuration space. Other approaches use the cost of linear quadratic regulators [5] and learning techniques [15,2]. In specific examples, these heuristics can significantly reduce the negative effects of finding nearest neighbors according to a metric inconsistent with the minimum cost path between configurations. However, the underlying inconsistency is not directly addressed. In contrast, a strong motivation to address it comes from recent research [8], which shows that a major speedup of sampling-based kynodinamic planners can be achieved by considering nonholonomy at the stage of nearest-neighbor search.

Specialized k-d tree algorithms have been proposed to account for non-standard topologies of some configuration manifolds [10,19,6]. However, no effort is known towards generalizing such algorithms to differential constraints.

In this work, we investigate the use of k-d trees for exact nearest-neighbor search in the presence of differential constraints. The main contributions can be summarized as follows: (i) we derive the expected complexity of nearest-neighbor queries with k-d trees built according to classic techniques and reveal that it is super-logarithmic (ii) we propose novel k-d tree build and query procedures tailored to sub-Riemannian metrics (iii) we provide numerical trials which verify our theoretical analysis and demonstrate the improvement afforded by the proposed algorithms as compared with popular open source software libraries, such as FLANN [14] and OMPL [18].

In Section 2, we review background material on sub-Riemannian geometries and show connections with nonholonomic systems, providing asymptotic bounds to their reachable sets. Based on these bounds, in Section 3 we propose a query procedure specialized for nonholonomic systems, after a brief review of the k-d tree algorithm. In Section 4, we study the expected complexity of m-nearest-neighbor queries on a classic k-d tree with sub-Riemannian metrics. Inspired by this analysis, in Section 5 we propose a novel incremental build procedure. Finally, in Section 6 we show positive experimental results for a nonholonomic mobile robot, which confirm our theoretical predictions and the effectiveness of the proposed algorithms.

2   Geometry of Nonholonomic Systems

Nonholonomic constraints are frequently encountered in robotics and describe mechanical systems whose local mobility is, in some sense, limited. Basic concepts from differential geometry, reviewed below, are used to clarify these limitations and discuss them quantitatively.

2.1   Elements of Differential Geometry

A subset $\mathcal{M}$ of $\mathbb{R}^n$ is a *smooth k-dimensional manifold* if for all $p \in \mathcal{M}$ there exists a neighborhood V of $\mathcal{M}$ such that $V \cap \mathcal{M}$ is diffeomorphic to an open subset of $\mathbb{R}^k$. A vector $v \in \mathbb{R}^n$ is said to be *tangent to* $\mathcal{M}$ *at point* $p \in \mathcal{M}$ if there exists a smooth curve $\gamma : [0,1] \to \mathcal{M}$, such that $\dot{\gamma}(0) = v$ and $\gamma(0) = p$. The *tangent space* of $\mathcal{M}$ at p, denoted $T_p\mathcal{M}$, is the subspace of vectors tangent to $\mathcal{M}$ at p. A map $Y : \mathcal{M} \to \mathbb{R}^n$ is a *vector field* on $\mathcal{M}$ if for each $p \in \mathcal{M}$, $Y(p) \in T_p\mathcal{M}$. A smooth euclidean *vector bundle* of rank r over $\mathcal{M}$ is defined as a set $E \subset \mathcal{M} \times \mathbb{R}^l$ such that the set $E_p : \{v \in \mathbb{R}^l, (p,v) \in E\}$ is an r-dimensional vector space for all $p \in \mathcal{M}$. E_p is called the *fiber* of bundle E at point p. Any set of h linearly independent smooth vector fields $Y_1, \dots Y_h$ such that for all $p \in \mathcal{M}$, $span(Y_1(p), ..., Y_h(p)) = E_p$ is called a *basis* of E and Y_i are called *generator vector fields* of E. The *tangent bundle* of $\mathcal{M}$ is defined as the vector bundle $T\mathcal{M}$ whose fiber at each point p is the tangent space at that point, $T_p\mathcal{M}$. A *distribution* $\mathcal{H}$ on a manifold is a subbundle of the tangent bundle, i.e., a vector bundle such that its fiber $\mathcal{H}_p$ at all points is a vector subspace of $T_p\mathcal{M}$.

Connection with nonholonomic systems. Consider a nonholonomic system described by the differential constraint:

$$\dot{x}(t) = f(x(t), u(t)), \tag{1}$$

where the configuration $x(t)$ and control $u(t)$ belong to the *smooth manifolds* $\mathcal{X}$ and $\mathcal{U}$ respectively. Each $\mathsf{u}_i \in \mathcal{U}$ defines a *vector field* $g_i(z) = f(z, \mathsf{u}_i)$ on $\mathcal{X}$. Therefore, for a fixed configuration z, $f(z, u)$ has values in a vector space $\mathcal{H}_z := Span(\{g_i(z)\})$. In other words, the dynamics described by equation (1) define a *distribution* $\mathcal{H}$ on the configuration manifold.

Throughout the paper, we will make use of the Reeds-Shepp vehicle [16] as an illustrative example.

Example (Configuration manifold of a Reeds-Shepp vehicle). The configuration manifold for this vehicle model is $\mathcal{X} = SE(2)$ with coordinates $x = (x_1, x_2, x_3)$. The mobility of the system is given by $\dot{x}_1 = u_1 \cos(x_3)$, $\dot{x}_2 = u_1 \sin(x_3)$, $\dot{x}_3 = u_1 u_2$, where $u_1, u_2 \in [-1,1]$. The inputs $\mathsf{u}_1 = (1,0)$ and $\mathsf{u}_2 = (1,1)$ define the vector fields $g_1(x) = (\cos(x_3), \sin(x_3), 0)$ and $g_2(x) = (\cos(x_3), \sin(x_3), 1)$. At each $z \in SE(2)$ the fiber of the Reeds-Shepp distribution $\mathcal{H}^{RS}$ is $Span\{g_1(z), g_2(z)\}$. Let $\hat{f}(x) = (\cos(x_3), \sin(x_3), 0)$, $\hat{l}(x) = (-\sin(x_3), \cos(x_3), 0)$ and $\hat{\theta}(x) = (0, 0, 1)$. These vector fields indicate the body frame of the vehicle, i.e. its *front* $\hat{f}$, *lateral* $\hat{l}$ and *rotation* $\hat{\theta}$ axes. The Reeds-Shepp distribution is then equivalently defined by the fibers $\mathcal{H}_z^{RS} = Span\{\hat{f}(z), \hat{\theta}(z)\}$.

2.2   Distances in a Sub-Riemannian Geometry

A *sub-Riemannian geometry* $\mathcal{G}$ on a manifold $\mathcal{M}$ is a tuple $\mathcal{G} = (\mathcal{M}, \mathcal{H}, \langle \cdot, \cdot \rangle_{\mathcal{H}})$, where $\mathcal{H}$ is a distribution on $\mathcal{M}$ whose fibers at all points p are equipped with the inner product $\langle \cdot, \cdot \rangle_{\mathcal{H}} : \mathcal{H}_p \times \mathcal{H}_p \to \mathbb{R}$. The distribution $\mathcal{H}$ is referred to as the *horizontal distribution*. A smooth curve $\gamma : [0, 1] \to \mathcal{M}$ is said to be *horizontal* if $\dot{\gamma}(t) \in \mathcal{H}_{\gamma(t)}$ for all $t \in [0, 1]$.

The length of smooth curves is defined $\ell_{\mathcal{G}}(\gamma) := \int_0^1 \sqrt{\langle \dot{\gamma}(t), \dot{\gamma}(t) \rangle_{\mathcal{H}}} \, dt$. If Γ_a^b denotes the set of horizontal curves between $a, b \in \mathcal{M}$, then $d_{\mathcal{G}}(a, b) := \inf_{\gamma \in \Gamma_a^b} \ell_{\mathcal{G}}(\gamma)$ is a *sub-Riemannian metric* defined by the geometry. The ball centered at p of radius r with respect to the metric $d_{\mathcal{G}}$ is denoted $\mathcal{B}(p, r)$. In control theory, the *attainable set* $\mathcal{A}(x_0, t)$ is the subset of $\mathcal{X}$ such that for each $p \in \mathcal{A}(x_0, t)$ there exists a $u : [0, \tau] \to \mathcal{U}$ for which the solution to (1) through $x(0) = x_0$ satisfies $x(\tau) = p$ for some $\tau \leq t$. Solutions to (1) are simply horizontal curves, so the attainable set $\mathcal{A}(x_0, t)$ is equivalent to the ball $\mathcal{B}(x_0, t)$ defined by the sub-Riemannian metric. This equivalence holds for systems with time-reversal symmetry, under controllability conditions stated in Theorem 1.

Example (Sub-Riemannian geometry of a Reeds-Shepp vehicle). The geometry associated with the Reeds-Shepp vehicle is $\mathcal{G}^{RS} = (SE(2), \mathcal{H}^{RS}, \langle \cdot, \cdot \rangle_{RS})$, with the standard inner product $\langle v, w \rangle_{RS} = v^T w$. Horizontal curves for this geometry are feasible paths satisfying the differential constraints. Geodesics correspond to minimum-time paths between two configurations and are known in closed form [16].

2.3   Iterated Lie Brackets and the Ball-box Theorem

A system is said to be *controllable* if any pair of configurations can be connected by a feasible (horizontal) path. Determining controllability of a nonholonomic system is nontrivial. For example, the Reeds-Shepp vehicle cannot move directly in the lateral direction, but intuition suggests that an appropriate sequence of motions can result in a lateral displacement (e.g., in parallel parking). Chow's Theorem and the Ball-box Theorem, reviewed below, are fundamental tools related to the controllability and the reachable sets of nonholonomic systems.

The *Lie derivative* of a vector field Y at $p \in \mathcal{M}$ in the direction $v \in T_p\mathcal{M}$ is defined as $dY(p)v = \frac{d}{dt} Y(\gamma(t))|_{t=0}$, where γ is a smooth curve starting in $p = \gamma(0)$ with velocity $v = \dot{\gamma}(0)$. Given two vector fields Y_1, Y_2 on $\mathcal{M}$, the *Lie bracket* $[Y_1, Y_2]$ is a vector field on $\mathcal{M}$ defined as $[Y_1, Y_2](p) = dY_2(p)Y_1(p) - dY_1(p)Y_2(p)$.

From the horizontal distribution $\mathcal{H}$, one can construct a sequence of distributions by iterating the Lie brackets of the generating vector fields, $Y_1, Y_2 \ldots Y_h$, with $h \leq k$. Recursively, this sequence is defined as: $\mathcal{H}^1 = \mathcal{H}$, $\mathcal{H}^{i+1} = \mathcal{H}^i \cup [\mathcal{H}, \mathcal{H}^i]$, where $[\mathcal{H}, \mathcal{H}^i]$ denotes the distribution given by the Lie brackets of each generating vector field of $\mathcal{H}$ with those of $\mathcal{H}^i$. Note that the Lie bracket of two vector fields can be linearly independent from the original fields, hence $\mathcal{H}^i \subseteq \mathcal{H}^{i+1}$. The *Lie hull* [13], denoted $Lie(H)$, is the limit of the sequence $\mathcal{H}^i$ as $i \to \infty$. A distribution $\mathcal{H}$ is said to be *bracket generating* if $Lie(\mathcal{H}) = T\mathcal{M}$.

Theorem 1. (Chow's Theorem *[13, p. 44]* **).** *If $Lie(\mathcal{H}) = T\mathcal{M}$ on a connected manifold $\mathcal{M}$, then any $a, b \in \mathcal{M}$ can be joined by a horizontal curve.*

Example (Lie hull of a Reeds-Shepp vehicle). Consider the Lie hull of $\mathcal{H}^{RS}$ generated by $\{\hat{f}(x), \hat{\theta}(x)\}$. For every $x \in SE(2)$, these vectors span a two-dimensional subspace of $T_x(SE(2))$. The first order Lie bracket is given by

$$[\hat{f}, \hat{\theta}](x) = \left(\frac{d}{dx}\hat{f}(x)\right)\hat{\theta}(x) - \left(\frac{d}{dx}\hat{\theta}(x)\right)\hat{f}(x) = (-\sin(x_3), \cos(x_3), 0). \quad (2)$$

This coincides with the body frame lateral axis $\hat{l}(x)$ of the vehicle. Therefore, the second order distribution $\mathcal{H}_x^2 = Span\{\hat{f}(x), \hat{\theta}(x), [\hat{f}, \hat{\theta}](x)\}$ spans the tangent bundle of $SE(2)$, and thus the Lie hull is obtained in the second step. By Theorem 1, there is a feasible motion connecting any two configurations which is consistent with one's intuition about the wheeled robot.

Above, we have shown that from a basis $Y_1 \ldots Y_h$ of a bracket-generating distribution $\mathcal{H}$, one can define a basis $y_1, \ldots y_k$ of TM. Specifically $y_i = Y_i$ for $i \leq h$, while the remaining $d - h$ fields are obtained with Lie brackets. The vector fields $\{y_i\}$ are called *privileged directions* [13]. Define the *weight* w_i of the privileged direction y_i as the smallest order of Lie brackets required to generate y_i from the original basis. More formally, w_i is such that $y_i \notin \mathcal{H}^{w_i-1}$ and $y_i \in \mathcal{H}^{w_i}$ for all i. The *weighted box* at p of size ϵ, weights $w \in \mathbb{N}_{>0}^k$ and multipliers $\mu \in \mathbb{R}_{>0}^k$ is defined as:

$$\mathrm{Box}^{w,\mu}(p,\epsilon) : \{y \in \mathbb{R}^n : |\langle y - p, y_i(p)\rangle_{\mathbb{R}^n}| < \mu_i \epsilon^{w_i} \text{ for all } i \in [1,k]\}. \quad (3)$$

Theorem 2 (The ball-box Theorem). *Let $\mathcal{H}$ be a distribution on a manifold M satisfying the assumptions of Chow's Theorem. Then, there exist constants $\epsilon_0 \in \mathbb{R}_{>0}$ and $c, C \in \mathbb{R}_{>0}^k$ such that for all $\epsilon < \epsilon_0$ and for all $p \in M$:*

$$\mathrm{Box}_i^w(p,\epsilon) := \mathrm{Box}^{w,c}(p,\epsilon) \subset \mathcal{B}(p,\epsilon) \subset \mathrm{Box}^{w,C}(p,\epsilon) := \mathrm{Box}_o^w(p,\epsilon), \quad (4)$$

where Box_i^w and Box_o^w are referred to as the inner and outer bounding boxes for the sub-Riemannian ball $\mathcal{B}$, respectively.

Example (Ball-Box Theorem visualized for a Reeds-Shepp vehicle). From the Lie hull construction in the previous example, we know that $\hat{f}(x), \hat{\theta}(x) \in \mathcal{H}^1$ so the corresponding weights are $w_{\hat{f}} = w_{\hat{\theta}} = 1$. Conversely, $\hat{l}(x)$ first appears in $\mathcal{H}^2$, so its weight is $w_{\hat{l}} = 2$. Theorem 2 states the existence of inner and outer bounding boxes for reachable sets as $t \to 0$ and predicts the infinitesimal order of each side of these boxes, as shown in figure 1. Higher order Lie brackets correspond to sides that approach zero at a faster asymptotic rate. The longitudinal and angular sides — along $\hat{f}$ and $\hat{\theta}$ — scale with $\Theta(t)$, while the lateral one — along $\hat{l}$ — with $\Theta(t^2)$. Therefore, both boxes become increasingly elongated along $\hat{f}$ and $\hat{\theta}$ and flattened along $\hat{l}$ as $t \to 0$. Intuitively, this geometric feature of boxes reflects the well known fact that a small lateral displacement of a car requires more time than an equivalent longitudinal one.

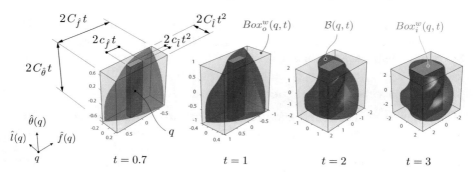

Fig. 1. Reachable sets and bounding boxes for a Reeds-Shepp vehicle around a configuration q and for different values of t. The lengths of the box sides are highlighted. For both the inner and outer boxes, as $t \to 0$, the sides along the front $\hat{f}$ and heading $\hat{\theta}$ axes are linear in t, while the side along the lateral axis $\hat{l}$ is quadratic in t.

For the Reeds-Shepp vehicle, the sides of both boxes can be computed explicitly with geometric considerations on special maneuvers that maximize or minimize the displacement along each axis. In particular, we get the values $C_{\hat{f}} = C_{\hat{\theta}} = c_{\hat{\theta}} = 1$, $C_{\hat{l}} = 1/2$, $c_{\hat{f}} = \sqrt{3/2} - 1$, $c_{\hat{l}} = 1/8$.

3   The k-d tree Data Structure

A k-d tree $\mathcal{T}$ is a binary tree organizing a *finite* subset $X \subset \mathcal{M}$, called a *database*, with its elements $x_i \in X$ called *data points*. We would like to find the m points in X closest to a given *query point* q on the manifold. Each point $x_i \in X$ is put in relation with a normal vector $n_i \in \mathbb{R}^n$. Together, the pair $v_i = (x_i, n_i)$ defines a *vertex* of the binary tree. The set of vertices is denoted with $\mathcal{V}$. A vertex defines a partition of $\mathbb{R}^n$ into two halfspaces, referred to as the *positive* and *negative* *halfspace*, and respectively described algebraically:

$$\begin{aligned} \mathfrak{h}^+(x_i, n_i) &:= \{x \in \mathbb{R}^n : \langle n_i, x \rangle_{\mathbb{R}^n} > \langle n_i, x_i \rangle_{\mathbb{R}^n}\}, \\ \mathfrak{h}^-(x_i, n_i) &:= \{x \in \mathbb{R}^n : \langle n_i, x \rangle_{\mathbb{R}^n} \le \langle n_i, x_i \rangle_{\mathbb{R}^n}\}. \end{aligned} \tag{5}$$

An *edge* is an ordered pair of vertices $e = (v_i, v_j)$. A binary tree is defined as $\mathcal{T} := (\mathcal{V}, \mathcal{E}^-, \mathcal{E}^+)$, with $\mathcal{V}$ set of vertices, $\mathcal{E}^-$ set of left edges and $\mathcal{E}^+$ set of right edges. Given one edge $e = (v_i, v_j)$, vertex v_j is referred to as the *left child* of v_i if $e \in \mathcal{E}^-$, or the *right child* if $e \in \mathcal{E}^+$. Let $\texttt{parent}(v_i \in \mathcal{V}) = v_j \in \mathcal{V}$ s.t. $(v_i, v_j) \in \mathcal{E}^- \cup \mathcal{E}^+$. By convention, $\texttt{parent}(v_i) = \emptyset$ if such v_j does not exist and v_i is called the *root* of $\mathcal{T}$, denoted $v_i = \texttt{root}(\mathcal{T})$. Let $\texttt{child}(v_i \in \mathcal{V}, s \in \{-, +\}) = v_j \in \mathcal{V}$ s.t. $(v_i, v_j) \in \mathcal{E}^s$, or otherwise $\emptyset$ if such v_j does not exist.

The fundamental property of k-d trees is that *left children belong to the negative halfspace defined by their parents and right children to the positive halfspace.* Recursively, a vertex belongs to the parent halfspaces of all its ancestors. As a result, a k-d tree defines a *partition* of $\mathcal{M}$ into non-overlapping polyhedra, called *buckets* [4], that cover the entire manifold. In the sequel, we let $\mathcal{B}_{\mathcal{T}}$ denote the set of buckets for a given k-d tree $\mathcal{T}$.

Buckets are associated with the leaves of $\mathcal{T}$ and with parents of only-child leaves (i.e., leaves without a sibling). For any given point $q \in \mathcal{M}$, we denote with $\mathfrak{b}_q$ the unique bucket containing q.

3.1 m-nearest-neighbor query algorithm

The computational efficiency afforded by the algorithm comes from the following observation: let q be the query point and suppose that among the distance evaluations computed thus far, the m closest points to q are within a distance d_m away. Now consider a vertex (x, n) and suppose the query point q is contained in $\mathfrak{h}^+(x, n)$ and $\mathcal{B}(q, d_m) \subset \mathfrak{h}^+(x, n)$. The data points represented by the sibling of (x, n) and all of its descendants are contained in $\mathfrak{h}^-(r, c)$ and are therefore at a distance greater than d_m. Thus, the corresponding subtree can be omitted from the search. The following primitives will be used to define the query procedure presented in Algorithm 1.

Side of hyperplane. A procedure $\mathtt{sideOf}(x, p, n) \to \{-, +\}$, with $x, p \in \mathcal{M}$, $n \in \mathbb{R}^n$. Returns $+$ iff $x \in \mathfrak{h}^+(p, n)$ and $-$ otherwise. For convenience, define $\mathtt{opposite}(+) = -$ and vice-versa.

Queue. For an m-nearest-neighbor query, the algorithm maintains a *Bounded Priority Queue*, Q of size m to collect the results. The queue can be thought of as a sequence $Q = [q_1, q_2, \ldots q_m]$, where each element q_i is defined as a distance-vertex pair $q_i : (d_i, v_i)$, $d_i \in \mathbb{R}_{\geq 0}$, $v_i \in \mathcal{V}$. The property $d_1 \leq d_2 \leq \cdots \leq d_m$ is an invariant of the data structure. When an element (d_{new}, n_{new}) is inserted in the queue, if $d_{new} < d_m$, then q_m is discarded and the indices of the remaining elements are rearranged to maintain the order.

Ball-Hyperplane Intersection. A procedure that determines whether a ball and a hyperplane intersect. Precisely, $\mathtt{ballHyperplane}(x, R, p, n)$, with $x \in \mathcal{M}$ and $R \geq 0$, returns true if $\mathcal{B}(x, R) \cap \mathfrak{h}(p, n) \neq \emptyset$. Note that it does not need to return false otherwise.

In the classic analysis of k-d trees [4, eq. 14] the distance is assumed to be a sum of component-wise terms, called *coordinate distance functions*. Namely:

$$d(x, y) = F\left(\sum_{i=1}^{k} d_i(x_i, y_i) \right). \tag{6}$$

When this holds, e.g., for L-p metrics, the ball-hyperplane intersection procedure reduces to the direct comparison of two numbers.

However, this property does not hold for other notions of distance and in general sub-Riemannian balls can have nontrivial shapes. For these cases, the procedure can be implemented by checking that $V_o(x, R) \cap \mathfrak{h}(p, n) \neq \emptyset$, where V_o is an "outer set" with a convenient geometry, i.e., a set such that $V_o(x, R) \supseteq \mathcal{B}(x, R)$ for all $x \in \mathcal{M}, R > 0$ and such that intersections with hyperplanes are easy to verify. Clearly, $\mathrm{vol}(V_o(x, R)) / \mathrm{vol}(\mathcal{B}(x, R)) \geq 1$ and it is desirable that this ratio stays as small as possible.

A crucial consequence of Theorem 2 is that the choice $V_o(x, R) = \mathrm{Box}_o^w(x, R)$ offers *optimal asymptotical behavior*, since the volume of the outer set scales with the smallest possible order, namely: $\mathrm{vol}(\mathrm{Box}_o^w(x,R)) \in \Theta[\mathrm{vol}(\mathcal{B}(x,R))]$. This fact is fundamental to devise efficient query algorithms for nonholonomic systems.

Example (Ball-hyperplane intersection for a Reeds-Shepp vehicle). The reachable sets shown in figure 1 have a nontrivial geometry. Let us consider two possible implementations of the ball-hyperplane intersection procedure:

1. *Euclidean bound* (EB). Define the set $\mathcal{C}(p,R) = \{x \in SE(2) \,|\, (x_1 - p_1)^2 + (x_2 - p_2)^2 \le R^2, |x_3 - p_3| \le 2R\}$, i.e., a cylinder in the configuration space with axis along $\hat{\theta}$. Since the Euclidean distance $\|a - b\|$ is a lower bound to $d_{RS}(a,b)$, then $\mathcal{B}(x,R) \subset \mathcal{C}(x,R)$ and `ballHyperplane` can be correctly implemented by checking $\mathcal{C}(x,R) \cap h(p,n) \ne \emptyset$

2. *Outer Box Bound* (BB). In this case, the procedure checks $\mathrm{Box}_o^w(x,R) \cap h(p,n) \ne \emptyset$. A simple implementation is to test whether all vertices of the box (8 in this case) are on the same halfspace defined by $\mathfrak{h}(p,n)$.

While Euclidean bounds (EB) are a simple choice, the Outer Box Bound (BB) method approximates the sub-Riemannian ball tightly. In fact, as $R \to 0$, $\mathrm{vol}(\mathcal{C}(p,R)) \in \Theta(R^3)$, while $\mathrm{vol}(\mathrm{Box}_o^w(p,R)) \in \Theta(R^4)$ and therefore the volume of the cylinder tends to be infinitely larger than the volume of the outer box. As a result, method (BB) yields an asymptotically unbounded speedup in the algorithm compared to (EB), as confirmed by our experimental results in Section 6. In addition, any other implementation of the procedure will at most provide a constant factor improvement with respect to method (BB).

Algorithm 1: k-d tree query

```
1  define query (q ∈ M, T = (V, E⁻, E⁺)):
2      Q ⟵ {q₁:ₘ = (∅, ∞)} ;                              // initialize queue
3      define querySubtree (q ∈ M, vᵢ = (xᵢ, nᵢ) ∈ V):
4          if vᵢ = ∅ then return;                          // caller reached leaf
5          s ⟵ sideOf (q, xᵢ, nᵢ);
6          querySubtree (q, child(vᵢ, s)) ;                // descend to child containing q
7          Q ⟵ Q ∪ {(dist(q, xᵢ), vᵢ)} ;                   // add vertex to the queue
8          if ballHyperplane (q, dₖ, xᵢ, nᵢ) then          // check for intersections
9              querySubtree (q, child(vᵢ, opposite(s))) ;  // visit sibling
10     querySubtree (q, root (T)) ;                        // start recursion from root
11     return Q;
```

3.2 k-d tree build algorithms

The performance of nearest-neighbor queries is heavily dependant on how the k-d tree is constructed. In the sequel, we describe two popular approaches to construct k-d trees: *batch* (or static) and *incremental* (or dynamic).

In the batch algorithm, all data points are processed at once and statistics of the database determine the vertices of the tree at each depth. This algorithm guarantees a balanced tree, but the insertion of new points is not possible without re-building the tree. Conversely, in the incremental version, the tree is updated on the fly as points are inserted, however, the tree balance guarantee is lost.

Both algorithms find applications in motion planning: the batch version can be used to efficiently build roadmaps for off-line, multiple-query techniques such as PRMs, while the incremental is suitable for anytime algorithms such as RRTs.

Batch k-d tree build algorithm

The following primitive procedures will be used to describe the batch construction of k-d trees, defined in Algorithm 2. The *range* of a set S along direction $\hat{d}$ is defined as $\operatorname{rng}_{\hat{d}}(S) = \sup_{x \in S} \langle x, \hat{d} \rangle - \inf_{x \in S} \langle x, \hat{d} \rangle$.

Maximum range. Let $\texttt{maxRange} : 2^X \to \mathbb{R}^n$. Given a subset $D \in 2^X$ of the data points, $\texttt{maxRange}$ determines a direction l along which the range of data is maximal. We consider the formulation in [4], where l is chosen among the cardinal directions $\{\hat{e}_i\}_{i \in [0,\ldots k-1]}$, so that $\texttt{maxRange}(D) = \arg\max_{\hat{e}_i}[\operatorname{rng}_{\hat{e}_i}(D)]$.

Median element. Let $\texttt{median} : 2^X \times \mathbb{R}^n \to D$. Given a subset $D \in 2^X$ of the data points and a direction l, $\texttt{median}$ returns the median element of D along direction l.

Algorithm 2: Batch k-d tree build

1 **define build** (X):
2 $\quad \mathcal{V} \longleftarrow \emptyset; \ \mathcal{E}^- \longleftarrow \emptyset; \ \mathcal{E}^+ \longleftarrow \emptyset$;   // Initialize tree
3 $\quad$ **define buildSubtree** $(D \in 2^X, \ v_i = (x_i, n_i) \in \mathcal{V}, \ s \in \{-, +\})$:
4 $\quad\quad$ **if** $D = \emptyset$ **then return**;   // caller reached leaf
5 $\quad\quad n_{new} \longleftarrow \texttt{maxRange}\ (D); \quad x_{new} \longleftarrow \texttt{median}\ (D, n_{new})$;
6 $\quad\quad v_{new} \longleftarrow (x_{new}, n_{new})$; $\quad \mathcal{V} \longleftarrow \mathcal{V} \cup v_{new}$;
7 $\quad\quad$ **if** $s \neq \emptyset$ **then** $\mathcal{E}^s \longleftarrow \mathcal{E}^s \cup (v_i, v_{new})$;
8 $\quad\quad L \longleftarrow D \cap \mathfrak{h}^-(x_{new}, n_{new}); \quad R \longleftarrow D \backslash (L \cup x_{new})$;
9 $\quad\quad$ buildSubtree $(L, v_{new}, -); \quad$ buildSubtree $(R, v_{new}, +)$;
10 $\quad$ buildSubtree $(X, \emptyset, \emptyset)$;
11 $\quad$ **return** $\mathcal{T} = (\mathcal{V}, \mathcal{E}^-, \mathcal{E}^+)$;

Incremental k-d tree build algorithm

A key operation to build a k-d tree incrementally is to pick splitting hyperplanes on the fly as new data points are inserted. This is achieved with a *splitting sequence*, which we define as:

Splitting sequence. A map $\mathcal{Z}_\mathcal{M} : \mathbb{N}_{\geq 0} \times \mathcal{M} \to \mathbb{R}^n$ that, given an integer $d \geq 0$ and a point $p \in \mathcal{M}$, returns a normal vector n associated with them.

When dealing with k-dimensional data, one usually chooses a basis of *cardinal axes* $\{\hat{e}_i\}_{i \in [0\ldots k-1]}$. The normal of the hyperplane associated with a vertex v is typically picked by cycling through the cardinal axes based on the depth of v in the binary tree. Let "$a \bmod b$" denote the remainder of the division of integer a by integer b. Then, the *classic splitting (CS) sequence* is defined as:

$$\mathcal{Z}_\mathcal{M}^{classic}(d, x) = \hat{e}_{[d \bmod k]}. \tag{7}$$

Algorithm 3: Incremental k-d tree build.
Insert in k-d tree $\mathcal{T} = (\mathcal{V}, \mathcal{E}^+, \mathcal{E}^-)$, start recursion with $\mathtt{insert}(x_{new}, \mathtt{root}(\mathcal{T}))$.

```
 1  define insert (x_new ∈ M, v_i = (x_i, n_i) ∈ V):
 2      if V = ∅ then                                          // if tree is empty...
 3          n_new ⟵ Z_M(0, x_new); V⟵(x_new, n_new) ;          //  ...add root
 4          return
 5      side ⟵ sideOf (x_new, x_i, n_i) ; c ⟵ child (v_i, side) ;
 6      if c = ∅ then
 7          n_new ⟵ Z_M(depth(v_i) + 1, x_new) ; v_new ⟵ (x_new, n_new);
 8          V ⟵ V ∪ v_new ; E^side ⟵ E^side ∪ (v_i, v_new) ;
 9          return
10      insert (x_new, c) ;
```

---

4   Complexity Analysis

In this section, we discuss the expected asymptotic complexity of m-nearest-neighbor queries in a k-d tree built according to Algorithm 2. The complexity of the nearest-neighbor query procedure (Algorithm 1) is measured in terms of the number n_v of vertices examined. Let $\mathcal{B}(q, d_m)$ be the ball containing the m nearest neighbors to q. For the soundness of the algorithm, all the vertices contained in this ball must be examined, and therefore, all the buckets overlapping with it. As we recall from Section 3, buckets are associated with leaves, and therefore the algorithm will visit as many leaves as the number of such buckets, formally: $n_l \in \Theta(|\beta|)$, where $\beta := \{b \in \mathscr{B}_\mathcal{T} \mid b \cap \mathcal{B}(q, d_m) \neq \emptyset\}$ and n_l denotes the number of leaves visited by the algorithm.

If the asymptotic order of visited leaves is known, proofs rely on the following fact: since the batch algorithm guarantees a balanced tree, descending into n_l leaves from the root requires visiting $n_v \in \Theta(n_l \log N)$ vertices, where N is the cardinality of the database X. Thus the expected query complexity is given by:

$$\mathbb{E}(n_v) \in \Theta[\mathbb{E}(n_l) \log N]. \tag{8}$$

Lemma 1 reviews the seminal result, originally from [4], that the expected asymptotic complexity for a Minkowski (i.e., L-p) distance is logarithmic. Our main theoretical contribution is the negative result described in Theorem 3, where we reveal that the expected complexity for sub-Riemannian metrics is in fact super-logarithmic. In the sequel, assume that data points are randomly sampled from $\mathcal{M}$ with probability distribution $p(x)$.

Lemma 1 ([4]). *If the distance function used for the nearest-neighbor query is induced by a p-norm, then the expected complexity of the nearest-neighbor search on a k-d tree built with Algorithm 2 is $O(\log(N))$ as $N \to \infty$.*

For the detailed proof, we suggest reading the original work by Friedman et al. [4], while here we report key facts used in our next result. In [4, pg. 214], it is shown that:

$$\mathbb{E}[\text{vol}(\mathcal{B}(q, d_m))] \approx \frac{m}{(N+1)} \frac{1}{p(q)}, \tag{9}$$

with no assumptions on the metric and the probability distribution $p(x)$.

In a batch-built k-d tree, the expected asymptotic shape of the bucket $\mathfrak{b}_q$ containing the query point q is assumed hyper-cubical. From equation (9) with $m = 1$, $\mathbb{E}(\text{vol}(\mathfrak{b}_q)) \approx \frac{1}{(N+1)} \frac{1}{p(q)}$. Thus, hyper-cubes in a neighborhood of q have sides of expected length $E(l_q) = [(N+1)p(q)]^{-1/k}$. Using these facts, it is then shown that the expected number of visited leaves is asymptotically constant with the size of the database, $E(n_l) \in \Theta(1)$ and therefore the average query complexity is logarithmic, as per equation (8).

Theorem 3. *Let the distance function used for the nearest-neighbor query be a sub-Riemannian metric on a smooth connected manifold $\mathcal{M}$ with a bracket-generating horizontal distribution of dimension h and weights $\{w_i\}$. Then the expected complexity of the nearest-neighbor query on a k-d tree built with Algorithm 2 is $\Theta(N^p \log(N))$ as $N \to \infty$, where the expression for p is:*

$$p = \sum_{w_i \leq W/k} \left(\frac{1}{k} - \frac{w_i}{W} \right), \quad \text{with } W := \sum_i w_i. \tag{10}$$

Proof. For N large enough, Theorem 2 ensures the existence of inner and outer bounding boxes to the sub-Riemannian ball $\mathcal{B}(q, d_m)$.

From equations (4) and (9), we get:

$$\mathbb{E}(d_m)^W \cdot \prod_{i=1}^{k} c_i \leq \frac{1}{N+1} \frac{1}{p(q)} \leq \mathbb{E}(d_m)^W \cdot \prod_{i=1}^{k} C_i, \tag{11}$$

where $W := \sum_i w_i$. Therefore, the expected distance to the m-th nearest neighbor scales asymptotically as $\mathbb{E}(d_m) \in \Theta(N^{-1/W})$ for $N \to \infty$. Recall that a bucket $\mathfrak{b}_q$ has sides of expected length $l_q = \sqrt[k]{1/[(N+1)p(q)]}$ or equivalently, $l_q \in \Theta(N^{-1/k})$. We are now interested in the asymptotic order of the number of buckets overlapping with a weighted box of size d_m. Let $n_h(d, l_q)$ the number of hypercubes intersected by a segment of length d embedded in a grid of $\mathbb{R}^k$ with side l_q. It can be shown that $\left\lceil (\sqrt{k})^{-1}(d/l_q) \right\rceil \leq n_h(d, l_q) \leq k + \sqrt{k}\lceil d/l_q \rceil$. Then, asymptotically, $n_h(d, l_q) \in \Theta(\lceil d/l_q \rceil)$ as $(d, l_q) \to 0$. Since a box has k orthogonal sides, each with expected length $\mathbb{E}(d_m)^{w_i}$, the expected number of visited buckets, is:

$$\mathbb{E}(n_l) \in \Theta \left(\prod_{i=1}^{k} \left\lceil \frac{\mathbb{E}(d_m)^{w_i}}{l_q} \right\rceil \right) = \Theta \left(\prod_{i=1}^{k} \left\lceil N^{\frac{1}{k} - \frac{w_i}{W}} \right\rceil \right). \tag{12}$$

In the latter product, only the factors with exponent > 0 contribute to the complexity of the algorithm, while the other terms tend to 1, as $N \to \infty$. In other words, the query complexity is determined only by low order Lie brackets up until $w_i \leq W/k$. In fact, along the direction of higher order Lie brackets, reachable sets shrink to zero faster than the side of a k-d tree bucket. Following up from equation (12), we get:

$$\mathbb{E}(n_l) \in \Theta\left(\prod_{w_i \leq W/k} N^{\frac{1}{k} - \frac{w_i}{W}}\right) = \Theta\left(N^p\right), \quad p = \sum_{w_i \leq W/k} \left(\frac{1}{k} - \frac{w_i}{W}\right). \qquad (13)$$

From equation (8) it follows that the expected complexity of the query algorithm is given by $\mathbb{E}(n_v) \in \Theta(N^p \log N)$ as $N \to \infty$. $\qquad\qquad\square$

Remark. For holonomic systems, the query complexity is logarithmic, in accordance with Lemma 1. In fact, for holonomic systems, $w_1 = w_2 = \cdots = w_k = 1$, then $W = k$ and $p = 0$ from equation (10).

Remark. The query complexity is always between logarithmic and linear. Since $W \geq k$ by definition, for all i such that $w_i \leq W/k$ one can state $0 < w_i/W \leq 1/k \leq 1$. It follows that $0 \leq \frac{1}{k} - \frac{w_i}{W} < \frac{1}{k}$. Then p can be bounded with:

$$0 \leq p = \sum_{w_i \leq W/k} \left(\frac{1}{k} - \frac{w_i}{W}\right) < \sum_{w_i \leq W/k} \left(\frac{1}{k}\right) \leq k \cdot \frac{1}{k} = 1 \quad \Rightarrow \quad p \in [0,1). \qquad (14)$$

Example. For the Reeds-Shepp car, $k = \dim[SE(2)] = 3$, $w_{\hat{f}} = w_{\hat{\theta}} = 1$, $w_{\hat{i}} = 2$ and $W = w_{\hat{f}} + w_{\hat{\theta}} + w_{\hat{i}} = 4$. Only $w_{\hat{f}}$ and $w_{\hat{\theta}}$ satisfy $w_i \leq W/k$, therefore, according to Theorem 3, the expected query complexity of a batch kd-tree is $\Theta\left(\sqrt[6]{N} \log N\right)$. We confirm this prediction with experiments in section 6.

5   The Lie splitting strategy

The basic working principle of k-d trees is the ability to discard subtrees without loss of information during a query. For each visited node $v = (x, n)$, the algorithm checks whether the current biggest ball in the queue, $\mathcal{B}(q, d_m)$ intersects $\mathfrak{h}(x, n)$. If such an intersection exists, the algorithm visits both children of v, otherwise one child is discarded and so is the entire subtree rooted in it. To reduce complexity, it is thus desirable that during query, a k-d tree presents as few ball-hyperplane intersections as possible.

In the classic splitting strategy, hyperplanes are chosen cycling through a globally defined set of cardinal axes. However, we have shown that nonholonomic systems have a set of locally-defined *privileged axes* and that the reachable sets have different infinitesimal orders along each of them. When the metric comes from a nonholonomic system, the hyperplanes in a classic k-d tree are not aligned with reachable sets and the buckets have different asymptotic properties than reachable sets. This can make intersections frequent, as shown in figure 2(a).

Ideally, to minimize the number of ball-hyperplane intersections, the buckets in a k-d tree should *approximate the bounding boxes for the reachable sets* of the

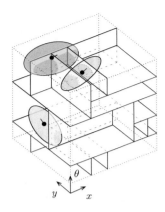
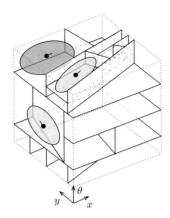

(a) Classic k-d tree construction   (b) Lie k-d tree splitting strategy

Fig. 2. Qualitative comparison of a classic k-d tree with its counterpart built with the proposed *Lie splitting strategy*. For nonholonomic systems, reachable sets (red, blue, green) are elongated along configuration-dependent *privileged directions*. The smaller the sets, the more pronounced their aspect ratios. The Lie splitting strategy adapts the hyperplanes locally to the balls and decreases the expected number of ball-hyperplane intersections, thus reducing the expected asymptotic query complexity.

dynamical system, as depicted in figure 2(b). To achieve this, we propose a novel splitting rule, named the *Lie splitting strategy*, which exploits the differential geometric properties of a system and the asymptotic scaling of its reachable sets. The Lie splitting strategy is based on the following two principles:

1. The splitting normal associated with each data point x_i is along one of the privileged axes in that point, i.e., $\mathcal{Z}_\mathcal{M}(d, x_i) \in \{y_1(x_i), y_2(x_i) \ldots y_k(x_i)\}$.
2. The buckets of the k-d tree, in expectation, scale asymptotically according to the weighted box along all privileged axes as $t \to 0$. Formally, for all $q \in \mathcal{M}$ and for all pairs of privileged axes $y_i(q), y_j(q)$,

$$\mathbb{E}\left[\text{rng}_{y_i(q)}(\mathfrak{b}_q)\right]^{w_j} \in \Theta\left(\mathbb{E}\left[\text{rng}_{y_j(q)}(\mathfrak{b}_q)\right]^{w_i}\right). \tag{15}$$

Requirement 2 prescribes the asymptotic behavior of the sequence $\mathcal{Z}_\mathcal{M}$ as $d \to \infty$, which can be formalized as follows: let $n_i(d) = |\{n < d : \mathcal{Z}_\mathcal{M}(n, x) = y_i(x)\}|$, i.e., the total number of splits in the sequence $\mathcal{Z}_\mathcal{M}$ along axis y_i before index d.

As $d \to \infty$, the expected bucket size along y_i after $n_i(d)$ splits has asymptotic order $\mathbb{E}\left[\text{rng}_{y_i(q)}(\mathfrak{b}_q)\right] \in \Theta[e^{-n_i(d)}]$. Then, in terms of the number of splits, equation (15) yields:

$$n_i(d) \cdot w_j \sim n_j(d) \cdot w_i \text{ for all } i, j \in [1, \ldots k] \text{ as } d \to \infty. \tag{16}$$

Simply put, each privileged direction should be picked as a splitting hyperplane with a frequency proportional to its weight. Note that this is only relevant *asymptotically*, i.e., as $d \to \infty$.

Example. For the Reeds-Shepp vehicle, a valid Lie splitting sequence is:

$$\mathcal{Z}_{RS}^{Lie}(d,x) := \begin{cases} \hat{l}(x), & d \equiv 0 \text{ or } d \equiv 1 \pmod 4 \\ \hat{f}(x), & d \equiv 2 \pmod 4 \\ \hat{\theta}(x), & d \equiv 3 \pmod 4 \end{cases} \tag{17}$$

It is easy to verify that this satisfies equation (16). In fact, lateral splits occur twice as often as longitudinal and angular splits. Thus, $n_{\hat{l}}(d) \sim 2n_{\hat{f}}(d) \sim 2n_{\hat{\theta}}(d)$.

6   Experimental results

In this section, we validate our results and compare our algorithms with other methods from widely used open-source libraries. Query performance is averaged over 1000 randomly drawn query points. The same insertion and query sequences are used across all analyzed algorithms and generated from a uniform distribution.

Experiment 1. In this experiment, we confirm the theoretical contributions presented in Section 4. In figure 3(a) we show the average number of leaves visited when querying a batch k-d tree with Euclidean (blue) and Reeds-Shepp metrics (solid red). While the blue curve settles to a constant value, in accordance with Lemma 1, the red curve exhibits exponential growth. When normalizing this curve by $\sqrt[6]{N}$ (dashed red), we observe a constant asymptotic behavior, consistent with the rate determined by Theorem 3. For comparison, we report the average time for Euclidean queries on an incremental k-d tree (black), which tends to visit more vertices than its batch, balanced counterpart (blue).

Experiment 2. In this experiment (figures 3(b-d)), we plot the total number of distance evaluations and the running times obseved with different combinations of build and query algorithms. Figure 3(d) reveals that the proposed outer Box Bound (BB) ball-hyperplane intersection (yellow, purple, green) reduces the query time significantly as compared to Euclidean bounds (EB) (blue, red), in accordance with our predictions in Section 3.1.

   Additionally, Lie splitting (green) further improves query time, as compared with the classic splitting (yellow). The corresponding incremental Lie k-d tree also outperforms a classic batch-built one (purple), guaranteed to be balanced.

   More important speedups emerge when comparing the proposed k-d trees with different techniques, such as Hierarchical Clustering from FLANN and Geometric Near-neighbor Access Tree (GNAT) from OMPL. Interestingly, off-the-shelf implementations of k-d trees offered by FLANN and other tools are unusable with nonholonomic metrics altogether, since they are limited to distances of the form of equation (6). Therefore a comparison is not possible.

   All the tested k-d trees visibly outperform Hierarchical Clustering (cyan) both in build time and query time. In contrast, GNAT offers competitive query times. However, its insertion is $\sim 100\times$ slower than incremental k-d trees, since a significant number of distances are evaluated in the build phase, while k-d trees

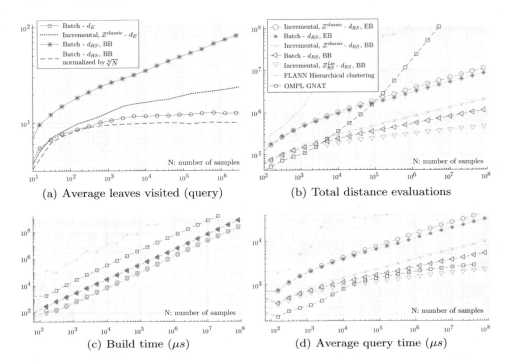

Fig. 3. (a) *Experiment 1*: Average number of leaves visited during Euclidean (blue) and Reeds-Shepp (red, solid) queries of a batch k-d tree. (b-d) *Experiment 2*: Performance of different algorithms. In the legends, d_E and d_{RS} indicate Euclidean and Reeds-Shepp queries, EB and BB indicate Euclidean Bound and outer Box Bound for ball-hyperplane intersections, $\mathcal{Z}^{classic}$ and $\mathcal{Z}^{Lie}_{RS}$ indicate the splitting sequence.

only evaluate distances during query. This is reflected in a noticeably higher asymptotic rate of distance evaluations, revealed in figure 3(b).

7   Conclusion

Motivated by applications in sampling-based motion planning, we investigated k-d trees for efficient nearest-neighbor search with distances defined by the length of paths of controllable nonholonomic systems. We have shown that for sub-Riemannian metrics, the query complexity of a classic batch-built k-d tree is $\Theta(N^p \log N)$, where $p \in [0, 1)$ depends on the properties of the system. In addition, we have proposed improved build and query algorithms for k-d trees that account for differential constraints. The proposed methods proved superior over classic ones in numerical experiments carried out with a Reeds-Shepp vehicle.

Future work will analyze whether logarithmic complexity is achieved for non-holonomic systems. In addition, the proposed algorithms are exact and rely on explicit distance evaluations. Since distances cannot be generally computed in closed form, we are interested in investigating approximate nearest-neighbor search algorithms with provable correctness bounds that do not require explicit distance computations.

References

1. J. L. Bentley. Multidimensional binary search trees used for associative searching. *Communications of the ACM*, 18(9):509–517, 1975.
2. M. Bharatheesha, W. Caarls, W. J. Wolfslag, and M. Wisse. Distance metric approximation for state-space RRTs using supervised learning. In *2014 IEEE/RSJ International Conference on Intelligent Robots and Systems*, pages 252–257, 2014.
3. B. Burns and O. Brock. Single-query motion planning with utility-guided random trees. In *Proceedings 2007 IEEE International Conference on Robotics and Automation*, pages 3307–3312, 2007.
4. J. H. Friedman, J. L. Bentley, and R. A. Finkel. An algorithm for finding best matches in logarithmic expected time. *ACM Transactions on Mathematical Software (TOMS)*, 3(3):209–226, 1977.
5. E. Glassman and R. Tedrake. A quadratic regulator-based heuristic for rapidly exploring state space. In *Robotics and Automation (ICRA), 2010 IEEE International Conference on*, pages 5021–5028, 2010.
6. J. Ichnowski and R. Alterovitz. Fast nearest neighbor search in SE(3) for sampling-based motion planning. In *Algorithmic Foundations of Robotics XI*, pages 197–214. Springer, 2015.
7. S. Karaman and E. Frazzoli. Sampling-based algorithms for optimal motion planning. *The International Journal of Robotics Research*, 30(7):846–894, 2011.
8. S. Karaman and E. Frazzoli. Sampling-based optimal motion planning for nonholonomic dynamical systems. In *Robotics and Automation (ICRA), 2013 IEEE International Conference on*, pages 5041–5047, 2013.
9. L. E. Kavraki, P. Svestka, J.-C. Latombe, and M. H. Overmars. Probabilistic roadmaps for path planning in high-dimensional configuration spaces. *IEEE transactions on Robotics and Automation*, 12(4):566–580, 1996.
10. J. J. Kuffner. Effective sampling and distance metrics for 3D rigid body path planning. In *Robotics and Automation, 2004. Proceedings. ICRA'04. 2004 IEEE International Conference on*, volume 4, pages 3993–3998, 2004.
11. S. M. LaValle. Rapidly-exploring random trees: A new tool for path planning. Citeseer, 1998.
12. S. M. LaValle and J. J. Kuffner. Randomized kinodynamic planning. *The International Journal of Robotics Research*, 20(5):378–400, 2001.
13. R. Montgomery. *A tour of subriemannian geometries, their geodesics and applications*. Number 91. American Mathematical Soc., 2006.
14. M. Muja and D. G. Lowe. Fast approximate nearest neighbors with automatic algorithm configuration. In *International Conference on Computer Vision Theory and Application VISSAPP'09)*, pages 331–340. INSTICC Press, 2009.
15. L. Palmieri and K. Arras. Distance metric learning for RRT-based motion planning for wheeled mobile robots. In *2014 IROS Machine Learning in Planning and Control of Robot Motion Workshop*, 2014.
16. J. Reeds and L. Shepp. Optimal paths for a car that goes both forwards and backwards. *Pacific journal of mathematics*, 145(2):367–393, 1990.
17. A. Shkolnik, M. Walter, and R. Tedrake. Reachability-guided sampling for planning under differential constraints. In *Robotics and Automation, 2009. ICRA'09. IEEE International Conference on*, pages 2859–2865, 2009.
18. I. A. Şucan, M. Moll, and L. E. Kavraki. The Open Motion Planning Library. *IEEE Robotics & Automation Magazine*, (4):72–82, December.
19. A. Yershova and S. M. LaValle. Improving motion-planning algorithms by efficient nearest-neighbor searching. *IEEE Transactions on Robotics*, 23(1):151–157, 2007.

Collision detection or nearest-neighbor search? On the computational bottleneck in sampling-based motion planning⋆

Michal Kleinbort[1]⋆⋆, Oren Salzman[2]⋆⋆, and Dan Halperin[1]

[1] Blavatnik School of Computer Science, Tel-Aviv University, Israel
[2] Carnegie Mellon University, Pittsburgh PA 15213, USA

Abstract. The complexity of nearest-neighbor search dominates the asymptotic running time of many sampling-based motion-planning algorithms. However, collision detection is often considered to be the computational bottleneck in practice. Examining various asymptotically optimal planning algorithms, we characterize settings, which we call *NN-sensitive*, in which the *practical* computational role of nearest-neighbor search is far from being negligible, i.e., the portion of running time taken up by nearest-neighbor search is comparable to, or sometimes even greater than the portion of time taken up by collision detection. This reinforces and substantiates the claim that motion-planning algorithms could significantly benefit from efficient and possibly specially-tailored nearest-neighbor data structures. The asymptotic (near) optimality of these algorithms relies on a prescribed connection radius, defining a ball around a configuration q, such that q needs to be connected to all other configurations in that ball. To facilitate our study, we show how to adapt this radius to non-Euclidean spaces, which are prevalent in motion planning. This technical result is of independent interest, as it enables to compare the radial-connection approach with the common alternative, namely, connecting each configuration to its k nearest neighbors (K-NN). Indeed, as we demonstrate, there are scenarios where using the radial connection scheme, a solution path of a specific cost is produced ten-fold (and more) faster than with K-NN.

1 Introduction

Given a robot $\mathcal{R}$ moving in a workspace $\mathcal{W}$ cluttered with obstacles, motion-planning (MP) algorithms are used to efficiently plan a path for $\mathcal{R}$, while avoiding collision with obstacles [9, 25]. Prevalent algorithms abstract $\mathcal{R}$ as a point in

⋆ This work has been supported in part by the Israel Science Foundation (grant no. 825/15), by the Blavatnik Computer Science Research Fund, and by the Hermann Minkowski–Minerva Center for Geometry at Tel Aviv University. O. Salzman has been also supported by the National Science Foundation IIS (#1409003), Toyota Motor Engineering & Manufacturing (TEMA), and the Office of Naval Research. Part of this work was carried out while O. Salzman was a student at Tel Aviv University.
⋆⋆ M. Kleinbort and O. Salzman contributed equally to this paper.

© Springer Nature Switzerland AG 2020
K. Goldberg et al. (Eds.): *Algorithmic Foundations of Robotics XII*, SPAR 13, pp. 624–639, 2020.
https://doi.org/10.1007/978-3-030-43089-4_40

a (possibly high-dimensional) space called the *configuration space* (C-space) $\mathcal{X}$ and plan a path (curve) in this space. A point, or a configuration, in $\mathcal{X}$ represents a placement of $\mathcal{R}$ that is either collision-free or not, subdividing $\mathcal{X}$ into the sets $\mathcal{X}_{\text{free}}$ and $\mathcal{X}_{\text{forb}}$, respectively. *Sampling-based* algorithms study the structure of $\mathcal{X}$ by constructing a graph, called a *roadmap*, which approximates the connectivity of $\mathcal{X}_{\text{free}}$. The nodes of the graph are collision-free configurations sampled at random. Two (nearby) nodes are connected by an edge if the straight line segment connecting their configurations is collision-free as well.

Sampling-based MP algorithms are typically implemented using two primitive operations: *Collision detection* (CD) [26], which is primarily used to determine whether a configuration is collision-free or not, and *Nearest-neighbor* (NN) search, which is used to efficiently return the nearest neighbor (or neighbors) of a given configuration. CD is also used to test if the straight line segment connecting two configurations lies in $\mathcal{X}_{\text{free}}$—a procedure referred to as *local planning* (LP). In this paper we consider both CD and LP calls when measuring the time spent on collision-detection operations.

Contribution The complexity of NN search dominates the asymptotic running time of many sampling-based MP algorithms. However, the main computational bottleneck in practical settings is typically considered to be LP [9, 25]. In this paper we argue that this may not always be the case. We describe settings, which we call *NN-sensitive*, where the (computational) role of NN search after finite running-time is far from negligible and merits the use of advanced and specially-tailored data structures; see Fig. 1 for a plot demonstrating this behavior. NN-sensitivity may be due to (i) planners that *algorithmically* shift the computational weight to NN search; (ii) scenarios in which certain planners perform mostly NN search; or (iii) parameters' values for which certain planners spend the same order of running time on NN and CD.

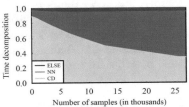

Fig. 1: Running-time breakdown of the main primitive operations used in MPLB [31] applied to the 3D-Grid scenario (Fig. 2a). For additional data, see Sec. 4. Best viewed in color.

Specifically, we focus on asymptotically (near) optimal MP algorithms. We study the ratio between the overall time spent on NN search and CD after N configurations were sampled. We observe situations where NN takes up to 100% more time than CD in scenarios based on the Open Motion Planning Library [11]; on synthetic high-dimensional C-spaces we even observe a ratio of 4500%.

We mostly concentrate on the *radial* version of MP algorithms, where the set of neighbors in the roadmap of a given configuration q includes all configurations of maximal distance r from q. To do so in *non-Euclidean* C-spaces, we derive closed-form expressions for the volume of a unit ball in several common C-spaces. This technical result is of independent interest, as the lack of such expressions seems to have thus far prevented the exploration and understanding of these types of algorithms in non-Euclidean settings—most experimental evaluation reported in the literature on the radial version of asymptotically-optimal planners is limited to Euclidean settings only. We show empirically that in certain scenarios, the radial version of an MP algorithm produces a solution of a specific cost more than *ten times faster* than the non-radial version, namely, where each node is connected to its k nearest neighbors.

We emphasize that we are not the first to claim that in certain cases NN may dominate the running time of MP algorithms, see, e.g., [5]. However, we take a systematic approach to characterize and analyze when this phenomenon occurs.

Throughout the paper we use the following notation: For an algorithm ALG, let $\chi_{\text{ALG}}(S)$ be the ratio between the overall time spent on NN search and CD for a specific motion-planning problem after a set S of configurations was sampled, where we assume that all other parameters of the problem, the workspace and the robot, are fixed—see details below. Let $\chi_{\text{ALG}}(N)$ be the expected value of $\chi_{\text{ALG}}(S)$ over all sample sets S of size N.

Organization We start with an overview of related work in Sec. 2 and continue in Sec. 3 to summarize, for several algorithms, the computational complexity in terms of NN search and CD. We show that asymptotically, as N tends to infinity, $\chi_{\text{ALG}}(N)$ tends to infinity as well. In Sec. 4 we point out several NN-sensitive settings together with simulations demonstrating how $\chi_{\text{ALG}}(N)$ behaves in such settings. These simulations make use of the closed-form expressions of the volume of unit balls, which are detailed in Sec. 5. Finally, Sec. 6 concludes with possible future work.

2   Background and related work

We start by giving an overview of asymptotically (near) optimal MP algorithms and continue with a description of CD and NN algorithms.

2.1   Asymptotically optimal sampling-based motion planning

A random geometric graph (RGG) $\mathcal{G}$ is a graph whose vertices are sampled at random from some space $\mathcal{X}$. Every two configurations are connected if their distance is less than a connection radius r_n (which is typically a function of the number of nodes n in the graph). We are interested in a connection radius such that, asymptotically, for any two vertices x, y, the cost of a path in the graph connecting x and y converges to the minimal-cost path connecting them in $\mathcal{X}$. A sufficient condition to ensure this property is that [20]

$$r_n \geq 2\eta \left(\frac{\mu(\mathcal{X}_{\text{free}})}{\zeta_d} \right)^{1/d} \left(\frac{1}{d} \right)^{1/d} \left(\frac{\log n}{n} \right)^{1/d}. \tag{1}$$

Here d is the dimension of $\mathcal{X}$, $\mu(\cdot)$ and ζ_d denote the Lebesgue measure (volume) of a set and of the d-dimensional unit ball, respectively, and $\eta \geq 1$ is a tuning parameter that allows to balance between exploring unvisited regions of the C-space and connecting visited regions. Alternatively, an RGG where every vertex is connected to its $k_n \geq e(1 + 1/d) \log n$ nearest neighbors will ensure similar convergence properties [21]. Unless stated otherwise, we focus on RGGs of the former type. For a survey on additional models of RGGs, their properties and their connection to sampling-based MP algorithms, see [33].

Most asymptotically-optimal planners sample a set of collision-free configurations (either incrementally or in batches). This set of configurations induces an RGG $\mathcal{G}$ or a sequence of increasingly dense RGGs $\{\mathcal{G}_n\}$ whose vertices are the sampled configurations. Set $\mathcal{G}' \subseteq \mathcal{G}$ to be the subgraph of $\mathcal{G}$ whose edges represent collision-free motions. These algorithms construct a roadmap $\mathcal{H} \subseteq \mathcal{G}'$.

PRM* and RRG [21] call the local planner for *all* the edges of $\mathcal{G}$. To increase the convergence rate to high-quality solutions, algorithms such as RRT* [21], RRT$^{\#}$ [1], LBT-RRT [32], FMT* [20], MPLB [31], Lazy-PRM* [14], and BIT* [13] call the local planner for a *subset* of the edges of $\mathcal{G}$.

Reducing the number of LP calls is typically done by constructing $\mathcal{G}$ (using nearest-neighbor operations only) and deciding for which edges to call the local planner. Many of the algorithms mentioned do so by using graph operations such as shortest-path computation. These operations often take a tiny fraction of the time required for LP computation. However, in more recent algorithms such as FMT* and BIT* this may not be true.

2.2   Collision detection

Most CD algorithms are bound to certain types of models, where rigid polyhedral models are the most common. They often allow answering proximity queries as well (i.e., separation-distance computation or penetration-depth estimation). Several software libraries for collision detection are publicly available [10, 24]. The most general of which is the Flexible Collision Library (FCL) [29] that integrates several techniques for fast and accurate collision checking and proximity computation. For polyhedral models, which are prevalent in MP settings, most commonly-used techniques are based on *bounding volume hierarchies* (BVH).

A collision query using BVHs may take $O(m^2)$ time in the worst case, where m is the complexity of the obstacle polyhedra and assuming that the robot is of constant-description complexity. However, tighter bounds may be obtained using methods tailored for large environments [10, 16]. Specifically, the time complexity is $O(m \log^{\delta-1} m + s)$, where $\delta \in \{2, 3\}$ is the dimension of the workspace $\mathcal{W}$ and s is the number of intersections between the bounding volumes. Other methods relevant to MP are mentioned in [23]. For a survey on the topic, see [26].

2.3   Nearest-neighbor methods: exact and approximate

Nearest-neighbor (NN) algorithms are frequently used in various domains. In the most basic form of the problem we are given a set P of n points in a metric space $M = (X, \rho)$, where X is a set and $\rho : X \times X \to \mathbb{R}$ is a distance metric. Given a query point $q \in X$, we wish to efficiently report the nearest point $p \in P$ to q. Immediate extensions include the k-nearest-neighbors (K-NN) and the r-near-neighbors (R-NN) problems. The former reports the k nearest points of P to q, whereas the latter reports all points of P within a distance r from q.

In the plane, the NN search problem can be efficiently solved by constructing a Voronoi diagram of P in $O(n \log n)$ time and preprocessing it to a linear-size point-location data structure in $O(n \log n)$ time. Queries are then answered in $O(\log n)$ time [4, 15]. However, for high-dimensional point sets this approach becomes infeasible, as it is exponential in the dimension d. This phenomenon is often termed "the curse of dimensionality" [19].

An efficient data structure for low dimensional spaces[3] is the kd-tree [3, 12], whose expected query complexity is logarithmic in n under certain assumptions. However, the constant factors hidden in the asymptotic query time depend exponentially on the dimension d [2]. Another structure suitable for low-dimensional spaces is the geometric near-neighbor access tree (GNAT); as claimed in [7], typically the construction time is $O(dn \log n)$ and only linear space is required. In the extended version [23] we also discuss methods that adapt to the intrinsic dimension of the subspace where the points lie.

All the aforementioned structures give an exact solution to the problem. However, many approximate algorithms exist, and often perform significantly faster than the exact ones, especially when d is high. Among the prominent approximate algorithms are *Balanaced box-decomposition trees* (BBD-trees) [2], and Locality-sensitive hashing (LSH) [19]. See [18] for a survey on approximate NN methods in high-dimensional spaces.

Finally, we note that in the context of MP, several specially-tailored exact [17, 34] and approximate [22, 30] techniques were previously described. A theoretical justification for using approximate NN methods rather than exact ones is proven in [33] for PRM*.

3   The asymptotic behavior of common MP algorithms

In this section we provide more background on the asymptotic complexity analysis of various sampling-based MP algorithms. We then show that for both PRM-type algorithms and RRT-type algorithms, the expected ratio between the time spent on NN search and the time spent on CD goes to infinity as $n \to \infty$.

We denote by N the total number of configurations sampled by the algorithm, and by n the number of collision-free configurations in the roadmap. Let m denote the complexity of the workspace obstacles and assume that the robot is of constant-description complexity[4].

3.1   Complexity of common motion-planning algorithms

We start by summarizing the computational complexity of the primitive operations and continue to detail the computational complexity of a selected set of algorithms. We assume familiarity with the planners that are discussed.

---

[3] We refer to a space as low dimensional when its dimension is at most a few dozens.

[4] The assumption that the robot is of constant-description complexity implies that testing for *self-collision* can be done in constant time.

Complexity of primitive operations The main primitive operations that we consider are (i) nearest-neigbhor operations (NN and R-NN) and (ii) collision-detection operations (CD and LP). Additionally, MP algorithms make use of priority queues and graph operations. We assume, as is typically the case, that the running time of these operations is negligible when compared to NN and CD.

Since many NN data structures require a preprocessing phase, the complexity of a single query should consider the amortized cost of preprocessing. However, since usually at least n NN or R-NN queries are performed, where n is the number of points stored in the NN data structure, this amortized preprocessing cost is asymptotically subsumed by the cost of a query.

A list of the common complexity bounds for the different types of NN queries can be found in the extended version of this paper [23]. As mentioned in Sec. 2.2, the complexity of a single CD operation for a robot of a constant-description complexity can be bounded by $O(m \log^{\delta-1} m + s)$, where $\delta \in \{2, 3\}$ is the dimension of the workspace and s is the number of intersections between the bounding volumes, which is $O(m^2)$ in the worst case. On the other hand, for a system with ℓ such robots, a CD operation is composed of ℓ single robot CD queries as well as $O(\ell^2)$ robot-robot collision checks. Local planning (LP) is often implemented using multiple CD operations along a densely-sampled C-space line-segment between two configurations. Specifically, we assume that the planner is endowed with a fixed parameter called STEP specifying the sampling density along edges. During LP, edges of maximal length r_n will be subdivided into $\lceil r_n/\text{STEP} \rceil$ collision-checked configurations (see also [25, p. 214]). Therefore, the complexity of a single LP query can be bounded by $O(r_n \cdot Q_{\text{CD}})$, where Q_{CD} is the complexity of a single CD query (here STEP is assumed to be constant).

Complexity of algorithms In order to choose which edges of $\mathcal{G}$ to explicitly check for being free, all algorithms need to determine (i) which of the N nodes are collision free and (ii) what are the neighbors of each node. Thus, these algorithms typically require N CD calls and n R-NN calls.

To quantify the number of LP calls performed by each algorithm, note that the expected number of neighbors of a node in $\mathcal{G}$ is $\Theta(\eta^d 2^d \log n)$ [33]. Therefore, if an algorithm calls the local planner for all (or for a constant fraction of) the edges of $\mathcal{G}$, then the expected number of LP calls will be $\Theta(\eta^d 2^d n \log n)$.

3.2 The asymptotic behavior of the ratio $\chi_{\text{ALG}}(N)$

Let $T_{\text{CD}}(S)$ be the overall time spent on CD for a specific motion-planning problem after a set S of configurations was sampled, where we assume, as before, that all other parameters of the problem are fixed. Let $T_{\text{CD}}(N)$ be the expected value of $T_{\text{CD}}(S)$ over all sample sets S of size N. We show here that the expected value $\chi_{\text{ALG}}(N)$ of the ratio over all sample sets of size N goes to infinty as $N \to \infty$ for both sPRM* and RRT*. Recall that we are interested in the expected value of the ratio. We do that by looking at the ratio between a lower bound on the time of NN and $T_{\text{CD}}(N)$, defined above.

To obtain a lower bound on the time of NN, we assume that the NN structure being used is a j-ary tree for a constant j, in which the data points are kept in the leaves. This is a reasonable assumption, as many standard NN structures are based on trees [2, 3, 7, 12]. Performing n queries of NN (or R-NN) using this structure, one for every data point, costs $\Omega(n \log n)$, as each query involves locating the leaf in which the query point lies. It is easy to show this both for sPRM*, in which the NN structure is constructed given a batch of all data points, and for RRT*, where the structure is constructed incrementally.

Additionally, we have the following lemma, whose proof is given in [23]:

Lemma 1 *If an algorithm uses a uniform set of samples and the C-space obstacles occupy a constant fraction of the C-space, then $n = \Theta(N)$ almost surely.*

For sPRM* it holds that $\mathrm{T_{CD}}(N) = \#_{\mathrm{CD}} \cdot \mathrm{Q_{CD}} + \#_{\mathrm{LP}} \cdot \mathrm{Q_{LP}}$. Clearly, $\#_{\mathrm{CD}} = N$ and $\mathrm{Q_{CD}} = O(m^2)$ (see Sec. 3.1). In expectation we have that $\#_{\mathrm{LP}} = O(n^2 \cdot r_n^d)$ and in addition $\mathrm{Q_{LP}} = O(r_n \cdot \mathrm{Q_{CD}})$. Finally, recall that $r_n = \Theta\left(2\eta \left(\log n/n\right)^{1/d}\right) = \Theta\left((\log n/n)^{1/d}\right)$. Therefore,

$$\begin{aligned}
\mathrm{T_{CD}}(N) &= N \cdot \mathrm{Q_{CD}} + O(\eta^d 2^d n \log n) \cdot \mathrm{Q_{LP}} \\
&= N \cdot \mathrm{Q_{CD}} + O(\eta^d 2^d n \log n) \cdot O(r_n \cdot \mathrm{Q_{CD}}) \\
&= O(m^2 N) + O(m^2 N^{1-1/d} \log^{1+1/d} N) = O(m^2 N).
\end{aligned} \tag{2}$$

As $\Omega(n \log n)$ is a valid lower bound on the overall complexity of NN, there exists a constant $c_2 > 0$ s.t. the time for NN for a roadmap with n nodes is at least $c_2 n \log n$. Moreover, since $\mathrm{T_{CD}}(N) = O(m^2 N)$ then there exists a constant $c_3 > 0$ s.t. the overall time for CD is at most $c_3 m^2 N$.

Thus, using Lemma 1, $\chi_{\mathrm{sPRM*}}(N) \geq \frac{c_2 n \log n}{c_3 m^2 N} \geq \frac{c' \log N}{m^2}$, where $c' > 0$ is a constant. Observing that the above fraction goes to infinity as N goes to infinity, we obtain that $\lim_{N \to \infty} \chi_{\mathrm{sPRM*}}(N) = \infty$, as anticipated.

We note that although m is assumed to be constant, we leave it in our analysis to emphasize its effect on $\chi_{\mathrm{ALG}}(N)$. In summary,

Proposition 2 *The values $\chi_{\mathrm{sPRM*}}(N)$ and $\chi_{\mathrm{RRT*}}(N)$ tend to infinity as $N \to \infty$.*

The proof for $\chi_{\mathrm{RRT*}}(N)$ can be found in the extended version [23].

From a theoretical standpoint, NN search determines the asymptotic running time of typical sampling-based MP algorithms. In contrast, the common experience is that CD dominates the running time in practice. However, we show in the remainder of the paper that in a variety of special situations NN search is a non-negligible factor in the running-time in practice.

4   Nearest-neighbor sensitive settings

In this section we describe settings where the computational role of NN search in practice is far from negligible, even for a relatively small number of samples. We call these settings *NN-sensitive*. For each such setting we empirically demonstrate this behavior. In Sec. 4.1 we describe our experimental methodology and outline properties common to all our experiments. Each of the subsequent sections is devoted to a specific type of NN-sensitivity.

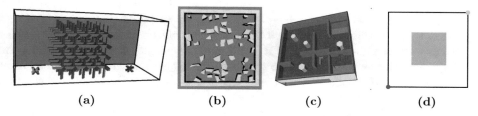

Fig. 2: Scenarios used in experiments. (a) 3D Grid, (b) 2D Random polygons, (c) 3D Cubicles and (d) dD Hypercube with a centered obstacle. Start and target configurations for a robot are depicted in green and red, respectively. Scenarios (b) and (c) are provided with the OMPL distribution. More details are provided in the body of the paper.

4.1 Experimental methodology

In our experiments, we ran the Open Motion Planning Library (OMPL 1.1) [11] on a 2.5GHz×4 Intel Core i5 processor with 8GB of memory. Each reported result is averaged over fifty (50) runs and includes error bars, which denote the 20'th and 80'th percentiles. The scenarios used are depicted in Fig. 2.

Several of our experiments are in non-Euclidean C-spaces, which in turn require a closed-form expression for ζ_d, the measure (volume) of the d-dimensional unit ball (see Eq. 1). In Sec. 5 we describe a general approach to compute this value together with a heuristic that makes the computed radius effective in practice. This heuristic is used in all the experiments presented in this section.

What do we measure? Recall that our main thesis is that while the folklore in MP is that the running time of sampling-based algorithms in practice is strongly dominated by CD, we (and others) observe that quite often the time taken up by NN-search is significant, and not rarely larger than the time taken up by CD. Therefore, our primary measure is *wall time*, namely the running time spent on the different primitives as gauged in standard clock time (to distinguish from CPU-time or other more system-specific measurements like number of floating point operations). The principal reason for doing that is that wall time is what matters most in practice. This, for example, will affect the response time of a planner used by a robot system. One may argue that this measurement may only be meaningful for a very limited suite of software tools used by motion planners. However, we use state-of-the-art tools that are used by many. There is not such an abundance of efficient stable software tools for this purpose, and most researchers in the field seem to use a fairly small set of tools for CD and NN. This said, we still provide additional measurements for each experiment— the average number of basic operations, which should allow people who come up with their own (or specialized) MP primitives to assess what will be the effect of their special primitives on the overall running time of the algorithms in practice.

4.2 NN-sensitive algorithms

In recent years, several planners were introduced, which *algorithmically* shift some of the computational cost from CD to NN search. Two such examples are

Lazy-PRM* [14] and MPLB [31], though lazy planners were described before (e.g., [6]). Both algorithms delay local planning by building an RGG $\mathcal{G}$ over a set of samples *without* checking if the edges are collision free. Then, they employ graph-search algorithms to find a solution. To construct $\mathcal{G}$ only NN queries are required. Moreover, using these graph-search algorithms dramatically reduces the number of LP calls. Thus, in many cases (especially as the number of samples grows) the weight of CD is almost negligible with respect to that of NN.

Specifically, Lazy-PRM* iteratively computes the shortest path in $\mathcal{G}$ between the start and target configurations. LP is called only for the edges of the path. If some are found to be in collision, they are removed from the graph. This process is repeated until a solution is found or until the source and target do not lie in the same connected component. We use a batch variant of Hauser's Lazy-PRM* algorithm [14], which we denote by Lazy-sPRM*. This variant constructs the roadmap in the same fashion as sPRM* does but delays LP to the query phase.

MPLB uses $\mathcal{G}$ to compute lower bounds on the cost between configurations to tightly estimate the cost-to-go [31]. These bounds are then used as a heuristic to guide the search of an anytime version of FMT* [20]. The bounds are computed by running a shortest-path algorithm over $\mathcal{G}$ from the target to the source. Fig. 1 (in Sec. 1) presents the amounts of time of NN, CD and other operations used by MPLB running on the 3D Grid scenario for two robots translating and rotating in space that need to exchange their positions (Fig. 2a). With several thousands of iterations, which are required for obtaining a high-quality solution, NN dominates the running time of the algorithm. For additional details see [23].

Additional experiments demonstrating the behavior of NN-sensitive algorithms can be found in [5, 14, 31].

4.3   NN-sensitive scenarios

A scenario $\mathcal{S} = (\mathcal{W}, \mathcal{R})$ is defined by a workspace $\mathcal{W}$ and a robot system $\mathcal{R}$. The robot system $\mathcal{R}$ may, in turn, be a set of ℓ single constant-description complexity robots operating simultaneously in $\mathcal{W}$. Let the dimension d of $\mathcal{S}$ be the dimension of the C-space induced by $\mathcal{R}$, and, hence, $d = \Theta(\ell)$. Let the complexity of $\mathcal{S}$ be the complexity m of the workspace obstacles. Note that CD is affected by ℓ, as both robot-obstacle and robot-robot collisions should be considered. Therefore, the bound on the complexity of a CD operation is: $O(\ell \cdot m^2 + \ell^2)$; see Sec. 3.1.

We next show how the role of NN may increase when (i) the dimension of $\mathcal{S}$ increases or (ii) the complexity of $\mathcal{S}$ decreases.

The effect of the dimension d Proposition 2 states that as the number of samples tends to infinity, NN dominates the running time of the algorithm. A natural question to ask is "what happens when we fix the number of samples and increase the dimension?" The different structure of RRT* and sPRM* merits a different answer for each algorithm.

*RRT** Here, we show that the NN sensitivity grows with the number of *unsuccessful iterations*[5]. This implies that if the number of unsuccessful iterations

---

[5] Here, an iteration is said to be unsuccessful when the RRT* tree is not extended.

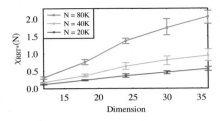

d	$\#_{\text{NN}}$	$\#_{\text{R-NN}}$	$\#_{\text{CD}}$	$\#_{\text{LP-A}}$	$\#_{\text{LP-B}}$	$\#_{\text{CD in LP}}$
12	80K	10.2K	80K	80K	325K	4,235K
24	80K	7.5K	80K	80K	724K	4,135K
36	80K	5.7K	80K	80K	886K	3,812K

Fig. 3: $\chi_{\text{RRT}^*}(\text{N})$ as a function of d in the 3D Cubicles scenario (Fig. 2c), when fixing the number N of iterations.

Table 1: Average number of calls to the main primitive operations for different values of d, for $N = 80K$ iterations.

grows with the dimension, so will $\chi_{\text{RRT}^*}(N)$. Indeed, we demonstrate this phenomenon in the 3D cubicles scenario (Fig. 2c). Note that in this situation the effect of d is indirect.

To better discuss our results we define two types of LP operations: the first is called when the algorithm attempts to grow the tree towards a random sample while the second is called during the rewiring step. We denote the former type of LP calls by LP-A and the latter by LP-B and note that LP-A will occur every iteration while LP-B will occur only in successful ones.

We use ℓ translating and rotating L-shaped robots. We gradually increase ℓ from two to six, resulting in a C-space of dimension $d = 6\ell$. Robots are placed in different sections of the workspace and can reach their target with little robot-robot interaction. We fix N and measure $\chi_{\text{RRT}^*}(N)$ as a function of d. The results for several values of N are depicted in Fig. 3. Additionally, Table 1 shows the average number of operation calls for various values of d.

As d grows, the number of unsuccessful iterations grows (see $\#_{\text{LP-A}}$ in Table 1). This growth, which is roughly linear with respect to d induces a linear increase in $\chi_{\text{RRT}^*}(N)$ for a given N (see Fig. 3). Furthermore, the slope of this line increases with N which further demonstrates the fact that for a fixed d, $\lim_{N\to\infty} \chi_{\text{RRT}^*}(N) = \infty$.

Finally, Fig. 4 depicts the time decomposition of the main primitives as a function of d, for $N = 80K$.

sPRM* Here, the NN sensitivity of the algorithm is more complex. The reason, roughly, is that for a *fixed* n, the expected value of the number κ of reported neighbors is $\Theta(2^d n \log n)$. Thus, in expectation, κ grows exponentially in d. How-

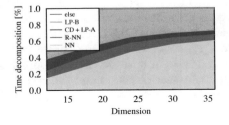

Fig. 4: Time breakdown of the main primitive operations in RRT* running on the 3D Cubicles scenario (Fig. 2c) as a function of d, for $N = 80K$ iterations. Best viewed in color.

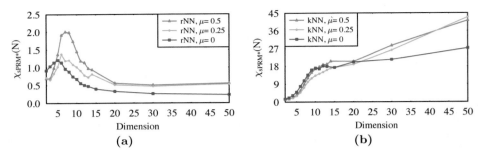

Fig. 5: $\chi_{\mathrm{sPRM*}}(N)$ as a function of d in the dD Hypercube scenario (Fig. 2d) for a roadmap with $n = 5000$ vertices. Experiments with obstacle measures μ of $0, 0.25$ and 0.5 are displayed. The connection strategies used in (a) and (b) are R-NN and K-NN, respectively.

ever, for large enough values of d, we have $\kappa = \Theta(n^2)$. Interestingly, this means that the computational cost of the overall NN time shifts from *finding* the set of nearest neighbors to *reporting* it.

According to our analysis, presented in [23], we expect to see an initial increase of $\chi_{\mathrm{sPRM*}}(N)$ followed by a convergence to a constant value. The increase in $\chi_{\mathrm{sPRM*}}(N)$ is due to the increasing complexity of finding the set of nearest neighbors, which grows with the dimension[6]. A possible decrease will occur as the computational weight "shifts" to reporting the set of neighbors followed by an asymptotic convergence to a constant value, for large values of d.

Aiming to test this conjecture, we solved a planning problem for a point robot moving in the d-dimensional unit hypercube containing a hyper-cubicle obstacle of measure μ (Fig. 2d). Indeed, this trend can be seen in Fig. 5a. For additional details see [23]. We repeated the experiment while using K-NN instead of R-NN queries. The results, depicted in Fig. 5b, show a growth in the ratio as a function of d by 2000%. This is not surprising, as the standard value of k that is commonly used is proportional to $\log n$, and is smaller by a factor of 2^d than the expected number of neighbors returned by an R-NN query.

The effect of the geometric complexity m of the obstacles Recall that a collision query may take $O(m^2)$ in the worst case. For small values of m, this becomes negligible with respect to other primitive operations, such as NN. To demonstrate this effect we ran the following experiment which is based on the 2D Random polygons scenario (Fig. 2b). We created two sequences of increasing geometric-complexity (growing m) environments. Each sequence was constructed as follows: we start with the empty environment and incrementally add random polygons from Fig. 2b until all the polygons have been added. We then placed *eight* robots that need to change their positions, and ran LBT-RRT (with approx. factor $\varepsilon = 0.4$) for a fixed N. Fig. 6 plots $\chi_{\mathrm{LBT\text{-}RRT}}(N)$ as a function of m for two sets of environments. As anticipated, the ratio in both sets of environments decays polynomially as m grows. See [23] for further discussion on the results.

---

[6] Here, we assume, as is common in the literature, that the cost of *finding* the set of NN grows with the dimension.

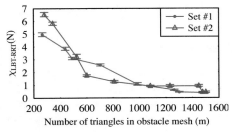

m	$\#_{\text{NN}}$	$\#_{\text{R-NN}}$	$\#_{\text{CD}}$	$\#_{\text{LP}}$	$\#_{\text{CD in LP}}$
266	4K	2.5K	4K	8.1K	496K
704	4K	1.7K	4K	7.7K	352K
1,262	4K	0.99K	4K	24.5K	271K
1,476	4K	0.7K	4K	28.7K	232K

Fig. 6: $\chi_{\text{LBT-RRT}}(N)$ as a function of m in workspaces of increasing obstacle complexity based on the 2D Random polygons scenario (Fig. 2b). The presented plots are for $N = 4$K iterations in two different randomly-generated experiment sets.

Table 2: Average number of calls to the main primitive operations for different values of m (chosen arbitrarily).

4.4 NN-sensitive parameters

In all planning algorithms, one of the critical user-defined parameters, is the step size (STEP); see Sec. 3. Using STEP which is too small may cause LP to be over-conservative and costly. Choosing larger values which are still appropriate for the scenario at hand allows to decrease the portion of time spent on CD checks.

We demonstrate how RRT* becomes NN-sensitive under certain step-size values. We ran RRT* for $N = 25$K iterations on the 3D Cubicles scenario (Fig. 2c). In order to modify the step size in OMPL, one needs to specify a state validity-checking resolution. This value, which we denote by RES, is specified as a fraction of the space's extent, that is, the maximal possible distance between two configurations in the C-space. Using larger values of RES may yield paths that are invalid. Thus, when increasing RES, we also used a model of the robot which was inflated accordingly to ensure that all paths are collision free (see [25, Ch.5.3.4]). A linear correlation between RES and $\chi_{\text{RRT}*}(N)$ was observed. The ratio has increased from 0.05 for the default OMPL value of 1% to 0.4 for RES=10%, which was the largest value yielding valid paths. See [23] for more details.

5 Asymptotically-optimal motion-planning using R-NN

In this section we address an existing gap in the literature of sampling-based MP algorithms: How to use Eq. 1 in non-Euclidean spaces, which are prevalent in motion planning. Specifically, we derive closed-form expressions for the volume of the unit ball in several common C-spaces and distance metrics and discuss how to effectively use this value.

Closing this gap allows to evaluate the connection scheme of an algorithm. Namely, should one choose connections using R-NN or K-NN. In NN-sensitive settings this choice may have a dramatic effect on the performance of the algorithm since (i) the number of reported neighbors may differ and (ii) the cost of the two query types for a certain NN data structure may be different. Indeed, we show empirically that there are scenarios where using R-NN, a solution path of a specific cost is produced ten-fold (and more) faster than with K-NN.

Due to lack of space, some of the technical details and experiments appear only in the extended version of this paper [23].

5.1   Well-behaved spaces and the volume of balls

Recall that $\mathcal{X}$ denotes a C-space and that given a set $A \subseteq \mathcal{X}$, $\mu(A)$ denotes the Lebesgue measure of A. Let $\rho : \mathcal{X} \times \mathcal{X} \rightarrow \mathbb{R}$ denote a distance metric and let $\mathcal{B}_{\mathcal{X}}^{\rho}(r, x) := \{y \in \mathcal{X} | \rho(x, y) \leq r\}$ and $\mathcal{S}_{\mathcal{X}}^{\rho}(r, x) := \{y \in \mathcal{X} | \rho(x, y) = r\}$ denote the ball and sphere of radius r (defined using ρ) centered at $x \in \mathcal{X}$, respectively. Finally, let $\mathbb{B}_{\mathcal{X}}^{\rho}(r) := \mu\left(\mathcal{B}_{\mathcal{X}}^{\rho}(r, 0)\right)$ and $\mathbb{S}_{\mathcal{X}}^{\rho}(r) := \mu\left(\mathcal{S}_{\mathcal{X}}^{\rho}(r, 0)\right)$. We will often omit the superscript ρ or the subscript $\mathcal{X}$ when they are clear from the context.

We now define the notion of a *well-behaved* space in the context of metrics; for a detailed discussion on well-behaved spaces see [27]. In such spaces there is a derivative relationship between $\mathbb{S}(r)$ and $\mathbb{B}(r)$. Formally,

Definition 3 *A space $\mathcal{X}$ is well behaved when $\frac{\partial \mathbb{B}_{\mathcal{X}}(r)}{\partial r} = \mathbb{S}_{\mathcal{X}}(r)$. Conversely, we say that $\mathcal{X}$ is well behaved when $\int_{\varrho \in [0,r]} \mathbb{S}_{\mathcal{X}}(\varrho) d\varrho = \mathbb{B}_{\mathcal{X}}(r)$.*

We continue with the definition of a *compound space* which is the Cartesian product of two spaces. Let $\mathcal{X}_1, \mathcal{X}_2$ be two C-spaces with distance metrics ρ_1, ρ_2, respectively. Define $\mathcal{X} = \mathcal{X}_1 \times \mathcal{X}_2$ to be their compound space. We adopt a common way to define the (weighted) distance metric over $\mathcal{X}$, when using weights $w_1, w_2 \in \mathbb{R}^+$ and some constant p [25, Chapter 5]: $\rho_{\mathcal{X}} = \left(w_1 \rho_1^p + w_2 \rho_2^p\right)^{1/p}$.[7]

The following Lemma states that the volume of balls in a compound space $\mathcal{X} = \mathcal{X}_1 \times \mathcal{X}_2$ where $\mathcal{X}_1$ is well behaved can be expressed analytically.

Lemma 4 *Following the above notation, if $\mathcal{X}_1$ is well behaved then*

$$\mathbb{B}_{\mathcal{X}_1 \times \mathcal{X}_2}(r) = \int_{\varrho \in [0, r/w_1^{1/p}]} \mathbb{S}_{\mathcal{X}_1}(\varrho) \cdot \mathbb{B}_{\mathcal{X}_2}\left(\left(\frac{r^p - w_1 \varrho^p}{w_2}\right)^{1/p}\right) d\varrho. \qquad (3)$$

Proof. By definition, $\mathbb{B}_{\mathcal{X}}(r) = \int_{x \in \mathcal{B}_{\mathcal{X}}(r)} dx$. Using Fubini's Theorem [28],

$$\mathbb{B}_{\mathcal{X}_1 \times \mathcal{X}_2}(r) = \int_{x_1 \in \mathcal{B}_{\mathcal{X}_1}(r/w_1^{1/p})} \left(\int_{x_2 \in \mathcal{B}_{\mathcal{X}_2}\left(\left(\frac{r^p - w_1 x_1^p}{w_2}\right)^{1/p}\right)} dx_2 \right) dx_1.$$

The inner integral is simply the volume of a ball of radius $\left(\frac{r^p - w_1 x_1^p}{w_2}\right)^{1/p}$ in $\mathcal{X}_2$. In addition, we know that $\mathcal{X}_1$ is well behaved, thus $\mathbb{B}_{\mathcal{X}_1}(r) = \int_{x_1 \in \mathcal{B}_{\mathcal{X}_1}(r)} dx = \int_{\varrho \in [0,r]} \mathbb{S}_{\mathcal{X}_1}(\varrho) d\varrho$. By changing the integration variable, substituting the inner integral and using the fact that $\mathcal{X}_1$ is well behaved we obtain Eq. 3. $\qquad \square$

For a full list of C-spaces for which a closed-form expression for $\mathbb{B}_{\mathcal{X}}(r)$ was derived, we refer the reader to the extended version of this paper [23].

---

[7] This distance metric is often used due to its computational efficiency and simplicity. However, alternative methods exist, which exhibit favorable properties such as invariance to rotation of the reference frame; see, e.g., [8].

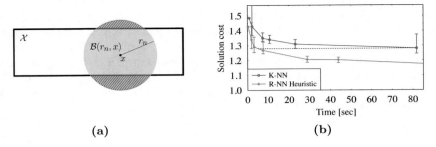

(a)   (b)

Fig. 7: R-NN heuristic (a) A two-dimensional C-space for which Assumption 5 does not hold. (b) Solution cost (optimal cost has a value of 1) as a function of time for Lazy-sPRM* running in the Cubicles scenario (Fig. 2c). The dashed line visualizes the difference in time for obtaining a solution of cost 1.28 between the R-NN heuristic and K-NN.

5.2   Effective use of R-NN in MP algorithms in practice

We now discuss how to effectively use the radial connection scheme of asymptotically (near) optimal MP algorithms. We first describe a common scenario for which the computed radii are practically useless, and continue by suggesting a simple heuristic to overcome this problem.

The proofs of asymptotic optimality provided by Karaman and Frazzoli [21] and by Janson et al. [20] rely on the following implicit assumption:

Assumption 5 *For $x \in \mathcal{X}$, w.h.p.* $\mathbb{B}_{\mathcal{X}}(r,x) = \mu(\mathcal{B}_{\mathcal{X}}(r,x) \cap \mathcal{X})$.

This assumption does not hold when the center of the ball is close to the boundary of the C-space $\mathcal{X}$. However, since the proofs consider balls of radii proportional to r_n which tends to zero as $n \to \infty$ (as in Eq. 1), there exists a value n_0 for which r_{n_0} is sufficiently small such that the sequence of balls of radius r_{n_0} covering a given solution path does not intersect the boundary of $\mathcal{X}$.

In many common settings, Assumption 5 does not hold for practical values of n. Consider, for example, Fig. 7a, which depicts a two-dimensional rectangular C-space where one dimension is significantly larger than the other. For small values of n, any ball of radius r_n intersects the boundary of $\mathcal{X}$ (as the ball $\mathcal{B}(r_n,x)$, drawn in orange, for which $\mathcal{B}(r_n,x) \setminus \mathcal{X} \neq \emptyset$). As a result, the number of configurations of distance at most r_n from a configuration x might be too small, and this, in turn, may cause the roadmap of n vertices to remain disconnected.

We start by formally describing the setting and propose a heuristic to choose a larger radius. Let $\mathcal{X} = \mathcal{X}_1 \times \mathcal{X}_2$ be a d-dimensional compound C-space and assume that $\mu(\mathcal{X}_1) \geq \mu(\mathcal{X}_2)$. Let d_1, d_2 denote the dimensions of $\mathcal{X}_1, \mathcal{X}_2$, respectively. Finally, let $\rho_{\max(\mathcal{X}_2)}$ be the maximal distance between any two points in $\mathcal{X}_2$ and assume that $\rho = w_1 \rho_1 + w_2 \rho_2$. When Assumption 5 does not hold, as in Fig. 7, the intuition is that the "effective dimension" of our C-space is closer to d_1 than to $d_1 + d_2$. If $r_n > w_2 \rho_{\max(\mathcal{X}_2)}$ then $\forall x \in \mathcal{X}$ $\mathbb{B}_{\mathcal{X}}(r_n,x) > \mu(\mathcal{B}_{\mathcal{X}}(r_n,x) \cap \mathcal{X})$. In such cases, we suggest to project all points to $\mathcal{X}_1$ and use the critical connection radius that we would have used had the planning occurred in $\mathcal{X}_1$.

To evaluate the proposed heuristic, we ran Lazy-sPRM* on the Cubicles scenario (Fig. 2c) using R-NN with and without the heuristic, and also using K-NN strategy (with the standard k_n value). We measured the cost of the solution path as a function of the running time. As depicted in Fig. 7b, the heuristic was able to find higher-quality solutions using less samples, resulting in a ten-fold speedup in obtaining a solution of a certain cost, when compared to K-NN. Moreover, R-NN without the heuristic was practically inferior, as it was not able to find a solution even for large values of n; results omitted.

6   Future work

An immediate extension of our work is to develop improved NN-search techniques tailored for MP. Moreover, it would be interesting to identify further NN-sensitive settings in which such tailored structures can be used to reduce the overall running time of the MP algorithm.

Another research question arising from our study is concerned with the choice of connection strategy. That is, which of R-NN or K-NN is preferable for a given setting in terms of NN-sensitivity and which other factors may affect this choice.

References

[1] Arslan, O., Tsiotras, P.: Use of relaxation methods in sampling-based algorithms for optimal motion planning. In: ICRA. pp. 2413–2420 (2013)
[2] Arya, S., Mount, D.M., Netanyahu, N.S., Silverman, R., Wu, A.Y.: An optimal algorithm for approximate nearest neighbor searching in fixed dimensions. Journal of the ACM 45(6), 891–923 (1998)
[3] Bentley, J.L.: Multidimensional binary search trees used for associative searching. Commun. ACM 18, 509–517 (1975)
[4] de Berg, M., Cheong, O., van Kreveld, M., Overmars, M.: Computational Geometry: Algorithms and Applications. Springer-Verlag, 3rd edn. (2008)
[5] Bialkowski, J., Otte, M.W., Karaman, S., Frazzoli, E.: Efficient collision checking in sampling-based motion planning via safety certificates. I. J. Robotic Res. 35(7), 767–796 (2016)
[6] Bohlin, R., Kavraki, L.E.: Path planning using lazy PRM. In: ICRA. pp. 521–528 (2000)
[7] Brin, S.: Near neighbor search in large metric spaces. In: VLDB. pp. 574–584 (1995)
[8] Chirikjian, G.S., Zhou, S.: Metrics on motion and deformation of solid models. J. Mech. Des. 120(2), 252–261 (1998)
[9] Choset, H., Lynch, K.M., Hutchinson, S., Kantor, G., Burgard, W., Kavraki, L.E., Thrun, S.: Principles of Robot Motion: Theory, Algorithms, and Implementation. MIT Press (June 2005)
[10] Cohen, J.D., Lin, M.C., Manocha, D., Ponamgi, M.: I-COLLIDE: an interactive and exact collision detection system for large-scale environments. In: Symposium on Interactive 3D Graphics. pp. 189–196, 218 (1995)
[11] Şucan, I.A., Moll, M., Kavraki, L.E.: The Open Motion Planning Library. IEEE Robotics & Automation Magazine 19(4), 72–82 (2012)
[12] Friedman, J.H., Bentley, J.L., Finkel, R.A.: An algorithm for finding best matches in logarithmic expected time. ACM Trans. Math. Softw. 3(3), 209–226 (1977)

[13] Gammell, J.D., Srinivasa, S.S., Barfoot, T.D.: Informed RRT*: Optimal sampling-based path planning focused via direct sampling of an admissible ellipsoidal heuristic. In: IROS. pp. 2997–3004 (2014)

[14] Hauser, K.: Lazy collision checking in asymptotically-optimal motion planning. In: ICRA. pp. 2951–2957 (2015)

[15] Hemmer, M., Kleinbort, M., Halperin, D.: Optimal randomized incremental construction for guaranteed logarithmic planar point location. Comput. Geom. 58, 110–123 (2016)

[16] Hubbard, P.M.: Approximating polyhedra with spheres for time-critical collision detection. ACM Trans. Graph. 15(3), 179–210 (1996)

[17] Ichnowski, J., Alterovitz, R.: Fast nearest neighbor search in SE(3) for sampling-based motion planning. In: WAFR. pp. 197–214 (2014)

[18] Indyk, P.: Nearest neighbors in high-dimensional spaces. In: Goodman, J.E., O'Rourke, J. (eds.) Handbook of Discrete and Computational Geometry, chap. 39, pp. 877–892. CRC Press LLC, Boca Raton, FL, 2nd edn. (2004)

[19] Indyk, P., Motwani, R.: Approximate nearest neighbors: Towards removing the curse of dimensionality. In: STOC. pp. 604–613 (1998)

[20] Janson, L., Schmerling, E., Clark, A.A., Pavone, M.: Fast marching tree: A fast marching sampling-based method for optimal motion planning in many dimensions. I. J. Robotic Res. 34(7), 883–921 (2015)

[21] Karaman, S., Frazzoli, E.: Sampling-based algorithms for optimal motion planning. I. J. Robotic Res. 30(7), 846–894 (2011)

[22] Kleinbort, M., Salzman, O., Halperin, D.: Efficient high-quality motion planning by fast all-pairs r-nearest-neighbors. In: ICRA. pp. 2985–2990 (2015)

[23] Kleinbort, M., Salzman, O., Halperin, D.: Collision detection or nearest-neighbor search? On the computational bottleneck in sampling-based motion planning. CoRR abs/1607.04800 (2016)

[24] Larsen, E., Gottschalk, S., Lin, M.C., Manocha, D.: Fast proximity queries with swept sphere volumes. Tech. rep., Department of Computer Science, University of North Carolina (1999), TR99-018

[25] LaValle, S.M.: Planning Algorithms. Cambridge University Press (2006)

[26] Lin, M.C., Manocha, D.: Collision and proximity queries. In: Goodman, J.E., O'Rourke, J. (eds.) Handbook of Discrete and Computational Geometry, chap. 35, pp. 767–786. CRC Press LLC, Boca Raton, FL, 2nd edn. (2004)

[27] Marichal, J.L., Dorff, M.: Derivative relationships between volume and surface area of compact regions in $\mathbb{R}^d$. Rocky Mountain J. Math. 37(2), 551–571 (2007)

[28] Mazzola, G., Milmeister, G., Weissmann, J.: Comprehensive Mathematics for Computer Scientists 2, chap. 31, pp. 87–95. Springer, Berlin, Heidelberg (2005)

[29] Pan, J., Chitta, S., Manocha, D.: FCL: A general purpose library for collision and proximity queries. In: ICRA. pp. 3859–3866 (2012)

[30] Plaku, E., Kavraki, L.E.: Quantitative analysis of nearest-neighbors search in high-dimensional sampling-based motion planning. In: WAFR. pp. 3–18 (2006)

[31] Salzman, O., Halperin, D.: Asymptotically-optimal motion planning using lower bounds on cost. In: ICRA. pp. 4167–4172 (2015)

[32] Salzman, O., Halperin, D.: Asymptotically near-optimal RRT for fast, high-quality, motion planning. IEEE Trans. Robotics 32(3), 473–483 (2016)

[33] Solovey, K., Salzman, O., Halperin, D.: New perspective on sampling-based motion planning via random geometric graphs. Robots Science and Systems (RSS) (2016)

[34] Yershova, A., LaValle, S.M.: Improving motion-planning algorithms by efficient nearest-neighbor searching. IEEE Trans. Robotics 23(1), 151–157 (2007)

Dynamic Region-biased Rapidly-exploring Random Trees

Jory Denny[1], Read Sandström[2], Andrew Bregger[2], and Nancy M. Amato[2]

[1] Department of Mathematics and Computer Science,
University of Richmond, Richmond, VA, USA,
jdenny@richmond.edu.
[2] Parasol Lab, Department of Computer Science and Engineering,
Texas A&M University, College Station, TX, USA,
{readamus, amato}@cse.tamu.edu.

Abstract. Current state-of-the-art motion planners rely on sampling-based planning to explore the problem space for a solution. However, sampling valid configurations in narrow or cluttered workspaces remains a challenge. If a valid path for the robot correlates to a path in the workspace, then the planning process can employ a representation of the workspace that captures its salient topological features. Prior approaches have investigated exploiting geometric decompositions of the workspace to bias sampling; while beneficial in some environments, complex narrow passages remain challenging to navigate.

In this work, we present Dynamic Region-biased RRT, a novel sampling-based planner that guides the exploration of a Rapidly-exploring Random Tree (RRT) by moving sampling regions along an embedded graph that captures the workspace topology. These sampling regions are dynamically created, manipulated, and destroyed to greedily bias sampling through unexplored passages that lead to the goal. We show that our approach reduces online planning time compared with related methods on a set of maze-like problems.

1 Introduction

State-of-the-art motion planners rely on randomized sampling [19] to construct an approximate model of the problem space that is then searched for a valid path. Planners for various types of robotic systems rely on this paradigm. However, constrained motion remains a fundamental challenge [13].

Motion planning applications such as robotics, virtual prototyping [34], and virtual reality [21] often require an object to move through narrow areas of an environment. Portions of the environment where the probability of sampling a valid object position is low are called "narrow passages". For problems with translational components, these "narrow passages" are often highly correlated with the environment geometry.

Many approaches extract information from the workspace to exploit the relationship between workspace geometry and allowable solutions. Examples include

© Springer Nature Switzerland AG 2020
K. Goldberg et al. (Eds.): *Algorithmic Foundations of Robotics XII*, SPAR 13, pp. 640–655, 2020.
https://doi.org/10.1007/978-3-030-43089-4_41

the use of collision information to adapt sampling [5,35], directing sampling with obstacle geometry [29,36], and decomposing the world into subproblems for the planner [18,24].

In this paper, we study how a representation of the workspace topology can be employed in a dynamic sampling strategy that leads a planner through the workspace and focuses its resources on unexplored areas. We present a novel planner called Dynamic Region-biased RRT based on Rapidly-exploring Random Trees (RRTs). Our method begins by pre-computing an embedded graph called a Workspace Skeleton that captures the topology of the free workspace (Figure 1(b)). At planning time, we compute a directed flow along the Workspace Skeleton called a Query Skeleton which extends outward from the start configuration (Figure 1(c)). Then, the method dynamically creates and moves sampling regions along the Query Skeleton to direct RRT exploration through the environment (Figure 1(d-f)). We use a Reeb Graph [27] as our Workspace Skeleton because it is a compact encoding of the changes in the topology of a volumetric space. Our contributions include:

- a novel workspace guidance for RRTs, called Dynamic Region-biased RRT, whereby sampling is biased toward sampling regions moving along an embedded skeleton of the workspace,
- the first sampling-based planner exploiting a Reeb Graph of the free workspace,
- and an experimental validation of our approach showing reduced online planning time in a set of maze-like environments.

2   Preliminaries and Related Work

In this section, we define important preliminaries of motion planning and describe relevant related work to our novel planner.

2.1   Motion Planning Preliminaries

In this paper, we discuss motion planning in the context of holonomic robots, i.e., robots whose *degrees of freedom* (DOFs) are fully controllable. The DOFs for a robot parameterize its position and orientation in the 2-d or 3-d world, or *workspace*. They include, for example, object position, orientation, joint angles, etc. A configuration is a single specification of the DOFs $q = \langle x_1, x_2, \ldots, x_n \rangle$, where x_i is the ith DOF and n is the total number of DOFs. The set of all possible parameterizations is called the *configuration space* ($\mathcal{C}_{space}$) [22].

$\mathcal{C}_{space}$ is often partitioned into two subsets, *free space* ($\mathcal{C}_{free}$) and *obstacle space* ($\mathcal{C}_{obst}$). In general, it is infeasible to explicitly compute a representation of $\mathcal{C}_{obst}$ [28], but we can often determine the validity of a configuration q quite efficiently by performing a workspace collision test between the robot placed at q and the environment. If the robot placed at q does not collide with itself or the environment, q is a valid configuration and $q \in \mathcal{C}_{free}$.

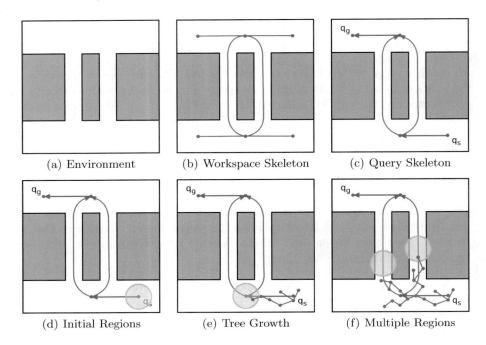

Fig. 1. Example execution of Dynamic Region-biased RRT: (a) environment; (b) pre-computed Workspace Skeleton (magenta) of the workspace; (c) Query Skeleton (magenta) computed from a query $\{q_s, q_g\}$; (d) initial region (green) placed at the source vertex of the Query Skeleton; (e) region-biased RRT growth (blue); and (f) multiple active regions (green) guiding the tree (blue) among multiple embedded arcs of the Query Skeleton (magenta) is shown.

Given a start configuration and a goal configuration or region, the motion planning problem is the problem of finding a continuous trajectory in $\mathcal{C}_{free}$ between the start and goal. We define a *query* as a start and goal pair.

While we describe our planner in terms of holonomic robots, i.e., planning in $\mathcal{C}_{space}$ only, it can be extended to nonholonomic robots, similar to [19].

Regions. We define a *region* as any bounded volume in the workspace, e.g., axis-aligned bounding boxes (AABBs) and bounding spheres (BSs). Each point p of a region R maps to a possibly infinite number of configurations in $\mathcal{C}_{space}$, e.g., by placing the center of mass of the robot at p and randomizing the remaining DOFs.

Homotopy. Two paths are defined to be *homotopy equivalent* if and only if one path can be continuously deformed to the other without transitioning through an obstacle region. A *homotopy class* is a set of paths such that any two are homotopy equivalent.

2.2   Sampling-based Planning

The high complexity of motion planning [28] has driven research toward developing sampling-based techniques that efficiently explore $\mathcal{C}_{free}$ in search of valid paths. Sampling-based planners, e.g., Rapidly-exploring Random Trees (RRTs) [19] attempt to construct a graph, called a *roadmap*, that is an approximate model of $\mathcal{C}_{free}$. We focus our related work on RRT approaches as they are most relevant to our planner.

Rapidly-exploring Random Trees (RRTs) Rapidly-exploring Random Trees (RRTs) [19] belong to the class of planning approaches tailored for solving single-query motion planning problems. RRTs iteratively grow a tree outwards from a root configuration q_{root}. In each expansion attempt, a random configuration q_{rand} is chosen, and the nearest configuration within the tree q_{near} is extended towards q_{rand} up to or at a fixed step size Δq [16]. From this extension, a new configuration q_{new} is computed and added to the tree if and only if there is a valid path from q_{near} to q_{new}.

RRTs have been quite useful for a broad range of robotic systems including dynamic environments [10], kinematically constrained robots [31], and nonholonomic robots [19]. Typically, RRTs are probabilistically complete and converge exponentially to the uniform sampling distribution over $\mathcal{C}_{free}$ due to Voronoi bias — RRTs are guaranteed to explore $\mathcal{C}_{free}$. Despite these important properties, RRTs lack efficiency in navigating many types of narrow passages. In this section, we highlight some RRT variants relevant to this research.

A few approaches attempt to limit needless RRT extensions. RRT with collision tendency [5] tracks the controls that have been tried when extending a node to reduce duplicated computations. Once all inputs have been tried, the node is excluded from nearest neighbor selection. Dynamic-Domain RRT [35] biases the random node selection to be within a radius r of q_{near} or expansion will not occur. The radius is dynamically determined from failed expansion attempts. RRT-Blossom [14] uses the idea of regression constraints to only add edges to a tree which explore new portions of the space and proposes a flood-fill approach when expanding from a node.

Obstacle-based RRT (OBRRT) [29] and Retraction-based RRT [36] exploit obstacle information to direct RRT growth along obstacle surfaces to improve RRT performance in narrow passages.

Recent approaches, e.g., RRT* [15], attempt to find optimal or near optimal motions. RRT* expands in the same way as RRT except after expansion the tree will locally "rewire" itself to ensure optimization of a cost function. RRT* has been shown to be quite effective in asymptotically finding shortest paths. In practice, RRT* requires many iterations to produce near optimal solutions. Recently proposed, Stable Sparse-RRT [20], a near-optimal planner, relaxes the optimality guarantee with a sparse tree to overcome this practical challenge.

Workspace-biased Planners Many planners use workspace information to aid in the planning process. Here we describe a few.

Feature Sensitive Motion Planning [24] recursively subdivides the space into "homogeneous" regions (regions of the environment containing similar properties, e.g., free or clutter), individually constructs a roadmap in each region, and merges them together to solve the aggregate problem. This framework adaptively decides the specific planning mechanism to map to each homogeneous region, e.g., use obstacle-based sampling [1] in cluttered regions and uniform sampling in open regions.

Other approaches utilize workspace decompositions to find narrow or difficult areas of the workspace to bias $\mathcal{C}_{space}$ sampling [2, 17, 18, 26]. These methods begin by decomposing the workspace using an adaptive cell decomposition [2] or a tetrahedralization [17, 18], and then they weight the decomposition to bias sampling. However, static determination of sampling regions often leads to oversampling in fully covered regions. Workspace Connectivity Oracle [18] mitigates this by preferring regions that bridge separate connected components.

SyCLoP [26] employs a graph-search over the cell decomposition to lead an RRT search and samples regions near the frontier of the resulting cell path. While similar in spirit to the method presented in this paper, it does not consider topology information in the biasing process.

In recent work, we proposed a planning approach that allows a user to define and manipulate regions of the workspace to bias probabilistic roadmap construction [7]. In this work, we utilize a similar concept of workspace regions, but do not rely on a human operator to direct region manipulation.

3   Dynamic Region RRT

In this section, we explore a fully automated planning algorithm that was inspired by the human collaboration techniques in [7].

3.1   Algorithm Overview

Our algorithm, Dynamic Region-biased RRT, is depicted in Algorithm 1 and Figure 1. The core idea is to compute a skeleton of the workspace topology that describes a path through each relevant homotopy class. We then exploit this representation by leading an RRT planner along the paths with dynamic sampling regions.

The workspace representation is a spatially embedded undirected graph called the *Workspace Skeleton* (Definition 1 and magenta in Figure 1(b)), which is generated during a pre-computation step in line 1. Once the query is known, we compute a directed version of the Workspace Skeleton in line 2, called the *Query Skeleton* (Definition 2 and magenta in Figure 1(c)), that encodes exploration directions leading through the workspace toward the query goal.

From the Query Skeleton, our planner creates a sampling region (green in Figure 1(d)) at the node that is nearest to the query start in line 4 and begins RRT growth. On each iteration (line 5), the planner chooses a region with uniform probability, generates the next growth target q_{rand} from within that volume

---

Algorithm 1 Dynamic Region-biased RRT

Input: Environment e and a Query $\{q_s, q_g\}$
Output: Tree T
 1: $W \leftarrow$ COMPUTEWORKSPACESKELETON(e) {Pre-computation}
 2: $S \leftarrow$ COMPUTEQUERYSKELETON(W, q_s, q_g)
 3: $T \leftarrow (\emptyset, \emptyset)$
 4: $R \leftarrow$ INITIALREGIONS(S, q_s)
 5: **while** $\neg done$ **do**
 6:     REGIONBIASEDRRTGROWTH(T, S, R)
 7: **return** T

---

of workspace, and extends the tree to a new node q_{new}. The entire environment is also considered to be a (static) region to ensure probabilistic completeness. When q_{new} is placed inside a dynamic sampling region, that region is advanced along the Query Skeleton edge until it no longer contains the new sample. The net effect is that the RRT tree in $\mathcal{C}_{free}$ is biased to follow these sampling regions along topologically distinct workspace paths that are encoded in the Query Skeleton. During this phase our planner also creates and destroys sampling regions in the workspace upon discovering new components of the Query Skeleton. Figure 1(e) shows how the regions bias tree growth (blue) and when they reach the next node of the Query Skeleton, two more regions are created to traverse the outgoing edges from that node as shown in Figure 1(f). The algorithm continues in this manner until a stopping criteria is met (line 5), e.g., query is solved or maximum number of iterations is met.

We note that this framework is amenable to many RRT-based planners because the only interface requirement is the ability to designate the sampling boundary for each iteration.

3.2   Representing Topology with a Workspace Skeleton

Dynamic Region-biased RRT begins by computing a *Workspace Skeleton* of the free workspace during pre-processing.

Definition 1. *A Workspace Skeleton is a spatially embedded 1-skeleton (undirected graph) of the free workspace satisfying the properties of a retraction [11].*

The Workspace Skeleton has certain requirements to maintain desirable properties. Namely, it must be a deformation retract [11] of the free workspace. This implies that every point in the free workspace can be mapped onto the skeleton, i.e., it represents the entire workspace and preserves its topology. This property allows the number of possible homotopy classes (distinct paths) in a workspace to be estimated.

There are a few examples of possible data structures that satisfy this property. One example, a Generalized Voronoi Graph (GVG) [6] is the set of points equidistant to m obstacles in a space of dimension m. The GVG, however, is not always guaranteed to be connected, even in 3-space. A few methods have proposed ideas to overcome this [6, 9].

Another example is a Reeb Graph [27]. It represents transitions in level sets of a real-valued function on a manifold, i.e., nodes of the graph are critical values of the function, referred to as a Morse function, and edges are the topological transitions between them. It has applications in various parts of computational geometry and computer graphics such as shape matching [12], iso-surface remeshing [33], and simplification [32].

For the implementation reported in this paper, we choose a Reeb Graph as the basis of our Workspace Skeleton because it satisfies our requirements and is sufficiently simple to compute [25]. We do not claim that this choice is optimal in any sense.

Our Workspace Skeleton algorithm, Algorithm 2, begins by computing a Delaunay tetrahedralization of the free workspace [30] in line 1. We then build a Reeb Graph from the resulting set of tetrahedra using the algorithm in [25] at line 2. In this implementation, the Morse function was the value of the principle axis of the environment boundary (e.g., y-value). The next step embeds the Reeb graph nodes in the free workspace by shifting each node to the center of its closest tetrahedron (lines 3–5). Finally, we embed each Reeb Graph arc by searching for a path through the dual of the tetrahedralization that connects the embedded start and end node (lines 6–7). We use a modified Dijkstra's algorithm that prefers the edges of the dual related to the specific Reeb Graph arc, and we ensure that the embedded arcs lie entirely within the free workspace by connecting adjacent tetrahedra through the center of their adjacent faces (as opposed to connecting the tetrahedra centers directly). In this way, the final Workspace Skeleton is a set of points and polygonal chains spatially contained in the free workspace.

The related tetrahedra for a Reeb Graph arc are a connected set of tetrahedra that are found between the topological critical points (Reeb Graph nodes). This is determined during the Reeb Graph construction: critical points occur whenever there are two or more adjacent tetrahedra that have a pair of vertices that cannot be connected by a straight-line through free space — the critical points represent a divide in the freespace volume. Thus, each Reeb Graph arc captures a distinct workspace homotopy class between two embedded Reeb nodes, and the embedded arc represents a specific (workspace) path in that class.

Additionally, we note that every workspace homotopy is a generalization of one or more $\mathcal{C}_{space}$ homotopies. This is easy to see by considering a projection of a $\mathcal{C}_{space}$ homotopy into workspace using only the positional DOFs. Thus, by representing all workspace homotopies, the Workspace Skeleton contains a partial representation of each $\mathcal{C}_{space}$ homotopy.

3.3   Specializing to a Query Skeleton

Once the Workspace Skeleton is computed and a query is posed, our planning algorithm computes a *Query Skeleton* from the Workspace Skeleton:

Algorithm 2 Compute Workspace Skeleton
___
Input: Environment e
Output: Workspace Skeleton W
1: $D \leftarrow$ TETRAHEDRALIZATION(e) {Using [30]}
2: $R = (R_V, R_E) \leftarrow$ CONSTRUCTREEBGRAPH(D) {Using [25]}
3: $W = (V, E) \leftarrow (\emptyset, \emptyset)$
4: **for all** $v \in R_V$ **do**
5:     $V \leftarrow V \cup \{D.\text{CLOSESTTETRAHEDRON}(v)\}$
6: **for all** $e \in R_E$ **do**
7:     $E \leftarrow E \cup (D.\text{FINDBIASEDPATH}(e))$
8: **return** W

Definition 2. *Let* $W = (W_V, W_E)$ *be an Workspace Skeleton and* $v_s, v_g \in W_V$ *be the nearest Workspace Skeleton nodes to the start and goal of a query respectively. A* Query Skeleton, $S = (S_V, S_E)$ *is a directed subgraph of* W *in which the edges point along possible paths from* v_s *toward* v_g.

To compute the Query Skeleton, we perform a single-source shortest-paths (SSSP) traversal of the Workspace Skeleton starting from the vertex closest to the query start configuration. We retain the (directed) shortest-path tree as well as forward and cross edges as the initial Query Skeleton. Then, we traverse the initial Query Skeleton with a reverse breadth-first search (BFS): starting from the Query Skeleton vertex that is closest to the goal, we back-track along the inbound edges to identify all of that vertex's ancestors. We then prune from the Query Skeleton any node that was not discovered as an ancestor in this second pass, thereby removing vertices and edges that do not lead from the start area toward the goal area (Figure 1(c)). The resulting Query Skeleton is used to coordinate region construction, modification, and deletion in our RRT planner.

3.4   Region-biased RRT Growth

Algorithm 3 shows our RRT growth strategy. The algorithm takes a tree T, a Query Skeleton S, and a set of regions R as input and combines four phases. In the first phase, the algorithm performs a region-biased RRT extension (lines 1–4). It selects a random region and picks a configuration q in the region by randomly selecting a point in the region to initialize the positional DOFs and randomizes the remaining DOFs of q. Note, an invisible region encompasses the entire environment to maintain probabilistic completeness as in the collaborative algorithm presented in [7]. In the second phase, we advance the regions along their associated Query Skeleton edges if needed (lines 5–9). To do this, we determine if the extended node q_{new} reached some portion of any region r. If so, we move each such region r to the next point on its embedded edge in the workspace until it no longer contains q_{new}. When a region reaches the end of its edge, it is deleted. The third phase deletes fruitless regions determined by some threshold τ of consecutive failed tree extensions (lines 10–11). In the final phase, we create new regions if q_{new} is within an ϵ distance of an unexplored skeleton vertex v

---

Algorithm 3 Region-biased RRTGrowth

Input: Tree T, Query Skeleton S, Regions R
Require: ϵ is a multiple of the robot radius, τ is a maximum for failed extension
{Region-biased RRT extension}
1: $r \leftarrow$ SELECTREGION(R)
2: $q_{rand} \leftarrow r$.GETRANDOMCFG$()$
3: $q_{near} \leftarrow$ NEARESTNEIGHBOR(T, q_{rand})
4: $q_{new} \leftarrow$ EXTEND$(q_{near}, q_{rand}, \Delta)$
{Advance regions along Query Skeleton edges}
5: **for** $r \in R$ **do**
6:    **while** r.INREGION(q_{new}) **do**
7:        r.ADVANCEALONGSKELETONEDGE$()$
8:        **if** r.ATENDOFSKELETONEDGE$()$ **then**
9:            $R \leftarrow R \setminus \{r\}$
{Delete useless regions}
10: **if** r.NUMFAILURES$() > \tau$ **then**
11:    $R \leftarrow R \setminus \{r\}$
{Create new regions}
12: **for all** $v \in S$.UNEXPLOREDVERTICES$()$ **do**
13:    **if** $\delta(v, q_{new}) < \epsilon$ **then**
14:        $R \leftarrow R \cup$ NEWREGION(v)
15:        S.MARKEXPLORED(v)

(lines 12–15). We create a region for each outgoing edge of v and mark v as explored on the Query Skeleton.

Our algorithm has little runtime overhead compared with standard RRT growth. It comes in the form of selecting, advancing, creating, and deleting regions. The number of regions is bounded by $O(|S_E|)$, where S_E is the set of edges in the Query Skeleton S. Our expansion step has an overhead of $O(|S_E| + |S_V|)$ where S_V is the set of vertices of the Query Skeleton S. The $O(|S_E|)$ term stems from checking if a new extension has reached any active region, while the $O(|S_V|)$ term comes from checking whether a new extension has reached an unvisited Query Skeleton node. Over whole problem, advancing the regions along the Query Skeleton edges S_e is bounded by the number of sub points $P_e \in S_e$, for a total cost of $O(|P_e|)$. However, this is an extemely pessimistic upper bound. In most scenarios, the number of active regions is much smaller than the number of Query Skeleton edges, and the number of unvisited vertices drops monotonically during execution.

3.5   Discussion

Our representation is able to capture the salient topological features of the free workspace. In problems where C_{space} is strongly related to the workspace, our algorithm provides a convenient methodology to explore the space.

Some problems exhibit only a partial correlation between C_{space} and workspace. One example is manipulation planning, where the end-effector's path is strongly correlated with the environment, but the movement of the rest of the robot may not be. In such cases, a Workspace Skeleton could be used to guide sampling for the correlated component. This would additionally require a sampling method that is able to place the correlated component first and subsequently sample the remaining DOFs, such as reachable volumes [23] or cyclic coordinate descent [4].

Unfortunately, not all problems exhibit a useful correlation between the workspace and C_{space} topology. In problems such as protein folding, the use of a Workspace Skeleton may not be sensible. However the underlying idea may still apply, which is to use regions and topological analysis in low-dimensional manifolds of C_{space} to explore C_{free}. Dynamic Region-biased RRT uses the workspace itself as such a manifold, but one could feasibly design alternatives for other classes of problems.

4   Experimental Evaluation

All methods were implemented in a C++ motion planning library developed in the Parasol Lab at Texas A&M University. It uses a distributed graph data structure from the Standard Template Adaptive Parallel Library (STAPL) [3], a C++ library designed for parallel computing.

All experiments were executed on a latop running Fedora 23 with an Intel® Core™ i7-6500U CPU at 2.5 GHz, 8 GB of RAM, and the GNU gcc compiler version 5.3.1.

We compared Dynamic Region-biased RRT (DRRRT) against RRT [19], Dynamic-Domain RRT [35], the Synergistic Combination of Layers of Planning framework (SyCLoP) with RRT [26], and the Probabilistic Roadmap (PRM) approach using Workspace Importance Sampling (WIS) [17]. We selected these representative methods for their similarities to our approach. Dynamic-Domain RRT, SyCLoP and WIS use adaptive sampling distributions. SyCLoP and WIS employ pre-computations to decompose the workspace. SyCLoP additionally uses a search over the decomposition graph to guide an RRT planner.

We demonstrate our algorithm on holonomic robots in five environments, including MazeTunnel, LTunnel, Garage, GridMaze4, and GridMaze8 as shown in Figure 2. We also evaluate performance on the LTunnel with nonholonomic robots, comparing against RRT and SyCLoP, as a proof-of-concept that the Workspace Skeleton can benefit nonholonomic systems. These environments contain narrow passages in the workspace that are correlated with valid paths in C_{space}, and are thus good exhibitions of our planner's ability to exploit that correlation. The difficulties in each environment vary. MazeTunnel contains false passages, LTunnel incorporates narrow entrances, Garage has multiple homotopy classes, and GridMaze environments have long, winding paths in a cramped tunnel that constrain the robot's rotational DOFs. The robots in all cases are 6 DOF rigid bodies.

We use a Δq value that is approximately 5-10% of the diagonal of each environment for the RRT planners. The nonholonomic experiment uses a fully

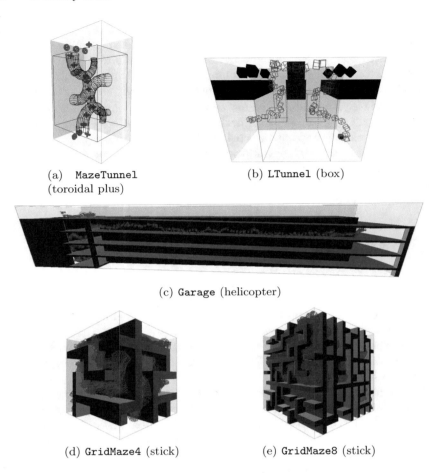

(a) MazeTunnel
(toroidal plus)

(b) LTunnel (box)

(c) Garage (helicopter)

(d) GridMaze4 (stick)

(e) GridMaze8 (stick)

Fig. 2. The experiment environments and (robot). The Garage contains multiple alternative paths. The GridMazes are 3D mazes with internal pathways and tunnels throughout. The robots in all cases are 6 DOF rigid bodies. In the GridMazes, the robot length is equal to the tunnel width to constrain rotational motion.

actuated robot with a 1:2 ratio of random vs. best control selection and a 1:2 ratio of fixed vs. variable timestep for extension [19].

Each experiment ran until the query was solved (success) or 20,000 nodes had been added to the roadmap or 20 minutes had elapsed (failure). We performed 35 trials for each experiment and removed the fastest and slowest run from each. Success rates are shown in Table 1, and average run times and standard deviations are shown in Figures 3 and 4. Times are reported over all runs.

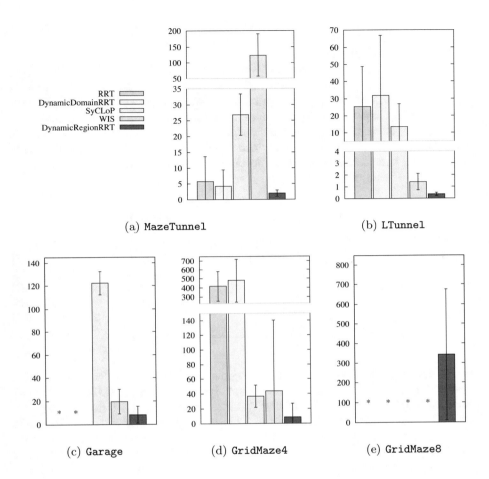

(a) MazeTunnel   (b) LTunnel

(c) Garage   (d) GridMaze4   (e) GridMaze8

Fig. 3. On-line planning time comparing Dynamic Region-biased RRT with RRT, Dynamic-Domain RRT, WIS, and SyCLoP in holonomic problems. The values are the average times over 33 trials. Error bars indicate standard deviation. A * indicates that none of the trials solved the query.

4.1 Discussion

As Figure 3 shows, our planner exhibits faster online planning time as compared with the other methods for the holonomic problems. The improvements over the closest competitors are significant, with p-values of .0083 against RRT and .0175 against Dynamic-Domain RRT in MazeTunnel (using the student's unpaired t-test). The closest competitor in the other environments were slower with p-values less than .0001 (WIS in LTunnel and Garage, SyCLoP in GridMaze4). Neither RRT nor Dynamic-Domain RRT was able to solve the Garage problem. WIS showed strong performance in LTunnel and Garage, but lacked efficiency

in `MazeTunnel` and consisitency in `GridMaze4`. Only DRRRT was able to solve
the large `GridMaze8`, which pushes the algorithm to the limits of what it can
handle with high reliability.

Compared with the other RRT methods, DRRRT's topological guidance
leads to more productive samples and thus smaller roadmaps. This results in
faster neighborhood finding tests and faster overall execution. Despite SyCLoP's
mechanism for expediting neighborhood finding, our guidance mechanism showed
significantly lower per-extension cost, which more than compensated for this in
the holonomic trials. WIS generally produced the smallest roadmaps, but often
took a long time to generate its samples and spent much more time connecting
them due to being a PRM method.

Table 1. Success rates in each experiment.

Planner	DRRRT	RRT	Dynamic-Domain RRT	SyCLoP-RRT	WIS
MazeTunnel	100%	100%	100%	100%	100%
LTunnel	100%	100%	100%	100%	100%
Garage	100%	0%	0%	100%	100%
GridMaze4	100%	100%	100%	100%	100%
GridMaze8	76%	0%	0%	0%	0%
LTunnel (nonholonomic)	90%	24%	-	97%	-

In the nonholonomic experiment (Figure 4), DRRRT was faster than RRT
(p-value less than .0001) and slower than SyCLoP (p-value .0002) in `LTunnel`
with high confidence. While DRRRT improves over RRT, it doesn't perform
as well as SyCLoP for this problem. This is because SyCLoP's neighborhood
finding expedition offers more benefit in problems with larger roadmaps (like
nonholonmic problems), and because it doesn't always expand from the front of
the tree, which helps prevent it from getting stuck when the leading nodes in
the tree are difficult to extend from. Emulating this aspect of SyCLoP leaves a
promising future refinement for DRRRT.

In terms of pre-computation time, our Workspace Skeleton took on average
3.5 seconds to build in `MazeTunnel`, 0.7 seconds in `LTunnel`, and 2.3 seconds in
`Garage`. It is important to note that the Reeb Graph computational efficiency
is $O(n^2)$ in the worst case where n is the number of points in the tetrahedralization [25]. In other words, the Reeb Graph computation is significantly more
efficient in simpler environments. It is possible to reduce this computation time
to a little worse than $O(n \log n)$ with a more sophisticated construction algorithm [8], which we leave to future work.

Beyond this, we would like to make a special note of the dynamic aspect of
our regions. In these test environments, there are false passages in the workspace.
Our algorithm may explore them while sampling from the whole environment,
but its region-guided efforts are concentrated on potentially viable paths in the
workspace and avoid the false passages. However, if a large number of consecutive
attempts to sample within a region fail, we delete the region as it is likely a path

that is only viable for a point in the workspace and not for the robot's $\mathcal{C}_{space}$. In this way, our algorithm is robust to very complex environments with multiple homotopy classes in the workspace.

5 Conclusion

In this paper, we introduced a new adaptive, workspace-based sampling strategy for RRTs and demonstrated its use in holonomic and nonholonomic problems. The key development is the use of dynamic sampling regions that exploit a embedded graph representing the workspace structure. While there are environmentally dependent trade-offs involved, this method demonstrates benefits in problems where valid paths are strongly correlated with the workspace geometry.

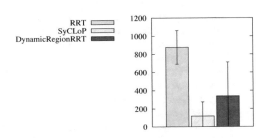

Fig. 4. On-line planning time comparing Dynamic Region-biased RRT with RRT and SyCLoP in the nonholonomic LTunnel. Error bars indicate standard deviation.

Future directions for this work include studying the effects of using a Workspace Skeleton with PRMs and optimal planners, enhancing the Workspace Skeleton construction, and improving performance for nonholonomic robots.

Acknowledgements

This research supported in part by NSF awards CNS-0551685, CCF-0833199, CCF-1423111, CCF-0830753, IIS-0916053, IIS-0917266, EFRI-1240483, RI-1217991, by NIH NCI R25 CA090301-11. J. Denny was supported in part by an NSF Graduate Research Fellowship during the main portion of this research conducted at Texas A&M University.

References

1. Amato, N.M., Bayazit, O.B., Dale, L.K., Jones, C., Vallejo, D.: OBPRM: an obstacle-based PRM for 3d workspaces. In: Proceedings of the third Workshop on the Algorithmic Foundations of Robotics. pp. 155–168. A. K. Peters, Ltd., Natick, MA, USA (1998), (WAFR '98)
2. van den Berg, J.P., Overmars, M.H.: Using workspace information as a guide to non-uniform sampling in probabilistic roadmap planners. Int. J. Robot. Res. 24(12), 1055–1071 (2005)

3. Buss, A.A., Harshvardhan, Papadopoulos, I., Pearce, O., Smith, T.G., Tanase, G., Thomas, N., Xu, X., Bianco, M., Amato, N.M., Rauchwerger, L.: STAPL: standard template adaptive parallel library. In: Proceedings of of SYSTOR 2010: The 3rd Annual Haifa Experimental Systems Conference, Haifa, Israel, May 24-26, 2010. pp. 1–10. ACM, New York, NY, USA (2010), http://doi.acm.org/10.1145/1815695. 1815713

4. Canutescu, A.A., Dunbrack, Jr., R.L.: Cyclic coordinate descent: A robotics algorithm for protein loop closure. Protein Sci. 12(5), 963–972 (2003)

5. Cheng, P., LaValle, S.: Reducing metric sensitivity in randomized trajectory design. In: Proc. IEEE Int. Conf. Intel. Rob. Syst. (IROS). vol. 1, pp. 43–48 vol.1 (2001)

6. Choset, H., Burdick, J.: Sensor-based exploration: The hierarchial generalized voronoi graph. Int. J. Robot. Res. 19(2), 96–125 (2000)

7. Denny, J., Sandstrom, R., Julian, N., Amato, N.M.: A region-based strategy for collaborative roadmap construction. In: Proc. Int. Workshop on Algorithmic Foundations of Robotics (WAFR). Istanbul, Turkey (August 2014)

8. Doraiswamy, H., Natarajan, V.: Efficient algorithms for computing reeb graphs. Comput. Geom. Theory Appl. 42(6-7), 606–616 (Aug 2009), http://dx.doi.org/ 10.1016/j.comgeo.2008.12.003

9. Foskey, M., Garber, M., Lin, M.C., Manocha, D.: A voronoi-based hybrid motion planner for rigid bodies. In: Proc. IEEE Int. Conf. Intel. Rob. Syst. (IROS). pp. 55–60 (2001)

10. Gayle, R., Sud, A., Lin, M.C., Manocha, D.: Reactive deformation roadmaps: Motion planning of multiple robots in dynamic environments. In: Proc. IEEE Int. Conf. Intel. Rob. Syst. (IROS) (2007)

11. Hatcher, A.: Algebraic Topology. Cambridge University Press (2001), http://www. math.cornell.edu/~hatcher/

12. Hilaga, M., Shinagawa, Y., Kohmura, T., Kunii, T.L.: Topology matching for fully automatic similarity estimation of 3d shapes. In: Proceedings of the 28th Annual Conference on Computer Graphics and Interactive Techniques. pp. 203–212. SIGGRAPH '01, ACM, New York, NY, USA (2001), http://doi.acm.org/10.1145/ 383259.383282

13. Hsu, D., Latombe, J.C., Kurniawati, H.: On the probabilistic foundations of probabilistic roadmap planning. Int. J. Robot. Res. 25, 627–643 (July 2006)

14. Kalisiak, M., van de Panne, M.: Rrt-blossom: Rrt with a local flood-fill behavior. In: Proceedings 2006 IEEE International Conference on Robotics and Automation, 2006. ICRA 2006. pp. 1237–1242 (May 2006)

15. Karaman, S., Frazzoli, E.: Sampling-based algorithms for optimal motion planning. International Journal of Robotics Research (IJRR) 30, 846–894 (2011)

16. Kuffner, J.J., LaValle, S.M.: RRT-connect: An efficient approach to single-query path planning. In: Proc. IEEE Int. Conf. Robot. Autom. (ICRA). pp. 995–1001 (2000)

17. Kurniawati, H., Hsu, D.: Workspace importance sampling for probabilistic roadmap planning. In: Proc. IEEE Int. Conf. Intel. Rob. Syst. (IROS). vol. 2, pp. 1618–1623 (sept-2 oct 2004)

18. Kurniawati, H., Hsu, D.: Workspace-based connectivity oracle - an adaptive sampling strategy for prm planning. In: Algorithmic Foundation of Robotics VII, pp. 35–51. Springer, Berlin/Heidelberg (2008), book contains the proceedings of the International Workshop on the Algorithmic Foundations of Robotics (WAFR), New York City, 2006

19. LaValle, S.M., Kuffner, J.J.: Randomized kinodynamic planning. Int. J. Robot. Res. 20(5), 378–400 (May 2001)

20. Li, Y., Littlefield, Z., Bekris, K.E.: Sparse Methods for Efficient Asymptotically Optimal Kinodynamic Planning, pp. 263–282. Springer International Publishing, Cham (2015), http://dx.doi.org/10.1007/978-3-319-16595-0_16
21. Lien, J.M., Pratt, E.: Interactive planning for shepherd motion (March 2009), the AAAI Spring Symposium
22. Lozano-Pérez, T., Wesley, M.A.: An algorithm for planning collision-free paths among polyhedral obstacles. Communications of the ACM 22(10), 560–570 (October 1979)
23. McMahon, T., Thomas, S.L., Amato, N.M.: Reachable volume RRT. In: Proc. IEEE Int. Conf. Robot. Autom. (ICRA). pp. 2977–2984. Seattle, Wa. (May 2015)
24. Morales, M., Tapia, L., Pearce, R., Rodriguez, S., Amato, N.M.: A machine learning approach for feature-sensitive motion planning. In: Algorithmic Foundations of Robotics VI, pp. 361–376. Springer Tracts in Advanced Robotics, Springer, Berlin/Heidelberg (2005), (WAFR '04)
25. Pascucci, V., Scorzelli, G., Bremer, P.T., Mascarenhas, A.: Robust on-line computation of reeb graphs: Simplicity and speed. ACM Trans. Graph. 26(3) (Jul 2007), http://doi.acm.org/10.1145/1276377.1276449
26. Plaku, E., Kavraki, L., Vardi, M.: Motion planning with dynamics by a synergistic combination of layers of planning 26(3), 469–482 (June 2010)
27. Reeb, G.: Sur les points singuliers d'une forme de pfaff complement integrable ou d'une fonction numerique. Comptes Rendus Acad. Sciences Paris 222, 847–849 (1946)
28. Reif, J.H.: Complexity of the mover's problem and generalizations. In: Proc. IEEE Symp. Foundations of Computer Science (FOCS). pp. 421–427. San Juan, Puerto Rico (October 1979)
29. Rodriguez, S., Tang, X., Lien, J.M., Amato, N.M.: An obstacle-based rapidly-exploring random tree. In: Proc. IEEE Int. Conf. Robot. Autom. (ICRA) (2006)
30. Si, H.: Tetgen, a delaunay-based quality tetrahedral mesh generator. ACM Trans. Math. Softw. 41(2), 11:1–11:36 (Feb 2015), http://doi.acm.org/10.1145/2629697
31. Suh, C., Kim, B., Park, F.C.: The tangent bundle RRT algorithms for constrained motion planning. In: 13th World Congress in Mechanism and Machine Science (2011)
32. Wood, Z., Hoppe, H., Desbrun, M., Schröder, P.: Removing excess topology from isosurfaces. ACM Trans. Graph. 23(2), 190–208 (Apr 2004), http://doi.acm.org/10.1145/990002.990007
33. Wood, Z.J., Schröder, P., Breen, D., Desbrun, M.: Semi-regular mesh extraction from volumes. In: Proceedings of the Conference on Visualization '00. pp. 275–282. VIS '00, IEEE Computer Society Press, Los Alamitos, CA, USA (2000), http://dl.acm.org/citation.cfm?id=375213.375254
34. Yan, Y., Poirson, E., Bennis, F.: Integrating user to minimize assembly path planning time in PLM. In: Product Lifecycle Management for Society, IFIP Advances in Information and Communication Technology, vol. 409, pp. 471–480. Springer Berlin Heidelberg (2013)
35. Yershova, A., Jaillet, L., Simeon, T., Lavalle, S.M.: Dynamic-domain RRTs: Efficient exploration by controlling the sampling domain. In: Proc. IEEE Int. Conf. Robot. Autom. (ICRA). pp. 3856–3861 (April 2005)
36. Zhang, L., Manocha, D.: An efficient retraction-based RRT planner. In: Proc. IEEE Int. Conf. Robot. Autom. (ICRA) (2008)

Motion Planning for Reconfigurable Mobile Robots Using Hierarchical Fast Marching Trees

William Reid[1], Robert Fitch[1,2], Ali Haydar Göktoğan[1] and Salah Sukkarieh[1]

[1]Australian Centre for Field Robotics, The University of Sydney, Australia
{w.reid,rfitch,a.goktogan,salah}@acfr.usyd.edu.au
[2]Centre for Autonomous Systems, University of Technology Sydney, Australia

Abstract. Reconfigurable mobile robots are versatile platforms that may safely traverse cluttered environments by morphing their physical geometry. However, planning paths for these robots is challenging due to their many degrees of freedom. We propose a novel hierarchical variant of the Fast Marching Tree (FMT*) algorithm. Our algorithm assumes a decomposition of the full state space into multiple sub-spaces, and begins by rapidly finding a set of paths through one such sub-space. This set of solutions is used to generate a biased sampling distribution, which is then explored to find a solution in the full state space. This technique provides a novel way to incorporate prior knowledge of sub-spaces to efficiently bias search within the existing FMT* framework. Importantly, probabilistic completeness and asymptotic optimality are preserved. Experimental results are provided for a reconfigurable wheel-on-leg platform that benchmark the algorithm against state-of-the-art sampling-based planners. In minimizing an energy objective that combines the mechanical work required for platform locomotion with that required for reconfiguration, the planner produces intuitive behaviors where the robot dynamically adjusts its footprint, varies its height, and clambers over obstacles using legged locomotion. These results illustrate the generality of the planner in exploiting the platform's mechanical ability to fluidly transition between various physical geometric configurations, and wheeled/legged locomotion modes.

1 Introduction

Reconfigurable mobile robots (RMRs) are platforms that gain versatility through the ability to dynamically alter certain elements of their geometric structure. An example RMR called the MAMMOTH rover [16] is shown in Fig. 1 in a variety of different poses. This wheel-on-leg platform is novel in how it combines wheeled and legged locomotion modes to traverse over, under or around challenging obstacles. It may drive omni-directionally while raising and lowering its body, packing or unpacking each of its legs between large and small footprints, as well as lifting or lowering individual legs to clamber over terrain. One benefit of this type of vehicle is in traversing unstructured terrain that conventional rovers with passive suspension systems may find inaccessible. Even though the

© Springer Nature Switzerland AG 2020
K. Goldberg et al. (Eds.): *Algorithmic Foundations of Robotics XII*, SPAR 13, pp. 656–671, 2020.
https://doi.org/10.1007/978-3-030-43089-4_42

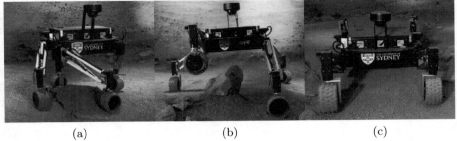

(a)   (b)   (c)

Fig. 1. The MAMMOTH Rover: an example of a reconfigurable mobile robot. The vehicle is capable of reconfiguring its footprint, changing its height, clambering over obstacles and driving omni-directionally.

rover may need to reconfigure to achieve a successful traversal, the dominant mode of motion remains the translation of its body frame relative to an inertial frame. In this work we propose a hierarchical motion planning technique that exploits this property by rapidly searching the dominant sub-space to find paths that are then used to bias search within the full state space.

The main challenge in planning for RMRs is to efficiently find a feasible low cost path given potentially many degrees of freedom. The problem is to generate a sequence of transitions between initial and terminal system states (configurations) while both avoiding environmental obstacles and satisfying internal kinematic constraints. The sequence of transitions is to be optimized with respect to a cost function that is used to define the objectives of the robot's overall motion. In this work, the energy expenditure of the robot is minimized.

For RMR planning, techniques such as discrete grid-based dynamic programming [5] are not suitable due to the infeasibility of explicitly constructing and searching occupancy grids over high-dimensional state spaces that include strict kinematic constraints. Sampling-based planners are also not immune from such plights, particularly because kinematic constraints may induce high-dimensional narrow passageways. *Hierarchical planning* addresses these difficulties by exploiting prior knowledge of task structure. Although automatically discovering effective hierarchical decompositions remains an open problem, using prior knowledge to inform the choice of hierarchy is known (in reinforcement learning) to be beneficial in taming high-DOF problems [11]. Typically, subroutine-like sequences of actions known as macro-actions, or *subtasks*, are predefined for a given level of the hierarchy and called by higher levels during planning. Hierarchical sampling-based planning is yet to be meaningfully explored.

We propose a novel hierarchical sampling-based planner for RMRs, the *Hierarchical Bidirectional Fast Marching Tree (HBFMT*)*. Our approach enables the use of prior knowledge of kinematic structure to inform the choice of hierarchical decomposition, but defined in a way that is more flexible than in other hierarchical approaches. The state-space decomposition must still be provided a priori, but the subtasks are continuous and generated probabilistically by the planner. The extent of the state space explored by subtasks is defined by a continuous parameter that may be automatically tuned.

A main technical challenge in developing a hierarchical sampling-based planner is to define the information flow between levels in a way that does not sacrifice completeness and optimality guarantees. In our planner, we decompose the state space as a hierarchy of increasingly larger sub-spaces, each searched by a Fast Marching Tree (FMT*) variant. Solutions from adjacent levels are used to bias search, thereby increasing path quality with equivalent computational resources. This technique is suited to platforms that have intuitively identifiable low-dimensional kinematic structure, such as the class of RMRs, but may also generalize to other high-DOF robots such as redundant manipulator arms.

We show that this approach retains the completeness and asymptotic optimality guarantees of FMT* [8], and evaluate its performance in path planning for RMRs. Experimental results in simulation show that HBFMT* outperforms FMT*, BMFT*, BIT*, RRT* and Informed RRT* in a variety of environments.

2   Related Work

Over the past two decades, sampling-based algorithms have become the go-to methodology for solving high-dimension planning problems. Ease of implementation, guarantees on probabilistic completeness, and more recently, asymptotic optimality are attractive properties of these techniques. Multi-query sampling-based techniques include the probabilistic roadmap (PRM, PRM*), and variants [4]. Single-query techniques include the rapidly-exploring random tree (RRT) and variants RRT* [10], RRT# [1] and Informed RRT* [6]. Additional asymptotically optimal single-query planners that, unlike RRT planners, rely on sampling batches of states prior to building a tree include (FMT*) [8], Batch Informed Tree (BIT*) [7] and MPLB [19].

FMT* is advantageous in that it uses *lazy* collision detection to reduce the number of expensive collision checking operations, and converges to optimal solutions more quickly than RRT* and PRM* (shown experimentally in [8, 21]). The most significant advantage of FMT* and its variants, however, is its asymptotic optimality guarantees [8], which are amenable to non-uniform sampling distributions.

Bidirectional search is a common method for improving performance of sampling-based planners and has been demonstrated with BFMT* [21], RRT-Connect [12] and bidirectional RRT* [9]. In our work, bidirectional search is employed for its performance boost as well as for finding alternative paths, which is a recent popular topic in motion planning [15] and [24].

When planning for RMRs, or a reconfigurable team of robots, it is common that a sub-space is initially searched to constrain the dimensionality of the problem. In [20], a plan for a high-DOF four-legged robot is found by first finding a footfall sequence and then finding paths between footfalls for the single swing leg using a modified RRT. Hierarchical methods that first find a path over a sub-space using discrete grid-based searches and then use this sub-space path to guide a search of the full space with a sampling-based planner have been demonstrated in various forms [18, 3, 14]. A similar state decomposition is also used in [2] for the multi-robot planning problem, where obstacle free polytope sub-regions are found within the configuration space and then route planning is performed for each robot within each region. For these hierarchical planners, a sub-space is selected by intuition. Automatic sub-space decomposition remains open in general [11]; a recent robotics example does this by finding principal cost variation components relative to an environment and robot model [23]. Our algorithm assumes a given state variable decomposition, but is the first principled hierarchical sampling-based planner that retains performance guarantees.

3 Problem Formulation and Background

In this section we formulate the path planning problem for RMRs. We have attempted to choose notation that agrees with previous work relevant to FMT*.

3.1 Problem Definition

Let $\mathcal{X} = [0, 1]^d$ be a state space with dimension $d \in \mathbb{N}$ constrained to $d \geq 2$. Let $\mathcal{X}_{obs}$ be the obstacle region such that $\mathcal{X} \setminus \mathcal{X}_{obs}$ is an open set. This implies that the free state space is a closed set $\mathcal{X}_{free} = cl(\mathcal{X} \setminus \mathcal{X}_{obs})$. A set of n points are randomly sampled from $\mathcal{X}_{free}$. Let $\mathcal{X}^{\#} = [0, 1]^f$ be a sub-space that is embedded within $\mathcal{X}$ where $2 \leq f \leq d$. Any variable with the "#" super-script is used to denote a variable related to the sub-space $\mathcal{X}^{\#}$.

A path planning problem is denoted by a triplet $(\mathcal{X}_{free}, x_{init}, \mathcal{X}_{goal})$ where x_{init} is the initial state and $\mathcal{X}_{goal}$ is the goal region. In this work, the goal region is considered as a single state, described by x_{goal}. A path is defined by a continuous mapping $\sigma : [0, 1] \rightarrow \mathbb{R}^d$ such that $0 \mapsto x_1$ and $1 \mapsto x_2$. Let Σ be the set of all paths in $\mathcal{X}$. A feasible path, π, in the planning problem $(\mathcal{X}_{free}, x_{init}, \mathcal{X}_{goal})$ is a path that is collision free with $\pi(0) = x_{init}$ and $\pi(1) = x_{goal}$. $\Pi = \{\pi_0, \pi_1, ..., \pi_k\}$ is a set of k feasible paths sorted according to path cost.

The optimal path planning problem is to find a path π^*, assuming problem $(\mathcal{X}_{free}, x_{init}, \mathcal{X}_{goal})$ and an arc cost function $c : \Sigma \rightarrow \mathbb{R}_{\geq 0}$ such that $c(\pi^*) = \min\{c(\pi)\}$, or to report failure. The optimal path π^* is δ-robustly feasible if every point along it is a minimum of δ away from $\mathcal{X}_{obs}$.

3.2   The Fast Marching Tree

HBFMT* builds on FMT*, which is presented in detail in [8] and summarized here for convenience. Tree $T = (V, E, H, W)$ is composed of states V, and edges E. The open set of nodes is denoted as H, and W is the unvisited set. Tree T is initialized with n nodes sampled from free space, $\mathcal{X}_{free}$, in addition to the start node, x_{start} and goal node, x_{goal}. The algorithm maintains an open set, H, and an unvisited set, W. Initially, x_{start} is placed in H, while every other node as well as the goal node is placed in W. The node set, V is initialized containing x_{start}, while the edge set, E is initialized as empty. The algorithm then performs a recursive dynamic programming procedure that propagates a wavefront of state transitions outwards from the start node. This iterative procedure is called **Expand**, and pseudocode can be found in [21].

Unvisited nodes that are within a radius r_n of the minimum cost node in the open set are considered as candidate transition nodes. For each candidate node, the locally optimal transition between it and the open set are found. This transition is tested for collision, and if it is collision free, then the transition is added to the edge set of the tree and the candidate node is placed in a temporary open set. After each candidate node has been considered all of the nodes in the temporary open set are placed in the open set. Tree expansion terminates when the goal node has been found or the open set is empty.

3.3   Bidirectional Fast Marching Tree

HBFMT* also builds on BFMT* [21], which is summarized here. BFMT* grows two trees, T and T', rooted at the start and goal states respectively. BFMT* starts by sampling a set of states, S, from $\mathcal{X}_{free}$. The two trees are then initialized. The first tree, T is initialized with S along with the start state, x_{start} at its root node and the goal state x_{goal}. The second tree T' is also initialized with S, x_{start} and x_{goal}, with x_{goal} at its root. Each tree is alternately expanded according to the FMT* expansion procedure until the two trees intersect.

The termination condition of BFMT* may occur in one of two separate conditions [21]. The first is when the two trees initially intersect, occurring when a node is in the open set of both trees. This is the "first path" termination condition, however it does not guarantee an optimal solution. The second termination condition occurs when a node is in the open set of one tree, while not inside the open or unvisited set of the other. This second "best path" condition results in a case where the node can no longer improve its cost-to-come value from either tree root and is therefore the optimal solution given the sample nodes.

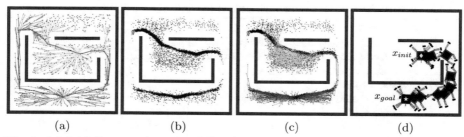

| (a) | (b) | (c) | (d) |

Fig. 2. (a) Initially, an exhaustive BFMT* search of a uniformly sampled sub-space returns a set of feasible paths. (b) States in the full space are sampled from a biased distribution focused around the sub-space paths. (c) A second BFMT* searches through the full state samples. (d) A full state-space path is returned.

4   Hierarchical Bidirectional Fast Marching Tree

In this section we present the HBFMT* algorithm. HBFMT* consists of several components, which are illustrated in Fig. 2.

Intuitively, HBFMT* biases search of the full state space based on knowledge gathered from an initial rapid search of a sub-space. To introduce HBFMT* we describe an example with an 8-DOF robot in a confined environment shown in Fig. 2. The robot starts in an open-stance state at the bottom of the map and its goal is to achieve a similar stance in the center of the map. Any feasible path requires reconfiguration of the robot to fit through narrow corridors. Initially, as in Fig. 2a, a sub-space is searched using BFMT*. The sub-space searched is in $\mathbb{R}^3$, the translation of the robot's body with respect to the inertial frame. Collision checking is performed between the environment and the translating body of the robot without its legs. A set composed of feasible paths is returned. The full 8-DOF state space of the robot is then sampled with a bias around the set of returned 3-DOF paths as shown in Fig. 2b. Another BFMT* instance then searches this biased distribution as in Fig. 2c. The bias is implemented using a tunable *tunnel radius* that defines how focussed the search of the solutions from the previous level of the hierarchy is. A final path through the shorter of the two narrow passageways is returned and shown in Fig. 2d.

Before we define HBFMT*, we first describe a modification to the termination condition of BFMT*. We introduce an "exhaustive" termination condition; pseudo-code is provided in Algorithm 1. When a "best path" termination condition is found at a tree collision node $x_{collision,i}$, BFMT* returns the feasible path π_i, where i is the path index, and adds it to the set of alternative paths Π. When a collision node is found it is *singed*. The collision node, all nodes within a distance r_{singe} of the collision node and their descendants are removed from the open and unvisited sets of both trees so that expansion may not continue through them. Descendant nodes are found via breadth first search of the

---

Algorithm 1: Bidirectional Fast Marching Tree (BFMT*)

---

    function BFMT* ($\mathcal{X}_{free}$,x_{init},x_{goal},Π,ℓ,r_{singe},r_{tunnel},tc)

1    $S \leftarrow x_{init} \cup x_{goal} \cup$ SampleFree $(n, \Pi, \ell, r_{tunnel})$

2    $T \leftarrow$ Initialize(S, x_{init}), $T' \leftarrow$ Initialize(S, x_{goal})

3    $\Pi \leftarrow \emptyset$, $z \leftarrow x_{init}$, $x_{collision} \leftarrow \emptyset$, $e \leftarrow false$

4    $success = false$

5   **while** $success = false$ **do**

6   Expand$(T, z, x_{collision})$

7   **if** ($tc =$ FIRST and $x_{collision} \neq \emptyset$) or ($tc =$ BEST and $z \in (V' \setminus H')$) **then**

8         $\pi \leftarrow$ Path$(x_{collision},T)$ $\cup$ Path$(x_{collision},T')$, $\Pi \leftarrow \Pi \cup \pi$

9         $success = true$

10   **else if** $tc =$ EXHAUSTIVE and $z \in (V' \setminus H')$ **then**

11         $\pi \leftarrow$ Path$(x_{collision},T)$ $\cup$ Path$(x_{collision},T')$, $\Pi \leftarrow \Pi \cup \pi$

12   SingeBranch$(x_{collision},T,r_{singe})$, SingeBranch$(x_{collision},T',r_{singe})$

13         $x_{collision} \leftarrow \emptyset$, $e \leftarrow true$

14   **if** $H = \emptyset$ and $H' = \emptyset$ and $e = false$ **then**

15   **return** Failure

16   **else if** $H = \emptyset$ and $H' = \emptyset$ and $e = true$ **then**

17         $success = true$

18      $z \leftarrow$ argmin$_{x' \in H'}\{$Cost$(x',T')\}$

19   Swap(T, T')

20   **return** Π

---

relevant sub-tree. This singeing procedure is performed in both FMT*s. Pseudo-code is presented in Alg. 2. BFMT* continues this search and singe procedure until both trees' open sets are exhausted.

We now formalize HBFMT*. Pseudocode is listed in Alg. 3. The first step is to plan a path through the sub-space $\mathcal{X}_{free}^{\#}$ using exhaustive BFMT* sampling uniformly from $\mathcal{X}_{free}^{\#}$. Depending on the space $\mathcal{X}_{free}^{\#}$ and the performance of the BFMT* search, there may be multiple unique paths $\Pi^{\#}$ returned. An example where two alternative paths are found is shown in Fig. 2 (a).

The set of paths $\Pi^{\#}$ is then passed to a second BFMT* instance along with a sampling variable ℓ. This second BFMT* search starts by sampling n states from $\mathcal{X}_{free}$ to create the tree nodes V. A collection of $n\ell$ samples are taken from a uniform distribution over $\mathcal{X}_{free}$ to comply with asymptotic optimality conditions discussed in Sec. 5. Every second path within $\Pi^{\#}$ is reversed and then all of the paths within $\Pi^{\#}$ are concatenated to form a path λ that goes back and forth between the goal and end nodes. The sub-space components of $n(1 - \ell)$ states are sampled from a Gaussian distribution $\mathcal{N}(d_{\lambda}, r_{tunnel})$ where d_{λ} is a uniformly sampled distance along λ. The parameter r_{tunnel}, the tunnel radius, determines how wide the biased search region around λ is. Once the biased set of samples

---

Algorithm 2: Singe Branch

---

1 function SingeBranch(z,T,r_{singe})

2 $A \leftarrow z \cup$ Near(z,T,r_{singe}), $C \leftarrow \emptyset$

3   **for** $a \in A$ **do**

4 $\lfloor$ $C \leftarrow C \cup$ BreadthFirstSearch(a,T) $\cup$ BreadthFirstSearch(a,T')

5 $A \leftarrow A \cup C$

6 $H \leftarrow H \setminus A$, $V \leftarrow V \setminus A$

7 $H' \leftarrow H' \setminus A$, $V' \leftarrow V' \setminus A$

Algorithm 3: Hierarchical Bidirectional Fast Marching Tree (HBFMT*)

Data: $\mathcal{X}_{free}$,x_{init},x_{goal},r_{singe},r_{tunnel}

Result: Π

1 begin

2 $\Pi^{\#} \leftarrow$ BFMT* ($\mathcal{X}_{free}^{\#}$,$x_{init}^{\#}$,$x_{goal}^{\#}$,$\emptyset$,1,r_{singe},r_{tunnel},$EXHAUSTIVE$)

3 $\Pi \leftarrow$ BFMT* ($\mathcal{X}_{free}$,x_{init},x_{goal},$\Pi^{\#}$,ℓ,$\emptyset$,r_{tunnel},$BEST$)

S has been found, the second BFMT* instance explores the set and returns the best path through the resulting tree as shown in Fig. 2d.

5 Analysis

In this section we show the asymptotic optimality (AO) of HBFMT* and discuss the effects of its tuning parameters. We prove AO (and thus, probabilistic completeness) of HBFMT* by observing that the terminal stage of the algorithm is BFMT* with a non-uniform sampling distribution and a metric cost function. The following theorem characterizes AO in terms of the number of sample nodes.

Theorem 1. *Let $\pi : [0,1]$ be a feasible path with strong δ-clearance, $\delta > 0$. ζ is the Lebesgue measure of the unit-cost ball and $\mu(\mathcal{X}_{free})$ is the Lebesgue measure of the free space. Consider running HBFMT* by sampling n nodes from a distribution φ and executing to completion using any termination criteria and a radius*

$$r_n = 2(1 + \eta) \left(\frac{1}{d}\right)^{\frac{1}{d}} \left(\frac{\mu(\mathcal{X}_{free})}{\zeta}\right)^{\frac{1}{d}} \left(\frac{\log(n\ell)}{n}\right)^{\frac{1}{d}} \left(\frac{1}{\ell}\right)^{\frac{1}{d}}, \qquad (1)$$

for a parameter $\eta \geq 0$ and $n\ell > 1$. Let c_n denote the cost of the path returned by HBFMT and c^* be the optimal path cost, then $\lim_{n \to \infty} \mathbb{P}(c_n > (1 + \varepsilon)c^*) = 0$ for all $\varepsilon > 0$ (which defines AO).*

Proof. The terminal stage of HBFMT* implements BFMT* operating in $\mathcal{X}_{free}$, using a metric cost function and samples from probability distribution φ that is lower bounded by ℓ. To prove AO of HBFMT* it therefore suffices to show that BFMT* is AO under these two conditions. Firstly, any metric cost function satisfies the AO conditions for FMT* (Sec. 5 of [8]). Secondly, we can represent φ as a mixture distribution, composed of a uniform distribution that is sampled with probability ℓ and a non-uniform distribution that is sampled with probability $1 - \ell$ [8]. This results in a uniform distribution being sampled $n\ell$ times, and thus the proof of AO of FMT* is preserved. Likewise, for BFMT*, $\lim_{n \to \infty} \mathbb{P}(c_n > (1 + \varepsilon)c^*) = 0$ for all $\varepsilon > 0$ [21].

The performance of HBFMT* compared with BFMT* and FMT* is largely dependent on the sub-space $\mathcal{X}_{free}^{\#}$ that is initially sampled. It is generally the case for reconfigurable mobile robots that $\mathcal{X}^{\#}$ is the workspace of the vehicle, which is in the translational $\mathbb{R}^3$ Euclidean space. However, if the platform requires a specific orientation or joint articulation motion that is not contained within the biased search region in $\mathcal{X}_{free}$, the algorithm may not perform as well as FMT* and BFMT* and will be dependent on samples taken from the uniform component of φ. Currently, the sub-space $\mathcal{X}_{free}^{\#}$ is chosen a priori, however a topic of future work is incorporation of autonomous selection of these spaces given a certain environment and robot model. Examples of such low-dimensional structure identification can be found in [23].

Additional tunable parameters that affect the performance of HBFMT* are the singe radius r_{singe} and the biased sampling distribution variance r_{tunnel}. The singe radius directly affects the number and spacing of sub-space paths. A large r_{singe} value relative to the dimensions of the robot body will create a larger space between alternate sub-space paths and result in a smaller number of sub-space paths. Decreasing the value of r_{tunnel} results in focused sampling around the nodes in $\Pi^{\#}$, thereby increasing the planner's reliance on the sub-state solutions $\Pi^{\#}$. Increasing r_{tunnel} results in a distribution resembling uniformity, thereby removing any advantage over BFMT* or FMT*. An experiment that investigates the effects of r_{singe} and r_{tunnel} is provided in Sec. 6.5.

6   Experiments

Our experimental evaluation has two aims. First, we benchmark HBFMT*'s performance against state-of-the-art sampling-based planners, including FMT*, BFMT*, BIT*, RRT* and Informed RRT*. Second, we showcase the algorithm's suitability to path planning for RMRs using the system described in Sec. 6.1 and the three distinct environments described in Sec. 6.2. The experimental setup is detailed in Sec. 6.3 and results are discussed in Sec. 6.4 and Sec 6.5.

6.1   The MAMMOTH Rover

The RMR used in all experiments is a simulated version of the MAMMOTH rover shown in Fig. 1. The simulated model has an 8-DOF state space. It may translate its body frame B relative to its inertial frame W along the x_B^W, y_B^W and z_B^W directions. Its roll, pitch and yaw positions are denoted by ϕ, θ and ψ respectively. In all experiments we hold ϕ and θ constant at $0°$, while ψ is free to rotate. Additionally, the four hip joints, $q_{H_1},...,q_{H_4}$ may rotate, thereby changing the robot's contact footprint. A randomly sampled state x contains values for each of the eight degrees of freedom. The state x also contains four thigh joint variables, $q_{U_1},..., q_{U_4}$ that are calculated as a function of the eight degrees of freedom as well as the local terrain profile. Additional dependent joints that are driven to meet a desired configuration are the wheel steering joints, $q_{S_1},...,q_{S_4}$ and wheel drive joints, $q_{A_1},...,q_{A_4}$. These joint positions and rates are determined as a function of the desired velocity of the platform.

The physical MAMMOTH rover has a mass of 75 kg and 16 points of actuation. Its maximum footprint is 1500 mm by 1500 mm, while its most compact footprint is 650 mm by 650 mm. It may drive and steer each of its wheels independently and continuously allowing for omnidirectional driving. Each hip actuator can be moved between $\pm270°$ to change the footprint of the vehicle. Lastly, each leg is a parallel structure. The legs can raise and lower between $-20°$ and $70°$ relative to the transverse plane of the rover's body.

The 8-DOF state space does not characterize the capability of the MAMMOTH rover to raise only one leg off the ground at a time. This ability is critical when the rover needs to clamber over obstacles. To allow this clambering behaviour to emerge, a sampling feedback method is used. If it is found that a sample from the 8-DOF space is deemed invalid due to leg i colliding with an obstacle, the sample is checked again with leg i raised. This is done by changing the leg's q_{U_i} value so that the leg is in its fully raised configuration. If this new leg-raised state is not in collision with obstacles and obeys static stability constraints, then it is inserted into $\mathcal{X}_{free}$. Alg. 4 provides pseudocode for the sampling implementation used for the MAMMOTH rover planning problem. As part of the collision detection procedure, the rover's static stability is checked according to the model described in [13] and [17]. Stability is characterized by the minimum angle between the rover weight vector and the four vectors that are normal to the sides of the rover's stability polygon and originate at the centre of mass. The stability polygon's vertices are the wheel/ground contact points. The minimum allowable stability margin is $11.5°$.

The function used to evaluate state transition costs approximates the amount of mechanical work in a state transition:

$$J(x_i, x_{i+1}) = \mu Mg\|\Delta x_B^W + \Delta y_B^W\| + (m_{body} + 2m_{leg})g\Delta z_B^W + \mu mgr_{leg}\Delta\psi$$
$$+ \mu m_{leg}gr_{leg}(\Delta q_{H_1} + \Delta q_{H_2} + \Delta q_{H_3} + \Delta q_{H_4}) + (r_{leg}/2)m_{leg}g\Delta q_{U_j}, \tag{2}$$

where Δ is the absolute difference over a single dimension. Additionally, $m_{body}=35$ kg is the mass of the rover's central chassis, $m_{leg}=10$ kg is the mass of a single leg,

---

Algorithm 4: Sample Mammoth State

---

1 function SampleFree(n,Π,l,r_{tunnel})

2    $m \leftarrow n$

3   **while** $m > 0$ **do**

4   **if** $m < (1-l)n$ **then**

5          $x \leftarrow$ SampleBiased(Π,r_{tunnel})

6   **else**

7          $x \leftarrow$ SampleUniform

8       $\mathbf{q_U} \leftarrow$ MammothIK(x)

9   **if** IsValid(x) **then**

10         $m \leftarrow m - 1,\ S \leftarrow S \cup x$

11   **else**

12   **for** $i \leftarrow 1$ **to** 4 **do**

13   **if** LegCollision(i) **then**

14              $x.q_{U_i} \leftarrow q_U.bounds.min$

15   **if** IsValid(x) **then**

16                $m \leftarrow m - 1,\ S \leftarrow S \cup x$

    return S

$M = (m_{body} + 4m_{leg})$ is the total mass of the rover, $r_{leg} = 0.7\ m$ is the average reach of the center of mass of a single rover leg, and $\mu = 0.1$ is an approximate rolling resistance coefficient for the rover's wheels. The constants used to generate J are approximations; an item of future work is to incorporate a high fidelity physical model of the rover into the planner to more accurately calculate energy expenditure. The traversal cost of a path is the sum of its state transition costs $c(\pi) = \sum_{a=0}^{b-1} J(x_a, x_{a+1})$, where b is the number of waypoints in the path.

6.2   Environments

The three environments used are shown in Fig. 3. In the "Gauntlet" environment, the rover must navigate around/over box obstacles, through a tunnel and through a narrow passageway to get to its goal. This environment contains a single homotopy class of paths within the workspace of the rover. HBFMT* will be able to focus its search on this single family of paths. The "Alternatives" environment is designed to contain two homotopy classes of paths within the translational sub-space. The robot may choose to clamber over the four step obstacles, or drive around/over the many block obstacles. Clambering over obstacles is energy intensive, so the optimal solution will lie in the region with the block obstacles. The "Corridor" environment contains a step obstacle with a narrow passageway on one side that the rover must clamber over. This space is designed to highlight the planner's ability to find clambering manoeuvres given the collision feedback sampling strategy described in Sec. 6.1.

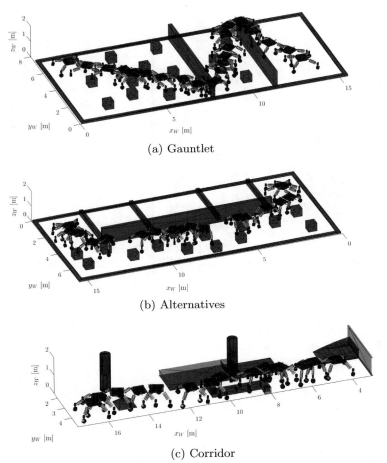

(a) Gauntlet

(b) Alternatives

(c) Corridor

Fig. 3. The three environments used in our experiments. The successful traversals shown are generated by HBFMT* with 10,000 nodes.

6.3 Experimental Setup

Paths are found for the three environments using a set of sampling-based planners: HBFMT*, BFMT*, FMT*, BIT*, RRT* and Informed RRT*. A batch of trials is composed of 50 planner executions. For FMT* and BFMT*, distinct batches are assigned between 1000 and 5000 nodes with 1000 node increments and between 5000 and 25,000 nodes with 5000 node increments. HBFMT* additionally uses 2500 nodes to sample $\mathcal{X}_{free}^{\#}$. Single batches of 50 planning trials, each with an expiry time of 350 s, are run for BIT*, RRT* and Informed RRT*. For the RRT* variants a goal-bias probability of 5% is used.

Tuning parameters for HBFMT* are set as ($r_{tunnel} = 0.1$ kJ, $\ell = 0.2$, $r_{singe} = 2$ m) for Gauntlet, ($r_{tunnel} = 0.75$ kJ, $\ell = 0.2$, $r_{singe} = 3$ m) for Alternatives

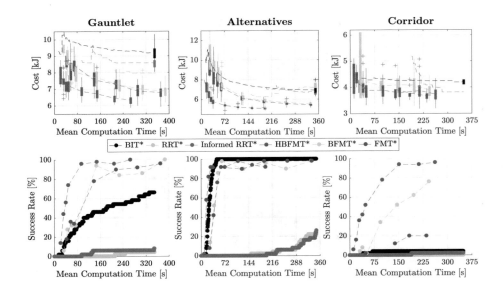

Fig. 4. Experimental results evaluating HBFMT* (red), BFMT* (green), FMT* (blue), BIT* (black), RRT* (orange) and Informed RRT* (pink) plans for the MAMMOTH rover in three distinct environments.

and ($r_{tunnel} = 0.3$ kJ, $\ell = 0.2$, $r_{singe} = 2$ m) for Corridor. HBFMT* begins by sampling the translational $\mathbb{R}^3$ state space of the rover and uses the bounding box of the rover's body (excluding the legs) for collision checking. Based on observations of best practice [8], k-nearest neighbour versions of all the planners are used. The value k_n, which denotes the number of nearest neighbour nodes, is calculated as $k_{n,FMT*} = 2^d e/d \log(n)$ for the FMT* variants and BIT*, and $k_{n,RRT*} = (e + e/d) \log(n)$ for RRT*.

All planners are implemented in simulation within the Open Motion Planning Library (OMPL) [22]. OMPL implementations for FMT*, RRT* and RRT-Connect have been used; the BFMT* implementation is our own. Experiments are run on a 3.4 GHz Intel i7 processor with 32 GB of RAM.

6.4   Benchmarking Results

The examples shown in Fig. 3 illustrate fluid transition between intuitively discrete locomotion modes such as driving, clambering and footprint reconfiguration. The experimental results shown in Fig. 4 characterize the performance, including computation time, of all of the planning algorithms operating on the three environments.

Generally, HBFMT* is observed to improve path quality with less computation time compared with its counterparts. HBFMT* returns a greater number of lower-cost feasible solutions sooner than the other planners for Gauntlet and Alternatives. On average, with 5000 nodes in the Gauntlet environment, HBFMT* returns a path that is approximately 1 kJ lower than those returned by FMT* and BFMT*, and 2 kJ lower in the Alternatives environment. In the Corridor environment, the cost difference between the planners is marginal, however it is noted that HBFMT* does a more effective job of finding feasible solutions through the narrow passageway introduced by the step obstacle.

The FMT* variants outperform the RRT* variants in terms of cost and success rate for all environments, and they outperform BIT* for Gauntlet and Corridor. BIT* returns more feasible solutions than any other planner in Alternatives. The feedback sampling strategy outlined in Sec. 6.1 is used by all planners in all environments, and removes asymptotic optimality and completeness guarantees given that the leg raise dimension is not being uniformly sampled. This explains why the FMT* variants do not converge to 100% success in any of the environments. When tested without feedback sampling, AO and completeness guarantees are preserved and HBFMT* converges to 100% success in the Gauntlet and Alternatives environments.

6.5   Parameter Tuning

An experiment investigating the robustness of the algorithm given varying parameter values of singe radius r_{singe} and tunnel radius r_{tunnel} is performed. In this experiment the HBFMT* planner is run for the MAMMOTH rover in each environment with the same start and goal configurations shown in Fig. 3. A total of 100 separate batches of trials are run, with 50 trials per batch. Each batch has a different r_{singe} and r_{tunnel} combination. The singe radius r_{singe} is varied between 0.5 m and 5 m, while r_{tunnel} is varied between 0.1 and 1.8 kJ.

The results of this experiment are shown in Fig. 5. Each mesh vertex represents the average of the 50 trials run for a specific r_{singe} and r_{tunnel} combination. The results for the Gauntlet environment demonstrate that lower cost is returned as the biased sampling gets more focused around the single $\mathcal{X}_{free}^{\#}$ homotopy class available in the environment. In Alternatives, it is seen that when r_{singe} is large and r_{tunnel} is small the number of successful traverses drops. This is caused by biased samples not being placed in the half of the environment filled with block obstacles and only being placed over the step obstacles. The planner has difficulty finding a path over the four step obstacles and therefore the success rate drops. Additionally, the Alternatives plot shows that if r_{tunnel} is too small then the biased sampling is too focused and cannot find configurations that successfully avoid the large number of block obstacles along the optimal path. A similar effect is seen with the Corridor environment.

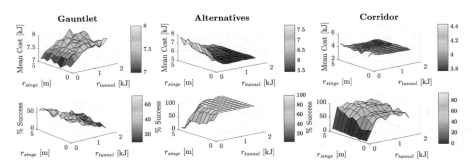

Fig. 5. Path cost and success rates for HBFMT* with 5000 nodes in each environment with varying r_{singe} and r_{tunnel} values.

7   Conclusions and Future Work

In this paper, we proposed a hierarchical sampling-based motion planner that biases search based on paths returned from a rapidly explored sub-space. The algorithm is based on the FMT* and BFMT* algorithms, and retains their asymptotic optimality guarantees. Numerical results showed that HBFMT* improves convergence to an energy-optimal path compared with state-of-the-art sampling-based planners. This approach opens up a new methodology for building efficient planners for RMRs and other high-DOF robots.

This work is motivated directly by a real-world RMR, and so the most important area of future work from our perspective is to use HBFMT* to plan energy efficient paths for the physical MAMMOTH rover in unstructured environments. One avenue of future algorithmic work is to expand the number of levels in the hierarchy to further reduce computational costs. Automatically tuning the tunnel radius parameter is another promising idea. Finally, it would be interesting to apply hierarchical planning to real-world systems with similar characteristics to RMRs, such as high-DOF manipulators and teams of mobile robots.

Acknowledgements. This work was supported by the Australian Centre for Field Robotics; the NSW Government; and the Faculty of Engineering & Information Technologies, The University of Sydney, under the Faculty Research Cluster Program. We would like to thank Oliver Cliff and Jack Umenberger for their helpful assistance.

References

1. Arslan, O., Tsiotras, P.: Use of relaxation methods in sampling-based algorithms for optimal motion planning. In: IEEE ICRA. pp. 2421–2428 (2013)
2. Ayanian, N., Kumar, V.: Decentralized feedback controllers for multiagent teams in environments with obstacles. IEEE Trans. Robot. 26(5), 878–887 (2010)
3. Belter, D., Labecki, P., Skrzypczynski, P.: Adaptive Motion Planning for Autonomous Rough Terrain Traversal with a Walking Robot. J. Field Robot. 33(3), 337–370 (2015)

4. Dobson, A., Bekris, K.E.: Sparse roadmap spanners for asymptotically near-optimal motion planning. Int. J. Robot. Res. 33(1), 18–47 (2014)
5. Ferguson, D., Stentz, A.: Using Interpolation to Improve Path Planning: The Field D* Algorithm. J. Field Robot. 23(2), 79–101 (2006)
6. Gammell, J.D., Srinivasa, S.S., Barfoot, T.D.: Informed RRT*: Optimal Sampling-based Path Planning focused via Direct Sampling of an Admissible Ellipsoidal Heuristic. In: IEEE/RSJ IROS. pp. 2997–3004 (2014)
7. Gammell, J.D., Srinivasa, S.S., Barfoot, T.D.: Batch Informed Trees (BIT*): Sampling-based Optimal Planning via the Heuristically Guided Search of Implicit Random Geometric Graphs. In: IEEE ICRA. pp. 3067–3074 (2015)
8. Janson, L., Schmerling, E., Clark, A., Pavone, M.: Fast marching tree: A fast marching sampling-based method for optimal motion planning in many dimensions. Int. J. Robot. Res. 34(7), 883–921 (2015)
9. Jordan, M., Perez, A.: Optimal Bidirectional Rapidly-Exploring Random Trees Random Trees. Tech. rep., MIT CSAIL, TR-2013-021 (2013)
10. Karaman, S., Frazzoli, E.: Sampling-based algorithms for optimal motion planning. Int. J. Robot. Res. 30(7), 846–894 (2011)
11. Kober, J., Bagnell, J.A., Peters, J.: Reinforcement learning in robotics :. Int. J. Robot. Res. 32(11), 1238–1274 (2013)
12. Kuffner, J.J., LaValle, S.M.: RRT-Connect: An Efficient Approach to Single-Query Path Planning. In: IEEE ICRA. pp. 995–1001 (2000)
13. Papadopoulos, E.G., Rey, D.A.: A New Measure of Tipover Stability Margin for Mobile Manipulators. In: IEEE ICRA. pp. 3111–3116 (1996)
14. Plaku, E., Kavraki, L.E., Vardi, M.Y.: Motion planning with dynamics by a synergistic combination of layers of planning. IEEE Trans. Robot. 26(3), 469–482 (2010)
15. Pokorny, F., Hawasly, M., Ramamoorthy, S.: Topological trajectory classification with filtrations of simplicial complexes and persistent homology. Int. J. Robot. Res. 35(1-3), 204–223 (2016)
16. Reid, W., Göktoğan, A.H., Pérez-Grau, F.J., Sukkarieh, S.: Actively Articulated Suspension for a Wheel-on-Leg Rover Operating on a Martian Analog Surface. In: IEEE ICRA. pp. 5596–5602 (2016)
17. Reid, W., Göktoğan, A.H., Sukkarieh, S.: Moving MAMMOTH : Stable Motion for a Reconfigurable Wheel-on-Leg Rover. In: ARAA ACRA (2014)
18. Rickert, M., Brock, O., Knoll, A.: Balancing exploration and exploitation in motion planning. IEEE T. Robot. 30(6), 2812–2817 (2014)
19. Salzman, O., Halperin, D.: Asymptotically-optimal motion planning using lower bounds on cost. In: IEEE ICRA. pp. 4167–4172 (2015)
20. Satzinger, B.W., Lau, C., Byl, M., Byl, K.: Tractable Locomotion Planning for RoboSimian. Int. J. Robot. Res. 34(13), 1541–1558 (2015)
21. Starek, J.A., Schmerling, E., Janson, L., Pavone, M.: Bidirectional Fast Marching Trees: An Optimal Sampling-Based Algorithm for Bidirectional Motion Planning. In: WAFR (2014)
22. Sucan, I., Moll, M., Kavraki, L.E.: The Open Motion Planning Library. IEEE Robot. Autom. Mag. pp. 72–82 (2012)
23. Vernaza, P., Lee, D.D.: Learning and exploiting low dimensional structure for efficient holonomic motion planning in high dimensional spaces. Int. J. Robot. Res. 31(14), 1739–1760 (2012)
24. Yi, D., Goodrich, M.A., Seppi, K.D.: Homotopy-Aware RRT* : Toward Human-Robot Topological Path-Planning. In: ACM/IEEE HRI. pp. 279–286 (2016)

SWIRL: A Sequential Windowed Inverse Reinforcement Learning Algorithm for Robot Tasks With Delayed Rewards

Sanjay Krishnan, Animesh Garg, Richard Liaw, Brijen Thananjeyan,
Lauren Miller, Florian T. Pokorny*, Ken Goldberg

The AUTOLAB at UC Berkeley (automation.berkeley.edu)
*CAS/CVAP, KTH Royal Institute of Technology, Stockholm, Sweden

Abstract. Inverse Reinforcement Learning (IRL) allows a robot to generalize from demonstrations to previously unseen scenarios by learning the demonstrator's reward function. However, in multi-step tasks, the learned rewards might be delayed and hard to directly optimize. We present Sequential Windowed Inverse Reinforcement Learning (SWIRL), a three-phase algorithm that partitions a complex task into shorter-horizon subtasks based on linear dynamics transitions that occur consistently across demonstrations. SWIRL then learns a sequence of local reward functions that describe the motion between transitions. Once these reward functions are learned, SWIRL applies Q-learning to compute a policy that maximizes the rewards. We compare SWIRL (demonstrations to segments to rewards) with Supervised Policy Learning (SPL - demonstrations to policies) and Maximum Entropy IRL (MaxEnt-IRL demonstrations to rewards) on standard Reinforcement Learning benchmarks: Parallel Parking with noisy dynamics, Two-Link acrobot, and a 2D GridWorld. We find that SWIRL converges to a policy with similar success rates (60%) in 3x fewer time-steps than MaxEnt-IRL, and requires 5x fewer demonstrations than SPL. In physical experiments using the da Vinci surgical robot, we evaluate the extent to which SWIRL generalizes from linear cutting demonstrations to cutting sequences of curved paths.

1 Introduction

One of the goals of learning from demonstrations (LfD) is to learn policies that generalize beyond the provided examples and are robust to perturbations in initial conditions, the environment, and sensing noise [1]. Inverse Reinforcement Learning (IRL) is a popular framework, where the objective is to infer an unknown reward function from a set of demonstrations [2, 3, 4]. Once a reward is learned, given novel instances of a task, a policy can be computed by optimizing for this reward function using an policy search approach like Reinforcement Learning (RL) [5, 4].

In IRL, a task is modeled as an MDP with a single unknown function that maps states and actions to scalar values. This model is limited in the way that it can represent *sequential tasks*, tasks where a robot must reach a sequence of intermediate state-space goals in a particular order. In such tasks, the inferred reward may be *delayed* and reflect a quantity observed after all of the goals are reached, and thus, making it very difficult to optimize directly. Furthermore, there may not exist a single stationary policy for a

© Springer Nature Switzerland AG 2020
K. Goldberg et al. (Eds.): *Algorithmic Foundations of Robotics XII*, SPAR 13, pp. 672–687, 2020.
https://doi.org/10.1007/978-3-030-43089-4_43

given state-space that achieves all of the goals in sequence, e.g., a figure-8 trajectory in the x,y plane.

To address this problem, one approach is to divide the task into segments with local reward functions that "build up" to the final goal. In existing work on multi-step IRL, this sequential structure is defined manually [2]. We propose an approach that automatically learns sequential structure and assigns local reward functions to segments. The combined problem is nontrivial because solving k independent problems neglects the shared structure in the value function during the policy learning phase (e.g., a common failure state). However, jointly optimizing over the segmented problem inherently introduces a dependence on history, namely, any policy must complete step i before step $i+1$. This potentially leads to an exponential overhead of additional states.

Sequential Windowed Inverse Reinforcement Learning (SWIRL) is based on a model for sequential tasks that represents them as a sequence of reward functions $\mathbf{R}_{seq} = [R_1, ..., R_k]$ and transition regions (subsets of the state-space) $G = [\rho_1, ..., \rho_k]$ such that R_1 is the reward function until ρ_1 is reached, after which R_2 becomes the reward and so on. SWIRL assumes that demonstrations are locally optimal (as in IRL), and all demonstrations reach each $\rho \in G$ in the same sequence. In the first phase of the algorithm, SWIRL segments the demonstrations and infers the transition regions using a kernelized variant of an algorithm proposed in our prior work [6, 7]. In the second phase, SWIRL uses the inferred transition regions to segment the set of demonstrations and applies IRL locally to each segment to construct the sequence of reward functions $\mathbf{R}_{seq}$. Once these rewards are learned, SWIRL computes a policy using an RL algorithm (Q-Learning) over an augmented state-space that indicates the sequence of previously reached reward transition regions. We show that this augmentation has an additional space complexity independent of the state-space and linear in the number of rewards.

Our contributions are:

1. A three-phase algorithm, SWIRL, to learn policies for sequential robot tasks.
2. We extend the Transition State Clustering algorithm [6, 7] with kernelization for segmentation that is robust to local non-linear (but smooth) dynamics.
3. A novel state-space augmentation to enforce sequential dependencies using binary indicators of the previously completed segments, which can be efficiently stored and computed based on the first phase of SWIRL.
4. Simulation and physical experiments comparing SWIRL with Supervised Learning and MaxEnt-IRL.

2 Related Work

The seminal work of Abbeel and Ng [4] explored learning from demonstrations using Inverse Reinforcement Learning. In [4], the authors used an IRL algorithm to infer the demonstrator's reward function and then an RL algorithm to optimize that reward. Our work re-visits this two-phase algorithm in the context of sequential tasks. It is well-established that RL problems often converge slowly in complex tasks when rewards are sparse and not "shaped" appropriately [8, 9]. These issues are exacerbated in sequential tasks where a sequence of goals must be reached. Related to this problem, Kolter et al. studied *Hierarchical Apprenticeship Learning* to learn bipedal locomotion [2], where the algorithm is provided with a hierarchy sub-tasks. These sub-tasks are not learned from data and assumed as given, but the algorithm infers a reward function from demonstrations. SWIRL applies to a restricted class of tasks defined by a sequence of reward functions and state-space goals.

There are have been some proposals in robotics to learn motion primitives from data. The approaches assume that reward functions are given (or the problem can be solved with planning-based methods). Motion primitives are example trajectories (or sub-trajectories) that bias search in planning towards paths constructed with these primitives [10, 11, 12]. Much of the initial work in motion primitives considered manually identified segments, but recently, Niekum et al. [13] proposed learning the set of primitives from demonstrations using the Beta-Process Autoregressive Hidden Markov Model (BP-AR-HMM). Calinon et al. [14] proposed the task-parametrized movement model with GMMs for action segmentation. Both Niekum and Calinon consider the motion planning setting in which analytical planning methods are used to solve a task and not RL. Konidaris et al. studied the primitives in the RL setting [15]. However, this approach assumed that the reward function was given and not learned from demonstrations as in SWIRL. Another relevant result is from Ranchod et al. [16], who use an IRL model to define the primitives, in contrast to the problem of learning a policy after IRL.

3   Problem Statement and Model

3.1   Notation

Consider a finite-horizon Markov Decision Process (MDP):

$$\mathcal{M} = \langle S, A, P(\cdot, \cdot), \mathbf{R}, T \rangle,$$

where S is the set of states (continuous or discrete), A is the set of actions (finite and discrete), $P : S \times A \mapsto Pr(S)$ is the dynamics model that maps states and actions to a probability density over subsequent states, T is the time-horizon, and $\mathbf{R}$ is a reward function that maps trajectories of length T to scalar values.

Sequential tasks are tasks composed of sequences of sub-tasks. There is a sequence $\mathbf{R}_{seq} = [R_1, ..., R_k]$, where each $R_i : S \times A \mapsto \mathbb{R}$. Associated with each R_i is a transition region $\rho_i \subseteq S$. Each trajectory accumulates a reward R_i until it reaches the transition ρ_i, then the robot switches to the next reward and transition pair. This process continues until ρ_k is reached. A robot is deemed *successful* when all of the $\rho_i \in G$ are reached in sequence. Further, a robot is *optimal* when it maximizes the expected cumulative reward and is successful. Given observations of an optimally acting robot through a set of demonstration trajectories $D = \{d_1, ..., d_k\}$, can we infer $\mathbf{R}_{seq}$ and G?

Assumptions: We make the following assumptions: (1) the changes in reward between transitions is non trivial (see next remark), (2) every demonstration is generated from k distinct stationary, locally optimal policies (maximized w.r.t R_i on the infinite horizon), (3) every demonstration visits each ρ_i in the same sequence, and (4) each R_i is a quadratic of the form.

Remarks: The key challenge in this problem is determining when a transition occurs—identifying the points in time in each trajectory at which the robot reaches a ρ_i and transitions the reward function. The natural first question is whether this is identifiable, that is, whether it is even theoretically possible to determine whether a transition

$\rho_i \to \rho_{i+1}$ has occurred after obtaining an infinite number of observations. Trivially, this is not guaranteed when $R_{i+1} = R_i$, where it would be impossible to identify a transition purely from the robot's behavior (i.e., no change in reward, implies no change in behavior). Perhaps surprisingly, this is still not guaranteed even if $R_{i+1} \neq R_i$ due to policy invariance classes [8]. Consider a reward function $R_{i+1} = 2R_i$, which functionally induce the same optimal behavior. Therefore, we consider a setting where all of the rewards in $\mathbf{R}_{seq}$ are distinct and are not equivalent w.r.t optimal policies. This formalism is a special case of the Hierarchical Reinforcement Learning [17], where each of the local rewards is a sub-goal and arrival at a ρ is a termination condition.

3.2   Target Tracking Controllers

As motivation for where such assumptions arise, consider the following system:

$$x_{t+1} = Ax_t + Bu_t + w_t \qquad w_t \sim N(0,\Sigma) \quad i.i.d$$

For quadratic rewards in the infinite horizon, the optimal policy is a linear state feedback controller $u_t = -Cx_t$ Given this model, suppose we wanted to control the robot to a final state μ_i with a linear state-feedback controller C_i, the dynamical system that would follow is:

$$\hat{x}_{t+1} = (A - BC_i)\hat{x}_t + w_t,$$

where $\hat{x}_t = x_t - \mu_i$. If this system is stable, it will converge to $x_t = \mu_i$ as $t \to \infty$. Now, suppose that the system has the following switching behavior: when $\|x_t - \mu_i\| \leq \varepsilon$, change the target state μ_i to μ_{i+1}. The resulting closed loop dynamics are:

$$A_i = (A - BC_i)$$

$$x_{t+1} = A_i \mathbf{x}_t + w_t : A_i \in \{A_1, ..., A_k\}.$$

The equation above defines an SLDS. This model maps back to the general case where the sequence $[\mu_1, ..., \mu_k]$ and their tolerances $[\varepsilon_1, ..., \varepsilon_k]$ define the regions $[\rho_1, ..., \rho_k]$. Each ρ_i corresponds to regions where transitions occur $A_i \neq A_j$. Intuitively, a change in the reward function results in a change of policy (C_i) for a locally optimal agent.

   This form of a target-tracking linear system inspires our approach, where a robot applies a closed-loop controller to reach a target within a tolerance. While the ultimate target is fixed, we still need to understand what cost function the robot is minimizing, i.e., are there directions in which the robot minimizes distance to the target faster? To learn these, we can leverage human demonstrations. A demonstration d is a trajectory of state and action tuples $[(s_0, a_0), ..., (s_T, a_T)]$. Let D be a set of demonstrations $\{d_1, ..., d_N\}$, and our objective is to infer a sequence of reward functions (or cost functions) corresponding to each of the task segments.

4   Sequential Windowed Inverse Reinforcement Learning

This section describes an algorithm to infer the parameters for the proposed model.

Algorithm Description Let D be a set of demonstration trajectories $\{d_1, ..., d_N\}$ of a task with a delayed reward. SWIRL can be described in terms of three sub-algorithms:

Inputs: Demonstrations D

1. **Sequence Learning:** Given D, SWIRL segments the task into k sub-tasks whose start and end are defined by arrival at a set of transitions $G = [\rho_1, ..., \rho_k]$.
2. **Reward Learning:** Given G and D, SWIRL associates a local reward function with the segment resulting in a sequence of rewards $\mathbf{R}_{seq}$.
3. **Policy Learning:** Given $\mathbf{R}_{seq}$ and G, SWIRL applies reinforcement learning for I iterations to learn a policy for the task π.

Outputs: Policy π

Phase I. Sequence Learning

The first phase of SWIRL is to segment the demonstrations into locally linear segments. We can then cluster the segment endpoints into k clusters to infer $[\rho_1, ..., \rho_k]$. This is an extension of our prior work on robust task segmentation [6, 7]. The overall procedure is summarized in Phase I.

Segmentation and Transition Clustering The first step is given a set of demonstration trajectories, decompose each trajectory into segments. A popular approach for segmentation is to use Gaussian Mixture Models [14], namely, cluster all state observations and identify times at which x_t is in a different cluster than x_{t+1}. For a given time t, we can define a window of length ℓ as:

$$\mathbf{n}_t^{(\ell)} = [x_{t-\ell}, ..., x_t]^\top$$

Then, for each demonstration trajectory we can also generate a trajectory of $T_i - \ell$ windowed states:

$$\mathbf{d}_i^{(\ell)} = [\mathbf{n}_\ell^{(\ell)}, ..., \mathbf{n}_{T_i}^{(\ell)}]$$

Over the entire set of windowed demonstrations, we collect a dataset of all of the $\mathbf{n}_t^{(\ell)}$ vectors. We fit GMM model to these vectors. The GMM model defines m multivariate Gaussian distributions and a probability that each observation $\mathbf{n}_t^{(\ell)}$ is sampled from each of the m distributions. We annotate each observation with the most likely mixture component. Times such that $\mathbf{n}_t^{(\ell)}$ and $\mathbf{n}_{t+1}^{(\ell)}$ have different most likely components are marked as transitions. This has the interpretation of fitting a locally linear regression to the data (refer to [18, 19, 20, 6, 7] for details). In typical GMM formulations, one must specify the number of mixture components k before hand. However, we apply results from Bayesian non-parametric statistics and jointly solve for the component parameters and the number of components using a Dirichlet Process [21]. Using a DP, the number of components grows with the complexity of the observed data (we denote this as DP-GMM).

In prior work, we noticed that motion-based segmentation algorithms can be unreliable when there is noise [6]. We, however, realized that applying a second level of cluster–clustering the segment endpoints found dense clusters of common transitions that occurred in all demonstrations–thus allowing us to reject spurious motions or observation noise. The insight of this work is that the same approach can be interpreted as identifying necessary sub-goals in a complex task.

Algorithm 1: Sequence Learning

Data: Demonstration $\mathcal{D}$

1 Fit a DP-GMM model to $\mathcal{D}$ and identify the set of transitions Θ, defined as all (x_t, t)
   where $(x_{t+1}, t+1)$ has a different cluster.
2 Fit a DP-GMM to the states in Θ.
3 Prune clusters that do not have one transition from all demonstrations.
4 The result of is $G = [\rho_1, \rho_2, ..., \rho_m]$ where each ρ is a disjoint ellipsoidal region of the
   state-space and time interval.

Result: G

We would like to be able to aggregate the transition times into state-space conditions for reward transitions $[\rho_1, ..., \rho_k]$. To each of these transition times, there is a corresponding *transition state*–the last state before the dynamics switched. We can model these regions againwith a Gaussian Mixture Model with k mixture components $\{m_1, ..., m_k\}$. As before, we use a DP to non-parametrically set k. Then, we prune clusters that do not have at least one transition from each demonstration. Thus, the result is the set of transition regions: $G = [\rho_1, \rho_2, ..., \rho_k]$, and segmentation of each demonstration trajectory into k segments.

Relaxing Local Linearity GMM's are a type of local Bayesian linear regression, but we can easily relax the linearity assumption. We relax the linear dynamics assumption with a kernel embedding of the trajectories. SWIRL does not require learning the exact regimes A_i, it only needs to detect changes in dynamics regime. The basic idea is to apply Kernelized PCA to the features before learning the transitions–a technique used in Computer Vision [22]. By changing the kernel function (i.e., the similarity metric between states), we can essentially change the definition of local linearity.

Let $\kappa(x_i, x_j)$ define a kernel function over the states. For example, if κ is the radial basis function (RBF), then: $\kappa(x_i, x_j) = e^{\frac{-\|x_i - x_j\|_2^2}{2\sigma}}$. κ naturally defines a matrix M where: $M_{ij} = \kappa(x_i, x_j)$. The top p' eigenvalues define a new embedded feature vector for each ω in $\mathbb{R}^{p'}$. We can now apply the algorithm above in this embedded feature space.

Phase II. Reward Learning

After Phase I, each demonstration is segmented into k sub-sequences. Phase II uses the learned $[\rho_1, ..., \rho_k]$ to construct the local rewards $[R_1, ..., R_k]$ for the task. Each R_i is a quadratic cost parametrized by a positive semi-definite matrix Q. SWIRL has two variants: a model-based variant which uses a locally estimated dynamics model to construct the reward, and a model-free variant which is a baseline which just estimates variation in different features. The Algorithm is summarized in Phase 2.

Model-based For the model-based approach, we use Maximum Entropy Inverse Reinforcement Learning (MaxEnt-IRL) [23]. The idea is to model every demonstration d_i as a noisy sample from an optimal policy. In other words, each d_i that is observed is a noisy observation of some hypothetical d^*.

Since each d_i is a path through a possibly discrete state and action space, we cannot simply average them to find d^*. Instead, we have to model trajectories that the system is likely to visit. This can be formalized with the following probability distribution:

$$P(d_i|R) \propto \exp\{\sum_{t=0}^{T} R(s_t)\}.$$

Paths with a higher cumulative reward are more likely.

MaxEnt-IRL uses the following linear parametrized representation:

$$R(s,a) = x^T \theta,$$

where x is the state vector. The resulting form is:

$$P(d_i|R) \propto \exp\{\sum_{t=0}^{T} x^T \theta\},$$

and MaxEnt-IRL proposes an algorithm to infer the θ that maximizes the posterior likelihood.

SWIRL applies MaxEnt-IRL to each segment of the task but with a small modification to learn quadratic rewards instead of linear ones. Let μ_i be the centroid of the next transition region. We want to learn a reward function of the form:

$$R_i(x) = -(x - \mu_i)^T Q(x - \mu_i).$$

for a positive semi-definite Q (negated since this is a negative quadratic cost). With some re-parametrization, this reward function can be written as:

$$R_i(x) = -\sum_{j=1}^{d}\sum_{l=1}^{d} q_{ij}x[j]x[l].$$

which is linear in the feature-space $y = x[j]x[l]$:

$$R_i(x) = \theta^T y.$$

This posterior inference procedure requires a dynamics model. We fit local linear models to each of the segments discovered in the previous section:

$$A_j = \arg\min_A \sum_{i=1}^{N} \sum_{\text{seg } j \text{ start}}^{\text{seg } j \text{ end}} \|Ax_t^{(i)} - x_{t+1}^{(i)}\|$$

In this form, the problem can be analytically solved with techniques proposed in [24]. SWIRL applies MaxEnt-IRL to the sub-sequences of demonstrations between 0 and ρ_1, and then from ρ_1 to ρ_2 and so on. The result is an estimated local reward function R_i modeled as a linear function of states that is associated with each ρ_i.

Model-free: Local Quadratic Rewards When the linearity assumption cannot be made, SWIRL can alternatively use a model-free approach for reward construction. The role of the reward function is to guide the robot to the next transition region ρ_i. A straight forward thing approach is for each segment i, we can define a reward function

---

Algorithm 2: Reward Learning

Data: Demonstration $\mathcal{D}$ and sub-goals $[\rho_1, ..., \rho_k]$

1 Based on the transition states, segment each demonstration d_i into k sub-sequences where the jth is denoted by $d_i[j]$.
2 If dynamics model is available, apply MaxEnt-IRL to each set of sub-sequences 1...k.
3 If the dynamics model is not available compute Equation 1 for each set of subsequences.

Result: $\mathbf{R}_{seq}$

---

as follows:

$$R_i(x) = -\|x - \mu_i\|_2^2,$$

which is just the Euclidean distance to the centroid.

A problem with using Euclidean distance directly is that it uniformly penalizes disagreement with μ in all dimensions. During different stages of a task, some features will likely naturally vary more than others–this is learned through IRL. To account for this, we derive a reasonable Q that is independent of the dynamics:

$$Q[j,l] = \Sigma_x^{-1},$$

which is the inverse of the covariance matrix of all of the state vectors in the segment:

$$Q[j,l] = (\sum_{t=start}^{end} xx^T)^{-1}, \tag{1}$$

which is a $p \times p$ matrix defined as the covariance of all of the states in the segment $i - 1$ to i. Intuitively, if a feature has low variance during this segment, deviation in that feature from the desired target it gets penalized. This is exactly the Mahalonabis distance to the next transition.

For example, suppose one of the features j measures the distance to a reference trajectory u_t. Further, suppose in step one of the task the demonstrator's actions are perfectly correlated with the trajectory ($Q_i[j,j]$ is low where variance is in the distance) and in step two the actions are uncorrelated with the reference trajectory ($Q_i[j,j]$ is high). Thus, Q will respectively penalize deviation from $\mu_i[j]$ more in step one than in step two.

Phase III. Policy Learning

In Phase III, SWIRL uses the learned transitions $[\rho_1, ..., \rho_k]$ and $\mathbf{R}_{seq}$ as rewards for a Reinforcement Learning algorithm. In this section, we describe learning a policy π given rewards $\mathbf{R}_{seq}$ and an ordered sequence of transitions G.

However, this problem is not trivial since solving k independent problems neglects potential shared value structure between the local problems (e.g., a common failure state). Furthermore, simply taking the aggregate of the rewards can lead to inconsistencies since there is nothing enforcing the order of operations. The key insight is that a single policy can be learned jointly over all segments over a modified problem where the state-space with additional variables that keep track of the previously achieved segments. To do so, we require an MDP model that also captures the history of the process.

Algorithm 3: Policy Learning

Data: Transition States G, Reward Sequence $\mathbf{R}_{seq}$, exploration parameter ε

1 Initialize $Q(\binom{s}{v}, a)$ randomly

2 **foreach** $iter \in 0, ..., I$ **do**

3 $\quad$ Draw s_0 from initial conditions

4 $\quad$ Initialize v to be $[0, ..., 0]$

5 $\quad$ Initialize j to be 1

6 $\quad$ **foreach** $t \in 0, ..., T$ **do**

7 $\quad\quad$ Choose best action a based on Q or random action w.p ε.

8 $\quad\quad$ Observe Reward R_j

9 $\quad\quad$ Update state to s' and Q via Q-Learning update

10 $\quad\quad$ If s' is $\in \rho_j$ update $v[j] = 1$ and $j = j + 1$

Result: Policy π

MDPs with Memory RL algorithms apply to problems that are specified as MDPs. The challenge is that some sequential tasks may not be MDPs. For example, attaining a reward at ρ_i depends on knowing that the reward at goal ρ_{i-1} was attained. In general, to model this dependence on the past requires MDPs whose state-space also includes history.

Given a finite-horizon MDP $\mathcal{M}$ as defined in Section 3, we can define an MDP $\mathcal{M}_H$ as follows. Let $\mathcal{H}$ denote set of all dynamically feasible sequences of length smaller than T comprised of the elements of S. Therefore, for an agent at any time t, there is a sequence of previously visited states $H_t \in \mathcal{H}$. The MDP $\mathcal{M}_H$ is defined as:

$$\mathcal{M}_H = \langle S \times \mathcal{H}, A, P'(\cdot, \cdot), R(\cdot, \cdot), T \rangle.$$

For this MDP, P' not only defines the transitions from the current state $s \mapsto s'$, but also increments the history sequence $H_{t+1} = H_t \sqcup s$. Accordingly, the parametrized reward function R is defined over S, A, and H_{t+1}.

$\mathcal{M}_H$ allows us to address the sequentiality problem since the reward is a function of the state and the history sequence. However, without some parametrization of H_t, directly solving this MDPs with RL is impractical since it adds an overhead of $\mathcal{O}(e^T)$ states.

Policy Learning Using our sequential task definition, we know that the reward transitions (R_i to R_{i+1}) only depend on an arrival at the transition state ρ_i and not any other aspect of the history. Therefore, we can store a vector v, a k dimensional binary vector ($v \in \{0, 1\}^k$) that indicates whether a transition state $i \in 0, ..., k$ has been reached. This vector can be efficiently incremented when the current state $s \in \rho_{i+1}$. Then, the additional complexity of representing the reward with history over $S \times \{0, 1\}^k$ is only $\mathcal{O}(k)$ instead of exponential in the time horizon.

The result is an augmented state-space $\binom{s}{v}$ to account for previous progress. Over this state-space, we can apply Reinforcement Learning algorithms to iteratively converge to a successful policy for a new task instance. SWIRL applies Q-Learning with a Radial Basis Function value function representation to learn a policy π over this state-space and the reward sequence $\mathbf{R}_{seq}$. This is summarized in Algorithm 3.

5   Experiments

We evaluate SWIRL with a series of standard RL benchmarks and in a physical experiment on the da Vinci surgical robot.

5.1   Methodology

All of the experimental scenarios followed a similar pattern: (1) start with an RL problem with a delayed reward, (2) generate N demonstration trajectories with motion planning in simulated scenarios and kinethestic demonstration in the physical experiments, (3) apply SWIRL, (4) evaluate the performance of the policy as a function of the I iterations. For all convergence curves presented, we show the probability of task success as a function of the number of RL iterations. For convergence rate, we measure the Area Under Curve of the learning curve (i.e., cumulative expected reward over the learning epoch).

The algorithms considered in the experiments are:

1. **Q-Learning:** This applies a Q-Learning algorithm with the same hyper-parameter setting as SWIRL.
2. **Pure MaxEnt-IRL:** Given N demonstrations this learns a reward using MaxEnt-IRL and no hierarchical structure. Then, it applies Q-Learning with the same hyper-parameter setting as SWIRL until convergence. (Only Phase II and III)
3. **SVM:** Given N demonstrations this learns a policy using a multi-class SVM classifier. There is no further learning after training. (Directly to Policy)
4. **SWIRL (Model-Based) and SWIRL (Model-Free)**

5.2   Parallel Parking

We constructed a parallel parking scenario for a robot with non-holonomic dynamics and two obstacles. The robot can control its speed ($\|\dot{x}\| + \|\dot{y}\|$) and heading ($\theta$), and observe its x position, y position, orientation, and speed in a global coordinate frame. If the robot parks between the obstacles, i.e., 0 velocity within a $15°$ tolerance, the task is a success and the robot receives a reward of 1. The robot's dynamics are noisy and with probability 0.1 will randomly add or subtract $5°$ degrees from the steering angle. If the robot collides with one of the obstacle or does not park in 200 timesteps the episode ends. The baseline approach is modeling the entire problem as an MDP with a quadratic reward function at the target state (where the robot parks). For comparison, we use this reward function for Q-Learning and infer a quadratic reward function using MaxEnt-IRL. We call this domain Parallel Parking with Full Observation (PP-FO) (see Figure 1).

Next, we made the Parallel Parking domain a little harder. We hid the velocity state from the robot, so the robot only sees (x, y, θ). As before, if the robot collides with one of the obstacle or does not park in 200 timesteps the episode ends. We call this domain Parallel Parking with Partial Observation (PP-PO).

We generated 5 demonstrations using an RRT motion planner (assuming deterministic dynamics) and applied SWIRL to learn the segments. Figure 1 illustrates the demonstrations and the learned segments. There are two intermediate goals corresponding to positioning the car and orienting the car correctly before reversing.

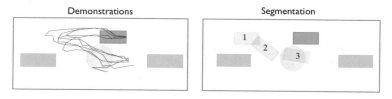

Fig. 1. This plot illustrates (left) the 5 demonstration trajectories for the parallel parking task, and (right) the sub-goals learned by SWIRL.

Performance In the fully observed problem, compared to MaxEnt-IRL, the model-based SWIRL converges to a policy with a 60% success rate with about 3x fewer time-steps. The gains for the model-free version are more modest with a 50% reduction. The supervised policy learning approach achieves a success rate of 47% and the baseline RL approach achieves a success rate of 36% after 250000 time-steps.

The baseline Q-Learning approach directly tries to learn a sequence of actions to minimize the quadratic cost around the target state. This leads to a lot of exploration since the robot must first make "negative" progress (pulling forward). SWIRL improves convergence since it structures the exploration through the segmentation. The local reward functions are better shaped to guide the car towards its short term goal. This focuses the exploration on solving the short term problem first. MaxEnt-IRL mitigates some of the problems since it rewards states based on their estimated cost-to-go, but as the time-horizon increases the estimates of this become nosier–leading to worse performance (see technical report for a characterization [25]).

In the partial observation problem (PP-PO), there is no longer a stationary policy that can achieve the reward. The techniques that model this problem with a single MDP all fail to converge. The learned segments in SWIRL help disambiguate dependence on history. After 250000 time-steps, the policy learned with model-based SWIRL has a 70% success rate in comparison to a <10% success rate for the baseline RL, MaxEnt-IRL, and 0% for the SVM.

Finally, we explore how well the constructed rewards transfer if the dynamics are perturbed in the fully observed setting. We expect MaxEnt-IRL to transfer well because it learns a delayed reward, which tends to encode success conditions and not task-specific details. After constructing the rewards, we randomly perturbed the system dynamics by introducing a bias towards turning left. We find that the model-based SWIRL technique transfers to this domain comparably to MaxEnt-IRL until the task is so different that the sub-goals learned with SWIRL are no longer informative. The model-free SWIRL algorithm converges more slowly; requiring 20% more time-steps to converge to the same success rate.

5.3   Acrobot

This domain consists of a two-link pendulum with gravity and with torque controls on the joint. The dynamics are noisy and there are limits on the applied torque. The robot has 1000 timesteps to raise the arm above horizontal ($y = 1$ in the images). If the task is successful and the robot receives a reward of 1. Thus, the expected reward is equivalent to the probability that the current policy will successfully raise the arm above horizontal. We generated $N = 5$ demonstrations for the Acrobot task and applied

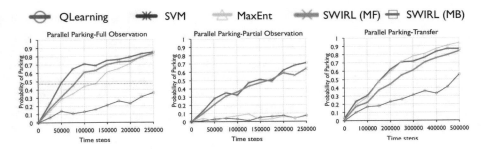

Fig. 2. Performance on a parallel parking task with noisy dynamics with full state observations (position, orientation, and velocity), partial observation (only position and orientation), and transfer (randomly permuting the action space). Success is measured in terms of the probability that the car successfully parked, and (M) denotes whether the approach used the dynamics model. In the fully observed case, both the model-based and model-free SWIRL algorithms converge faster than MaxEnt-IRL and quickly outperforms the SVM. In the partially observed case, MaxEnt-IRL, Q-Learning, and the SVM fail–while SWIRL succeeds. Both techniques also demonstrate comparable transferability to MaxEnt-IRL when the domain's dynamics are perturbed.

segmentation. These demonstrations were generated by training the Q-Learning baseline to convergence and then sampling from the learned policy. In Figure 3, we plot the performance of the all of the approaches. We include a comparison between a Linear Multiclass SVM and a Kernelized Multiclass SVM for the policy learning alternative. In this example, we find that applying MaxEnt-IRL does not improve the convergence rate. For this state-space, MaxEnt-IRL merely recovers the reward used in the original RL problem. On the other hand, added segments using SWIRL improve convergence rates.

We also vary the number of input demonstrations to SWIRL and find that it requires fewer demonstrations than policy learning and MaxEnt-IRL to converge to a more reliable policy. It takes about 10x more demonstrations for the supervised learning approach to reach comparable reliability. Finally, we find that SWIRL does not sacrifice much transferability. We learn the rewards on the standard pendulum, and then during learning we vary the size of the second link in the pendulum. We plot the success rate (after a fixed 50000 steps) as a function of the increase link size. SWIRL is significantly more robust than supervised policy learning to the increase in link size and has a significantly higher success rate than IRL for small perturbations in link size.

5.4 Summary of Simulated Experiments

Table 1 summarizes the results of our experiments in terms of convergence rate and maximum attained reward on the Parallel Parking domain (with and without partial observation), Acrobot domain, and additional experiments using variants of GridWorld. GridWorld is a two-room map where the robot has to reach two target states in sequence to get the full reward. GridWorld-2 is a substantially harder map with "pits" (i.e., instant failure if reached). The Two-Bridges domain is another GridWorld based environment in which there is a short "unsafe" path between start and goal and a longer "safe" path (which is actually the optimal solution). Please refer to the arXiv report [25] for more details.

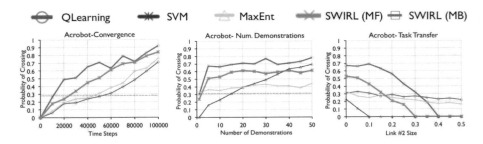

Fig. 3. Acrobot: We measured the performance of rewards constructed with SWIRL and the alternatives. We find that SWIRL (model-based and model-free) converges faster than MaxEnt-IRL, Q-Learning, and the SVM. Furthermore, SWIRL requires less demonstrations, which we measure by comparing the performance of the alternatives after a fixed 50000 time-steps and with varied input demonstrations. We also vary the task parameters by changing the size of the second link of the pendulum and find that the learned rewards are robust to this variation (as before comparing the performance of the alternatives after a fixed 50000 time steps). MaxEnt-IRL shows improved transfer performance since once the task has changed enough the segments learned during the demonstrations may not be informative and may even hurt performance if they are misleading.

Table 1. This table summarizes the convergence rate (**AUC**) and max reward (**MAX**) attained by a Q-learning robot using the alternatives after a fixed number of iterations.

	GridWorld		GridWorld-2		Two-Bridges		PP(FO)		PP(PO)		Acrobot	
	Max	AUC	Max	AUC	Max	AUC	Max	AUC	Max	AUC	Max	AUC
Q-Learning	0.984	10.976	0.861	15.440	1.090	16.270	0.911	109.76	0.311	27.419	0.944	3.447
MaxEnt-IRL	0.987	299.556	0.861	16.956	0.759	16.270	0.950	299.556	0.444	33.128	0.920	44.111
SWIRL (MF)	1.830	322.125	1.764	14.070	**1.751**	**18.953**	**0.991**	164.127	0.934	123.115	0.906	20.935
SWIRL (MB)	**1.835**	**514.113**	**1.827**	**28.632**	1.577	17.141	0.965	**514.113**	**0.958**	**333.897**	**0.987**	**65.512**

5.5   Physical Experiments with the da Vinci Surgical Robot

In physical experiments, we apply SWIRL to learn to cut along a marked line in gauze similar to Murali et al. [26]. This is a multi-step problem where the robot starts from a random initial state, has to move to a position that allows it to start the cut, and then cut along the marked line. We provide the robot 5 kinesthetic demonstrations by positioning the end-effector and then following various marked straight lines. The state-space of the robot included the end-effector position (x, y) as well as a visual feature indicating its pixel distance to the marked line (pix). This visual feature is constructed using OpenCV thresholding for the black line. Since the gauze is planar, the robot's actions are unit steps in the $\pm x, \pm y$ axes. Figure 4 illustrates the training and test scenarios.

As expected, the algorithm identifies two consistent changes in local linearity corresponding to the positioning step and the termination. The learned reward function for the position step minimizes x, y, pix distance to the starting point and for the cutting step the reward function is more heavily weighted to minimize the pix distance. We defined task success as positioning within 1 cm of the starting position of the line and during the following stage, missing the line by no more than 1 cm (estimated from pixel distance). Since we did not have a dynamics model, we evaluated the model-free version of SWIRL, Q-Learning, and the SVM. SWIRL was the only technique able to achieve

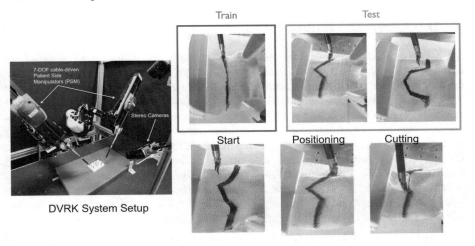

Fig. 4. We collected demonstrations on the da Vinci surgical robot kinesthetically. The task was to cut a marked line on gauze. We demonstrated the location of the line without actually cutting it. The goal is to infer that the demonstrator's reward function has two steps: position at a start position before the line, and then following the line. We applied this same reward to curved lines that started in different positions.

the combined task. This is because the policy for this task is non-stationary, and SWIRL is the only approach of the alternatives that can learn such a policy.

We evaluated the learned tracking policy to cut gauze. We ran trials on different sequences of curves and straight lines. Out of the 15 trials, 11 were successful. 2 failed due to SWIRL errors (tracking or position was imprecise) and 2 failed due to cutting errors (gauze deformed causing the task to fail). 1 of the failures was on the 4.5 cm curvature line and 3 were one the 3.5 cm curvature line.

Table 2. With 5 kinesthetic demonstrations of following marked straight lines on gauze, we applied SWIRL to learn to follow lines of various curvature. After 25 episodes of exploration, we evaluated the policies on the ability to position in the correct cutting location and track the line. We compare to SVM on each individual segment. SVM is comparably accurate on the straight line (training set) but does not generalize well to the curved lines.

Curvature Radius (cm)	SVM Pos. Error (cm)	SVM Tracking Error (cm)	SWIRL Pos. Error (cm)	SWIRL Tracking Error (cm)
straight	0.46	0.23	0.42	0.21
4.0	0.43	0.59	0.45	0.33
3.5	0.51	1.21	0.56	0.38
3.0	0.86	3.03	0.66	0.57
2.5	1.43	-	0.74	0.87
2.0	-	-	0.87	1.45
1.5	-	-	1.12	2.44

Next, we characterized the repeatability of the learned policy. We applied SWIRL to lines of various curvature spanning from straight lines to a curvature radius of 1.5 cm. Table 2 summarizes the results on lines of various curvature. While the SVM approach did not work on the combined task, we evaluated its accuracy on each individual step to illustrate the benefits of SWIRL. On following straight lines, SVM was comparable to SWIRL in terms of accuracy. However, as the lines become increasingly curved, SWIRL generalizes more robustly than the SVM.

6   Discussion and Future Work

SWIRL is a three-phase algorithm that first segments a task, learns local rewards, and then learns a policy. Experimental results suggest that sequential segmentation can indeed improve convergence in RL problems with delayed rewards. Results suggest that SWIRL is robust to perturbations in initial conditions, the environment, and sensing noise. This paper formalizes the interaction and composability of the three phases (sequence, reward, and policy learning). In future work, we will explore extensions to each of the phases and quantify the degree of generalization. We will explore how the Q-Learning step could be replaced with Guided Policy Search, Policy Gradients, and optimal control. We will modify the segmentation algorithm to incorporate more complex transition conditions and allow for sub-optimal demonstrations. We will explore more robotic tasks including suturing, surgical knot tying, and assembly. Another avenue for future work is modeling complex tasks as hierarchies of MDPs.

Acknowledgements: This research was performed at the AUTOLAB at UC Berkeley in affiliation with the AMP Lab, BAIR, and the CITRIS "People and Robots" (CPAR) Initiative in affiliation with UC Berkeley's Center for Automation and Learning for Medical Robotics (Cal-MR). The authors were supported in part by the U.S. National Science Foundation under NRI Award IIS-1227536: Multilateral Manipulation by Human-Robot Collaborative Systems, and by Google, UC Berkeley's Algorithms, Machines, and People Lab, Knut & Alice Wallenberg Foundation, and by a major equipment grant from Intuitive Surgical and by generous donations from Andy Chou and Susan and Deepak Lim. We thank our colleagues and the anonymous WAFR reviewers who provided valuable feedback and suggestions, in particular, Pieter Abbeel, Anca Dragan, and Roy Fox.

References

1. Argall, B.D., Chernova, S., Veloso, M., Browning, B.: A Survey of Robot Learning from Demonstration. Robotics and Autonomous Systems **57**(5) (2009) 469–483
2. Kolter, J.Z., Abbeel, P., Ng, A.Y.: Hierarchical apprenticeship learning with application to quadruped locomotion. In: NIPS. (2007) 769–776
3. Coates, A., Abbeel, P., Ng, A.Y.: Learning for control from multiple demonstrations. In: ICML, ACM (2008)
4. Abbeel, P., Ng, A.Y.: Apprenticeship learning via inverse reinforcement learning. In: ICML, ACM (2004) 1
5. Ng, A.Y., Russell, S.J., et al.: Algorithms for inverse reinforcement learning. In: Icml. (2000) 663–670
6. Krishnan*, S., Garg*, A., Patil, S., Lea, C., Hager, G., Abbeel, P., Goldberg, K., (*denotes equal contribution): Transition State Clustering: Unsupervised Surgical Trajectory Segmentation For Robot Learning. In: International Symposium of Robotics Research, Springer STAR (2015)
7. Murali*, A., Garg*, A., Krishnan*, S., Pokorny, F.T., Abbeel, P., Darrell, T., Goldberg, K., (*denotes equal contribution): TSC-DL: Unsupervised Trajectory Segmentation of Multi-Modal Surgical Demonstrations with Deep Learning. In: IEEE Int. Conf. on Robotics and Automation (ICRA). (2016)
8. Ng, A.Y., Harada, D., Russell, S.J.: Policy invariance under reward transformations: Theory and application to reward shaping. In: ICML. (1999) 278–287
9. Judah, K., Fern, A.P., Tadepalli, P., Goetschalckx, R.: Imitation learning with demonstrations and shaping rewards. In: AAAI. (2014) 1890–1896

10. Ijspeert, A., Nakanishi, J., Schaal, S.: Learning attractor landscapes for learning motor primitives. In: Neural Information Processing Systems (NIPS). (2002) 1523–1530

11. Pastor, P., Hoffmann, H., Asfour, T., Schaal, S.: Learning and generalization of motor skills by learning from demonstration. In: IEEE ICRA. (2009)

12. Manschitz, S., Kober, J., Gienger, M., Peters, J.: Learning movement primitive attractor goals and sequential skills from kinesthetic demonstrations. Robotics and Autonomous Systems (2015)

13. Niekum, S., Osentoski, S., Konidaris, G., Barto, A.: Learning and generalization of complex tasks from unstructured demonstrations. In: Int. Conf. on Intelligent Robots and Systems (IROS), IEEE (2012)

14. Calinon, S.: Skills learning in robots by interaction with users and environment. In: IEEE Int. Conf. on Ubiquitous Robots and Ambient Intelligence (URAI). (2014)

15. Konidaris, G., Kuindersma, S., Grupen, R., Barto, A.: Robot Learning from Demonstration by Constructing Skill Trees. Int. Journal of Robotics Research **31**(3) (2011) 360–375

16. Ranchod, P., Rosman, B., Konidaris, G.: Nonparametric bayesian reward segmentation for skill discovery using inverse reinforcement learning. In: IEEE/RSJ Int. Conf. on Intelligent Robots and Systems (IROS), IEEE (2015)

17. Dietterich, T.G.: Hierarchical reinforcement learning with the maxq value function decomposition. J. Artif. Intell. Res.(JAIR) **13** (2000) 227–303

18. Moldovan, T., Levine, S., Jordan, M., Abbeel, P.: Optimism-driven exploration for nonlinear systems. In: Int. Conf. on Robotics and Automation (ICRA). (2015)

19. Khansari-Zadeh, S.M., Billard, A.: Learning stable nonlinear dynamical systems with gaussian mixture models. Robotics, IEEE Transactions on **27**(5) (2011) 943–957

20. Kruger, V., Herzog, D., Baby, S., Ude, A., Kragic, D.: Learning actions from observations. Robotics & Automation Magazine, IEEE **17**(2) (2010) 30–43

21. Kulis, B., Jordan, M.I.: Revisiting k-means: New algorithms via bayesian nonparametrics. arXiv preprint arXiv:1111.0352 (2011)

22. Mika, S., Schölkopf, B., Smola, A.J., Müller, K., Scholz, M., Rätsch, G.: Kernel PCA and de-noising in feature spaces. In: NIPS. (1998) 536–542

23. Ziebart, B.D., Maas, A.L., Bagnell, J.A., Dey, A.K.: Maximum entropy inverse reinforcement learning. In: AAAI. (2008)

24. Ziebart, B., Dey, A., Bagnell, J.A.: Probabilistic pointing target prediction via inverse optimal control. In: UIST, ACM (2012) 1–10

25. Krishnan, S., Garg, A., Liaw, R., Miller, L., Pokorny, F.T., Goldberg, K.: Hirl: Hierarchical inverse reinforcement learning for long-horizon tasks with delayed rewards. arXiv preprint arXiv:1604.06508 (2016)

26. Murali*, A., Sen*, S., Kehoe, B., Garg, A., McFarland, S., Patil, S., Boyd, W., Lim, S., Abbeel, P., Goldberg, K., (*denotes equal contribution): Learning by Observation for Surgical Subtasks: Multilateral Cutting of 3D Viscoelastic and 2D Orthotropic Tissue Phantoms. In: IEEE Int. Conf. on Robotics and Automation (ICRA). (2015)

Adapting Deep Visuomotor Representations with Weak Pairwise Constraints

Eric Tzeng[*][1], Coline Devin[*][1], Judy Hoffman[1], Chelsea Finn[1],
Pieter Abbeel[1], Sergey Levine[1], Kate Saenko[2], Trevor Darrell[1]

[1] University of California, Berkeley
[2] Boston University

Abstract. Real-world robotics problems often occur in domains that differ significantly from the robot's prior training environment. For many robotic control tasks, real world experience is expensive to obtain, but data is easy to collect in either an instrumented environment or in simulation. We propose a novel domain adaptation approach for robot perception that adapts visual representations learned on a large easy-to-obtain source dataset (e.g. synthetic images) to a target real-world domain, without requiring expensive manual data annotation of real world data before policy search. Supervised domain adaptation methods minimize cross-domain differences using pairs of aligned images that contain the same object or scene in both the source and target domains, thus learning a domain-invariant representation. However, they require manual alignment of such image pairs. Fully unsupervised adaptation methods rely on minimizing the discrepancy between the feature distributions across domains. We propose a novel, more powerful combination of both distribution and pairwise image alignment, and remove the requirement for expensive annotation by using weakly aligned pairs of images in the source and target domains. Focusing on adapting from simulation to real world data using a PR2 robot, we evaluate our approach on a manipulation task and show that by using weakly paired images, our method compensates for domain shift more effectively than previous techniques, enabling better robot performance in the real world.

1   Introduction

Transfer and domain shift are major challenges in learning-based robotic perception and control. Perception systems built using offline datasets often fail when deployed on a robot, robots trained to perceive and act in a laboratory setting might fail outside of the lab, and robots trained in simulation often fail in the real world. However, accurate data annotations (such as the state of the world) are often only available in simulated or instrumented environments, which usually look too different from the real world to use directly. To enable adaptation

---

[*] Authors contributed equally.

© Springer Nature Switzerland AG 2020
K. Goldberg et al. (Eds.): *Algorithmic Foundations of Robotics XII*, SPAR 13, pp. 688–703, 2020.
https://doi.org/10.1007/978-3-030-43089-4_44

of robotic perception between domains, we present a deep learning architecture that learns to map images from each domain into a common feature space.

We propose a novel framework with losses for both pairwise alignment and distribution-level alignment. We also introduce a new algorithm for aligning source and target images without labels in the target domain. This method is general and can be applied to many perception tasks, and we show that it increases performance on adapting pose estimation (predicting object keypoints) from synthetic images to real images. Furthermore, this technique can be used to pretrain visual features for visuomotor policy search. Recently proposed end-to-end visuomotor networks [1] can learn both image representations and the control policy for a particular task directly from visual data. In particular, the method in [1] first learns a convolutional network to predict keypoint locations from raw images, then fine-tunes the representation with guided policy search to map keypoints to actions. However, this previous method uses 1000 pose annotated images to train the keypoint predictor. We show that pretraining using our framework allows us to construct effective vision-based manipulation policies without any pose annotated real images.

Existing deep domain adaptation methods have focused on the category-level domain invariance task, and used optimization to generally reduce the discrepancy, or maximize confusion, between domains [2,3]; this is valuable, but misses a significant opportunity in the setting of synthetic to real image adaptation. It is often feasible to generate a large enough variety of synthetic images such that for each unlabeled real image, there exists a matching synthetic image. This can provide instance level training constraints for a deep domain adaptation architecture that minimizes the distance between features of the instance pair. Previous work has not tackled the problem of learning these pairs in settings where explicit annotations are unavailable. Additionally, while such constraints have been explored in earlier adaptation schemes [4], to our knowledge they have not been combined with contemporary deep discrepancy or deep confusion models.

Fig. 1. A pair of corresponding synthetic (left) and real-world (right) images used for our pose estimation evaluation. Our method finds pairs without real-world supervision

We report experiments with our framework on the pose pretraining stage of the visuomotor model of [1], using a real and simulated PR2, as shown in Figure 1. We also evaluate the learned representations by using them as input for training a visuomotor policy. Our results confirm (1) there can be a significant domain shift in visuomotor task learning, (2) that domain adaptation methods specialized to the deep spatial feature point architecture introduced in [1] can learn to be relatively invariant to such shifts and improve performance, (3) that inclusion of pairwise constraints provides a performance boost relative to previous deep domain adaptation approaches based solely on discrepancy minimization or domain confusion maximization, and (4) that, even in settings where pose annotations are unavailable for target domain imagery, annotations can be transferred from a source domain dataset (e.g. generated by a low-fidelity renderer). We validate our method by training a visuomotor policy on the PR2 robot to perform a simple manipulation task.

2   Related work

In both vision and robotics, it has long been a desirable goal to use easily obtainable data (such as synthetic rendered images) to train models that are effective in real environments. In robotics, past work has used domain adaptation and simulated data to reduce the need for labeled target domain examples. Lai and Fox used a variant of feature augmentation [5] to use human-made 3D models for laser scan classification [6]. Saxena et al. used rendered objects to learn to grasp from vision [7].

Classically, in computer vision, hand-engineered features were designed to be invariant to the domain shift between synthetic and real worlds, e.g., efforts dating from the earliest model alignment methods in computer vision using edge detection-based representations [8]. One of the earliest visuomotor neural network learning methods, ALVINN [9], exploited simulated training data of observed road shapes when training a multi-layer perceptron for an autonomous driving task. Many approaches to pose estimation in the recent decade were trained using rendered scenes from POSER and other human form rendering systems [10,11,12]; reliance on fixed feature representations limited their performance, however, and state-of-the-art pose estimation methods generally train exclusively on real imagery [13,14].

Traditional visual domain adaptation methods tackled the problem where a fixed representation extraction algorithm was used for both visual domains, and adaptation took the form of learning a transformation between the two spaces [4,15,16] or regularizing the target domain model based on the source domain [17,18]. Later models improved upon this by proposing adaptation which both transformed the representation spaces and regularized the target model using the source data [19,20]. Since the resurgence in the popularity of convolutional networks for visual representation learning, adaptation approaches have been proposed to optimize the full target representation and model to better align with the source, for example by minimizing the maximum mean discrepancy [21,22] or by minimizing the a-distance (specific form of discrepancy distance [23]) between the two distributions [3,2].

Recently, a method has been proposed to use 3D object models to render synthetic training examples for training visual models with limited human annotations needed [24]. It was shown that there is a specific domain shift problem that arises when applying a synthetically trained visual model to the real world data. This paradigm of synthetic to real was further used to study deep representations and the types of invariances they learn by [25].

While classic robotic perception already provides ample motivation for exploring scalable and effective domain adaptation methods, recent progress in deep reinforcement learning (RL) raises another intriguing possibility. Deep RL methods have shown remarkable performance and generality on tasks ranging from simulated locomotion to playing Atari games [26,27,28], but often at the cost of very high sample complexity. Other than the method in [1], many of these methods are impractical to use directly on real physical systems due to the sample requirements, and a key question is whether policies learned with deep reinforcement learning in simulation could be extended for use in the real world. In this paper, we present an initial step in this direction by showing that vision systems trained on simulated data and adapted using our technique can be used to initialize deep visuomotor policies that achieve superior performance on real-world tasks, when compared to policies trained using small amounts of real-world data.

Previous attempts to learn transformations from source to target domains for visual domain adaptation such as [29] and [4] have used a contrastive metric learning loss. In these methods the learned adaptation was a kernelized transformation over a fixed representation. Earlier work introduced Siamese networks [30,31], for which a shared representation is directly optimized using the contrastive loss for signature and face verification. These were later used for dimensionality reduction [32] and person hand and head pose alignment [12]. Taylor et al. [12] further explored combining synthetic data along with real data to improve representation invariance and overall performance. However, this method used the synthetic data to regularize the learning of the real model and found that performance suffered once the amount of simulated data overwhelmed the amount of real world data. In contrast, our approach uses synthetic data to learn a complete model and uses a very limited number of real examples for refining and adapting that model.

Recently, there has been considerable interest in learning visuomotor policies directly from visual imagery using deep networks [33,1,34,28]. This tight coupling between perception and control simplifies both the vision and control aspects of the problem, but suffers from the major limitation that each new task requires collection, annotation, and training on real world visual data in order to successfully learn a policy. To overcome this issue, we explore how simulated imagery can be adapted for robotic tasks in the real world. Directly applying models learned in simulation to the real world typically does not succeed [35], due to systematic discrepancies between real and simulated data. We demonstrate that our domain adaptation method can successfully perform pose estimation for a real robotic task using minimal real world data, suggesting that adaption from simulation to the real world can be effective for robotic learning. In an earlier version of this paper [36], we demonstrated initial results using domain confusion constraints on PR2 visuomotor policies but without the pairwise constraint reported below. Contemporaneously to our work, [37] also reported success with a domain confusion-style regularizer on a domain adaptive visual behavior task on autonomous MAV flight.

3   Preliminaries

We address the problem of adapting visual representations for robotic learning from a source domain where labeled data is easily accessible (such as simulation) to a target domain without labels. Domain adapation is often necessary because of *domain shift*: a discrepancy in the data distributions between domains that prevents a model trained only on source data to perform well on target data. We define the problem as finding image features $f(x; \theta_{\mathrm{repr}})$ such that this representation allows learning visuomotor policies from a large dataset x_S of labeled source images and a small dataset x_T of unlabeled target images.

When training models for regression, we generally seek to take input images x and directly output some label ϕ. This involves learning a representation θ_{repr} and a regressor θ_ϕ that minimizes the following loss:

$$\mathcal{L}_\phi(x, \phi; \theta_\phi, \theta_{\mathrm{repr}}) = \frac{1}{2K} \sum_{i=1}^{K} ||\theta_\phi^T f(x^{(i)}; \theta_{\mathrm{repr}}) - \phi^{(i)}||_2^2 \tag{1}$$

where $f(x^{(i)}; \theta_{\mathrm{repr}})$ denotes the feature vector corresponding to $x^{(i)}$ under the representation defined by θ_{repr}.

However, collecting ground truth labels in the real world can be impractical, often requiring expensive instrumented setups. As a result, it is difficult to gather enough training data to properly train models from scratch. We instead rely on the existence of a simulator that can render synthetic versions of the task environment. This enables us to quickly generate an unlimited amount of training data with full annotations by simply changing the environment configuration, recording the ground truth label, and rendering a view.

Ideally, we would be able to simply train on our rendered data and have the learned model transfer to the real world. However, because they are acquired independently, our synthetic and real-world images differ significantly in appearance. This discrepancy between the two domains is referred to as *domain shift*, and generally results in reduced performance when attempting to directly transfer source models to the target domain.

To combat the negative effects of domain shift, we model this as a domain adaptation problem, with synthetic renders serving as our source domain, and real-world images serving as our target domain. We propose a model that augments the task loss with two additional adaptation loss functions designed to specifically align the two domains in feature space. This ensures we learn a model that successfully performs the task and transfers robustly between domains.

4   Domain alignment with weakly supervised pairwise constraints

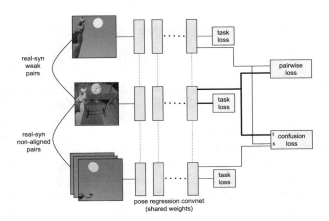

Fig. 2. After determining a weak pairing between source and target images, optimization proceeds via backpropagation on our model architecture. Our model combines a task loss, a domain confusion loss for aligning domains at the distribution level, and a pairwise loss for aligning specific pairs of source and target images. Together, these three losses ensure that our model learns to accurately perform the task while remaining robust to domain shift.

Our method attempts to solve the domain shift problem via two approaches. The first is a distribution-based approach, which seeks to align the source domain with the target domain in feature space. By ensuring that all images lie in the same general neighborhood in representation space, we better facilitate the transfer of task-relevant features from source to target. The second approach incorporates weakly supervised pairwise constraints, and seeks to ensure that images with identical labels are treated identically by the network, regardless of their originating domain. This encourages the network to disregard domain-specific features in favor of features that are relevant to the perception task. Together with the task loss, these approaches ensure that we learn a representation that is meaningful to the chosen visual task while remaining robust to the source-target domain shift.

Domain confusion loss. To align the source and target domains at the overall distribution level, we adopt the domain confusion loss introduced by [2,3]. The model trains a domain classifier θ_D that attempts to correctly classify each image into the domain it originates from. In parallel, the loss $\mathcal{L}_{\mathrm{conf}}$ tries to learn a representation θ_{repr} such that the domain classifier cannot distinguish the two domains in feature space. This loss is the negative cross entropy loss between the predicted domain label of each image x and a uniform distribution over the D domains, which is minimized the domain classifier is maximally confused:

$$\mathcal{L}_{\text{conf}}(x_S, x_T, \theta_D; \theta_{\text{repr}}) = - \sum_{x \in (x_S \cup x_T)} \sum_d \frac{1}{D} \log q_d(x, \theta_D, \theta_{\text{repr}}). \tag{2}$$

Here, q corresponds to the domain classifier activations:

$$q(x, \theta_D; \theta_{\text{repr}}) = \text{softmax}(\theta_D^T f(x; \theta_{\text{repr}})) \tag{3}$$

Pairwise loss. While the confusion loss ensures that the source and target domains as a whole are treated similarly by the model, it does not make use of the task labels. Thus, we include an additional term that seeks to find specific pairs of source and target images with similar labels and align them in representation space. By explicitly aligning images with similar labels, we can optimize the representation to focus only on task-relevant features. However, we assume that task labels are unavailable in the target domain. Thus, we need to determine a pairing P of the target images x_T with the target images x_S so that we can ensure that their distances in the feature space defined by θ_{repr} lie close together. We write this objective as the loss function

$$\mathcal{L}_{\text{pairwise}}(x_S, x_T; P, \theta_{\text{repr}}) = \sum_{(i,j) \in P} \left[\frac{1}{2} \rho \left(x_S^{(i)}, x_T^{(j)}; \theta_{\text{repr}} \right)^2 \right], \tag{4}$$

where we define our distance function ρ as the Euclidean distance in the feature space corresponding to θ_{repr}:

$$\rho \left(x_S^{(i)}, x_T^{(j)}; \theta_{\text{repr}} \right) = \left\| f(x_S^{(i)}; \theta_{\text{repr}}) - f(x_T^{(j)}; \theta_{\text{repr}}) \right\|_2. \tag{5}$$

Intuitively, this objective encourages a pairing P that correctly matches target and source images, as well as a representation θ_{repr} that is task-sensitive while disregarding domain-specific features. However, because the source-target pairing P and the feature representation θ_{repr} depend on each other, it is not immediately clear how to directly optimize for both simultaneously. Thus, we propose an iterative approach.

First, we minimize $\mathcal{L}_{\text{pairwise}}$ with respect to the source-target pairing P. We begin by finding an initial representation θ_{repr} that minimizes the task loss $\mathcal{L}_\phi$ and optionally $\mathcal{L}_{\text{conf}}$ on only the source imagery. Once this source-only model has been trained, we extract a feature representation for every image in our dataset, both source and target. These representations are used to find a source image nearest-neighbor for each target image, thereby determining a weak pairing P. Finding such a pairing additionally enables us to transfer task labels between each pair of images, thus annotating the target images using the labels from their corresponding source images. These transferred weak labels can then be used to minimize the task loss $\mathcal{L}_\phi$ over the target images as well.

Once P has been determined, we keep it fixed and minimize $\mathcal{L}_{\mathrm{pairwise}}$ with respect to the representation θ_{repr} to ensure that pairs lie close in feature space. We note that when used to optimize θ_{repr}, this loss function is similar to the contrastive loss function introduced by [32]. As typically formulated, the contrastive loss function seeks to draw paired images closer together in feature space while pushing unpaired images apart. However, our source dataset has many examples similar to any particular target image, which means there are often many other valid source-target pairs in the dataset that are not explicitly identified. The dissimilarity term in the contrastive loss function would force these unlabeled similar pairs apart, making the optimization poorly conditioned, so our pairwise loss omits this dissimilarity term.

Complete objective. Our full model thus minimizes the joint loss function

$$
\begin{aligned}
\mathcal{L}(x_S, &\phi_S, x_T, \phi_T, P, \theta_D; \theta_\phi, \theta_{\mathrm{repr}}) = \\
&\mathcal{L}_\phi(x_S, \phi_S; \theta_\phi, \theta_{\mathrm{repr}}) + \mathcal{L}_\phi(x_T, \phi_T; \theta_\phi, \theta_{\mathrm{repr}}) \\
&+ \lambda \mathcal{L}_{\mathrm{conf}}(x_S, x_T, \theta_D; \theta_{\mathrm{repr}}) \\
&+ \nu \mathcal{L}_{\mathrm{pairwise}}(x_S, x_T; P, \theta_{\mathrm{repr}})
\end{aligned}
\tag{6}
$$

where the hyperparameters λ and ν trade off how strongly we enforce domain confusion and weakly supervised pairwise constraints.

The feature used to form P is a low-level convolutional feature of a network trained to perform the visual task on the source data. In this feature space, we match each target image with its nearest neighbor in the source domain. Because the feature used to determine P is from a network trained to perform the perception task, it focuses primarily on task-relevant features of the image. After the pairing P has been determined, we can then minimize the complete loss function outlined in Equation 6 via backpropagation. This procedure of determining a weak alignment P and using it to learn a domain-invariant representation is summarized in Algorithm 1.

We depict the architecture setup for a given sampled target image in Figure 2. The task loss is applied to all images the network sees, regardless of whether they came from the source or target environment. Because the target examples do not have labels, we use the labels transferred from the source using the pairing P. Each pair is input to the pairwise loss which pushes the feature representations of the explicitly paired images closer together. Finally, all images are additionally optimized by the confusion loss, which seeks to make the representation agnostic to the overall differences between the two domains.

The combination of losses presented here is architecture-agnostic, thereby making our method applicable to many different visual tasks. We implement our networks using the Caffe framework [38], and plan to release the code and datasets from our experiments upon acceptance of this paper.

5   Adapting visuomotor control policies

As mentioned above, our domain adaptation approach is general and can be applied to many visual tasks. Here we use it to directly adapt deep visual rep-

---

Algorithm 1 Learning domain-invariant image features

1: Collect x_S source domain images with labeled object pose
2: Collect x_T target domain images
3: Minimize $\mathcal{L}_\phi(x_S, \phi_S; \theta_\phi, \theta_{\mathrm{repr}}) + \lambda\mathcal{L}_{\mathrm{conf}}(x_S, x_T, \theta_D; \theta_{\mathrm{repr}})$ with respect to $\theta_\phi, \theta_{\mathrm{repr}}$
4: **for** $x_T^{(j)}$ in x_T **do**
5:     $i^* = \arg\min_i ||f_{\mathrm{conv1}}(x_S^{(i)}; \theta_{\mathrm{repr}}) - f_{\mathrm{conv1}}(x_T^{(j)}; \theta_{\mathrm{repr}})||_2$
6:     Add (i^*, j) to P
7: **end for**
8: Minimize $\mathcal{L}(x_S, \phi_S, x_T, \phi_T, P, \theta_D; \theta_\phi, \theta_{\mathrm{repr}})$ with respect to $\theta_\phi, \theta_{\mathrm{repr}}$

---

resentations for pose estimation and visual policy learning. We build upon the end-to-end architecture presented by [1] for training deep visuomotor policies that can learn to accomplish tasks such as screwing a cap onto a bottle or placing a coat hanger on a rack. The method first pretrains a convolutional neural network on a pose estimation task, then finetunes this network with guided policy search to map from input image to action. Guided policy search is initialized with trajectories from a fully observed state (where the locations of both the manipulated and target object are known), but once learned, the policy only requires visual input at test time.

Once we have learned a visual representation that is robust to the synthetic-real domain shift and can effectively locate salient objects in a scene, we use guided policy search (GPS) with these features to train a parametrized controller θ_{ctrl}. GPS turns reinforcement learning into a supervised learning problem by using time-varying linear controllers to collect (observation, control) data that is used to train a neural network policy. During training, the position of the target object is known, but the neural network policy is trained to act based on the visual feature points; at test time, this policy can succeed solely from vision without being provided the location of the target.

Like in [39], we fit time-varying linear models to the robot joint angles and velocities and use these to collect a dataset of feature points, feature point velocities, joint angles, and joint efforts. We use this dataset to train a neural network policy θ_{ctrl}. The feature points are generated by the θ_{repr} trained with our method, and we do not backpropagate gradients from θ_{ctrl} through θ_{repr} during policy learning. As in [1], we used BADMM to jointly optimize the controllers and neural network with a penalty on the KL divergence between them. θ_{ctrl} is 2 layer network with 40 hidden units per layer that takes the learned feature points and joint state as input and outputs joint efforts. Unlike in [39], we do not apply any filtering or smoothing to the feature points. We refer the reader to [1] for a more in depth explanation of the BADMM GPS algorithm. The final result is a visuomotor control policy from images features pretrained solely on unannotated real imagery and low-fidelity synthetic renderings, while the policy itself is trained in the real world.

We empirically evaluate our method in a variety of experimental settings. We begin with an evaluation on a simple pose estimation task in Section 5.1.

Next, we investigate the quality of synthetic-real pairings produced by our unsupervised alignment method. Finally, we use the learned pairings to train a representation via our method, then use this representation to train a full visuomotor control policy on a "hook loop" manipulation task in Section 5.3. These experiments demonstrate the effectiveness of incorporating synthetic imagery into the pretraining of visuomotor policies.

5.1   Supervised robotic pose estimation evaluation

As a self-contained evaluation of our visual adaptation method, we first evaluate our method in a supervised setting, using a pose estimation task that is representative of the visual estimation required for robotic visuomotor control. This is intended as a toy task to evaluate the use of known pairs for simulation to real world adaptation. By using the gripper, we are able to generate images that are exactly paired between the domains.

We first obtain real world images with gripper pose annotations using the PR2's forward kinematics. We also collected pose labeled images from the Gazebo simulator, where we know the exact location of all objects, and we can specifically obtain paired images by replaying the joint angles used in the real world data collection. With this data, we train a model to regress to the 3D gripper pose from an image. We adopt the deep spatial feature point architecture introduced by Levine et al. [1]. Both the domain confusion loss ($\lambda = 0.1$) and pairwise loss ($\nu = 0.01$) are applied at the third convolutional layer, after the ReLU nonlinearity. As before, when both losses are employed simultaneously, we further halve each of their weights. Results from this experimental setting are presented in Table 1.

Table 1.   Using pairwise constraints improves pose estimation. We report supervised evaluation results averaged over 3 trials on PR2 gripper pose estimation using 5 labeled and paired real examples. Each real example is paired with a corresponding synthetic image. Minibatches are sampled such that an equal number of real and synthetic images are present. We report the average error of the prediction in centimeters. We find that, through combining both a domain confusion loss and a pair alignment loss, we are able to improve performance by 20% (relative).

Method	#Sim	#Real	Error (cm)
Synthetic only	1005	0	25.37 ± 1.18
Real only	0	5	4.43 ± 0.23
Synthetic and real	1005	5	7.74 ± 3.90
Domain confusion [2]	1005	0	6.68 ± 0.01
Pairwise loss	1005	5	5.21 ± 2.48
Domain alignment with strong pairwise constraints	1005	5	3.98 ± 0.02
Oracle	0	1000	0.90 ± 0.13

The results indicate that adaptation with paired examples yields improved performance. We find that incorporating synthetic imagery during training is nontrivial, confirming our hypothesis that simulation to real world has a significant domain shift. Simply combining synthetic and real imagery into one large training set negatively impacts performance, due to slight variations in appearance and viewpoint. We see that domain confusion alone does not help either, since domain confusion does not offer a way to learn the specific viewpoint variations between the real and synthetic domains. Nonetheless, by exploiting the presence of pairs, our method is able to account for these differences, performing better than all other baselines. Comparing against the "Oracle" setting, in which we train on 1000 labeled real examples, we see that our method is able to remove most of the negative effects of domain shift despite training on relatively few real examples. (For additional results on vision-only adaptation from CAD models to real PASCAL images, we refer the reader to our earier report [36].)

5.2   Unsupervised synthetic-real alignment evaluation

To evaluate the effectiveness of our alignment method, we transfer pose annotations from paired synthetic images to their corresponding real images, then compute the error relative to the real-world ground truth pose annotations. In order to test on a real control setting, we perform this experiment on the "hook loop" task introduced in [39], where the robot is expected to place a loop of rope on a hook, as depicted in Figure 3. We generate low-fidelity renderings of the PR2 and a hook in 4000 different configurations and attempt to align these with 100 real-world images of the task without hook pose annotations. As the goal is to learn a policy that can place the loop on an arbitrarily located hook, the policy must locate the hook from visual input.

Table 2. Comparing the pairing error for different strategies of learning f_{conv1} for weak alignment. We compare using only the task loss during pretraining against combining both the task loss and the domain confusion images and report the average error between the object positions within each pair. We see that both the task loss as well as the task loss with confusion do significantly better than random, and in simpler settings their performance approaches that of the optimal alignment (reported as Oracle) if the real labels were known.

Error of hook pose in weak pairings (cm)		
Method	Static camera	Head motion
Random pairs	22.7 ± 0.4	23.9 ± 0.6
Task loss	5.9 ± 0.2	10.9 ± 2.0
Task loss + confusion	6.1 ± 0.4	10.6 ± 2.0
Oracle (known real labels)	4.1	4.9

To learn the representation used for producing the alignment in this setting, we attempt to estimate the 3D pose of the target hook. We evaluate both the

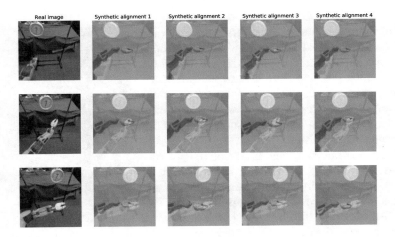

Fig. 4. Example alignments generated by our unsupervised synthetic-real alignment method in the static camera setting. The first column shows an example real image, and the next four columns show the top four corresponding images from our rendered dataset. **The goal is to match the hook position, with the arm position being irrelevant**, because the policy needs to be conditioned on the hook position. We overlay a translucent version the real image on the synthetic images to better show the quality of our alignment.

alignment produced using the simple synthetic-only model, as well as a model trained with an additional domain confusion loss. Table 2 shows the resuls of this experiment on two experimental settings: one with a fixed camera, and one in which the head of the robot (and the camera as well) moves around slightly. The relatively low error in the results indicates that the alignments are generally of high quality.

Visual inspection of the results also indicates that our method produces high-quality pairings. Figure 4 shows example results of our unsupervised alignment method in the static camera setting using the representation trained only on the synthetic data. The hooks in the synthetic renderings match quite closely with the hooks in the corresponding real images. As expected, the position of the arm is largely ignored as desired, and the alignment focuses primarily on the portion of the image that is relevant for the pose estimation task.

5.3   Visuomotor policies for manipulation tasks

After determining the synthetic-real pairings using our method, we retrain the pose predictor on the combined data to learn the final feature points θ_{repr}. To evaluate these feature points, we set up the "hook loop" task from [39]. This task requires a PR2 to bring a loop of rope to the hook of a supermarket scale, as depicted in Figure 3. As the location of the scale is not instrumented, the robot must adjust its actions by visually perceiving the location of the hook/scale.

We used four target hook positions along a bar to learn the linear dynamics and generate trajectories. GPS was run for 13 iterations, where each iteration obtained 5 sample trajectories for 4 training hook position. The linear-quadratic controller was given only the arm joint state, while the neural network policy was given the arm joint state as well as the learned feature point (x, y) positions and velocities.

The performance of the final policy θ_{ctrl} was measured by testing it 14 times: twice at each of 7 positions (including the 4 training positions). Success was defined as the loop being on the hook. As shown in Table 3, the features learned with our method allowed GPS to learn a much more accurate policy than the other methods not using labeled real images.

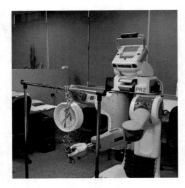

Fig. 3. In the "hook loop" task, the PR2 must position a loop of rope over the hook of a supermarket scale.

We also compared against the deep spatial autoencoder from [39]. Trained on either 100 or 500 images, this method did not perform well, as the feature points tended to model the robot arm's position rather than the hook. Without the simulated hook pose supervision that our method has, the network has no incentive to model the hook over the much more varied positions of the arm and gripper. We also trained an "Oracle" controller. The feature points used were from a pose estimation model trained directly on 500 real images with ground truth data. This controller performed equally well to the one trained with adapted features on only 100 unlabeled real images.

Table 3. Performance of visuomotor tasks trained using domain alignment with weakly supervised pairwise constraints. We report the percentage of successful attempts at placing a loop of rope on a hook after training with 12 iterations of GPS. Each experiment was repeated 3 times.

Method	# Sim	# Real (unlabeled)	Success rate
Synthetic only	4000	0	38.1% ± 8%
Autoencoder (100)	0	100	28.6% ± 25%
Autoencoder (500)	0	500	33.2% ±15%
Domain alignment with randomly assigned pairs	4000	100	33.3% ±16%
Domain alignment with weakly supervised pairwise constraints	4000	100	**76.2% ± 16%**
Oracle	0	500 (labeled)	71.4% ± 14%

Because of the optimization that happens during guided policy search, the performance of the final controller is dependent on the quality of the feature points that are passed in: if the feature points give θ_{ctrl} enough information about the position of the hook, then the controller will learn to use it. However, if the feature points are not consistent enough in where they activate (such as in many of our baselines), the controller cannot learn a policy that takes the hook location into account. For example, when the controller failed a trial it put the loop at a possible hook position, but not at the current hook position. These results show that we can successfully learn visual features that are sufficient for control from synthetic data and a small number of unlabeled real images.

In contrast to our prior work, which required either ground truth pose labels for the real-world images [1] or fifty 100-frame videos for a total of 5000 images for unsupervised learning [39], our method only uses 100 unlabeled real-world images. Being able to use unlabeled images is important for for practical real-world robotic applications, where determining the ground truth pose of movable objects in the world with a high degree of precision typically requires specialized equipment such as motion capture.

6   Conclusion

In this paper, we present a novel model for domain adaptation that is able to exploit the presence of weakly paired source-target examples. Our model extends existing adaptation architectures by combining pairwise and distribution alignment loss functions, and optimizaing over weak label assignments. Because of its generality, our method is applicable to a wide variety of deep adaptation architectures and tasks. Through a pose estimation task, we experimentally validate the importance of using image pairs and show that they are integral to achieving strong adaptation performance. We demonstrate the ability to adapt in settings where pose annotations on real-world data is unavailable.

We address domain adaptation for visual inputs in the context of robotic state estimation. The tasks used in our robotic evaluation involve estimating information that is highly relevant for robotic control [40], as well as for pretraining visuomotor control policies [1]. While we show successful transfer of simulated data for learning real-world visual tasks, training full control policies entirely in simulation will also require tackling the question of physical adaptation, to account for the mismatch between simulated and real-world physics. Addressing this question in future work would pave the way for large-scale training of robotic control policies in simulation.

References

1. S. Levine, C. Finn, T. Darrell, and P. Abbeel, "End-to-end training of deep visuomotor policies," *Journal of Machine Learning Research*, vol. 17, 2016.
2. E. Tzeng, J. Hoffman, T. Darrell, and K. Saenko, "Simultaneous deep transfer across domains and tasks," in *International Conference in Computer Vision (ICCV)*, 2015.
3. Y. Ganin and V. Lempitsky, "Unsupervised domain adaptation by backpropagation," in *International Conference in Machine Learning (ICML)*, 2015.

4. K. Saenko, B. Kulis, M. Fritz, and T. Darrell, "Adapting visual category models to new domains," in *Proc. ECCV*, 2010.
5. H. D. III, "Frustratingly easy domain adaptation," *ACL*, vol. 45, pp. 256–263, 2007.
6. K. Lai and D. Fox, "3d laser scan classification using web data and domain adaptation." in *Robotics: Science and Systems, 2009*, 2009.
7. A. Saxena, J. Driemeyer, and A. Y. Ng, "Robotic grasping of novel objects using vision," *The International Journal of Robotics Research*, vol. 27, no. 2, pp. 157–173, 2008.
8. R. Brooks, R. Greiner, and T. Binford, "The acronym model-based vision system," in *International Joint Conference on Artificial Intelligence 6*, 1979, pp. 105–113.
9. D. Pomerleau, "ALVINN: an autonomous land vehicle in a neural network," in *Advances in Neural Information Processing Systems (NIPS)*, 1989.
10. G. Shakhnarovich, P. Viola, and T. Darrell, "Fast pose estimation with parameter-sensitive hashing," in *Computer Vision, 2003. Proceedings. Ninth IEEE International Conference on*. IEEE, 2003, pp. 750–757.
11. R. Urtasun and T. Darrell, "Sparse probabilistic regression for activity-independent human pose inference," in *Computer Vision and Pattern Recognition, 2008. CVPR 2008. IEEE Conference on*, June 2008, pp. 1–8.
12. G. W. Taylor, R. Fergus, G. Williams, I. Spiro, and C. Bregler, "Pose-sensitive embedding by nonlinear nca regression," in *Advances in Neural Information Processing Systems 23*, J. Lafferty, C. Williams, J. Shawe-Taylor, R. Zemel, and A. Culotta, Eds. Curran Associates, Inc., 2010, pp. 2280–2288.
13. A. Toshev and C. Szegedy, "Deeppose: Human pose estimation via deep neural networks," *CoRR*, vol. abs/1312.4659, 2013.
14. J. J. Tompson, A. Jain, Y. Lecun, and C. Bregler, "Joint training of a convolutional network and a graphical model for human pose estimation," in *Advances in Neural Information Processing Systems 27*, Z. Ghahramani, M. Welling, C. Cortes, N. Lawrence, and K. Weinberger, Eds. Curran Associates, Inc., 2014, pp. 1799–1807.
15. R. Gopalan, R. Li, and R. Chellappa, "Domain adaptation for object recognition: An unsupervised approach," in *Proc. ICCV*, 2011.
16. B. Gong, Y. Shi, F. Sha, and K. Grauman, "Geodesic flow kernel for unsupervised domain adaptation," in *Proc. CVPR*, 2012.
17. J. Yang, R. Yan, and A. G. Hauptmann, "Cross-domain video concept detection using adaptive svms," *ACM Multimedia*, 2007.
18. Y. Aytar and A. Zisserman, "Tabula rasa: Model transfer for object category detection," in *IEEE International Conference on Computer Vision*, 2011.
19. L. Duan, D. Xu, and I. W. Tsang, "Learning with augmented features for heterogeneous domain adaptation," in *Proc. ICML*, 2012.
20. J. Hoffman, E. Rodner, J. Donahue, K. Saenko, and T. Darrell, "Efficient learning of domain-invariant image representations," in *International Conference on Learning Representations*, 2013.
21. E. Tzeng, J. Hoffman, N. Zhang, K. Saenko, and T. Darrell, "Deep domain confusion: Maximizing for domain invariance," *CoRR*, vol. abs/1412.3474, 2014.
22. M. Long, Y. Cao, J. Wang, and M. I. Jordan, "Learning transferable features with deep adaptation networks," in *International Conference in Machine Learning (ICML)*, 2015.
23. Y. Mansour, M. Mohri, and A. Rostamizadeh, "Domain adaptation: Learning bounds and algorithms," in *COLT*, 2009.

24. B. Sun and K. Saenko, "From virtual to reality: Fast adaptation of virtual object detectors to real domains," in *British Machine Vision Conference (BMVC)*, 2014.
25. X. Peng, B. Sun, K. Ali, and K. Saenko, "Exploring invariances in deep convolutional neural networks using synthetic images," *CoRR*, vol. abs/1412.7122, 2014. [Online]. Available: http://arxiv.org/abs/1412.7122
26. V. Mnih, K. Kavukcuoglu, D. Silver, A. Graves, I. Antonoglou, D. Wierstra, and M. Riedmiller, "Playing Atari with deep reinforcement learning," *NIPS '13 Workshop on Deep Learning*, 2013.
27. J. Schulman, S. Levine, P. Moritz, M. Jordan, and P. Abbeel, "Trust region policy optimization," in *International Conference on Machine Learning (ICML)*, 2015.
28. T. P. Lillicrap, J. J. Hunt, A. Pritzel, N. Heess, T. Erez, Y. Tassa, D. Silver, and D. Wierstra, "Continuous control with deep reinforcement learning," *arXiv preprint arXiv:1509.02971*, 2015.
29. B. Kulis, K. Saenko, and T. Darrell, "What you saw is not what you get: Domain adaptation using asymmetric kernel transforms," in *Proc. CVPR*, 2011.
30. J. Bromley, I. Guyon, Y. LeCun, E. Säckinger, and R. Shah, "Signature verification using a "siamese" time delay neural network," in *Advances in Neural Information Processing Systems 6*, J. Cowan, G. Tesauro, and J. Alspector, Eds.   Morgan-Kaufmann, 1994, pp. 737–744.
31. S. Chopra, R. Hadsell, and Y. LeCun, "Learning a similarity metric discriminatively, with application to face verification," in *Computer Vision and Pattern Recognition, 2005. CVPR 2005. IEEE Computer Society Conference on*, vol. 1. IEEE, 2005, pp. 539–546.
32. R. Hadsell, S. Chopra, and Y. LeCun, "Dimensionality reduction by learning an invariant mapping," in *Proc. Computer Vision and Pattern Recognition Conference (CVPR'06)*.   IEEE Press, 2006.
33. M. Riedmiller, S. Lange, and A. Voigtlaender, "Autonomous reinforcement learning on raw visual input data in a real world application," in *International Joint Conference on Neural Networks*, 2012.
34. M. Watter, J. Springenberg, J. Boedecker, and M. Riedmiller, "Embed to control: a locally linear latent dynamics model for control from raw images," in *Advances in Neural Information Processing Systems (NIPS)*, 2015.
35. F. Zhang, J. Leitner, M. Milford, B. Upcroft, and P. Corke, "Towards Vision-Based Deep Reinforcement Learning for Robotic Motion Control," *ArXiv e-prints*, Nov. 2015.
36. E. Tzeng, C. Devin, J. Hoffman, C. Finn, X. Peng, S. Levine, K. Saenko, and T. Darrell, "Towards adapting deep visuomotor representations from simulated to real environments," *CoRR*, vol. abs/1511.07111, 2015. [Online]. Available: http://arxiv.org/abs/1511.07111
37. S. Daftry, J. A. Bagnell, and M. Hebert, "Learning transferable policies for monocular reactive mav control," in *International Symposium on Experimental Robotics*, 2016.
38. Y. Jia, E. Shelhamer, J. Donahue, S. Karayev, J. Long, R. Girshick, S. Guadarrama, and T. Darrell, "Caffe: Convolutional architecture for fast feature embedding," *arXiv preprint arXiv:1408.5093*, 2014.
39. C. Finn, X. Tan, Y. Duan, T. Darrell, S. Levine, and P. Abbeel, "Deep spatial autoencoders for visuomotor learning," in *International Conference on Robotics and Automation (ICRA)*, 2016.
40. P. Pastor, M. Kalakrishnan, J. Binney, J. Kelly, L. Righetti, G. Sukhatme, and S. Schaal, "Learning task error models for manipulation," in *IEEE International Conference on Robotics and Automation*, 2013.

Bandit-Based Model Selection for Deformable Object Manipulation

Dale McConachie and Dmitry Berenson

University of Michigan, Ann Arbor MI 48109, USA
dmcconac@umich.edu, berenson@eecs.umich.edu

Abstract. We present a novel approach to deformable object manipulation that does not rely on highly-accurate modeling. The key contribution of this paper is to formulate the task as a Multi-Armed Bandit problem, with each arm representing a model of the deformable object. To "pull" an arm and evaluate its utility, we use the arm's model to generate a velocity command for the gripper(s) holding the object and execute it. As the task proceeds and the object deforms, the utility of each model can change. Our framework estimates these changes and balances exploration of the model set with exploitation of high-utility models. We also propose an approach based on Kalman Filtering for Non-stationary Multi-armed Normal Bandits (KF-MANB) to leverage the coupling between models to learn more from each arm pull. We demonstrate that our method outperforms previous methods on synthetic trials, and performs competitively on several manipulation tasks in simulation.

1 Introduction

One of the primary challenges in manipulating deformable objects is the difficulty of modeling and simulating them. The most common simulation methods use Mass-Spring models [1, 2], which are generally not accurate for large deformations [3], and Finite-Element models [4, 5], which require significant tuning and are very sensitive to the discretization of the object. Approaches like [6, 7] bypass this challenge by using offline demonstrations to teach the robot specific manipulation tasks; however, when a new task is attempted a new training set needs to be generated. In our application we are interested in a way to manipulate a deformable object without a high-fidelity model or training set available *a priori*. For instance, imagine a robot encountering a new piece of clothing for a new task. While it may have models for previously-seen clothes or training sets for previous tasks, there is no guarantee that those models or training sets are appropriate for the new task. Also, depending on the state of the clothing different models may be most useful at different times in the manipulation task.

Rather than assuming we have a high-fidelity model of a deformable object interacting with its environment, our approach is to have multiple models available for use, any one of which may be useful at a given time. We do not assume these models are correct, we simply treat the models as having some measurable *utility* to the task. The *utility* of a given model is the expected reduction in task

© Springer Nature Switzerland AG 2020
K. Goldberg et al. (Eds.): *Algorithmic Foundations of Robotics XII*, SPAR 13, pp. 704–719, 2020.
https://doi.org/10.1007/978-3-030-43089-4_45

error when using this model to generate robot motion. As the task proceeds, the utility of a given model may change, making other models more suitable for the current part of the task. However, without testing a model's prediction, we do not know its true utility. Testing every model in the set is impractical, as all models would need to be tested at every step, and performing a test changes the state of the object and may drive it into a local minimum. The key question is then which model should be selected for testing at a given time.

The central contribution of this paper is framing the model selection problem as a Multi-Armed Bandit (MAB) problem where the goal is to find the model that has the highest utility for a given task. An arm represents a single model of the deformable object; to "pull" an arm is to use the arm's model to generate and execute a velocity command for the robot. The reward received is the reduction in task error after executing the command. In order to determine which model has the highest utility we need to explore the model space, however we also want to exploit the information we have gained by using models that we estimate to have high utility. One of the primary challenges in performing this exploration versus exploitation trade-off is that our models are inherently coupled and non-stationary; performing an action changes the state of the system which can change the utility of every model, as well as the reward of pulling each arm. While there is work that frames robust trajectory selection as a MAB problem [8], we are not aware of any previous work which either 1) frames model selection for deformable objects as a MAB problem; or 2) addresses the coupling between arms for non-stationary MAB problems.

In our experiments, we show how to formulate a MAB problem with coupled arms for Jacobian-based models. We perform our experiments on three synthetic systems, and on three deformable object manipulation tasks in the Bullet [9] simulator. We demonstrate that formulating model selection as a MAB problem is able to successfully perform all three manipulation tasks. We also show that our proposed MAB algorithm outperforms previous MAB methods on synthetic trials, and performs competitively on the manipulation tasks.

2 Related Work

Deformable Object Modeling: One of the key challenges in manipulating deformable objects is the difficulty inherent in modeling and simulating them. While there has been some progress towards online modeling of deformable objects [10, 11] these methods rely on a time consuming training phase for each object to be modeled. Of particular interest are Jacobian-based models such as [12] and [13]. In these models we assume that there is some function $F : SE(3)^G \rightarrow \mathbb{R}^N$ which maps a configuration of G robot grippers $q \in SE(3)^G$ to a parametrization of the deformable object $\mathcal{P} \in \mathbb{R}^N$, where N is the dimensionality of the parametrization of the deformable object. These models are then linearized by calculating an approximation of the the Jacobian of F:

$$\mathcal{P} = F(q)$$
$$\frac{\partial \mathcal{P}}{\partial t} = \frac{\partial F(q)}{\partial q} \frac{\partial q}{\partial t}$$
$$\dot{\mathcal{P}} = J(q)\dot{q} \ . \tag{1}$$

Computation of an exact Jacobian $J(q)$ at a given configuration q is often computationally intractable and requires high-fidelity models and simulators, so instead approximations are frequently used. A shared characteristic of these approximations is some reliance on tuned parameters. This tuning process can be tedious, and in some cases needs to be done on a per-task basis.

In this paper we consider two types of approximate Jacobian models. The first approximation we use is a *diminishing-rigidity Jacobian* [12] which assumes that points on the deformable object that are near a gripper move "almost rigidly" with respect to the gripper while points that are further away move "less rigidly". This approximation uses deformability parameters to control how quickly the rigidity decreases with distance. The second approximation we use is an *adaptive Jacobian* [13] which uses online estimation to approximate $J(q)$. Adaptive Jacobian models rely on a learning rate to control how quickly the estimation changes from one timestep to the next.

Model Selection: In order to accomplish a given manipulation task, we need to determine which type of model to use at the current time to compute the next velocity command, as well as how to set the model parameters. Frequently this selection is done manually, however, there are methods designed to make these determinations automatically. Machine learning techniques such as [14, 15] rely on supervised training data in order to intelligently search for the best regression or classification model, however, it is unclear how to acquire such training data for the task at hand without having already performed the task. The most directly applicable methods come from the Multi-Armed Bandit (MAB) literature [16–18]. In this framework there are multiple actions we can take, each of which provides us with some reward according to an unknown probability distribution. The problem then is to determine which action to take (which arm to pull) at each time step in order to maximize reward.

The MAB approach is well-studied for problems where the reward distributions are *stationary*; i.e. the distributions do not change over time [16, 19]. This is not the case for deformable object manipulation; consider the situation where the object is far away from the goal versus the object being at the goal. In the first case there is a possibility of an action moving the object closer to the goal and thus achieving a positive reward; however, in the second case any motion would, at best, give zero reward.

Recent work [20] on non-stationary MAB problems offer promising results that utilize independent Kalman filters as the basis for the estimation of a non-stationary reward distribution for each arm. This algorithm (KF-MANB) provides a Bayesian estimate of the reward distribution at each timestep, assuming that the reward is normally distributed. KF-MANB then performs Thompson sampling [19] to select which arm to pull, choosing each in proportion to the belief that it is the optimal arm. We build on this approach in this paper to produce a method that also accounts for dependencies between arms by approximating the coupling between arms at each timestep.

For the tasks we address, the reward distributions are both non-stationary as well as *dependent*. Because all arms are operating on the same physical system, pulling one arm both gives us information about the distributions over other arms, as well as changing the future reward distributions of all arms. While work has been done on dependent bandits [21, 22], we are not aware of any work addressing the combination of non-stationary and dependent bandits. Our method for model selection is inspired by KF-MANB, however we directly use coupling between models in order to form a joint reward distribution over all models. This enables a pull of a single arm to provide information about all arms, and thus we spend less time exploring the model space and more time exploiting useful models to perform the manipulation task.

3 Problem Statement

Let the robot be represented by a set of G grippers with configuration $q \in SE(3)^G$. We assume that the robot configuration can be measured exactly; in this work we assume the robot to be a set of free floating grippers; in practice we can track the motion of these with inverse kinematics on a real robot. We use the Lie algebra [23] of $SE(3)$ to represent robot gripper velocities. This is the tangent space of $SE(3)$, denoted as $\mathfrak{se}(3)$. The velocity of a single gripper g is then $\dot{q}_g = \begin{bmatrix} v_g^T & \omega_g^T \end{bmatrix}^T \in \mathfrak{se}(3)$ where v_g and ω_g are the translational and rotational components of the gripper velocity. We define the velocity of the entire robot to be $\dot{q} = \begin{bmatrix} \dot{q}_1^T & \dots & \dot{q}_G^T \end{bmatrix}^T \in \mathfrak{se}(3)^G$. We define the inner product of two gripper velocities $\dot{q}_1, \dot{q}_2 \in \mathfrak{se}(3)$ to be $\langle \dot{q}_1, \dot{q}_2 \rangle = \langle \dot{q}_1, \dot{q}_1 \rangle_c = v_1^T v_2 + c \omega_1^T \omega_2$, where c is a non-negative scaling factor relating rotational and translational velocities.

The configuration of a deformable object is a set $\mathcal{P} \subset \mathbb{R}^3$ of P points. We assume that we have a method of sensing $\mathcal{P}$. To measure the norm of a deformable object velocity $\dot{\mathcal{P}} = \begin{bmatrix} \dot{\mathcal{P}}_1^T & \dots & \dot{\mathcal{P}}_P^T \end{bmatrix}^T \in \mathbb{R}^{3P}$ we will use a weighted Euclidean norm

$$\|\dot{\mathcal{P}}\|_W^2 = \sum_{i=1}^{P} w_i \dot{\mathcal{P}}_i^T \dot{\mathcal{P}}_i = \dot{\mathcal{P}}^T \operatorname{diag}(W) \dot{\mathcal{P}} \tag{2}$$

where $W = \begin{bmatrix} w_1 \dots w_P \end{bmatrix}^T \in \mathbb{R}^P$ is a set of non-negative weights. The rest of the environment is denoted $\mathcal{O}$ and is assumed to be both static, and known exactly.

Let a *deformation model* be defined as a function $\phi : \mathfrak{se}(3)^G \to \mathbb{R}^{3P}$ which maps a change in robot configuration $\dot{q}$ to a change in object configuration $\dot{\mathcal{P}}$. Let $\mathcal{M}$ be a set of M deformable models which satisfy this definition. Each model is associated with a robot command function $\psi : \mathbb{R}^{3P} \times \mathbb{R}^P \to \mathfrak{se}(3)^G$ which maps a desired deformable object velocity $\dot{\mathcal{P}}$ and weight W (Sec. 5.2) to a robot velocity command $\dot{q}$. ϕ and ψ also take the object and robot configuration $(\mathcal{P}, q)$ as additional input, however this is omitted for clarity. When a model m

is selected for testing, the model generates a gripper command

$$\dot{q}_m(t) = \psi_m(\dot{\mathcal{P}}(t), W(t)) \tag{3}$$

which is then executed for one unit of time, moving the deformable object to configuration $\mathcal{P}(t+1)$.

The problem we address in this paper is which model $m \in \mathcal{M}$ to select in order to to move G grippers such that the points in $\mathcal{P}$ align as closely as possible with some task-defined set of T target points $\mathcal{T} \subset \mathbb{R}^3$, while avoiding gripper collision and excessive stretching of the deformable object. Each task defines a function ρ which measures the alignment error between $\mathcal{P}$ and $\mathcal{T}$. The method we present is a local method which picks a single model m_* at each timestep to treat as the true model. This model is then used to reduce error as much as possible while avoiding collision and excessive stretching.

$$m_* = \underset{m \in \mathcal{M}}{\operatorname{argmin}} \, \rho(\mathcal{T}, \mathcal{P}(t+1)) \tag{4}$$

We show that this problem can be treated as an instance of the multi-arm non-stationary dependent bandit problem.

4   Bandit-Based Model Selection

The primary difficulty with solving (4) directly is that the effectiveness of a particular model in minimizing error is unknown. It may be the case that no model in the set produces the optimal option, however, this does not prevent a model from being useful. In particular the *utility* of a model may change from one task to another, and from one configuration to another as the deformable object changes shape, and moves in and out of contact with the environment. We start by defining the utility $u_m(t) \in \mathbb{R}$ of a model as the expected improvement in task error ρ if model m is used to generate a robot command at time t. If we know which model has the highest utility then we can solve (4). This leads to a classic exploration versus exploitation trade-off where we need to explore the space of models in order to learn which one is the most useful, while also exploiting the knowledge we have already gained. The multi-armed bandit framework is explicitly designed to handle this trade-off.

In the MAB framework, each arm represents a model in $\mathcal{M}$; to pull arm m is to command the grippers with velocity $\dot{q}_m(t)$ (Eq. 3) for 1 unit of time. We then define the *reward* $r_m(t+1)$ after taking action $\dot{q}_m(t)$ as the improvement in error

$$r_m(t+1) = \rho(t) - \rho(t+1) = u_m(t) + w \tag{5}$$

where w is a zero-mean noise term. The goal is to pick a sequence of arm pulls to minimize total expected regret $R(T_f)$ over some (possibly infinite) horizon T_f

$$E[R(T_f)] = \sum_{t=1}^{T_f} (E[r^*(t)] - E[r(t)]) \tag{6}$$

where $r^*(t)$ is the reward of the best model at time t. The next section describes how to use bandit-based model selection for deformable object manipulation.

5   MAB Formulation for Deformable Object Manipulation

Our algorithm (Alg. 1) can be broken down into four major sections and an initialization block. In the initialization block we pre-compute the geodesic distance between every pair of points in $\mathcal{P}$ when the deformable object is in its "natural" or "relaxed" state and store the result in D. These distances are used to construct the deformation models (Sec. 5.3), as well as to avoid overstretching the object (Sec. 5.2). At each iteration we: 1) pick a model to use to achieve the desired direction (Sec. 5.1); 2) compute the task-defined desired direction to move the

---

Algorithm 1 MainLoop($\mathcal{O}, \beta, \lambda$)

1: $t \leftarrow 0$
2: $D \leftarrow$ GeodesicDistanceMatrix($\mathcal{P}_{relaxed}$)
3: $\mathcal{M} \leftarrow$ InitializeModels(D)
4: InitialzeBanditAlgorithm()
5: $\mathcal{P}(0) \leftarrow$ SensePoints()
6: $q(0) \leftarrow$ SenseRobotConfig()
7: **while** true **do**
8:     $m \leftarrow$ SelectArmUsingBanditAlgorithm()
9:     $\mathcal{T} \leftarrow$ GetTargets()
10:    $\dot{\mathcal{P}}_e, W_e \leftarrow$ ErrorCorrection($\mathcal{P}(t), \mathcal{T}$)
11:    $\dot{\mathcal{P}}_s, W_s \leftarrow$ StretchingCorrection($D, \lambda, \mathcal{P}(t)$)
12:    $\dot{\mathcal{P}}_d, W_d \leftarrow$ CombineTerms($\dot{\mathcal{P}}_e, W_e, \dot{\mathcal{P}}_s, W_s$)
13:    $\dot{q}_d \leftarrow \psi_m(\dot{\mathcal{P}}_d, W_d)$
14:    $\dot{q} \leftarrow$ ObstacleRepulsion($\dot{q}_d, \mathcal{O}, \beta$)
15:    CommandConfiguration($q(t) + \dot{q}$)
16:    $\mathcal{P}(t+1) \leftarrow$ SensePoints()
17:    $q(t+1) \leftarrow$ SenseRobotConfig()
18:    UpdateBanditAlgorithm()
19:    $t \leftarrow t + 1$
20: **end while**

---

formable object (Sec. 5.2); 3) generate a velocity command using the chosen model (Sec. 5.3); 4) modify the command to avoid obstacles (Sec. 5.2); and 5) update bandit algorithm parameters (Sec. 5.1).

5.1   Algorithms for MAB

Previous solutions [16, 20] to minimizing (6) assume that rewards for each arm are normally and independently distributed and then estimate the mean and variance of each Gaussian distribution. We test three algorithms in our experiments: Upper Confidence Bound for normally distributed bandits (UCB1-Normal), Kalman Filter Based Solution to Non-Stationary Multi-arm Normal Bandits (KF-MANB), and our extension of KF-MANB, Kalman Filter Based Solution to Non-Stationary Multi-arm Normal Dependent Bandit (KF-MANDB).

UCB1-Normal: The UCB1-Normal algorithm [16] treats each arm (model) as independent, estimating an optimistic Upper Confidence Bound (UCB) for the utility of each model. The model with the highest UCB is used to command the robot at each timestep. This algorithm assumes that the utility of each model is stationary and independent, shifting from exploration to exploitation as more information is gained. While our problem is non-stationary and dependant, we use UCB1-Normal as a baseline algorithm to compare against due to its prevalence in previous work.

KF-MANB: The Kalman Filter Based Solution to Non-Stationary Multi-arm Bandit (KF-MANB) algorithm [20] uses independent Kalman filters to estimate the utility distribution of each model, and then uses Thompson sampling [19] to chose which model to use at each timestep. Because this algorithm explicitly allows for non-stationary reward distributions, it is able to "switch" between models much faster than UCB1-Normal.

KF-MANDB: We also propose a variant of KF-MANB, replacing the independent Kalman filters with a single joint Kalman filter. This enables us to capture the correlations between models, allowing us to learn more from each pull. We start by defining utility as a linear system with Gaussian noise with process model $u(t+1) = u(t) + v$ and observation model $r(t) = Cu(t) + w$ where $u(t)$ is our current estimate of the relative utility of each model, while v and w are zero-mean Gaussian noise terms. C is a row vector with a 1 in the column of the model we used and zeros elsewhere. The variance on w is defined as $\sigma_{obs}^2\eta^2$. η is a tuning parameter to scale the covariance to match the reward scale of the specific task, while σ_{obs} controls how much we believe each new observation.

To define the process noise v we want to leverage correlations between models; if two model predictions are similar, the utility of these models is likely correlated. To measure the similarity between two models i and j we use the angle between their gripper velocity commands $\dot{q}_i$ and $\dot{q}_j$. This similarity is then used to directly construct a covariance matrix for each arm pull:

$$v \sim \mathcal{N}\left(0, \sigma_{tr}^2\eta^2(\xi\Sigma + (1-\xi)\,\mathbf{I})\right)$$
$$\Sigma_{i,j} = \frac{\langle\dot{q}_i, \dot{q}_j\rangle}{\|\dot{q}_i\|\|\dot{q}_j\|} = \cos\theta_{i,j}\ . \tag{7}$$

σ_{tr} is the standard Kalman Filter transition noise factor tuning parameter. $\xi \in [0,1]$ is the correlation strength factor; larger ξ gives more weight to the arm correlation, while smaller ξ gives lower weight. When ξ is zero then KF-MANDB will have the same update rule as KF-MANB, thus we can view KF-MANDB as a generalizion of KF-MANB, allowing for correlation between arms.

After estimating the utility of each model and the noise parameters at the current timestep, these values are then passed into a Kalman filter which estimates a new joint distribution. The next step is the same as KF-MANB; we draw a sample from the resulting distribution, then use the model that yields the largest sample to generate the next robot command. In this way we automatically switch between exploration and exploitation as the system evolves; if we are uncertain of the utility of our models then we are more likely to choose different models from one timestep to the next. If we believe that we have accurate estimates of utility, then we are more likely to choose the model with the highest utility.

5.2   Determining $\dot{q}$

Error Correction We build on previous work [12], splitting the desired deformable object movement into two parts: an error correction part and a stretching correction part. When defining the direction we want to move the deformable object to minimize error we calculate two values; which direction to move the deformable object points $\dot{\mathcal{P}}_e$ and the importance of moving each deformable object point W_e. This is analogous to computing the gradient of error, as well as an "importance factor" for each part of the gradient. We need these weights to be able to differentiate between points of the object where the error function is a plateau versus points where the error function is at a local minimum (Fig. 1). Typically this is achieved using a Hessian, however our error function does not have a second derivative at many points. We use the `ErrorCorrection` (Alg. 2) function to calculate these values. Each target point $\mathcal{T}_i \in \mathcal{T}$ defines a potential field, pulling the nearest point on the deformable object $\mathcal{P}_k$ towards $\mathcal{T}_i$. W_e is set to the maximum distance $\mathcal{P}_k$ is being pulled by any target point. This allows W_e to be insensitive to changes in discretization.

Stretching Correction Our algorithm for stretching correction is similar to that found in [12], with the addition of a weighting term W_s, and a change in how we combine the two terms. We use the `StretchingCorrection` function (Alg. 3) to compute $\dot{\mathcal{P}}_s$ and W_s based on a task-defined stretching threshold $\lambda \geq 0$. First we compute the distance between every two points on the object and store the result in E. We then compare E to D which contains the relaxed lengths between every pair of points. If any two points are stretched by more than λ, we attempt to move the points closer to each other. We use the same strategy for setting the importance of this stretching correction W_s as we use for error correction. When combining stretching correction and error correction terms (Alg. 4) we prioritize stretching correction, accepting only the portion of the error correction that is orthogonal to the stretching correction term for each point.

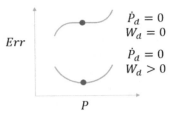

$$\dot{P}_d = 0$$
$$W_d = 0$$

$$\dot{P}_d = 0$$
$$W_d > 0$$

Fig. 1. Top Line: moving the point does not change the error, thus the desired movement is zero, however, it is not important to achieve zero movement, thus $W_d = 0$. Bottom Line: error is at a local minimum; thus moving the point increases error.

Algorithm 2 ErrorCorrection($\mathcal{P}, \mathcal{T}$)

1: $\dot{P}_e \leftarrow \mathbf{0}_{3P \times 1}$, $W_e \leftarrow \mathbf{0}_{P \times 1}$
2: **for** $i \in \{1, 2, \ldots, T\}$ **do**
3:     $k \leftarrow \operatorname{argmin}_{j \in \{1,2,\ldots,P\}} \|\mathcal{T}_i - \mathcal{P}_j\|$
4:     $\dot{P}_{e,k} \leftarrow \dot{P}_{e,k} + \mathcal{T}_i - \mathcal{P}_k$
5:     $W_{e,k} \leftarrow \max(W_{e,k}, \|\mathcal{T}_i - \mathcal{P}_k\|)$
6: **end for**
7: **return** $\{\dot{P}_e, W_e\}$

Algorithm   3 StretchingCorrection($D, \lambda, \mathcal{P}$)

1: $E \leftarrow$ EuclidianDistanceMatrix($\mathcal{P}$)
2: $\dot{P}_s \leftarrow \mathbf{0}_{3P \times 1}$, $W_s \leftarrow \mathbf{0}_{P \times 1}$
3: $\Delta \leftarrow E - D$
4: **for** $i \in \{1, 2, \ldots, P\}$ **do**
5:     **for** $j \in \{i+1, \ldots, P\}$ **do**
6:         **if** $\Delta_{i,j} > \lambda$ **then**
7:             $v \leftarrow \Delta_{i,j}(\mathcal{P}_j - \mathcal{P}_i)$
8:             $\dot{P}_{s,i} \leftarrow \dot{P}_{s,i} + \frac{1}{2}v$
9:             $\dot{P}_{s,j} \leftarrow \dot{P}_{s,j} - \frac{1}{2}v$
10:            $W_{s,i} \leftarrow \max(W_{s,i}, \Delta_{i,j})$
11:            $W_{s,j} \leftarrow \max(W_{s,j}, \Delta_{i,j})$
12:        **end if**
13:    **end for**
14: **end for**
15: **return** $\{\dot{P}_s, W_s\}$

Obstacle Avoidance In order to guarantee that the grippers do not collide with any obstacles, we use the same strategy from [12], smoothly switching between collision avoidance and other objectives (see Alg. 5). For every gripper g and an obstacle set $\mathcal{O}$ we find the distance d_g to the nearest obstacle, a unit vector $\dot{x}_{p_g}$ pointing from the obstacle to the nearest point on the gripper, and a Jacobian J_{p_g} between the gripper's DOF and the point on the gripper. The `Proximity` function can be found in the expanded version of this paper [24]. $\beta > 0$ sets the rate at which we change between servoing and collision avoidance objectives. $\dot{q}_{\max,o} > 0$ is an internal parameter that sets how quickly we move the robot away from obstacles.

5.3   Jacobian Models

Algorithm   **4**
CombineTerms($\dot{\mathcal{P}}_e, W_e, \dot{\mathcal{P}}_s, W_s$)

1: **for** $i \in \{1, 2, \ldots, P\}$ **do**
2:      $\dot{\mathcal{P}}_{d,i} \leftarrow \dot{\mathcal{P}}_{s,i} + \left(\dot{\mathcal{P}}_{e,i} - \mathrm{Proj}_{\dot{\mathcal{P}}_{s,i}} \dot{\mathcal{P}}_{e,i} \right)$
3:      $W_{d,i} \leftarrow W_{s,i} + W_{e,i}$
4: **end for**
5: **return** $\{\dot{\mathcal{P}}_d, W_d\}$

Algorithm 5 ObstacleRepulsion($\mathcal{O}, \beta$)

1: **for** $g \in \{1, 2, \ldots, G\}$ **do**
2:      $J_{p_g}, \dot{x}_{p_g}, d_g \leftarrow \mathrm{Proximity}(\mathcal{O}, g)$
3:      $\gamma \leftarrow e^{-\beta d_g}$
4:      $\dot{q}_{c,g} \leftarrow J_{p_g}^+ \dot{x}_{p_g}$
5:      $\dot{q}_{c,g} \leftarrow \frac{\dot{q}_{\max,o}}{\|\dot{q}_{c,g}\|} \dot{q}_{c,g}$
6:      $\dot{q}_g \leftarrow \gamma \left(\dot{q}_{c,g} + \left(\mathbf{I} - J_{p_g}^+ J_{p_g} \right) \dot{q}_g \right) + (1 - \gamma) \dot{q}_g$
7: **end for**
8: **return** $\dot{q}$

Every model must define a prediction function $\phi(\dot{q})$ and has an associated robot command function $\psi(\dot{\mathcal{P}}, W)$. This paper focuses on Jacobian-based models whose basic formulation Eq. (1) directly defines the deformation model ϕ

$$\phi(\dot{q}) = J\dot{q}. \tag{8}$$

When defining the robot command function ψ, we use the weights W to focus the robot motion on the important part of $\dot{\mathcal{P}}$. This is done by using a weighted norm in a standard minimization problem

$$\psi(\dot{\mathcal{P}}, W) = \underset{\dot{q}}{\mathrm{argmin}} \, \|J\dot{q} - \dot{\mathcal{P}}\|_W^2 \text{ s.t. } \|\dot{q}\|^2 < \dot{q}_{\max,e}^2. \tag{9}$$

We also need to ensure that the grippers do not move too quickly, so we add the constraint that the robot moves no more than $\dot{q}_{\max,e} > 0$. To solve (9) we use the Gurobi [25] optimizer. We use two different Jacobian approximation methods in our model set; a diminishing rigidity Jacobian, and an adaptive Jacobian, which are described below.

Diminishing Rigidity Jacobian The key assumption used by this method [12] is *diminishing rigidity*: the closer a gripper is to a particular part of the deformable object, the more that part of the object moves in the same way that the gripper does (i.e. more "rigidly"). The further away a given point on the object is, the less rigidly it behaves; the less it moves when the gripper moves. Details of how to construct a diminishing rigidity Jacobian are shown in the expanded version of this paper [24]. This approximation depends on two parameters k_{trans} and k_{rot} which control how the translational and rotational rigidity scales with distance. Small values entail very rigid objects; high values entail very deformable objects.

Table 1. Controller parameters

		Synthetic Trials	Rope Winding	Table Coverage	Two Stage Coverage
$\mathfrak{se}(3)$ inner product constant	c	-	0.0025	0.0025	0.0025
Servoing max gripper velocity	$\dot{q}_{max,e}$	0.1	0.2	0.2	0.2
Obstacle avoidance max gripper velocity	$\dot{q}_{max,o}$	-	0.2	0.2	0.2
Obstacle avoidance scale factor	β	-	200	1000	1000
Stretching correction scale factor	λ	-	0.005	0.03	0.03

Table 2. KF-MANB and KF-MANDB parameters

		Synthetic Trials	Rope Winding	Table Coverage	Two Stage Coverage
Correlation strength factor (KF-MANDB only)	ξ	0.9	0.9	0.9	0.9
Transition noise factor	σ_{tr}^2	1	0.1	0.1	0.1
Observation noise factor	σ_{obs}^2	1	0.01	0.01	0.01

Adaptive Jacobian A different approach is taken in [13], instead using online estimation to approximate $J(q)$. In this formulation we start with some estimate of the Jacobian $\tilde{J}(0)$ at time $t = 0$ and then use the Broyden update rule [26] to update $\tilde{J}(t)$ at each timestep t

$$\tilde{J}(t) = \tilde{J}(t-1) + \Gamma \frac{\left(\dot{\mathcal{P}}(t) - \tilde{J}(t-1)\dot{q}(t)\right)}{\dot{q}(t)^T \dot{q}(t)} \dot{q}(t)^T \ . \tag{10}$$

This update rule depends on a update rate $\Gamma \in (0, 1]$ which controls how quickly the estimate shifts between timesteps.

6   Experiments and Results

We test our method on three synthetic tests and three deformable object manipulation tasks in simulation. The synthetic tasks show that the principles we use to estimate the coupling between models are reasonable; while the simulated tasks show that our method is effective at performing deformable object manipulation tasks. Table 1 shows the parameters used by the Jacobian-based controller, while Table 2 shows the parameters used by the the bandit algorithms for all experiments. η is set dynamically and discussed in Sec. 6.1.

6.1   Synthetic Tests

For the synthetic tests, we set up an underactuated system that is representative of manipulating a deformable object with configuration $y \in \mathbb{R}^n$ and control input $\dot{x} \in \mathbb{R}^m$ such that $m < n$ and $\dot{y} = J\dot{x}$. To construct the Jacobian of this system we start with $J = \begin{bmatrix} \mathbf{I}_{m \times m} \\ \mathbf{0}_{(n-m) \times m} \end{bmatrix}$ and add uniform noise drawn from $[-0.1, 0.1]$ to each element of J. The system configuration starts at $\begin{bmatrix} 10 \dots 10 \end{bmatrix}^T$ with the target configuration set to the origin. Error is defined as $\rho(t) = \|y(t)\|$, and the desired direction to move the system at each timestep is $\dot{y}_d(t) = -y(t)$. These

Table 3. Synthetic trial results showing total regret with standard deviation in brackets for all bandit algorithms for 100 runs of each setup.

# of Models	n	m	UCB1-Normal	KF-MANB	KF-MANDB
10	3	2	4.41 [1.65]	3.62 [1.73]	2.99 [1.40]
60	147	6	5.57 [1.37]	4.89 [1.32]	4.53 [1.42]
60	6075	12	4.21 [0.64]	3.30 [0.56]	2.56 [0.54]

tasks have no obstacles or stretching, thus β, λ, and $\dot{q}_{max,o}$ are unused. Rather than setting the utility noise scale η *a priori*, we use an annealing filter

$$\eta(t+1) = \max(10^{-10}, 0.9\eta(t) + 0.1|r(t+1)|) \ . \tag{11}$$

This enables us to track the changing available reward as the task evolves.

To generate a model for the model set we start with the true Jacobian J and add uniform noise drawn from $[-0.025, 0.025]$ to each element of J. For an individual trial, each bandit algorithm uses the same J and the same model set. Each bandit algorithm receives the same random number stream during a trial, ensuring that a more favourable stream doesn't bias results. We ran one small test using a 3×2 Jacobian with 10 arms in order to yield results that are easily visualised. The second and third tests are representative of the scale of the simulation experiments, using the same number of models and similar sizes of Jacobian as are used in simulation. A single trial consists of 1000 pulls (1000 commanded actions); each test was performed 100 times to generate statistically significant results. Our results in Table 3 show that KF-MANDB clearly performs the best for all three tests.

6.2   Simulation Trials

We now demonstrate the effectiveness of multi-arm bandit techniques on three example tasks, show how to encode those tasks for use in our framework, and discuss experimental results. The first task shows how our method can be applied to a rope, with the goal of winding the rope around a cylinder in the environment. The second and third tasks show the method applied to cloth. In the second task, two grippers manipulate the cloth so that it covers a table. In the third task, we perform a two-stage coverage task, covering portions of two different cylinders. In all three tasks, the alignment error $\rho(\mathcal{P}, \mathcal{T})$ is measured as the sum of the distances between every point in $\mathcal{T}$ and the closest point in $\mathcal{P}$ in meters. Figure 2 shows the target points in red, and the deformable object in green.[1]

All experiments were conducted in the open-source Bullet simulator [9], with additional wrapper code developed at UC Berkeley. The rope is modeled as a series of 49 small capsules linked together by springs and is 1.225m long. The cloth is modeled as a triangle mesh of size 0.5m $\times$ 0.5m for the table coverage task, and size 0.5m $\times$ 0.625m for the two-stage coverage task. We emphasize that our method does not have access to the model of the deformable object or the simulation parameters. The simulator is used as a "black box" for testing.

---

[1] The video accompanying this paper (https://youtu.be/r86O7PTSVlY) shows the task executions.

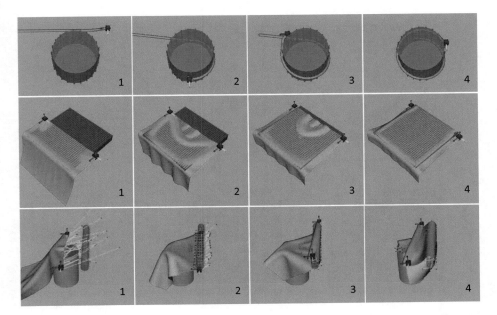

Fig. 2. Sequence of snapshots showing the execution of the simulated experiments using the KF-MANDB algorithm. The rope and cloth are shown in green, the grippers is shown in blue, and the target points are shown in red. The bottom row additionally shows $\dot{\mathcal{P}}_d$ as green rays with red tips.

We use models generated using the same parameters for all three tasks with a total of 60 models: 49 diminishing rigidity models with rotation and translational deformability values k_{trans} and k_{rot} ranging from 0 to 24 in steps of 4, as well as 11 adaptive Jacobian models with learning rates Γ ranging from 1 to 10^{-10} in multiples of 10. All adaptive Jacobian models are initialized with the same starting values; we use the diminishing rigidity Jacobian for this seed with $k_{trans} = k_{rot} = 10$ for the rope experiment and $k_{trans} = k_{rot} = 14$ for the cloth experiments to match the best model found in [12]. We use the same strategy for setting η as we use for the synthetic tests.

We evaluate results for the MAB algorithms as well as using each of the models in the set for the entire task. To calculate regret for each MAB algorithm, we create copies of the simulator at every timestep and simulate the gripper command, then measure the resulting reward $r_m(t)$ for each model. The reward of the best model $r^*(t)$ is then the maximum of individual rewards. As KF-MANB and KF-MANDB are not deterministic algorithms, each task is performed 10 times for these methods. All tests are run on an Intel Xeon E5-2683 v4 processor with 64 GB of RAM. UCB1-Normal and KF-MANB solve Eq. (9) once per timestep, while KF-MANDB solves it for every model in $\mathcal{M}$. Computation times for each test are shown in their respective sections.

Winding a Rope Around a Cylinder: In the first example task, a single gripper holds a rope that is lying on a table. The task is to wind the rope around a cylinder which is also on the table (see Fig. 2). Our results (Fig. 3) show that at

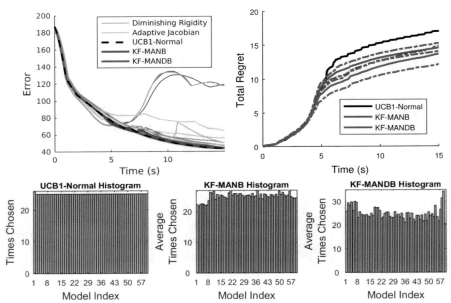

Fig. 3. Experimental results for the rope-winding task. Top Left: alignment error for 10 trials for each MAB algorithm, and each model when used in isolation. UCB1-Normal, KF-MANB, KF-MANDB lines overlap in the figure for all trials. Top Right: Total regret averaged across 10 trials for each MAB algorithm with the minimum and maximum drawn in dashed lines. Bottom row: histograms of the number of times each model was selected by each MAB algorithm.

the start of the task all the individual models perform nearly identically, starting to split at 2 seconds (when the gripper first approaches the cylinder) and again at 6 seconds. Despite our model set containing models that are unable to perform the task, our formulation is able to successfully perform the task using all three bandit algorithms. Interestingly, while KF-MANDB outperforms UCB1-Normal and KF-MANB in terms of regret, all three algorithms produce very similar results. Solving Eq. (9) at each iteration requires an average of 17.3 ms (std. dev. 5.5 ms) for a single model, and 239.5 ms (std. dev. 153.7 ms) for 60 models.

Spreading a Cloth Across a Table: The second scenario we consider is spreading a cloth across a table. In this scenario two grippers hold the rectangular cloth at two corners and the task is to cover the top of the table with the cloth. All of the models are able to perform the task (see Fig. 4), however, many single-model runs are slower than the bandit methods at completing the task, showing the advantage of the bandit methods. When comparing between the bandit methods, both error and total regret indicate no performance difference between the methods. Solving Eq. (9) at each iteration requires an average of 89.5 ms (std. dev. 82.4 ms) for a single model, and 605.1 ms (std. dev. 514.3 ms) for 60 models.

Two-Part Coverage Task: In this experiment, we consider a two-part task. The first part of the task is to cover the top of a cylinder similar to our second scenario. The second part of the task is to cover the far side of a second cylinder.

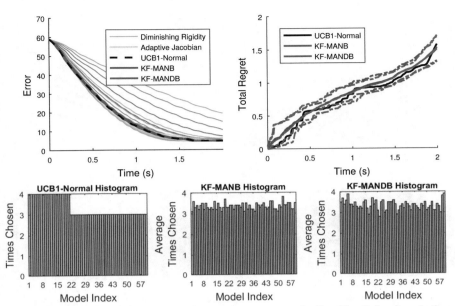

Fig. 4. Experimental results for the table coverage task. See Fig. 3 for description.

For this task the `GetTargets` function used previously pulls the cloth directly into the second cylinder. The collision avoidance term then negates any motion in that direction causing the grippers to stop moving. To deal with this, we discretize the free space using a voxel grid, and then use Dijkstra's algorithm to find a collision free path between each cover point and every point in free space. We use the result from Dijkstra's algorithm to define a vector field that pulls the nearest (as defined by Dijkstra's) deformable object point p_k along the shortest collision free path to the target point. This task is the most complex of the three (see Fig. 5); many models are unable to perform the task at all, becoming stuck early in the task. We also observe that both KF-MANB and KF-MANDB show a preference for some models over others. Two interesting trials using KF-MANDB stand out; in the first the grippers end up on opposite sides of the second cylinder, in this configuration the physics engine has difficulty resolving the scene and allows the cloth to be pulled straight through the second cylinder. In the other trial the cloth is pulled off of the first cylinder, however KF-MANDB is able to recover, moving the cloth back onto the first cylinder. KF-MANDB and UCB1-Normal are able to perform the task significantly faster than KF-MANB, though all MAB methods complete the task using our formulation. Solving Eq. (9) at each iteration requires an average of 102.6 ms (std. dev. 30.6 ms) for a single model, and 565.5 ms (std. dev. 389.8 ms) for 60 models.

7   Conclusion

We have formulated model selection for deformable object manipulation as a MAB problem. Our formulation enables the application of existing MAB algorithms to deformable object manipulation as well as introduces a novel *utility*

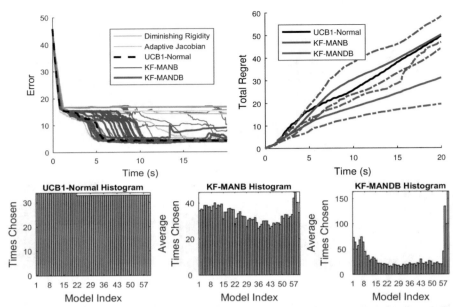

Fig. 5. Experimental results for the two-part coverage task. See Fig. 3 for description.

metric to measure how useful a model is at performing a given task. We have also presented KF-MANDB to leverage coupling between dependent bandits to learn more from each arm pull. Our experiments show how to perform several interesting tasks for rope and cloth using our method.

During our experiments we observed that finding and exploiting the best model is less important than avoiding poor models for extended periods of time; UCB1-Normal never leaves its initial exploration phase, yet it is able to success-fully perform each task. This may be due to many models being able to provide commands that have a positive dot-product with the correct direction of motion.

One limitation of KF-MANDB is handling bifurcations; when very small differences in command sent to the robot cause large differences in the result the assumption of coupling between models in KF-MANDB does not hold. In future work we seek to explore how to overcome this limitation, as well as using the predictive accuracy of each model as an additional measure of model coupling.

8   Acknowledgements

This work was supported in part by NSF grants IIS-1656101 and IIS-1551219. We gratefully acknowledge Calder Phillips-Grafflin for his assistance with Bullet.

References

1. Gibson, S.F.F., Mirtich, B.: A survey of deformable modeling in computer graphics. Technical report, Mitsubishi Electric Research Laboratories (1997)
2. Essahbi, N., Bouzgarrou, B.C., Gogu, G.: Soft Material Modeling for Robotic Manipulation. In: Applied Mechanics and Materials. (April 2012)

3. Maris, B., Botturi, D., Fiorini, P.: Trajectory planning with task constraints in densely filled environments. In: IROS. (2010)
4. Müller, M., Dorsey, J., McMillan, L., Jagnow, R., Cutler, B.: Stable real-time deformations. In: SIGGRAPH. (2002)
5. Bathe, K.J.: Finite Element Procedures. Klaus-Jurgen Bathe (2006)
6. Schulman, J., Ho, J., Lee, C., Abbeel, P.: Learning from demonstrations through the use of non-rigid registration. In: Springer Tracts in Advanced Robotics. Volume 114., Springer International Publishing (2016) 339–354
7. Huang, S.H., Pan, J., Mulcaire, G., Abbeel, P.: Leveraging appearance priors in non-rigid registration, with application to manipulation of deformable objects. In: IROS. (2015)
8. Koval, M.C., King, J.E., Pollard, N.S., Srinivasa, S.S.: Robust trajectory selection for rearrangement planning as a multi-armed bandit problem. In: IROS. (2015)
9. Coumans, E.: Bullet physics library. Open source: bulletphysics.org (2010)
10. Jochen Lang, Pai, D.K., Woodham, R.J.: Acquisition of Elastic Models for Interactive Simulation. IJRR **21**(8) (aug 2002) 713–733
11. Cretu, A.M., Payeur, P., Petriu, E.: Neural Network Mapping and Clustering of Elastic Behavior From Tactile and Range Imaging for Virtualized Reality Applications. IEEE TIM **57**(9) (sep 2008) 1918–1928
12. Berenson, D.: Manipulation of deformable objects without modeling and simulating deformation. In: IROS. (2013)
13. Navarro-Alarcon, D., Romero, J.G.: Visually servoed deformation control by robot manipulators. In: ICRA. (2013)
14. Maron, O., Moore, A.W.: Hoeffding Races: Accelerating Model Selection Search for Classification and Function Approximation. In: NIPS. (1994)
15. Sparks, E.R., Talwalkar, A., Haas, D., Franklin, M.J., Jordan, M.I., Kraska, T.: Automating model search for large scale machine learning. In: SoCC. (2015)
16. Auer, P., Cesa-Bianchi, N., Fischer, P.: Finite-time Analysis of the Multiarmed Bandit Problem. Machine Learning **47**(2/3) (2002) 235–256
17. Gittins, J., Glazebrook, K., Weber, R.: Multi-armed Bandit Allocation Indices. John Wiley & Sons (2011)
18. Whittle, P.: Restless Bandits: Activity Allocation in a Changing World. Journal of Applied Probability **25** (1988) 287
19. Agrawal, S., Goyal, N.: Analysis of Thompson Sampling for the multi-armed bandit problem. Conference on Learning Theory (2012)
20. Granmo, O.C., Berg, S.: Solving Non-Stationary Bandit Problems by Random Sampling from Sibling Kalman Filters (2010)
21. Pandey, S., Chakrabarti, D., Agarwal, D.: Multi-armed bandit problems with dependent arms. In: ICML, New York, New York, USA (2007)
22. Langford, J., Zhang, T.: The Epoch-Greedy Algorithm for Multi-armed Bandits with Side Information. In: NIPS. (2008)
23. Murray, R.M., Li, Z., Sastry, S.S.: A Mathematical Introduction to Robotic Manipulation. Volume 29. CRC Press (1994)
24. McConachie, D., Berenson, D.: Bandit-Based Methods for Deformable Object Manipulation. arXiv preprint arXiv:1703.10254 (2017)
25. Gurobi: Gurobi optimization library. Proprietary: gurobi.com (2016)
26. Broyden, C.G.: A class of methods for solving nonlinear simultaneous equations. Mathematics of Computation **19**(92) (1965) 577–593

Matrix Completion as a Post-Processing Technique for Probabilistic Roadmaps

Joel M. Esposito[1] and John N. Wright[2]

[1] United States Naval Academy, Annapolis, MD, USA
esposito@usna.edu,
[2] Columbia University, New York, NY, USA
jw2966@columbia.edu

Abstract. This paper describes a novel post-processing algorithm for probabilistic roadmaps (PRMs), inspired by the recent literature on matrix completion. We argue that the adjacency matrix associated with real roadmaps can be decomposed into the sum of low-rank and sparse matrices. Our method, based on Robust Principal Component Analysis (RPCA), numerically computes a relaxation of this decomposition by solving a convex optimization problem–even when most of the entries in the adjacency matrix are unknown. Given a PRM with n vertices and only $O(n \log^2 n)$ collision-checked candidate edges, the algorithm estimates the status of all $n(n-1)/2$ possible edges in the full road map with high accuracy, without performing any additional collision checks. Numerical experiments on problems from the Open Motion Planning Library indicate that they posses the requisite low-rank plus sparse structure; and that after checking 5% of the possible edges, the algorithm estimates the full visibility graph with 96% accuracy. Based on numerical experiments, we propose sharper estimates of the error as a function of number of edge checks than what is previously reported. The practical utility of the algorithm is that average path length across the resulting denser edge set is significantly shorter (at the cost of somewhat increased spatial complexity and query times). An ancillary benefit is that the resulting low-rank plus sparse decomposition readily reveals information about that would be otherwise difficult to compute, such as the number of convex cells in free configuration space and the number of vertices in each. We believe that this novel connection between motion planning and matrix completion–two previously disparate lines of research–provides a new perspective on sampling-based motion planning and may guide future algorithm development.

1 Introduction

Sampling-based motion planning techniques, such as the *probabilistic road map* (PRM) method [1], efficiently generate a graph, G_{prm}, which approximates the free configuration space by randomly sampling a discrete set of configurations (vertices) and collision checking only a small sub-set of the possible edges between them. This paper describes a novel post-processing step for PRMs, inspired by the success of so-called matrix completion techniques [2] which recover

© Springer Nature Switzerland AG 2020
K. Goldberg et al. (Eds.): *Algorithmic Foundations of Robotics XII*, SPAR 13, pp. 720–735, 2020.
https://doi.org/10.1007/978-3-030-43089-4_46

large, structured, matrices from a small set of randomly observed entries. Specifically, given a PRM with n vertices and $O(n \log^2 n)$ collision-checked candidate edges, we are able to estimate, with high probability, nearly all of the remaining $n(n-1)/2$ possible edges in the full visibility graph without additional collision checks. From a computational perspective, the method involves solving a convex optimization problem, whose unique optimum can be found efficiently, without the need for parameter tuning or a priori knowledge of the solution. The practical benefit of post-processed denser edge set is that it contains shorter paths–at the cost of increased spatial complexity and query times.

At a more fundamental level we believe that the primary contribution of this work is the previously unexploited insight that the adjacency matrix can be decomposed into sparse and low-rank terms, which encode information about problem difficulty. The empirical successes of PRM and other sampling-based planners, such as RRT [3] or EST [4], at estimating the connectivity of realistic environments, using very few collision checks, has outstripped our theoretical understanding. Several analyses of PRM completeness [5,6], convergence [7,8], and optimality [9] have been offered. Yet most require knowing the geometric parameters of a particular solution path, such as obstacle clearance [9] or path length [7]. In [10] it was hypothesized that the apparent success of PRM can be attributed to the favorable properties possessed by most real-world environments, called expansiveness. While this notion does not require knowledge of a solution path, its parameters are difficult to compute for high dimensional environments. However, some easily computable properties of our low-rank plus sparse decomposition provide a similar characterization of problem difficulty. Moreover, we hope this novel application of matrix completion to the motion planning problem will inspire cross-pollination between these previously disparate fields and provide new perspective on sampling-based motion planning.

The remainder of this paper is organized as follows. Section 2.1 reviews results on matrix completion with a focus on robust principal component analysis; Section 2.2 reviews the PRM algorithm; and Section 2.3 discusses connections between these concepts. Section 3 and 4 describe our post-processing technique along with some performance analysis in Sect. 5. Section 6 presents the results of our computational experiments on a set of benchmark problems. Finally, Section 7 places our results in context and presents a program for future work.

2   Background

2.1   Matrix recovery via convex optimization

In many areas of engineering, we are confronted with problems of missing data. Examples include predicting how users will rate novel items on a website, determining the relative position of sensors in a large network, and filling in missing values in a degraded digital image. We can formalize these problems as follows: in each, we observe a small subset $\Omega \subseteq [n] \times [n]$ of the entries of a large matrix $A \in \mathbb{R}^{n \times n}$, and the goal is to fill in the remaining entries of A. Clearly, without prior knowledge about the structure of A, this problem is ill-posed.

Fortunately, in many applications, the target matrix $\boldsymbol{A}$ is known to be approximately *low-rank*. In this situation, the problem becomes well-posed: provided $\boldsymbol{A}$ is not too concentrated on any row or column[3], only about $nr\log^2 n$ observations are needed to uniquely determine the matrix $\boldsymbol{A}$. This is nearly the same order as the number of degrees of freedom in a rank-r matrix– $O(nr)$. Moreover, not only is the problem well-posed but seminal results in theory of *matrix completion* show that it can be solved using efficient, well-structured algorithms based on convex optimization [2, 11–14].

Many problems in imaging and statistics pose an additional challenge: gross errors due to sensor noise, partial object occlusions, etc. This situation can be modeled as follows. Instead of assuming that $\boldsymbol{A}$ is low-rank, we assume that $\boldsymbol{A}$ is a *superposition* of a low-rank matrix $\boldsymbol{L}$ and a *sparse* error matrix $\boldsymbol{S}$: $\boldsymbol{A} = \boldsymbol{L} + \boldsymbol{S}$.

Unfortunately, the rank and sparsity (as measured by the vector ℓ^0 norm) are not convex functions. Instead, ℓ^1 is known to be the tightest convex relation of the sparsity criteria; and the *nuclear norm* $\|\boldsymbol{L}\|_* = \sum_i \sigma_i(\boldsymbol{L})$ –the sum of the singular values–is used as a convex relaxation of the rank function. Both $\boldsymbol{L}$ and $\boldsymbol{S}$ can be recovered by solving a convex optimization problem [15, 16]:

$$\min_{\boldsymbol{L},\boldsymbol{S}} \|\boldsymbol{L}\|_* + \lambda\|\boldsymbol{S}\|_1 \quad \text{such that} \quad P_{\Omega_{obs}}(\boldsymbol{S} + \boldsymbol{L}) = P_{\Omega_{obs}}(\boldsymbol{A}). \tag{1}$$

The parameter λ balances between rank and sparsity. Fortunately, theory provides good guidance for choosing λ: $O(1/\sqrt{n})$ is shown to work under broad conditions in [16]. $P_{\Omega_{obs}}$ is the orthoprojector onto the set of observed entries:

$$[P_{\Omega_{obs}}(\boldsymbol{A})]_{ij} = \begin{cases} \boldsymbol{A}_{ij} & (i,j) \in \Omega_{obs} \\ 0 & \text{else} \end{cases} \tag{2}$$

In words, this constraint forces $\boldsymbol{L} + \boldsymbol{S}$ to agree with the observed entries in $\boldsymbol{A}$. The past ten years have seen a tremendous development of the theory [14–18] and algorithms [19, 20]–see [21] for an overview.

2.2   Motion planning with probabilistic roadmaps

Some of the earliest work on motion planning [22] introduced the idea of a *roadmap*, which abstracts the robot's free configuration space, C_{free}, as a graph, G, whose vertices $V = \{v_1, \ldots, v_n\}$ are configurations in C_{free} and whose edges $E = \{e_{ij}\}$ are simple paths connecting vertices. Good roadmaps satisfy two properties [23]: *accessibility* means any point in C_{free} can be easily connected to a vertex; and *connectivity* means the graph has the same number of connected component as the free configuration space. Unfortunately computing an exact solution to the motion planning problem is known to be PSPACE-hard [24] and

---

[3] This condition is needed to rule out very sparse $\boldsymbol{A}$. For example, if $\boldsymbol{A}$ has only one nonzero entry, we would need to sample exhaustively to locate it. When $\boldsymbol{A}$ is not too concentrated on any row or column, every entry carries information about the global structure of $\boldsymbol{A}$, and only a few entries are needed to correctly complete it.

the complexity of complete algorithms, such as [25], makes them impractical. Sampling-based motion planning methods, such as PRM [1], trade a reduction in computation time for weaker (probabilistic) completeness. A generalization of the PRM algorithm is outlined in Fig. 1, and a sample PRM is shown Fig. 2.

We refer to several graphs throughout this paper, all of which share the same randomly generated vertex set V but differ in their edges sets. G_{full}, termed the full visibility graph on V, is the graph created by the algorithm in Fig. 1 when the function $CandidateVertices$ returns all other vertices. This is identical to the simplified-PRM algorithm [7], which results in $O(n^2)$ calls to $CollisionFree$— known to be the most computationally expensive step [9].

G_{prm} is the graph returned by other variants of PRM, when more restrictive heuristics are used for $CandidateVertices$. For example, the range-limited variant [26] only checks vertices in a ball of (possibly varying) radius R about v_i; another uses the K-nearest vertices [26]. The motivation is to limit the number of calls to $CollisionFree$ to $O(n \log n)$ [9]. However, in doing so they create a sparse graph whose edges are a subset of G_{full}; and therefore G_{prm} has at least as many connected components and the length of the resulting paths are at least as large G_{full}. In fact, under many heuristics, the resulting path lengths are not even asymptotically optimal [9].

Generic Probabilistic Road Map (g-PRM): Preprocessing Phase

$GenerateVertices$ $V = \{v_1, \ldots, v_n\} \in C_{free}$
Initialize edges $E = \emptyset$
for $i = 1, \ldots, n$ **do**
  $U \leftarrow CandidateVertices(\ v_i\ )$
  for $\forall v_j \in U$ **do**
    if $CollisionFree\ (v_i, v_j)$ **then**
      add edge e_{ij} to E
    end if
  end for
end for
RETURN $G = \{V, E\}$

Fig. 1. A generalization of the PRM pre-processing algorithm.

2.3   Connections: matrix completion and PRM

Our goal is to compute a graph $\hat{G}$ that estimates the edges in G_{full} based on the limited edge information in G_{prm}, using the low-rank plus sparse matrix completion technique outlined in Sect. 2.1. One way to represent a graph is an $n \times n$ adjacency matrix, $\boldsymbol{A}$, where $A_{ij} = 1$ if $e_{ij} \in E$, and $A_{ij} = 0$ otherwise. Figure 2 (left) depicts G_{full}; in the right panel, the shading indicates the corresponding

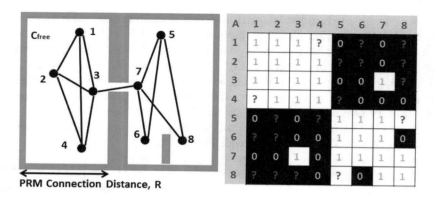

Fig. 2. The graph generated by simplified-PRM for a point robot in a simple 2-D environment (left) and its adjacency matrix (right). The question marks represent edges that the range-limited variant of PRM would not collision check.

adjacency matrix, $\boldsymbol{A}_{full}$, with black cells indicating 0 entries and white cells indicating 1 entries. The entries marked "?" will be discussed later.

We argue that in general $\boldsymbol{A}_{full}$ can be expressed as the sum of low rank and sparse matrices. To see this, consider Figure 2. Within each convex cell of C_{free} the vertices form a connected cluster. Ignoring entries $(3, 7)$ and $(6, 8)$, for a moment, $\boldsymbol{A}_{full}$ has an underlying block diagonal structure shaded in white (under a particular choice of vertex labeling), whose rank is equal to the number of convex cells of C_{free} that contain vertices. Note that this low-rank property requires adopting the convention $A_{ii} = 1$. The sparse matrix can be used to model the connecting edges *between* the convex cells, such as entry $(3, 7)$, as well as to capture the effect of small occlusions, as in the case of $(6, 8)$. Loosely, the accessibility property of roadmaps results in a low-rank structure, while the connectivity property necessitates the sparse component.

Now consider a G_{prm} produced when $CandidateVertices(q_i) = \{q_j| \ \|q_i - q_j\| \le R\}$ (where R is shown in Fig. 2-left). The matrix entries of A_{prm} in the lower panel marked with "?" are unknown, e.g. the distance between vertex 1 and 4 is greater than R so $CollisionFree$ will not be called for this pair. G_{prm} still satisfies the accessibility and connectivity criteria yet the resulting paths would be longer in some cases. For example, since $e_{1,4}$ is not included in the sub-graph, a query to connect 1 to 4 would be routed via 2.

The RPCA method in Sect. 2.1 can be used to compute an estimate, $\hat{\boldsymbol{A}}$, of $\boldsymbol{A}_{full}$ using a limited set of observed entries such as those in $\boldsymbol{A}_{prm}$. Of course there are two notable differences between matrix completion and PRM. First is the way in which negative answers to $CollisionFree$ are used. In most PRM algorithms if $CollisionFree(q_i, q_j)$ is false, that information is essentially discarded; yet for matrix completion, this observation represents an additional constraint for the optimization problem. Second, proofs of the completeness of RPCA assume that the set of observations, Ω_{obs}, is selected at random, uniformly over $[n] \times [n]$ rather than using the geometric heuristic employed in PRM.

3   Algorithm overview

Given a motion planning problem, we run g-PRM where *CandidateVertices* will simply be $F \cdot n(n-1)/2$ vertex pairs selected uniformly at random. $F \in (0,1]$ is called the *sample fraction*. Using the edge information observed during PRM's *CollisionFree* step, including negative results, we assemble the constraint (1). Since the graph is undirected, we add the corresponding symmetric entries of A to the set of observations Ω_{obs}. We then use the numerical implementation described in Sect. 4 to solve (1) with two additional constraints:

$$L_{ii} = 1 \qquad \forall\, i \in [n], \tag{3}$$

$$-1 \leq S_{ij} \leq 1 \qquad \forall\, i,j \in [n]. \tag{4}$$

We round the final estimate $\hat{A} = round(L + S)$ to ensure it is a binary matrix.

The additional constraints are motivated by the fact that the objective function in (1) is merely a convex relaxation of the low-rank plus sparse model. Constraint (3) allows on-block observed "1"s to be included in L for "free" rather than pay a penalty for including them in S, since the nuclear norm of a block diagonal matrix is the same as that of the identity matrix (n). It also incentivizes the algorithm to extrapolate by completing the blocks when possible. It is worth noting that while a true rank minimizing L would have as few blocks as possible, the nuclear norm of a binary block diagonal matrix is n regardless of the number of blocks it contains so there is less incentive to "over generalize" by, for example, setting L to the all ones matrix. Finally, combined with the objective function, it effectively constrains $0 \leq L_{ij} \leq 1$. Constraint (4) is simply a convex relaxation of the property that $S_{ij} \in \{-1, 0, 1\}$.

4   Numerical implementation

We developed a numerical solver for (1) based on the Alternating Directions Method of Multipliers (ADMM) [27, 28]. ADMM is a flexible framework for constrained optimization, which is especially effective for problems in which the objective is separable–it can be decomposed into a sum of functions on blocks of variables [29]. ADMM has been successfully applied for RPCA and related problems in [19, 20]. A virtue of ADMM in this setting is that it allows us to easily handle constraints such as (3) and (4). This is accomplished by *splitting*: we introduce auxiliary variables $\bar{L}, \tilde{L}, \bar{S}, \tilde{S}$, and solve the equivalent problem

$$\min \|L\|_* + \lambda \|S\|_1 \quad \text{s.t.} \quad \begin{aligned} & L = \bar{L}, \ S = \bar{S}, \ \mathcal{P}_\Omega[\bar{L} + \bar{S}] = A, \\ & \tilde{L} = \bar{L}, \ \tilde{S} = \bar{S}, \ \tilde{L} \in \mathfrak{L}, \ \tilde{S} \in \mathfrak{S} \end{aligned} \tag{5}$$

Here, $\mathfrak{L}$ and $\mathfrak{S}$ are convex sets which encode any additional constraints on $\boldsymbol{L}$ and $\boldsymbol{S}$, such as the constraint that the diagonal elements of $\boldsymbol{L}$ equal one, interval constraints, etc. While splitting increases the number of variables, it makes it much easier to develop efficient numerical methods, by grouping the "difficult" parts of the objective and the constraint in a way that allows them to be handled independently. To accomplish this, we introduce Lagrange multipliers $\boldsymbol{\Lambda}$, $\boldsymbol{\Gamma}_L, \boldsymbol{\Gamma}_S, \boldsymbol{\zeta}_L, \boldsymbol{\zeta}_S$ corresponding to the equality constraints in (5), and write the *Augmented Lagrangan* for this modified problem as

$$
\begin{aligned}
&\mathcal{L}_\nu(\boldsymbol{L}, \boldsymbol{S}, \bar{\boldsymbol{L}}, \bar{\boldsymbol{S}}, \boldsymbol{\Lambda}, \boldsymbol{\Gamma}_L, \boldsymbol{\Gamma}_S, \boldsymbol{\zeta}_L, \boldsymbol{\zeta}_S) \\
&= \|\boldsymbol{L}\|_* + \lambda\|\boldsymbol{S}\|_1 + \langle \mathcal{P}_\Omega[\bar{\boldsymbol{L}} + \bar{\boldsymbol{S}}] - \boldsymbol{A}, \boldsymbol{\Lambda}\rangle + \langle \boldsymbol{L} - \bar{\boldsymbol{L}}, \boldsymbol{\Gamma}_L\rangle + \langle \boldsymbol{S} - \bar{\boldsymbol{S}}, \boldsymbol{\Gamma}_S\rangle \\
&\quad + \tfrac{\nu}{2}\|\boldsymbol{L} - \bar{\boldsymbol{L}}\|_F^2 + \tfrac{\nu}{2}\|\boldsymbol{S} - \bar{\boldsymbol{S}}\|_F^2 + \tfrac{\nu}{2}\left\|\tilde{\boldsymbol{L}} - \bar{\boldsymbol{L}}\right\|_F^2 + \tfrac{\nu}{2}\left\|\tilde{\boldsymbol{S}} - \bar{\boldsymbol{S}}\right\|_F^2 \\
&\quad + \tfrac{\nu}{2}\left\|\mathcal{P}_\Omega[\bar{\boldsymbol{L}} + \bar{\boldsymbol{S}}] - \boldsymbol{A}\right\|_F^2.
\end{aligned}
\tag{6}
$$

The basic iteration in ADMM then consists in minimizing $\mathcal{L}$ with respect to each block $(\boldsymbol{L}, \boldsymbol{S}, \tilde{\boldsymbol{L}}, \tilde{\boldsymbol{S}})$ and $(\bar{\boldsymbol{L}}, \bar{\boldsymbol{S}})$ of primal variables individually, and then updating the Lagrange multipliers $(\boldsymbol{\Lambda}, \boldsymbol{\Gamma}_L, \boldsymbol{\Gamma}_S)$ by one step of gradient ascent, with a very particular choice of the step size:

$$
\begin{aligned}
&(\boldsymbol{L}^{(k+1)}, \boldsymbol{S}^{(k+1)}, \tilde{\boldsymbol{L}}^{(k+1)}, \tilde{\boldsymbol{S}}^{(k+1)}) \\
&= \arg\min_{\boldsymbol{L}, \boldsymbol{S}} \mathcal{L}_\nu\left(\boldsymbol{L}, \boldsymbol{S}, \tilde{\boldsymbol{L}}, \tilde{\boldsymbol{S}}, \bar{\boldsymbol{L}}^{(k)}, \bar{\boldsymbol{S}}^{(k)}, \boldsymbol{\Lambda}^{(k)}, \boldsymbol{\Gamma}_L^{(k)}, \boldsymbol{\Gamma}_S^{(k)}, \boldsymbol{\zeta}_L^{(k)}, \boldsymbol{\zeta}_S^{(k)}\right),
\end{aligned}
\tag{7}
$$

$$
\begin{aligned}
&(\bar{\boldsymbol{L}}^{(k+1)}, \bar{\boldsymbol{S}}^{(k+1)}) \\
&= \arg\min_{\bar{\boldsymbol{L}}, \bar{\boldsymbol{S}}} \mathcal{L}_\nu\left(\boldsymbol{L}^{(k+1)}, \boldsymbol{S}^{(k+1)}, \tilde{\boldsymbol{L}}^{(k+1)}, \tilde{\boldsymbol{S}}^{(k+1)}, \bar{\boldsymbol{L}}, \bar{\boldsymbol{S}}, \boldsymbol{\Lambda}^{(k)}, \boldsymbol{\Gamma}_L^{(k)}, \boldsymbol{\Gamma}_S^{(k)}, \boldsymbol{\zeta}_L^{(k)}, \boldsymbol{\zeta}_S^{(k)}\right) \\
&\boldsymbol{\Lambda}^{(k+1)} = \boldsymbol{\Lambda}^{(k)} + \nu\left(\mathcal{P}_\Omega\left[\bar{\boldsymbol{L}}^{(k+1)} + \bar{\boldsymbol{S}}^{(k+1)}\right] - \boldsymbol{A}\right) \\
&\boldsymbol{\Gamma}_L^{(k+1)} = \boldsymbol{\Gamma}_L^{(k)} + \nu\left(\boldsymbol{L}^{(k+1)} - \bar{\boldsymbol{L}}^{(k+1)}\right), \quad \boldsymbol{\zeta}_L^{(k+1)} = \boldsymbol{\zeta}_L^{(k)} + \nu\left(\tilde{\boldsymbol{L}}^{(k+1)} - \bar{\boldsymbol{L}}^{(k+1)}\right) \\
&\boldsymbol{\Gamma}_S^{(k+1)} = \boldsymbol{\Gamma}_S^{(k)} + \nu\left(\boldsymbol{S}^{(k+1)} - \bar{\boldsymbol{S}}^{(k+1)}\right), \quad \boldsymbol{\zeta}_S^{(k+1)} = \boldsymbol{\zeta}_S^{(k)} + \nu\left(\tilde{\boldsymbol{S}}^{(k+1)} - \bar{\boldsymbol{S}}^{(k+1)}\right).
\end{aligned}
$$

With these choices, it can be shown that the Lagrange multipliers converge to a dual optimal solution; under mild conditions the primal variables also converge to a primal optimal solution for (5) [29, 30]. The main utility in introducing the variables $\bar{\boldsymbol{L}}$ and $\bar{\boldsymbol{S}}$ is that it makes the solutions to subproblem (7) computable in nearly closed form. For example, (7) decouples into four independent minimizations, each of which has a closed form solution.

The major computation at each iteration consists in computing one partial singular value decomposition (SVD). The overall cost of the algorithm is proportional to the number of iterations times the cost of performing one partial SVD. It can be proved that ADMM converges at a rate of $O(1/k)$; for well-structured instances, it is often observed to outperform its worst case convergence theory.

5   Solution Properties and Errors

In general when only a portion of the entries in $\boldsymbol{A}_{full}$ are observed (i.e. $F < 1$), some subset of the entries in $\hat{\boldsymbol{A}}$ will be erroneous. Define $\boldsymbol{E} = \boldsymbol{A}_{full} - \hat{\boldsymbol{A}}$. The non-zero elements of $\boldsymbol{E}$ fall into two categories. *False negatives* are edges in G_{full} that are not present in $\hat{G}$. However, this is not great cause for concern since $\hat{G}$ still has fewer false negatives than G_{prm}, since the constraint in (1) ensures that $\hat{G}$ is a super-graph of G_{prm}. This implies that $\hat{G}$ has no more connected components than G_{prm}, inheriting its connectivity property. It also implies that the average path length in $\hat{G}$ is at least as small as that of G_{prm}. Obviously, since they share the same vertex set, $\hat{G}$ inherits the accessibility property. On the other hand *false positives*–fictitious edges in $\hat{G}$–are of greater concern, since a robot executing such a path blindly will experience a collision. In other applications this can be corrected for at runtime using range or bump sensors.

There are a few general properties of the errors worth noting. First, there are no errors on the observation set Ω_{obs}, by virtue of the constraint in (1). From this it follows that there can be no errors on the support of $\boldsymbol{S}$ since $Supp(\boldsymbol{S}) \subset \Omega_{obs}$. To see this, consider that if S had a nonzero entry off of the observation set, setting the entry to zero would reduce the objective function without violating a constraint. These imply that: 1.the false negatives are a subset of $\neg Supp(\boldsymbol{L}) - \Omega_{obs}$; and 2. the false positives are a subset of $Supp(\boldsymbol{L}) - \Omega_{obs}$.

Of course, the structure of L and Ω_{obs} are influenced by the sample fraction F. There are three regimes: *under-sampled, transition* and *over-sampled*. In the *under-sampled regime* $(F < 2/n)$, no extrapolation occurs–*i.e.* $\hat{\boldsymbol{A}} = P_{\Omega_{obs}}(\boldsymbol{A}) = \boldsymbol{A}_{prm}$. The diagonal constraint (3) effectively prevents extrapolation since the nuclear norm of the identity matrix is n. Up to one additional "1" can be added to each row (column) of $\boldsymbol{L}$ without increasing the nuclear norm so it is possible to satisfy the constraint (1) with $\boldsymbol{L} = P_{\Omega_{obs}}(\boldsymbol{A}_{full})$ and $\boldsymbol{S} = \boldsymbol{0}$. Just as in G_{prm}, there are no false positives but the fraction of false negatives is quite high:

$$Err = (1 - F)\frac{\|\boldsymbol{A}_{full}\|_0}{n^2}. \tag{8}$$

The *transition regime* $(F > 2/n)$ begins after there are 3 or more observed "1"'s in any row (including the diagonal), causing L to begin extrapolating. When the extrapolations contradict an observed entry, S begins populating as well. The errors decrease monotonically from (8) to (9). The number of false negatives drops sharply while the number of false positives rises slightly.

During the *over-sampled regime* $(F > C\log^2 n/n)$, the remaining errors come from the unsampled sparse entries. Let $\boldsymbol{L}_{full}$ and $\boldsymbol{S}_{full}$ refer to the decomposition when $F = 1$. Once F is above the transition threshold $\boldsymbol{L} = \boldsymbol{L}_{full}$ with high probability while $\boldsymbol{S} = P_{\Omega_{obs}}(\boldsymbol{S}_{full})$. Therefore in expectation

$$Err = (1 - F)\frac{\|\boldsymbol{S}_{full}\|_0}{n^2}, \tag{9}$$

with false positives and negatives on the unsampled negative and positive entries, respectively, of S_{full}. Since no extrapolation occurs in the sparse term, the error only vanishes when $F = 1$.

6   Results

6.1   Examples

We consider four motion planning problems in this paper. For each problem G_{prm} was computed 10 times and for each of the resulting vertex sets, a ground truth visibility graph G_{full} and the corresponding adjacency matrix, A_{full}, were computed by collision checking all possible edges using a resolution of 1% of the extent of C_{free}. All the results reported in this paper for a given problem are averaged over the 10 instances.

The first problem, *Simple 2-D* involves a point robot navigating in the environment shown in Fig. 2 and is primarily included because the results are easy to visualize. Next, three more realistic examples were generated using the Open Motion Planning Library (OMPL) [31]. *Twisty Cool* (Fig. 3 - left) involves an L-shaped object (red) traveling through a narrow passageway connecting the left and the right open regions. The passage can only be entered with a limited set of orientations and requires the object to rotate when partially though to exit the passage. Such problems are known to be challenging for sampling-base planners. *Twisty Cooler* (Fig. 3 - center) is an even more challenging scenario with multiple narrow passages. *Cubicles* (Fig. 3-right) involves a "T"-shaped robot (gold) moving through a multi-floor office building.

Fig. 3. Examples: Twisty Cool (left); Twisty Cooler (center) and Cubicles (right).

6.2   Model Applicability

First we support our claim that A_{full} has the low-rank plus structure posited in Sect. 2.3. In order to do so we set the sample fraction $F = 1$ and solve the optimization problem. Naturally $\hat{A} = A_{full}$ and $E = 0$; however we are interested in the properties of the resulting L and S.

In the case of *Simple 2-D* with $n = 200$, the results can be easily visualized. L is a rank-2 matrix. Figure 4 (left) illustrates the principle components of L, which correspond to the edges within each large convex regions of C_{free}. Of course the edges connecting the two regions are missing from L, and there are a few erroneous edges going through the small protrusion in the lower right corner. The *relative sparsity* of S, defined as $\|S\|_0/n^2$, is 0.0191, where 0.0121 of the entries are positive and 0.007 are negative. Fig. 4 (center) depicts the positive entries–edges that are added to L. Many connect the two large convex regions. Fig. 4 (right) illustrates the negative entries, which are deleted edges that correct the false positives in L near the protrusion on the lower right.

The non-zero singular values of L (126 and 74) tell us there are two convex clusters of vertices in C_{free} containing 126 and 74 vertices. In higher dimensional examples this provides insight into the problem structure that is otherwise hard to glean. For Twisty Cool with 500 vertices, Figure 5 shows the original adjacency matrix on the left from which little can be discerned. The center panel shows the support pattern of L after the vertex labeling has been permuted so that identical row/columns are consecutively numbered. There are two clusters of size 247 and 229 (the two largest singular values). Another set of 20 vertices form the small interconnected group in the lower right corner. Their non-integer singular values (12.3 and 7.7) indicate that they cannot be successfully partitioned into clusters by the algorithm. They likely lie in the narrow passageway. Four additional vertices belong to no cluster (v_{490}, v_{494}, v_{499} and v_{500}); since their singular values are all 1. Their inclusion in L is an artifact of the diagonal constraint. The right panel shows S under the same permutation, where white entries are 1 and black entries are -1. We can see that most of the edges added by S connect v_{499} and v_{500} to the second large block (along outter edge of bottom right quadrant). v_{494} has a degree of 3–the smallest of any vertex.

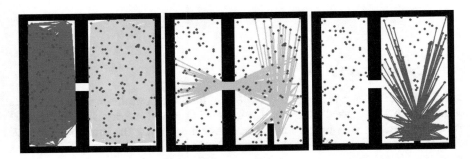

Fig. 4. The edges corresponding to the two principle components of L (left); the positive entries in S (center) and negative entries in S (right) for problem in Fig. 2 with 100% of the entries observed.

A similar analysis was conducted for the remaining example problems, the results of which are summarized in Table 1. The effective rank (eliminating components whose singular values are ≤ 1) is around 5% of full and the sparsity

Fig. 5. The low-rank plus sparse decomposition for Twisty Cool (Fig.3-left). A_{full} (left) is difficult to interpret. While the permuted L (center) and S (right) provide insight into the the location of the vertices in C_{free}.

is around 4%; suggesting that they all fit the low-rank plus sparse assumption. Our experiments suggest their relative values are not sensitive to n.

Example	n	Eff. Rank	smallest s.v.	$\lVert S > 0 \rVert / n^2$	$\lVert S < 0 \rVert / n^2$
Simple 2-D	200	2	74	0.0121	0.0071
Simple 2-D	500	2	151	0.0124	0.0068
Twisty Cool	250	4	6	0.0012	0.0399
Twisty Cool	500	4	7.7	0.0013	0.0370
Twisty Cooler	250	23	2	0.0152	0.0227
Twisty Cooler	500	29	2	0.0168	0.0238
Cubicles	250	16	2	0.0081	0.0312
Cubicles	500	23	2	0.0091	0.0323

Table 1. Properties of the decomposition when all entries in A_{full} are observed.

6.3   Completion Errors

Next we turn to our primary objective–completing the unknown entries in the adjacency matrix based on a small sample of observed entries. In these experiments we ran the post-processing algorithm while varying the sample fraction F. Figure 6 shows the results for the Twisty Cool example with $n = 400$. The mean error rate is the solid blue trend line. The dashed blue line shows the predicted mean error rate during the under-sampled, transition and over-sampled phases from Sect. 5. The predicted end of the transition region is $F = 0.0508$; after this point the edge structure of $\hat{G}$ is better than 96% correct. Note that the mean false positive rate (shown in red), peaks at 0.0303 when $F = 0.0508$. Error plots are shown for Twisty Cooler (Fig. 8) and Cubicles (Fig. 9).

The dramatic reduction in path length for the Twisty Cool example is shown in Fig. 7 and is representative of the results for the other examples. At the end of the transition region, $F = 0.0503$, the average path length is 53% shorter in

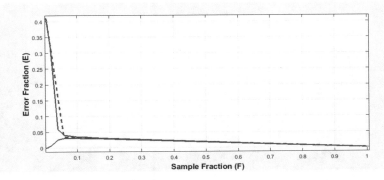

Fig. 6. The mean fraction of incorrect edges for Twisty Cool ($n = 400$) vs. the sample fraction F: experimental (solid blue); predicted (dashed blue); and false positives (red).

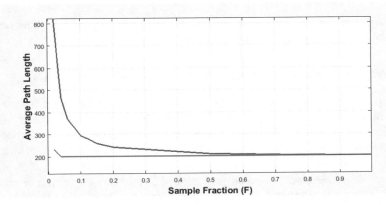

Fig. 7. The mean path length for Twistycool ($n = 400$): G_{prm} (upper red line) and $\hat{G}$ (lower blue line).

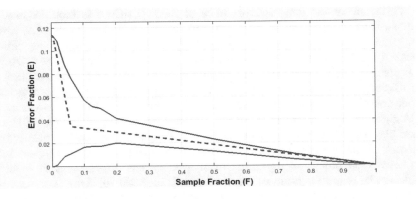

Fig. 8. The mean fraction of incorrect edges for Twisty Cooler ($n = 400$) vs. the sample fraction F: experimental (solid blue); predicted (dashed blue); and false positives (red).

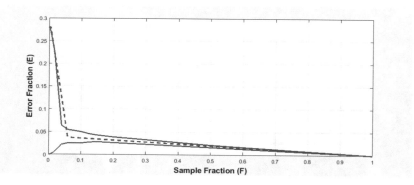

Fig. 9. The mean fraction of incorrect edges for Cubicles ($n = 400$) vs. the sample fraction F: experimental (solid blue); predicted (dashed blue); and false positives (red).

the post-processed graph, even after excluding the false positives. The number of connected components is identical in both graphs.

7   Connections to the Literature and Future Work

This paper introduced a post-processing technique for PRMs that uses convex optimization to estimate the status of unobserved graph edges without any additional collision checks. The post-processed graph, $\hat{G}$, inherits the connectivity and accessibility of G_{prm}. In a typical experiment, after directly checking only 5% of the possible edges, the structure of $\hat{G}$ was 96% accurate, with 53% shorter paths and 3% false positives. The practical utility of our algorithm is that the paths in the post-processed graph, $\hat{G}$, are dramatically shorter than those in G_{prm}– almost as short as in the full visibility graph which is prohibitively expensive to compute directly. A forthcoming study will extend the comparison to the path lengths produced by PRM*.

While few in number, the presence of false positives is troublesome. The ℓ^1 norm is symmetric, and hence does not introduce a clear preference towards false positives or false negatives. Different types of errors may have very different practical implications. A preference for one type or the other can be introduced by replacing the ℓ^1 norm with an asymmetric penalty, in a manner similar to quantile regression [32]. Determining how to choose the penalty as a function of a user specified risk tolerance is another interesting topic for future work.

A rigorous assessment of the comparative computational savings of the proposed algorithm is a topic of future work. Our proposed method essentially trades the cost of many additional collision checks for the cost of solving an interative convex optimization problem– motivated by the observation that collision checking the edges is the most expensive step in PRM [9]. Our post-processing algorithm is dominated by the time to compute a partial singular value decomposition. This cost can be further reduced by replacing ADMM with the Frank-Wolfe method, which only computes a single singular value/vector pair at each iteration [33]. Order analysis suggests that our approach is favorable. However, rigorously comparing the computation time of two radically different operations (collision checking vs. the optimization) is difficult because each is implemented in different programming languages, packages and machines; and depends on the complexity of the environment, the termination criteria and tolerances.

Overall, we believe the primary contribution of this work is in the novel application of tools from matrix completion to the application domain of robot motion planning. Our algorithm exploits the structure of the adjacency matrix which we hypothesize can be expressed as the sum of low-rank and sparse components. In essence this means that the adjacency matrix is highly redundant and can be encoded with very few bits of information, which can be captured with high probability by purely random sampling. We believe this provides an alternate explanation for how sampling-based planning methods are able to capture the connectivity of C_{free} with relatively few samples. Of course it is impossible to establish that this structure is representative of the set of *all possible* motion planning problems. However, in the four examples considered here, the rank and sparsity of the resulting decomposition were less than 5% of their maximum values. In the same way that [10] argues that most motion planning problems of practical interest are expansive, we argue that the low-rank plus sparse structure is representative of common environments. In fact one area of future work is to relate the concept of $(\epsilon, \alpha, \beta)$-goodness to the properties of the low-rank and sparse matrices. If the examples here are indeed typical, it suggests motion planning problems are amenable to other techniques from that field as well. Results from the graph clustering literature concede that problems become difficult when the size of the smallest cluster approaches $\sqrt{n}$ and the sparsity approaches 1/3 [34]. By that measure, the graphs here are well outside of the hard regime (though the smallest cluster size is clearly the limiting factor); suggesting the use of convexified maximum-likelihood estimation.

The analysis of RPCA assumes the sparse entries are i.i.d Bernoulli random variables. However, we conducted preliminary experiments where we used the non-uniform sampling pattern induced by the common range limited variant of PRM and the results were not markedly different. However, adaptive sampling schemes are a promising area of future work.

Finally, a fundamental topic of interest in the matrix completion literature is *sample complexity*–the number of observations required to recover the low-rank term with high probability. The tightness of the $O(n \log^2 n)$ transition estimate of RPCA is an open question–the structure of adjacency matrices is not accounted for in the generic analysis [16]. Eliminating the second log factor would suggest a common range-limited PRM (requiring $O(n \log n)$ edge checks [9]) contains enough information, up to constant factors, to complete the low-rank term–eliminating the need for additional edge checks.

Acknowledgment

Joel Esposito is supported by ONR-N00014-13-0489; and John Wright is supported by ONR-N00014-13-0492. The authors would like to thank Lydia Kavraki and Mark Moll for providing the data sets used in this paper.

References

1. L. E. Kavraki, P. Svestka, J.-C. Latombe, and M. H. Overmars. Probabilistic roadmaps for path planning in high-dimensional configuration spaces. *IEEE Transactions on Robotics and Automation*, 12(4):566580, June 1996.

2. E. Candès and B. Recht. Exact matrix completion via convex optimization. *Foundations of Computational Mathematics*, 9(6):717–772, 2009.

3. S. M. LaValle and J. J. Kuffner. Randomized kinodynamic planning. *International Journal of Robotics Research*, 20(5):378–400, May 2001.

4. D. Hsu, J.-C. Latombe, and R. Motwani. Path planning in expansive configuration spaces. *International Journal Computational Geometry & Applications*, 4:495–512, 1999.

5. J. Barraquand, L. Kavraki, J.-C. Latombe, T.-Y. Li, R. Motwani, and P. Raghavan. A random sampling scheme for robot path planning. In G. Giralt and G. Hirzinger, editors, *Proceedings International Symposium on Robotics Research*, pages 249–264. Springer-Verlag, New York, 1996.

6. A. Ladd and L. E. Kavraki. Measure theoretic analysis of probabilistic path planning. *IEEE Transactions on Robotics & Automation*, 20(2):229–242, 2004.

7. L. E. Kavraki, M. N. Kolountzakis, and J.-C. Latombe. Analysis of probabilistic roadmaps for path planning. *IEEE Transactions on Robotics and Automation*, 14(1):166–171, 1998.

8. S. M. LaValle, M. S. Branicky, and S. R. Lindemann. On the relationship between classical grid search and probabilistic roadmaps. *International Journal of Robotics Research*, 23(7/8):673–692, July/August 2004.

9. S. Karaman and E. Frazzoli. Sampling-based algorithms for optimal motion planning. *International Journal of Robotics Research*, 30(7):846–894, 2011.

10. D. Hsu, J.-C. Latombe, and H. Kurniawati. On the probabilistic foundations of probabilistic roadmap planning. *International Journal of Robotics Research*, 25(7):627–643, July 2006.

11. E. Candès and T. Tao. The power of convex relaxation: Near-optimal matrix completion. *IEEE Transactions on Information Theory*, 56(5):2053–2080, 2010.

12. D. Gross. Recovering low-rank matrices from few coefficients in any basis. *IEEE Transactions on Information Theory*, 2010.

13. S. Neghaban and M. Wainwright. Estimation of (near) low-rank matrices with noise and high-dimensional scaling. *Annals of Statistics*, 39(2):1069–1097, 2011.

14. Y. Chen. Incoherence optimal matrix completion. *IEEE Transactions on Information Theory*, 61(5):2909–2923, 2015.

15. V. Chandrasekaran, S. Sanghavi, P. Parillo, and A. Willsky. Rank-sparsity incoherence for matrix decomposition. *SIAM Journal on Optimization*, 21(2):572–596, 2011.

16. E. Candès, X. Li, Y. Ma, and J. Wright. Robust principal component analysis? *Journal of the ACM*, 58(3), May 2011.

17. Daniel Hsu, Sham M Kakade, and Tong Zhang. Robust matrix decomposition with sparse corruptions. *IEEE Transactions on Information Theory*, 57(11):7221–7234, 2011.

18. X. Li. Compressed sensing and matrix completion with constant proportion of corruptions. *Constructive Approximation*, 37(1), 2013.

19. X. Yuan and M. Tao. Recovering low-rank and sparse components of matrices from incomplete and noisy observations. *SIAM Journal on Optimization*, 21(1):57–81, 2011.

20. Z. Lin, M. Chen, L. Wu, and Y. Ma. The augmented Lagrange multiplier method for exact recovery of corrupted low-rank matrices. Technical report, arXiv:1009.5055, 2009.

21. M. Davenport and J. Romberg. An overview of low-rank matrix recovery from incomplete observations. *IEEE Journal on Selected Topics in Signal Processing*, 2016.

22. N. Nilsson. A mobile automaton: An application of artificial intelligence techniques. In *Proceedings International Joint Conference on Artificial Intelligence*, May 1969.

23. B. Siciliano and O. Khatib, editors. *Springer Handbook of Robotics*. Springer, 2008.

24. J. H. Reif. Complexity of the mover's problem and generalizations. In *Proceedings IEEE Symposium on Foundations of Computer Science*, pages 421–427, 1979.

25. J. F. Canny. *The Complexity of Robot Motion Planning*. MIT Press, Cambridge, MA, 1988.

26. S. M. LaValle. *Planning Algorithms*. Cambridge University Press, Cambridge, U.K., 2006. Available at http://planning.cs.uiuc.edu/.

27. R. Glowinski and A. Marrocco. Sur l'approximation par él'ements finis d'ordre un, et la r'esolution, par p'enalisation-dualit'e d'une classe de probl'emes de dirichlet nonlin'eaires. *Revue Francaise d'Automatique, Informatique, Recherche Opérationnelle*, 9(2):41–76, 1975.

28. D. Gabay and B. Mercier. A dual algorithm for the solution of nonlinear variational problems via finite-element approximations. *Computers and Mathematics with Applications*, 2:17–40, 1976.

29. S. Boyd, N. Parikh, E. Chu, B. Peleato, and J. Eckstein. Distributed optimization and statistical learning via the alternating direction method of multipliers. *Foundations and Trends in Machine Learning*, 3(1):1–122, 2011.

30. J. Eckstein. Augmented lagrangian and alternating direction methods for convex optimization: A tutorial and some illustrative computational results. *RUTCOR Technical Report*, 2012.

31. I. A. Sucan, M. Moll, and L. E. Kavraki. The Open Motion Planning Library. *IEEE Robotics and Automation Magazine*, 19(4):72–82, 2012.

32. R. Koenker and K. Hallock. Quantile regression: An introduction. *Journal of Economic Perspectives*, 15(4):43–56.

33. C. Mu, Y. Zhang, J. Wright, and D. Goldfarb. Scalable robust matrix recovery: Frank-wolfe meets proximal methods. *SIAM Journal on Scientific Computing*, 2016.

34. Y. Chen, A. Jalali, S. Sanghavi, and H. Xu. Clustering partially observed graphs via convex optimization. *Journal of Machine Learning Research*, 15:2213–2238, 2014.

Persistent Surveillance of Events with Unknown Rate Statistics

Cenk Baykal[1], Guy Rosman[1], Kyle Kotowick[1], Mark Donahue[2], and Daniela Rus[1]

[1] Massachusetts Institute of Technology (MIT), Cambridge MA 02139, USA,
{baykal, rosman, kotowick, rus}@csail.mit.edu,
[2] MIT Lincoln Laboratory, Lexington MA 02421, USA,
mark.donahue@ll.mit.edu

Abstract. We present a novel algorithm for persistent monitoring of stochastic events that occur at discrete locations in the environment with unknown event rates. Prior research on persistent monitoring assumes knowledge of event rates, which is often not the case in robotics applications. We consider the multi-objective optimization of maximizing the total number of events observed in a balanced manner subject to real-world autonomous system constraints. We formulate an algorithm that quantifies and leverages uncertainty over events' statistics to greedily generate adaptive policies that simultaneously consider learning and monitoring objectives. We analyze the favorable properties of our algorithm as a function of monitoring cycles and provide simulation results demonstrating our method's effectiveness in real-world inspired monitoring applications.

Keywords: persistent monitoring, optimization and optimal control, probabilistic reasoning, mobile robots, sensor planning, machine learning

1 Introduction

Robotic surveillance missions often require a mobile robot to navigate an unknown environment and monitor stochastic events of interest over a long period of time. Equipped with limited a priori knowledge, the agent is tasked with exploring the environment by traveling from one landmark to the other and identifying regions of importance in an efficient way. The overarching monitoring objective is to observe as many events as possible in a uniform, balanced manner so that sufficient heterogeneous information can be collected across different parts of the environment. Applications include people and vehicle surveillance of friendly and unfriendly activity and environmental monitoring of natural phenomena.

In our formulation, we consider monitoring stochastic, instantaneous events of interest that occur at discrete stations, i.e., locations, over an infinite time horizon. Our robot is equipped with a limited-range sensor that can only record accurate measurements when the robot is stationary, e.g., a microphone on a

K. Goldberg et al. (Eds.): *Algorithmic Foundations of Robotics XII*, SPAR 13, pp. 736–751, 2020.
https://doi.org/10.1007/978-3-030-43089-4_47

UAV. Hence, the robot must travel to each location and listen for events for a predetermined amount of time before traveling to another location.

We are given a fixed, cyclic path for the robot to traverse, but do not know the dwell time at each station. The persistent monitoring problem is to compute the optimal *observation time* for each station with respect to problem-specific optimality criteria. In particular, we consider maximizing the number of events observed in a maximally balanced way to be the overarching monitoring objective. Our multi-objective problem formulation extends previous work by relaxing the assumption that the rates of events are known, which introduces the notorious exploration and exploitation trade-off.

This paper contributes the following:

1. A novel per-cycle monitoring problem that hinges on an *uncertainty constraint*, a hard constraint that enables computation of policies conducive to balancing the exploration and exploitation trade-off.
2. A persistent monitoring algorithm with provable per-cycle guarantees that quantifies and employs the uncertainty over events' statistics to generate per-cycle optimal policies.
3. An analysis proving probabilistic bounds on the accuracy of rate approximations and the per-cycle optimality of the generated policies as a function of monitoring cycles.
4. Simulation results that characterize our algorithm's effectiveness in robotic surveillance scenarios and compare its performance to state-of-the-art monitoring algorithms.

1.1   Related Work

We build on important prior work in persistent surveillance, sensor scheduling, and machine learning. Robotic surveillance missions have been considered for a variety of applications and objectives such as ecological monitoring, underwater marine surveillance, and detection of natural phenomena [1–10]. Examples of monitoring objectives include facilitating high-value data collection for autonomous underwater vehicles [2], keeping a growing spatio-temporal field bounded using speed controllers [5], and generating the shortest watchman routes along which every point in a given space is visible [11].

Surveillance of discrete landmarks is of particular relevance to our work. Monitoring discrete locations such as buildings, windows, doors using a team of autonomous micro-aerial vehicles (MAVs) is considered in [3]. [12] presents different approaches to the min-max latency walk problem in a discrete setting. [13] extends this work to include multiple objectives, i.e. [13] considers the objective of minimizing the maximum latency and maximizing balance of events across stations using a single mobile robot. The authors show a reduction of the optimization problem to a quasi-convex problem and prove that a globally optimal solution can be computed in $O(\text{poly}(n))$ time where n is the number of discrete landmarks. Persistent surveillance in a discrete setting can be extended to the case of reasoning over different trajectories as shown in [5,9,14]. However, most prior work assumes that the rates of events are known prior to the surveillance mission, which is very often not the case in real world robotics applications. In this paper, we relax the assumption of known rates and present an algorithm with provable guarantees to generate policies conducive to learning event rates and optimizing the monitoring objectives.

[10] considers controlling multiple agents to minimize an uncertainty metric in the context of a 1D spatial domain. Decentralized approaches to controlling a network of robots for purposes of sensory coverage are investigated in [9], where a control law to drive a network of mobile robots to an optimal sensing configuration is presented. Persistent monitoring of dynamic environments has studied in [4,5,7]. For instance, [7] considers optimal sensing in a time-changing Gaussian Random Field and proposes a new randomized path planning algorithm to find the optimal infinite horizon trajectory. [15] presents a surveillance method based on Partially Observable Markov Decision Processes (POMDPs), however, POMDP-based approaches are often computationally intractable, especially when the action set includes continuous parameters, as in our case.

Persistent surveillance is closely related to sensor scheduling [16], sensor positioning [17], and coverage [18]. Previous approaches have considered persistent monitoring in the context of a mobile sensor [19]. Other related work includes variants and applications of the Orienteering Problem (OP) to generate informative paths that are constrained to a fixed length or time budget [20]. Yu et al. present an extension of OP to monitor spatially-correlated locations within a predetermined time [21]. In [22] and [23] the authors consider the OP problem in which the reward is a known function of the time spent at each point of interest. In contrast to our work, approaches in OP predominantly consider known environments and budget-constrained policies that visit each location at most once and optimize only a single objective.

The main challenge for the problem we address in this paper stems from the inherent exploration and exploitation trade-off, which has been rigorously analyzed in the form of regret bounds in Reinforcement Learning [24–26] and more relevantly, in Multi-armed Bandit (MAB) literature [27–29]. However, the traditional MAB problem considers minimizing regret with respect to the accumulated reward by appropriately pulling one of the $K \in \mathbb{N}_+$ levers at each discrete time step to obtain a stochastic reward that is generally assumed to bounded or subgaussian.

Our work differs from the canonical MAB formulation in that we consider a multi-objective optimization problem, i.e. we consider both the number and balance of event observations, in the face of travel costs, distributions with infinite support, cyclic policy structure, and continuous state and parameter space. To the best of our knowledge, this paper presents the first treatment of a MAB variant exhibiting all of the aforementioned complexities and an adaptive algorithm with provable guarantees as a function of monitoring cycles.

2   Problem Definition

Let there be $n \in \mathbb{N}_+$ spatially-distributed stations in the environment whose locations are known. At each station $i \in [n]$, stochastic events of interest occur according to a Poisson process with an unknown, station-specific rate parameter λ_i that is independent of other stations' rates. We assume that the robot executes a given cyclic path, taking $d_{i,j} > 0$ time to travel from station i to station j and let $D := \sum_{i=1}^{n-1} d_{i,i+1} + d_{n,1}$ denote the total travel time per cycle. The robot can only observe events at one station at any given time and cannot make observations while traveling.

We denote each complete traversal of the cyclic path as a *monitoring cycle*, indexed by $k \in \mathbb{N}_+$. We denote the observations times for all stations $\pi_k := (t_{1,k}, \ldots, t_{n,k})$ as the *monitoring policy* at cycle k. Our monitoring objective is to generate policies that maximize the number of events observed in a balanced manner across all stations within the allotted monitoring time $T_{\max}$ that is assumed to be unknown and unbounded. We introduce the function $f_{\text{obs}}(\Pi)$ that computes the total number of expected observations for a sequence of policies $\Pi := (\pi_k)_{k \in \mathbb{N}_+}$: $f_{\text{obs}}(\Pi) := \sum_{\pi_k} \sum_{i \in [n]} \mathbb{E}[N_i(\pi_k)]$, where $N_i(\pi_k)$ is the Poisson random variable, with realization $n_{i,k}$, denoting the number of events observed at station i under policy π_k and $\mathbb{E}[N_i(\pi_k)] := \lambda_i t_{i,k}$ by definition.

To reason about balanced attention, we let $f_{\text{bal}}(\Pi)$ denote as in [13] the expected observations ratio taken over the sequence of policies Π:

$$f_{\text{bal}}(\Pi) := \min_{i \in [n]} \frac{\sum_{\pi_k} \mathbb{E}[N_i(\pi_k)]}{\sum_{\pi_k} \sum_{j=1}^{n} \mathbb{E}[N_j(\pi_k)]}. \tag{1}$$

The *idealized* persistent surveillance problem is then:

Problem 1 (Idealized Persistent Surveillance Problem). Generate the optimal sequence of policies $\Pi^* = \operatorname{argmax}_{\Pi \in S} f_{\text{obs}}(\Pi)$ where S is the set of all possible policies that can be executed within the allotted monitoring time $T_{\max}$.

Generating the optimal solution Π^* at the beginning of the monitoring process is challenging due to the lack of knowledge regarding both the upper bound $T_{\max}$ and the station-specific rates. Hence, instead of optimizing the entire sequence of policies at once, we take a greedy approach and opt to subdivide the problem into multiple, *per-cycle* optimization problems. For each cycle $k \in \mathbb{N}_+$, our goal is to adaptively generate the policy π_k^* that optimizes the monitoring objectives with respect to the most up-to-date knowledge of event statistics. We let $\hat{f}_{\text{bal}}$ represent the per-cycle counterpart of f_{bal}

$$\hat{f}_{\text{bal}}(\pi_k) := \min_{i \in [n]} \frac{\mathbb{E}[N_i(\pi_k)]}{\sum_{j=1}^{n} \mathbb{E}[N_j(\pi_k)]}.$$

We note that the set of policies that optimize $\hat{f}_{\text{bal}}$ is uncountably infinite and policies of all possible lengths belong to this set [13]. To generate observation times that are conducive to exploration, we impose the hard constraint $t_{i,k} \geq t_{i,k}^{\text{low}}$

on each observation time, where $t_{i,k}^{\text{low}}$ is a lower bound that is a function of our uncertainty of the rate parameter λ_i (see Sec. 3). The optimization problem that we address in this paper is then of the following form:

Problem 2 (Per-cycle Monitoring Optimization Problem). At each cycle $k \in \mathbb{N}_+$, generate a per-cycle optimal policy π_k^* satisfying

$$\pi_k^* \in \operatorname*{argmax}_{\pi_k} \hat{f}_{\text{bal}}(\pi_k) \quad \text{s.t.} \quad \forall i \in [n] \quad t_{i,k} \geq t_{i,k}^{\text{low}}. \tag{2}$$

3 Methods

In this section, we present our monitoring algorithm and detail the main subroutines employed by our method to generate dynamic, adaptive policies and interleave learning and approximating of event statistics with policy execution.

3.1 Algorithm for Monitoring Under Unknown Event Rates

The entirety of our persistent surveillance method appears as Alg. 1 and employs Alg. 2 as a subprocedure to generate adaptive, uncertainty-reducing policies for each monitoring cycle.

Algorithm 1: Core monitoring algorithm	**Algorithm 2:** Generates a per-cycle optimal policy π^*
1 $\alpha_i \leftarrow \alpha_{i,0}; \quad \beta_i \leftarrow \beta_{i,0};$ $\hat{\lambda}_i \leftarrow \alpha_i / \beta_i;$ 2 **Loop** 3 $\pi^* \leftarrow \text{Algorithm2}(\alpha_i, \beta_i);$ 4 **for** $i \in [n]$ **do** 5 Observe for t_i^* time to obtain n_i observations; 6 $\alpha_i \leftarrow \alpha_i + n_i; \beta_i \leftarrow \beta_i + t_i^*;$ 7 $\hat{\lambda}_i \leftarrow \alpha_i / \beta_i;$	1 **for** $i \in [n]$ **do** 2 $\lfloor$ Compute t_i^{low} using (8); 3 $\pi_{\text{low}} \leftarrow (t_1^{\text{low}}, \ldots, t_n^{\text{low}});$ 4 $N_{\max} \leftarrow \max_{i \in [n]} t_i^{\text{low}} \alpha_i / \beta_i;$ 5 **for** $i \in [n]$ **do** 6 Compute t_i^* using $N_{\max}$ according to (10); 7 **return** $\pi^* = (t_1^*, \ldots, t_n^*);$

3.2 Learning and Approximating Event Statistics

We use the Gamma distribution as the conjugate prior for each rate parameter because it provides a closed-form expression for updating the posterior distribution after observing events. We let $\text{Gamma}(\alpha_i, \beta_i)$ denote the Gamma distribution with hyper-parameters $\alpha_i, \beta_i \in \mathbb{R}_+$ that are initialized to user-specified values $\alpha_{i,0}, \beta_{i,0}$ for all stations i and are updated as new events are observed.

For any arbitrary number of events $n_{i,k} \in \mathbb{N}$ observed in $t_{i,k}$ time, the posterior distribution is given by $\text{Gamma}(\alpha_i + n_{i,k}, \beta_i + t_{i,k})$ for any arbitrary station $i \in [n]$ and cycle $k \in \mathbb{N}_+$. For notational convenience, we let $X_i^k := (n_{i,k}, t_{i,k})$

represent the summary of observations for cycle $k \in \mathbb{N}_+$ and define the aggregated set of observations up to any arbitrary cycle as $X_i^{1:k} := \{X_i^1, X_i^2, \ldots, X_i^k\}$ for all stations $i \in [n]$. After updating the posterior distribution using the hyperparameters, i.e. $\alpha_i \leftarrow \alpha_i + n_{i,k}$, $\beta_i \leftarrow \beta_i + t_{i,k}$, we use the maximum probabiliy estimate of the rate parameter λ_i, denoted by $\hat{\lambda}_{i,k}$ for any arbitrary station i:

$$\hat{\lambda}_{i,k} := E[\lambda_i | X_i^{1:k}] = \frac{\alpha_{i,0} + \sum_{k=1}^{n} n_{i,k}}{\beta_{i,0} + \sum_{k=1}^{n} t_{i,k}} = \frac{\alpha_i}{\beta_i}. \tag{3}$$

3.3   Per-cycle Optimization and the Uncertainty Constraint

Inspired by confidence-based MAB approaches [28–30], our algorithm adaptively computes policies by reasoning about the uncertainty of our rate approximations. We introduce the *uncertainty-constraint*, an optimization constraint that enables the generating a station-specific observation time based on uncertainty of each station's parameter. The constraint helps bound the policy lengths adaptively over the course of the monitoring process so that approximation uncertainty decreases uniformly across all stations. We use the posterior variance of the rate parameter λ_i, $\mathrm{Var}(\lambda_i | X_i^{1:k})$, as our uncertainty measure of each station i after executing k cycles. We note that in our Gamma-Poisson model, $\mathrm{Var}(\lambda_i | X_i^{1:k}) := \frac{\alpha_i}{\beta_i^2}$ by definition of the Gamma distribution.

Uncertainty constraint For a given $\delta \in (0,1), \epsilon \in \left(0, 2(1 + 2e^{1/\pi})^{-1}\right)$ and arbitrary cycle $k \in \mathbb{N}_+$, π_k must satisfy the following

$$\forall i \in [n] \quad \mathbb{P}\left(\mathrm{Var}(\lambda_i | X_i^{1:k}, \pi_k) \le \delta \mathrm{Var}(\lambda_i | X_i^{1:k-1}) \big| X_i^{1:k-1}\right) > 1 - \epsilon. \tag{4}$$

We incorporate the uncertainty constraint as a hard constraint and recast the per-cycle optimization problem from Sec. 2 in terms of the optimization constraint.

Problem 3 (Recast Per-cycle Monitoring Optimization Problem). For each monitoring cycle $k \in \mathbb{N}_+$ generate a per-cycle optimal policy π_k^* that simultaneously satisfies the uncertainty constraint (4) and maximizes the balance of observations, i.e.,

$$\pi_k^* \in \underset{\pi_k}{\mathrm{argmax}} \, \hat{f}_{\mathrm{bal}}(\pi_k) \tag{5}$$

$$\text{s.t.} \quad \forall i \in [n] \quad \mathbb{P}\left(\mathrm{Var}(\lambda_i | X_i^{1:k}, \pi_k) \le \delta \mathrm{Var}(\lambda_i | X_i^{1:k-1}) \big| X_i^{1:k-1}\right) > 1 - \epsilon.$$

3.4   Controlling Approximation Uncertainty

We outline an efficient method for generating observation times that satisfy the uncertainty constraint and induce uncertainty reduction at each monitoring cycle. We begin by simplifying (4) to obtain

$$\mathbb{P}\left(N_i(t_{i,k}) \le \delta k(t_{i,k}) | X_i^{1:k-1}\right) > 1 - \epsilon \tag{6}$$

where $N_i(t_{i,k}) \sim \text{Pois}(\lambda_i t_{i,k})$ by definition of Poisson process and $k(t_{i,k}) := \delta \alpha_i (\beta_i + t_{i,k})^2 / \beta_i^2 - \alpha_i$. Given that the distribution of the random variable $N_i(t_{i,k})$ is a function of the unknown parameter λ_i, we use interval estimation to reason about the cumulative probability distribution of $N_i(t_{i,k})$.

For each monitoring cycle $k \in \mathbb{N}_+$ we utilize previously obtained observations $X_i^{1:k-1}$ to construct the equal-tail credible interval for each parameter λ_i, $i \in [n]$ defined by the open set $(\lambda_i^l, \lambda_i^u)$ such that

$$\forall \lambda_i \in \mathbb{R}_+ \quad \mathbb{P}\left(\lambda_i \in (\lambda_i^l, \lambda_i^u) \mid X_i^{1:k-1}\right) = 1 - \epsilon$$

where $\epsilon \in (0, 2(1 + 2e^{1/\pi})^{-1})$. By leveraging the relation between the Poisson and Gamma distributions, we compute the end-points of the equal-tailed credible interval:

$$\lambda_i^l := \frac{Q^{-1}(\alpha_i, \frac{\epsilon}{2})}{\beta_i} \quad \lambda_i^u := \frac{Q^{-1}(\beta_i, 1 - \frac{\epsilon}{2})}{\beta_i}$$

where $Q^{-1}(a, s)$ is the Gamma quantile function and α_i and β_i are the posterior hyper-parameters after observations $X_i^{1:k-1}$. Given that we desire our algorithm to be *cycle-adaptive* (Sect. 2), we seek to generate the minimum feasible observation time satisfying the uncertainty constraint for each station $i \in [n]$, i.e.,

$$t_{i,k}^{\text{low}} = \inf_{t_{i,k} \in \mathbb{R}_+} t_{i,k} \quad \text{s.t.} \quad \mathbb{P}\left(N_{i,k}(t_{i,k}) \leq \delta k(t_{i,k}) \mid X_i^{1:k-1}\right) > 1 - \epsilon. \tag{7}$$

For computational efficiency in the optimization above, we opt to use a tight and efficiently-computable lower bound for approximating the Poisson cumulative distribution function that improves upon the Chernoff-Hoeffding inequalities by a factor of at least two [31]. As demonstrated rigorously in Lemma 1, the expression for an approximately-minimal observation time satisfying constraint (4) is given by

$$t_{i,k}^{\text{low}} := t \in \mathbb{R}_+ \mid D_{\text{KL}}\left(\text{Pois}(\lambda_i^u t) \,\|\, \text{Pois}(k(t))\right) - W_\epsilon = 0 \tag{8}$$

where $D_{\text{KL}}\left(\text{Pois}(\lambda_1) \,\|\, \text{Pois}(\lambda_2)\right)$ is the Kullback-Leibler (KL) divergence between two Poisson distributions with mean λ_1 and λ_2 respectively and W_ϵ is defined using the Lambert W function [32]: $W_\epsilon = \frac{1}{2}W\left(\frac{(\epsilon-2)^2}{2e^2\pi}\right)$. An appropriate value for $t_{i,k}^{\text{low}}$ can be obtained by invoking a root-finding algorithm such as Brent's method on the equation above [33].

The constant factor $\delta \in (0, 1)$ is the exploration parameter that influences the rate of uncertainty decay. Low values of δ lead to lengthy, and hence less cycle-adaptive policies, whereas high values lead to shorter, but also less efficient policies due to incurred travel time. We found that values generated by a logistic function with respect to problem-specific parameters as input worked well in practice for up to 50 stations: $\delta(n) := (1 + \exp(-n/D))^{-1}$ where D is the total travel time per cycle.

3.5   Generating Balanced Policies that Consider Approximation Uncertainty

We build upon the method introduced in the previous section to generate a policy π_k^* that simultaneously satisfies the uncertainty constraint and balances attention given to all stations in approximately the minimum time possible. The key insight is that the value of $t_{i,k}^{\text{low}}$ given by (8) acts as a lower bound on the observation time for each station $i \in [n]$ for satisfying the uncertainty constraint (see Lemma 2). We also leverage the following fact from [13] regarding the optimality of the balance objective for a policy π_k:

$$\mathbb{E}[N_1(\pi_k)] = \cdots = \mathbb{E}[N_n(\pi_k)] \Leftrightarrow \pi_k \in \underset{\pi}{\text{argmax}}\ \hat{f}_{\text{bal}}(\pi). \tag{9}$$

We use a combination of this result and the fact that any observation time satisfying $t_{i,k} \geq t_{i,k}^{\text{low}}$ also satisfies the uncertainty constraint to arrive at an expression for the optimal observation time for each station. In constructing the optimal policy $\pi_k^* = (t_{1,k}^*, \ldots, t_{n,k}^*)$, we first identify the "bottleneck" value, $N_{\max}$, which is computed using the lower bounds for each $t_{i,k}$, i.e., $N_{\max} := \max_{i \in [n]} \hat{\lambda}_{i,k} t_{i,k}^{\text{low}}$. Given (9), we use the bottleneck value $N_{\max}$ to set the value of each observation time $t_{i,k}^*$ appropriately so that each $t_{i,k}^* \geq t_{i,k}^{\text{low}}$ and the policy defined by $\pi_k^* := (t_{1,k}^*, \ldots, t_{n,k}^*)$ maximizes the balance objective function. Namely, the optimal observation times for all stations which constitute the per-cycle optimal policy $\pi_k^* = (t_{1,k}^*, \ldots, t_{n,k}^*)$ are computed individually:

$$\forall k \in \mathbb{N}_+ \quad \forall i \in [n] \quad t_{i,k}^* := \frac{N_{\max}}{\hat{\lambda}_{i,k}} = N_{\max} \frac{\beta_i}{\alpha_i}. \tag{10}$$

4   Analysis

The outline of results in this section is as follows: we begin by proving the uncertainty-reducing property and per-cycle optimality of policies generated by Alg. 2 with respect to the rate approximations. We present a probabilistic bound on posterior variance and error of our rate approximations with respect to the ground-truth rates by leveraging the properties of each policy. We use the previous results to establish a probabilistic bound on the per-cycle optimality of any arbitrary policy generated by Alg. 2 with respect to the ground-truth optimal solution of Problem 3.

We impose the following assumption on user-specified input.

Assumption 1. *The parameters ϵ and δ are confined to the intervals $(0, 2(1 + 2e^{1/\pi})^{-1})$ and $(0, 1)$ respectively, i.e., $\epsilon \in (0, 2(1 + 2e^{1/\pi})^{-1})$, $\delta \in (0, 1)$.*

A policy π_k is said to be *approximately-optimal* at cycle $k \in \mathbb{N}_+$ if π_k is an optimal solution to Problem 3 with respect to the rate approximations $\hat{\lambda}_{1,k}, \ldots, \hat{\lambda}_{n,k}$, i.e., if it is optimal under the approximation of expectation: $\mathbb{E}[N_i(\pi_k)] \approx \hat{\lambda}_{i,k} t_{i,k} \ \forall i \in [n]$. In contrast, a policy π_k is *ground-truth optimal* if it is an optimal solution to Problem 3 with respect to the ground-truth rates $\lambda_1, \ldots, \lambda_n$. For sake of notational brevity, we introduce the function $g : \mathbb{R} \to \mathbb{R}$ denoting

$$g(x) := 1 - \frac{e^{-x}}{\max\left\{2, 2\sqrt{\pi x}\right\}},$$

and note the bound established by [31] for a Poisson random variable Y with mean m and $k \in \mathbb{R}_+$ such that $k \geq m$

$$\mathbb{P}\left(Y \leq k\right) > g\big(D_{\mathrm{KL}}(\mathrm{Pois}(m) \,\|\, \mathrm{Pois}(k))\big). \tag{11}$$

We begin by proving that each policy generated by Alg. 2 is optimal with respect to the per-cycle optimization problem (Problem 3).

Lemma 1 (Satisfaction of the uncertainty constraint). *The observation time $t_{i,k}^{low}$ given by (8) satisfies the uncertainty constraint (4) for any arbitrary station $i \in [n]$ and iteration $k \in \mathbb{N}_+$.*

Proof. We consider the left-hand side of (6) from Sect. 3 and marginalize over the unknown parameter $\lambda_i \in \mathbb{R}_+$:

$$\mathbb{P}\left(N_i(t_{i,k}) \leq k(t_{i,k})|X_i^{1:k-1}\right) = \int_0^\infty \mathbb{P}\left(N_i(t_{i,k}) \leq k(t_{i,k})|X_i^{1:k-1}, \lambda\right) \mathbb{P}\left(\lambda|X_i^{1:k-1}\right) d\lambda$$

where the probability is with respect to the random variable $N_i(t_{i,k}) \sim \mathrm{Pois}(\lambda t_{i,k})$ $\forall \lambda \in \mathbb{R}_+$ by definition of a Poisson process with parameter λ. Using the equal-tails credible interval constructed in Alg. 2, i.e. the interval $(\lambda_i^l, \lambda_i^u)$ satisfying

$$\forall i \in [n] \ \ \forall \lambda_i \in \mathbb{R}_+ \ \mathbb{P}\left(\lambda_i^l > \lambda_i|X_i^{1:k-1}\right) = \mathbb{P}\left(\lambda_i^u < \lambda_i|X_i^{1:k-1}\right) = \frac{\epsilon}{2},$$

we establish the inequalities:

$$\mathbb{P}\left(N_i(t_{i,k}) \leq k(t_{i,k})|X_i^{1:k-1}\right) > \int_0^{\lambda_i^u} \mathbb{P}\left(N_i(t_{i,k}) \leq k(t_{i,k})|X_i^{1:k-1}, \lambda\right) \mathbb{P}\left(\lambda|X_i^{1:k-1}\right) d\lambda$$

$$\geq \mathbb{P}\left(N_i(t_{i,k}) \leq k(t_{i,k})|X_i^{1:k-1}, \lambda_i^u\right) \int_0^{\lambda_i^u} \mathbb{P}\left(\lambda|X_i^{1:k-1}\right) d\lambda$$

$$= \left(1 - \frac{\epsilon}{2}\right) \mathbb{P}\left(N_i(t_{i,k}) \leq k(t_{i,k})|X_i^{1:k-1}, \lambda_i^u\right). \tag{12}$$

where we utilized the fact that $\mathbb{P}\left(N_i(t_{i,k}) \leq k(t_{i,k})|X_i^{1:k-1}, \lambda_i^u\right)$ is monotonically decreasing with respect to λ. By construction, $t_{i,k}^{low}$ satisfies $D_{\mathrm{KL}}(\mathrm{Pois}(\lambda_i^u t_{i,k}^{low}) \,\|\, \mathrm{Pois}(k(t_{i,k}^{low}))) = W_\epsilon$ which yields $1 - g(W_\epsilon) = 1 - \frac{\epsilon}{2-\epsilon}$ by definition and thus by (11) we have:

$$\mathbb{P}\left(N_i(t_{i,k}^{low}) \leq k(t_{i,k}^{low})|X_i^{1:k-1}, \lambda_i^u\right) > 1 - g(W_\epsilon) = 1 - \frac{\epsilon}{2-\epsilon}.$$

Combining this inequality with the expression of (12) establishes the result. $\qquad\square$

Lemma 2 (Monotonicity of solutions satisfying (4)). *For any arbitrary station $i \in [n]$ and monitoring cycle $k \in \mathbb{N}_+$, the observation time $t_{i,k}$ satisfying $t_{i,k} \geq t_{i,k}^{low}$, where $t_{i,k}^{low}$ is given by (8), satisfies the uncertainty constraint.*

Theorem 1 (Per-cycle approximate-optimality of solutions). *For any arbitrary cycle $k \in \mathbb{N}_+$, the policy $\pi_k^* := (t_{1,k}^*, \ldots, t_{n,k}^*)$ generated by Alg. 2 is an approximately-optimal solution with respect to Problem 3.*

Proof. By definition of (10), we have for any arbitrary cycle $k \in \mathbb{N}_+$ and station $i \in [n]$, $t_{i,k}^* = N_{\max}/\hat{\lambda}_{i,k} \geq t_{i,k}^{\text{low}}$ by definition of $N_{\max} := \max_{i \in [n]} \hat{\lambda}_{i,k} t_{i,k}^{\text{low}}$. Applying Lemma 2 and observing that

$$\hat{\lambda}_{1,k} t_{1,k}^* = N_{\max},\ \hat{\lambda}_{2,k} t_{2,k}^* = N_{\max},\ \ldots,\ \hat{\lambda}_{n,k} t_{n,k}^* = N_{\max}$$

implies that the uncertainty constraint is satisfied for all stations $i \in [n]$ and that $\pi_k^* \in \text{argmax}_{\pi_k} \hat{f}_{\text{bal}}(\pi_k)$, which establishes the optimality of π_k with respect to Problem 3. □

Using the fact that each policy satisfies the uncertainty constraint, we establish probabilistic bounds on uncertainty, i.e. posterior variance, and rate approximations.

Lemma 3 (Bound on posterior variance). *After executing an arbitrary number of cycles $k \in \mathbb{N}_+$, the posterior variance $Var(\lambda_i|X_i^{1:k})$ is bounded above by $\delta^k Var(\lambda_i)$ with probability at least $(1-\epsilon)^k$, i.e.,*

$$\forall i \in [n]\ \ \forall k \in \mathbb{N}_+\ \ \mathbb{P}\left(Var(\lambda_i|X_i^{1:k}) \leq \delta^k Var(\lambda_i)|X_i^{1:k}\right) > (1-\epsilon)^k$$

for all stations $i \in [n]$ where $Var(\lambda_i) := \alpha_{i,0}/\beta_{i,0}^2$ is the prior variance.

Proof. Iterative application of the inequality $\text{Var}(\lambda_i|X_i^{1:k}) \leq \delta \text{Var}(\lambda_i|X_i^{1:k-1})$ each with probability $1-\epsilon$ by the uncertainty constraint (4) yields the result. □

Corollary 1 (Bound on variance of the posterior mean). *After executing an arbitrary number of cycles $k \in \mathbb{N}_+$, the variance of our approximation $Var(\hat{\lambda}_{i,k}|X_i^{1:k-1})$ is bounded above by $\delta^{k-1} Var(\lambda_i)$ with probability greater than $(1-\epsilon)^{k-1}$, i.e.,*

$$\forall i \in [n]\ \ \mathbb{P}\left(Var(\hat{\lambda}_{i,k}|X_i^{1:k-1}) \leq \delta^{k-1} Var(\lambda_i)|X_i^{1:k-1}\right) > (1-\epsilon)^{k-1}.$$

Proof. Application of the law of total conditional variance and invoking Lemma 3 yields the result. □

Theorem 2 (ξ-bound on approximation error). *For all $\xi \in \mathbb{R}_+$ and cycles $k \in \mathbb{N}_+$, the inequality $|\hat{\lambda}_{i,k} - \lambda_i| < \xi$ holds with probability at least $(1-\epsilon)^{k-1}(1 - \frac{\delta^{k-1} Var(\lambda_i)}{\xi^2})$, i.e.,*

$$\forall i \in [n]\ \ \mathbb{P}\left(|\hat{\lambda}_{i,k} - \lambda_i| < \xi|X_i^{1:k-1}\right) > (1-\epsilon)^{k-1}\left(1 - \frac{\delta^{k-1} Var(\lambda_i)}{\xi^2}\right).$$

Proof. Applying Corollary 1 and using Chebyshev's inequality gives the result. □

Theorem 3 (Δ-bound on optimality with respect to Problem 3). *For any $\xi_i \in \mathbb{R}_+$, $i \in [n], k \in \mathbb{N}_+$, given that $|\hat{\lambda}_{i,k} - \lambda_i| \in (0, \xi_i)$ with probability as given in Theorem 2, let $\sigma_{min} := \sum_{i=1}^n (\lambda_i - \xi_i)^{-1}$ and $\sigma_{max} := \sum_{i=1}^n (\lambda_i + \xi_i)^{-1}$. Then, the objective value of the policy π_k^* at iteration k is within a factor of Δ of the ground-truth optimal solution, where $\Delta := \frac{\sigma_{min}}{\sigma_{max}}$ with probability greater than $(1-\epsilon)^{n(k-1)}\left(1 - \frac{\delta^{k-1} Var(\lambda_i)}{\xi^2}\right)^n$.*

Proof. Note that for any arbitrary total observation time $T \in \mathbb{R}_+$, a policy $\pi_k = (t_{1,k}^*, \ldots, t_{n,k}^*)$ satisfying

$$\forall i \in [n] \quad t_{i,k}^* := \frac{T}{\lambda_i \sum_{l=1}^n \frac{1}{\lambda_l}}. \tag{13}$$

optimizes the balance objective function $\hat{f}_{\text{bal}}$ [13]. Using the fact that $|\hat{\lambda}_{i,k} - \lambda_i| < \xi_i$ with probability given by Theorem 2, we arrive at the following inequality for $\hat{f}_{\text{bal}}(\pi_k^*)$

$$\hat{f}_{\text{bal}}(\pi_k^*) > \frac{\frac{T}{\sum_{l=1}^n (\lambda_i + \xi_l)^{-1}}}{\frac{nT}{\sum_{l=1}^n (\lambda_l - \xi_l)^{-1}}} = \frac{\sum_{l=1}^n (\lambda_l - \xi_l)^{-1}}{n \sum_{l=1}^n (\lambda_l + \xi_l)^{-1}}$$

with probability at least $(1 - \epsilon)^{n(k-1)} \left(1 - \frac{\delta^{k-1} \text{Var}(\lambda_i)}{\xi^2}\right)^n$. $\qquad \square$

5   Results

We evaluate the performance of Alg. 1 in two simulated scenarios modeled after real-world inspired monitoring tasks: (i) a synthetic simulation in which events at each station precisely follow a station-specific Poisson process and (ii) a scenario simulated in Armed Assault (ARMA) [34], a military simulation game, involving detections of suspicious agents. We note the statistics do not match our assumed Poisson model, and yet our algorithm performs well compared to other approaches. We compare Alg. 1 to the following monitoring algorithms:

1. Equal Time, Min. Delay (ETMD): computes the total cycle time to minimize latency T_{obs} [13] and partitions T_{obs} evenly across all stations.
2. Bal. Events, Min. Delay (BEMD): the algorithm introduced by [13] which generates policies that minimize latency and maximize observation balance.
3. Incremental Search, Bal. Events (ISBE): generates policies to maximize balance that increase in length by a fixed amount $\Delta_{\text{obs}} \in \mathbb{R}_+$ after each cycle.
4. Oracle Algorithm (Oracle Alg.): an omniscient algorithm assuming perfect knowledge of ground-truth rates and monitoring time T_{max} where each observation time is generated according to (13).

5.1   Synthetic Scenario

We consider the monitoring scenario involving the surveillance of events in three discrete stations over a monitoring period of 10 hours. We characterize the average performance of each monitoring algorithm with respect to 10,000 randomly generated problem instances with the following statistics:

1. Prior hyper-parameters: $\alpha_{i,0} \sim \text{Uniform}(1, 20)$ and $\beta_{i,0} \sim \text{Uniform}(0.75, 1.50)$.
2. Rate parameter of each station: $\mu_{\lambda_i} = 2.23$ and $\sigma_{\lambda_i} = 1.02$ events per minute.
3. Initial percentage error of the rate estimate $\lambda_{i,0}$, denoted by ρ_i: $\mu_{\rho_i} = 358.29\%$ and $\sigma_{\rho_i} = 221.32\%$.

4. Travel cost from station i to another j: $\mu_{d_{i,j}} = 9.97$ and $\sigma_{d_{i,j}} = 2.90$ minutes.

where μ and σ refer to standard deviation and variance of each parameter respectively and the transient events at each station $i \in [n]$ are simulated precisely according to $\text{Pois}(\lambda_i)$.

The performance of each algorithm with respect to the the monitoring objectives defined in Sect. 2 is shown in Figs. 1a and 1b respectively. The figures show that our algorithm is able to generate efficient policies that enable the robot

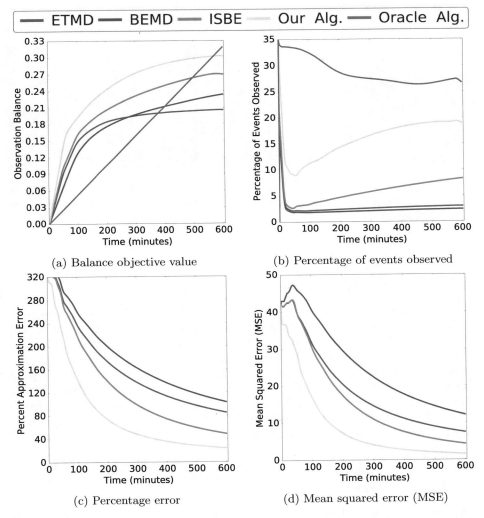

(a) Balance objective value

(b) Percentage of events observed

(c) Percentage error

(d) Mean squared error (MSE)

Fig. 1: Results of the synthetic simulation averaged over 10,000 trials that characterize and compare our algorithm to the four monitoring algorithms in randomized environments containing three discrete stations.

to observe significantly more events that achieve a higher balance in comparison to those computed by other algorithms (with exception of Oracle Alg.) at all times of the monitoring process. Figs. 1c and 1d depict the efficiency of each monitoring algorithm in rapidly learning the events' statistics and generating accurate approximations. The error plots show that our algorithm achieves lower measures of error at any given time in comparison to those of other algorithms and supports our method's practical efficiency in generating exploratory policies conducive to rapidly obtaining accurate approximations of event statistics. Figs. 1a-1d show our algorithm's dexterity in balancing the inherent trade-off between exploration vs. exploitation.

Fig. 2: Viewpoints from two stations in the ARMA simulation of the yellow backpack scenario. Agents wearing yellow backpacks whose detections are of interest appear in both figures.

5.2   Yellow Backpack Scenario

In this subsection, we consider the evaluation of our monitoring algorithm in a real-world inspired scenario, labeled the *yellow backpack scenario*, that entails monitoring of suspicious events that do not adhere to the assumed Poisson model (Sect. 2). Using the military strategy game ARMA, we simulate human agents that wander around randomly in a simulated town. A subset of the agents wear yellow backpacks (see Fig. 2). Under this setting, our objective is to optimally monitor the yellow backpack-wearing agents using three predesignated viewpoints, i.e. stations. We considered a monitoring duration of 5 hours under the following simulation configuration:

1. Environment dimensions: 250 meters x 250 meters $(62, 500$ meters$^2)$.
2. Number of agents with a yellow backpack: 10 out of 140 ($\approx 7.1\%$ of agents).
3. Travel cost (minutes): $d_{1,2} = 3$, $d_{2,3} = 2$, $d_{3,1} = 12$.

We used the Faster Region-based Convolutional Neural Network (Faster R-CNN, [35]) for recognizing yellow backpack-wearing agents in real-time at a frequency of 1 Hertz. We ran the simulation for a sufficiently long time in order to obtain estimates for the respective ground-truth rates of 23.3, 20.3, and

18.5 yellow backpack recognitions per minute, which were used to generate Figs. 3c and 3d. The results of the yellow backpack scenario, shown in Figs. 3a-3d, tell the same story as did the results of the synthetic simulation. We note that at all instances of the monitoring process, our approach that leverages uncertainty estimates outperforms others in generating balanced policies conducive to efficiently observing more events and obtaining accurate rate approximations.

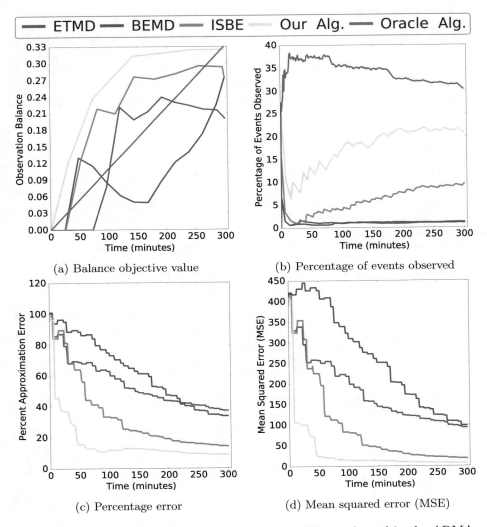

Fig. 3: The performance of each monitoring algorithm evaluated in the ARMA-simulated yellow backpack scenario.

6   Conclusion

In this paper, we presented a novel algorithm with provable guarantees for monitoring stochastic, transient events that occur at discrete stations over a long period of time. The algorithm developed in this paper advances the state of the art in persistent surveillance by removing the assumption of known event rates. Our simulation experiments show that our approach has potential applications to important real world scenarios such as detection and tracking efforts at a large scale. We conjecture that our algorithm can be extended to persistent surveillance of events in dynamic environments where event statistics are both unknown and time-varying.

Acknowledgments This material is based upon work supported by the Assistant Secretary of Defense for Research and Engineering under Air Force Contract No. FA8721-05-C-0002 and/or FA8702-15-D-0001. Any opinions, findings, conclusions or recommendations expressed in this material are those of the author(s) and do not necessarily reflect the views of the Assistant Secretary of Defense for Research and Engineering.

References

1. Jonathan Binney, Andreas Krause, and Gaurav S Sukhatme. Informative path planning for an autonomous underwater vehicle. In *ICRA*, pages 4791–4796, 2010.
2. Ryan N Smith, Mac Schwager, Stephen L Smith, et al. Persistent ocean monitoring with underwater gliders: Adapting sampling resolution. *Journal of Field Robotics*, 28(5):714–741, 2011.
3. Nathan Michael, Ethan Stump, and Kartik Mohta. Persistent surveillance with a team of mavs. In *IROS*, 2011.
4. Stephen L Smith, Mac Schwager, and Daniela Rus. Persistent robotic tasks: Monitoring and sweeping in changing environments. *T-RO*, 28(2):410–426, 2012.
5. Daniel E. Soltero, Mac Schwager, and Daniela Rus. Generating informative paths for persistent sensing in unknown environments. In *IROS*, pages 2172–2179, 2012.
6. Geoffrey A Hollinger, Sunav Choudhary, Parastoo Qarabaqi, et al. Underwater data collection using robotic sensor networks. *JSAC*, 30(5):899–911, 2012.
7. Xiaodong Lan and Mac Schwager. Planning periodic persistent monitoring trajectories for sensing robots in gaussian random fields. In *ICRA*, pages 2415–2420. IEEE, 2013.
8. Geoffrey A Hollinger and Gaurav S Sukhatme. Sampling-based motion planning for robotic information gathering. In *RSS*, pages 72–983. Citeseer, 2013.
9. Mac Schwager, Daniela Rus, and Jean-Jacques E. Slotine. Decentralized, adaptive coverage control for networked robots. *IJRR*, 28(3):357–375, 2009.
10. Christos Cassandras, Xuchao Lin, and Xuchu Ding. An optimal control approach to the multi-agent persistent monitoring problem. *Automatic Control, IEEE Transactions on*, 58(4):947–961, 2013.
11. Wei-pang Chin and Simeon Ntafos. Optimum watchman routes. *ISO*, 28(1):39–44, 1988.
12. Soroush Alamdari, Elaheh Fata, and Stephen L Smith. Persistent monitoring in discrete environments: Minimizing the maximum weighted latency between observations. *IJRR*, 33(1):138–154, 2014.

13. J. Yu, S. Karaman, and D. Rus. Persistent monitoring of events with stochastic arrivals at multiple stations. *T-RO*, 31(3):521–535, 2015.
14. Daniel E Soltero, Mac Schwager, and Daniela Rus. Decentralized path planning for coverage tasks using gradient descent adaptive control. *IJRR*, page 0278364913497241, 2013.
15. Jesus Capitan, Luis Merino, and Anibal Ollero. Decentralized cooperation of multiple uas for multi-target surveillance under uncertainties. In *ICUAS*, pages 1196–1202, 2014.
16. Ying He and Edwin KP Chong. Sensor scheduling for target tracking in sensor networks. In *ICDC*, volume 1, pages 743–748. IEEE, 2004.
17. Alfred O Hero III, Christopher M Kreucher, and Doron Blatt. Information theoretic approaches to sensor management. In *Foundations and applications of sensor management*, pages 33–57. Springer, 2008.
18. Yoav Gabriely and Elon Rimon. Competitive on-line coverage of grid environments by a mobile robot. *Computational Geometry*, 24(3):197–224, 2003.
19. Jerome Le Ny, Munther A Dahleh, Eric Feron, et al. Continuous path planning for a data harvesting mobile server. In *ICDC*, pages 1489–1494. IEEE, 2008.
20. Aldy Gunawan, Hoong Chuin Lau, and Pieter Vansteenwegen. Orienteering problem: A survey of recent variants, solution approaches and applications. *European Journal of Operational Research*, 2016.
21. Jingjin Yu, Mac Schwager, and Daniela Rus. Correlated orienteering problem and its application to informative path planning for persistent monitoring tasks. In *IROS*, pages 342–349. IEEE, 2014.
22. Güneş Erdoğan and Gilbert Laporte. The orienteering problem with variable profits. *Networks*, 61(2):104–116, 2013.
23. Jingjin Yu, Javed Aslam, Sertac Karaman, et al. Anytime planning of optimal schedules for a mobile sensing robot. In *IROS*, pages 5279–5286. IEEE, 2015.
24. P Ortner and R Auer. Logarithmic online regret bounds for undiscounted reinforcement learning. *NIPS*, 19:49, 2007.
25. Thomas Jaksch, Ronald Ortner, and Peter Auer. Near-optimal regret bounds for reinforcement learning. *JMLR*, 11(Apr):1563–1600, 2010.
26. Kenji Kawaguchi. Bounded optimal exploration in MDP. In *AAAI*, pages 1758–1764. AAAI Press, 2016.
27. Sébastien Bubeck and Nicolo Cesa-Bianchi. Regret analysis of stochastic and nonstochastic multi-armed bandit problems. *arXiv preprint arXiv:1204.5721*, 2012.
28. Peter Auer, Nicolo Cesa-Bianchi, and Paul Fischer. Finite-time analysis of the multiarmed bandit problem. *JMLR*, 47(2-3):235–256, 2002.
29. Peter Auer and Ronald Ortner. UCB revisited: Improved regret bounds for the stochastic multi-armed bandit problem. *Periodica Mathematica Hungarica*, 61(1-2):55–65, 2010.
30. Peter Auer. Using confidence bounds for exploitation-exploration trade-offs. *JMLR*, 3(Nov):397–422, 2002.
31. Michael Short. Improved inequalities for the Poisson and binomial distribution and upper tail quantile functions. *ISRN Probability and Statistics*, 2013.
32. Robert M Corless, Gaston H Gonnet, David EG Hare, et al. On the Lambert W function. *Advances in Computational mathematics*, 5(1):329–359, 1996.
33. Richard P Brent. *Algorithms for minimization without derivatives*. Courier Corporation, 2013.
34. Bohemia Interactive. ARMA 3. http://arma3.com/.
35. Shaoqing Ren, Kaiming He, Ross Girshick, et al. Faster R-CNN: Towards real-time object detection with region proposal networks. In *NIPS*, pages 91–99, 2015.

Planning and Resilient Execution of Policies For Manipulation in Contact with Actuation Uncertainty

Calder Phillips-Grafflin[1] and Dmitry Berenson[2]

[1]Worcester Polytechnic Institute, [2]University of Michigan

Abstract. We propose a method for planning motion for robots with actuation uncertainty that incorporates contact with the environment and the compliance of the robot to reliably perform manipulation tasks. Our approach consists of two stages: (1) Generating partial policies using a sampling-based motion planner that uses particle-based models of uncertainty and simulation of contact and compliance; and (2) Resilient execution that updates the planned policies to account for unexpected behavior in execution which may arise from model or environment inaccuracy. We have tested our planner and policy execution in simulated $SE(2)$ and $SE(3)$ environments and Baxter robot. We show that our methods efficiently generate policies to perform manipulation tasks involving significant contact and compare against several simpler methods. Additionally, we show that our policy adaptation is resilient to significant changes during execution; e.g. adding a new obstacle to the environment.

1 Introduction

Many real-world tasks are characterized by uncertainty: actuators and sensors may be noisy, and often the robot's environment is poorly modelled. Unlike robots, humans effortlessly perform everyday tasks, like inserting a key into a lock, which require fine manipulation despite limited sensing and imprecise actuation. We observe that humans often perform these tasks by exploiting *contact*, *compliance*, and *resilience*. Using compliance to safely make contact and move while in contact allows us to reduce uncertainty. We also exhibit resilience: when an action fails to produce the desired result, we may withdraw and try again. Seminal motion planning work by Lozano-Pérez et al. [1] shows that incorporating contact and compliance is critical to performing fine motions like inserting a peg into a hole. Building from this work and our observations of human motions, we have developed a motion planner that incorporates contact, compliance, and resilience to generate behavior for robots with actuation uncertainty.

Motion in the presence of actuation uncertainty is an example of a continuous Markov Decision Process (MDP), adding in sensor uncertainty, the problem becomes a Partially-Observable Markov Decision Process (POMDP). Solving an MDP or POMDP is often framed as the problem of computing an optimal policy

© Springer Nature Switzerland AG 2020
K. Goldberg et al. (Eds.): *Algorithmic Foundations of Robotics XII*, SPAR 13, pp. 752–767, 2020.
https://doi.org/10.1007/978-3-030-43089-4_48

π^* that maps each state to an action a that maximizes the expected reward (e.g. the probability of reaching the goal). This paper focuses on motion planning with actuation uncertainty, and thus we frame the problem as an MDP. This MDP formulation is representative of the challenges face by low-cost and compliant robots such as Baxter or Raven, which have accurate sensing but noisy actuators.

Instead of planning in the configuration or state-space of the robot, we represent the uncertainty of the state of the robot as a probability distribution over possible configurations, and plan in the space of these distributions—the *belief space*[1]. The computational expense of optimal motion planning leads us to adopt a thresholding approach from *conformant* planning [2]. Instead of attempting to find a global optimal policy, we seek to generate a *partial* policy that allows a robot with actuation uncertainty to move from start configuration q_{start} to reach goal q_{goal} within tolerance ϵ_{goal} with at least planning threshold P_{goal} probability. A partial policy, which maps a subset of possible states to actions rather than a global policy that maps all states to actions, simplifies the problem and is appropriate for the single-query planning problems we seek to solve.

The complexity of robot kinematics and dynamics preclude analytical modeling of compliance and contact for practical, high-dimensional problems, and thus we rely on the ability to forward simulate the state of the robot given a starting state and action. In the presence of uncertainty, individual actions may have multiple distinct outcomes: for example, when trying to insert a key into a lock, some attempts will succeed in inserting the key, while some will miss the keyhole. In advance of performing such an action, we cannot *select* between desired outcomes (as is assumed in [3]). However, we can *distinguish* between the outcomes after the action is executed. We directly incorporate this behavior into our planner using *splits* and *reversibility*. Splits are single actions that produce multiple distinct outcomes, which we distinguish between using a series of clustering algorithms. Reversibility is the ability of a specific action and outcome to be "undone" and return to the previous state, which allows the robot to attempt the action again. Of course, the planner may not accurately model the outcomes of every action, so we incorporate an online adaptation process to update the planned policy during execution to reflect the results of actions.

Our primary contributions are thus 1) incorporating contact and compliance into policy generation, thus allowing contacts that other planners would discard but that, in fact, can be used to reduce uncertainty; and 2) introducing resilience into policy execution and thus significantly increasing the probability of successfully completing the task. Our experiments with simulated test environments suggest that our planner efficiently generates policies to reliably perform motion for robots with actuation uncertainty. We apply our methods to problems in $SE(2)$, $SE(3)$, and a simulated Baxter robot ($\mathbb{R}^7$) and show performance improvements over simpler methods and the ability to recover from an unanticipated blockage.

---

[1] The term *belief* is borrowed from POMDP literature, which assumes that the state is partially-observable. Though this paper considers only MDPs, we nevertheless use "belief" as it is a convenient and widely-used term for a distribution over states.

2   Related Work

Planning motion in the presence of actuation uncertainty dates back to the seminal work of Lozano-Pérez et al. [1], which introduced pre-image backchaining. A pre-image, i.e. a region of configuration space from which a motion command attains a certain goal recognizably, was used in a planner that produced actions guaranteed to succeed despite pose and action uncertainty. However, constructing such pre-images is prohibitively computationally expensive [4, 5].

In its general form, belief-space planning is formulated as a Partially-Observable Markov Decision Process (POMDP), which are widely known to be intractable for high-dimensional problems. However, recent developments of general approximate point-based solvers such as SARSOP [6] and MCVI [7] have made considerable progress in generating policies for complex POMDP problems. For some lower-dimensional robot motion problems like [8], the POMDP can be simplified by extracting the part of the task that incorporates uncertainty (e.g. the position of an item to be grasped) and applying off-the-shelf solvers to the problem. Others have investigated learning approaches [9] for similar problems; however, we are interested in planning because we want our methods to generalize to a broad range of tasks without collecting new training data.

Several sampling-based belief-space planners have been developed [3, 10–13]. Others have evaluated the belief-space distance functions [14] and show that the selection of distance function greatly impacts the performance of the planner. Additionally, approaches using LQG and LQR controllers [12, 15, 16] and trajectory optimizers [17, 18] have been proposed. These approaches use Gaussian distributions to model uncertainty, but such a simple distribution cannot accurately represent the belief of a robot moving in contact with obstacles, where belief may lose support in one or more dimensions, or the state may become trans-dimensional. Other approaches like [3] use a set of particles to model belief like a particle filter; while we also use a particle-based representation, our approach more accurately captures the behavior of splits and also includes resilience during execution.

The importance of compliance has long been known, with [1] demonstrating the important role of compliance in performing precise motion tasks. Sampling-based motion planning for compliant robots has been previously explored in [19], albeit limited to disc robots with simplified contact behavior. We draw from these methods, but our approach differs significantly from previous work by incorporating contact and compliance directly into the planning process by using forward simulation like the kinodynamic RRT [20]. A major advantage over existing methods is that the policies we generate are not fixed; instead, we update them online during execution, which allows us to reduce the impact of differences between our planning models and real-world execution conditions.

3   Problem Statement

We consider the problem of planning motion for a *controlled compliant* robot R with configuration space Q in an environment with obstacles E. For given start

(q_{start}) and goal (q_{goal}), we seek to produce motion which allows the robot to reach q_{goal} within tolerance ϵ_{goal} with at least P_{goal} probability.

The robot is assumed to have actuation uncertainty modelled by $q_{t+1} = q_t + (\varDelta q_t + r_{\varDelta q})$ in which the next configuration q_{t+1} is the result of the previous configuration q_t, control input $\varDelta q$ and actuation error $r_{\varDelta q}$. We assume that a function $\mathbf{F}$, which models the probability distribution of the uncertainty, is available from which to sample $r_{\varDelta q} \sim \mathbf{F}(\varDelta q)$ for a given $\varDelta q$. Due to this actuation uncertainty, when executing actions in our planner the result is a belief distribution b. The robot is compliant, meaning that for a motion from collision-free $q_{current}$ to colliding $q_{desired}$, the resulting configuration q_{result} will be in contact and the robot will not damage itself or the environment.

Since the motion of the robot is uncertain, a path τ that is a discrete sequence of configurations may not be robust to errors. Instead, we wish to produce a partial policy $\pi : \mathcal{Q}' \to A$ that maps $\mathcal{Q}' \subseteq \mathcal{Q}$ to actions A such that for a configuration $q \in \mathcal{Q}'$, the policy returns an action to perform. Even π may not always be robust to unexpected errors, therefore during execution we wish to detect actions that do not reach their expected results; i.e. when an action produces $q_{result} \notin \mathcal{Q}'$. In such an event, we wish to adapt $\mathcal{Q}'$ and π such that $q_{result} \in \mathcal{Q}'$ and continue attempting to complete the task.

4   Methods

We have developed a motion planner consisting of an anytime RRT-based global planner and a local planner that uses a kinematic simulator to model robot behavior. Together, they produce a set of solution paths S, where each solution $s \in S$ is a sequence of nodes $n_i = (b_i, a_i)$, in which b_i is the belief distribution for n_i and a_i is the action that produced b_i. Using this set of solution paths, we construct a single partial policy π. As π is queried during execution, we update the policy to reflect the "true" state observed during the execution process.

Because it is difficult to model the belief state in contact using a parametric distribution, we use a particle-based approach similar to [3] in which we represent the belief b_i of node n_i with a set of configurations $q_1, q_2, ..., q_n$ that are forward-simulated by the local planner. Like previous work [3], we expect that performing some actions will result in multiple qualitatively different states as illustrated in Figure 1 (e.g. in contact with an obstacle some particles will become stuck on the obstacle while others slide along the surface). These distinct parts of the belief state, which we refer to as *splits*, are distinguished in our planner by a series of clustering operations. To ensure that all actions are adequately modeled, a fixed number of particles $N_{particles}$ is used to simulate every action; since splits reduce the number of particles at a given state, a new set of particles must be resampled for these states to avoid particle starvation.

It is important to understand that we cannot select between the different result states of a split when performing the action; however, we can *distinguish* using our clustering methods if we have reached an undesirable result. To be *resilient* to such errors, we incorporate the ability to reverse the action back

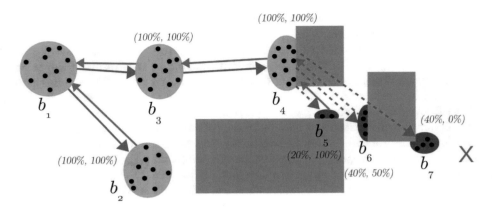

Fig. 1. Our belief-space RRT extending toward a random target (red X) from b_4. Due to compliance, the particles (dots) can slide along the obstacles (gray). Solid blue edges denote 100% probability edges, dashed edges denote a split resulting in multiple states; solid magenta edges denote 100% reversible edges, while dashed edges denote lower reversibility. Because the extension is attempting to move through a narrow passage, particles separate and a split occurs, resulting in three distinct states (b_5, b_6, b_7).

to the previous state and try the action again. Clearly, not all actions will be reversible, so we perform additional simulation to estimate the ability to reverse each action after identifying the resulting states.

We first introduce our RRT-based global planner that uses a simulation-based local planner and a series of particle clustering methods to generate policies incorporating actuation uncertainty, and then discuss our online policy execution and adaptation that enables resiliency to unexpected behavior encountered during execution.

4.1   Global planner

Until it reaches time limit $t_{planning}$, our global planner iteratively grows a tree T using the local planner to extend the tree towards a sampled configuration q_{target}. Like the RRT, q_{target} is either a uniformly sampled $q_{rand} \in \mathcal{C}$, or with some probability, the goal q_{goal}. Each time we sample a q_{target}, we select the closest node in the tree $n_{near} = \text{argmin}_{n_i \in T} \text{PROXIMITY}(b_i, q_{target})$. The local planner plans from n_{near} towards q_{target} and returns new nodes $\mathcal{N}_{new}$ and edges $\mathcal{E}_{new}$ that grow the tree. We check each new node $n_{new} \in \mathcal{N}_{new}$ to see if it meets the goal conditions, and if so, add the new solution path to S.

We also incorporate several features distinct from the RRT. First, using PROXIMITY we consider more than distance when selecting the nearest neighbor node n_{near}. We want to bias the growth of the tree toward nodes that can be reached with higher probability and have more concentrated b_i, so we incorporate weighting using $P(n_{start} \rightarrow n_i)$, the probability the entire path from n_{start}

to n_i succeeds, and var(n_i), the variance of b_i. The proximity of a node n_i to a configuration q is given by the following equation:

$$\textsc{Proximity}(n_i, q) = \text{dist}(\text{expect}(b_i), q)$$
$$* [(1 - P(n_{start} \to n_i)) * \alpha_P + (1 - \alpha_P)][\text{erf}(|\text{var}(b_i)|_1) * \alpha_V + (1 - \alpha_V)] \quad (1)$$

Here, $\text{expect}(b_i) = q_{expected}$ is the expected value of the belief distribution b_i and $\text{dist}(q_{expected}, q)$ is the $\mathcal{C}$-space distance function. Two weights α_P and α_V control the effect of the probability and variance weighting, respectively. Values of α_P and α_V closer to 1 increase the effect of the weighting, while values closer to 0 increase the effect of the $\mathcal{C}$-space distance. Using the error function $\text{erf}(x) = 2/\sqrt{\pi} \int_0^x e^{-t^2} dt$ maps variance in the range $[0, \inf]$ to the range $[0, 1)$ to simplify computation. Previous work in belief-space planning has used a range of distance functions, such as L1, Kullback-Leibler divergence, Hausdorff distance, or Earth Mover's Distance (EMD) [14]; however, many of these choices only provide useful distances between belief states with overlapping support. While EMD encompasses both the $\mathcal{C}$-space distance and probability mass of two beliefs, it is expensive to compute. Since most of our distance computations are between beliefs with non-overlapping support, the $\mathcal{C}$-space distance between expected configurations is an efficient approximation [14].

Second, we cannot simply test if $n_{new} = q_{goal}$, since the $P(n_{start} \to n_{new})$ may be low; instead, we check if a new solution has been found. To be a solution, the probability n_{new} reaches the goal must be greater than P_{goal}, i.e. the product of $P(n_{start} \to n_{new})$ and $|q \in b_{new}|\text{dist}(q, q_{goal}) \leq \epsilon_{goal}|$. Finally, once a path to the goal has been found, we continue planning to find alternative paths. We want to encourage a diverse range of solutions, so once a solution path has been found, we remove nodes on solution branches from consideration for nearest neighbor lookups. This process recurses towards the root of the planner's tree T until it either reaches the root node n_{start} or a node n_i which is the result of a split. Once the base of the solution branch is found, we remove the branch from nearest neighbors consideration and continue planning until reaching $t_{planning}$.

4.2   Local planner

Our local planner grows the planner tree T from nearest neighbor node n_{near} towards a target configuration q_{target} by forward-propagating belief using EXTEND to produce one or more result nodes $n_{new} \in \mathcal{N}_{new}$ and edges $e_{new} \in \mathcal{E}_{new}$ (recall that splits may occur). To improve the time-to-first-solution, the local planner operates like RRT-Connect, repeatedly calling EXTEND, until a solution is found, whereupon it switches to RRT-Extend, calling EXTEND only once, to improve coverage of the space and encourage solution diversity. Note that the RRT-Connect behavior is stopped if an extension results in a split.

EXTEND forward-simulates particles $Q_{initial}$ from node n_{near} towards q_{target}, clusters the resulting particles $Q_{results}$ into new nodes $\mathcal{N}_{new}$, and computes the transition probabilities. As previously discussed, we simulate every action with the same number of particles. If node n_{near} is *not* the result of a split, and thus b_{near} contains a full set of particles, then we simply copy b_{near} to use for simulation. If, instead, n_{near} is the result of a split, then we uniformly resample $N_{particles}$ particles from b_i. We then simulate the extension toward q_{target} for each particle. Any simulation engine that simulates contact and compliance could be used, but the simulation should be as fast as possible to minimize planning time. In our experiments, we used an approximate kinematic simulator described in the expanded version of this paper [21]. The resulting particles are then grouped into one or more clusters using CLUSTERPARTICLES, which we describe in Section 4.3. For each cluster $Q_{cluster}$, we form a new node $n_{new} = (b_{new}, a_{new})$ with belief $b_{new} = Q_{cluster}$ and action $a_{new} = q_{target}$. In the case of splits, where multiple nodes are formed, we assign $P(n_{near} \to n_{new,i}) = |b_{new,i}|/N_{particles}$. We then estimate the probability that action a_{new} can be reversed from node n_{new} by simulating $N_{particles}$ particles back towards node n_{near}. Note that some particles may become stuck while reversing, and thus the probability of reversing the action may not be 1.

The ability to reverse an action allows us to detect an undesired outcome, reverse to the parent node, and retry the action until we either reach the desired outcome or become stuck. Thus, we estimate the *effective* probability $P(n_{near} \to n_{new})_{effective}$ for each node n_{new} by estimating the probability that a particle has reached n_{new} after $N_{attempt}$ attempts, where at each attempt, particles that have not reached the n_{new} try to return to n_{near} and try again.

Analysis – The planner always stores $N_{stored} = N_{actions} N_{particles}$ particles. For every action $N_{particles}$ particles are forward-simulated, and all of them are stored in $\mathcal{N}_{new}$. In the worst case, where every action produces $N_{particles}$ distinct nodes, the number of particles that must be simulated $N_{simulated} = N_{nodes}(N_{particles} + 1)$, as each node itself is the product of one initial simulation and $N_{particles}$ simulations are required to estimate reverse probability. In practice, as we discuss in Section 5.1, the space requirements to perform complex tasks are low as most actions produce a small number of nodes, and the time cost can be reduced by simulating particles in parallel.

4.3   Particle clustering

Intuitively, we want every configuration in a cluster to be reachable from every other one using the local planner. However, testing this directly is computationally expensive, so we also consider two approximate methods. All clustering methods use two successive passes to cluster the configurations resulting from forward simulation: first, a spatial-feature-based pass that groups configurations based on their relationship to different parts of the workspace, and second, a distance-based pass that refines the initial clusters. All of our clustering methods use complete-link hierarchical clustering, as it produces smaller, more dense clusters, while not requiring the number of clusters to be known in advance [22, 3]. Below we discuss the ideal approach and our two approximations, shown in Figure 2. We compare the performance of these methods in Section 5.1.

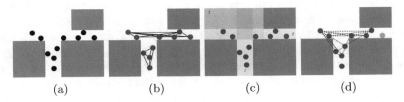

Fig. 2. Our proposed spatial-feature particle clustering methods. (a) The positions of particles after an extension of the planner. (b) Actuation center clustering, with clusters (red, blue) and the straight-line paths for each cluster. (c) Weakly Convex Region Signature clustering, with the three convex regions shown and labeled. (d) Particle movement clustering, with successful particle-to-particle motions shown dashed for the main cluster (red) and two unconnected particles (blue, green).

Particle Connectivity (PC) Clustering We run the local planner from every configuration to every other configuration and record which simulations reach within ϵ_{goal} of the target. For a pair of configurations q_1, q_2, going from q_1 to q_2 may fail while the opposite succeeds; however, to be conservative, we only record success if both executions succeed. We then perform clustering using the complete-link clustering method with distance threshold 0, where successful simulations correspond to distance 0 and unsuccessful simulations correspond to distance 1. Note that this method is very expensive, since it requires simulating $N^2 - N$ particles for N configurations considered.

Weakly Convex Region Signature (WCR) Clustering Intuitively, in many environments a robot can move freely from q_1 to q_2 if both configurations reside entirely in the same convex region of the workspace. This is also true for some slight concave features, so long as the features do not block the robot. Conversely, for configurations in clearly distinct regions, it is less likely that the robot can move from one configuration to the other.

Illustrated in Figure 2c, we capture this intuition by recording the position of the robot relative to *weakly convex regions* of the free workspace, to form what we call the *convex region signature*. These regions form a weakly convex covering: individual regions may contain slight concavity, and multiple regions overlap. Techniques such as [23] exist to automatically compute these regions, but for simple environments these regions can be directly encoded. The convex region signature of a configuration q, $WCR(q)$, records the region(s) occupied by every point of the robot at q. Distance metric D_{WCR} between two region signatures $WCR(q_1)$ and $WCR(q_2)$ is the percentage of points in the robot that do *not* share a common region between the signatures. Using this metric, we perform complete-link clustering. We test different thresholds for D_{WCR} in Section 5.1. This method allows configurations with some points in a shared region to be clustered together, while separating configurations that share no regions. At runtime, this method requires N computations of $WCR(q)$ and $(N^2-N)/2$ evaluations of D_{WCR} to compute all pairwise distances.

Actuation Center (AC) clustering We observe that many successful motions in contact occur when the actuation (or joint) centers of the starting and ending configuration can be connected by collision-free straight lines, so this method checks the straight-line path from the joint centers of one configuration to those of the other configuration. As with the particle movement clustering approach, configurations with successful (collision-free) paths have distance 0, while those with unsuccessful (colliding) paths have distance 1. Like the previous approach, clusters are then produced using complete-link clustering with threshold 0. At runtime, this method requires $(N^2-N)/2$ checks of the straight-line paths.

4.4   Partial policy construction

Once the global planner has produced a set of solution paths S, we construct a partial policy π. Policy construction consists of the following steps:

1. Graph construction – An explicit graph is formed, in which the vertices of the graph are nodes $n_i \in S$, and the edges correspond to the edges forming the paths in S. An edge $n_i \to n_{i+1}$ is assigned an initial cost $1/P(n_i \to n_{i+1}))$. This means that likely edges receive low cost, which is necessary to compute maximum-probability paths through the graph.

2. Edge cost updating – The edge costs are updated to reflect the estimated number of attempts needed to successfully traverse the edge by multiplying the cost of the edge by the estimated number of attempts required to reach $P(n_i \to n_{i+1}) \geq P_{goal}$. This estimate is the complement of the effective probability discussed in Section 4.2; instead of computing the probability of reaching a node after a fixed number of attempts, we compute the number of attempts needed to reach the node with P_{goal} probability. The fewer attempts necessary to traverse the edge, the faster the policy can be executed, and thus this cost represents an expected execution time.

3. Dijkstra's search – The optimal path from every vertex in the graph to the goal state is computed using Dijkstra's algorithm. This determines the optimal next state (and thus action to perform) for every state in the graph.

4.5   Partial policy execution and adaptation

At every step during execution, the partial policy π is queried for the next action to perform. While we could simply find the "closest" node in the policy using a distance function like Equation 1, doing so would discard important information. Not only do we know the configuration $q_{current}$ that results from executing an action, but we also know the action $a_{performed}$ we attempted to perform. Using this information, we know exactly which nodes(s) in π the robot should have reached. As shown in Algorithm 1, we first collect all potential result nodes (i.e. those nodes n_i with actions $a_i = a_{performed}$). We then use our particle clustering method to cluster $q_{current}$ with the belief b_i of each n_i. This clustering tells us if the robot reached a given state (if a single cluster is formed) or not (multiple clusters). In the unlikely (but possible) event that $q_{current}$ clusters with multiple

---

Algorithm 1 Partial policy query algorithm

---

procedure POLICYQUERY$(S, \pi, q_{current}, a_{performed})$
    $\mathcal{N}_{potential} \leftarrow \{n_i \in S \mid a_i = a_{performed}\}$
    $\mathcal{N}_{matching} \leftarrow \{n_i \in \mathcal{N}_{potential} \mid |ClusterParticles(b_i \cup q_{current})| = 1\}$
    if $\mathcal{N}_{matching} \neq \emptyset$ **then**
        $n_{reached} \leftarrow argmin_{n_i \in \mathcal{N}_{matching}} \text{DijkstraDistance}(n_i)$
        INCREASEPROBABILITY$(n_{reached}, a_{performed})$;
        for $n_i \in \mathcal{N}_{potential} \mid n_i \neq n_{reached}$ **do**
            REDUCEPROBABILITY$(n_i, a_{performed})$
        $\pi \leftarrow$ CONSTRUCTPOLICY(π)
        if $P(n_{reached} \rightarrow q_{goal}) \geq P_{goal}$ **then**
            $a_{next} \leftarrow \pi(n_{reached})$
            return a_{next}
        else
            return failure
    else
        $n_{observed} \leftarrow \{\{q_{current}\}, a_{performed}\}$
        $S \leftarrow S \cup n_{observed}$
        return POLICYQUERY$(S, \pi, q_{current}, a_{performed})$

---

potential result nodes, we select the "best" matching node $n_{reached}$ using the distance-to-goal computed via Dijkstra's algorithm.

The key contribution of our policy execution is that we adapt the policy π to reflect the results of actual execution. If a matching node $n_{reached}$ is found, we then update π to increase the probability that $n_{reached}$ is the result of $a_{performed}$. We assign a constant $A_{importance} \in \mathbb{N}$ that reflects how much we value the results of executing an action compared to the results of simulating a particle during planning. To update the probability, we increase the counts of attempted $N_{attempts}$ actions and successful $N_{successful}$ actions, then recompute probability:

$$P(n_{previous} \rightarrow n_{reached}|a) = \frac{N_{successful} + A_{importance}}{N_{attempts} + A_{importance}} \qquad (2)$$

Likewise, we reduce the probability for other potential result states:

$$P(n_{previous} \rightarrow n_{other}|a) = \frac{N_{successful}}{N_{attempts} + A_{importance}} \qquad (3)$$

This update process allows us to learn online, during execution, the true probabilities of reaching states given an action. In effect, the probabilities computed by the global planner serve as an initialization for this online learning. Once updated, we rebuild policy π to reflect the new probabilities. If the probability of reaching the goal $P(n_{reached} \rightarrow q_{goal})$ is at least P_{goal}, we query π for the next action to take. If the probability of reaching the goal has dropped below P_{goal}, policy execution terminates.

However, sometimes no matching node $n_{reached}$ exists. This means a split occurred during execution that was not captured in S during planning (e.g. an obstacle that is not accurately modelled in E, or where the behavior of the simulator diverges from the true robot). To handle this case, we insert a new node $n_{observed}$ with belief $b_{observed} = \{q_{current}\}$ into S, and then retry the policy query (which will now have an exactly matching state). To incorporate reversibility, we initially assign new nodes a reverse probability $N_{attempts} = N_{successful} = 1$. Thus, the next action selected by the policy will be to return to the previous node. Together with updating probabilities by inserting new states in this manner, we can thus extend the policy to reflect behavior observed during execution that was not captured during the planning process.

Analysis – In the worst case, a policy π cannot be executed successfully, and performing every action a results in a new node $n_{observed}$. For any $A_{importance} \in \mathbb{N}, P_{goal} > 0$, adapting the policy will detect failure and terminate in this case.

Proof – For every action $a_{i+1}, ...,$ node $n_{observed}$ will be created with a reverse prior $P(n_{observed} \rightarrow n_{previous}) = 1/1$. If reversing to $n_{previous}$ fails, we update $P(n_{observed} \rightarrow n_{previous}) = 1/1+A_{importance}$. For the ith successive failed reverse and $n_{observed,i}$ generated, $P(n_{observed,i} \rightarrow n_{previous}) = \Pi_i \frac{1}{1+A_{importance}}$. As the number of failed actions increases $P(n_{observed,i} \rightarrow n_{previous}) \rightarrow 0$, and thus $P(n_{observed,i} \rightarrow q_{goal}) \leq P(n_{observed,i} \rightarrow n_{previous}) \rightarrow 0$. Thus eventually $P(n_{observed,i} \rightarrow q_{goal})$ will fall below $P_{goal} > 0$ and execution will terminate. $\square$

5   RESULTS

We present results of testing our planner in simulated $SE(2)$ and $SE(3)$ environments and a simulated Baxter robot. For dynamic simulation during execution, we use the Gazebo simulator. As our kinematic simulator does not consider friction, we use *contact motion controllers* to reduce contact forces (see the expanded paper [21] for more details). We present statistical results over a range of actuation uncertainty and clustering methods and show that our planner produces policies that allow execution of tasks incorporating contact and robot compliance. We also present statistical results showing that our online policy updating adapts to unexpected behavior during execution. All planning and simulation testing was performed using 2.4 GHz Xeon E5-2673v3 processors. Likewise, all planning was performed with PROXIMITY weights $\alpha_P = \alpha_V = 0.75$ (see Equation 1), $N_{attempt} = 50$ attempted reverse/repeats of each action, and planning threshold $P_{goal} = 0.51$, such that solutions must be more likely than not to reach the goal.

5.1 $SE(3)$ simulation

Peg-in-hole In $SE(3)PegInHole$, a version of the classical peg-in-hole task [1] shown in Figure 3, the free-flying 6-DoF robot "peg" must reach the bottom of the hole. This task is difficult for robots with actuation uncertainty, as the hole is only 30% wider than the peg. Even without uncertainty, attempting to avoid

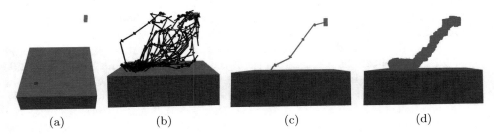

(a)   (b)   (c)   (d)

Fig. 3. (a) The $SE(3)PegInHole$ task involves moving from the start (red) to the bottom of the hole. (b) An example policy produced from 296 solutions, the (c) initial action sequence (blue arrows), actions the policy will return if every action is successful, and (d) the swept volume of the peg executing the policy. Note that the peg makes contact with the environment to reduce uncertainty, then slides into the hole.

	AC	PC	WCR with $D_{WCR} =$				
			0.125	*0.25*	*0.5*	*0.75*	*0.99*
P_{plan}	1.0	1.0	1.0	0.97	0.97	0.97	0.97
P_{exec}	0.97 [0.17]	0.89 [0.19]	0.73 [0.42]	0.95 [0.18]	0.84 [0.34]	0.99 [0.02]	0.96 [0.18]

Table 1. $SE(3)PegInHole$ particle clustering performance comparison (mean [std.dev.]) of P_{exec}, the probability of reaching the goal with 300 seconds, between policies produced using our planner with different clustering methods. P_{plan} is the probability that a policy is planned within 5 minutes, averaged over 30 plans, and P_{exec} is averaged over 40 executions on each successfully-planned policy.

	Simplified			Planned policies (WCR, $D_{WCR} = 0.75$)			
	Simple RRT		Contact RRT	24 particles		48 particles	
γ	P_{plan}	P_{exec}	P_{plan}	P_{exec}	P_{plan}	P_{exec}	P_{plan}	P_{exec}
0	0	0 [0]	1	0.78 [0.38]	1	0.42 [0.48]	0.97	0.59 [0.47]
$1/16$	0	0 [0]	1	0.78 [0.39]	1	0.60 [0.43]	1	0.625 [0.43]
$1/8$	0	0 [0]	1	0.79 [0.38]	1	0.99 [0.18]	0.93	0.81 [0.37]
$1/4$	0	0 [0]	1	0.50 [0.37]	1	0.90 [0.28]	0.86	0.72 [0.41]

Table 2. $SE(3)PegInHole$ policy performance comparison between simplified planners and our planner with 24 and 48 particles and actuation uncertainty γ.

contact greatly restricts the motion of the robot entering the hole. Instead, as shown in [1], the best strategy is to use contact with the environment and the compliance of the robot to guide the peg into the hole. We assess the performance of a policy approach in terms of P_{exec}, the probability that executing the policy reaches the goal within a time limit of 300 seconds. For a given value of γ, linear velocity uncertainty $\gamma_v = \gamma$ (m/s) and angular velocity uncertainty $\gamma_\omega = 1/4\gamma$ (rad/s). Linear and angular velocity noise is sample from a zero-mean truncated normal distribution with bounds $[-\gamma_{v,\omega}, \gamma_{v,\omega}]$ and standard deviation $1/2\gamma_{v,\omega}$. While this differs from zero-mean normal distributions conventionally used to model uncertainty, we believe the bounded truncated distribution better reflects the reality of robot actuators, which do not exhibit unbounded velocity error. Goal distance threshold ϵ_{goal} was set to $1/2$ the length of the peg.

We first compared the performance of our planner at a fixed $\gamma = 1/8$ and $N_{particles} = 24$ using the clustering approaches introduced in Section 4.3, including several thresholds for $D_{WCR} = 0.125, 0.25, 0.5, 0.75, 0.99$, with 30 plans per approach (5 minutes planning time) and 40 executions of each planned policy. As seen in Table 1, WCR clustering with $D_{WCR} = 0.75$ clearly outperformed the others in terms of policy success, reaching the goal in 99% of executions. Planning time is overwhelmingly dominated by simulation, accounting for approximately 99.9% of the allotted time. Using WCR and $D_{WCR} = 0.75$, we then compared the performance of our planner against two simplified RRT-based approaches:

1. Simple RRT – Does not model uncertainty or allow contact, but like our planner produces multiple solutions in the allotted planning time.
2. Contact RRT – Incorporates contact and compliance but does not model uncertainty. Equivalent to planning with $\gamma = 0$ and one particle.

In addition, we tested our planner with both 24 and 48 particles to show the effects of increasing the number of particles used. As before, we planned 30 policies for each, and executed each planned policy 40 times. Note that the Simple RRT was unable to produce solutions in 5 minutes due to the confined narrow passage. Results are shown in Table 2. With low actuator error the Contact RRT performs better, as it does not expend planning time on simulating multiple particles and instead produces more solutions. As error increases, our planner clearly outperforms the alternatives. Note that increasing particles does not improve performance, indicating that 24 particles is sufficient without requiring unnecessary simulation. Low P_{exec} overall, in particular when planning with low γ, is due to the mismatch between the planning simulator and the dynamics of Gazebo (i.e. motions that are possible in the planner, but not in Gazebo) which disproportionally affects motions near the entrance to the hole. In particular, the planner at low values of γ overestimates how successfully motions at the edge of the hole can be performed and thus results in a lower-than-expected P_{exec}.

In terms of the number of particles stored, the worst case was $N_{particles} = 48, \gamma = 0$, with an average of 148894 (std.dev. 25015) particles stored. The worst case for simulated particles was $N_{particles} = 24, \gamma = 1/4$, with an average 268634 (std.dev. 16856) particles simulated. In practice, the storage and computational expense is limited; the worst-case for particles stored requires a mere 15 megabytes, while for a planning time of 300 seconds and using eight threads, the planner evaluated more than 100 particles per second per thread.

5.2   Baxter simulation

In addition to $SE(3)$ and $SE(2)$ tests, we apply our planner and policy execution to a simulated Baxter robot shown in Figure 4, with the robot reaching into a confined space. We compare the performance of our planner with $N_{particles} = 24$

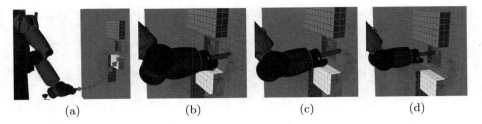

(a)   (b)   (c)   (d)

Fig. 4. An execution of the Baxter task, from start (a) to goal (d), and environment with confined space around the goal. The planned policy is shown in blue. Note the use of contact with the environment to reduce uncertainty and reach the target passage.

and WCR clustering method with $D_{WCR} = 0.1$ with the simplified Contact RRT in terms of success probability P_{exec} for uncertainty $\gamma = 0.1$. To simulate Baxter's actuation uncertainty γ defines a truncated normal distribution with $\sigma = 1/2\gamma\dot{q}_i$ and bounds $[-\gamma\dot{q}_i, \gamma\dot{q}_i]$ for each joint i with velocity $\dot{q}_i$. Goal distance threshold $\epsilon_{goal} = 0.15$ radians. We generated 10 policies using each approach with a planning time of 10 minutes to ensure both approaches would produce multiple solutions, then executed each 8 times for up to 5 minutes. As expected, Contact RRT finds solutions faster; on average 8.42s (std.dev. 2.61) versus 65.4s (std.dev. 32.9) and policies contain more solutions; on average 17.2 (std.dev. 16.5) versus 6.33 (std.dev. 3.97) since each solution requires less simulation time. However, our planner incorporating uncertainty outperforms the Contact RRT baseline with $P_{exec} = 0.79$ (std.dev. 0.30) versus $P_{exec} = 0.70$ (std.dev. 0.30). This suggests that, while planning with uncertainty does help in this environment, our approach to policy execution and resilience also works well when uncertainty is not accounted for in the planner, but we have a diverse policy.

5.3   Policy adaptation

We performed tests in a planar $SE(2)$ (3-DoF) environment to show that our policy adaptation recovers from unexpected behavior during execution. As shown in Figure 5, the L-shaped robot attempts to move from the start (upper left) to the goal (lower right). Due to the obstacles present, there are three distinct horizontal passages. Using the same controllers and uncertainty models as the $SE(3)$ tests with uncertainty $\gamma = 0.125$ and WCR clustering method with $D_{WCR} = 0.75$, 24 particles, and a planning time of 2 minutes, we generated 30 policies using our planner. Goal distance threshold ϵ_{goal} was set to 1/8 the length of the robot.

We evaluated the performance of the planned policies in the unmodified environment and an environment in which we blocked the horizontal passage used by the initial path of each policy. Each was executed 8 times for a total of 240 policy executions, for a maximum of 600 seconds. In the unmodified environment, 97% of policies were executed successfully, with an average of 15.4 actions

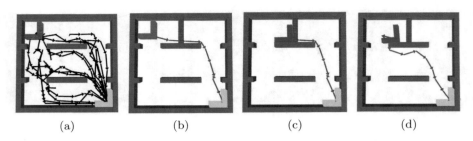

(a)   (b)   (c)   (d)

Fig. 5. (a) Our planar test environment, in which the robot must move from upper left (red) to lower right (green), with an example policy produced by our planner, with solutions through each of the horizontal passages. (b) The initial action sequence (blue arrows), showing actions the policy will return if every action is successful. (c) Following the policy, the robot becomes stuck on the new obstacle (gray). (d) Once the policy detects the failed action, it adapts to avoid the obstacle.

(std.dev. 9.62). In the modified environment with policy execution importance $A_{importance} = 500$ (this high value results in rapid policy adaptation), 73% of policies were executed successfully, with an average of 26.7 actions (std.dev. 12.3). This result shows that adapting the policy using our methods allows us to circumvent the new obstacle, however, doing so may result in following a path that is less likely to succeed.

6   Conclusion

We have developed a method for planning motion for robots with actuation uncertainty that incorporates environment contact and compliance of the robot to reliably perform manipulation tasks. First, we generate partial policies using an RRT-based motion planner that uses particle-based models of uncertainty and kinematic simulation of contact and compliance. Second, we adapt planned policies online during execution to account for unexpected behavior that arises from model or environment inaccuracy. We have tested our planner and policy execution in simulated $SE(2)$ and $SE(3)$ environments and on the simulated Baxter robot and show that our methods generate policies that perform manipulation tasks involving significant contact and compare against two simpler methods. Additionally, we show that our policy adaptation is resilient to significant changes during execution; e.g. adding a new obstacle to the environment.

Acknowledgements This work was supported in part by NSF grants IIS-1656101 and IIS-1551219.

References

1. Lozano-Prez, T., Mason, M.T., Taylor, R.H.: Automatic synthesis of fine-motion strategies for robots. IJRR **3**(1) (1984) 3–24

2. Goldman, R.P., Boddy, M.S.: Expressive planning and explicit knowledge. In: Artificial Intelligence Planning Systems. (May 1996)

3. Melchior, N.A., Simmons, R.: Particle rrt for path planning with uncertainty. In: ICRA. (April 2007)

4. Canny, J.: On computability of fine motion plans. In: ICRA. (May 1989)

5. Erdmann, M.: Using backprojections for fine motion planning with uncertainty. The International Journal of Robotics Research **5**(1) (1986) 19–45

6. Kurniawati, H., Hsu, D., Lee, W.S.: Sarsop: Efficient point-based pomdp planning by approximating optimally reachable belief spaces. In: RSS. (2008)

7. Bai, H., Hsu, D., Kochenderfer, M., Lee, W.S.: Unmanned aircraft collision avoidance using continuous-state pomdps. In: RSS. (June 2011)

8. Koval, M., Pollard, N., Srinivasa, S.: Pre- and post-contact policy decomposition for planar contact manipulation under uncertainty. In: RSS. (July 2014)

9. Levine, S., Wagener, N., Abbeel, P.: Learning contact-rich manipulation skills with guided policy search. In: ICRA. (May 2015)

10. Roy, N., Prentice, S.: The belief roadmap: Efficient planning in belief space by factoring the covariance. IJRR **28**(11-12) (2009) 1448–1465

11. Bry, A., Roy, N.: Rapidly-exploring random belief trees for motion planning under uncertainty. In: ICRA. (May 2011)

12. Agha-mohammadi, A.a., Chakravorty, S., Amato, N.M.: Firm: Sampling-based feedback motion planning under motion uncertainty and imperfect measurements. The International Journal of Robotics Research (2013)

13. Alterovitz, R., Simon, T., Goldberg, K.: The stochastic motion roadmap: A sampling framework for planning with markov motion uncertainty. In: RSS. (June 2007)

14. Littlefield, Z., Klimenko, D., Kurniawati, H., Bekris, K.E.: The importance of a suitable distance function in belief-space planning. In: ISRR. (September 2015)

15. Berg, J.V.D., Abbeel, P., Goldberg, K.: Lqg-mp: Optimized path planning for robots with motion uncertainty and imperfect state information. In: RSS. (June 2010)

16. Huynh, V.A., Karaman, S., Frazzoli, E.: An incremental sampling-based algorithm for stochastic optimal contro. In: ICRA. (May 2012)

17. Davis, B., Karamouzas, I., Guy, S.J.: C-opt: Coverage-aware trajectory optimization under uncertainty. IEEE Robotics and Automation Letters **1**(2) (July 2016) 1020–1027

18. Lee, A., Duan, Y., Patil, S., Schulman, J., McCarthy, Z., van den Berg, J., Goldberg, K., Abbeel, P.: Sigma hulls for gaussian belief space planning for imprecise articulated robots amid obstacles. In: IROS. (Nov 2013)

19. Nieuwenhuisen, D., van der Stappen, A.F., Overmars, M.H.: Pushing using compliance. In: ICRA. (May 2006)

20. LaValle, S.M., Kuffner, J.J.: Randomized kinodynamic planning. IJRR **20**(5) (2001) 378–400

21. Phillips-Grafflin, C., Berenson, D.: Planning and Resilient Execution of Policies For Manipulation in Contact with Actuation Uncertainty. arXiv preprint arXiv:1703.10261 (2017)

22. Sneath, P.H.A., Sokal, R.R.: Numerical taxonomy: the principles and practice of numerical classification. Freeman (1973)

23. Asafi, S., Goren, A., Cohen-Or, D.: Weak convex decomposition by lines-of-sight. Computer Graphics Forum **32**(5) (2013) 23–31

Configuration Lattices for Planar Contact Manipulation Under Uncertainty

Michael Koval[1], David Hsu[2], Nancy Pollard[1], and Siddhartha S. Srinivasa[1]

[1] The Robotics Institute, Carnegie Mellon University
[2] Department of Computer Science, National University of Singapore

Abstract. This work addresses the challenge of a robot using real-time feedback from contact sensors to reliably manipulate a movable object on a cluttered tabletop. We formulate this task as a partially observable Markov decision process (POMDP) in the joint space of robot configurations and object poses. This formulation enables the robot to explicitly reason about uncertainty and all major types of kinematic constraints: reachability, joint limits, and collision. We solve the POMDP using DESPOT, a state-of-the-art online POMDP solver, by leveraging two key ideas for computational efficiency. First, we lazily construct a discrete lattice in the robot's configuration space. Second, we guide the search with heuristics derived from an unconstrained relaxation of the problem. We empirically show that our approach outperforms several baselines on a simulated seven degree-of-freedom manipulator.

1 Introduction

Our goal is to enable robots to use real-time contact sensing to reliably manipulate their environments. We focus on contact sensing because it is an ideal source of feedback for manipulation: the sense of touch directly observes the forces that the robot imparts on its environment and is unaffected by occlusion.

Consider a robot trying to push a bottle into its palm (Figure 1). The robot is uncertain about the initial pose of the bottle, has only an approximate model of physics, and receives feedback from noisy contact sensors on its fingertips. To localize the object and complete the task, the robot must take *information-gathering actions* to force the bottle into contact with one of its sensors. Simultaneously, the robot must avoid *kinematic constraints* such as colliding with an obstacle or pushing the bottle outside of its workspace.

Prior work focuses on planning under kinematic constraints and under uncertainty in isolation (Section 2). Work that considers kinematic constraints, but ignores uncertainty, often produces brittle policies that fail when executed. Work that considers uncertainty, but ignores kinematic constraints, often produces policies that are infeasible to execute in clutter.

In this paper, we formulate planar manipulation as a *partially observable Markov decision process* (POMDP) [42] in the joint space of robot configurations and object poses (Section 3). This formulation enables the robot to explicitly reason about uncertainty and, when necessary, plan information-gathering

© Springer Nature Switzerland AG 2020
K. Goldberg et al. (Eds.): *Algorithmic Foundations of Robotics XII*, SPAR 13, pp. 768–783, 2020.
https://doi.org/10.1007/978-3-030-43089-4_49

(a) Problem Definition   (b) Online Search   (c) Relaxed Problem

Fig. 1: HERB [44] uses real-time feedback from contact sensors to manipulate a bottle on a cluttered table. (a) We represent the robot's configuration as a point in a lattice that models the reachable (—) and unreachable (—) parts of the robot's configuration space. (b) We use an online POMDP solver guided by powerful heuristics to construct only the portions of the lattice relevant to completing the task. These heuristics are derived from (c) a relaxation of the problem that considers only the motion of the object relative to the end-effector.

actions to reduce it. Unlike in our prior work [25], this model also includes all major types of kinematic constraints: reachability, joint limits, and collision.

While exactly solving a POMDP is intractable in the worst case, recent advances in approximate online [41, 43] and point-based [26] solvers have enabled successful POMDP planning for manipulation tasks [17, 18, 25]. We use DESPOT [43], a state-of-the-art online POMDP solver, and leverage two key ideas for computational efficiency. First, we lazily discretize the robot's configuration space into a lattice of configurations connected by constrained trajectories (Section 4). Second, we leverage heuristics computed from an unconstrained relaxation of the problem to guide the online search (Section 5). We prove that the optimal policy will not take infeasible actions and our heuristics do not compromise the optimality of the solution.

We validate the algorithm on a simulation of HERB [44], a robot equipped with a 7-DOF Barrett WAM arm [39], manipulating an object on a cluttered tabletop (Figure 1, Section 6). Our results show that the proposed algorithm succeeds more often than approaches that consider either uncertainty or kinematic constraints in isolation.

2   Related Work

Our work builds on a long history of research on manipulating objects under uncertainty. Early work focused on planning open-loop trajectories that successfully reconfigure an object despite non-deterministic uncertainty [27] in its pose [5, 14]. Recently, the same type of worst-case analysis has been used to plan robust open-loop trajectories for grasping [9, 10] and rearrangement planning [11, 23]. Our approach makes the quasistatic assumption [32], similar to this prior work, but differs in two key ways: it (1) considers probabilistic uncertainty [27] and (2) produces a closed-loop policy that uses real-time feedback.

Another line of research aims to incorporate feedback from contact sensors into manipulator control policies, e.g. to optimize the quality of a grasp [37], servo towards a desired contact sensor observation [28, 47], or learn a feedback policy to achieve a grasp [34, 45]. These approaches achieve impressive real-time performance, but require a higher-level planner to specify the goal.

One common strategy is to plan a sequence of move-until-touch actions that localize an object, then execute an open-loop trajectory to complete the task [16, 21, 35]. Other approaches formulate the problem as a POMDP [42] and find a policy that takes information-gathering actions only when they are necessary to achieve the goal [17, 19]. Unfortunately, most of this work assumes that the end-effector can move freely in the workspace and that objects do not significantly move when touched.

Recent work, including our own [25], has relaxed the latter assumption by incorporating a stochastic physics model into the POMDP [17] and using SAR-SOP [26], an offline point-based POMDP solver, to find a policy for manipulating an object relative to the hand. Unfortunately, hand-relative policies often fail when executed on a manipulator due to kinematic reachability or collision with obstacles. We use a hand-relative policy to guide DESPOT [43], an online POMDP solver [38], in a search that explicitly models these constraints.

Our approach represents the robot's configuration space as a state lattice [36], a concept that we borrow from mobile robot navigation [29] and search-based planning [8]. Similar to many randomized motion planners [3, 15], we use lazy evaluation to defer collision checking until an action is queried by the planner.

3   Problem Formulation

We formulate planar contact manipulation as a partially observable Markov decision process (POMDP) [42]. A POMDP is a tuple (S, A, O, T, Ω, R) where S is the set of states, A is the set of actions, O is the set of observations, $T(s, a, s') = p(s'|s, a)$ is the transition model, $\Omega(s, a, o) = p(o|s, a)$ is the observation model, and $R(s, a) : S \times A \to \mathbb{R}$ is the reward function (Figure 2).

The robot does not know the true state s_t. Instead, the robot tracks the *belief state* $b(s_t) = p(s_t|a_{1:t}, o_{1:t})$, a probability distribution over the current state s_t conditioned on the history of actions $a_{1:t} = \{a_1, \ldots, a_t\}$ and observations $o_{1:t} = \{o_1, \ldots, o_t\}$. The set of all belief states is known as *belief space* Δ.

Our goal is to find a policy $\pi : \Delta \to A$ that optimizes the *value function*

$$V^\pi [b] = E \left[\sum_{t=0}^{\infty} \gamma^t R(s_t, a_t) \right]$$

where the expectation $E[\cdot]$ is taken over the sequence of states visited by π. The *discount factor* $\gamma \in [0, 1)$ adjusts the value of present versus future reward.

In our problem, *state* $s = (q, x_o) \in S$ is the configuration of the robot $q \in Q$ and the pose of the movable object $x_o \in X_o = \mathrm{SE}(3)$ (Figure 2a). An *action*

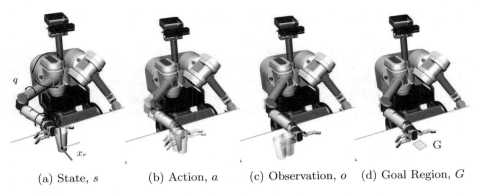

(a) State, s (b) Action, a (c) Observation, o (d) Goal Region, G

Fig. 2: We formulate planar contact manipulation as a POMDP with a state space (a) that contains the robot configuration q and the pose of the movable object x_r. The environment contains one movable object (white glass) and static obstacles. (b) action is a short joint-space trajectory ξ of duration T. After executing an action, the robot receives feedback from (c) binary contact sensors on its end-effector. The goal of pushing the object into (d) the hand-relative goal region G is encoded in the reward function $R(s, a)$.

$a = (\xi, T) \in A$ is a trajectory $\xi : [0, T] \to Q$ that starts in configuration $\xi(0)$ and ends configuration $\xi(T)$ at time T (Figure 2b). We assume that uncertainty over object pose dominates controller and proprioceptive error. Therefore, we treat q as fully-observable and assume that the manipulator is perfectly position controlled.

The robot executes a quasistatic push if it comes in contact with the movable object. The *quasistatic assumption* states that friction is high enough to neglect acceleration of the object, i.e. the object stops moving as soon as it leaves contact [32], and is accurate for many tabletop manipulation tasks [10, 11]. We define the stochastic *transition model* $p(s'|s, a) = T(s, a, s')$ in terms of a deterministic quasistatic physics model [32] by introducing noise into its parameters [12]. We do not attempt to refine our estimate of these parameters during execution.

After executing an action, the robot receives an *observation* $o \in \{0, 1\}^{n_o} = O$ from its n_o binary contact sensors (Figure 2c). We assume that an *observation model* $p(o|s', a) = \Omega(s', a, o)$ is available, but make no assumptions about its form.

The robot's goal is to push the movable object into a hand-relative *goal region* $X_{\text{goal}} \subseteq X$ (Figure 2d). We encode this in the *reward function* that assigns $R(s, a) = 0$ for states $[T_{\text{ee}}(q)]^{-1}x_o \in X_{\text{goal}}$ in the goal region and $R(s, a) = -1$ otherwise. In this expression, $T_{\text{ee}} : Q \to \text{SE}(3)$ is the forward kinematics of the end effector. Note that the choice of -1 reward is arbitrary: any negative reward would suffice.

4   Configuration Lattice POMDP

Solving this POMDP is challenging. We simplify the problem by constraining the end effector to a fixed transformation relative to the support surface and build a lattice in configuration space (Section 4.1). Configurations in the lattice are connected by action templates that start and end on lattice points (Section 4.2) and are penalized if rendered infeasible by kinematic constraints (Section 4.3).

4.1   Configuration Lattice

We assume that the robot's end effector is constrained to have a fixed transformation $^{\mathrm{sup}}T_{\mathrm{ee}} \in \mathrm{SE}(3)$ relative to the support surface $T_{\mathrm{sup}} \in \mathrm{SE}(3)$. A configuration q satisfies this constraint iff

$$T_{\mathrm{ee}}(q) = T_{\mathrm{sup}}\,^{\mathrm{sup}}T_{\mathrm{ee}}\,\mathrm{Trans}([x_{\mathrm{r}}, y_{\mathrm{r}}, 0])\,\mathrm{Rot}(\theta_{\mathrm{r}}, \hat{e}_z) \tag{1}$$

where $(x_{\mathrm{r}}, y_{\mathrm{r}}, \theta_{\mathrm{r}}) \in X_{\mathrm{r}} = SE(2)$ is the pose of the end effector in the plane, $\mathrm{Rot}(\theta, \hat{v})$ is a rotation about $\hat{v}$ by angle θ, $\mathrm{Trans}(v)$ is a translation by v.

We also assume that the movable object also has a fixed transformation $^{\mathrm{sup}}T_{\mathrm{o}} \in SE(3)$ relative to the support surface. We parameterize its pose as

$$x_{\mathrm{o}} = T_{\mathrm{sup}}\,^{\mathrm{sup}}T_{\mathrm{o}}\,\mathrm{Trans}([x_{\mathrm{o}}, y_{\mathrm{o}}, 0])\,\mathrm{Rot}(\theta_{\mathrm{o}}, \hat{e}_z)$$

where $(x_{\mathrm{o}}, y_{\mathrm{o}}, \theta_{\mathrm{o}}) \in X_{\mathrm{o}} = \mathrm{SE}(2)$ is the pose of the object in the plane.

We discretize the space of the end effector poses X_{r} by constructing a *state lattice* $X_{\mathrm{r,lat}} \subseteq X_{\mathrm{r}}$ with a translational resolution of $\Delta x_{\mathrm{r}}, \Delta y_{\mathrm{r}} \in \mathbb{R}^+$ and an angular resolution of $\Delta\theta_{\mathrm{r}} = 2\pi/n_\theta$ for some integer value of $n_\theta \in \mathbb{N}$ [36]. The lattice consists of the discrete set of points

$$X_{\mathrm{r,lat}} = \{(i_x \Delta x_{\mathrm{r}}, i_y \Delta y_{\mathrm{r}}, i_\theta \Delta\theta_{\mathrm{r}}) : i_x, i_y, i_\theta \in \mathbb{Z}\}.$$

Each *point* $x_{\mathrm{r}} \in X_{\mathrm{r,lat}}$ may be reachable from multiple configurations. We assume that we have access to an *inverse kinematics function* $q_{\mathrm{lat}}(x_{\mathrm{r}})$ that returns a single solution $\{q\}$ that satisfies $T_{\mathrm{ee}}(q) = x_{\mathrm{r}}$ or $\emptyset$ if no such solution exists. A solution may not exist if x_{r} is not reachable, the end effector is in collision, or the robot collides with the environment in all possible inverse kinematic solutions.

4.2   Action Templates

Most actions do not transition between states in the lattice S_{lat}. Therefore, we restrict ourselves to actions that are instantiated from one of a finite set A_{lat} of action templates. An *action template* $a_{\mathrm{lat}} = (\xi_{\mathrm{lat}}, T) \in A_{\mathrm{lat}}$ is a Cartesian trajectory $\xi_{\mathrm{lat}} : [0, T] \to SE(3)$ that specifies the relative motion of the end effector. The template starts at the origin $\xi_{\mathrm{lat}}(0) = I$ and ends at some lattice point $\xi_{\mathrm{lat}}(T) \in X_{\mathrm{r,lat}}$. It is acceptable for multiple actions templates in A_{lat} to end at the same lattice point or have different durations.

An action $a = (\xi, T) \in A$ instantiates template a_{lat} at lattice point $x_{\text{r}} \in X_{\text{r,lat}}$ if it satisfies three conditions: (1) starts in configuration $\xi(0) = q_{\text{lat}}(x_{\text{r}})$, (2) ends in configuration $\xi(T) = q_{\text{lat}}(x_{\text{r}}\xi_{\text{lat}}(T))$, and (3) satisfies $T_{\text{ee}}(\xi(\tau)) = x_{\text{r}}\xi_{\text{lat}}(\tau)$ for all $0 \leq \tau \leq T$. These conditions are satisfied if ξ moves between two configurations in Q_{lat} and produces the same end effector motion as ξ_{lat}.

We define the function $\text{Proj}(x_{\text{r}}, a) \mapsto a_{\text{lat}}$ to map an action a to the template a_{lat} it instantiates. The pre-image $\text{Proj}^{-1}(x_{\text{r}}, a_{\text{lat}})$ contains the set of all possible instantiations of action template a_{lat} at x_{r}. We assume that we have access to a *local planner* $\phi(q, a_{\text{lat}})$ that returns a singleton action $\{a\} \subseteq \text{Proj}^{-1}(q, a_{\text{lat}})$ from this set or $\emptyset$ to indicate failure. The local planner may fail due to kinematic constraints, end effector collision, or robot collision.

4.3   Configuration Lattice POMDP

We use the lattice to define a *configuration lattice POMDP*, Lat-POMDP, $(S_{\text{lat}}, A_{\text{lat}}, O, T_{\text{lat}}, \Omega_{\text{lat}}, R_{\text{lat}})$ in state space $S_{\text{lat}} = X_{\text{r,lat}} \times X_{\text{o}}$. The structure of the lattice guarantees that that all instantiations $\text{Proj}^{-1}(q, a_{\text{lat}})$ of the action template a_{lat} execute the same motion ξ_{lat} of the end effector. This motion is independent of the starting pose of the end effector x_{r} and configuration $q_{\text{lat}}(x_{\text{r}})$ of the robot.

If the movable object only contacts the end effector—not other parts of the robot or the environment—then the motion of the object is also independent of these variables. We refer to a violation of this assumption as *un-modelled contact*. The lattice transition model $T_{\text{lat}}(s_{\text{lat}}, a_{\text{lat}}, s'_{\text{lat}})$ is identical to $T(s, a, s')$ when a_{lat} is feasible and no un-modelled contact occurs. If either condition is violated, the robot transitions to the absorbing state s_{invalid}. Similarly, the lattice observation model $\Omega_{\text{lat}}(s_{\text{lat}}, a_{\text{lat}}, o)$ is identical to $\Omega(s, a, o)$ for valid states and is uniform over O for $s_{\text{lat}} = s_{\text{invalid}}$.

We penalize invalid states $s_{\text{lat}} = s_{\text{invalid}}$, infeasible actions $\phi(q_{\text{lat}}(x_{\text{r}}), a_{\text{lat}}) = \emptyset$, and un-modelled contact in the reward function R_{lat} by assigning them $\min_{s \in S, a \in A} R(s, a) = -1$ reward. Otherwise, we define $R_{\text{lat}}(s, a) = R(s, a)$. This choice guarantees that an optimal policy π^*_{lat} will never take an infeasible action:

Theorem 1. *An optimal policy π^*_{lat} of Lat-POMDP will not execute an infeasible action in belief b if $V^*_{lat}[b] > \frac{-1}{1-\gamma}$.*

Proof. Suppose $V^*_{\text{lat}}[b] > \frac{-1}{1-\gamma}$ and an optimal policy π_{lat} executes the invalid action in belief state b. The robot receives a reward of -1 and transitions to s_{invalid}. For all time after that, regardless of the actions that π_{lat} takes, the robot receives a reward of $R_{\text{lat}}(s_{\text{invalid}}, \cdot) = -1$ at each timestep. This yields a total reward of $V^{\pi_{\text{lat}}}_{\text{lat}} = \frac{-1}{1-\gamma}$, which is the minimum reward possible to achieve.

The value function of the optimal policy satisfies the Bellman equation $V^*_{\text{lat}}[b] = \arg\max_{a_{\text{lat}} \in A_{\text{lat}}} Q^*[b, a_{\text{lat}}]$, where $Q^*[b, a]$ denotes the value of taking action a in belief state b, then following the optimal policy for all time. This contradicts the fact that $V^*_{\text{lat}}[b] > \frac{-1}{1-\gamma}$ and $V^{\pi_{\text{lat}}}_{\text{lat}}[b] = \frac{-1}{1-\gamma}$. Therefore, π_{lat} must not be the optimal policy. □

We can strengthen our claim if we guarantee that every lattice point reachable from q_0 has at least one feasible action and it is possible to achieve the goal with non-zero probability. Under those assumptions we know that $V_{\mathrm{lat}}^*[b] > \frac{-1}{1-\gamma}$ and Theorem 1 guarantees that π_{lat}^* will never take an infeasible action. One simple way to satisfy this condition is to require that all actions be reversible.

5   Online POMDP Planner

Lat-POMDP has a large state space that changes whenever obstacles are added to, removed from, or moved within the environment. We use DESPOT [43], an online POMDP solver, to efficiently plan in this space. DESPOT incrementally explores the action-observation tree rooted at $b(s_0)$ by performing a series of trials. Each *trial* starts at the root node, descends the tree, and terminates by adding a new leaf node to the tree.

In each step, DESPOT chooses the action that maximizes the upper bound $\bar{V}[b]$ and the observation that maximizes *weighted excess uncertainty*, a regularized version of the gap $\bar{V}[b] - \underline{V}[b]$ between the bounds. This strategy heuristically focuses exploration on the optimally reachable belief space [26]. Finally, DESPOT backs up the upper and lower bounds of all nodes visited by the trial.

We leverage two key ideas for computational efficiency. First, we interleave lattice construction with planning to evaluate only the parts of the lattice that are visited by DESPOT (Section 5.1). Second, we guide DESPOT with upper (Section 5.2) and lower (Section 5.3) bounds derived from a relaxation of the problem that considers only the pose of the movable object relative to the hand.

5.1   Configuration Lattice Construction

DESPOT uses upper and lower bounds to focus its search on belief states that are likely to be visited by the optimal policy. We exploit this fact to avoid constructing the entire lattice. Instead, we interleave lattice construction with planning and only instantiate the lattice edges visited by the search, similar to the concept of *lazy evaluation* used in motion planning [3, 15].

We begin with no pre-computation and run DESPOT until it queries the transition model T_{lat}, observation model Ω_{lat}, or reward function R_{lat} for a state-action pair $(x_{\mathrm{r}}, a_{\mathrm{lat}})$ that has not yet been evaluated. When this occurs, we pause the search and check the feasibility of the action by running the local planner $\phi(x_{\mathrm{r}}, a_{\mathrm{lat}})$. We use the outcome of the local planner to update the Lat-POMDP model and resume the search. Figure 1 shows the (a) full lattice and (b) subset evaluated by DESPOT, only a small fraction of the full lattice.

It is also possible to use a hybrid approach by evaluating some parts of the lattice offline and deferring others to be computed online. For example, we may compute inverse kinematics solutions, kinematic feasibility checks, and self-collision checks in an offline pre-computation step. These values are fixed for a given support surface and, thus, can be used across multiple problem instances.

5.2   Hand-Relative Upper Bound

Recall from Section 4 that the motion of the end effector and the object is independent of the pose of the end effector x_r or the robot configuration q. We use this insight to define a *hand-relative POMDP*, Rel-POMDP, $(S_{rel}, A_{lat}, O, T_{rel}, \Omega_{rel}, R_{rel})$ with a state space that only includes the pose $x_{o,rel} = x_r^{-1} x_o \in S_{rel}$ of the movable object relative to the hand. The hand-relative transition model T_{rel}, observation model Ω_{rel}, and reward function R_{rel} are identical to the original model when no un-modelled contact occurs.

Rel-POMDP is identical to the hand-relative POMDP models used in prior work [17, 24, 25] and is equivalent to assuming that environment is empty and the robot is a lone end effector actuated by an incorporeal planar joint. As a result, Rel-POMDP is a relaxation of Lat-POMDP:

Theorem 2. *The optimal value function V_{rel}^* of Rel-POMDP is an upper bound on the optimal value function V_{lat}^* of Lat-POMDP: $V_{rel}^*[b] \geq V_{lat}^*[b]$ for all $b \in \Delta$.*

Proof. We define a *scenario* $\psi = (s_0, \psi_1, \psi_2, \dots)$ as an abstract simulation trajectory that captures all uncertainty in our POMDP model [33, 43]. A scenario is generated by drawing the initial state $s_o \sim b(s_0)$ from the initial belief state and each random number $\psi_1 \sim \text{uniform}[0, 1]$ uniformly from the unit interval. Given a scenario ψ, we assume that the outcome of executing a sequence of actions is deterministic; i.e. all stochasticity is captured in the initial state s_0 and the sequence of random numbers $\psi_1, \psi_2, \dots$.

Suppose we have a policy π for Rel-POMDP that executes the sequence of actions $a_{lat,1}, a_{lat,2}, \dots$ in scenario ψ. The policy visits the sequence of states $s_{rel,1}, s_{rel,2}, \dots$ and receives the sequence of rewards $R_1, R_2, \dots$.

Now consider executing π in the same scenario ψ on Lat-POMDP. Without loss of generality, assume that π first takes an infeasible action or makes un-modelled contact with the environment at timestep H. The policy receives the same sequence of rewards $R_1, R_2, \dots, R_2, \dots, R_{H-1}, -1, -1, \dots$ as it did on Rel-POMDP until timestep H. Then, it receives -1 reward for taking an infeasible action, transitions to absorbing state $s_{invalid}$, and receives -1 reward for all time.

Policy π achieves value $V_{rel,\psi}^\pi = \sum_{t=0}^{\infty} \gamma^t R_t$ on Rel-POMDP and $V_{lat,\psi}^\pi = \sum_{t=0}^{H-1} \gamma^t R_t - \frac{\gamma^H}{1-\gamma}$ on Lat-POMDP in scenario ψ. Since $R_t \geq -1$, we know that $V_{rel,\psi}^\pi \geq V_{lat,\psi}^\pi$. The value of a policy $V^\pi = E_\psi[V_\psi^\pi]$ is the expected value of π over all scenarios.

Consider the optimal policy π_{lat}^* of Lat-POMDP. There exists some Rel-POMDP policy π_{mimic} that executes the same sequence of actions as π_{lat}^* in all scenarios. From the reasoning above, we know that $V_{rel}^{\pi_{mimic}} \geq V_{lat}^*$. We also know that $V_{rel}^* \geq V_{rel}^{\pi_{mimic}}$ because the value of any policy is a lower bound on the optimal value function. Therefore, $V_{rel}^* \geq V_{rel}^{\pi_{mimic}} \geq V_{lat}^*$. □

This result implies that *any* upper bound $\bar{V}_{rel}$ is an upper bound on the value of the optimal value function $\bar{V}_{rel} \geq V_{rel}^* \geq V_{lat}^*$. Therefore, we may also use $\bar{V}_{rel}$ as an upper bound on Lat-POMDP. The key advantage of doing so is that the $\bar{V}_{rel}$ may be pre-computed once per hand-object pair. In contrast, the same upper bound on $\bar{V}_{lat}$ must be re-computed for each problem instance.

5.3   Hand-Relative Lower Bound

We exploit the fact that the value function of any policy is a lower bound on the optimal value function to define $\underline{V}_{lat}$. We use offline pre-computation to compute a rollout policy on $\pi_{rollout}$ for Rel-POMDP once per hand-object pair, e.g. using MDP [31] or point-based [17, 25] value iteration.

Given $\pi_{rollout}$, we construct an approximate lower bound $\underline{V}_{lat}$ for Lat-POMDP by estimating the value $V_{lat}^{\pi_{rollout}}$ of executing $\pi_{rollout}$ on Lat-POMDP via Monte Carlo rollouts. Approximating a lower bound with a *rollout policy* is commonly used in POMCP [41], DESPOT [43], and other online POMDP solvers.

6   Experimental Results

We validated the efficacy of the proposed algorithm by running simulation experiments on HERB [44], a robot equipped with a 7-DOF Barrett WAM arm [39] and the BarrettHand [46] end-effector. The robot attempts to push a bottle into the center of its palm on a table littered with obstacles.

6.1   Problem Definition

The state space of the problem consists of the configuration space $Q = \mathbb{R}^7$ of the robot and the pose of the object X_o relative to the end effector. The robot begins in known configuration q_0 and x_o is drawn from a Gaussian distribution centered in front of the palm with a covariance matrix of $\Sigma^{1/2} = \mathrm{diag}[5 \text{ mm}, 10 \text{ cm}]$.

Transitions Model. At each timestep, the simulated robot chooses an action a_{lat} that moves 1 cm at a constant Cartesian velocity in the xy-plane. The motion of the object is simulated using the Box2D physics simulator [7]. We simulate uncertainty in the model by sampling the hand-object friction coefficient and center of the object-table pressure distribution at each timestep [12, 24].

Configuration Lattice. These actions define a lattice centered at $T_{ee}(q_0)$ with a resolution of $\Delta x_r = \Delta y_r = 1$ cm. To construct this lattice, we select a configuration $q_{lat}(x_r)$ using an iterative inverse kinematics solver initialized with the solution of an adjacent lattice point. Then, we use a Cartesian motion planner to find a trajectory that connects adjacent points while satisfying the action template. As described in Section 5.1, the kinematic structure of the lattice is computed offline, but all collision checking is deferred until runtime. Forward kinematics, inverse kinematics, and collision detection is provided by the Dynamic Animation and Robotics Toolkit (DART) [1].

Observation Model. The simulated robot receives binary observations from contact sensors on its fingertips. We assume that the sensors perfectly discriminate between contact and no-contact [24, 25], but provide no information about where contact occurred on the sensor. The robot must take information-gathering actions, by moving side-to-side, to localize the object.

Discretization. We discretize S_{rel}, as in prior work [17, 25], to speed up evaluation of the model and to enable calculation of the QMDP [31] and SARSOP [26] policies. We discretize a region of size 20 cm × 44 cm centered around the palm at a 1 cm resolution (Figure 3a). To do so, we compute a discrete transition model, observation model, and reward function by taking an expectation over the underlying continuous state. States outside of this region are considered to be $s_{invalid}$.

6.2   Baseline Policies

We compare Lat-DESPOT against several baseline policies:

Rel-QMDP chooses the action at each timestep that greedily optimizes the Q-value of the MDP value function defined by Rel-POMDP [31]. QMDP does not perform multi-step information-gathering actions, but has been shown to perform well in domains where information is easily gathered [13, 20].

Rel-SARSOP uses SARSOP [26], a point-based method, to compute an offline policy for Rel-POMDP that is capable of taking information-gathering actions. SARSOP has been shown to perform well on Rel-POMDP in prior work [17, 25]. As in that work, we used the implementation of SARSOP provided by the APPL toolkit and allowed it to run for 10 minutes offline.

Rel-DESPOT uses DESPOT [43] to plan for Rel-POMDP using Rel-QMDP as an upper bound and rollouts of Rel-QMDP as a lower bound. We use the implementation of DESPOT provided by the APPL toolkit and tuned its parameters on a set of training problem instances distinct from these results.

Lift-QMDP and Lift-SARSOP use the state lattice to evaluate the feasibility of the action returned by Rel-QMDP and Rel-SARSOP, respectively, before executing it. If the desired action is infeasible, instead execute the feasible action with the next highest Q-value. This represents a heuristic solution for modifying a Rel-POMDP policy to avoid taking infeasible actions.

Lat-DESPOT, the proposed algorithm, uses DESPOT [43] to plan for Lat-POMDP using Rel-QMDP as the upper bound and rollouts of Lift-QMDP as the lower bound. This algorithm considers both kinematic constraints and uncertainty during planning.

6.3   Rel-POMDP Experiments

We begin by considering Rel-POMDP to isolate the effect of uncertainty from that of kinematic constraints. First, we confirm that our POMDP formulation faithfully encodes our goal. Next, we demonstrate information-gathering is necessary to achieve good performance. Finally, we verify that DESPOT—an online method—does not sacrifice the solution quality achieved by offline methods.

Figure 3b shows the value achieved by each policy in a 100 timestep simulation of the discretized Rel-POMDP problem. Figure 3c shows the probability

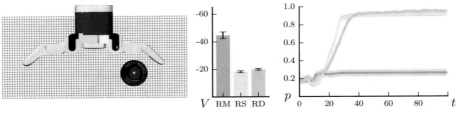

(a) Hand-Relative Discretization   (b) Discrete V (c) Continuous Success Prob.

Fig. 3: Performance of Rel-QMDP (RM ■ —), Rel-SARSOP (RS □ —), and Rel-DESPOT (RD □ —) on Rel-POMDP. (a) Discretization of Rel-POMDP used during planning. (b) Value V_{rel} achieved by each policy after 100 timesteps on the discretized Rel-POMDP problem. Note that the y-axis is inverted; lower (less negative) is better. (c) Probability $p = \Pr(s_t \in X_{\mathrm{goal}})$ that the movable object is in the goal region at each timestep on the continuous Rel-POMDP problem. Results are averaged over 500 trials and error bars denote a 95% confidence interval. Best viewed in color.

that the movable object is in X_{goal} at each timestep when simulated using the continuous model. As expected, the higher value achieved by Rel-SARSOP (□ —) and Rel-DESPOT (■ —) on the discretized problem translates to those algorithms achieving a higher success rate than Rel-QMDP (■ —) on the continuous problem. This result suggests that *discretizing the state space does not harm a policy's performance on the continuous problem.*

Rel-QMDP (■ —) performs poorly on this problem, achieving $< 30\%$ success probability, because QMDP does not take multi-step information-gathering actions [31]: the robot pushes straight without localizing the object.

Rel-SARSOP (□ —) and Rel-DESPOT (■ —) execute information-gathering actions by moving the hand laterally to drive the movable object into one of the fingertip contact sensors, then push the object into the goal region. These results replicate those in prior work [17, 25] by confirming that *information-gathering is necessary to perform well on this problem.* Our POMDP formulation provides a principled method of automatically constructing policies that gather information when necessary to complete the task.

Our intuition is that it is more difficult to solve Lat-POMDP than Rel-POMDP. Therefore, it is important that we verify that DESPOT solves Rel-POMDP before applying it to Lat-POMDP. Our results confirm *Rel-DESPOT (□ —) achieves comparable value and success probability to Rel-SARSOP (□ —).*

6.4   Lat-POMDP Experiments

We evaluate the proposed approach (Lat-DESPOT) on Lat-POMDP in four different environments: (a) an empty table, (b) obstacles on the right, (c) obstacles on the left, and (d) more complex obstacles on the right. Unlike in the

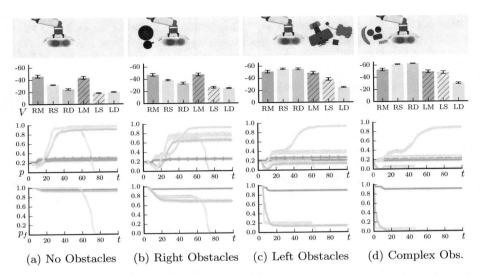

Fig. 4: Performance of Rel-QMDP (RM ■ —), Rel-SARSOP (RS □ —), Rel-DESPOT (RD □ —), Lift-QMDP (LM ▨ ✦), Lift-SARSOP (LS ▨ ✦), and Lat-DESPOT (LD □ —), the proposed algorithm, on four Lat-POMDP environments. (Top) Value achieved by each policy after 100 timesteps on the discretized problem (less negative is better). (Middle) Probability $p = \Pr(s_t \in X_{\text{goal}})$ that the movable object is in the goal region at each timestep on the continuous problem. (Bottom) Probability that execution is feasible. Lift-QMDP, Lift-SARSOP, and Lat-DESPOT are omitted because they do not take infeasible actions. Results are averaged over 500 trials and error bars denote a 95% confidence interval and axis labels are shared across the plots in each row. Best viewed in color.

Rel-POMDP experiments, kinematic constraints are present in the form of reachability limits, self-collision, and collision between the arm and the table. Scenes (b), (c), and (d) are constructed out of objects selected from the YCB dataset [6].

Figure 4 shows results for each scene. Figure 4-Top shows the value V_{lat} achieved by each policy on the discretized Lat-POMDP. Figure 4-Middle shows the probability that the movable object is in X_{goal} at each timestep, treating instances that have terminated as zero probability. Figure 4-Bottom shows the proportion of Rel-QMDP, Rel-SARSOP, and Rel-DESPOT policies that are active at each timestep; i.e. have not yet terminated by taking an infeasible action.

Rel-QMDP (RM ■ —) and Lift-QMDP (LM ▨ ✦) perform poorly across all environments, achieving < 30% success probability, because they do not take multi-step information-gathering actions. Figure 4 confirms this: both QMDP policies perform poorly on all four environments. This result demonstrates that *it is important to gather information even when kinematic constraints are present.*

Rel-SARSOP (RS □ —) and Rel-DESPOT (RD □ —) perform well on environments (a) and (b) because they hit obstacles late in execution. The converse is true on environments (c) and (d): both policies hit obstacles so quickly that they perform worse than Rel-QMDP! This result highlights that *it is important to consider kinematic constraints even when uncertainty is present.*

Lift-SARSOP (LS ▨ ↦) performs near-optimally on environments (a) and (b) because it does not take infeasible actions and gathers information. However, it performs no better than Rel-QMDP on problem (d). This occurs because Lift-SARSOP myopically considers obstacles in a one-step lookahead and may oscillate when blocked. Small changes in the environment are sufficient to induce this behavior: the key difference between environments (b) and (d) is the introduction of a red box that creates a cul-de-sac in the lattice.

Our approach, Lat-DESPOT (□ ↦), avoids myopic behavior by considering action feasibility during planning. Lat-DESPOT performs no worse than Lift-SARSOP on environments (a) and (b) and outperforms it on environments (c) and (d). Unlike Rel-SARSOP, Lat-DESPOT identifies the cul-de-sac in (d) during planning and avoids becoming trapped in it. In summary, *Lat-DESPOT is the only policy that performs near-optimally on all four environments because it considers both uncertainty and kinematic constraints during planning.*

Our unoptimized implementation of Lat-DESPOT took between 200 μs and 2.4 s to select an action on a single core of a 4 GHz Intel Core i7 CPU. The policy was slowest to evaluate early in execution, when information-gathering is necessary, and fastest once the movable object is localized because the upper and lower bounds become tighter. The QMDP and SARSOP policies, which are computed offline, took an average of 1.6 μs and 218 μs to evaluate respectively.

We are optimistic about achieving real-time performance from Lat-DESPOT by optimizing our implementation of the algorithm in future work. Since DESPOT is an anytime algorithm, speeding up the search will both improve the quality of a solution given a fixed time budget and reduce the time required to find a solution of a desired quality.

6.5 Upper Bound Validation

Finally, we combine the data in Figure 3-Left and Figure 4-Top to empirically verify the bound we proved in Theorem 2. The value of Rel-SARSOP (□ ↦) and Rel-DESPOT (□ ↦) on Rel-POMDP (Figure 3-Left) are greater (i.e. less negative) than the value of all policies we evaluated on Lat-POMDP (Figure 4-Top). The data supports our theory: the optimal value achieved on Rel-POMDP is no worse than the highest value achieved on Lat-POMDP in environment (a) and greater than the highest value achieved in environments (b), (c), and (d).

7   Discussion

In this paper, we formulated the problem of planar contact manipulation under uncertainty as a POMDP in the joint space of robot configurations and poses of the movable object (Section 3). For computational efficiency, we simplify the problem by constructing a lattice in the robot's configuration space and prove that, under mild assumptions, the optimal policy of Lat-POMDP will never take an infeasible action (Section 4). We find a near-optimal policy for Lat-POMDP using DESPOT [43] guided by upper and lower bounds derived from Rel-POMDP (Section 5).

Our simulation results show that Lat-DESPOT outperforms five baseline algorithms on cluttered environments: it achieves a $> 90\%$ success rate on all environments, compared to the best baseline (Lift-SARSOP) that achieves only a $\sim 20\%$ success rate on difficult problems. This highlights the importance of reasoning about both object pose uncertainty and kinematic constraints during planning. However, Lat-DESPOT has several limitations that we plan to address in future work.

First, our approach assumes that the robot has perfect proprioception and operates in an environment with known obstacles. In practice, robots often have imperfect proprioception [4, 22] and uncertainty about the pose of *all* objects in the environment. We hope to relax both of these assumptions by replacing the deterministic transition model for robot configuration with a stochastic model that considers the probability of hitting an obstacle. This extension should not significantly affect computational complexity because DESPOT—as with most online solvers—does not scale directly with the size of the state space.

Second, we are excited to scale our approach up a larger repertoire of action templates (including non-planar motion), solving more complex tasks, and planning in environments that contain multiple movable objects. Solving these more complex problems will require more informative heuristics. We are optimistic that more sophisticated Rel-POMDP policies, e.g. computed by Monte Carlo Value Iteration [2], could be used to guide the search. Additionally, we are interested in using macro actions [30] consisting of the repeated execution of an action template to reduce the effective horizon of the search and methods that operate on a continuous action space to incrementally densify the lattice [40].

Third, our approach commits to a single inverse kinematics solution $q_{lat}(x_r)$ for each lattice point. This prevents robots from using redundancy to avoid kinematic constraints. We plan to relax this assumption in future work by generating multiple inverse kinematic solutions for each lattice point and instantiating an action template for each. Our intuition is that many solutions share the same connectivity and, thus, may be treated identically during planning.

Finally, we plan to implement Lat-DESPOT on a real robotic manipulator and evaluate the performance of our approach on real-world manipulation tasks.

Acknowledgements

This work was supported by a NASA Space Technology Research Fellowship (award NNX13AL62H), the National Science Foundation (awards IIS-1218182 and IIS-1409003), the U.S. Office of Naval Research, and the Toyota Motor Corporation. We would like to thank Rachel Holladay, Shervin Javdani, Jennifer King, Stefanos Nikolaidis, and the members of the Personal Robotics Lab for their input. We would also like to thank Nan Ye for assistance with APPL.

Bibliography

[1] Dynamic Animation and Robotics Toolkit. http://dartsim.github.io (2013)

[2] Bai, H., Hsu, D., Lee, W., Ngo, V.: Monte Carlo value iteration for continuous-state POMDPs. In: WAFR (2011)

[3] Bohlin, R., Kavraki, L.: Path planning using lazy PRM. In: IEEE ICRA. pp. 521–528 (2000)

[4] Boots, B., Byravan, A., Fox, D.: Learning predictive models of a depth camera & manipulator from raw execution traces. In: IEEE ICRA (2014)

[5] Brokowski, M., Peshkin, M., Goldberg, K.: Curved fences for part alignment. In: IEEE ICRA (1993)

[6] Calli, B., Singh, A., Walsman, A., Srinivasa, S., Abbeel, P., Dollar, A.: The YCB object and model set: Towards common benchmarks for manipulation research. In: ICAR (2015)

[7] Catto, E.: Box2D. http://box2d.org (2010)

[8] Cohen, B., Chitta, S., Likhachev, M.: Single-and dual-arm motion planning with heuristic search. IJRR (2013)

[9] Dogar, M., Hsiao, K., Ciocarlie, M., Srinivasa, S.: Physics-based grasp planning through clutter. In: R:SS (2012)

[10] Dogar, M., Srinivasa, S.: Push-grasping with dexterous hands: Mechanics and a method. In: IEEE/RSJ IROS (2010)

[11] Dogar, M., Srinivasa, S.: A planning framework for non-prehensile manipulation under clutter and uncertainty. AuRo 33(3), 217–236 (2012)

[12] Duff, D., Wyatt, J., Stolkin, R.: Motion estimation using physical simulation. In: IEEE ICRA (2010)

[13] Emery-Montemerlo, R., Gordon, G., Schneider, J., Thrun, S.: Approximate solutions for partially observable stochastic games with common payoffs. In: AAMAS (2004)

[14] Erdmann, M., Mason, M.: An exploration of sensorless manipulation. IEEE T-RA (1988)

[15] Hauser, K.: Lazy collision checking in asymptotically-optimal motion planning. In: IEEE ICRA (2015)

[16] Hebert, P., Howard, T., Hudson, N., Ma, J., Burdick, J.: The next best touch for model-based localization. In: IEEE ICRA (2013)

[17] Horowitz, M., Burdick, J.: Interactive non-prehensile manipulation for grasping via POMDPs. In: IEEE ICRA (2013)

[18] Hsiao, K.: Relatively robust grasping. Ph.D. thesis, MIT (2009)

[19] Hsiao, K., Lozano-Pérez, T., Kaelbling, L.: Robust belief-based execution of manipulation programs. In: WAFR (2008)

[20] Javdani, S., Bagnell, J., Srinivasa, S.: Shared autonomy via hindsight optimization. In: R:SS (2015)

[21] Javdani, S., Klingensmith, M., Bagnell, J., Pollard, N., Srinivasa, S.: Efficient touch based localization through submodularity. In: IEEE ICRA (2013)

[22] Klingensmith, M., Galluzzo, T., Dellin, C., Kazemi, M., Bagnell, J., Pollard, N.: Closed-loop servoing using real-time markerless arm tracking. In: IEEE ICRA Humanoids Workshop (2013)

[23] Koval, M., King, J., Pollard, N., Srinivasa, S.: Robust trajectory selection for rearrangement planning as a multi-armed bandit problem. In: IEEE/RSJ IROS (2015)

[24] Koval, M., Pollard, N., Srinivasa, S.: Pose estimation for planar contact manipulation with manifold particle filters. IJRR 34(7), 922–945 (2015)

[25] Koval, M., Pollard, N., Srinivasa, S.: Pre- and post-contact policy decomposition for planar contact manipulation under uncertainty. IJRR (2015), in press

[26] Kurniawati, H., Hsu, D., Lee, W.: SARSOP: Efficient point-based POMDP planning by approximating optimally reachable belief spaces. In: R:SS (2008)

[27] LaValle, S., Hutchinson, S.: An objective-based framework for motion planning under sensing and control uncertainties. IJRR (1998)

[28] Li, Q., Schürmann, C., Haschke, R., Ritter, H.: A control framework for tactile servoing. In: R:SS (2013)

[29] Likhachev, M., Ferguson, D.: Planning long dynamically feasible maneuvers for autonomous vehicles. IJRR 28(8), 933–945 (2009)

[30] Lim, Z., Hsu, D., Sun, L.: Monte Carlo value iteration with macro-actions. In: NIPS (2011)

[31] Littman, M., Cassandra, A., Kaelbling, L.: Learning policies for partially observable environments: Scaling up. ICML (1995)

[32] Lynch, K., Maekawa, H., Tanie, K.: Manipulation and active sensing by pushing using tactile feedback. In: IEEE/RSJ IROS (1992)

[33] Ng, A., Jordan, M.: PEGASUS: A policy search method for large MDPs and POMDPs. In: UAI (2000)

[34] Pastor, P., Righetti, L., Kalakrishnan, M., Schaal, S.: Online movement adaptation based on previous sensor experiences. In: IEEE/RSJ IROS (2011)

[35] Petrovskaya, A., Khatib, O.: Global localization of objects via touch. IEEE T-RO 27(3), 569–585 (2011)

[36] Pivtoraiko, M., Kelly, A.: Efficient constrained path planning via search in state lattices. In: i-SAIRAS (2005)

[37] Platt, R., Fagg, A., Grupen, R.: Nullspace grasp control: theory and experiments. IEEE T-RO 26(2), 282–295 (2010)

[38] Ross, S., Pineau, J., Paquet, S., Chaib-Draa, B.: Online planning algorithms for POMDPs. JAIR (2008)

[39] Salisbury, K., Townsend, W., Eberman, B., DiPietro, D.: Preliminary design of a whole-arm manipulation system (WAMS). In: IEEE ICRA (1988)

[40] Seiler, K., Kurniawati, H., Singh, S.: GPS-ABT: An online and approximate solver for POMDPs with continuous action space. In: IEEE ICRA (2015)

[41] Silver, D., Veness, J.: Monte-Carlo planning in large POMDPs. In: NIPS (2010)

[42] Smallwood, R., Sondik, E.: The optimal control of partially observable Markov processes over a finite horizon. Operations Research 21(5), 1071–1088 (1973)

[43] Somani, A., Ye, N., Hsu, D., Lee, W.: DESPOT: Online POMDP planning with regularization. In: NIPS (2013)

[44] Srinivasa, S., Berenson, D., Cakmak, M., Collet, A., Dogar, M., Dragan, A., Knepper, R., Niemueller, T., Strabala, K., Vande Weghe, M.: HERB 2.0: Lessons learned from developing a mobile manipulator for the home. Proc. IEEE 100(8), 1–19 (2012)

[45] Stulp, F., Theodorou, E., Buchli, J., Schaal, S.: Learning to grasp under uncertainty. In: IEEE ICRA. pp. 5703–5708 (2011)

[46] Townsend, W.: The BarrettHand grasper–programmably flexible part handling and assembly. Industrial Robot: An International Journal 27(3), 181–188 (2000)

[47] Zhang, H., Chen, N.: Control of contact via tactile sensing. IEEE T-RA 16(5), 482–495 (2000)

On the Effects of Measurement Uncertainty in Optimal Control of Contact Interactions

Brahayam Pontón[1], Stefan Schaal[12], and Ludovic Righetti[1]

[1] Max-Planck Institute for Intelligent Systems, Tuebingen-Germany
firstname.lastname@tuebingen.mpg.de
[2] University of Southern California, Los Angeles-USA
sschaal@usc.edu

Abstract. Stochastic Optimal Control (SOC) typically considers noise only in the process model, i.e. unknown disturbances. However, in many robotic applications involving interaction with the environment, such as locomotion and manipulation, uncertainty also comes from lack of precise knowledge of the world, which is not an actual disturbance. We analyze the effects of also considering noise in the measurement model, by developing a SOC algorithm based on risk-sensitive control, that includes the dynamics of an observer in such a way that the control law explicitly depends on the current measurement uncertainty. In simulation results on a simple 2D manipulator, we have observed that measurement uncertainty leads to low impedance behaviors, a result in contrast with the effects of process noise that creates stiff behaviors. This suggests that taking into account measurement uncertainty could be a potentially very interesting way to approach problems involving uncertain contact interactions.

1 Introduction

In a not distant future, personal robots will be a common part of our daily lives, with a broad range of applications going from industrial and service applications to common household scenarios [1]. Being able to safely operate among humans by *optimally adapting to uncertainty* in a dynamic environment is a key ingredient for this to happen. In this contribution, we address this aspect by studying the effects of measurement uncertainty in stochastic optimal control problems[3].

For instance, we would like to understand the effects of considering uncertainty information upon optimal control solutions in problems that involve contact interactions. We distinguish between two sources of uncertainty: one due to external forces that physically perturb the robot, and the other due to uncertain knowledge of the robot's state, infered from noisy measurements. Figure 1 shows a schematic. On the one hand, external disturbances (process model's noise) directly affect the dynamic evolution of the robot and a stiff behavior is usually adopted to control the robot in their presence. On the other hand, our belief about the robot state, e.g. distance to a contact location, can be thought of as a noisy sensor signal. This, however, does not affect our actuation directly;

---

[3] An unpublished preliminary version of this work is available at https://arxiv.org/abs/1605.04344.

K. Goldberg et al. (Eds.): *Algorithmic Foundations of Robotics XII*, SPAR 13, pp. 784–799, 2020.
https://doi.org/10.1007/978-3-030-43089-4_50

instead, it influences the way in which decisions over control signals are taken. For example, if we are walking down a stair in the dark and we are uncertain about the floor location because we cannot see it, we reach for it with a gentle touch in order not to harm ourselves or fall. We observe that in this case, compliance was used to handle measurement uncertainty.

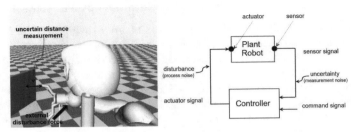

Fig. 1: *To the left*: Schematic of two different sources of uncertainty. *To the right*: Simplified control diagram showing where and how they enter the system.

Why is it important to consider measurement noise effects? First of all, properly addressing robustness issues due to uncertainty is important in robotics [1]. More concretely, imagine a reaching motion: in general, increasing the magnitude of an external disturbance implies one must increase the stiffness of the robot to maintain a desired tracking accuracy. Now, consider the experiment under measurement noise, e.g. you would like to grasp an object using no vision information, or you try to reach a wall to orient yourself in a dark corridor. Under these conditions, compliance is key to carefully reach an object. This distinction is important, because it suggests that modeling interactions of a robot with its environment (a fundamental problem in robotics) as an optimal control problem with measurement uncertainty, one could naturally get optimal compliant behaviors as a function of the uncertainty level.

Optimal control techniques based on *applying Bellman's Principle of Optimality around nominal trajectories*, e.g. techniques such as DDP or iLQG, have been very successfully used in robotics for large degrees of freedom system [2–6]. They maintain a single trajectory as a local method; and improve it iteratively based on dynamic programming along a neighborhood of the trajectory. This allows them to overcome to some extent the curse of dimensionality while remaining computationally efficient. Our algorithm belongs to this family, but distinguishes itself by the ability to explicitly consider measurement uncertainty.

Typically these methods find a solution for a nonlinear problem by iteratively solving a first- or second-order Taylor approximation of it [7,8]. This, however, imposes limitations: it only considers the mean of the objective function (expectation of a quadratic form) and systems under purely additive noise. As a result, the optimal control for the stochastic and deterministic problems is the same, i.e. uncertainty is dealt with by the estimator separately, and the control design is independent from noise. While it is reasonable for systems with small noise intensity, intuitively, one would not expect the same to be true for systems with large noise intensity, where a *control strategy capable of reasoning about noise statistics and cost of uncertainty* would be more appropriate. While there exists other equally valid methods for dealing with uncertainty, e.g. avoiding uncertain regions of the state space [9], we are interested in understanding how to control desired contact interactions with an inherently uncertain environment. This is the reason why we prefer a risk-sensitive over a risk-neutral optimal control approach. In this way, our algorithm is sensitive to the statistical properties of the noise and can incorporate this information into the optimal control law.

Risk-sensitive optimal control algorithms usually invalidate the assumptions of the Certainty Equivalence Principle, e.g. by using a nonlinear state equation, multiplicative noise or a non-quadratic objective function. Care is taken though at preserving the computational efficiency. The problem of considering multiplicative process noise was studied in [6], and extended to multiplicative measurement noise in [10, 11]. These methods construct an affine control law dependent on noise statistics and apply it to control a two-DOF model of a biomechanical arm. Another appealing alternative is to capture noise effects *on higher order statistics of the objective function by using non-quadratic costs*.

Jacobson [12] introduced a Linear-Exponential-Gaussian (LEG) algorithm to consider higher order statistics *using as cost an exponential transformation of the original objective function*. He derived feedback controllers for a linear system with additive process noise that explicitly depend on noise statistics. At low noise levels, the LEG control law is similar to the Linear-Quadratic-Gaussian (LQG) controller; but the larger the noise, the more they differ. This idea was extended in [13] for continuous-time nonlinear SOC problems using an iterative algorithm and they illustrated both risk-seeking and risk-averse behaviors in a continuous-time cliff problem. A more comprehensive review on the risk-sensitive literature for systems under process noise can be found in [14], where they present a unified theory of linearly solvable optimal control problems that includes both standard and risk-sensitive optimal control problems.

It has been shown [15], that due to the multiplicative nature of the exponential cost, it is not straightforward to extend the results for the case of measurement noise (where the control law is not a linear functional of the current state, but of the whole smoothed history of states). As solution, they proposed to define a state that grows every timestep to comprise the entire history of states seen so far. Because of this increasingly growing computational complexity, only two cases, where simplifications occur, are practical: when the objective function is a functional only of the final state, and when there is no process noise.

In this paper, we use recent results in risk-sensitive control [13] and extend them to incorporate measurement uncertainty. We will then show how different types of noise can significantly change the optimal controls. In our approach, as in [13], we sequentially approximate the nonlinear problem and design risk-sensitive controllers. However, in order to include measurement uncertainty effects, instead of using a growing state composed of the entire history of states, we *use an enlarged dynamical system composed of the control and estimation problems* [16, 17], where the number of states only doubles. We reduce the amount of information for constructing the optimal control to statistics that can be captured in the state estimate (i.e. expectation and variance). But by doing this, we gain increased flexibility at designing the objective function and are able to simultaneously capture process and measurement noise to compare their effects. The most important contributions of this work are as follows:

- We propose a theoretical contribution, where we extend recent work on risk-sensitive control to the measurement noise case by incorporating a state observer. This makes the optimal control explicitly dependent on statistical properties of process and measurement uncertainty.
- By applying our algorithm in a contact interaction experiment of a 2D manipulator with an uncertain wall, we show that our approach produces optimal impedance behaviors for contact interaction, that differ from the usual stiff behaviors in that compliance is encouraged under measurement noise.

In the following, we present the problem formulation and background material. Then, we show the algorithm derivation and illustrate its performance in two simple robotic tasks: a viapoint and a contact interaction task.

2   Problem Formulation and Background Material

In this section, we present the stochastic optimal control problem under measurement noise. The following stochastic differential equations (SDE's) define the dynamical evolution of the state and measurement models respectively

$$dx = m(x, u)dt + M(x, u)d\omega \ , \tag{1}$$

$$dy = n(x, u)dt + N(x, u)d\gamma \ . \tag{2}$$

Let $x \in \mathbb{R}^n$, $u \in \mathbb{R}^m$ and $y \in \mathbb{R}^p$ be the system states, control and measured outputs. $d\omega$ and $d\gamma$ are zero-mean Brownian motions with covariance Ωdt, Γdt. $m(x, u)$ and $n(x, u)$ are the drift coefficients representing the deterministic components of the dynamics and measurement models. $M(x, u)$ and $N(x, u)$ are the diffusion coefficients that encode the stochasticity of the problem.

In optimal control, we are interested in minimizing an objective function J^π, which is a functional of the control policy $u = \pi(x)$, defined as

$$J^\pi(x, t) = \Phi_f(x_{t_f}) + \int_t^{t_f} L(x_t, u_t, t)dt \ , \tag{3}$$

where $L(x_t, u_t, t)$ is the rate at which cost increases. $\Phi_f(x_{t_f})$ is the cost at the final time t_f. In standard optimal control, the mean of the objective function $\mathbb{E}[J^\pi]$ would typically be minimized. However, in order to analyze the effects of uncertainty on the optimal controls, this is not sufficient. We need to include the notion of the cost of uncertainty into the objective function. For this purpose, we use tools from the risk-sensitive control literature, that allow us to include higher order statistics of the objective function J^π in the minimization. This is done by reformulating the objective function as an exponential transformation of the original objective function [12]. The risk-sensitive cost is then given by

$$J = \min_\pi \mathbb{E}\{\exp[\sigma J^\pi]\} \ . \tag{4}$$

J^π is a random variable functional of the policy $u = \pi(x)$. $\sigma \in \mathbb{R}$ is the risk-sensitive parameter, $\mathbb{E}$ is the expectation over J^π. J is therefore the risk-sensitive cost and corresponds to the moment generating function, an alternative specification of the probability distribution of the random variable J^π [18].

In the following, we recall two previous results elaborated in [13, 12, 15] that we will use for the development of our approach: the meaning of the transformed cost and the form of the Hamilton Jacobi Bellman (HJB) equation under the exponential transformation.

2.1   Meaning of the Exponential Transformation of the Cost

It has been shown [13] that the cumulant generating function (logarithmic transformation of the moment generating function) of the risk-sensitive cost J can be rewritten as a linear combination of the moments of the objective function J^π

$$\frac{1}{\sigma} \log [J] = \mathbb{E}[J^\pi] + \frac{\sigma}{2}\mu_2[J^\pi] + \frac{\sigma^2}{6}\mu_3[J^\pi] + \cdots \ , \tag{5}$$

μ_2, μ_3 denote the variance and skewness of J^π. Therefore, the risk-sensitive cost is a linear combination of all the moments of the original objective function. It provides an additional degree of freedom, namely the risk-sensitive parameter σ, which allows us to define if the higher order moments act as a penalty or a reward in the cost, giving rise to risk-averse or risk-seeking behaviors respectively. Depending on σ, there will be a compromise between increasing control effort and narrowing confidence intervals. The lower the values of σ, the less weight is given to higher order moments. When it is negative, they even act as a reward, therefore leading to risk-seeking solutions.

2.2   HJB Equation under the Exponential Transformation

From [13], we recall the form of the HJB equation under the exponential transformation (the dynamics are given only by (1) and the cost by (3)-(4))

$$-\partial_t \Psi = \min_u \underbrace{\left\{ L + \nabla_x \Psi^T \mathbf{m} + \frac{1}{2}Tr\left(\nabla_{xx}\Psi \mathbf{M}\mathbf{M}^T\right) \right.}_{\text{Usual HJB equation}} + \underbrace{\left. \frac{\sigma}{2}\nabla_x \Psi^T \mathbf{M}\mathbf{M}^T \nabla_x \Psi \right\}}_{\text{Term due to uncertainty}}, \tag{6}$$

where the value function Ψ is a function of x and t. The terms without σ represent the usual HJB equation for a stochastic dynamical system with cost rate L due to the current state and control, the free drift and control benefit costs, and the diffusion cost. The interesting term is the last one which captures noise effects on statistical properties of the cost (higher moments). When σ is zero, the problem reduces to the minimization of the usual expected value of the cost $E[J^\pi]$. It is worth highlighting that these two results presented as background material model only process noise and do not include measurement noise.

2.3   Problem Formulation

Finally, we conclude the problem definition. The cost to minimize is given by (3)-(4). Our goal is to find a risk-sensitive optimal control law π^* that minimizes the cost $J^{\pi}(\boldsymbol{x}_0, t_0)$ for the stochastic system in the presence of additive process and measurement noise (1)-(2). The globally optimal control law $\pi^*(\boldsymbol{x}, t)$ does not depend on an initial state. However, finding it is in general intractable. Instead, we are interested in a locally-optimal feedback control law that approximates the globally optimal solution in the vicinity of a nominal trajectory $\boldsymbol{x}_t^n$. Since this nominal trajectory depends on the initial state of the system, so does the optimal feedback control law. As can be noted, our formulation of the problem differs from the results in the background material, because we include measurement noise. However, as we will show in detail in the next section, these results can still be used in our case, after a certain reformulation of the problem.

3   Algorithm Derivation

As mentioned in the Introduction, there have been numerous contributions on risk-sensitive control with process noise [14, 13]. Therefore, our goal in this section is to reformulate our problem including the stochastic dynamics of the measurement model, such that we can use some of these previous results. The algorithmic idea is to extend the state dynamics (1), with the dynamics of a state estimator. As will be seen later in detail, the key element then is to include a forward propagation of measurement uncertainty along a nominal trajectory and to precompute optimal estimation gains. This allows for the use of standard techniques to compute backwards in time optimal feedback controllers [7]. This idea, however, is not particular to the algorithm we present in the following, but could be used in combination with other methods, such as the ones presented in [14]. In the following, we derive a continuous time algorithm[4].

At each iteration, the algorithm begins with a nominal control sequence $\boldsymbol{u}_t^n$ and the corresponding zero-noise trajectory $\boldsymbol{x}_t^n$, obtained by applying the control sequence to the dynamics $\dot{\boldsymbol{x}} = \boldsymbol{m}(\boldsymbol{x}, \boldsymbol{u})$ with initial state $\boldsymbol{x}(0) = \boldsymbol{x}_0$. Next, we follow a standard approach in iterative optimal control [7, 8] to form a linear approximation of the dynamics and a quadratic approximation of the cost along the nominal trajectories $\boldsymbol{x}_t^n$ and $\boldsymbol{u}_t^n$, in terms of state and control deviations $\delta\boldsymbol{x}_t = \boldsymbol{x}_t - \boldsymbol{x}_t^n$, $\delta\boldsymbol{u}_t = \boldsymbol{u}_t - \boldsymbol{u}_t^n$. Dynamics and measurement models then become

$$d(\delta\boldsymbol{x}_t) = (\boldsymbol{A}_t\delta\boldsymbol{x}_t + \boldsymbol{B}_t\delta\boldsymbol{u}_t)dt + \boldsymbol{C}_t d\boldsymbol{\omega}_t \ , \tag{7}$$

$$d(\delta\boldsymbol{y}_t) = (\boldsymbol{F}_t\delta\boldsymbol{x}_t + \boldsymbol{E}_t\delta\boldsymbol{u}_t)dt + \boldsymbol{D}_t d\boldsymbol{\gamma}_t \ . \tag{8}$$

Evaluated along the nominal trajectories $\boldsymbol{x}_t^n$, $\boldsymbol{u}_t^n$, the matrices are given by

$$\boldsymbol{A}_t = \partial\boldsymbol{m}(\boldsymbol{x}, \boldsymbol{u})/\partial\boldsymbol{x}^T|_{\boldsymbol{x}_t^n, \boldsymbol{u}_t^n}, \ \boldsymbol{B}_t = \partial\boldsymbol{m}(\boldsymbol{x}, \boldsymbol{u})/\partial\boldsymbol{u}^T|_{\boldsymbol{x}_t^n, \boldsymbol{u}_t^n}, \ \boldsymbol{C}_t = \boldsymbol{M}(\boldsymbol{x}_t^n, \boldsymbol{u}_t^n) \ .$$

$$\boldsymbol{F}_t = \partial\boldsymbol{n}(\boldsymbol{x}, \boldsymbol{u})/\partial\boldsymbol{x}^T|_{\boldsymbol{x}_t^n, \boldsymbol{u}_t^n}, \ \boldsymbol{E}_t = \partial\boldsymbol{n}(\boldsymbol{x}, \boldsymbol{u})/\partial\boldsymbol{u}^T|_{\boldsymbol{x}_t^n, \boldsymbol{u}_t^n}, \ \boldsymbol{D}_t = \boldsymbol{N}(\boldsymbol{x}_t^n, \boldsymbol{u}_t^n) \ .$$

---

[4] The derivation of the discrete-time version of the algorithm is presented in the accompanying appendix.

In the same way, the quadratic approximation of the cost $\mathcal{J}$ is given by

$$\tilde{\ell}(\boldsymbol{x}, \boldsymbol{u}) = q_t + \boldsymbol{q}_t^T \delta \boldsymbol{x}_t + \boldsymbol{r}_t^T \delta \boldsymbol{u}_t + \frac{1}{2} \delta \boldsymbol{x}_t^T \boldsymbol{Q}_t \delta \boldsymbol{x}_t + \delta \boldsymbol{x}_t^T \boldsymbol{P}_t \delta \boldsymbol{u}_t + \frac{1}{2} \delta \boldsymbol{u}_t^T \boldsymbol{R}_t \delta \boldsymbol{u}_t , \quad (9)$$

$$\tilde{\ell}_f(\boldsymbol{x}) = q_f + \boldsymbol{q}_f^T \delta \boldsymbol{x}_t + \frac{1}{2} \delta \boldsymbol{x}_t^T \boldsymbol{Q}_f \delta \boldsymbol{x}_t . \tag{10}$$

In order to explicitly take into account noise present in our measurement model (2), we include the dynamics of a state observer. This is an important step, that allows us to define an enlarged dynamical system composed of the control and estimation problems. By using this enlarged dynamical system, we are able to include measurement noise and extend previous results, while remaining computationally efficient. The state observer could in principle be of any type, but it is required that out of it, we can obtain a sequence of estimation gains. Therefore, we use an Extended Kalman filter (EKF), whose dynamics are given by

$$d(\delta \hat{\boldsymbol{x}}_t) = (\boldsymbol{A}_t \delta \hat{\boldsymbol{x}}_t + \boldsymbol{B}_t \delta \boldsymbol{u}_t) dt + \boldsymbol{K}_t [d(\delta \boldsymbol{y}_t) - d(\delta \hat{\boldsymbol{y}}_t)] . \tag{11}$$

The dynamics of the control-estimation problem can be compactly written as

$$\underbrace{\begin{bmatrix} d(\delta \boldsymbol{x}_t) \\ d(\delta \hat{\boldsymbol{x}}_t) \end{bmatrix}}_{d(\delta \tilde{\boldsymbol{x}}_t)} = \underbrace{\begin{bmatrix} \boldsymbol{A}_t \delta \boldsymbol{x}_t + \boldsymbol{B}_t \delta \boldsymbol{u}_t \\ \boldsymbol{A}_t \delta \hat{\boldsymbol{x}}_t + \boldsymbol{B}_t \delta \boldsymbol{u}_t + \boldsymbol{K}_t \boldsymbol{F}_t (\delta \boldsymbol{x}_t - \delta \hat{\boldsymbol{x}}_t) \end{bmatrix}}_{\boldsymbol{f}(\delta \tilde{\boldsymbol{x}}_t, \delta \boldsymbol{u}_t)} dt + \underbrace{\begin{bmatrix} \boldsymbol{C}_t & 0 \\ 0 & \boldsymbol{K}_t \boldsymbol{D}_t \end{bmatrix} \begin{bmatrix} d\boldsymbol{\omega}_t \\ d\boldsymbol{\gamma}_t \end{bmatrix}}_{\boldsymbol{g}(t)} .$$

$$(12)$$

$\delta \hat{\boldsymbol{x}}_t$ is the estimate of $\delta \boldsymbol{x}_t$, and $\delta \tilde{\boldsymbol{x}}_t$ represents the vector $[\delta \boldsymbol{x}_t, \delta \hat{\boldsymbol{x}}_t]^T$. Equation (12) is a bilinear system in $\delta \tilde{\boldsymbol{x}}_t$, $\delta \boldsymbol{u}_t$ and $\boldsymbol{K}_t$. Below, we show in detail the derivation. However, the algorithm's main idea is to use this special problem structure to iteratively find a solution. We forward propagate measurement noise and compute estimation gains $\boldsymbol{K}_t$ along the nominal trajectory. Then, with fixed estimation gains, we use a usual backward pass to compute feedback controllers [7, 8]. This eases the design of a locally optimal estimator and controller, while still being able to consider the effects of process and measurement noise. As can be easily noticed, $\boldsymbol{f}(\delta \tilde{\boldsymbol{x}}_t, \delta \boldsymbol{u}_t)$ and $\boldsymbol{g}(t)$ correspond to what in (6), we called $\boldsymbol{m}$ and $\boldsymbol{M}$ respectively. However, now they include measurement noise by incorporating the dynamics of a state estimator.

Estimator Design. We use an EKF; however, other estimators could be used as long as we can extract a sequence of estimation gains. The optimal estimation gains that minimize the error dynamics

$$\dot{\boldsymbol{\Sigma}}_t^e = (\boldsymbol{A}_t - \boldsymbol{K}_t \boldsymbol{F}_t) \boldsymbol{\Sigma}_t^e + \boldsymbol{\Sigma}_t^e (\boldsymbol{A}_t - \boldsymbol{K}_t \boldsymbol{F}_t)^T + \boldsymbol{K}_t \boldsymbol{D}_t \boldsymbol{\Gamma}_t \boldsymbol{D}_t^T \boldsymbol{K}_t^T + \boldsymbol{C}_t \boldsymbol{\Omega}_t \boldsymbol{C}_t^T$$

$$(13)$$

are given by

$$\boldsymbol{K}_t = \boldsymbol{\Sigma}_t^e \boldsymbol{F}_t^T (\boldsymbol{D}_t \boldsymbol{\Gamma}_t \boldsymbol{D}_t^T)^{-1} . \tag{14}$$

They are updated at each iteration in a forward pass along the nominal trajectories, and are then fixed for the backward pass. In this way, the estimation-control system (12), is linear in $\delta \tilde{\boldsymbol{x}}_t$ and $\delta \boldsymbol{u}_t$. This allows us to make use of the HJB Eq. (6) to compute a control law $\boldsymbol{\pi}$ sensitive to both process and measurement noise of the original system.

Controller Design. The locally-optimal control law is affine, of the form $\delta\boldsymbol{u}_t = \boldsymbol{l}_t + \boldsymbol{L}_t\delta\hat{\boldsymbol{x}}_t$. Notice that, we assume it to be a functional only of the state estimate. The HJB equation for this system has the same form as (6) (remember that $\boldsymbol{m}$ and $\boldsymbol{M}$ correspond now to $\boldsymbol{f}$ and $\boldsymbol{g}$ respectively), the cost is given by (9)-(10) (remember that we use the HJB equation under the exponential transformation; therefore, the cost need not to be exponentiated), and the dynamics by (12). The Ansatz for the value function $\Psi(\delta\tilde{\boldsymbol{x}}_t, t)$ is quadratic of the form

$$\Psi(\delta\tilde{\boldsymbol{x}}_t, t) = \frac{1}{2}\begin{bmatrix}\delta\boldsymbol{x}_t \\ \delta\hat{\boldsymbol{x}}_t\end{bmatrix}^T \begin{bmatrix}\boldsymbol{S}_t^x & \boldsymbol{S}_t^{x\hat{x}} \\ (\boldsymbol{S}_t^{x\hat{x}})^T & \boldsymbol{S}_t^{\hat{x}}\end{bmatrix}\begin{bmatrix}\delta\boldsymbol{x}_t \\ \delta\hat{\boldsymbol{x}}_t\end{bmatrix} + \begin{bmatrix}\delta\boldsymbol{x}_t \\ \delta\hat{\boldsymbol{x}}_t\end{bmatrix}^T \begin{bmatrix}\boldsymbol{s}_t^x \\ \boldsymbol{s}_t^{\hat{x}}\end{bmatrix} + s_t \ .$$

and the partial derivatives of the Ansatz Ψ are given by

$$\partial_t\Psi = \frac{1}{2}\begin{bmatrix}\delta\boldsymbol{x}_t \\ \delta\hat{\boldsymbol{x}}_t\end{bmatrix}^T \begin{bmatrix}\dot{\boldsymbol{S}}_t^x & \dot{\boldsymbol{S}}_t^{x\hat{x}} \\ \dot{\boldsymbol{S}}_t^{\hat{x}x} & \dot{\boldsymbol{S}}_t^{\hat{x}}\end{bmatrix}\begin{bmatrix}\delta\boldsymbol{x}_t \\ \delta\hat{\boldsymbol{x}}_t\end{bmatrix} + \begin{bmatrix}\delta\boldsymbol{x}_t \\ \delta\hat{\boldsymbol{x}}_t\end{bmatrix}^T \begin{bmatrix}\dot{\boldsymbol{s}}_t^x \\ \dot{\boldsymbol{s}}_t^{\hat{x}}\end{bmatrix} + \dot{s}_t \ .$$

$$\nabla_{\delta\tilde{\boldsymbol{x}}}\Psi = \begin{bmatrix}\boldsymbol{S}_t^x & \boldsymbol{S}_t^{x\hat{x}} \\ \boldsymbol{S}_t^{\hat{x}x} & \boldsymbol{S}_t^{\hat{x}}\end{bmatrix}\begin{bmatrix}\delta\boldsymbol{x}_t \\ \delta\hat{\boldsymbol{x}}_t\end{bmatrix} + \begin{bmatrix}\boldsymbol{s}_t^x \\ \boldsymbol{s}_t^{\hat{x}}\end{bmatrix} \ .$$

$$\nabla_{\delta\tilde{\boldsymbol{x}}\delta\tilde{\boldsymbol{x}}}\Psi = \begin{bmatrix}\boldsymbol{S}_t^x & \boldsymbol{S}_t^{x\hat{x}} \\ \boldsymbol{S}_t^{\hat{x}x} & \boldsymbol{S}_t^{\hat{x}}\end{bmatrix} \ .$$

The right super-scripts x and $\hat{x}$ for $\boldsymbol{S}$ and $\boldsymbol{s}$ denote that they are sub-blocks that multiply x and $\hat{x}$, respectively. Under the assumed linear dynamics and quadratic cost and value function, the HJB eq. can be written as follows. The LHS corresponds to the time derivative of the value function and is given by

$$-\frac{1}{2}\delta\boldsymbol{x}_t^T \dot{\boldsymbol{S}}_t^x\delta\boldsymbol{x}_t - \frac{1}{2}\delta\hat{\boldsymbol{x}}_t^T \dot{\boldsymbol{S}}_t^{\hat{x}}\delta\hat{\boldsymbol{x}}_t - \delta\boldsymbol{x}_t^T \dot{\boldsymbol{S}}_t^{x\hat{x}}\delta\hat{\boldsymbol{x}}_t - \delta\boldsymbol{x}_t^T \dot{\boldsymbol{s}}_t^x - \delta\hat{\boldsymbol{x}}_t^T \dot{\boldsymbol{s}}_t^{\hat{x}} - \dot{s}_t \ , \qquad (15)$$

and the RHS corresponds to the following minimization (where for presentation clarity, we call $\alpha_t = \boldsymbol{C}_t\Omega_t\boldsymbol{C}_t^T$ and $\beta_t = \boldsymbol{K}_t\boldsymbol{D}_t\Gamma_t\boldsymbol{D}_t^T\boldsymbol{K}_t^T$):

$$= \min_{\delta\boldsymbol{u}_t} \left\{ q_t + \boldsymbol{q}_t^T\delta\boldsymbol{x}_t + \boldsymbol{r}_t^T\delta\boldsymbol{u}_t + \frac{1}{2}\delta\boldsymbol{x}_t^T\boldsymbol{Q}_t\delta\boldsymbol{x}_t + \delta\boldsymbol{x}_t^T\boldsymbol{P}_t\delta\boldsymbol{u}_t + \frac{1}{2}\delta\boldsymbol{u}_t^T\boldsymbol{R}_t\delta\boldsymbol{u}_t + \right.$$
$$(\boldsymbol{S}_t^x\delta\boldsymbol{x}_t + \boldsymbol{S}_t^{x\hat{x}}\delta\hat{\boldsymbol{x}}_t + \boldsymbol{s}_t^x)^T(\boldsymbol{A}_t\delta\boldsymbol{x}_t + \boldsymbol{B}_t\delta\boldsymbol{u}_t) + (\boldsymbol{S}_t^{\hat{x}x}\delta\boldsymbol{x}_t + \boldsymbol{S}_t^{\hat{x}}\delta\hat{\boldsymbol{x}}_t + $$
$$\boldsymbol{s}_t^{\hat{x}})^T(\boldsymbol{A}_t\delta\hat{\boldsymbol{x}}_t + \boldsymbol{B}_t\delta\boldsymbol{u}_t + \boldsymbol{K}_t\boldsymbol{F}_t(\delta\boldsymbol{x}_t - \delta\hat{\boldsymbol{x}}_t)) + \frac{\sigma}{2}(\boldsymbol{S}_t^x\delta\boldsymbol{x}_t + \boldsymbol{S}_t^{x\hat{x}}\delta\hat{\boldsymbol{x}}_t$$
$$+ \boldsymbol{s}_t^x)^T\alpha_t(\boldsymbol{S}_t^x\delta\boldsymbol{x}_t + \boldsymbol{S}_t^{x\hat{x}}\delta\hat{\boldsymbol{x}}_t + \boldsymbol{s}_t^x) + \frac{\sigma}{2}(\boldsymbol{S}_t^{\hat{x}x}\delta\boldsymbol{x}_t + \boldsymbol{S}_t^{\hat{x}}\delta\hat{\boldsymbol{x}}_t + $$
$$\left. \boldsymbol{s}_t^{\hat{x}})^T\beta_t(\boldsymbol{S}_t^{\hat{x}x}\delta\boldsymbol{x}_t + \boldsymbol{S}_t^{\hat{x}}\delta\hat{\boldsymbol{x}}_t + \boldsymbol{s}_t^{\hat{x}}) + \frac{1}{2}\text{Tr}\left(\boldsymbol{S}_t^x\alpha_t\right) + \frac{1}{2}\text{Tr}\left(\boldsymbol{S}_t^{\hat{x}}\beta_t\right) \right\} \ . \quad (16)$$

To perform the minimization of the RHS, we analyze its control dependent terms, corresponding to the part of the cost to go that is control dependent:

$$V_{\delta\boldsymbol{u}_t} = \frac{1}{2}\delta\boldsymbol{u}_t^T\underbrace{\boldsymbol{R}_t}_{\boldsymbol{H}_t}\delta\boldsymbol{u}_t + \delta\boldsymbol{u}_t^T\underbrace{\left(\boldsymbol{r}_t + \boldsymbol{B}_t^T\left(\boldsymbol{s}_t^x + \boldsymbol{s}_t^{\hat{x}}\right) + \right.}_{\boldsymbol{g}_t}$$
$$\underbrace{\left(\boldsymbol{P}_t^T + \boldsymbol{B}_t^T\left(\boldsymbol{S}_t^x + \boldsymbol{S}_t^{\hat{x}x}\right)\right)}_{\boldsymbol{G}_t^x}\delta\boldsymbol{x}_t + \underbrace{\boldsymbol{B}_t^T\left(\boldsymbol{S}_t^{x\hat{x}} + \boldsymbol{S}_t^{\hat{x}}\right)}_{\boldsymbol{G}_t^{\hat{x}}}\delta\hat{\boldsymbol{x}}_t) \ . \qquad (17)$$

The above expression is quadratic in $\delta\boldsymbol{u}_t$ and is easy to minimize. However, the minimum is a functional not only of $\delta\hat{\boldsymbol{x}}_t$, but also of $\delta\boldsymbol{x}_t$. Here, we use the assumption that we do not have access to full state information, only a statistical description of it, given by the state estimate. Therefore, in order to perform the minimization, we take an expectation of $V_{\delta\boldsymbol{u}_t}$ over $\delta\boldsymbol{x}_t$ conditioned on $\delta\hat{\boldsymbol{x}}_t$

$$\mathbb{E}_{\delta\mathbf{x}_t|\delta\hat{\mathbf{x}}_t}\left[V_{\delta\boldsymbol{u}_t}\right] = \frac{1}{2}\delta\boldsymbol{u}_t^T \boldsymbol{H}_t \delta\boldsymbol{u}_t + \delta\boldsymbol{u}_t^T(\boldsymbol{g}_t + (\boldsymbol{G}_t^x + \boldsymbol{G}_t^{\hat{x}})\delta\hat{\boldsymbol{x}}_t) \ .$$

This means that the cost of uncertainty due to measurement noise, considers only the effects of mean and variance of the measurement (captured by the EKF) when evaluating noise effects on the statistical properties of the performance criteria. Consequently, the risk-sensitive control law, considers only as cost of measurement uncertainty the one that can be computed by means of the state estimate, in other words, the one that can be extracted from using mean and variance of the state estimate and neglecting higher order terms.

From the above expression, the minimizer can be analytically computed. In case of control constraints, a quadratic program can be used to solve for the constrained minimizer [5]. In both cases, the minimizer is an affine functional of the state-estimate. For the unconstrained case, it is given by

$$\delta\boldsymbol{u}_t = \boldsymbol{l}_t + \boldsymbol{L}_t\delta\hat{\boldsymbol{x}}_t = -\boldsymbol{H}_t^{-1}\boldsymbol{g}_t - \boldsymbol{H}_t^{-1}(\boldsymbol{G}_t^x + \boldsymbol{G}_t^{\hat{x}})\delta\hat{\boldsymbol{x}}_t \ . \tag{18}$$

$V_{\delta\boldsymbol{u}_t}$ can then be written in terms of the optimal control as

$$V_{\delta\boldsymbol{u}_t^*} = \frac{1}{2}\delta\hat{\boldsymbol{x}}_t^T\left((\boldsymbol{G}_t^x)^T\boldsymbol{H}_t^{-1}\boldsymbol{G}_t^x - (\boldsymbol{G}_t^{\hat{x}})^T\boldsymbol{H}_t^{-1}\boldsymbol{G}_t^{\hat{x}}\right)\delta\hat{\boldsymbol{x}}_t - \delta\boldsymbol{x}_t^T(\boldsymbol{G}_t^x)^T\boldsymbol{H}_t^{-1}(\boldsymbol{G}_t^x+$$
$$\boldsymbol{G}_t^{\hat{x}})\delta\hat{\boldsymbol{x}}_t - \frac{1}{2}\boldsymbol{g}_t^T\boldsymbol{H}_t^{-1}\boldsymbol{g}_t - \delta\boldsymbol{x}_t^T(\boldsymbol{G}_t^x)^T\boldsymbol{H}_t^{-1}\boldsymbol{g}_t - \delta\hat{\boldsymbol{x}}_t^T(\boldsymbol{G}_t^{\hat{x}})^T\boldsymbol{H}_t^{-1}\boldsymbol{g}_t \ . \tag{19}$$

The negative coefficients in the terms of $V_{\delta\boldsymbol{u}_t^*}$ are the benefit of control at reducing the cost. It should be noted that even setting measurement noise to zero does not give a control law equivalent to what was found in [13]. It should be clear from (18) that mathematically they are not the same. However, it is worth pointing out that, [13] considers neither measurement noise, nor the combined effect of process and measurement noise over optimal controls. Here, we do, and setting measurement noise to zero has the specific meaning that we are absolutely sure about our state, and because of it, this control law allows the use of more control authority. In the presence of measurement noise, our control law has more conservative gains than [13], in order to remain compliant enough for the measurement noise level. Writing these terms back into the RHS of the HJB, we can drop the minimization and verify that the quadratic Ansatz for the value function remains quadratic and is therefore valid. Finally, matching terms in LHS and RHS of the HJB eq., we write the backward pass recursion eqns. as:

$$-\dot{\boldsymbol{S}}_t^x = \boldsymbol{Q}_t + \boldsymbol{A}_t^T \boldsymbol{S}_t^x + (\boldsymbol{S}_t^x)^T \boldsymbol{A}_t + \boldsymbol{S}_t^{x\hat{x}} \boldsymbol{K}_t \boldsymbol{F}_t + \boldsymbol{F}_t^T \boldsymbol{K}_t^T (\boldsymbol{S}_t^{x\hat{x}})^T +$$
$$\sigma(\boldsymbol{S}_t^x)^T \alpha_t \boldsymbol{S}_t^x + \sigma \boldsymbol{S}_t^{x\hat{x}} \beta_t (\boldsymbol{S}_t^{x\hat{x}})^T \ .$$

$$-\dot{\boldsymbol{S}}_t^{\hat{x}} = (\boldsymbol{A}_t - \boldsymbol{K}_t \boldsymbol{F}_t)^T \boldsymbol{S}_t^{\hat{x}} + (\boldsymbol{S}_t^{\hat{x}})^T (\boldsymbol{A}_t - \boldsymbol{K}_t \boldsymbol{F}_t) + (\boldsymbol{G}_t^x)^T \boldsymbol{H}_t^{-1} \boldsymbol{G}_t^x -$$
$$(\boldsymbol{G}_t^{\hat{x}})^T \boldsymbol{H}_t^{-1} \boldsymbol{G}_t^{\hat{x}} + \sigma(\boldsymbol{S}_t^{x\hat{x}})^T \alpha_t \boldsymbol{S}_t^{x\hat{x}} + \sigma(\boldsymbol{S}_t^{\hat{x}})^T \beta_t \boldsymbol{S}_t^{\hat{x}} \ .$$

$$-\dot{\boldsymbol{S}}_t^{x\hat{x}} = \boldsymbol{A}_t^T \boldsymbol{S}_t^{x\hat{x}} + \boldsymbol{S}_t^{x\hat{x}} (\boldsymbol{A}_t - \boldsymbol{K}_t \boldsymbol{F}_t) + \boldsymbol{F}_t^T \boldsymbol{K}_t^T \boldsymbol{S}_t^{\hat{x}} - (\boldsymbol{G}_t^x)^T \boldsymbol{H}_t^{-1} (\boldsymbol{G}_t^x + \boldsymbol{G}_t^{\hat{x}}) +$$
$$\sigma(\boldsymbol{S}_t^x)^T \alpha_t \boldsymbol{S}_t^{x\hat{x}} + \sigma \boldsymbol{S}_t^{x\hat{x}} \beta_t \boldsymbol{S}_t^{\hat{x}} \ .$$

$$-\dot{\boldsymbol{s}}_t^x = \boldsymbol{q}_t + \boldsymbol{A}_t^T \boldsymbol{s}_t^x + \boldsymbol{F}_t^T \boldsymbol{K}_t^T \boldsymbol{s}_t^{\hat{x}} - (\boldsymbol{G}_t^x)^T \boldsymbol{H}_t^{-1} \boldsymbol{g}_t + \sigma(\boldsymbol{S}_t^x)^T \alpha_t \boldsymbol{s}_t^x + \sigma \boldsymbol{S}_t^{x\hat{x}} \beta_t \boldsymbol{s}_t^{\hat{x}} \ .$$

$$-\dot{\boldsymbol{s}}_t^{\hat{x}} = (\boldsymbol{A}_t - \boldsymbol{K}_t \boldsymbol{F}_t)^T \boldsymbol{s}_t^{\hat{x}} - (\boldsymbol{G}_t^{\hat{x}})^T \boldsymbol{H}_t^{-1} \boldsymbol{g}_t + \sigma(\boldsymbol{S}_t^{x\hat{x}})^T \alpha_t \boldsymbol{s}_t^x + \sigma(\boldsymbol{S}_t^{\hat{x}})^T \beta_t \boldsymbol{s}_t^{\hat{x}} \ .$$

$$-\dot{s}_t = q_t - \frac{1}{2} \boldsymbol{g}_t^T \boldsymbol{H}_t^{-1} \boldsymbol{g}_t + \frac{1}{2} \operatorname{Tr}\left((\boldsymbol{S}_t^x)\,\alpha_t\right) + \frac{1}{2} \operatorname{Tr}\left(\boldsymbol{S}_t^{\hat{x}} \beta_t\right) +$$
$$\frac{\sigma}{2}(\boldsymbol{s}_t^x)^T \alpha_t \boldsymbol{s}_t^x + \frac{\sigma}{2}(\boldsymbol{s}_t^{\hat{x}})^T \beta_t \boldsymbol{s}_t^{\hat{x}} \ . \tag{20}$$

The integration runs backward in time with $\boldsymbol{S}_t^x = \boldsymbol{Q}_f$, $\boldsymbol{S}_t^{\hat{x}} = 0$, $\boldsymbol{S}_t^{x\hat{x}} = 0$, $\boldsymbol{s}_t^x = \boldsymbol{q}_f$, $\boldsymbol{s}_t^{\hat{x}} = 0$ and $s_t = q_f$. Despite being long, it is a very simple to implement solution, similar to any other LQR-style recursion.

Remark 1. The effects of process and measurement noise appear in pairs due to the fact that we assumed their Brownian motions to be uncorrelated (see $\boldsymbol{g}(t)$ in (12)). However, their combined effect is not just as having higher process noise. Estimation couples their effects, and this can be seen in the recursion equation, where we do not only have costs for the state and its estimate $\boldsymbol{S}_t^x$ and $\boldsymbol{S}_t^{\hat{x}}$, but also the coupling cost $\boldsymbol{S}_t^{x\hat{x}}$; whose products with the covariances of process noise and estimation error determine how process noise and measurement uncertainty affect the value function and therefore the control law.

4 Experimental Results

In this section, we use the control algorithm on a 2-DOF manipulator on 2 different tasks: a viapoint task and a contact task. This setup allows us to analyze in a simple setting the important properties of the algorithm. The equations of motion are given by

$$\boldsymbol{H}(\boldsymbol{q})\ddot{\boldsymbol{q}} + \boldsymbol{C}(\boldsymbol{q}, \dot{\boldsymbol{q}}) = \boldsymbol{\tau} + \boldsymbol{J}(\boldsymbol{q})^T \boldsymbol{\lambda} \ . \tag{21}$$

The vector $\boldsymbol{q} = [q_1, \ q_2]^T$ contains the joints positions. $\boldsymbol{H}(\boldsymbol{q})$ is the inertia matrix, $\boldsymbol{C}(\boldsymbol{q}, \dot{\boldsymbol{q}})$ the vector of Coriolis and centrifugal forces, $\boldsymbol{J}(\boldsymbol{q})$ is the end-effector Jacobian, $\boldsymbol{\lambda} \in \mathbb{R}^2$ the external forces and $\boldsymbol{\tau} \in \mathbb{R}^2$ the input torques. The system dynamics can be easily written in the form given by (1), with additive process noise $d\omega$ and state $\boldsymbol{x} = [\boldsymbol{q}^T, \ \dot{\boldsymbol{q}}^T]^T$. The measurement model can also easily be written in the form given by (2) ($d\boldsymbol{y} = d\boldsymbol{x} + d\gamma$), with Brownian motion γ with variance Γdt.

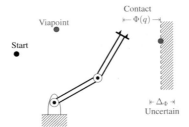

Fig. 2: Schematic: pass through a viapoint and establish contact with a wall with uncertain location $\Phi(q) + \Delta_\Phi$.

4.1   Experiment 1: Process Noise vs. Measurement Uncertainty

We compare the effect of process and measurement noise in the control law in a motion task between two points with two viapoints. The objective function

$$
\mathcal{J} = \sum_0^{t_f} c_u \boldsymbol{\tau}^T \boldsymbol{\tau} + \sum_{i=1}^{N_{via}} c_i \log(\cosh(||\boldsymbol{x} - \boldsymbol{x}_i||_2)) + c_{t_f} \log(\cosh(||\boldsymbol{x} - \boldsymbol{x}_{t_f}||_2))
$$

measures task performance. $\boldsymbol{x}$, $\boldsymbol{x}_i$, $\boldsymbol{x}_{t_f} \in \mathbb{R}^4$ are current, viapoints and final desired end-effector positions and velocities, respectively. c_u, c_i, c_{t_f} are cost weights. The nonlinear cost $\log(\cosh(\cdot))$ is a soft absolute value to demonstrate that general nonlinear costs functions can be used.

We first evaluate the effects of increasing process noise under no measurement uncertainty (Fig. 3 - left). Feedback gains for the motion task under several noise intensities are shown. In general, they are higher for regulating behavior at the viapoints and goal position. As process noise increases, the cost of uncertainty does too, because we might miss the viapoints or the goal due to disturbances. This can be seen in sample trajectories, where the variance of the trajectories due to noise has increased. In this case, the trade-off between cost of uncertainty and control-effort involves feedback gains proportional to the process noise, **the higher the process noise, the higher the feedback gains**.

In a second set of simulations, we test the effect of increasing measurement uncertainty under no process noise (Fig. 3 - right). Feedback gains and sample trajectories for different values of measurement noise are shown. Feedback gains are also higher near viapoints and goal position, and sample trajectories are similar to the ones with process noise. The big difference is that the optimal control solution for this case is to trust feedback proportionally to the information content of the measurements, namely, **the higher the measurement uncertainty, the lower the feedback gains**. It shows how under low measurement noise, feedback control with higher gains is possible and optimal. Under high measurement noise, lower impedance is better. We note that during the evaluation of the controller online estimation is used as it achieves better performance than using the precomputed sequence of estimation gains. In these experiments, we kept the risk sensitive parameter constant as it is not the focus of this paper (see for example [13]). However, the effects of process and measurement noise are qualitatively similar for all allowed values of σ (data not shown).

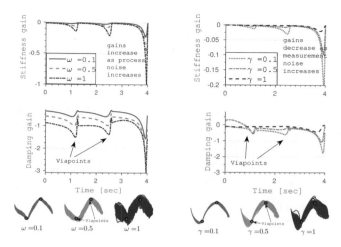

Fig. 3: Comparison between process noise and measurement uncertainty. The control gains for various level of noise are shown in the upper graphs. Sample trajectories are shown in the lower graphs for both varying process noise (left) and measurement uncertainty (right). The red dots represent the viapoints. In all the experiments the risk sensitive parameter $\sigma = 2.5$. Stiffness gains are normalized to 1 corresponding to 100 N/m.

4.2   Experiment 2: Establishing Contact with the Environment

In this experiment, the robot needs to pass through a viapoint and then make contact with a wall at an uncertain location (Δ_Φ), as shown in Fig. 2. While simple enough to be carefully analyzed, the experiment addresses the role of measurement uncertainty when interacting with an uncertain environment, which is important for manipulation and locomotion tasks. Performance is measured by

$$\mathcal{J} = \sum_0^{t_f} c_u \boldsymbol{\tau}^T \boldsymbol{\tau} + c_{\text{via}} \log(\cosh(||\boldsymbol{x}_{t_{\text{via}}} - \boldsymbol{x}_{\text{via}}||_2)) +$$

$$\sum_{t_{\text{cnt}_0}}^{t_{\text{cnt}_f}} c_{\text{cnt}} \log(\cosh(\Phi(\boldsymbol{q})) \cosh(||\boldsymbol{\lambda} - \boldsymbol{\lambda}_{\text{des}}||_2)) \ .$$

$\boldsymbol{\lambda}_{\text{des}} \in \mathbb{R}^2$ is the desired contact force at contact; c_{cnt}, c_{via}, c_u are cost weights. This cost rewards low torques, passing a viapoint x_{via} at time t_{via}, making contact $\Phi(\boldsymbol{q}) = 0$ and exerting a desired force from t_{cnt_0} to t_{cnt_f} (shown as shaded areas in Figs. 4-5). The external force $\boldsymbol{\lambda}$ is modeled as a stiff spring and is part of the dynamic model such that its effect is known to the optimizer. Given that its value depends on the uncertain position of the wall, it is also an uncertain variable. There are two possible ways to encode the uncertainty in the distance to the contact $\Phi(\boldsymbol{q})$. On the one hand, it is a function of the joint positions q, and therefore, we could model measurement noise directly into these components. The other alternative is to add a new state x' to the state vector $\boldsymbol{x}$. This new state would be defined as the distance to the contact $\Phi(\boldsymbol{q})$. For the dynamic model, we need its derivative, which is given by $\nabla_q \Phi(q)^T \dot{q}$, and in the measurement model we model directly our uncertainty in the value of $\Phi(q)$.

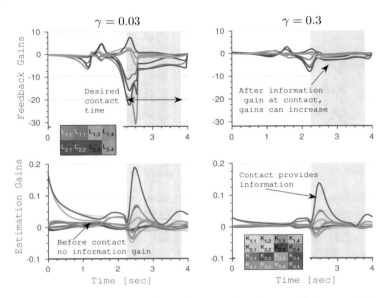

Fig. 4: Feedback and Estimation Gains for a motion-force task for two different values of measurement noise γ, that encodes the uncertainty in the distance to the contact.

Fig. 4 shows feedback and estimation gains for two measurement noise values γ that encode uncertainty in the state (distance to contact). Feedback gains show two peaks around 1 and 2.2 sec, when passing the viapoint and when contact happens. Feedback gains for $\gamma = 0.03$ are higher than for $\gamma = 0.3$, where control is more cautious. Estimation gains are qualitatively similar. Interestingly, passing the viapoint does not affect them but when the contact is expected, they are higher because contact provides location information. As we expected, under increasing measurement noise, feedback gains decrease, which allows us to have a compliant interaction in the presence of measurement uncertainty.

Fig. 5 shows force profiles of contact interactions with the wall. Black dashed lines are the reference forces. Dashed blue lines show the interaction force profiles using a controller not sensitive to measurement uncertainty, optimized for process noise ($\omega = 0.2$) but very low measurement uncertainty ($\gamma = 0.003$). The distribution of force profiles under the sensitive control law and the stochastic dynamics is shown in green. It was optimized for process noise ($\omega = 0.2$) and measurement uncertainty ($\gamma = 0.3$). We see that with the controller using measurement uncertainty when contact happens before it was expected ($\Delta_\Phi = 1.5$ or $\Delta_\Phi = 3.0$ cm), forces are higher than the reference, but the interaction is not as aggressive as it would be with the higher feedback gains of a usual non-sensitive optimal controller. In the case of the controller sensitive only to process noise, since the feedback gains are higher we see much higher contact forces (blue lines) and even a loss of contact (Fig. 5 - right).

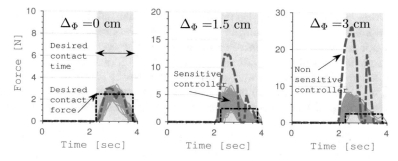

Fig. 5: Contact forces given a perturbation $\Delta\Phi$ at the contact location. Black dashed lines show reference desired forces and dashed blue lines show the interaction profile of a typical optimal feedback controller that does not take into account measurement uncertainty. The controller sensitive to measurement uncertainty (green lines) robustly deal with the uncertain dynamic interaction.

These results illustrate a behavior relevant for robotic applications involving contact interactions: policies sensitive to measurement uncertainty lead to low impedance behavior in face of too high uncertainty. In a receding horizon setting, the impedance behavior would then be adapted as the robot gains more information about the state of the environment (e.g. after making a contact). While the execution does not exploit sensed contact forces, which could improve further the dynamic interaction, it is still able to find a feedback control policy that can safely interact with the environment, despite uncertainty in the position of the wall. This example illustrates that taking into account measurement uncertainty in the control law can lead to more robust behaviors for contact interaction, that cannot be achieved with an approach taking only into account process noise. It is worth noting that our noise description is not limited to state transition due to actuators noise or sensor measurements noise, but can include uncertainty on process and measurement models, which are only an approximation of the true underlying dynamics.

5   Discussion

In our experiments, we have seen that process noise is fundamentally different from measurement noise. While the first one is a dynamics disturbance that requires control using high feedback gains; the second one represents uncertainty in the state information, and requires compliance proportional to the uncertainty to dynamically interact with the world given our limited knowledge of it.

The fundamental difference between process and measurement noise effects on the control law comes from the cost they penalize. Cost of uncertainty due to process noise increases with terms of the form $C_t \Omega_t C_t^T$. If there is no control action, the process noise increases the cost. Therefore, regulation with high gains is optimal. For measurement noise, cost increases with terms $K_t D_t \Gamma_t D_t^T K_t^T$ and estimation gains are inversely proportional to measurement noise. Therefore, not using informative measurements is costly and requires high feedback gains. For poorly-informative measurements, we incur very low cost and control with lower gains is optimal. This behavior can be exploited in robotic tasks with dynamic interactions. For example when making a contact, behaving compliant under poor contact-information is robust. Once the contact is established and position certainty is higher, feedback gains would then be increased. In a receding horizon setup, gain in information about the current state of the world after contact would allow to online adapt the feedback policy, which would improve the performance of the controller compared to only executing a plan, as shown in Fig. 5.

In this work, we have looked into a problem with simple geometry and unidirectional contact interaction. Besides, the uncertainty in the distance to contact has been modelled as a Gaussian distribution. While the first two assumptions, using a point contact model, are common and therefore transferable to more general cases of multi-body systems interacting with a more geometrically rich environment, it is not clear if the same holds for the uncertainty model as a Gaussian distribution. Despite of this fact, the value of this work resides in the insight provided about what is important to consider for controlling the dynamic interaction of a robot with its environment.

From a computational point of view, the algorithm should scale to more complex systems. We can approximate the complexity of a call to the dynamics with its heaviest computation (factorization and back-substitution of $H(q)$) as roughly $O(n^3)$, n being the number of states. The most expensive computation is that of first derivatives $O(Nn^4)$, N being the number of timesteps in the horizon. This is in the same order of complexity as other iterative approaches that show very good performance on more complicated robotic tasks [4],[5], although those examples did not exploit measurement uncertainty for control. While our approach requires using twice the number of actual states, which increases the solving time a small amount, this should still be fine given the impressive results of recent papers on high dimensional robotics problems with contacts [19].

6   Conclusion

We have presented an iterative algorithm for finding locally-optimal feedback controllers for nonlinear systems with additive measurement uncertainty. In particular we showed that measurement uncertainty leads to very different behaviors than process noise and it can be exploited to create low impedance behaviors in uncertain environments (e.g. during contact interaction). This opens the possibility for planning and controlling contact interactions robustly based on controllers sensitive to measurement noise. In a receding horizon setting, it could be possible to regulate impedance in a meaningful way depending on the current uncertainty about the environment.

Acknowledgments. This research was mainly supported by the Max-Planck-Society and the European Research Council under the European Unions Horizon 2020 research and innovation programme (grant No 637935). It was also supported by National Science Foundation grants IIS-1205249, IIS-1017134, EECS-0926052, the Office of Naval Research, the Okawa Foundation, and the Max-Planck ETH Center for Learning Systems.

References

1. Stefan Schaal and Atkeson C. Learning control in robotics. *Robotics and Automation Magazine*, 17:20–29, 2010.
2. Situan Feng, Xinjilefu X., Weiwei Huang, and Christopher G. Atkeson. 3D walking based on online optimization. *Humanoids Atlanta*, pages 21–27, 2013.
3. Michael Neunert, Cedric de Crousaz, Fadri Furrer, Mina Kamel, Farbod Farshidian, Roland Siegwart, and Jonas Buchli. Fast Nonlinear Model Predictive Control for Unified Trajectory Optimization and Tracking. *ICRA*, 2016.
4. Yuval Tassa, Tom Erez, and E Todorov. Fast model predictive control for reactive robotic swimming.
5. Yuval Tassa, Nicolas Mansard, and Emo Todorov. Control-limited differential dynamic programming. In *ICRA, Hong Kong, China*, pages 1168–1175, 2014.
6. Emmanuel Todorov and Weiwei Li. A generalized iterative LQG method for locally-optimal feedback control of constrained nonlinear stochastic systems. *ACC 2005*, 1:300 – 306, 2005.
7. David Mayne. A Second-order Gradient Method for Determining Optimal Trajectories of Non-linear Discrete-time Systems. *Int. Journal of Control*, 3(1):85–95, 1966.
8. Athanasios Sideris and James Bobrow. An efficient sequential linear quadratic algorithm for solving nonlinear optimal control problems. *IEEE TAC*, 50(12):2043–2047, 2005.
9. Aaron M. Johnson, Jennifer E. King, and Siddhartha Srinivasa. Convergent Planning. *IEEE Robotics and Automation Letters*, 1(2):1044–1051, 2016.
10. Weiwei Li and Emmanuel Todorov. Iterative linearization methods for approximately optimal control and estimation of non-linear stochastic system. *International Journal of Control*, 80(9):1439 – 1453, 2007.
11. Emmanuel Todorov. Stochastic optimal control and estimation methods adapted to the noise characteristics of the sensorimotor system. *Neural Computation*, 17(5), 2005.
12. D. Jacobson. Optimal stochastic linear systems with exponential performance criteria and their relation to deterministic differential games. *IEEE Transactions on Automatic Control*, 18(2):124–131, 1973.
13. Farbod Farshidian and Jonas Buchli. Risk Sensitive, Nonlinear Optimal Control: Iterative Linear Exponential-Quadratic Optimal Control with Gaussian Noise. 2015.
14. Krishnamurthy Dvijotham and Emanuel Todorov. A Unifying Framework for Linearly Solvable Control. *CoRR*, abs/1202.3715, 2012.
15. Jason Speyer, John Deyst, and D. Jacobson. Optimization of stochastic linear systems with additive measurement and process noise using exponential performance criteria. *IEEE Transactions on Automatic Control*, 19(4):358–366, 1974.
16. MR. James, JS. Baras, and LJ. Elliot. Risk-sensitive control and dynamic games for partially observed discrete-time nonlinear systems. *IEEE Transactions on Automatic Control*, 39(4):780–792, 1994.
17. P. Whittle and J. Kuhn. A hamiltonian formulation of risk-sensitive linear quadratic gaussian control. *International Journal on Control*, 43:1–12, 1986.
18. Horacio Wio. *Path Integrals for Stochastic Processes*. World Scientific, 2013.
19. Sergey Levine and Vladlen Koltun. Guided Policy Search. In *Proceedings of the 30th International Conference in Machine Learning*, pages 1–9, 2013.

Feedback Control of the Pusher-Slider System: A Story of Hybrid and Underactuated Contact Dynamics

François Robert Hogan and Alberto Rodriguez

Department of Mechanical Engineering, Massachussetts Institute of Technology,
77 Massachusetts Avenue, Cambridge, MA, USA

Abstract. This paper investigates real-time control strategies for dynamical systems that involve frictional contact interactions. Hybridness and underactuation are key characteristics of these systems that complicate the design of feedback controllers. In this research, we examine and test a novel feedback controller design on a planar pushing system, where the purpose is to control the motion of a sliding object on a flat surface using a point robotic pusher. The pusher-slider is a simple dynamical system that retains many of the challenges that are typical of robotic manipulation tasks.

Our results show that a model predictive control approach used in tandem with integer programming offers a powerful solution to capture the dynamic constraints associated with the friction cone as well as the hybrid nature of the contact. In order to achieve real-time control, simplifications are proposed to speed up the integer program. The concept of *Family of Modes* (FOM) is introduced to solve an online convex optimization problem by selecting a set of contact mode schedules that spans a large set of dynamic behaviors that can occur during the prediction horizon. The controller design is applied to stabilize the motion of a sliding object about a nominal trajectory, and to re-plan its trajectory in real-time to follow a moving target. We validate the controller design through numerical simulations and experimental results on an industrial ABB IRB 120 robotic arm.

1   INTRODUCTION

Humans manipulate objects within their hands with impressive agility and ease. While doing so, they also make many and frequent mistakes from which they recover seamlessly. The mechanical complexity of the human hand along with its array of sensors sure play an important role. However, despite recent advances in the design of complex robotic hands [1–3] and sensory equipment (tactile sensors, vision markers, proximity sensors, etc. [4, 5]), autonomous robotic manipulation remains far from human skill at manipulating with their hands or teleoperating robotic interfaces.

We argue that this gap in performance can largely be attributed to robots' inability to use sensor information for real-time control purposes. Whereas humans effectively process and react to information from tactile and vision sensing, robot manipulators are most often programmed in an open-loop fashion, incapable of adapting or correcting

© Springer Nature Switzerland AG 2020
K. Goldberg et al. (Eds.): *Algorithmic Foundations of Robotics XII*, SPAR 13, pp. 800–815, 2020.
https://doi.org/10.1007/978-3-030-43089-4_51

their motion. With the recent development of sensing equipment, the question remains: how should robots use sensed information?

This work is concerned with the challenges involved in closing the loop through contact in robotic manipulation. To the knowledge of the authors, a general feedback controller design methodology is still lacking in the field of robotic manipulation, which is essential for robots to be aware and reactive to contact. In this article, we focus our attention on dexterous manipulation tasks, where the manipulated object moves relative to the robot's end effector.

In this article we examine and test a feedback controller design for the pusher-slider system, where the purpose is to control the motion of a sliding object on a flat surface using a point pusher. The pusher-slider system is a simple dynamical system that incorporates several of the challenges that are typical of robotic manipulation tasks. In particular, we are concerned with two main challenges:

1. It is a **hybrid dynamical system** that exhibits different contact modes between the pusher and slider (e.g. separation, sticking, sliding up, and sliding down). Transitions between these modes result in discontinuities in the dynamics, which complicate controller design.
2. It is an **underactuated** system where the contact forces from the pusher acting on the sliding object are constrained to remain inside the friction cone. These constraints on the control inputs lead to a dynamical system where the velocity control of the pusher is not sufficient to produce an arbitrary acceleration of the slider. Ultimately, the controller must reason about finite horizon trajectories and not just instantaneous actuation.

The purpose of this article is to develop a feedback controller design that can handle both challenges described above. Our results show that a model predictive control approach used in tandem with integer programming offers a powerful solution to capture the dynamic constraints associated with the friction cone as well as the hybrid nature of contact. In order to achieve real-time control, simplifications are proposed to speed up the integer program. The concept of *Family of Modes* (FOM) is proposed to solve an online convex optimization problem by simulating the dynamical system forward using a set (i.e., family) of mode schedules that are identified as being key. Numerical simulations and experimental results, performed using an industrial robotic manipulator to push the sliding object and a Vicon system to track its pose, show that the FOM methodology yields a feedback controller design that can be implemented in real-time and stabilize the motion of a slider through a single contact point about a nominal trajectory.

2 RELATED WORK

Historically, grasping and in-hand dexterous manipulation have been two main focuses of robotic manipulation research. In grasping, conventional control approaches first search for the location of grasp contact points that yield some form of geometric closure on the object [6], then close the gripper either blindly or with force control, and finally treat the object as a rigid extension of the robotic arm. In-hand dexterous manipulation was first explored by Salisbury and Craig [7] for dynamic in-hand motion. The controller design techniques presented in the in-hand dexterous literature typically apply to complex robotic hands and rely on the gripper and the manipulated objects to stick together at the contact points.

Dafle et al. [8, 9] demonstrated that simple robotic hands can be used to perform fast and effective regrasp strategies by using the robotic arm and the environment as an external source of actuation. When performing these regrasp strategies, the contact interactions are not limited to sticking contact but also necessarily exploit sliding motion. The control actions proposed by [8, 9] require offline trajectory planning and rely on accurate contact models.

Recently, Posa et al. [10] proposed to apply trajectory optimization tools to determine the motion of the robot and the manipulated object by including contact reaction forces along with motions as decision variables in a large optimization program. This method has been shown to be effective for path planning of high degree of freedom systems undergoing contact interactions. This paper shares the motivation of including contact forces as decision variables as part of an optimization program.

The application of feedback control strategies to discontinuous contact dynamical systems is a relatively unexplored field of research. Tassa and al. [11, 12] have achieved remarkable simulation results by using smoothed contact models in an optimal control framework, and a similar approach has been proposed by Stewart and Anitescu [13]. These methods contrast to the approach explored in this paper, where hybridness of contact is considered explicitly.

With regard to the pusher-slider system, Mason [14] presented an early study of the mechanics of planar pushing. This theory has been applied to the design of controllers that achieve stable pushing [15], which offers the advantage of operating without sensor feedback by acting as an effective grasp. In [16], this theory is expanded to a tactile feedback based controller for the case of a point pusher-slider system.

3   CHALLENGES OF CONTROL THROUGH CONTACT

The aim of this work is to accurately control the motion of objects through contact. Two major difficulties are typical of these systems: hybridness and underactuation.

3.1   Hybridness

When in contact, object and manipulator can interact in different manners. For example, the object can slip within the fingers of the gripper, the gripper can throw the object in the air or perform pick and place maneuvers, etc. These manipulation actions correspond to different contact interaction modes, namely sliding, separation, and sticking. The hybridness associated with the transitions between modes can result in a non-smooth dynamical system. This complicates the design of feedback controllers as the vast majority of standard control techniques rely on smoothness of the dynamical model.

In many applications involving hybrid dynamical systems, this difficulty is overcome by setting the mode scheduling of the controller offline or using on-board sensing to detect mode transitions. For example, in the locomotion community, it is common to transition between two feedback controllers as the robot switches from a stance phase to an aerial phase [17], as in Fig. 1. For robotic manipulation tasks, the mode schedul-

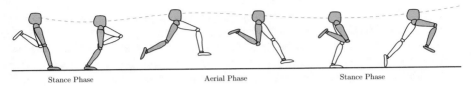

Stance Phase   Aerial Phase   Stance Phase

Fig. 1: Human running gait adapted from Decker et al. [18]. The periodic nature of human gait permits to use control strategies that rely on offline mode scheduling.

ing is often not known a priori and can be challenging to predict. In such cases, we must rely on the controller to decide during execution what interaction mode is most beneficial to the task. Figure 2 illustrates the example of picking a book from a shelf. The hand interacts with the book in a complex manner. It is difficult to say when fingers and palm stick or slide, but those transitions not only happen, but are necessary to pick the book. Likely the hand initially sticks to the book and drags it backwards exploiting friction. Then, the thumb and fingers swiftly slide to regrasp the book. Finally, the book is retrieved from the shelf using a stable grasp. For such manipulation tasks where the motion is not periodic, determining a fixed mode sequencing strategy is not obvious and likely impractical. Errors during execution will surely require that the mode sequencing be altered.

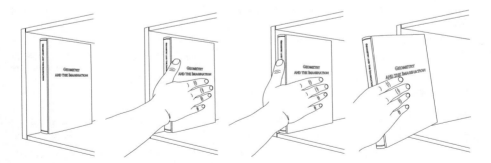

Fig. 2: Animation of a simple manipulation task that exploits multiple contact modalities. First, the hand sticks to the book and drags it backwards exploiting friction. Second, thumb and fingers slide to perform a regrasp maneuver. Finally, the book is retrieved from the shelf using a stable grasp.

3.2 Underactuation

Underactuation is due to the fact that contact interactions can only transmit a limited set of forces and torques to the object. As such, the controller must reason only among the forces that can physically be realized. For example, the normal forces commanded should be positive, as contact interactions can only "push" and cannot "pull." In order to achieve this, it is required to explicitly integrate the physical constraints associated with contact interactions in the controller design. A second important consequence of underactuation is that the controller must be capable of reasoning about future not just instantaneous actuation since the forces required to drive the task in the direction of the goal might not be feasible at the current instant. The controller must reason on a finite horizon.

4   PUSHER-SLIDER SYSTEM

In this article, we study the pusher-slider system, a simple nonprehensile manipulation task where the goal is to control the motion of a sliding object (slider) through a single frictional contact point (pusher). The pusher-slider system is a useful test case dynamical system for controller design where actuation arises from friction.

4.1   Kinematics

Consider the system in Fig. 3. The pose of the slider is $\mathbf{q}_s = \begin{bmatrix} x & y & \theta \end{bmatrix}^\mathsf{T}$ where x and y denote the cartesian coordinates of the center of mass of the slider and θ its orientation relative to the inertial reference frame $\mathcal{F}_a$. The position of the pusher relative to point b resolved in $\mathcal{F}_b$ is $\mathbf{r}_b^{pb} = \begin{bmatrix} p_x & p_y \end{bmatrix}^\mathsf{T}$. Figure 3 shows the kinematics of the slider subject to contact interactions with a single point of contact robotic pusher.

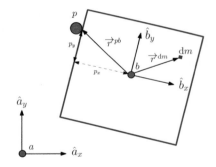

Fig. 3: Kinematics of a slider subject to contact interactions with a single point of contact robotic pusher.

4.2   Quasi-Static Approximation

During contact interactions, two external forces are exerted on the slider: the generalized frictional force applied by the pusher on the slider (denoted $\mathbf{f}^P$) and the generalized frictional force applied by the ground on the sliding object (denoted $\mathbf{f}^G$). Applying Newton's second law in the $\hat{a}_x$ - $\hat{a}_y$ plane yields the motion equations

$$\mathbf{H}\ddot{\mathbf{q}}_s = \mathbf{f}^G + \mathbf{f}^P, \tag{1}$$

where $\mathbf{H}$ denotes the inertia matrix of the system. The quasi-static assumption suggests that at low velocities, frictional contact forces dominate and inertial forces do not have a decisive role in determining the motion of the slider. Under this assumption, the applied frictional force by the pusher is of equal magnitude and opposite direction to the ground planar frictional force (i.e., $\mathbf{f}^P = -\mathbf{f}^G$). This quasi-static assumption leads to a simplified analysis of the motion of a sliding object using a single point of contact robotic pusher. Note that including the term $\mathbf{H}\ddot{\mathbf{q}}_s$ does not complicate the controller design and could easily be integrated into the control formulation presented in Section 5. The resulting controller from a dynamic analysis yields a mapping between the motion of the slider to the reaction forces applied on the object. In contrast, the quasi-static assumption leads to a direct mapping between the motion of the slider and the motion of the pusher. This proved desirable from an experimental implementation standpoint using a position controlled robotic manipulator.

The motion equations of the pusher-slider system are formulated in [16] assuming a quasi-static formulation with a uniform pressure distribution. Prior to presenting these motion equations, it is necessary to review two important concepts of frictional contact interactions: the limit surface and the motion cone.

4.3   Limit Surface

The limit surface is a useful geometric representation which, under the quasi-static assumption, maps the applied frictional force on an object to its resulting velocity. First introduced in [19], the limit surface is defined as a convex surface which bounds the set of all possible frictional forces and moments that can be sustained by frictional interface. In this paper, we use the ellipsoidal approximation to the limit surface [20], where the semi-principal axes are given by f_{max}, f_{max}, and m_{max} defined by $f_{max} = \mu_g mg$ and $m_{max} = \frac{\mu_g mg}{A} \int_A \left\| \vec{r}^{dm\,b} \right\| dA$, where μ_g is the coefficient of friction between the object and the ground, m is the mass of the object, g is the gravitational acceleration, A is the surface area of the object exposed to friction, and $\vec{r}^{dm\,b}$ denotes the position of dm relative to the origin of $\mathcal{F}_b$.

4.4   Motion Cone

Depending on the direction of motion of the pusher, different contact interaction modes can arise between the pusher and the slider. The motion cone [14], shown in Fig. 4, is useful to determine if a given velocity of the pusher will result in sticking or sliding behavior between the pusher and the slider. Each boundary of the motion cone is constructed by mapping the resulting velocity of the slider at the contact point p when subject to a frictional force that lies on a boundary of the friction cone. It can be shown using the ellipsoidal approximation to the limit surface, that for flat faced objects, the two boundaries of the motion cone are given as $\vec{v}_t^{\mathcal{MC}} = 1\hat{b}_x + \gamma_t \hat{b}_y$ and $\vec{v}_b^{\mathcal{MC}} = 1\hat{b}_x + \gamma_b \hat{b}_y$, with

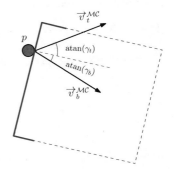

Fig. 4: Motion cone ($\mathcal{MC}$) at contact point p. If the pusher velocity lies within the two boundaries of the motion cone, the pusher will stick to the slider, else, it will slide.

$$\gamma_t = \frac{\mu c^2 - p_x p_y + \mu p_x^2}{c^2 + p_y^2 - \mu p_x p_y} \qquad (2)$$

and

$$\gamma_b = \frac{-\mu c^2 - p_x p_y - \mu p_x^2}{c^2 + p_y^2 + \mu p_x p_y}, \tag{3}$$

where p_x and p_y are shown in Fig. 3 and $c = \frac{f_{max}}{m_{max}}$. Given the velocity of the pusher resolved in $\mathcal{F}_b$, denoted $\mathbf{u} = [v_n \ v_t]^\mathsf{T}$, the conditions stated below determine the resulting contact interaction mode that will arise between the pusher and the slider.

Sticking When the sliding object is sticking to the pusher, the relative tangential velocity between the pusher and the object is zero. In order to have this behavior, the velocity vector must lie within the boundaries of the motion cone in Fig. 4. This constraint, denoted as $\mathbf{u} \in \mathcal{MC}$, is defined as

$$\mathbf{u} \in \mathcal{MC} : \quad \begin{cases} v_t \leq \gamma_t v_n, & (4) \\ v_t \geq \gamma_b v_n. & (5) \end{cases}$$

Sliding Up When the pusher is sliding up in the tangential direction relative to the object, velocity of the pusher must lie above the upper boundary of the motion cone. This constraint is expressed by

$$\mathbf{u} > \mathcal{MC} : \quad \{ \ v_t > \gamma_t v_n. \tag{6}$$

Sliding Down When the pusher is sliding down in the tangential direction relative to the object, the velocity of the pusher must lie below the lower boundary of the motion cone . This condition is enforced as

$$\mathbf{u} < \mathcal{MC} : \quad \{ \ v_t < \gamma_b v_n. \tag{7}$$

4.5 Motion Equations

The motion equations of the pusher-slider system are formulated in [16] and stated below. The equations presented in Eq. (8) describe hybrid dynamics, where the contact interaction mode depends upon the direction of the pusher velocity.

$$\dot{\mathbf{x}} = \begin{cases} \mathbf{f}_1(\mathbf{x}, \mathbf{u}) & \text{if } \mathbf{u} \in \mathcal{MC}, \\ \mathbf{f}_2(\mathbf{x}, \mathbf{u}) & \text{if } \mathbf{u} > \mathcal{MC}, \\ \mathbf{f}_3(\mathbf{x}, \mathbf{u}) & \text{if } \mathbf{u} < \mathcal{MC}, \end{cases} \tag{8}$$

with $\mathbf{x} = [\mathbf{q}_s^\mathsf{T} \ p_y]^\mathsf{T}$, $\mathbf{u} = [v_n \ v_t]^\mathsf{T}$ denotes the velocity of the pusher resolved in $\mathcal{F}_b$, and

$$\mathbf{f}_j(\mathbf{x}, \mathbf{u}) = \begin{bmatrix} \mathbf{C}^\mathsf{T} \mathbf{Q} \mathbf{P}_j \\ \mathbf{b}_j \\ \mathbf{c}_j \end{bmatrix} \mathbf{u}, \quad \mathbf{C} = \begin{bmatrix} \cos\theta & \sin\theta \\ -\sin\theta & \cos\theta \end{bmatrix}, \quad \mathbf{Q} = \frac{1}{c^2 + p_x^2 + p_y^2} \begin{bmatrix} c^2 + p_x^2 & p_x p_y \\ p_x p_y & c^2 + p_y^2 \end{bmatrix},$$

$$\mathbf{b}_1 = \begin{bmatrix} \frac{-p_y}{c^2 + p_x^2 + p_y^2} & \frac{p_x}{c^2 + p_x^2 + p_y^2} \end{bmatrix}, \quad \mathbf{b}_2 = \begin{bmatrix} \frac{-p_y + \gamma_t p_x}{c^2 + p_x^2 + p_y^2} & 0 \end{bmatrix}, \quad \mathbf{b}_3 = \begin{bmatrix} \frac{-p_y + \gamma_b p_x}{c^2 + p_x^2 + p_y^2} & 0 \end{bmatrix},$$

$$\mathbf{c}_1 = \begin{bmatrix} 0 \\ 0 \end{bmatrix}^{\mathsf{T}}, \; \mathbf{c}_2 = \begin{bmatrix} -\gamma_t \\ 1 \end{bmatrix}^{\mathsf{T}}, \; \mathbf{c}_3 = \begin{bmatrix} -\gamma_b \\ 1 \end{bmatrix}^{\mathsf{T}}, \; \mathbf{P}_1 = \mathbf{I}_{2\times2}, \; \mathbf{P}_2 = \begin{bmatrix} 1 & 0 \\ \gamma_t & 0 \end{bmatrix}, \; \mathbf{P}_3 = \begin{bmatrix} 1 & 0 \\ \gamma_b & 0 \end{bmatrix},$$

where $j = 1, 2, 3$ correspond to sticking, sliding up, and sliding down contact interaction modes, respectively. For simplicity, we do not consider the case of separation when the pusher is not in contact with the object. Under the assumption of small forces with low impact, separation is the least relevant mode to the pusher-slider system.

4.6 Linearization

This section develops the linearized motion equations and motion cone constraints developed in Sections 4.4 and 4.5 about a given nominal trajectory. This linearization yields linear equations, which can be enforced as linear matrix inequalities in an optimization program and are computationally tractable for real-time execution of the controller design presented in Section 5. Consider a feasible nominal trajectory $\mathbf{x}^\star(t)$ of the sliding object with nominal control input $\mathbf{u}^\star(t)$ of the pusher. The notation $(\cdot)^\star$ is used to evaluate a term at the equilibrium state and $(\bar{\cdot})$ is used to denoted a perturbation about the equilibrium state. The linearization of motion equations Eq. (8) about a nominal trajectory yields

$$\dot{\bar{\mathbf{x}}} = \mathbf{A}_j(t)\bar{\mathbf{x}} + \mathbf{B}_j(t)\bar{\mathbf{u}}, \quad \text{with} \quad \bar{\mathbf{x}} = \mathbf{x} - \mathbf{x}^\star, \quad \bar{\mathbf{u}} = \mathbf{u} - \mathbf{u}^\star,$$

and

$$\mathbf{A}_j(t) = \left.\frac{\partial \mathbf{f}_j(\mathbf{x}, \mathbf{u})}{\partial \mathbf{x}}\right|_{\mathbf{x}^\star(t), \mathbf{u}^\star(t)}, \quad \mathbf{B}_j(t) = \left.\frac{\partial \mathbf{f}_j(\mathbf{x}, \mathbf{u})}{\partial \mathbf{u}}\right|_{\mathbf{x}^\star(t), \mathbf{u}^\star(t)}. \tag{9}$$

Similarly, the constraints enforcing a sticking interaction between the pusher and the slider presented in Eqs. (4) and (5) are perturbed about the nominal trajectory as

$$(v_t^\star + \bar{v}_t) \le (\gamma_t^\star + \bar{\gamma}_t)(v_n^\star + \bar{v}_n), \quad \text{and} \quad (v_t^\star + \bar{v}_t) \ge (\gamma_b^\star + \bar{\gamma}_b)(v_n^\star + \bar{v}_n), \tag{10}$$

respectively. Expanding the perturbations $\bar{\gamma}_t$ and $\bar{\gamma}_b$ in terms of $\bar{\mathbf{x}}$ as

$$\bar{\gamma}_t = \mathbf{C}_t\bar{\mathbf{x}}, \quad \bar{\gamma}_b = \mathbf{C}_b\bar{\mathbf{x}}, \quad \mathbf{C}_t = \left.\frac{\partial \gamma_t}{\partial \bar{\mathbf{x}}}\right|_{\mathbf{x}^\star(t), \mathbf{u}^\star(t)}, \quad \mathbf{C}_b = \left.\frac{\partial \gamma_b}{\partial \bar{\mathbf{x}}}\right|_{\mathbf{x}^\star(t), \mathbf{u}^\star(t)}, \tag{11}$$

permits to write Eq. (10) in matrix form as

$$\mathbf{E}_1(t)\bar{\mathbf{x}} + \mathbf{D}_1(t)\bar{\mathbf{u}} \le \mathbf{g}_1(t), \tag{12}$$

where $\mathbf{E}_1 = v_n^\star \begin{bmatrix} -\mathbf{C}_t \\ \mathbf{C}_b \end{bmatrix}$, $\mathbf{D}_1 = \begin{bmatrix} -\gamma_t^\star & 1 \\ \gamma_b^\star & -1 \end{bmatrix}$, $\mathbf{g}_1 = \begin{bmatrix} -v_t^\star + \gamma_t^\star v_n^\star \\ v_t^\star - \gamma_b^\star v_n^\star \end{bmatrix}$, where higher order perturbations are neglected. Similarly, the sliding up and sliding down constraints in Eqs. (6) and (7) are perturbed about the nominal trajectory as

$$(v_t^\star + \bar{v}_t) > (\gamma_t^\star + \bar{\gamma}_t)(v_n^\star + \bar{v}_n), \quad \text{and} \quad (v_t^\star + \bar{v}_t) < (\gamma_b^\star + \bar{\gamma}_b)(v_n^\star + \bar{v}_n), \tag{13}$$

respectively and can be rearranged in matrix form as

$$\mathbf{E}_2(t)\bar{\mathbf{x}} + \mathbf{D}_2(t)\bar{\mathbf{u}} \le \mathbf{g}_2(t), \quad \text{and} \tag{14}$$

$$\mathbf{E}_3(t)\bar{\mathbf{x}} + \mathbf{D}_3(t)\bar{\mathbf{u}} \le \mathbf{g}_3(t), \tag{15}$$

with $\mathbf{E}_2 = v_n^\star \mathbf{C}_t, \mathbf{E}_3 = -v_n^\star \mathbf{C}_b, \mathbf{D}_2 = \begin{bmatrix} \gamma_t^\star & -1 \end{bmatrix}, \mathbf{D}_3 = \begin{bmatrix} -\gamma_b^\star & 1 \end{bmatrix}, \mathbf{g}_2 = \begin{bmatrix} v_t^\star - \gamma_t^\star v_n^\star - \epsilon \end{bmatrix}$,
and $\mathbf{g}_3 = \begin{bmatrix} -v_t^\star + \gamma_b^\star v_n^\star - \epsilon \end{bmatrix}$, where higher order perturbations are neglected, ϵ is a
small scalar value, and $\mathbf{C}_t$ and $\mathbf{C}_b$ are given by Eq. (11).

5   MODEL PREDICTIVE CONTROL

In this section, we present a feedback controller design for the motion of the slider that
minimizes perturbations from a desired trajectory. The proposed controller determines
the desired pusher velocity at each time step based on the sensed pose of the slider. A
successful feedback controller must: 1) address hybridness and underactuation 2) allow
for sliding at contact 3) be fast enough to solve online 4) drive perturbations from the
nominal trajectory to zero. To satisfy these requirements, we use a Model Predictive
Control (MPC) formulation, which takes the form of an optimization program over
the control inputs during a finite time horizon $t_0, \ldots, t_N$. The decision variables of
the optimization program include the perturbed states of the system for N time steps
$\bar{\mathbf{x}}_1, \ldots, \bar{\mathbf{x}}_N$ and the perturbed control inputs $\bar{\mathbf{u}}_0, \ldots, \bar{\mathbf{u}}_{N-1}$. The goal is represented by
a finite-horizon cost-to-go function that we will minimize subject to the constraints on
the control inputs and the dynamics of the system detailed in Section 4. We express the
cost-to-go for N time steps as:

$$J(\bar{\mathbf{x}}_n, \bar{\mathbf{u}}_n) = \bar{\mathbf{x}}_N^\mathsf{T} \mathbf{Q}_N \bar{\mathbf{x}}_N + \sum_{n=0}^{N-1} \left(\bar{\mathbf{x}}_{n+1}^\mathsf{T} \mathbf{Q} \bar{\mathbf{x}}_{n+1} + \bar{\mathbf{u}}_n^\mathsf{T} \mathbf{R} \bar{\mathbf{u}}_n \right). \tag{16}$$

The terms $\mathbf{Q}$, $\mathbf{Q}_N$, and $\mathbf{R}$ denote weights matrices associated with the error state, final
error state, and control input, respectively. We subject the search for optimal control
inputs to the constraints describing the motion equations and contact dynamics of the
system. Due to the hybridness of the dynamical equations, the constraints to be enforced
depend on the contact mode j at play at each iteration n of the prediction finite horizon:

$$\text{if } n = 0 : \quad \begin{cases} \bar{\mathbf{x}}_1 = \mathbf{f}_{j0} + h\mathbf{B}_{j0}\bar{\mathbf{u}}_0, & \text{(Dynamics)} \tag{17} \\ \mathbf{D}_{j0}\bar{\mathbf{u}}_0 \leq \mathbf{g}_{j0}, & \text{(Motion Cone)} \tag{18} \end{cases}$$

$$\text{if } n > 0 : \quad \begin{cases} \bar{\mathbf{x}}_{n+1} = [\mathbf{I} + h\mathbf{A}_{jn}]\,\bar{\mathbf{x}}_n + h\mathbf{B}_{jn}\bar{\mathbf{u}}_n, & \text{(Linearized Dynamics)} \tag{19} \\ \mathbf{E}_{jn}\bar{\mathbf{x}}_n + \mathbf{D}_{jn}\bar{\mathbf{u}}_n \leq \mathbf{g}_{jn}, & \text{(Linearized Motion Cone)} \tag{20} \end{cases}$$

where the terms $\mathbf{A}_{jn}, \mathbf{B}_{jn}, \mathbf{D}_{jn}, \mathbf{E}_{jn}$ are developed in Eqs. (9), (12), (14), and (15), and
the subscript n is used to denote the time stamp at which each expression is evaluated
(e.g. $\mathbf{A}_{jn} = \mathbf{A}_j(t_n)$). The constraints Eqns. (17) and (19) describe the dynamical motion equations while Eqns. (18) and (20) represent the motion cone constraints. Note
that for the special case ($n = 0$), the nonlinear dynamical and contact constraint equations given by Eqs. (4), (5), (6), (7), and (8) reduce to linear form due to the knowledge
of the state $\mathbf{x}_0$. In this case, the expressions $\mathbf{B}_{j0}, \mathbf{D}_{j0}$, and $\mathbf{g}_{j0}$ are evaluated at $\mathbf{x}_0$ rather
than $\mathbf{x}^\star(t_0)$, and the term $\mathbf{f}_{j0}$ is defined as

$$\mathbf{f}_{j0} = \bar{\mathbf{x}}_0 + h\left[\mathbf{B}_{j0}\mathbf{u}^\star(t_0) - \mathbf{f}(\mathbf{x}^\star(t_0), \mathbf{u}^\star(t_0))\right].$$

The constraints in Eqs. (17), (18), and (19), and (20) depend on the contact mode i, which complicates the search for optimal control inputs. Contact modes and control inputs must be chosen simultaneously. As illustrated in Fig. 5, this problem takes the form of a tree of optimization programs with 3^N possible contact schedules, each yielding a convex optimization program, which is too computationally expensive to solve online.

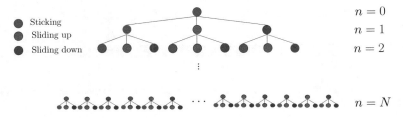

Sticking
Sliding up
Sliding down

$n = 0$
$n = 1$
$n = 2$

$n = N$

Fig. 5: Tree of optimization programs for a MPC program with N prediction steps. Scales exponentially due to contact hybridness.

5.1 Mixed-Integer Quadratic Program

The combinatorial hybrid nature of the pusher-slider dynamics can be modeled by adding integer decision variables into the optimization program, as is commonly done in Mixed-Integer programming. The resulting Mixed-Integer Quadratic Program (MIQP) can be solved rather efficiently using numerical tools, such as Gurobi [21]. In the case of the pusher-slider system, we introduce the integer variables: $z_{1n} \in \{0, 1\}$, $z_{2n} \in \{0, 1\}$, and $z_{3n} \in \{0, 1\}$, where $z_{1n} = 1$, $z_{2n} = 1$, or $z_{3n} = 1$ indicate that the contact interaction mode at step n is either sticking, sliding up, or sliding down, respectively. We will use the big-M formulation [22] to write down the problem, where M is a large scalar value used to activate and deactivate the contact mode dependent constraints, through a set of linear equations. The mode dependent constraints are reformulated as

$$\text{if } n = 0: \begin{cases} \begin{bmatrix} 1 \\ -1 \end{bmatrix} \bar{\mathbf{x}}_1 \leq \begin{bmatrix} 1 \\ -1 \end{bmatrix} \{\mathbf{f}_{j0} + h\mathbf{B}_{j0}\bar{\mathbf{u}}_0\} + \mathbf{1}_{8 \times 1} M(1 - z_{jn}) \\ \mathbf{D}_{j0}\bar{\mathbf{u}}_0 \leq \mathbf{g}_{j0} + \mathbf{1}_{2 \times 1} M(1 - z_{jn}) \end{cases}$$

$$\text{if } n > 0: \begin{cases} \begin{bmatrix} 1 \\ -1 \end{bmatrix} \bar{\mathbf{x}}_{n+1} \leq \begin{bmatrix} 1 \\ -1 \end{bmatrix} \{[\mathbf{1} + h\mathbf{A}_j] \bar{\mathbf{x}}_n + h\mathbf{B}_j\bar{\mathbf{u}}_n\} + \mathbf{1}_{8 \times 1} M(1 - z_{jn}) \\ \mathbf{E}_{jn}\bar{\mathbf{x}}_n + \mathbf{D}_{jn}\bar{\mathbf{u}}_n \leq \mathbf{g}_{jn} + \mathbf{1}_{2 \times 1} M(1 - z_{jn}), \end{cases}$$

where $\mathbf{1}_{m \times 1} = [1 \ 1 \ \dots \ 1]^\mathsf{T}$. Finally, the constraint $z_{1n} + z_{2n} + z_{3n} = 1$ is enforced to ensure that only one mode can be activated at a time.

5.2 Family of Modes (FOM) Scheduling

The MIQP formulation greatly reduces the computational cost associated with the optimization program in Eq. (16). With an efficient implementation, it can be solved in almost real-time for the low dimensional pusher-slider system. However it does not scale

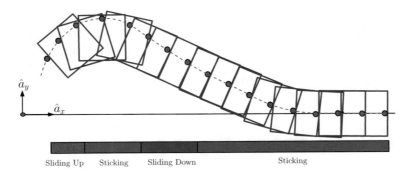

$\hat{a}_y$

$\hat{a}_x$

| Sliding Up | Sticking | Sliding Down | Sticking |

Fig. 6: Example of optimal mode schedule for the pusher-slider system converging to a straight horizontal trajectory.

well for systems with more degrees of freedom or additional contact points. The method presented in this section is motivated by the observation that many of the branches of the tree in Fig. 5 will give very good solutions, even if not exactly optimal. For the pusher-slider system, it is reasonable to expect the optimal mode schedule will follow a certain predictable structure. For example, we can intuitively expect that when the object is located in the y direction above the reference straight line trajectory, as in Fig. 6, the pusher will likely slide up to correct the orientation of the object followed by a downward sliding motion as the object converges to the desired trajectory. This pushing strategy represents one possible mode schedule. The family of modes algorithm consists in determining a fixed set of probable mode schedules that span a large range of primitive behaviors of the system. Each mode schedule in the family specifies a sequence of N contact modes to be imposed during the finite prediction horizon in the MPC formulation. By doing so, the combinatorial problem reduces to solving m convex optimization programs, where m is the number of mode sequences in the family. A key challenge is in determining a small number of mode schedules that spans a "significant" set of dynamic behaviors. For the pusher-slider system, one could consider a family of three mode sequences:

$\mathcal{M}_1$: the pusher slides up relative to the object followed by a sticking phase,
$\mathcal{M}_2$: the pusher slides down relative to the object followed by a sticking phase,
$\mathcal{M}_3$: the pusher sticks to the object for the full length of the prediction horizon.

Even though this family of mode sequences only contains a very small fraction of all the possible contact mode combinations in the tree in Fig. 5, it spans a very large set of dynamic behaviors between the pusher and the slider. Part of the reason is that the controller will re-optimize the selection of optimal modes in real-time. Solving Eq. (16) for each mode schedule leads to the finite horizon costs $J_1, \ldots, J_m$. Given that all possible contact modes are predetermined, all combinatorial aspects disappear, and each mode schedule in the family leads to a computationally solvable quadratic problem. The controller then chooses the optimal among the "m" mode schedules. The control input is selected at each time step by choosing the first element of the sequence of control inputs as $\mathbf{u} = \mathbf{u}_0^\star + \bar{\mathbf{u}}_0$, where the term $\bar{\mathbf{u}}_0$ is obtained from the optimization program with minimum cost and $\mathbf{u}^\star$ denotes the nominal control input.

Table 1: Physical parameters of pusher-slider system.

Property	Symbol	Value
coefficient of friction (pusher-slider)	μ_p	0.3
coefficient of friction (slider-table)	μ_g	0.35
mass of slider, kg	m	1.05
length of slider, m	a	0.09
width of slider, m	b	0.09

6 RESULTS

The controller design based on the Family of Modes approach is implemented in this section as described in Section 5. Two test scenarios are considered. Section 6.1 considers a trajectory tracking problem, where an external perturbation is applied to the system. Section 6.2 adapts the controller design to a target tracking problem, where the pusher guides the slider through 3 successive targets. We evaluate the performance of the controller design in both test scenarios through numerical simulations and experiments. The experiments are conducted using an ABB IRB 120 industrial robotic manipulator along with a Vicon system to track the pose of the slider. The experimental setup is depicted in Fig. 7, where a metallic rod (pusher) attached to the robot is used to push an aluminum object (slider) on a flat surface (plywood). The physical parameters of the system are reported in Table 1.

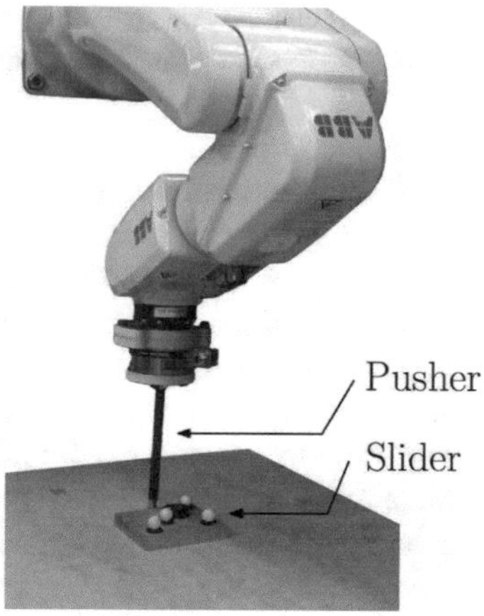

Fig. 7: Experimental setup. A metallic rod (pusher) is attached to an ABB IRB 120 industrial robotic arm to push an aluminum object (slider). The pose of the slider is tracked using a Vicon camera system.

6.1 Straight Line Tracking

Consider the problem of tracking a straight line nominal trajectory at a constant velocity, defined by $\mathbf{x}^\star(t) = [0.05t \ 0 \ 0 \ 0]^\mathsf{T}$ and $\mathbf{u}^\star = [0.05 \ 0]^\mathsf{T}$.

$$\mathcal{M}_1 := \begin{cases} \text{Slide up} & \text{if } n = 0 \\ \text{Stick} & \text{if } n > 0 \end{cases}, \quad \mathcal{M}_2 := \begin{cases} \text{Slide down} & \text{if } n = 0 \\ \text{Stick} & \text{if } n > 0 \end{cases}, \quad \mathcal{M}_3 := \text{Stick},$$

as detailed in Section 5.2. The controller design parameters used in the numerical simulations in Fig. 8(a) and the experiments in Fig. 8(b) are set to $N = 35$ steps, $h = 0.03$ seconds, $\mathbf{Q} = 10 \ \text{diag}\{1, 3, .1, 0\}$, $200 \ \text{diag}\{1, 3, .1, 0\}$, and $\mathbf{R} = 0.5 \ \text{diag}\{1, 1\}$. To

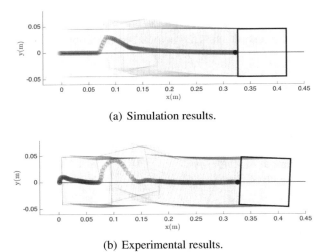

(a) Simulation results.

(b) Experimental results.

Fig. 8: Tracking of a straight line trajectory at a constant velocity. Both simulated and experimental trajectories of the sliding object recover from an external lateral perturbation. The pusher quickly reacts to stabilize the slider about the nominal trajectory.

limit the maximum velocity of the pusher, we include the constraints $|v_n| \le 0.1$ m/s and $|v_t| \le 0.1$ m/s to the optimization program.

In Fig. 8, an impulsive force is applied in the y direction to perturb the system about its nominal trajectory and evaluate the performance of the feedback controller. In Fig. 8(a), the simulated response of the slider to a state perturbation $\delta = [0 \ 0.01 \ \frac{15\pi}{180} \ 0]^{\mathsf{T}}$ applied at $x = 0.075$ m is compared to the experimental response of the slider in Fig. 8(b) to an external impulsive force applied using a hand held poker. Both simulated and experimental responses show that the feedback controller is successful in driving the perturbations from the nominal trajectory to zero. The initial motion of the pusher in Fig. 8(b) is trying to correct the non-zero initial conditions of the slider, which was imperfectly placed by hand. The experimental performance of the feedback controller design to a variety of external perturbations can be visualized at https://mcube.mit.edu/videos.

6.2 Trajectory Tracking

In this section, the controller design developed in Sections 5 and 6 is adapted to the problem of tracking a moving target position. The objective is to control the motion of the robotic pusher such that the sliding object reaches a target (x_c, y_c). At each instant, an intermediate reference frame $\mathcal{F}_c$ is defined where the unit vector $\hat{c}_x$ points from the center of mass of the sliding object to the target position. The angle θ_c in Fig. 9 is the orientation of $\hat{c}_x$ relative to the horizontal $\hat{a}_x$ and the angle $\theta_{rel} = \theta - \theta_c$ is the orientation of the slider relative to $\hat{c}_x$. The tracking of a target position is converted into a trajectory tracking problem where the objective is to track the straight line between the current position of the slider and the target. Assuming a constant desired velocity of the

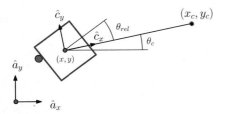

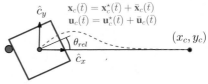

Fig. 9: The intermediate reference frame $\mathcal{F}_c$ is defined such that $\hat{c}_x$ points in the direction of the target.

Fig. 10: A desired trajectory $\mathbf{x}_c^\star(\bar{t})$ is defined from the center of the object to the goal position along the direction $\hat{c}_x$ with nominal control input sequence $\mathbf{u}_c^\star(\bar{t})$.

sliding object, the nominal trajectory $\mathbf{x}_c^\star(\bar{t})$ and control input $\mathbf{u}_c^\star(\bar{t})$, defined relative to the intermediate reference frame $\mathcal{F}_c$, are $\mathbf{x}_c^\star(\bar{t}) = \begin{bmatrix} v_x\bar{t} \ 0 \ 0 \ 0 \end{bmatrix}^\mathsf{T}$ and $\mathbf{u}_c^\star(\bar{t}) = \begin{bmatrix} v_x \ 0 \end{bmatrix}^\mathsf{T}$, with v_x the desired velocity. The term $\bar{t}$ denotes the prediction horizon time, which is reinitialized at each time step as $\bar{t} = 0$. The controller design parameters used in Fig. 11 are identical to those presented in Section 8, with the exception that the max tangential velocity is set to $|v_t| \leq 0.3$ m/s. The target tracking results are performed from zero initial conditions with the target positions $[x_c \ y_c]^\mathsf{T}$ given by

$$\text{Target 1: } \begin{bmatrix} 0.23 \\ -0.11 \end{bmatrix} (m), \quad \text{Target 2: } \begin{bmatrix} 0.23 \\ 0.11 \end{bmatrix} (m), \quad \text{Target 3: } \begin{bmatrix} 03 \\ 0.08 \end{bmatrix} (m)$$

The simulation begins with the position tracking of Target 1. When the slider falls within a distance of 0.01 meters of a target, the position is updated to the next target position and so on until the final position is reached. Both simulated and experimental trajectories achieve target tracking within the specified tolerance (video of experiments available at https://mcube.mit.edu/videos). The trajectories in Fig. 11 depict the robotic pusher favoring a sliding behavior when the relative angle of the sliding object is large relative to the target position. The controller elects to slide the pusher relative to the object to rotate it and then favors a sticking contact mode to push the object in a straight line towards the target position. This control strategy is intuitive and is in line with the way in which humans manipulate objects using a single finger. It is observed that the simulated trajectory results in Fig. 11(a) achieves slightly more aggressive turns than the experimental trajectory results in Fig. 11(b). Despite the unmodeled aspects of the problem, such as delay in the robot position control, quasistatic assumption, imprecise nature of friction, etc., the feedback controller makes good control decisions which drive the system in the right direction towards the desired trajectory.

7 CONCLUSION

In this work we present a feedback controller design for manipulation tasks where control acts through contact. Using a model predictive control approach combined with an integer programming formulation permits for the integration of the physical constraints associated with the dynamics of contact as well as explicitly consider the hybrid nature of the interactions. We describe two methodologies to solve the optimization program,

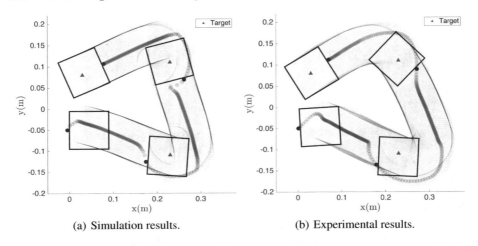

(a) Simulation results.   (b) Experimental results.

Fig. 11: Control of a sliding object through a single point robotic pusher. Both simulation and experimental results successfully reach all 3 targets within the specified tolerance.

namely Mixed-Integer Quadratic Programming and Family of Modes, which we propose to speed up the integer program for real-time control purposes. The Family of Modes method is validated through numerical simulations and experiments, where the feedback controller successfully stabilizes the motion of a sliding object through single point contact.

Following the success of the Family of Modes approach in the pusher-slider system, a natural question raised is: "How does the Family of Modes approach extend to more complex manipulation tasks?" For large set of manipulation problems involving multiple contact locations, key mode sequences can be readily identified either using physical insight of by investigating solutions to offline planning algorithms. Moreover, for many manipulation tasks, such as those involving parallel jaw grippers, there is a natural symmetry of the problem that can be leveraged to deduce important mode sequences. The application of the Family of Modes approach to multiple contact problems, such as in-hand manipulation with extrinsic dexterity [8, 9], is a direction of future work.

References

1. Kawasaki, H., Komatsu, T., Uchiyama, K.: Dexterous Anthropomorphic Robot Hand With Distributed Tactile Sensor: Gifu Hand II. IEEE/ASME Transactions on Mechatronics **7**(3) (2002) 296 – 303
2. Mouri, T., Kawasaki, H., Yoshikawa, K., Takai, J., Ito, S.: Anthropomorphic Robot Hand: Gifu Hand III. International Conference on Control, Automation and Systems, Jeonbuk, Korea, October 16 – 19, 2002, pp. 1288 – 1293
3. Gaiser, I., Schulz, S., Kargov, A., Klosek, H., Bierbaum, A., Pylatiuk, C., Oberle, R., Werner, T., Asfour, T., Bretthauer, G., and, R.D.: A New Anthropomorphic Robotic Hand. IEEE-RAS International Conference on Humanoid Robots, Daejon, Korea, December 1 – 03, 2008, pp. 418 – 422
4. Yousef, H., Boukallel, M., Althoefer, K.: Tactile Sensing for Dexterous In-Hand Manipulation in Robotics–A Review. Sensors and Actuators A: Physical **167**(2) (2011) 171 – 187

5. Li, R., Jr., R.P., Yuan, W., Andreas ten Pas, N.R., Srinivasan, M.A., Adelson, E.: Localization and Manipulation of Small Parts Using GelSight Tactile Sensing. IEEE/RSJ International Conference on Intelligent Robots and Systems, Chicago, Illinois, September 14 – 18, 2014, pp. 3988 – 3993

6. Bicchi, A., Kumar, V.: Robotic Grasping and Contact: A Review. IEEE International Conference on Robotics and Automation, San Francisco, California, April 24 – 28, 2000, pp. 348 – 353

7. Salisbury, J.K., Craig, J.J.: Articulated Hands: Force Control and Kinematic Issues. The International Journal of Robotics Research 1(1) (1982) 4 – 17

8. Chavan-Dafle, N., Rodriguez, A., Robert Paolini, B.T., Srinivasa, S., Erdmann, M., Mason, M.T., Lundberg, I., Staab, H., Fuhlbrigge, T.: Extrinsic Dexterity: In-Hand Manipulation with External Forces. IEEE International Conference on Robotics and Automation, Hong Kong, China, May 31 – June 5, 2014, pp. 1578 – 1585

9. Chavan-Dafle, N., Rodriguez, A.: Prehensile Pushing: In-Hand Manipulation with Push Primitives. IEEE/RSJ International Conference on Intelligent Robots and Systems, Hamburg, Germany, September 28 – October 02, 2015

10. Posa, M., Cantu, C., Tedrake, R.: A Direct Method for Trajectory Optimization of Rigid Bodies Through Contact. The International Journal of Robotics Research 33(1) (2014) 69 – 81

11. Tassa, Y., Todorov, E.: Stochastic Complementarity for Local Control of Discontinuous Dynamics. Proceedings of Robotics: Science and Systems, Zaragoza, Spain, June 27 – 30, 2010

12. Tassa, Y., Todorov, E.: Synthesis and Stabilization of Complex Behaviors through Online Trajectory Optimization. IEEE/RSJ International Conference on Intelligent Robots and Systems, Vilamoura, Portugal, October 7 – 12, 2012

13. Stewart, D.E., Anitescu, M.: Optimal Control of Systems with Discontinuous Differential Equations. Numerische Mathematik 114(4) (2009) 653 – 695

14. Mason, M.T.: Mechanics and Planning of Manipulator Pushing Operations. The International Journal of Robotics Research 5(3) (1996) 53 – 71

15. Lynch, K.M., Mason, M.T.: Stable Pushing: Mechanics, Controllability, and Planning. The International Journal of Robotics Research 15(6) (1986) 533 – 556

16. Lynch, K., Maekawa, H., Tanie, K.: Manipulation and Active Sensing by Pushing Using Tactile Feedback. IEEE/RSJ International Conference on Intelligent Robots and Systems, Raleigh, USA, July 7 – 10, 1992, pp. 416–421

17. Poulakakis, I., Grizzle, J.W.: The Spring Loaded Inverted Pendulum as the Hybrid Zero Dynamics of an Asymmetric Hopper. IEEE Transactions on Automatic Control 54(8) (2009) 1779 – 1793

18. Decker, L., Berge, C., Renous, S., Penin, X.: An Alternative Approach to Normalization and Evaluation for Gait Patterns: Procrustes Analysis Applied to the Cyclograms of Sprinters and middle-Distance runners. Journal of Biomechanics 40(9) (2007) 2078 – 2087

19. S. Goyal, A.R., Papadopoulos, J.: Wear. Planar Sliding with Dry Friction Part 1. Limit Surface and Moment Function 143 (1991) 307 – 330

20. Lee, S.H., Cutkosky, M.: Journal of Manufacturing Science and Engineering. Fixture Planning with Friction 113(3) (1991) 320 – 327

21. Gurobi Optimization, I.: Gurobi optimizer reference manual (2015)

22. Nemhauser, G.L., Wolsey, L.A.: Integer Programming and Combinatorial Optimization. Wiley, Chichester, England (1988)

Assembling and disassembling planar structures with divisible and atomic components

Yinan Zhang, Emily Whiting, Devin Balkcom

Dartmouth Computer Science

Abstract. This paper considers an assembly problem. Let there be two interlocking parts, only one of which may be cut into pieces. How many pieces should we cut the divisible part into to separate the parts using a sequence of rigid-body motions? In this initial exploration, we primarily consider 2D polygonal parts. The paper presents an algorithm that computes a lower bound on the number of pieces that the divisible part must be cut into. The paper also presents a complete algorithm that constructs a set of cuts and a motion plan for disassembly, yielding an upper bound on the required number of pieces. Applications of the future extension of this work to three dimensions may include robot self-assembly, interlocking 3D model design, search-and-rescue, packaging, and robotic surgery.

1 Introduction

Assembly of parts is one of the oldest problems in robotics. This paper considers a variant of the assembly problem in which some parts can be cut into pieces, and others cannot be. How many pieces must an amber fossil be cut into to extract a fly? How many pieces must a model ship be broken into in order to construct a ship-in-a-bottle? How many pieces must rubble be cut into to rescue an injured person? How should styrofoam packaging be assembled to support a delicate object for transport?

As an initial exploration, we consider planar devices composed of one polygon of each of the two material types. We allow arbitrary rigid body motion of the parts after cutting; the cuts may be along arbitrary curves. Figure 1a shows an example of a mammoth in a ice cube. The gray material may not be cut, but the white material may be cut to be separated from the indivisible, or *atomic*, part.

We provide algorithms to find lower and upper bounds on the number of pieces that the device must be cut into. We also provide a prove-ably complete algorithm for design and for determining an assembly sequence. Figure 1b shows three rotation centers that could be used to locally separate ice edges of corresponding colors from the mammoth; the ice must be cut into at least three pieces. However, there is no guarantee that three is a sufficient number, because

© Springer Nature Switzerland AG 2020
K. Goldberg et al. (Eds.): *Algorithmic Foundations of Robotics XII*, SPAR 13, pp. 816–830, 2020.
https://doi.org/10.1007/978-3-030-43089-4_52

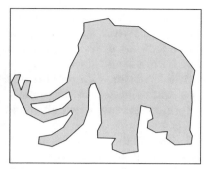

(a) A mammoth body inside an ice cube. We want to separate the mammoth and the ice (except the hole) by breaking the ice into pieces.

(b) Edges with the same color might be locally separated from the atomic part using a single rotation about the corresponding rotation center.

Fig. 1. A mammoth in the ice and one necessary solution to remove ice without damaging the mammoth body.

global properties of the geometry also matter; we find that cutting into 61 pieces is sufficient (Figure 8c).

We believe that extension to three dimensions would have practical value for many problems in 3D printing and prototyping for robotics. Many structures cannot be printed out of a single material; robots may contain atomic components such as motors, wires, micro-controllers, and batteries that cannot be cut into pieces, supported by a rigid but divisible structure that fits around and supports the atomic components. We can also imagine applications in other areas, including search-and-rescue and robotic surgery.

However, the focus of the current paper is not on applications, but on the exploration of a fundamental robotics problem: the difficulty of assembly, measured by the number of required pieces, if some of the parts can themselves be disassembled. Such analysis of lower and upper bounds on *physical complexity* forms a useful basis for thinking about robotics problems, just as bounds on computational complexity are useful in computer science. Because the difficulty of assembly appears to depend on the shapes of the parts in non-trivial ways, we cannot simply provide an interesting bound directly; however, we can devise algorithms that compute bounds for an input shape. These bounds may give insights into the question of which shapes are hard, which are easy, and how to design appropriate shapes.

There are both local and global properties of the geometry that may cause a part to need to be cut into many pieces. Locally, two contacting rigid bodies may be interlocked, in the same way that a robot grasp may interlock with a part, effectively forming a single rigid body that must be cut to allow separation. Globally, there may be narrow openings though which only small parts can fit, the classic furniture-mover's problem.

When computing a *necessary* number of pieces, we focus on the local geometry, and relax or ignore the global geometric constraints. We consider cutting the interlocking parts into sufficiently many pieces such that there is no single immobilizing grasp, using a geometric method inspired by Reuleaux's method [20, 24] to formulate the problem as a minimum-set-cover problem. Computationally, minimum-set-cover is NP-complete, but can be approximated in polynomial-time to within a logarithmic factor. In future work, better lower bounds might be found by also considering global constraints.

We also give an algorithm to compute a *sufficient* number of pieces, by constructing cuts and an assembly order that respect global and local constraints. We prove that as long as atomic components do not contain voids, the parts can be cut into a finite set of pieces and disassembled using only translation; rotations are not required.

1.1   Related work

The current paper is closest in spirit to work on k-moldability. In the k-moldability problem, a separable k-piece mold is taken apart using a single translation per piece to expose a molded atomic part [16, 22]. Ravi and Srinivasan [23] give a list of criteria to aid the engineer in making decisions of parting surfaces. Pryadarshi and Gupta [22] used accessible directions to decompose molds into a small number of pieces. Exact-cast-mold design methods require models to be moldable or result in a large number of mold pieces. Herholz *et al.* [14] deform a model into an approximate but moldable shape, and then decompose mold pieces.

The primary contribution of the current paper is the relaxation of the requirement that the mold be separated using single translations; this allows study of the fundamental theoretical limits of dis-assembly. Most of the structures studied in this paper are not k-moldable, because there is no set of directions from which all of the divisible structure is visible; some portions of the structure are occluded. We show that in fact, any structure without inaccessible voids can be dis-assembled using only sequences of translations. The lower and upper bounds that we study are true physical bounds — they hold over *any* sequence of rigid body motions, not just single translations or rotations.

This paper is also inspired by Snoeyink's work on the number of hands required to dis-assemble a collection of rigid parts [31]. Because we allow parts to be cut, simultaneous motions are not required for dis-assembly. In the *Carpenter's Rule* problem studied by Rote, Demaine, Connelly [9], Streinu [35], and others, the rigid pieces are also all atomic, and connected by joints, typically requiring many simultaneous motions.

This paper is therefore somewhat closer in spirit to work by Wang [37] that studies the number of fingers needed to tie a knot – in that work, the string is treated as a collection of rigid bodies, but the joints may be placed arbitrarily, essentially cutting the carpenter's rule, without disconnecting the pieces. The current work is also close in spirit to work by Bell and coauthors [5] that studies the number of pieces that a mechanical knotting device must be cut into to extract the knotted string.

There has been significant work in the graphics community in computational fabrication. Song et al. [32]'s approach to fabricating large 3D objects was to break the shell of the object into pieces and assemble after fabrication. Hu et al. [15] presented a method to decompose 3D object into a set of pyramidal shapes such that no support material is needed when 3D printing the model. Fu and Song [13, 33] studied computational interlocking furniture design.

Our approach to the lower-bounds problem in particular grows out of seminal work on immobilizing rigid bodies, or *grasping*. Traditional geometric approaches to grasping attempt to prevent all possible sliding and rotational motions of a polygonal object by placing fingers around the object. Reuleaux [24] is credited with the concept of *form closure*. Mishra et al. [21] proved the sufficiency of four fingers to immobilize any polygonal object. Czyzowicz et al. [10, 11] showed that polygons without parallel edges can be immobilized using three fingers. Rimon and Burdick [26, 27] showed how two-finger grasps can be analyzed using *second-order immobility*. Cheong [7] provided an algorithm to compute all immobilizing grasps of a simple polygon. Cheong et al. [8] also showed that $n + 3$ contacts suffice to immobilize a chain of n hinged polygons.

A polygonal object can be *caged* by surrounding an object with fingers, such that the object has some freedom locally but cannot escape the cage. Some of the earliest work on caging was by Rimon and Blake [25]. Vahedi et al. and Allen et al. [36, 1] proposed algorithms to find all caging grasps of two disk fingers. Erickson et al. [12] studied the case of three-finger caging for arbitrary convex polygon. Makita et al. [19] extend the caging problem from 2D to 3D with multi-fingers. Our work, instead of caging an object, can be viewed as removing contacting pieces to uncage a polygon in the plane.

A classic problem of *self-assembly* is to move a set of small robots to specified target positions; some of the challenges with narrow corridors and coordination are similar to those faced in the current work. Kotay et al. [17], for example, designed a robotic module, groups of which aggregate into 3D structures. Rus and Vona's early work on the Crystalline robot [30] presented an algorithm to do self-reconfiguration. Recently, working on the problem of scale, Rubenstein et al. [29] provided an algorithm for moving kilo-bots one-by-one to form certain planar shapes. Arbuckle et al. [2] allowed identical memoryless agents to construct and repair arbitrary shapes in the plane. In the authors' own work [38] on assembly of interlocking structures, 9 kinds of blocks are used to build large-scale voxelized models such that all blocks are interlocked and the whole structure is rigid as a whole. Self-assembly and modular robots in the presence of obstacles has also been extensively studied. Becker et al. [4] proposed an algorithm to efficiently control a large population of robots in this scenario. Rubenstein et al. [28] used multiple robotic units to manipulate the positions of obstacles.

2   Computing a lower bound on the number of pieces

Let parts A and B be interlocked. In this section, we show how to find a lower bound on the number of pieces that part B must be cut into to separate the

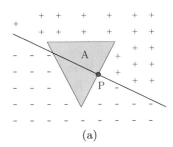

(a)

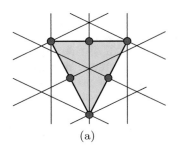

(a)

Fig. 2. Contacting point P can rotate in positive direction about centers in + area and in negative direction about centers in − area.

Fig. 3. The normal lines through the contact points form a set of cells, such that the number of points that may be separated from the gray part is maximized by choosing a rotation center at a cell vertex.

parts. The approach is inspired by analysis of contact modes for contacting 2D rigid bodies [3, 20], as well as by Reuleaux's graphical method for analyzing whether a collection of points fully constrains (or in our case, is constrained by) the motion of a rigid body [24].

We first replace B with a collection of points P from B along the boundary of A; whichever points we choose, *at least* these points must be separated from A using a collection of rigid-body motions. For simplicity, we place three vertices per edge: one in the center, and one at each endpoint. Choosing more points may allow a larger lower bound to be computed, at the cost of some additional computation.

Now consider any subset of these points. Can this subset be contained in a single rigid piece after cutting, in such a way that the rigid piece may be separated from A? In order to separate this piece from A, there must at least instantaneously be a single rigid body motion that at least does not cause collision for any of the points, and ideally, simultaneously separates all of the points from A. Every motion of a planar rigid body is instantaneously a rotation or a translation. Does there exist a translation direction or a rotation center that allows separation? How do we compute good (large) subsets without considering the power set over P?

Theorem 1. *If n points P of a planar rigid body B are in contact with a polygonal rigid body A, then there are at most $n(n-1)/2 + 2n$ maximal subsets of P, such that each subset may be moved together as a rigid body without colliding with A, and any other non-colliding subset is a contained within one of the maximal subsets.*

Proof. We would like to group subsets of points in P and see if they can be separated from A as a group. Let us first consider rotations. To find a compatible group, we might choose a particular rotation center and a positive or negative direction for rotation. Then find all points in P that separate from (or at least do not collide with) A under that motion; we have found a compatible subset. This is our basic approach; but how many potential rotations must we consider, and how many compatible subsets may be generated?

Consider a rotation center r somewhere in the plane, with an associated direction (either positive or negative rotation). Reuleaux's method makes use of the fact that for a particular point P_i, for most choices of rotation center, one direction of rotation (either positive or negative) is permissible, while the other causes collision of P_i with A. Along the normal to the edge of A at P_i, either negative or positive rotation is possible. Let $P_r \in P$ be the set of points compatible with rotation center and direction r. If we move r along a continuous trajectory, membership in P_r only changes as r crosses one of the normals through one of the points in P. The constraint is least restrictive along the normals themselves, so to compute the *maximal* subsets of compatible points, such that any other compatible subset is either a singleton or a subset of a computed subset, we need only consider rotation centers at the intersections of the normals, as shown in Figure 3. The normals form an *arrangement* [34], and there are $n(n-1)/2$ possible intersections, each with at most one corresponding maximal subset(Figure 3). Once candidate maximal subsets have been generated, discard any that are contained within other computed subsets.

Translation directions may be analyzed similarly. Each point P_i, if included in a subset, forbids an open half-plane of translation directions. So for each point, it is sufficient to test two translations directions, each corresponding to sliding in one direction or another along the point. Collect all $2n$ directions, and for each direction, test the remaining points against that direction to generate candidate maximal subsets.

The proof of Theorem 1 implies an algorithm for computing a lower bound on the number of pieces that the divisible part must be cut into to allow assembly or disassembly. Compute the maximal subsets as suggested; then solve the minimum-set-cover problem to find the minimum number of such sets needed to separate all points in P from A. If the number of maximal subsets is small, minimum-set-cover may be solved exactly. The simple examples presented in this paper were solved exactly using integer-linear programming. If there is a large number of subsets, then a greedy approach yields a solution in polynomial time, with guaranteed logarithmic approximation quality.

Figure 1b shows a solution of the necessary number of pieces needed to extract the planar mammoth from the ice. If two points on the same edge are in the same set, the segments between the points are considered able to rotate about centers in the same region as the two points. In this case, three sets cover all edges.

Edges containing points in the same set are not necessarily connected. Whether there exist cuts to separate edges exactly into the derived sets as connected rigid bodies is an open question, as is whether those bodies can be extracted after initial separation. This technique thus yields only a lower bound, and we expect that the lower bound might be significantly improved in future work.

Results for other shapes can be found in Figure 4. A few statistics are shown in Table 2. Time costs were measured on a 2016-model MacBook Pro with a 2.6 GHz Intel processor and 8 GB 1600 MHz DDR3 memory, and are intended only to give a sense of the practicality of analysis of analyzing structures with varying numbers of edges. From the table, we can see that with the increase of maximum number of rotation sets, the time cost increases dramatically, as we would expect for a $O(n^3)$ checking of rotations centers and a linear-integer program optimal solution to minimum-set-cover; we expect that much larger problems could be solved with good approximation by using greedy minimum-set-cover techniques.

shape	# edges	largest set size	set cover size	time cost
cavity	8	144	2	0.2018 s
spiral	14	612	3	1.8764 s
dumbbell	12	1104	4	7.2543 s
mammoth	64	12012	3	540.327 s

Table 1. Lower-bound analysis examples.

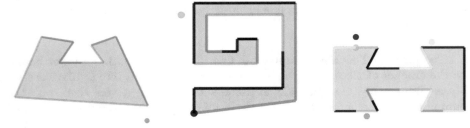

(a) Cavity. Orange edges move to the left, and green edges rotate about the green point.

(b) Spiral. Black and orange edges rotate about black and orange points respectively. Green edges move to the right.

(c) Dumbbell. 4 sets of edges in 4 colors. Each set of edges rotates about centers with the same color.

Fig. 4. Three examples of analyzing necessary number of pieces the divisible part should be cut into. In each example, edges with the same color are in the same set.

3   Computing a sufficient number of pieces

In this section, we present a complete algorithm that chooses where to cut the divisible part B, and finds a motion plan to achieve separation from the indivisible part A. We want to find an upper bound on the minimum number of pieces B can be cut into to move each piece out of a planar box that contains both A and B.

The algorithm first decomposes the part into a small number of convex polygons, and moves pieces within these convex shapes; the number of pieces depends on the number and size of polygons. In our implementation, we used a Delaunay triangulation, but better bounds could be achieved by finding larger convex components.

After decomposition of B into convex polygons, consider two adjacent convex polygons that share an edge. Flipping one polygon about the shared edge and intersecting with the other polygon gives a new convex polygon. Inside the new polygon we compute a largest inner square with one edge on the shared edge. This square, called the *transit square*, can move freely between the two convex polygons without leaving the interior.

We would like to find an axis-aligned grid such that at least one complete grid cell fits entirely within the transit squares for each pair of adjacent convex polygons. For each transit square, we find the largest axis-aligned inscribed square, divide the width by two (shown in Figure 5c), and take the minimum over all such values as the width of cells in the grid.

To find the size of the largest axis-aligned square, assume the width of the outer square is L and the small angle between x-axis and edges of the square is α. There exists another square of edge length $l = L\sqrt{1 - 2 \cdot \tan\alpha/(1 + \tan\alpha)^2}$ inside the outer square.

We intersect the grid cells with the convex polygons to create small pieces that we will call *components*. We will extract the material from each convex polygon; we will say that a polygon that has already had its material extracted is empty, and one that has not is full. We will prove that components can move from one full convex polygon to an adjacent empty polygon without leaving either convex polygon (and thus without collision with atomic parts or uncleared divisible parts), assuming each component disappears once it has completely entered the empty polygon. This is sufficient to prove inductively that the entire structure can be disassembled.

Lemma 1. *In a planar grid of square cells with a designated target square, continuous translation of each grid cell to the target will not cause collision, if the cells are translated in order of L_2 distance from the target.*

Proof. Let A be a square whose bottom-left point is at position (x_a, y'_a), moving in one direction towards the target square O, with bottom-left point at (x_o, y_o). Without loss of generality, assume $x_a > 0, y_a > 0$ and $x_o = 0, y_o = 0$, and that the width of each cell is 1. We know that, during the motion, every point of square A is bounded by the rectangle R defined by its bottom-left point $(0,0)$ and top-right point $(x_a + 1, y_a + 1)$.

Assume A collides with a square B whose bottom-left point is at (x_b, y_b), $x_b, y_b \in \mathbb{Z}^+$. Then $0 \leq x_b \leq x_a$ and $0 \leq y_b \leq y_a$; otherwise, no point in the square is in R. Because OA is along the diagonal of R, OA is the longest edge in

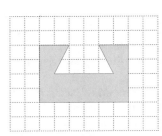

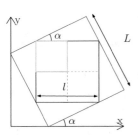

(a) Two interlocked parts where the gray part is atomic and the rest is divisible. Gridding the divisible part into small enough cells and removing them will separate both parts.

(b) Between two adjacent convex shapes, there exists a square that can move freely from one to another without leaving either shape.

(c) Partitioning the space using a (green) square of width $l/2$ guarantees there exists at least one square in the transition square.

Fig. 5. Gridding the divisible part to find sufficient pieces to separate parts.

triangle $\triangle OAB$. So $|OB| < |OA|$, and square B would have been moved before A using the proposed order.

Although we claim and prove Lemma 1 in the plane, it extends easily to similar results in arbitrary dimensions. We now come to the main result of this section:

Theorem 2. *Given an intersection of a grid with a pair of adjacent convex polygons, such that at least one complete grid cell is completely contained within each of the transit squares of the polygons, one polygon can be emptied into the other without collisions, using the intersections of the cells with the polygons as components, assuming each component disappears once it has completely entered the empty polygon.*

Proof. Lemma 1 indicates that cells in a grid can be translated to a target in order of distance without collision. Since all motions of all points in cells are along straight lines during translation, the result extends trivially to a case where the space is constrained to a convex polygon, and the cells are clipped by the convex polygon, as are the previously-defined *components*.

We therefore have the following approach. First, empty the transit square in the full polygon into the transit square in the empty polygon, by sorting the grid cells in order of distance from an arbitrarily chosen complete grid cell in the empty polygon, and translating those cells to the target in that order; since the pair of transit squares is together a convex polygon, there are no collisions. Then choose a target cell in the now-empty transit square in the first polygon.

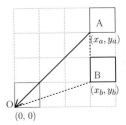 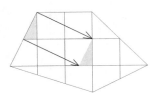 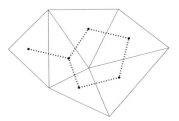

(a)   Squares   in   a grid space can move to any target position with no collision following the order of their distances to the target.

(b)   Cells can also move to any target square with no collision in the grid space bounded by a convex polygon using the same method.

(c)   The convex polygons forms   a   graph that   may be weighted by the size of the transit squares connecting each pair of polygons. Removing the smallest edges but keeping the connection between vertices improves the grid square size.

Fig. 6. Moving cells inside a convex shape, and between multiple convex shapes.

Sort the components in the polygon based on their distance from the target cell. In this order, first translate each component first into the transit square, and then into the adjacent empty polygon.

3.1   Algorithm 1: simple separation

The previous theorem suggests a simple, but complete, algorithm for cutting and separating the interlocked parts. First, using triangulation or some other means, decompose B and its containing square (from which we would like to remove B) into convex polygons. Polygons within B will be full, and polygons outside of B will be empty. Define a *boundary* polygon as a polygon that is connected by a sequence of adjacent empty polygons (the *exit sequence*) to the outside of the containing square, the *exit*.

Figure 7 shows the approach to extraction. Choose a boundary polygon, and an exit sequence. Extract components from that polygon one at a time in the order suggested in the proof of the theorem. As each component enters the exit sequence, move it through the sequence of empty polygons to the exit, using translation first to and then through each pair of transit squares along the exit sequence.

3.2   Algorithm 2: greedy separation with grouped components

There are many possible ways to improve Algorithm 1. For example, if there is a tiny shape in the convex decomposition of the divisible part, the cell size will be very small. In this section, we propose an algorithm to plan a path for every cell generated by the decomposition algorithm. Based on the paths, we

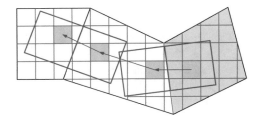

Fig. 7. Moving components through a chain of convex polygons. White polygons are empty and the gray one is filled. Using the pairs of transit squares, cells can move between any two adjacent polygons, or through a chain of connected polygons.

aggregate components into *pieces* and cut only along piece divisions; this will greatly reduce the number of pieces needed.

Let the *path* of a component be $P = [p_0, p_1, p_2, \ldots, p_n]$ where $p_i \in P$ is the intermediate position after the i-th control, and p_0 is the initial position of the component, where a *control* is a single translation. Define the *control sequence* for the component as $C = [c_1, c_2, \ldots, c_n]$ where $c_i = p_i - p_{i-1}$ is the i-th control.

A simple greedy approach to grouping components is as follows. Components may be *whole* (entire grid cells) or *partial*. We first deal only with whole components. Simulation motion of all of the whole components in each of the four cardinal directions. For each direction, count the number of components that reach the exit area, and may be grouped together based on 4-connectivity; greedily choose the direction that gives the fewest such grouped pieces.

Add the resulting cleared squares as a target area, and attempt motion in each of the four cardinal translation directions, potentially creating new piece groups; attempt to then move these new groups to the exit.

Table 2 shows some statistics for the sufficient decomposition for some example shapes, and a few illustrative run times. The main time cost is spent on testing collisions. Decomposition solutions are shown in Figure 8.

shape	# components	# pieces	decomposition time	planning time
cavity	144	6	0.862 s	0.822 s
spiral	462	26	0.786 s	1.500 s
dumbbell	269	8	0.529 s	0.496 s
mammoth	61	9954	16.816 s	149.192 s

Table 2. Analysis of sufficient number of pieces for a few example shapes.

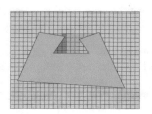 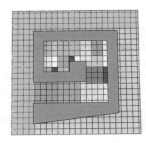 

(a)   Cavity. The divisible part is divided into 6 pieces.   (b)   Spiral. The divisible part is divided into 26 pieces.   (c) Mammoth. The divisible part is divided into 61 pieces.

Fig. 8. Decompositions of separable parts into pieces. Colors are reused as needed; each set of same-color components is a single piece.

4   Conclusions, limitations, and future work

We proposed the problem of separating a divisible polygon from an interlocked atomic polygon by cutting. We explored lower and upper bounds on the number of pieces the divisible part must be cut into, and presented algorithms to make the cuts and achieve the separation using a sequence of translations. Both bounds are very conservative; the main contribution of this work is the proposal of the problem and an initial exploration of solutions.

Improving the quality of bounds is of great interest for future work. In the section of computing the *necessary* number of pieces the divisible part must be cut into, we only considered some sets of points immediately adjacent to the edges of the atomic part, ignoring global properties. For example, extremely small exit corridors through the atomic part should increase the lower bound. Such properties might be considered by how large a component might be before it 'plugs' a hole in configuration space.

While computing a *sufficient* number of pieces, the algorithm firstly decomposes the divisible part into convex polygons, then computes a grid resolution using all pairs of adjacent polygons. The algorithm can generate a large number of pieces if there exists a single narrow corridor anywhere in the divisible part. Different convex decomposition methods might also yield better decompositions, as might the selection of good disassembly sequences [18, 6].

In future work, we would also like to expand the problem scope. We expect that most practical problems in this area are 3D, rather than planar. Also, we can imagine situations where there are several atomic components, and several divisible components. How should such multi-component 3D puzzles be designed, assembled, or disassembled?

References

1. Thomas F Allen, Joel W Burdick, and Elon Rimon. Two-finger caging of polygonal objects using contact space search. *IEEE Transactions on Robotics*, 31(5):1164–1179, 2015.
2. DJ Arbuckle and Aristides AG Requicha. Self-assembly and self-repair of arbitrary shapes by a swarm of reactive robots: algorithms and simulations. *Autonomous Robots*, 28(2):197–211, 2010.
3. Devin J Balkcom and Jeffrey C Trinkle. Computing wrench cones for planar rigid body contact tasks. *The International Journal of Robotics Research*, 21(12):1053–1066, 2002.
4. Aaron Becker, Golnaz Habibi, Justin Werfel, Michael Rubenstein, and James McLurkin. Massive uniform manipulation: Controlling large populations of simple robots with a common input signal. In *2013 IEEE/RSJ International Conference on Intelligent Robots and Systems*, pages 520–527. IEEE, 2013.
5. Matthew P Bell, Weifu Wang, Jordan Kunzika, and Devin Balkcom. Knot-tying with four-piece fixtures. *The International Journal of Robotics Research*, 33(11):1481–1489, 2014.
6. Lukas Beyeler, Jean-Charles Bazin, and Emily Whiting. A graph-based approach for discovery of stable deconstruction sequences. In *Advances in Architectural Geometry 2014*, pages 145–157. Springer, 2015.
7. Jae-Sook Cheong, Herman J Haverkort, and A Frank van der Stappen. Computing all immobilizing grasps of a simple polygon with few contacts. *Algorithmica*, 44(2):117–136, 2006.
8. Jae-Sook Cheong, A Frank Van Der Stappen, Ken Goldberg, Mark H Overmars, and Elon Rimon. Immobilizing hinged polygons. *International Journal of Computational Geometry & Applications*, 17(01):45–69, 2007.
9. Robert Connelly, Erik D Demaine, and Günter Rote. Straightening polygonal arcs and convexifying polygonal cycles. In *Foundations of Computer Science, 2000. Proceedings. 41st Annual Symposium on*, pages 432–442. IEEE, 2000.
10. Jurek Czyzowicz, Ivan Stojmenovic, and Jorge Urrutia. Immobilizing a polytope. In *Workshop on Algorithms and Data Structures*, pages 214–227. Springer, 1991.
11. Jurek Czyzowicz, Ivan Stojmenovic, and Jorge Urrutia. Immobilizing a shape. *International Journal of Computational Geometry & Applications*, 9(02):181–206, 1999.
12. Jeff Erickson, Shripad Thite, Fred Rothganger, and Jean Ponce. Capturing a convex object with three discs. In *Robotics and Automation, 2003. Proceedings. ICRA'03. IEEE International Conference on*, volume 2, pages 2242–2247. IEEE, 2003.
13. Chi-Wing Fu, Peng Song, Xiaoqi Yan, Lee Wei Yang, Pradeep Kumar Jayaraman, and Daniel Cohen-Or. Computational interlocking furniture assembly. *ACM Transactions on Graphics (TOG)*, 34(4):91, 2015.
14. Philipp Herholz, Wojciech Matusik, and Marc Alexa. Approximating Free-form Geometry with Height Fields for Manufacturing. *Computer Graphics Forum (Proc. of Eurographics)*, 34(2):239–251, 2015.
15. Ruizhen Hu, Honghua Li, Hao Zhang, and Daniel Cohen-Or. Approximate pyramidal shape decomposition. *ACM Trans. Graph.*, 33(6):213–1, 2014.
16. Jun Huang, Satyandra K Gupta, and Klaus Stoppel. Generating sacrificial multi-piece molds using accessibility driven spatial partitioning. *Computer-Aided Design*, 35(13):1147–1160, 2003.

17. Keith Kotay, Daniela Rus, Marsette Vona, and Craig McGray. The self-reconfiguring robotic molecule: Design and control algorithms. In *Workshop on Algorithmic Foundations of Robotics*, pages 376–386. Citeseer, 1998.

18. Anne Loomis. {Computation reuse in stacking and unstacking}. 2005.

19. Satoshi Makita and Yusuke Maeda. 3d multifingered caging: Basic formulation and planning. In *2008 IEEE/RSJ International Conference on Intelligent Robots and Systems*, pages 2697–2702. IEEE, 2008.

20. Matthew T Mason. *Mechanics of robotic manipulation*. MIT press, 2001.

21. Bhubaneswar Mishra, Jacob T Schwartz, and Micha Sharir. On the existence and synthesis of multifinger positive grips. *Algorithmica*, 2(1-4):541–558, 1987.

22. Alok K Priyadarshi and Satyandra K Gupta. Geometric algorithms for automated design of multi-piece permanent molds. *Computer-Aided Design*, 36(3):241–260, 2004.

23. R. Ravi and M. N. Srinivasan. Decision criteria for computer-aided parting surface design. *Comput. Aided Des.*, 22(1):11–17, January 1990.

24. Franz Reuleaux. *Theoretische Kinematik: Grundzüge einer Theorie des Maschinenwesens*, volume 1. F. Vieweg und Sohn, 1875.

25. Elon Rimon and Andrew Blake. Caging 2d bodies by 1-parameter two-fingered gripping systems. In *Robotics and Automation, 1996. Proceedings., 1996 IEEE International Conference on*, volume 2, pages 1458–1464. IEEE, 1996.

26. Elon Rimon and Joel W Burdick. Mobility of bodies in contact. i. a 2nd-order mobility index for multiple-finger grasps. *IEEE transactions on Robotics and Automation*, 14(5):696–708, 1998.

27. Elon Rimon and Joel W Burdick. Mobility of bodies in contact. ii. how forces are generated by curvature effects. *IEEE Transactions on Robotics and Automation*, 14(5):709–717, 1998.

28. Michael Rubenstein, Adrian Cabrera, Justin Werfel, Golnaz Habibi, James McLurkin, and Radhika Nagpal. Collective transport of complex objects by simple robots: theory and experiments. In *Proceedings of the 2013 international conference on Autonomous agents and multi-agent systems*, pages 47–54. International Foundation for Autonomous Agents and Multiagent Systems, 2013.

29. Michael Rubenstein, Alejandro Cornejo, and Radhika Nagpal. Programmable self-assembly in a thousand-robot swarm. *Science*, 345(6198):795–799, 2014.

30. Daniela Rus and Marsette Vona. Crystalline robots: Self-reconfiguration with compressible unit modules. *Autonomous Robots*, 10(1):107–124, 2001.

31. Jack Snoeyink and Jorge Stolfi. Objects that cannot be taken apart with two hands. In *Proceedings of the ninth annual symposium on Computational geometry*, pages 247–256. ACM, 1993.

32. Peng Song, Bailin Deng, Ziqi Wang, Zhichao Dong, Wei Li, Chi-Wing Fu, and Ligang Liu. Cofifab: Coarse-to-fine fabrication of large 3d objects. *ACM Transactions on Graphics*.

33. Peng Song, Zhongqi Fu, Ligang Liu, and Chi-Wing Fu. Printing 3d objects with interlocking parts. *Computer Aided Geometric design (Proc. of GMP 2015)*, 35-36:137–148, 2015.

34. Jacob Steiner. Einige gesetze über die theilung der ebene und des raumes. *Journal für die reine und angewandte Mathematik*, 1:349–364, 1826.

35. Ileana Streinu. A combinatorial approach to planar non-colliding robot arm motion planning. In *Foundations of Computer Science, 2000. Proceedings. 41st Annual Symposium on*, pages 443–453. IEEE, 2000.

36. Mostafa Vahedi and A Frank van der Stappen. Caging polygons with two and three fingers. *The International Journal of Robotics Research*, 27(11-12):1308–1324, 2008.

37. Weifu Wang and Devin Balkcom. Grasping and folding knots. In *2016 IEEE International Conference on Robotics and Automation (ICRA)*, pages 3647–3654. IEEE, 2016.

38. Yinan Zhang and Devin Balkcom. Interlocking structure assembly with voxels. In *Intelligent Robots and Systems (IROS), 2016 IEEE/RSJ International Conference on*. IEEE, 2016.

A BCMP Network Approach to Modeling and Controlling Autonomous Mobility-on-Demand Systems

Ramon Iglesias, Federico Rossi, Rick Zhang, and Marco Pavone

Abstract In this paper we present a queuing network approach to the problem of routing and rebalancing a fleet of self-driving vehicles providing on-demand mobility within a *capacitated* road network. We refer to such systems as autonomous mobility-on-demand systems, or AMoD. We first cast an AMoD system into a closed, multi-class BCMP queuing network model. Second, we present analysis tools that allow the characterization of performance metrics for a given routing policy, in terms, e.g., of vehicle availabilities, and first and second order moments of vehicle throughput. Third, we propose a scalable method for the synthesis of routing policies, with performance guarantees in the limit of large fleet sizes. Finally, we validate the theoretical results on a case study of New York City. Collectively, this paper provides a unifying framework for the analysis and control of AMoD systems, which subsumes earlier Jackson and network flow models, provides a quite large set of modeling options (e.g., the inclusion of road capacities and general travel time distributions), and allows the analysis of second and higher-order moments for the performance metrics.

1 Introduction

Personal mobility in the form of privately owned automobiles contributes to increasing levels of traffic congestion, pollution, and under-utilization of vehicles (on average 5% in the US [17]) – clearly unsustainable trends for the future. The pressing need to reverse these trends has spurred the creation of cost competitive, on-demand personal mobility solutions such as car-sharing (e.g. Car2Go, ZipCar) and ride-sharing (e.g. Uber, Lyft). However, without proper fleet management, car-sharing and, to some extent, ride-sharing systems lead to vehicle imbalances: vehicles aggregate in some areas while becoming depleted in others, due to the asymmetry be-

Ramon Iglesias · Federico Rossi · Rick Zhang · Marco Pavone
Autonomous Systems Laboratory, Stanford University, 496 Lomita Mall, Stanford, CA 94305,
e-mail: {rdit, frossi2, rickz, pavone}@stanford.edu

© Springer Nature Switzerland AG 2020
K. Goldberg et al. (Eds.): *Algorithmic Foundations of Robotics XII*, SPAR 13, pp. 831–847, 2020.
https://doi.org/10.1007/978-3-030-43089-4_53

tween trip origins and destinations [24]. This issue has been addressed in a variety of ways in the literature. For example, in the context of bike-sharing, [5] proposes rearranging the stock of bicycles between stations using trucks. The works in [18], [4], and [1] investigate using paid drivers to move vehicles between car-sharing stations where cars are parked, while [2] studies the merits of dynamic pricing for incentivizing drivers to move to underserved areas.

Self-driving vehicles offer the distinctive advantage of being able to rebalance themselves, in addition to the convenience, cost savings, and possibly safety of not requiring a driver. Indeed, it has been shown that one-way vehicle sharing systems with self-driving vehicles (referred to as autonomous mobility-on-demand systems, or AMoD) have the potential to significantly reduce passenger cost-per-mile-traveled, while keeping the advantages and convenience of personal mobility [22]. Accordingly, a number of works have recently investigated the potential of AMoD systems, with a specific focus on the synthesis and analysis of coordination algorithms. Within this context, the goal of this paper is to provide a principled framework for the analysis and synthesis of routing policies for AMoD systems.

Literature Review: Previous work on AMoD systems can be categorized into two main classes: heuristic methods and analytical methods. Heuristic routing strategies are extensively investigated in [7, 8, 16] by leveraging a traffic simulator and, in [25], by leveraging a model predictive control framework. Analytical models of AMoD systems are proposed in [20], [24], and [26], by using fluidic, Jackson queuing network, and capacitated flow frameworks, respectively. Analytical methods have the advantage of providing structural insights (e.g., [26]), and provide guidelines for the synthesis of control policies. The problem of controlling AMoD systems is similar to the System Optimal Dynamic Traffic Assignment (SO-DTA) problem (see, e.g., [6, 19]) where the objective is to find optimal routes for all vehicles within congested or capacitated networks such that the total cost is minimized. The main differences between the AMoD control problem and the SO-DTA problem is that SO-DTA only optimizes customer routes, and *not* rebalancing routes.

This paper aims at devising a general, unifying analytical framework for analysis and control of AMoD systems, which subsumes many of the analytical models recently presented in the literature, chiefly, [20], [24], and [26]. Specifically, this paper extends our earlier Jackson network approach in [24] by adopting a BCMP queuing-theoretical framework [3, 15]. BCMP networks significantly extend Jackson networks by allowing almost arbitrary customer routing and service time distributions, while still admitting a convenient product-form distribution solution for the equilibrium distribution [15]. Such generality allows one to take into account several real-world constraints, in particular road capacities (that is, congestion). Indeed, the impact of AMoD systems on congestion has been a hot topic of debate. For example, [16] notes that empty-traveling rebalancing vehicles may increase congestion and total in-vehicle travel time for customers, but [26] shows that, with congestion-aware routing and rebalancing, the increase in congestion can be avoided. The proposed BCMP model recovers the results in [26], with the additional benefits of taking into account the stochasticity of transportation networks and providing estimates for performance metrics.

Statement of Contributions: The contribution of this paper is fourfold. First, we show how an AMoD system can be cast within the framework of closed, multi-class BCMP queuing networks. The framework captures stochastic passenger arrivals, vehicle routing on a road network, and congestion effects. Second, we present analysis tools that allow the characterization of performance metrics for a given routing policy, in terms, e.g., of vehicle availabilities and second-order moments of vehicle throughput. Third, we propose a scalable method for the synthesis of routing policies: namely, we show that, for large fleet sizes, the stochastic optimal routing strategy can be found by solving a linear program. Finally, we validate the theoretical results on a case study of New York City.

Organization: The rest of the paper is organized as follows. In Section 2, we cover the basic properties of BCMP networks and, in Section 3, we describe the AMoD model, cast it into a BCMP network, and formally present the routing and rebalancing problem. Section 4 presents the mathematical foundations and assumptions required to reach our proposed solution. We validate our approach in Section 5 using a model of Manhattan. Finally, in Section 6, we state our concluding remarks and discuss potential avenues for future research.

2 Background Material

In this section we review some basic definitions and properties of BCMP networks, on which we will rely extensively later in the paper.

2.1 Closed, Multi-Class BCMP Networks

Let $\mathscr{X}$ be a network consisting of N independent queues (or nodes). A set of agents move within the network according to a stochastic process, i.e. after receiving service at queue i they proceed to queue j with a given probability. No agent enters or leaves the network from the outside, so the number of agents is fixed and equal to m. Such a network is also referred to as a *closed* queuing network. Agents belong to one of $K \in \mathbb{N}_{>0}$ classes, and they can switch between classes upon leaving a node.

Let $x_{i,k}$ denote the number of agents of class $k \in \{1,\dots,K\}$ at node $i \in \{1,\dots,N\}$. The state of node i, denoted by x_i, is given by $x_i = (x_{i,1},\dots,x_{i,K}) \in \mathbb{N}^K$. The state space of the network is [10]:

$$\Omega_m := \{(x_1,\dots,x_N) : x_i \in \mathbb{N}^K, \ \sum_{i=1}^{N} \|x_i\|_1 = m\},$$

where $\|\cdot\|_1$ denotes the standard 1-norm (i.e., $\|x\|_1 = \sum_i |x_i|$). The relative frequency of visits (also known as relative throughput) to node i by agents of class k, denoted as $\pi_{i,k}$, is given by the traffic equations [10]:

$$\pi_{i,k} = \sum_{k'=1}^{K} \sum_{j=1}^{N} \pi_{j,k'} p_{j,k';i,k}, \quad \text{for all } i \in \{1,\dots,N\}, \tag{1}$$

where $p_{j,k';i,k}$ is the probability that upon leaving node j, an agent of class k' goes to node i and becomes an agent of class k. Equation (1) does not have a unique solution (a typical feature of closed networks), and $\pi = \{\pi_{i,k}\}_{i,k}$ only determines frequencies up to a constant factor (hence the name "relative" frequency). It is customary to express frequencies in terms of a chosen reference node, e.g., so that $\pi_{1,1} = 1$.

Queues are allowed to be one of four types: First Come First Serve (FCFS), Processor Sharing, Infinite Server, and Last Arrived, First Served. FCFS nodes have exponentially distributed service times, while the other three queue types may follow any Cox distribution [10]. Such a queuing network model is referred to as a closed, multi-class BCMP queuing network [10].

Let $\mathcal{N}$ represent the set of nodes in the network and N its cardinality. For the remainder of the paper, we will restrict networks to have only two types of nodes: FCFS queues with a single server (for short, SS queues), forming a set $\mathcal{S} \subset \mathcal{N}$, and infinite server queues (for short, IS queues), forming a set $\mathcal{I} \subset \mathcal{N}$. Furthermore, we consider class-independent and load-independent nodes, whereby at each node $i \in \{1, \ldots, N\}$ the service rate is given by:

$$\mu_i(x_i) = c_i(x_i)\mu_i^o,$$

where $x_i := \|\boldsymbol{x}_i\|_1$ is the number of agents at node i, μ_i^o is the (class-independent) base service rate, and $c_i(x_i)$ is the (load-independent) *capacity* function

$$c_i(x_i) = \begin{cases} x_i & \text{if } x_i \leq c_i^o, \\ c_i^0 & \text{if } x_i > c_i^o, \end{cases}$$

which depends on the number of servers c_i^o at the queue. In the case considered in this paper, $c_i^o = 1$ for all $i \in \mathcal{S}$ and $c_i^o = \infty$ for all $i \in \mathcal{I}$.

Under the assumption of class-independent service rates, the multi-class network $\mathcal{L}$ can be "compressed" into a single-class network $\mathcal{L}^*$ with state-space $\Omega_m^* := \{(x_1, \ldots, x_N) : x_i \in \mathbb{N}, \ \sum_{i=1}^N x_i = m\}$ [14]. Performance metrics for the original, multi-class network $\mathcal{L}$ can be found by first analyzing the compressed network $\mathcal{L}^*$, and then applying suitable scalings for each class. Specifically, let $\pi_i = \sum_{k=1}^K \pi_{i,k}$ and $\gamma_i = \sum_{k=1}^K \frac{\pi_{i,k}}{\mu_i^o}$, be the total relative throughput and relative utilization at a node i, respectively. Then, the stationary distribution of the compressed, single-class network $\mathcal{L}^*$ is given by

$$\mathbb{P}(x_1, \ldots, x_N) = \frac{1}{G(m)} \prod_{i=1}^N \frac{\gamma_i^{x_i}}{\prod_{a=1}^{x_i} c_i(a)}, \quad \text{where} \quad G(m) = \sum_{\Omega_m^*} \prod_{i=1}^N \frac{\gamma_i^{x_i}}{\prod_{a=1}^{x_i} c_i(a)}$$

is a normalizing constant. Remarkably, the stationary distribution has a product form, a key feature of BCMP networks.

Three performance metrics that are of interest at each node are throughput, expected queue length, and availability. First, the throughput at a node (i.e., the number of agents processed by a node per unit of time) is given by

$$\Lambda_i(m) = \pi_i \frac{G(m-1)}{G(m)}. \tag{2}$$

Second, let $\mathbb{P}_i(x_i; m)$ be the probability of finding x_i agents at node i; then the expected queue length at node i is given by $L_i(m) = \sum_{x_i=1}^{m} x_i \mathbb{P}_i(x_i; m)$.

In the case of IS nodes (i.e., nodes in $\mathscr{I}$), the expected queue length can be more easily derived via Little's Law as [11]

$$L_i(m) = \Lambda_i(m)/\mu_i^o \quad \text{for all } i \in \mathscr{I}. \tag{3}$$

Finally, the availability of single-server, FCFS nodes (i.e., nodes in $\mathscr{S}$) is defined as the probability that the node has at least 1 agent, and is given by [11]

$$A_i(m) = \gamma_i \frac{G(m-1)}{G(m)} \quad \text{for all } i \in \mathscr{S}.$$

The throughputs and the expected queue lengths for the original, multi-class network $\mathscr{X}^*$ can be found via scaling [14], specifically, $\Lambda_{i,k}(m) = (\pi_{i,k}/\pi_i)\Lambda_i(m)$ and $L_{i,k}(m) = (\pi_{i,k}/\pi_i)L_i(m)$.

It is worth noting that evaluating the three performance metrics above requires computation of the normalization constant $G(m)$, which is computationally expensive. However, several techniques are available to avoid the direct computation of $G(m)$. In particular, in this paper we use the Mean Value Analysis method, which, remarkably, can be also used to compute higher moments (e.g., variance) [23]. Details are provided in the extended version of this paper [13].

2.2 Asymptotic Behavior of Closed BCMP Networks

In this section we describe the asymptotic behavior of closed BCMP networks as the number of agents m goes to infinity. The results described in this section are taken from [11], and are detailed for a single-class network; however, as stated in the previous section, results found for a single-class network can easily be ported to the multi-class equivalent in the case of class-independent service rates.

Let $\rho_i := \gamma_i/c_i^o$ be the utilization factor of node $i \in \mathscr{N}$, where c_i^o is the number of servers at node i. Assume that the relative throughputs $\{\pi_i\}_i$ are normalized so that $\max_{i \in \mathscr{S}} \rho_i = 1$; furthermore, assume that nodes are ordered by their utilization factors so that $1 = \rho_1 \geq \rho_2 \geq \ldots \geq \rho_N$, and define the set of bottleneck nodes as $\mathscr{B} := \{i \in \mathscr{S} : \rho_i = 1\}$.

It can be shown [11, p. 14] that, as the number of agents m in the system approaches infinity, the availability at all bottleneck nodes converges to 1 while the availability at non-bottleneck nodes is strictly less than 1, that is

$$\lim_{m \to \infty} A_i(m) \begin{cases} = 1 & \forall i \in \mathscr{B}. \\ < 1 & \forall i \notin \mathscr{B}. \end{cases} \tag{4}$$

Additionally, the queue lengths at the non-bottleneck nodes have a limiting distribution given by

$$\lim_{m \to \infty} \mathbb{P}_i(x_i; m) = \begin{cases} (1-\rho_i)\rho_i^{x_i} & i \in S, i \notin \mathscr{B}, \\ e^{-\gamma_i} \frac{\gamma_i^{x_i}}{x_i!} & i \in I. \end{cases} \tag{5}$$

Together, (4) and (5) have strong implications for the operation of queuing networks with a large number of agents, and in particular for the operation of AMoD systems. Intuitively, (4) shows that as we increase the number of agents in the network, they will be increasingly queued at bottleneck nodes, driving availability in those queues to one. Alternatively, non-bottleneck nodes will converge to an availability strictly less than one, implying that there is always a non-zero probability of having an empty queue. In other words, agents will aggregate at the bottlenecks and become depleted elsewhere. Additionally, (5) shows that, as the number of agents goes to infinity, non-bottleneck nodes tend to behave like queues in an equivalent open BCMP network with the bottleneck nodes removed, i.e., individual performance metrics can be calculated in isolation.

3 Model Description and Problem Formulation

In this section, we introduce a BCMP network model for AMoD systems, and formalize the problem of routing and rebalancing such systems under stochastic conditions. Casting an AMoD system as a queuing network allows us to characterize and compute key performance metrics including the distribution of the number of vehicles on each road link (a key metric to characterize traffic congestion) and the probability of servicing a passenger request. To emphasize the relationship with the theory presented in the previous section, we reuse the same notation whenever concepts are equivalent.

3.1 Autonomous Mobility-on-Demand Model

Consider a set of stations[1] $\mathscr{S}$ distributed within an urban area connected by a network of individual road links $\mathscr{I}$, and m autonomous vehicles providing one-way transportation between these stations for incoming customers. Customers arrive to a station $s \in \mathscr{S}$ with a target destination $t \in \mathscr{S}$ according to a time-invariant Poisson process with rate $\lambda \in \mathbb{R}_{>0}$. The arrival process for all origin-destination pairs is summarized by the set of tuples $\mathscr{Q} = \{(s^{(q)}, t^{(q)}, \lambda^{(q)})\}_q$.

If on customer arrival there is an available vehicle, the vehicle drives the customer towards its destination. Alternatively, if there are no vehicles, the customer leaves the system (i.e., chooses an alternative transportation system). Thus, we adopt a *passenger loss* model. Such model is appropriate for systems where high quality-of-service is desired; from a technical standpoint, this modeling assumption decouples the passenger queuing process from the vehicle queuing process.

A vehicle driving a passenger through the road network follows a routing policy $\alpha^{(q)}$ (defined in Section 3.2) from origin to destination, where q indicates the origin-destination-rate tuple. Once it reaches its destination, the vehicle joins the station first-come, first-serve queue and waits for an incoming trip request.

A known problem of such systems is that vehicles will inevitably accumulate at one or more of the stations and reduce the number of vehicles servicing the rest of

---

[1] Stations are not necessarily physical locations: they can also be interpreted as a set of geographical regions.

the system [11] if no corrective action is taken. To control this problem, we introduce a set of "virtual rebalancing demands" or "virtual passengers" whose objective is to balance the system, i.e., to move empty vehicles to stations experiencing higher passenger loss. Similar to passenger demands, rebalancing demands are defined by a set of origin, destination and arrival rate tuples $\mathscr{R} = \{(s^{(r)}, t^{(r)}, \lambda^{(r)})\}_r$, and a corresponding routing policy $\alpha^{(r)}$. Therefore, the objective is to find a set of routing policies $\alpha^{(q)}, \alpha^{(r)}$, for all $q \in \mathscr{Q}$, $r \in \mathscr{R}$, and rebalancing rates $\lambda^{(r)}$, for all $r \in \mathscr{R}$, that balances the system while minimizing the number of vehicles on the road, and thus reducing the impact of the AMoD system on overall traffic congestion.

3.2 Casting an AMoD System into a BCMP Network

We are now in a position to frame the AMoD system in terms of a BCMP network model. To this end, we represent the vehicles, the road network and the passenger demands in the BCMP framework.

First, the passenger loss assumption allows the model to be characterized as a queuing network with respect only to the *vehicles*. Thus, we will henceforth use the term "vehicles" to refer to the queuing agents. From this perspective, the stations $\mathscr{S}$ are equivalent to SS queues, and the road links $\mathscr{I}$ are modeled as IS queues.

Second, we map the underlying road network to a directed graph with the queues as *edges*, and introduce the set of road intersections $\mathscr{V}$ to function as graph vertices. As in Section 2, the set of all queues is given by $\mathscr{N} = \{\mathscr{S} \cup \mathscr{I}\}$. Let Parent($i$) and Child($i$) be the origin and destination vertices of edge i. Then, a road that goes from intersection j to intersection l is represented by a queue $i \in \mathscr{I}$ such that Parent(i)= j and Child(i)= l. Note that the road may not have lanes in the opposite direction, in which case a queue i' with Parent(i')= l and Child(i')= j would not exist. For example, in Figure 1, queue 14 starts at vertex 1 and ends at vertex 5. However, there is no queue that connect the vertices in the opposite direction. Similarly, we assume that stations are adjacent to road intersections, and therefore stations are modeled as edges with the same parent and child vertex. An intersection may have access to either one station (e.g., vertex 2 in Figure 1), or zero stations (e.g., vertex 5 in Figure 1).

Fig. 1 BCMP network model of an AMoD system. Diamonds represent infinite-server road links, squares represent the single-server vehicle stations, and dotted circles represent road intersections.

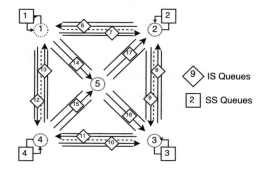

Third, we introduce classes to represent the process of choosing destinations. We map the set of tuples $\mathscr{Q}$ and $\mathscr{R}$ defined in Section 3.1 to a set of classes $\mathscr{K}$ such that $\mathscr{K} = \{\mathscr{Q} \cup \mathscr{R}\}$. Moreover, let $\mathscr{O}_i$ be the subset of classes whose origin $s^{(k)}$ is the station i, such that $\mathscr{O}_i := \{k \in \mathscr{K} : s^{(k)} = i\}$ and $\mathscr{D}_i$ be the subset whose destination $t^{(k)}$ is the station i, such that $\mathscr{D}_i := \{k \in \mathscr{K} : t^{(k)} = i\}$. Thus, the probability that a vehicle at station i will leave for station j with a (real or virtual) passenger is the ratio between the respective (real or virtual) arrival rate $\lambda^{(k)}$, with $s^{(k)} = i, t^{(k)} = j$, and the sum of all arrival rates at station i. Formally, the probability that a vehicle of class k switches to class k' upon arrival to its destination $t^{(k)}$ is $\widetilde{p}_{t^{(k)}}^{(k')} = \left(\lambda^{(k')} / \widetilde{\lambda}_{t^{(k)}} \right)$, where $\widetilde{\lambda}_i = \sum_{k \in \mathscr{O}_i} \lambda^{(k)}$ is the sum of all arrival rates at station $t^{(k)}$. Consequently, at any instant in time a vehicle belongs to a class $k \in \mathscr{K}$, regardless of whether it is waiting at a station or traveling along the road network. By switching class on vehicle arrival, the vehicles's transition probabilities $\widetilde{p}_{t^{(k)}}^{(k')}$ encode the passenger and rebalancing demands defined in Section 3.1.

As mentioned in the previous section, the traversal of a vehicle from its source $s^{(k)}$ to its destination $t^{(k)}$ is guided by a routing policy $\alpha^{(k)}$. This routing policy, in queuing terms, consists of a matrix of transition probabilities. Let $\mathscr{W}_i = \{j \in \mathscr{N} : \text{Parent}(j) = i\}$ be the set of queues that begin in vertex i, and $\mathscr{U}_i = \{j \in \mathscr{N} : \text{Child}(j) = i\}$ the set of queues that end in vertex i. A vehicle of class k leaves the station $s^{(k)}$ via one of the adjacent roads $j \in \mathscr{W}_{\text{Child}(s^{(k)})}$ with probability $\alpha^{(k)}_{s^{(k)}, j}$. It continues traversing the road network via these adjacency relationships following the routing probabilities $\alpha^{(k)}_{i,j}$ until it is adjacent to its goal $t^{(k)}$. At this point, the vehicle proceeds to the destination and changes its class to $k' \in \mathscr{O}_{t^{(k)}}$ with probability $\widetilde{p}_{t^{(k)}}^{(k')}$. This behavior is encapsulated by the routing matrix

$$
p_{i,k;j,k'} = \begin{cases} \alpha^{(k)}_{i,j} & \text{if } k = k', \, j \in \mathscr{W}_{\text{Child}(i)}, \, t^{(k)} \notin \mathscr{W}_{\text{Child}(i)}, \\ \widetilde{p}_j^{(k')} & \text{if } j = t^{(k)}, \, t^{(k)} \in \mathscr{W}_{\text{Child}(i)}, \, k' \in \mathscr{O}_j, \\ 0 & \text{otherwise}, \end{cases} \tag{6}
$$

such that $\sum_{j,k'} p_{i,k;j,k'} = 1$. Thus, the relative throughput $\pi_{i,k}$, total relative throughput π_i, and utilization γ_i have the same definition as in Section 2.

As stated before, the queuing process at the stations is modeled as a SS queue where the service rate of the vehicles $\mu_i(a)$ is equal to the sum of real and virtual passenger arrival rates, i.e. $\mu_i(a) = \widetilde{\lambda}_i$ for any station i and queue length a. Additionally, by modeling road links as IS queues, we assume that their service rates follow a Cox distribution with mean $\mu_i(a) = \frac{c_i(a)}{T_i}$, where T_i is the expected time required to cross link i in absence of congestion, and $c_i(a)$ is the capacity factor when there are a vehicles in the queue. In this paper, we only consider the case of load-independent travel times, therefore $c_i(a) = a$ for all a, i.e., the service rate is the same regardless of the number of vehicles on the road. We do not make further assumptions on the distribution of the service times. The assumption of load-independent travel times is representative of uncongested traffic [21]: in Section 3.3 we discuss how to incorporate probabilistic constraints for congestion on road links.

3.3 Problem formulation

As stated in Equation (4), vehicles tend to accumulate in bottleneck stations driving their availability to 1 as the fleet size increases, while the rest of the stations have availability strictly smaller than 1. In other words, for unbalanced systems, availability at most stations is capped regardless of fleet size. Therefore, it is desirable to make all stations "bottleneck" stations, i.e., set the constraint $\gamma_i = \gamma_j$ for all $i, j \in \mathcal{S}$, so as to (i) enforce a natural notion of "service fairness," and (ii) prevent needless accumulation of empty vehicles at the stations.

However, it is desirable to minimize the impact that the rebalancing vehicles have on the road network. We achieve this by minimizing the expected number of vehicles on the road serving customer and rebalancing demands. Using Equation (3), the expected amount of vehicles on a given road link i is given by $\Lambda_i(m)T_i$.

Lastly, we wish to avoid congestion on the individual road links. Traditionally, the relation between vehicle flow and congestion is parametrized by two basic quantities: the *free-flow travel time* T_i, i.e., the time it takes to traverse a link in absence of other traffic; and the *nominal capacity* C_i, i.e., the measure of traffic flow beyond which travel time increases very rapidly [19]. Assuming that travel time remains approximately constant when traffic is below the nominal capacity (an assumption typical of many state-of-the-art traffic models [19]), our approach is to keep the expected traffic $\Lambda_i(m)T_i$ below the nominal capacity C_i and thus avoid congestion effects. Note that by constraining in expectation there is a non-zero probability of exceeding the constraint; however, in Section 4.2, we show that, asymptotically, it is also possible to constrain the *probability* of exceeding the congestion constraint.

Accordingly, the routing problem we wish to study in this paper (henceforth referred to as the *Optimal Stochastic Capacitated AMoD Routing and Rebalancing problem*, or OSCARR) can now be formulated as follows:

$$\underset{\lambda^{(r\in\mathcal{R})},\,\alpha_{ij}^{(k\in\mathcal{K})}}{\text{minimize}} \quad \sum_{i\in\mathcal{S}} \Lambda_i(m)T_i,$$

$$\text{subject to} \quad \gamma_i = \gamma_j, \qquad\qquad\qquad\qquad\qquad i,j \in \mathcal{S}, \tag{7a}$$

$$\Lambda_i(m)T_i \le C_i, \qquad\qquad\qquad\qquad i \in \mathcal{S}, \tag{7b}$$

$$\pi_{s^{(k)},k} = \sum_{k'\in\mathcal{K}} \sum_{j\in\mathcal{N}} \pi_{j,k} P_{j,k;t^{(k)},k'}, \qquad k \in \mathcal{K}, \tag{7c}$$

$$\pi_{i,k} = \sum_{k'=1}^{K} \sum_{j=1}^{N} \pi_{j,k'} P_{j,k';i,k} \qquad i \in \{\mathcal{S} \cup \mathcal{I}\}, \tag{7d}$$

$$\sum_{j} \alpha_{ij}^{(k)} = 1, \quad \alpha_{ij}^{(k)} \ge 0, \qquad\qquad i,j \in \{\mathcal{S} \cup \mathcal{I}\}, \tag{7e}$$

$$\lambda_r \ge 0, \qquad\qquad\qquad\qquad\qquad\qquad r \in \mathcal{R}. \tag{7f}$$

Constraint (7a) enforces equal availability at all stations, while constraint (7b) ensures that all road links are (on average) uncongested. Constraints (7c)–(7f) enforce consistency in the model. Namely, (7c) ensures that all traffic leaving the source $s^{(k)}$ of class k arrives at its destination $t^{(k)}$, (7d) enforces the traffic equations (1), (7e) ensures that $\alpha_{ij}^{(k)}$ is a valid probability measure, and (7f) guarantees nonnegative rebalancing rates.

At this point, we would like to reiterate some assumptions built into the model. First, the proposed model is time-invariant. That is, we assume that customer and rebalancing rates remain constant for the segment of time under analysis, and that the network is able to reach its equilibrium distribution. An option for including the variation of customer demand over time is to discretize a period of time into smaller segments, each with its own arrival parameters and resulting rebalancing rates. Second, the passenger loss model assumes impatient customers and is well suited for cases where high level of service is required. This allows us to simplify the model by focusing only on the vehicle process; however, it disregards the fact that customers may have different waiting thresholds and, consequently, the queuing process of waiting customers. Third, we focus on keeping traffic within the nominal road capacities in expectation, allowing us to assume load-independent travel times and to model exogenous traffic as a reduction in road capacity. Finally, we make no assumptions on the distribution of travel times on the road links: the analysis proposed in this paper captures arbitrary distributions of travel times and only depends on the *mean* travel time.

4 Asymptotically Optimal Algorithms for AMoD routing

In this section we show that, as the fleet size goes to infinity, the solution to OS-CARR can be found by solving a linear program. This insight allows the efficient computation of asymptotically optimal routing and rebalancing policies and of the resulting performance parameters for AMoD systems with very large numbers of customers, vehicles and road links.

First, we introduce simplifications possible due to the nature of the routing matrix $\{\alpha_{i,j}^{(k)}\}_{(i,j),k}$. Then, we express the problem from a flow conservation perspective. Finally, we show that the problem allows an asymptotically optimal solution with bounds on the probability of exceeding road capacities. The solution we find is equivalent to the one presented in [26]: thus, we show that the network flow model in [26] also captures the asymptotic behavior of a stochastic AMoD routing and rebalancing problem.

4.1 Folding of traffic equations

The next two lemmas show that the traffic equations (1) at the SS queues can be expressed in terms of other SS queues, and that the balanced network constraint can be expressed in terms of real and virtual passenger arrivals. The proof of Lemmas 1 and 2 are omitted for space reasons and can be found in the extended version of this paper [13].

Lemma 1 (Folding of traffic equations). *Let $\mathscr{Z}$ be a feasible solution to OSCARR. Then, the relative throughputs of the single server stations can be expressed in terms of the relative throughputs of the other single server stations, that is*

$$\pi_i = \sum_{k \in \mathscr{D}_i} \widetilde{p}_{s(k)}^{(k)} \pi_{s(k)}, \quad \text{for all } i \in S. \tag{8}$$

Lemma 2 (Balanced system in terms of arrival rates). *Let $\mathscr{L}$ be a feasible solution to OSCARR, then the constraint $\gamma_i = \gamma_j \ \forall i, j$, is equivalent to*

$$\widetilde{\lambda}_i = \sum_{k \in \mathscr{D}_i} \lambda_{s(k)}^{(k)}. \tag{9}$$

4.2 Asymptotically Optimal Solution

As discussed in Section 2.1, relative throughputs are computed up to a constant multiplicative factor. Thus, without loss of generality, we can set the additional constraint $\pi_{s(1)} = \widetilde{\lambda}_1$, which, along with (7a), implies that

$$\pi_i = \widetilde{\lambda}_i, \quad \pi_{s(k),k} = \lambda^{(k)}, \quad \text{and} \quad \gamma_i = 1, \quad \text{for all } i \in \mathscr{S}. \tag{10}$$

As seen in Section 2.2, the availabilities of stations with the highest relative utilization tend to one as the fleet size goes to infinity. Since the stations are modeled as single-server queues, $\rho_i = \gamma_i$ for all $i \in \mathscr{S}$. Therefore, if the system is balanced, $\gamma_i = \gamma_S^{\max} = \gamma = 1$ for all $i \in \mathscr{S}$. That is, the set of bottleneck stations $\mathscr{B}$ includes all stations in $\mathscr{S}$ and $\lim_{m \to \infty} \frac{G(m-1)}{G(m)} = 1$ by Equation (4) .

As $m \to \infty$ and $\frac{G(m-1)}{G(m)} \to 1$, the throughput at every station $\Lambda_i(m)$ becomes a linear function of the relative frequency of visits to that station, according to Equation (2). Thus, the objective function and the constraints in (7) are reduced to linear functions. We define the resulting problem (i.e., Problem (7) with $G(m-1)/G(m) = 1$) as the *Asymptotically Optimal Stochastic Capacitated AMoD Routing and Rebalancing problem*, or A-OSCARR. The following lemma shows that the optimal solution to OSCARR approaches the optimal solution to A-OSCARR as m increases.

Lemma 3 (Asymptotic behavior of OSCARR). *Let $\{\pi_{i,k}^*(m)\}_{i,k}$ be the set of relative throughputs corresponding to an optimal solution to OSCARR with a given set of customer demands $\{\lambda_i\}_i$ and a fleet size m. Also, let $\{\hat{\pi}_{i,k}\}_{i,k}$ be the set of relative throughputs corresponding to an optimal solution to A-OSCARR for the same set of customer demands. Then,*

$$\lim_{m \to \infty} \frac{G(m-1)}{G(m)} \sum_{i \in I} T_i \sum_{k \in \mathscr{K}} \pi_{i,k}^* = \sum_{i \in I} T_i \sum_{k \in \mathscr{K}} \hat{\pi}_{i,k}. \tag{11}$$

Proof. We arrive to the proof by contradiction. Recall that $\pi_i = \sum_{k \in \mathscr{K}} \pi_{i,k}$. Assume Equation (11) did not hold. By definition, the following equations hold for all m and $\{\pi_{i,k}\}_{i,k}$:

$$\frac{G(m-1)}{G(m)} \sum_{i \in I} T_i \pi_i^* \leq \frac{G(m-1)}{G(m)} \sum_{i \in I} T_i \pi_i, \quad (12) \qquad \sum_{i \in I} T_i \hat{\pi}_i \leq \sum_{i \in I} T_i \pi_i. \tag{13}$$

Applying the limit to (12) and using (4), we obtain $\sum_{i \in I} T_i \lim_{m \to \infty} (\pi_i^*) \leq \sum_{i \in I} T_i \pi_i$. However, according to our assumption, either $\sum_{i \in I} T_i \lim_{m \to \infty} (\pi_i^*) > \sum_{i \in I} T_i \hat{\pi}_i$ or $\sum_{i \in I} T_i \lim_{m \to \infty} (\pi_i^*) < \sum_{i \in I} T_i \hat{\pi}_i$ but the former violates Equation (12), and the latter (13).

As discussed in Section 3.3, constraint 7b only enforces an upper bound on the expected number of vehicles traversing a link. However, in the asymptotic regime, it is possible to enforce an analytical upper bound on the *probability* of exceeding the nominal capacity of any given road link. As seen in Equation (5), as the fleet size increases, the distribution of the number of vehicles on a road link i converges to a Poisson distribution with mean $T_i\pi_i$. The cumulative density function of a Poisson distribution is given by $Pr(X < \bar{x}) = Q(\lfloor\bar{x}+1\rfloor, \tilde{C})$, where $\tilde{C}$ is the mean of the distribution and Q is the regularized upper incomplete gamma function. Let ε be the maximum tolerable probability of exceeding the nominal capacity. Set $\widehat{C}_i = Q^{-1}(1-\varepsilon; \lfloor C_i+1\rfloor)$, i.e. $Q(\lfloor C_i+1\rfloor, \widehat{C}_i) = 1-\varepsilon$. Then the constraint $\Lambda_i(m)T_i \leq \widehat{C}_i$ is equivalent to $\lim_{m\to\infty} \mathbb{P}_i(x_i < C_i; m) \geq 1-\varepsilon$.

4.3 Linear programming formulation and multi-commodity flow equivalence

We now show that an asymptotically optimal routing and rebalancing problem can be framed as a multi-commodity flow problem. Specifically, we show that A-OSCARR is equivalent to the Congestion-Free Routing and Rebalancing problem presented in [26]: thus, (i) A-OSCARR can be solved efficiently by ad-hoc algorithms for multi-commodity flow (e.g. [12]) and (ii) the theoretical results presented in [26] (namely, the finding that rebalancing trips do not increase congestion) extend, in expectation, to *stochastic* systems.

First, we show that the problem can be solved exclusively for the relative throughputs on the road links, and then we show that the resulting equations are equivalent to a minimum cost, multi-commodity flow problem.

The relative throughput going from an intersection i into adjacent roads is $\sum_{j\in\mathscr{W}_i'} \pi_{j,k}$, where $\mathscr{W}_i' = \{\mathscr{W}_i \cap \mathscr{I}\}$ is the set of road links that begin in node i. Similarly, the relative throughput entering the intersection i from the road network is $\sum_{j\in\mathscr{U}_i'} \pi_{j,k}$, where $\mathscr{U}_i' = \{\mathscr{U}_i \cap \mathscr{I}\}$ is the set of road links terminating in i. Additionally, define $d_i^{(k)}$ as the difference between the relative throughput leaving the intersection and the relative throughput entering the intersection. From (7d), (7c), and (10), we see that for customer classes

$$\sum_{j\in\mathscr{W}_i'} \pi_{j,q} - \sum_{j\in\mathscr{U}_i'} \pi_{j,q} = d_i^{(q)}, \quad \text{where} \quad d_i^{(q)} = \begin{cases} \lambda^{(q)} & \text{if } i = s^{(q)}, \\ -\lambda^{(q)} & \text{if } i = t^{(q)}, \\ 0 & \text{otherwise.} \end{cases}$$

While the rebalancing arrival rates $\lambda^{(r)}$ are not fixed, we do know from Equation (7c) and from the definition of $d_i^{(q)}$ that $d_{s^{(r)}}^{(r)} = -d_{t^{(r)}}^{(r)}$. Thus,

$$\sum_{j\in\mathscr{W}_{s^{(r)}}'} \pi_{j,r} - \sum_{j\in\mathscr{U}_{s^{(r)}}'} \pi_{j,r} = -\sum_{j\in\mathscr{W}_{t^{(r)}}'} \pi_{j,r} + \sum_{j\in\mathscr{U}_{t^{(r)}}'} \pi_{j,r}.$$

Finally, we can rewrite Lemma 2 as

$$\sum_{q \in \mathcal{Q}} d_i^{(q)} + \sum_{r \in \mathcal{R}} \sum_{j \in \mathcal{W}_i'} \pi_{j,r} - \sum_{j \in \mathcal{U}_i'} \pi_{j,r} = 0.$$

Thus, in the asymptotic regime Problem (7) can be restated as

$$\underset{\pi_{i,k}, i \in \mathcal{I}, k \in \mathcal{K}}{\text{minimize}} \quad \sum_{i \in I} T_i \sum_{k \in \mathcal{K}} \pi_{i,k},$$

subject to

$$\sum_{q \in \mathcal{Q}} d_i^{(q)} + \sum_{r \in \mathcal{R}} \sum_{j \in \mathcal{W}_i'} \pi_{j,r} - \sum_{j \in \mathcal{U}_i'} \pi_{j,r} = 0 \qquad \forall i \in \mathcal{S}, \tag{14a}$$

$$T_i \sum_{k \in \mathcal{K}} \pi_{j,k} \leq \widehat{C}_i \qquad \forall i \in \mathcal{I}, \tag{14b}$$

$$\sum_{j \in \mathcal{W}_i'} \pi_{j,q} - \sum_{j \in \mathcal{U}_i'} \pi_{j,q} = d_i^{(q)} \qquad \forall i \in \mathcal{S}, \tag{14c}$$

$$\sum_{j \in \mathcal{W}_{s(r)}'} \pi_{j,r} - \sum_{j \in \mathcal{U}_{s(r)}'} \pi_{j,r} = \sum_{j \in \mathcal{U}_{t(r)}'} \pi_{j,r} - \sum_{j \in \mathcal{W}_{t(r)}'} \pi_{j,r} \qquad \forall r \in \mathcal{R}, \tag{14d}$$

$$\sum_{j \in \mathcal{W}_i'} \pi_{j,r} - \sum_{j \in \mathcal{U}_i'} \pi_{j,r} = 0 \qquad \forall i \in \mathcal{S} \setminus \{s^{(r)}, t^{(r)}\}, \tag{14e}$$

$$\sum_{j \in \mathcal{W}_{s(r)}'} \pi_{j,r} - \sum_{j \in \mathcal{U}_{s(r)}'} \pi_{j,r} \geq 0 \qquad \forall r \in \mathcal{R}, \tag{14f}$$

$$\pi_{i,k} \geq 0, \qquad \forall i \in \mathcal{I}, k \in \mathcal{K}. \tag{14g}$$

Here, constraints (14a) and (14b) are direct equivalents to (7a) and (7b), respectively. By keeping traffic continuity and equating throughputs at source and target stations, (14c) enforces (7c) and (7d) for the customer classes. For the rebalancing classes, (14d) is equivalent to (7c) and (14e) to (7d). Non-negativity of rebalancing rates (7f) is kept by (14f).

Thus, A-OSCARR can be solved efficiently as a linear program. Note that this formulation is very similar to the multi-commodity flow found in [26]. The formulation in this paper prescribes specific routing policies for distinct rebalancing origin-destination pairs, while [26] only computes a single "rebalancing flow": however, stochastic routing policies can be computed from the rebalancing flow in [26] with a flow decomposition algorithm [9].

5 Numerical Experiments

To illustrate a real-life application of the results in this paper, we applied our model to a case study of Manhattan, and computed the system performance metrics as a function of fleet size using the Mean Value Analysis. Results show that the solution correctly balances vehicle availability across stations while keeping road traffic within the capacity constraints, and that the assumption of load-independent travel times is relatively well founded.

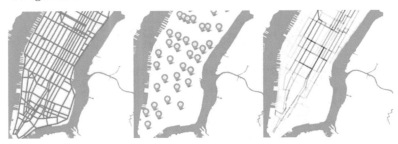

Fig. 2: Manhattan scenario (detail). Left: modeled road network. Center: Station locations. Right: Resulting vehicular flow (darker flows show higher vehicular presence).

The model used for this case study consists of a simplified version of Manhattan's real road network, with 1005 road links and 357 nodes. To select station positions and compute the rates $\lambda^{(q)}$ of the origin-destination flows $\mathcal{Q}$, we used the taxi trips within Manhattan that took place between 7:00AM and 8:00AM on March 1, 2012 (22,416 trips) from the New York City Taxi and Limousine Commission dataset[2]. We clustered the pickup locations into 50 different groups with K-means clustering, and placed a station at the road intersection closest to each cluster centroid. We fitted an origin-destination model with exponential distributions to describe the customer trip demands between the stations. Road capacities were reduced to ensure that the model reaches maximum utilization in some road links; in the real world, a qualitatively similar reduction in road capacity is caused by traffic exogenous to the taxi system.

We considered two scenarios: the "baseline" scenario where traffic constraints on each road link are based on expectation, i.e., on average the number of vehicles on a road link is below its nominal capacity; and the "conservative" scenario where the constraints are based on the asymptotic probability of exceeding the nominal capacity (specifically, the asymptotic probability of exceeding the nominal capacity is constrained to be lower than 10%). Figure 2 shows the station locations, the road network, and the resulting traffic flow, and Figure 3 shows our results.

We see from Figure 3a that, as intended, the station availabilities are balanced and approach one as the fleet size increases. However, Figure 3b shows that there is a trade off between availability and vehicle utilization. For example, for a fleet size of 4,000 vehicles, half of the vehicles are expected to be waiting at the stations. In contrast, a fleet of 2,400 vehicles results in availability of 91% and only 516 vehicles are expected to be at the stations. Not shown in the figures, 34% of the trips are for rebalancing purposes; in contrast, only about 18% of the traveling vehicles are rebalancing. This shows that rebalancing trips are significantly shorter than passenger trips, in line with the goal of minimizing the number of empty vehicles on the road and thus road congestion.

Although Figures 3a and 3b show only the results for the baseline case, for the conservative scenario the difference in availabilities is less than 0.1%, and the differ-

---

[2] http://www.nyc.gov/html/tlc/html/about/trip_record_data.shtml

ence in the total number of vehicles on the road is less than 7, regardless of fleet size. However, road utilization is significantly different in the two scenarios we considered. In Figure 3c, we see that, as the fleet size increases, the likelihood of exceeding the nominal capacity approaches 50%. In contrast, in the conservative scenario, the probability of exceeding the capacity is never more than 10% –by design– regardless of fleet size.

Lastly, we evaluated how much the assumption of load-independent travel times deviates from the more realistic case where travel time depends on traffic. Assuming asymptotic conditions (i.e., the number of vehicles on each road follows a Poisson distribution), we computed for both scenarios the expected travel time between each origin-destination pair by using the Bureau of Public Roads (BPR) delay model [21], and estimated the difference with respect to the load-independent travel time used in this paper. The results, depicted in Figure 3d, show that the maximum difference for the baseline and conservative scenarios are an increase of around 8% and 4%, respectively, and the difference tends to be smaller for higher trip times. Thus, for this specific case study, our assumption is relatively well founded.

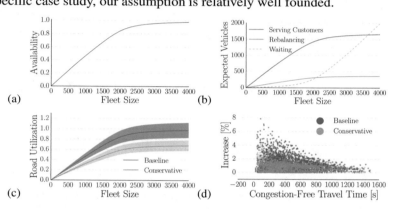

Fig. 3: (a) Station availabilities as a function of fleet size for the baseline case. (b) Expected number of vehicles by usage as a function of fleet size for the baseline case. (c) Utilization as a function of fleet size for the most utilized road. The colored band denotes ±1 standard deviation from the mean. (d) Increase in expected travel time for each O-D pair when considering the BPR delay model.

6 Conclusions

In this paper, we presented a novel queuing theoretic framework for modeling AMoD systems within capacitated road networks. We showed that, for the routing and rebalancing problem, the stochastic model we propose asymptotically recovers existing models based on the network flow approximation. The model enables the analysis and control of the probabilistic distribution of the vehicles, and not only of its expectation: in particular (i) it enables the computation of higher moments of the vehicle distribution on road links and at stations and (ii) it allows to establish an arbitrary bound on the asymptotic probability of exceeding the capacity of individual road links.

The flexibility of the model presented will be further exploited in future work. First, we would like to incorporate a more accurate congestion model, using load-dependent IS queues as roads, in order to study heavily congested scenarios. Second, we currently consider the system in isolation from other transportation modes, whereas, in reality, customer demand depends on the perceived quality of the different transportation alternatives. Future research will explore the effect of AMoD systems on customer behavior and how to optimally integrate fleets of self-driving vehicles with existing public transit. Third, we would like to examine scenarios where the vehicle fleet is electric-powered and explore the relationship between the constraints imposed by battery charging and the electric grid. Fourth, the current model assumes that each customer travels alone: future research will address the problem of *ride-sharing*, where multiple customers may share the same vehicle. Lastly, the control policy proposed in this paper is open-loop, and thus sensitive to modeling errors (e.g., incorrect estimation of customer demand). Future research will characterize the stability, persistent feasibility and performance of *closed-loop* model predictive control schemes based on a receding-horizon implementation of the controller presented in this paper.

Acknowledgements The authors would like to thank the National Science Foundation for funding this work via the NSF CAREER award.

References

1. Acquaviva, F., Di Paola, D., Rizzo, A.: A novel formulation for the distributed solution of load balancing problems in mobility on-demand systems. In: Connected Vehicles and Expo (ICCVE), 2014 International Conference on. pp. 906–911 (2014)
2. Banerjee, S., Johari, R., Riquelme, C.: Pricing in ride-sharing platforms: A queueing-theoretic approach. In: Proceedings of the Sixteenth ACM Conference on Economics and Computation. pp. 639–639. ACM (2015)
3. Baskett, F., Chandy, K.M., Muntz, R.R., Palacios, F.G.: Open, closed, and mixed networks of queues with different classes of customers. Journal of the Association for Computing Machinery 22(2), 248–260 (Apr 1975)
4. Boyacı, B., Zografos, K.G., Geroliminis, N.: An optimization framework for the development of efficient one-way car-sharing systems. European Journal of Operational Research 240(3), 718–733 (2015)
5. Chemla, D., Meunier, F., Calvo, R.W.: Bike sharing systems: Solving the static rebalancing problem. Discrete Optimization 10(2), 120–146 (2013)
6. Chiu, Y.C., Bottom, J., Mahut, M., Paz, A., Balakrishna, R., Waller, T., Hicks, J.: Dynamic traffic assignment: A primer. Transportation Research E-Circular (E-C153) (2011)
7. Fagnant, D.J., Kockelman, K.M.: The travel and environmental implications of shared autonomous vehicles, using agent-based model scenarios. Transportation Research Part C: Emerging Technologies 40, 1–13 (2014)
8. Fagnant, D.J., Kockelman, K.M., Bansal, P.: Operations of shared autonomous vehicle fleet for austin, texas, market. Transportation Research Record: Journal of the Transportation Research Board (2536), 98–106 (2015)

9. Ford, L.R., Fulkerson, D.R.: Flows in Networks. Princeton University Press (1962)
10. Gelenbe, E., Pujolle, G., Nelson, J.: Introduction to queueing networks, vol. 2. Wiley Chichester (1998)
11. George, D.K.: Stochastic Modeling and Decentralized Control Policies for Large-Scale Vehicle Sharing Systems via Closed Queueing Networks. Ph.D. thesis, The Ohio State University (2012)
12. Goldberg, A.V., Oldham, J.D., Plotkin, S., Stein, C.: An implementation of a combinatorial approximation algorithm for minimum-cost multicommodity flow. In: Bixby, R., Boyd, E., Ros-Mercado, R. (eds.) Integer Programming and Combinatorial Optimization, Lecture Notes in Computer Science, vol. 1412, pp. 338–352. Springer Berlin Heidelberg (1998)
13. Iglesias, R., Rossi, F., Zhang, R., Pavone, M.: A BCMP network approach to modeling and controlling autonomous mobility-on-demand systems (extended version) (2016), available at http://arxiv.org/abs/1607.04357
14. Kant, K., Srinivasan, M.: Introduction to computer system performance evaluation. McGraw-Hill College (1992)
15. Kobayashi, H., Gerla, M.: Optimal routing in closed queueing networks. In: ACM SIGCOMM Computer Communication Review. vol. 13, pp. 26–26. ACM (1983)
16. Levin, M.W., Li, T., Boyles, S.D., Kockelman, K.M.: A general framework for modeling shared autonomous vehicles. In: 95th Annual Meeting of the Transportation Research Board (2016)
17. Neil, D.: Could self-driving cars spell the end of ownership? wsj.com (2015)
18. Nourinejad, M., Zhu, S., Bahrami, S., Roorda, M.J.: Vehicle relocation and staff rebalancing in one-way carsharing systems. Transportation Research Part E: Logistics and Transportation Review 81, 98–113 (2015)
19. Patriksson, M.: The traffic assignment problem: models and methods. Courier Dover Publications (2015)
20. Pavone, M., Smith, S.L., Frazzoli, E., Rus, D.: Robotic load balancing for mobility-on-demand systems. International Journal of Robotics Research 31(7), 839–854 (Jun 2012)
21. Bureau of Public Roads: Traffic assignment manual. Tech. rep., U.S. Department of Commerce, Urban Planning Division, Washington, D.C (1964) (1964)
22. Spieser, K., Treleaven, K., Zhang, R., Frazzoli, E., Morton, D., Pavone, M.: Toward a systematic approach to the design and evaluation of automated mobility-on-demand systems: A case study in Singapore. In: Lecture Notes in Mobility, pp. 229–245. Springer (Jun 2014)
23. Strelen, J.: A generalization of mean value analysis to higher moments: moment analysis. In: ACM Sigmetrics Performance Evaluation Review. vol. 14, pp. 129–140. ACM (1986)
24. Zhang, R., Pavone, M.: Control of robotic mobility-on-demand systems: A queueing-theoretical perspective. International Journal of Robotics Research 35(1-3), 186–203 (Jan 2016)
25. Zhang, R., Rossi, F., Pavone, M.: Model predictive control of autonomous mobility-on-demand systems. In: Proc. IEEE Conf. on Robotics and Automation. pp. 1382 – 1389. Stockholm, Sweden (May 2016)
26. Zhang, R., Rossi, F., Pavone, M.: Routing autonomous vehicles in congested transportation networks: Structural properties and coordination algorithms. In: Robotics: Science and Systems (Mar 2016)

Optimal Policies for Platooning and Ride Sharing in Autonomy-Enabled Transportation

Aviv Adler, David Miculescu, and Sertac Karaman**

Massachusetts Institute of Technology,
77 Massachusetts Avenue, Cambridge, MA, 02139
{adlera, dmicul, sertac}@mit.edu,

Abstract. Rapid advances in autonomous-vehicle technology may soon allow vehicles to platoon on highways, leading to substantial fuel savings through reduced aerodynamic drag. While these aerodynamic effects have been widely studied, the systems aspects of platooning have received little attention. In this paper, we consider a class of problems, applicable to vehicle platooning and passenger ride-sharing, from the systems perspective. We consider a system in which vehicles arrive at a station according to a stochastic process. If they wait for each other, they can form platoons to save energy, but at the cost of incurring transportation delays. Our analysis explores this tradeoff between energy consumption and transportation delays. Among our results is the derivation of the Pareto-optimal boundary and characterization of the optimal polices in both the open-loop and feedback regimes. Surprisingly, the addition of feedback improves the energy-delay curve very little when compared to open-loop policies.

Keywords: Logistics, Autonomy-Enabled Transportation Systems, Platooning, Optimization

1 Introduction

Autonomous vehicles hold the potential to revolutionize transportation. Autonomous cars may enable effective mobility-on-demand systems [1]. Autonomous drones may allow us to deploy the next generation urban logistics infrastructure [2, 3]. Even though tremendous effort has been invested in developing autonomous vehicles themselves, it is still unclear how teams of autonomous vehicles may operate as an efficient and sustainable transportation system.

We draw inspiration from the development of technology that allows autonomous vehicles to follow each other very closely [4], beyond what human drivers and pilots can safely do. This technology may allow road/air vehicles to drive/fly in platoons/formation in order to save energy. It is known that platooning trucks [4–9] and formation flying airplanes [10, 11] can save a substantial amount of fuel, simply by mitigating the effects of aerodynamic drag. In fact, many bird species frequently utilize formation flight precisely for this purpose

---

** The first two authors contributed equally, and they are listed in alphabetical order.

© Springer Nature Switzerland AG 2020
K. Goldberg et al. (Eds.): *Algorithmic Foundations of Robotics XII*, SPAR 13, pp. 848–863, 2020.
https://doi.org/10.1007/978-3-030-43089-4_54

on long-distance flights [12, 13]. Similarly, bikers and race car drivers often pack behind a leader in long-distance competitions [14], a behavior called *drafting*.

Motivated by this potential, we study the systems aspects of this technology from the perspective of sustainability-efficiency tradeoffs, in a mathematically rigorous way. Even though aerodynamics of platooning and control policies for best platooning results have been studied, to our knowledge this paper is the first to study its systems aspects from a queueing perspective as in [15, 16].

Specifically, we consider vehicles arriving at an initial location, from here on called the station, and all headed to the same destination. If the vehicles wait at the station for each other, then they can travel in platoons to reduce energy consumption per vehicle. However, clearly, forming long platoons will require some vehicles to wait for others, which will impact the average delay. In this paper, we explore this fundamental "energy-delay tradeoff."

Our contributions include the following: formalizing the problem, including formal definitions for station control policies and the energy-delay tradeoff; several general results, greatly narrowing down the range of policies we need to find optimal policies; a full characterization of the optimal delay-energy combinations, as well as a description of a class of policies which achieve them; and a description and analysis of a class of open-loop policies, which are potentially much simpler to implement. Surprisingly, in many cases, feedback policies show little improvement compared to open-loop policies, which we show rigorously.

Our results apply directly to truck platooning, air vehicle formation flight, and bicycle and race car drafting. These results also apply to ride-sharing (or pooling) systems [17, 18]. Consider passengers arriving at a bus station. When should each bus head out? What policies guarantee minimize delays and energy consumption (and the negative environmental impact of transportation) at the same time? Even though we utilize terminology from platooning, drafting, and formation flight throughout the paper, our results also apply to ride sharing systesm, which may become prominent with the introduction of autonomous vehicles [18].

Furthermore, while most recent rigorous work in transportation is on efficiency metrics, for instance focusing on capacity (throughput) and delay tradeoffs [19], our study involves the tradeoff between efficiency and sustainability, in this case, energy and delay. The study of the tradeoff between efficiency and sustainability in autonomy-enabled transportation systems may become more prominent in the future, as autonomous vehicles are deployed broadly.

This paper is organized as follows. In Section 2, we introduce our model and formalize the metrics of evaluation. In Section 3, we provide a set of general results characterizing optimal policies. Then, we focus on Poisson arrivals and restrict our attention to certain classes of energy usage models. In Section 4, we analyze optimal feedback policies. In Section 5, we analyze optimal open-loop policies. In Section 6, we compare these two policies, and conclude with remarks.

Due to space constraints some of our proofs are omitted and replaced either with sketches or the underlying intuition. The complete proofs are available in our full online paper [20].

2   Problem Setup

Even though the problem setup is relevant for both platooning and pooling, we motivate the rest of the paper with the truck platooning problem and use terminology from platooning. We remind the reader that all results reported in this paper can be applied to similar pooling problems as well.

We are interested in identifying the *energy-delay tradeoff* for platooning vehicles running between two locations connected by a highway. The vehicles arrive at an initial location, called the station, at random time intervals. All vehicles are headed to the same final location. In this case, if the vehicles wait for each other, then they can platoon to reduce energy consumption. Platoons with many vehicles are clearly better in terms of energy consumption; however, to form large platoons, the vehicles need to wait long periods of time for each other, which increases delay. Our aim is to analyze this fundamental energy-delay tradeoff.

The model includes two essential components: the vehicle arrival model and the energy consumption of a platoon; there are two metrics of performance: average delay and average energy per vehicle. We consider two types of policies: time-table policies and feedback optimal policies. In what follows, we formalize these six notions, leading to a formal problem definition.

Vehicle Arrival Model: Generically, the vehicles arrive at the station according to the stochastic process $\{t_i : i \in \mathbb{Z}_{>0}\}$, where t_i is the (random) arrival time for the ith vehicle (so $t_1 \leq t_2 \leq \dots$). In this paper, we assume that the arrival process $\{t_i : i \in \mathbb{Z}_{>0}\}$ is a Poisson process with intensity λ. That is, the inter-arrivals $a_{i+1} = t_{i+1} - t_i$ are independent random variables whose distributions are exponential with parameter λ.

Energy Consumption Model: The energy usage of a platoon of vehicles is a function of the number of vehicles in the platoon. This function is called the *total energy function*, and denoted by $\mathrm{TE}: \mathbb{Z}_{\geq 0} \to \mathbb{R}_{\geq 0}$, where $\mathrm{TE}(k)$ denotes the combined energy usage of a platoon consisting of k vehicles, which is called a k-platoon. By convention, we define $\mathrm{TE}(0) = 0$.

In the case of vehicle platooning, these functions are determined by the aerodynamics of drafting for the particular vehicles and speeds involved. In the case of pooling, these relate to the energy characteristics of the engine that powers the vehicles carrying the passengers.

We also define a particular class of total energy functions, which are simple but capture the intuition that while the first and last vehicles in a platoon pay higher energy costs, vehicles in the middle of a platoon spand roughly the same energy as each other. Thus, each vehicle will have an energy cost c, with an additional cost of c^* which is not influenced significantly by the number of vehicles in the platoon. We call this the class of *affine energy functions*:

Definition 1. *The* affine energy function *with parameters $c \geq 0$ and $c^* > 0$ is* $\mathrm{TE}^{\mathrm{aff}}(k) = ck + c^*$ *for $k > 0$, and* $\mathrm{TE}^{\mathrm{aff}}(0) = 0$.

This has been previously used in studying platooning optimization problems [5].

Control policies: We formally define the state of the system at time t as $x(t) :=$ $\left(t, (t_i^{\mathrm{arr}}, t_i^{\mathrm{dep}})_{i=1}^n\right)$, where t denotes the current time in the system, and $(t_i^{\mathrm{arr}}, t_i^{\mathrm{dep}})$ is the arrival time and departure time of the ith vehicle to enter the system. If a vehicle has arrived but not departed, then we formally define the departure time of such a vehicle as ∞. Note that the state of the system incorporates the entire history of arrivals and departure in the system, but only up until the current time t. We denote the collection of all states as the state space $\mathcal{X}$:

$$\mathcal{X} := \left\{ \left(t, (t_i^{\mathrm{arr}}, t_i^{\mathrm{dep}})_{i=1}^n\right) : 0 \le t_i^{\mathrm{arr}} \le t, \, t_i^{\mathrm{arr}} \le t_i^{\mathrm{dep}}, \, t_i^{\mathrm{arr}} \le t_{i+1}^{\mathrm{arr}}, n \in \mathbb{N} \right\}.$$

Let $\mathcal{Y}$ be a predetermined sequence of random variables. We define a *control policy* $\pi \colon (\mathcal{X}, \mathcal{Y}) \to 2^{\mathbb{N}}$ (the $\mathcal{Y}$ is there to accommodate control policies with internal randomness), identifying which vehicles must leave the station as a platoon at the current time t. Note that a control policy can control a vehicle at the instant of its arrival in the station, but has no knowledge of future arrivals.

We also define the notion of *regularity* of a policy. This is to capture the intuition that whatever is optimal now should be optimal also in the future, and to avoid the possibility of a policy which changes permanently over time and prevents average delay and/or energy consumption from converging. In order to properly define it, we first assume that any reasonable policy will *sometimes* send all vehicles out of the station; otherwise, there will be always be at least one vehicle at the station, and we can immediately improve the policy by always having one fewer vehicle at the station (without altering when platoons leave).

Definition 2 (Reset Points and Regular Policies). *Suppose we have a policy π and it has been running up until time t^*. Then, t^* is a* reset point *if:*

- *no vehicles are present at the station at time t^**
- *the distribution of the future behavior of the system is identical to the distribution of future behavior at $t = 0$*

A policy π is regular *if the probability of getting a reset point at some time in the future is 1, and the expected time between reset points is finite.*

The fundamental problem considered here is the tradeoff between *energy consumption* and *delay* on the vehicles. The station seeks to minimize energy usage by combining vehicles into platoons. However, in order to do this, it will have to hold some vehicles at the station to wait for additional vehicles to arrive. The amount of time a vehicle spends at the station is its *delay*. Both energy consumption and delay for a policy are calculated as *averages per vehicle* as the policy runs over long time horizons.

Definition 3 (Average Energy Consumption Metric). *Suppose we run policy π on a Poisson-distributed sequence of vehicles (indexed by order of arrival). Let k_i be the size of the platoon in which vehicle i leaves. We define $E(i) = \frac{\mathrm{TE}(k(i))}{k(i)}$. If there is some E such that*

$$\lim_{n \to \infty} \frac{\sum_{i=1}^n E(i)}{n} = E \text{ with probability 1}$$

then we refer to this as the average energy *of policy π and denote it by $\mathbb{E}[E(\pi)]$.*

Note that in this definition, we are using $E(i) = \frac{\text{TE}(k(i))}{k(i)}$ – the average energy consumption of a vehicle in vehicle i's platoon – as a stand-in for the amount of energy actually consumed by vehicle i. Individually, vehicle i may consume a different amount of energy, but this averages out correctly when considering all the vehicles in vehicle i's platoon.

Definition 4 (Average Delay Metric). *Suppose we run policy π on a Poisson-distributed sequence of vehicles (indexed by order of arrival). If there is some D such that*

$$\lim_{n \to \infty} \frac{\sum_{i=1}^{n}(t_i^{\text{dep}} - t_i^{\text{arr}})}{n} = D \text{ with probability 1,}$$

then we refer to this as the average delay *of policy π and denote it by $\mathbb{E}[D(\pi)]$.*

Note that these definitions require that the average energy or delay per vehicle converges to a predictable value with probability 1 as the policy is run over an infinite time. Although this may seem strict, it is necessary since otherwise there is no clear way to measure a policy's performance in terms of saving energy and delay. We restricted ourselves to regular policies because, in addition to being intuitively the only reasonable kind of policy, the regularity condition guarantees that the above metrics are well-defined:

Lemma 1. *If π is regular, then $\mathbb{E}[E(\pi)]$ and $\mathbb{E}[D(\pi)]$ exist.*

The proof is straightforward, but due to space constraints we omit it.

Now that we have a rigorous notion of the energy consumption and delay incurred by different policies, we can discuss what it means for a policy to be 'efficient' with respect to these metrics. We use the notion of *Pareto optimality*, in which a policy is optimal for a class Π if no other policy in Π can improve one metric without performing worse on the other.

Definition 5 (Pareto-optimality). *A policy π is* Pareto-optimal *among a class of policies Π if there is no other policy $\pi' \in \Pi$ such that $\mathbb{E}[E(\pi')] \leq \mathbb{E}[E(\pi)]$ and $\mathbb{E}[D(\pi')] \leq \mathbb{E}[D(\pi)]$ and at least one of the two inequalities is strict.*

When implementing a policy, one should always choose from the set of Pareto-optimal policies, if possible. This then leads to the notions of the *profile* of policy π and the *energy/delay tradeoff curve*:

Definition 6 (Profile and energy-delay curve). *The* profile *of policy π is the point $(\mathbb{E}[D(\pi)], \mathbb{E}[E(\pi)]) \in \mathbb{R}^2$. The* energy-delay curve *is the set of profiles of Pareto-optimal policies.*

Equivalent ways of viewing the energy-delay curve are: (1) as the lower convex hull of the set of profile points of policies; (2) the curve representing solutions of the optimization problem, "minimize average energy among all policies with average delay at most D".

All-or-Nothing Policies: We also note that, intuitively, keeping vehicles at the station when a platoon leaves is wasteful – if we wanted to send a k-platoon, we should do it when there are k vehicles rather than wait for additional vehicles to arrive. This leads to the notion of an *All-or-Nothing policy*, which sends every vehicle present with every platoon.

Definition 7. *An* all-or-nothing policy *is a mapping* $\pi : (\mathcal{X}, \mathcal{Y}) \to \{\emptyset, \mathbb{N}\}$.

In fact, we will show later that *only* All-or-Nothing policies (and others that behave similarly) are optimal (see Theorem 2). Hence, we can restrict ourselves to considering only these policies.

We now introduce the two specific kinds of policies we will analyze: *time-table* and *mixed-threshold* policies. We are interested in time-table policies because they require almost no information on the state of the system, and hence can be much easier and quicker to implement. Mixed-threshold policies are also relatively simple (though they require more information on the state of the system) and simple to analyze, and, as it turns out, there is always a Pareto-optimal policy of this form. Both are by definition All-or-Nothing policies.

Time-Table Policies: General policies are allowed to use a significant range of information and internal randomization to make their decisions, while time-table policies are restricted to have no randomization and make decisions based only on the elapsed time since the beginning of the process. Formally:

Definition 8. *A* time-table policy *is a policy* $\pi : \mathbb{R}_+ \to \{0, \mathbb{N}\}$.

Thus, a time-table policy is really just a list of predetermined times for which it sends all vehicles out of the station – hence the name. Because of the regularity condition, time-table policies must have reset points after which their behavior repeats. Thus, a time-table policy can be thought of in terms of a repeating cycle of intervals during which vehicles collect at the station; at the end of each interval, the vehicles exit.

We define the (ordered) set of *waiting times* $(T_1, T_2, \ldots, T_M)$ so that $T_m > 0$ for all m, and denote $C := \sum_{m=1}^{M} T_m$ as the cycle time, *i.e.*, the waiting times repeat every C time units in the same order. In other words, let $\mathcal{T}$ be the set of times when vehicles exit the station. Then, $\mathcal{T}$ has the following structure:

$$\mathcal{T} := \left\{ kC + \sum_{m=1}^{n} T_m : \ k \in \mathbb{Z}_{\geq 0}, \ n \in \{1, 2, \ldots, M\} \right\},$$

where $k \geq 0$ counts the number of cycles and $n \in \{1, \ldots, M\}$ tracks the current interval T_n. Thus, for the remainder of the paper, we will use the simpler but equivalent representation of time-table policies:

$$\Pi := \left\{ (T_1, T_2, \ldots, T_M) \in \mathbb{R}_{>0}^{M} : \ M \in \mathbb{Z}_{>0} \right\}$$

We note that a time-table policy has a reset point every time the cycle ends.

Mixed-Threshold Policies: Another simple type of policy is one which determines what distribution of platoon sizes it wants to send and directly produces them.

A *mixed-threshold* policy π is determined by a probability distribution $p(\pi) = (p_1, p_2, \dots)$ over platoon sizes; p_k is the probability that any given platoon will have k vehicles. Every time it sends out a platoon, some k is drawn according to $p(\pi)$ (independently of everything that has happened so far) and the next platoon leaves when the kth vehicle arrives at the station.

3   General Results

We present some general results pertaining to this optimization problem.

Theorem 1 (Mixing Regular Policies). *Given two regular policies π, π' and some $\alpha \in [0, 1]$, it is possible to create a 'mixed' policy π^* such that*

$$\mathbb{E}[D(\pi^*)] = \alpha\mathbb{E}[D(\pi)] + (1-\alpha)\mathbb{E}[D(\pi')] \ and \ \mathbb{E}[E(\pi^*)] = \alpha\mathbb{E}[E(\pi)] + (1-\alpha)\mathbb{E}[E(\pi')]$$

Proof. π^* is constructed by using the "cycles" of π and π'. A *cycle* of a policy is an interval between consecutive reset points. By the definition of regularity and because vehicle arrivals are Poisson, a policy can be viewed as a sequence of independent cycles. Furthermore, each cycle has finite expected length (by definition of regularity) and therefore a finite expected number of vehicle arrivals. Let $L(\pi)$ be the expected number of vehicle arrivals in a cycle of π, and $L(\pi')$

We then construct π^* as follows: randomly decide between π and π', and run one cycle; then repeat. The random decisions are i.i.d. (so every time a cycle ends, we have a reset point of π^*). The probability of picking π is:

$$\mathbb{P}[\text{pick } \pi \text{ to control next cycle}] = \frac{\alpha L(\pi')}{\alpha L(\pi') + (1 - \alpha)L(\pi)}$$

This is because we want each *vehicle* to be controlled under π with probability α, and under π' with probability $(1 - \alpha)$ – and thus, we pick more often from the policy with shorter cycles to address this imbalance. Thus, by linearity of expectations, the average delay and energy per vehicle under π^* are $\alpha\mathbb{E}[D(\pi)] + (1 - \alpha)\mathbb{E}[D(\pi')]$ and $\alpha\mathbb{E}[E(\pi)] + (1 - \alpha)\mathbb{E}[E(\pi')]$ respectively.   ∎

Corollary 1. *The energy-delay curve for regular policies is convex.*

Proof. This is a simple consequence of Theorem 1, since it shows that any point on the line segment between the profiles of π and π' is the profile of some π^*. Thus, the set of profile points is a convex set and so its lower boundary (which is the energy-delay curve) is a convex function.   ∎

We now define the *platoon distribution* of a given policy π, which is a table of what fraction of platoons have k vehicles and leave behind j at the station.

Definition 9 (Platoon Distribution). *Consider running a policy π. Let $\rho_{j,k}^{(m)}$ denote the number of platoons of the first m which leave with k vehicles while j stay behind at the station. Then, if there is some set of values $\{\rho_{j,k}\}$ (for $j \in \mathbb{Z}_{\geq 0}$ and $k \in \mathbb{Z}_{>0}$) such that $\lim_{m \to \infty} \frac{\rho_{j,k}^{(m)}}{m} = \rho_{j,k}$ with probability 1 for all j, k, we call $\{\rho_{j,k}\}$ the platoon distribution of π, and denote it $\rho(\pi)$.*

Note that $\rho(\pi)$ is really a probability distribution over $\mathbb{Z}_{\geq 0} \times \mathbb{Z}_{>0}$. For this notion to be useful, we want to make sure that it exists for regular policies.

Lemma 2. *For any regular policy* π, $\rho(\pi)$ *exists.*

This result is similar to Lemma 1.

Note that for an all-or-nothing policy, $\rho_{j,k} = 0$ for all $j > 0$ since no vehicles are ever left behind. The platoon distribution is important because it turns out that the platoon distribution of a policy π can be used to calculate the average energy and average delay.

Theorem 2 (Policy Equivalence from Platoon Dist.). *Let* π, π' *be policies such that* $\rho(\pi) = \rho(\pi')$. *Then* $\mathbb{E}[D(\pi)] = \mathbb{E}[D(\pi')]$ *and* $\mathbb{E}[E(\pi)] = \mathbb{E}[E(\pi')]$.

Proof (sketch). Since $\rho(\pi) = \rho(\pi')$, we can use $\rho_{j,k}$ to refer to elements of both; we also refer to the time between consecutive platoons as a *platoon interval.*

First, we note that by definition, the probability of getting a platoon of size k is $p_k = \sum_{j=0}^{\infty} \rho_{j,k}$ (this is of course the same for both policies). Since energy consumption of a platoon depends only on its size, $\mathbb{E}[E(\pi)] = \mathbb{E}[E(\pi')]$.

Now we turn to the delay. The basic intuition for why the average delay is the same in both policies is that waiting with m vehicles at the station has the same effect no matter what happened in the past, because future vehicle arrivals are Poisson-distributed and hence not dependent on past events. Formally, the proof works by computing $\mathbb{E}[D(\pi)]$ using only $\rho(\pi)$ – which automatically shows that $\mathbb{E}[D(\pi')]$ can be computed the same way and is therefore the same.

The basic method is to consider running π, and look at the set S_m of times where exactly m vehicles are at the station. If we allow empty segments, each platoon interval has exactly one segment of time with m vehicles. Furthermore, each segment can end either because (i) a new vehicle arrived or (ii) a platoon leaves. We will compute the probability of (i) from $\rho(\pi)$.

Note that if we piece together the segments of S_m, including where new vehicles arrived, we get a Poisson distribution of new vehicle arrivals (as Poisson arrivals are independent of the past, and π is only dependent on the past).

But we also note that a segment of S_m ends with a new vehicle if and only if the platoon interval starts with m or fewer vehicles and ends with more than m vehicles, which can be calculated from $\rho(\pi)$. This, and the insight that new vehicle arrivals on S_m are Poisson, gives us the average length (empty segments included) of a segment of S_m, which gives us the average amount of time in a platoon interval spent with exactly m vehicles at the station. Since we can compute this for all m, we also get the average length of a platoon interval.

This then gives us the fraction of total time with m vehicles at the station (when m vehicles are accruing delay) – call this fraction α_m. Thus, the average amount of delay being accrued at any moment as $\sum_{m=0}^{\infty} m\alpha_m$. Since the average number of vehicles which show up in a unit of time is λ, this means if we divide the total amount of delay accrued over a long period of time by the number of vehicles which arrived during that time, we get (with cancellation of T):

$$\mathbb{E}[D(\pi)] = \lim_{T \to \infty} \left(\frac{\text{total delay in } T \text{ time}}{\# \text{ vehicles in } T \text{ time}} \right) = \frac{\sum_{m=0}^{\infty} m\alpha_m}{\lambda}$$

with the approximation converging to the right-hand side as $T \to \infty$.

Thus, we have shown that we can compute $\mathbb{E}[D(\pi)]$ using only $\rho(\pi)$.   ∎

This proof has a very useful consequence: it can be used to show that for every policy π, there is an all-or-nothing policy π^* which performs at least as well; and if it has a positive probability of leaving behind vehicles when sending a (nonempty) platoon, then π^* has strictly lower average delay.

Theorem 3 (Optimality of All-or-Nothing). *Let π be a regular policy, and π^* be the mixed-threshold policy whose probability of sending a k-platoon is $p_k^* = \sum_{j=0}^{\infty} \rho_{j,k}$ for all k (that is, the mixed-threshold policy with the same proportion of k-platoons as π). Then, $\mathbb{E}[E(\pi)] = \mathbb{E}[E(\pi^*)]$ and $\mathbb{E}[D(\pi)] \geq \mathbb{E}[D(\pi^*)]$ with a strict inequality if and only if there is some (j', k') such that $j' > 0$ and $\rho_{j',k'} > 0$.*

The proof of this relies on the fact that π^* has the same distribution of platoon sizes as π. This immediately gives $\mathbb{E}[E(\pi)] = \mathbb{E}[E(\pi^*)]$.

For the delay, we use the method from the proof of Theorem 2 to compute the average delay of π and π^*. If there is no (j', k') such that $j' > 0$ and $\rho_{j',k'} > 0$, this means that π and π^* have exactly the same platoon distribution (and π is effectively all-or-nothing since it never leaves vehicles behind at the station) and so we apply Theorem 2 to get $\mathbb{E}[D(\pi)] = \mathbb{E}[D(\pi')]$. If there is such a (j', k'), then we can show algebraically (through the notion of stochastic dominance) that $\mathbb{E}[D(\pi)] > \mathbb{E}[D(\pi^*)]$.

4   On Pareto-optimal Regular Policies

We now consider the problem of finding policies which are Pareto-optimal over all regular policies. We do this by using the machinery from the previous section. In particular, we know from Theorem 3 that we do not need to consider policies which are not all-or-nothing. We then analyze the average delay and energy of these policies. First, we define the class of *hard-threshold* policies:

Definition 10. *The* hard-threshold *policy $\bar{\pi}_k$ is the policy which sends all vehicles out of the station whenever there are k vehicles (or more) waiting.*

The "or more" is just for completeness, as such a policy would prevent the queue from becoming longer than k vehicles. Furthermore, a hard-threshold policy has a reset point every time a platoon is sent out.

Theorem 4 (Mixed-Threshold is Optimal). *Any point on the energy-delay curve can be achieved by a mixed-threshold policy with at most two platoon sizes.*

Proof. Recall that a mixed-threshold policy is one which pre-selects the distribution of platoon sizes it produces with a probability distribution $(p_1, p_2, \dots)$, with p_k representing the probability of a platoon having k vehicles; after every platoon is sent, it chooses the size k of the next platoon from this distribution (independently) and sends the next platoon when the kth vehicle arrives.

If (D, E) is on the energy-delay curve, then there must be a policy π for which it is the profile. But by Theorem 3, we can use $\rho(\pi)$ to construct the mixed-threshold policy π^* where $p_k = \sum_{j=0}^{\infty} \rho_{j,k}$, and π^* will then by definition achieve (D, E) as well.

We now want to show why at most two unique platoon sizes are necessary. We note that mixed-threshold policies can be expressed as probabilistic mixtures of the set of policies $\{\bar{\pi}_k\}$, as in Theorem 1. Thus, the set of all profiles of mixed-threshold policies is just the convex hull of the set of profiles of $\{\bar{\pi}_k\}$.

But the energy-delay curve lies on the boundary of this set, and thus any point on the curve is on a line between at most two profile points – and thus is the profile of a policy that mixes at most two elements of $\{\bar{\pi}_k\}$. ∎

In order to gain a complete characterization of the energy-delay curve, we compute the average energy and delay of hard-threshold policies:

Lemma 3 (Delay and Energy of Threshold Policies). *If the vehicles arrive with intensity λ, then* $\mathbb{E}[D(\bar{\pi}_k)] = \frac{k-1}{2\lambda}$ *and* $\mathbb{E}[E(\bar{\pi}_k)] = \frac{\mathrm{TE}(k)}{k}$

Proof. Because $\bar{\pi}_k$ has a reset point every time it sends a platoon, we can examine one platoon in isolation. By definition, every platoon it sends has k vehicles and hence $\mathbb{E}[E(\bar{\pi}_k)] = \frac{\mathrm{TE}(k)}{k}$ is immediate.

Recall that $a_{i+1} = t_{i+1}^{\mathrm{arr}} - t_i^{\mathrm{arr}}$ (the amount of time between the arrivals of vehicles i and $i+1$). Because arrivals are Poisson with intensity λ, we know that a_{i+1} follows an exponential distribution with parameter λ, which in turn means $\mathbb{E}[a_{i+1}] = 1/\lambda$ for all i. But the amount of time that vehicle i waits before the platoon leaves is $t_k^{\mathrm{arr}} - t_i^{\mathrm{arr}} = \sum_{j=i+1}^{k} a_j$, which means that $\mathbb{E}[t_k^{\mathrm{arr}} - t_i^{\mathrm{arr}}] = \frac{k-i}{\lambda}$. Averaging this over vehicles $i = 1, 2, \ldots, k$ then gives $\mathbb{E}[E(\bar{\pi}_k)] = \frac{k-1}{2\lambda}$. ∎

We can now give the form of the energy-delay curve:

Corollary 2. *The energy-delay curve is piecewise-linear, with its vertices at* $\left(\frac{k_i-1}{2\lambda}, \frac{\mathrm{TE}(k_i)}{k_i}\right)$ *for some (increasing) sequence of integers $k_1, k_2, \ldots$.*

Proof. This follows from Lemma 3 and Theorem 4. The sequence $\{k_i\}$ just represents the profiles of vertices the lower convex hull. ∎

Finally, we look at the optimal policies for the affine total energy function:

Corollary 3. *When the total energy function is $\mathrm{TE}^{\mathrm{aff}}$, the Pareto-optimal policy achieving delay at most x is the mixed-threshold policy which, letting $k^* = \lfloor 2\lambda x + 1 \rfloor$ and $p^* = (2\lambda x + 1) - k^*$, sends a platoon of size $k^* + 1$ with probability p^* and k^* with probability $1 - p^*$.*

Also, the minimum possible energy with delay x is $y = \left(\frac{c^*}{k^*} - \frac{p^*}{k^*(k^*+1)}\right) + c$, which is just the piecewise-linear interpolation of the function $f(x) = \frac{c^*}{2\lambda x + 1} + c$ with vertices at $\{x = \frac{k-1}{2\lambda} : k \in \mathbb{Z}_{>0}\}$.

Proof. This follows simply from the fact that $\frac{\mathrm{TE}^{\mathrm{aff}}(k)}{k} = c + \frac{c^*}{k}$ for all $k > 0$, which is decreasing and convex – so all points in the sequence $(x_k, y_k) = \left(\frac{k-1}{2\lambda}, \frac{\mathrm{TE}(k)}{k}\right) = \left(\frac{k-1}{2\lambda}, c + \frac{c^*}{k}\right)$ are on the lower hull. The exact definition of k^* is derived from the platoon-mixing construction in Theorem 1's proof. ∎

5   On Pareto-optimal Time-table Policies

We now analyze the important class of *time-table* policies. Recall that a time-table policy is a control policy that releases all currently present vehicles as a platoon at a set of predetermined and regular departure times, as described in Section 2. These policies are *open-loop* policies, in the sense that no information regarding the number or arrival times of the vehicles is required to realize them. In practice, there may be difficulties in obtaining state information and/or sharing it throughout the system, making time-table policies important to consider.

The average delay of policy π can also be expressed as the expected cumulative delay in one cycle divided by the expected number of arrivals:

$$\mathbb{E}[D(\pi)] = \frac{\frac{\lambda}{2} \sum_{m=1}^{M} T_m^2}{\lambda \sum_{m=1}^{M} T_m} = \frac{\sum_{m=1}^{M} T_m^2}{2 \sum_{m=1}^{M} T_m}.$$

Note that the average delay of policy π is not a function of the arrival rate λ. Additionally, for any time-table policy $\pi \in \Pi$, there exists a time-table policy with only one interval size that yields the same average delay, namely, the policy $(T) \in \Pi$ satisfying $T = 2\mathbb{E}[D(\pi)]$. We call such a policy *fixed-interval*.

Similarly, we can express the average energy of a time-table policy by dividing the expected cumulative energy in a cycle by the expected number of arrivals:

$$\mathbb{E}[E(\pi)] = \frac{\sum_{m=1}^{M} \mathbb{E}[\mathrm{TE}(N_m)]}{\lambda \sum_{m=1}^{M} T_m},$$

where N_m is the number of arrivals in interval T_m, *i.e.*, N_m is distributed as a Poisson random variable with parameter λT_m.

We now look at the *fixed-interval* time-table policies, *i.e.*, $\mathcal{T} = \{kT_1 : k \in \mathbb{N}_{\geq 0}\}$, which releases the accumulated vehicles at the station every T_1 time units. To simplify notation, define $\epsilon(\lambda T)$ as the average accumulated energy in time interval T with arrival rate λ:

$$\epsilon(\lambda T) := \mathbb{E}[\mathrm{TE}(N)] = e^{-\lambda T} \sum_{k>0} \mathrm{TE}(k) \frac{(\lambda T)^k}{k!}, \tag{1}$$

where $N \sim Pois(\lambda T)$. Note that $\epsilon(0) = 0$, and $\epsilon(T) > 0$ for all $T > 0$. We assume that ϵ is infinitely differentiable on the open interval $(0, \infty)$. The function $\epsilon(x)$ is not necessarily infinitely differentiable in general, *e.g.*, $\mathrm{TE}(k) = k^k$; however, such TE are pathological and not meaningful to consider.[1] We can express the average energy of a fixed-interval policy $(T) \in \Pi$ as $\mathbb{E}[E(T)] = \frac{\epsilon(\lambda T)}{\lambda T}$.

Another important class of time-table policies is the class of *alternating time-table policies*, *i.e.*, $(T_1, T_2) \in \Pi$. Notice that the average energy of an alternating time-table policy $(T_1, T_2) \in \Pi$ can be expressed as $\mathbb{E}[E(T_1, T_2)] = \frac{\epsilon(\lambda T_1) + \epsilon(\lambda T_2)}{\lambda T_1 + \lambda T_2}$.

---

[1] We note that a sub-exponential, monotonically increasing TE guarantees that ϵ is infinitely differentiable. It is sufficient to check that $\left| k \frac{\mathrm{TE}(k-1)}{\mathrm{TE}(k)} \right|$ goes to ∞ in order to establish that ϵ is infinitely differentiable.

In Lemma 4 below, we characterize the class of TE functions for which fixed-interval policies are Pareto-optimal among all time-table policies. Consider a general time-table policy $\pi := (x_1, x_2, \ldots, x_M) \in \Pi$ and its corresponding fixed-interval policy $\pi_f := (x) \in \Pi$, where $x = 2\mathbb{E}[D(\pi)]$. Item 1 of Lemma 4 essentially states that the fixed-interval policy π_f never has larger average energy than the original time-table policy π, i.e., $\mathbb{E}[E(\pi)] \geq \mathbb{E}[E(\pi_f)]$, i.e., Pareto-optimality of fixed-interval policies among all *time-table* policies. Item 2 is similarly states that fixed-interval policies are Pareto-optimal among all *alternating* time-table policies, i.e., $M = 2$. Item 3 gives us a key insight into the general structure of the Pareto-optimal energy-delay curve among time-table policies, since $\frac{\epsilon(x)}{x}$ is the energy-delay curve up to horizontal scaling.

Lemma 4. [2] *Given some TE function, suppose the corresponding $\epsilon(x)$ (see Equation (1)) is infinitely differentiable. Then, the following are equivalent:*

1. $\dfrac{\sum_m \epsilon(x_m)}{\sum_m x_m} \geq \dfrac{\epsilon\left(\frac{\sum_m x_m^2}{\sum_m x_m}\right)}{\frac{\sum_m x_m^2}{\sum_m x_m}}$, *for all $x_m \geq 0$, such that $\sum_{m=1}^{M} x_m > 0$;*

2. $\dfrac{\epsilon(x)+\epsilon(y)}{x+y} \geq \dfrac{\epsilon\left(\frac{x^2+y^2}{x+y}\right)}{\frac{x^2+y^2}{x+y}}$, *for all $x, y \geq 0$, such that $x + y > 0$;*

3. $\dfrac{\epsilon(x)}{x}$ *is convex on $(0, \infty)$.*

Note that the special case of Item 1 for $M = 2$ yields Item 2. Also, item 3 implies Item 1, via a simple application of Jensen's inequality. The main difficulty of this lemma is in showing Item 2 implies Item 3. Notice that Item 2 is similar to the notion of *midpoint convexity* of the function $\frac{\epsilon(x)}{x}$. The key insight behind the proof is the following. A particular TE function uniquely determines ϵ, from which one obtains the function $\epsilon(\lambda T)/(\lambda T)$ describing the (not necessarily optimal) energy-delay curve of all fixed-interval time-table policies. The previous lemma says that if the given TE function is such that the energy-delay curve of all fixed-interval time-table policies is convex, then fixed-interval time-table policies are Pareto-optimal among all other time-table policies.

Next, we present a more meaningful sufficient condition on the TE function to ensure the Pareto-optimality of fixed-interval time-table policies (among all time-table policies). This sufficient condition has the class of affine TE functions as a special case.

Theorem 5. *Let TE be given such that ϵ as defined in Equation (1) is infinitely differentiable. Furthermore, suppose $\left\{\frac{\mathrm{TE}(k)}{k}\right\}_{k>0}$ is a convex sequence. Then, fixed-interval time-table policies are Pareto-optimal among all time-table policies.*

Proof. We show that TE satisfies Item 3 of Lemma 4. Then, by Lemma 4, Item 3 implies Item 1; hence, Pareto-optimality of fixed-interval time-table policies follows directly. Since ϵ is infinitely differentiable, we prove convexity of ϵ by showing $(\epsilon(x)/x)'' \geq 0$. Expanding $(\epsilon(x)/x)''$ as a power series, one has

---

[2] For ease of presentation, we write $\sum_m$ in the place of $\sum_{m=1}^{M}$ in Item 1.

$$\left(\frac{\epsilon(x)}{x}\right)'' = e^{-x} \sum_{k \geq 0} \left(\frac{\text{TE}(k+1)}{k+1} - 2\frac{\text{TE}(k+2)}{k+2} + \frac{\text{TE}(k+3)}{k+3}\right) \frac{x^k}{k!}.$$

Since $\text{TE}(k)/k$ is convex, all the coefficients of the series are nonnegative, from which the series is never negative for any $x > 0$.   ∎

In particular, for an affine TE function, we have the following result:

Theorem 6. *Consider the affine total energy function,* $\text{TE}^{\text{aff}}(k) = ck + c^*\mathbf{1}(k > 0)$ *with* $c \geq 0$, $c^* > 0$. *Then, fixed-interval policies are Pareto-optimal among all time-table policies. Furthermore, the Pareto-optimal energy-delay curve is given by* $c + c^*\left(\frac{1-e^{-2\lambda D}}{2\lambda D}\right)$, *where* D *is the average delay and* λ *is the arrival rate.*

Proof. We begin by deriving the energy-delay curve for the fixed-interval time-table policies, and then verify its convexity. We compute $\frac{\epsilon(\lambda T)}{\lambda T} = c + c^*\frac{1-e^{-\lambda T}}{\lambda T}$. Note that $c \geq 0$, $c^* > 0$ guarantees convexity. By invoking Lemma 4, this energy-delay curve is Pareto-optimal among all time-table policies (equivalence of Item 3 and Item 1). Note that the energy-delay curve in terms of delay is given by tracing out the curve: $(\frac{T}{2}, \frac{\epsilon(\lambda T)}{\lambda T})$. By changing variables from time interval T to delay D, *i.e.*, $D = T/2$, we arrive at the optimal energy-delay curve. ∎

Example 1. Consider Figure 1. The simplest nontrivial TE function is $\text{TE}^{\text{aff}}(k) = ck + c^*\mathbf{1}(k > 0)$ (given in Section 2) with $c \geq 0$ and $c^* > 0$. Note that the sufficient condition of Theorem 5 is satisfied; thus, fixed-interval policies are Pareto-optimal among all time-table policies.

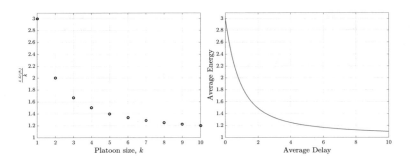

Fig. 1. Pareto-optimal (among time-table policies) energy-delay curve for affine TE function: $\text{TE}(k) = 2 + k$. Note convexity of both the energy-delay curve (Lemma 4) and $\frac{\text{TE}(k)}{k}$ (Theorem 5).

Example 2. Consider Figure 2. Note that Theorem 5 is not a necessary condition for optimality of fixed-interval time-table policies. Consider the following function $\text{TE}(k) = 2 \cdot \mathbf{1}(k > 0) + \lfloor k/2 \rfloor$. This TE function does not meet the convexity condition of Theorem 5, while its energy-delay curve is indeed Pareto-optimal among all time-table policies according to Lemma 4.   .

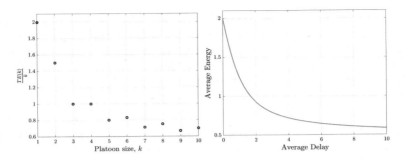

Fig. 2. Pareto-optimal (among time-table policies) energy-delay curve with non-convex $\frac{TE(k)}{k}$; $TE(k) = 2 + \lfloor k/2 \rfloor$. Note convexity of energy-delay curve (Lemma 4).

Example 3. Now, we show that if one violates Items 1, 2, or 3 of the Lemma 4, then a fixed-interval time-table policy can be arbitrarily worse than some alternating time-table policy. Consider the almost affine TE function: $TE(k) = k+2$, for all $k \neq 4$, and $TE(4) = K$. Consider the alternating policy $(1,5)$, and its delay-equivalent fixed-interval policy $(\frac{13}{3})$. Then, the difference in average energy between these two policies grows linearly with K, with the alternating policy having lower average energy.

6   Discussion and Conclusion

In this section, we compare time-table and threshold policies. We assume that vehicles arrive at the station according to a Poisson arrival, and we assume an affine total energy function. See Figure 3. In the plot on the left, we show the

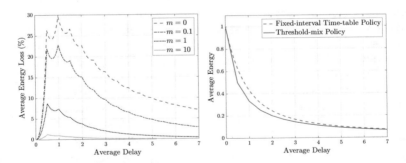

Fig. 3. On the left, average energy loss versus average delay plot for various total energy curves, $TE(k) = mk + 1$. On the right, Pareto-optimal energy-delay tradeoff for the class of time-table (fixed-interval) and feedback policies (mixed-threshold) for $TE(k) = 1$; Poisson arrival process, intensity $\lambda = 1$.

loss in average energy by implementing a time-table policy in lieu of a feedback policy, for various energy functions. We only consider energy functions of the form $TE(k) = mk + 1$, since the plot of average energy loss is the same for any other total energy function $\mathrm{TE}(k) = cmk + c$, for any $c > 0$. We note that the energy loss is no greater than thirty percent, and corresponds to a flat total energy function, $m = 0$, $i.e.$, the total energy of a platoon is the same regardless of how many vehicles are in the platoon. Also, we fixed the intensity of Poisson arrivals λ to 1, since changing the arrival rate λ simply corresponds to a horizontal scaling of λ in the plots of Figure 3.

One interesting insight is that hard-threshold policies – the building blocks of the Pareto-optimal mixed-threshold policies – derive much of their advantage over fixed-interval time-table policies by getting (on average) one more vehicle into each platoon. Specifically, the hard-threshold policy $\bar{\pi}_k$ gets k-vehicle platoons for an average delay of $\frac{k-1}{2\lambda}$ (as shown in Lemma 3); a time-table policy with an average delay of $\frac{k-1}{2\lambda}$ would require an interval of length $\frac{k-1}{\lambda}$, which gives it an average platoon size of $k - 1$.

Another interesting point is the 'spikiness' of the curves on the left of Figure 3. These curves plot the relative energy usage of timetable and mixed-threshold policies. The spikes occur at average delays which match up exactly to $\frac{k-1}{2\lambda}$ for some k – at these points, the Pareto-optimal policies are hard-threshold. This happens because mixing two hard-threshold policies is in some sense a compromise; ideally, one would like to have a hard-threshold policy for non-integer sized platoons. However, of course, that is impossible, and mixing hard-threshold policies is thus a stand-in for this. On the other hand, fixed-interval timetable policies smoothly increase their intervals, and hence don't have to make this compromise.

In this paper, we considered a class of decision-making problems which model a number of logistical scenarios – including platooning, drafting, vortex riding, ride sharing, and many others – that are relevant in transportation systems involving autonomous vehicles. In our problem, vehicles arrive at a station following a Poisson process; the station then decides when to send the vehicles currently present to their destination. Sending more vehicles at once allows for more energy-efficient transportation, but to do so consistently the station must delay some vehicles to wait for others to arrive. We explored the energy-delay tradeoff this problem presents, focusing on open loop (time-table) and feedback policies, and we characterized both the Pareto-optimal policies and the optimal energy-delay curves for each case. We also proved some general results concerning the form that the optimal policies can take, and showed that the performance of any policy is uniquely determined by the sizes of the platoons it produces.

Future work will include studies involving multiple stations, complex transportation networks, and unknown statistics, among other generalizations. In this setting, we hope to identify the tradeoffs between efficiency metrics, $e.g.$, capacity and delay, and sustainability metrics, $e.g.$, energy consumption and environmental impact.

Acknowledgements This work was supported by NSF under grant #1544413.

References

1. J. B. Greenblatt and S. Saxena. Autonomous taxis could greatly reduce greenhouse-gas emissions of US light-duty vehicles. *Nature Climate Change*, 5(9):860–863, July 2015.
2. K. B. Sandvik and K. Lohne. The Rise of the Humanitarian Drone: Giving Content to an Emerging Concept. *Millennium - Journal of International Studies*, 43(1):145–164, September 2014.
3. R. D'Andrea. Guest Editorial Can Drones Deliver? *IEEE Transactions on Automation Science and Engineering*, 11(3):647–648, July 2014.
4. B. Besselink, V. Turri, S. H. van de Hoef, K.-Y. Liang, A. Alam, J. Martensson, and K. H Johansson. Cyber–Physical Control of Road Freight Transport. *Proceedings of the IEEE*, 104(5):1128–1141, April 2016.
5. S. H. van de Hoef, K. H. Johansson, and D. V. Dimarogonas. Fuel-Optimal Centralized Coordination of Truck Platooning Based on Shortest Paths. In *American Control Conference*, July 2015.
6. U. Franke, F. Bottiger, Z. Zomotor, and D. Seeberger. Truck Platooning in Mixed Traffic. In *IEEE Intelligent Vehicles Symposium*, 1995.
7. S. Tsugawa, S. Kato, and K. Aoki. An Automated Truck Platoon for Energy Saving. In *IEEE/RSJ IROS Conference*, September 2011.
8. Overview of Platooning Systems. In *ITS World Congress*, October 2012.
9. O. Gehring and H. Fritz. Practical results of a longitudinal control concept for truck platooning with vehicle to vehicle communication. In *IEEE Conference on Intelligent Transportation Systems*, November 2003.
10. W. B. Blake, S. R. Bieniawski, and T. C. Flanzer. Surfing aircraft vortices for energy. *The Journal of Defense Modeling and Simulation: Applications, Methodology, Technology*, 12(1):31–39, December 2014.
11. M. J. Vachon, R. J. Ray, K. R. Walsh, and K. Ennix. F/A-18 Performance Benefits Measured During Autonomous Formation Flight Project. Technical report, 2003.
12. H. Weimershirsh, J. Martin, Y. Clerquin, P. Alexandre, and S. Jiraskova. Energy saving in flight formation. *Nature*, pages 697–698, October 2001.
13. M. T. Florian and M. H. Dickinson. Fly with a little flap from your friends. *Nature*, pages 295–296, January 2014.
14. J. Brisswalter and C. Hausswirth. Consequences of Drafting on Human Locomotion: Bene ts on Sports Performance. *International Journal of Sports Physiology and Performance*, 3:3–15, 2008.
15. D. Gross and C. Harris. *Fundamentals of Queueing Theory*. Wiley, 1998.
16. R. C. Larson and A. R. Odoni. *Urban Operations Research*. MIT Press, 1999.
17. N. A. H. Agatz, A. L. Erera, M. W. P. Savelsbergh, and X. Wang. Optimization for dynamic ride-sharing: A review. *European Journal of Operational Research*, 223(2):295–303, December 2012.
18. D. J. Fagnant and K. M. Kockelman. Dynamic ride-sharing and fleet sizing for a system of shared autonomous vehicles in Austin, Texas. In *Transportation Research Board Annual Meeting*, September 2015.
19. Z. Tian, T. Urbanik, and R. Engelbrecht. Variations in capacity and delay estimates from microscopic traffic simulation models. *Transportation Research Record: Journal of the Transportation Research Board*, 1802:23–31, 2002.
20. A. Adler, D. Miculescu, and S. Karaman. Optimal policies for platooning and ride sharing in autonomy-enabled transportation. http://peris.mit.edu/AdlerMiculescuKaramanWAFR16.pdf, July 2016.

Decentralised Monte Carlo Tree Search
for Active Perception

Graeme Best[1], Oliver M. Cliff[1], Timothy Patten[1],
Ramgopal R. Mettu[2], and Robert Fitch[1,3]

[1]Australian Centre for Field Robotics (ACFR), The University of Sydney, Australia,
{g.best,o.cliff,t.patten,rfitch}@acfr.usyd.edu.au
[2]Department of Computer Science, Tulane University, New Orleans, LA, USA
rmettu@tulane.edu
[3]Centre for Autonomous Systems, University of Technology Sydney, Australia

Abstract. We propose a decentralised variant of Monte Carlo tree search (MCTS) that is suitable for a variety of tasks in multi-robot active perception. Our algorithm allows each robot to optimise its own individual action space by maintaining a probability distribution over plans in the joint-action space. Robots periodically communicate a compressed form of these search trees, which are used to update the locally-stored joint distributions using an optimisation approach inspired by variational methods. Our method admits any objective function defined over robot actions, assumes intermittent communication, and is anytime. We extend the analysis of the standard MCTS for our algorithm and characterise asymptotic convergence under reasonable assumptions. We evaluate the practical performance of our method for generalised team orienteering and active object recognition using real data, and show that it compares favourably to centralised MCTS even with severely degraded communication. These examples support the relevance of our algorithm for real-world active perception with multi-robot systems.

1   Introduction

Information gathering is a fundamentally important family of problems in robotics that plays a primary role in a wide variety of tasks, ranging from scene understanding to manipulation. Although the idea of exploiting robot motion to improve the quality of information gathering has been studied for nearly three decades [3], most real robot systems today (both single- and multi-robot) still gather information passively. The motivation for an active approach is that sensor data quality (and hence, perception quality) relies critically on an appropriate choice of viewpoints [17]. One way to efficiently achieve an improved set of viewpoints is through teams of robots, where concurrency allows for scaling up the number of observations in time and space. The key challenge, however, is to coordinate the behaviour of robots as they actively gather information, ideally in a decentralised manner. This paper presents a new, decentralised approach

© Springer Nature Switzerland AG 2020
K. Goldberg et al. (Eds.): *Algorithmic Foundations of Robotics XII*, SPAR 13, pp. 864–879, 2020.
https://doi.org/10.1007/978-3-030-43089-4_55

for active perception that allows a team of robots to perform complex information gathering tasks using physically feasible sensor and motion models, and reasonable communication assumptions.

Decentralised active information gathering can be viewed, in general, as a partially observable Markov decision process (POMDP) in decentralised form (Dec-POMDP) [1]. There are several known techniques for solving this class of problems, but policies are usually computed in advance and executed in a distributed fashion. There are powerful approximation algorithms for special cases where the objective function is monotone submodular [21]. Unfortunately this is not always the case, particularly in field robotics applications.

Our approach provides convergence guarantees but does not require submodularity assumptions, and is essentially a novel decentralised variant of the Monte Carlo tree search (MCTS) algorithm [5]. At a high level, our method alternates between exploring each robot's individual action space and optimising a probability distribution over the joint-action space. In any particular round, we first use MCTS to find locally favourable sequences of actions for each robot, given a probabilistic estimate of other robots' actions. Then, robots periodically attempt to communicate a highly compressed version of their local search trees which, together, correspond to a product distribution approximation. These communicated distributions are used to estimate the underlying joint distribution. The estimates are probabilistic, unlike the deterministic representation of joint actions typically used in multi-robot coordination algorithms. Optimising a product distribution is similar in spirit to the mean-field approximation from variational inference, and also has a natural game-theoretic interpretation [19].

Our algorithm is a powerful new method of decentralised coordination for any objective function defined over the robot action sequences. Notably, this implies that our method is suitable for complex perception tasks such as object classification, which is known to be highly viewpoint-dependent [17]. Further, communication is assumed to be intermittent, and the amount of data sent over the network is small in comparison to the raw data generated by typical range sensors and cameras. Our method also inherits important properties from MCTS, such as the ability to compute anytime solutions and to incorporate prior knowledge about the environment. Moreover, our method is suitable for online replanning to adapt to changes in the objective function or team behaviour.

We evaluate our algorithm in two scenarios: generalised team orienteering and active object recognition. These experiments are run in simulation, and the second scenario uses range sensor data collected a priori by real robots. We show that our decentralised approach performs as well as or better than centralised MCTS even with a significant rate of message loss. We also show the benefits of our algorithm in performing long-horizon and online planning.

2   Related work

Information gathering problems can be viewed as sequential decision processes in which actions are chosen to maximise an objective function. Decentralised coordination in these problems is typically solved myopically by maximising the objective function over a limited time horizon [25, 10]. Unfortunately, the quality of solutions produced by myopic methods can be arbitrarily poor in the general case. Recently, however, analysis of submodularity [16] has shown that myopic methods can achieve near-optimal performance [14], which has led to considerable interest in their application to information gathering with multiple robots [21]. While these methods provide theoretical guarantees, they require a submodular objective function, which is not applicable in all cases.

Efficient non-myopic solutions can be designed by exploiting problem-specific characteristics [4, 6]. But in general, the problem is a POMDP, which is notoriously difficult to solve. The difficulty of solving Dec-POMDPs is compounded because the search space grows exponentially with the number of robots. For our problem, we focus our attention on reasoning over the unknown plans of the other robots, while assuming other aspects of the problem are fully observable. The problems we consider here are therefore not Dec-POMDPs, but our algorithm is general enough to be extended to problems with partial observability.

MCTS is a promising approach for online planning because it efficiently searches over long planning horizons and is anytime [5]. MCTS has also been extended to partially observable environments [20]. However, MCTS has not been extended for decentralised multi-agent planning, and that is our focus here.

MCTS is parallelisable [7], and various techniques have been proposed that split the search tree across multiple processors and combine their results. In the multi-robot case, the joint search tree interleaves actions of individual robots and it remains a challenge to effectively partition this tree. A related case is multi-player games, where a separate tree may be maintained for each player [2]; however, a single simulation traverses all of the trees and therefore this approach would be difficult to decentralise. We propose a similar approach, except that each robot performs independent simulations while sampling from a locally stored probability distribution that represents the other robots' action sequences.

Coordination between robots is achieved in our method by combining MCTS with a framework that optimises a product distribution over the joint action space in a decentralised manner. Our approach is analogous to the classic mean-field approximation and related variational approaches [26, 19]. Variational methods seek to approximate the underlying global likelihood with a collection of structurally simpler distributions that can be evaluated efficiently and independently. These methods characterise convergence based on the choice of product distribution, and work best when it is possible to strike a balance between the convergence properties of the product distribution and the KL-divergence between the product and joint distributions. As discussed in the body of work on *probability collectives* [22, 24, 23], such variational methods can also be viewed under a game theoretic interpretation, where the goal is to optimise each agent's choice of actions based on examples of the global reward/utility function. The latter method has been used for solving the multiple-TSP in a decentralised manner [15]; we propose a similar approach, but we leverage the power of the MCTS to select an effective and compact sample space of action sequences.

3   Problem statement

We consider a team of R robots $\{1, 2, ..., R\}$, where each robot i plans its own sequence of future actions $\boldsymbol{x}^i = (x_1^i, x_2^i, ...)$. Each action x_j^i has an associated cost c_j^i and each robot has a cost budget B^i such that the sum of the costs must be less than the budget, i.e., $\sum_{x_j^i \in \boldsymbol{x}^i} c_j^i \leq B^i$. This cost budget may be an energy or time constraint defined by the application, or it may be used to enforce a planning horizon. The feasible set of actions and associated costs at each step n are a function of the previous actions $(x_1^i, x_2^i, ..., x_{n-1}^i)$. Thus, there is a predefined set of feasible action sequences for each robot $\boldsymbol{x}^i \in \mathcal{X}^i$. Further, we denote $\boldsymbol{x}$ as the set of action sequences for all robots $\boldsymbol{x} := \{\boldsymbol{x}^1, \boldsymbol{x}^2, ..., \boldsymbol{x}^R\}$ and $\boldsymbol{x}^{(i)}$ as the set of action sequences for all robots except i, i.e., $\boldsymbol{x}^{(i)} := \boldsymbol{x} \setminus \boldsymbol{x}^i$.

The aim is to maximise a global objective function $g(\boldsymbol{x})$ which is a function of the action sequences of all robots. We assume each robot i knows the global objective function $g(\boldsymbol{x})$, but does not know the actions $\boldsymbol{x}^{(i)}$ selected by the others. Moreover, the problem must be solved in an online setting.

We assume that robots can communicate to improve coordination. The communication channel may be unpredictable and intermittent, and all communication is asynchronous. Therefore, each robot will plan based on the information it has available locally. Bandwidth may be constrained and therefore message sizes should remain small, even as the plans grow. Although we do not consider explicitly planning to maintain communication connectivity, this may be encoded in the objective function $g(\boldsymbol{x})$ if a reliable communication model is available.

4   Dec-MCTS

In this section we present the Dec-MCTS algorithm. Dec-MCTS runs simultaneously and asynchronously on all robots; here we present the algorithm from the perspective of robot i. The algorithm cycles between three phases (Alg. 1): 1) grow a search tree using MCTS, while taking into account information about the other robots, 2) update the probability distribution over possible action sequences, and 3) communicate probability distributions with the other robots. These three phases continue regardless of whether or not the communication was successful, until a computation budget is met.

4.1   Local utility function

The global objective function g is optimised by each robot i using a local utility function f^i. We define f^i as the difference in utility between robot i taking actions $\boldsymbol{x}^i$ and a default "no reward" sequence $\boldsymbol{x}_\emptyset^i$, assuming fixed actions $\boldsymbol{x}^{(i)}$ for the other robots, i.e., $f^i(\boldsymbol{x}) := g(\boldsymbol{x}^i \cup \boldsymbol{x}^{(i)}) - g(\boldsymbol{x}_\emptyset^i \cup \boldsymbol{x}^{(i)})$. In practice, this improves the performance compared to optimising g directly since f^i is more sensitive to robot i's plan and the variance of f^i is less affected by the uncertainty of the other robots' plans [23].

Algorithm 1 Overview of Dec-MCTS for robot i.

input: global objective function g, budget B^i, feasible action sequences and costs
output: sequence of actions $\boldsymbol{x}^i$ for robot i
1: $\mathcal{T} \leftarrow$ initialise MCTS tree
2: **while** not converged or computation budget not met **do**
3:     $\hat{\mathcal{X}}_n^i \leftarrow$ SELECTSETOFSEQUENCES$(\mathcal{T})$
4:     **for** fixed number of iterations **do**
5:         $\mathcal{T} \leftarrow$ GROWTREE$(\mathcal{T}, \hat{\mathcal{X}}_n^{(i)}, q^{(i)}, B^i)$
6:         $q^i \leftarrow$ UPDATEDISTRIBUTION$(\hat{\mathcal{X}}_n^i, q^i, \hat{\mathcal{X}}_n^{(i)}, q^{(i)}, \beta)$
7:         COMMUNICATIONTRANSMIT$(\hat{\mathcal{X}}_n^i, q^i)$
8:         $(\hat{\mathcal{X}}_n^{(i)}, q^{(i)}) \leftarrow$ COMMUNICATIONRECEIVE
9:         $\beta \leftarrow$ COOL(β)
10: **return** $\boldsymbol{x}^i \leftarrow \arg\max_{\boldsymbol{x}^i \in \hat{\mathcal{X}}_n^i} \left[q^i(\boldsymbol{x}^i) \right]$

4.2   Monte Carlo tree search

The first phase of the algorithm is the MCTS update shown in Alg. 2. A single tree is maintained by robot i which only contains the actions of robot i. Coordination occurs implicitly by considering the plans of the other robots when performing the rollout policy and evaluation of the global objective function. This information about the other robots' plans comes from the distributed optimisation of probability distributions detailed in the following subsection. We use MCTS with a novel bandit-based node selection policy.

Standard MCTS incrementally grows a tree by iterating through four phases: selection, expansion, simulation and backprogation [5]. In each iteration t, a new leaf node is added, where each node represents a sequence of actions and contains statistics about the expected reward. The selection phase begins at the root node of the tree and recursively selects child nodes s_j until an expandable node is reached. For selecting the next child, we propose an extension of the UCT policy [13], detailed later, to balance exploration and exploitation. In the expansion phase, a new child is added to the selected node, which extends the parent's action sequence with an additional action.

In the simulation phase, the expected utility $\mathbb{E}[g_j]$ of the expanded node s_j is estimated by performing and evaluating a rollout policy that extends the action sequence represented by the node until a terminal state is reached. This rollout policy could be a random policy or a heuristic for the problem. The objective is evaluated for this sequence of actions and this result is saved as $\mathbb{E}[g_j]$.

For our problem, the objective is a function of the action sequence $\boldsymbol{x}^i$ as well as the unknown plans of the other robots $\boldsymbol{x}^{(i)}$. To compute the rollout score, we first sample $\boldsymbol{x}^{(i)}$ from a probability distribution $q_n^{(i)}$ over the plans of the other agents (defined later). A heuristic rollout policy extended from s_j defines $\boldsymbol{x}^i$, which should be a function of $\boldsymbol{x}^{(i)}$ to simulate coordination. Additionally, we optimise $\boldsymbol{x}^i$ using the local utility f^i (defined in Sec. 4.1) rather than g. The rollout score is then computed as the utility of this joint sample $f^i(\boldsymbol{x}^i \cup \boldsymbol{x}^{(i)})$,

---

Algorithm 2 Grow the tree using Monte Carlo Tree Search for robot i.

1: **function** GROWTREE($\mathcal{T}, \hat{\mathcal{X}}_n^{(i)}, q^{(i)}, B^i$)

    input: partial tree $\mathcal{T}$, distributions for other robots $(\hat{\mathcal{X}}_n^{(i)}, q^{(i)})$

    output: updated partial tree $\mathcal{T}$

2:     **for** fixed number of samples τ **do**

3:         $s_j \leftarrow$ NODESELECTIOND-UCT($\mathcal{T}$)   ▷ Find the next node to expand

4:         $s_j^+ \leftarrow$ EXPANDTREE(s_j)   ▷ Add a new child to s_j

5:         $\boldsymbol{x}^{(i)} \leftarrow$ SAMPLE($\hat{\mathcal{X}}_n^{(i)}, q^{(i)}$)   ▷ Sample the policies of the other robots

6:         $\boldsymbol{x}^i \leftarrow$ PERFORMROLLOUTPOLICY($s_j^+, \boldsymbol{x}^{(i)}, B^i$)

7:         $score \leftarrow f^i(\boldsymbol{x}^i \cup \boldsymbol{x}^{(i)})$ ▷ Local utility function

8:         $\mathcal{T} \leftarrow$ BACKPROPAGATION($s_j^+, score$)   ▷ Update scores

9:     **return** $\mathcal{T}$

---

which is an estimate for $\mathbb{E}_{q_n}[f^i \mid \boldsymbol{x}^i]$. Thus, we define the reward $F_t(s, s_j)$ as the rollout score when child s_j was expanded from node s at sample t.

In the backpropagation phase, the rollout evaluation is added to the statistics of all nodes along the path from the expanded node back to the root of the tree. Typically, each rollout t is treated as equally relevant and therefore $\mathbb{E}[g_j]$ is an unbiased average of the rollout evaluations. However, our algorithm alternates between growing the tree for a fixed number of rollouts τ and updating the probability distributions for other robots at each iteration n, where $n = \lfloor t/\tau \rfloor$. Therefore the most recent rollouts are more relevant since they are obtained by sampling the most recent distributions. Thus, we use a variation on the standard UCT algorithm, which we term D-UCT after discounted UCB [11]. This policy accounts for non-stationary reward distributions by biasing each sample by a weight γ which increases at each rollout.

Specifically, we propose a node selection policy (Alg. 2 line 3) that maximises a UCB $\bar{F}_t(\cdot) + c_t(\cdot)$ for the discounted expected reward, i.e., for parent node s and sample t, D-UCT selects the child node $s_t^+ = \arg\max_{s_j}[\bar{F}_t(\gamma, s, s_j) + c_t(\gamma, s, s_j)]$. This continues recursively until a node with unvisited children is reached. The discounted empirical average $\bar{F}_t$ is given by

$$\bar{F}_t(\gamma, s, s_j) = \frac{1}{N_t(\gamma, s, s_j)} \sum_{u=1}^{t} \gamma^{t-u} F_u(s, s_j) \mathbb{1}_{\{s_u^+ = s_j\}}, \tag{1}$$

$$N_t(\gamma, s, s_j) = \sum_{u=1}^{t} \gamma^{t-u} \mathbb{1}_{\{s_u^+ = s_j\}},$$

and the discounted exploration bonus c_t is given by

$$c_t(\gamma, s, s_j) = 2C_p \sqrt{\frac{\log N_t(\gamma, s)}{N_t(\gamma, s, s_j)}}, \quad N_t(\gamma, s) = \sum_{j=1}^{K} N_t(\gamma, s, s_j). \tag{2}$$

In this context, $N_t(\gamma, s)$ is the discounted number of times the current (parent) node s has been visited, and $N_t(\gamma, s, s_j)$ is the discounted number of times child node s_j has been visited. The discount factor $\gamma \in (1/2, 1)$ and constant $C_p > 1/\sqrt{8}$. This D-UCT selection policy guarantees a rate of regret in the bandit case with abruptly changing distributions, which we discuss in Sec. 5.

4.3   Decentralised product distribution optimisation

The second phase of the algorithm updates a probability distribution q^i over the set of possible action sequences for robot i (Alg. 3). These distributions are communicated between robots and used for performing rollouts during MCTS. To optimise these distributions in a decentralised manner for improving global utility, we adapt a type of variational method originally proposed in [22]. This method can be viewed as a game between independent robots, where each robot selects their action sequence by sampling from a distribution.

One challenge is that the set of possible action sequences is typically of exponential size. We obtain a sparse representation by selecting the sample space $\hat{\mathcal{X}}_n^i \subset \mathcal{X}^i$ as the most promising action sequences $\{x_1^i, x_2^i, ...\}$ found by MCTS. We select a fixed number of nodes with the highest $\mathbb{E}[f^i]$ obtained so far. $\hat{\mathcal{X}}_n^i$ is the action sequences used during the initial rollouts when the selected nodes were first expanded.

The set $\hat{\mathcal{X}}_n^i$ has an associated probability distribution q^i such that $q^i(x^i)$ defines the probability that robot i will select $x^i \in \hat{\mathcal{X}}_n^i$. The distributions for different robots are independent and therefore define a product distribution, such that the probability of a joint action sequence selection x is $q(x) := \prod_{i \in \{1:R\}} q^i(x^i)$. The advantage of defining q as a product distribution is so that each robot selects its action sequence independently, and therefore allows decentralised execution.

Consider the general class of joint probability distributions p that are not restricted to product distributions. Define the expected global objective function for a joint distribution p as $\mathbb{E}_p[g]$, and let Γ be a desired value for $\mathbb{E}_p[g]$. According to information theory, the most likely p that satisfies $\mathbb{E}[g] = \Gamma$ is the p that maximises entropy. The most likely p can be found by minimising the maxent Lagrangian: $L(p) := \lambda (\Gamma - \mathbb{E}_p[g]) - \mathrm{H}(p)$, where $\mathrm{H}(p)$ is the Shannon entropy and λ is a Lagrange multiplier. The intuition is to iteratively increase Γ and optimise p. A descent scheme for p can be formulated with Newton's method.

For decentralised planning and execution, we are interested in optimising the product distribution q rather than a more general joint distribution p. We can approximate q by finding the q with the minimum pq KL-divergence $D_{\mathrm{KL}}(p \parallel q)$. This formulates a descent scheme with the update policy for q^i shown in Alg. 3 line 5, and where we use f^i rather than g. Intuitively, this update rule increases the probability that robot i selects x^i if this results in an improved global utility, while also ensuring the entropy of q^i does not decrease too rapidly.

Pseudocode for this approach is in Alg. 3. We require computing two expectations (lines 3 and 4) to evaluate the update equation (line 5). It is often necessary to approximate these expectations by sampling x, since it is intractable to sum over the enumeration of all $x \in \hat{\mathcal{X}}_n$. Parameter β should slowly decrease and α remain fixed. For efficiency purposes, in our implementation q^i is set to a uniform distribution when $\hat{\mathcal{X}}_n^i$ changes (Alg. 1 line 3).

Algorithm 3 Probability distribution optimisation for robot i.

1: **function** UPDATEDISTRIBUTION($\hat{\mathcal{X}}_n^i, q^i, \hat{\mathcal{X}}_n^{(i)}, q^{(i)}, \beta$)

    input: action sequence set for each robot $\hat{\mathcal{X}}_n := \{\hat{\mathcal{X}}_n^1, \hat{\mathcal{X}}_n^2, ..., \hat{\mathcal{X}}_n^R\}$
                      with associated probability distributions $\{q^1, q^2, ..., q^R\}$,
                      update parameter β

    output: updated probability distribution q^i for robot i

2:    **for each** $\boldsymbol{x}^i \in \hat{\mathcal{X}}_n^i$ **do**

3:        $\mathbb{E}_q[f^i] \leftarrow \sum_{\boldsymbol{x} \in \hat{\mathcal{X}}_n} \left[f^i(\boldsymbol{x}) \prod_{i' \in \{1:R\}} q^{i'}(\boldsymbol{x}^{i'}) \right]$

4:        $\mathbb{E}_q[f^i \mid \boldsymbol{x}^i] \leftarrow \sum_{\boldsymbol{x}^{(i)} \in \hat{\mathcal{X}}_n^{(i)}} \left[f^i(\boldsymbol{x}^i \cup \boldsymbol{x}^{(i)}) \prod_{i' \in \{1:R\} \setminus i} q^{i'}(\boldsymbol{x}^{i'}) \right]$

5:        $q^i(\boldsymbol{x}^i) \leftarrow q^i(\boldsymbol{x}^i) - \alpha q^i(\boldsymbol{x}^i) \left[\dfrac{\mathbb{E}_q[f^i] - \mathbb{E}_q[f^i \mid \boldsymbol{x}^i]}{\beta} + \mathrm{H}(q^i) + \ln\left(q^i(\boldsymbol{x}^i)\right) \right]$

6:        $q^i \leftarrow$ NORMALISE(q^i)

7:    **return** q^i

4.4 Message passing

At each iteration n, robot i communicates its current probability distribution $(\hat{\mathcal{X}}_n^i, q_n^i)$ to the other robots. If robot i receives any updated distributions then this replaces its stored distribution. The updated distribution is used during the next iteration. If no new messages are received from a robot, then robot i continues to plan based on its most recently received distribution. If robot i is yet to receive any messages from a robot then it may assume a default policy.

4.5 Online replanning

The best action is selected as the first action in the highest probability action sequence in $\hat{\mathcal{X}}_n^i$. The search tree may then be pruned by removing all children of the root except the selected action. Planning may then continue while using the sub-tree's previous results. If the objective function changes, e.g. as a result of a new observation, then the tree should be restarted. In practice, if the change is minor then it may be appropriate to continue planning with the current tree.

5 Analysis

We characterise the convergence properties of the two key algorithmic components of our approach: tree search (Sec. 4.2) and sample space contraction (Sec. 4.3). Our first aim is to show that, with the D-UCT algorithm (Alg. 2), we maintain an exploration-exploitation trade-off for child selection while the distributions q_n^i are changing (and converging). Then we characterise the convergence of Alg. 3 given a contracted sample space of distributions $\hat{\mathcal{X}}_n^i \subset \mathcal{X}^i$.

We analyse the D-UCT algorithm by considering a particular type of non-stationary (adversarial) multi-armed bandit (MAB) problem [13, 11]. That is, we consider a unit-depth tree where the reward distributions can change abruptly at a given *breakpoint*, such that a previously optimal arm becomes suboptimal. We denote the number of breakpoints before time t as Υ_t.

Theorem 1. *Suppose we restrict $\mathcal{T}$ to a unit-depth tree in Alg. 1. Let $\tilde{N}(\boldsymbol{x}^i)$ denote the number of times action $\boldsymbol{x}^i$ was taken when it was suboptimal w.r.t. $\mathbb{E}_{q_n}[f^i]$ and fix $C_p > 1/\sqrt{8}$ and $\gamma_t = 1 - \sqrt{\Upsilon_t/16t}$ for known Υ_t. Then, running Alg. 2, where the bias sequence $c(\gamma, s, s_j)$ is given by (2), yields*

$$\mathbb{E}[\tilde{N}(\boldsymbol{x}^i)] \leq O(\sqrt{t\Upsilon_t} \log t),$$

for any action and sample index $t > 1$.

Proof. Observe that the problem of action selection for a unit-depth tree is equivalent to an MAB problem [13]. Denote by Υ_t the number of breakpoints of the reward distribution up until the sample index t. In the context of Sec. 4.2, these are sample rounds when a previously optimal child s_j becomes suboptimal (i.e., $F_t(s, s_j)$ changes abruptly). Under the D-UCB policy applied to the MAB problem, given a known number of breakpoints Υ_t, we can minimise the number of times a suboptimal action is taken by setting the discount factor to $\gamma_t = 1 - \sqrt{\Upsilon_t/16t}$. Selecting this discount factor upper bounds the expected number of times a suboptimal action is taken to $O(\sqrt{t\Upsilon_t} \log t)$ [11]. □

Remark 1. Although Theorem 1 analyses Alg. 1 in the MAB setting, we believe it is possible to extend to arbitrary depth trees. This could be achieved in a similar way to [13] by showing drift and breakpoint conditions are satisfied at all levels of the tree and proving regret bounds by induction.

As a consequence of Theorem 1 and Remark 1, during abrupt changes to q_n, the child selection policy in the tree search balances exploration and exploitation.

Depending on the rate of increasing β, the expected utility will converge asymptotically. As a consequence, the number of breakpoints Υ_t are known, monotonic and bounded. Importantly, note that nearer to the root node of the tree the breakpoints will decay faster than the leaf nodes. Thus, toward the root node, as n becomes large, $\gamma \to 1$ and the number of breakpoints Υ_t remain constant with high probability. For these nodes, the D-UCT algorithm becomes equivalent to UCT. As per Theorem 2 of [13], the bias of the estimated utility then converges polynomially. This is reasonable since, given that we will typically want to adaptively replan (Sec. 4.5), it should be sufficient to guarantee optimal short-term action sequences. Moreover, note that γ can be optimally set given bounds on the convergence *rate* of the distributions.

We now consider the effect of contracting the sample space $\hat{\mathcal{X}}_n \subset \mathcal{X}$ on the convergence of Alg. 3. Recall that the pq KL-divergence is the divergence from a product distribution q to the optimal joint distribution p. We then have the following proposition:

Proposition 1. *Alg. 3 asymptotically converges to a distribution that locally minimises the pq KL-divergence, given an appropriate subset $\hat{\mathcal{X}}_n \subset \mathcal{X}$.*

Proposition 1 relies on growing β slowly. Consider the situation where, at each iteration n, we randomly choose a subset $\hat{\mathcal{X}}_n^i \subset \mathcal{X}^i$ for each robot. This approach is equivalent to Monte Carlo sampling of the expected utility and thus the biased estimator is consistent (asymptotically converges to $\mathbb{E}[f^i]$). For tractable computation, in our algorithm we modify the random selection by choosing a sparse set of strategies $\hat{\mathcal{X}}_n$ with the highest expected utility (Sec. 4.3). Although this does not ensure we sample the entire domain $\mathcal{X}$ asymptotically, in practice $q_n(\hat{\mathcal{X}}_n)$ is a reasonably accurate representation of $q_n(\mathcal{X})$, and therefore this gives us an approximation to importance sampling [24].

The analyses above show separately that the tree search of Alg. 2 balances exploration and exploitation and that, under reasonable assumptions, Alg. 3 converges to the product distribution that best optimises the joint action sequence. These two components do not immediately yield a characterisation of optimality for Alg. 1. To prove global convergence rates, we would need to characterise the co-dependence between the evolution of the reward distributions $\mathbb{E}_{q_n}[f^i \mid x^i]$ and the contraction of the sample space $\hat{\mathcal{X}}_n$. The following experiments show that the algorithm converges rapidly to a high-quality solution.

6   Experiments: Generalised team orienteering

In this section we evaluate the performance of our algorithm in an abstract multi-robot information gathering problem (Fig. 1). We show convergence, robustness to intermittent communication and a comparison to centralised MCTS.

The problem is motivated by tasks where a team of Dubins robots maximally observes a set of features of interest in an environment, given a travel budget [4]. Each feature can be viewed from multiple viewpoints and each viewpoint may be in observation range of multiple features. This formulation generalises the orienteering problem [12] by combining the set structure of the generalised travelling salesman problem with the budget constraints of the orienteering problem with neighbourhoods [9] extended for multi-agent scenarios [4].

Robots navigate within a graph representation of an environment with vertices $v_i \in \mathcal{V}$, edges $e_{ij} := \langle v_i, v_j \rangle \in \mathcal{E}$ and edge traversal costs c_{ij}. Each vertex v_i represents a location and orientation (x, y, θ) within a square workspace with randomly placed obstacles. The action sequences of each robot are defined as paths through the graph from a fixed start vertex unique to each robot to a free destination vertex. The edge costs are defined as the distance of the Dubins path between the two configurations. All edges are connected within a fixed distance.

For the objective function, we have a collection of sets $\mathcal{S} = (S_1, S_2, ...)$, where each $S_i \subseteq \mathcal{V}$. These sets may represent a set of features of interest, where a vertex is an element of a set only if the associated feature can be observed from the vertex location. We assume each set is a disc, however the formulation could extend to more complex models [4]. The vertices $v_j \in \mathcal{V}$ are randomly placed within the sets. A set S_i is visited if $\exists v_j \in x, v_j \in S_i$ and each visited set yields an associated reward w_i. There is no additional reward for revisiting a set. The objective is defined as the sum of the rewards of all visited sets.

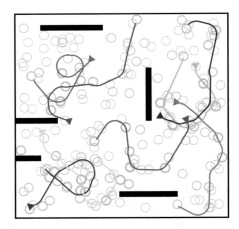

Fig. 1. The generalised team orienteering problem. The 8 robots (coloured paths) aim to collectively visit a maximal number of goal regions (green circles, weighted by importance). The robots follow Dubins paths, are constrained by distance budgets and must avoid obstacles (black).

6.1 Experiment setup

We compare our algorithm (Dec-MCTS) to centralised MCTS (Cen-MCTS), which consists of a single tree where robot i's actions appear at tree depths $(i, i+R, i+2R, ...)$. Intermittent communication is modelled by randomly dropping messages. Messages are broadcast by each robot at 4 Hz and a message has a probability of being received by each individual robot.

Experiments were performed with 8 simulated robots running in separate ROS nodes on an 8-core computer. Each random problem instance (Fig. 1) consisted of 200 discs with rewards between 1 and 10, 4000 graph vertices and 5 obstacles. Each iteration of Alg. 1 performs 10 MCTS rollouts, and $\hat{\mathcal{X}}_n^i$ consists of 10 paths that are resampled every 10 iterations. The MCTS rollout policy recursively selects the next edge that does not exceed the travel budget and maximises the ratio of the increase of the weighted set cover to the edge cost.

6.2 Results

The first experiments (Fig. 2(a)) show that Dec-MCTS achieved a median 7 % reward improvement over Cen-MCTS after 120 s, and a higher reward in 91 % of the environments. Dec-MCTS typically converged after ∼60 s. A paired single-tailed t-test supports the hypothesis ($p < 0.01$) that Dec-MCTS achieves a higher reward than Cen-MCTS for time > 7 s. Cen-MCTS performs well initially since it performs a centralised greedy rollout that finds reasonable solutions quickly. Dec-MCTS soon reaches deeper levels of the search trees, though, which allows it to outperform Cen-MCTS. Dec-MCTS uses a collection of search trees with smaller branching factors than Cen-MCTS, but still successfully optimises over the joint-action space.

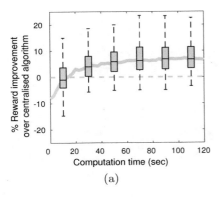 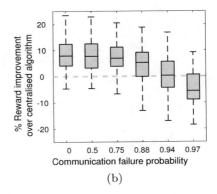

Fig. 2. (a) Comparison of Dec-MCTS with varying computation time to Cen-MCTS (120 s). (b) Performance of Dec-MCTS with intermittent communication (60 s computation time). (a,b) Vertical axes show percentage additional reward achieved by Dec-MCTS compared to Cen-MCTS. Error bars show 0, 25, 50, 75 and 100 percentiles (excluding outliers) of 100 random problem instances.

The second experiments analysed the effect of communication degradation. When the robots did not communicate, the algorithm achieved a median 31 % worse than Cen-MCTS, but with full communication achieves 7 % better than centralised, which shows the robots can successfully cooperate by using our proposed communication algorithm. Fig. 2(b) shows the results for partial communication degradation. When half of the packets are lost, there is no significant degradation of performance. When 97 % of packets are lost the performance is degraded but still performs significantly better than with no communication.

7   Experiments: Active object recognition

This section describes experiments in the context of online active object recognition, using point cloud data collected from an outdoor mobile robot in an urban scene (Fig. 3). We first outline the problem and experiment setup, and then present results that analyse the value of online replanning and compare Dec-MCTS to a greedy planner.

A team of robots aim to determine the identity of a set of static objects in an unknown environment. Each robot asynchronously executes the following cycle: 1) plan a path that is expected to improve the perception quality, 2) execute the first planned action, 3) make a point cloud observation using onboard sensors, and then 4) update the belief of the object identities. The robots have the same motion model and navigation graph as Sec. 6. Each graph edge has an additional constant cost for the observation processing time.

The robots maintain a belief of the identity of each observed object, represented as the probability that each object is an instance of a particular object from a given database. The aim is to improve this belief, which is achieved by

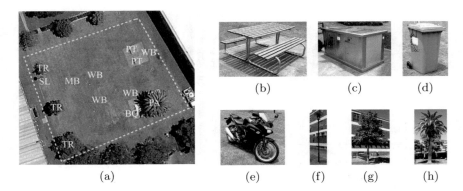

Fig. 3. Experiment setup for the point cloud dataset. (a) Environment with labelled locations, (b) picnic table (PT), (c) barbecue (BQ), (d) wheelie bin (WB), (e) motorbike (MB), (f) street light (ST), (g) tree (TR), (h) palm tree (PT).

maximising the mutual information objective proposed in [18]. The posterior probability distribution for each object after a set of observations is computed using recursive Bayes' rule. The observation likelihood is computed as the symmetric residual error of an iterative closest point (ICP) alignment [8] with each model in the database. Objects may merge or split after each observation if the segmentation changes. Observations are fused using decentralised data fusion or a central processor and shared between all robots. While planning, the value of future observations are estimated by simulating observations of objects in the database for all possible object identities, weighted by the belief.

7.1   Experiment setup

The experiments use a point cloud dataset [18] of Velodyne scans of outdoor objects in a $30 \times 30\,\mathrm{m}^2$ park (Fig. 3(a)). The environment consisted of 13 objects from 7 different model types as shown in Figs. 3(b)-(h). The dataset consists of single scans from 50 locations and each scan was split into 8 overlapping observations with different orientations. Each observation had a 180° field of view and 8 m range. These locations and orientations form the roadmap vertices with associated observations. Each object was analysed from separate data to generate the model database. The robots are given a single long-range observation from the start location to create an initial belief of most object locations.

The experiments simulate an online mission where each robot asynchronously alternates between planning and perceiving. Three planners were trialled: our Dec-MCTS algorithm with 120 s replanning after each action, Dec-MCTS without replanning, and a decentralised greedy planner that selects the next action that maximises the mutual information divided by the edge cost. The *recognition score* of an executed path was calculated as the belief probability that each object matched the ground-truth object type, averaged over all objects.

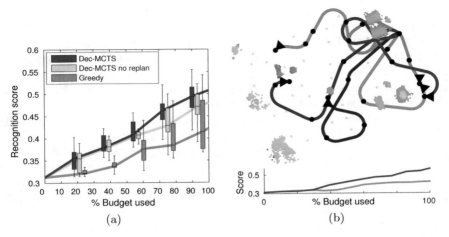

Fig. 4. (a) Task performance over mission duration for 10 trials (maximum possible score is 0.62). (b) Overlay of 2 example missions with 3 robots. Blue paths denote online Dec-MCTS (score 0.53). Orange paths denote greedy policy (score 0.42). Objects are green point clouds where shading indicates height. Robots observe at black dots in direction of travel. Start location top right.

7.2   Results

Overall, results validate the coordination performance of Dec-MCTS. Fig. 4(a) shows the recognition score (task performance) over the duration of the mission for 10 trials with 3 robots. The maximum possible recognition score subject to the perception algorithm and dataset was 0.62. Dec-MCTS outperformed greedy halfway through the missions since some early greedy decisions and poor coordination reduced the possibility of making subsequent valuable observations. By the end of the missions some greedy plans successfully made valuable observations, but less often than Dec-MCTS. The no-replanning scenario achieved a similar score as the online planner in the first half, showing that the initial plans are robust to changes in the belief. For the second half, replanning improved the recognition score since the belief had changed considerably since the start. This shows that while the generated plans are reasonable for many steps into the future, there is also value in replanning as new information becomes available.

Fig. 4(b) shows two example missions using online Dec-MCTS (blue) and greedy (orange) planners, and their score over the mission duration. Greedy stayed closer to the start location to improve the recognition of nearby objects, and consequently observed objects on the left less often; reaching this part of the environment would require making high cost/low immediate value actions. On the other hand, Dec-MCTS achieved a higher score since the longer planning horizon enabled finding the high value observations on the left, and was better able to coordinate to jointly observe most of the environment.

8   Discussion and future work

We have presented a novel approach to decentralised coordination with convergence properties. The performance (i.e., solution quality) of our approach is as good or better than its centralised counterpart in real-world applications with sensor data, and shows that our approach can effectively optimise a joint-action space even with intermittent communication. A key conceptual feature of our approach is its generality in representing joint actions probabilistically rather than deterministically. This facilitates the decentralisation of a variety of tasks while maintaining convergence properties drawn from statistics and game theory.

One interesting aspect of our work is that it is straightforward to extend it to problems that have partial observability. That is, we can replace MCTS with POMCP [20] and apply the same general framework. For example, this provides a convenient decentralised approach to problems in active perception, such as active classification. An interesting question is whether the same or similar convergence properties to the fully observable case can be maintained.

Another interesting line of inquiry is to incorporate coalition forming into our approach. As formulated, static coalitions of agents can be formed by generalising the product distributions in our framework to be partial joint distributions. The product distribution described in Sec. 4.3 would be defined over *groups* of robots rather than individuals. Each group acts jointly, with a single distribution modelling the joint actions of its members, and coordination between groups is conducted as in our algorithm. A natural field robotics application would be mobile robots, each with multiple manipulators, for weeding and harvesting in agriculture. Just as our current approach corresponds to mean-field methods, this approach maps nicely to region-based variational methods [26] and guarantees from these approaches may be applicable. It would also be interesting to study *dynamic* coalition forming, where the mapping between agents and robots is allowed to change, and to develop convergence guarantees for this case.

Acknowledgements. This work was supported by the Australian Centre for Field Robotics; the NSW Government; the Australian Research Council's Discovery Project funding scheme (No. DP140104203); the Faculty of Engineering & Information Technologies, The University of Sydney, under the Faculty Research Cluster Program; and the University's International Research Collaboration Award.

References

1. Amato, C.: Decision Making Under Uncertainty: Theory and Application, chap. Cooperative Decision Making. MIT Press, Cambridge and London (2015)
2. Auger, D.: Multiple tree for partially observable Monte-Carlo tree search. In: Proc. of EvoApplications. pp. 53–62 (2011)
3. Bajcsy, R.: Active perception. Proceedings of the IEEE 76(8), 966–1005 (1988)
4. Best, G., Faigl, J., Fitch, R.: Multi-robot path planning for budgeted active perception with self-organising maps. In: Proc. of IEEE/RSJ IROS (2016)
5. Browne, C., Powley, E., Whitehouse, D., Lucas, S., Cowling, P., Rohlfshagen, P., Tavener, S., Perez, D., Samothrakis, S., Colton, S.: A survey of Monte Carlo tree search methods. IEEE T. Comput. Int. AI Games 4(1), 1–43 (2012)

6. Charrow, B.: Information-theoretic active perception for multi-robot teams. Ph.D. thesis, University of Pennsylvania (2015)
7. Chaslot, G.M.J.B., Winands, M.H.M., van den Herik, H.J.: Parallel Monte-Carlo tree search. In: Proc. of CG. pp. 60–71 (2008)
8. Douillard, B., Quadros, A., Morton, P., Underwood, J.P., Deuge, M.D.: A 3D classifier trained without field samples. In: Proc. of ICARCV. pp. 805 – 810 (2012)
9. Faigl, J., Pěnička, R., Best, G.: Self-organizing map-based solution for the orienteering problem with neighborhoods. In: Proc. of IEEE SMC (2016)
10. Gan, S.K., Fitch, R., Sukkarieh, S.: Online decentralized information gathering with spatial–temporal constraints. Auton. Robots 37(1), 1 – 25 (2014)
11. Garivier, A., Moulines, E.: On upper-confidence bound policies for switching bandit problems. In: Proc. of AMA ALT. pp. 174–188 (2011)
12. Gunawan, A., Lau, H.C., Vansteenwegen, P.: Orienteering problem: A survey of recent variants, solution approaches and applications. Eur. J. Oper. Res. 255(2), 315 – 332 (2016)
13. Kocsis, L., Szepesvári, C.: Bandit based Monte-Carlo planning. In: Proc. of ECML. pp. 282–293 (2006)
14. Krause, A., Singh, A., Guestrin, C.: Near-optimal sensor placements in Gaussian processes: Theory, efficient algorithms and empirical studies. J. Mach. Learn. Res. 9, 235 – 284 (2008)
15. Kulkarni, A.J., Tai, K.: Probability collectives: A multi-agent approach for solving combinatorial optimization problems. Appl. Soft Comput. 10(3), 759 – 771 (2010)
16. Nemhauser, G.L., Wolsey, L.A., Fisher, M.L.: An analysis of approximations for maximizing submodular set functions–I. Math. Program. 14(1), 265 – 294 (1978)
17. Patten, T., Zillich, M., Fitch, R., Vincze, M., Sukkarieh, S.: Viewpoint evaluation for online 3-D active object classification. IEEE Robot. Autom. Lett. 1(1), 73–81 (2016)
18. Patten, T., Kassir, A., Martens, W., Douillard, B., Fitch, R., Sukkarieh, S.: A Bayesian approach for time-constrained 3D outdoor object recognition. In: ICRA 2015 Workshop on Scaling Up Active Perception (2015)
19. Rezek, I., Leslie, D.S., Reece, S., Roberts, S.J., Rogers, A., Dash, R.K., Jennings, N.R.: On similarities between inference in game theory and machine learning. J. Artif. Intell. Res. 33, 259–283 (2008)
20. Silver, D., Veness, J.: Monte-Carlo planning in large POMDPs. In: Lafferty, J.D., Williams, C.K.I., Shawe-Taylor, J., Zemel, R.S., Culotta, A. (eds.) Advances in Neural Information Processing Systems 23, pp. 2164–2172. Curran Inc. (2010)
21. Singh, A., Krause, A., Guestrin, C., Kaiser, W.J.: Efficient informative sensing using multiple robots. J. Artif. Int. Res. 34(1), 707 – 755 (2009)
22. Wolpert, D.H., Bieniawski, S.: Distributed control by Lagrangian steepest descent. In: Proc. of IEEE CDC. pp. 1562–1567 (2004)
23. Wolpert, D.H., Bieniawski, S.R., Rajnarayan, D.G.: Handbook of Statistics 31: Machine Learning: Theory and Applications, chap. Probability collectives in optimization, pp. 61 – 99. Elsevier (2013)
24. Wolpert, D.H., Strauss, C.E.M., Rajnarayan, D.: Advances in distributed optimization using probability collectives. Adv. Complex Syst. 09(04), 383–436 (2006)
25. Xu, Z., Fitch, R., Underwood, J., Sukkarieh, S.: Decentralized coordinated tracking with mixed discrete-continuous decisions. J. Field Robot. 30(5), 717 – 740 (2013)
26. Yedidia, J.S., Freeman, W.T., Weiss, Y.: Constructing free-energy approximations and generalized belief propagation algorithms. IEEE Trans. Inf. Theor. 51(7), 2282–2312 (2005)

Decentralized Multi-Agent Navigation Planning with Braids

Christoforos I. Mavrogiannis[†] and Ross A. Knepper[‡]

[†]Sibley School of Mechanical & Aerospace Engineering, Cornell University
cm694@cornell.edu
[‡]Department of Computer Science, Cornell University
rak@cs.cornell.edu

Abstract. We present a novel planning framework for navigation in dynamic, multi-agent environments with no explicit communication among agents, such as pedestrian scenes. Inspired by the collaborative nature of human navigation, our approach treats the problem as a coordination game, in which players coordinate to avoid each other as they move towards their destinations. We explicitly encode the concept of coordination into the agents' decision making process through a novel inference mechanism about future joint strategies of avoidance. We represent joint strategies as equivalence classes of topological trajectory patterns using the formalism of braids. This topological representation naturally generalizes to any number of agents and provides the advantage of adaptability to different environments, in contrast to the majority of existing approaches. At every round, the agents simultaneously decide on their next action that contributes collision-free progress towards their destination but also towards a global joint strategy that appears to be in compliance with all agents' preferences, as inferred from their past behaviors. This policy leads to a smooth and rapid uncertainty decrease regarding the emerging joint strategy that is promising for real world scenarios. Simulation results highlight the importance of reasoning about joint strategies and demonstrate the efficacy of our approach.

Keywords: motion planning, navigation, topology, braids

1 Introduction

Human environments, such as crowded hallways, sidewalks, and rooms are often characterized by un-structured motion, imposed by the lack of formal rules to control traffic and the lack of explicit communication among agents. Nonetheless, humans are capable of traversing such environments with remarkable efficiency and without hindering each other's paths. Human navigation not only achieves collision avoidance; it does so while respecting several social considerations, such as the passing preference of others and their personal space, ensuring smooth co-navigation. Cooperation has been identified to be the key for the generation of such a complex behavior (see e.g. Wolfinger [26]). In the absence of explicit communication, cooperation relies on intention inference, which in turn is based on trust. Pedestrians infer others' intentions and preferences by observing their motion, while communicating their own, essentially negotiating a joint

© Springer Nature Switzerland AG 2020
K. Goldberg et al. (Eds.): *Algorithmic Foundations of Robotics XII*, SPAR 13, pp. 880–895, 2020.
https://doi.org/10.1007/978-3-030-43089-4_56

strategy of avoidance. Past experiences of cooperatively resolved pedestrian encounters build and reinforce a form of trust among pedestrians that allows them to relax uncertainty and agree on a joint strategy of avoidance.

Inspired by the efficiency of humans in resolving pedestrian encounters, we explicitly employ the concept of cooperation into the design of an online navigation algorithm for multi-agent environments. Our approach treats navigation as a cooperative game in which agents make decisions by compromising between a notion of personal efficiency and a concept of joint efficiency. The concept of joint efficiency concerns the joint strategy that all of the agents will follow to reach their destinations, while avoiding others. We model joint strategies as topological patterns of agents' trajectories using *braids* [3].

In contrast to the majority of existing approaches that are either too myopic, only focusing on local collision avoidance resolution, or too specific, reproducing demonstrated behaviors in specific contexts, we contribute: (1) a topological model of a multi-agent scene, based on braids, that allows us to reduce the problem of planning joint strategies to a graph search problem that can be efficiently solved with existing techniques; (2) a framework for collaborative motion planning that generalizes across environments and numbers of agents; (3) a probabilistic intent inference mechanism for cooperative navigation that accelerates the rate of convergence among agents' plans; and (4) simulation results demonstrating the importance of incorporating a collective, global topological understanding in the planning process. Our framework was designed according to the insights of sociology studies on pedestrian behavior and psychology studies on action interpretation, reflecting our goal to employ it on a mobile robot platform navigating in crowded human environments. The topological structure that our model offers to the motion planning process is expected to reduce the emergence of undesired situations such as deadlocks and livelocks that are frequently observed in human-robot pedestrian encounters.

2   Related Work

2.1   Navigation

Navigation has been the focus of various diverse scientific communities, ranging from sociology and cognitive science to computer vision and robotics, aiming at understanding and simulating human navigation but also at reproducing robotic navigation.

Several works have proposed models for crowd dynamics that have been validated in simulation of various scenarios in different contexts. The social force model [9] and its variants [12, 19, 24] introduced a physics-inspired way of modeling pedestrian interactions: pedestrians are attracted to their destination and repulsed by obstacles or other agents. Hoogendoorn and Bovy [11] modeled the problem as a differential game in which the agents are cost-minimizing predictive controllers. Moussaïd et al. [17] looked at the problem from a cognitive science perspective, proposing a set of behavioral heuristics that guide human walking behaviors. Bonneaud and Warren [4] proposed a decomposition of locomotion into a set of elementary behaviors, each modeled as an experimentally tuned nonlinear dynamical system. Finally, Zhou et al. [28] presented a data-driven approach for learning macroscopic collective crowd behaviors.

Designing artificial agents, capable of seamlessly navigating dynamic human environments typically requires a predictive framework and a planning framework. Over the past two decades, the robotics community has made a number of significant contributions related to both components.

In human motion prediction, roboticists have employed learning techniques to derive models of human behavior. Bennewitz et al. [2] clustered human behavior into typical motion patterns which they used to perform online trajectory predictions on a mobile robot platform. Ziebart et al. [29] and Henry et al. [10] presented data-driven approaches, based on Inverse Reinforcement Learning for learning context-specific humanlike navigation behaviors for static and dynamic environments respectively.

In the area of planning and control, emphasis has been given to the design of strategies that would enable robots to integrate smoothly in human environments. To this end, Sisbot et al. [20] presented a cost-based planner that incorporates considerations of human comfort and context-specific social conventions, whereas Park et al. [18] proposed an online model-predictive control framework that generates locally optimal collision-free smooth trajectories for autonomous robotic wheelchairs. Another class of works have focused on modeling the interactions among multiple agents. The reactive multi-robot planning framework of van den Berg et al. [25] made explicit use of the assumption that the responsibility for collision avoidance is shared among interacting agents. Under the same assumption, Knepper and Rus [13], inspired by human navigation, contributed a sampling-based planner that also incorporates predictions about other agents' trajectories in the planning process. Kuderer et al. [14] and Trautman et al. [23] presented learning frameworks for predicting the trajectories of interacting pedestrians, which they used to plan socially compliant robot motion.

2.2 Human Behavior

One of the central principles guiding human decision making in pedestrian environments appears to be *cooperation*. Wolfinger [26] concluded to a concise, high-level protocol that captures the essence of the cooperative nature of human navigation: the *pedestrian bargain*. The pedestrian bargain is a set of foundational social rules that regulate pedestrian cooperation: (1) *people must behave like competent pedestrians* and (2) *people must trust copresent others to behave like competent pedestrians*. Pedestrians' trust to the rules of the bargain constitutes the basis of smooth co-navigation in shared environments.

In the absence of explicit communication, pedestrians rely on inference mechanisms for both prediction of others' behaviors and for generation of their own behaviors. As a result, building an autonomous system capable of seamlessly navigating human environments requires the design of a realistic, human-like inference mechanism. To this end, under the assumption of rational action, we design a goal-driven probabilistic model for action understanding and generation, that is in line with the insights of Csibra and Gergely [5, 6] regarding teleological action interpretation of humans. The concept of teleological reasoning, describing the tendency of humans to attribute potential *goals* to observed *actions*, has recently been employed in human robot-interaction by Dragan and Srinivasa [8], who formalized a framework for intent-expressive robot motion.

2.3 Topology

Finally, the foundational inspiration for this work is the topological concept of braids. The formalism of braids, first formulated by Artin [1] and extensively studied by Birman [3] has been an inspiration for applications in various disciplines, including robotics. Diaz-Mercado and Egerstedt [7] were the first to develop a framework for centralized multi-robot mixing, in which the agents are assigned trajectories that contribute to a specified topological pattern corresponding to a given braid. Although we are also making use of braids to model multi-robot behaviors, the scope of our approach is inherently different, since our target application concerns navigation in dynamic environments where no explicit communication takes place. In our case the agents do not follow a pre-specified braid pattern, but rather employ a braid-based probabilistic reasoning to reach a topological consensus that best complies with everyone's intentions or preferences. For our purposes, braids provide a basis for reasoning about uncertainty in a principled fashion, as their dual geometric and algebraic representation enables us to symbolically enumerate diverse distinct topological scene evolutions. As a result, our algorithm generates *socially competent* behaviors, i.e., behaviors that explicitly take into consideration the *social welfare* of the whole system of agents.

It should be noted that this work solidifies and extends the concepts first presented in our previous works [15, 16], where we made use of braids in a planning framework based on trajectory optimization.

3   Foundations

Consider n agents navigating a workspace $\mathcal{W}$. Each agent i starts from an initial configuration $q_i \in \mathcal{W}$ and aims at reaching a destination $d_i \in \mathcal{W}$ by following a trajectory $\zeta_i : I \to \mathbb{R}^2$, with $I = [0, 1]$ being a normalized time parametrization. The agents do not explicitly exchange information regarding their planned paths and are assumed to be acting rationally, which in our context means that (1) they always aim at making progress towards their destinations and (2) they have no motive for acting adversarially against other agents (e.g. blocking their paths or colliding with them). The notion of rationality is in line with the concept of *competence* as described by Wolfinger [26] in his definition of the *Pedestrian Bargain*.

3.1   Game-Theoretic Formulation

Inspired by Wolfinger's observations on the cooperative nature of human navigation, we approach the problem of robotic navigation in multi-agent, dynamic environments as a finitely repeated coordination game of imperfect information and perfect recall. The game is repeated a finite number of rounds (until all agents reach their destinations). At every round t, each agent decides on an action a_i from a set of available[1] actions A_i. All agents are simultaneously selecting their actions and therefore they have no access to other agents' plans (imperfect information); we assume however that they maintain a history of all previous rounds (perfect recall).

Based on our assumption of rationality, we can model agents' decision making to be the result of optimizing a utility function u_i. Agents generally aim at reaching their destination by spending low energy. However, in a multi-agent, uncertain environment, each agent's decisions are not independent of the decisions of others. Many decisions might lead to collisions, whereas others, greedily serving one agent's own interests

---

[1] Actions that could probabilistically lead to collisions or actions that violate the agent's dynamics are excluded.

might not be able to guarantee long term efficiency in such a complex context. In particular, the latter might actually lead to undesired outcomes such as longer paths, antisocial hindering of others' paths or even deadlocks and livelocks. For this reason, it is important that each agent's utility function incorporates a term reflecting the *social welfare*, i.e., the "common good", besides its own efficiency.

The set of actions selected by all players at a round t, $A = \{a_1, ..., a_n\}$ constitute a *strategy profile*. The sequence of strategy profiles of all rounds from the beginning to the end of the game form a global *joint strategy s*. In this paper, we make use of the concept of joint strategies to imbue artificial agents with an understanding of how their own actions affect the actions of others.

3.2   Modeling Joint Strategies

As the agents move from their initial configurations $Q = \langle q_1, ..., q_n \rangle$ to their intended destinations $D = \langle d_1, ..., d_n \rangle$, their trajectories $Z : I \to (\mathbb{R}^2)^n$ form a 3-dimensional pattern in space-time. This pattern corresponds to the global *joint strategy s* that the agents engaged in, to avoid each other and reach their destinations, from the beginning to the end of the game. Its topological properties are particularly interesting, as they can provide a qualitative characterization of the strategy and hence of the agents' interactions. For this reason, we represent joint strategies as equivalence classes of topological trajectory patterns. Thus, a joint strategy s is an equivalence class of trajectory patterns from a set of classes $\mathcal{S}$. In this paper, we model this set as the braid group [3].

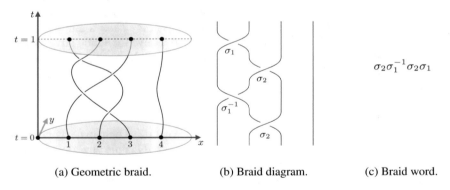

(a) Geometric braid.   (b) Braid diagram.   (c) Braid word.

Fig. 1: Alternative braid representations.

3.2.1   Background on Braids

Consider the finite set $N = \{1, 2, ..., n\}$, where $n \in \mathbb{N}^+$ and denote by $Perm(N)$ the set of permutations on N. A permutation in $Perm(N)$ is a bijection $b : N \to N$, often represented as:

$$b = \begin{pmatrix} 1 & 2 & ... & n \\ b(1) & b(2) & ... & b(n) \end{pmatrix}, \tag{1}$$

where $b(i)$ is the image of element $i \in N$, through the permutation b.

From a geometric perspective, a braid on $n \geq 1$ strands can be described as a system of n curves in $\mathbb{R}^3$, called the strands of the braid, such that each strand i connects the point $(i, 0, 0)$ with the point $(b(i), 0, 1)$ and intersects each plane $\mathbb{R}^2 \times t$ exactly once for any $t \in I$ (see Fig. 1a). A braid is usually represented with a *braid diagram*, a projection of the braid to the plane $\mathbb{R} \times 0 \times I$ with indications of the strand crossings (see Fig. 1b).

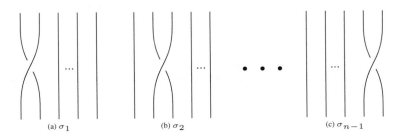

(a) σ_1 (b) σ_2 (c) σ_{n-1}

Fig. 2: The generators of the Braid Group B_n.

The set of all braids on n strands, along with the composition operation, form a group B_n that can be generated from a set of $n-1$ elementary braids $\sigma_1, \sigma_2, \ldots, \sigma_{n-1}$, depicted in Fig. 2, called the generators of B_n, that satisfy the following *relations*:

$$\sigma_j \sigma_k = \sigma_k \sigma_j, \qquad |j - k| > 1; \qquad \sigma_j \sigma_k \sigma_j = \sigma_k \sigma_j \sigma_k, \qquad |j - k| = 1. \qquad (2)$$

Intuitively, a generator σ_i, $i \in \{1, 2, ..., n - 1\}$, can be described as the pattern that emerges upon exchanging the ith strand (counted from left to right) with the $(i + 1)$th strand, such that the left strand passes over the right, whereas the inverse element σ_i^{-1} implements the same strand exchange, with the difference that the left strand passes under the right (see Fig. 1b). A trivial element is a braid where no strand exchanges occur. The group operation can be described as a concatenation of braids: given two braids $b_1, b_2 \in B_n$, the product $b_1 \cdot b_2$ results in b_2 being placed on the top of b_1, by attaching the top endpoints of b_1 to the bottom endpoints of b_2 and shrinking each braid by a factor of 2, along the t axis (e.g. see Fig. 1b). Any braid can be written as a product of generators and their inverses. This product is commonly referred to as a *braid word* (Fig. 1c).

3.2.2 Representing a Trajectory Collection as a Braid

We use the aforementioned braid representations to characterize topologically a collection of trajectories, based on the method of Thiffeault [21]. Given a trajectory collection Z and a line $\epsilon \in \mathcal{W}$, we can extract the braid word that corresponds to the scene evolution by (1) projecting the states of all agents at every point in time, $Z(t)$ to ϵ, (2) labeling any emerged projected trajectory intersections as generators (or their inverses) according to the intersection pattern, and (3) arranging them into temporal order. This word describes the topological properties of agents' trajectories Z, from the beginning to the end of time. Fig. 3 depicts an example of transitioning from a collection of trajectories to a braid diagram and finally a braid word (considering ϵ to be the x axis).

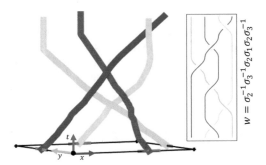

Fig. 3: Projecting a trajectory collection (left) to the x axis to derive a braid diagram and a braid word (right). The visualization of the braid diagram and the extraction of the braid word was done using BraidLab [22].

3.2.3   Using Braids to Represent Joint Strategies

In this paper, we make use of the braid group B_n to represent the set of classes of *joint strategies* S that a set of n agents could follow in a common workspace to avoid each other and reach their destinations. Instead of explicitly planning geometric representations of agents' potential future behaviors, i.e., predicting their future trajectories Z', the agents are reasoning symbolically about possible emerging collective topologies-braids. A planning agent i, headed towards a destination d_i, determines a set of joint strategies S by considering an appropriate braid definition, i.e., selecting an appropriate projection line ϵ to define braids with respect to.

3.3   Modeling Agents' Inference Mechanism

The braid group B_n, for $n > 1$, is infinite; therefore infinitely many alternative joint strategies could be mathematically possible. However, under a context M and observations of past collective behaviors Z, agents may form a belief over the set of emerging joint strategies S. For a planning agent, this inference process serves as a form of a context-specific social understanding, in the sense that it enables the agent to understand how its decision over a navigation strategy is coupled with the decisions of others.

In this paper, we design agents' inference mechanism in a human-inspired fashion, reflecting our goal to employ our framework on an autonomous social robot. In particular, we follow the main insight of Csibra and Gergely [5, 6] regarding the teleological nature of human action interpretation: humans tend to attribute potential context-specific *goals* to observed *actions*. In our framework, a joint strategy $s \in S$ is a goal, whereas agents' state history Z is the action and M is a variable that models the context of the scene (encoding the understanding of the static environment, such as obstacles, points of interest but also secondary inferences such as predictions about the destinations of others, agents' groupings etc). Formally, we model an agent i's inference as a belief distribution $P(s|Z, M)$ over a future joint strategy $s \in S$, given past trajectories Z and the context M.

3.4   Modeling Agents' Utilities

We model the interests of an agent i with a utility function $u_i : \mathcal{A}_i \to \mathbb{R}$ that maps an action $a_i \in \mathcal{A}_i$ to a real number, representing a reward for selecting it (higher rewards are better), whereas the *Social Welfare* is defined as mean of the individual utilities:

$$W(A) = \frac{1}{n} \sum_{i=1}^{n} u_i(a_i). \tag{3}$$

As we discussed earlier, in a multi-agent environment where no explicit communication takes place among agents, it is important that agents take into consideration their scene understanding in their decision making process. For this reason, we model u_i to be a weighted sum, compromising between personal efficiency and social compliance:

$$u_i(a_i) = \lambda E_i(a_i) - (1 - \lambda) H_i(a_i). \tag{4}$$

$E_i = \exp(-C_i(a_i))$ represents the efficiency of an action $a_i \in \mathcal{A}_i$ with respect to a geometric cost to destination $C_i : \mathcal{A}_i \to \mathbb{R}$, whereas H_i is the information entropy of the agent's belief regarding the emerging strategy s, defined as:

$$H_i(a_i) = - \sum_{s \in \mathcal{S}} P(s|Z^+, M) \log_2 P(s|Z^+, M) \tag{5}$$

where Z^+ is Z (the state history so far), augmented with the action in consideration a_i and λ is a weighting factor.

The Efficiency represents agents' intention of reaching their destinations by spending low energy and is in line with the principle of rational action as highlighted in the definitions of the pedestrian bargain [26] and the teleological reasoning [5]. The entropy reflects the state of the global consensus among pedestrians regarding the joint strategy to be followed and therefore, it directly incorporates a form of social understanding in an agent's decision making policy. The lower the entropy, the lower the uncertainty regarding the emerging joint strategy. Thus, by consistently picking actions that contribute to entropy reduction, an agent communicates its intention of complying with a subset of joint strategies that appear to be more likely or "social" according to the model $P(s|Z, M)$. Another interpretation of the functionality of this policy is that it implicitly biases others towards complying with the same strategy. As a result, the agents are expected to reach a consensus over s easier and faster, avoiding ambiguous situations such as livelocks or deadlocks and reach their destinations more *comfortably*; not necessarily faster or with less energy, but with a higher degree of cooperation, requiring a lower planning cognitive load.

The superposition of the specifications for Efficiency and Entropy reduction, represent what -to our interpretation of the pedestrian bargain [26]- constitutes *competent* behavior in the pedestrian context: (1) rationality and (2) social understanding.

4   Planning Global Joint Strategies with Braids

As discussed in Sec. 3.3, the braid group B_n is infinite; however, in practice, only a subset of joint strategies-braids are meaningful under the context of a scene and given observations of agents' past behaviors. In particular, given predictions of agents' destinations D, we can determine a set $\mathcal{S}$, comprising only strategies that take agents from their current configurations to their predicted destinations. Making use of the observation that all braids in B_n describe transitions between permutations of the set $N = \{1, 2, ..., n\}$,

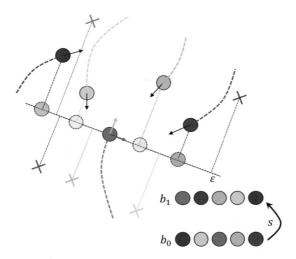

Fig. 4: A multi-agent scene from the perspective the planning agent (blue color). The robot arranges all agents according to the projections of their current configurations on the line ϵ (coincident to the x-axis of its body frame) and derives a corresponding permutation b_0. Given agents' past trajectories (dashed lines) and the context it also makes a prediction of their intended destinations (denoted with crosses) and derives a corresponding permutation b_1. Transitioning from b_0 to b_1 is implemented with a joint strategy $s \in \mathcal{S}$.

we convert the problem of planning joint strategies to a graph search in a permutation graph.

Fig. 4 illustrates the concept of our method. Assume that at planning time, the agents have already followed trajectories Z, denoted with thick dashed lines and are located at positions Q (circles in vivid colors, thick lines). The planning agent (blue color) has predicted that they are aiming for the destinations contained in the tuple D (denoted with crosses). From its perspective, Q and D correspond to the permutations b_0 and b_1 respectively, derived upon their projection on the line ϵ (parallel to the x-axis of its body frame).

4.1   Permutation Graph Construction

The set of all permutations on N, $Perm(N)$, along with the composition operation, form the symmetric group S_n. S_n is a group of order $n!$, that can be generated by a set of adjacent transpositions $\beta_k = \begin{bmatrix} k & k+1 \end{bmatrix}$, with $1 \leq k < n-1$ (i.e., the set of permutations that implement exactly one swap of a pair of adjacent elements in the set). It should be noted that these transpositions/generators just implement swaps of adjacent pairs, whereas *braid* generators, besides implementing adjacent swaps, also prescribe a swapping *quality* (which strand passes over or under, as discussed in Sec. 3.2.1).

We construct a permutation graph $G = (V, E)$, comprising a set of vertices that correspond to the elements of S_n. A pair of nodes $v_i, v_j \in V$ is connected iff there exists a permutation β_k (from the set of the generating transpositions described in the previous paragraph) that permutes v_i into v_j. Our graph can be graphically represented

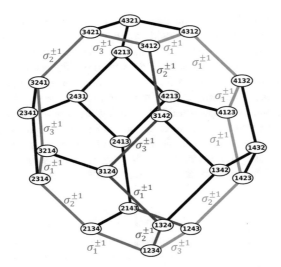

Fig. 5: A permutation graph for a scene with four agents, represented as a permutohedron of order 4. Three alternative paths implementing the transition from the permutation 1234 to the permutation 3412 are depicted in different colors. Each path consists of a sequence of transitions, each of which can be implemented topologically with a braid generator or its inverse.

as a *permutohedron* [30]. Fig. 5 depicts a permutohedron of 4th order (for a scene with four agents) along with example paths and indications of braid transitions.

4.2 Searching in the Permutation Graph

Two vertices $v_i, v_j \in V$ corresponding to the permutations $p_i, p_j \in S_n$ respectively can be connected with a path of vertices P_{ij}, corresponding to a sequence of permutations that transitions p_i into p_j. Each edge in the graph corresponds to a generator of S_n that can only be implemented topologically with a generator of B_n or its inverse. Therefore, given a path P_{ij}, a corresponding joint strategy (braid) s, transitioning p_i into p_j, can be derived by assigning a braid at each transition between consecutive waypoints (vertices) in the path. For each transition, there are always two candidate braid generators, i.e., a σ_i^+ and its inverse, σ_i^-. Fig. 5 schematically demonstrates the assignment of braids to permutation transitions in three different paths implementing a symbolic plan with the same starting and ending permutations.

5 Algorithm Design

Algorithm 1 describes our online algorithm for socially competent navigation (SCN) in multi-agent environments. The Algorithm starts by updating the context[2] M with function `UpdateContext`. Next, the function `DetermineReactiveAgents` returns a subset of all observed agents that should be taken into consideration in the motion plan (e.g. ignoring agents that are behind the planning agent or agents that are too far

---

[2] The context M comprises a static component (map, obstacles, points of interest in the scene etc) and a dynamic component that depends on agents' behaviors.

ahead). Subsequently, the algorithm determines the set of actions $\mathcal{A}$ that are available to the planning agent, taking into consideration its dynamics and the positions and intentions of others (function `GetAvailableActions`). In case there are not other agents to which the planning agent should be reacting, the algorithm returns the most efficient action towards the agent's destination (function `MaximizeEfficiency`). Otherwise, the algorithm continues with the function `GetStrategies` that derives a set of topological joint strategies/braids $\mathcal{S}$. Finally, the function `MaximizeUtility` returns a control command a that corresponds to the action that both makes progress towards the planning agent's destination and communicates compliance with the most likely joint strategies at the given time. The algorithm runs until the planning agent reaches its destination, i.e., until the boolean variable $AtGoal$ becomes 1.

Algorithm 1 SCN($q, d, map, N, Z, AtGoal, a$)

Input: q − agent's current state; d − agent's intended destination; map; Z − state history of all agents; $AtGoal$ − boolean variable signifying arrival at agent's destination; M − context

Output: a − action selected for execution

1: **while** ¬AtGoal **do**
2:      $M \leftarrow UpdateContext(Z, M)$
3:      $\mathcal{R} \leftarrow DetermineReactiveAgents(M)$
4:      $\mathcal{A} \leftarrow GetAvailableActions(M)$
5:      **if** $\mathcal{R}$ **then**
6:          $\mathcal{S} \leftarrow GetStrategies(M, \mathcal{R})$
7:          $a \leftarrow MaximizeUtility(\mathcal{A}, \mathcal{S}, d, M, \mathcal{R})$
8:      **else**
9:          $a \rightarrow MaximizeEfficiency(\mathcal{A}, d)$
10: **return** a

---

6   Application

We tested our algorithm in simulation in the following game. Consider a workspace $\mathcal{W}$, partitioned into a set of m polytopes (Fig. 6). A set of n agents navigate the workspace, each starting from an initial configuration $q_i \in \mathcal{W}$ and aiming at reaching a final configuration $d_i \in \mathcal{W}$. The game is played in rounds until all agents reach their destinations. At every round, the players simultaneously pick an action, i.e., a neighboring square. Forward, backward, left, right and diagonal, collision-free transitions are allowed. Since at planning time each agent has no access to others' plans, in order to ensure collision avoidance, transitioning to a square that is adjacent to a square currently occupied by another agent is not allowed.

To demonstrate the importance of considering the emerging joint strategy in the decision making stage, we compare the performance of our algorithm against a simple baseline that only plans actions that seek to maximize the efficiency (the progress to the agent's destination) at every round. This baseline is conceptually similar to the widely used *social force* algorithm [9]. We show that explicitly reasoning about the emerging joint strategy when planning an action, benefits everyone in the scene, as it leads to a rapid uncertainty decrease that simplifies everyone's decision making. This allows agents to avoid ambiguous situations that could lead to livelocks or deadlocks.

6.1 Implementation Details

For the simulations we made the assumption that the agents were aware of others' destinations. The agents were starting from one side of the board and aiming at reaching a destination in the opposite side of the board. As a geometric cost C_i we selected the Manhattan distance to destination. For each agent, the projection line, with respect to which braids were defined, was selected to be constantly parallel to the line defining their starting board side (and coinciding with agent's body frame x-axis). The belief over strategies was modeled as:

$$P(s|Z, M) = \frac{1}{\Lambda} \prod_{j=1:l} \exp(-(l - j)\Delta x_j) \tag{6}$$

where l is the length of the word representing a strategy s, Δx_j is the current distance along the x-axis between a pair of swapping agents, corresponding to the jth generator in the braid, and Λ is an appropriate normalizer. This distribution is a simplified approximation that cannot guarantee robust performance and generalization. We are using it in this paper to provide a proof of our concept. We plan on approximating it using human pedestrian data. Finally, for deriving a set of candidate paths in the permutation graph, we use the algorithm of Yen [27] for finding K *shortest paths*.

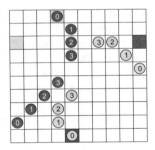
(a) Greedy agents.
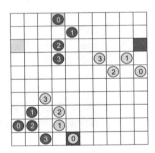
(b) Social agents.

Fig. 6: Partial Execution: 4 agents play the same game (same initial configurations and destinations) running the baseline (Fig. 6a) and our algorithm (Fig. 6b). The destination of each agent corresponds to the square of the same color. The actions taken by each agent at every round so far are noted with a corresponding round number.

6.2 Simulation Results

Figs 6a and 6b depict partial executions of the same game, after three rounds, for the case of the baseline and our algorithm respectively. It can be observed that the agents running our algorithm ("social agents") have achieved a better status, as all of their encounters are essentially resolved by the end of the third round. On the contrary, the agents running the baseline ("greedy agents") are about to engage in an ambiguous encounter involving three of them aiming to pass from the same region of the board.

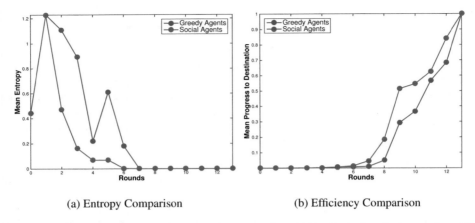

(a) Entropy Comparison   (b) Efficiency Comparison

Fig. 7: Comparative Performance plots for the game of Fig. 6: Fig. 7a depicts the comparison of the mean entropy progression between the agents running our algorithm (red color) and the agents running the baseline (blue color), whereas Fig. 7b depicts the mean progress to destinations (mean efficiency) per round.

Fig. 7 depicts comparative performance diagrams for the game of Fig. 6. The progression of the quality of agents' decision making is demonstrated by plots of the mean entropy across all agents (Fig. 7a) and the mean progress to the destination (Fig. 7b) per round of the game. Comparative plots of the complete trajectories of the game are shown in Fig. 8. It can be noticed that the *social agents* achieve a rapid decrease in uncertainty, expressed by the smooth convergence of the entropy, whereas in the case of the *greedy agents*, the entropy fluctuates before the agents reach consensus, reflecting the ambiguity of agents' actions. At the same time, it appears that the social agents' actions are also ensuring faster progress to their destinations compared to the greedy ones.

Fig 9 depicts performance diagrams extracted from a similar scenario that besides four agents involves a static obstacle that blocks the agents' way to their destinations. The obstacle is treated as an extra static agent. For our braid model an obstacle is not different than an agent, as it can be represented with a stationary strand. Fig. 9a demonstrates again an improved entropy progression, reflecting a faster consensus, while Fig. 9b shows that our algorithm stays quite close to the baseline in terms of efficiency. The complete trajectories for the two cases are depicted in Fig. 10.

7   Discussion and Future Work

We presented an online framework for navigation in multi-agent environments with no explicit communication, inspired by the insights of recent studies on the cooperative nature of pedestrian behavior [26] and the goal-directed inference of humans [6]. Our framework explicitly incorporates the concept of cooperation by modeling multi-agent collective behaviors as topological global joint strategies, using the formalism of braids [3]. Our topological model forms the basis of an inference mechanism that associates

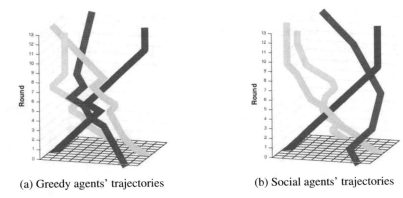

(a) Greedy agents' trajectories   (b) Social agents' trajectories

Fig. 8: Comparative plot of the trajectories followed by agents running the baseline (Fig. 8a) and our algorithm (Fig. 8b).

observed behaviors with future collective topologies. In the decision making stage, each agent decides on an action that corresponds to a compromise between its personal efficiency (progress towards destination) and a form of joint efficiency (the status of a consensus on a joint strategy of avoidance). Simulation results on a discretized workspace demonstrated the benefit of incorporating this joint efficiency in the decision making stage, as opposed to picking actions that only contribute progress to one's destination. Our approach was shown to lead to a rapid drop in uncertainty that allows agents to efficiently cooperate towards avoiding each other and reaching their destinations.

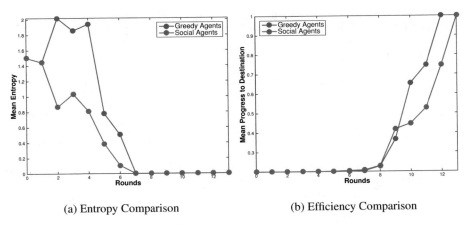

(a) Entropy Comparison   (b) Efficiency Comparison

Fig. 9: Performance comparisons in a workspace with four agents and an obstacle.

Ongoing work involves the development of a continuous implementation of our algorithm and learning a distribution over topologies from human data. This will enable us to design a more realistic and accurate belief distribution over joint strategies and test our algorithm experimentally, in real world scenarios involving human agents in a variety of pedestrian environments. Finally, we plan on conducting a user study to get feedback from humans and improve our design.

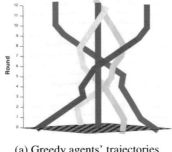 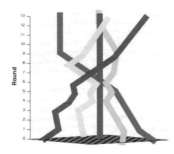

(a) Greedy agents' trajectories   (b) Social agents' trajectories

Fig. 10: Comparative plots of trajectories for a scenario involving an obstacle (purple trajectory) for greedy (Fig. 10a) and social agents (Fig. 10b).

Acknowledgement

This material is based upon work supported by the National Science Foundation under Grant IIS-1526035. We are grateful for this support.

Bibliography

[1] E. Artin. Theory of braids. Annals of Mathematics, 48(1):pp. 101–126, 1947.

[2] M. Bennewitz, W. Burgard, G. Cielniak, and S. Thrun. Learning motion patterns of people for compliant robot motion. Int. J. of Robotics Res., 24:31–48, 2005.

[3] J. S. Birman. Braids Links And Mapping Class Groups. Princeton University Press, 1975.

[4] S. Bonneaud and W. H. Warren. An empirically-grounded emergent approach to modeling pedestrian behavior. In Pedestrian and Evacuation Dynamics 2012, pages 625–638. Springer International Publishing, 2014.

[5] G. Csibra and G. Gergely. The teleological origins of mentalistic action explanations: A developmental hypothesis. Developmental Science, 1(2):255–259, 1998.

[6] G. Csibra and G. Gergely. 'Obsessed with goals': Functions and mechanisms of teleological interpretation of actions in humans. Acta Psychologica, 124(1):60–78, Jan. 2007.

[7] Y. Diaz-Mercado and M. Egerstedt. Interactions in multi-robot systems using braids. Robotics, IEEE Transactions on, 2016. Under Review.

[8] A. D. Dragan and S. Srinivasa. Integrating human observer inferences into robot motion planning. Auton. Robots, 37(4):351–368, 2014.

[9] D. Helbing and P. Molnár. Social force model for pedestrian dynamics. Phys. Rev. E, 51: 4282–4286, May 1995.

[10] P. Henry, C. Vollmer, B. Ferris, and D. Fox. Learning to navigate through crowded environments. In IEEE International Conference on Robotics and Automation (ICRA), 2010.

[11] S. Hoogendoorn and P. H. Bovy. Simulation of pedestrian flows by optimal control and differential games. Optimal Control Applications and Methods, 24(3):153–172, 2003.

[12] I. Karamouzas, B. Skinner, and S. J. Guy. Universal power law governing pedestrian inter-actions. Phys. Rev. Lett., 113:238701, Dec 2014.

[13] R. A. Knepper and D. Rus. Pedestrian-inspired sampling-based multi-robot collision avoid-ance. In RO-MAN, pages 94–100. IEEE, 2012.

[14] M. Kuderer, H. Kretzschmar, C. Sprunk, and W. Burgard. Feature-based prediction of trajectories for socially compliant navigation. In Proc. of Robotics: Science and Systems (RSS), Sydney, Australia, 2012.

[15] C. I. Mavrogiannis and R. A. Knepper. Towards socially competent navigation of pedes-trian environments. In Robotics: Science and Systems 2016 Workshop on Social Trust in Autonomous Robots, 2016.

[16] C. I. Mavrogiannis and R. A. Knepper. Interpretation and communication of pedestrian intentions using braid groups. In Workshop on Intention Recognition in Human-Robot Interaction, 11th ACM/IEEE International Conference on Human Robot Interaction (HRI), 2016.

[17] M. Moussaïd, D. Helbing, and G. Theraulaz. How simple rules determine pedestrian be-havior and crowd disasters. Proceedings of the National Academy of Sciences, 108(17): 6884–6888, Apr. 2011.

[18] J. J. Park, C. Johnson, and B. Kuipers. Robot navigation with model predictive equilibrium point control. In Intelligent Robots and Systems (IROS), 2012 IEEE/RSJ International Conference on, pages 4945–4952. IEEE, 2012.

[19] S. Pellegrini, A. Ess, K. Schindler, and L. J. V. Gool. You'll never walk alone: Modeling social behavior for multi-target tracking. In ICCV, pages 261–268. IEEE Computer Society, 2009.

[20] E. A. Sisbot, L. F. Marin-Urias, R. Alami, and T. Siméon. A human aware mobile robot motion planner. IEEE Transactions on Robotics, 23(5):874–883, 2007.

[21] J.-L. Thiffeault. Braids of entangled particle trajectories. Chaos, 20(1), 2010.

[22] J.-L. Thiffeault and M. Budišić. Braidlab: A software package for braids and loops, 2013–2016. URL http://arXiv.org/abs/1410.0849. Version 3.2.1.

[23] P. Trautman, J. Ma, R. M. Murray, and A. Krause. Robot navigation in dense human crowds: Statistical models and experimental studies of human-robot cooperation. Int. J. of Robotics Res., 34(3):335–356, 2015.

[24] A. Treuille, S. Cooper, and Z. Popović. Continuum crowds. ACM Transactions on Graphics, 25(3):1160–1168, July 2006.

[25] J. van den Berg, S. J. Guy, M. C. Lin, and D. Manocha. Reciprocal n-body collision avoid-ance. In Robotics Research - The 14th International Symposium, ISRR 2009, August 31 - September 3, 2009, Lucerne, Switzerland, pages 3–19, 2009.

[26] N. H. Wolfinger. Passing Moments: Some Social Dynamics of Pedestrian Interaction. Journal of Contemporary Ethnography, 24(3):323–340, 1995.

[27] J. Yen. Finding the k shortest loopless paths in a network. management Science, pages 712–716, 1971.

[28] B. Zhou, X. Tang, and X. Wang. Learning collective crowd behaviors with dynamic pedestrian-agents. International Journal of Computer Vision, 111(1):50–68, 2015.

[29] B. D. Ziebart, N. Ratliff, G. Gallagher, C. Mertz, K. Peterson, J. A. Bagnell, M. Hebert, A. K. Dey, and S. Srinivasa. Planning-based prediction for pedestrians. In Proc. of the International Conference on Intelligent Robots and Systems, 2009.

[30] G. M. Ziegler. Lectures on polytopes. Springer-Verlag, New York, 1995.

A Geometric Approach for Multi-Robot Exploration in Orthogonal Polygons

Aravind Preshant Premkumar, Kevin Yu, and Pratap Tokekar[*]

Department of Electrical & Computer Engineering, Virginia Tech, USA.
{aravindp, klyu, tokekar}@vt.edu

Abstract. We present an algorithm to explore an orthogonal polygon using a team of p robots. Our algorithm is based on a single-robot polygon exploration algorithm and a tree exploration algorithm. We show that the exploration time of our algorithm is competitive (as a function of p) with respect to the offline optimal exploration algorithm. In addition to theoretical analysis, we discuss how this strategy can be adapted to real-world settings. We investigate the performance of our algorithm through simulations for multiple robots and experiments with a single robot. We conclude with a discussion of our ongoing work.

1 Introduction

Exploration of unknown environments using a single robot has been a well studied problem [1, 2]. The task can be performed much faster if multiple robots are used. The challenge is to come up with an algorithm to efficiently coordinate multiple robots to explore the environment in the minimum amount of time. Broadly speaking, there have been two types of approaches towards solving the exploration problem: geometric and information-theoretic. In geometric approaches (e.g., [3]), it is typical to assume that the robots have perfect, unrestricted and omni-directional sensing. Geometric algorithms typically give global guarantees on distance traveled at the expense of restrictive assumptions about the environment and sensor models. On the other hand, information-theoretic approaches (e.g., [4]) explicitly take into account practical constraints such as noisy sensors and complex environments. However, these approaches are often greedy (e.g., frontier-based [5]) and typically do not yield any guarantees on the total time taken. In this paper we investigate the challenges in applying geometric exploration algorithms to practical settings.

We use competitive analysis [6] in order to analyze the cost of exploration. Competitive ratio of an online algorithm is defined as the worst-case ratio (over all inputs) of the cost of the online algorithm and the optimal offline algorithm. The optimal offline algorithm corresponds to the case when the input (i.e., polygon) itself is known. The goal is to find online algorithms with low, constant competitive ratios. That is, algorithms whose online performance is comparable

---

[*] This material is based upon work supported by the National Science Foundation under Grant No. 1566247.

K. Goldberg et al. (Eds.): *Algorithmic Foundations of Robotics XII*, SPAR 13, pp. 896–911, 2020.
https://doi.org/10.1007/978-3-030-43089-4_57

to algorithms who know the input a priori. We focus on the case of exploring unknown orthogonal polygons[1] without any holes. Deng et al. [7] showed that there is an algorithm with a constant competitive ratio for exploring orthogonal polygons with a single robot. Our main result is a constant competitive ratio algorithm for exploring with p robots, when p is fixed (Section 3).

The analysis of this algorithm requires certain assumptions that do not necessarily hold in practice. Our second contribution is to show how to adapt this purely geometric algorithm for real-world constraints to incorporate sensing limitations and uncertainty (Section 4). We evaluate our algorithm through simulations and experiments on a mobile robot (Sections 5 and 6). We begin with a discussion of related work in computational geometry and robotics literature.

2   Related Work

In this section, we present the existing work related to the exploration problem. We organize the related work into three broad categories: polygon exploration, graph exploration, and information-theoretic exploration.

2.1   Exploration of Polygons

The study of geometric problems that are based on visibility is a well-established field within computational geometry. The classic problems are the art gallery problem [8], watchman route problem [9], and target search [10] and shortest path planning [11] in unknown environments.

Using a fixed set of positions for guarding a known polygonal region, i.e., a set of points from which the entire polygon is visible, is known as the classical art gallery problem. If the gallery is represented by a polygon (having n vertices) and the guards are points in the polygon, then visibility problems can be equivalently stated as problems of covering the gallery with star-shaped polygons. Chvatal [12] and Fisk [13] proved that $\lfloor n/3 \rfloor$ guards are always sufficient and sometimes necessary to cover a polygon of n edges. The minimum number of guards required for a specific polygon may be much smaller than this upper bound. However, Schuchardt and Hecker [14] showed that finding a minimum set of guards is NP-hard, even for the special case of an orthogonal polygon.

Finding the shortest tour along which one mobile guard can see the polygon completely is the watchman route problem. Chin and Ntafos[9] showed that the watchman route can be found in polynomial time in a simple rectilinear polygon. Exploring an unknown polygon is the online watchman route problem. Bhattacharya et al. [3] and Ghosh et al. [15] approached the exploration problem with discrete vision, i.e., they assume that the robot doesn't have continuous visibility and has to stop at different scan points in order to sense the environment. They focus on the worst-case number of necessary scan points.

---

[1] An orthogonal polygon is one in which all edges are aligned with either the X or the Y axes.

Their algorithm results in a competitive ratio of $(r + 1)/2$, where r is the number of reflex vertices in the polygon. For the case of a limited range of visibility they give an algorithm where the competitive ratio in a polygon P can be limited by $\lfloor \frac{8\pi}{3} + \frac{\pi R \times Perimeter(P)}{Area(P)} + \frac{(r+h+1)\pi R^2}{Area(P)} \rfloor$, where h is the number of holes in the polygon.

For a simple polygon, Hoffmann et al. [16] presented an algorithm which achieves a constant competitive ratio of $c = 26.5$. For the special case of an orthogonal polygon, Deng et al. [7] presented a $\sqrt{2}$ competitive exploration strategy for the case of a single robot. We show how to extend the single robot exploration algorithm by Deng et al. [7] to the case of p robots. The resulting algorithm has a competitive ratio that is a function of p.

2.2   Graph Exploration

The problem of visiting all the nodes in a graph in the least amount of time is known as the Traveling Salesperson Problem (TSP). Here, all nodes of the graph are known before-hand and the objective is to determine the shortest path visiting all the nodes in the graph exactly once. Finding the optimal TSP tour for a given graph is known to be NP-hard, even for the special case where the nodes in the graph represent points on the Euclidean plane [17]. For the Euclidean version of the problem, there exist polynomial time approximation schemes [17, 18], i.e., for any $\epsilon > 0$, there exists a polynomial time algorithm which guarantees an approximation factor of $1 + \epsilon$.

In the graph exploration version of the problem nodes are revealed in an online fashion. The objective is to minimize the total distance (or time) traveled. Fraigniaud et al. [19] presented an algorithm for exploration of trees using p robots with a competitive ratio of $\mathcal{O}(p/\log p)$ and a lower bound of $\Omega(2 + 1/p)$. This lower bound was improved by Dynia et al. [20] to $\Omega(\log p/\log \log p)$. They modeled the cost as the maximal number of edges traversed by a robot and presented a $(4 - 2/p)$-competitive online algorithm.

Higashikawa et al. [21] showed that greedy exploration strategies have an even stronger lower bound of $\Omega(p/\log p)$ and presented a $(p + \log p/1 + \log p)$ competitive algorithm.

Better bounds have been achieved for restricted graphs. Dynia et al. [22] presented an algorithm that achieves faster exploration for trees restricted by a density parameter k which forces a minimum depth for any subtree depending on its size. Trees embeddable in d-dimensional grids can be explored with a competitive ratio of $\mathcal{O}(d^{1-1/k})$. For 2-dimensional grids with only convex obstacles, Ortolf et al. improved the competitive ratio to $\mathcal{O}(\log^2 d)$ [23]. Despite these strong restrictions on the graph, the same lower bound of $\Omega(\log p/\log \log p)$ holds for all trees.

We show that the problem of exploring a polygon can be formulated as a multi-robot tree exploration problem.

2.3   Information-Theoretic Exploration

The geometric and graph-based approaches typically assume perfect sensing with no noise. In practice, however, we observe that measurements are not perfect and we have to account for noise in these measurements while employing exploration strategies. Exploration strategies can be broadly classified into two types: frontier based and information-gain based. We refer the reader to [2] for a comprehensive survey of exploration strategies.

Frontier based exploration strategies are primarily greedy in nature and drive the robots to a deterministic boundary between known and unknown spaces in the map. This strategy has been extended and used to perform exploration of unknown 2D [5, 24, 25] and 2.5D environments [26]. For a comprehensive survey please refer to Holz et al. [27].

Information-gain based strategies seek to optimize some information-theoretic measures such as minimizing the entropy of the map [28, 29], or maximizing mutual information [30, 31]. While these algorithms produce better maps, the planner is typically one-step greedy algorithms that do not have global guarantees on the total distance traveled. Charrow et al.[4] attempted to resolve this by combining a global planner for a single robot to determine goals whilst locally following a path which locally maximizes mutual information. Here, we present a global planner for multi-robot teams that has strong performance guarantees in the form of constant competitive ratio.

3   Multi-Robot Exploration Algorithm

In this section, we present the details of our algorithm for exploring an orthogonal polygon without any holes, P, using a team of p robots. Our algorithm builds on the algorithm by Deng et al. [7] for exploring an orthogonal polygon with a single robot and extends it to the case of multiple robots using the graph exploration strategy from [21]. Our main insight is to show that the path followed by the robot using the algorithm in [7] can be used to construct a tree, denoted by $\mathcal{T}$, in P. That is, exploring the polygon is equivalent to visiting all nodes in this tree. We show that a multi-robot tree exploration algorithm from [21] can be used to explore and visit every node in this tree. Furthermore, we show that the competitive ratio of our algorithm with respect to the optimal offline algorithm is bounded (as a function of p).

We assume all robots start at a common location. The cost of exploration is defined as the time taken for all the robots to return to the starting location having explored the polygon. We say that a polygon P is explored if all points in its interior and on the boundary were seen from at least one robot. For the purpose of the analysis, we assume that the sensor on the robot is an omni-directional camera with infinite sensing range which returns the exact coordinates of any object in its field of view. We also assume that the robots move at unit speeds and can communicate at all times. In the next section, we show how to adapt our algorithm to realistic sensing models and evaluate it through experiments.

We introduce some terminology used in our algorithm (refer to Figure 1) before presenting the details. The sub-polygon that is visible from a point x is known as the *visibility polygon* of x and is denoted by $VP(x)$. Some of the edges in the visibility polygon are part of the boundary of P where as others are chords in the interior of P (e.g., segment $\overline{gb}$ in Figure 1). A reflex vertex of P which breaks the continuity of the part of the boundary of P visible from x is known as a *blocking vertex*. Vertex b in Figure 1 is a blocking vertex. Let $\overline{bc}$ be the edge incident to b which is (partly) visible from x. The line segment perpendicular to $\overline{bc}$ drawn from b till the boundary of P is known as the *extension* of the blocking vertex b for an orthogonal polygon. The robot

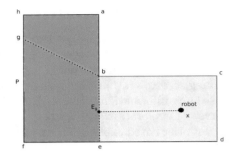

Fig. 1. Vertex b blocks the visibility of the robot and is known as a *blocking vertex*. The robot has to cross the segment $\overline{be}$ (known as the *extension*) to increase the visible boundary of the polygon P. The *extension* divides P into two sub-polygons. The sub-polygon not containing the robot (shaded in dark gray) is known as the *foreign polygon* w.r.t. b and is denoted as $FP(b)$.

must cross the extension in order to "look beyond" the blocking vertex and explore the polygon. Draw the line segment starting from the robot's position x perpendicular to the extension $\overline{be}$. The point at which these two line segments intersect (E_g) is known as the *extension goal* corresponding to the blocking vertex b. An extension divides P into two sub-polygons. The one which contains the robot is known as the *home polygon* of the robot with respect to b. The other sub-polygon is known as the *foreign polygon* of the robot with respect to b and is denoted as $FP(b)$.

The algorithm starts by creating a tree with the initial position of the robots as the root. All robots start in one cluster located at the root node. We use three labels to keep track of the status of any node in the tree: *unexplored*, *under-exploration*, and *explored*. Whenever a cluster of robots reach a node (using a subroutine goto not shown), we call the subroutine shown in Algorithm 1. While navigating using goto, if two clusters run into each other, then they merge and travel up the tree together.

The root is initially marked as *under-exploration*. We then check to determine any blocking vertices visible from the current node. We add the *extension goals* corresponding to any blocking vertices visible from the current node as its children. All of the corresponding extension goals are added as children by sorting them in the clockwise direction. These new children of the current node are adjusted according to the conditions mentioned. These conditions define an ordering over the nodes, by rewiring the tree. The cluster of robots at the current node is then divided as equally as possible and sent to visit the children of the current node. If the current node doesn't have any children, the current node

```
 1 Function explore()
        Data: Cluster of robots, C, located at some node, A, in the tree.
 2  if A is marked under-exploration then
 3   │  A ← children of A not marked as explored;
 4   │  if A == ∅ then
 5   │   │  if A == root of the tree then
 6   │   │   │  Terminate exploration;
 7   │   │  else
 8   │   │   │  goto(C,parent(A));
 9   │   │  end
10  │  else
11  │   │  Divide C equally among A;
12  │   │  When any cluster reaches a child of A, call explore;
13  │  end
14  else
15   │  Mark A as under-exploration;
16   │  if blocking vertices detected from A then
17   │   │  Sort in clockwise direction and add as children of A;
18   │   │  for p and q are distinct blocking vertices do
19   │   │   │  if FP(p) ⊆ FP(q) then
20   │   │   │   │  add p as child of q;
21   │   │   │  else
22   │   │   │   │  if FP(p) and FP(q) intersect then
23   │   │   │   │   │  add the vertex that appears first in the clockwise order
        │   │   │   │   │     as the parent of the vertex that appears later;
24   │   │   │   │  end
25   │   │   │  end
26   │   │  end
27   │   │  Divide C equally among children of A;
28   │   │  When any cluster reaches a child of A, call explore;
29   │  else
30   │   │  Mark A as explored;
31   │   │  goto(C,parent(A));
32   │  end
33  end
```

Algorithm 1: Multi-Robot Exploration Subroutine

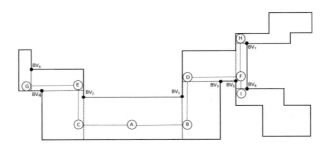

Fig. 2. An intermediate stage of exploration of a polygon by a team of 4 robots. Two robots located at the location G and the other two are located at F.

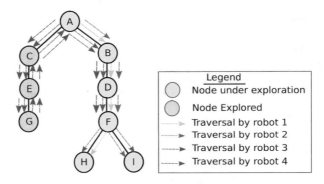

Fig. 3. The tree corresponding to the stage of exploration shown in Figure 2.

is marked as *explored*. The cluster moves to the parent of the current node to explore any of its other children which are *unexplored* or *under-exploration*. If a node does not have any children that are *unexplored* or *under-exploration*, then that node and its sub-tree is said to be *explored*. The exploration is said to be completed if the root of the tree is marked as *explored*.

Consider an example where P has been explored partially by a team of four robots as shown in Figure 2. All four robots $\{r_1, r_2, r_3, r_4\}$ start off at the location A which is added as the root of the tree as shown in Figure 3. Node A is marked as *under-exploration*. The robots observe two blocking vertices, BV_1 and BV_2. The extension goals B and C corresponding to BV_1 and BV_2, respectively, are added as the children of A in the tree. The robots split into two clusters $\{r_1, r_2\}$ and $\{r_3, r_4\}$. Cluster $\{r_1, r_2\}$ moves towards B and cluster $\{r_3, r_4\}$ moves towards C. The corresponding nodes in the tree are marked as *under-exploration*. At B, cluster $\{r_1, r_2\}$ observes a new blocking vertex, BV_3, and the corresponding extension goal D is added as the child of B. Since there is only one child, the cluster does not split and both robots move towards D. The corresponding node, D, in the tree is marked as *under-exploration*.

At C, the cluster $\{r_3, r_4\}$ observes a blocking vertex, BV_3, and the corresponding extension goal E is added as the child of C. Similarly, F is added as the child of D and G is added as the child of E. At F, cluster $\{r_1, r_2\}$ observes two blocking vertices, the cluster splits into two clusters $\{r_1\}$ and $\{r_2\}$. Cluster $\{r_1\}$ moves towards H and cluster $\{r_2\}$ moves towards I. At G, cluster $\{r_3, r_4\}$ observes that there are no more blocking vertices. Hence, no children are added to G. G is marked as *explored* and the algorithm checks the predecessor in the tree, E. Since E does not have any other children for exploration, E is marked as *explored* as well.

Similarly, the algorithm checks its predecessor, C, and marks it as *explored*. Now the algorithm checks A, it has a child B which is *under-exploration*. Since there is no blocking vertex, the algorithm proceeds to check B. Similarly, the algorithm proceeds to check D and subsequently F. At F, there are two blocking vertices since neither of the two clusters , $\{r_1\}$ or $\{r_2\}$, have reached their goals. Thus, cluster splits into two $\{r_3\}$ and $\{r_4\}$. H is assigned to $\{r_3\}$ and I is assigned to $\{r_4\}$. The exploration algorithm proceeds in this manner up until the root is marked as *explored*.

3.1   Competitive Ratio Analysis

In this section, we prove that the competitive ratio of our algorithm is bounded with respect to the offline optimal algorithm. We divide our analysis into three steps. First, we show that the paths followed by all the robots can be mapped to navigating on a tree. Next we bound the sum of the costs of edges in the the tree with respect to the offline optimal cost. Finally, we bound the cost of our algorithm with respect to the cost of the tree.

Lemma 1. *The graph created in the exploration algorithm is a tree.*

Proof. We add an edge between two nodes when a new *extension goal* is observed due to a blocking vertex visible from the current location of the robot. Suppose for contradiction that graph created by the robots during exploration is not a tree. Hence, there exists a cycle in the graph. Draw the closed polygon corresponding to this cyclic path inside P. This polygon contains at least one blocking vertex (in fact, all blocking vertices corresponding to the cycle) in its interior. Furthermore, the boundary of this polygon does not intersect with the boundary of P. Therefore, there is an obstacle (hole) in the interior of P. This contradicts with our assumption that P is a simple, simply-connected orthogonal polygon. Hence, this graph is in fact a tree.

When $p = 1$, the proposed algorithm is the same as the one given by Deng et al. [7] for orthogonal polygons without holes. They showed that the competitive ratio of this algorithm is $\sqrt{2}$. Let C_{OPT} denote the time taken by the optimal algorithm for a single robot to explore P. Let C_{RECT} denote the time taken by the strategy from [7] for a single robot to explore P. We have $C_{\text{RECT}} \leq \sqrt{2} C_{\text{OPT}}$ from Theorem 3 in [7] and the assumption that the robots travel with unit speeds. Let C_{OPT}^p be the time taken by the optimal p robot exploration algorithm, and C_{ALG} be the time taken by the proposed algorithm. Our goal is to show an upper bound for $C_{\text{ALG}}/C_{\text{OPT}}$. We will show this by relating both quantities with C_{TREE} which is the sum of the lengths of edges in the tree.

Lemma 2. $C_{TREE} \leq C_{RECT} \leq \sqrt{2}C_{OPT}$.

Proof. The inequality $C_{\text{RECT}} \leq \sqrt{2}C_{\text{OPT}}$ holds from Theorem 3 of [7] for simple rectilinear (orthogonal) polygons. The inequality $C_{\text{TREE}} \leq C_{\text{RECT}}$ holds because any robot will have to back track on its path to reach previously unexplored areas and return back to the root.

Lemma 3. *If p robots are used to explore P and C_{OPT}^p is the cost of the optimal offline algorithm, then $C_{TREE} \leq \sqrt{2}C_{OPT} \leq \sqrt{2}pC_{OPT}^p$.*

Proof. Given the optimal algorithm for p robots, one can construct a tour for a single robot that executes each of the p tours. The length of such a tour is upper bounded by pC_{OPT}^p. Since C_{OPT} is the optimal cost for a single robot's tour, we have $C_{\text{OPT}} \leq C_{\text{OPT}}^p$. The other inequality follows from the previous lemma.

Theorem 1. *If C_{ALG} is the cost of exploring the polygon using the proposed strategy and C_{OPT} is the cost of exploring the polygon using an optimal offline strategy, then we have,*

$$\frac{C_{ALG}}{C_{OPT}} \leq \frac{2(\sqrt{2}p + \log p)}{1 + \log p} \tag{1}$$

Proof. The cost of exploring a tree with a recursive depth-first strategy used in the proposed algorithms is given by:

$$C_{\text{ALG}} \leq \frac{2(C_{\text{TREE}} + d_{\max}\log p)}{1 + \log p}, \tag{2}$$

where $d_{\max}$ is the maximum distance of a leaf node from the root in the tree. This bound comes directly from the result in [21]. In our case, $d_{\max}$ corresponds to the maximum distance of any extension goal from the starting position of the robots. It is easy to see that $d_{\max} \leq C_{\text{OPT}}^p$.

From Lemma 3, we have:

$$C_{\text{ALG}} \leq \frac{2(\sqrt{2}pC_{\text{OPT}}^p + C_{\text{OPT}}^p\log p)}{1 + \log p}$$

which yields

$$\frac{C_{\text{ALG}}}{C_{\text{OPT}}} \leq \frac{2(\sqrt{2}p + \log p)}{1 + \log p} \tag{3}$$

Thus, we show that the competitive ratio is bounded as a function of p. Figure 4 shows a plot of this bound as a function of p. We note that while the analysis only holds for the case of an orthogonal polygon without holes, the resulting algorithm can also be applied for polygons with holes. However, in such a case, the underlying graph created by the robots is not guaranteed to be a tree. Consequently, we would have to apply a bound for exploring general graphs with multiple robots to yield a similar competitive ratio. In the simulations, we show the empirical performance of our algorithm in environments with holes.

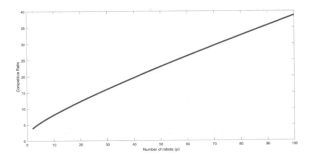

Fig. 4. Plot of the growth of competitive ratio as a function of the number of robots.

4 Adapting to Real-World Scenarios

Some of the assumptions made for the analysis do not hold in most practical scenarios. In this section, we show how to extend our basic algorithm framework in order to incorporate real-world constraints.

The main assumption is that of unlimited sensing range. In practice, the robot has a limited sensing range. For example, the robot cannot sense a long corridor with a single observation. Thus, in addition to blocking vertices, the robot has to move to *frontiers* at the end of its sensing range to sense more of the environment. This increases the distance the robot would have to cover compared to the distance it would have to cover if it had an infinite sensor. The robot thus has to detect two types of frontiers, one due to blocking vertices and the other due to sensing range. The only change to the algorithm is in Line 1 where we check for blocking vertices as well as frontiers due to sensing range.

We first detect blocking vertices in a given scan (as described below). Then frontier cells are clustered together to form frontiers due to sensing range. Any frontier which has a constituent frontier cell neighboring a blocking vertex is then discarded. For a blocking vertex, the *extension goal* is added on its *extension* with a slight offset. For frontiers due to sensing range, the middle frontier cell, after clustering, is chosen as the *frontier goal*.

Due to the sensing uncertainty, we represent the map built by the robots as a 2D occupancy grid (OG) as opposed to a geometric map. An OG is a discretized representation of the environment where each cell represents the probability of that space being occupied. Figure 5-right shows a representative OG. Cells with a lower probability of occupancy (< 0.5) are designated as free cells (represented as white in the OG) and cells with a higher probability of occupancy (> 0.5) are designated as occupied cells (represented as black in the OG). Cells which have not been observed are marked as unknown (represented as gray in the OG).

Typical sensors such as cameras and laser range finders have finite angular resolution. Consider three rays from the laser as shown in Figure 5-left. The rays intersect obstacles at the cells marked as black and all the cells the ray intersect between the robot (marked in blue) and the cell are marked as free. Due to the finite resolution of the laser ($0.395°$ for the hokuyo laser used in our system), the gray cells, even though they are in the field of the laser, are unobserved and this leads to gaps in observations. This leads to false frontiers being detected and hence such erroneous frontiers have to be discarded. We employ a simple heuristic of checking the size of a candidate frontier and discard those below a threshold.

Consider the robot (and the laser) located at the blue circle in Figure 5-right. The green ray represents one of the laser rays. In order to check for blocking vertices, occupied cells with a neighboring frontier cell are shortlisted first. In the figure, the cells marked with yellow and red 'X' are identified. A blocking vertex, as defined earlier, is a reflex vertex. We can detect a reflex vertex in an OG by checking its four neighbors. If the four neighbors are an occupied cell, an unknown cell, a free cell which a frontier, and a free cell which is not a frontier we mark it as a blocking vertex. In Figure 5-right, the cell marked with the yellow 'X' is detected as a blocking vertex.

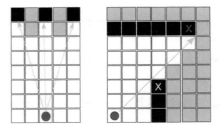

Fig. 5. (Left) Finite resolution of the laser leads to gaps in observations. **(Right)** A blocking vertex (marked with yellow 'X') has four distinctive neighbors: an occupied cell, an unknown cell, a free cell which is a frontier, one which is not.

Our algorithm also works for environments which are not orthogonal when occupancy grids are used as the underlying representation. Occupancy grids, typically, are orthogonal polygons by construction. Furthermore, for environments with holes (i.e., obstacles) the algorithm would create a tree and explore this tree. While this is correct, the distance traveled by the robots can be much higher than the optimal cost and as such the competitive ratio does not hold.

In the next section, we evaluate the empirical performance of our algorithm through simulations in such scenarios.

5   Simulations

We implemented our algorithm using ROS [32] and carried out simulations in Gazebo [33] in order to verify the correctness of the exploration algorithm in realistic environments. The five gazebo simulation environments used (Figure 6) are not all orthogonal and simply-connected – assumptions required for the analysis. The simulated robot is a differential-drive Pioneer P3-DX robot with a 2D Hokuyo laser range finder with a maximum range of 5 meters. The robot is localized in the environment using the *amcl* [34] package using a predefined map. Mapping during exploration is done using the octomap ROS package. The laser scan is converted into a point cloud which is then fed as input to the *octomap* [35] node. Our implementation is available online at `https://github.com/raaslab/Exploration`.

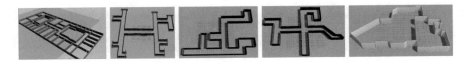

Fig. 6. Simulation environments 1 − 5 from left to right.

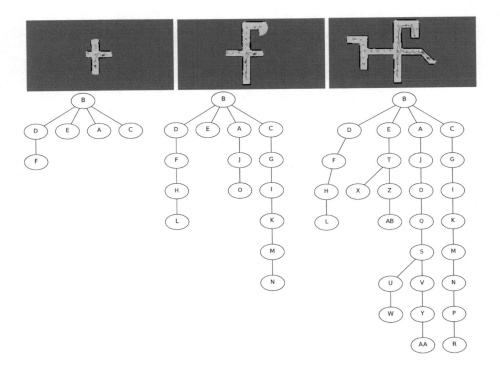

Fig. 7. Various stages while exploring environment 3 using four robots. The map explored and built along with the corresponding trees are shown.

Figure 7 shows various stages of exploration with four robots. The figure also shows the partial exploration tree built by the algorithm. The final trees produced while exploring all the environments is given in Figure 8. Table 1 shows the maximum distance traveled by a robot (in meters) during exploration for all the environments.

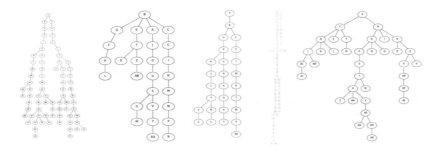

Fig. 8. Final tree produced after exploring the five environments in Figure 6 with four robots. The maximum distance traveled by a robot during exploration is given in Table 1.

	Env. 1	Env. 2	Env. 3	Env. 4	Env. 5
1 Robot	214.11	362.29	124.77	135.62	156.03
2 Robots	121.95	223.50	76.87	82.69	74.50
4 Robots	127.78	152.18	83.39	63.51	64.36

Table 1. Maximum distance traveled by a robot to explore environments in Fig. 6.

The cost of exploring environments 1, 3, 4, and 5 remains almost the same when the number of robots is increased from two to four. This is due to the fact that the exploration tree is not a balanced tree (Figure 8). On the other hand, in environment 2, the tree contains four or more *under-exploration* branches at all times. Consequently, the cost of exploration decreases significantly when four robots are used as opposed to just two.

6   Experiments

We carried out experiments using a Pioneer P3-DX robot mounted with a Hokuyo URG-04LX-UG01 2D laser and a Kinect2 RGBD sensor (Figure 9-left). The laser was configured to use 180° field of view and a resolution of 0.395°. During the exploration experiments, the robot used the 2D laser for localization using on a pre-built map. The map was generated using RTAB-map [36, 37] and the localization was carried out using the amcl package from ROS. Note that having a pre-built map is not a requirement for our algorithm. The amcl localization component could be replaced by, for example, any SLAM implementation. Furthermore, the robots generated a new map during the exploration process using octomapping [38] with the Kinect2 sensor. This map (and not the pre-built map) was used as the basis for finding blocking vertices in the proposed exploration algorithm.

The mapping and localization was performed on the pioneer P-3DX robot with two onboard computers in a master/slave configuration. The slave i7 Intel

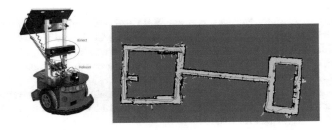

Fig. 9. (Left) Pioneer P3-DX with the Hokuyo laser used for the experiments. The 2D Hokuyo laser was used for robot localization and the Kinect2 sensor was used for mapping during exploration. **(Right)** Final map as well as the exploration path during the corridor experiment.

NUC was dedicated to processing the Kinect2 data and the master i7Intel NUC was dedicated to running the localization and exploration algorithm. Figure 9-right shows the results of an exploration experiment in a corridor environment along with the path followed by the robot. A video of the system in operation is available online at `https://github.com/raaslab/Exploration`

7   Conclusion

We presented an algorithm for exploring an unknown polygonal environment using a team of p robots in the least amount of time. Our main theoretical contribution was to show that if the underlying environment is an orthogonal polygon without holes then our algorithm yields a constant competitive ratio for fixed p. Next, we showed how to adapt our algorithm so that it can extend to real-world sensing constraints. We verified the behavior of our algorithm through simulations in five representative environments with up to four robots and experiments with a single robot. We are currently working on experiments with multiple robots.

Future work includes extending our analysis to more general environments. Handling general polygons without holes, not necessarily orthogonal, is a direct extension of the algorithm presented here. The notion of blocking vertices remains the same and the underlying graph will still be a tree. However, the *extension goal* corresponding to the blocking vertex needs to be carefully defined. An immediate avenue of future work is to leverage the algorithm from [7] that allows for obstacles in orthogonal environments. For polygons with holes, the underlying graph is no longer a tree. Hence, a general graph exploration algorithm would have to be used.

References

1. Rao, N.S., Kareti, S., Shi, W., Iyengar, S.S.: Robot navigation in unknown terrains: Introductory survey of non-heuristic algorithms. Technical report, Citeseer (1993)
2. Juliá, M., Gil, A., Reinoso, O.: A comparison of path planning strategies for autonomous exploration and mapping of unknown environments. Autonomous Robots **33**(4) (2012) 427–444
3. Bhattacharya, A., Ghosh, S.K., Sarkar, S.: Exploring an unknown polygonal environment with bounded visibility. In: International Conference on Computational Science, Springer (2001) 640–648
4. Charrow, B., Kahn, G., Patil, S., Liu, S., Goldberg, K., Abbeel, P., Michael, N., Kumar, V.: Information-theoretic planning with trajectory optimization for dense 3d mapping. In: Proceedings of Robotics: Science and Systems. (2015)
5. Yamauchi, B.: A frontier-based approach for autonomous exploration. In: Computational Intelligence in Robotics and Automation, 1997. CIRA'97., Proceedings., 1997 IEEE International Symposium on, IEEE (1997) 146–151
6. Motwani, R., Raghavan, P.: Randomized algorithms. Chapman & Hall/CRC (2010)
7. Deng, X., Kameda, T., Papadimitriou, C.: How to learn an unknown environment. i: the rectilinear case. Journal of the ACM (JACM) **45**(2) (1998) 215–245
8. O'Rourke, J.: Art gallery theorems and algorithms. Oxford University Press Oxford (1987)
9. Chin, W.p., Ntafos, S.: Optimum watchman routes. Information Processing Letters **28**(1) (1988) 39–44
10. Fleischer, R., Kamphans, T., Klein, R., Langetepe, E., Trippen, G.: Competitive online approximation of the optimal search ratio. SIAM Journal on Computing **38**(3) (2008) 881–898
11. Papadimitriou, C.H., Yannakakis, M.: Shortest paths without a map. Theoretical Computer Science **84**(1) (1991) 127–150
12. Chvatal, V.: A combinatorial theorem in plane geometry. Journal of Combinatorial Theory, Series B **18**(1) (1975) 39–41
13. Fisk, S.: A short proof of chvátal's watchman theorem. Journal of Combinatorial Theory, Series B **24**(3) (1978) 374
14. Schuchardt, D., Hecker, H.D.: Two np-hard art-gallery problems for ortho-polygons. Mathematical Logic Quarterly **41**(2) (1995) 261–267
15. Ghosh, S.K., Burdick, J.W., Bhattacharya, A., Sarkar, S.: Online algorithms with discrete visibility-exploring unknown polygonal environments. IEEE robotics & automation magazine **15**(2) (2008) 67–76
16. Hoffmann, F., Icking, C., Klein, R., Kriegel, K.: The polygon exploration problem. SIAM Journal on Computing **31**(2) (2001) 577–600
17. Mitchell, J.S.: Guillotine subdivisions approximate polygonal subdivisions: A simple polynomial-time approximation scheme for geometric tsp, k-mst, and related problems. SIAM Journal on Computing **28**(4) (1999) 1298–1309
18. Arora, S.: Polynomial time approximation schemes for euclidean traveling salesman and other geometric problems. Journal of the ACM (JACM) **45**(5) (1998) 753–782
19. Fraigniaud, P., Gasieniec, L., Kowalski, D.R., Pelc, A.: Collective tree exploration. Networks **48**(3) (2006) 166–177
20. Dynia, M., Lopuszański, J., Schindelhauer, C.: Why robots need maps. In: International Colloquium on Structural Information and Communication Complexity, Springer (2007) 41–50

21. Higashikawa, Y., Katoh, N., Langerman, S., Tanigawa, S.i.: Online graph exploration algorithms for cycles and trees by multiple searchers. Journal of Combinatorial Optimization **28**(2) (2014) 480–495
22. Dynia, M., Kutyłowski, J., auf der Heide, F.M., Schindelhauer, C.: Smart robot teams exploring sparse trees. In: International Symposium on Mathematical Foundations of Computer Science, Springer (2006) 327–338
23. Ortolf, C., Schindelhauer, C.: Online multi-robot exploration of grid graphs with rectangular obstacles. In: Proceedings of the twenty-fourth annual ACM symposium on Parallelism in algorithms and architectures, ACM (2012) 27–36
24. Burgard, W., Moors, M., Stachniss, C., Schneider, F.E.: Coordinated multi-robot exploration. IEEE Transactions on robotics **21**(3) (2005) 376–386
25. Schwager, M., Dames, P., Rus, D., Kumar, V.: A multi-robot control policy for information gathering in the presence of unknown hazards. In: Proceedings of International Symposium on Robotics Research, Aug. (2011)
26. Cesare, K., Skeele, R., Yoo, S.H., Zhang, Y., Hollinger, G.: Multi-uav exploration with limited communication and battery. In: 2015 IEEE International Conference on Robotics and Automation (ICRA), IEEE (2015) 2230–2235
27. Holz, D., Basilico, N., Amigoni, F., Behnke, S.: A comparative evaluation of exploration strategies and heuristics to improve them. In: ECMR. (2011) 25–30
28. Whaite, P., Ferrie, F.P.: Autonomous exploration: Driven by uncertainty. IEEE Transactions on Pattern Analysis and Machine Intelligence **19**(3) (1997) 193–205
29. Moorehead, S.J., Simmons, R., Whittaker, W.L.: Autonomous exploration using multiple sources of information. In: Robotics and Automation, 2001. Proceedings 2001 ICRA. IEEE International Conference on. Volume 3., IEEE (2001) 3098–3103
30. Julian, B.J., Karaman, S., Rus, D.: On mutual information-based control of range sensing robots for mapping applications. The International Journal of Robotics Research (2014) 0278364914526288
31. Amigoni, F., Caglioti, V.: An information-based exploration strategy for environment mapping with mobile robots. Robotics and Autonomous Systems **58**(5) (2010) 684–699
32. Quigley, M., Gerkey, B., Conley, K., Faust, J., Foote, T., Leibs, J., Berger, E., Wheeler, R., Ng, A.: ROS: an open-source Robot Operating System. In: ICRA Workshop on Open Source Software. (2009)
33. Koenig, N., Howard, A.: Design and use paradigms for gazebo, an open-source multi-robot simulator. In: Intelligent Robots and Systems, 2004.(IROS 2004). Proceedings. 2004 IEEE/RSJ International Conference on. Volume 3., IEEE (2004) 2149–2154
34. : AMCL ROS Packagel. http://wiki.ros.org/amcl Accessed: 2016-10-30.
35. Hornung, A., Wurm, K.M., Bennewitz, M., Stachniss, C., Burgard, W.: Octomap: An efficient probabilistic 3d mapping framework based on octrees. Autonomous Robots **34**(3) (2013) 189–206
36. Labbé, M., Michaud, F.: Online global loop closure detection for large-scale multi-session graph-based slam. In: 2014 IEEE/RSJ International Conference on Intelligent Robots and Systems, IEEE (2014) 2661–2666
37. Labbe, M., Michaud, F.: Appearance-based loop closure detection for online large-scale and long-term operation. IEEE Transactions on Robotics **29**(3) (2013) 734–745
38. Hornung, A., Wurm, K.M., Bennewitz, M., Stachniss, C., Burgard, W.: Octomap: An efficient probabilistic 3d mapping framework based on octrees. Autonomous Robots **34**(3) (2013) 189–206

Assignment Algorithms for Variable Robot Formations

Srinivas Akella

Department of Computer Science
University of North Carolina at Charlotte
Charlotte, NC 28223, USA

Abstract. This paper describes algorithms to perform optimal assignment of teams of robots translating in the plane from an initial formation to a variable goal formation. We consider the case when each robot is to be assigned a goal position, the individual robots are interchangeable, and the goal formation can be scaled or translated. We compute the costs for all candidate pairs of initial, goal robot assignments as functions of the parameters of the goal formation, and partition the parameter space into equivalence classes invariant to the cost order using computational geometry techniques. We compute a minimum completion time assignment for an equivalence class by formulating it as a linear bottleneck assignment problem (LBAP). To improve efficiency, we solve the LBAP problem for each equivalence class by incrementally updating the solution as the formation parameters are varied. This work is motivated by applications that include the motion of droplet formations in digital microfluidic lab-on-a-chip devices, and of robot and drone formations in the plane.

Keywords: Robot formations, linear bottleneck assignment, multiple robots

1   Introduction

The assignment and motion of teams of robots from one formation to another is a problem that arises in multiple applications ranging from robot and drone formation planning ([13, 30, 31, 34]) to the motion of droplet formations in digital microfluidic lab-on-a-chip devices ([19]). Most previous work on multiple robot assignment has either assumed fixed initial and goal formations when computing robot assignments, or has assumed a fixed assignment while varying the scale, location, or orientation of the goal formations.

In this paper, we introduce the problem of performing robot assignment while simultaneously considering variable goal formations that can be either scaled or translated. Since energy expenditure and completion time are important for many tasks, we wish to miminize the maximum distance traveled or the maximum time taken by any of the robots. We introduce the scaled goal formation problem and the translated goal formation problem, where the goal

© Springer Nature Switzerland AG 2020
K. Goldberg et al. (Eds.): *Algorithmic Foundations of Robotics XII*, SPAR 13, pp. 912–927, 2020.
https://doi.org/10.1007/978-3-030-43089-4_58

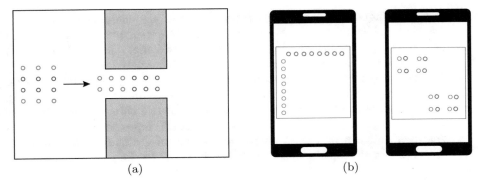

Fig. 1. Example motivating applications. (a) A robot formation on the left that has to change its shape to pass through the passage between two obstacles. Each circle represents a robot. (b) Schematic figure showing overhead view of an LADM chip (indicated by a square) on a smartphone, with circles indicating chemical droplets. (Left) Initial formation of droplets. (Right) Goal formation of droplets for mixing. LADM chip and droplets are shown enlarged and chemicals are color coded for clarity.

formation may be scaled or the goal formation may be translated in a region respectively. Given an initial formation of the robots and a specified shape formation, we wish to determine the assignment of each robot in the initial formation to a configuration in the variable goal formation that minimizes the maximum distance traveled by the robots. We formulate these assignment problems with variable goal formations as linear bottleneck assignment problems (LBAP) with geometric constraints where the size or location of the goal formation, which influences the LBAP costs, must be computed. Our approach exploits the relative cost ordering property of LBAPs and the geometric structure captured by arrangements of cost curves and surfaces. We thus partition the goal formation parameter space into equivalence classes invariant to the cost order. This enables a comprehensive evaluation to identify the globally optimum solution.

Motivating applications: The multi-robot assignment problem for variable goal formations is motivated by applications including the following examples.

1. Multiple-robot assignment: The assignment and planning of teams of robots, drones, and spacecraft often requires motion between formations. Example applications include search and rescue operations, package delivery, construction, surveillance, and sensor network monitoring [30, 31, 33, 9, 3]. These tasks can be subject to spatial constraints (e.g., the team of robots must squeeze through a passage while maintaining a minimum separation distance, as in Figure 1(a)) or communication range constraints (e.g., a team of UAVs must stay within a maximum distance of team members), which can be represented by the allowable scaling and/or translation of the goal formation.

2. Digital microfluidic lab-on-chip systems: Low-cost, portable lab-on-a-chip systems can impact a wide variety of applications including point-of-care

medical diagnostics. Recent hardware advances are enabling lab-on-a-chip devices that can be optically actuated using smartphone and tablet LCD screens (e.g., [23], [22], [25]). In such *light-actuated digital microfluidic* (LADM) chips, discrete droplets of chemicals are optically actuated using moving patterns of projected light to perform chemical reactions by repeatedly moving and mixing droplets (Figure 1(b)). A key issue is the automated planning and coordination of droplet formations on the LADM chip, which we address by modeling the droplets as robots. An important step is computing goal formations that can fit within a specified region on the chip, and identifying assignments of droplets that can efficiently travel to the goal formation.

2   Background: Assignment Problems

Assignment problems deal with how to assign n items (robots, tasks) to n other items (locations, resources). Consider a problem where there are n tasks, indexed by i, that must be assigned to n resources, indexed by j. The cost of performing task i on resource j is c_{ij}. The cost matrix C consists of entries c_{ij} corresponding to the cost of assigning task i to resource j. x_{ij} is a binary variable that is 1 if task i is assigned to resource j, and 0 otherwise.

In the linear sum assignment problem (LSAP) [7], which is the standard form of the assignment problem, the objective function to be minimized is the sum of the assigned costs $\sum_{1 \leq i,j \leq n} c_{ij} x_{ij}$. LSAP is the dominant form of assignment that has been considered in the robotics literature (for example, [30, 17]), and can be solved using the Hungarian algorithm [21, 7].

2.1   Linear Bottleneck Assignment Problem

Since we want to minimize the maximum completion time or the maximum distance for any robot, we consider the linear bottleneck assignment problem [7]. The linear bottleneck assignment problem (LBAP) formulation is:

$\min_{1 \leq i,j \leq n} \max c_{ij} x_{ij}$

$\sum_{j=1}^{n} x_{ij} = 1 \ \forall i = 1, \ldots, n$

$\sum_{i=1}^{n} x_{ij} = 1 \ \forall j = 1, \ldots, n$

$x_{ij} \in \{0, 1\} \ \forall i, j = 1, \ldots, n$

LBAPs occur in connection with assigning jobs to parallel machines to minimize the latest completion time. The LBAP minimizes the maximum cost of completing any task, whereas the LSAP minimizes the total cost of completing all the tasks. When the objective function is the time taken or distance traveled, the LBAP minimizes the maximum completion time or maximum distance traveled by a robot. We use the *threshold algorithm* [7], outlined in Section 4, for solving LBAPs.

3   Finding Optimal Assignments and Scaled Goal Formations: An LBAP Formulation

We consider the problem of simultaneously computing both the optimal robot assignment and scaled goal formation while minimizing the maximum robot travel distance. The *Scaled Goal Formation problem* is: Given n robots in an initial formation $P = \{p_i\}$ where $p_i \in \mathbb{R}^2$ is the initial configuration of robot i, and a specified shape formation $S = \{s_j\}$, where $s_j \in \mathbb{R}^2$, assign the robots to configurations in the goal formation $Q = \{q_j\}$, where $q_j \in \mathbb{R}^2$ is the jth scaled goal configuration, such that the maximum distance traveled by any robot is minimized and Q is equivalent up to a scale factor $\alpha \in \mathbb{R}_+$ of S. Here $q_j = \alpha s_j + d_0$, where $d_0 \in \mathbb{R}^2$ is a user specified translation. We assume that only scale is varied and that all robots move in straight lines to their goals with an equal and constant speed.

We formulate this as a linear bottleneck assignment problem (LBAP) [7]. For a given scale value α, the cost c_{ij} of assigning robot i in the initial formation to a location j in the goal formation is the distance from p_i to q_j, yielding cost matrix $C = [c_{ij}]$. x_{ij} is a binary variable that is 1 if robot i in the initial formation is assigned to robot j in the goal formation, and 0 otherwise.

The *scaled goal formation LBAP formulation* is:

$$\min \max c_{ij} x_{ij}$$
$$\sum_{j=1}^{n} x_{ij} = 1 \ \forall i = 1, \ldots, n$$
$$\sum_{i=1}^{n} x_{ij} = 1 \ \forall j = 1, \ldots, n$$
$$x_{ij} \in \{0, 1\} \ \forall i, j = 1, \ldots, n$$
where cost c_{ij} is a function of scale α.
$$c_{ij} = \|p_i - q_j\|_2 = [a_i + b_{ij}\alpha + e_j\alpha^2]^{1/2} \text{ where}$$
$$a_i = (p_{ix} - d_{0x})^2 + (p_{iy} - d_{0y})^2, \ b_{ij} = -2((p_{ix} - d_{0x})s_{jx} + (p_{iy} - d_{0y})s_{jy}),$$
and $e_j = s_{jx}^2 + s_{jy}^2$.

We must find the value of α that minimizes the objective function. An interesting property of LBAP is that the optimal assignment depends only on the relative order of the costs c_{ij} and not on their actual values (Lemma 6.1 [7]). The key idea is to exploit this property and identify intervals of α over which the relative order is invariant. Over each such interval, the value of the optimal LBAP assignment is determined by the cost of one particular initial and goal position combination. We call this the *critical cost*. The value of this cost varies as α is varied over the interval, but the cost element that determines the value of the optimal assignment does not change.

The cost c_{ij} of each assignment is a function of α. So we have a set of curves c_{ij} and the relative order of the costs changes only at the intersection points of the curves. See Figure 2. If we can compute all the intersection points as α is varied, for each interval we can compute the LBAP assignment at a specified value (say at the midpoint of the interval α_{mid}) and the associated critical cost.

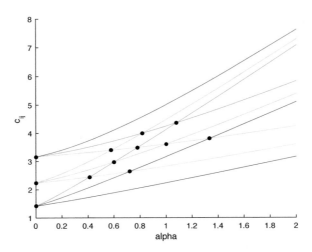

Fig. 2. Cost curves, showing cost c_{ij} as a function of scale factor α, with curve intersections indicated by dark points. There is an invariant cost order between every pair of consecutive intersections.

By minimizing the critical cost in the interval, we find the α value in the interval that gives the optimal assignment and cost for the interval. We compare the optimum costs over all intervals to find the globally optimal assignment and cost. See the example in Figure 3.

Algorithm for scaled goal formation LBAP:

1. From the given initial formation P and specified shape formation S, and range of permitted α values, compute the c_{ij} cost curves.
2. Compute all intersections of the cost curves by a line sweep algorithm. For each interval, compute the ordering of the costs using α_{mid}, the alpha value at the midpoint of the interval. See Figure 2.
3. Compute the cost matrix at α_{mid} in the first α interval, and compute an LBAP solution for this cost matrix.
4. Compute the optimal LBAP solution for each α interval using the optimal LBAP solution from the previous interval (to avoid solving the problem from scratch). Using the known optimal assignment for the previous cost ordering, the new optimal assignment when two consecutive elements of the ordering swap places can be efficiently computed (as will be described in Section 4).
5. From the computed LBAP solution, identify the critical cost c_{ij}, the c_{ij} that determines the objective function value, and optimal assignment for the interval. Then compute the minimum value of the critical cost c_{ij} over the interval to obtain the optimal objective value and corresponding optimal alpha value α_{opt}.
6. Compare the LBAP solutions for all intervals at their respective α_{opt}, and select the minimum among them as the global optimal assignment and opti-

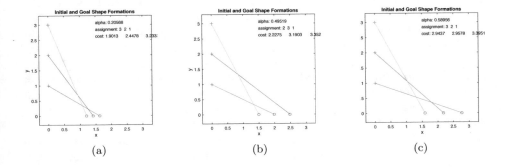

Fig. 3. A three robot example using the cost curves of Figure 2. (a) The optimal cost assignment, at $\alpha = 0.2057$. (c) An example assignment for $\alpha = 0.49519$. (d) A different assignment for $\alpha = 0.58956$. The start positions are indicated by plus symbols, and the goal positions are indicated by circles.

mal scale factor. (This global optimal solution can be updated as the optimal assignment for each interval is computed.)

There are $O(n^2)$ costs c_{ij}. Finding all pairwise intersections of c_{ij} curves by the sweep line algorithm will yield $O(n^4)$ intersections in $O(n^4 \log n)$ time [11]. There are $O(n^4)$ intervals, and explicitly writing out the cost order for an interval takes $O(n^2)$ time. As we will discuss in Section 4, solving a new instance of the LBAP problem takes $O(n^{2.5}/\sqrt{\log n})$ time [7] while using a solution to a closely related LBAP takes $O(n^2)$ time. So the overall running time is $O(n^6)$.

4   Incremental Updates to Compute the Optimal LBAP

We now present an approach to update the optimal solution to an LBAP from the optimal solution of a closely related LBAP. We use the *threshold algorithm* [7] for solving an LBAP instance. The threshold algorithm alternates between two phases. In the first phase, a cost element c^*, the threshold value, is chosen and a threshold matrix $\overline{C}$ is defined by $\overline{c}_{ij} = 1$ if $c_{ij} > c^*$ and 0 otherwise. In the second phase, we check whether there exists an assignment with total cost 0 for the cost matrix $\overline{C}$. For this, we construct a bipartite graph $\overline{G} = (U, V; E)$ with $|U| = |V| = n$ and edges $[i, j] \in E$ if and only if $\overline{c}_{ij} = 0$. In other words, we check whether the bipartite graph with threshold matrix $\overline{C}$ contains a perfect matching or not. A perfect matching exists for a bipartite graph G with n vertices if there exists a set of n edges in G such that each vertex is incident on exactly one edge. Let the *threshold graph* $\overline{G}(c)$ be the bipartite graph such that an edge $e_{ij} = [i, j]$ exists if and only if $c_{ij} \leq c$; the edge has weight 0. The edge e_{ij} has a cost c_{ij} associated with it in the cost matrix C. The smallest value c^* for which the threshold graph $\overline{G}(c^*)$ contains a perfect matching is the optimal value of the LBAP; this c^* is the *critical cost*. For the case of dense graphs, there is a version of the threshold algorithm with a total time complexity of $O(n^{2.5}/\sqrt{\log n})$ [7].

To explain and analyze our incremental LBAP update algorithm, we use the LBAP property that the optimal solution of an LBAP depends only on the relative order of the cost coefficients and not on their numerical value [7]. This enables us to simplify the cost structure of the cost matrix so it consists of consecutive integers starting at 0. In the remainder of this section, c_{ij} will be a representative integer cost (and not the cost from the actual matrix C). Now the questions are: If two entries in the cost matrix swap their cost values, is there an efficient way of telling whether the optimal assignment has changed? If the optimal assignment has changed, is there an efficient way of computing it from the previously computed optimal assignment?

When two entries in the cost matrix swap their cost values due to the intersection of their cost curves, they are represented by consecutive integers (we assume here that the costs are all distinct). Let the two edges that are swapping costs be e_{ij} and e_{kl} with costs c_{ij} and c_{kl} respectively before the swap (where c_{ij} and c_{kl} are consecutive integers). Then edge e_{ij} will have cost c_{kl} and e_{kl} will have cost c_{ij} after the swap. Let the optimum objective value of the LBAP prior to the swap be c_{opt}. Therefore there is a perfect matching in the threshold graph $\overline{G}(c_{opt})$. Using the threshold algorithm for solving the LBAP after the swap, the four possible cases that can occur are:

1. If $c_{opt} > \max(c_{ij}, c_{kl})$, then both edges e_{ij} and e_{kl} are present in the threshold graph $\overline{G}(c_{opt})$ before the swap. The threshold graph after the swap $\overline{G}_{swap}(c_{opt})$ will be unchanged because the only edges affected are e_{ij} and e_{kl}, and they will both be present. Therefore the optimum value and the perfect matching will be unchanged after the swap.

2. If $c_{opt} < \min(c_{ij}, c_{kl})$, then both edges e_{ij} and e_{kl} are not present in the threshold graph $\overline{G}(c_{opt})$ before the swap. The threshold graph after the swap $\overline{G}_{swap}(c_{opt})$ will be unchanged because the only edges affected are e_{ij} and e_{kl}, and they will both still be absent. Therefore the optimum value and the perfect matching will be unchanged after the swap.

3. If $c_{opt} = \max(c_{ij}, c_{kl})$, assume without loss of generality that $c_{opt} = c_{ij}$, that is, $c_{ij} > c_{kl}$ before the swap. Therefore there is a perfect matching in the threshold graph $\overline{G}(c_{ij})$. We have two cases to consider since the LBAP optimum after the swap must be greater than or equal to c_{kl}[1].

   (a) If we consider the threshold graph $\overline{G}_{swap}(c_{ij})$ after the swap of c_{ij} and c_{kl}, it will be identical to $\overline{G}(c_{ij})$ before the swap. Therefore a perfect matching is present, and the objective function value c_{ij} and the matching will be unchanged after the swap.

---

[1] Proof that optimum value of the LBAP after the swap, c_{opt}^{swap}, cannot be less than c_{kl}: Assume the LBAP after the swap has an optimum value $c_{opt}^{swap} = c_* < c_{kl} < c_{ij}$. This implies that $\overline{G}_{swap}(c_*)$ has a perfect matching without edges e_{ij} and e_{kl}. Since $\overline{G}_{swap}(c_*) = \overline{G}(c_*)$, this implies the optimum LBAP value before the swap is c_*, which leads to a contradiction.

(b) The optimum value after the swap may decrease at most to c_{kl}. So we need to additionally evaluate $\overline{G}_{swap}(c_{kl})$, which is a subgraph of $\overline{G}(c_{ij})$. Here edge e_{ij} is present, but edge e_{kl} is not present. A new matching must be computed, and if the graph $\overline{G}_{swap}(c_{kl})$ contains a perfect matching, the optimum objective will be c_{kl} ($= c_{ij} - 1$). We can avoid recomputing the matching from scratch in this graph $\overline{G}_{swap}(c_{kl})$ by using the matching in $\overline{G}(c_{ij})$ before the swap to efficiently compute the new matching. To do this, we use the standard reduction of the maximum cardinality matching problem to the maximum flow problem [21], and use a procedure linear in the size of the graph to compute the new maximum flow for $\overline{G}_{swap}(c_{kl})$ with one edge (e_{kl}) whose capacity has decreased to 0 from 1 in $\overline{G}(c_{ij})$, for which we have the maximum flow from its perfect matching. If this corresponds to a perfect matching, it represents the optimum value of the LBAP.

4. If $c_{opt} = \min(c_{ij}, c_{kl})$, assume without loss of generality that $c_{opt} = c_{kl}$, that is, $c_{kl} < c_{ij}$ before the swap. Therefore there is a perfect matching in the threshold graph $\overline{G}(c_{kl})$. We have two cases to consider.

(a) If we consider the threshold graph $\overline{G}_{swap}(c_{ij})$ after the swap, it will include the edge e_{ij} in addition to all edges in $\overline{G}(c_{kl})$ before the swap. Therefore a perfect matching is present, with an optimum value of c_{ij}.

(b) We need to additionally evaluate $\overline{G}_{swap}(c_{kl})$. Here edge e_{ij} is present, but edge e_{kl} is not present. A new matching must be computed, and if this graph contains a perfect matching, the optimum objective will be c_{kl}. We can use the matching before the swap to efficiently compute the new matching. We use the reduction from the maximum cardinality matching problem to the maximum flow problem, and use a procedure linear in the size of the graph to compute the new maximum flow. To do this, we use the perfect matching from $\overline{G}(c_{kl})$ in $\overline{G}_{swap}(c_{ij})$, and then reduce the capacity of edge e_{kl} from 1 to 0 in the flow problem corresponding to $\overline{G}_{swap}(c_{kl})$. If this corresponds to a perfect matching, it represents the optimum value of the LBAP.

The threshold graph $\overline{G}$ has $2n$ vertices and up to n^2 edges, where n is the number of robots. Computing the new maximum flow takes time linear in the size of this graph, and so takes $O(n^2)$ time. Therefore computing the optimal LBAP by an incremental update takes $O(n^2)$ time.

5   Finding Optimal Translated Goal Formations

Selecting an optimal location for the goal formation Q can minimize the maximum distance traveled by any robot (or maximum completion time). We therefore introduce the *Translated Goal Formation problem*: Given an initial formation P with n robots and a specified shape formation S, assign the robots to configurations in the goal formation Q such that the maximum distance traveled by any robot is minimized and Q is equivalent to S up to translation d, which is to be computed. As before, $P = \{p_i\}$, $S = \{s_j\}$, $Q = \{q_j\}$, and $p_i, s_j, q_j, d \in \mathbb{R}^2$. Here we assume that the scale is fixed, and that the origin of the shape formation can be translated by $d = (d_x, d_y)$ in the plane. So $q_j = s_j + d$, where d is measured with respect to the same reference frame as P.

We formulate this problem as an LBAP. We now need to compute the relative order of the costs c_{ij} as functions of two variables: d_x, d_y. Our goal is to partition the $d_x d_y$-plane into cells with invariant cost orderings. We can then apply the algorithm of Section 3, appropriately modified, to identify the optimal (d_x, d_y) value.

The *translated goal formation LBAP formulation* is:

min max $c_{ij} x_{ij}$

$\sum_{j=1}^{n} x_{ij} = 1 \ \forall i = 1, \ldots, n$

$\sum_{i=1}^{n} x_{ij} = 1 \ \forall j = 1, \ldots, n$

$x_{ij} \in \{0, 1\} \ \forall i, j = 1, \ldots, n$

where cost c_{ij} can be written as a function of translation (d_x, d_y).

$c_{ij} = ||p_i - q_j||_2 = [(p_{ix} - (s_{jx} + d_x))^2 + (p_{iy} - (s_{jy} + d_y))^2]^{1/2}$
$= [(d_x - (p_{ix} - s_{jx}))^2 + (d_y - (p_{iy} - s_{jy}))^2]^{1/2}$

Each $c_{ij}(d_x, d_y)$ describes a surface, and the obvious approach is to determine the intersections of all the surfaces for all pairs i, j, and use this to determine the d_x, d_y regions over which the cost ordering is invariant. We instead characterize the d_x, d_y regions with invariant cost order using a different approach. It will be helpful to first introduce new variables $r_{ij} = p_i - s_j$ and rewrite the cost equation as: $c_{ij} = [(d_x - r_{ijx})^2 + (d_y - r_{ijy})^2]^{1/2}$.

Each c_{ij} can now be viewed as describing the distance of the point (d_x, d_y) from a fixed site r_{ij}. We have a set $R = \{r_{ij}\}$ of n^2 sites in the plane, and we use the reformulated costs to compute the regions of $d_x d_y$-space that are invariant in their cost ordering. For this, we use some beautiful connections between Voronoi diagrams, unit paraboloids, and arrangements from computational geometry [6]. Consider the unit paraboloid $z = d_x^2 + d_y^2$. For each site r_{ij}, lift it by projecting it vertically upwards to the paraboloid, to the point $(r_{ijx}, r_{ijy}, r_{ijz})$ where $r_{ijz} = r_{ijx}^2 + r_{ijy}^2$. Next construct the plane H_{ij} tangent to the paraboloid at the point $(r_{ijx}, r_{ijy}, r_{ijz})$. The equation of this plane is $z = 2r_{ijx}d_x + 2r_{ijy}d_y - (r_{ijx}^2 + r_{ijy}^2)$. Given any point (d_x, d_y), the upward vertical ray from it intersects the set of planes $H = \{H_{ij}\}$. The order in which the ray intersects these planes corresponds to their decreasing cost order, where the cost associated with a plane H_{ij} is the distance of its site r_{ij} to the point d. The highest plane identifies the site closest to the point[2], and the lowest plane identifies the site farthest from the point,

---

[2] To understand the connection between the height ordering of the planes and the point's distances to the sites, consider two planes H_{ij} and H_{kl}. Let the equation of plane H_{ij} be $z_{ij} = 2r_{ijx}d_x + 2r_{ijy}d_y - (r_{ijx}^2 + r_{ijy}^2)$ and the equation of plane H_{kl} be $z_{kl} = 2r_{klx}d_x + 2r_{kly}d_y - (r_{klx}^2 + r_{kly}^2)$. When plane H_{ij} is higher than plane H_{kl}, $z_{ij} > z_{kl}$, which implies that $2r_{ijx}d_x + 2r_{ijy}d_y - (r_{ijx}^2 + r_{ijy}^2) > 2r_{klx}d_x +$

and the planes from highest to lowest are ordered in increasing order of distance of their sites to the point (d_x, d_y).

An *arrangement* $A(H)$ induced by the set of planes H is the convex subdivision of space defined by the set of planes H [6]. A *cell* lies at *depth* i if there are exactly i planes above the cell. *Level* i refers to the boundary of the union of cells at depths zero, one, up to $i - 1$. Here a level is a piecewise-linear surface formed by pieces of the planes H_{ij}.

The arrangement $A(H)$ of the planes H provides useful geometric structure. The projection of the upper envelope of these planes (i.e., level 1 of $A(H)$) onto the $d_x d_y$-plane gives the Voronoi diagram of the set of sites R [6]. This is in fact the order-1 Voronoi diagram, where each cell contains the points closest to one site. More generally, we can partition the space according to the k closest sites of N sites, for some $1 \leq k \leq N - 1$. The resulting diagrams are called higher-order Voronoi diagrams, and for a given k, the diagram is called the order-k Voronoi diagram. An order-2 Voronoi diagram is one where each cell contains the points closest to an unordered pair of sites. This order-2 Voronoi diagram can be computed by projecting level 2 of $A(H)$. We can similarly compute the order-3 through order-$(N - 1)$ Voronoi diagrams. Conceptually, the overlay of these order-1 through order-$(N - 1)$ Voronoi diagrams on the $d_x d_y$-plane partitions the plane into convex cells, and each cell has an invariant ordering of sites based on the distances of points in the cell to the sites. In other words, within each cell, the cost order is invariant.

There are $N = n^2$ sites in R, which implies $O(n^2)$ planes that intersect at $O(n^4)$ intersection curves. When projected onto the $d_x d_y$-plane, we obtain a planar arrangement with at most $O(n^8)$ faces, edges, and vertices [11, 10]. This arrangement can be computed in $O(n^8)$ time, and the cost order for each cell can be enumerated in $O(n^2)$ time.

At each cell of the planar arrangement, we compute an LBAP at an interior point to identify the critical cost, say c_{ij}. Since each cell is a convex polygon, we can compute the optimal d_x, d_y values that minimize the LBAP by formulating a convex quadratic program (QP) that minimizes c_{ij}^2 given the linear boundary constraints of the cell. The QP formulation is:

$$\text{minimize } c_{ij}^2 = (d_x - r_{ijx})^2 + (d_y - r_{ijy})^2 \text{ subject to}$$
$$2(r_{vwx} - r_{tux})d_x + 2(r_{vwy} - r_{tuy})d_y + (r_{tux}^2 + r_{tuy}^2) - (r_{vwx}^2 + r_{vwy}^2) \leq 0 \ \forall r_{tu}, r_{vw}$$
$$l_x \leq d_x \leq u_x$$
$$l_y \leq d_y \leq u_y$$

where sites $r_{tu}, r_{vw} \in R$ correspond to consecutive distances in the cost order in the cell, and l_x, l_y and u_x, u_y are lower and upper bounds for d_x and d_y.

Since this is a convex QP in just two variables, it can be solved in time that is linear in the number of edges of the cell. Each cell has an invariant cost ordering

---

$2r_{kly}d_y - (r_{klx}^2 + r_{kly}^2)$. Adding $d_x^2 + d_y^2$ to both sides and rearranging terms shows that $(d_x - r_{klx})^2 + (d_x - r_{kly})^2 > (d_x - r_{ijx})^2 + (d_x - r_{ijy})^2$, thus establishing that point (d_x, d_y) is closer to site r_{ij} than site r_{kl}.

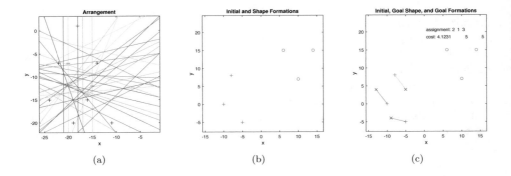

Fig. 4. A translation LBAP example for 3 robots. (a) Arrangement that results from overlay of order-1 through order-$(N-1)$ Voronoi diagrams, for $N = 9$ sites indicated by black plus symbols. (b) Initial formation and specified shape formation. (c) Initial formation, shape formation, and optimal goal formation, with the lines indicating the optimal assignment. The initial positions are indicated by red plus symbols, the shape positions are indicated by blue circles, and the goal positions are indicated by black x symbols.

of N sites, with each edge corresponding to a bisector of two sites that defines the relative ordering of a successive pair of distance costs. At each cell edge, we can interchange only consecutive cost elements. Since there is a maximum of $N - 1$ swaps between consecutive elements, this implies a maximum of $N - 1$ edges for each cell. A cell has less than $N - 1$ edges if it is completely on one side of a bisector corresponding to two of its consecutive distances (i.e., it is completely contained in the corresponding halfplane). Therefore the QP can be solved in $O(N)$ time, that is, $O(n^2)$ time.

We find the globally optimal solution by comparing the optimal LBAP solutions over all the cells in the $d_x d_y$-plane. See Figure 4 for an example. The overall running time of the algorithm is $O(n^{10})$, since there are $O(n^8)$ cells with $O(n^2)$ computation at each cell.

6   Related Work

Multi-robot Assignment and Path Planning: Coordinating multiple robots in a shared workspace has attracted much attention in the robotics community [16], [26], [2], [33]. A typical goal of coordination is to achieve collision-free and time-optimal robot motions. There has been extensive research on motion planning for multiple robots; see [14, 15, 8] for overviews and [16, 26, 24, 33] for example approaches related to our work. There have been many exciting recent developments in multiple robot motion planning and formation planning, some of which are summarized below.

Derenick and Spletzer[9] coordinated a large-scale robot team to change the shape of formation, for a fixed robot assignment, by modifying scale, translation, and rotation through second-order cone programming techniques. The solution is explored by minimizing the total distance or minimizing the maximum distance robots travel while geometric shape constraints are satisfied.

Kloder and Hutchinson [13] developed a representation for collision-free path planning of multiple unlabeled robots translating in the plane from one formation to another formation. They represent a formation by the coefficients of a complex polynomial whose roots represent the robot configurations.

The CAPT algorithm [30] developed by Turpin et al. performs concurrent assignment and planning of trajectories for unlabeled robots when there are N robots and M goals, and each goal is visited by one robot. They used the Hungarian assignment algorithm to minimize the sum of individual robot-goal costs. They developed synchronized trajectories that are collision-free when the start and goal locations are at least $2\sqrt{2}R$ from each other. They also present an online decentralized version of the algorithm.

Yu and LaValle [34, 35] demonstrated that multi-robot path planning on unit distance graphs can be modelled as multi-commodity dynamic network flow problems. They provide fast and complete algorithms to efficiently solve permutation-invariant versions of these problems. They also studied time and distance optimality of the feasible solutions. They [35] use integer linear program formulations to solve time and distance problems. Katsev, Yu, and LaValle [12] perform path planning for large robot formations of indistinguishable robots using a hierarchical approach. The algorithm provides paths with total distance within a constant multiple of the optimal total distance.

Solovey and Halperin [27] developed a sampling-based algorithm for the k-color multi-robot motion planning problem, where the robots are partitioned into groups such that the robots are interchangeable within each group. The algorithm reduces the k-color problem to several discrete pebble problems. Adler et al. [1] present an efficient algorithm for multi-robot motion planning for unlabeled discs in simple polygons. They transform a continuous problem into a discrete pebble motion on a graph problem.

Solovey, Yu, Zamir, Halperin [29] present a polynomial-time complete algorithm for unlabeled disc robots in the plane. It minimizes the total path length for the set of robots, shown to be at most $4m$ longer than the optimal solution, where m is the number of robots, assuming certain robot-obstacle and start-goal separation constraints are satisfied. Solovey and Halperin [28] show that the problem of unlabeled unit-square robots translating among obstacles is PSPACE-hard.

Luna and Bekris [18] developed a method for cooperative path-finding that is polynomial in the number of robots and is complete for all problem instances with at least two empty vertices in the graph. It uses push-and-swap primitives to restrict unnecessary exploration of search space. Liu and Shell [17] developed a hierarchical dynamic partitioning and distribution scheme for large-scale multi-robot task allocation. Nam and Shell [20] analyzed cost uncertainties in multi-robot task allocation problems, and the sensitivity of optimal assignments to variations in the cost matrix. Wagner and Choset [32] develop an efficient multiple robots path planning algorithm, where paths for individual robots are initially generated and coordination among robots is performed when needed due to collisions.

van den Berg et al. [4] presented an efficient method for reciprocal n-body collision avoidance that provides a sufficient condition for multiple robots to select an action that avoids collisions with other robots, though each acts independently without communication with others. van den Berg et al. [5] study path planning for multiple robots in the presence of obstacles. They apply optimal decoupling techniques to problems with low degrees of coupling. They decompose a multi-robot problem into a sequence of subproblems with a minimum degree of coupled control. The arrival times of the robots are not optimal as the plans are executed sequentially.

Light-actuated digital microfluidics: Lab-on-a-chip technology scales down multiple laboratory processes to miniature chips capable of performing automated chemical analyses.

Light-actuated digital microfluidics (LADM) systems use moving patterns of projected light on a continuous photoconductive surface to move droplets [22],[23]. The projected pattern of light effectively creates virtual electrodes on the lower substrate. By moving the virtual electrodes, droplets can be moved in parallel on the microfluidic chips to perform multiple biochemical reactions (Figure 1(a)).

By modeling droplets as robots, we can achieve collision-free motions optimized to reduce completion time. We have explored the problem of coordinating multiple droplets in light-actuated digital microfluidic systems intended for use as lab-on-a-chip systems. We focused primarily on creating matrix formations of droplets for biological applications. To achieve collision-free droplet coordination while optimizing completion times, we applied multiple robot coordination techniques. We used a mixed integer linear programming (MILP) approach to schedule coordination of both individual droplets and batches of droplets given their paths [19]. We also developed a linear time coordination algorithm for batch coordination of droplet matrix layouts.

7   Conclusion

This paper described algorithms to perform optimal assignment of teams of robots translating in the plane from an initial formation to a variable goal formation. We considered the case when each robot is to be assigned a goal position, the individual robots are interchangeable, and the goal formation can be either scaled or translated. We computed the costs for all pairs of initial, goal robot assignments as functions of the parameters of the goal formation, and partitioned the parameter space into equivalence classes using computational geometry techniques. We compute a minimum completion time assignment for an equivalence class by formulating it as a linear bottleneck assignment problem (LBAP). To improve efficiency, we solved the LBAP problem for each equivalence class by incrementally updating the solution as the formation parameters are varied. This work is motivated by applications that include the motion of droplet formations in digital microfluidic lab-on-a-chip devices, and of robot and drone formations in the plane.

Our emphasis has been on robot assignment with variable goal formations. In the event that two or more robots have assigned paths that lead to collisions, we must take additional steps to avoid collisions. These could include coordination using MILP solutions [2], path coordination [26], or collision avoidance [4].

With the LBAP solution, once the critical edge that determines the objective function value has been determined, some of the lower cost edges can have their assignments swapped without affecting the LBAP cost. This can potentially result in multiple optimal assignments. One way to enforce a consistent assignment for each cell of the arrangement is to solve the Lexicographic LBAP [7] instead of the standard LBAP. The advantage of using the Lexicographic LBAP (LexL-BAP) is that for the optimum LBAP value, it minimizes the distance traveled by each robot. We need to evaluate the LexLBAP only once, for the optimal LBAP found.

This paper describes initial steps towards addressing the problem of simultaneously optimizing robot assignments and variable goal formations. There are many aspects to be explored in future work. We are working to solve the assignment problem with combined scaling and translation. The relatively high computational complexity of the approach will make application to very large formations (with hundreds of robots) challenging. We are therefore interested in approaches to reduce the effective complexity. For example, methods to prune the set of cells at which an LBAP has to be solved will be useful. Another approach is to partition the set of robots into smaller subsets. Since UAVs and spacecraft operate in a 3D workspace, another useful direction is to extend the approach to 3D formations.

Acknowledgments: This work benefited from discussions with Dan Halperin, Zhiqiang Ma, and Erik Saule. This work was supported in part by NSF Award IIS-1547175.

References

1. Adler, A., de Berg, M., Halperin, D., Solovey, K.: Efficient multi-robot motion planning for unlabeled discs in simple polygons. In: 11th International Workshop on the Algorithmic Foundations of Robotics (WAFR) (2014)
2. Akella, S., Hutchinson, S.: Coordinating the motions of multiple robots with specified trajectories. In: IEEE International Conference on Robotics and Automation. pp. 624–631. Washington, DC (May 2002)
3. Alonso-Mora, J., Baker, S., Rus, D.: Multi-robot navigation in formation via sequential convex programming. In: 2015 IEEE/RSJ International Conference on Intelligent Robots and Systems (IROS). pp. 4634–4641 (2015)

4. van den Berg, J., Guy, S.J., Lin, M., Manocha, D.: Reciprocal n-body collision avoidance. In: Pradalier, C., Siegwart, R., Hirzinger, G. (eds.) Robotics Research: The 14th International Symposium ISRR, pp. 3–19. Springer, Berlin, Heidelberg (2011)

5. van den Berg, J., Snoeyink, J., Lin, M., Manocha, D.: Centralized path planning for multiple robots: Optimal decoupling into sequential plans. In: Robotics: Science and Systems. pp. 137–144 (2010)

6. de Berg, M., Cheong, O., van Kreveld, M., Overmars, M.: Computational Geometry: Algorithms and Applications. Springer-Verlag, Berlin, third edn. (2008)

7. Burkard, R., Dell'Amico, M., Martello, S.: Assignment Problems. SIAM, revised reprint edn. (2012)

8. Choset, H., Lynch, K.M., Hutchinson, S., Kantor, G.A., Burgard, W., Kavraki, L.E., Thrun, S.: Principles of Robot Motion: Theory, Algorithms, and Implementations. MIT Press (2005)

9. Derenick, J.C., Spletzer, J.R.: Convex optimization strategies for coordinating large-scale robot formations. IEEE Transactions on Robotics 23(6), 1252–1259 (2007)

10. Halperin, D.: Personal communication (2016)

11. Halperin, D., Sharir, M.: Arrangements. In: Goodman, J.E., O'Rourke, J., Tóth, C.D. (eds.) Handbook of Discrete and Computational Geometry. CRC Press, Boca Raton, FL, third edn. (2017), to appear

12. Katsev, M., Yu, J., LaValle, S.M.: Efficient formation path planning on large graphs. In: 2013 IEEE International Conference on Robotics and Automation (ICRA) (2013)

13. Kloder, S., Hutchinson, S.: Path planning for permutation-invariant multirobot formations. IEEE Transactions on Robotics 22(4), 650–665 (2006)

14. Latombe, J.C.: Robot Motion Planning. Kluwer Academic Publishers, Norwell, MA (1991)

15. LaValle, S.M.: Planning Algorithms. Cambridge University Press, Cambridge, U.K. (2006), available at http://planning.cs.uiuc.edu/

16. LaValle, S.M., Hutchinson, S.A.: Optimal motion planning for multiple robots having independent goals. IEEE Transactions on Robotics and Automation 14(6), 912–925 (Dec 1998)

17. Liu, L., Shell, D.: Large-scale multi-robot task allocation via dynamic partitioning and distribution. Autonomous Robots 33(3), 291–307 (2012)

18. Luna, R., Bekris, K.E.: Efficient and complete centralized multi-robot path planning. In: IEEE/RSJ International Conference on Intelligent Robots and Systems (IROS-11). San Francisco, CA (25-30 Sept 2011)

19. Ma, Z., Akella, S.: Coordination of droplets on light-actuated digital microfluidic systems. In: IEEE International Conference on Robotics and Automation. pp. 2510–2516. St. Paul, MN (May 2012)

20. Nam, C., Shell, D.A.: When to do your own thing: Analysis of cost uncertainties in multi-robot task allocation at run-time. In: 2015 IEEE International Conference on Robotics and Automation (ICRA). Seattle, Washington (May 2015)

21. Papadimitriou, C.H., Steiglitz, K.: Combinatorial Optimization: Algorithms and Complexity. Prentice-Hall, Englewood Cliffs, New Jersey (1982)

22. Park, S.Y., Teitell, M.A., Chiou, E.P.Y.: Single-sided continuous optoelectrowetting (SCOEW) for droplet manipulation with light patterns. Lab Chip 10, 1655–1661 (2010), http://dx.doi.org/10.1039/C001324B

23. Pei, S.N., Valley, J.K., Neale, S.L., Jamshidi, A., Hsu, H., Wu, M.C.: Light-actuated digital microfluidics for large-scale, parallel manipulation of arbitrarily sized droplets. In: 23rd IEEE International Conference on Micro Electro Mechanical Systems. pp. 252–255. Wanchai, Hong Kong (Jan 2010)

24. Peng, J., Akella, S.: Coordinating multiple robots with kinodynamic constraints along specified paths. International Journal of Robotics Research 24(4), 295–310 (Apr 2005)

25. Shekar, V., Campbell, M., Akella, S.: Towards automated optoelectrowetting on dielectric devices for multi-axis droplet manipulation. In: IEEE International Conference on Robotics and Automation. pp. 1431–1437. Karlsruhe, Germany (May 2013)

26. Simeon, T., Leroy, S., Laumond, J.P.: Path coordination for multiple mobile robots: A resolution-complete algorithm. IEEE Transactions on Robotics and Automation 18(1), 42–49 (Feb 2002)

27. Solovey, K., Halperin, D.: k-color multi-robot motion planning. International Journal of Robotics Research 33(1), 82–97 (2014)

28. Solovey, K., Halperin, D.: On the hardness of unlabeled multi-robot motion planning. In: Robotics: Science and Systems. Rome, Italy (Jul 2015)

29. Solovey, K., Yu, J., Zamir, O., Halperin, D.: Motion planning for unlabeled discs with optimality guarantees. In: Robotics: Science and Systems. Rome, Italy (Jul 2015)

30. Turpin, M., Michael, N., Kumar, V.: CAPT: Concurrent assignment and planning of trajectories for multiple robots. The International Journal of Robotics Research 33(1), 98–112 (2014)

31. Turpin, M., Mohta, K., Michael, N., Kumar, V.: Goal assignment and trajectory planning for large teams of interchangeable robots. Autonomous Robots 37(4), 401–415 (Dec 2014)

32. Wagner, G., Choset, H.: Subdimensional expansion for multirobot path planning. Artificial Intelligence 219, 1–24 (Feb 2015)

33. Wurman, P., D'Andrea, R., Mountz, M.: Coordinating hundreds of cooperative, autonomous vehicles in warehouses. AI Magazine 29(1), 9–19 (2008)

34. Yu, J., LaValle, S.M.: Multi-agent path planning and network flow. In: Frazzoli, E., Lozano-Perez, T., Roy, N., Rus, D. (eds.) Algorithmic Foundations of Robotics X, Springer Tracts in Advanced Robotics, vol. 86, pp. 157–173. Springer Berlin Heidelberg (2013)

35. Yu, J., LaValle, S.M.: Planning optimal paths for multiple robots on graphs. In: 2013 IEEE International Conference on Robotics and Automation (ICRA) (2013)

Author Index

© Springer Nature Switzerland AG 2020
K. Goldberg et al. (eds.), *Algorithmic Foundations of Robotics XII*, SPAR 13, pp. 929–931, 2020.
https://doi.org/10.1007/978-3-030-43089-4

Printed in the United States
by Baker & Taylor Publisher Services